W9-ADB-446

United States Map

World Map

North America
South America
Europe
Africa
Asia
Australia
Arctic Ocean
Atlantic Ocean
Atlantic Ocean
Pacific Ocean
Pacific Ocean
Indian Ocean
Greenland
Canada
United States
Mexico
Cuba
Haiti
Dominican Republic
Puerto Rico
Belize
Jamaica
Honduras
Nicaragua
Guatemala
El Salvador
Costa Rica
Panama
Venezuela
Guyana
Suriname
French Guiana
Colombia
Ecuador
Peru
Brazil
Bolivia
Paraguay
Chile
Argentina
Uruguay
Iceland
Norway
Sweden
Finland
United Kingdom
Ireland
Denmark
Estonia
Latvia
Lithuania
Belarus
Poland
Germany
Ukraine
Moldova
Romania
Bulgaria
France
Spain
Portugal
Italy
Greece
Russia
Kazakhstan
Mongolia
Georgia
Armenia
Azerbaijan
Turkey
Uzbekistan
Turkmenistan
Kyrgyzstan
Tajikistan
Afghanistan
Pakistan
Iran
Iraq
Syria
Cyprus
Lebanon
Israel
Jordan
Kuwait
Saudi Arabia
Bahrain
Qatar
U.A.E.
Oman
Yemen
Nepal
Bhutan
Bangladesh
India
China
North Korea
South Korea
Japan
Taiwan
Myanmar (Burma)
Laos
Thailand
Vietnam
Cambodia
Philippines
Brunei
Malaysia
Singapore
Sumatra
Borneo
Indonesia
Papua New Guinea
New Zealand
Morocco
Algeria
Libya
Egypt
Mauritania
Mali
Niger
Chad
Sudan
Eritrea
Ethiopia
Somalia
Senegal
Gambia
Guinea Bissau
Guinea
Sierra Leone
Liberia
Ivory Coast
Burkina Faso
Ghana
Benin
Togo
Nigeria
Cameroon
Central African Rep.
Guinea
Gabon
Congo
Democratic Republic of the Congo
Cabinda (Angola)
Uganda
Kenya
Rwanda
Burundi
Tanzania
Malawi
Angola
Zambia
Zimbabwe
Mozambique
Namibia
Botswana
South Africa
Swaziland
Lesotho
Madagascar
1. Netherlands
2. Belgium
3. Luxemburg
4. Switzerland
5. Czech. Rep.
6. Austria
7. Slovakia
8. Hungary
9. Slovenia
10. Croatia
11. Bosnia and Herzegovina
12. Yugoslavia
13. Albania
14. Macedonia

SOCIOLOGY

NINTH EDITION

Jon M. Shepard
Virginia Polytechnic Institute
and State University

Australia • Canada • Mexico • Singapore • Spain
United Kingdom • United States

Sociology Editor: Robert Jucha
Development Editor: Natalie Cornelison
Assistant Editor: Stephanie Monzon
Editorial Assistant: Melissa Walter
Technology Project Manager: DeeDee Zobian
Marketing Manager: Matthew Wright
Advertising Project Manager: Linda Yip
Project Manager, Editorial Production: Emily Smith
Print/Media Buyer: Doreen Suruki
Permissions Editor: Sarah Harkrader

Production Service: Hespenheide Design
Text Designer: Gary Hespenheide/Hespenheide Design
Photo Researcher: John Kelty/Hespenheide Design
Copy Editor: Christianne Thillen
Illustrator: Randy Miyake/Hespenheide Design
Compositor: Hespenheide Design
Cover Designer: Yvo Riezebos
Cover Image: *November 19, 1863* by Christopher Brown: Collection of the Modern Art Museum of Fort Worth. Museum purchase made possible by a grant from the Burnett Foundation
Cover Printer: The Lehigh Press, Inc.
Text Printer: Quebecor World/Dubuque

Printed in the United States of America
2 3 4 5 6 7 07 06 05

For more information about our products, contact us at:
Thomson Learning Academic Resource Center
1-800-423-0563
For permission to use material from this text, contact us by:
Phone: 1-800-730-2214 **Fax**: 1-800-730-2215
Web: http://www.thomsonrights.com

Library of Congress Control Number: 2003117164

Student Edition: ISBN 0-534-62073-6

Instructor's Edition: ISBN 0-534-62075-2

Wadsworth/Thomson Learning
10 Davis Drive
Belmont, CA 94002-3098
USA

Asia
Thomson Learning
5 Shenton Way #01-01
UIC Building
Singapore 068808

Australia/New Zealand
Thomson Learning
102 Dodds Street
Southbank, Victoria 3006
Australia

Canada
Nelson
1120 Birchmount Road
Toronto, Ontario M1K 5G4
Canada

Europe/Middle East/Africa
Thomson Learning
High Holborn House
50/51 Bedford Row
London WC1R 4LR
United Kingdom

Latin America
Thomson Learning
Seneca, 53
Colonia Polanco
11560 Mexico D.F.
Mexico

Spain/Portugal
Paraninfo
Calle/Magallanes, 25
28015 Madrid, Spain

About the Author

While an undergraduate student, Jon Shepard was inspired and nurtured by his sociology professor, Richard Scudder. After graduating from Michigan State University with a Ph.D. in sociology, Shepard taught introductory sociology and the sociology of organizations at the University of Kentucky. For the past fourteen years, he has been Head of the Virginia Tech Department of Management. He is the author of ten books and more than forty professional journal articles. He has received teaching awards at both universities, including the University of Kentucky Great Teacher Award. He lives with his wife, Kay Vogel Shepard, in Blacksburg, Virginia. Just retired from college administration, Jon will now be able to devote even more time to writing and teaching.

Enjoying four new friends:

Mocha, Muffin, Tiny, and Pumpkin

Contents in Brief

Contents

BOXES

Doing Research

Sociology and the News Media

Technology and Society

MAPS

Preface

A Note to Students from the Author

The issues discussed in my freshman introductory sociology class were not the sort of things I usually spent time thinking about: Is divorce more likely or less likely when people have the same social-class background? Are some races inferior to others? What is the social significance of Darwinism?

Suddenly, I began to see human behavior in a different light. I discovered that Richard Wright's classics, *Native Son and Black Boy*, are not merely stories about black youth but rather autobiographical reflections of the black experience in America. Prejudice and discrimination are not just characteristics of individuals; they are part of society as a whole. I learned that football is actually as much a business as a sport. It became apparent that the fraternity I was about to join was not only a brotherhood but also part of the campus social hierarchy.

I began to see social relationships as essential for human survival. And if the world is a stage and all its men and women merely players, these players generally deliver their lines and act out their parts as if they were rehearsed, and with a definite flair for mimicry. Yet, the action that sociologists label "social structure" depends less on the conscious learning of appropriate attitudes, beliefs, and behavior than on unreflective acceptance of our culture and society. In this sense, we are like puppets responding to tugs on the strings that bind us to essential social relationships—relationships in which people, I came to understand, do not usually behave randomly and do not always behave only as individuals. People often think, feel, and behave in rather predictable ways because of what they have been taught and because of the many social pressures to which they have been exposed. At the same time, however, individuals interacting with others create their own understandings of situations. In this sense, human beings are not like puppets, because they have the potential to buck tradition; they are active, thinking creatures even when they are conforming.

Society was demystified for me. I came to value sociology as a tool for understanding the world. In fact, this experience led me to major in sociology and subsequently to obtain my Ph.D. in the field. I have never regretted those choices.

You might not major in sociology. You can, however, enjoy this course and take lifelong benefits from the slant on social life that sociology provides.

A Note to Instructors from the Author

Several specific goals continue to energize me as we introduce the significance and excitement of sociology to students. First, sociology—with its perspectives, concepts, theories, and research findings—offers a window on the social forces that affect us all daily. This perspective is vital for students as they grapple to understand the social factors that promote patterned behavior in themselves and others. Second, the material must be readable and at the same time theoretically and empirically sound. Third, students deserve a textbook design that is dynamic and demonstrates, clearly, the application of a new perspective to their personal lives. Fourth, I want a presentation from which students can not only learn the basics of sociology, but can acquire the ability to pose their own questions about social life. Finally, the catalyst: stimulate students to become more active learners.

Unifying Themes and Features

Sociological Imagination

The study of sociology encourages critical thinking about conventional wisdom through the development of the *sociological imagination*—the mind-set that enables individuals to see the relationship between events in their personal lives and events in their society. To this end, each chapter opens with a question about some aspect of social life. The answer to each question contradicts a popular or commonsense belief. Sometimes the question will focus on a result that even sociologists may have doubted until a sufficient amount of convincing research had been done. The correct answer to each question is given at the beginning of the chapter and is elaborated on within the chapter itself. Topics covered include the following:

- Suicide (Chapter 1, "The Sociological Perspective")
- Television and violence (Chapter 4, "Socialization Over the Life Course")
- Selfishness and human nature (Chapter 5, "Social Structure and Society")
- Date rape (Chapter 7, "Deviance and Social Control")
- Gender income inequality (Chapter 10, "Inequalities of Gender")
- National health care (Chapter 16, "Health and Health Care")
- Revolutions and social change (Chapter 18, "Social Change and Collective Behavior")

Of course, opportunities to build a sociological imagination permeate the entire text.

Doing Research: Sociologists as Scientists

A boxed feature within each chapter, entitled *Doing Research*, presents the theory, methods, conclusions, and implications of significant sociological studies. This feature is intended to demonstrate the ways sociologists use the scientific method in their work.

Several criteria guided the selection of these research studies. Some studies, such as Emile Durkheim's work on the social antecedents of suicide, are sociological landmarks. Others, such as Philip Zimbardo's experiment involving a simulated prison and George Ritzer's analysis of the McDonaldization of higher education, are included because they reinforce a major point in a chapter. Still other studies are included because they illustrate the imaginative use of a major research method. Durkheim's use of existing sources in the study of suicide and Donna Eder's use of a variety of research methods in her study of popularity in middle school are examples.

Given these criteria, it is hardly surprising that many of the studies included are sociological "classics"; they have had a lasting influence on the field and are continuously being cited by other researchers. Like classics in all fields, these pieces of research generally have high interest value. They are innovative in approach and explore important topics in ingenious ways. If read carefully, these detailed accounts of significant sociological studies cannot fail to pique interest in social research and stimulate the sociological imagination. Here are some examples:

- "Teenagers in a Cultural Bind" (Chapter 3, "Culture")
- "High School Reunions" (Chapter 4, "Socialization Over the Life Course")
- "Adopting Statuses in a Simulated Prison" (Chapter 5, "Culture and Society")
- "Who's Popular, Who's Not" (Chapter 8, "Social Stratification")
- "The McDonaldization of Higher Education" (Chapter 13, "Education")
- "The Secularization of Religion" (Chapter 15, "Religion")
- "Gang Violence" (Chapter 17, "Population and Urbanization")
- "The Withering of the American Dream" (Chapter 18, "Social Change and Collective Behavior")

Technology and Society

Technology is a major engine of social change in modern society. Because our present students grew up with telephones, television, and computers, it may be difficult for them to understand the social effects of these technologies. But think of the changes they have seen with the emergence of the Internet and wireless phones. Each chapter has a boxed feature entitled *Technology and Society*. Topics new to this edition are privacy in the workplace, aging in the mass media, terrorism, and "smart mobs." Other topics include

- "Are Researchers Peeping Toms?" (Chapter 2, "Social Research")
- "*Star Wars* and the Internet" (Chapter 3, "Culture")
- "Can the Internet Stunt Your Growth?" (Chapter 4, "Socialization Over the Life Course")
- "Look Out for Identity Thieves!" (Chapter 7, "Deviance and Social Control")
- "Gender Equality and the Internet" (Chapter 10, "Inequalities of Gender")
- "Technology and Protection Against Terrorism" (Chapter 14, "Political and Economic Institutions")
- "Is Human Cloning Ethical?" (Chapter 15, "Religion")
- "Virtual Communities" (Chapter 17, "Population and Urbanization")

Sociology and the News Media

C. Wright Mills, creator of the term sociological imagination, valued sociology for its potential in developing a better understanding of news events. His point is only stronger with the addition of cable television and the Internet. A feature called *Sociology and the News Media* stimulates students' newly acquired sociological perspective as they view political and social events in the news. Each chapter contains a boxed feature keyed to a recently broadcast story that appeared on the major news network, CNN. Four of the *Sociology and the News Media* boxes are new to this edition. Each is keyed to a CNN video (available to instructors); each contains brief news stories; and up to one-half are cross-cultural. Here are examples:

- "Being Gay in America" (Chapter 5, "Social Structure and Society")
- "Gang Violence" (Chapter 7, "Deviance and Social Control")
- "Hate Crimes" (Chapter 9, "Inequalities of Race and Ethnicity")
- "Marry for Love or Money" (Chapter 12, "Family")
- "Are School Vouchers a Good Thing?" (Chapter 13, "Education")
- "The Politics of Smoking" (Chapter 14, "Political and Economic Institutions")
- "Modernization and the American Family" (Chapter 18, "Social Change and Collective Behavior")

Accent on Theory

Each chapter contains a prominent section on the distinctive views of three major theoretical perspectives—functionalism, conflict theory, and symbolic interactionism. Each chapter also presents a table entitled "Focus on Theoretical Perspectives," in which succinct illustrations are offered for the three theoretical perspectives. For example, the following table appears in Chapter 11 ("Family"):

Cross-Cultural Perspective

Never has the world been as interconnected as it is now. Gone are the comfort and simplicity of geographically based social isolation. It is now apparent that events in

TABLE 11.2

FOCUS ON THEORETICAL PERSPECTIVES: Perspectives on the Family

Both functionalism and conflict theory are more concerned with the ways social norms affect the nature of the family. Symbolic interactionism tends to examine the relationship of the self to the family. If functionalism and conflict theory were used to focus on the self, what examples would you give?

Theoretical Perspective	Social Arrangement	Example
Functionalism	• Sex norms	• Children are taught that sexual activity should be reserved for married couples.
Conflict theory	• Male dominance	• Husbands use their economic power to control how money is spent.
Symbolic interactionism	• Developing self-esteem	• A child is abused by her parents learns to dislike herself.

cities, states, and nations have repercussions for each other. Thus, it is imperative that stimulating sociological phenomena be viewed in a larger geographic frame.

To this end, cross-cultural examples and research are presented throughout the text. This cross-cultural emphasis alerts students to our tendency to accept our own culture while rejecting others. By interacting mentally with other cultures within the context of sociological concepts, theories, and research findings, students are better able to apply the sociological perspective to their daily lives. This cross-cultural emphasis is intended to encourage in students a more self-conscious awareness of their own society and a better understanding of other cultures.

In addition, maps of the United States and the world are included in each chapter. Each map permits the geographic comparison of a particular social phenomenon. Here are some examples from the *Snapshot of America* feature:

- "Gun Control" (Chapter 3, "Culture")
- "High School Exit Exams" (Chapter 9, "Inequalities of Race and Ethnicity")
- "Americans Without Health Insurance" (Chapter 16, "Health and Health Care")

The world map feature, entitled *World View*, includes these social phenomena:

- "The Wired World" (Chapter 2, "Social Research")
- "National Death Penalty Policy" (Chapter 7, "Deviance and Social Control")
- "Political Freedom" (Chapter 14, "Political and Economic Institutions")

What is New in the Ninth Edition?

All New Full-Color Design

The Ninth Edition of *Sociology* marks a return to the use of full color in the design, figures, and photographs. Long-time users of *Sociology* may recall that from the first through the sixth editions the text was in full color. Then, in order to save students money in their textbook purchases, starting with the seventh and continuing with the eighth editions we published the text using only two colors. Now with the Ninth Edition I am pleased that we are able to offer the text in rich, full color. In the animated world of today's students, visual learning is even more important. The use of full color thus enhances the learning process. I am, however, most pleased that this use of full color has been achieved without having to resort to a large price increase for the text. *Sociology*, Ninth Edition, continues to be offered as one of the most economical college sociology textbooks available today.

Text Additions

This edition of *Sociology* has undergone visibly significant revision throughout. Perhaps the most noticeable change is the replacement of a single chapter on inequalities of gender and age with separate chapters on gender stratification and age stratification. Sociological interest in gender and age stratification has grown to the point that sufficient detail on each topic cannot be presented in a single chapter. New separate chapters provide the space needed for more complete elaboration.

Other important revisions come in the form of expanded coverage and new topics in existing chapters. These revisions are highlighted in the following descriptions.

Chapter 1 introduces students to the practical uses of the sociological perspective and the history of sociology. Also, the three major sociological theories (functionalism, conflict theory, and symbolic interactionism) are illustrated with sports examples. *Liberation sociology* is introduced as a new concept in this edition. Conflict theory is now illustrated with new information on corporate manipulation of tax laws, including Enron. Emile Durkheim's study of suicide has been moved to this chapter.

Chapter 2 introduces the major research methods used by sociologists. It concludes with a discussion of ethics in social research. An interesting new field experiment by Marianne Bertrand and Sendhil Mullainathan ("What's in a Name?") on racial discrimination is introduced in *Doing Research*. New data are presented regarding the relationship between sex, race, and education.

Chapter 3 is thoroughly reorganized—promoting a broader understanding of the pivotal concept of culture. The chapter offers many new contemporary examples, including the Taliban's maltreatment of women in Afghanistan, corporate scandals in Enron, Arthur Anderson, and WorldCom, and attitudes in Arab countries. New to this edition are considerations of *gestures* as an aspect of language and the concept of *multiculturalism*. The sections on subculture and counterculture are expanded and revised. New data on premarital sexual intercourse among U.S. teenage women are presented. Also included are two new tables: one on do's and taboos around the world, the other on symbols created for Internet communication.

Chapter 4 explores our learning to participate in society through the acquisition of culture (socialization). Socialization from infancy to old age is investigated within the context of functionalism, conflict theory, and symbolic interactionism. A concept, *continuity theory*, is introduced in a new discussion of adjustment among aging adults.

Chapter 5 removes, for students, the mystery of the meaning of social structure. It explains and illustrates the major sociological concepts underlying the central concept of social structure. A new table compares the nature of different types of societies from hunting and gathering to postindustrial. A new *World View* displays the extent of employment in service industries around the globe.

Chapter 6 distinguishes among various types of groups and the basic interactions within them. A new table compares primary and secondary groups, and *social networks* is presented as a new concept. The last half of the chapter is devoted to formal organizations or bureaucracies. A description of formal organization in Japan closes the chapter. New maps include a *Snapshot of America* on the size of state bureaucracies and a revised *World View* on the size of military budgets in major countries.

Chapter 7 The first part of this chapter focuses on biological, psychological, and sociological explanations of deviant behavior. An expanded discussion of deviance in high places includes contemporary examples from college sports and the Catholic Church. An almost entirely new discussion of white-collar deviance emphasizes the recent corporate scandals. The last half of the chapter deals with a more detailed discussion of crime in the United States. The crime control section has a completely updated exploration of capital punishment and imprisonment in the United States, including new data from the FBI. A new *Sociology and the News Media*, "Capital Punishment," has been added to this edition.

Chapter 8 is a pivotal chapter on one of the most important concepts in sociology—social stratification. Material on poverty in America is thoroughly updated and revised, including new information on the outcomes of welfare reform. An updated *Snapshot of America* shows the percentage of the U.S. population in poverty by state. A completely new *World View* displays the nature of global inequality. A heavily revised section on global poverty closes the chapter.

Chapter 9 explores the significance of race and ethnicity and the operation of prejudice and discrimination in the United States. *Scapegoat* is included as a new concept. A subsequent large section, "Institutionalized Discrimination," details the nature and extent of inequality among American minorities. Discussions of inequality among African Americans and Latinos are completely revised and updated. A new *Snapshot of America* focuses on high school exit exams and their impact on minority and poor children. A new *World View* illustrates the extent of global ethnic diversity.

Chapter 10 concentrates on gender inequality. As noted earlier, this area of study has grown in such importance that it requires a stand-alone chapter. This expanded coverage is reflected in the sections on "Women as a Minority Group," "Occupational and Economic Inequality," "Sexism in Sports," and "Legal and Political Inequality." All tables and graphs from the eighth edition are updated and revised. A new figure shows the persistent gender inequality in college sports. A new *World View* displays women's earnings as a proportion of men's earnings around the globe.

Chapter 11 covers age inequality with sections added on "Aging and Stratification" and "The Graying of America." Five new concepts are included: *age cohort, age structure, feminization of poverty, geronticide,* and *gerontocracy*. Each feature in this chapter is new. Technology and Society deals with the unfavorable depiction of aging in the mass media. *Doing Research* documents the consequences of aging in rural Ireland. *Sociology and the News Media* has a new topic entitled "Baby Boomer Marketing." Three new tables and four new figures are incorporated. A new *Snapshot of America* displays the percentage of the U.S. population 65 years and over by state. Gender differences in life expectancy around the world are shown in a new *World View*.

Chapter 12 explores the nature of the family as viewed by sociologists. There are also sections on "Family and Marriage in the United States" and "Lifestyle Variations." Expanded coverage is given to the divorce rate in the United States. An added section entitled "Family Resiliency" presents two new concepts: *family resiliency* and *public policy*. Nearly all graphs have been updated, and a new bar graph on international divorce rates is included. "Cyber Sperm" is a *Sociology and the News Media* new to this edition.

Chapter 13 examines the organization of schools, the functions of education, educational inequality, and the transmission of culture in schools. The section on the back-to-basics movement is enlarged. New material is presented in the section on "Competitors to the Traditional Public School," including vouchers, charter schools, and for-profit schools. The section on "Promoting Equality in Education" is beefed up. "Higher Education" is expanded. Data in graphs and U.S. maps are updated. A new *Sociology and the News Media*, entitled "School Test Trouble," is included in this edition.

Chapter 14 covers political and economic institutions with a significant revision of "Political Power in American Society," particularly the discussions about voting and political action committees. The concept of *corporate welfare* is introduced in this edition. A new *Technology and Society* explores the uses of technology in the fight against terrorism.

Chapter 15 presents sociology's unique perspective on the institution of religion. This edition includes a new section entitled "Gender and Religion." The Raelians have been added to the discussion of "New Religious Movements." A study challenging the secularization of religion argument is presented in a new *Doing Research*.

Chapter 16 explores health care in the United States. The relationship between illness and age is significantly expanded. A completely new section on mental illness is introduced. Treatment of HMOs is expanded. All graphs and maps are updated.

Chapter 17 is devoted to the dual and related topics of population and urbanization. In an entirely new section, "Population Growth in the United States," the concept *natural increase* is introduced. *Boomburgs* is a new concept included in the section on "Suburbanization." This chapter contains new *Snapshot of America* and *World View* maps. All data on population and urbanization have been updated.

Chapter 18 combines the areas of social change and collective behavior. The section on "World-System Theory" is increased. Most of the discussion on "Mass Hysteria and Panics" is new. A new section entitled "The Future Direction of Social Movement Theory" closes this chapter.

Distinctive Study Aids

Joseph Butler, an eighteenth-century English minister and moral philosopher, wrote:

> *[P]eople habituate themselves to let things pass through their minds, as one may speak, rather than to think of them. Thus by use they become satisfied merely with seeing what is said, without going any further. Review and attention, and even forming a judgment, becomes fatigue; and to lay anything before them that requires it, is putting them quite out of their way. (Butler 1983:11; originally published in 1726)*

Butler eloquently expresses the elements underlying our approach to learning in this textbook: the SQ3R method and critical thinking. Students continue to express enthusiasm for this approach. It helps them combat passivity and become better active learners, comprehending the material more fully as they become progressively more mentally involved.

SQ3R: A Format for Study

This text is designed with the "SQ3R" study format at its core. Research tested, this approach helps students identify significant ideas, understand these ideas rapidly, remember important points, and review effectively for exams. As a result, students learn more about sociology more easily while performing better on tests. *I recommend that instructors and students go over these steps together.*

The letters in SQ3R symbolize five steps in effective reading and learning: survey (S); question (Q); read, recite, and review (3R). The steps in the SQ3R method are built into each chapter.

1. *Survey*. Before reading the chapter, students should read the outline, the introduction, and the summary. This survey, which will give them an overall picture of the chapter content, should take only a few minutes.
2. *Question*. Third-order headings are phrased as questions to help students select and concentrate on the important points. For example, instead of seeing subtopic headings such as "Working Women," they will find such questions as "Have men and women reached financial equality?" and "How do American women fare globally?"
3. *Read*. For increased comprehension, students should focus on each third-order question as they read the material that answers it.
4. *Recite*. Students should answer each third-order heading immediately after reading the relevant material. If they are unable to answer a question, they should examine the material until they find the answer. Also, at the end of every major topic is a final recitation check called *Feedback*. (See sample *Feedback* on Page xvii.) If students cannot answer one of these self-test questions, they should note the correct answer given under the questions and look back at the text material to find out why this is the right answer. The recitation dimension of the SQ3R method will prevent them from deluding themselves into believing that they understand material when, in fact, they do not. This process is designed to replace surface recognition with a more thorough comprehension.

FEEDBACK

1. _____ is the scientific study of social structure.
2. Match the social sciences listed below with the examples of research projects beside them.

___ a. sociology	(1) a study of how children learn to talk
___ b. anthropology	(2) a study of the impact of taxation on consumer spending
___ c. psychology	(3) a study of African American family structure during the slavery era
___ d. economics	(4) a study of village ruins
___ e. political science	(5) a study of presidential power
___ f. history	(6) a study of drug use patterns among high school students

Answers: 1. Sociology 2. a. (6) b. (4) c. (1) d. (2) e. (5) f. (3)

5. *Review.* After completing a chapter, students should once again briefly answer the questions posed in the third-order headings. Even better, they should have another person ask them the questions. They can complete their review of a chapter by using the *Review Guide* that appears at the end of each chapter.

Critical Thinking

Critical thinking—questioning commonly held assumptions—is traced by intellectual historians to fifth-century-B.C. Greece, particularly to the Athenians (Brinton 1963). Full-fledged Western interest in critical analysis did not appear until the eighteenth century, the period known as the Enlightenment (Gay 1966). Respect for critical and reasoned analysis, rather than judgments based on emotion, has been a key element of the Western world ever since.

Critical thinking is crucial for today's college students. First, the tradition of liberal education is the tradition of critical thought (Pelikan 1992). Second, as the nature of work continues to move from physical labor to cerebral activities, the facility for critical reasoning becomes an increasingly valuable asset on the job. Third, it doesn't take a rocket scientist to figure out the need for critical thought among all American voters; behavior of our political leaders makes this point daily. Finally, not least, critical thinking is vital in personal contexts such as family life, decision making outside of work, and personal enrichment.

Critical thinking is incorporated in this edition in three ways:

1. **Critical Thinking within the SQ3R Method.** The promotion of critical thinking lies at the heart of the "question" step of the SQ3R method. This point can be made by contrasting two approaches to learning: the *informational* approach and the *critical thinking* approach. The objective in the informational approach is knowledge acquisition, important when one is exposed to new material. Following are some sample SQ3R informational questions.

- Do extremely isolated children develop human characteristics?
- To what extent do Americans exercise power through the ballot box?
- Do American workers like their jobs?

Although gaining information is essential, it is only the beginning. Critical thinking emphasizes further interaction with knowledge as it is being acquired. Following are some sample critical-thinking SQ3R questions:

- If prisons do not rehabilitate, what are some alternatives?
- Is the negative image of the poor in America accurate?
- What is required to prove the existence of the power elite in American society?

2. **Critical-Thinking Questions.** Critical-thinking questions are liberally interspersed throughout each chapter. Questions follow the *Doing Research, Technology and Society,* and *Sociology and the News Media* features. A critical-thinking question is also included in each table, figure, and map. A set of four to six critical-thinking questions appears in the *Review Guide* at the end of each chapter. Consequently, a critical-thinking opportunity is offered at least every few pages. These wide-ranging questions encourage students to think critically and creatively about the ideas within a chapter. Sometimes students will apply these ideas to a particular aspect of society. At other times students will use sociological ideas to analyze and understand events and experiences in their own lives.

3. **Critical Feedback.** Each *Doing Research* closes with a series of critical-thinking questions under the heading, "Thinking About the Research," designed to help students better understand the piece of sociological research involved and to probe below the surface. For example, these questions are posed following the description of Stanley Milgram's study of group pressure and conformity:

- Discuss the ethical implications of Milgram's experiment.
- If the researcher had not been present as an authority figure during the experiment to approve the use

of all shock levels, do you think group pressure would have been as effective? Explain.

4. **The Sociological Imagination.** It is easy to fall into a pattern of nonreflection about prevailing ideas that are passed from generation to generation. The feature *Using the Sociological Imagination*, described earlier, opens each chapter with a question designed to challenge some aspect of a social myth.

Review Guide

Each *Review Guide* begins with a chapter summary, followed by the chapter learning objectives. Next is a concept review of approximately 50 percent of the concepts introduced in the chapter. Students can test their grasp of key concepts by matching concepts with definitions. Several critical-thinking questions follow the concept review. These broad questions provide practice for essay tests. A set of multiple-choice questions then acts as a mini self-test. The feedback review consists of a sample of questions taken directly from the *Feedback* questions throughout the chapter. In most chapters a graphic review feature tests understanding of a particular table or figure in the chapter. An answer key closes each *Review Guide*.

Supplements for the Ninth Edition

Supplements for the Instructor

Instructor's Edition of *Sociology: The Essentials*. An Instructor's Edition (IE) of this text containing several features useful to instructors is available. The IE contains the Visual Preface, a walk-through of the several themes and many features of *Sociology* along with a complete listing of available bundles for this text. To obtain a copy of the Instructor's Edition, contact your Thomson Sales Representative.

Instructor's Resource Manual with Test Bank (with the MultiMedia Manager CD-ROM). This manual offers the instructor chapter outlines, discussion topics, and lecture suggestions to facilitate in-class discussion, and innovative class activities for each chapter. The test bank includes 50–75 multiple-choice questions and true/false questions with answers and page references, as well as essay questions for each chapter. A concise user guide for InfoTrac and WebTutor is included as an appendix. A new MultiMedia Manager CD-ROM is now located in the print Instructor's Resource Manual. The CD also includes book-specific PowerPoint lecture slides, graphics from the book itself, the Instructor's Resource Manual and Test Bank as Word documents, CNN video clips, and links to many of Wadsworth's important sociology resources.

ExamView Computerized Testing. Create, deliver, and customize tests and study guides (both print and online) in minutes with this easy-to-use assessment and tutorial system. ExamView offers both a Quick Test Wizard and an Online Test Wizard that guide you step-by-step through the process of creating tests. The test appears on screen exactly as it will print or display online. Using ExamView's complete word processing capabilities, you can enter an unlimited number of new questions or edit existing questions included with ExamView.

Wadsworth's Introduction to Sociology 2005 Transparency Acetates. A set of four-color acetates consisting of tables and figures from Wadsworth's introductory sociology texts is available to help prepare lecture presentations. Free to qualified adopters.

Videos. Adopters of *Sociology* have several different video options available with the text.

Wadsworth's Lecture Launchers for Introductory Sociology An exclusive offering jointly created by Wadsworth/Thomson Learning and Dallas TeleLearning, this video contains a collection of video highlights taken from the "Exploring Society: An Introduction to Sociology" Telecourse (formerly "The Sociological Imagination"). Each 3- to 6-minute-long video segment has been especially chosen to enhance and enliven class lectures and discussion of 20 key topics covered in any introductory sociology text. Accompanying the video is a brief written description of each clip, along with suggested discussion questions to help effectively incorporate the material into the classroom.

Sociology: Core Concepts. An exclusive offering jointly created by Wads-worth/Thomson Learning and Dallas TeleLearning, this video contains a collection of video highlights taken from the "Exploring Society: An Introduction to Sociology" Telecourse (formerly "The Sociological Imagination"). Each 15- to 20-minute video segment will enhance student learning of the essential concepts in the introductory course and can be used to initiate class lectures, discussion, and review. The video covers topics such as the sociological imagination, stratification, race and ethnic relations, social change, and more.

CNN® Today Sociology Video Series (Volumes I–VII). Illustrate the relevance of sociology to everyday life with this exclusive series of videos for the Introduction to Sociology course. Jointly created by Wadsworth and CNN, each video consists of approximately 45 minutes of footage originally broadcast on CNN and specifically selected to illustrate important sociological concepts. The videos are broken into short 2- to 7-minute segments, perfect for use as lecture launchers or as illustrations of key sociological concepts. Each video includes an annotated table of contents, descriptions of the segments, and suggestions on their use within the course.

Wadsworth Sociology Video Library. Bring sociological concepts to life with videos from Wadsworth's Sociology Video Library, which includes thought-provoking offerings from Films for Humanities, as well as other excellent educational video sources. This extensive collection illustrates important sociological concepts covered in many sociology courses. Certain adoption conditions apply.

Supplements for the Student

Wadsworth's Sociology Online Resources and Writing Companion, First Edition. This valuable guide shows students how they can use Wadsworth's exclusive online resources—InfoTrac College Edition, the Opposing Viewpoints Resource Center (OVRC), and MicroCase Online—to assist them in their study of sociology and to build essential research and writing skills. Part One provides informative user guides that introduce each of these powerful research tools. Part Two contains directed exercises designed to develop research and critical-thinking proficiency for each of the core topics in sociology. Part Three provides an overview of some of the research and writing tools available online, such as InfoWrite and the OVRC Research Guide, and shows students how they can effectively integrate their research findings into class assignments.

SocCoach CD-ROM. The new, interactive SocCoach CD-ROM can be packaged for free with each new copy of the text. This tutorially driven CD-ROM is firmly grounded in sociology. It enables students to review chapter content, conduct online research, think critically about sociology statistics, watch well-known sociologists discussing important concepts, and complete book-specific quizzes, all on one, easy-to-use CD-ROM. The new Study Plan feature prompts students to take a diagnostic chapter quiz, and then generates a personalized Study Plan that shows students exactly what they need to review further. Students can then access study material for each concept, including material from the book itself, illustrative graphs, videos, and statistics that help students to better understand each concept.

Readers

***Classic Readings in Sociology,* Third Edition (edited by Eve Howard).** This series of classic articles written by key sociologists will complement any introductory sociology textbook. This reader serves as a touchstone for students, where they can read original works that teach the fundamental ideas of sociology.

***Understanding Society: An Introductory Reader,* Second Edition (edited by Margaret Andersen, University of Delaware, Kim Logio, St. Joseph's University, and Howard Taylor, Princeton University).** This reader includes articles with a variety of styles and perspectives, with a balance of the classic and contemporary. The editors selected readings that students will find accessible yet intriguing, and have maximized the instructional value of each selection by prefacing each with an introduction and following each with discussion questions. The articles center on the following five themes: classical sociological theory, contemporary research, diversity, globalization, and application of the sociological perspective.

Online Resources

Wadsworth's Virtual Society: The Wadsworth Sociology Resource Center. *www.wadsworth.com/sociology*
Here you will find a wealth of sociology resources, such as Census 2000: A Student Guide for Sociology, Breaking News in Sociology, a Guide to Researching Sociology on the Internet, Sociology in Action, and much more. Contained on the home page is the text-specific site for *Sociology: The Essentials*, Third Edition.

Shepard, *Sociology*, Ninth Edition Companion Web Site. sociology.wadsworth.com/shepard/soc9e
Access useful learning resources for each chapter of the book. Here are some of the many resources available:

- Tutorial practice quizzes that can be scored and emailed to the instructor
- Internet exercises and Web links
- Video exercises
- Periodical exercises via InfoTrac College Edition
- Flashcards of the text's glossary
- Crossword puzzles
- Essay questions
- Learning objectives
- MicroCase online data exercises
- Virtual explorations

And much more!

WebTutor™ ToolBox for WebCT or Blackboard. Preloaded with content and available free via pincode when packaged with this text, WebTutor ToolBox pairs all the content of this text's rich Book Companion Web Site with all the sophisticated course management functionality of a WebCT or Blackboard product. You can assign materials (including online quizzes) and have the results flow automatically to your gradebook. ToolBox is ready to use as soon as you log on—or, you can customize its preloaded content by uploading images and other resources, adding Web links, or creating your own practice materials. Students have access only to student resources on the Web site. Instructors can enter a pincode for access to password-protected Instructor Resources.

InfoTrac® College Edition. With each purchase of a new copy of the text comes a free 4-month pincode to InfoTrac College Edition, the online library that gives students anytime, anywhere access to reliable resources.

This fully searchable database offers twenty years' worth of full-text articles from almost five thousand diverse sources, such as academic journals, newsletters, and up-to-the-minute periodicals including *Time, Newsweek, Science, Forbes,* and *USA Today*. This incredible depth and breadth of material—available 24 hours a day from any computer with Internet access—makes conducting research so easy, your students will want to use it to enhance their work in every course. Through InfoTrac's InfoWrite, students now also have instant access to critical-thinking and paper writing tools. Both adopters and their students receive unlimited access for 4 months.

Opposing Viewpoints Resource Center (OVRC). Newly available from Wadsworth, this online center presents varying perspectives on today's most compelling issues. OVRC draws on Greenhaven Press's acclaimed Social Issues Series, as well as core reference content from other Gale and Macmillan Reference USA sources. The result is a dynamic online library of current event topics—the facts as well as the arguments of each topic's proponents and detractors. Special sections focus on critical thinking—walking students through the steps involved in critically evaluating point-counterpoint arguments—and on researching and writing papers. OVRC is also available through Wadsworth's Sociology Online Resources and Writing Companion.

Acknowledgments

Several colleagues have provided thoughtful and helpful reviews for the Ninth Edition. Many thanks to the following individuals:

Reba Chaisson
Purdue University

Michael Granata
Community College of South Nevada

Laura Gruntmeir
Redlands Community College

Justine Gueno
Xavier University of Louisiana

Amy Holzgang
Cerritos College

Elaine McDuff
Truman State University

Charles Mulford
Iowa State University

Roksana Rahman
Rutgers University

Edward Silva
El Paso Community College

La Fleur Small
University of Miami

Some of my colleagues listed above have also reviewed earlier editions. In addition to those individuals, the following colleagues have provided critiques of previous editions: G. William Anderson, Robert Anwyl, Paul J. Baker, Melvin W. Barber, Jerry Bode, Patricia Bradley, Ruth Murray Brown, Brent Bruton, Victor Burke, Bruce Bylund, David Caddell, Albert Chabot, Stephen Childs, William T. Chute, Carolie Coffey, Kenneth Colburn Jr., William F. Coston, Jerome Crane, Ray Darville, Alline DeVore, Mary Van DeWalker, Mary Donahy, Susan B. Donohue, M. Gilbert Dunn, Lois Easterday, Mark G. Eckel, Irving Elan, Ralph England, K. Peter Etzkorn, Mark Evers, Susan Farrell, Joseph Faulkner, Kevin M. Fitzpatrick, John W. Fox, Jesse Frankel, Larry Frye, Roberta Goldberg, Ramona Grimes, James W. Grimm, Rebecca Guy, Penelope J. Hanke, Thomas Harlach, Cynthia Hawkins, Kenneth E. Hinze, Carla Howery, Emmit Hunt, Gary Kiger, James A. Kitchens, Joseph Kotarba, Irving Krauss, Mark LaGory, Raymond P. LeBlanc, Jerry M. Lewis, Jieli Li, Roger Little, Richard Loper, Roy Lotz, Scott Magnuson-Martinson, Judy Maynard, R. Lee McNair, Doris Miga, Edward V. Morse, Charles Mulford, Bill Mullin, Daniel F. O'Connor, Jon Olson, Charles Osborn, Thomas R. Panko, Margaret Poloma, Carol Axtell Ray, Ellen Rosengarten, William Roy, Josephine A. Ruggiero, Steven Schada, Paul M. Sharp, James Skellinger, Robert P. Snow, Mary Steward, Ron Stewart, Robert F. Szafran, David Terry, Ralph Thomlinson, Charles M. Tolbert, David Waller, Carrol W. Waymon, Michael E. Weissbuch, Dorether M. Welch, Carol S. Wharton, Douglas L. White, Paul Wozniak, David Zaret, Wayne Zatopek.

It takes a community to produce a textbook. Foremost among our community is Sandy Crigger, a multitalented wonder posing as a staff assistant. Saying that I could not have done this edition without her will not surprise those familiar with Sandy's spirit and work ethic. Professor Craig VanSandt, now at Augustana College, once again ensured the accuracy, completeness, and timeliness of text material. Bob Jucha, my Wadsworth editor, was invaluable in guiding this revision, while giving me maximum freedom as an author. Gary Hespenheide and his staff at Hespenheide Design once again did a sterling production and design job.

1

The Sociological Perspective

© Deborah Davis/Photoedit

OUTLINE

LEARNING OBJECTIVES

- Illustrate the unique sociological perspective from both the micro and macro levels of analysis.
- Describe three uses of the sociological perspective.
- Distinguish sociology from other social sciences.
- Outline the contributions of the major pioneers of sociology.
- Summarize the development of sociology in the United States.
- Identify the three major theoretical perspectives in sociology today.
- Discuss feminist theory.
- Compare two recent developments in symbolic interactionism.
- Apply the dominant theoretical perspectives to sport.

USING THE SOCIOLOGICAL IMAGINATION

Why do people commit suicide? Answers immediately come to mind: prolonged illness, loss of a lover, public disgrace, depression, heavy financial loss. Each of these explanations assumes personal crisis as the sole motivation. Yet, suicide is affected not only by personal trauma but also by social forces. Specifically, sociologist Emile Durkheim revealed that suicide rate varies with social characteristics. Highly socially integrated people—married, females, Catholics—exhibit lower rates of suicide. More socially isolated persons—unmarried, males, Protestants—show higher suicide rates. After 100 years, research continues to support Durkheim's findings and conclusions.

Durkheim is one of the pioneers of sociology who will be profiled in this chapter. Before turning to these pioneers, however, we will discuss the unique sociological perspective that Durkheim identified and developed (Jones 1999).

The Sociological Perspective

The Importance of Perspective

Why begin your study of sociology with a discussion of perspective? Perspective matters. Babies are usually brighter and better looking to their parents than to others. Newlyweds nearly always find their spouses infinitely more attractive than do their friends.

We all interpret what is happening around us through our own perspective—our own particular point of view. It can hardly be any other way because we normally do not even realize the extent to which our beliefs, attitudes, and views of reality are determined by our perspective. Sometimes, though, when our outlook is challenged, we may be jarred into realizing how much we take it for granted (see Worldview 1.1).

Perspective is reflected in a Native American's reply to white men in eighteenth-century Virginia. The Virginia colonists had offered to "properly educate" some young Indian boys at the College of William and Mary in Williamsburg. Here is the Native American's response:

WORLDVIEW 1.1

A World Turned Upside Down

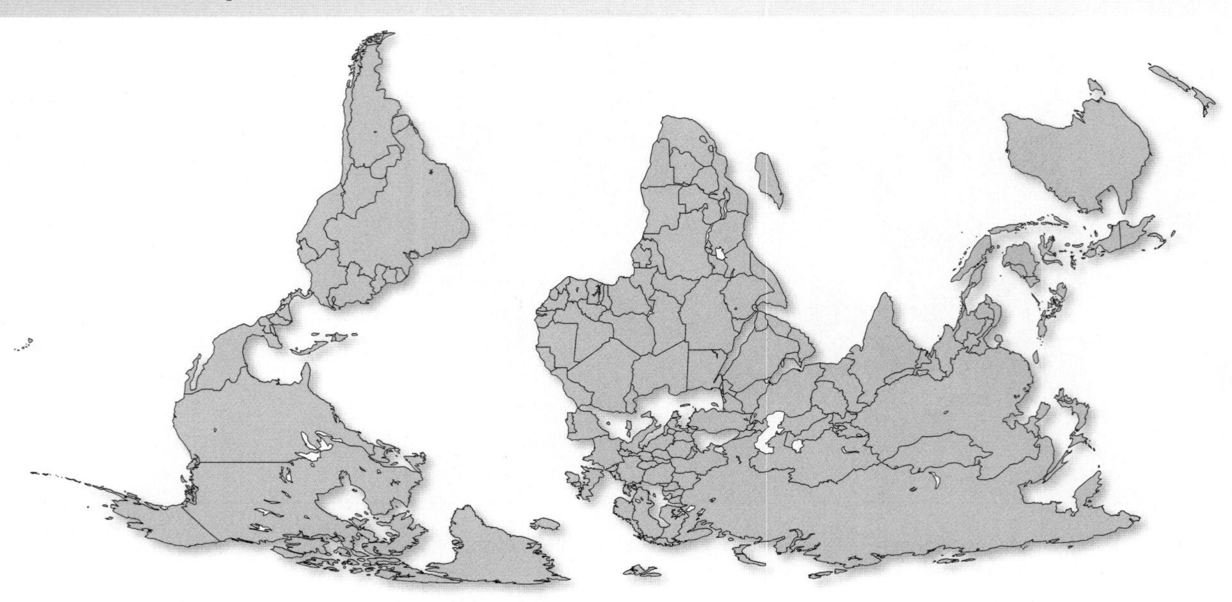

Without turning this map upside down, locate the United States. Do you find it more difficult than usual? Do you find this view of the world disorienting? Because you are so used to the conventional representation of the Earth, you may reject this worldview. So it is with any perspective.

In this book, you will be asked to abandon the typical American psychological perspective in favor of the sociological perspective.

1. How would you feel if this map were universally substituted for the one you know? Describe your feelings, and explain why you feel this way.
2. What does your reaction to this map tell you about the power of the perspective you bring to a situation?

We know that you highly esteem the kind of learning taught in . . . [your] colleges. . . . But you, who are wise, must know that different nations have different conceptions of things; and you will not therefore take it amiss, if our ideas of this kind of education happen not to be the same with yours. We have had some experience of it; several of our young people were formerly brought up at the colleges of the northern provinces; they were instructed in all your sciences; but, when they came back to us, they were bad runners, ignorant of every means of living in the woods, unable to bear either cold or hunger, knew neither how to build a cabin, take a deer, nor kill an enemy, spoke our language imperfectly, were therefore neither fit for hunters, warriors, nor councellors; they were totally good for nothing.

We are however not the less obligated by your kind offer, though we decline accepting it; and, to show our grateful sense of it, if the gentlemen of Virginia will send us a dozen of their sons, we will take care of their education, instruct them in all we know, and make men of them. (Anonymous)

To the colonists' surprise, the benefits of a William and Mary education were not desired by the tribal elders. In fact, from the Native American perspective, the gentlemen of Virginia would be much better off to leave William and Mary and pursue an education in the tribe. Because of their own perspective, no William and Mary students transferred.

© Deborah Davis/Photoedit

The decision to attend college is not merely a choice made by an isolated individual. Social relationships, with teachers, parents, peers, and others, influence a person's decision to become a college student. Sociologists study these patterned social relationships referred to as social structure.

Sociology has a unique perspective. The first segment of this chapter is devoted to this perspective—a focus on groups, not individuals, as an object of investigation.

Defining Sociology

At this point, how would you define sociology? As a novice to the field, you may at first view sociology as the study of human groups. This inclination has the advantage of tapping ideas you already possess. As you progress, however, you will acquire a more precise understanding of **sociology** as the scientific study of social structure.

In the meantime, your understanding of the concept of social structure will be understandably vague. For now it is sufficient for you to become familiar with two key aspects of the sociological perspective: (1) the interaction between social structure and the individual and (2) the idea of two levels of analysis.

Social Structure and the Individual: A Two-Way Street

What is unique about the perspective of sociology? The starting point for sociology is the predictability and recurrence of social behavior. Sociologists do not concentrate on the idiosyncratic behavior of individuals, but on the patterns of behavior individuals share with others in their group or society. Sociologists are interested in the patterns of social relationships referred to as **social structures**. The average person tends to explain human behavior in individualistic or personal terms: A young man goes to war to prove his patriotism; a woman divorces her husband to develop her potential; a college student commits suicide to escape depression. Sociologists attempt to explain these same events without relying on the personal motivations of individuals. They look for social rather than personal explanations when they examine war, divorce, or suicide: Young men go to war because they have been taught by their society to be patriotic; more women divorce because of the social trend toward sexual equality; college students commit suicide because of pervasive societal expectations of academic performance. Sociologists do not speak of a young man, a married woman, or a college student. They concentrate on categories of people—young men, married women, college students.

Can you think of an example of "patterned" social behavior? As you know, college students in a classroom do not all behave exactly alike. Some attempt to write down everything their professors say, some just listen to the lecture, some use tape recorders, and some tune in and out. Yet, if you visit almost any college or university, you will find patterned relationships. Professors lecture, students remain in their seats; professors give examinations, students take them. Although the individual characteristics of students and professors may vary from school to school, students and professors relate in similar patterned ways. It is the *recurrent* patterned interaction of people and the social structures created by such interaction that capture the attention of sociologists.

How does group behavior differ from individual behavior? Sociologists assume that social relationships are not solely determined by the particular characteristics of the individuals involved. Emile Durkheim (1966; originally published in 1895), a pioneering nineteenth-century sociologist, argued that we do not attempt to explain bronze in terms of its component parts (lead, copper, and tin). Instead, we consider bronze an alloy, a unique metal produced by the synthesis of several distinct metals. Even the consistency of bronze is not predictable from its components: bronze is hard, whereas lead, copper, and tin are soft and malleable. Durkheim reasoned that if a combination of certain metals produces a unique metal, some similar process might happen in groups of people. Indeed, people's behavior within a group setting cannot be predicted from the characteristics of individual group members. Something new is created when individuals come together as a collective. For example, when the Denver Broncos won the 1999 Super Bowl, some exuberant and rowdy Bronco fans tipped over trash cans and cars, spilled newspaper boxes, broke windows in buildings, tore down street signs, and started bonfires—behavior they would not have exhibited as lone individuals. An interesting contrast to the Denver celebration occurred when Brazil, an ethnically diverse and often conflicted country, won the men's soccer World Cup in 2002. Nearly fifty thousand Brazilians peacefully and joyfully cheered in São Paulo as their national team beat Germany in the championship. Even if only temporarily, many Brazilians put aside their differences to share in the national pride generated by their soccer team. You may recall other well-publicized examples from 1999. Some normally law-abiding participants at the 1999 Woodstock music festival engaged in arson, looting, property destruction, and rape. When the World Trade Organization met in Seattle in late 1999, many protestors peacefully demonstrated their opposition to its global economic effects, while others broke shop windows and looted stores. On the positive side, the 1999 collapse of the wooden structure built for the pregame bonfire for the annual Texas A&M (Aggies) and University of Texas (Longhorns) football game, killing twelve A&M students, led to unprecedented behavior. The bitter rivalry between the teams was temporarily laid aside when at halftime the Longhorn band played "Amazing Grace" and taps and then saluted the deceased Aggies by removing their cowboy hats. At a joint Aggie-Longhorn candlelight vigil two nights before the game, the A&M student body president said that the relationship between the two schools will be changed forever.

Group members generally tend to act like each other even when their personal preferences are not the same as their group's. Some fraternity and sorority members, for example, drink alcoholic beverages even though, as individuals, they would prefer not to. Similarly, because of peer pressure, members of campus religious groups may not drink alcohol despite their preference for it. A study of helping behavior illustrates the power of social structure. Social psychologist Robert Levine (2003) found the best predictor of people's willingness to aid others in trouble to be population density. Helping behavior declines as population density increases. However, when individuals from large cities are in less densely populated places, they are as helpful as the residents. Due to the high degree of conformity within a group, similarities in social behavior exist despite extensive differences among individual members.

How is conformity related to group behavior? We live in groups ranging in size from a family to an entire society, and they all encourage conformity—often, conformity promoted by social forces that individuals do not create and cannot control. Because of this accommodation, people who belong to similar groups tend to think, feel, and behave in similar ways. American, Russian, and Chinese citizens, for instance, have distinctive eating habits, types of dress, religious beliefs, and attitudes toward family life. Groups of teenagers tend to listen to the same music, dress alike, and follow similar dating customs.

Why do people conform? Conformity within a group occurs in part because most of its members believe that their group's ways of thinking, feeling, and behaving are the best; they have been successfully taught to value their group's ways. And as already noted, some group members tend to comply even against their personal inclinations. Whether because members value their group's ways or because they yield to social pressures of the moment, the behavior of a group cannot be predicted simply from knowledge about the individual members. A group is not equal to the sum of its parts any more than is bronze.

Why is the existence of conformity important to sociology? Because conformity within societies exists, sociologists can attempt to understand, explain, and predict the often invisible social processes that permit successive generations to live predictable and orderly lives without each generation creating its own new guidelines for social living. Because a particular generation is spared this trouble, its members usually fail to ask why things are the way they are, or why things are changing. Sociologists, in contrast, constantly wrestle with these questions.

But don't people, in turn, affect society? Yes. If our discussion of sociology to this point causes you to believe that sociology is interested only in the effects of social structures on individuals, it has been misleading. It is necessary, however, to begin with this emphasis because without some considerable degree of conformity, there can be no group life. Still, this necessary nonindividualistic emphasis should not obscure the effect that individuals can have on social structures. Individuals are active, thinking beings who do not always follow social scripts. The interplay between individuals and their social structures is a two-way street: People are affected by social structures, and people change their social structures. Consequently, you should not conclude that all human behavior in groups is determined by preexisting social structures. For example, historian Joyce Appleby (2000) documents how the first generation born after the American Revolution fashioned a national culture based on an interpretation of their parents' creation of a new nation.

In summary, the sociological perspective focuses on the group, examines patterns of behavior, isolates patterns of conformity, and recognizes the effects of people on society. Table 1.1 illustrates these unique aspects of the sociological perspective using suicide rates among Catholics and Protestants.

TABLE 1.1

WHAT SOCIOLOGISTS SEE AND HOW THEY STUDY SUICIDE: Comparing Protestants and Catholics

SOCIOLOGISTS FOCUS . . .	SOCIOLOGISTS MIGHT INVESTIGATE . . .
on the group, not the individual.	distinct characteristics of both Protestants and Catholics that affect their differential suicide rates.
on patterns of social behavior.	data to determine if Protestants and Catholics actually do consistently commit suicide at different rates.
on social forces that encourage conformity.	the ways that greater social integration among Catholics promotes a lower rate of suicide compared to more socially isolated Protestants.
on the effects people have on social structure.	means that Protestants may use to lower their suicide rate, such as the promotion of social involvement through church activities.

Levels of Analysis: Microsociology and Macrosociology

An important distinction is implied in our discussion of the sociological perspective. A careful reading reveals the interest of sociology in both the interaction of people "within" social structures (*microsociology*) and the "intersection" of social structures per se (*macrosociology*).

Macrosociology and microsociology should be thought of as levels of analysis, because they determine which aspects of social structure a sociologist wishes to analyze. *Macro* means large; *micro* means small. Macro research focuses on groups as a whole. Micro analysis investigates the relationships *within* groups. Macro views from a distance, micro from close range (Helle and Eisentadt 1985a, 1985b; Scheff 1990; Hawley 1992).

Suppose for the moment that social structures are directly observable as tangible objects and that we are cruising at 20,000 feet at the controls of *Sociologist One*. On a clear day at this altitude we would be able to observe definite social structures, just as we can observe the lay of the land from an airplane. At 20,000 feet we will see no movement in the form of people interacting. We are at the macro level. Suppose we wish to get a closer look and start descending. As we descend, we will begin to see people interacting. When we focus on these relationships, we are at the micro level. Formal definitions of microsociology and macrosociology will make more sense against this general background.

Microsociology is concerned with the study of people as they interact in daily life. Consider the practice of knife fighting among street gangs. At the micro level, a sociologist attempts to explain participation in gang knife fighting based on the social relationships involved. For example, gang leaders, to validate their right to leadership, may feel that they must either fight members of their own gang who challenge their position or fight leaders of other gangs.

Macrosociology focuses on groups without regard to the interaction of the people involved. Some sociologists examine entire societies (Tilly 1978, 1986; Wallerstein 1979, 1984; Skocpol 1985; Lenski 1988; Nolan and Lenski 1999). We will use the term *macrosociology* in referring to the study of societies as a whole as well as to the relationships between social structures within societies. For example, sociologists might study the patterned relationships between the defense industry and the federal government, or the effect of the economy on the stability of the family.

FEEDBACK

1. Sociology is the scientific study of _________.
2. _________ explanations of group behavior are inadequate because human activities are influenced by social forces that individuals have not created and cannot control.
3. Microsociology focuses on relationships between social structures without reference to the interaction of the people involved. T or F?

Answers: 1. social structure 2. Individualistic 3. F

The macro and micro levels of analysis are complementary; their combined use tells us more about social behavior than either one alone (Alexander et al. 1987; Sawyer 2001). To understand gang warfare, a microsociologist would want to know about the social relationships involved—the relationships between gang leaders and followers or between gang members and police on the beat. To supplement this understanding of gangs obtained at the micro level, the macro level can be used to examine the larger social structures of which gangs are a part. A macrosociologist might look for the aspects of a society or social structure that produce the poverty promoting delinquency in the first place—such as lack of education and joblessness.

Uses of the Sociological Perspective

Why study sociology? The sociological perspective offers you three personal benefits. First, it enables you to develop the *sociological imagination*. Second, sociological theory and research can be applied to important public issues. Third, the study of sociology can sharpen skills useful in many occupations.

The Sociological Imagination

How does the self interplay with society? In his famous essay "On Liberty," nineteenth-century philosopher John Stuart Mill wrote:

> *It will probably be conceded that it is desirable [that] people should exercise their understandings, and that an intelligent following of custom, or even occasionally an intelligent deviation from custom, is better than blind and simple mechanical adhesion to it.* (Mill 1993:68–69; originally published in 1859)

Knowing how social forces affect our lives can help prevent us from being prisoners of those forces. C. Wright Mills called this personal use of sociology the **sociological imagination**—the set of mind that enables individuals to see the relationship between events in their personal lives and events in their society. Events affecting us as individuals, Mills pointed out, are closely related to the ebb and flow of society (Mills and Etzioni 1999). Decisions, both minor and momentous, are not isolated, individual matters. Historically, for example, American society has shown a bias against childless marriages and only children. Couples without children have been considered selfish, and only children have often been labeled "spoiled" (Benokraitis 2001). These values date back to a time when there was a societal need for large families because of the high infant mortality rate and the need for labor on family farms. Only now, as the need for large families is disappearing, are we reading of the benefits of one-child families—to the child, to the family, and to society. The sociological imagination enables us to understand the effects of such social forces on our lives. With this understanding, we are in a stronger position to make autonomous decisions rather than merely conform (Game and Metcalfe 1996; Peck and Hollingsworth 1996; Erikson 1997).

This broadened social awareness permits us to read the newspaper with a more complete understanding of the implications of social events. Instead of interpreting an editorial opposing welfare as merely a selfish expression, we might see the letter as a reflection of the importance Americans place on independence and self-help (A. M. Lee 1990; Straus 2002).

Ashley Montagu has coined the term *psychosclerosis* to describe a mental condition that deprives people of the ability to accept new ideas. What are the symptoms of psychosclerosis, or "hardening of the mind"?

> *Well, a certain lack of ability to see 360 degrees, to see all the way behind, sideways, as well as forward, to pick up the small and subliminal signals. To be wedded to fixed, stereotyped ideas that this is the right way and the only way.* (Montagu 1977:50)

Because the sociological imagination expands our horizons, it may be considered an antidote for psychosclerosis.

How will the debunking theme stimulate your sociological imagination? Despite the availability of accurate information and explanations, people tend to cling to myths and false ideas about social life—impressions that are passed from generation to generation. Plato's "Allegory of the Cave" comes to mind. In *The Republic*, Plato describes a cave in which people have been chained, from childhood, so they cannot move. Forced to look only straight ahead, these prisoners cannot see behind them the blazing fire and a raised walkway in front of the fire. Men walk along the wall carrying all types of objects. The shadows on the wall in front of the prisoners, made by the walking men and the fire, are the only things the prisoners have ever seen. Their interpretations of these shadows constitute reality to them. Sociology attempts to replace common misconceptions about social life (shadows on the cave wall) with accurate information and explanations (Ruggiero 1996).

Because the task of sociology is to reveal the nature of human social behavior, it often leads us to question what we usually take for granted. What people take to be unassailable truth may, under scientific examination, be proven false. The sociological imagination, then, replaces common misconceptions about social life with accurate information and explanations.

How is the sociological imagination intellectually liberating? Like all liberal arts courses—anthropology, history, literature, and philosophy—sociology encourages intellectual liberation (Bierstedt 1963; Brouilette 1985). You can, for example, learn that the sacrificial murder of infants has been a normal and legitimate aspect of some societies and that those who participate in this practice are not evil and despicable. Among the Winnebago Indians of Wisconsin, for example, it was socially acceptable for parents to kill children who became too great a liability (Radin 1953). Among the Toda of southern India, female infants were frequently killed at birth; and if twins of different sexes were born, the female was always killed (Murdock 1935). The Winnebago and Toda, like all humans, had been taught the ways of their group. Through the sociological imagination, you can come to understand why a society under extreme population pressure, living on the margin of subsistence, might adopt infanticide.

Sociology provides a window to the social world that lies outside ourselves. It allows us to see the many social forces that shape our own lives. Sociology thus complements rather than replaces other ways of viewing human behavior. The more we seek to apply a range of different perspectives in attempting to interpret social forces and their meanings, the greater will be our understanding of our own behavior as well as the behavior of others. With such understanding comes the potential for greater personal freedom from social pressures. Consider divorce, for example. Social research reveals many factors in modern society that promote marital failure. Divorce is more likely to occur among couples who share particular characteristics: low levels of education, early marriages, premarital pregnancies, or living arrangements with parents. As predictive as these characteristics are, they cannot be used to force couples not to marry. However, awareness of the inevitable pressures of such marriages may convince couples under these circumstances either not to marry, or at least to anticipate and prepare more effectively for such problems. The successful application of the sociological imagination enables us to share Somerset Maugham's insight that "tradition is a guide and not a jailer."

For this couple, their marriage and its particular ceremonial form are the result of their decision as individuals. To sociologists, however, the decision to marry and to do so in a particular way is socially channeled.

Applied Sociology

Do sociologists have a moral responsibility to speak against and attempt to change aspects of social life they believe to be wrong? From the time it first appeared on the American academic scene in the late 1800s, sociology has steadily attempted to move from its origins as a social problem-solving discipline to a nonsocially involved science. During the intervening years, disagreement has periodically surfaced on the compatibility of these two viewpoints (A. M. Lee 1978; Weinstein 2000; Hamilton and Thompson 2002). Those arguing against the involvement of social scientists in the eradication of social ills view science as "value neutral"; that is, scientific research is supposed to be directed at discovering what actually exists, with no room for personal value judgments as to what *ought* to exist. The idea of value neutrality has dominated sociological thought for a long time. And although the idea of science as a value-neutral enterprise remains strong among sociologists, those favoring the interjection of standards of good and bad into scientific activity are on the increase (Bickman and Rog 1997). This issue has gained considerable prominence in the form of **humanist sociology**, which places human needs and goals at the center (A. M. Lee 1978; Scimecca 1987; Giddens 1997), and **liberation sociology**, whose objective is to replace human oppression with greater democracy and social justice (Feagin and Vera 2001).

Humanist sociologists apply sociological theories and research findings to problems such as those that occupied Nobel Peace Prize winner Jane Addams—the urban poor in general and the plight of poor working women in particular.

Does sociological research influence public policies and programs? Yes. Social scientists, for instance, contributed to the 1954 Supreme Court decision outlawing separate but equal schools for African Americans and whites. This is one of the earliest examples of the U.S. court system accepting social science research as a basis for legal decisions. Subsequent research has explored a variety of related topics, including the effects of school quality on student performance, the impact of social environment on measured IQ, the influence of school desegregation on the performance of African American and white students, and the effects of busing in desegregating schools. Referring to the application of sociology to public policy, William Julius Wilson, in his 1990 presidential address to the American Sociological Society, offered these closing words:

> *With the reemergence of poverty on the nation's public agenda, researchers have to recognize that they have the political and social responsibility as social scientists to ensure that their findings and theories are interpreted accurately by those in the public who use their ideas. They also have the intellectual responsibility to do more than simply react to trends or currents of public thinking. They have to provide intellectual leadership with arguments based on systematic research and theoretical analyses that confront ideologically driven and short-sighted public views.* (W. J. Wilson 1991:12)

Some sociologists advocate intervention beyond policy-related research. Following a long-neglected lead (Wirth 1931), clinical sociologists are now using sociological theories, principles, and research to promote social change. Clinical sociologists provide help for individuals or serve as agents for change in organizations, communities, and even entire societies (Black 1984; Erickson and Simon 1998;R. H. Hall 2001; Du Bois and Wright 2000; Koppel 2002). Clinical sociologists may work as marriage and family therapists, design intervention programs to reduce juvenile delinquency, or redesign the social environment of cancer patients.

Sociology and Occupational Skills

How will the study of sociology contribute to the development of work skills? Most employers are interested in four types of skills: the ability to work well with others, the ability to write and speak fluently, the ability to solve problems, and the ability to analyze information (see Snapshot 1.1). Because computers have revolution-

SNAPSHOT OF AMERICA 1.1

Literacy Rates

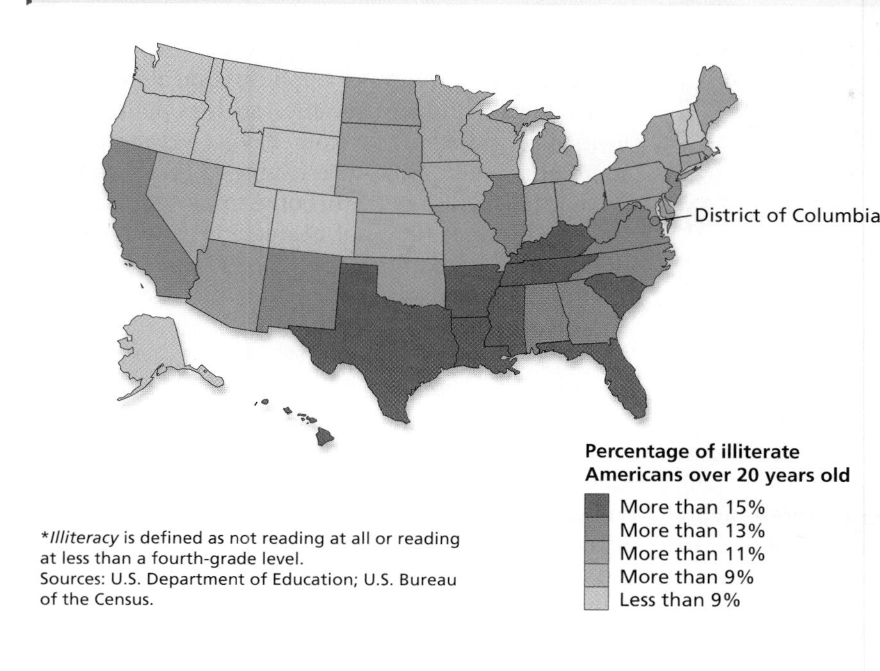

**Illiteracy* is defined as not reading at all or reading at less than a fourth-grade level.
Sources: U.S. Department of Education; U.S. Bureau of the Census.

Employers look for functional skills such as the ability to write and speak well, capacity for solving problems, and competency in information analysis. One of the basic requirements for these skills is literacy, defined as the ability to read at a fourth-grade level. This map shows, by state, the percentage of Americans over twenty years old who are illiterate.

1. List the states with highest and lowest literacy rates.
2. Where does your state stand on literacy?
3. How does the United States compare with other Western nations in relation to literacy?

ized the office, information analysis skills are becoming much more important to managers in all types of organizations. In addition, the increasing complexity of work demands greater critical analysis and problem-solving skills. The levels of each of these skills can be improved through the broad liberal arts foundation of sociology (Billson and Huber 1993; D. C. Miller 1994; American Sociological Association 1997; Stephens 2002).

What about more specific preparation for employment? In addition to general skills, specific sociology subfields offer preparation for fairly specialized jobs. Consider these examples:

- Training in race relations is an asset for working in human resources (personnel) departments, hospitals, or day-care centers.
- Background in urban sociology can be put to good use in urban planning, law enforcement, and social work.
- Courses focusing on gender and race serve as valuable background for work in community planning, arbitration, and sexual harassment cases.
- Training in criminology is sought by agencies dealing with criminal justice, probation, and juvenile delinquency.
- Courses in social psychology are valuable for sales, marketing, and advertising, as well as for counseling.

These jobs only scratch the surface; students of sociology are prepared to pursue many other careers. Consider this selected list: manager, executive, college placement officer, community planner, employment counselor, foreign service worker, environmental

FEEDBACK

1. The __________ is the set of mind that enables individuals to see the relationship between events in their personal lives and events in their society.
2. Which of the following is *not* one of the uses of the sociological imagination?
   a. seeing the interplay of self and society
   b. creating new aspects of culture not thought of by others
   c. questioning aspects of social life most people take for granted
   d. understanding the social forces that shape daily life
3. __________ sociology places human needs and goals at its center.

Answers: 1. sociological imagination 2. b 3. Humanist

© David Longstreath/AP/Wide World Photos

The modern world has entered the computer age. Job requirements faced by this woman at her computer, however, go beyond technical skills. She will be required to exercise higher-order thinking skills such as handling complex ideas, recognizing pattern in apparent disorder, and developing alternative solutions to problems.

specialist, guidance counselor, health planner, journalist, labor relations specialist, marketing researcher, public relations supervisor, research analyst, writer, and editor (American Sociological Association, 1997).

The Social Sciences

Your initial understanding of sociology includes its unique focus and the uses of the sociological perspective. You can further clarify your grasp of sociology by distinguishing it from the other social sciences.

It was not until the nineteenth century that Auguste Comte coined the term *sociology*. Comte's ideas about sociology were embedded in his ideas about psychology, economics, and political theory. He viewed sociology as the one comprehensive social science.

Gradually, scholars developed the various social sciences to the point that they could be legitimately distinguished. Nevertheless, the social sciences have enough in common to overlap with one another at many points.

Anthropology

Because anthropology investigates culture, it is the social science most closely related to sociology. Historically, anthropologists have concentrated on the study of "primitive" or nonliterate societies, whereas sociologists have focused on the modern, complex, industrial societies of Western civilization. Because anthropologists are interested primarily in small nonindustrial societies, they tend to study entire societies. Sociologists, on the other hand, investigate more limited aspects of modern societies, such as political revolutions or the status of women vis-à-vis men. Today some anthropologists have extended their studies to include research in modern, complex societies, examining, for instance, cultural characteristics of neighborhoods and communities.

Psychology

Psychology focuses on the development and function of mental-emotional processes in human beings. Whereas sociology and anthropology concentrate on human groups, psychology takes as its domain the individual. Sociology and psychology have some common interest in an aspect of behavioral science called social psychology, a field that focuses on the interactions between individuals and groups. A social psychologist, for example, is interested in the ways in which groups promote conformity among their individual members.

Economics

Economics is the study of the production, distribution, and consumption of goods and services. It deals with such matters as the impact of interest rates on the flow of money and the effects of taxation on consumption. Economics uses highly developed mathematical models to predict changes in economic indicators. Although these economic models are not always correct, they are sophisticated, and they give economists greater predictive power than that of most other social scientists. Sociology and economics merge in an area of study called economic sociology, which concentrates on the interrelationships between the economic and noneconomic aspects of social life. An economic sociologist, for example, would be interested in the relationship between the extent of industrialization in a particular society and the degree of workers' involvement in union activity.

Political Science

Political science studies the organization, administration, history, and theory of government. Political scientists are concerned, for example, with voting patterns and political party participation. In recent years, political science has moved beyond its earlier preoccupation with the workings of government toward the study of political behavior in its broadest sense. This trend has inevitably bound closer together the interests of political scientists and political sociologists. Both political scientists and political sociologists share an interest in the social interaction that occurs within the political institution as well as the interaction between the political institution and other institutions, including economic and educational institutions. Both disciplines conduct research on the location and use of power, the process of political socialization, the functioning of special interest groups, and the workings of political protest.

History

History examines past events in human societies. As a field of study, history currently is sharply divided into two camps (Landes and Tilly 1991). Humanistic historians believe that history cannot use the methods of the social sciences. They view any attempts at scientific generalization as the simplification of complex, unique human experiences rooted in specific times and places. Description, they argue, is the only legitimate function of the historian. Social scientific historians, in contrast, assume the existence of uniform, recurring patterns of human behavior that transcend time and place. The role of the historian, they contend, is to identify and verify these patterns. In recent years, the relationship between sociology and history has become more mutually beneficial. Sociology is currently more open to the use of historians' work in their attempts to document recurring patterns and provide explanations of historical reality; and more historians are now committed to the collection and analysis of primary sources with an eye to establishing recurrent and universal historical patterns.

Social Sciences and the Family

Here, by way of illustration, are distinctive social science approaches to the specific study of the family:

- *Sociologists*. Some sociologists investigating the family focus on the social causes of divorce. Factors thought to contribute to higher divorce rates in American society include early marriage, poor mate selection, economic prosperity, increased participation of women in the labor force, and the weakening of the stigma formerly associated with divorce (see Chapter 12, "Family").
- *Anthropologists*. An interest of some anthropological researchers is whether the nuclear family of married biological parents and their legitimate offspring exists universally in human societies. This topic can be studied by examining documented evidence on as many societies as possible and by attempting to find one or more societies where the nuclear family is not the norm (Leach 1982).
- *Psychologists*. Some psychologists are interested in studying familial retardation—cases of mental retardation that tend to appear in poor families with no known organic basis. Familial retardation is traced largely to an impoverished environment, including inferior nutrition and medical care and the relative absence of early emotional support and intellectual stimulation (Coon 2002).
- *Economists*. Economists use the family as the unit of analysis for the distribution of income. For example, currently, the bottom 60 percent of American families receive less than one-third of the nation's total income (Baumol and Blinder 2001).
- *Political scientists*. The relationship between voting and the family is a topic of research for political scientists. Political scientists have found, for example, that although voting on election day is done by private ballot, voting is primarily a group phenomenon. The family is the most important influence in forming political party identification and determining the voting behavior of its members. Political attitudes and behaviors are shaped, often unintentionally, by family members who also exist in similar economic, religious, social class, and geographical environments (Patterson 2002).
- *Historians*. The relationship between slavery and the family has been the subject of research by historians. Using slavery in the United States as a case study, a controversial study by historian Henry Gutman (1983) presents evidence from twenty-one urban and rural communities in the South between 1865 and 1880 to challenge the long-standing assumption that slavery destroys the family structure. For example,

FEEDBACK

Match the following fields of study and aspects of family life.

____ a. anthropology	(1) distribution of income and the family
____ b. sociology	(2) effects of slavery on family stability
____ c. history	(3) relationship between voting and the family
____ d. political science	(4) effects of early marriage on divorce
____ e. economics	(5) link between early childhood emotional support and familial retardation
____ f. psychology	(6) universality of the nuclear family

Answers: a. (6) b. (4) c. (2) d. (3) e. (1) f. (5)

Gutman concluded that most African American households during the slavery era had a father and mother present. In short, African American slave families tended to be stable and intact.

Founders of Sociology

European Origins

Sociology is a relatively new science, emerging as a distinct area of study in the late nineteenth century (Ray, 1999; Turner, 2001; J. H. Turner, Beeghley, and Powers, 2002). In fact, it is only during the past half century that sociology has taken its place as a legitimate social science. The origins of sociology, however, can be seen in the work of eighteenth-century philosopher Adam Smith. A man of the Scottish Enlightenment, Smith believed that humans possess a natural affection for others. This led him to seek hidden forces of social harmony to match unseen physical forces like gravity and magnetism. In Smith's desire to do for the understanding of social life what Newton had done for the physical world lay the seeds of modern social science (Wood 2002).

Who was Adam Smith? Upon first meeting Scotsman Adam Smith (1723–1790), a female French novelist described him as harsh in speech, big in the teeth, and "ugly as the devil." Realizing that outward beauty is only skin deep, she later wrote a friend: "I wish that the devil would carry off all of our men of letters, all of our philosophers, and bring Mr. Smith to me" (Rae 1965:211–212). Famous for his absentmindedness, Smith was revered by other Enlightenment scholars for his intellectual gifts and practical nature. Smith began as a moral philosopher at the University of Glasgow, spent years as a tutor to the stepson of one of the men whose tax policies provoked the American Revolution, and during his last years held the position of Commissioner of Customs for Scotland (Muller 1993).

The Smith influence is most often associated with economics. This is small wonder; Smith's most famous book, *The Wealth of Nations* (Smith and Cannan 2000; originally published in 1776), is considered foundational to capitalism. Even sociologists themselves, though, often fail to place him among the founders of their field, despite the reflection of his ideas by the founders of sociology in the nineteenth century.

What were Smith's contributions to sociology? Smith made two major contributions to the development of sociology. His first achievement was to attribute much of human behavior to the influence of society. People learn honesty, for example, through participation in the family, church, community, and government. The roots of the human personality lie in our association with others.

In addition to emphasizing the sociological perspective, Smith laid the foundation for a sociological theory known as *symbolic interactionism* (discussed later in this chapter). He wanted to explain the development of the ability to follow society's rules. This ability, reasoned Smith, develops from the human need for the approval of others (Herman 2001). The concepts Smith used to express this idea in his second book, *The Theory of Moral Sentiments* (1976; originally published in 1759), inspired later sociologists to explore the relationship between human personality and society.

Smith thought like a sociologist and used a sociological perspective in his moral philosophy and economic analysis. The term *sociology*, however, had not yet been coined, and the field had not been established. Sociology was not formally introduced until the nineteenth century, some forty years after the death of Adam Smith.

What was occurring in the nineteenth century that sparked the emergence of sociology? Nineteenth-century Europe was torn by social change and controversy. The old social order of the Middle Ages—based on social position, land ownership, church, kinship, community, and autocratic political leadership—was crumbling under the social and economic influences of the Industrial and French Revolutions. The move from farm to factory resulted in the loss of a stabilizing community. This social disharmony led Auguste Comte, Harriet Martineau, and other intellectuals to grapple with the restoration of community, social order, and predictability (Nisbet 1966; Mazlish 1993). So in an important sense, the emergence of sociology as a distinct area of study was a conservative reaction to the social chaos of the nineteenth century. The central ideas of the major pioneers of sociology, all of whom were European, will shed further light on the origins of the field (J. A. Hughes, Martin, and Sharrock 1995; Camic 1998).

Who was Auguste Comte? Auguste Comte (1798–1857) was a Frenchman born to a government bureaucrat. A small child, often ill, Comte proved early on to be an excellent student, although he had difficulty balancing his genuine interest in school and his rebellious and stubborn nature. In fact, his protest against the examination procedures at the Ecole Polytechnique (of MIT stature) led to his dismissal. A year later, Comte became secretary to the famous philosopher Henri Saint-Simon. Most of Comte's significant ideas were reflected in the writings he and Saint-Simon did together. Differences between Comte and his mentor slowly eroded their relationship, which finally ended in an argument over whose name was to appear on a publication.

What were Comte's major ideas? Comte is recognized as the father of sociology. The improvement of society was his main concern. If societies were to advance, Comte believed, social behavior had to be studied scientifically. Because no science of society existed, Comte attempted to create one himself and coined the term *sociology* to describe this science. Sociology, he asserted, should rely on **positivism**—the use of scientific observation and experimentation in the study of social behavior. When Comte wrote that sociology should rely on positivism, he meant that sociology should be a science based on knowledge of which we can be "positive," or sure. He also distinguished between **social statics**, the study of social stability and order, and **social dynamics**, the study of social change. This distinction between social stability and social change remains at the center of modern sociology.

As a result of his bitter breakup with Saint-Simon, Comte was almost prevented from writing his masterwork, *Positive Philosophy*. (It was in this work that Comte elucidated his principle of "cerebral hygiene." To prevent others from polluting his thoughts, Comte stopped reading.) After parting ways with Saint-Simon, Comte never found another well-paid position. He was forced to rely on intermittent activities—tutoring and testing in mathematics, for example. It was only after a few of his followers invited him to prepare some private lectures that Comte presented his ideas to the world.

Comte died before very many beyond his private circle came to appreciate his work. His belief that sociology could apply scientific procedures and promote social progress was later recognized and widely adopted by European scholars. Many nineteenth-century scholars came to embrace Comte's ideas; some did not.

Who was Harriet Martineau? Harriet Martineau (1802–1876) was an Englishwoman born into a solidly middle-class home. Her writing career, which included fiction as well as sociology, began in 1825 after the Martineau family textile mill was lost to a business depression (Webb 1960; Fletcher 1974; Pichanik 1980). Without family income, and with few immediate prospects for marriage following a broken engagement, Martineau was forced to seek a dependable source of income to support herself. She became a popular writer of celebrity stature, whose work outsold that of Charles Dickens.

What did Martineau contribute to the early development of sociology? Harriet Martineau, an important figure in the founding of sociology, is best known for her translation of Comte's *Positive Philosophy*. Done with Comte's approval, Martineau's translation remains the most readable. Despite being deaf, she also made original contributions in the areas of research methods, political economy, and feminist theory (Lengermann and Niebrugge-Brantley 1998; Hill and Hoecker-Drysdale 2001; Turner, Beeghley, and Powers 2002).

Harriet Martineau (1802–1876)

Martineau's book *How to Observe Manners and Morals* (1838) was the first methodology book in sociology. In it she emphasized several crucial research principles: the use of a theoretical framework to guide social observation; the development of predetermined questions in gathering information; objectivity; and representative sampling (Lipset 1962).

Martineau followed these research methods while conducting a comparative study of European and American society. Although not as well known as Alexis de Tocqueville's *Democracy in America*, which was published at about the same time (2000; originally published in 1835), Martineau's *Society in America* (1837) remains her most widely read book today. Martineau compared America favorably with England, but believed that America had not lived up to its promise of democracy and freedom for all its people. Slavery and the domination of women, she wrote, revealed a significant gap between American ideals and social practices (Terry 1983).

In *Society in America*, Martineau (1837) established herself as a pioneering feminist theorist who, like other early feminists, saw a link between slavery and the oppression of women. Consequently, she was a strong and outspoken supporter for the emancipation of both women and slaves.

Through her penetrating analysis of the subordination of American women, Martineau stands as an important envoy of contemporary feminist theory. Doors to the economic institution, she observed, were closed to women. Martineau thus identified economic dependency as the linchpin of female subordination. The oppression of women, in short, was due to sociological forces; subordination of women was part of the structure of society.

Who was Herbert Spencer? Herbert Spencer (1820–1903), the sole survivor of nine children, was born to an English schoolteacher. Due to continual ill health, Spencer was taught exclusively by his father and uncle, mostly in mathematics and the natural sciences. Because of his poor background in Latin, Greek, English, and history, Spencer did not feel qualified to enter Cambridge University, his father's alma mater. His subsequent career became a mixture of engineering, drafting, inventing, journalism, and writing.

How did Spencer view society? To explain social stability, Spencer offered an analogy based on biology. Like a human body, a society is composed of interrelated parts working together to promote its well-being and survival. People have brains, stomachs, nervous systems, limbs. Societies have economies, religions, governments, families. Just as the eyes and the heart make essential contributions to the functioning of the human body, legal and educational institutions are crucial for a society's functioning.

Why did Spencer oppose social reform? Spencer introduced a theory of social change called social Darwinism, based on Charles Darwin's theory of evolution. Spencer believed that evolutionary social change led to progress, provided that people did not interfere. If left alone, natural social selection would ensure the survival of the fittest. On these grounds, Spencer opposed social reform. The poor deserve to be poor, the rich to be rich. Society profits from allowing individuals to find their own social class level without outside help or hindrance. To interfere with poverty—or the result of any other natural process—is to harm society.

When Spencer visited America in 1882, he was warmly greeted, particularly by captains of industry. After all, his ideas of noninterference provided a moral justification for their economic practices, whether they be respectable or ruthless. Later, public support for government intervention increased, and Spencer's ideas began to slip out of fashion. He reportedly died with a sense of having failed.

Spencer's theories have crumbled over time. But he brought to the table for the science of sociology a discussion of how societies are structured.

Who was Karl Marx? The most influential of the nineteenth-century intellectuals was the German scholar Karl Marx (1818–1883). Although he did not consider himself a sociologist (his doctorate was in philosophy), his ideas have had a significant influence in shaping the field. Marx was deeply troubled by the long hours, low pay, and harsh working conditions in the capitalist system of his day. Preferring social activism to the abstractness of philosophy, Marx was guided by his conviction that social scientists should seek to change the world rather than merely observe it. In part through long association with his friend and collaborator Friedrich Engels, the son of an industrialist, Marx's commitment to democracy and humanism was channeled toward a concern for the poverty and inequality suffered by the working class. Forced by political pressures, Marx moved from Germany to France and finally settled in London, where he devoted the remaining decades of his life to a systematic analysis of capitalism.

Unfortunately, in the minds of many today, Marxism is equated with communism. Although some of his writings were later used as a basis for communism, it is likely that he would be as discouraged with the practice of communism in the twentieth century as he was with the utopian communists of his time. In fact, at one point Marx was so unhappy with others' erroneous interpretations of his work that he disavowed being a Marxist himself.

What is the legacy of Marx? Herbert Spencer and Karl Marx had different conceptions of society and social change. Spencer depicted society as a set of interrelated parts that promoted its own welfare. Marx described

Karl Marx (1818–1883)

society as a set of conflicting groups with different values and interests; the selfishness and ruthlessness of capitalists harmed society. Spencer saw progress coming only from noninterference with natural, evolutionary processes. Marx disagreed.

Marx did, however, believe in an unfolding, evolutionary pattern of social change. He envisioned a linear progression of modes of production from primitive communism through slavery, feudalism, capitalism, and communism. However, he was also convinced that the transformation from capitalism to a classless society could be accelerated through planned revolution. Although his political objective in interpreting capitalism was to hasten its fall through revolution, he believed that capitalism would eventually self-destruct anyway because of its inherent contradictions.

Although recognizing the presence of several social classes in nineteenth-century industrial society—farmers, servants, factory workers, craftspeople, owners of small businesses, moneyed capitalists—Marx predicted that ultimately all industrial societies would contain only two social classes: the **bourgeoisie**, those who owned the means for producing wealth in industrial society, and the **proletariat**, those who labored for the bourgeoisie at subsistence wages. For Marx, the key to the unfolding of history was **class conflict**—conflict between those controlling the means for producing wealth and those laboring for them. Just as slaves rebelled against slave owners and peasants revolted against the landed aristocracy, wage workers would overtake the capitalists. Out of this conflict would emerge a classless society—one without exploitation of the powerless by the powerful.

Which capitalist institution did Marx consider primary? According to the principle of **economic determinism** (an idea often associated with Marx), the nature of a society is based on the society's economy. A society's economic structure determines its other systems: legal, religious, cultural, political. Marx himself did not use the term economic determinism. The term was applied to his ideas by others, no doubt a consequence of his concentration on the economic sphere in capitalist society. Some interpreters mistakenly assume that because Marx perceived the economic institution as primary in capitalist society, he believed that all societies operated according to the same principles. Actually, Marx recognized that even in capitalist society economic and noneconomic institutions affect each other. Not only that, but he wrote that sometimes the economy "conditions" rather than "determines" the historical evolution of capitalist society.

Who was Emile Durkheim? Emile Durkheim (1858–1917) was the son of a French rabbi. Durkheim originally intended to follow this family rabbinical tradition; but later experiences led him to become an agnostic, although he retained an intellectual interest in religion throughout his life. In fact, one of Durkheim's major concerns was social and moral order, an emphasis undoubtedly related to his upbringing in the home of a rabbi:

> *From Judaic family training and an intimate environment Durkheim gained a deep and permanent concern for universal moral law and the problems of ethics, a concern that was not combined with any indulgent sense of humor. Indeed, he was eminently without humor and somewhat "heavy-handed."* (Simpson 1963:1)

Durkheim was a brilliant student even during his early school years. In college, he was so intensely studious that his schoolmates nicknamed him "the metaphysician." He eventually became the first French academic sociologist.

What were Durkheim's foremost contributions? According to Durkheim, social order exists because of a broad consensus among members of a society. This consensus is especially characteristic of preindustrial, nonliterate, simple societies based on **mechanical solidarity**—social unity that comes from a consensus of values and beliefs, strong social pressures for conformity, and dependence on tradition and family. Witnessing the social upheaval brought on by industrial and democratic revolutions, Durkheim attempted to describe social order in complex, industrial society. In this modern, more complicated society, he contended, social order is based on **organic solidarity**—social unity based on a complex of highly specialized roles. These specialized roles render members of a society dependent on one another for goods and services.

Although Comte, Martineau, and other early sociologists emphasized the need to make sociology scientific, they did not have the necessary research tools. They did, however, influence later sociologists to develop more scientific methods: to replace armchair speculation with careful observation, to engage in the collection and classification of data, and to use data for formulating and testing social theories. Durkheim was one of the most prominent of these later sociologists. He first introduced the use of statistical techniques in his groundbreaking research on suicide. (See "Using the Sociological Imagination," at the beginning of this chapter, and Doing Research.) In that study, Durkheim demonstrated that suicide involves more than an individual process. By revealing that suicide rates vary according to group characteristics—the suicide rate is lower among Catholics than Protestants and lower among married than single persons—Durkheim documented that human social behavior must be explained by social factors in addition to just psychological ones.

Who was Max Weber? Max Weber (1864–1920) was the eldest son of a well-to-do German lawyer and politician. Weber's father was a man rooted in the pleasures of this world. His mother, in stark contrast, was a strongly devout Calvinist who rejected the pleasure seeking of her husband. While Weber's father was preoccupied with life on earth, Weber's mother concentrated on her future salvation. Weber was affected psychologically by the conflicting values of his parents. Even though he suffered a complete mental breakdown, he eventually recovered to do some of his best work. His mother's influence is clearly reflected in his work, especially in his regard for the sociology of religion and in one of his most famous books, *The Protestant Ethic and the Spirit of Capitalism* (Ritzer 2000a).

How did Weber contribute to the development of sociology? As a university professor trained in law and economics, Weber wrote on a wide variety of topics, including the relationship between capitalism and Protestantism, the nature of power, the development and nature of bureaucracy, the religions of the world, and the nature of social classes. Thanks to the quality of his work and the diversity of his interests, Weber is credited with being the single most important influence on the development of sociological theory.

Max Weber (1864–1920)

Unlike Durkheim, Weber did not favor studying human beings as though they were physical things. Weber's approach was a subjective one: humans act on the basis of their own understanding of a situation. Consequently, sociologists must discover the personal meanings, values, beliefs, and attitudes that people bring to social situations. Understanding the subjective intentions of human social behavior could be accomplished through what Weber called the method of ***verstehen***—understanding social behavior by putting oneself in the place of others. This empathetic perspective was apparently rooted in Weber's personal nature. According to Weber's wife, he "showed throughout his life an extraordinary appreciation for the problems other people faced and for the shades of mood and meaning that characterized their outlook on life" (Bendix 1962:8).

Since Weber and Marx were not intellectual contemporaries, and since Weber disagreed with Marx on many points, it is often said that Weber was debating with Marx's ghost (Wiley 1987). It was out of honest differences on issues, not disrespect, that Weber opposed some of Marx's ideas. Although Marx was, at times, openly political in advocating his ideas, Weber stressed objectivity. He strongly counseled sociologists to conduct **value-free research**—research in which personal biases are not allowed to affect the research process and its outcome. Marx saw religion as retarding social change, whereas Weber believed that religion could promote change. Also, whereas Marx emphasized the role of the economy in social stratification, Weber's analysis involved several dimensions.

One should not conclude from these differences, however, that Weber's approach to sociology was antagonistic to Marx's. Actually, Weber was the most prominent of the second-generation German scholars who were concerned with power and conflict in society. Although there are significant differences in detail, both Marx and Weber took the problems of capitalism as their unifying theme (R. Collins 1994).

One of Weber's most notable contributions is the identification of rationalization as a key influence in the transition from preindustrial society. **Rationalization**—the use of knowledge, reason, planning, and objectivity—in industrial society marked a change from the tradition, emotion, superstition, and personal relationships of preindustrial society. With rationalization, for example, businesses were to be run on proven economic principles rather than custom. One of the most familiar effects of rationalization is the expansion of bureaucracy as the dominant type of organization. (See Doing Research, "The Study of Suicide," p. 25.) According to Weber, rationalization, with its emphasis on social control, would ultimately imprison us all in an "iron cage," robbing us of our freedom, autonomy, and individuality.

Sociology in America

Although the early development of sociology occurred in Europe, the maturation of sociology has taken place primarily in the United States. Because sociology has become a science largely through the efforts of American sociologists, it is not surprising that worldwide, most sociologists are American. Sociological writings in English are used by sociologists throughout the world, reflecting the tremendous global influence American sociologists have had since World War II. But of course, American sociologists have been heavily influenced by the European originators of sociology. This is partly because early American sociologists had to rely on the writings of Smith, Comte, Martineau, Spencer, Marx, Durkheim, Weber, and others. Where would they have turned but to the existing body of literature published by the Europeans? There is, however, another reason for the strong European influence. Just as sociology in Europe was born during a time of rapid, disruptive change, sociology emerged in the United States during the period of rapid urbanization and industrialization following the Civil War. It is little wonder that Lester Ward (1841–1913), the founder of American sociology, picked up on Comte's conviction that sociology could promote social progress. Like Comte, Ward believed that industrial, urban society could be improved through sociological analysis.

What is a brief history of American sociology? From its founding in 1892 to World War II, the first department of sociology at the University of Chicago stood at the forefront of American sociology. Housed at the University of Chicago have been a number of exceptional minds. George Herbert Mead, John Dewey, William I. Thomas, and Dorthony Swane Thomas pioneered the study of human nature and personality. Urban social problems such as prostitution, slums, and crime were studied extensively by Robert E. Park and Ernest Burgess. The study of social and cultural change was championed by William F. Ogburn. On the whole, the early Chicago School became closely linked with the idea of social reform. Emphasis on seeking solutions to problems of race, crime, and poverty continued at the University of Chicago after World War II with scholars such as Howard Becker, Joseph Gusfield, Herbert Blumer, David Riesman, and Erving Goffman (Abbott 1999).

Edith Abbott, Sophinista Breckenridge, Marion Talbott, and other female sociology researchers collaborated with the men at the Chicago School. They contributed to many studies of such urban problems as housing, child rearing, education, and poverty. Some of these women were professors in the Department of Sociology; others worked as university administrators (Deegan 1991, 2000). Many more women, having been denied university appointments, contributed to the development of social work rather than academic sociology.

After World War II, eastern universities such as Harvard and Columbia, midwestern universities such as Wisconsin and Michigan, and western universities such as Stanford and the University of California at Berkeley emerged as leading institutions fostering a diversity of sociological emphases. Sociologists such as George A. Lundberg underscored the application of the research methods of physical scientists; they wanted to make sociological research more sophisticated and scientific. Lundberg was known as a neo-positivist for his revival of Comte's scientific approach to sociology. A sociologist at Harvard, Talcott Parsons, returned to the development of general theories of society à la Weber, Durkheim, and other European scholars. Parsons was one of the leading proponents of functionalism (discussed later), which is one of several dominant theoretical perspectives in sociology today. Robert K. Merton, a Columbia University sociologist and a student of Parsons, advocated applying more specific theories, rather than general ones. Merton also stressed the importance of empirical research. By the 1970s, the number of practicing sociologists had grown dramatically. As a result, talented and influential sociologists were no longer located in only a few elite colleges and universities.

In their quest toward developing general sociological theories and establishing sociology as a science, sociologists between World War II and the 1960s had nearly lost sight of the idea that sociology could help solve social problems. Social reform emerged again during the turbulent sixties and remains an important part of sociology today. "Humanist" sociologists, who owe much to sociologist C. Wright Mills, believe that sociology cannot be ethically neutral about important social issues. Rather, they argue, sociologists have an inherent obligation to question social arrangements and work for social transformation (A. M. Lee 1978; Giddens 1987, 2000; Scimecca 1987).

Two Early Contributors

Two early contributors who are often neglected in the history of American sociology are Jane Addams and W. E. B. Du Bois, close friends and coworkers. Although neither was a researcher or scientist, both were concerned about applying sociology to solve social problems.

How did Jane Addams contribute to American sociology? The best known and most influential of the early female social reformers in the United States was Jane Addams (1860–1935). Although her mother died when she was two years old, Addams's father provided a loving and comfortable home for her and her eight brothers and sisters. In addition to the physical comforts he provided, her father set an example with his love of learning.

Addams was an excellent student. Her earlier education emphasized practical knowledge and the improvement of "the organizations of human society." She attended The Women's Medical College of Philadelphia, but was compelled to drop out because of illness (Addams 1981; originally published in 1910).

During her childhood, Addams was exposed to many episodes of government corruption and to business practices that enriched owners and harmed workers. This was the beginning of her social conscience. While on one of her European trips, she saw the work being done to help the poor in London. With this example of social action and with her earlier exposure in the United States to corruption and exploitation, Addams began her life's work seeking social justice.

She cofounded Hull House in Chicago's slums. There, people who needed help—immigrants, the sick, the poor, the aged—could find some relief. Addams focused on the problems caused by the imbalance of power among the social classes. She invited sociologists from the University of Chicago to Hull House to witness firsthand the effects of the exploitation of the lower class. In addition to her work with the underclass, Addams was active in the women's suffrage and peace movements (Deegan 2000).

As a result of her tireless work for social reform, Addams was awarded the Nobel Peace Prize in 1931—the only sociologist to receive this honor. The irony is that Addams herself suffered a sort of class discrimination. She was not considered a sociologist during her lifetime, in part because she did not teach at a university. She was considered a social worker (a less prestigious career) because she was a woman and because she worked directly with the poor.

How did W. E. B. Du Bois contribute to an understanding of black communities? Together with other early African American sociologists, W. E. B. Du Bois (1868–1963), an African American educator and social activist, influenced the early development of sociology in the United States (Young and Deskins 2001). Although Du Bois's father was part white and part black (known then as a mulatto) and his mother was a direct descendant of a freedman Dutch slave, Du Bois was considered black by his associates. He attended an integrated high school in his native Great Barrington, Massachusetts, and was the first black to receive a diploma there. Du Bois earned a doctorate degree from Harvard University in 1895 and taught at a number of predominantly black universities during his career.

Image Bank

W. E. B. Du Bois (1868–1963)

Du Bois learned firsthand about racial discrimination and segregation when he attended Fisk University in Nashville, Tennessee. From this experience and from teaching in rural, all-black schools around Nashville, Du Bois decided to fight what was then called "the Negro problem"—a racist policy based on the assumption that blacks were an inferior race. He did this by scientifically studying the social structure of black communities, first in Philadelphia and later in other places. In collaboration with Isabel Eaton, Du Bois published *The Philadelphia Negro*, a classic work on urban African Americans.

Besides contributing to the understanding of black communities, Du Bois worked for civil rights. He was the only black member of the Board of Directors of the National Association for the Advancement of Colored People (NAACP) when it was founded in 1910, and he was the editor of its journal *Crisis* for the next twenty-four years.

Du Bois's concern for his race did not stop at the borders of the United States. He was also active in the Pan African movement, which was concerned with the rights of all African descendants, no matter where they lived. While continuing to document the experience and contributions of African people throughout the world, Du Bois died in the African country of Ghana, at the age of ninety-five (D. Lewis 1993, 2000).

FEEDBACK

1. __________ is the idea that sociology should use observation and experimentation in the study of social life.
2. Which of the following sociologists contended that social change leads to progress, provided that people do not interfere?
   a. Durkheim b. Weber c. Martineau d. Spencer e. Marx f. Du Bois
3. According to the principle of __________ a society's economic system molds the society's legal system, religion, art, literature, political structure, and other social arrangements.
4. According to Durkheim, the United States is a social order based on __________ solidarity.
5. According to Weber, the method of __________ involves an attempt to understand the behavior of others by putting oneself mentally in their place.
6. Du Bois focused only on the American race question. T or F?

Answers: 1. Positivism 2. d 3. economic determinism 4. organic 5. verstehen 6. F

Theoretical Perspectives

Theory and Perspective

Do scientific theories compete? It is normal in science for competing, even conflicting, theories to exist simultaneously. This may be because of insufficient evidence to determine which theory is accurate. Or it may be that different theories explain different aspects of reality. This is true even for the "hard" sciences, such as modern physics. Einstein's theory of general relativity, for example, was inconsistent with the widely accepted Big Bang theory of the origin of the universe. And Einstein was deeply bothered by quantum theory, whose laws have become the foundation of modern developments in such fields as chemistry and molecular biology (Hawking 1998). Just as theories in physics oppose one another, several major theoretical perspectives compete in sociology.

How does perspective affect perception? In *Beyond Good and Evil*, nineteenth-century philosopher Friedrich Nietzsche wrote:

> *It is more comfortable for our eye to react to a particular object by producing again an image it has often produced before than by retaining what is new and different in an impression.* (Nietzsche 1987:97; originally published in 1886)

As implied at the beginning of this chapter, a person's perspective draws attention to some things but blinds him or her to other possibilities. This fundamental

FIGURE 1.1

The Importance of Perspectives

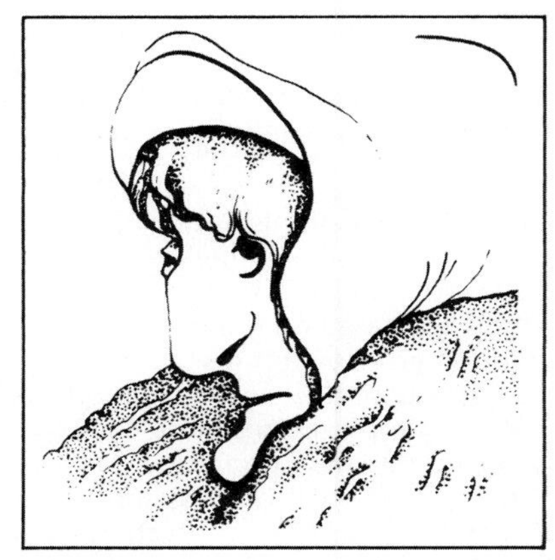

View I

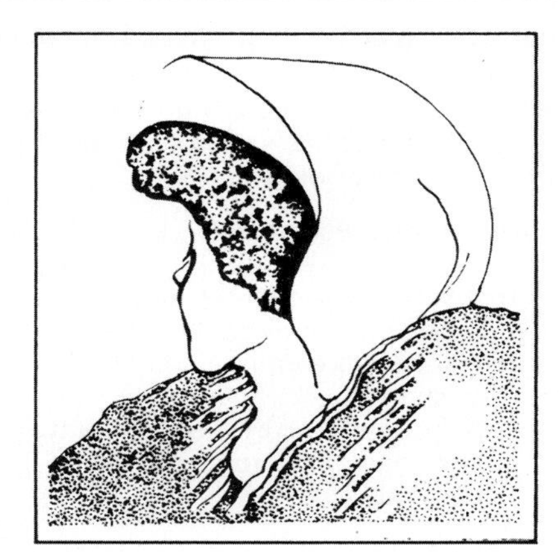

View II

View III

Our perception of things is heavily influenced by our perspective. Decide whether you want to look at View I or View II (cover the other view). After examining View I or View II, look at View III. If you decided to look at View I, you should see the old woman in View III. If you decided to look at View II, you should see the young woman in View III. Can you see both? Try it on your friends.

power of perspective is graphically illustrated in two drawings that psychologists often use to illustrate the concept of perception. One, shown in Figure 1.1, is a picture of an old woman—or is it? If you stare at the old woman long enough, she becomes a beautiful young woman with a feather boa around her neck. The other drawing (not shown in Figure 1.1) is a picture of a rabbit. If you stare at the rabbit long enough, a duck appears. If you continue to look, both drawings return to their original form. You cannot, however, see the old woman and the young woman or the rabbit and the duck at the same time.

Which image is real, the old woman or the young woman? The rabbit or the duck? It depends on your focus; your perspective influences what you see. So it is with any perspective. One perspective emphasizes certain aspects of an event; another perspective accents different aspects of the same event. When a perspective highlights certain parts of something, it necessarily places other parts in the background.

Sociology has three overarching theoretical perspectives: *functionalism*, *conflict theory*, and *symbolic interactionism*. Each of these perspectives provides a different slant on human social behavior (see Table 1.2). The exclusive use of any one of them prevents our seeing other aspects of social behavior, just as one cannot see the old woman and the young woman at the same time. All three perspectives together, however, allow us to see most of the important dimensions of social life that are of interest in sociology. These three perspectives can be placed within the context of macrosociology or microsociology. Functionalism and conflict theory are best viewed from the macro level, symbolic interactionism from the micro level.

Macrosociology: Functionalism

Functionalism emphasizes the contributions (functions) performed by each part of a society. For example, family, economy, and religion are all parts of a society. The family contributes to society by providing for the reproduction and care of its new members. The economy contributes by dealing with the production, distribution, and consumption of goods and services. Religion contributes by emphasizing beliefs and practices related to sacred things. Functionalism focuses on social integration, stability, order, and cooperation.

Who were the originators? Among the major pioneers of this perspective was, of course, Herbert Spencer, who compared societies to living organisms. Although sociologists no longer use the organism metaphor, they still view society as a system of interrelated parts. Emile Durkheim, a contemporary of Spencer's, greatly contributed to the development of this theoretical perspective. Two of the strongest supporters of functionalism have been American sociologists Talcott Parsons and Robert Merton.

What are the assumptions of functionalism? First, functionalists see the parts of a society as an integrated whole. A change in one part of a society leads to changes in other parts. A major change in the economy, for example, may change the family. This is precisely

TABLE 1.2

FOCUS ON THEORETICAL PERSPECTIVES:
Assumptions of Functionalism, the Conflict Perspective, and Symbolic Interactionism

This table contains the major assumptions of the three major theoretical perspectives. Although the assumptions of functionalism and conflict theory are direct opposites, each focuses on the nature of society. Symbolic interactionism uniquely addresses interpersonal issues in sociology.

Functionalism	The Conflict Perspective	Symbolic Interactionism
1. A society is a relatively integrated whole. 2. A society tends to seek relative stability. 3. Most aspects of a society contribute to the society's well-being and survival. 4. A society rests on the consensus of its members.	1. A society experiences inconsistency and conflict everywhere. 2. A society is continually subjected to change. 3. A society involves the constraint and coercion of some members by others.	1. People's interpretations of symbols are based on the meanings they learn from others. 2. People base their interaction on their interpretations of symbols. 3. Because symbols permit people to have internal conversations, they can gear their interaction to the behavior they think others expect of them and they expect of others.

Source: The assumptions of functionalism and the conflict perspective are based on the work of Ralf Dahrendorf, "Toward a Theory of Social Conflict," *Journal of Conflict Resolution* 2 (June 1958):174. Adapted by permission of Sage Publications, Inc., © 1958.

what happened as a result of the Industrial Revolution. Before the Industrial Revolution, when most people made their living by farming, the need for a large farm labor force was fulfilled by having many children. The need for a large family declined dramatically as industrialization substituted machines for manual labor. Smaller families, then, became the norm.

Functionalists realize that societies are not perfectly integrated. Although the actual degree of integration varies, a certain degree of integration is necessary for the survival of the society.

Second, functionalists assume that societies tend to return to a state of stability after some upheaval has occurred. A society may change over time, but functionalists believe that it will return to a stable state by incorporating change in such a way that society will be similar to what it was before any change occurred.

The idea that a society changes yet maintains most of its original structure over time is referred to as **dynamic equilibrium**—a constantly changing balance among the parts. The student unrest on college and university campuses during the late 1960s is an illustration of dynamic equilibrium. The activities of student protestors did create some changes. The public no longer accepts all American wars as legitimate; universities are now more responsive to students' needs and goals; and the public is more aware of the importance of environmental protection. These changes, however, have not revolutionized American society. They have been absorbed into it, leaving it only somewhat different from the way it was before the student unrest. In fact, many of the student "radicals" are now part of the middle-class society they once rejected. It will be interesting to see if the widespread concern over corporate misbehavior that erupted in 2002 as a result of the Enron, WorldCom, and Arthur Anderson scandals will bring about any significant changes in American society.

Third, most aspects of a society exist to promote the society's survival and welfare. It is for this reason that all complex societies have economies, families, governments, and religions. If these elements did not contribute to a society's well-being and survival, they would disappear.

According to Robert Merton (1996), **manifest functions** are intended and recognized at the time; **latent functions** are unintended and unrecognized until later. In 1696, Scotland's Parliament established free public schools in all parishes. These schools were to enable the young to read Holy Scripture. Unintended consequences included the establishment of education as a way of life for middle-class Scots and the intellectual flowering of the Scottish Enlightenment in the eighteenth century (Herman 2001). More recently, the automakers' attempt to make it more difficult to hot-wire expensive cars had the desired effect of reducing car theft. An unintended consequence, however, was the emergence of a new and more dangerous crime, carjacking (Mann 2002). One of the manifest functions of secondary schools, for example, is to teach math skills. A latent (and positive) function of schools is the development of close friendships.

Not all elements of a society make a positive contribution. Those that have negative consequences result in **dysfunction**. Bureaucratic rules, for example, often result in rigidity and impersonality. If a secretary in the sociology department treats you like a "number" rather than an individual, you don't like the inflexibility and impersonality.

Finally, according to functionalism, there is among most members of a society a consensus on values. Most Americans, for example, agree on the desirability of democracy, success, and equal opportunity. This consensus of values, say the functionalists, accounts for the high degree of cooperation found in any society.

There are legitimate criticisms of the functionalist perspective: It tends to legitimatize the status quo, and it neglects social change. Conflict theory, in contrast, takes social change as a focal point.

Macrosociology: Conflict Theory

Conflict theory emphasizes conflict, competition, change, and constraint within a society (Schellenberg 1982; Giddens 1979, 1987). Although this theoretical perspective was not very popular among most American sociologists until the 1960s, its roots go back as far as those of functionalism.

Who were the originators? Karl Marx, Max Weber, and Georg Simmel (1858–1918) are three German sociologists whose work underlies the conflict perspective. According to Marx, class conflict is inevitable in capitalist economies. Modern conflict advocates such as C. Wright Mills, Ralf Dahrendorf, and Randall Collins do not limit themselves to class conflict. They broaden Marx's insights to include conflict between any segments of a society, a point Weber had made much earlier (R. Collins 1994). For example, conflict exists between Republicans and Democrats, unions and management, industrialists and environmentalists.

How are the assumptions of conflict theory often the reverse of the functionalist perspective? Functionalism emphasizes a basic agreement on values within a society, concentrating on the ways people cooperate to reach common goals. Conflict theory, in contrast, focuses on the disagreements between various groups in a society or between societies. Groups and societies compete and conflict as they attempt to preserve and promote their own special values and interests.

© Bettmann/CORBIS

Conflict theorists can see no function for rural poverty such as this. Poor people, they contend, end up at the bottom of the stratification structure in part because they lack power to influence social policy.

How do conflict advocates view power? Because advocates of conflict theory see social living as a contest, their central question is, Who gets what? Their answer: Those with the most **power**—the ability to control the behavior of others, even against their will—get the largest share of whatever is considered valuable in a society. Those with the most power have the most wealth, the most prestige, and the most privileges. Because some segments have more power than others, they are able to constrain the less powerful.

The wealthy and powerful, for example, can often manipulate income tax laws (Ehrenreich 1994; Barlett and Steele 2000; Huffington 2003). In 1998, U.S. oil and pipeline companies were required to pay only 5.7 percent of their domestic profits to the U.S. Treasury even though the official tax rate was 35 percent. Actually, half of the oil and pipeline firms surveyed were entitled to tax rebates (McIntyre 2000). According to a recent study by Citizens for Tax Justice, 250 of America's biggest corporations were taxed at only 20.1 percent in 1998. Moreover, forty-one of these companies received tax rebates for at least one year from 1996 to 1998. That is, they made additional money after filing their tax forms on the profits they reported to the Internal Revenue Service (IRS). For example, just before the accounting scandals erupted in 2002, Enron had paid no income taxes in four of the previous five years. Instead, the corporation was eligible for $382 million in tax refunds (Johnston 2002; Behr 2003). Tax evasion or avoidance is not limited to corporations. Millions of Americans pay less than their fair share of taxes annually, including more than one thousand American taxpayers with incomes over $200,000 who in a given year pay no federal income tax whatever (Ehrenreich 1994; Lewis and Allison 2001). The lowest 20 percent family income category in the United States gained 19 percent in after-tax income between 1980 and 1999, while the top 5 percent gained 241 percent (see Table 1.3). And the wealthiest 1 percent of Americans will receive 52 percent of President Bush's 2001 individual tax cuts, totaling some $477 billion (Citizens for Tax Justice 2002).

Other examples of wealthy and powerful interests getting their way are numerous. In 1985, Congress passed legislation continuing a federal sugar subsidy

TABLE 1.3

Average Income Before and After Tax, 1980, 1998, AND 1999 (in constant 1998 dollars, adjusted for changes in family size)

	Pretax Incomes			After-Tax Incomes		
Family Income Category	1980	1998	% Change	1980	1999 Projected	% Change
Lowest 20%	12,697	12,526	−1%	7,382	8,800	19%
Second 20%	27,641	29,482	7%	17,477	20,000	14%
Middle 20%	41,756	46,662	12%	26,968	31,400	16%
Fourth 20%	57,931	68,430	18%	37,816	45,100	19%
Highest 20%	97,539	140,846	44%	61,168	102,300	67%
Top 5%	138,659	246,520	78%	84,972	289,700	241%

As this table reveals, people in certain income categories profited more between 1980 and 1999 than others. Relate these data to the conflict perspective.

Sources: U.S. Bureau of the Census. "Historical Income Tables—Families; Table F-3, Mean Income Received by Each Fifth and Top 5 Percent of Families (All Races): 1966 to 1998," October, 1999. At http://www.census.gov/hhes/income/histinc/f03.html; and Isaac Shapiro, and Robert Greenstein. "The Widening Income Gulf." Washington, DC: Center on Budget and Policy Priorities, 1999.

that placed an average of $250,000 in the pockets of some 12,000 American sugar growers. As a result, American consumers had to pay among the highest prices in the world for their sugar: A quantity of sugar costing 85 cents in Washington, D.C., sold for half that amount in Ottawa, Canada. According to the evidence, members of the House of Representatives were likely to vote for the bill in direct proportion to the amount of campaign contributions they received from the sugar lobby (Birnbaum and Murray 1988; Stern 1988). The 1999 Ticket to Work and Work Incentives Improvement Act for the disabled contains a provision unrelated to disabilities that saves corporations $1.5 billion in taxes every two years (Barlett and Steele 2000). And President Bush's proposed budget for fiscal 2003 introduced new corporate farm subsidies that are projected to total $70 billion over the next decade (Gleckman 2002).

How does conflict theory explain social change? Because many conflicting groups exist and the balance of power among these groups may shift, conflict theory assumes that social change is continual. For example, the women's movement is attempting to change the balance of power between men and women. As this movement progresses, we see larger numbers of women in occupations once limited to men. More women are either making or influencing decisions in business, politics, medicine, and law. Gender relations are changing in other ways as well. More women are choosing to remain single, to marry later in life, and to have fewer children—or none. According to the conflict perspective, these changes are the result of increasing power among women. This brings up feminist theory.

What is feminist theory? **Feminist theory**, a branch of conflict theory, links the lives of women (and men) to the structure of gender relationships within society. Three frameworks within the broad umbrella of feminist theory can be isolated: liberal feminism, radical feminism, and socialist feminism (Freedman 2001; Anderson 2002). Each of these frameworks attempts to explain the relationship between gender, social class, and minority status.

Advocates of *liberal feminism* focus on equal opportunity for women and heightened public awareness of women's rights. Operating within the existing structure of society, liberal feminists look to legal and social change to rectify the abridgment of women's rights. If women are to receive the same treatment accorded men, long-standing laws, social policies, and social structures must be altered. Equal pay for equal work, abortion rights, and civil rights are among the issues uniting the liberal feminist perspective.

Liberal feminism attributes the subordination of women to unequal rights and to the ways in which males and females are taught by society to view themselves. *Radical feminists* trace the oppression of women to the fact that societies are dominated by men. A *patriarchal society* is controlled by males who use their power, prestige, and economic advantage to dominate females, including their sexuality. Male-controlled institutions—family, religion, economy, education, government—define women as subordinate to men and are used to ensure the perpetuation of female subordination. Radical feminists are especially interested in male supremacy and the ways men gain and maintain the power to control all social institutions (Crow 1998).

What does socialist feminism add to radical feminism? *Socialist feminism*, which is actually more radical than radical feminism, sees capitalism as a source of female oppression. This perspective on capitalism, however, goes beyond the Marxist interpretation. Socialist feminists contend that the problem of female oppression extends beyond the fact that under capitalist societies women are the property of men. Because female oppression exists in societies without a capitalist economic base, socialist feminists reason that more than capitalism is at work in female oppression. That something else is patriarchy. Patriarchy is the means for male domination of females in preindustrial and authoritarian societies (Khouri 2003). In capitalist society, the power relations of the class structure combine with the force of patriarchy to create and maintain male oppression of women.

Despite some significant differences in theoretical perspectives, feminists share at least two common themes. First, they believe that sociology carries a bias from years of theory and research shared by white middle-class males from Western Europe and North America. Second, they believe that the nature of gender and gender relationships is sociological (rather than psychological), embedded as they are in social structures.

Which is the better perspective, functionalism or conflict theory? There is no "better" theoretical perspective. Each perspective emphasizes certain aspects of social life. The advantages of one perspective highlight the disadvantages of the other. Functionalism explains much of the consensus, stability, and cooperation within a society. Conflict theory explains much of the constraint, conflict, and change.

Because each perspective has captured an essential facet of society's nature, their combination, or synthesis, is a reasonable next step. Some attempts to combine functionalism and conflict theory have already been made (Dahrendorf 1958b; van den Berghe 1963, 1978). One of the most promising is the attempt to specify the conditions under which conflict and cooperation occur. Gerhard Lenski (1984; Nolan and Lenski 1999 contends that people cooperate—even share the fruits of their labors—when scarcity threatens their survival. But conflict, competition, and constraint are likely to occur when there is more than enough for everyone. Thus, as a society moves from a subsistence economy to an affluent one, conflict, competition, and constraint increase. And the more a deprived group questions the legitimacy of its condition, the more likely it is to conflict with privileged groups (Coser 1998), as illustrated by the civil rights and women's movements.

Microsociology: Symbolic Interactionism

Who were the originators, and what were they thinking? Both functionalism and conflict theory deal with large social units and broad social processes—the state, the economy, evolution, class conflict. At the close of the nineteenth century, some sociologists began to introduce a new focus on human social behavior. Instead of being preoccupied with larger structures per se (such as society and social class), they began to recognize the importance of people's interactions within social structures. Max Weber and Georg Simmel were among the earliest contributors to the theoretical perspective of interactionism, which holds that groups exist only because their members influence one another's behavior. Later sociologists—namely, Charles Horton Cooley, George Herbert Mead, W. I. Thomas, Erving Goffman, Harold Garfinkel, and Herbert Blumer—expanded the insight of interactionism. They created symbolic interactionism, the most influential approach to interactionism.

What is symbolic interactionism? Symbolic interactionism targets a less abstract view of human social behavior than either functionalism or conflict theory. This perspective focuses on the actual interaction between people (Reynolds and Herman-Kinney 2003).

Basic to the symbolic interactionist perspective is the concept of a symbol. A **symbol** is something chosen to represent something else. It may be an object, a word, a gesture, a facial expression, or a sound. A symbol is something that is observable, something concrete; but it may represent something that is not observable, something abstract. A frown can be a symbol of either disapproval or concentration. Your school mascot, like the Hokie Bird of Virginia Tech, is a symbol of your college or university:

> *One cannot tell by looking at an X in an algebraic equation what it stands for; one cannot ascertain with the ears alone the symbolic value of the phonetic compound si; one cannot tell merely by weighing a pig how much gold he will exchange for; one cannot tell from the wave length of a color whether it stands for courage or cowardice, "stop" or "go."*
> (L. A. White 1969:26)

The meaning of a symbol is not determined by its own physical characteristics. Those who create and use the symbol assign a meaning to it. If people in a group do not share the same meaning for a given symbol, confusion results. For example, if some people interpret the red light of a traffic signal to mean "go," while others see the green light to mean "stop," chaos will result.

The importance of shared symbols is reflected in the formal definition of **symbolic interactionism**: the theoretical perspective that focuses on interaction among people—interaction based on mutually understood symbols.

What are the basic assumptions of symbolic interactionism? Herbert Blumer (1969a, 1969b), who coined the term *symbolic interactionism*, outlined three assumptions central to this perspective (see Table 1.2). First, according to symbolic interactionism, we learn the meaning of a symbol from the way we see others reacting to it. For example, American musicians in Latin America soon learn that when audience members whistle at the end of a performance, they are unhappy; their whistling is a symbol of disapproval, as is the symbol of booing in the United States.

Second, once we learn the meanings of these symbols, we base our behavior (interactions) on them. Now that the musicians have learned that whistling symbolizes a negative response, they will, if the crowd begins whistling, decide against an encore. This is opposite the response they would have in the United States, where whistling has a very different meaning.

Finally, symbols are central to interaction in yet another way. We use the meanings of symbols to imagine how others will respond to our behavior. Through this capability, we can have "internal conversations" with ourselves. These internal conversations enable us to visualize how others will respond to us *before* we act. This is crucial because we guide our interactions with people according to the behavior we think others expect of us and we expect of others. Meanwhile, these others are also having internal conversations. The interaction (acting on each other) that follows is therefore *symbolic* interaction.

What approach has Erving Goffman added to the recent developments in symbolic interactionism? In an attempt to understand human interaction, Erving Goffman introduced **dramaturgy**, an approach that depicts human interaction as theatrical performance (Goffman 1961a, 1963, 1974, 1979, 1983; C. Lemert and Branaman 1997). Like actors on a stage, people (the performers) present themselves—by their dress, gestures, tone of voice—in such a way as to enhance their performance and create in others a favorable evaluation. Goffman labels this effort **presentation of self**. Through this kind of "impression management," we

Emile Durkheim— The Study of Suicide

Emile Durkheim, the first person to be formally recognized as a sociologist and the most scientific of the pioneers, conducted a study that stands as a research model for sociologists today. His investigation of suicide was, in fact, the first sociological study to use statistics. In *Suicide* (1964b; originally published in 1897), Durkheim documented his contention that some aspects of social behavior—even something as allegedly individualistic as suicide—can be explained without reference to individuals.

Like all of Durkheim's work, his study of suicide is best considered within the context of his concern for social integration (R. Collins 1994; Pickering and Walford 2000). Durkheim wanted to see if suicide rates within a social entity (for example, a group, organization, or society) are related to the degree to which individuals are socially involved (integrated and regulated). In his study, Durkheim described three types of suicide: egoistic, altruistic, and anomic. He hypothesized that *egoistic suicide* increases when individuals do not have sufficient social ties. Because single (never married) adults, for example, are not heavily involved with family life, they are more likely to commit suicide than are married adults. On the other hand, he predicted *altruistic suicide* as more likely to occur when social integration is extremely strong. The Al Qaeda agents who slammed jet liners into the World Trade Center and the Pentagon in 2001 are one example, as are suicide bombers. Altruistic suicide need not be this extreme. Military personnel who lay down their lives for their country are another illustration.

Durkheim forecasted his third type of suicide—*anomic suicide*—to increase when existing social ties are broken. For example, suicide rates increase during economic depressions. People suddenly without jobs, or hope of finding any, are more prone to kill themselves. Suicide may also increase during periods of prosperity. People may loosen their social ties by taking new jobs, moving to new communities, or finding new mates.

Using pre-collected data from government population reports of several countries (much of it from the French government statistical office), Durkheim found strong support for his predictions. Suicide rates were, in fact, higher among unmarried than married people and among military personnel than civilians. They were also higher among people involved in nationwide economic crises.

Durkheim's primary interest, however, was not in the empirical (observable) indicators he used, such as suicide rates among military personnel, married people, and so forth. Rather, Durkheim used the results of his study to support several of his broader contentions: (1) Social behavior can be explained by social rather than psychological factors; (2) suicide is affected by the degree of integration and regulation within social entities; and (3) because society can be studied scientifically, sociology is worthy of recognition in the academic world (Ritzer 2000a). Durkheim was successful on all three counts. If Auguste Comte told us that sociology *could* be a science, Durkheim showed us *how* it could be a science.

Thinking about the Research

1. Do you believe that Durkheim's study of suicide supported his idea that much social behavior cannot be explained psychologically? Why or why not?
2. Which approach do you think Durkheim followed in his study of suicide: functionalist, conflict, or symbolic interactionist? Support your choice by relating his study to the assumptions of the perspective you chose.

attempt to create a favorable evaluation of ourselves in the minds of others.

Think about your own impression management. Most students come to college concerned about acceptance by new peers. Consequently, you make decisions (consciously or not) about how you want others to view you. Choices in clothing are made—some students wear blue jeans and T-shirts; others choose a preppie look. Posters and magazines may be prominently displayed to present the desired image. Personal information such as athletic, musical, or sexual prowess will be appropriately leaked.

Performances, according to Goffman's theater analogy, may have a front and a back stage. College students may behave one way with dates and let down when alone with other campus classmates. Students may conduct themselves maturely when their parents are visiting campus, only to revert to old habits with their roommates. Behavior changes as it is managed in various settings.

Goffman's theoretical approach has stimulated the research of other sociologists. Dramaturgical symbolic interactionists have documented diverse arenas of American drama and ritual: social life, including singles

Theoretical Focus on the Internet

The number of Americans paying for an online service is skyrocketing. First invented as a way for military and scientific users to share information after a nuclear war, the Internet was formed in 1969 with only four hosts (a host is a single computer connected to the Internet). The number of hosts expanded to 213 by 1981 and exploded to more than 153 million in the year 2002 (Wright 2002).

Because of its rapid spread throughout American society, cyberspace technology is a timely example showcasing the value of the three theoretical perspectives. The viewpoints of functionalism, conflict theory, and symbolic interactionism may converge at times. But they also contribute, in very different ways, to an understanding of the social implications of this new technology. Some brief examples will make this clear.

Functionalism

Functionalists view cyberspace technology as having both positive and negative consequences. Through computer links to others, many parents can work at home, spending more time with their children. Handicapped individuals can do jobs at home that would be denied them otherwise, thus integrating them more fully into society. On the other hand, there are dysfunctions. Young people have unprecedented access to pornographic material that can distort their view of the opposite sex. Hate groups can be formed by strangers who live hundreds of miles apart. Their anonymity may encourage them to engage in violent behavior they would otherwise avoid.

Conflict Theory

The Internet is clearly changing American society, contributing to an unprecedented speed of technological change. An advocate of conflict theory might investigate the social instability created by this situation. Workers in increasing numbers may be released as jobs formerly requiring a human touch are performed by computers. Conflict theorists could investigate the discrepancy in computer allocation to school districts of varying socioeconomic levels; students who attend wealthy schools with access to computers have an advantage over students in poorer schools.

Symbolic Interactionism

Symbolic interactionists are interested, for example, in the ease with which the Internet can affect a child's socialization process. The popularity of cartoon characters on television is reinforced by Web pages that allow children to join fan clubs, interact with other fans, and view on demand video clips of their favorite cartoon characters. The popular cartoons "Beavis and Butt-Head," "Simpsons," and "South Park" feature children behaving in ways deemed unacceptable in most American homes. Instead of the more limited exposure to these characters that occurs with television alone, the Internet allows cartoon characters to become an important part of a child's daily life. What children come to accept as desirable behavior is being based increasingly on their interpretations of the symbols and behaviors represented by these characters. Symbolic interactionists might conclude that to the extent that this occurs, the Internet diminishes adult influence on children.

Analyzing the Trends

Do you think that the Internet will have some dysfunctions that Americans should consider? Discuss two possible negative consequences for society.

bars; work, including nurses; athletics, including college football; politics, including effects of corruption; and religion, including the public's view of their social world (Deegan and Stein 1978; T. R. Young 1990; Hochschild 2003).

Although the feminist perspective grew most directly out of conflict theory, feminist sociologists also use the micro perspective. They have done very important research within the tradition of symbolic interactionism (West and Zimmerman 1987; Gardner 1994).

What are the limitations of symbolic interactionism? By virtue of being a microsociological theoretical perspective, symbolic interactionism sometimes fails to take the larger social picture into account. Social interaction in everyday life is sometimes affected by societal forces beyond the control (or even awareness) of individuals. Anthony Giddens (1979, 2000), in his *structuration theory*, attempts to overcome this potential limitation. Although Giddens recognizes that social structures influence the individual—a central point of both functionalism and conflict theory—he emphasizes the effects individuals have on social structures. He looks at individual recognition of social forces and their subsequent actions based on this recognition. Despite the micro limitations, symbolic interactionism is one of the most influential theoretical perspectives in sociology today (Fine 1993).

What are the contributions of the theoretical perspectives? Because the three theoretical perspectives have some contradictory assumptions and each is incomplete in itself, you may be somewhat confused. Where do perspectives in sociology now stand? The answer is easy: These three perspectives tend to complement one another because they focus on different aspects of social life.

Each of the perspectives could be used, for example, to better understand such deviant behavior as crime, juvenile delinquency, and mental illness. Functionalism emphasizes integration, stability, and consensus, so it accents the negative consequences of deviance. It underscores the cost of crime, including the money spent on police departments and prisons, violence in the streets, and the threat to public safety in inner cities and suburbs. The conflict perspective views deviance quite differently. Because various groups compete to influence the passage and enforcement of legislation serving their interests, many laws are created and enforced by those with political power. Consequently, those with the least influence—such as prostitutes and drug addicts—are the most likely to be classified as deviants. Interactionism focuses on the relationships within a group of deviants. It looks, for example, at the processes by which members of a drug-using group influence others to join them. It studies the ways drugs are bought from pushers, the methods for raising money to pay the pushers, and the ways people who use drugs interpret the user's activities.

The unique contribution of each of the three perspectives can be illustrated by examining another aspect of group life. American women currently earn about 76 cents for every dollar earned by men. An earning gap exists even between men and women in the same occupations with the same educational qualifications. How would functionalists, conflict advocates, and symbolic interactionists view this piece of social reality? Functionalists claim that income inequality between men and women is a dysfunction, or negative consequence, of the broader social condition of sexual inequality. Conflict advocates attribute income inequality to men's power in controlling women's income. Symbolic interactionists would point out that income inequality has existed for a very long time. It is only in recent years that many Americans have come to define income inequality as an unfair condition in need of eradication. In other words, income inequality now has a different symbolic meaning than it has had in the past.

Because each of these perspectives emphasizes different aspects of social life, one perspective cannot be said to be inherently superior to the others. Except for sheer personal preference, the old witchlike lady is no worse than the beautiful young woman, the rabbit no better than the duck. By the same token, no theoretical perspective is innately superior to another. Each of the three major perspectives has its own worth, because each tells us something different about group life. The remaining chapters will illustrate each perspective in the context of the chapter topic.

FEEDBACK

1. A __________ perspective is a set of assumptions accepted as true by its advocates.
2. Indicate whether the following statements represent functionalism (F), conflict theory (C), or symbolic interactionism (S).
   ____ a. Societies are in dynamic equilibrium.
   ____ b. Power is one of the most important elements in social life.
   ____ c. Religion helps hold a society together morally.
   ____ d. Symbols are crucial to social life.
   ____ e. A change in the economy leads to a change in the family structure.
   ____ f. People conduct themselves according to their subjective interpretations of reality.
   ____ g. Many elements of a society exist to benefit the powerful.
   ____ h. Different segments of a society compete to achieve their own self-interests rather than cooperate to benefit others.
   ____ i. Social life should be understood from the viewpoint of the individuals involved.
   ____ j. Most members of a society agree that democracy is desirable.
   ____ k. Social change is constantly occurring.
   ____ l. Conflict is harmful and disruptive to society.
3. According to the __________ approach, humans use impression management to control the attitudes and responses of others toward them.
4. Match the three feminist theoretical frameworks with the following words and phrases.
   ____ a. liberal feminism
   ____ b. radical feminism
   ____ c. socialist feminism
   (1) patriarchy
   (2) capitalism and patriarchy
   (3) equality of opportunity
   (4) social and legal reform

Answers: 1. theoretical 2. a. (F) b. (C) c. (F) d. (S) e. (F) f. (S) g. (C) h. (C) i. (S) j. (F) k. (C) l. (F) 3. dramaturgical 4. a. (3),(4) b. (1) c. (2)

Theoretical Perspectives and Sport

Because sport is a major avenue through which culture is created and reinforced, it is ideal for differentiating the three major theoretical perspectives. Sociologists working within the functionalist, conflict, and symbolic interactionist perspectives have recognized this important aspect of sport, but they differ regarding the implications.

Functionalism

How do functionalists perceive the role of sport in society? Functionalists see sport as an important aspect of society primarily because of its contribution to the smooth functioning of society's other parts (Leonard 1998; Coakley 1998). Consider these four functions:

1. *Sport socializes people to the basic beliefs, norms, and values of society*. Sport readies us for adult roles (Kleiber and Kelly 1980). Games, for instance, prepare participating athletes for work in an organizational setting. In fact, young people who are exposed to competitive sport become more achievement motivated than those who are not. And the earlier the exposure, the higher the achievement orientation. This is important because achievement motivation is considered essential to the maintenance of productivity in the modern economy (Lüschen 1968; Lüschen and Sage 1981).
2. *Sport promotes a sense of social identification.* Teams bind people to their community and nation. Clevelanders are united in their love of the Browns, Indians, and Cavaliers. Around the middle of the twentieth century, the United States seemed to be divided into Dodger and Yankee fans. And the Atlanta Braves are trying to be "America's team." Higher social integration results.
3. *Sport offers a safe and controlled release of aggressive feelings generated by the frustrations, anxieties, and strains of modern life*. It is socially acceptable to yell and scream for an athletic team. Similar behavior directed at a professor, employer, parent, or sorority sister can have negative consequences.
4. *Sport promotes the development of physical fitness and sound character*. With its emphasis on hard work, discipline, and exercise, sport encourages people to live responsible lives. Sport contributes to the development of motor skills needed for living in the modern world (Wohl 1979).

TABLE 1.4

Paradoxes In Sport

Stanley Eitzen (2003) argues that sports in America is inherently contradictory. Existing alongside examples of self-sacrifice, teamwork, courage, and hard work are displays of selfishness, greed, violence, and exploitation. Here are a few of the paradoxes Eitzen identifies and dissects.

Effects of Sports	Positive Side	Negative Side
Social integration	• Sports can unite different social classes and racial/ethnic groups.	• Sports can heighten barriers that separate groups.
Fair play	• Sports promote fair play through adherence to rules.	• Sports' emphasis on winning induces some cheating.
Physical fitness	• Sports promote muscle strength, weight control, endurance, and coordination.	• Sports can lead to the use of drugs, excessive weight loss or gain, and injuries.
Academics	• Sports contribute to higher education through scholarships and fund-raising.	• Sports siphon money away from academics toward athletics and emphasize athletic performance over the classroom and graduation.
Social mobility	• Sports allow athletes to obtain an education who might otherwise not attend college.	• Sports' promise of fame and wealth in the professional ranks after graduation can be kept only for a few.

Do you agree with Eitzen that these paradoxes exist? Which theoretical perspective best explains these paradoxes?

Source: D. Stanley Eitzen, *Fair and Foul: Beyond the Myths and Paradoxes of Sport*, 2nd ed., Lanham, MD: Rowman & Littlefield Publishers, Inc., 2003, pp. 4–7.

The functionalist perspective provides some important understandings about the role of sport in society. The perspective does, however, have some shortcomings.

What are the shortcomings of the functionalist perspective on sport? In the first place, functionalism tends to emphasize the positive consequences of sport while overlooking its negative effects. For example, sport can distort values, as when the pressure to win leads to cheating or when top professional golfers playing in the annually televised "Skins Game" can walk away with hundreds of thousands of dollars after eighteen holes of golf while countless children are sick, hungry, and dying. Second, by focusing on how the parts of the system contribute to the whole system, functionalists neglect the conflicts of interests between different social groups. High school athletes, for example, enjoy high prestige at the expense of nonathletes.

Some sociologists have raised disturbing questions about the effects of sport on society (Eitzen 2003). These questions are best understood through the conflict perspective.

Conflict Theory

How do conflict theorists perceive the role of sport in society? Conflict theorists are interested in who has the power and how the elite use power to satisfy their own interests. To conflict theorists, sport is a social institution in which the most powerful oppress, manipulate, coerce, and exploit others. Conflict theorists view sport as a reflection of the unequal distribution of power and money, and they emphasize the role of sport in maintaining inequality (Leonard 1998).

Modern economies are based on highly efficient systems of production. These large bureaucracies, whether white-collar or blue-collar, place most workers in highly fragmented and alienating jobs. Sport, contend conflict theorists, prepares people for a world full of stopwatches, time schedules, and production quotas.

While functionalists see sport as contributing to the integration of society, conflict theorists do not. People from all major segments of the community or society may join in cheering for the same team, but this is only a temporary occurrence:

> *When the game is over, the enthusiasm dies, the solidarity runs short, and disharmony in other relations reasserts itself. Much as one hour a week cannot answer to the religious impulse, one game a week cannot answer to the solidarity needs of a racist, sexist, or elitist society.* (T. R. Young 1986:9)

Basic social class divisions, in other words, will continue to exist and to affect social relationships in a community even if the local team has just won the World Cup or the Super Bowl.

The contribution of sport to character formation is questioned by conflict theorists. Among college athletes, the degree of sportsmanship apparently declines as athletes become more involved in the sports system. Nonscholarship athletes display greater sportsmanship than those with athletic scholarships, and those who have not earned letters exhibit more sportsmanship than letter winners (Eitzen 2000).

Moreover, sport itself becomes a product to be bought and sold. For example, rather than exercise alone or with groups of friends, many people join athletic clubs that charge large fees for the use of their facilities and instructors. The equipment associated with sport has become big business as well. Equipment that was unavailable or considered a luxury a few years ago is increasingly defined as necessary for successful involvement in a particular sport.

Athletes, according to the conflict perspective, are also commodities traded on the market. Professional athletes are purchased and sold just like any other item. Market value, for example, is reflected in the draft pick of college football and basketball players—the higher the draft pick, the higher the value of the first contract. And use of free agency is merely the professional athlete's way of determining market value.

Conflict theorists can point to any number of scandals in both the amateur and professional ranks. Articles in the sports section of the daily newspaper report on athletes—from high school to the professional level—who are taking drugs, cheating in school, or accepting illegitimate cash "gifts." One university after another is being investigated and penalized by the National Collegiate Athletic Association (NCAA). Coaches as well as players are involved in misconduct:

> *The enormous pressures to win result sometimes in scandalous behaviors. Sometimes there are illegal payments to athletes. Education is mocked by recruiting athletes unprepared for college studies, by altering transcripts, by having surrogate test-takers, by providing phantom courses, and by not moving the athletes toward graduation. Florida State University provides an example of the wanton disregard for following the rules. FindLaw, an Internet site that provides legal news and commentary, annually presents a list of football powers with legal problems, called "Football's Tarnished Twenty." In November 2001, it gave the yearly title to the Florida State University Seminoles (called "Creminoles" by FindLaw) for the second consecutive year* (Eitzen 2003:128)

What are the limitations of the conflict perspective on sport? In the first place, conflict theory tends to overlook forces other than economic ones (such as entertainment value, affluence, and the availability of leisure) in attempting to explain the role of sport in modern society. Second, some critics believe that conflict theory places too much emphasis on the degree to which sport is manipulated and controlled by the elite. Third, conflict theory is charged by some with ignoring the personal satisfaction derived from participation in sport (Leonard 1998).

Both the functionalist and conflict perspectives make important points regarding the role of sport in society. Sport does, as functionalists suggest, serve several important functions for society. On the other hand, as conflict theorists suggest, many sports have become so closely tied to elite interests that they contribute more to private profit than to the general well-being of society. The conflict perspective reflects major social concerns in American sport, such as racism and sexism.

Symbolic interactionism also contributes to our understanding of sport as a social institution. This theoretical perspective concentrates on the subjective meanings, social relationships, and self-identity processes associated with sport.

Symbolic Interactionism

What is the unique view that symbolic interactionists have of sport? Symbolic interactionists are concerned with the meanings assigned to the symbols of sports activities and the interpretations made by those involved in these activities. The meanings and interpretations are important in large part because they affect the self-concepts of the participants as well as the relationships among those involved.

The social context of Little League baseball illustrates this perspective. For three years, Gary Alan Fine (1987) studied American adolescent suburban males who played Little League baseball. He discovered and documented a variety of ways in which the boys assigned meanings and interpretations to their team activities. In addition, he described how these meanings and interpretations influenced the boys' social interaction and affected their self-definitions.

What were the meanings and interpretations of the participants? Much of the activity of coaches and parents centered on teaching the rules of the game and imparting moral lessons on matters such as the value of team play, hard work, fair play, competition, and winning. But consistent with the perspective of symbolic interactionism, these ten- to twelve-year-old boys formed their own subjective interpretations of the social messages to which they were being exposed. According to Fine, the boys filtered the adult meanings and symbols through their own shared need for social acceptance among their male peers. Because popularity was directly related to their traditional view of masculinity, the boys tended to interpret their own behavior and the behavior of teammates through this social lens. Consequently, they misinterpreted the adult value of hard work, competition, and so on, as "masculine" and then placed high value on dominance, "toughness," and risky behavior.

How were social interaction and self-concepts affected? The boys acted in ways that convinced coaches and parents of their understanding and acceptance of the intended adult lessons. For example, the aggressive behavior that the boys considered as evidence of their

SOCIOLOGY AND THE NEWS MEDIA

Effects of Divorce

This CNN report focuses on the many sides of divorce in modern American society. One of the most devastating effects of divorce is the psychological pain often experienced by children. This pain can extend into adulthood. Divorce is an excellent topic on which to assess your understanding of the major theoretical perspectives in sociology.

1. State how functionalists would view the divorce rate in the United States.

____

____

2. Focusing on self-esteem, identify some problems faced by children of divorce as a symbolic interactionist would.

____

____

Divorce is usually painful for everyone involved. It can be particularly difficult for children, who are least able emotionally to handle the situation.

masculinity was seen by the coaches and parents as evidence of "hustle," dedication to team competition, and a desire to win. The boys were praised for this behavior, which encouraged them to continue. Moreover, the emphasis on displaying characteristics associated with traditional masculinity led many of the boys to behave toward each other in ways reflecting the U.S. Marine cliché "Real men don't bond." This accent on emotional independence led to social distance and a devaluation of anyone showing opposite traits, such as weakness and submissiveness. "Weaker" peers, younger children, and girls in general frequently experienced the disdain of Little Leaguers attempting to fashion themselves in the image of professional players who pride themselves on sacrificing whatever part of their body it takes to win. From Fine's study, it is easy to imagine the loss of self-esteem suffered by some children from this drive for a traditional masculine identity and social acceptance.

What does the symbolic interactionist perspective leave out? Symbolic interactionism contributes to our understanding of attributed meanings, social relationships, and the resulting development, maintenance, and change in self-concept. Exclusive reliance on symbolic interactionism and its microprocesses of subjective interpretation, interpersonal interaction, and self-development, however, would lead to a failure to consider the ways in which sport is linked to the broader society. Symbolic interactionism, for example, does not directly address the role of sport in society or explore sport within the context of power and social inequality.

FEEDBACK

1. According to the __________ perspective, sport prepares people to function successfully in society, promotes social integration, provides a safe outlet for pent-up feelings, and contributes to the development of character.
2. Which of the following is *not* one of the criticisms of the functionalist perspective?
   a. It underemphasizes the negative consequences of sport.
   b. It overlooks conflicts of interests in sport.
   c. It is blinded by concern for sport and social change.
3. Which of the following *best* describes conflict theorists' views about the relationship between sport and social integration?
   a. Sport promotes social integration by providing a common set of symbols with which people can identify.
   b. Because very few people are interested in sport, it has no real effect on social integration.
   c. Because sport involves competition, it is destructive to social integration.
   d. Although sport may promote temporary social integration, major divisions continue to exist in society.
4. __________ is the theoretical perspective concerned with the meanings assigned and interpretations made by those involved in sport activities.
5. Which of the following does symbolic interactionism fail to explore?
   a. effects of sport on the self-concept
   b. the importance of subjective interpretation of cultural symbols
   c. the ways in which gender stereotypes can be perpetuated by sport activity
   d. the positive correlation between social class and involvement in sport in the United States

Answers: 1. functionalist 2. c 3. d 4. Symbolic interactionism 5. d

SUMMARY

1. Sociology is the scientific study of social structure. It maintains a group rather than an individual focus. It emphasizes the patterned and recurrent social relationships between group members and uses social factors to explain human social behavior. Macrosociology and microsociology are levels of analysis crucial to understanding the sociological perspective.
2. Sociology benefits both the individual and the public. First, through the sociological imagination, individuals can better understand the relationship between what is happening in their personal lives and the social events occurring in their society. The sociological imagination promotes the questioning of conventional, and often misleading, ways of thinking, and it provides a vision of social life that extends far beyond the often narrow confines of one's limited personal experience. Second, sociological research contributes to public policies and programs. Third, sociology enhances the development of occupational skills.
3. It was not until the nineteenth century that the process of differentiating the various social sciences seriously began. Five of the major social sciences—anthropology, psychology, economics, political science, and history—are clearly distinguishable from sociology. Although each is a separate field in its own right, each shares some areas of interest with sociology.

4. Sociology is rather young. It was born out of the social upheaval created by the French and Industrial Revolutions. In an attempt to understand the social chaos of their time, early sociologists emphasized social stability and social change. Sociology received its start primarily from the writings of European scholars Adam Smith, Auguste Comte, Harriet Martineau, Herbert Spencer, Karl Marx, Emile Durkheim, and Max Weber.
5. Auguste Comte, the generally acknowledged father of sociology, believed that society could advance only if studied scientifically. Harriet Martineau contributed to research methods, political theory, and feminism. According to Herbert Spencer, social progress occurs only if people do not interfere with natural processes. Karl Marx argued that history unfolds according to the outcome of conflict between social classes. In capitalist societies, the conflict is between the ruling bourgeoisie and the ruled proletariat.
6. Emile Durkheim shared with the earlier pioneers a concern for social order. Two of his major contributions were the nonpsychological explanation of social life and the introduction of statistical techniques in social research. One of Max Weber's major contributions was also methodological. His method of verstehen assumed an understanding of human social behavior based on mentally putting oneself in the place of others. He also explored the process of rationalization as it existed in the transition from traditional to industrial society.
7. American sociology has been heavily influenced by the early European sociologists, in part because it, too, was born during a time of social upheaval (following the Civil War). From its founding in the 1800s to World War II, the hotbed of American sociology was the University of Chicago. After World War II, sociology departments in the East and Midwest rose to prominence.
8. A theoretical perspective is a set of assumptions considered true by its advocates. Functionalism and conflict theory are the province of macrosociology. Symbolic interactionism is part of microsociology.
9. According to functionalism a society is an integrated whole, seeks a dynamic equilibrium, is composed of elements promoting its well-being, and is based on the consensus of its members. The conflict perspective contradicts these assumptions. A society experiences conflict at all times, is constantly changing, and rests on the domination of some of its members by other members.
10. Feminist theory, a form of conflict theory, can be divided into three frameworks—liberal feminism, radical feminism, and socialist feminism. Although there are important differences between these theoretical frameworks, they each link the lives of women and men to the structure of gender relationships within society.
11. According to symbolic interactionism, people's interpretations of symbols are based on the meanings they learn from others. People base their interaction on their interpretations of symbols. Because symbols permit people to have internal conversations, they can gear their interaction to the behavior they think others expect of them and they expect of others.
12. No single sociological perspective is all-encompassing. Each perspective makes a unique contribution; combined, these perspectives shed considerable light on social life.
13. Functionalists see sport as contributing to society: socializing young people, promoting social integration, providing a release for tensions, and developing sound character.
14. Conflict theorists view sport as a tool of the elite. Not only are athletes used as commodities, contend conflict theorists, but athletics are actually harmful to character development.
15. Whereas both functionalism and conflict theory yield insights on the macro level, symbolic interactionism informs us about micro processes. Through this perspective, we can learn about the meanings, social relationships, and self-identity processes involved in sport.

INFOTRAC® COLLEGE EDITION

http://www.infotrac-college.com/wadsworth

Another unique option available to you at the Student Resources section of the companion Web site is Infotrac College Edition, an online library with access to hundreds of scholarly and popular periodicals. Below are suggested search terms for this chapter. Results from these and other searches are found at the site.

Search keywords: **social conformity**. Conformity is an essential element of group behavior, and we all conform to group expectations (perhaps more than we would like to acknowledge). Find articles that discuss social conformity and relate them to your introduction of sociology.

Search keywords: **positivism**. Find two articles that discuss the use of positivism in a scientific experiment. Was this research method appropriate for the studies?

Search keywords: **sociological perspective**. Look up articles that use the term *sociological perspective*. Try to determine what the authors meant by this term. Does the authors' use of *sociological perspective* match the textbook's use?

LEARNING OBJECTIVES REVIEW

- Illustrate the unique sociological perspective from both the micro and macro levels of analysis.
- Describe three uses of the sociological perspective.
- Distinguish sociology from other social sciences.
- Outline the contributions of the major pioneers of sociology.
- Summarize the development of sociology in the United States.
- Identify the three major theoretical perspectives in sociology today.
- Discuss feminist theory.
- Compare two recent developments in symbolic interactionism.
- Apply the dominant theoretical perspectives to sport.

CONCEPT REVIEW

____ a. economic determinism
____ b. mechanical solidarity
____ c. positivism
____ d. dynamic equilibrium
____ e. social structure
____ f. bourgeoisie
____ g. macrosociology
____ h. sociology
____ i. *verstehen*
____ j. symbol
____ k. latent function
____ l. conflict theory
____ m. presentation of self
____ n. social dynamics
____ o. theoretical perspective

1. a set of assumptions accepted as true by its advocates
2. the theoretical perspective that emphasizes conflict, competition, change, and constraint within a society
3. an unintended and unrecognized consequence of some element of a society
4. the ways people attempt to create a favorable evaluation of themselves in the minds of others
5. the study of social change
6. the method of understanding the behavior of others by putting oneself mentally in another's place
7. patterned, recurring social relationships
8. the scientific study of social structure
9. the use of observation, experimentation, and other methods of the physical sciences in the study of social life
10. something that stands for or represents something else
11. the idea that the nature of a society is based on the society's economy
12. social unity based on a consensus of values and norms, strong social pressure for conformity, and dependence on tradition and family
13. the assumption by functionalists that a society both changes and maintains most of its original structure over time
14. members of industrial society who own the means for producing wealth
15. the level of analysis that focuses on relationships between social structures without reference to the interaction of the people involved

CRITICAL-THINKING QUESTIONS

1. Think of a recent time you conformed. Were you responding to group pressure? Explain.

____

____

2. Apply the sociological imagination to your college choice. Identify the social forces (for example, family, friends, school, and the media) that you believe had the most influence on your decision.

____

____

3. Discuss the possible benefits of sociology to your future personal and work life.

____

____

4. Sociology, psychology, anthropology, economics, political science, and history are all considered social sciences. What do they have in common that justifies this grouping? Explain.

____

____

5. Max Weber introduced the concept of verstehen. How would you use this approach to social research if you wanted to investigate the importance of money to your peers?

____

____

6. Think of an aspect of human social behavior (for example, college sports or fraternities and sororities) that you

would like to know more about. Which of the three theoretical perspectives would you use? Explain your choice.

____

____

7. Think about some social change (for example, increased drug use or later age at first marriage) that you think is particularly significant. Is this change better explained by conflict theory or functionalism?

____

____

8. Could any of your behavior over the past week be described by the dramaturgical perspective? Explain, using personal examples.

____

____

9. Has your self-concept been affected by sports? Has a friend's? Explain the effects from the symbolic interactionist viewpoint.

____

____

MULTIPLE-CHOICE QUESTIONS

1. According to C. Wright Mills, the _________ is the set of mind that allows individuals to see the relationship between events in their personal lives and events in their society.
   a. sociological awareness
   b. macrosociology
   c. sociological imagination
   d. microsociology
   e. moral imperative

2. Shepard defines sociology as
   a. the scientific study of socialism and other political-economic systems.
   b. the opposite of psychology.
   c. the scientific arm of social work.
   d. the study of how culture affects the personality.
   e. the scientific study of social structure.

3. Microsociology focuses on
   a. class conflicts.
   b. relationships among social structures.
   c. the way individuals interact with each other in social relationships.
   d. material very similar to psychology.
   e. emotional-mental development.

4. Macrosociology focuses on
   a. relationships between individuals within groups.
   b. political systems.
   c. various systems of producing, distributing, and consuming goods and services.
   d. how individuals interpret and react to their own culture.
   e. relationships between social structures without concern for the interaction of the individuals involved.

5. The sociological perspective that places human needs and goals as the primary concern of sociology is referred to as
   a. the sociological imagination.
   b. qualitative sociology.
   c. the ministerial perspective.
   d. humanist sociology.
   e. pure sociology.

6. Positivism is defined as
   a. the approach that should be used in dealing with the soft sciences.
   b. a religion that started in the early 1900s in France.
   c. a mental attitude designed to ensure success in life.
   d. the use of observation and experimentation in research.
   e. the balance of power between groups.

7. Karl Marx saw class conflict as
   a. one of the major economic consequences of World War I.
   b. the product of insecurities of the Cold War.
   c. one of the social and economic consequences of the social chaos of nineteenth-century Europe.
   d. conflict between those controlling the means for producing wealth and those laboring for them.
   e. conflict among those controlling the means for producing wealth.

8. Karl Marx's political objective was to
   a. end the British Monarchy.
   b. obtain a seat in the British Parliament.
   c. explain the workings of capitalism to hasten its fall through revolution.
   d. promote the use of religion as the opium of the masses.
   e. identify the distinctive characteristics of monogamy to enhance family stability.

9. According to Emile Durkheim, the social unity based on a complex of highly specialized roles that make members of a society dependent on one another is
   a. mechanical solidarity.
   b. organic solidarity.
   c. positivism.
   d. value freedom
   e. dialectic materialism.

10. The theorist who wrote *The Protestant Ethic and the Spirit of Capitalism* was
   a. Emile Durkheim.
   b. Alexis de Tocqueville.
   c. Max Weber.
   d. Lester Ward.
   e. Robert Merton.

11. Functionalism is defined as

a. the theoretical perspective that emphasizes the contributions made by each part of society.
b. the type of society characterized by weak family ties, competition, and impersonal social relationships.
c. the form of control by which authority is split evenly between husband and wife.
d. a system in which strategic industries are owned and operated by the state.
e. a sense of identification with the goals and interests of the members of one's social class.

12. _________ emphasizes conflict, competition, change, and constraint within a society.

a. The functionalist perspective
b. Critical theory
c. The symbolic interactionist perspective
d. Social revolutionist theory
e. Conflict theory

13. The concept of patriarchy is most closely associated with

a. liberal feminism.
b. socialist feminism.
c. conservative feminism.
d. Marxist feminism.
e. radical feminism.

14. Impression management is a central concept of which of the following theoretical perspectives?

a. functionalism
b. conflict theory
c. symbolic interactionism
d. critical theory
e. exchange theory

15. Functionalists view all of the following as functions of sport *except*

a. encouraging an uncritical acceptance of the worst aspects of society.
b. socializing people to the basic beliefs, norms, and values of society.
c. providing people with a sense of social identification that increases their feelings of attachment to the larger society and to one another.
d. serving a socioemotional function for society by providing a safe and controlled release of aggressive feelings generated by the frustrations, anxieties, and strains of modern life.
e. promoting the development of physical fitness and sound character.

FEEDBACK REVIEW

True-False

1. Microsociology focuses on the relationships between social structures without reference to the interaction of the people involved. T or F?
2. W. E. B. Du Bois focused only on the American race question. T or F?

Fill in the Blank

3. _________ explanations of group behavior are inadequate because human activities are influenced by social forces that individuals have not created and cannot control.
4. According to Durkheim, the United States is a social order based on _________ solidarity.
5. The _________ is the set of mind that enables individuals to see the relationship between events in their personal lives and events in their society.
6. A _________ perspective is a set of assumptions accepted as true by its advocates.

Matching

7. Match the following fields of study and aspects of family life.

____ a. anthropology
____ b. sociology
____ c. history
____ d. political science
____ e. economics
____ f. psychology

(1) distribution of income and the family
(2) effects of slavery on family stability
(3) relationship between voting and the family
(4) effects of early marriage on divorce
(5) link between early childhood emotional support and familial retardation
(6) universality of the nuclear family

8. Indicate whether the following statements represent functionalism (F), conflict theory (C), or symbolic interactionism (S).

____ a. Religion helps hold a society together morally.
____ b. Symbols are crucial to social life.
____ c. A change in the economy leads to a change in the family structure.
____ d. Many elements of a society exist to benefit the powerful.
____ e. Social life should be understood from the viewpoint of the individuals involved.
____ f. Social change is constantly occurring.

9. Match the three feminist theoretical frameworks with the words or phrases.

____ a. liberal feminism
____ b. radical feminism
____ c. socialist feminism

(1) patriarchy
(2) capitalism and patriarchy
(3) equality of opportunity

MULTIPLE-CHOICE QUESTIONS

10. Which of the following is not one of the uses of the sociological imagination?
a. seeing the interplay of self and society
b. capacity for creating new aspects of culture no thought of by others
c. ability to question aspects of social life most people take for granted
d. capability of understanding the social forces that shape daily life

11. Which of the following sociologists contended that social change leads to progress, provided that people do not interfere?
a. Durkheim
b. Weber
c. Martineau
d. Spencer
e. Marx
f. Du Bois

GRAPHIC REVIEW

Table 1.3 contains data on average income (after taxes) in America — by income category. Answering the following questions will test your understanding of this table.

1. What is the most important generalization you can make from these data?

____
____
____

2. How would conflict theorists interpret the data?

____
____
____

3. Would functionalists agree with the interpretation of conflict theorists? Why or why not?

____
____
____

ANSWER KEY

Concept Review
a. 11
b. 12
c. 9
d. 13
e. 7
f. 14
g. 15
h. 8
i. 6
j. 10
k. 3
l. 2
m. 4
n. 5
o. 1

Multiple Choice
1. c
2. e
3. c
4. e
5. d
6. d
7. d
8. c
9. b
10. c
11. a
12. e
13. e
14. c
15. a

Feedback Review
1. F
2. F
3. Individualistic
4. organic
5. sociological imagination
6. theoretical
7. a. 6
   b. 4
   c. 2
   d. 3
   e. 1
   f. 5
8. a. F
   b. S
   c. F
   d. C
   e. S
   f. C
9. a. 3, 4
   b. 1
   c. 2
10. b
11. d

2

Social Research

© Mary Kate Denny/PhotoEdit

OUTLINE

LEARNING OBJECTIVES

- Identify major nonscientific sources of knowledge; then explain why science is a superior source of knowledge.
- Discuss cause-and-effect concepts and apply the concept of causation to the logic of science.
- Differentiate the major quantitative research methods used by sociologists.
- Describe the major qualitative research methods used by sociologists.
- Explain the steps sociologists use to guide their research.
- Describe the role of ethics in research.
- State the importance of reliability, validity, and replication in social research.

USING THE SOCIOLOGICAL IMAGINATION

Does lower church attendance result in a rise in juvenile delinquency? The finding of a statistical link between church attendance and delinquency (delinquency increasing as church attendance decreases) meets the test of common sense. One can easily speculate on why this would be the case. An observed relationship between these two events, however, does not necessarily mean that one causes the other. In fact, delinquency increases as church attendance decreases because of a third factor—age. Age is related to both delinquency and church attendance. Older adolescents not only go to church less often, but are more likely to be delinquents. The apparent relationship between church attendance and delinquency, then, is actually produced by a third factor—age—that affects both of the original two factors.

Mistaken ideas can survive when people rely on sources of knowledge not grounded in the use of reason and not stimulated by the search for truth. A major benefit of sociological research lies in its replacement of false beliefs with more accurate knowledge. Before we turn to the logic of scientific research and the methods of sociological research, it will be helpful to consider some sources of knowledge.

Sources of Knowledge

Nonscientific Sources of Knowledge

How do we know what we think we know? Four major nonscientific sources of knowledge are *intuition*, common sense, authority, and tradition. Intuition is quick and ready insight that is not based on rational thought. To intuit is to have the feeling of immediately understanding something because of insight from an unknown inner source. For example, the decision against dating a particular person because "it feels wrong" is a decision based on intuition.

Common sense refers to opinions that are widely held because they seem so obviously correct. The problem with commonsense ideas is that they are often wrong. Some people take as common sense, for example, that property values will decline when African Americans move into a white middle-class neighborhood. Yet, the research doesn't support this idea. As philosopher Alasdair MacIntyre writes, "Because common sense is never more than an inherited amalgam of past clarities and past confusions, the defenders of common sense are unlikely to enlighten us" (MacIntyre 1997:117).

An *authority* is someone who is supposed to have special knowledge that we do not have. A king believed to be ruling by divine right is an example of an authority. Reliance on authority is often appropriate. It is more reasonable to accept a doctor's diagnosis of an illness than to rely on information from a neighbor whose friend had the same symptoms (although even a single doctor's diagnosis should not be accepted uncritically). In other instances, however, authority can obscure the truth. Astrologers who advise people to guide their lives by the stars are an example of a misleading authority. Advocates of science and supporters of authority are currently debating an approach to teaching science: evolution versus creation.

The fourth major nonscientific source of knowledge is *tradition*. Despite evidence to the contrary, it is traditional to believe that an only child will be self-centered and socially inept. In fact, to avoid these alleged personality traits, most Americans still wish to have two or more children (Sifford 1989). And barriers to equal opportunity for women persist in industrial societies despite evidence that traditional negative ideas about the capabilities of women are fallacious.

Nonscientific sources of knowledge often provide false or misleading information (Adler and Clark 2003), sometimes leading to completely opposite conclusions. One person's intuition tells him to buy oil stocks, whereas another person's intuition tells her to avoid all energy stocks. One person's commonsense conclusion may deem the feminist movement as harmful to the family, while it seems perfectly obvious to someone else that feminism promotes family values.

Because nonscientific sources of knowledge are often accepted at face value, most people do not challenge the information. Consequently, reality can be distorted for an extended time. Science is a more reliable method for obtaining knowledge because it is based on the principles of objectivity and verifiability.

Science as a Source of Knowledge

What is objectivity? According to the principle of **objectivity**, scientists are expected to prevent their personal biases from influencing the interpretation of their results. A male, antifeminist biologist investigating aptitudes, for example, is supposed to guard against any unwarranted tendency to conclude that males make better scientists than females. Researchers must interpret their data solely on the basis of merit; the outcome they personally prefer is irrelevant. This is what Max Weber (1918–1946) meant by value-free research (see Chapter 1).

SNAPSHOT OF AMERICA 2.1

Suicide Rates, 1997 (deaths per 100,000 population)

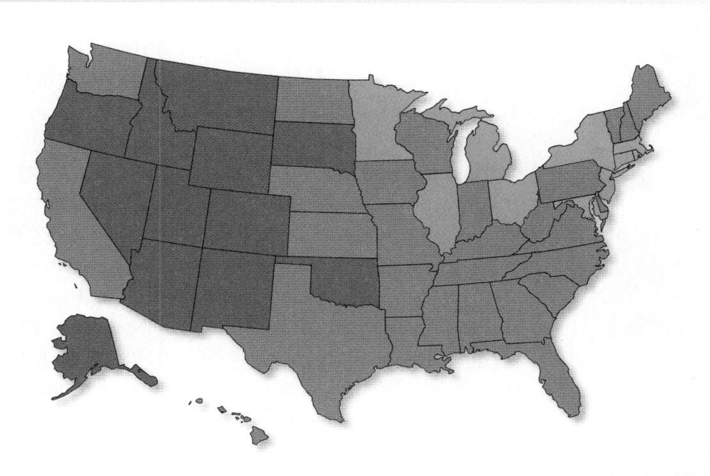

Source: From the Center for Disease Control and Prevention, *National Vital Statistics Reports*, Vol. 47, No. 19, Table 26 (June 30, 1999), p. 85.

Above average 14.4 or more
Average: 10.5–14.3
Below average: 10.4 or fewer

Refer to Snapshot 2.2.

SNAPSHOT OF AMERICA 2.2

People per Square Mile

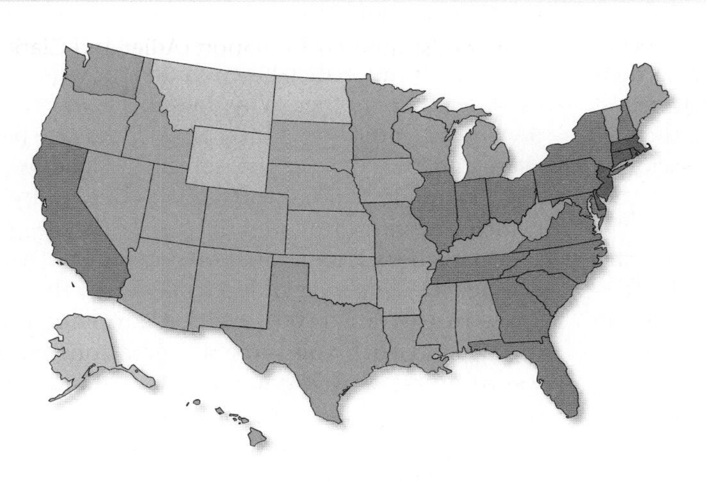

Above 1,300
260–1,299
130–259
26–129
Below 25

Source: Lisa Thomas, *Student Atlas* (New York: DK Publishing, 1998), 39.

Population densities in the United States are very low compared with those in other countries. For example, if New York City were as densely populated as Lagos, Nigeria, it would contain three-quarters of the total U.S. population. This map shows the number of people per square mile in each state.

1. Emile Durkheim's study of suicide suggested that one factor in the suicide rate is the degree to which an individual has group ties. One indication of social ties is population density. Based on this map, where would you expect to find the highest suicide rates in the United States?
2. Does Snapshot 2.1 agree with your predictions?
3. What other factors might account for the variations in suicide rates?

Can scientists really be objective? Sometimes, scientists unintentionally let their personal biases influence their work. For example, in the 1950s, pioneering sex researcher Alfred Kinsey revolutionized popular thinking about sex in America. He has been accused of being both a homosexual and a masochist, characteristics said to unduly influence his research. James Jones (1997) makes this charge and presents evidence that Kinsey was a man with an ideological agenda whose research methods undermine his claim to objectivity. Kinsey's generalizations about the American population were based, according to Jones, on data gathered largely from volunteers, including disproportionate numbers of male prostitutes, gays, and prison inmates.

© Bettmann/CORBIS

When forming opinions about the risk of AIDS, many Americans rely on nonscientific sources. The protestors in this picture are concerned about the harm done to children in an environment void of scientifically based knowledge.

Historian Howard Zinn (2201, 2003) emphasizes the impossibility of being completely neutral with respect to the outcomes of research. Since researchers bring their own interests to work, they must decide what to report and what to omit from a wide array of possible theories and "facts." This process of choice, whether consciously or unconsciously, reflects personal preferences.

How can subjectivity be reduced? Scientists, then, cannot possibly be completely objective. But if subjectivity in research cannot be eliminated, it can be reduced. According to Zinn, the best approximation of objectivity exists when researchers strive for the truth and follow specific safeguards: continually reexamine their thinking, permit contradicting evidence to alter their view, and make public any evidence that runs counter to their view of the truth. If researchers are aware of their biases, they can consciously take them into account. They can be more careful in designing research instruments, selecting samples, choosing statistical techniques, and interpreting results. According to Swedish economist Gunnar Myrdal (1983), personal recognition of biases is insufficient; public exposure of them is essential. Personal values, Myrdal contends, should be explicitly stated, so that those who read a research report can be aware of the author's biases.

What is verifiability? **Verifiability** means that a study can be repeated by other scientists. This is possible because scientists report in detail on their research methods. Verifiability is important because it exposes scientific work to critical analysis, retesting, and revision by colleagues. If researchers repeating a study produce results at odds with the original study, the original findings will be questioned. Under these circumstances, erroneous theories, findings, and conclusions will not survive (Begley 1997).

Causation and the Logic of Science

The Nature of Causation

What is causation? Scientists assume that an event occurs for a reason. According to the concept of **causation**, events occur in predictable, nonrandom ways, and one event leads to another. Why does this book remain stationary rather than rise slowly off your desk, go past your eyes, and rest against the ceiling? Why does a ball thrown into the air return to the ground? Why do the planets stay in orbit around the sun?

More than two thousand years ago, Aristotle claimed that heavier objects fall faster than lighter ones. In the late 1500s, Galileo contended that all objects fall with the same acceleration (change of speed) unless slowed down by air resistance or some other force. But it was not until the late 1600s that Sir Isaac Newton developed the theory of gravity. We now know that objects fall because the Earth has a gravitational attraction for objects near it. The planets remain in orbit around the sun because of the gravitational force created by the sun.

FEEDBACK

1. Intuition is quick and ready insight based on rational thought. T or F?
2. The major problem with nonscientific sources of knowledge is that such sources often provide erroneous information. T or F?
3. Define objectivity and verifiability as used in science.
4. According to Gunnar Myrdal, it is enough that scientists themselves recognize their biases. T or F?

Answers: 1. F 2. T 3. Objectivity exists when an effort is made to prevent personal biases from distorting research. Verifiability means that any given piece of scientific research done by one scientist can be duplicated by other scientists. 4. F

Because scientists assume causation, one of their main goals is to discover cause-and-effect relationships. They attempt to discover the factors—there is usually more than one—that cause events to happen.

Why multiple causation? Leo Rosten, noted author and political scientist, once wrote, "If an explanation relies on a single cause, it is surely wrong." Events in the physical or social world are generally too complex to be explained by any single factor. For this reason, scientists are guided by the principle of **multiple causation**, which states that an event occurs as a result of several factors operating in combination. What, for example, causes crime? Cesare Lombroso, a nineteenth-century Italian criminologist, believed that the predisposition to commit crimes was inherited and that criminals could be identified by certain physical traits (large jaws, receding foreheads). Modern criminologists reject Lombroso's (or anyone else's) one-factor explanation of crime. They cite numerous factors that contribute to crime, including drugs, excessive materialism, peer pressure, hopeless poverty in slums, and overly lax, overly strict, or erratic child-rearing practices. Each of these factors is a variable.

Causation and Variables

How, in a research project, are variables related to causation? A **variable** is a characteristic (such as age, education, or social class) that is subject to change. It occurs in different (varying) degrees. Some materials have greater density than others; some people have higher incomes than others; the average literacy rate is higher in developed countries than in developing ones. Each of these is a **quantitative variable**, a variable that can be measured and given a numerical value. Because differences can be measured numerically, individuals, groups, objects, or events can be pinpointed at some specific point along a continuum. In contrast, a **qualitative variable** consists of variation in kind rather than in number. It denotes differences in distinguishing characteristics of a category rather than differences in numerical degree. Sex, marital status, and group membership are three qualitative variables often used by sociologists.

TABLE 2.1

Forming Quantitative and Qualitative Variables

Please indicate whether you strongly agree, agree, disagree, or strongly disagree with each of the following statements:

	Strongly Agree	Agree	Disagree	Strongly Disagree
a. Most schoolteachers really know what they are talking about.	1	2	3	4
b. To get ahead in life, you have to get a good education.	1	2	3	4
c. My parents encouraged me to get a good education.	1	2	3	4
d. Children should not be allowed to quit school at sixteen.	1	2	3	4
e. Too little emphasis is put on education these days.	1	2	3	4
f. My parents think that going to college is essential.	1	2	3	4

These questions are designed to measure the value people place on education. The total score on this measure can range from 6 for those strongly agreeing with all six statements (highest value on education) to 24 for those strongly disagreeing with all six statements (lowest value on education). Like any quantitative variable, most actual scores will fall somewhere between these two extremes.

In your own words, describe your views on the importance of education.

____

____

____

Answers to this question could be used by a sociologist to classify people into one of two categories: those who place a high value on education or those who place a low value on education.

This is an either/or or yes/no variable: People are either male or female; they are either single or married; they are either Greek or they are not (see Table 2.1.).

Once scientists identify the variables (qualitative or quantitative) to investigate, they define these variables as either independent or dependent. Variables that cause something to occur are **independent variables**. Variables in which a change (or effect) can be observed are **dependent variables**. Marital infidelity is an independent variable (although, of course, not the only one) that may cause the dependent variable of divorce. The independent variable of poverty is one of several independent variables that can produce a change in the dependent variable of hunger. Whether a variable is dependent or independent varies with the context. Hunger may be a dependent variable in a study of poverty; it may be an independent variable in a study of crime.

An **intervening variable** influences the relationship between an independent variable and a dependent variable. A government support program, for example, may intervene between poverty and hunger. With a strong safety net, very poor parents and their children may experience no more hunger than working-class families. Poverty is the cause of hunger, but it does not have to be if there is government intervention in the form of income and food. The poor *without* a safety net will experience more hunger; the poor *with* a safety net will not.

What is a correlation? After identifying which independent and dependent variables they will study, sociologists attempt to find out how the variables are related. They describe these relationships in terms of correlations. A **correlation** exists when a change in one variable is associated with a change (either positively or negatively) in the other.

A correlation may be positive or negative. A **positive correlation** exists if both the independent variable and the dependent variable change in the same direction. A positive correlation exists if we find that grades (dependent variable) improve as study time (independent variable) increases (see Figure 2.1). In a **negative correlation**, the variables change in opposite directions. For example, an increase in the inde-

WORLDVIEW 2.1

The Wired World

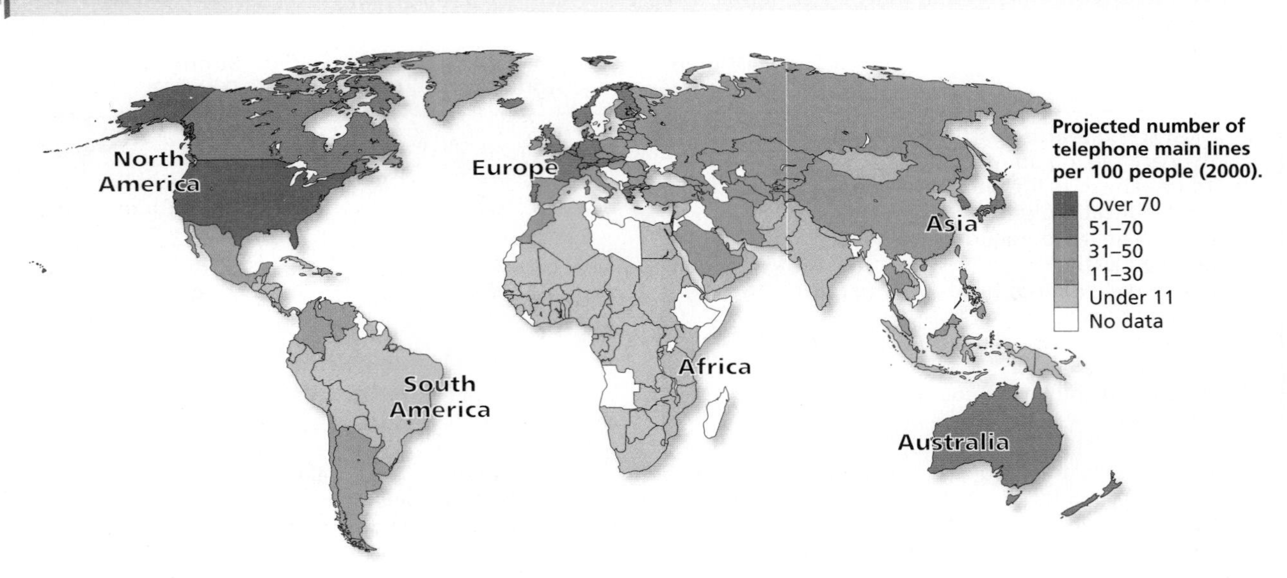

Source: Ian Pearson (ed), *Atlas of the Future* (New York: Macmillan Books, 1998), pp. 56–57.

This world map shows the number of telephone main lines per 100 people projected for 2000. It illustrates the creation of a quantitative variable that could be used in research. Some questions below will allow you to apply your understanding of the nature of variables to this example.

1. Explain why the data in this map constitutes a quantitative variable.
2. What would need to be done with the data to make it a qualitative variable?
3. If you were to use the number of telephone main lines per 100 people as a research variable, to which sociological variable would you most like to relate it? Would it be a dependent or independent variable? Explain.

FIGURE 2.1

Positive and Negative Correlations

Examine the graphs below. Choose two examples of your own to illustrate a positive and a negative correlation. Briefly explain why you would expect the type of correlation in each case.

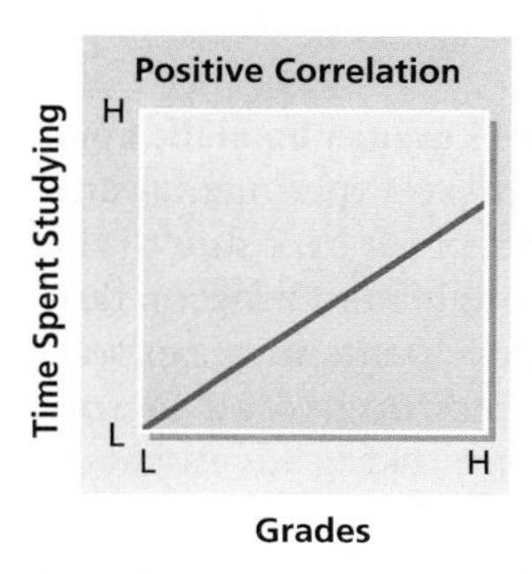

In a positive correlation, increases in one variable are associated with increases in the other. Grades (dependent variable) improve with time spent studying (independent variable).

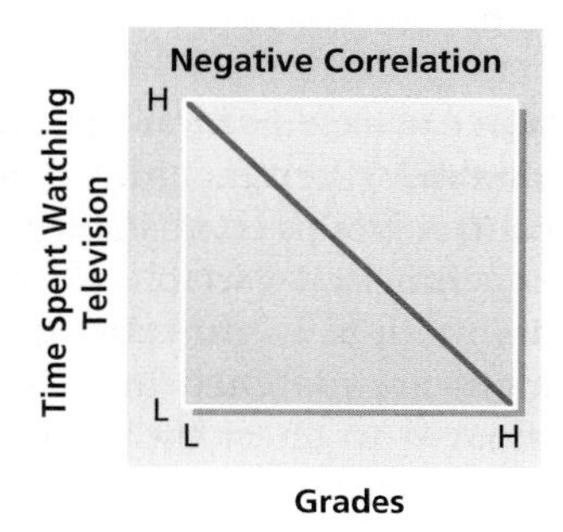

In a negative correlation, increases in one variable are associated with decreases in the other. Grades (dependent variable) decrease as time spent watching television (independent variable) increases.

pendent variable is linked to a decrease in the dependent variable. A negative correlation exists if we find that grades (dependent variable) go down as time spent watching television (independent variable) increases.

What are the criteria for establishing a causal relationship? Establishing causation is much more complicated than establishing a correlation between two variables. In a causal relationship, one variable actually causes the other to occur. Three standards are commonly used for establishing causality (Lazarsfeld 1955; Hirschi and Selvin 1973). These standards can be illustrated in the mistaken assumption that lower church attendance causes higher juvenile delinquency. (See "Using the Sociological Imagination.")

1. *Two variables must be correlated.* Some researchers found that juvenile delinquency increases as church attendance declines (R. Stark, Kent, and Doyle 1982). Does this negative correlation mean that lower church attendance causes higher delinquency? To answer this question, the second criterion of causality must be met.
2. *All possible contaminating factors must be taken into account.* A major problem in establishing causality lies in the control of all relevant variables. Holding contaminating factors constant (ruling out their influence) is one of the greatest challenges in science.

   Although all cause-and-effect relationships involve a correlation, the existence of a correlation does not necessarily indicate a causal relationship. Just because two events vary together (are correlated) does not mean that one causes the other. Two totally unrelated variables may have a high correlation. In fact, the correlation between lower church attendance and delinquency is known as a **spurious correlation**—an apparent relationship between two variables, which is actually produced by a third variable that affects both of the original two variables.

   The negative relationship between church attendance and delinquency occurs because age is related to both church attendance and delinquency (older adolescents attend church less frequently, and older adolescents are also more likely to be delinquents). Thus, before we can predict that a causal relationship exists between church attendance and delinquency, we need to take into account all variables relevant to the relationship. In this instance, the age variable reveals that the relationship between church attendance and delinquency is not a causal one. Church attendance is the contaminating factor here.
3. *A change in the independent variable must occur before a change in the dependent variable can occur.* The cause must occur before the effect. Does lack of church attendance precede delinquency, or vice versa? Logically, either one can precede the other, or they can occur simultaneously. Thus, even if the original correlation between church attendance and delinquency is maintained after holding age and other possible contaminating factors constant, causality between these two variables still cannot be established. Why? Because it cannot be determined which occurred first.

The Experiment as a Model

Describing an experiment is an excellent way to illustrate causation. Though used infrequently by sociologists, experiments provide insight into the nature of all scientific research because they are grounded in the concept of causation.

An **experiment** takes place in a laboratory in an attempt to eliminate all possible contaminating influences. By ruling out extraneous factors, a researcher can determine the effects (if any) of an independent variable on a dependent variable. According to the logic of experiments, if the dependent variable changes when the experimental (independent) variable is introduced but does not change when it is absent, the change must have been caused by the independent variable.

The basic ingredients of an experiment are a *pretest*, a *posttest*, an *experimental variable*, an *experimental group*, and a *control group*. Suppose a researcher wants to study, experimentally, the effects of providing information on drug use (the experimental, or independent, variable) on student attitudes toward drug use (the dependent variable). After selecting a class of eighth graders, the researcher could first measure the teenagers' attitudes toward drug use (pretest). Then, at a later time, a film demonstrating the harmful effects of drugs might be shown to the class. After the movie, the students could again be questioned about their attitudes toward drug use (posttest). Any changes in their attitudes toward drug use that took place between the pretest and the posttest could be attributed to the experimental variable. Such a conclusion might be wrong, however, because the change could have been due to factors other than the experimental variable: A student in the school might have died from an overdose, a rock star might have publicly endorsed drug use, or a pusher might have begun selling drugs to the students.

The conventional method for controlling the influence of contaminating variables is to select a control group as well as an experimental group. In the preceding example, half of the eighth-grade class could have been assigned to the **experimental group**, the group exposed to the experimental variable, and half to the **control group**, the group not exposed to the experimental variable. Assuming that the members of each group had similar characteristics and that their experiences between the pretest and the posttest were the same, any difference in attitudes toward drug use between the two groups could reasonably be attributed to the students' exposure or lack of exposure to the film.

How can experimental and control groups be made comparable? The standard way to make experimental and control groups comparable—except for exposure to the experimental variable—is through matching or randomization. In **matching**, participants in an experiment are matched in pairs according to all factors thought to affect the relationship being investigated. Members of each pair are then assigned to one group or the other. In **randomization**, which is preferable to matching, subjects are assigned to the experimental or control group on a random (chance) basis. Assignment to one group or the other can be determined by flipping a coin or by having subjects draw numbers from a container. Whether matching or randomization is used, the goal is the same—to form experimental and control groups that are alike with respect to all relevant characteristics except the experimental variable. If this requirement has been met, any significant change in the experimental group, as compared to the control group, can be attributed with considerable confidence to the experimental variable. That is, a causal link will have been established between the independent and dependent variables.

FEEDBACK

1. Match the following concepts and statements.

____ a. causation
____ b. multiple causation
____ c. variable
____ d. quantitative variable
____ e. qualitative variable
____ f. independent variable
____ g. dependent variable
____ h. correlation
____ i. spurious correlation

(1) something that occurs in varying degrees
(2) the variable in which a change or effect is observed
(3) a change in one variable associated with a change in another variable
(4) the idea that an event occurs as a result of several factors operating in combination
(5) a factor that causes something to happen
(6) the idea that the occurrence of one event leads to the occurrence of another event
(7) a factor consisting of nonnumerical categories
(8) when a relationship between two variables is actually the result of a third variable
(9) a variable consisting of numerical units

2. An ________ attempts to eliminate all possible contaminating influences on the variables being studied.
3. The group in an experiment that is not exposed to the experimental variable is the ________ group.
4. Experimental and control groups are made comparable in all respects except for exposure to the experimental variable through ________ or ________.

Answers: 1. a.(6) b.(4) c.(1) d.(9) e.(7) f.(5) g.(2) h.(3) i.(8) 2. experiment 3. control 4. matching; randomization

Quantitative Research Methods

Because sociologists find it difficult to create experimental and control groups, they tend to rely more on other research methods, classified as either quantitative or qualitative. Quantitative research uses numerical data. Qualitative research rests on descriptive data.

Quantitative research methods include survey research and precollected data. About 90 percent of the research published in major sociological journals is based on surveys, so this approach is discussed first.

Survey Research

The **survey**, in which people are asked to answer a series of questions, is the most widely used research method among sociologists. It is ideal for studying large numbers of people.

How are effective surveys conducted? In survey research, care must be taken in selecting respondents and in formulating the questions to be asked (T. R. Black 2001). Researchers describe the people surveyed based on populations and samples.

A **population** consists of all those people with the characteristics a researcher wants to study. A population can be all college sophomores in the United States, all former drug addicts living in Connecticut, or all current inmates of the Ohio State Penitentiary. Most populations are too large and inaccessible to collect information on all members. For this reason, a sample is drawn. A **sample** is a limited number of cases drawn from the larger population. The sample must be selected carefully if it is to have the same basic characteristics as the population. If a sample is not representative of the population from which it is drawn, the survey findings cannot be used to make generalizations about the entire population (Winship and Mare 1992). The U.S. Census Bureau regularly uses sample surveys in its highly accurate work. The Gallup and Harris polls are also recognized as reliable sample indicators of national trends and public opinion.

How are representative samples selected? A **random sample** is selected on the basis of chance, so that each member of a population has an equal opportunity of being selected. A random sample can be selected by assigning each member of the population a number and then drawing numbers from a container after they have been thoroughly scrambled. An easier and more practical method, particularly with large samples, uses a table of random numbers in which numbers appear without pattern. After each member of the population has been assigned a number, the researcher begins with any number in the table and goes down the list until enough subjects have been selected.

If greater precision is desired, a **stratified random sample** can be drawn. With this method, the population is divided into categories such as sex, race, or age. Subjects are then selected randomly from each category. In this way, the proportion of persons in a given category reflects the population at large.

How is information gathered in surveys? In surveys, information is obtained through either a questionnaire or an interview. A **questionnaire** is a written set of questions that survey participants answer by themselves. In an **interview**, a trained interviewer asks questions and records answers. Questionnaires or interviews may be composed of either closed-ended or open-ended questions (May 2001).

Closed-ended questions are those for which a limited, predetermined set of answers is possible—multiple-choice questions, for example. Because participants must choose from rigidly predetermined responses, closed-ended questions sometimes fail to elicit the participants' underlying attitudes and opinions. On the positive side, closed-ended questions make answers easier to quantify and compare. **Open-ended questions** ask the respondent to answer in his or her own words. Answers to open-ended questions, however, are not easy to quantify, and the interviewer may

© John A. Rizzo/Getty Images

These employees of the U.S. Census Bureau are entering data from one of this government agency's many surveys. The results of these surveys, considered to be representative of the United States population, are widely utilized for decision making by private individuals, business organizations, and political leaders.

BASIC STATISTICAL MEASURES

Statistics are methods used for tabulating, analyzing, and presenting quantitative data (Levin and Fox 2002). Sociologists, like all scientists, use statistical measures. The statistics you will encounter in this textbook and in the sources you are likely to read later, such as *The Wall Street Journal*, *Time*, *Newsweek*, and *The Economist*, are easily comprehended. Among the most basic statistical measures are averages (modes, means, medians) and correlations.

An *average* is a single number representing the distribution of several figures: Suppose the following annual salary figures were those of nine highly paid Major League Baseball players:

$3,300,000 (Catcher)
$4,500,000 (First Base)
$3,600,000 (Second Base)
$4,900,000 (Starting Pitcher)
$3,600,000 (Third Base)
$5,300,000 (Left Field)
$4,200,000 (Center Field)
$6,100,000 (Right Field)
$4,300,000 (Shortstop)

There are three kinds of averages that can make these numerical values more meaningful (see Figure 2.2). Each gives a different picture.

The *mode*—in this case $3,600,000—is the number that occurs most frequently. The mode is appropriate only when the objective is to indicate the most popular number. If a researcher were to rely on the mode alone, though, in a report of these Major League Baseball salaries, readers would be misled, because no mention is made of the wide range of salaries ($3,300,000 to $6,100,000).

The *mean* is the measure closest to the everyday meaning of the term *average*. It lies somewhere in the middle of a range. The mean of the salary figures listed earlier—$4,422,222—is calculated by adding all the salaries together ($39,800,000) and dividing by the number of salaries (9). The mean, unlike the mode, takes all of the figures into account, but it is distorted by the extreme figure of $6,100,000. Although one player earns $6,100,000, most players make considerably less—the highest-paid player earns nearly twice as much as the lowest-paid player. The mean is distorted when there are extreme values at either the high or low end of a scale; it is more accurate when extremes are not widely separated.

The *median* is the number that divides a series of values in half; that is, half of the values lie above it, half below. In this example, the median is $4,300,000—half of the salaries are above $4,300,000, half are below. What is the advantage here? The median is not distorted by extremes.

FIGURE 2.2

Mode, Mean, and Median of Salaries

Suppose that you were negotiating your salary with a Major League Baseball team. Also suppose that your agent presented these average salaries as currently accurate with the news that management would give you only one chance to declare your bottom-line salary. Which type of average would you choose? Explain why.

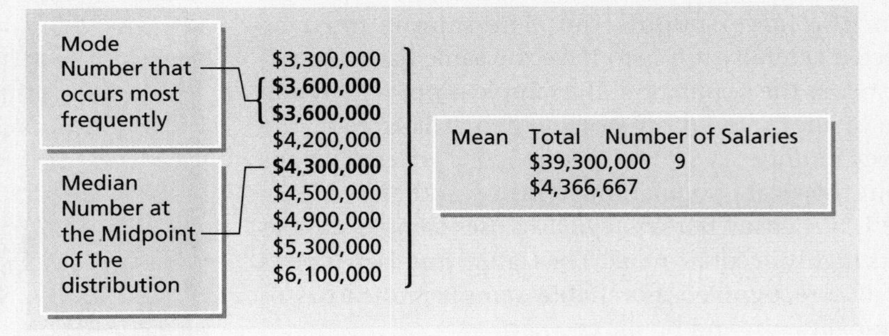

inadvertently change the meaning of the respondent's answers by rephrasing them.

What are the advantages and disadvantages of closed-ended survey research? Surveys based on closed-ended questions have the advantage of precision and comparability of responses. They permit the use of statistical techniques, a feature they have in common with experiments. Statistical techniques can be used because of still other advantages in survey research. For example, surveys permit the collection of large samples, which in turn permit more detailed analysis; surveys include a large number of variables; and variables in surveys can be quantified.

The survey research method has several disadvantages, however (Schuman 2002). First, surveys are expensive due to the large samples usually involved. Second, because survey questions are predetermined, interviewers cannot always include important unanticipated information, although they are encouraged to write such information in the margin or on the back of the interview form. Third, the response rate—particularly for mailed questionnaires—is often low. A respectable return rate is about 50 percent, although researchers make an effort to obtain a return rate of 80 percent or higher. Even in interviews, some people are not available and some refuse to answer the questions. Because nonresponses can make the sample unrepresentative, surveys may be biased. Fourth, the phrasing of survey questions may also introduce bias. For example, negatively phrased questions are more likely to receive a no answer than neutrally phrased questions. It is better to ask, "Are you in favor of abortion?" than "You aren't in favor of abortion, are you?" Respondents also interpret the same question differently. If asked about the extent of their drug use, some respondents may include alcohol in their answers; others may not. Fifth, surveys cannot probe deeply into the context of the social behavior being studied; they draw specific bits of information from respondents, but they cannot capture the total social situation. Finally, survey researchers must be on guard for the *Hawthorne effect,* when unintentional behavior on the part of researchers influences the results they obtain from those they are studying (Roethlisberger and Dickson 1964; originally published in 1939). As researchers and survey participants interact, participants detect cues regarding what the researchers are trying to find. The participants, depending on the circumstances, may attempt to please the researcher, frustrate the researcher's goals, or give "socially acceptable" answers.

Precollected Data

The use of previously collected information is a well-respected method of obtaining data. This is known as **secondary analysis**. In fact, the first sociologist to use statistics in a sociological study—Emile Durkheim—relied on precollected data (see Doing Research).

What are the major types of precollected data? The sources for precollected data are as varied as government reports, company records, voting records, prison records, and reports of research done by other social scientists. The U.S. Census Bureau is one of the most important sources of precollected data for sociologists. The Census Bureau collects information on the total population every ten years and conducts countless specific surveys each year. Consequently, detailed information exists on such topics as income, education, race, sex, age, marital status, occupation, death rates, and birth rates. Other government agencies also collect information. The U.S. Department of Labor regularly collects information on the nation's income and unemployment levels across a variety of jobs. The U.S. Department of Commerce issues monthly reports on various aspects of the economy.

What are the advantages and disadvantages of precollected data? Precollected data provide sociologists with inexpensive, quality information. Existing sources of information also permit the study of a topic over a long period of time. With census data, for example, we can begin in 1960, with the onset of the War on Poverty, and trace the changes in the relative income levels of African Americans and whites. Also, because the data have been collected by others, the researcher cannot influence the answers.

Because the information was collected for different purposes, there is also a downside to precollected data. The existing information may not exactly suit the current researcher's needs. Also, the people who originally collected the data may have been biased. Finally, sometimes precollected data are simply too old to be valid.

FEEDBACK

1. Match the following terms and statements:
   ____ a. population
   ____ b. representative sample
   ____ c. random sample
   ____ d. sample
   ____ e. survey
   (1) selected on the basis of chance so that each member of a population has an equal opportunity of being selected
   (2) all those people with the characteristics the researcher wants to study within the context of a particular research question
   (3) a limited number of cases drawn from the larger population
   (4) a sample that has basically the same relevant characteristics as the population
   (5) the research method in which people are asked to answer a series of questions
2. Use of company records would be an example of using ________ data.

Answers: 1. a. (2) b. (4) c. (1) d. (3) e. (5) 2. precollected

READING TABLES AND GRAPHS

Tables and graphs present information concisely. They are chock-full of fascinating findings, which you can easily decode. Just follow the steps outlined here. The steps are keyed to Table 2.2 and Figure 2.3.

1. Begin by reading the title of the table or graph carefully; it will tell you what information is being presented. Table 2.2 shows median annual incomes in the United States by sex, race, and education.
2. Find out the source of the information. You will want to know whether the source is reliable, whether its techniques for gathering and presenting data are sound. The figures originated from the U.S. Bureau of the Census, a highly trusted source.
3. Read any notes accompanying the table or graph. Not all tables and graphs have notes, but if notes are present, they offer further information about the data. The notes in Table 2.2 and in Figure 2.3 explain that all the data refer to the total money income of full-time and part-time workers, ages 25 and over, in a March 2001 survey.
4. Examine any footnotes. Footnotes in Table 2.2 and Figure 2.3 indicate that the data are categorized by the highest grade actually completed. Although you may have assumed this, "years of schooling" could have referred to the total number of years in school, regardless of the grade level attained.
5. Look at the headings across the top and down the left-hand side of the table or graph. To observe any pattern in the data, it is usually necessary to keep both types of headings in mind. Table 2.2 and Figure 2.3 show the median annual income of African American and white males and females for several levels of education.

TABLE 2.2

Median Annual Income by Sex, Race, and Education

Examine the following data on median annual income by sex, race, and education. State the conclusions you can make from these data. Which findings most surprise you?

Demographic Group	Overall Median Income	Years of Schooling*				
		Less than 9	9–11	12	13–15	16 or more
White males	$40,151	$20,353	$24,711	$31,937	$37,265	$56,788
African American males	$28,554	$15,806	$20,557	$25,183	$30,297	$43,274
White females	$25,251	$12,229	$13,614	$19,667	$22,979	$36,885
African American females	$22,851	$12,208	$13,280	$19,200	$24,189	$36,463

*Based on highest grade completed.
Note: These figures include the total money income of full-time and part-time workers, ages 25 and over, as of March 2001.
Source: U.S. Bureau of the Census, "PINC-03, Educational Attainment—People 25 Years Old and Over, by Total Money Earnings in 2001, Work Experience in 2001, Age, Race, Hispanic Origin, and Sex"; available at http://ferret.bls.census.gov/macro/032002/perinc/new03_000.htm

Qualitative Research Methods

Some aspects of social reality can be revealed best through qualitative (nonquantitative) research methods. Most qualitative research methods include field research and the subjective approach (Strauss and Corbin 1998; Neuman 2002; Lincoln and Denzin 2003).

Field Research

Field research investigates aspects of social life that cannot be measured quantitatively and that are best understood within a natural setting. The world of prosti-

6. Find out what units are being used. Data can be expressed in percentages, hundreds, thousands, millions, billions, means, and so forth. In Table 2.2 and Figure 2.3, the units are dollars and years of schooling.
7. Check for trends in the data. For tables, look down the columns (vertically) and across the rows (horizontally) for the highest figures, lowest figures, trends, irregularities, and sudden shifts. If you read Table 2.2 vertically, you can see how income varies by race and sex within each level of education. If you read the table horizontally, you can see how income varies with educational attainment for white males, African American males, white females, and African American females. A major advantage of presenting these data as a graph (Figure 2.3) is that the sudden shifts, trends, irregularities, and extremes are easier to spot than they are in tables.
8. Draw conclusions from your own observations. Again, look at Table 2.2 and Figure 2.3. Although income tends to rise with educational level for both African Americans and whites, it increases much less for African American men and for women of both races than for white men. At each level of schooling, African American men earn less than white men. In fact, white male high school dropouts have incomes only $1,118 below African American male high school graduates; white male high school graduates earn over $1,400 more than African American males with some college but no degree. White women and African American women both appear to improve their earning power through high school and college education.

FIGURE 2.3

Median Annual Income by Gender, Race, and Education

Clearly, this table documents the income advantage that white males in the United States have over white females and African Americans of both sexes. Explain how this situation can be used to challenge the existence of a true meritocracy.

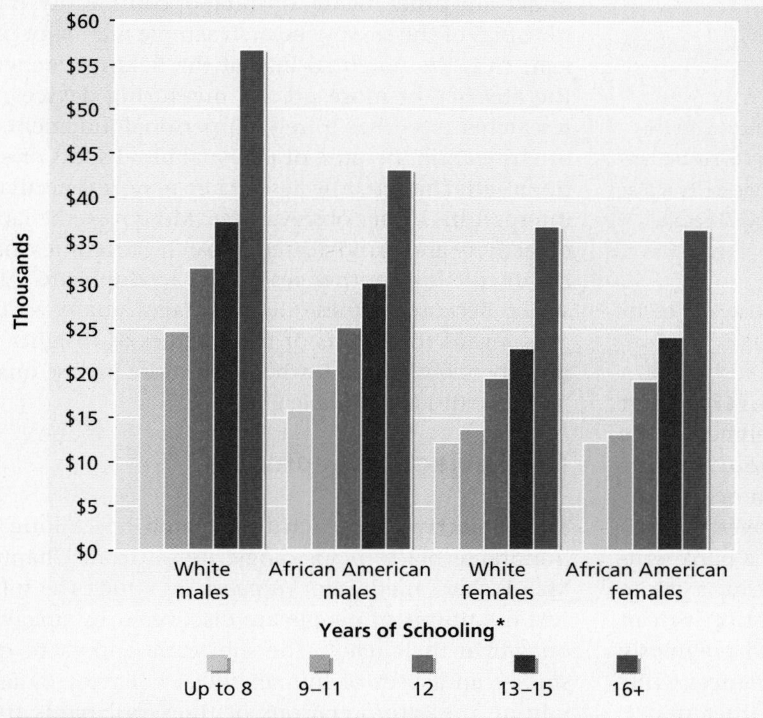

*Based on highest grade completed.
Note: These figures include the total money income of full-time and part-time workers, ages 25 and over, as of March 2001.
Source: U.S. Bureau of the Census, "PINC-03, Educational Attainment—People 25 Years Old and Over, by Total Money Earnings in 2001, Work Experience in 2001, Age, Race, Hispanic Origin, and Sex"; available at http://ferret.bls.census.gov/macro/032002/perinc/new03_000.htm.

tution, the inner workings of a Mafia family, and events during a riot are all topics best studied by field research.

The most popular approach to field research is the **case study**—a thorough investigation of a single group, incident, or community. Case studies are accomplished primarily through intensive observation, information obtained from informants, and informal interviews. Newspaper files, formal interviews, official records, and surveys can be used to supplement these techniques.

This method assumes that the findings in one case can be generalized to similar situations. The conclusions of a study on drug use in Chicago, for example, should apply to other large cities as well. It is the researcher's responsibility to indicate factors that are unique and that would not apply to other situations.

The survey is the most widely used research method for collecting data in sociology. Surveys are usually conducted in person, although use of the telephone is becoming much more common. An advantage of the survey is that it permits the gathering of information about a large number of people.

Researchers conducting case studies often use the technique of participant observation.

What is participant observation? In **participant observation**, a researcher becomes a member of the group being studied. A researcher may join a group with or without informing its members that he or she is a sociologist. A compelling account of covert participant observation appears in *Black Like Me*, a book written by John Howard Griffin (1961). Griffin, a white journalist, dyed his skin to study the life of African Americans in the South. Although he had previously visited the South as a white man, his experiences while posing as an African American were quite different.

Sociologists sometimes reveal themselves as participant researchers. Elliot Liebow studied two dozen disadvantaged African American men who hung around a particular corner in Washington, D.C. His study illustrates the open approach to participant observation. Even though he was a white outsider, Liebow was allowed to participate in the daily activities of the men: "The people I was observing knew that I was observing them, yet they allowed me to participate in their activities and take part in their lives to a degree that continues to surprise me" (Liebow 1998:253).

What are the advantages and disadvantages of field studies? Field studies can produce a depth and breadth of understanding unattainable with quantitative research methods. Uniquely, they can reveal insights into a social situation from the experiences of the people involved. Adaptability is another advantage. Even after a field study has begun, it is easily altered when new insights or oversights are discovered. (This is in contrast to survey research, where changes are not practical once research is in progress.) In addition, field studies are valuable where survey research would be either impossible or biased—as in a study of skid-row derelicts or organized crime. (Imagine studying a delinquent gang with a set of questions about their activities.) Because of these advantages, field studies may produce insights and explanations not likely to be unearthed through quantitative research.

Disadvantages do exist, however. The findings from one case may not be generalizable to similar situations. One mental hospital or community may be quite unlike any other mental hospital or community. If possible bias of the survey research sample is a major problem, so is the potential bias of the field researcher. In the absence of more precise measuring devices, the researcher often has to rely on personal judgment and interpretation. Because of personal blind spots or emotional attachment, the researcher may not accurately interpret his or her observations. Moreover, the lack of objectivity and standardized research procedures makes it difficult for another researcher to duplicate a field study. Because of these disadvantages, many sociologists regard the results of field studies as insights that must be investigated further with more precise quantitative methods (see Table 2.3).

The Subjective Approach

The subjective approach to research has a long and honorable place in sociology. Recall from Chapter 1 Max Weber's method of *verstehen*, in which the subjective intentions of people are discovered by imagining oneself in their place. The subjective approach, then, studies an aspect of human social behavior by ascertaining the interpretations of the participants themselves. A prominent example of the **subjective approach** is ethnomethodology, a development in microsociology that attempts to uncover taken-for-granted social routines.

How does ethnomethodology work? **Ethnomethodology** is the study of processes people develop and use in understanding the routine behaviors expected of themselves and others in everyday life. Ethnomethodologists assume that people share the meanings that underlie much of their everyday behavior. Through observing others and through a process of trial and error in social situations, people develop a sense of appropriate behavior. This understanding prevents them from making silly or serious social errors and saves them from having to continually decide the fitting behavior for particular sit-

Marianne Bertrand and Sendhil Mullainathan—What's in a Name?

Social researchers infrequently use the experiment as a method of study. And when they do use the experimental method, it is usually a field or "natural" experiment—an experiment done outside the laboratory in the context of normal social life (Babbie 2002). Marianne Bertrand and Sendhil Mullainathan (2002) conducted a field experiment on racial discrimination in the labor market.

Bertrand and Mullainathan began their research with the knowledge that racial inequality exists in the U.S. labor market. This chapter has documented the presence of this inequality. Using race as their experimental variable, the authors asked two research questions: (1) Do employers actively discriminate against African Americans in hiring by rejecting them in favor of whites who are only equally qualified; and (2) Does improving the credentials of African Americans compared with whites overcome this discrimination?

With their field experiment design, Bertrand and Mullainathan sent resumes in response to a series of Chicago and Boston help-wanted ads and used the rate of callback for interviews as the measure of success for each resume. To manipulate perception of race, the researchers randomly assigned half of the resumes very white-sounding names such as Emily Walsh or Brendan Baker. The other half of the resumes were randomly given very African American–sounding names like Lakisha Washington or Jamal Jones. The researchers responded to each employment ad by sending two higher-quality resumes (one with a white-sounding name, the other with an African American–sounding name) and two lower-quality resumes (one with a white-sounding name, one with an African American–sounding name). They responded to over 1,300 ads for a wide variety of white-collar jobs, sending out almost 5,000 resumes.

Even though race was assigned randomly, callback rates for African Americans and whites were quite different. Applicants with white names received one callback for every 10 or so resumes. Fifteen resumes were required for resumes with African American names. In other words, white names attracted 50 percent more callbacks than African American names. Since applicants' names were randomly assigned, the researchers could safety attribute the differential callback rate to racial discrimination.

Race, they found, also influenced the likelihood of higher-quality resumes being rewarded with invitations to interviews. Whites with better resumes received 30 percent more invitations to interviews than whites with lower-quality resumes. In contrast, African American applicants did not benefit from appearing better qualified. That is, African Americans saw little benefit from upgrading their credentials.

Sendhil and Mullainathan conclude that discrimination is an important reason for poor African American performance in the U.S. labor market. African Americans get fewer interview opportunities and are not highly rewarded for improving their observable qualifications. As long as these conditions exists, racial inequality will persist.

Thinking About the Research

1. Discuss the implications of these findings for African American incentives.
2. Relate this research to the tendency to blame the victim.
3. Identify some public policy implications of these findings.

uations. Predictable, patterned behavior is a result of this process (Livingston 1987; Atkinson 1988; Hilbert 1990; Pollner 1991).

How can ethnomethodologists discover what is going on in the minds of individuals as they construct a mental sense of social reality? Because they are not mind readers, ethnomethodologists have had to be inventive. Harold Garfinkel (1984) is a prominent advocate of ethnomethodology. He believes the best course to understanding people's construction of social reality is to deprive them momentarily of their mental maps of daily routines. If people are deprived of their definitions of expected behaviors, they reconstruct a coherent picture of social reality. Ethnomethodologists can learn by observing this process of reconstruction.

Garfinkel writes of situations that his students have created. Here the researchers can observe what people do when deprived of their taken-for-granted social routines. The following passage describes a situation in which an experimenter (E) is attempting to deprive a subject (S) of his sense of expected routine by asking for more detailed information than is normally required in everyday situations. In the context of watching television, the experimenter first asks, "How are you tired? Physically, mentally, or just bored?"

TABLE 2.3

Research Methods: Advantages and Disadvantages

A summary of the advantages and disadvantages of several basic research methods is presented here. Suppose that you wished to do a study of the relationship between dating and self-esteem in high school. From the table, select the method that you think is best suited to your research. Explain the advantages and disadvantages of this method to your research.

Research Method	Definition	Advantages	Disadvantages
Quantitative Methods Survey research	People answer a series of questions, usually predetermined.	• Precision and comparability of answers. • Use of statistical techniques. • Information on large numbers of people. • Detailed analysis.	• Expensive due to large numbers. • Low response rate. • Phrasing of questions introduces bias in favor of certain answers. • Researchers' behavior can affect answers given.
Secondary analysis	Information gathered by one researcher is used by another researcher for a different purpose.	• Inexpensive. • Can study a topic over a long period of time. • Researchers' influence on subjects avoided.	• Information collected for a different reason may not suit another researcher's needs. • Original researcher may already have introduced biases. • Information may be outdated.
Experiment	Occurs in a laboratory setting with a minimum of contaminating influences.	• Can be replicated with precision. • Variables can be manipulated. • Can be relatively inexpensive. • Permits the establishment of causation (rather than just correlation).	• Laboratory environment is artificial. • Not suited to most. sociological research. • Number of variables studied is limited.
Qualitative Methods Case study	Thoroughly investigates a small group, incident, or community.	• Provides depth of understanding from group members' viewpoint. • Unexpected discoveries and new insights can be incorporated into the research. • Permits the study of social behavior not feasible with quantitative methods.	• Difficult to generalize findings from one group to another group. • Presence of researcher can influence results. • Hard to duplicate. • Takes lots of time. • Difficult to be accepted as a group member (in case of participant observation).

FEEDBACK

1. Field studies are best suited for situations in which _________ measurement cannot be used.
2. A _________ is a thorough investigation of a small group, an incident, or a community.
3. In _________, a researcher becomes a member of the group being studied.
4. According to the _________ approach, some aspect of social structure is best studied through an attempt to ascertain the interpretations of the participants themselves.
5. _________ is the study of the processes people develop and use in understanding the routine behaviors expected of themselves and others in everyday life.

Answers: 1. quantitative 2. case study 3. participant observation 4. subjective 5. Ethnomethodology

SCRUTINIZING POPULAR REPORTS

We are living in the "age of instant information," and unfortunately the media often report these data without adequate "processing." Here are some suggestions that can help us become more savvy consumers in the information marketplace.

Maintain a Skeptical Attitude. Be skeptical, because the media "sound bite treatment" tends to sensationalize and distort information. For example, the media may report that a university researcher spent $500,000 to find out that love keeps families together when, in fact, this may have been only one small aspect of the larger research project. Moreover, chances are the media have oversimplified even this part of the researcher's conclusions.

Consider the Source of Information. The credibility of a study may be affected by the organization funding the research. For example, find out whether a study on the relationship between cancer and tobacco has been sponsored by the tobacco industry or by the American Cancer Society. Representatives of tobacco companies deny the existence of any research linking throat and mouth cancer with snuff dipping. An impartial medical researcher concluded that putting a "pinch between your cheek and gum" has, in the long run, led to cancer in humans. Which report would you believe? At the very least you want to know the source of information before making a judgment about scientific conclu- sions. This caution is especially relevant to the Internet, which is now a major source of information. Because this information varies widely in its accuracy and reliability, sources must be evaluated with particular care.

Determine Whether a Control Group Has Been Used. Knowing whether a control group (a group not exposed to the experimental variable) has been used may be important. For instance, increases in self-esteem and physical energy may be reported in a study of participants in a meditation program. Was this because of the meditation techniques themselves or because of the respect and attention the participants were given during the training period? Or a study may report that the productivity of a group of workers in an office increased dramatically because the workers were allowed to participate in work-related decisions. Did the employees' productivity increase because of their participation in decision making or because they were involved in something new and exciting? Without one or more control groups, you cannot be certain of what caused the changes in the meditation participants or in the office workers.

Do Not Mistake Correlation for Causation. A correlation between two variables does not necessarily mean that one caused the other. For example, at one time, the percentage of Americans who smoked was increasing at the same time life expectancy was increasing. Did this mean that smoking caused people to live longer? Actually, a third factor—improved health care—accounts for the increased life expectancy. Do not assume that two events are related causally just because they occur together.

Critical Thinking

Bring to class a report of a sociological study from a popular magazine or news weekly. Be prepared to share how these four safeguards can be applied.

S: I don't know, I guess physically, mainly.
E: You mean that your muscles ache or your bones?
S: I guess so. Don't be so technical. (After more watching)
S: All these old movies have the same kind of old iron bedstead in them.
E: What do you mean? Do you mean all old movies, or some of them, or just the ones you have seen?
S: What's the matter with you? You know what I mean.
E: I wish you would be more specific.
S: You know what I mean! Drop dead! (Garfinkel 1984:43)

The researcher continues this type of conversation until the subject is disoriented and can no longer respond within a previously developed frame of reference. The researcher can then observe the subject's creation of a new definition for expected or "normal" social interaction.

A Model for Doing Research

In an effort to obtain accurate knowledge, social scientists use a research model known as the *scientific method*. This model follows several distinct steps: identifying a problem, reviewing the literature, formulating hypotheses, developing a research design, collecting data, analyzing data, and stating findings and conclusions (Schutt 2001).

Identifying the Problem

Researchers begin by choosing an object or topic for investigation. A research question may be chosen for a variety of reasons—because it interests the researcher, addresses a social problem, tests a major theory, or responds to a government agency's needs.

Reviewing the Literature

Once the object of study has been identified, the researcher must examine the literature for relevant theories and previous research findings. For example, a sociologist investigating suicide will probably develop an approach that makes use of the classic study of suicide by Emile Durkheim as well as other studies on the topic.

Formulating Hypotheses

From a careful examination of relevant theory and previous research findings, a sociologist can state one or more **hypotheses**—tentative, testable statements of relationships among variables. These variables must be defined precisely enough to be measurable. One hypothesis might be "The longer couples are married, the less likely they are to divorce." The independent variable (length of marriage) and the dependent variable (divorce) must be defined and measured. Scientists measure variables through the use of **operational definitions**—definitions of abstract concepts in terms of simpler, observable procedures. Divorce could be defined operationally as the legal termination of marriage. If so, the measurement of divorce would be qualitative—the couple is either legally married or not. Length of marriage could be measured quantitatively—the number of years a couple has been legally married.

Developing a Research Design

A research design describes the procedures the researcher will follow for collecting and analyzing data. Will the study be a survey or a case study? If it is a survey, will data be collected from a cross section of an entire population, as with the Harris and Gallup polls, or will a sample be selected from only one city? Will simple percentages or more sophisticated statistical methods be used? These and many other questions must be answered while the research design is being developed.

Collecting Data

There are three basic ways of gathering data in sociological research: asking people questions, observing behavior, and analyzing existing materials and records. Sociologists studying the harmony in interracial marriages could question couples about their communication skills and compatibility. They could locate an organization with a large number of interracially married couples and observe couples' behavior. Or they could compare the divorce rate among interracially married couples with the divorce rate of the population as a whole.

Analyzing Data

Once the data have been collected and classified, they can be analyzed to determine whether the hypotheses

TABLE 2.4

FOCUS ON THEORETICAL PERSPECTIVES: Investigating School Violence and School Funding

This table illustrates the research method a sociologist of a particular theoretical persuasion would *most likely* choose to investigate school violence and school funding. Any of the three sociologists, of course, could use *any* of the three research methods.

Theoretical Perspective	Research Method	Approach to the Research Question
Functionalism	Survey	• A questionnaire on violence in high school is sent to a national, random sample of principals. The survey examines a possible relationship between incidence of school violence and level of school funding.
Conflict theory	Case study	• A particular high school with low funding is studied with respect to a relationship between school violence and school funding. Researchers interview administrators, teachers, and students.
Symbolic interactionism	Participant observation	• Concealing her identity, a researcher takes a temporary job at a high school with low funding. She attempts to covertly observe a possible link between school violence and school funding.

FEEDBACK

Listed below are the steps in the research model. Following are some concrete examples related to the sociability of the only child. Indicate the appropriate example for each step number.

____ Step 1: identifying the problem
____ Step 2: reviewing the literature
____ Step 3: formulating hypotheses
____ Step 4: developing a research design
____ Step 5: collecting data
____ Step 6: analyzing data
____ Step 7: stating findings and conclusions

a. The researcher reads past theory and research on the sociability of only children.
b. From previous research and existing theory, the researcher states that only children appear to be more intelligent than children with siblings.
c. The researcher collects data on only children from a high school in a large city.
d. The researcher writes a report giving evidence that only children are more intelligent than children with brothers or sisters.
e. The researcher decides to study the intelligence level of only children.
f. The researcher classifies and processes the data collected to test a hypothesis.
g. The researcher decides on the data needed to test a hypothesis, the methods for data collection, and the techniques for data analysis.

Answers: Step 1: e Step 2: a Step 3: b Step 4: g Step 5: c Step 6: f Step 7: d

are supported. This is not as easy or automatic as it sounds, because results are not always obvious. Because the same data can be interpreted in several ways, judgments have to be made. Guarding against personal biases is especially important in this phase of research.

Stating Findings and Conclusions

After analyzing the data, a researcher is ready to state the conclusions of the study. It is during this phase that the methods are described and the hypotheses are formally accepted, rejected, or modified. The conclusions of the study are then related to the theory and research findings on which the hypotheses are based, and directions for further research are suggested. Depending on the findings, the original theory itself may have to be altered. Whether the statement of conclusions appears in a scientific journal, a book, or online, it includes a description of the methods used. By making the research procedures public, scientists make it possible for others to either duplicate the research, conduct a slightly different study, or proceed in a very different direction.

Using the Research Model

Realistically, do sociologists follow these steps? Some sociologists believe that this research model is too rigid to capture spontaneous, subjective, and changeable social behavior. They may prefer to discover their findings without preconceived ideas of outcome, without hypothesis-biased observations, and without an inflexible research design. Even though most sociologists do follow the model, they do not do so mechanically. They may conduct exploratory studies prior to stating hypotheses or before developing research designs. Or they may alter their hypotheses and research designs as their investigations proceed.

This does not mean that these steps can be ignored in conducting sociological research. Most sociologists more or less go through this process anyway. And all researchers, even those who say they do not follow this procedure, have the model in mind as they do their work. Moreover, research reports are evaluated with this method in mind. If researchers seriously violate the research process, their findings and conclusions may not be viewed as credible.

Ethics in Social Research

The Issue of Ethics

Research is a distinctly human activity. Although there are principles for conducting research, such as objectivity and verifiability, scientists sometimes forgo them. At times, the ethics of research is not honored by researchers (Greenberg 2001; K. Chang 2002; Goodstein 2002; Spotts 2002).

Unfortunately, there is a long list of examples of ethical lapses in medical research. During the Nuremberg trials, twenty Nazi doctors were convicted of conducting sadistic experiments on concentration camp inmates. From 1932 to 1972, the Public Health Service of the U.S.

government deliberately did not treat approximately 400 syphilitic African American sharecroppers and day laborers, so that biomedical researchers could study the full evolution of the disease (J. M. Jones 1993). For twenty years, researchers at Germany's University of Heidelberg used human corpses, those of adults and children, in high-speed automobile crash tests (Fedarko 1993). Federal investigators in the United States have documented over ten years of fraud in some of the most important breast cancer research ever conducted (Crewdson 1994). In a Cincinnati hospital, from 1960 to 1972, approximately 100 women and men with cancer were subjected to experimental radiation over their entire bodies in U.S. military–funded research on the effects of nuclear war. Twenty-one died within a month of exposure and nearly all of them died within months of this "therapy" (Stephens 2002).

Plagiarism is an ethical lapse that can occur in any area of research. For example, in 2001 three prominent historians (Stephen Ambrose, Doris Kearns Goodwin, and Joseph Ellis) faced charges of including in their writings the work of others without citation.

Several eminent social scientists have been criticized for conducting research that many scientists consider unethical. In each case, subjects were placed in stressful situations without being informed of the true nature of the experiment (Milgram 1963, 1965, 1974; Zimbardo, Anderson, and Kabat 1981). These and other studies have created great interest in a code of ethics. There is, in fact, a formal code of ethics for professional sociologists (American Sociological Association 1997).

A Code of Ethics in Sociological Research

The formal code of ethics for sociologists covers important areas beyond research, including relationships with students, employees, and employers (American Sociological Association 1997). In broad terms, the code is concerned with maximizing the benefits of sociology to society and minimizing the harm that sociological work might create. Of importance in the present context are the research-related aspects of the code.

Sociologists are committed to objectivity, the highest technical research standards, accurate reporting of their methods and findings, and protection of the rights, privacy, integrity, dignity, and autonomy of the subjects of their research. Because the first several topics have already been covered in this chapter, the focus in this section is on the rights, privacy, integrity, dignity, and autonomy of participants in sociological research.

Occasionally, adherence to the code is documented. Mario Brajuha, a graduate student at a major American university, kept detailed field notes while engaging in a participant observation study of restaurant work (Brajuha and Hallowell 1986). Because of suspected arson after a fire at the restaurant where he was employed as a waiter, his field notes became the object of interest by the police, the district attorney, the courts, and some suspects. By refusing to reveal the contents of his field notes, Brajuha protected the rights of those individuals described in his notes. He did so in the face of a subpoena, threats of imprisonment, and the specter of personal harm to himself, his wife, and his children. The case was finally dropped after two difficult years.

Though lapses in ethics are rare in sociological research, much can be learned from such cases. Laud Humphreys (1979) studied homosexual activities in men's public bathrooms ("tearooms"). By acting as a lookout to warn the men of approaching police officers,

SOCIOLOGY AND THE NEWS MEDIA

Counting Americans

Politics entered the 2000 U.S. Census. Democrats generally favored sampling the population, while Republicans argued for a total count. In this CNN segment, you are introduced to this controversy.

The Census is important because it affects the number of seats in the House of Representatives held by each state. Do you agree with the Democrats or Republicans?

Using material in this chapter, make a case for or against sampling in the 2000 census.

____

____

Discuss the ethics of undercounting minority populations.

____

____

Are Researchers Peeping Toms?

A recent episode of NBC's *Today* show featured a segment about a Louisiana woman whose male neighbor had secretly installed video cameras in her bedroom and bathroom. Because of the cameras, the neighbor was able to secretly observe this woman in her most private moments. Obviously, this is an immoral and extremely illegal use of technology. It is alarming that this type of "peeping" is now within the financial range and technological ability of many people.

Consider the lawsuits that have been filed in some states by workers who discovered that their employers, in an effort to reduce high levels of employee theft, installed cameras in restrooms or changing rooms. Management claims that dishonest employees often use these areas to hide company products in purses or bags. And besides, management asserts, if workers know the cameras are present, they won't do anything for which they can be caught. But workers object; they feel that they are entitled to expect a minimum level of privacy and that cameras violate this expectation.

Not surprisingly, this technology is available to social scientists. While some sociologists and psychologists see distinct advantages to observing subjects through video cameras, many others are concerned about the ethics of this practice. In fact, one requirement of the Code of Ethics of the American Sociological Association is to protect the privacy of research subjects. Suppose a sociologist came to your university and requested permission to place video cameras in the hallways, classrooms, and dining halls. Would permission from university administrators be enough to protect the privacy of research subjects as required by the code of ethics? Even if every student in the university gave permission for the cameras, someone could do something absurd and embarrassing.

There are other concerns as well. If researchers begin videotaping with the consent of their subjects, will they get a true record of behavior? And what happens if a criminal act is recorded? Do the researchers have an obligation to release the tape to the authorities? Such complicated issues will continue to arise as technology allows investigators to invade areas where custom and culture have traditionally prevented them from going.

Analyzing the Trends
Develop an argument for or against the use of video equipment in a sociological research project. Be sure to use logical arguments that relate to the maximizing of benefits to society while minimizing the harms that sociological work might create.

© Spencer Grant/PhotoEdit

he observed their activities closely. As participants left the tearooms, Humphreys recorded their license plate numbers to obtain their addresses for personal interviews. After waiting a year, Humphreys falsely presented himself as a survey researcher to obtain additional information.

Did Humphreys violate the code of ethics as a covert participant observer? Yes, Humphreys violated the privacy of these people. Most did not want their sexual activities known, and Humphreys did not give them the opportunity to refuse to participate in the study. Humphreys also deceived the men by misrepresenting himself in both the tearooms and their homes. Finally, by recording his observations, Humphreys placed these people in jeopardy of public exposure, arrest, or loss of employment. (Actually, because of his precautions, none of the subjects was injured as a result of his research. In fact, to protect their identities, Humphreys even allowed himself to be arrested.)

Do ethical concerns make research harder? Yes, but it is the researcher's responsibility to decide when a particular action crosses an ethical line—a decision not always easy to make, because moral lines are often blurred. Moreover, the researcher must balance the interests of those being studied against the need for accurate, updated data.

Although Kai Erikson is one of the most sensitive and outspoken critics of disguised observation, he has defended it on grounds that it is—on occasion—the only way to obtain relevant information:

FEEDBACK

1. Situations a, b, and c describe three research situations involving possible ethical violations (Babbie 2003). Match each situation with the appropriate aspect of the social science code of ethics for research on human subjects.
   (1) concern for participants' privacy
   (2) avoidance of deception
   (3) obligation not to harm participants
   a. After a field study of deviant behavior during a riot, law enforcement officials demand that the researcher identify those people who were observed looting. Rather than risk arrest as an accomplice after the fact, the researcher complies.
   b. A research questionnaire is circulated among students as part of their university registration packet. Although students are not told they must complete the questionnaire, the hope is that they will believe they must, thus ensuring a higher completion rate.
   c. Researchers obtain a list of gay men they wish to study. They contact them with the explanation that each has been selected "at random" from among the general population to take a sampling of "public opinion."
2. Match the concepts on the left side with the definitions on the right side.

____ a. reliability	(1) when a measurement technique yields consistent results on repeated applications
____ b. validity	(2) the duplication of the same study to ascertain its accuracy
____ c. replication	(3) when a measurement technique actually measures what it is designed to measure

Answers: 1. a. (3) b. (2) c. (1) 2. a. (1) b. (3) c. (2)

> *Some of the richest material in the social sciences has been gathered by sociologists who were true participants in the group under study but who did not announce to other members that they were employing this opportunity to collect research data. . . . It would be absurd, then, to insist as a point of ethics that sociologists should always introduce themselves as investigators everywhere they go and should inform every person who figures in their thinking exactly what their research is about.* (K. T. Erikson 1967:368)

Balance is the key to the issue of ethics. At the very least, subjects—whether in experiments, surveys, or field studies—should be protected from social, financial, psychological, or legal damage (M. Hunt 1999).

A Final Note

Reliability, Validity, and Replication

Researchers can be guided by all the important research considerations we have discussed in this chapter and still not conduct a good study. They can be mindful of objectivity, sensitive to the criteria of causation, and careful in the selection of the most appropriate method (survey, precollected data, field study). Still, they may fail to produce knowledge superior to that yielded by intuition, common sense, authority, or tradition.

What else must a researcher do? Sociologists must pay careful attention to the quality of measurement (Babbie 2002). Consequently, they must emphasize reliability and validity in the creation and evaluation of the measuring devices they use for the variables they wish to investigate.

What is meant by reliability? A measurement technique must yield consistent results on repeated applications—a requirement called **reliability**. Reliability is tested by repeated administration of a measurement technique, such as a questionnaire, to the same subjects to ascertain whether the same results occur each time. Suppose a researcher, after deciding to study satisfaction with day care among parents, designs a questionnaire. If, on repeated applications of the questionnaire to a sample of parents, the level of satisfaction remains consistent, then confidence in the reliability of the measurement device rises. Should, on the other hand, the level of satisfaction—from one administration of the questionnaire to the next—vary over a period of time, then we would doubt that satisfaction with child care is actually being measured.

The problem of reliability is also an issue in qualitative research. Suppose that our field researcher is also interested in satisfaction with day care among the children. If the level of satisfaction among the children seems different each day to the researcher, then doubt is raised about the reliability of the measurement technique being used.

Even when a measurement technique is reliable, it still may not produce scientifically sound results. This is because a measurement technique must be not only reliable, but valid.

What is validity? Validity exists when a measurement technique actually measures what it is designed to measure. Thus, a technique intended to measure parental satisfaction with day care may yield consistent results on repeated applications to a sample of parents, but not really be measuring satisfaction at all. The measurement device might be tapping parental need to view day care positively (to mask guilt feelings about permitting someone else to be the care provider during working hours). Children at a day-care center may appear satisfied to the visiting researcher because they are neglected during the day and welcome his or her attention, or because the day-care provider has coached them to appear satisfied. A measurement technique, in short, may be consistently measuring something very different from what it purports to measure.

How does replication contribute to the self-corrective nature of research? **Replication**—the duplication of the same study to ascertain its accuracy—is closely linked to both reliability and validity in that reliability and validity problems unknown to original researchers are likely to be revealed as subsequent social scientists repeat the research. It is partially through replication that scientific knowledge accumulates and changes over time.

A major goal of scientific research is to generate knowledge that is more reliable than can be obtained from such nonscientific sources as intuition, common sense, authority, and tradition. Through efforts to be objective and to make their research subject to replication by others, researchers attempt to portray reality as accurately as possible. The methods of research presented in this chapter are the specific tools sociologists use to create knowledge of social life that is as accurate as possible.

However, empirical results obtained through the use of research methods are not the final goal of science. As Gerhard Lenski has stated, "Science is more than method: its ultimate aim is the development of a body of 'verified' general theory" (Lenski 1988:163). For this reason, there is constant interaction between sociological theory and research methods. Theory is used to develop hypotheses capable of being supported or falsified through testing. These results, in turn, may support existing theory, alter it, or lead to its ultimate rejection and the creation of a new theory. Lenski points out that divorced from research methods, "sociological theory has more in common with seminary instruction in theology and biblical studies" (1988:165) than it does with the natural sciences model that sociology is emulating. Theory is trustworthy and useful only to the extent that it has been tested and found to be valid.

SUMMARY

1. People tend to get information from such nonscientific sources as intuition, common sense, authority, and tradition. Generally speaking, these sources are inadequate for obtaining accurate knowledge about social life. The advantage of scientific knowledge is grounded in its objectivity and verifiability.
2. Objectivity is the aim of social scientists. Subjectivity can be minimized if researchers make themselves aware of their biases and make their biases public when presenting their findings. Through verification of past research, erroneous theories, findings, and conclusions can be exposed to other scientists.
3. The concept of causation—the idea that the occurrence of one event leads to the occurrence of another event—is central to science. All events have causes, and scientists attempt to discover the factors causing those events.
4. Three criteria must be met before a cause-and-effect relationship can be said to exist. First, two variables must be correlated. That is, change in the independent variable (the causal factor) must be associated with a change in the dependent variable (effect). Second, the correlation must not be spurious, that is, due to the effects of a third variable. Third, it must be shown that the independent variable always occurs before the dependent variable. Scientists think in terms of multiple causation because events are usually caused by several factors, not simply by a single factor.
5. The experiment is a good example of a research method that is based on the idea of causation. Sociologists, however, do not generally use the experimental research method, because it is often impossible to control the relevant social variables.
6. Two major quantitative research methods are the survey and precollected data. Surveys can draw on large samples, are quantitative, include many variables, are relatively precise, and permit the comparison of responses. Researchers must take care to collect representative samples for their surveys. Precollected data permits high-quality research at reasonably low cost and reveals changes in variables over an extended period of time.
7. Field studies are best used when some aspect of social structure cannot be measured quantitatively, when interaction should be observed in a natural setting, and when in-depth analysis is needed. The case study is the popular approach to field research. Some sociologists have adopted a subjective approach in which emphasis is on ascertaining the subjective interpretations of the participants themselves.
8. A research model involves several distinct steps: identifying the problem, reviewing the literature, formulating hypotheses, developing a research design, collecting data, analyzing data, and stating findings and conclusions. These steps form a model for scientific research.

9. Researchers have an ethical obligation to protect participants' privacy and to avoid deceiving or harming participants. Preserving the rights of subjects is sometimes weighed against the value of the knowledge to be gained. While usually harmless, these compromises sometimes place the subjects in jeopardy.
10. Demands on researchers go even further. They must design measurement devices that give consistent results each time they are used (reliability), and measurement techniques must actually measure what they intend (validity).

INFOTRAC® COLLEGE EDITION

http://www.infotrac-college.com/wadsworth

Another unique option available to you at the Student Resources section of the companion Web site is Infotrac College Edition, an online library with access to hundreds of scholarly and popular periodicals. Below are suggested search terms for this chapter. Results from these and other searches are found at the site.

Search keyword: **causation**. The textbook lists three criteria for establishing a causal relationship. Review those criteria and then search Infotrac for articles related to causation. Do the articles use the same criteria?

Search keywords: **research ethics**. Ethics is a popular topic in many fields of research. Using these keywords, look for different areas in which the ethics of research is a concern. How many different fields did you find?

Search keywords: **research validity**. Validity is the ability of a measurement technique to measure what it is actually supposed to. Find at least two articles that discuss the validity of scientific measurements. What did you learn from these articles?

LEARNING OBJECTIVES REVIEW

- Identify major nonscientific sources of knowledge; then explain why science is a superior source of knowledge.
- Discuss cause-and-effect concepts and apply the concept of causation to the logic of science.
- Differentiate the major quantitative research methods used by sociologists.
- Describe the major qualitative research methods used by sociologists.
- Explain the steps sociologists use to guide their research.
- Describe the role of ethics in research.
- State the importance of reliability, validity, and replication in social research.

CONCEPT REVIEW

Match the following concepts with the definitions below them.

____ a. participant observation
____ b. experiment
____ c. verifiability
____ d. subjective approach
____ e. experimental group
____ f. independent variable
____ g. objectivity
____ h. correlation
____ i. population
____ j. sample
____ k. field research
____ l. case study
____ m. survey
____ n. replication

1. the group in an experiment exposed to the experimental variable
2. a statistical measure in which a change in one variable is associated with a change in another variable
3. a research approach for studying aspects of social life that cannot be measured quantitatively and that are best understood within a natural setting
4. a thorough, recorded investigation of a small group, incident, or community
5. all those people with the characteristics a researcher wants to study within the context of a particular research question
6. the principle of science stating that scientists are expected to prevent their personal biases from influencing their results and their interpretation of the results
7. a variable that causes something to happen
8. the type of field research technique in which a researcher becomes a member of the group being studied
9. a principle of science by which any given piece of research can be duplicated (replicated) by other scientists
10. a research method in which people are asked to answer a series of questions
11. the duplication of the same study to ascertain its accuracy
12. a limited number of cases drawn from a population
13. laboratory research that attempts to eliminate all possible contaminating influences on the variables being studied

14. a research method in which the aim is to understand some aspect of social reality through the study of the subjective interpretations of the participants themselves

CRITICAL-THINKING QUESTIONS

1. Suppose a noncollege friend insists that you are wasting your time in college because the experience gained from the "university of hard knocks" is all a person needs to know the truth. What arguments would you use to defend science as a better source of knowledge?

___

___

2. Suppose your sociology professor reports on his recent study showing that men are generally better business managers than women. If you are concerned about a possible lack of objectivity on his part, what questions could you ask him to boost your confidence in his results?

___

___

3. The experiment is the research model for investigating causal relationships. What is there about the nature of causation and the design of experiments that supports this claim?

___

___

4. Do you think that a selected sample of three thousand individuals could yield an accurate picture of the leisure habits of Americans? Why or why not?

___

___

5. Pretend that you are a sociologist studying the relationship between the receipt of welfare payments and commitment to working. Describe the research method you would use, and show why it is the most appropriate to this topic.

___

___

MULTIPLE-CHOICE QUESTIONS

1. **The concept of intuition refers to**
   a. quick and ready insight that is not based on rational thought.
   b. opinions that are widely held because they seem so obviously correct.
   c. someone who is supposed to have special knowledge that we do not have.
   d. the fourth major nonscientific source of knowledge.
   e. a variable that causes something to happen.
2. **According to your text, causation can be asserted when**
   a. going from particular instances to general principles.
   b. there are only a limited number of cases taken from society.
   c. events occur in a predictable, nonrandom way, and one event leads to another.
   d. people develop and use routine behaviors expected of themselves and others in everyday life.
   e. a change in one variable is often accompanied by a change in another variable.
3. **Several factors have been shown to influence crime rates in poor neighborhoods. This illustrates the principle of**
   a. the poverty/crime hypothesis.
   b. multiple causation.
   c. verifiability.
   d. criminology.
   e. variance.
4. **A variable that causes something else to occur is called**
   a. a dependent variable.
   b. a correlation variable.
   c. a causation variable.
   d. a independent variable.
   e. a qualitative variable.
5. **The term *correlation* is defined as**
   a. a change in one variable associated with a change in the other.
   b. an apparent relationship between two variables, which is actually produced by a third variable that affects both of the original two variables.
   c. an event that occurs as a result of several factors operating in combination.
   d. something that occurs in different degrees among individuals, groups, objects, and events.
   e. a research method in which people are asked to answer a series of questions.
6. **All of the following are criteria for establishing a causal relationship *except*:**
   a. All possible contaminating factors must be controlled.
   b. A relationship representing a spurious relationship must exist.
   c. The independent variable must occur before the dependent variable.
   d. Two variables must be correlated.

7. All of the following statements about experiments are true *except*:
a. A description of the experiment provides an excellent means to illustrate causation.
b. A controlled experiment provides insight into the nature of all scientific research.
c. Controlled experiments take place in a laboratory.
d. The basic idea of the controlled experiment is to rule out the effects of extraneous factors to see the effects of an independent variable on a dependent variable.
e. Controlled experiments do not need a control group, because of the controlled atmosphere the laboratory provides.

8. The experimental group is exposed to the experimental variable; the group that is not exposed to the experimental variable is the
a. natural group.
b. experiential group.
c. control group.
d. dependent group.
e. independent group.

9. A standard way of making experimental and control groups comparable in all respects, except for exposure to the experimental variable, is through
a. qualifying or quantifying.
b. matching or randomizing.
c. pretesting or posttesting.
d. testing or retesting.
e. verification or replication.

10. A written set of questions that survey participants are asked to fill out by themselves is called
a. a survey.
b. an interview.
c. a questionnaire.
d. a survey research.
e. an independent variable.

11. Use of data from the U.S. Bureau of the Census is an example of
a. primary analysis.
b. population sampling.
c. the Hawthorne effect.
d. secondary analysis.
e. a case study.

12. The type of research used for studying aspects of social life that cannot be measured quantitatively and that are best understood in a natural setting is called
a. field research.
b. survey research.
c. participant observation.
d. analysis of precollected data.
e. content analysis.

13. Ethnomethodologists assume that
a. the subjective approach relies too much on intuition.
b. the behavior of people is random.
c. the underlying factor explaining human behavior is ethnicity.
d. questionnaires need to be tightly structured.
e. people share the meanings underlying much of their everyday behavior.

14. A tentative, testable statement of a relationship among variables is called
a. a hypothesis.
b. an operational definition.
c. a formal argument.
d. a correlation.
e. a conclusion.

15. In Laud Humphreys's study of homosexual activities occurring in men's public bathrooms ("tearooms"), what ethical standard did he violate?
a. He studied homosexuals.
b. He acted as a participant observer.
c. He violated the privacy of the participants.
d. He used the research to become famous.
e. He did not violate any ethical standards.

FEEDBACK REVIEW

True-False

1. The major problem with nonscientific sources of knowledge is that such sources often provide erroneous information. T or F?
2. According to Gunnar Myrdal, it is enough that scientists themselves recognize their biases. T or F?

Fill in the Blank

3. The group in an experiment that is not exposed to the experimental variable is the _________ group.
4. Field studies are best suited for situations in which _________ measurement cannot be used.
5. In _________, a researcher becomes a member of the group being studied.
6. An _________ attempts to eliminate all possible contaminating influences on the variables being studied.
7. Use of company records would be an example of using _________ data.
8. According to the _________ approach, some aspects of social structure are best studied through an attempt to ascertain the interpretations of the participants themselves.

Matching

9. Listed below are the steps in the research model. Following are some concrete examples related to the sociability of the only child. Indicate the appropriate example for each step number.

____ Step 1: identifying the problem
____ Step 2: reviewing the literature
____ Step 3: formulating hypotheses
____ Step 4: developing a research design
____ Step 5: collecting data
____ Step 6: analyzing data
____ Step 7: stating findings and conclusions

a. Read past theory and research on the sociability of only children.
b. From previous research and existing theory, a researcher states that only children appear to be more intelligent than children with siblings.
c. A researcher collects data on only children from a high school in a large city.
d. A researcher writes a report giving evidence that only children are more intelligent than children with brothers or sisters.
e. A researcher decides to study the intelligence level of only children.
f. A researcher classifies and processes the data collected in order to test a hypothesis.
g. A researcher decides on the data needed to test a hypothesis, the methods for data collection, and the techniques for data analysis.

10. Situations a, b, and c describe three research situations involving possible ethical violations (Babbie 2003). Match each situation with the appropriate aspect of the social science Code of Ethics for research on human subjects.
____ a. After a field study of deviant behavior during a riot, law enforcement officials demand that the researcher identify those people who were observed looting. Rather than risk arrest as an accomplice after the fact, the researcher complies.
____ b. A research questionnaire is circulated among students as part of their university registration packet. Although students are not told they must complete the questionnaire, the hope is that they will believe they must, thus ensuring a higher completion rate.
____ c. Researchers obtain a list of right-wing radicals they wish to study. They contact the radicals with the explanation that each has been selected "at random," from among the general population, to take a sampling of "public opinion."
(1) concern for participants' privacy
(2) avoidance of deception
(3) obligation not to harm participants

GRAPHIC REVIEW

Table 2.2 displays the median annual income in the United States by sex, race, and education. Demonstrate your understanding of the information in this table by answering the following questions.

1. What does this table tell us about the relationship among sex, race, and education in the United States?

____

____

2. Identify the demographic group that enjoys the greatest economic benefits of education.

____

3. Identify the demographic group that benefits the least, economically, from higher levels of education.

____

ANSWER KEY

Concept Review
a. 8
b. 13
c. 9
d. 14
e. 1
f. 7
g. 6
h. 2
i. 5
j. 12
k. 3
l. 4
m. 10
n. 11

Multiple Choice
1. a
2. c
3. b
4. d
5. a
6. b
7. e
8. c
9. b
10. c
11. d
12. a
13. e
14. a
15. c

Feedback Review
1. T
2. F
3. control
4. quantitative
5. participant observation
6. experiment
7. precollected
8. subjective
9. Step 1: e
Step 2: a
Step 3: b
Step 4: g
Step 5: c
Step 6: f
Step 7: d
10. a. 3
b. 2
c. 1

3 Culture

© Amy Etra/PhotoEdit

OUTLINE

LEARNING OBJECTIVES

- Identify the three major dimensions of culture.
- Describe and illustrate the interplay between language and culture.
- Discuss cultural diversity and its promotion within a society.
- Describe and illustrate the relationship between cultural diversity and ethnocentrism.
- Outline the advantages and disadvantages of ethnocentrism and discuss the role of cultural relativism in combating ethnocentrism.
- Explain the existence of cultural similarities that are shared around the world.
- Explain the relationship between culture and heredity.

USING THE SOCIOLOGICAL IMAGINATION

Does the existence of cultural diversity rule out cultural similarities? In other words, what, if anything, could the United States have in common with Afghanistan? What could we Americans have in common with a society that does not approve of women appearing in public with uncovered faces? What do we share with the citizens of Bhutan, who believe that gods assume human form to direct normal day-to-day affairs? How do we relate to the culture of Ingushetia, a republic inside Russia, where it is legal for men to have several wives—all at the same time?

Many Americans, in enthusiastic endorsement of multiculturalism, ignore the similarities among cultures. In reality, however, despite the surface differences, social scientists document a wide variety of "cultural universals" shared by all cultures. Like the United States, for instance, Iraq (or Bhutan or Nigeria) has families, schools, houses of worship, economics, and governments. Later in this chapter we will elaborate on this evidence. First, however, we will examine the nature of culture and society.

© Robert Caputo/Stock, Boston

Because a people's ways of thinking, feeling, and behaving are based on learning rather than biology, there is much variation in human expression. These Shilluk (in the Sudan) learned to celebrate the thatching of their king's huts through this unique ritual.

Dimensions of Culture

Culture and Society

What is the difference between culture and society? **Culture** is a people's way of life that is passed on from generation to generation. It consists of physical objects as well as patterns of thinking, feeling, and behaving. On the material side, American culture includes such physical objects as skyscrapers, fast-food restaurants, video games, and cars. On the nonmaterial side, American culture includes various beliefs, rules, customs, a family system, and a capitalist economy. Although culture and society are tightly interwoven and cannot exist without each other, they are not identical. A **society** is a group of people living in a defined territory and participating in a common culture. Culture is that society's total way of life.

Why is culture important? Culture underlies human social behavior. What people do and don't do, what they like and dislike, what they believe and don't believe, and what they value and discount are all based, in large part, on culture. Culture provides the blueprints people in a society use to guide their relationships with others. It is from culture that teenage girls are stimulated to compete for positions on the women's basketball team, and it is from culture that teenage boys come to believe that "pumping iron" is a gateway to popularity.

Human social behavior, then, is based on culture. And because culture is not innate, human behavior must be learned. The concepts presented in this chapter will help substantiate this point.

If you wanted to describe and analyze a culture, what would you look for? How could you begin to classify the components of a people's way of life? The sociological classification system consists of three major dimensions of culture: the normative (standards for behavior), the cognitive (knowledge and beliefs), and the material (tangible objects). An elaboration of each dimension will help you better understand the nature of culture.

The Normative Dimension

The normative dimension of culture, which consists of the standards for appropriate behavior of a group or society, is heavily tied into functionalism with its emphasis on social integration, stability, and consensus.

The most important aspects of the normative dimension are norms, sanctions, and values.

What are norms? **Norms** are rules defining appropriate and inappropriate behavior (Hechter and Opp 2001). A Hindi peasant in India can be found lying dead of starvation beside perfectly healthy and edible cattle, because in India cows are sacred. A Far Eastern woman may have her head severed for going to bed with a man before marriage. Roman emperors sent relatives to small isolated islands for life for disgracing the family. Each of these instances is a result of cultural norms—rules that specify ways of behaving for specific situations. Norms help to explain why people in a society or group behave similarly in similar circumstances.

William Graham Sumner (1906), an early sociologist, wrote perceptively about norms. Anything, he stated, can be considered appropriate when norms exist that approve of it. This is because once norms are learned, members of a society use them to guide their social behavior. Norms are so ingrained that they guide our social behavior without our awareness. In fact, we may not even be consciously aware of a norm until it has been violated. For instance, we do not think about standing in line for concert tickets as a norm until someone attempts to step in front of us. Then it immediately registers. Taking one's turn in line is expected behavior and cutting in line is violating that norm.

FIGURE 3.1

Premarital Sexual Experience Among Teenage Women in the United States

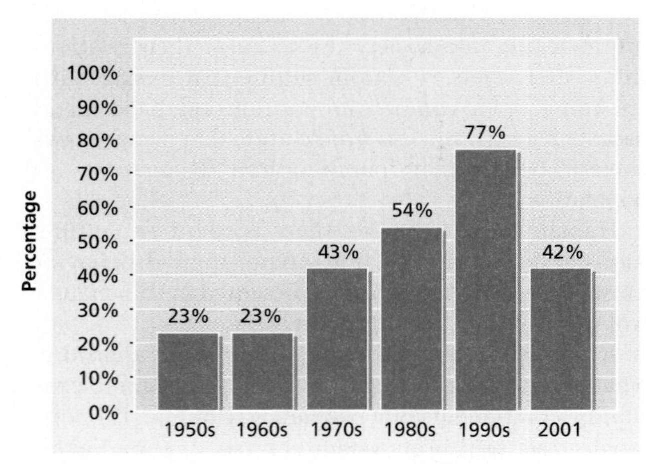

Note: Studies in each decade covered a two-year period.
Source: Based on Sandra L. Hofferth, John R. Kahn, and Wendy Baldwin, "Premarital Sexual Activity Among U.S. Teenage Women over the Past Three Decades," *Family Planning Perspectives* 19 (1987):49; Planned Parenthood Federation of America, Inc. "Planned Parenthood Fact Sheet: Sexuality Education in the U.S"; www.ppfa.org/ppfa/sex-ed.html/(June, 1995); and "Assessing Health Risk Behaviors Among Young People," U.S. Department of Health and Human Services, Centers for Disease Control and Prevention, 2002.

INTERNET LINK The Internet has an extensive list of sources that pertain to premarital sexual activity. Several of these provide useful data for the sociologist researching in this field. To begin, go to Campaign for Our Children at www.cfoc.org/

Norms vary widely from society to society. While visiting the Persian King Gelese in the nineteenth century, Sir Richard Francis Burton witnessed the "Ku to Man" ceremony performed in his honor. The dramatic climax of this ceremony was the ritual execution of twenty-three resigned and terrified slaves (Kruschwitz and Roberts 1987). Norms also change within the same society. Premarital sexual intercourse among teenage American women has increased dramatically since the 1950s (see Figure 3.1). Interestingly, the rate of premarital intercourse among teenagers has declined rather sharply since the 1990s, indicating that sexual norms may be changing once again (Risman and Schwartz 2002).

Norms range from relatively minor rules, such as applause after a performance, to extremely important ones, such as laws against stealing. Sumner identified three basic types of norms: folkways, mores, and laws. These three types of norms vary in their importance within a society, and their violation is tolerated to different degrees.

What are folkways? Rules that cover customary ways of thinking, feeling, and behaving but lack moral overtones are called **folkways**. For example, whether one uses a cell phone in a restaurant is not a moral issue; it qualifies as a folkway. Folkways in the United States include shaking hands when introduced, opening doors for older persons, and, if you are a male, removing your hat when you enter a house.

Because folkways are not considered vital to the welfare of the group, disapproval of violators is not very great. Those who consistently violate folkways—say, by persistently talking loudly in quiet places, wearing shorts with a suit coat and tie, or wearing a different-colored sock on each foot—may appear odd. We may avoid these people, but we do not consider them wicked or immoral.

Some folkways are more important than others, and the social reaction to their violation is more intense. Failure to offer a woman a seat on a crowded bus draws little reaction, but obnoxious behavior at a party after excessive drinking may bring a severe negative response from others (see Table 3.1).

What are mores? **Mores** (pronounced MOR-ays; singular, *mos*) are norms of great moral significance. They are thought to be vital to the well-being of a society. Conformity to mores elicits significant social approval; violation of this type of norm evokes strong disapproval. For example, Americans subscribe to a long-

TABLE 3.1

Do's and Taboos Around the World

Bulgaria	• A nod means no, and a shake of the head from side to side means yes.
Great Britain	• Never touch the Queen, not even to gently guide her, and don't offer to shake her hand unless she extends her hand to you first.
Germany	• Shaking hands while your other hand is in your pocket is considered impolite.
Greece	• Unlike the British, the Greeks do not respect lines, or queues, in public places.
Italy	• Italians consider it unfeminine for a woman to pour wine.
Russia	• Whistling at public gatherings is a sign of disagreement and disapproval.
Turkey	• It is considered rude to cross your arms over your chest or to put your hands in your pockets while talking to someone.
Oman	• It is an insult to sit in such a way as to face your host with the soles of your shoes showing. Do not place your feet on a desk, table, or chair.
Iran	• Shaking hands with a child shows respect of his parents.
Saudi Arabia	• At political events, it is customary for men to greet elders and dignitaries by kissing their right front shoulder.
The People's Republic of China	• Some pushing and shoving in stores or when boarding public transportation is common and not considered rude.
India	• Since the head is considered a sacred part of the body in India, you should not pat children on the head or touch an older person's head.
Japan	• Women should avoid wearing high heels in order not to risk towering over Japanese counterparts.
South Korea	• Koreans, especially women, cover their mouths when laughing to avoid impolitely showing the inside of their mouths.

Source: Roger E. Axtell, *Do's and Taboos Around the World*, revised and expanded ed. (New York: Wiley, 1998).

standing mos requiring able-bodied men to work for a living. Able-bodied men who do not work are stigmatized (Waxman 1983).

Although conformity to folkways is generally a matter of personal choice, conformity to mores is a social requirement. As in the case of folkways, some mores are more vital to a society than others. Failure to stand while the national anthem is being played is not as serious a violation of American mores as using loud profanity during a church service. Choosing, unnecessarily, to live on welfare is a more serious breach of appropriate conduct than either.

Some mores are more serious than others. A **taboo** is a mos so important that its violation is considered repugnant. Although definitions of incest vary from society to society, the incest taboo is generally regarded as the only taboo existing in all societies. There are, however, taboos other than the prohibition against having sexual relations with relatives. Hindus in India have a taboo forbidding the killing of cows. The mother-in-law taboo in some societies prohibits or severely restricts social contact between a husband and his wife's mother. Taboos in Western societies prohibit transvestism and bestiality (having sexual relations with animals). As Christie Davies (1982) points out, taboos such as those against transvestism and bestiality cannot have a psychological or biological origin, because some societies do not have such taboos. The world generally shares a taboo against the use of nuclear weapons. And it has been taboo in the United States to torture political prisoners.

Norms, which are rules specifying appropriate behavior, vary from the optional to the mandatory and from society to society. To bow in greeting is the Japanese norm equivalent of the American handshake.

How do laws differ from mores? **Laws**, the third type of norm, are norms that are formally defined and enforced by officials. Folkways and mores emerge slowly and are often unconsciously created, but laws are consciously created and enforced.

The Taliban militia in Afghanistan created a strict set of gender-apartheid laws. Women were not permitted to work outside the home, and females were banned from schools. Women could go outside only if accompanied by a male relative; women had to hide themselves behind painted windows; and women could not be examined by male physicians.

Mores are an important source of laws, as was the case in many of the Taliban laws. At one time in human history, the norm against murder was not written down. But as civilization advanced, the norm prohibiting murder became formally defined and was enforced by public officials. But not all mores become laws. For example, it is not against the law to cheat on a college examination (although one can be expelled or otherwise punished). Nor have all laws been mores at one time. Fines for overtime parking and laws against littering have never been mores.

Laws often remain on the books for a long time after the mores of a society have changed (Koon, Powell, and Schumaker 2002). It is illegal in Minnesota to hang male and female undergarments on the same clothesline; card playing on trains is prohibited in New York; elephants in Natchez, Mississippi, cannot legally drink beer; and in Portland, Oregon, it is against the law to roller-skate in public bathrooms. Mores and laws sometimes overlap. Although private citizens may show strong disapproval of a father who fails to support his children, public officials can do little unless the mother is willing to start legal action.

Because norms must be learned and accepted by individuals, conformity to them is not automatic. For this reason, groups must have some means for teaching norms and encouraging conformity to them. They do this, in part, through sanctions.

How are norms enforced? **Sanctions** are rewards and punishments used to encourage conformity to norms. They can be formal or informal. **Formal sanctions** are sanctions that may be given only by officially designated persons, such as judges and college professors. Formal sanctions range widely in their severity. From the Middle Ages to the Protestant Reformation, it was an unpardonable sin for lenders to charge interest on money. (This practice, called usury, was condemned in the Bible.) This civil crime was punishable on the third offense by public humiliation and social and economic ruin (Cockburn 1994). More recently, under Taliban rule in Afghanistan women were stoned to death if found outside the home with a nonrelative male; women were beaten for dressing "improperly"; women were shot for going to a doctor unescorted by a male relative. In 2002 a northern Nigerian woman was sentenced by an Islamic court to die by stoning for having an illegitimate child (G. Robinson 2002). Less severely, courts across the United States have handed down sentences involving public shaming. For example, some courts have required child molesters to place, in front of their residences, signs describing their crimes (El Nasser 1996). In 1997, Latrell Sprewell, National Basketball Association star player for the Golden State Warriors, physically attacked his coach, P. J. Carlesimo. The NBA revoked his $32 million, four-year contract and suspended him for one year before he joined the New York Knicks. Jack Kevorkian ("Dr. Death") was sentenced to ten to twenty-five years in prison for euthanasia (McGinn 1999). In response to the multiple business scandals involving companies such as Enron, Arthur Anderson, and WorldCom in the early 2000s, Congress approved and the president signed legislation to stiffen regulation of corporate auditors and hold executives more accountable for their behavior.

Formal sanctions can also be positive. Soldiers earn the Congressional Medal of Honor for heroism, and some students receive an A for academic performance.

Informal sanctions are sanctions that can be applied by most members of a group. They, too, can be positive or negative. Informal sanctions include thanking someone for pushing your car out of a snowbank to glaring harshly at someone who is cheating on a test. The star of the syndicated television show *Roseanne*, then Roseanne Barr Arnold, was booed and heavily criticized for her shrill rendition of the national anthem and her parody of baseball players (at the conclusion of the song, she grabbed herself between the legs and spat) at a major-league baseball game.

Sanctions are not used randomly or without reason. Specific sanctions are associated with specific norms. Teenagers who violate their parents' curfew are not supposed to be beaten or locked in a closet. Convicted murderers are not invited to the White House for special celebrations. Instead, a society or group develops appropriate sanctions for following or failing to follow specific norms (A. V. Horwitz 1990).

Sanctions, most often, do not have to be used. After we reach a certain age, most of us conform without the threat of sanctions. We may conform because we believe that the behavior expected of us is appropriate, because we wish to avoid guilt feelings, or because we fear social disapproval. In other words, if we have been properly socialized, we will mentally sanction ourselves before breaking a norm.

The severity of the sanction for a deviant act varies from one society to another and from one time to another. A convicted thief in the United States might be sentenced to only a few months in jail. A convicted thief in North Yemen might have to pick up his amputated hand and salute the judge with it. In 1993, five Somalian women accused of adultery were stoned to death and a sixth woman was lashed 100 times by a mob of Muslim fundamentalists. During the 1960s, possession of an ounce of marijuana could result in several years in prison. Today, possession of a small amount of marijuana in most states is a misdemeanor, and in several states it is equivalent to a minor traffic violation.

Sometimes, informal sanctions are illegally imposed. A soccer sportscaster was shot in the knees in Avellino, Italy, by irate fans. The rash of "dowry deaths" in India, not too long ago, is also an example of illegal sanctioning. According to police statistics, almost 400 women were set afire in New Delhi in 1981. Although some of these deaths were accidental and some were suicides, it is suspected that many of them were murders. Some of the mothers-in-law, upset over the small size of their son's wife's dowry, poured gasoline over their daughters-in-law and threw matches on them. A Jordanian father killed his daughter recently upon learning of her premarital relations with her new husband. The dishonor she brought to the family could not be excused despite her pregnancy (Khouri 2003).

So far, we have discussed norms and sanctions as aspects of normative culture. Though norms and sanctions are relatively specific and concrete, the next major component of the normative dimension of culture—values—is rather broad and abstract. Values are much more general than norms.

© Grant LeDuc/Stock, Boston

Achievement is one of America's most basic values. The expressions on these parents' faces reflect happiness and pride over the significant accomplishment of their daughter.

What are values? **Values** are broad cultural principles that most people in a society consider desirable. Values are so general that they do not specify precise ways of thinking, feeling, and behaving. Thus, different societies or different groups within the same society can have quite different norms based on the same value. For instance, the value of freedom has been expressed differently in America and in the former Soviet Union. In the Soviet Union, as Robin Williams (1970) notes, freedom was expressed in the right to such things as employment, medical care, and education. Americans have different norms based on the value of freedom—the right to free speech and assembly, the right to engage in private enterprise, the right to a representative government, the right to change where they live and work. And patriotism for some citizens might involve death for their country. For others, refusing to fight in a war they consider immoral might be patriotic. Identical values do not result in identical norms (Hurley 2000).

Why are values important? Values are important because they have a tremendous influence on human social behavior—mainly because norms are based on them. A society that values democracy will have norms ensuring personal freedom; a society that values humanitarianism will have norms providing for its most unfortunate members; a society that values hard work will have norms against laziness. Values are also important because they are so general that they tend to permeate most aspects of daily life. In America, for example, the value of freedom affects more than a person's political life. It affects such diverse areas as relationships in the family, treatment within the legal system, the operation of organizations, and the choice of religious affiliation.

What are the basic values in American society? Any attempt to state the basic values of American society is risky. Because America has so many diverse groups and because it is constantly changing, any one set of values is unlikely to receive unanimous support (Etzioni 2001). Despite these problems, Robin Williams (1970) has done an excellent job of outlining fifteen major values guiding the daily lives of most Americans. Whether or not his classification is complete, it provides a picture of the major values influencing Americans. These values include achievement and success, activity and work, humanitarianism, efficiency and practicality, progress, material comfort, equality, freedom, democracy, individuality, science and rationality, external conformity, group (racial, ethnic, religious) superiority, morality, and patriotism. Following is a sample of the values identified by Williams.

- *Achievement and success*. Americans emphasize achievement, especially in the world of work. Success is supposed to be based on effort and competition, and is viewed as a reward for performance. Wealth is a symbol of success and personal worth, and quantity often represents quality.
- *Activity and work*. We tend to stress action over inaction in almost all instances. For most Americans, continuous and regular work is an end in itself, advancement is to be based on merit rather than favoritism, and all citizens are supposed to have the opportunity to perform at their best.
- *Efficiency and practicality*. Americans pride themselves on getting things done by the most rational means. We search for better, faster ways of doing things, praise good workmanship, judge work performance by its practical consequences, and depend on science and technology.
- *Equality*. Americans advocate equality for all citizens. In interpersonal activities, we tend to treat one another as equals, to defend everyone's legal rights, and to favor equal opportunity for everyone.
- *Democracy*. Americans emphasize that all citizens are entitled to equal rights. All Americans should have equal rights and equal opportunity under the law, and power should not be concentrated in the hands of an elite.
- *Group superiority*. Despite their concern for equality of opportunity, Americans tend to place greater value on people of their own race, ethnic group, social class, or religious group.

These values, clearly, are interrelated. The values of achievement, success, activity, and work are related to the values of efficiency and practicality. Equally obvious is the conflict between some values: Americans value group superiority while stressing equality and democracy.

Are these values prominent in American society today? Williams identified fifteen major American values approximately thirty years ago. Although these values have remained remarkably stable over the years, some have changed somewhat. For example, today there is less emphasis on group superiority than in the past, as can be seen in the apparent decline of openly racist attitudes and behavior (R. Farley 1998; Rochon 1998). However, it is usually norms and behavior, rather than underlying values, that change radically. In other words, it is probably because of civil rights laws that many Americans are now less likely to make overt racist statements and discriminate against minority members. They simply behave in a way that is consistent with changed norms. So, as yet, racism remains part of the fabric of American culture.

Although Americans stress equality of opportunity and individual merit, they also evaluate others on the basis of ascribed characteristics such as social class, race, and ethnic group. Members of this Chinese subculture in the United States experience prejudice and discrimination based on their physical and cultural characteristics.

The norms related to hard work and activity have also changed in recent years. According to Lionel Lewis (1982), many Americans now work as hard at their leisure (for example, long-distance running and mountain climbing) as they do at their jobs.

Although Williams's analysis of major American values remains basically sound today, it is not surprising that some, quite validly, believe that his list is incomplete. George and Louise Spindler (1983), for example, wish to add optimism, honesty, and sociability to the list of major American values.

Although functionalism is the basis of the normative dimension of culture, symbolic interactionism is related to the cognitive and material dimensions. Be alert to these theoretical relationships.

The Cognitive Dimension

Do beliefs matter? Cognition is the process of thinking, knowing, or processing information. The cognitive dimension of culture, then, refers to its complex of ideas and knowledge. The most important aspect of the cognitive dimension of culture is **beliefs**—ideas concerning the nature of reality. Actually, beliefs are influential whether they be true or false. The Romans believed that Caesar Augustus was a god; the Tanala, a hill tribe of Madagascar, believed that the souls of their kings passed into snakes; and many Germans believed that pictures of Hitler on their walls would prevent the walls from crumbling during bombing raids. In 1953, Edmund Hillary, a New Zealander, became the first man to climb to the top of Mount Everest (29,035 feet).

When he descended, he was greeted with hatred by the Sherpas of Nepal, who "believed that Buddhist gods resided up there, in the clouds, and they did not want to accept that the first human foot there had not been one of their own" (Hoffer 1999:122). On the American frontier, it was believed that a dog howling in the distance foretold a death in one's family, that anyone who brought a shovel into a cabin would carry out a coffin, and that a bird on one's windowsill or inside one's home meant sorrow in the future. While each of these beliefs is false, other beliefs—such as the belief that the human eye can distinguish over 7 million colors or the belief that no life exists on Mars (Begley 1994; Lemonick 1994)—are factually supported by evidence. Regardless of their truth, beliefs are important, because people tend to base their behavior on what they believe.

The Material Dimension

How do a society's physical objects reflect its culture? **Material culture** consists of the concrete, tangible objects within a culture—automobiles, basketballs, chairs, highways, birth-control pills, art, jeans. Artifacts, or physical objects, have no meaning or use apart from the meanings people give them. Consider newspaper and pepper as physical objects. Of course, each of these two things separately has some meaning for you, but can you think of a use for them in combination? Some midwives have used pepper and newspaper in a process known as nettling. An elderly medical doctor tells the story of his first encounter with nettling:

> *The ink of my medical license was hardly dry, and as I was soon to find out, my ears would not be dry for some time. I had never delivered a baby on my own and faced my maiden voyage with some fear.*
>
> *Upon entering Mrs. Williamson's house, I found a local midwife and several neighbors busily at work preparing for the delivery. My fear caused me to move rather slowly and my happiness over my reprieve prompted me to tell the women that they were doing just fine and to proceed without my services.*
>
> *Having gotten myself off the hook, I watched the ladies with a fascination that soon turned to horror.*
>
> *At the height of Mrs. Williamson's labor pains, one of the neighbors rolled a piece of newspaper into a funnel shape. Holding the bottom end of the cone she poured a liberal amount of pepper into it. Her next move was to insert the sharp end of the cone into Mrs. Williamson's nose. With the cone in its "proper" place, the neighbor inhaled deeply and blew the pepper from the cone into the inner recesses of Mrs. Williamson's nose—if not her mind.*
>
> *Suddenly alert, Mrs. Williamson's eyes widened as her senses rebelled against the pepper. With a mighty sneeze, I was introduced to nettling. The violence of that sneeze reverberated through her body to force the baby from her womb in a skittering flight across the bed. An appropriately positioned assistant fielded the baby in midflight and only minor details of Orville's rite of birth remained.*

Before this doctor was introduced to nettling, this particular combination of newspaper and pepper had no meaning for him. And until nettling was devised, the combination was without meaning for anyone, even though the separate physical objects existed as part of the culture.

Physical objects do not have the same meanings and uses in all societies. Although it is conventional to use a 747 jet for traveling, it is possible that a 747 downed in a remote jungle region of the world could be used as a place of worship, a storage bin, or a home. In the

TABLE 3.2

FOCUS ON THEORETICAL PERSPECTIVES: Culture

This table illustrates the unique slant that each of the three major theoretical perspectives takes on culture. Select a different aspect of culture for each perspective and make up your own table.

Theoretical Perspective	Aspect of Culture	Example
Functionalism	• The norm requiring students to listen to their teachers on request	• This norm promotes better education for everyone because of a lack of disorder.
Conflict theory	• Drug use by a member of a drug subculture	• Because of a lack of power, he or she is more likely to go to jail than an upper-class user.
Symbolic interactionism	• Confederate flag on the back window of a truck	• To the driver, the flag may be a symbol of cultural heritage; to an African American, it may represent slavery.

Star Wars and the Internet

When *Star Wars* first appeared at theaters in the late 1970s, director George Lucas probably thought he had made a good movie. Little did he realize that he had almost single-handedly created a full-fledged cultural phenomenon! Virtually everyone in the United States now recognizes Luke Skywalker, Darth Vader, Yoda, and Jar Jar Binks. We know what "may the Force be with you" means.

The five movies in the *Star Wars* series have certainly been extremely popular in their own right, but the Internet has also been instrumental in their penetration into popular culture. In 1999, *Star Wars* fans kept in touch over the Internet as they eagerly awaited the first new *Star Wars* installment in sixteen years:

> *[H]undreds of thousands of the* Star Wars *faithful have chugged along for years without a new movie to gawk at, but now this scattered community is enjoying a glorious moment of solidarity, coalescing on the Internet and preparing to burst forth as one from their far-flung basements and rec rooms.* (Hamilton and Gordon 1999:62)

Anticipation of the first "prequel," *The Phantom Menace*, was incredibly intense, and pirated footage spread to more than sixty Web sites within hours of first being posted. In response, Lucasfilm's official Web site posted the film's trailer and was promptly overwhelmed with 340 "hits" per second. In 2002, there were more than 1.3 million Web sites related to *Star Wars*. The impact of the Internet on this bit of American culture is undeniable:

> *"Everyone said this was the most top-secret movie ever made, that it was tighter than Fort Knox, no leaks whatsoever," says Scott Chitwood, 25, who's the emperor of TheForce.net. Well, most Website operators knew the plot a year ago. That's all because of the Internet.* (Hamilton and Gordon 1999:63)

Of course, the cultural effects of *Star Wars* are not limited to the box office. The movie has become a subculture unto itself, and like all subcultures, it has its own icons, symbols, and language. And elements of the subculture have even entered the larger culture. For instance, between 1977, when the first film was released, and 2002, merchandise related to the *Star Wars* movies totaled over $7 billion in sales. That alone amounts to more than six times the revenues generated from the films themselves. These items include toys, soundtracks, costumes, and licensing fees. Experts predicted that the toy maker Hasbro could ring up sales exceeding $5 billion from the three "prequels" to be released. Pepsico made sure that almost everyone in the United States knew about the new movies by placing their images on soft drinks and packages of Frito-Lay chips and in Taco Bell, Kentucky Fried Chicken, and Pizza Hut restaurants (Rogers 1999). With the increased popularity of e-commerce, *Star Wars* merchandise will, for some time, continue to permeate American culture.

Analyzing the Trends

What other recent events have become part of the popular culture in the United States? Name one that meets the criteria for being included as culture, and tell what aspects of this event have made their way into our thinking, feeling, and behaving.

United States, out-of-service buses, trains, and trolley cars have been converted to restaurants.

Clearly, the cultural meaning of a physical object is not determined by the physical characteristics of the object. People use newspaper and pepper during childbirth or make a temple of a downed 747 jet because of cognitive and normative definitions. The meanings of physical objects are based on the beliefs, norms, and values people hold with regard to them. This is readily apparent when new meanings of a physical object are considered. At one time, only pianos and organs were used in church services; guitars, drums, and trumpets were not "holy" enough to accompany a choir. Yet, many churches today use these "worldly" instruments regularly in their worship activities. The instruments have not changed, but the cultural meanings placed on them have.

Ideal and Real Culture

There is sometimes a gap between cultural guidelines and actual behavior. This gap is captured in the concepts of ideal and real culture. **Ideal culture** refers to cultural guidelines publicly embraced by members of a society; these are the guidelines we *claim* to accept. **Real culture** refers to *actual* behavior patterns. Sometimes subterranean real cultural patterns are publicly denied because they conflict with the ideal culture. For example, one aspect of America's ideal culture is honesty. Yet, in real culture, some taxpayers annually violate both the letter and spirit of existing tax laws, some businesspeople engage in dishonest business practices, some students cheat on exams, and some college athletes do the "high $500" handshake (during which a team booster leaves illegal money in their palms). These are not isolated instances. These real cultural patterns are often passed from generation to generation. Keep in mind that we are not referring to cases of individual deviance, such as people who murder, rape, and rob. These types of antisocial behavior violate even real culture.

Does the fact that we sometimes ignore cultural guidelines make ideal culture meaningless? Absolutely not. In an imperfect world, ideal culture provides high standards. These ideals are targets that most people attempt to reach most of the time. Otherwise, social chaos would prevail. Ideal culture also permits the detection of deviant behavior. Individuals who deviate too far from the ideal cultural pattern are sanctioned. This helps to preserve the ideal culture.

Culture as a Tool Kit

According to the dominant view, culture guides behavior by providing the values or ends toward which behavior is directed. Ann Swidler, a contemporary critic of the culture-as-a-way-of-life approach, thinks that culture should be viewed as a "'tool kit' of symbols, stories, rituals, and world views, which people may use in varying configurations to solve different kinds of problems" (Swidler 1986:273). In this view, culture provides a range of choices to be applied in defining and solving the problems of living. Because the content of any culture is not fully consistent, the "tools" that are chosen vary among individuals and situations. This view of culture is gaining visibility. Its eventual prominence in the sociological approach to culture remains to be seen (Hays 1994).

Language and Culture

Culture is the social heritage of humans. This heritage is altered by each generation and must be learned by new members of society. Both the creation and the transmission of culture depend heavily on the human capacity to develop and use symbols, the most significant of which make up language. The following discussion relies on concepts central to symbolic interactionism. Symbolic interactionism, you will recall,

FEEDBACK

1. __________ consists of all the material objects as well as the patterns of thinking, feeling, and behaving that are passed from generation to generation among members of a society.
2. A __________ is composed of a people living within defined territorial borders who share a common culture.
3. __________ are rules defining appropriate and inappropriate ways of behaving.
4. __________ are rewards and punishments used to encourage desired behavior.
5. Indicate whether the following are formal sanctions (F) or informal sanctions (I).
   ____ a. A mother spanks her child.
   ____ b. A professor fails a student for cheating on an exam.
   ____ c. A jury sentences a person to life in prison for espionage.
   ____ d. A husband separates from his wife after she has an affair.
6. __________ are broad cultural principles embodying ideas about what most people in a society consider desirable.
7. __________ are ideas concerning the nature of reality.
8. Indicate whether each of the following best reflects a belief (B), folkway (F), mos (M), law (L), or value (V).
   ____ a. conception that God exists
   ____ b. norm against cursing aloud in church
   ____ c. norm encouraging the eating of three meals daily
   ____ d. idea of progress
   ____ e. norm against burning a national flag
   ____ f. norm encouraging sleeping in a bed
   ____ g. norm prohibiting murder
   ____ h. norm against overtime parking
   ____ i. idea that the Earth's orbit is elliptical
   ____ j. idea of freedom
9. __________ culture consists of the concrete, tangible objects within a culture.
10. __________ culture refers to aspects of culture publicly embraced by members of a society.

Answers: 1. Culture 2. society 3. Norms 4. Sanctions 5. a. (I) b. (F) c. (F) d. (I) 6. Values 7. Beliefs 8. a. (B) b. (M) c. (F) d. (V) e. (M) f. (F) g. (L) h. (L) i. (B) j. (V) 9. Material 10. Ideal

SOCIOLOGY AND THE NEWS MEDIA

Cultural Change in America

© Cable News Network, Inc.

The media revealed more about the personal life of President Clinton than about any other president in our history.

You don't have to have been around during John Kennedy's presidency to know that political norms have changed. Reporters, editors, and news anchors overlooked sexual behavior on Kennedy's part that would have shocked the nation. In today's cultural environment, it is considered appropriate among the mass media to reveal for public consumption almost all personal presidential behavior. This CNN segment explores this cultural change.

1. Relate the changes identified in this CNN segment to the three dimensions of culture.

2. Describe these changes within the context of ideal and real culture.

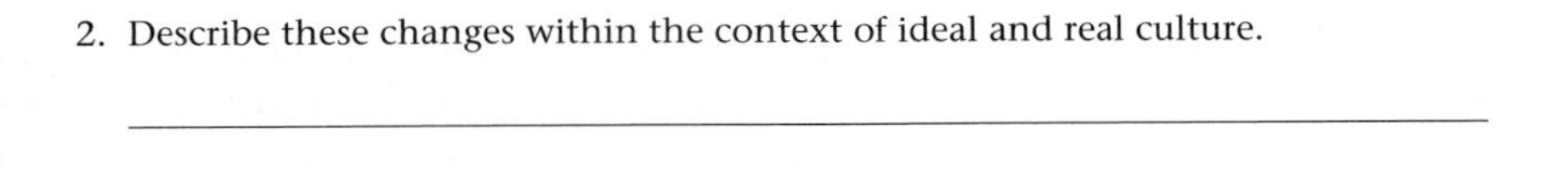

emphasizes our use of symbols in subjective interpretations of human social behavior.

Symbols, Language, and Culture

Once again, what are symbols? **Symbols**—things that stand for, or represent, something else—can range from physical objects to words, sounds, smells, and tastes. As you saw in Chapter 1, however, the meaning of a symbol is not dictated by its characteristics. There is nothing intrinsically good about the sound created by applause. In the United States, for example, applause warms the heart of an entertainer, politician, or professor, but in Latin America it symbolizes disapproval. Of course, once meaning has been assigned to applause and we learn to associate it with approval or disapproval, it seems as though the appropriate meaning is determined by the applause itself. A more recently defined symbol, a red ribbon insignia, identifies AIDS supporters (Peyser 1993). Tiger Woods is worth $100,000,000 to Nike as a symbol for the firm's new golf division. The baseball that Barry Bonds hit for his record-breaking seventy-third home run in 2001 is estimated to be a $1 million symbol. Because it symbolized to the Arab world a takeover, the American flag first draped over the head of a Saddam Hussein statute in 2003 was quickly replaced by an Iraqi flag.

In Lewis Carroll's *Through the Looking Glass*, Humpty Dumpty says to Alice with some finality, "When I use a word, it means just what I choose it to mean—neither more nor less." So it is with symbols. Of course, the same word can symbolize very different things. Shortly after September 11, American teenagers appropriated newly current words for their own purposes. Ground-zero bedrooms are a total mess; mean teachers are terrorists; a disciplined student experienced a jihad; and out-of-style clothes are burqas (Wax 2002). Even the symbol "nine-eleven" was expanded from its conventional meanings as a numerical value and an emergency phone number to include the 2001 terrorist attacks in New York, Washington, and Pennsylvania.

Symbolic culture is not limited to vocal, written, or material symbols. **Gestures**, whether in the form of facial expression, body movement, or posture, also carry culturally defined and shared symbolic meanings. Most Americans know that shaking one's fist at another driver, forming a circle with one's forefinger and thumb, and extending one's thumb in the direction of traffic while walking along a highway each convey a commonly held meaning to other Americans. Sometimes, though not frequently, gestures may change within a culture. For example, it used to be that men publicly holding hands could only suggest a homosexual relationship. When NFL players hold hands in the huddle on national television today, it symbolizes team solidarity. And different gestures transmit different meanings in various cultures. When Americans hold a thumb upright, either strong approval or the desire for a ride is assumed. The same gesture in Nigeria is taken as an insult (Axtell 1998).

What is the relationship between language and culture? Language frees humans from the limits of time and

TABLE 3.3

Symbols Created for Internet Communication

The symbols in the left column were created to stand for the symbols in the right column. Why are both columns called symbols?

:-)	Happy
;-)	Winking
:-(	Sad
:-D	Laughing
:-O	Surprised
:-@	Screaming
:-P	Tongue in cheek
R	Are
UOK?	Are you OK?
B	Be
B4	Before
BBL	Be back later
BRB	Be right back
BTW	By the way
XLNT	Excellent
4	For, Four
GR8	Great
L8R	Later
LOL	Laughing out loud
LUV	Love
NO1	No one
OIC	Oh, I see
PLS	Please
PCM	Please call me
C	See
THX	Thanks
TTUL	Talk to you later
2	Too, to, two
2DAY	Today
2MORO	Tomorrow
WAN2	Want to
?	What
U	You
YR	Your

place. It allows us to create culture. The Wright brothers' victory over gravity did not come just from their own solitary efforts. They constructed their airplane according to aerodynamic principles known for some time before their first successful flight. They could read, discuss, and recombine existing ideas and technology.

Equipped with the symbols of language, humans can transmit their experiences, ideas, and knowledge to others. Children can be taught things without any actual experience on their part. Although it may take some time and repetition, children can be taught the dangers of fire and heights without being burned or toppling from stairs. This process of social learning, of course, applies to other cultural patterns as well, such as exhibiting patriotism, consuming food, or staying awake in class.

The Sapir-Whorf Hypothesis

According to Edward Sapir (1929) and Benjamin Whorf (1956), language is our guide to reality; our view of the world depends on the particular language we have learned. Our perception of reality is at the mercy of the words and grammatical rules of our language. And because our perceptions are different, our worlds are different. This is known as the **hypothesis of linguistic relativity**.

What can vocabulary reveal about a culture? When something is important to a society, its language will contain many words to describe that entity. The Agta of the Philippines have thirty-one different verbs meaning "to fish" (M. Harris 1990). The importance of time in American culture is reflected in many words—*era, moment, interim, recurrent, century, afternoon, semester, eternal, annual, meanwhile, regularly,* and *semester*, just to name a few. When something is unimportant to people, they may not even have a word for it. When Christian missionaries first went to Asia, they were dismayed because the Chinese language contained no word for—and therefore no concept of—sin. Other missionaries were no less dismayed to learn that Africans and Polynesians had no way to express the idea of a single, all-powerful God. The Tzeltal language in southern Mexico recognizes five basic colors—white, black, red, green, and yellow. The language of the Jalé of New Guinea has terms only for black and white (Berlin and Kay 1969).

Does the hypothesis of linguistic relativity mean people are forever prisoners of their language? As far as we know, all people have the genetic capacity to learn any language. Thus, although the principle of linguistic relativity states that a person's view of the world is colored by his or her language, it does not state that people are forever trapped by their language. Exposure to another language can alter a person's perception of the world.

People can begin to view the world differently as they learn a new language. Most people, however, confine themselves to their native language. Consequently, they tend not to change their view of the world. Language, then, so necessary for the creation and transmission of culture, is also constricting.

What other factors help to shape our perception of reality? Application of the Sapir-Whorf hypothesis of lin-

FEEDBACK

1. ________ are things that stand for, or represent, something else.
2. Because of ________, culture can be created and transmitted.
3. According to the hypothesis of linguistic relativity, words and the structure of a language cause people to live in a distinct world. T or F?
4. The language we learn determines forever how we see the world. T or F?

Answers: 1. Symbols 2. language 3. T 4. F

guistic relativity is almost limitless; it extends far beyond vocabulary into areas as diverse as visual perception, audio perception, and ideas about privacy and space. The Japanese use paper walls as sound barriers and are not bothered by noise from adjacent rooms. Americans staying at a hotel in Japan may complain that they are being bombarded with noise because Westerners have not been conditioned to screen out sound (E. T. Hall 1990). At the other extreme, Germans feel very protective about their privacy. German executives, for example, have a closed-door policy. Problems arise, then, in subsidiaries of American firms in Germany because American executives leave their doors open. The German preference for physical barriers to protect their privacy is reflected in the prevalence of double doors in German hotels and other public and private buildings (E. T. Hall 1976, 1979).

Cultural Diversity and Similarity

Inherent in the discussion thus far is the relationship between sociology's major theoretical perspectives and other central cultural concepts. Symbolic interactionism is intimately connected with the role of language in culture and with the cognitive and material dimensions of culture. The assumptions of functionalism underlie the normative dimension of culture. As you view cultural diversity and similarity, you will again be aware of the appropriate theories. Conflict theory, with its emphasis on inconsistency, conflict, and change, undergirds any discussion of cultural diversity, whereas functionalism is the basis for a sociological analysis of cultural similarity.

Cultural Diversity

In ancient Rome, adultery was an accepted practice among the upper classes, and divorce was granted upon agreement of the couple. In modern Italy, adultery is frowned on, and divorce has only recently been legalized. Under Saudi Arabian religious laws, adulterers are stoned to death in public. Some American married couples advertise their "swinging" behavior in newspapers and magazines. Eskimos used to offer their wives as sexual favors to visitors—even strangers—and felt offended if the offer was refused. These examples reflect the almost endless cultural variations in the world. Not only do different societies have different beliefs, norms, values, and sanctions, but various groups within the same society have their own cultural patterns. Even in a simple society, cultural diversity makes it impossible for all members to participate in all aspects of culture. In modern societies, cultural diversity is staggering. Given the biological similarity among humans, this diversity must be explained by nongenetic factors.

How is cultural diversity promoted? Cultural diversity exists in all societies in part due to the presence of social categories. A **social category** is a group of persons who share a social characteristic such as age, sex, or religion. Members of social categories are expected to participate in aspects of culture unique to them. A particular age, gender, or religion, for example, often has distinctive activities associated with it.

Cultural diversity also stems from the existence of groups—known as *subcultures* and *countercultures*—that are somewhat different from the larger culture. Although these groups do participate in the dominant culture—they may speak the language, work at regular jobs, eat and dress like most others, and attend recognized houses of worship—they also think, feel, and behave in ways that set them apart. Subcultures and countercultures are more prevalent in large, complex societies.

What is a subculture? A **subculture** is a group that is part of the dominant culture but differs from it in some important respects. By tradition, Americans like to see themselves as part of a large, single culture. This view is fueled by the long-standing conception of the United States as a "melting pot" that absorbs peoples from around the world into "American" culture. This viewpoint minimizes the presence of cultural diversity in the United States. Since the 1960s there has been a movement toward an emphasis on the cultural uniqueness of subgroups. This movement, known as **multiculturalism**, accents the viewpoints, experiences, and contributions of minorities (women as well as ethnic and racial minorities). In fact, many scholars are concerned that the current emphasis on cultural differences is unduly fragmenting American society (Schlesinger 1998).

SNAPSHOT OF AMERICA 3.1

Gun Control

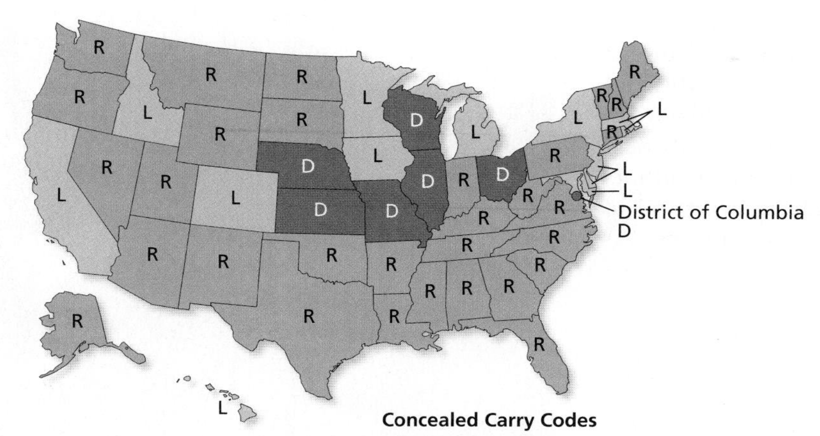

Concealed Carry Codes

- R Right-to-Carry Permitted: Less restrictive discretionary permit system.
- L Right-to-Carry Permitted: Limited by local authority's discretion over permit issuance.
- D Right-to-Carry Denied: No permit system exists; concealed carry is prohibited.

Source: National Rifle Association of America Institute for Legislative Action, "Compendium of State Laws Governing Firearms 1998," March 1999. www.nra.org/research/riflaws.html

Some observers believe groups that promote gun ownership form a subculture. For example, the National Rifle Association (NRA) brings together people who share an interest in guns and the right to own them. The map shown here displays the states that permit citizens to carry concealed guns.

1. What code marks the states with the most liberal gun control laws?
2. Can you find a relationship between gun control and regions in the United States?
3. Do you believe that members of the NRA form a subculture? Why or why not?

Sociologists have routinely studied subcultures, both before and after the rise of multiculturalism. The residents of southern Appalachia are an illustration of early sociological research on subcultures. According to some sociologists in the early 1960s, these people subscribe to ways of thinking, feeling, and behaving quite different from those of middle-class America. Middle-class Americans believe in progress, planning ahead, status seeking, education, social participation, and freedom to determine one's destiny. In contrast, Jack Weller (1980) contended, southern Appalachians are fatalistic, present oriented, unambitious, and nonparticipative. Though they are different from middle-class Americans, Weller notes that southern Appalachians are not opposed to the dominant culture. Rather, Weller argues, the southern Appalachian subculture constitutes a cultural adaptation to the reality of living a long-standing deprived and frustrating existence.

Another example of a subculture in the United States is the Old Order Amish, a group that has consciously attempted to retain its unique cultural identity ever since entering the United States in the early 1700s. The highly religious 100,000-strong Amish, who live primarily in the Midwest, exert tremendous efforts to maintain physical and cultural separation from the dominant culture. You may not have seen any Amish in person, but you probably have seen them depicted in the movies or on television as very traditionally dressed farmers traveling by horse and buggy. The Amish are also culturally conservative in their view of women (subservient to men), birth control (prohibited), education (send their children to Amish schools), and technology (only recently beginning to welcome machines, electricity, and indoor plumbing) (Janofsky 1997; Zellner and Kephart 1997).

The Amish generally reject living and participating in the dominant culture, but members of another group—devotees of the Grateful Dead band—both live and participate in the dominant culture. "Deadheads," as they call themselves, subscribe to a subculture constructed around the band. They share a unique language, mode of dress (featuring tie-dyed clothing), and liberal attitudes toward drug use. Deadheads tend to have mainstream jobs and lives. For them, participation in the subculture revolves around hearing the band when it is close enough to attend. Many Deadheads, however, take off for months to follow the band on tour around the country (R. G. Adams 1995).

A less extreme subculture is that of San Francisco's Chinatown. Early Chinese immigrants brought much of their native culture with them to America and have attempted to retain it by passing it from generation to generation. Although Chinese residents of Chinatown have been greatly affected by American culture, they have retained many cultural patterns of their own, such as language and family structure.

Other subcultures in America are those formed by athletes, actors, and surfers, as well as by people in mental hospitals, convents, and universities. (For other excellent examples, see Fine 1996; Zellner and Kephart 1997; Redhead 1997; Meuller 1999.)

What is a counterculture? A **counterculture** is a subculture that is *deliberately* and *consciously* opposed to certain central aspects of the dominant culture. A counterculture can be understood only within the context of its underlying opposition to the dominant culture (see Doing Research). Countercultures openly defy the norms, values, and beliefs of the dominant culture. Rebelling against the dominant culture is central to their members. For example, a counterculture was formed on American college and university campuses in the 1960s. What Edward Suchman (1968) called the "hang-loose" ethic rejected such central aspects of American culture as material success, achievement, hard work, efficiency, authority, premarital chastity, and the nuclear family. A more recent counterculture, the militia movement, is based on political opposition. Members of the militia counterculture congregate in isolated compounds, sharing a way of life fashioned to resist the authority of the federal government. Fully armed, members of the militia organize their lives around a future revolution against their oppressor (Stern 1996; Ferber 1998). Many Americans view the 1995 Oklahoma City federal building bombing that killed 168 people as either part of the militia movement or an outgrowth of it. The skinheads, a neo-Nazi counterculture founded in the United States at the beginning of the 1980s, is a counterculture based on racial, ethnic, and sexual orientation intolerance. This intolerance appears in violent acts against gay men and lesbians, ethnic minorities (particularly Jews), and blacks. Skinheads draw their members from among young, white working-class males. Prime marks of membership are a shaved head and a quasi-military outfit. Skinheads see themselves as patriotic defenders of a belief in racial superiority and "sexual correctness" who have a right to use violence against their targets (Wooden 1995).

Prison counterculture surfaced at the 1999 trial of John King. King was convicted of the gruesome truck-dragging murder of James Byrd Jr. During an earlier prison stretch, King had become a member of a prison gang named the Confederate Knights of America's Texas Rebel Soldiers. Dedicated to white supremacy, this prison counterculture advocates violence in many forms. For example, the norm "blood in, blood out" involves the entrance requirement of a violent act and the threat of losing one's own blood for getting "out," or leaving the group (Galloway 1999). Certain types of delinquent gangs, motorcycle gangs, drug groups, and religious groups may also form countercultures (Cushman 1995; Roszak 1995; Zellner 1995; Zellner and Kephart 1997; Heineman 2002).

These members of a street gang are participants in a counterculture. They are aware of the beliefs, values, and norms of the larger culture, but they consciously participate in a lifestyle that opposes it.

How do cultural differences affect attitudes toward others? Once members of a society learn their culture, they tend to become strongly committed to it. In fact, they tend to become so committed that they cannot conceive of any other way of life. If members of other groups seem different, they may be considered strange. Cultural differences may even become signs of alleged inferiority. This tendency to judge others in relation to one's own cultural standards is referred to as **ethnocentrism**.

People who spend most of their lives with others culturally similar to themselves—who hardly ever deal with people different from themselves—will almost inevitably use their own cultural standards to judge others. The ethnocentric eye may see those who are different as inferior, ignorant, crazy, or immoral. We are shocked by a people such as the Ik, who willingly abandon their children at three years of age (Turnbull 1994). And it is hard to understand the Yanomamö women of Venezuela and Brazil who measure the affection of their husbands by the number of beatings they are given (Chagnon 1997).

Less exotic examples of ethnocentrism are plentiful—extreme nationalism being one. Members of all societies tend to offer themselves as exemplary models. The Olympic Games are much more than an arena for young men and women to engage in healthy and exuberant competition; they are also an expression of ethnocentrism. Influential political and nationalistic undercurrents run through the Olympics. A country's final ranking in this athletic competition for gold, silver, and bronze medals is frequently taken as a reflection of the country's worth and status on the world stage.

Ethnocentrism also exists within societies. Regional rivalries in the United States are a source of many humorous stories, but these jokes reflect an underlying ethnocentrism. Boston is said by some (mostly Bostonians) to be the hub of the universe. Some inhabitants of the eastern seaboard believe that only a wasteland separates them from the Pacific Ocean, while some more enlightened easterners are willing to concede the addition of California in describing all that is good in the United States. Of course, this long-standing ethnocentric viewpoint has been subject to attack by those in Middle America, who have always felt that virtue resides with them. And finally, the members of country clubs, churches, and schools all over America typically feel that their particular ways of living should be adopted by others.

Are there advantages of ethnocentrism? Up to this point, ethnocentrism has been portrayed as either politically incorrect or ridiculous. It may be both. But to some degree it is inevitable. And its inevitability is rooted partly in the advantages it offers to social life. Imagine the expense and effort required to create the integration, high morale, and loyalty that ethnocentrism provides a society or group. Few things draw people closer together than shared loyalty or conviction that what they are doing is right and superior vis-à-vis the actions and beliefs of others. Such commitment makes people feel good about themselves and their fellow group members. Fires of nationalism and patriotism are not fanned by declaring that your nation is slightly below average! Finally, ethnocentrism promotes social stability because people who are convinced that truth and beauty are theirs seldom entertain the need for change.

People are ethnocentric not simply because they recognize the benefits of ethnocentricity for social life. No. Ethnocentrism would be pervasive even if it did not offer these advantages. We are taught the rightness of our culture from a variety of sources (home, church, peers, schools, mass media), so it is only natural that this "rightness" becomes the yardstick for evaluating other cultures and how their members think, feel, and behave. Judgments of all kinds are nearly always alloyed with socially derived convictions regarding right and wrong. Ethnocentrism, then, is a predictable by-product of the transmission of culture.

WORLDVIEW 3.1

Patterns of Tourism

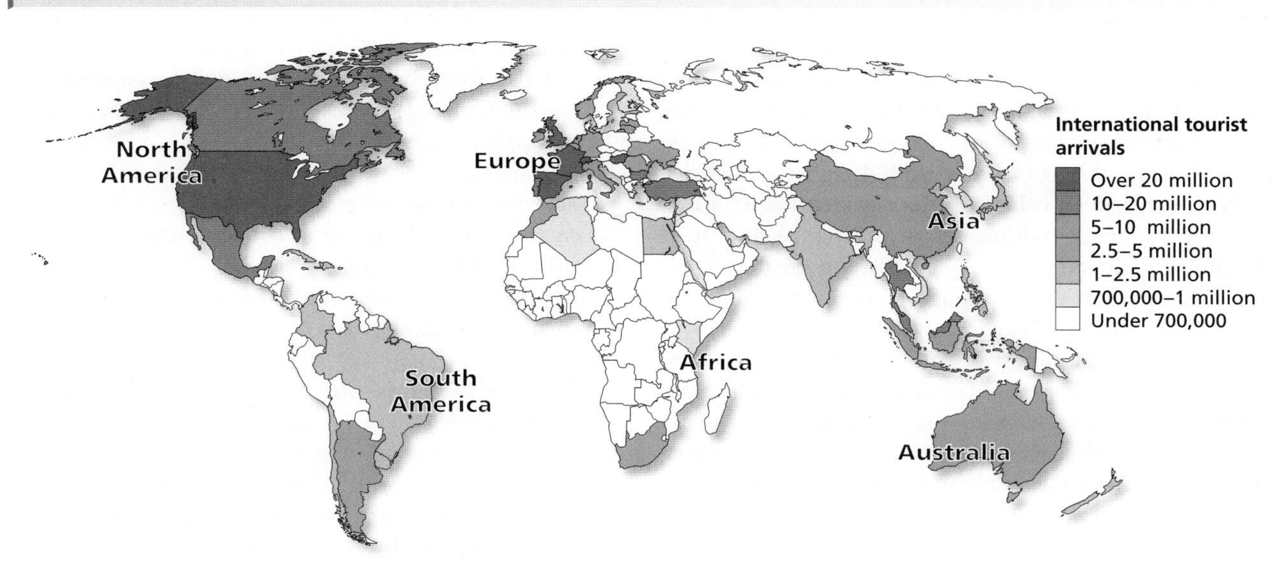

Source: David Lambert, *Student Atlas* (DK Publishing, 1998), p. 23.

While people often want to observe and experience cultures different from their own, exposure to cultural diversity can be uncomfortable. It is not surprising, therefore, that most international tourist travel occurs among countries sharing common cultural traditions and languages.

1. Can you identify world regions that might share cultural traditions and language?
2. Are there any reasons to believe that these travel patterns might change?

© AP/Wide World Photos

The effects of ethnocentrism can be disastrous. The recent genocide in Bosnia is only one contemporary example.

What are the disadvantages of ethnocentrism? A price may be paid for the integration, morale, loyalty, and stability that ethnocentrism provides. Extreme ethnocentrism has ill effects within societies as well as between them. On the intrasocietal side, extreme ethnocentrism may create such a high degree of integration and stability that innovation is hampered. Societies whose members are too firmly convinced of their righteousness may choke off exploration for new solutions to persistent problems. Further, they may reject, without examination, solutions that might be gleaned from other societies. And if, heightened ethnocentrism impedes both the internal creation of ideas and the external influence for solutions to long-standing problems, it may create an even greater disadvantage in facing the challenge of new problems and unfamiliar circumstances.

On the intersocietal front, social and political conflict is nearly always a result of extreme ethnocentrism. Global peace and welfare become secondary goals in a world in which ethnocentrism intensifies intersocietal conflicts based on power struggles for economic superiority and/or military supremacy.

We have already seen some of the negative consequences of ethnocentrism for a society. Ethnocentrism has harmful effects on individuals as well. Culture shock—the psychological and social stress we may experience when confronted with a radically different cultural environment—is one such negative consequence. Even cultural anthropologists, who are trained professionals, may experience culture shock. Napoleon Chagnon gives this account of his reaction to his first contact with the Yanomamö:

> *My heart began to pound as we approached the village and heard the buzz of activity within the circular compound. Mr. Barker commented that he was anxious to see if any changes had taken place while he was away and wondered how many of them had died during his absence. I felt into my back pocket to make sure that my notebook was still there and felt personally more secure when I touched it.*
>
> *The entrance to the village was covered over with brush and dry palm leaves. We pushed them aside to expose the low opening to the village. The excitement of meeting my first Yanomamö was almost unbearable as I duck-waddled through the low passage into the village clearing.*
>
> *I looked up and gasped when I saw a dozen burly, naked, sweaty, hideous men staring at us down the shafts of their drawn arrows! Immense wads of green tobacco were stuck between their lower teeth and lips making them look even more hideous, and strands of dark-green slime dropped or hung from their nostrils—strands so long that they clung to their pectoral muscles or drizzled down their chins. We arrived at the village while the men were blowing a hallucinogenic drug up their noses. One of the side effects of the drug is a runny nose. The mucus is always saturated with the green powder and they usually let it run freely from their nostrils. . . . My next discovery was that there were a dozen or so vicious, underfed dogs snapping at my legs, circling me as if I were to be their next meal. I just stood there holding*

Jacquelynne Eccles—Teenagers in a Cultural Bind

Adolescence is often marked by drama and difficulty. Jacquelynne Eccles and colleagues (1993) investigated the experience of American teenagers entering a midwestern junior high school and discovered that some teenage troubles are more than hormonal—they are cultural as well. The relationships between seventh graders, teachers, and parents are embedded in the normative and cognitive aspects of culture.

Eccles conducted a two-year study of students from twelve school districts in middle-class Michigan communities. She studied 1,500 early adolescents moving from sixth grade in an elementary school to seventh grade in a junior high school. Students filled out questionnaires at school in two consecutive years—the sixth and seventh grades. This procedure permitted Eccles to document changes the teenagers experienced after the first year of their transition.

The findings were not encouraging. The relationships between students and teachers deteriorated over the year, a deterioration Eccles found related to the culture of junior high school. At the very time when young adolescents especially needed supportive relationships outside the home, personal and positive relationships with teachers fell victim to cultural and organizational changes in junior high school. There was increased grouping based on academic achievement and increased public evaluation comparing students with one another. This heightened emphasis on student ranking comes just when young adolescents are at the peak of concern about their status relative to their peers. In addition, in the junior high culture, the students experienced less opportunity to participate in classroom decision making. The resulting diminished school motivation and lessened academic self-confidence is as predictable as it is unfortunate. The culturally defined educational structure of junior high works directly against the fact that adolescents develop better in emotionally supportive environments.

Eccles's news was no better on the home front. In fact, the results in the family paralleled those reported for the school. Excessive parental control over teenagers went up at the same time school motivation and self-esteem of the junior high students went down.

As a check on these general findings, Eccles compared students in more culturally supportive schools and families with those in less supportive ones. In both the school and the family settings, she found more positive results in culturally supportive environments. Students who were able to participate in school and family decision making showed higher levels of academic motivation and self-esteem than their peers with less opportunity to participate.

The solution, Eccles concludes, exists in a cultural change within both the schools and the family. Schools and the family need to develop balanced cultural expectations of young adolescents based on their developmental needs. At this age, young people have a growing need for autonomy. Neither cracking down on them nor relinquishing control strikes the proper balance. The task is for the family and the school to provide "an environment that changes in the right way and at the right pace" (Eccles et al. 1993:99).

Thinking About the Research

1. Do you recall your first year in junior high school? Analyze your experience based on the cultural definitions that existed in your school and home at the time.
2. Do adolescent students belong to a subculture? A counterculture? Explain.
3. Which of the three theoretical perspectives is most helpful in understanding the social relationships Eccles describes? Apply this perspective to explain her findings.

my notebook, helpless and pathetic. Then the stench of the decaying vegetation and filth hit me and I almost got sick. I was horrified. What kind of welcome was this for the person who came here to live with you and learn your way of life, to become friends with you? They put their weapons down when they recognized Barker and returned to their chanting, keeping a nervous eye on the village entrances. (Chagnon 1997:11–12)

If trained anthropologists can experience culture shock, it is easy to understand why immigrants are stunned by cultural practices foreign to their own. As a U.S. native, you might be surprised at the shocks, large and small, that internationals face in this country. When one newcomer was asked by a checkout clerk if he wished paper or plastic, he got very confused and upset. How could he choose paper or plastic when he thought payment required money? More seriously, imagine the difficulties of dating if you came from a culture in which interaction with the opposite sex among the young is limited to relatives (Pipher 2002). You will probably better understand culture shock if you take an overseas job assignment (Wagster 1993).

Cultural Relativism

Ethnocentrism can injure others when we attempt to impose our ways of thinking, feeling, and behaving on them. Despite good intentions, outsiders have, in many instances, harmed members of other cultures.

What can be done to reduce the negative personal effects of ethnocentrism? Awareness of ethnocentrism and the harm it can cause is a necessary first step to its control. An important second step lies in a perspective known as **cultural relativism**. According to this perspective, values, norms, beliefs, and attitudes are not in themselves correct or incorrect, desirable or undesirable; they simply exist within the total cultural framework of a people and should be evaluated in relation to their place within the larger cultural context of which they are a part rather than according to some alleged universal standard that applies across all cultures. Cultural relativism gives us a unique window through which to observe cultural variations.

Offering one's mate for sexual activity with an overnight guest is not allowed in most societies. Hans Ruesch, in his novel *Top of the World* (1959), however, reveals a society in which this practice was not only acceptable but expected. In traditional Eskimo society, it was a serious personal affront to the husband if a guest refused to "laugh" (have sexual intercourse) with his wife.

In the following passage from Ruesch's novel, an Eskimo named Ernenek tries to explain how his rage caused him to accidentally kill a guest who had refused to have sexual relations with his wife Asiak. Ernenek and Asiak are talking with a potential ally in the matter, a white man whose life Ernenek has saved:

> *"You have saved my life, Ernenek," the white man said, "and I wish to straighten things out so that you need no longer fear my companions. But you will have to stand before a judge. I will help you explain things."*
> *"You are very kind," said Ernenek, happy.*
> *"You said the fellow you killed provoked you?"*
> *"So it was."*
> *"He insulted Asiak?"*
> *"Terribly."*
> *"Presumably he was killed as you tried to defend her from his advances?"*
> *Ernenek and Asiak looked at each other and burst out laughing.*
> *"It wasn't so at all," Asiak said at last.*
> *"Here's how it was," said Ernenek. "He kept snubbing all our offers although he was our guest. He scorned even the oldest meat we had."*
> *"You see, Ernenek, many of us white men are not fond of old meat."*
> *"But the worms were fresh!" said Asiak.*
> *"It happens, Asiak, that we are used to foods of a quite different kind."*
> *"So we noticed," Ernenek went on, "and that's why, hoping to offer him at last a thing he might relish, somebody proposed him Asiak to laugh with."*
> *"Let a woman explain," Asiak broke in. "A woman washed her hair to make it smooth, rubbed tallow into it, greased her face with blubber and scraped herself clean with the knife, to be polite."*
> *"Yes," cried Ernenek, rising. "She had purposely groomed herself! And what did the white man do? He turned his back to her! That was too much! Should a man let his wife be so insulted? So somebody grabbed the scoundrel by his miserable little shoulders and beat him a few times against the wall—not in order to kill him, just wanting to crack his head a little. It was unfortunate it cracked a lot."*
> *"Ernenek has done the same to other men," Asiak put in helpfully, "but it was always the wall that went to pieces first."*
> *The white man winced, "Our judges would show no understanding for such an explanation. Offering your wife to other men!"*
> *"Why not? The men like it and Asiak says it's good for her. It makes her eyes sparkle and her cheeks glow."*
> *"Don't you people borrow other men's wives?" Asiak inquired.*
> *"Never mind that! It isn't fitting, that's all."*
> *"Refusing isn't fitting for a man!" Ernenek said indignantly.* (Ruesch 1959:87–88)

Now apply cultural relativism. How did this norm fit with other aspects of Eskimo culture? Under conditions at that time, all Eskimo possessions were handmade, difficult to replace, subject to hard use, and easily destroyed. Within this cultural context, lending wives made sense, as Ernenek explains:

> *Anybody would much rather lend out his wife than something else. Lend out your sled and you'll get it back cracked, lend out your saw and some teeth will be missing, lend out your dogs and they'll come home crawling, tired—but no matter how often you lend out your wife she'll always stay like new.* (Ruesch 1959:88)

Cultural relativism does not require our accepting other cultural ways as our own, nor does it require engagement in alien cultural practices. But because it is impossible, as a cultural relativist, to view aspects of another person's culture ethnocentrically, cultural relativism removes barriers between ourselves and those who are culturally different; it is essentially an antidote to ethnocentrism.

Cultural relativism can help us adjust more easily when we are meeting new people or entering new situations. For example, your adjustment to a college roommate from another country, another region of the United States, or even another social class will be smoother if you attempt to understand your roommate from the viewpoint of his or her own cultural background. And if significant cultural differences exist between you and your spouse, cultural relativism can enhance your marital relationship. It is also useful at work, where you will need to cooperate with people from other cultural environments. And given the growing importance of America's international business ties, someday you may well be practicing cultural relativism with people from Japan, Germany, Egypt, or Saudi Arabia.

Cultural Similarity

The world today seems to display an overwhelming diversity of social and cultural behaviors: In 1999, the president of Ingushetia, a republic inside Russia, signed a bill into law making it legal for a man to have multiple wives. German husbands and wives wear their wedding band on their right hand. In Bulgaria, a nod means no and a shake of the head means yes. Upon entering a home in India, you may be adorned with a garland of flowers—which you should remove immediately as a sign of humility. It is not polite to accept food in Iran until your host has offered it to you a number of times. Despite these surface differences, sociologists and anthropologists have identified many behaviors that are shared by all cultures. All societies have families, schools, houses of worship, economies, governments, and systems of prestige.

A recent poll compared Arabs (eight countries) and Americans with regard to the values they consider most important to be taught to their children (Zogby 2002). As Table 3.4 reveals, Arabs and Americans generally subscribe to common values. The same values (self-respect, good health and hygiene, responsibility) are among the top three for both Arabs and Americans. In fact, convergence is very high on five of the top six values. The biggest differences are on a couple of values lower on both lists. Americans place greater emphasis on teaching their children respect for authority. Arabs give the teaching of religious faith a more prominent place. For both, shared values focus on personal and family concerns rather than on external matters such as political issues.

George Murdock (1945) identified about seventy **cultural universals**, general cultural traits thought to exist in all known cultures. These universals included athletic sports, cooking, courtship, division of labor, education, etiquette, funeral rites, family, government, hospitality, housing, incest taboos, inheritance rules, joking, language, law, medicine, marriage, mourning, music, obstetrics, property rights, religious rituals, sexual restrictions, status differences, and toolmaking. When each of these universals is examined more closely, the similarity among cultures becomes even more apparent (see Figure 3.2).

How are cultural universals expressed? Just because societies share a cultural universal does not mean that they

TABLE 3.4

A Comparison of Values Among Arabs and Americans

This table contains a comparison of the ranking of values to be taught to children among Arabs and Americans. Do the data support the existence of cultural diversity or cultural similarity? Explain.

Value	Arab Rank	American Rank
Self-respect	1	2
Good health and hygiene	2	3
Responsibility	3	1
Respect for elders	4	5
Achieve a better life	5	8
Self-reliance	6	4
Religion faith	7	12
Serious work habits	8	6
Obedience	9	11
Creativity	10	9
Tolerance of others	11	10
Respect for authority	12	7

Source: James J. Zogby, *What Arabs Think: Values, Beliefs, and Concerns* (Utica, NY: Zogby International, 2002), 94.

FIGURE 3.2

Cultural Universals

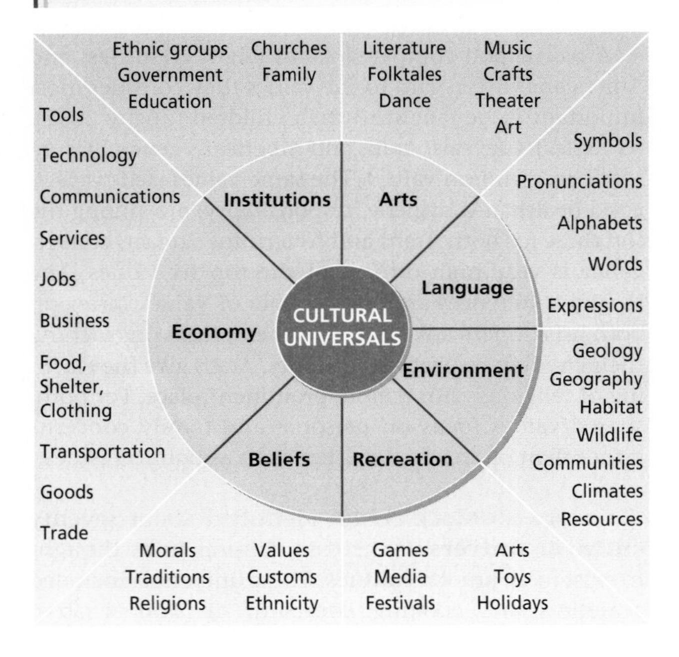

express it in the same way. Usually, cultures develop different ways of demonstrating the same universal trait. These may be thought of as **cultural particulars**. Traditionally, in the United States (though it has changed in recent years), the division of labor was based on gender roles. Women performed the domestic tasks and men worked outside the home. The Manus of New Guinea, on the other hand, place the man completely in charge of child rearing. Among the Mbuti pygmies, the Lovedu of Africa, and the Navajo and Iroquois Indians, men and women share equally in domestic and economic tasks (Little 1975).

Why do cultural universals exist? The biological similarity shared by all humans helps to account for the presence of cultural universals. If a society is to survive, children must be born and cared for and some type of family structure must exist. Groups that have deliberately eliminated the family—such as the Shakers—have disappeared. The existence of two sexes limits the type of marital relationships that can exist: one male and one female (monogamy), several females and one male (polygyny), one female and several males (polyandry), several males and females (group marriage), and two people of the same sex (homosexual marriage). Because people become ill, there must be some sort of medical care. Because people die, there must be funeral rites, mourning, and inheritance rules. Because babies are born, some form of obstetrics is required. Because food is necessary, cooking must exist. So goes the list of human biological similarities and their influence on culture.

A second source of cultural universals is the physical environment. For example, the awesomeness of nature and people's inability to explain physical phenomena—such as the eclipse of the sun or the creation of the universe—helped to encourage the development of religion. Because humans cannot survive in extreme climates without artificial protection, some form of housing must be created. Conflicts often occur over natural territorial borders such as rivers and mountains, so some provision must be made for maintaining order within as well as between societies.

Finally, cultural universals exist because societies face many of the same problems. If a society is to survive, certain social provisions must be made. New members must be socialized; goods and services must be produced and distributed; means of dealing with the supernatural must be devised; tasks must be assigned; and work must be accomplished.

FEEDBACK

1. Indicate which of the following are social categories (SC), subcultures (S), or countercultures (C).
   ____ a. Chinatown in New York City
   ____ b. motorcycle gang
   ____ c. Catholics
   ____ d. females
   ____ e. revolutionary political group
   ____ f. the superrich
2. The tendency to judge other societies or groups according to one's own cultural standards is known as _________.
3. The psychological and social stress one may experience when confronted with a radically different cultural environment is called _________.
4. _________ is the idea that any given aspect of a particular culture should be evaluated in terms of its place within the larger cultural context of which it is a part rather than in terms of some alleged universal standard that applies across all cultures.
5. _________ are general cultural traits thought to exist in all known cultures.
6. Which of the following is not one of the reasons that cultural universals exist?
   a. physical environment
   b. biological similarity of humans
   c. cultural predetermination
   d. problems of maintaining social life

Answers: 1. a. (S) b. (C) c. (SC) d. (SC) e. (C) f. (SC) 2. ethnocentrism 3. culture shock 4. Cultural relativism 5. Cultural universals 6. c

Culture, Society, and Heredity

Culture and Heredity

What is the relationship between culture and biology? **Instincts** are genetically inherited, complex patterns of behavior that always appear among members of a particular species under appropriate environmental conditions. Although nonhuman animals are heavily hardwired for action through instincts, human infants cannot go very far on the basis of their genetic heritage alone. They lack immediate and automatic solutions to the problems they face. And later, without instincts to determine the type of shelter to build, the type of food to eat, the time of year to have children, or the type of mating pattern to follow, humans are forced to create and learn their own ways of thinking, feeling, and behaving. That is, humans are forced to rely on the culture they have created—even for such basic needs as eating and reproducing.

If humans were controlled by instincts, they would all behave in the same way. If, for example, women had an instinct for mothering, then all women would want children, and all women would love and protect their children. In fact, some women do not want to have children, and some women who give birth abuse or abandon, or kill their children (Hrdy 2000).

Is genetic heritage without influence on human behavior? Although humans lack instincts, they are nevertheless affected by their genetic nature. Research on twins reared together and apart has examined the relative influence of heredity and environment on personality traits. It is estimated that about 50 percent of the diversity in measured personality characteristics is attributable to genetic heritage (Tellegen et al. 1993; Messner 1997).

In addition, humans have **reflexes**—simple, biologically inherited, automatic reactions to physical stimuli. A human baby, for example, cries when pinched; its eyes blink in response to a foreign particle; and the pupils of its eyes contract in bright light. Humans also have biologically inherited **drives**, or impulses, to reduce discomfort. They want to eat, drink, sleep, associate with others, and have sexual relations.

Genetically inherited personality traits, reflexes, and drives, however, do not "determine" human social behavior. Instead, culture channels the expression of these biological characteristics (Good et al. 1994). Most boys in traditional American culture, for example, are taught not to cry in response to pain, whereas boys in Jewish and Italian cultures learn to pay more attention to physical discomfort and express it more obviously (Zborowski 1952, 1969). To cite another example, humans have inherited the capacity for love. Awareness of this fact, however, does not allow us to predict the precise ways that different groups of people will express the ability to love. The people of one culture may believe that being one of several husbands is the most natural form of marriage, whereas members of another culture may endorse monogamous marriage.

Some social and biological scientists are challenging the traditional sociological perspective on human nature. Their divergent perspective is called *sociobiology*.

Sociobiology

Sociobiology is the study of the biological basis of human behavior. Sociobiologists argue that like physical characteristics, human social behavior is shaped through the evolutionary process. Thus, sociobiology is the application of Darwinian natural selection and modern genetics to human social behavior (Pinker 2002).

How do sociobiologists view human behavior? According to Darwin's theory of evolution, all organisms evolve through the process of natural selection. Those organisms best suited to an environment survive and reproduce themselves, and the rest perish. If human behavior is based on millions of years of evolution, say the sociobiologists, then most human behavior is basically self-protective (Dawkins 1990). Those behaviors that improve our chances for survival are genetically retained and reproduced. Parental affection and care, friendship, sexual reproduction, and the education of children must therefore be biologically based because they contribute to the survival of the human species. Some sociobiologists even contend that there may be specific genes for such things as aggression, religion, homosexuality, group loyalty or altruism, the creation of hierarchies, and the incest taboo.

Sociobiologists believe that the sharp line usually drawn between human and nonhuman animals is inappropriate. DNA testing shows that humans and apes (chimpanzees, gorillas, orangutans, and gibbons) are more closely related than either are to monkeys. Moreover, the chimpanzees' closest relative is not the gorilla but humans; we differ in only 1.6 percent of our DNA (Diamond 1993, 1999).

Nonhuman animals learn and transmit their knowledge—as when a group of chimpanzees spontaneously begin to use long sticks to ferret ants from an anthill for a meal (De Waal 1999; Whiten et al. 1999). Many nonhuman animals, assert sociobiologists, exhibit intelligence of a kind formerly thought to be unique to humans, including use of language (Begley 1993; Linden 1993a; Fagot, Wasserman, and Young 2001; Hauser 2001). For the first time among wild animals, chimps have been observed deliberately, consistently, and repeatedly using a tool as a weapon against one of their own kind (Linden 2002a, 2002b). These findings tend to support pioneer sociobiologist Edward Wilson's (1978, 1986) contention that the study of human behavior must begin with the genetic heritage of humans.

Contemporary evolutionary social scientists report some aspects of social behavior they believe are genetically based. Women, it is reported, look for different characteristics in men they prefer to date and marry than men value in women (Buss, Malamuth, and Winstead 1998). And stepfathers are more likely to abuse their stepchildren than biological fathers their own offspring (Daly and Wilson 1997).

What do the critics of sociobiology say? Some critics of sociobiology fear that this perspective will be used as a justification to label specific races inferior; others fear that it will be used to argue for the superiority of the male (Andersen 2002). Still other critics fear that the sociobiological perspective will be used to bolster the assumption of innate human selfishness. Marxists object to sociobiology on the grounds that it justifies the existence of Western capitalism.

Critics join in common objection to a genetic explanation of human social behavior. There is, they contend, too much social diversity around the world for human behavior to be explained on strictly biological grounds (Parker and Easton 1998). The human brain and the unique human capacity for using language and creating social life have allowed humans to overcome any contribution to behavior that might come from their genes. Humans, liberated from the confines of their genes by a large, well-developed cerebral cortex—which permits, among other things, abstract thinking—create and transmit a dazzling array of ways for thinking, feeling, and behaving. Nonhuman species, which either have no cerebral cortex or have an undeveloped one, behave as they do because of a strict genetic code. Birds do not walk south for the winter; salmon do not fly upstream; and lions do not prefer ferns to fresh meat. Critics believe that sociobiology should not sidetrack sociologists from their efforts to understand and explain human behavior from the sociological perspective. They agree with the nineteenth-century English philosopher John Stuart Mill: "Of all the vulgar modes of escaping from consideration of the effect of social and moral influence upon the human mind, the most vulgar is that of attributing the diversities of conduct and character to inherent natural differences" (Zinn 2003:33).

Since the early days of sociobiology, some common ground has emerged between sociologists and sociobiologists. Anthropologist Marvin Harris (1990) agrees that there is some biological continuity between human and nonhuman animals and that there is a basic human nature. Evolutionary psychologists and some sociologists now contend that human biology and the human capacity for creating a nearly infinite variety of ways for thinking, feeling, and behaving are two sides of the same coin. The genetic heritage of humans, they argue, shapes and limits human nature and social life. Consequently, they contend, sociologists should not overlook the complex relationship between genetic heritage and the human capacity for creating social life (Lopreato 1990; Weingart et al. 1997; Konner 1999; Pinker 2002; Ridley, 2003).

FEEDBACK

1. A genetically inherited, complex pattern of behavior that always appears among members of a particular species under appropriate environmental conditions is a(n)
   a. reflex
   b. instinct
   c. drive
   d. need
2. Scientists agree that humans have instincts for self-preservation, motherhood, and war. T or F?
3. Indicate which of the following are drives (D), which are reflexes (R), which are instincts (I), and which are human creations (H).
   ____ a. eye blinking in dust storm
   ____ b. need for sleep
   ____ c. reaction to a loud noise
   ____ d. socialism
   ____ e. sex
   ____ f. racial inequality
4. According to sociology, culture totally determines the nature of human behavior. T or F?
5. __________ is the study of the biological basis of human behavior.
6. Sociologists now agree that genetic heritage plays no part in the shaping and limiting of social life. T or F?

Answers: 1. b 2. F 3. a. (R) b. (D) c. (R) d. (H) e. (D) f. (H) 4. F 5. Sociobiology 6. F

SUMMARY

1. Culture consists of all the material objects as well as the patterns of thinking, feeling, and behaving passed from generation to generation within a society. A society is a group of people living within defined territorial borders who share a culture.
2. The three broad categories of culture are the normative, cognitive, and material dimensions. The normative dimension of culture is composed of norms, sanctions, and values. Norms are rules defining appropriate and inappropriate ways of behaving. There are several types of norms. Folkways are not considered vital to a group and may be violated without significant consequences. When mores—norms considered essential—are violated, social disapproval is strong. Some norms become laws. Sanctions—positive and negative, formal and informal—are used to encourage conformity to norms. Values are broad cultural principles defining the desirable. Norms and values are not the same thing; the same value can be expressed through radically different norms. In complex societies, some values conflict.
3. Language is an important aspect of the cognitive dimension of culture. Beliefs, another important aspect of the cognitive dimension, are ideas concerning the nature of reality. Whether or not they are actually true, beliefs have a great influence on the members of a society.
4. Material culture is composed of the concrete, tangible aspects of a culture. Aspects of material culture—desks, trucks, cups, money—have no inherent meanings. Material objects have meanings only when people assign meanings to them.
5. Cultural guidelines and actual behavior do not always match. Sociologists distinguish between the cultural guidelines a society claims to accept (ideal culture) and the behavior patterns actually practiced (real culture).
6. Because humans are capable of creating and communicating arbitrary symbols such as language, they have the ability to create and transmit culture. Gestures, such as facial expressions or body movements, also carry culturally defined and shared symbolic meanings. Language is also important because it organizes people's view of reality. According to the hypothesis of linguistic relativity, people actually live in different worlds because their languages make them aware of different aspects of their environment. People are not forever trapped by their language, but they are limited by it unless they learn to see the world from another linguistic viewpoint.
7. Wide cultural variation exists from one society to another as well as within a single society. Because all humans are basically the same biologically, cultural diversity must be explained by nongenetic factors. Cultural diversity within societies is promoted by the existence of social categories, subcultures, and countercultures.
8. Ethnocentrism—judgment of another group in relation to one's own cultural standards—is a major consequence of cultural diversity. The consequences of ethnocentrism are both positive and negative. The principle of cultural relativism—the idea that any given aspect of a culture should be evaluated according to its place within the larger cultural context of which it is a part—helps to combat ethnocentrism.
9. Some cultural universals are found in all societies. Their expression in specific practices, however, varies widely from one society to another. Cultural universals exist because of the biological similarity of humans, common limitations of the physical environment, and the common problems of sustaining social life.
10. Contrary to popular belief, humans do not have instincts. Most behavior among nonhuman animals is instinctual, but human behavior is not the sole product of genetic heritage. Human behavior is learned. Even genetically inherited reflexes and drives do not determine how humans will behave, because people are heavily influenced by culture. Although culture does not determine human nature, it does significantly condition it. Sociobiologists are now arguing for recognition of the role of biology in human behavior.

INFOTRAC® COLLEGE EDITION

http://www.infotrac-college.com/wadsworth

Another unique option available to you at the Student Resources section of the companion Web site is Infotrac College Edition, an online library with access to hundreds of scholarly and popular periodicals. Below are suggested search terms for this chapter. Results from these and other searches are found at the site.

Search keyword: **counterculture**. The "hippie" movement in the 1960s was often referred to as a counterculture. Using Infotrac, find out what types of activities are being called countercultures today. Based on your understanding from the textbook, do you agree that the activities you found are truly countercultures?

Search keyword: **culture**. The textbook defines culture as "a people's way of life, consisting of material objects as well as the patterns of thinking, feeling, and behaving, that is passed from generation to generation among members of a society." Find several articles that refer to culture and determine how those authors use the term *culture*. Based on your findings, can you assume that different individuals have the same concept of culture?

Search keywords: **Stockholm syndrome**. In 1973 during a robbery, four Swedes were held in a bank vault for six days. They became attached to their captors, a phenomenon called the Stockholm syndrome. Search for articles related to this topic. What did you learn? How might this phenomenon relate to the study of sociology?

LEARNING OBJECTIVES REVIEW

- Identify the three major dimensions of culture.
- Describe and illustrate the interplay between language and culture.
- Discuss cultural diversity and its promotion within a society.
- Describe and illustrate the relationship between cultural diversity and ethnocentrism.
- Outline the advantages and disadvantages of ethnocentrism and discuss the role of cultural relativism in combating ethnocentrism.
- Explain the existence of cultural similarities that are shared around the world.
- Explain the relationship between culture and heredity.

CONCEPT REVIEW

Match the following concepts with the definitions listed below them.

____ a. cultural universals
____ b. sociobiology
____ c. sanctions
____ d. informal sanctions
____ e. real culture
____ f. cognition
____ g. society
____ h. cultural particulars
____ i. laws
____ j. taboos
____ k. ethnocentrism
____ l. mores
____ m. subculture
____ n. formal sanctions
____ o. culture shock
____ p. gestures

1. the study of the biological basis of human behavior
2. general cultural traits thought to exist in all known cultures
3. subterranean patterns for thinking, feeling, and behaving
4. rewards and punishments that may be applied by most members of a group
5. norms that are formally defined and enforced by designated persons
6. norms so strong that their violation is thought to be punishable by the group or society or even by some supernatural force
7. the widely varying, often distinctive ways societies demonstrate cultural universals
8. the psychological and social stress one may experience when confronted with a radically different cultural environment
9. the process of human thinking
10. norms of great moral significance, thought to be vital to the well-being of a society
11. rewards and punishments used to encourage desired behavior
12. a group of people who live within defined territorial borders and who share a common culture
13. rewards and punishments that may be given only by officially designated persons
14. the tendency to judge others in terms of one's own cultural standards
15. a group that is part of the dominant culture but differs from it in some important respects
16. facial expressions, posture, and body movements that carry culturally defined and shared symbolic meanings

CRITICAL-THINKING QUESTIONS

1. Distinguish the normative, cognitive, and material aspects of culture. Cite illustrations to show an understanding of the differences.

   ____

   ____

2. The ability to use language is cited by sociologists as a characteristic separating humans from nonhuman animals. Discuss the relationship between language and culture. Provide examples.

   ____

   ____

3. Functionalists and conflict theorists would be expected to have different views of countercultures. Identify these differences and defend the position you prefer in relation to

the theoretical perspective you support.

4. Discuss the relationship between cultural relativism and ethnocentrism. Give one or more examples to make the connection.

5. Sometimes you hear it said that people are just "naturally" selfish. Do you agree or disagree that humans are capable only of pursuing their own self-interest? Defend your position.

MULTIPLE-CHOICE QUESTIONS

1. **In your text, *culture* is defined as**
   a. a number of persons who share a social goal.
   b. rules defining appropriate and inappropriate ways of behaving for specific situations.
   c. the material objects as well as the patterns for thinking, feeling, and behaving that are passed from generation to generation among members of a society.
   d. the idea that any given aspect of a particular culture should be evaluated in relation to its place within the larger cultural context of which it is a part.
   e. people who live within defined territorial borders and who share common ties.
2. **This chapter distinguishes between three major dimensions of culture: the normative, the material, and the**
   a. cognitive.
   b. ecological.
   c. ethical.
   d. genetic.
   e. noncognitive.
3. **_________ are rules defining appropriate and inappropriate ways of behaving for specific situations.**
   a. Folkways
   b. Values
   c. Mores
   d. Norms
   e. Laws
4. **_________ are applied only by officially designated persons, such as judges, executioners, or college professors.**
   a. Sanctions
   b. Formal sanctions
   c. Laws
   d. Mores
   e. Statutes
5. **The example, in the text, of a doctor's introduction to nettling demonstrates that**
   a. material aspects of culture have the same meanings and uses in all cultures.
   b. it is conventional to use a 747 jet for traveling.
   c. separate physical objects existing as part of a culture may be combined to have meaning for some people and no meaning for others.
   d. newspaper and pepper are an ineffective way of delivering babies.
   e. laws regulating midwifery are unnecessary.
6. **Which of the following statements is *not* consistent with the "tool kit" view of culture?**
   a. Culture provides a range of choices to be applied in the definition and solution of the problems of living.
   b. Culture is a means for "fixing" social problems.
   c. Culture should be viewed as a group of symbols, stories, rituals, and worldviews.
   d. Because the content of any culture is not fully consistent, the symbols, stories, rituals, and worldviews that are chosen vary among individuals and situations.
7. **The Sapir-Whorf hypothesis holds that**
   a. people are forever prisoners of their language.
   b. our view of the world depends on the particular language we have learned.
   c. Germans are less concerned about privacy than are Americans.
   d. exposure to another language can cause a person to become marginalized in their own culture.
   e. functionalism is more plausible than conflict theory.
8. **Societies experience cultural diversity in part because of**
   a. biological variation.
   b. social policy.
   c. social categories.
   d. social determinism.
   e. social disaggregration.
9. **In your text, a counterculture is referred to as**
   a. a subculture that is deliberately and consciously opposed to certain central aspects of the dominant culture.
   b. the material objects as well as the patterns for thinking, feeling, and behaving that are passed from generation to generation among members of a society.
   c. general cultural traits thought to exist in all known cultures.
   d. aspects of culture publicly embraced by members of a society.
   e. a group that is part of the dominant culture but differs from it in some important respects.
10. **Negative consequences of extreme ethnocentrism include all of the following except**

a. culture shock.
b. intrasocietal fragmentation.
c. intersocial conflict.
d. resistance to innovation.
e. feelings of cultural relativism.

11. **The idea that a given aspect of a particular culture should be evaluated in relation to its place within the larger cultural context of which it is a part, rather than according to some alleged universal standard that applies across all cultures, is known as**
    a. cultural universalism.
    b. ethnocentrism.
    c. ecological adaptation.
    d. cultural relativism.
    e. pragmatism.
12. **__________ are general traits thought to exist in all cultures.**
    a. Cultural particulars
    b. Cultural universals
    c. Ecological adaptations
    d. Material adaptations
    e. Cultural adaptations
13. **Simple, biologically inherited, automatic reactions to physical stimuli are known as**
    a. reflexes.
    b. impulses.
    c. instincts.
    d. drives.
    e. needs.
14. **__________ is the study of the biological basis of human behavior.**
    a. Anthropology
    b. Ethnomethodology
    c. Sociolinguistics
    d. Sociobiology
    e. Social psychology

FEEDBACK REVIEW

True-False

1. According to the hypothesis of linguistic relativity, words and the structure of a language cause people to live in a distinct world. T or F?
2. Sociologists now agree that genetic heritage plays no part in the shaping and limiting of social life. T or F?

Fill in the Blank

3. __________ consists of all the material objects as well as the patterns for thinking, feeling, and behaving that are passed from generation to generation among members of a society.
4. Because of __________, culture can be created and transmitted.
5. The tendency to judge other societies or groups according to one's own cultural standards is known as __________.
6. __________ is the idea that any given aspect of a particular culture should be evaluated in relation to its place within the larger cultural context of which it is a part rather than according to some alleged universal standard that applies across all cultures.
7. __________ are broad cultural principles embodying ideas about what most people in a society consider to be desirable.

Matching

8. Indicate whether these statements best reflect a belief (B), folkway (F), mos (M), law (L), or value (V).
    ____ a. conception that God exists
    ____ b. norm against cursing aloud in church
    ____ c. norm encouraging the eating of three meals daily
    ____ d. idea of progress
    ____ e. norm against burning a national flag
    ____ f. norm encouraging sleeping in a bed
    ____ g. norm prohibiting murder
    ____ h. norm against overtime parking
    ____ i. idea that the Earth's orbit is elliptical
    ____ j. idea of freedom
9. Indicate whether the following are formal sanctions (F) or informal sanctions (I).
    ____ a. A mother spanks her child.
    ____ b. A professor fails a student for cheating on an exam.
    ____ c. A jury sentences a person to life in prison for espionage.
    ____ d. A husband separates from his wife after she has an affair.
10. Indicate which of the following are social categories (SC), subcultures (S), or countercultures (C).
    ____ a. Chinatown in New York City
    ____ b. motorcycle gang
    ____ c. Catholics
    ____ d. females
    ____ e. revolutionary political group
    ____ f. the superrich

GRAPHIC REVIEW

The trend toward increased premarital sexual experience among teenage females in the United States is graphically illustrated in Figure 3.1. The following questions ask you to relate this trend to the normative dimension of culture.

1. During the 1950s, in the United States, what type of norm did the prohibition against premarital sex among teenage females represent?

   ____
   ____
   ____

2. What type of norm regarding premarital sex among teenage females exists in America today?

   ____
   ____
   ____

3. Compare the types of formal and informal sanctions used to discourage premarital sexual activity among American female teenagers during the 1950s and today.

   ____
   ____
   ____

ANSWER KEY

Concept Review

a. 2
b. 1
c. 11
d. 4
e. 3
f. 9
g. 12
h. 7
i. 5
j. 6
k. 14
l. 10
m. 15
n. 13
o. 8
p. 16

Multiple Choice

1. c
2. a
3. d
4. b
5. c
6. b
7. b
8. c
9. a
10. e
11. d
12. b
13. a
14. d

Feedback Review

1. T
2. F
3. Culture
4. language
5. ethnocentrism
6. Cultural relativism
7. Values
8. a. B
   b. M
   c. F
   d. V
   e. M
   f. F
   g. L
   h. L
   i. B
   j. V
9. a. I
   b. F
   c. F
   d. I
10. a. S
    b. C
    c. SC
    d. SC
    e. C
    f. SC

4 Socialization Over the Life Course

OUTLINE

LEARNING OBJECTIVES

- Discuss the contribution of socialization to the process of human development.
- Describe the contribution of symbolic interactionism to our understanding of socialization, including the concepts of the self, the looking-glass self, significant others, and role taking.
- Compare and contrast the life course theories of Freud, Erikson, and Piaget.
- Distinguish among the concepts of desocialization, resocialization, and anticipatory socialization.
- Better understand the socialization process of young people.
- Describe the stages of adult development.
- Discuss the unique demands of socialization encountered in late adulthood.
- Compare and contrast the application of functionalism and conflict theory to the socializing effects of mass media.

USING THE SOCIOLOGICAL IMAGINATION

Does violence on television lead to real-life violent acts? Until fairly recently, Americans strongly disagreed about the spillover effect of television violence. Moreover, social scientists hesitated to claim a causal link between television violence and actual violence. No longer. After hundreds of studies, researchers now find common ground in citing the connection between televised aggression and personal aggressiveness.

This link between televised violence and real violence is another example of a culturally transmitted human social behavior. As humans learn the culture around them, they adopt certain patterns of behavior. This learning process begins at birth and continues into old age. It is called *socialization*.

The first section of this chapter documents the importance of socialization for human development. A discussion of theoretical perspectives, with an emphasis on symbolic interactionism, then follows. The next two sections portray socialization as it unfolds throughout the life cycle. The chapter closes with an examination of mass media in light of functionalism and conflict theory.

Social Participation or Social Deprivation

According to an enduring popular belief, human nature is spoiled by civilization. Society corrupts. Jean-Jacques Rousseau's "noble savage" has been a frequent theme in romantic novels. Underlying this idea is the belief in a fundamental human nature that exists prior to social contact. Whether this is true or not, the nature of humans is shaped by socialization. Human beings at birth are helpless and without knowledge of their society's ways of thinking, feeling, and behaving. If a human infant is to learn how to participate in social life, much cultural learning has to take place.

Nearly all the human social behavior we consider natural and normal is learned. How, for example, is one supposed to behave at a cocktail party? Certainly, an individual is expected to have a drink. But what will the drink be—white wine or sparkling water? Asking seriously for a Bloody Mary with fresh cow's blood would result in signs of disgust and social disapproval. Yet, in some African tribes, a glass of cow's blood would be relished (Douglas 1979). Children in some Palestinian refugee camps aspire to becoming suicide bombers just as other children wish to be policemen or athletes. These statements from children will seem shocking to most people:

> *"Everyone here* [residents of a refugee camp near Bethlehem] *wants to be a Shahid* [martyr], *everyone has a friend who died and went to heaven,"' says Salem, a genial 10-year-old. "Everything I could wish for is in paradise," says Abdullah, who is 13. "I would have green gardens and fresh fruit trees and we would have freedom. Not like here."* (Miller and Penniman 2002:49)

One source of these aspirations is the family. Palestinian mothers have been observed celebrating their children's success in blowing up other children. Their infants are sometimes outfitted as suicide bombers (Will 2002). Preference for cow's blood or admiration for a suicide martyr are, like nearly all aspects of social life, acquired through **socialization**—the process of learning to participate in group life through the acquisition of culture.

Learning about the countless aspects of social life begins at birth and continues throughout life. For example, infants in most American homes are taught to eat certain foods, to sleep at certain times, and to smile at certain facial expressions. But socialization is not limited to the early years; it is a lifelong process enabling people to fit into all kinds of social groups. Socialization must occur if freshmen are to adjust to their new environment, if plebes are to survive at West Point, if presidents of the United States are to govern, and if nursing-home residents are to adapt to their unfamiliar setting. Would-be executives who prefer warm cow's blood to dry white wine will have to keep their tastes a secret or abandon their occupational aspirations.

How is socialization related to personality? Although socialization is a lifelong process, the most important learning occurs early in life. Childhood cases of extreme social isolation reveal that without early prolonged and intensive social contact, children do not learn such basics as walking, talking, and loving. Without socialization, a human infant cannot develop a **personality**—the relatively organized complex of attitudes, beliefs, values, and behaviors associated with an individual.

The Importance of Socialization

How can the effects of socialization be assessed? An accurate assessment of socialization would require an experiment comparing a control group of normally

© Bob Krist/CORBIS

Socialization is the process of learning to participate in group life, through the influence of culture. These alums are visiting the college where they spent formative years learning aspects of their culture in preparation for the adult lives they now lead.

socialized infants with an experimental group of socially isolated infants. Assuming that these two groups of children were biologically the same, the differences between them at the end of the experimental period could be attributed to socialization. For obvious reasons, there is no such evidence on human infants. There has, however, been some experimental research with monkeys and some nonexperimental evidence from studies of socially isolated children.

How do monkeys react to social isolation? Experiments by Harry Harlow (Harlow and Zimmerman 1959; Harlow and Harlow 1962; Harlow 1967; Blum 2002) have shown the negative effects of social isolation among rhesus monkeys. In one experiment, infant monkeys were separated from their mothers at birth and then exposed to two artificial mothers—wire dummies of the same approximate size and shape as real adult monkeys. One of the surrogate (substitute) mothers had an exposed wire body; the other was covered with soft terry cloth. Free to choose between them, the infant monkeys consistently spent more time with the soft, warm mother. Even when the exposed wire surrogate became the only source of food, the terry cloth mother remained their preference. Clearly, closeness and comfort were more important to these monkeys than food. When agitated by a mechanical toy bear or a rubber snake, these infant monkeys consistently ran to their cloth mothers for security and protection.

Apparently, infant monkeys need intimacy, warmth, physical contact, and comfort. Indeed, Harlow has shown that infant monkeys raised in isolation become distressed, apathetic, withdrawn, hostile adult animals. They never exhibit normal sexual patterns; and as mothers they either reject or ignore any babies they may have, sometimes even abusing them physically.

Can we generalize from monkeys to humans? Generalizing research findings from nonhumans to humans is risky, because people are not monkeys. Nevertheless, many experts on human development believe that a human infant's emotional need for affection, intimacy, and warmth—like Harlow's monkeys—is as important as the infant's physiological need for food, water, and physical protection. Babies denied close human contact usually have difficulty forming emotional ties with others. Touching, holding, stroking, and communicating appear to be essential to normal human development. Although Harlow's findings with monkeys cannot be applied directly to humans, similar findings on human children have given them considerable credibility. According to a Lawrence Casler study (1965), for example, the developmental growth rate of institutionalized children can be improved by only 20 minutes of extra touching a day. Cases of socially isolated children provide additional support.

Social Isolation Among Humans

There is dramatic evidence that children deprived of social contact do not develop all of the characteristics associated with being human. Prominent among this evidence are cases of children who have been deprived—both socially and emotionally. Fortunately, documentation includes observations of the changes that occurred when such unfortunate children were placed in an environment designed to socialize them. We will look at the histories of three children—Anna, Isabelle, and Genie—who were socially and emotionally neglected or abused (K. Davis 1947; Curtiss 1977; Pines 1981).

Who was Anna? Anna was the second child born to her unmarried mother. They lived with Anna's grandfather, who was incensed by the latest evidence that his daughter did not measure up to his strict moral code. Anna and her mother were forced to move out of the house. After several months of moving from one place to another, repeated failures at adoption, and no further alternatives, Anna and her mother returned to the grandfather's house. Anna's mother so feared that the sight of the child would anger her father that she confined Anna to an attic-like room on the second floor of their farmhouse. Anna was kept alive on milk alone until she was discovered at the age of five. Barely alive, she was extremely emaciated and undernourished. Her legs were skeleton-like, her stomach bloated.

Apparently, Anna had seldom been moved from one position to another, and her clothes and bed were filthy. Positive emotional attention was unfamiliar to her. When she was found, Anna could exhibit no signs of intelligence, nor could she walk or talk.

During the first year and a half after being found, Anna was in a county home for children. Among other things, she learned to walk, understand simple commands, eat by herself, tend to personal neatness somewhat, and recall people she had seen. But her speech was that of a one-year-old. Anna was then transferred to a school for retarded children, where she made some further progress. But still, at the age of seven, her mental age was only nineteen months, and her social maturity was that of a two-year-old. A year later she could bounce and catch a ball, participate as a follower in group activities, eat normally (although with a spoon only), attend to her toilet needs, and dress herself (except for handling buttons and snaps). Significantly, she had acquired the speech level of a two-year-old. By the time of her death at age ten, Anna had made some additional progress. She could carry out instructions, identify a few colors, build with blocks, wash her hands, brush her teeth, and try to help other children. Her developing capacity for emotional attachment was reflected in the love she had developed for a doll.

Who was Isabelle? Nine months after Anna was found, Isabelle was discovered. She, too, was the child of a single mother and was kept in isolation for fear of social disapproval. Isabelle's mother, a deaf-mute since the age of two, stayed with the child in a dark room, secluded from the rest of the family. When found at the age of six and a half, Isabelle was physically ill from an inadequate diet and lack of sunshine. Her legs were so bowed that when she stood, the soles of her shoes rested against each other, and her walk was a skittering movement. Unable to talk except for a strange croaking sound, Isabelle communicated with her mother by means of gestures. Like an animal in the wild, she reacted with fear and hostility to strangers, especially men. Some of her actions were those of a six-month-old infant.

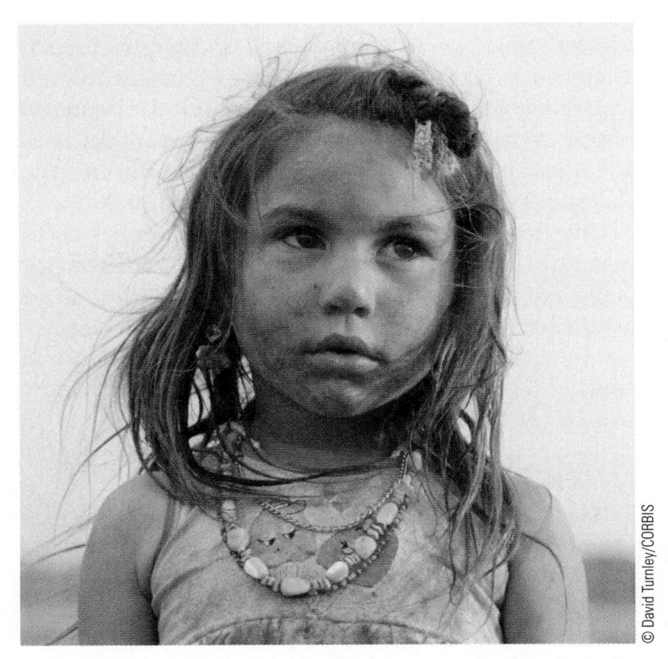
© David Turnley/CORBIS

Normal human social and personality development requires intensive and prolonged human interaction in the formative years. Lack of adult warmth and care can produce many enduring social and psychological problems. Loneliness and difficulty in interacting with others, which are among these problems, are easy to sense in this young girl.

Isabelle's first IQ score was near the zero point, and her social maturity was at the level of a two-and-a-half-year-old. Despite the belief that Isabelle was feebleminded, an intensive program of rehabilitation was begun. After a slow start, Isabelle progressed through the usual stages of learning and development, much like any normal child progressing from ages one to six. Although the pace was faster than normal, the stages of development were in their proper order. It took her only two years to acquire the skills mastered by a normal six-year-old. By the time she was eight and a half, Isabelle was on an educational par with children her age. By outward appearances, she was an intelligent, happy, energetic child. At age fourteen, she participated in all the school activities normal for other children in her grade. To Isabelle's good fortune, she, unlike Anna, benefited from intensive instruction at the hands of trained professionals. Her ability to progress may also have been due to the presence of her mother during the period of isolation.

Who was Genie? A more recent case did not end as happily as Isabelle's. Genie, from the time she was nearly two, had been kept isolated in a locked room by her father. When she was found, at the age of thirteen, much of her behavior was subhuman. Because Genie's father severely punished her for making any vocal sounds whatever, she was completely silent. In fact, she neither sobbed when she cried nor spoke when in a fit of rage. Because she had hardly ever worn clothing, Genie did not notice changes in temperature. Never having been given solid food, she could not chew. Because she had spent her entire life strapped in a potty chair, Genie could not stand erect, straighten her arms or legs, or run. Her social behavior was primitive. She blew her nose on whatever was handy or into the air when nothing was available. Without asking, she would take from people things that attracted her attention.

Attempts to socialize Genie over a four-year period were not successful. At the end of efforts to rehabilitate her, Genie could not read, could speak only in short phrases, and had just begun to control some of her feelings and behavior. Genie paid a high price—her full development as a human being—for the isolation, abuse, and lack of human warmth she experienced.

Anna, Isabelle, and Genie are the most fully documented cases of socially isolated children, but other cases continue to surface. For example, six-year-old Betty Topper was kept chained to a bedpost (by a two-foot-long chain and a harness normally used on horses) since she was a year old in a home littered with trash and debris. Early reports state that when found, 30-pound Betty was covered with her own waste and could only make moaning sounds (McNary 1999). Reports of her social development should be available in the future.

What other disruptions in social contact retard human development? The implication from case studies such as Anna, Isabelle, and Genie is unmistakable: The personal and social development associated with being human is acquired through intensive and prolonged social contact with others. But cases of extreme social isolation are not the only evidence for this generalization. Children can be affected adversely when the degree of contact with others is limited or when emotional attachments are not formed (Blum 2002).

In pioneering research, René Spitz (1946a, 1946b) compared the infants in an orphanage with those in a women's prison nursery. After two years, some of the children in the orphanage were retarded, and all were psychologically and socially underdeveloped for their age. By the age of four, a third of them had died. No such problems were observed among the prison nursery infants. Not one of them died during this period, even though the physical environment was not as clean as that of the orphanage. Spitz traced the difference between the two groups of infants to the emotional, physical, and mental stimulation offered in each setting. In the prison nursery, the infants' mothers were with them for the first year of their lives. The mothers were not present in the foundling home, and the infant-nurse ratio there was seven or eight to one. They were isolated from the other children as well. Also, in contrast to the infants in the prison nursery, the infants in the orphanage lacked the stimulation normally provided by toys. Social isolation, then, need not be extreme—as it was for Anna, Isabelle, and Genie—to cripple social and personality development.

Further research supports Spitz's conclusions. William Goldfarb (1945), for example, investigated a group of children reared in an institutionalized infant home. He compared these children, who were in the infant home from shortly after birth to three years of age, with children who had spent nearly all their lives in foster homes. He found that the institutionalized children suffered personality defects from their early years of deprivation that persisted even after they had been placed in foster homes. Compared with the children who had known only foster homes, the institutionalized children had lower IQ scores, were more emotionally and socially immature, had more difficulty caring about others, and were more passive and apathetic.

Lytt Gardner (1972) attributed a condition known as deprivation dwarfism to emotional deprivation. He conducted an intensive study of six "thin dwarfs"—children who were underweight and short for their age and who had retarded skeletal growth. Gardner found that these children had come from families missing the normal emotional attachment between parents and children. When such children were removed from their hostile family environments, they began to grow physically; on returning home their growth was stunted once again. According to Gardner, deprivation dwarfism is a concrete example—an experiment of nature, so to speak—demonstrating "the delicacy, complexity and crucial importance of infant-parent interactions" (Gardner 1972).

In another study, some infants adopted from a Lebanese orphanage were compared with children who remained in the orphanage. Initially, all of these children existed in stark cribs, lying on their backs in a barren room. Human touch was limited to diaper changing. After one year, the children's intellectual, motor, and social development was on a six-month-old level. Those children left in the orphanage continued to be underdeveloped, whereas the adopted infants made many developmental strides. In another orphanage study, some preschool-age children (average IQ 64) were placed in an institution for retarded adults, where each was "adopted" by an older woman and received much attention from patients and staff members. Children who remained in the orphanage (or another orphanage) lost an average of 20 IQ points over three years. An average gain of 28 points was experienced among the "adopted" orphans (Ornstein and Ehrlich 1991).

The process of socialization, then, permits us to develop the basic characteristics we associate with being human. It is also through socialization that we learn culture and learn how to participate in society.

Theoretical Perspectives and Socialization

Functionalism, Conflict Theory, and Symbolic Interactionism

How does each theoretical perspective view the socialization process? Each of the three major theoretical perspectives sheds light on the processes of socialization, although the symbolic interactionist perspective facilitates a more complete view than the other two. The

FEEDBACK

1. __________ is the process through which people learn to participate in group life through the acquisition of culture.
2. According to sociologists, no fundamental human nature exists prior to social contact. T or F?
3. Thanks to recent breakthroughs, research findings on the need of infant monkeys for warmth and affection can easily be applied to humans. T or F?
4. The cases of Anna, Isabelle, and Genie indicate that the personal and social development associated with being human is acquired through intensive and prolonged social contact with others. T or F?
5. Social isolation need not be extreme to damage social and personality development. T or F?

Answers: 1. Socialization 2. T 3. F 4. T 5. T

contribution of functionalism is more implied than explicit. That is, the very concept of socialization is based on the idea that people fit into groups by virtue of learning the culture of their society. People take their place in society by learning and accepting what society expects of them. The process of socialization assumes continuity and stability—as does functionalism. In fact, if it were otherwise, the existence of society would not be possible.

The conflict perspective views socialization (the learning of roles and acceptance of statuses) as a way of perpetuating the status quo. When people are socialized to accept their family's social class, for example, they help perpetuate the existing class structure. People are socialized into accepting their social fate before they have enough self-awareness to realize what is happening. Once social class socialization has taken place, it is very difficult to overcome. Consequently, socialization maintains the social, political, and economic advantages of the higher social classses. People who do not challenge their lot in life can never mount a revolution against the class structure.

The contribution of symbolic interactionism to our understanding of the socialization process is more precise than either functionalism or conflict theory. Symbolic interactionism helps us appreciate the subtleties of socialization: the development of the self-concept, the role of symbols and language in interpreting the social environment, the process of learning and assuming roles, and the social antecedents of human nature (Hewitt 2002).

Symbolic Interactionism and Socialization

As indicated in Chapter One, Adam Smith, in the eighteenth century, laid the foundation for symbolic interactionism. This passage from Smith's *The Theory of Moral Sentiments* suggests the nature of symbolic interactionism:

> *When I endeavor to examine my own conduct . . . and either to approve or condemn it . . ., I divide myself, as it were, into two persons. . . . The first is the spectator, whose sentiments with regard to my own conduct I endeavor to enter into, by placing myself in his situation, and by considering how it would appear to me, when seen from that particular point of view. The second is the agent, the person whom I properly call myself, and of whose conduct. . . . I was endeavoring to form some opinion. The first is the judge; the second the person judged of.* (Smith 1976/1759:113)

Charles Horton Cooley (1864–1929) and George Herbert Mead (1863–1931) were the originators of symbolic interactionism. At the turn of the twentieth century they challenged the belief, prominent in their day, that human nature is biologically determined. To them, like Adam Smith, human nature is a social product (Herman 2001).

Symbolic interactionism, when applied to socialization, involves a number of key concepts, including the self-concept, the looking-glass self, significant others, role taking (the imitation stage, the play stage, the game stage), and the generalized other.

What is the looking-glass self? From watching his own children at play, Cooley formulated insights about the development of the **self-concept**—an image of oneself as an entity separate from other people—that still stand today. Cooley and Schubert (1998) noted the many ways that children interpret the reactions of others toward them. From such insights, children learn to judge themselves in relation to how they imagine others will react to them. Thus, others serve as mirrors for the development of the self.

Cooley called this way of learning the **looking-glass self**—a self-concept based on our perception of others' judgments of us. We use others as mirrors reflecting back our imagined reactions of them to us. According to Cooley, the looking-glass self is the product of a three-stage process that is constantly taking place. First, we imagine how we appear to others. Next, we imagine the reaction of others to our imagined appearance. Finally, we evaluate ourselves according to how we imagine others have judged us. The result of this process is a positive or negative self-evaluation. Of course, this is not a conscious process, and the three stages occur in rapid succession in any given instance.

Suppose that you have a new professor you want to impress. To do so, you prepare especially well for the next day's class. As you participate in the class discussion the next day, you have an image of your performance (stage 1). After finishing your comments, you think your professor is disappointed (stage 2). Because you wished your professor to be impressed, you feel bad about yourself (stage 3).

Can the looking glass be distorted? Because the looking-glass self comes from our imagination, the mirrors we use may be distorted. The looking glass may not accurately reflect others' opinions of us. In the preceding example, your professor may have been so impressed that he could not show his genuine reaction. If so, you misread her lack of expression and silence. Parents, to take another example, may punish a child for some deed that in fact endears the child to them. A child sent to his room for using obscene language may not know how hard it was for his parents to conceal their amusement. An anorexic teenager may continue to threaten her health by not eating because she believes that others see her as fat.

Although we may misinterpret others' perceptions of us, this does not diminish the effectiveness of the looking-glass process. Even though we incorrectly believe that a professor dislikes us, the consequences to us are just as real as if our interpretation were true. E. L. Quarantelli and Joseph Cooper (1966) found the self-concept of dental students to be nearer to the evaluations they *thought* their instructors had given them than to the evaluation their instructors had actually given them. Despite the possibility of distortion in the looking glass, the relationship between self and others is well established.

Can we go beyond the looking glass? Although not denying that our self-evaluations are heavily influenced by our perception of others' evaluations of us, Viktor Gecas and Michael Schwalbe (1983) think that other factors also affect self-concept. They suggest going beyond the looking glass to consider other sources of influence on the development of self-concept.

According to Gecas and Schwalbe, the looking-glass self-orientation depicts human beings as oversocialized, passive conformists. That is, self-concepts are assumed to be based solely on the real or imagined opinions of others. (This is ironic, because Cooley and other symbolic interactionists view individuals as active and creative.) To correct this passive view of human beings, Gecas and Schwalbe developed the argument that individuals' self-concepts are also derived from personal judgments regarding how effectively they control their environments. Individuals with power on the job, for example, tend to have higher self-esteem than powerless individuals (Kanter 1993). True, part of the response is due to social recognition (or lack of it) from others; part of this effect, however, is attributable to the presence or absence of control over one's fate. A feeling of self-efficacy, in short, contributes to self-esteem.

Are all people in our looking glass equally significant? One's self-concept is not equally influenced by everyone. As George Herbert Mead pointed out, some people are more important to us than others (Mead 1934). Those whose judgments are most important to our self-concept are called **significant others**. For a child, significant others are likely to include mother, father, grandparents, teachers, and playmates. Teenagers place heavy reliance on their peers. The variety of significant others is greater for adults, ranging from spouses, parents, and friends to ministers and employers.

What is role taking? Because humans have language and the capacity for thinking, we can carry on silent conversations. That is, we can think something to ourselves and respond internally to it. This facility is crucial for anticipating the behavior of others. Through internal conversation we can imagine the thoughts, emotions, and behavior of others in all social situations. This ability enables us to engage in **role taking**—the process which allows us to take the viewpoint of another individual and then respond to ourselves from that imagined viewpoint.

Role taking is a cognitive process that permits us to play out scenes in our mind and anticipate what others will say or do. Thus, we avoid a trial-and-error method, which would be necessary if we could not mentally anticipate the behavior of others. If, for example, you wanted to ask your employer for a raise, and if you could not mentally put yourself in your boss's place, you would have no idea of the objections that she might raise. But by role-taking her reaction, you can be ready to justify your request.

How does the ability for role taking develop? According to Mead, the ability for role taking is the product of a three-stage process: the imitation stage, the play stage, and the game stage. In the **imitation stage**, which begins around age one and a half to two years, the child imitates (without understanding) the physical and verbal behavior of a significant other. This is the first step in developing the capacity for role taking. At about the age of three or four, a young child can be seen playing at being mother, father, police officer, teacher, or astronaut. This play involves acting and thinking as a child imagines another person would. This is what Mead called the **play stage**—the stage during which children take on roles of others one at a time.

The third phase in the development of role taking Mead labeled the **game stage**, the stage in which children learn to engage in more sophisticated role taking.

TABLE 4.1

FOCUS ON THEORETICAL PERSPECTIVES: Socialization

Each theoretical perspective has a unique view of the socialization process. This table identifies these views and illustrates the unique interpretation of each view with respect to the influence of the mass media on the socialization process.

Theoretical Perspective	View of Socialization	How the Media Influences Socialization
Functionalism	• Stresses how socialization contributes to a stable society	• Network television programs encourage social integration by exposing the entire society to shared beliefs, values, and norms.
Conflict theory	• Views socialization as a way for the powerful to keep things the same	• Newspaper owners and editors exercise power by setting the political agenda for a community.
Symbolic interactionism	• Holds that socialization is the major determinant of human nature	• Through words and pictures, children's books expose the young to the meaning of love, manners, and motherhood.

After a few years in the play stage, children are able to consider the roles of several people simultaneously. Games they play involve several participants, and there are specific rules designed to ensure that the behavior of the participants fits together. All participants in a game must know what they are supposed to do and what is expected of others in the game. Imagine the confusion in a baseball game if young first-base players have not yet mastered the idea that the ball hit to a teammate infielder will usually be thrown to them. In the play stage, a child may pretend to be a first-base player one moment and pretend to be a base runner the next. In the game stage, however, first-base players who drop their glove and run to second base when an opposition player hits the ball will not remain in the game for very long. It is during the game stage that children learn to gear their behavior to the norms of the group.

© Gary Conner/PhotoEdit

Others serve as mirrors for the development of our self-concepts. The looking-glass self is based on our preceptions of others' judgments of us. Part of what this young woman is thinking involves her imagined reactions of others to her appearance.

When do we start acting out of principle? During the game stage, a child's self-concept, attitudes, beliefs, and values gradually come to depend less on individuals and more on generalized referents. Being an honest person is no longer merely a matter of pleasing significant others such as one's mother, father, or minister. Rather, it begins to seem wrong *in principle* to be dishonest. As this change takes place, a **generalized other**—an integrated conception of the norms, values, and beliefs of one's community or society—emerges.

What is the self? According to Mead, the self is composed of two analytically separable parts: the "me" and the "I." The "**me**" is the part of the self formed through socialization. Because it is socially derived, the "me" accounts for predictability and conformity. Yet, much human behavior is spontaneous and unpredictable. An angry brother may, for example, spontaneously and unaccountably yell hurtful words at a sister he loves. Afterward, his reaction may be, "I don't know what came over me. How could I do that to a person I care for?" To account for this spontaneous and unpredictable part of the self, Mead wrote of another dimension of the self—the "**I**."

The "I" doesn't operate just in extreme situations of rage or excitement but interacts constantly with the "me" as we conduct ourselves in social situations. According to Mead, the first reaction of the self comes

© Felicia Martinez/PhotoEdit

The ability for role taking begins with children imitating significant others. As this young child "helps" her father sweep the sidewalk, she is laying the foundation for her ability to assume mentally the viewpoint of another person and then respond to herself from that imagined viewpoint

from the "I"; but before we act, the initial impulse is directed in socially acceptable channels by the "me." Thus, the "I" normally takes the "me" into account before acting. However, the uniqueness and unpredictability of much human behavior demonstrates that the "me" does not always control the innovative, unpredictable dimension of the self.

Psychology and Life Course Theories

Because socialization involves our entire life course and because the development of human personality is part of that life course, we will veer temporarily from sociology and present several psychologically oriented life course theories. The theories range from the psychoanalytic theory of Sigmund Freud to the psychosocial perspective of Erik Erikson and the cognitive perspective of Jean Piaget.

A Psychoanalytic Perspective

Aside from his theories of the unconscious, Sigmund Freud's (1856–1939) greatest contribution was his work regarding the influence of early childhood experiences on personality development. Experiences within the family during the first few years of life, Freud contended, largely shape future psychological and social functioning.

What is the composition of the human personality? According to Freud, the personality has three parts: the id, the ego, and the superego. The *id* is made up of biologically inherited urges, impulses, and desires. It is selfish, irrational, impulsive, antisocial, and unconscious. The id operates on the pleasure principle—the principle of having whatever feels good. Newborn infants are said to be totally controlled by the id; they want their every desire fulfilled without delay. Very early in life, members of society—usually parents—interfere with the pleasure of infants. Infants gradually learn to wait until it is time to eat, to control their bowels and bladder, and to hold their temper. In many ways, they must face the fact that others are not around merely to satisfy their impulses.

To cope with this denial of pleasure, children begin to develop an *ego*—the conscious, rational part of the personality that thinks, plans, and decides. The ego is ruled by the reality principle, which allows us to delay action until a time when the gratification of our desires is more likely. Thus, the ego mediates between the biological, unconscious impulses of the id and the denying social environment. But the ego is not itself sufficient to control the id. Nor need it be.

At about four or five years of age, the *superego*—roughly the conscience—begins to develop. It contains all the "right" and "wrong" ideas that we have learned

FEEDBACK

1. In its approach to socialization, _________ emphasizes social interaction based on symbols.
2. According to the looking-glass process, our _________ is based on how we think others judge the way we look and act.
3. Those individuals whose judgments of us are the most important to our self-concept are _________.
4. _________ is the process of mentally assuming the viewpoint of another individual and then responding to oneself from that imagined viewpoint.
5. According to George Herbert Mead, children learn to take on the roles of individuals, one at a time, during the _________ stage.
6. In Mead's theory of the self, the _________ is the unsocialized side, and the _________ is the socialized side.

Answers: 1. symbolic interactionism 2. self-concept 3. significant others 4. Role taking 5. play 6. "I"; "me"

from those close to us, particularly our parents. By incorporating their ideals into our personality, we develop what may be thought of as an internal parent. Parents are no longer the only source of punishment for wrongdoing; we punish ourselves through guilt feelings. At the same time, we feel good about ourselves when we live up to the standards contained in our superego. Through this internal monitoring system, we learn to channel our behavior in socially acceptable ways and to repress socially undesirable thoughts and actions.

What is the relationship between the id, the ego, and the superego? Freud did not see the id, ego, and superego as separate regions of the brain or as little people battling and negotiating inside our heads. Rather, he saw them as separate, interacting, and conflicting processes within the mind. The id demands satisfaction; the superego prohibits it. The ego supplies rational information in this conflict; it attempts to gain satisfaction within the limits set by the superego and the social environment. The following example, although an oversimplification, should clarify the relationship between these three parts of the personality:

> *Let's say you are sexually attracted to an acquaintance. The id clamors for immediate satisfaction of its sexual desires, but is opposed by the superego (which finds the very thought of sexual behavior shocking). The id says, "Go for it!" The superego icily replies, "Never even think that again!" And what does the ego say? The ego says, "I have a plan!"* (Coon 2002:369)

If the ego has difficulty controlling the id, then some sexually related experience—sexual intercourse, masturbation, even rape—may follow. Should the superego be dominant, sexual desires may be sublimated into other activities, such as working, dancing, studying, or stamp collecting. If the ego is not overwhelmed by the id or superego, it will encourage more socially acceptable behavior, such as dating or marriage.

Psychosocial Development

Psychoanalyst Erik Erikson (1902–1994) extended the human development research of his mentor, Sigmund Freud. Like his teacher, Erikson (1973) emphasized the role of the ego as the mediator between the individual and society and wrote of "crises" accompanying each stage in human development. Erikson, however, differed from Freud in his underlying assumptions and in his specific ideas on the relationship between self and socialization. To Freud, the human personality is almost totally determined in early childhood. Erikson, in contrast, believed that personality may change at any time in life. Whereas Freud attributed most human behavior to universal instincts, Erikson emphasized cultural variations. Although Freud contended that parents are the most influential forces in personality development, Erikson considered the impact of peers, friends, and spouses to be significant as well.

According to Erikson, all individuals pass through a series of eight developmental stages from infancy to old age. Each of these developmental stages involves a psychosocial crisis or developmental task (see Table 4.2). Personal identity is firmly established or impaired, depending on how successfully an individual handles each turning point. The effects of successfully or unsuccessfully meeting these crises are cumulative: Successful management of a crisis at an earlier stage

TABLE 4.2

Erik Erikson's Stages of Psychosocial Development

Examine Erik Erikson's eight stages of psychosocial development. Do any or all of these stages ring true in your life? Explain.

Theoretical Perspective	Crisis	Favorable Outcome
1. First year of life	• Trust versus mistrust	• Faith in the environment and in others
2. Ages two to three	• Autonomy versus shame and doubt	• Feelings of self-control and adequacy
3. Ages four to five	• Initiative versus guilt	• Ability to begin one's own activities
4. Ages six to twelve	• Industry versus inferiority	• Confidence in productive skills; learning how to work
5. Adolescence (ages twelve to eighteen)	• Identity versus role confusion	• Integrated image of oneself as a unique person
6. Early adulthood (ages eighteen to thirty-five)	• Intimacy versus isolation	• Ability to form bonds of love and friendship with others
7. Middle adulthood (ages thirty-five to sixty)	• Generativity versus stagnation	• Concern for family, society, and future generations
8. Late adulthood (over age sixty)	• Integrity versus despair	• Sense of dignity and fulfillment; willingness to face death

increases the chances of mastering later developmental tasks; those who fall behind in one stage will have increasingly greater difficulty in the following stages. Those individuals whose egos adequately meet the psychosocial crisis at each developmental stage are the most mature and the happiest and have the most stable personal identities.

Several additional points about Erikson's eight stages of psychosocial development should be kept in mind. First, Erikson does not use the word crisis to mean a threat or catastrophe. Rather, it is a turning point, a crucial period during which growth or maladjustment may occur, depending on how the crisis is handled. Second, although the way a crisis in one developmental period is handled affects the management of later crises, no crisis is necessarily solved permanently. Thus, a sense of trust developed early in life may, because of social experiences, turn to distrust later; and a sense of self-doubt may, through relationships with others, later change to self-confidence. Third, as already implied, each crisis must be solved via interaction with others, and characteristics developed at earlier stages of development (trust, autonomy, industry, and so forth) require the support of others if they are to be maintained. Fourth, each crisis is solved within a particular social setting—family group, school, peer group, neighborhood, or workplace.

Cognitive Development

Jean Piaget (1896–1980) was concerned primarily with the development of intelligence, or cognitive abilities—thinking, knowing, perceiving, judging, and reasoning. According to Piaget (1981; Piaget and Inhelder 1973; Piaget and Valsiner 1999), children gradually develop cognitive abilities through interaction with their social setting—through the process of socialization. Piaget contended that children are not merely passive recipients of social stimuli; they are actively engaged in interpreting their environment as they attempt to adjust to it. Cognitive ability, in Piaget's theory, advances in stages. Learning to solve problems in one stage must precede problem solving at later stages. Piaget has isolated four stages of cognitive development: the sensorimotor stage, the preoperational stage, the concrete operational stage, and the formal operational stage.

What cognitive development occurs in the sensorimotor stage? The *sensorimotor stage*, where the basis for thought is laid, begins at birth and lasts until the age of eighteen months to two years. It is called the sensorimotor stage precisely because most of its activities are associated with learning to coordinate body movements with information obtained through the senses—touching, hearing, seeing, feeling. Only gradually do children come to realize that they are an object separate from other things. Initially, they do not realize that they can cause things to happen. A baby shaking bells on its crib does not realize who has caused the sound.

One of the most significant developments during this stage is the development of a sense of "object permanence"—a sense that objects exist even when they cannot be seen. Once this level of thought is reached, a child will, for example, pursue a ball that has rolled under a bed. Later a child can anticipate the reappearance of an object. Instead of looking into the entrance of a tunnel after an electric train has entered, a child will watch for the train's reappearance at the other end of the tunnel. During this stage, children come to see their world as an understandable and predictable place.

How do children think during the preoperational stage? Between the ages of two and seven, during the *preoperational stage*, children learn to think symbolically and to use language. Initially, they have difficulty distinguishing between what they call something (a symbol) and the object itself; an object and the symbol used to represent it are the same to them. If you throw away a milk carton that has been used as a child's toy truck, you will have a difficult time convincing the child that it was only a milk carton.

Another dominant characteristic of children during the preoperational stage is self-centeredness. Children's extreme egocentrism is clearly reflected in their inability to see things from others' points of view. Many children believe that the sun and the moon follow them when they are taking a walk. In one type of experiment,

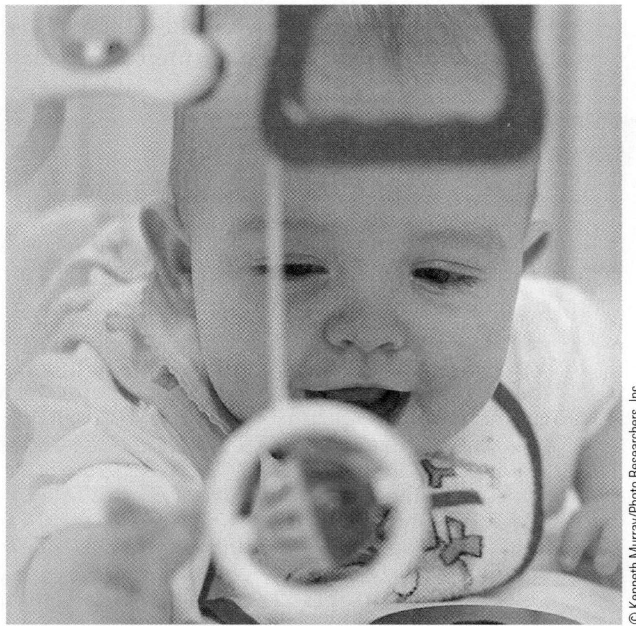

This toy is helping this baby develop the capacity to distinguish itself from other objects. If the toy makes a sound, the infant is not aware of having caused it.

children have been placed opposite adults with a two-sided mirror between them. When asked what the adults see in their side of the mirror, the children answer "me!"

In the preoperational stage, thoughts and operations are not reversible. Consider this conversation researcher John Phillips (1975) had with a four-year-old boy:

> *"Do you have a brother?"*
> *"Yes."*
> *"What's his name?"*
> *"Jim."*
> *"Does Jim have a brother?"*
> *"No."*

What about the stage of concrete operations? Increasing abstractness dominates the stage of *concrete operations*, which spans the ages of seven to eleven. A child begins to think logically about time, quantity, and space; handle arithmetic operations; and place things logically in categories. In addition, children can reverse thoughts and operations during this stage. The reversibility of thought allows an older child to recognize that A-B-C is the reverse of C-B-A and to recognize that if he has a brother, then his brother also has one. A certain degree of abstract thinking is reflected in the children's ability to imagine themselves in the place of others and to gear their behavior to others.

Despite the movement toward abstract thinking that occurs during the stage of concrete operations, the child still has difficulty if concrete objects are not involved. Logical thinking is tied to action, to the observation or manipulation of concrete symbols and objects. For example, children at this stage of development can master the concept of the conservation of quantity by watching someone pour liquids into different-sized glasses. In the preoperational stage, if children see water poured from a slender glass into a wide glass, they assume that the wider glass has less water in it because the level of the water appears to be lower. During the stage of concrete operations, however, children come to realize that the amount of water is the same in both glasses. They could not grasp this idea without an understanding of concrete operations. But it is only in the next stage that completely abstract thought becomes possible.

What is unique about thinking in the stage of formal operations? A fundamental change in cognitive ability begins to occur sometime after the age of eleven. Children learn to think without the aid of concrete objects and manipulations; they can begin to think in terms of abstract ideas and principles. Gradually, the adolescent can reason about hypothetical matters. For example, if you ask younger children, "What would life be like if people had wings?" they would say something like "That's silly" or "People don't have wings." But during the stage of formal operations, children become

FEEDBACK

1. Listed below are the three major parts of the personality according to Freud. Match these parts with the examples.
   ____ a. id
   ____ b. ego
   ____ c. superego
   (1) A student decides not to cheat on an examination because the professor seems to be watching the class closely.
   (2) A student does not cheat on an examination because she feels too guilty about it.
   (3) A man kills his girlfriend in a fit of jealous rage.
2. The major role of the __________ in Freud's theory of the personality is to mediate between innate impulses and the conscience.
3. Whereas Freud believed that the human personality is almost totally determined in early childhood, Erikson held that personality can change __________.
4. According to Erikson, the successful management of a psychosocial crisis at an earlier stage increases the chances of mastering later crises. This means that the effects of success or lack of success at earlier crisis points are __________.
5. Match Jean Piaget's stages of cognitive growth with the statements below them.
   ____ a. sensorimotor stage
   ____ b. concrete operational stage
   ____ c. formal operational stage
   ____ d. preoperational stage
   (1) A child thinks without the aid of concrete objects and manipulations.
   (2) A child recognizes that 3 + 2 is the same as 2 + 3.
   (3) A child learns that a cat that runs under a couch still exists.
   (4) A child believes that a box is a spaceship.
6. The ability to imagine oneself in the place of others develops during the stage of __________.

Answers: 1. a.(3) b.(1) c.(2) 2. ego 3. at any time 4. cumulative 5. a.(3) b.(2) c.(1) d.(4) 6. concrete operations

willing to speculate on such matters. Children also learn during the *formal operations* stage to consider relationships that are logical, even if they are ridiculous. If younger children are asked to complete the sentence "If day is dark, night is __________," they will argue that day is not dark. Older children will answer that night is light. It is during this stage, then, that the capacity for adult thinking is developed.

Socialization and the Life Course

Desocialization, Resocialization, and Anticipatory Socialization

Over our life course, old ways of life must be abandoned, new ways adopted, and preparation made for transitions from one period of life to another. *Desocialization, resocialization,* and *anticipatory socialization,* concepts that describe these processes, are associated with symbolic interactionism (Lachman et al. 1994; Lachman and James, 1997). Erving Goffman's work on "total institutions" provides an excellent, albeit extreme, illustration of these concepts.

Goffman believes that people are not always free to manage their own lives but are under the direction of others. In *Asylums*, Goffman (1961b) writes about places such as mental hospitals and prisons as **total institutions**—places in which residents are separated from the rest of society. These residents are controlled and manipulated by those in charge, the purpose being to change the residents. The first step toward change is **desocialization**—the process of relinquishing old norms, values, attitudes, and behaviors. In extreme situations, such as mental hospitals and prisons, desocialization involves an attempt to obliterate the resident's old self-concept. Along with the self-concept go the norms, values, attitudes, and behaviors associated with the personal identity the resident had upon entering the institution. This is accomplished in many ways. Replacing personal possessions with standard-issue items promotes sameness among the residents. It deprives them of the personal effects (long hair, hair brushes, baseball caps, T-shirts) that identify them as unique individuals. Serial numbers and the loss of privacy also contribute to the breakdown of past identity.

Once the self-concept has been fractured, **resocialization**—the process of learning to adopt new norms, values, attitudes, and behaviors—can begin. The staff, using an elaborate system of rewards and punishments, attempts to instill a new self-concept in the residents. Rewards for conformity to these new identities include extra food or special periods of privacy. Punishments for nonconformity involve loss of special privileges, physical punishment, or physical isolation.

The concepts of desocialization and resocialization were developed by Goffman to analyze social processes in extreme situations. They still apply to other social settings, including basic training in the U.S. Marine Corps and plebe (freshman) year at the United States Military Academy. In much less extreme form, these concepts illuminate changes in our normal life course. Desocialization and resocialization occur as a child makes the transition into the adolescent subculture, when young adults begin their occupational careers, and as the elderly move into retirement or widowhood.

Can socialization begin prior to joining a group? **Anticipatory socialization**—the process of preparing oneself for learning new norms, values, attitudes, and behaviors—does not generally occur in the extreme social settings represented by total institutions. This is because anticipatory socialization involves voluntary change. Teenagers, because they want to resemble those their own age, may willingly abandon many of the norms, values, attitudes, and behaviors learned previously. Consequently, preteens begin very early to observe the ways of teenagers, their new **reference group**—a group used to evaluate oneself and from which to acquire attitudes, values, beliefs, and norms. University seniors, normally seen on campus only in jeans and oversized sweatshirts, suddenly, as graduation nears, are wearing tailored suits and more serious expressions. Initiating their transition into the business world, they are talking with friends who have graduated as well as with company recruiters. By anticipating the new environment they are about to enter, graduating seniors are, in effect, preparing themselves for the resocialization they know awaits them (Atchley 2000).

Are these life stages culturally universal? Stages of the life course as we know them were unknown in previous societies. A "midlife" crisis could not have been imagined in preindustrial society, because that is when people died. Even as late as 1900, life expectancy in the United States was just over forty-nine years (Atchley 2001). The concept of childhood did not exist in medieval society. The problems of adolescence are a relatively recent development.

Among the Truks of Micronesia, the specter of death looms at forty.

> *In Truk, they rely basically on bread, fruit and fish. In order to get fish a young man has to be strong and agile; he has to be a good paddler of a canoe, a good navigator to go out on the reef and gather fish. When the Trukese reaches about forty years of age, his strength begins to decline; he also does not climb trees as well as he used to. And when his*

SNAPSHOT OF AMERICA 4.1

Rates of Imprisonment

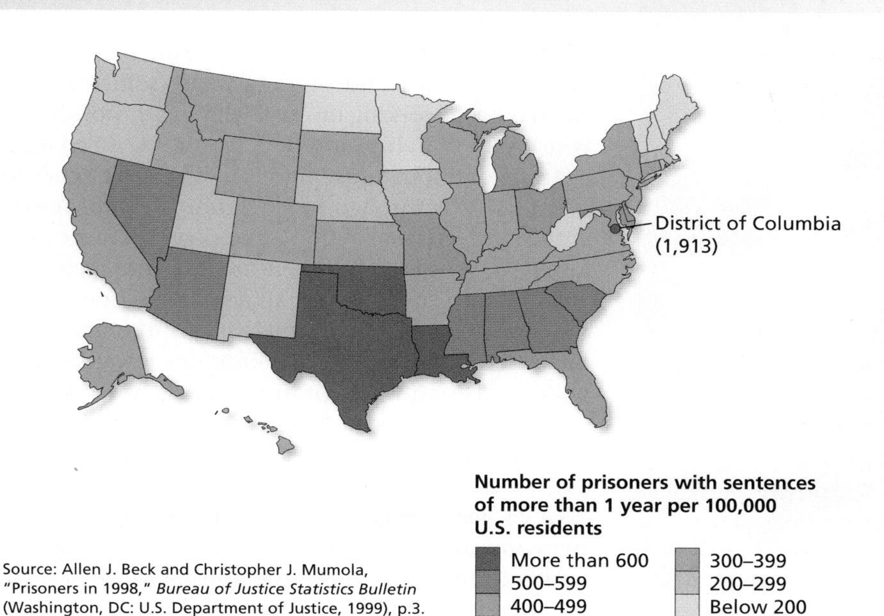

Source: Allen J. Beck and Christopher J. Mumola, "Prisoners in 1998," *Bureau of Justice Statistics Bulletin* (Washington, DC: U.S. Department of Justice, 1999), p.3.

The U.S. has one of the highest rates of imprisonment in the industrialized world. The imprisonment rate in the U.S. is over four times that of any Western European Country.

Critics depict prisons as total institutions functioning as "schools for crime" because of the negative influence of the inmate code. If prisons do first desocialize and then resocialize inmates toward a criminal identity, then the U.S. prison system is unintentionally increasing the criminal portion of the population. It is at least interesting that one-third to two-thirds of those released from prison return within two to five years.

1. Where does your state rank in imprisonment rate? Explain why this might be the case.
2. Describe the type of desocialization and resocialization that might be occurring through the prison inmate code.

strength begins to wane, he begins to feel that his life is ebbing away and he begins to prepare himself for death. (Kübler-Ross 1986:28)

Thus, the stages in the life cycle are not strictly defined by biology or psychology. They are socially and culturally defined (Kübler-Ross 1997a, 1997b).

Socialization of the Young

What are the major effects of the family? The child's first exposure to the world occurs within the family. Being dependent and highly impressionable, the child is virtually defenseless during the first few years of life. Through close interaction with a small number of people—none of whom the child has selected—the child learns to think and speak; internalizes norms, beliefs, and values; forms some basic attitudes; develops a capacity (or incapacity) for intimate and personal relationships; and acquires a self-image. By the time the child develops some independence and judgment, much of the socializing work of the family has been accomplished. Development does not end at age five, but a foundation for later development has been firmly established (Santrock 2002).

The impact of the family reaches far beyond its direct effects on the personal and social development of the child. Our family's social class significantly affects how others treat us and what we think of ourselves. Our family of birth largely determines our place in society. Jean Evans offers an illustration of this in the case of Johnny Rocco, a twenty-year-old living in a city slum:

Johnny hadn't been running the streets long when the knowledge was borne in on him that being a Rocco made him "something special"; the reputation of the notorious Roccos, known to neighbors, schools, police, and welfare agencies as "chiselers, thieves, and trouble-makers" preceded him. The cop on the beat, Johnny says, always had some cynical smart crack to make. . . . Certain children were not permitted to play with him. Wherever he went—on the streets, in the neighborhood, settlement house, at the welfare agency's penny milk station, at school, where other Roccos had been before him—he recognized himself by a gesture, an oblique remark, a wrong laugh. (J. Evans 1954:11)

The effects of social class usually extend into adulthood. For example, in a major study of occupational attainment, Peter Blau and Otis Duncan make this observation about the relationship between one's family and subsequent work life:

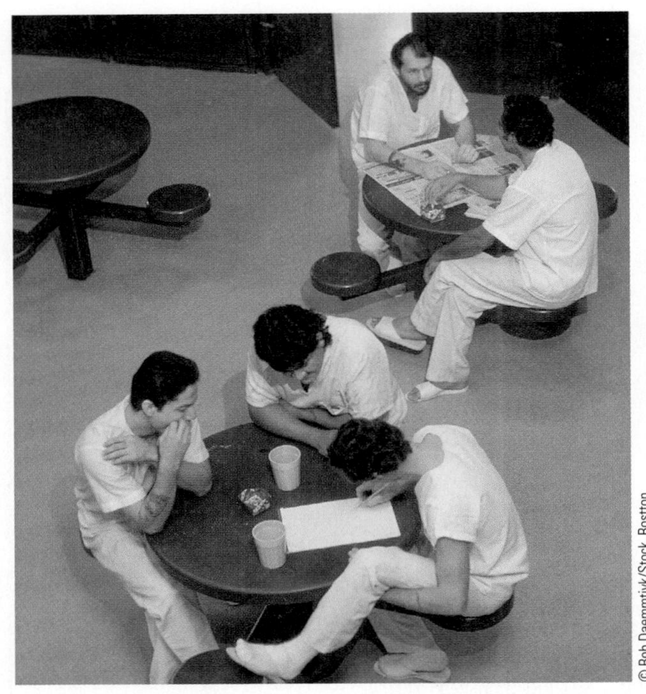

© Bob Daemmrich/Stock, Boston

Total institutions are places in which residents are separated from the rest of society and are controlled and manipulated by those in charge. These inmates in a Texas jail share the experience of life in a total institution.

> *The family into which a man is born exerts a profound influence on his career, because his occupational life is conditioned by his education, and his education depends to a considerable extent on his family.* (Blau and Duncan 1967:330)

What about school? In school, children are under the care and supervision of adults who are not relatives. The first year of school involves a transition from an environment saturated with personal relationships to an impersonal environment. Rewards and punishments are based on performance rather than affection. Although a mother may cherish any picture that her child creates, a teacher evaluates her students by more objective standards and informs those who are not meeting these standards. Slowly, children are taught to be less dependent emotionally on their parents. In addition, the school links children to the broader society. It creates feelings of loyalty and allegiance to something beyond the children's families.

The socialization process in school involves more than reading, writing, and arithmetic. Underlying the formal goals of the school is the **hidden curriculum**—the subterranean informal and unofficial aspects of culture that children are taught as preparation for life in the larger society. The hidden curriculum teaches children discipline, order, cooperation, and conformity—characteristics thought to be needed for success in modern bureaucratic society, whether the child becomes a doctor, college president, secretary, assembly-line worker, or professional athlete.

Is socialization in the public school always functional? Not according to educational critic John Holt (1995). Life in schools, for example, is run by the clock. Whether or not a student understands something he or she has been working on, and whether or not the child is ready to switch to a different subject, a bell signals that all children must move to the next scheduled event. Getting through a predetermined set of activities within a given time period often becomes more important than learning. And there are rules and regulations, Holt says, to cover almost all activities—how to dress, how to wear one's hair, which side of the hall to walk on, when to speak in class. Teachers reward children with praise and acceptance when they recite the "right" answers, behave "properly," or exhibit "desirable" attitudes.

How do peer groups contribute to socialization? The **peer group**—composed of individuals of roughly the same age and interests—is the only agency of socialization that is not controlled primarily by adults. Children usually belong to several peer groups. A child may belong to groups composed of neighborhood children, schoolmates, service club members, and religiously affiliated friends.

The peer group provides children with experiences they are unlikely to capture elsewhere (Adler and Adler 1998). In the family and at school, children are subordinated to adults, but through the peer group, young people have an opportunity to engage in give-and-take relationships. Children experience conflict, competi-

© Kevin Radford/SuperStock

In addition to teaching academic subjects, schools offer a "hidden curriculum"—informal and unofficial aspects of culture taught to prepare children for life in the larger society. The children in the photograph are learning the values of order and waiting in turn.

Keiko Ikeda—High School Reunions

If you asked most Americans to talk about their experience at a recent high school reunion, what would they say? "It was great seeing old friends." "I was curious about how things turned out for people I loved and hated as a teenager." "I plan to get together with some old friends in the near future." High school reunions are generally thought to be playful amusement, a time to recapture fond memories of youth.

One researcher wished to investigate the meaning of high school reunions. Keiko Ikeda (1998) studied eight reunions in the American Midwest. He observed these reunions armed with a camera, tape recorder, and notebook. After each reunion, he conducted in-depth, life-story interviews with samples of participants.

Like all in-depth observational studies, the results are too complex and varied for comprehensive reporting. One aspect of Ikeda's study, however, reveals the cultural nature of high school reunions. He compared several reunions of one high school—tenth, fifteenth, twentieth, thirtieth, fortieth, and fiftieth. Ikeda focused on the relative emphasis on the past and the present. As you can see from the following passage, the past becomes more important as age increases:

> *In the earlier reunions (the tenth and fifteenth years), a concern with relative status and a sense of competitiveness is expressed, often blatantly, through award-giving ceremonies. . . . The hall was decorated in the school colors, and images of the high school mascot were present, but beyond this no high school memorabilia were displayed. The music, too, was current, and not the rock 'n' roll of the late sixties and early seventies.*
>
> *The twentieth-year reunion of the Class of '62 is typical of a transitional phase in which elements from the past begin to assume an important role. The past is expressed in high school memorabilia . . . in . . . films and slides taken during high school, and in . . . high school anecdotes that are playfully interwoven throughout the ceremonial events.*
>
> *In the thirtieth-year reunion of the Class of '52, the past firmly occupied center stage. A carefully crafted, chronological narrative of the senior year, entitled "The Way We Were," was read, in which major class activities were recalled month by month. . . .*
>
> *In the fiftieth-year reunion, we find a dramatic disappearance of all ritual activities. According to the president of the Class of '32, his class had held reunions every ten years since graduation, and in earlier ceremonies they had given awards, but this time, "none of the folks in the reunion committee felt like doing that kind of thing." It seemed that attendees at the fiftieth-year reunion, for the most part, had risen above concerns of past and present and were content to celebrate together the simple fact that they all still had the vigor to attend a reunion.*

Source: Keiko Ikeda, *A Room Full of Mirrors* (Stanford, CA: Stanford University Press, 1998), 143–145.

Thinking About the Research

1. Ask several adults to describe the activities at one or more high school reunions they have attended. Compare their descriptions with Ikeda's findings.
2. Suppose you had a class assignment to study an upcoming reunion at your school. Select a research question you would want to ask. Identify the research methods you would use.

tion, and cooperation. The peer group also gives children experience in self-direction. Children can begin to make their own decisions; experiment with new ways of thinking, feeling, and behaving; and engage in activities that involve self-expression.

The peer group further promotes independence from adults by introducing the child to a social milieu that is often in conflict with the adult world. Children learn to be different from their parents in ways that contribute to the development of self-sufficiency.

The capacity for intimacy is enhanced because the peer group provides an opportunity for children to develop close ties with friends outside the family, including members of the opposite sex. While children are making close friends with a few individuals, they are also learning to get along with large numbers of people, many of whom are quite different from themselves. This helps develop the social flexibility needed in a mobile, rapidly changing society.

The rise of formal education has contributed immensely to the emergence of a peer world that is not only separate from adults but also beyond their control. Children are isolated from adult society by being set apart, in school, for most of their preadult lives.

While these teenagers are eating pizza and bantering with peers, they are also engaged in the serious process of learning to participate in group life. In a rapidly changing world, these young people will continue to be "socialized" throughout their lives.

Because they are separated from the adult world for such a long time, young people are forced to depend on one another for much of their social life *(The State of Our Nation's Youth 2000)*.

Who has more influence on young people—friends or family? An important factor contributing to the dependence of adolescents on each other is the distribution of the population in advanced industrial societies. Most Americans now live in either urban or suburban areas. In addition to the prevalence of the dual-employed family in American society, both parents may commute many miles to work. If so, they spend much of their time away from home. Consequently, once children reach the upper levels of grade school, they may be spending more time with their peers than they do with their parents.

Urie Bronfenbrenner (1970) notes that peer groups fill the vacuum created in the lives of children who receive an insufficient amount of attention from their parents. And according to Judith Harris (1998), peers are more important than parents in socializing children. Even though most sociologists do not agree with this author's extreme conclusion, they do believe that the peer group is having a growing effect on social development.

What role do the mass media play in socialization? Mass media are means of communication designed to reach the general population. It is not technology per se that makes these media "mass." Newspapers existed in the American colonies, but they were limited in circulation and aimed at more local audiences (DeFleur and Ball-Rokeach 1982). Technologies become mass media only when their audiences include any average person who wishes exposure to them. Sociologists agree that the mass media are powerful socializing agency. A primary function of the mass media is informing children about their culture (Riesman 2001).

From the mass media—television, radio, newspapers, magazines, movies, books, the Internet, tapes, and disks—children learn the behavior expected of individuals in certain social statuses. Though these popular images may be distorted (detective and police work are not as exciting and glamorous as depicted) or stereotyped, the media display role models for children to imitate (Cavender and Fishman 1998). The media often present characters in such simple, one-sided forms that it is easy to recognize behavior suitable for men, women, heroes, and villains. Learning these role models helps to integrate the young into society:

> *The mass media, by their content alone, teach many of the ways of the society. This is evident in the behavior we take for granted—the duties of the detective, waitress, or sheriff; the functions of the hospital, advertising agency, and police court; behavior in hotel, airplane, or cruise ship; the language of the prison, army, or courtroom; the relationship between nurses and doctors or secretaries and their bosses. Such settings and relationships are portrayed time and again in films, television shows, and comic strips; and all "teach"—however misleadingly—norms, status positions, and institutional functions.* (Elkin and Handel 2001:189)

Mass media also offer children ideas (sometimes real, sometimes idealized) about values in their society: achievement and success, activity and work, democracy and equality. Mass media are also purveyors of consumerism and violence (Bagdikian 2000).

Is there a relationship between television violence and real-life violence? The only activity that American children engage in more than watching television is sleep-

WORLDVIEW 4.1

Availability of Television

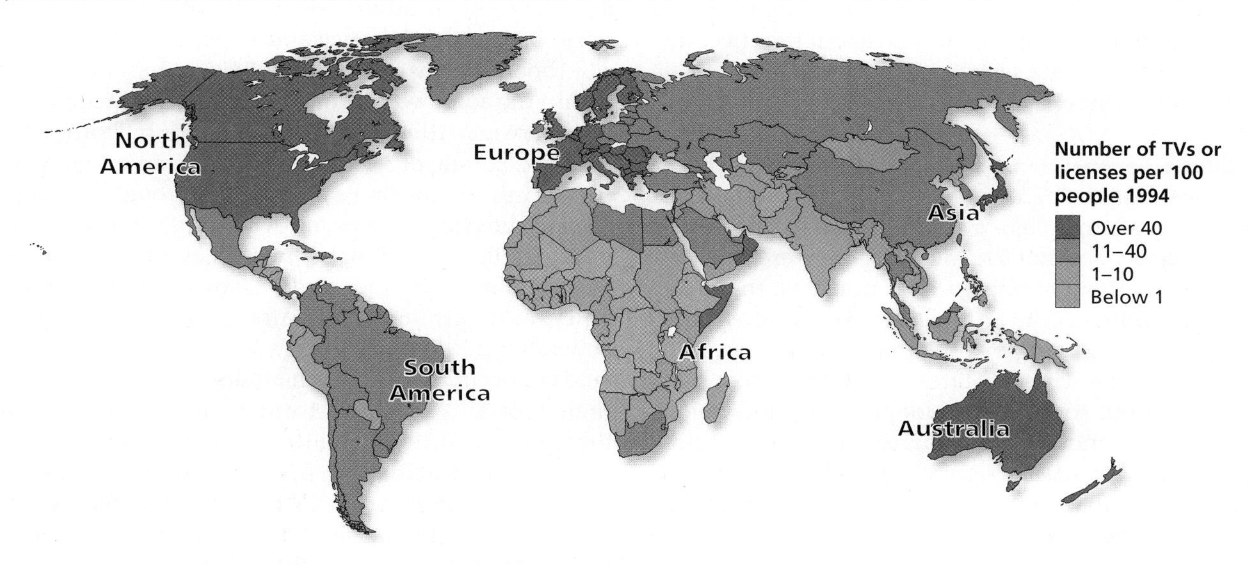

Source: Ian Pearson, (ed.), *Atlas of the Future* (New York: Macmillan Books, 1998), 60.

The mass media play a key role in the socialization process. Since nearly every U.S. home has at least one television (and most have more), this medium is one of the most influential in the United States. This map shows that ownership of televisions varies widely around the world.

1. What geographical factor(s) might contribute to the density of TV households in South America?
2. Do you think members of societies with few televisions are as well socialized as members of societies with more sets? Why or why not? (Refer to *Tech Trends* in this chapter before answering.)

ing. There is at least one television set in 98% of U.S. hoseholds; 76 percent have more than one set. And an average American watches television more than 7.5 hours each day *(2000 Report on Television 2000)*. Any medium with which people spend so much time must contribute significantly to the socialization process.

Consider the relationship between violence on television and real-life violence. By age sixteen, the average American child will have seen 20,000 homicides on television (Leonard 1998). As noted in "Using the Sociological Imagination" at the beginning of this chapter, social scientists have, in the past, been reluctant to recognize a causal connection between media violence and real-life violence. However, after hundreds of studies involving over 10,000 children, most now conclude that watching aggressive behavior on television significantly increases aggression (Strasburger 1995; W. Dudley, 1999; Anderson and Bushman 2002; Carter and Weaver 2003).

Researchers now generally agree that at times the effect of television is direct, concrete, and dramatic. A nine-year-old girl, for example, was gang-raped with a beer bottle by three teenage girls and a boy three days after they had seen a similar incident on a television movie. At least twenty-nine Americans have shot themselves while imitating the Russian roulette scene from *The Deer Hunter*. A two-year-old girl died when her older brother, age five, set the house on fire with matches while imitating behavior he had seen on *Beavis and Butt-Head*. A sixteen-year-old murdered his mother to kick off an intended killing spree modeled by the *Scream* movies. A rash of would-be copycats followed the massacre at Columbine High School, just on the basis of the news reports alone. Television's effects, of course, are usually more hidden, subtle, and long term:

> *[N]ot every child who watches a lot of violence or plays a lot of violent games will grow up to be violent. Other forces must converge, as they did* [at Columbine]. . . . *But just as every cigarette increases the chance that someday you will get lung cancer, every exposure to violence increases the chances that some day a child will behave more violently than they would otherwise.* (To Establish Justice, 1999:vi)

Extreme pressure from a variety of sources has caused the major television networks to institute ratings for their programs (Farhi 1997). Most critics of violence on television have seen this self-policing move as an evasion tactic by an industry under public attack. Among other things, the critics strongly recommend the "V chip," which a parent can preset to block any program rated V-for-violent. More will be said about the effects of the mass media in the socialization process in the closing section of this chapter.

Are there other agencies of socialization during childhood and adolescence? The family, school, peer group, and mass media are the major agencies of socialization during childhood and adolescence, but they are not the only ones. Although religion does not have the same degree of influence in all societies, it can affect the moral outlook of young people, even in secular societies such as the United States. Athletic teams teach young children to compete, cooperate, follow rules, make friends, and handle disagreeable peer relationships. Other potentially significant agencies of socialization are youth organizations such as the Cub Scouts, Girl Scouts, and YMCA.

Do these agencies of socialization ever conflict? Yes. In complex societies, conflict between agencies of socialization is inevitable. Some families work at cross-purposes with the schools; the church and the peer group may place conflicting demands on adolescents; parents may believe that the mass media are undermining the values they have taught their children. Conflict may even occur within a single agency of socialization. One of a child's parents may emphasize materialism, whereas the other parent minimizes the importance of possessions. Little League baseball coaches may talk of fair play but encourage their players to win at all costs.

Early and Middle Adulthood Socialization

While early socialization lays an important foundation, the socialization process is not over at age eighteen. Although indispensable, childhood and adolescent socialization does not completely prepare us to meet the demands of adult life. If adults are to participate successfully in society, they must continue to undergo socialization (Lachman et al. 1994; Lachman and James 1997; Santrock 2002). And with the extension of the life span, the elderly confront new situations and roles (Clausen 1986; Hogan and Astone 1986; Atchley 2000; Hillier and Barrow 1999).

There are several models of adult developmental stages (van Gennep 1960; E. H. Erikson 1982; Neugarten 1968; R. Gould 1975; Levinson 1978, 1986). Despite differences in detail, it is possible to paint a general portrait of adult socialization.

What are the stages of development in adulthood? *Early adulthood* begins toward the end of the teen years and extends into the late thirties. (See Figure 4.1 for a graphic presentation of one model of the stages of development in adulthood.) This period involves a move beyond adolescence and a preliminary step into adulthood; it ends when the individual has made a life within the adult world. Young people forge a temporary link between themselves and the adult world that involves, among other things, choosing an initial occupation and establishing a new family through marriage. Toward the end of the twenties and early thirties, some of the earlier provisional choices may be reevaluated. Marriages may be in jeopardy, divorce may be a possibility, and extramarital affairs are not uncommon. Between ages thirty and forty, a settling-down period tends to occur, during which adults, more conscious of their mortality, attempt to make a place in life for themselves and their families. Central to this period is an interest in achieving occupational success, contributing to society, and establishing a solid family life.

Next comes the transition from early to *middle adulthood*. This is the period when new questions about one's place in the world may arise. Adults may wonder what they have done with their lives and what they wish to do with the rest of their working years. Sometime during the forties, choices are made either to continue the path already taken or to establish a somewhat different life. In the case of the latter, a number of things may occur, including such drastic steps as an occupational change, a divorce, an extramarital affair, or a move to a new community. Typically, more subtle changes take place. For example, family life improves or deteriorates; work becomes more important, or thoughts turn toward retirement. Between the mid-forties and retirement age, adults complete the tasks of middle adulthood. Acceptance of one's level of occupational achievement occurs. Emphasis on success is replaced by concern with personal relationships and the "small pleasures of life," including children and grandchildren.

Does this model apply to women? Research supporting the preceding model is primarily based on the male experience. Because insufficient research has been done on females, it remains to be seen how well these developmental stages fit women. Preliminary work, however, indicates that the model is less applicable to women than to men. Women have unique socialization experiences (Rosenfeld and Stark 1987).

What are some of the socialization experiences unique to women? In early adulthood, women may experience conflict between the ideal wife-mother model they learned as children and the modern roles they may now

FIGURE 4.1

Developmental Periods in the Eras of Early and Middle Adulthood

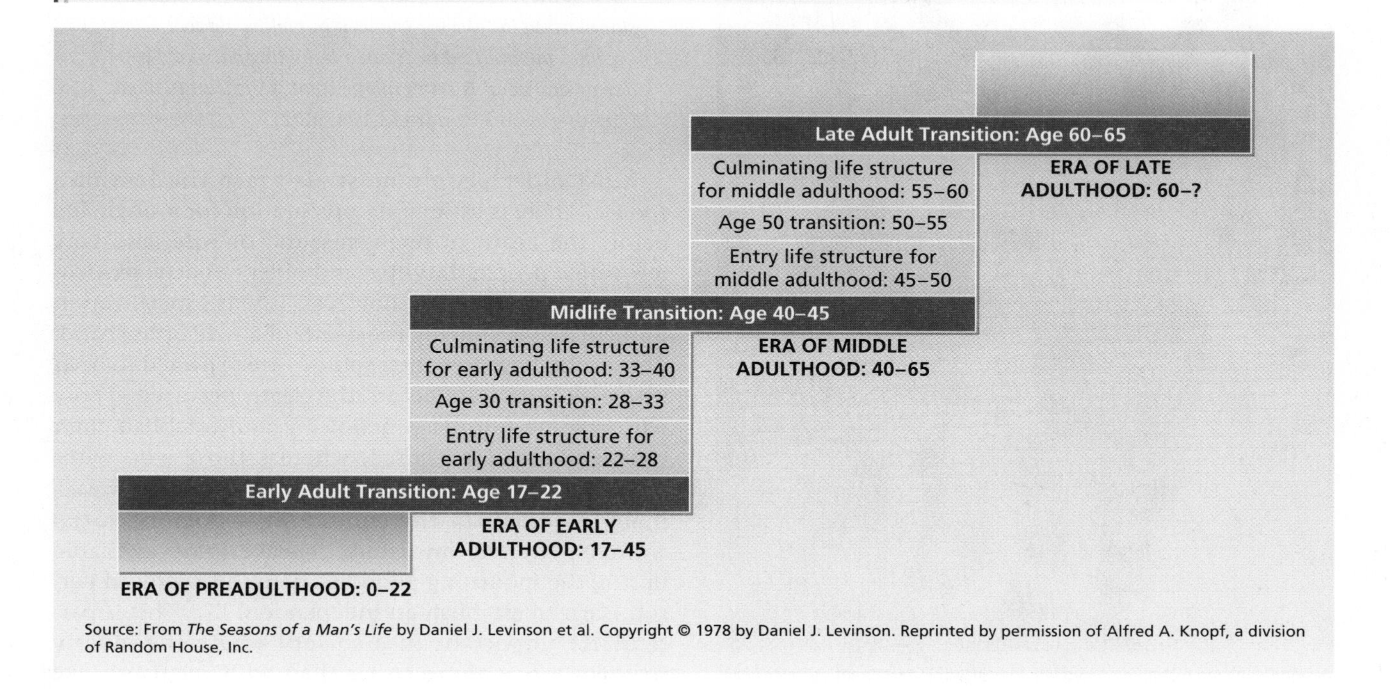

Source: From *The Seasons of a Man's Life* by Daniel J. Levinson et al. Copyright © 1978 by Daniel J. Levinson. Reprinted by permission of Alfred A. Knopf, a division of Random House, Inc.

value. Men often reinforce the traditional female domestic roles by expecting their mates, who have demanding jobs outside the home, to perform most of the household work (Pendleton, Poloma, and Garland 1980; Skinner 1983; Hochschild 1997, 2001, 2003). Many women now experience added pressure in a society that expects more of them than it does of men.

For women who spend most of their adult lives as full-time wives and mothers, middle adulthood presents some special problems. Those women who have based their identity and self-worth on their home, husbands, and children may experience a void when grown children don't require as much attention and husbands at the peak of their careers are preoccupied with work. Also, economically dependent women who experience the death of a husband or a divorce may experience multiple shocks—declining level of living, breaking into the job market, returning to school.

Although signs of early aging, such as wrinkles, weight gains, and loss of body shape, come to men as well as women, it is females who feel the greatest negative repercussions. This is in large part due to the double standard of aging. Many of the characteristics associated with masculinity—aggressiveness, competitiveness, ambitiousness, and decisiveness—are heightened by physical maturity. In contrast, traditionally held images of femininity are diminished in midlife.

The current generation of more highly educated and occupationally successful young women will be in a much stronger position to handle the socialization difficulties of middle adulthood. Actually, some evidence indicates that women in middle adulthood are faring better than the popular image suggests. The loss of family responsibilities is opening new doors of opportunity for many women at the midpoint of their lives (Baruch and Brooks-Gunn 1984; Mirowsky and Ross 1986; L. M. Coleman et al. 1987; Rosenfield 1989). Many women in midlife are initially entering the labor force, resuming earlier careers, or pursuing higher education.

Late Adulthood Socialization

What are the major demands of late adulthood? Length of life is closely linked with the economic base of a society. A study of the Bushmen, a hunting and gathering people, shows that 40 percent of them die before the age of fifteen. Even a century ago, the life expectancy of most people did not exceed thirty-five to forty years. In modern societies, adults can now expect to live beyond seventy. This longer life span exposes aging people to demands unique in human history. The major challenge in American society during *late adulthood* is the withdrawal from participation in certain major aspects

© Brian Bailey/Getty Images

A significant proportion of Americans live into the retirement years. While not all older people adjust well to separation from meaningful work roles, this couple appears to be adjusting nicely to the role requirements of late adulthood.

of social life (Mitford 1998; Cavanaugh and Blanchard-Fields 2001; Benokraitis 2001). Roles are lost because statuses are lost. Two of the most important lost statuses are those of worker and spouse. One cannot act like a worker if no coworker or bosses are present; one cannot act like a wife if no husband exists.

Older people are expected to retire from work. There are many cues. Workmates begin to encourage retirement; a high level of job performance is no longer expected; deserved promotions are given to younger people; rewards for work diminish; duties are reduced; demotions occur; retraining programs are not made available. Moreover, retirement is considered by most people to be the right thing to do.

Although older workers are expected to retire, they receive little preparation for life without work (Seale 1998). They are simply cut off at an arbitrary age—to be creative and active or to stagnate and be passive. They are forced to assume what has been called a "roleless status" (Shanas 1980; Hooyman and Kiyak 2001). The transition from work to retirement may carry negative consequences for social identity and self-esteem:

> *An individual may intellectually apprehend the processes of compulsory retirement policies and yet be emotionally unable to accept them as beyond control. If the older worker incorporates the negative societal stereotypes, a gradual shift in self-image will occur, one that is difficult to overcome in the retirement years. When the negative appraisal summarized in "We see you as a bumbling old fool" becomes internalized to "I am a bumbling old fool," a downward spiral is set in motion that is difficult to break.* (Hendricks and Hendricks 1987:332)

Most older people must also face the loss of a spouse. There is little or no preparation for a single life before the death of one's husband or wife, and very few single people sixty-five and older remarry, particularly women. There is, however, intense socialization immediately following the death of a wife or husband. Those who have lost their spouses are expected to be as active as they were before the death occurred. Those who resume normal functioning and establish their independence are praised, whereas those who withdraw and mourn too long are criticized. However, many members of the community—doctors, social workers, friends, ministers, relatives—are available during the mourning period to help the widowed person learn to establish an independent life. This is particularly important in a society where the elderly frequently live alone rather than with relatives. Like retirement, there is scant cultural definition of roles for a widowed person. Widowhood is another roleless status. (For a further discussion of the problems faced by the elderly, see Chapter 17, "Population and Urbanization.")

Why, then, do aging adults tend to be well adjusted? Gerontologists have long puzzled over the consistent research finding that a large majority of aging adults adjust relatively well in a youth-oriented society—even one that pushes them aside. To solve this puzzle, noted gerontologist Robert Atchley (2000) studied 1,271 residents of a small Midwest town of 25,000 over a twenty-year period. In the process, he developed **continuity theory** to explain how people adjust psychologically and socially to the aging process. Continuity theory presumes that most aging people maintain consistency with their past lives and use their life experiences to intentionally continue developing in self-determined channels. Because change and continuity are not mutually exclusive, Atchley contends, details can change as long as basic life themes and patterns are maintained. Moreover, adaptation via continuity can vary from one area of a person's life to another. Atchley found that the highest degree of continuity occurs in people's values, beliefs, and ideas, followed by continuity in social relationships. Not surprisingly, given their changes in status (retirement, widowhood, health), aging people tend to be less successful in maintaining continuity in activities.

SOCIOLOGY AND THE NEWS MEDIA

Aging in America

Adult socialization in American society is about to change. The combination of aging baby boomers and older people of previous generations will alter the age structure. This, in turn, will affect various aspects of American culture. This CNN report raises the specter of change.

1. Identify and discuss the most important changes the aging of America will bring.

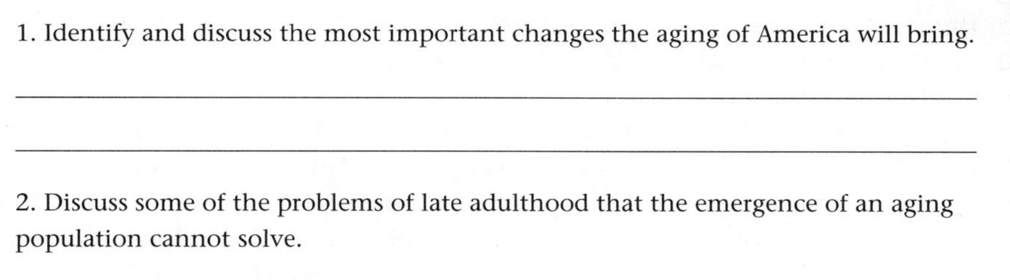

2. Discuss some of the problems of late adulthood that the emergence of an aging population cannot solve.

Even at mid-twentieth century it was common for people to work until age 65 and die soon after. Today senior citizens contribute to the economy for many years after retirement.

How do older people face death? Americans are not very well socialized when it comes to handling death (Moller 1995a, 1995b). Death is simply not a subject very much discussed in the United States. Americans do not even die. They "pass on," "go to their just reward," "go to a better world," "leave us," "buy the farm," or just "go." In fact, sixty-six euphemisms for dying have been identified (DeSpelder and Strickland 1991). Americans do not like to talk about death, particularly to the dying. Relatives often attempt to keep the dying person from knowing that he or she is dying, and dying people sometimes keep their own prognosis secret (Atchley 2001; Leming and Dickinson 2002). Consequently, dying Americans do not know how to fulfill the status of a dying person. The status of a dying person is, like retirement and widowhood, a status almost devoid of roles. Dying Americans face conflicting advice. One person may advise a dying person to accept death after a long and fruitful life. Another person may offer Dylan Thomas's advice:

> *Do not go gentle into that good night,*
> *Old age should burn and rave at close of day;*
> *Rage, rage against the dying of the light.*
> (D. Jones 1971:207)

Death is handled differently in other cultures (Kübler-Ross 1997a; Seale 1998). Among Alaskan Indians, the dying participate in the timing of and planning for death; they play a very active role in their own dying and death. According to Lois Grace, a deacon of the Kawaiahao Church in Hawaii, the Hawaiian people believe that death is a constant companion, whether they are hunting, partying, or staying at home. When someone dies, a big luau among friends and relatives follows a wailing procession. Children are prominently included in the funeral so that they can learn early that death is a part of life. Among the Japanese, death is not something to fear. A dead person "goes to the pure land, the other shore, a place often described as beautifully decorated pools and silver, gold and *lapis lazuli*" (Kübler-Ross 1986:30). Memorial services held by ministers are more for the family than for the deceased.

The American aversion to the topic of death has, until lately, prevented us from gaining much knowledge about death and dying. Partly because of the gradual aging of the American population, death and dying are topics of greater interest to researchers. Consequently, we are beginning to learn how dying people face death. We know, for example, that people are more accepting of death if they are permitted to talk openly and freely about it (Martocchio 1985; Sanders 1992). Those in **hospices**—organizations designed to provide support for the dying and their families—appear to be better socialized in part because they can share their feelings with others who are dying (Mor et al. 1988; DeSpelder and Strickland 1991).

Elisabeth Kübler-Ross (1997b) has identified five stages of the terminally ill. In the first stage, *denial and isolation*, there is a refusal to accept the fact that one is going to die; some individuals even attempt to convince themselves that they have been suddenly cured. In the second stage, *anger* is the dominant emotion. After the dying person has accepted the fact that this is actually happening to him or her, the reaction becomes "Why me?" Patients in this stage may express anger in various ways. The third stage involves *bargaining*. Most bargains are made with God and involve asking for more time. Postponement is the basis of bargaining: "If only I could see my son married" or "If only I could see my grandchild born." The fourth stage, *depression*, begins when the terminally ill patient can no longer deny the reality of dying. The predominant emotion during this stage is a deep sense of loss for oneself and for one's family. In the *acceptance*, or last, phase, the dying person is able to think about death with a degree of quiet expectation. Although this is not a happy stage, it is, as one patient expressed it to Kübler-Ross, a time for "the final rest before the long journey."

Whether these five stages apply to cultures other than the United States has yet to be determined. And as some sociologists have pointed out, even for American society, these stages are not necessarily followed in all cases. People may move back and forth among the stages and may exhibit characteristics of more than one stage at the same time. It is dangerous to view these stages in the dying process as an inevitable progression.

The Sociology of the Life Course

Some qualifications to this description of early, middle, and late adulthood need to be noted. First, as you have already seen, the model of early and middle adulthood is based primarily on samples of males. Second, although this model may describe a general pattern, not all people go through each of these stages. Some individuals handle developmental tasks in unique ways, and some skip periods of growth and change. Third, the ages given for each period are only approximate; individuals may enter stages at different ages. Although these qualifications do nothing to diminish the value of examining adulthood from a developmental perspective, they do alert us to the existence of sociological factors that differentially affect the socialization experiences of both women and men. Thus, complete reliance on this model for predicting what will happen to people and how they will behave masks variations traceable to sociological factors that need to be discussed (Rossi 1974; George, 1993). The model omits, for example, important social influences of gender, social class, race, and ethnicity. For purposes of illustration, let's consider the effects of gender and social class on retirement and widowhood, the two major late adulthood adjustments.

How does social class influence retirement? **Social class** refers to a segment of a population whose members have a relatively similar share of society's desirable things and who share attitudes, values, norms, and an identifiable lifestyle. (Social class is explored in detail in Chapter 8, "Social Stratification.") Two desirable things associated with social class are economic resources and occupational status. Each of these sociological factors affects the experience of retirement.

Financial security is one of the most important determinants of satisfaction with life among retirees (Atchley and Miller 1983; Calasanti 1988). Occupational status also ranks high as a predictor of satisfaction with retirement. First, retirees who held higher-status jobs are generally more financially secure than those with lower-status jobs. Second, higher-status occupations have other characteristics promoting better adjustment to retirement. For example, higher-status occupations involve more interpersonal contact and greater participation in a variety of structured activities. The social skills and the more positive attitudes toward group activities can be transferred into more satisfying nonwork retirement activities (Kohn and Schooler 1983). One positive consequence is less social isolation and loneliness.

How do gender differences affect widowhood? Widowhood negatively influences the quality of life of

FEEDBACK

1. The informal and unofficial things that are taught to children in school to prepare them for life in the larger society are called the __________.
2. The __________ provides children with needed experiences they are unlikely to obtain within the family.
3. Because there is almost no cultural definition of what retired and widowed persons should do, they are in what has been called a __________ status.
4. Match the following stages of dying with the accompanying statements.

____ a. denial and isolation ____ c. bargaining ____ e. acceptance
____ b. anger ____ d. depression

(1) "Why me?"
(2) "What will Martha do without me?"
(3) "I would be fine if I could just see Julia graduate."
(4) "I am going to be fine, you'll see; just leave me alone for a while."
(5) "My marriage has been a good one. There is little left for me to do."

Answers: 1. hidden curriculum 2. peer group 3. roleless 4. a.(4) b.(1) c.(3) d.(2) e.(5)

both elderly men and elderly women. Gender, however, appears to exercise differential effects. The major sociological factors associated with gender are the degree of financial security (gender and social class interact) and the extent of social involvement.

Adjustment to widowhood appears to be related to financial resources (Brubaker 1991). And older widows are usually at a disadvantage financially, especially if they have been economically dependent on their husband and are not covered by their husband's pension plan (Choi 1992). The situation is even worse for economically dependent women who are under sixty years of age, because they do not qualify for Social Security benefits. Even if they qualify for Social Security benefits, they live well below the poverty line (Watterson 1990). Thus, poverty or near poverty is the fate of a substantial proportion of widows in the United States (S. Thomas 1994).

Although women are at a disadvantage in adjusting to widowhood financially, they benefit from their greater capabilities for emotional expression and for attaining and maintaining a network of social support involving friends, relatives, and other widows. Widowers, on the other hand, are more socially isolated, have more difficulty displaying grief, express greater loneliness, and recover more slowly from the loss of their spouse (T. B. Anderson 1984; J. K. Burgess 1988; Morgan 1989).

Socialization and the Mass Media: Functionalist and Conflict Theories

Because the mass media are playing an increasingly important role in the socialization process in American society, additional coverage of them is appropriate. This coverage involves application of the functionalist and conflict theories to the mass media. Socialization through the family, schools, and religion are covered later in separate chapters.

Functions of the Mass Media

A major problem in identifying the functions of the mass media is separating their unique functions from those of other institutions that use the mass media for their own purposes. Although isolating the societal functions of the mass media is difficult, some successful attempts have been made (Lasswell 1948; C. R. Wright 1960, 1974; McQuail 1987).

What are the functions of the mass media? The media have consequences for societies, groups, and individuals. We focus here only on societal consequences:

1. *The mass media provide valuable information regarding events inside and outside a society.* Warnings of natural disasters or military invasions are delivered by the mass media. The media also provide data regarding political candidates, stock market trends, or consumer products. In addition, by publicizing information about deviant behavior, the media enhance social control.
2. *The mass media promote social continuity and integration.* The mass media expose an entire population to the society's dominant beliefs, values, and norms. The media socialize the young, continue its process of socialization into adulthood, and transmit new cultural developments to the population.
3. *The mass media provide entertainment.* By providing amusement, diversion, and relaxation, the media contribute to the reduction of social tension. People are better able to perform their social functions, and society is more stable.
4. *The mass media explain and interpret meanings of events and information.* This process tends to support established cultural forms and to enhance the building of consensus among members of a society. Obviously, the media are involved in the process of socialization as they attempt to translate meanings of events and information for the population.
5. *The mass media can help mobilize society.* The mass media can galvanize a society for war, motivate great numbers of people to join a humanitarian movement, or move citizens to protest government policies.

The functions just outlined are positive ones. Functional analysis, as you know, also involves the identification of negative consequences, or dysfunctions.

What are some dysfunctions of the mass media? The media may foster panic while delivering information; increase social conformity; legitimate the status quo and impede social change while promoting social continuity and integration; divert the public from serious long-term social issues through a constant menu of trivial entertainment and short-term emphasis; shape public views through editorializing as they "interpret" the meanings of events and information; and create violence via public mobilization (Rosenwein 2000).

Conflict Theory and the Media

As you have come to expect, conflict theory underscores more fundamental drawbacks to the mass media. In this section we examine the mass media from the conflict perspective, beginning with the neo-Marxian model and concluding with the power elite model.

Can the Internet Stunt Your Social Growth?

We have seen the effects of extreme social isolation in Anna, Isabelle, and Genie. Although no one expects the results to be nearly as detrimental, many sociologists are concerned about computers and the Internet and the effects that their increased use might have on the socialization of young people.

Mary Pipher (1997) writes about the current generation of children. Because of all the time spent alone on the computer, these children may be the first to grow up with inadequate social skills. Sherry Turkle also claims that the social isolation brought about by heavy use of the Internet leads to the destruction of meaningful community and social integration (Katz and Aspden 1997). In a similar vein, Cliff Stoll says that excessive Internet activity reduces people's commitment to real friendships (Stoll 1995). Many parents see their offspring using computers inappropriately. Betsy Fein's seven-year-old son is using the family's home computer as a private world into which he can escape (Van Gelder 1989). Stories appear in the news quite often about young children who have run away from home to meet adults they have met through the Internet. These children are often lured to these meetings because they do not have the social skills and experience to make sound judgments about their actions.

According to an important nationwide study, the Internet is promoting social isolation (Nie and Erbring 2000). As people spend more time on the Internet (55 percent of Americans have access), they experience less meaningful social contact. Impersonal electronic relationships are replacing face-to-face interaction with family and friends. According to the author of this study, political scientist Norman Nie, "When you spend time on the Internet, you don't hear a human voice and you never get a hug."

Some critics contend that vital social skills are not developing because computer activity is replacing childhood games. Whereas computer use is a solitary activity, traditional games such as sandlot ballgames are socially oriented. These games require interaction and negotiation, encourage sensitivity to others' viewpoints, help establish mutual understanding, promote the delay of gratification, and increase cooperative behavior (Casbergue and Kieff 1998). These social skills are a by-product of traditional activities and are lost to children who spend considerable time socially isolated. Other critics are concerned that extensive computer use will shorten the attention span of children. This could cause them to grow up requiring a continuous flow of outside stimulation ("Lego: Fighting the Video Monsters" 1999).

Defenders of computers and the Internet point to a survey based on 2,500 Americans that showed Internet users as likely as non-Internet users to join religious, leisure, or community groups (Katz and Aspden 1997). This finding, however, does not settle the matter, because critics claim that the researchers who conducted this survey failed to ask some important questions. They did not distinguish between heavy users of the Internet and more moderate users. Nor did they focus on children; those surveyed were adults who had already gone through the important early years of socialization. Future research will enable us to better understand the effects of new technologies on childhood social growth.

Analyzing the Trends

Will children who substitute solitary computer use for traditional childhood games grow up to be fully socialized adults who can interact effectively with their peers? Base your answer on some of the information in this chapter.

What is the Marxian view of the media in America? Marxists consider the media to be monopolized by the ruling class for the purposes of earning a profit and bolstering their power. In the process, workers are exploited by being paid less than they deserve, consumers are overcharged, and the ruling class receives excessive profits. They view the media as necessary to disseminate the **ideology** of the ruling class (that is, the set of ideas they use to justify and defend their interests and actions) and to reduce the impact of competing ideas. By controlling the marketplace of ideas, the ruling class can dilute the development of class consciousness and subsequent political action on the part of the workers. The media, then, are seen as a tool of manipulation by which the ruling class maintains its power.

There is a non-Marxian theory that also sees a few elite individuals in control of American society and, of course, in control of the mass media. This is the theory of the **power elite**—a unified coalition of top military, corporate, and government leaders (the executive branch in particular). The power elite theory is covered in greater depth in Chapter 14 ("Political and Economic Institutions"). For now, we will concentrate on this theory as it applies to the mass media.

Do members of the power elite control the mass media? Thomas Dye (2001) explores three lines of indirect evi-

This NBC news studio contributes to the functioning of society. Among other things, its newscasts contribute to the socialization of the population.

© Najlah Feanny/CORBIS SABA

dence: the organizational concentration of the media, the agenda-setting power of the media, and the media's ability to socialize the population.

The mass media are increasingly becoming concentrated in fewer corporate hands. Likening the mass media to a near totalitarian state, Ben Bagdikian (2000) documents the decline, since 1983, in the number of corporations controlling most of America's print publications, television, radio, and movies—from fifty to six. Despite the growth of cable television, American television is still controlled primarily by three major networks: the American Broadcasting Companies (ABC); Columbia Broadcasting System (CBS); and the National Broadcasting Company (NBC). This control is still intact despite the recent competition of the Fox network, a relatively new network making some bold moves to become a big player. Most of America's local television stations must affiliate with one of these four networks because they cannot afford to produce all of their own programming. Each of the networks is permitted to own local stations, a combination that covers over one-third of households in the United States. Although some ground has been lost to cable television—over 69 percent of all households in the United States are cable subscribers—the four networks held about 44 percent of all television viewing in 2000, down from 61 percent in 1988 *(The World Almanac and Book of Facts 2002).* Moreover, all four networks are now part of larger corporations, a move that greatly magnifies their power. ABC, for example, is now part of The Walt Disney Company, a corporation that owns, in addition to its famous movie studio and theme parks, the ABC television network, eight television stations, six radio networks, nineteen radio stations, four daily and thirty-five weekly newspapers, numerous magazine and book publications, and such cable television holdings as ESPN (an all-sports network), the Arts and Entertainment Network, and Lifetime Cable Network *(Directory of Corporate Affiliations 2001).*

Of the four networks, Fox is the only one located outside the United States. News Corp. Ltd, an Australian-based corporation, in 2000 also owned Fox Television Stations, HarperCollins Publishers, *TV Guide,* the *Boston Herald,* News American Publishing Company, News Ltd. of Australia, and Twentieth Century-Fox Film Corporation *(Directory of Corporate Affiliations 2001).* News Corp. Ltd.'s principal activities are the printing and publishing of newspapers, magazines, and books; commercial printing; television broadcasting; film production and distribution; and motion picture studio operations.

Newspapers and radio stations receive most of their national and international news from two wire services: the Associated Press (AP) and United Press International (UPI). Newspaper ownership is increasingly becoming monopolized as major newspaper chains absorb local papers. The New York Times Company owns not only the *New York Times* but also many other daily newspapers along with radio and television stations. Consider the concentration of power created by the merger of TimeWarner and AOL:

> *When AOL took over TimeWarner, it also took over: Warner Brothers Pictures, Morgan Creek, New Regency, Warner Brothers Animation, a partial stake in Savoy Pictures, Little Brown & Co., Bullfinch, Back Bay, Time-Life Books, Oxmoor House, Sunset Books, Warner Books, the Book-of-the-Month Club, Warner/Chappel Music, Atlantic Records, Warner Audio Books, Elektra, Warner Brothers Records, Time-Life Music, Columbia House, a 40 percent stake in Seattle's Sub-Pop records,* Time *magazine,* Fortune, Life, Sports Illustrated, Vibe, People,

> Entertainment Weekly, Money, In Style, Martha Stewart Living, Sunset, Asia Week, Parenting, *Weight Watchers,* Cooking Light, *DC Comics, 49 percent of the Six Flags theme parks, Movie World and Warner Brothers parks, HBO, Cinemax, Warner Brothers Television, partial ownership of Comedy Central, E!, Black Entertainment Television, Court TV, the Sega channel, the Home Shopping Network, Turner Broadcasting, the Atlanta Braves and Atlanta Hawks, World Championship Wrestling, Hanna-Barbera Cartoons, New Line Cinema, Fine Line Cinema, Turner Classic Movies, Turner Pictures, Castle Rock productions, CNN, CNN Headline News, CNN International, CNN/SI, CNNN Airport Network, CNNfi, CNN radio, TNT, WTBS, and the Cartoon Network.* (Alterman 2003:22–23)

Similar lists can be made for each of the other five media giants *(Directory of Corporate Affiliations* 2001*)*. In 1996, the federal government loosened regulations on radio station ownership. Scooping up hundreds of local stations, six chains currently broadcast to 42 percent of the national radio audience (Fonda 2003). Although subsequently overturned by Congress, the Federal Communications Commission attempted in 2003 to loosen media ownership restrictions, eliminating the long-standing ban on the cross-ownership of newspapers, television stations, and radio stations in the same cities. The commission ruling also permitted television broadcast networks to own more stations at the local and national level, up to 45 percent of the national audience (Ahrens 2003).

The ability of the media to set the political agenda is also underscored by advocates of the power elite model. Through agenda setting the media select for the public which issues are important for consideration (R. H. Turner and Killian 1987). The media can create, publicize, and dramatize an issue to the point that it becomes a topic for debate among both political leaders and the public. Or the media can downplay an issue or an event so that it disappears from public view. Although agenda setting is not accomplished only in the television arena, television is the most powerful force: "TV is the Great Legitimator. TV confers reality. Nothing happens in America, practically everyone seems to agree, until it happens on television" (W. A. Henry 1981:134). Not only can the media attempt to tell us *what* to think, as the functionalists suggest; they are also said to be extremely successful in determining what we think *about*.

Conflict theorists also underscore the media's role in socializing the population. It is through television that most Americans stay in touch with what is happening in their country and in the world. Ninety-eight percent of all households in the United States have at least one television set; the average number of sets in a household is 2.4; those in an average household view television over 29 hours per week *(The World Almanac and Book of Facts 2002)*. America's favorite television shows are viewed by an average of almost 20 percent of households each week.

Because television is the most widely shared experience in the United States, it has the dominant role in portraying what the media consider to be the most important beliefs, norms, attitudes, and values for people to hold. Through entertainment programming, news coverage, and advertising, Americans are constantly being exposed to a particular view of the most desirable and appropriate ways to think, feel, and behave.

The foregoing analysis does not prove that the mass media primarily serve the interests of the elite. It does, however, suggest the potential for the media to be doing so. Conflict theorists, of course, contend that this potential is being realized.

FEEDBACK

1. The __________ are those means of communication that reach large audiences without any personal interaction between the senders and the receivers of messages.
2. Which of the following was *not* discussed in the text as a function of the mass media?
   a. provision of information
   b. creation of consensus
   c. mobilization of society
   d. promotion of social discontinuity and disintegration
3. Which of the following was not presented as indirect evidence for the power elite model of the mass media?
   a. foreign infiltration into the media industry
   b. organizational concentration of the media industry
   c. agenda setting
   d. socialization of the population

Answers: 1. mass media 2. d. 3. a.

SUMMARY

1. Socialization, the process of learning to participate in group life through the acquisition of culture, is one of the most important social processes in human society. Without it, we would not be able to participate in group life and would not develop many of the characteristics we associate with being human. Evidence of the importance of socialization has been shown in studies of isolated monkeys and children. Socially isolated primates—including humans—do not develop as we would expect.
2. Several major theories emphasize the importance of socialization for human development. Although functionalism and conflict theory bear on the socialization process, the most fully developed sociological approach to socialization comes from symbolic interactionism. Key concepts include the self-concept; the looking-glass self; significant others; role taking; the imitation, play, and game stages; and the generalized other.
3. Several major life course theories emphasize the importance of socialization for human development from a psychological viewpoint. Sigmund Freud accentuated the interaction between the biological nature of human beings and the social environment. Erik Erikson, a student of Freud's, described a series of developmental stages that occur from infancy to old age. Each of these developmental stages is accompanied by a psychosocial crisis, or developmental task. In Erikson's view, socialization and personality development are not confined to childhood; they are lifelong processes. According to Jean Piaget, the ability to think, know, and reason develops through interaction with others. If we are to mature cognitively, Piaget contends, each of us must pass through four identifiable stages in the proper developmental sequence.
4. Symbolic interactionism contributes to our understanding that socialization is a lifelong process. The concepts of total institutions, desocialization, resocialization, anticipatory socialization, and the reference group are particularly important in this regard.
5. During childhood and adolescence, the family, school, peer group, and mass media are the major agencies of socialization. The family makes a tremendous impression on the child both because it is the first agency of socialization and because the child is dependent and highly impressionable.
6. The school is generally the first agency of socialization that is controlled by nonrelatives. In school, children are exposed to new standards of performance applied to everyone, are encouraged to develop loyalties beyond their own families, and are prepared in a number of ways for adult life. In addition, by exposing children to such skills as reading and writing and such subject matter as mathematics and English, schools train children to be disciplined, orderly, cooperative, and conforming.
7. The peer group is the first agency of socialization that is not controlled by adults and consequently provides young people with experiences they cannot easily obtain elsewhere. In peer groups, young people learn to deal with others as equals, gain experience in self-direction, establish some degree of independence from their parents and other adults, develop close relationships with friends of their own choosing, and learn to associate with various types of people.
8. Specific effects of the mass media are difficult to document. Still, we know that television, radio, newspapers, and other mass media play a significant part in the socialization process by introducing children to various aspects of their culture.
9. Socialization in infancy, childhood, and adolescence makes a profound contribution to personality development and one's ability to participate in social life. There is, however, an increasing recognition that socialization continues throughout life, even to old age. There is much interest now in viewing adulthood from a developmental perspective.
10. The extension of life expectancy through medical science has created problems of adjustment unparalleled in human history. At an arbitrary age, most older workers are deprived of a meaningful work life. They are forced to retire whether or not they are still productive. Just as there is little preparation for retirement, there is hardly any prior socialization for losing a husband or wife. Like retirement, widowhood is a roleless status. Despite these changes, aging adults tend to be psychologically well adjusted. Unlike people in some other cultures, Americans are not well socialized to face the prospect of dying. Despite this relative lack of socialization, Americans tend to go through several stages in the process of dying.
11. The mass media encompass those means of communication that reach large audiences without any personal interaction between the senders and the receivers of messages. The major forms of mass communication are television, radio, newspapers, magazines, movies, books, tapes, disks, and now the Internet.
12. Although the functions of the media are intertwined with other institutions, it is possible to isolate their major functions. The media provide information, promote social continuity and integration, supply entertainment, explain and interpret events and information, and can mobilize the society when necessary. Some negative consequences, or dysfunctions, are also associated with the mass media.
13. Both the neo-Marxist and power elite views have been applied to the mass media. Proponents of the power elite model point to concentration of power in the media, the agenda-setting power of the media, and the media's ability to socialize the population as at least indirect evidence of a ruling class in America.

INFOTRAC® COLLEGE EDITION

http://www.infotrac-college.com/wadsworth

Another unique option available to you at the Student Resources section of the companion Web site is Infotrac College Edition, an online library with access to hundreds of scholarly and popular periodicals. Below are suggested search terms for this chapter. Results from these and other searches are found at the site.

Search keywords: **mass media**. Find several sources discussing the role of mass media in society. What viewpoints did the authors express about this subject? How would you characterize the publications in which these articles appeared?

Search keyword: **socialization**. As the textbook discussed, socialization is a lifelong process. For many people one of the significant steps in this process is attending college. Find an article about the socialization process of college. You may want to focus your search on a more specific type of socialization, for example, the sexual socialization of college students.

Search keywords: **total institution**. Find three articles that use the term *total institution*. To what specific institutions do these articles refer? Using the sociological definition for this term, do you concur with the authors' characterization?

LEARNING OBJECTIVES REVIEW

- Discuss the contribution of socialization to the process of human development.
- Describe the contribution of symbolic interactionism to our understanding of socialization, including the concepts of the self, the looking-glass self, significant others, and role taking.
- Compare and contrast the life course theories of Freud, Erikson, and Piaget.
- Distinguish among the concepts of desocialization, resocialization, and anticipatory socialization.
- Better understand the socialization process of young people.
- Describe the stages of adult development.
- Discuss the unique demands of socialization encountered in late adulthood.
- Compare and contrast the application of functionalism and conflict theory to the socializing effects of mass media.

CONCEPT REVIEW

Match the following concepts with the definitions listed below them.

____ a. significant others
____ b. ideology
____ c. looking-glass self
____ d. personality
____ e. total institution
____ f. anticipatory socialization
____ g. resocialization
____ h. hospices
____ i. social class
____ j. peer group
____ k. "I"
____ l. reference group
____ m. ego

1. the process of preparing oneself for learning new norms, values, attitudes, and behaviors
2. the conscious, rational part of the personality
3. organizations designed to provide support for the dying and their families
4. the spontaneous and unpredictable part of the self
5. the set of ideas used to justify and defend the interests of those in power in a society
6. one's self-concept based on perceptions of others' judgments
7. a group composed of individuals of roughly the same age and interests
8. a relatively organized complex of attitudes, beliefs, values, and behaviors associated with an individual
9. a group we use to evaluate ourselves and to acquire attitudes, beliefs, values, and norms
10. the process of learning to adopt new norms, values, attitudes, and behaviors
11. those persons whose judgments are the most important to an individual's self-concept
12. a segment of the population whose members have a relatively similar share of the desirable things and who share attitudes, values, norms, and an identifiable lifestyle
13. a place in which residents separated from the rest of society are controlled and manipulated by those in charge

CRITICAL-THINKING QUESTIONS

1. Defend the proposition that human nature is more a matter of nurture than nature. Cite evidence in your argument.
2. What do you think is the greatest contribution that symbolic interactionism has made to our understanding of the socialization process? Why?
3. You are now a college student. Have you undergone (or are you currently undergoing) desocialization, resocialization, or anticipatory socialization? Provide examples from your own experience.
4. Evaluate the statement that socialization ends after childhood. Be specific in your line of argument.
5. Discuss the pros and cons of the mass media from the viewpoint of sociologists. Use specifics in developing your thoughts.

MULTIPLE-CHOICE QUESTIONS

1. **_________ is the process through which people learn to participate in group life through the acquisition of their culture.**
   a. Imprinting
   b. Behavior modification
   c. Socialization
   d. Imitation
   e. Acculturation
2. **The relatively organized system of attitudes, beliefs, values, and behaviors associated with an individual is known as**
   a. personality.
   b. internalization.
   c. humanization.
   d. socialization.
   e. dualism.
3. **René Spitz's study of infants in an orphanage and infants in a women's prison nursery found that**
   a. the infants in both the orphanage and the prison nursery developed normally.
   b. social isolation need not be extreme to damage social and personality development.
   c. social isolation must be extreme to damage social and personality development.
   d. infants in both the orphanage and the women's prison died at early ages.
   e. children in prison nurseries are worse off than children in orphanages.
4. **_________ allows for the greatest understanding of the socialization process.**
   a. Functionalism
   b. Conflict theory
   c. Symbolic interactionism
   d. Sociobiology
   e. Game theory
5. **The _________ is based on our perceptions of others' judgments of us.**
   a. significant other self
   b. alternative self
   c. distorted self-concept
   d. other-directed self
   e. looking-glass self
6. **The process of mentally assuming the viewpoint of another individual and then responding to oneself from that imagined viewpoint is referred to as**
   a. role reversal.
   b. play acting.
   c. gaming.
   d. role taking.
   e. role distancing.
7. **According to George Herbert Mead, children learn to take on the roles of other individuals, one at a time, during the _________ stage.**
   a. game
   b. imitation
   c. work
   d. play
   e. oedipal
8. **According to Freud, the id, ego, and superego should be thought of as**
   a. separate regions of the brain.
   b. separate, interacting, and conflicting processes within the mind.
   c. a mass of impulses, urges, and desires.
   d. multiple personalities.
   e. little people negotiating inside our heads.
9. **According to Erikson, all individuals pass through a series of eight developmental stages from infancy to old age, each of which involves a ________ crisis or ________ task.**
   a. psychosocial; developmental
   b. sociobiological; cognitive
   c. developmental; psychopathic
   d. psychosocial; cognitive
   e. developmental; cognitive

10. Jean Piaget was concerned primarily with the development of _________, which is the ability to think, know, perceive, judge, and reason.
a. psychosocial development
b. cognitive ability
c. self-concept
d. socialization
e. role taking

11. _________ are most likely to participate in anticipatory socialization.
a. New prison inmates
b. Teenagers
c. Office workers
d. New mental patients
e. Cab drivers

12. A group composed of individuals of roughly the same age and interests is known as a
a. subculture.
b. culture cohort.
c. peer group.
d. resocialization group.
e. society.

13. Because there is almost no cultural definition of what retired and widowed persons should do, they occupy what has been called a/an _________ status.
a. undefinable
b. casual
c. master
d. limbo
e. roleless

14. According to the Marxian view, the media in America are
a. a threat to social conformity.
b. the "mouthpiece" of organized labor.
c. a means of production.
d. a tool of manipulation by which the ruling class maintains its power.
e. a threat to the proletarian revolution.

15. The power elite is composed of
a. top university, corporate, and government elite.
b. those in direct control of the mass media.
c. top military, corporate, and government leaders.
d. top religious, university, and corporate leaders.
e. those who place controlling others over economic success.

FEEDBACK REVIEW

True-False

1. According to sociologists, no fundamental human nature exists prior to social contact. T or F?
2. Thanks to recent breakthroughs, research findings on the need of infant monkeys for warmth and affection can easily be applied to humans. T or F?

Fill in the Blank

3. In its approach to socialization, _________ emphasizes social interaction based on symbols.
4. According to the looking-glass process, our _________ is based on how we think others judge the way we look and act.
5. Whereas Freud believed that the human personality is almost totally determined in early childhood, Erikson held that personality can change _________.
6. According to Erikson, the successful management of a psychosocial crisis at an earlier stage increases the chances for mastering later crises. This means that the effects of success or lack of success at earlier crisis points are _________.
7. The informal and unofficial things that are taught to children in school to prepare them for life in the larger society are called the _________.
8. Because there is almost no cultural definition of what retired and widowed persons should do, they are in what has been called a _________ status.

Matching

9. Match Jean Piaget's stages of cognitive growth with the statements.
____ a. sensorimotor stage
____ b. concrete operational stage
____ c. formal operational stage
____ d. preoperational stage
(1) A child thinks without the aid of concrete objects and manipulations.
(2) A child recognizes that 3 + 2 is the same as 2 + 3.
(3) A child learns that a cat that runs under a couch still exists.
(4) A child believes that a box is a spaceship.

10. Match the following stages of dying with the accompanying statements.
____ a. denial and isolation
____ b. anger
____ c. bargaining
____ d. depression
____ e. acceptance
(1) "Why me?"
(2) "What will Martha do without me?"
(3) "I would be fine if I could just see Julia graduate."
(4) "I am going to be fine, you'll see; just leave me alone for a while."
(5) "My marriage has been a good one. There is little left for me to do."

GRAPHIC REVIEW

Snapshot of America 4.1 shows the rate of imprisonment by state.

1. Why was a rate used as the measure rather than the total number of inmates?

____

____

____

2. Why would a high number like 100,000 U.S. residents be used in the calculation?

____

____

____

ANSWER KEY

Concept Review

a. 11
b. 5
c. 6
d. 8
e. 13
f. 1
g. 10
h. 3
i. 12
j. 7
k. 4
l. 9
m. 2

Multiple Choice

1. c
2. a
3. b
4. c
5. e
6. d
7. d
8. b
9. a
10. b
11. b
12. c
13. e
14. d
15. c

Feedback Review

1. T
2. F
3. symbolic interactionism
4. self-concept
5. at any time
6. cumulative
7. hidden curriculum
8. roleless
9. a. 3
   b. 2
   c. 1
   d. 4
10. a. 4
    b. 1
    c. 3
    d. 2
    e. 5

5 Social Structure and Society

© Bob Daemmrich/Stock, Boston

OUTLINE

LEARNING OBJECTIVES

- Explain what sociologists mean by social structure.
- Distinguish and illustrate the basic building blocks of social structure.
- Define the concept of society.
- Describe the means of subsistence practiced by the major types of preindustrial societies.
- Discuss the characteristics of industrial society.
- Compare and contrast preindustrial, industrial, and postindustrial societies.

USING THE SOCIOLOGICAL IMAGINATION

Are humans genetically selfish? Many would say that, by their very nature, people are basically selfish. Evidence indicates otherwise. If humans, as a species, shared the trait of selfishness, we would not expect to find a society populated by cooperative members. Yet, the economic relationships in hunting and gathering societies are grounded in the practice of sharing. Lacking a conception of private property or ownership, members of hunting and gathering societies view even thrift as an indication of selfishness. If genetically programmed to be selfish, these people could not exist without a sense of personal ownership. They would be incapable of such communal concern.

Members of hunting and gathering societies, like members of most groups, know what is expected of them and what they can expect from others, and they engage in the same basic social patterns time after time. All groups have patterned and predictable social relationships that are passed from generation to generation. These patterned social relationships are referred to as social structure, which is the basic subject matter of sociology.

Social Structure and Status

Your understanding of sociology as the study of social structure will jell in this chapter. You learned in Chapter 3 that culture shapes human social behavior—that in the absence of biological preprogramming, culture has to provide the raw material for thinking, feeling, and behaving. Without culture, humans would have no blueprint for social living. This chapter will lead you through a set of sociological concepts that demonstrate the flow between culture and social structure.

In *As You Like It,* William Shakespeare wrote:

> *All the world's a stage. And all the men and women merely players; They have their exits and their entrances; And one man in his time plays many parts.*

All members of a group have parts they are expected to play. Students are expected to attend class, listen to the instructor, fulfill class requirements, and take examinations. Professors are expected to teach class, make assignments, and grade examinations. On any American college campus, you will find similar relationships between students and faculty members. Here, each student-faculty interaction is orderly and predictable. If, however, you found yourself in a class where students shouted the professor down, the custodian presented lessons, and the students did all the teaching, you would miss the presence of structure. Without the order and predictability you expected, you would become apprehensive about the educational process in general and about relating to those unfamiliar professors. To fit in, you would need some awareness of the underlying social structure.

Fortunately, we are usually spared such confusion because groups are relatively predictable. Upon entering a new group, we bring some knowledge of the ways people normally relate to one another. In other words, in our minds, we carry a social map for various group situations. We have mental images—however unconscious and hazy—of the group structure in which we want to participate. This awareness permits us to engage in patterned social relationships within groups without personal embarrassment or social disruption. This underlying pattern of social relationships is called **social structure**.

The mental maps of social structure are not part of our genetic heritage; they must be learned from others. In the process, we learn statuses and roles—major elements of social structure.

What do sociologists mean by status? People may refer to themselves as students, doctors, welders, secretaries, mothers, or sons. Each of these labels refers to a **status**—a position a person occupies within a social structure. Status helps us define who and what we are in relation to others within the same social structure. Some social statuses are acquired at birth. For example, a newly born female instantly becomes a child and a daughter. From then on, she assumes an increasingly larger number and variety of statuses.

How do people acquire social statuses? There are two basic types of social statuses—ascribed and achieved statuses. An **ascribed status** is neither earned nor chosen; it is assigned to us. At birth, an infant is either a male or a female. Except in instances of sex-change operations, we cannot select our gender. Age is another prominent example of an ascribed social status. In some societies, religion and social class are ascribed by the family of birth. If you were born into a lower-class home in India, for example, you would not be permitted to rise to a higher social class.

An **achieved status** is earned or chosen because people have some degree of control and choice. In most modern societies, for example, an individual can decide to become a spouse or a parent. Occupations are also achieved statuses in modern societies; people have

Sociologists often use the stage as a metaphor for social structure. Just as the actors play parts in a play, we occupy social statuses in real life.

latitude to choose their work. Plumber, electrician, sales representative, nurse, executive, lawyer, and doctor are achieved statuses.

Sociologists are interested in the relationships among social statuses. A sociologist investigating college athletics, for example, may focus on the status of the athlete in relation to the statuses of teammate, coach, parent, and professor (see Figure 5.1).

What is a status set? In addition to these relationships, an athlete will occupy various other statuses that may be totally unrelated to that of athletics. A **status set** is all of the statuses that an individual occupies at any particular time. One athlete may be an architect major, an artist, and a church choir member; another may be a service club leader, a motivational speaker, and a mother. Each of these statuses is part of another network of statuses. Assume, for example, that in addition to being an athlete, an individual is also a part-time jazz musician. In this status, she would interact with the statuses of nightclub owner, dancer, and fellow musician, among others.

Are all of a person's statuses equal? Some statuses are more important than others. **Master statuses** are important because they influence most other aspects of a person's life. Master statuses may be achieved or ascribed. In industrial societies, occupations—achieved for the most part—are master statuses. Our occupation strongly influences such matters as where we live, how well we live, and how long we live. "Criminal" can be an achieved master status because its effects permeate the rest of that person's life.

Ascribed master statuses are no less important to a person's life than achieved master statuses. A person who acquires immune deficiency syndrome through a blood transfusion in a hospital becomes an AIDS vic-

FIGURE 5.1

The Interrelationships of Social Statuses

Social statuses cannot exist in isolation. Without other statuses (parent, administrator, referee), the status of college athlete would have no meaning. All statuses, then, are interrelated with other statuses.

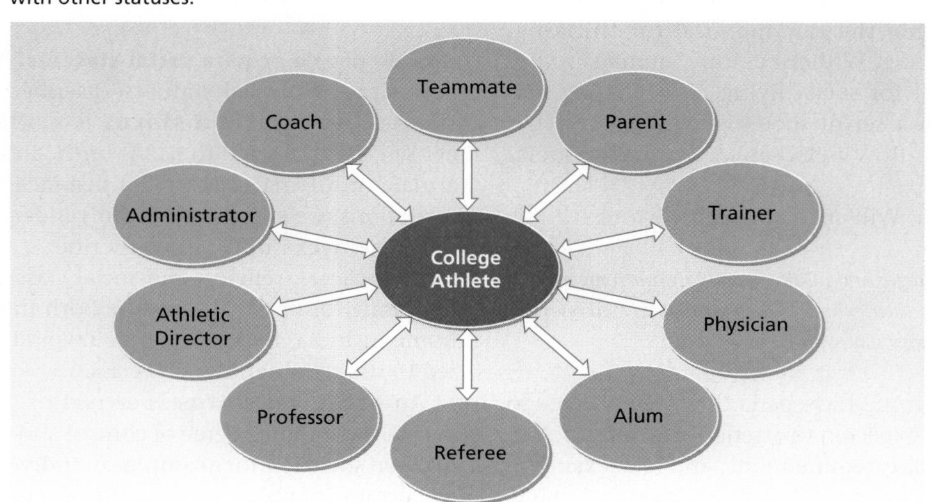

As the occupant of the achieved status of talk-show host, Oprah Winfrey has a variety of roles. For example, she is expected to prepare for the show, ask the questions, and entertain the audience.

tim who will likely suffer prejudice and discrimination in employment, housing, and social relationships. The physically handicapped often have similar experiences.

Age, gender, race, and ethnicity are also ascribed master statuses because they significantly affect the likelihood of achieving other social statuses. Age categories, gender, and various racial and ethnic minorities do not normally constitute social classes, because all social classes contain males, females, young people, retired persons, racial categories, and ethnic people. But these statuses are especially influential in social class placement because they affect other social statuses that a person may occupy. When will the United States have a female president? Would you let a nineteen-year-old or a ninety-year-old handle your case in court—or remove your appendix?

Social statuses are similar to the parts performers play on the stage. Prostitute, pimp, police officer, and judge are parts that individuals play in real life. The behavior of an individual depends largely on the part, or status, that individual holds. This is because parts in real life are based on culturally defined roles.

Social Structure and Roles

Roles

What are roles? **Roles** are the culturally defined rights and obligations attached to a status; they indicate the behavior expected of an individual holding that particular status. Any status carries with it a variety of roles. For example, the roles of a modern doctor include keeping abreast of new medical developments, scheduling office appointments, diagnosing illnesses, and prescribing treatment.

Roles can be thought of as the glue that holds a network of social statuses together; the roles of one status are matched with the roles of other statuses through rights and obligations. **Rights** inform individuals of behavior they can expect from others. **Obligations** inform individuals of the behavior others expect from them. The rights of one status correspond to the obligations of another. Doctors, for example, are obligated to diagnose their patients' illnesses. Correspondingly, patients have the right to expect a diagnosis. Patients have an obligation to keep their office appointments, and doctors have a right to expect that they will do so.

To continue the stage metaphor, roles are the script that indicates to the actors (status holders) what beliefs, feelings, and actions are expected of them (see Table 5.1). Just as a playwright or screenwriter specifies the content of a performer's part, culture determines the parts played in real life. Mothers, for instance, follow dissimilar maternal "scripts" in different cultures. Most

FEEDBACK

1. __________ refers to the patterned relationships among individuals and groups.
2. A __________ is a position a person occupies within a group.
3. Match the type of status with the examples that best illustrate it.
   ____ a. ascribed status
   ____ b. achieved status
   ____ c. master status
   ____ d. status set
   (1) basketball coach, mother, author, daughter, professor
   (2) occupation, gender, race, ethnicity
   (3) president of the United States
   (4) sex

Answers: 1. Social structure 2. status 3. a. (4) b. (3) c. (2) d. (1)

American mothers tend to emphasize independence more than most German mothers do.

As noted in Chapter 1, Erving Goffman, with his dramaturgical approach, deserves the most credit for developing the analogy between the stage and social life. In a sense, Goffman's life's work was devoted to the distinction between culturally prescribed roles and the fulfillment of those roles.

Role Performance and Social Interaction

How is role performance tied to social interaction? Statuses and roles provide the basis for group life. But roles are activated only when people in statuses engage in role performance. This occurs through the process of social interaction.

Role performance is the actual conduct, or behavior, involved in activating a role. Although some role performance can occur in isolation, as when a student studies alone for an examination, most of it involves social interaction. **Social interaction** is the process of two or more persons influencing each other's behavior. For example, before two boys begin to fight, they have probably gone through a process of insult, counter-insult, and challenge. Fortunately, most social interaction is not as negative and violent, but the same process of influence and reaction is involved (Turner 2002).

Not all social interaction follows expected role performance. Spectators at a professional golf tournament are not supposed to cheer bad shots or boo good ones. When some American fans at the 1999 Ryder Cup competition did both, they insulted European players and fans.

If statuses are analogous to the parts of a play and roles are the script, then social interaction represents the way actors respond to cues given by other actors, and role performance is the actual performance. Table 5.1 illustrates this metaphor.

TABLE 5.1

The Stage Analogy

This table draws an analogy between rehearsed behavior on the stage and real social behavior. In your own words, describe the parallel between the sociological concepts on the right and the stage terms on the left.

Stage	Social Life
Parts	Statuses
Script (lines)	Roles
Cues	Social interaction
Actual performances	Role performance

FIGURE 5.2

The Links Between Culture and Social Structure

Sociologists concentrate on the study of social structure. They have developed a set of concepts and a conception of their relationship in order to understand the basic nature of social structure.

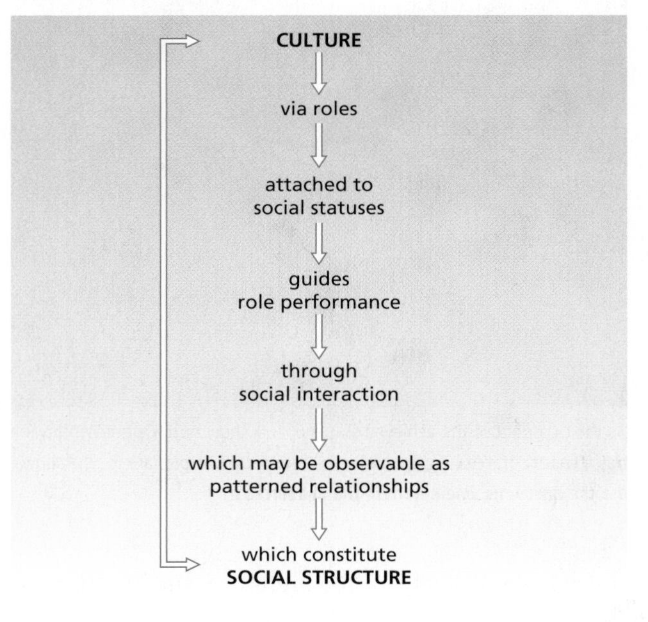

Relationship Between Culture and Social Structure

How are culture and social structure related? Figure 5.2 outlines the links that connect culture and social structure. Starting at the top of this figure, the first major bond between culture and social structure is the concept of role, culturally defined rights and obligations attached to a social status. The second link is a status, a position that a person occupies within a group. It is through the roles attached to each status that culture enters the picture. Yet people do not always follow roles exactly. The manner in which roles are actually carried out is role performance, the third link in the conceptual chain. Role performance occurs through social interaction. This is the fourth link between culture and social structure. Social interaction based on roles is observable as patterned relationships, which constitute social structure. In turn, existing social structure affects the creation of—and changes in—culture. (Note the two-headed arrow in Figure 5.2.)

Are all these terms necessary? Concepts are as necessary for sociology as they are for physics. Both sociologists and physicists deal with abstractions, with unobservable phenomena whose existence must be assumed from the observation of things that can be seen. Before 1970, physicists had never directly observed individual atoms. Instead, they used indirect methods, such as photographs of X rays beamed at a small crystal of rock salt, to infer the existence of atoms. Just as physicists have used indirect methods, sociologists use the observation of patterned relationships (something concrete) to infer the existence of social structure (something abstract). If the concepts and relationships outlined in Figure 5.2 were not known, it would be as difficult for sociologists to establish the existence of social structure as it would have been for the ancient Greeks to document the existence of atoms. Concepts and their relationships are the necessary building blocks of any area of scientific study.

How does social interaction differ from playacting? Although the stage metaphor is a valid one, there is danger in taking it literally. First, "delivery of the lines" in life is not the conscious process practiced by stage performers. Most real-life role performance occurs without much forethought because we act in ways that have been unconsciously adopted through observation, imitation, and socialization.

Second, there is more of a discrepancy between roles and role performance in real life than between a stage script and its performance. Although performers may sometimes ad-lib, change lines to suit themselves, or introduce a little "business," overall they adhere to the script. Differences between a role and role performance

Many people experience role conflict—they find themselves performing a role in one status that clashes with a role in another status. The smile on this waiter's face belies the conflict he feels from attending college and having to earn money at the same time.

SNAPSHOT OF AMERICA 5.1

Students Expelled for Bringing Guns to School, 1998–1999

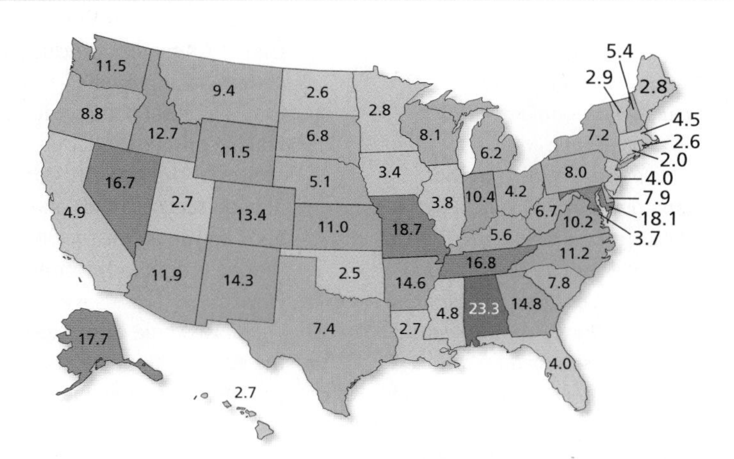

Student gun-related school expulsions, 1998–99
Number of students expelled per 100,000

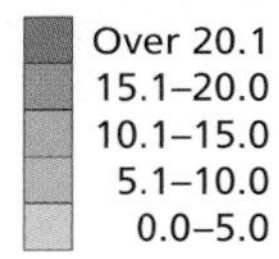

Over 20.1
15.1–20.0
10.1–15.0
5.1–10.0
0.0–5.0

Source: U.S. Department of Education; available at www.ed.gov/offices/OESE/SDFS/GFSA/report_2000/Table1.html

It is illegal for students to bring guns to public school. Bringing firearms to school is also a violation of the student role. Teachers expect students to come to school unarmed (a teacher's right). Students are expected not to bring weapons to school (a student's obligation). This U.S. map gives us some idea of the relative extent to which this role is being violated and punished in various states. In total, 3,523 students were expelled during the 1998–99 school year for carrying guns to school. The good news—this number is down roughly 30 percent from 1996–1997.

1. Was your high school more like your state or more like some other state? Explain.
2. Explain why the violation of a law can also be a role violation.

Philip Zimbardo—Adopting Statuses in a Simulated Prison

Social psychologist Philip Zimbardo and his colleagues designed an experiment to observe the behavior of people without criminal records in a mock "prison." They were amazed at the rapidity with which statuses were adopted and roles fulfilled by the college students playing "prisoners" and "guards." This experiment reveals the ease with which people can be socialized to statuses and roles. Zimbardo's own words describe the design and results of this experiment.

> *In an attempt to understand just what it means . . . to be a prisoner or a prison guard, Craig Haney, Curt Banks, Dave Jaffe and I created our own prison. We carefully screened over 70 volunteers who answered an ad in a Palo Alto city newspaper and ended up with about two dozen young men who were selected to be part of this study. They were mature, emotionally stable, normal, intelligent college students from middle-class homes. . . . They appeared to represent the cream of the crop of this generation. None had any criminal record. . . .*
>
> *Half were arbitrarily designated as prisoners by a flip of a coin, the others as guards. These were the roles they were to play in our simulated prison. The guards . . . made up their own formal rules for maintaining law, order and respect, and were generally free to improvise new ones during their eight-hour, three-man shifts. The prisoners were unexpectedly picked up at their homes by a city policeman in a squad car, searched, handcuffed, fingerprinted, booked at the Palo Alto station house and taken blindfolded to our jail. There they were stripped, deloused, put into a uniform, given a number and put into a cell with two other prisoners where they expected to live for the next two weeks. . . .*
>
> *At the end of only six days we had to close down our mock prison because what we saw was frightening. It was no longer apparent to most of the subjects (or to us) where reality ended and their roles began. The majority had indeed become prisoners or guards, no longer able to clearly differentiate between role playing and self. There were dramatic changes in virtually every aspect of their behavior, thinking and feeling. . . . We were horrified because we saw some boys (guards) treat others as if they were despicable animals, taking pleasure in cruelty, while other boys (prisoners) became servile, dehumanized robots who thought only of escape, of their own individual survival and of their mounting hatred for the guards. We had to release three prisoners in the first four days because they had such acute situational traumatic reactions as hysterical crying, confusion in thinking, and severe depression. Others begged to be paroled, and all but three were willing to forfeit all the money they had earned [$15 per day] if they could be paroled. By then (the fifth day) they had been so programmed to think of themselves as prisoners that when their request for parole was denied they returned docilely to their cells. . . .*
>
> *About a third of the guards became tyrannical in their arbitrary use of power, in enjoying their control over other people. They were corrupted by the power of their roles and became quite inventive in their techniques of breaking the spirit of the prisoners and making them feel they were worthless. . . . By the end of the week the experiment had become a reality.*

Thinking About the Research

1. If you were asked to discuss Zimbardo's experiment within the context of one of the three major theoretical perspectives, which would you choose? Why?
2. One of Zimbardo's conclusions, not stated in the preceding account, is that the brutal behavior found in real-life prisons is not due to the antisocial characteristics or personality defects of guards and prisoners. Can you argue, sociologically, that he is right in this conclusion? How?
3. There was some controversy over the ethics of this experiment. Do you think this experiment could be carried out today under the sociological code of ethics discussed in Chapter 2? Why or why not?

in real life are neither as easy to detect nor as easy to control as departures from a script.

Third, on the stage, there is a programmed and predictable relationship between cues and responses. One performer's line is a cue for a specific response from another actor. In life, we can choose our own cues and responses. A student may decide to tell a professor that her examinations are the worst he has ever encountered. On hearing this, the professor may tell the student that it is not his place to judge, or she may ask for further explanation so that improvement may be made. In effect, the professor can choose from among several roles the one she wants to "play" at that time. Likewise, the student can choose from a variety of responses to the professor's behavior. If the professor tells the student he is out of line, the student may report the matter to the head of the department, or he may decide to forget about it altogether. This process of role selection prior to behavior occurs in nearly all instances of social interaction.

Keep in mind, however, that the range of acceptable responses is not limitless. Only certain responses are considered culturally legitimate. It is not an appropriate response for the professor to bodily eject the student from her office, and the student would be foolish to pound the professor's desk in protest.

Role Conflict and Role Strain

The existence of statuses and roles permits social life to be predictable, orderly, and recurrent. At the same time, each person holds many statuses, and each status involves many roles. This diversity invites conflict and strain.

What is the difference between role conflict and role strain? **Role conflict** occurs when the performance of a role in one status clashes with the performance of a role in another status. College students who hold the statuses of student and employee often find it difficult to balance study and work demands.

Because groups and subcultures have unique role prescriptions, they also exhibit unique role conflicts. As young, affluent, successful individuals, athletes attract a large number of "friends." Because the athletes pay all the bills, these friends often evolve into employees. Role conflict for athletes inevitably flows from the mixing of business and personal relationships.

Role strain occurs when some of the roles of a single status clash. College basketball coaches, for example, have to recruit for the next season while attempting to win games in the current season. University professors teach classes, spend time with students, and publish their research findings. Each of these roles is time consuming, and the fulfillment of one role may interfere with the performance of others. If expectations for you as a college student pressure you to perform well academically, join a fraternity or sorority, pursue a sport, date, and participate in other college activities, you will probably experience some degree of role strain.

SOCIOLOGY AND THE NEWS MEDIA

Being Gay in America

It doesn't take the 1999 murder of Matthew Shepard, then a college student at the University of Wyoming, to remind us that gays are stigmatized in American society. This CNN report explores the treatment of gays in public schools. The personal consequences for the victims of homophobia are enormous.

1 Discuss being gay as a master status in American society.

____

____

2. Do known gays in public schools experience role conflict or role strain? Explain and illustrate.

____

____

Though not often murder victims like Matthew Shepard, gays are nonetheless targets in less obvious ways.

How can role conflict and role strain be managed? Role conflict and strain may lead to discomfort and confusion. To reduce anxiety and have smoother relationships with others, we often solve role dilemmas by setting priorities. When roles clash, we decide which role is more important to us and act accordingly. A student who misses work frequently because of school requirements can eliminate the role conflict completely by quitting work and assigning priority to school. If she remains in both statuses, she can reduce work hours or cut down on extracurricular school activities. A university professor may have to decide to emphasize either teaching or research. By ranking incompatible roles based on their importance, we reduce role conflict and strain.

We can also segregate roles. This is especially effective for reducing the negative effects of conflicting roles. An organized crime member may reduce role conflict by segregating his criminal activities from his role as a loving father. A college coach experiencing the role strain associated with coaching and recruiting simultaneously can decide to give priority to one of those roles. He may, for example, let his assistant coach do most of the recruiting until the season ends.

Can conformity to roles ever be complete? No. Complete conformity to all roles is impossible. This poses no problem as long as role performance occurs within accepted limits. Professors at research-oriented universities may be permitted to emphasize teaching over research. Coaches may accent fair play, character building, and scholarship. But professors at research universities who do too little publishing or coaches who win too few games usually will not be rewarded for very long. At some point, they will have exceeded the acceptable limits for deviation from expected role performance.

Theoretical Perspectives and Social Structure

If you stop to think about it for a moment, you will realize that the concepts of role, social status, and patterned relationships reflect the concerns of functionalism with stability, order, and consensus (see Table 5.2). The concepts of role behavior and social interaction have the strong flavor of symbolic interactionism. Also present in the study of social structure, but not as apparent, are the ideas of conflict theory. To illustrate conflict, consider the struggles between street gangs and law enforcement officials and the clashes between teenagers and their parents. To illustrate change, consider women's more prominent role in sports and the economy, and ponder the effects of emerging globalization on your career choices.

Society

A society is the largest and most self-sufficient social structure in existence. As you may recall from Chapter 3, a **society** is composed of a people living within defined territorial borders, sharing a common culture. Theoretically, a society is independent of all outsiders. It contains enough smaller social structures—family, economy, government, religion—to meet most of the needs of its members. As you will see, preindustrial societies met this criteria; they truly could be independent and self-sufficient. But modern societies, although capable of caring for most members' needs, must have political, military, economic, cultural, and technological ties with other societies. In fact, as we move toward a global society, the fate of all people becomes linked.

FEEDBACK

1. _________ are culturally defined rights and obligations attached to statuses.
2. The rights of one social status correspond to the _________ of another one.
3. _________ occurs when the performance of a role in one status clashes with the performance of a role in another status.
4. _________ occurs when the roles of a single status are inconsistent.
5. Match the following concepts with the examples listed beside them.
   ___ a. role (1) A mother is expected to take care of her children.
   ___ b. role performance (2) A husband and wife discuss disciplining one of their children.
   ___ c. social interaction (3) A university president hands out diplomas at a graduation ceremony.
6. Which of the following is *not* one of the differences between the stage and social life?
   a. There is considerably more discrepancy between roles and role performance in social life than between a stage script and its performance.
   b. Unlike the stage, there are no cues and responses in real life.
   c. Role performance in real life is not the conscious process that actors go through on the stage.
   d. In social life, the cues and responses are not as programmed and predictable as on the stage.

Answers: 1. Roles 2. obligations 3. Role conflict 4. Role strain 5. a. (1) b. (3) c. (2) 6. b

TABLE 5.2

FOCUS ON THEORETICAL PERSPECTIVES: Illustrating Social Structure Concepts

This table illustrates how each theoretical perspective might view a major concept. The concepts could be switched to any other theoretical perspective and illustrated from that perspective. Associate each concept with a different theoretical perspective and provide your own example.

Theoretical Perspective	Social Structure Concept	Example
Functionalism	Role	• Social integration is promoted by culturally defined rights and obligations honored by group members.
Conflict theory	Ascribed master status	• Ascribed master statuses such as gender and race empower some to subjugate others.
Symbolic interactionism	Social interaction	• Roles are carried out by individuals based on the symbols and meanings they share.

Anthropologists and sociologists have classified societies in various ways. Because all societies must fulfill the need for food, one classification system (the one we will use in this chapter) is based on the way subsistence is addressed. This classification system is based on the evolution from the four types of preindustrial societies (hunting and gathering, horticultural, pastoral, and agricultural societies) to industrial societies and finally to postindustrial societies (Haviland 2002). A society's solution to the subsistence problem is clearly reflected in its culture and social structures (see Table 5.3.).

Preindustrial Societies

What is unique about a hunting and gathering society? The **hunting and gathering society** survives by hunting animals and gathering edible foods such as wild fruits and vegetables. This is the oldest solution to the subsistence problem. In fact, it was only about nine thousand years ago that other methods of solving the subsistence problem emerged.

Although not all hunting and gathering societies are exactly alike, they do share some basic features (Beals, Hoijer, and Beals 1977; Service 1979). Hunting and gathering societies are usually nomadic; that is, they must move from place to place as the food supply and seasons change. Because nomads must carry all their possessions with them, they accumulate few material goods. Hunting and gathering societies also tend to be very small—usually fewer than fifty people—with members scattered over a wide area. Because the family is the only institution in hunting and gathering societies, it tends to all the needs of its members. On a day-to-day basis, association is typically limited to the immediate family. In fact, these societies are organized either as self-sufficient family groups or as loose combinations of families or bands. Thus, hunting and gathering societies are tied together by kinship. Most members are related by blood or marriage, although marriage is usually limited to those outside the family or band.

Cooperation or competition? Which is more valued in preindustrial society? Economic relationships within hunting and gathering societies are based on cooperation—members sharing their surplus with other members. Most members of this type of society are kin, and sharing takes place without the expectation of reciprocation. Members of hunting and gathering societies seem simply to give things to one another. In fact, the scarcer something is, the more freely it is shared. Generosity and hospitality are valued; thrift is considered a reflection of selfishness. Because the obligation to share goods is one of the most binding aspects of their culture, members of hunting and gathering societies have little or no conception of private property or ownership.

Without a sense of private ownership and with few possessions for anyone to own, hunting and gathering societies have no social classes, no rich or poor. These societies also lack status differences based on political authority because they have no political institutions; there is no one to organize and control activities. According to E. Adamson Hoebel, when the Inuits wanted to settle disputes, they had dueling songs. The people involved in the dispute prepared and sang songs to express their sides of the issue. Their families, as choruses, accompanied them. Those listening to the duel applauded their choice for the victor (Hoebel 1983).

The division of labor in hunting and gathering societies is limited to the sex and age distinctions found in most families. Men and women are assigned separate tasks, and certain tasks are given to the old, the young, and young adults. This scant division of labor exists because there are no institutions beyond the family.

TABLE 5.3

Comparison of Major Types of Society

Carefully examine this comparison of major types of society. Choose one type of society and explain why the nature of its culture and social structure fits with its means of subsistence and technological base.

Type of Society	Time of Origin	Means of Subsistence	Technological Base	Culture and Social Structure
Hunting and Gathering	First type of society to emerge among humans	Hunting animals and gathering edible foods in nature	Simple handmade tools	• Small nomadic bands based on kinship and cooperation; common ownership of property; scant division of labor along sex and age lines within families
Horticultural	About 9,000 years ago	Domesticating plants	Slightly more advanced handmade tools (digging sticks, hoes, spades)	• Less nomadic living in more permanent settlements; bands organized around families; more conflict among bands; neglible division of labor based on sex and age
Pastoral	About 9.000 years ago	Domesticating animals (goats, sheep, cattle)	Handmade horticultural tools; meat-cutting tools; knowledge in such areas as grazing land carrying capacity, breeding, size and composition of herds, weather, water supply	• Live in long-term villages; some trade emerges; women remain at home while males attend herds; greater economic surplus leads to more complex division of labor; greater competition over surplus
Agricultural	About 5,000–6,000 years ago	Permanent land cultivation and harnessing of animal energy	Plow and animal energy	• Increased productivity releases people from land; more complex division of labor branching outside economic roles; establishment of smaller cities; emergence of separate political, economic, and religious institutions; state replaces family as prime mover; appearance of distinct social classes; emergence of trade and money economy
Industrial	About 250 years ago	Application of science and technology to production	Power-driven machines	• Economy shifts from subsistence to an open market due to greater food surplus; growth of large cities; importance of kinship declines; education moves out of homes and increases in importance; women become less subordinate to husbands; influence of religion less pervasive; social institutions become more complex, specialized, and independent; social relationships more impersonal
Postindustrial	Around 1970	Development of service industries	Intellectual	• Greater social instability in the form of divorce, crime, and rapid social change; less social and cultural consensus; reduced gender inequality; individualism increases; urban population moves out of large cities

Also, more leisure time is available in hunting and gathering societies than in any other. Today, few true hunting and gathering societies remain other than the Khoi-San (Bushmen) in southern Africa, the Kaska Indians in Canada, and the Yanomamö of Brazil.

What is a horticultural society? According to archaeological evidence, **horticultural societies** solved the subsistence problem primarily through the domestication of plants. This type of society came into being about nine thousand years ago. The transition from hunting and gathering to horticultural societies was not abrupt. It occurred over several centuries. In fact, some hunters and gatherers in the Middle East were harvesting wild grains with stone sickles 10,000 years ago, and early horticultural societies also hunted and gathered (Nolan and Lenski 1999).

The shift from hunting and gathering to horticulture led to more permanent settlements. Subsistence no longer required people to move frequently in order to take advantage of available food. Instead, people could work a land area for extended periods of time before moving on to more fertile soil. This relative stability in simple horticultural societies—and particularly in the permanent settlements in the Middle East and southeastern Europe—permitted the growth of large societies with greater population densities.

The family is even more basic to social life in horticultural societies than in hunting and gathering societies. In hunting and gathering societies, the survival of the band usually has uppermost priority. In horticultural societies, primary emphasis is on providing for household members. This is because producing food in horticultural societies can be accomplished through the labor of family members. With the labor necessary for survival, households can be more self-sufficient and independent: "The family is to the . . . . [horticultural] economy as the manor to medieval European economy, or the industrial corporation to modern capitalism: each of these is the central production-institution of its time" (Sahlins 1968:75).

While hunters and gatherers seldom engage in interband raids, there is considerable intervillage conflict in horticultural societies. Much of the fighting, however, resembles modern contact sports more than warfare. Satisfaction is often gained more from defeating the enemy in games than from killing him. Although conflict does sometimes lead to battle and death, religious beliefs and rituals keep slaughter within bounds (Plog and Bates 1990). Still, Napoleon Chagnon (1997) estimates that at least 24 percent of all male deaths among the Yanomamö (South American preindustrial people) are due to warfare—a figure consistent with those for other tribes that constantly feud with their neighbors.

How is the subsistence problem addressed in pastoral societies? Most horticultural societies keep domesticated animals such as pigs and chickens. They do not, however, depend economically on the products of these animals the way *pastoralists* (herders) do. In **pastoral societies**, food is obtained primarily by raising and taking care of domesticated animals. For the most part, these are gregarious (herd) animals such as cattle, camels, goats, and sheep, all of which provide both milk and meat. Because food grains are needed to supplement the food obtained from animals, pastoralists must also either cultivate some land or trade with people who do (Peoples and Bailey 2000; Nanda and Warms 2002).

There is more physical mobility in pastoral societies than in those based more fully on the cultivation of land, although the degree of mobility may vary greatly. Permanent (or at least long-term) villages can be maintained if, as seasons change, herd animals are simply moved to different pastures within a given area. In such societies, the women remain at home while the males take the herds to different pastures. A permanent village, however, cannot exist if, as the weather changes, an entire population moves with the herd (nomads).

Because both horticultural and pastoral societies can produce a surplus of food, they usher in important social changes unknown in hunting and gathering societies. With a surplus food supply, some members of the community are freed to create a more complex division of labor. People can become political and religious leaders or makers of surplus goods such as pottery, spears,

Horticultural societies solve the subsistence problem primarily through the domestication of plants. These Indonesians are threshing grain by hand in the rice terrace at harvest season.

and clothing. Because nonedible goods are produced, an incentive to trade with other peoples emerges.

The creation of a surplus also permits the development of social inequality, though it is limited by the size and dependability of the surplus. Even a relatively small surplus, however, means that some families (or villages or clans) have more wealth than others. Competition over resources can take the form of warfare, which in turn leads to the emergence of political leaders or possibly the creation of slavery.

What are the characteristics of an agricultural society? The transition from horticultural to **agricultural society** was made possible largely through the invention of the plow (Nolan and Lenski 1999). The longer the horticulturalists worked the soil with their digging sticks, hoes, and spades, the greater their disadvantage. With these crude implements, horticulturalists could not control the weeds, and they could not dig deeply enough to reach the soil's nutrients, which recede farther below the root line each year of cultivation. The plow, which appeared about 5,000–6,000 years ago, did not solve these problems completely, but it was effective enough to permit the permanent cultivation of land. The plow not only controlled the weeds but also turned the weeds into fertilizer by burying them under the soil. By digging more deeply into the ground, the plow was able to reach nutrient-rich dirt that had sunk below root level. The result was more productivity—more food per unit of land.

Moreover, the plow permitted a shift from human to animal energy. Only humans could use a hoe or spade, but oxen could pull a plow. Using draft animals also increased productivity because larger areas could be cultivated with fewer people. As a result, more people were released from the land, freed to engage in noneconomic activities such as formal education, concerts, and political rallies. This led, in turn, to other significant changes: the introduction of occupations not directly tied to farming (such as politician, blacksmith, and haberdasher); the establishment of cities; the development of more complex economic specialization; and the emergence of separate political, economic, and religious institutions.

Although family ties remained important, the state replaced the kinship group as the guiding force for agricultural societies. Advanced agricultural societies were monarchical, headed by a king or an emperor who desired as much control as possible. Distinct social

WORLDVIEW 5.1

Employment in Service Industries

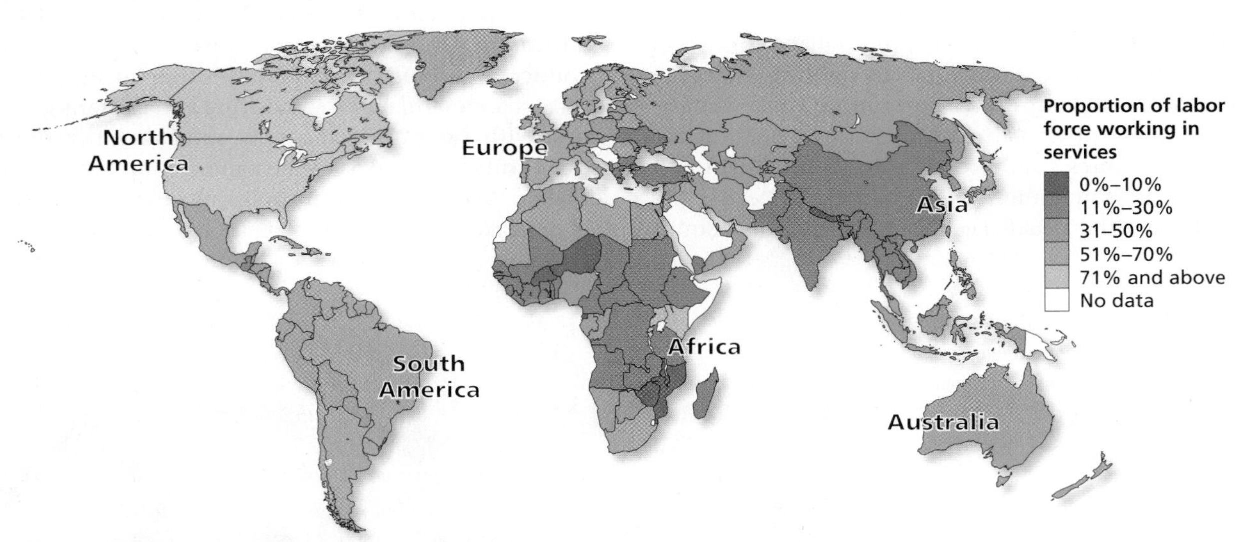

Source: *Human Development Report, 1998*, United Nations Development Report (New York: Oxford University Press, 1998).

As societies move from the industrial to the postindustrial stage, more people are needed in service occupations. This map shows the percentage of each country's population involved in the service industries.

1. After examining this map, what generalizations about types of societies around the world would you make? Explain.
2. Which countries do you think could be ready to move from one type of society to another? Be specific on countries and types of societies.
3. What parts of the world are least likely to change in the near future? Explain your answer.

classes appeared for the first time. Wealth and power were based on land ownership, which was controlled by the governing elite. While the elite enjoyed the benefits of the economic surplus, the peasants did most of the work.

Urban merchants held less prestige than the governing class. They were better off than the peasants; but they, too, worked hard for their living. An economy, based on trade and a monetary system, began to emerge as an identifiable institution. The agricultural economy developed two basic sectors: a rural, agricultural sector and an urban, commercial sector. Institutional specialization was also reflected in the increasing separation of the political and religious elite. Although rulers were believed to be divinely chosen, few of them doubled as religious leaders.

Industrial Societies

The Industrial Revolution was "a kind of discontinuous leap in human history; a leap as important as that which had lifted the first . . . [horticultural] settlements above the earlier hunting communities" (Heilbroner 1993:102). The Industrial Revolution created **industrial society**—a society whose subsistence is based primarily on the application of science and technology to the production of goods and services.

The problem of subsistence is solved in industrial society primarily through the application of science and technology to the production of goods and provision of services. The technology available earlier in the development of industrial society was crude, as shown in this cotton mill in North Carolina in 1908. Still, the power looms used to weave cotton cloth were an important technological advance in the rise of industrial society.

What happens to societies as new technologies appear? Alberto Martinelli and Neil Smelser (1990) have identified some basic structural changes that occur in societies shifting from an agricultural to an industrial economic base. Industrialism brings with it a change away from simple, traditional technology (plows, hammers, harnesses) toward the application of scientific knowledge. With this change comes the discovery and adoption of new, complex technological devices. Early examples of industrial technology include the steam engine and the use of electrical power in manufacturing. More recent technological developments include nuclear energy, aerospace-related inventions, and the computer. These technological devices are used in factories and in agriculture.

In industrial societies, the trend is away from human and animal power toward power-driven machines. These machines are operated by wage earners who produce goods for sale on the market. With this mechanization, the shift is away from subsistence farming toward the selling of agricultural goods on the open market. Farmers are able to produce a surplus sufficient to support themselves and many others. This food surplus supports urbanization, the movement of people from farms and villages to gigantic urban concentrations. Thus, industrial society is also an urban society, with vast concentrations of people located in and around many urban centers. In fact, the Population Division of the United Nations now predicts that by the year 2007 half of the world's population will, for the first time, live in cities (Howard 2002).

The concept of structural differentiation also helps to describe industrial society. **Structural differentiation** occurs when a single social structure divides into two or more social structures. Under the new circumstances, these new separate social structures operate more efficiently than the one alone could. Under the domestic form of production, for example, the family performs a variety of economic roles. In the change from domestic to factory production, economic activities are moved from the family to the factory. Because economic and familial functions are now separated into two institutions, these activities become increasingly segregated.

Other kinds of changes occur through structural differentiation. The education of the young moves from the home to the school. Because an industrial society requires an educated labor force, education is extended from the elite to the masses. Kinship declines in importance as the immediate family begins to separate socially and physically from the extended family. Personal choice and love replace arranged marriages. Women, through their entrance into the workforce, become less subordinate to their husbands. Religion no longer pervades all aspects of social life as various institutions—political, economic, familial, educational—become more and more independent of religious influence. Individual mobility increases dramatically, and social class is based more on occupational achievement than on the social class of one's parents.

Because the United States has been an industrial society for so long, its characteristics are taken as a given. The effects of industrialization are easier to observe in societies currently moving from an agricultural to an industrial economic base. For example, Vietnam and Malaysia, two countries currently undergoing industrialization, are experiencing these kinds of changes at the beginning of the twenty-first century (Phu 1998; S. Singh 1998).

How do preindustrial and industrial societies differ? Ferdinand Tönnies, Emile Durkheim, and Robert Redfield were three early sociologists who, each with his own unique approach, compared preindustrial and industrial societies. Contemporary sociologists generally agree that each of these three men were successful at isolating central features differentiating the two types of societies.

Ferdinand Tönnies (1957; originally published in 1887), an early German sociologist, distinguished between **gemeinschaft** (German for "community") and **gesellschaft** (German for "society"). The former—closely approximating preindustrial society—is based on tradition, kinship, and intimate social relationships. The latter—representing industrial society—is characterized by weak family ties, competition, and less personal social relationships.

Shortly after Tönnies formulated his distinction, Emile Durkheim (1997; originally published in 1893) made a similar observation. He distinguished the two types of societies by the nature of their social solidarity—the degree to which a society is unified.

Social solidarity, Durkheim contended, is predicated on a society's division of labor. In societies in which the division of labor is simple—in which most people are doing the same type of work—**mechanical solidarity** is the foundation for social unity. A society based on mechanical solidarity achieves social unity through a consensus of beliefs, values, and norms; strong social pressures for conformity; and dependence on tradition and family. In this type of society, which is best observed in small, nonliterate societies, people tend to behave, think, and feel in much the same ways, to place the group above the individual, and to emphasize tradition and family.

Modern industrial societies, in contrast, involve a very complex and differentiated division of labor. Because people perform very specialized jobs, they are dependent on others and must cooperate with them. Whereas farmers in a simple society are largely self-sufficient, members of an industrial society depend on a variety of people to fulfill their needs—barbers, bakers, manufacturers, and other suppliers of services. This modern industrial society is based on **organic solidarity**. It achieves social unity through a complex of specialized statuses that makes members of a society interdependent. Dependence and need for cooperation replace the homogeneity of beliefs, values, and norms characteristic of simpler societies.

The term *organic solidarity* is based on an analogy with biological organisms. If a biological organism composed of highly specialized parts is to survive, its parts must work together. Similarly, the parts of a society based on organic solidarity must cooperate if the society is to survive. In modern societies, individualism tends to replace strict group conformity; people do not share all ways of thinking, feeling, and behaving; and the importance of kinship declines.

In this same vein, anthropologist Robert Redfield (1941) distinguished between a folk society and an urban society. A **folk society** rests on tradition, cultural and social consensus, family, personal ties, little division of labor, and an emphasis on the sacred. In an **urban society**, social relationships are impersonal and contractual; the importance of the family declines; cultural and social consensus is diminished; economic specialization becomes even more complex; and secular concerns outweigh sacred ones.

Postindustrial Society

Some societies, including the United States, have passed beyond industrial society into **postindustrial society**. The postindustrial economic base is grounded more in service industries than in manufacturing industries and relies on expertise in production, consumption, and government (Kumar 1995).

Theoretical knowledge is the key organizing feature in postindustrial society. Contrast the relatively high theoretical knowledge requirements of this plant worker at his computer with the minimal theoretical knowledge required of the young girl and woman in the early twentieth-century North Carolina cotton mill shown in the preceding photograph.

The Dark Side of a Bright Technology

Computers and the Internet are vital to the development of postindustrial society. According to one business visionary, one of the most important changes that will occur in the postindustrial workplace is the "virtual organization." It will be fueled by electronic technology, including digital technologies, wireless transfer of information, computer networks, and telecommuting (Barner 1996). Most of the writing about technology in postindustrial society has been distinctly utopian, forecasting a new social order with such benefits for employees as higher job satisfaction and greater social equality (Nelson and Cooperman 1998).

If this optimistic view of new technology is correct, workers in high-tech jobs should be much happier than employees doing low-tech work. However, in a survey of 1,509 workers in California's Silicon Valley (an area where high-tech industry is concentrated), researchers found no difference in job satisfaction between employees in high-tech companies and those in more traditional manufacturing firms. They also found that there are still large social class differences within the workplace. These findings challenge the belief that work in high-tech society will be more satisfying and economically fair (Gamst and Otten 1992).

Nelson and Cooperman (1998) go further. They cite the destabilizing effect of innovations on organizations. When a new method of production is invented, for example, risk and uncertainty increase. To avoid risk and uncertainty, organizations attempt to transfer any potential burden to others. To protect themselves from financial loss, for example, businesses often release employees from permanent, full-time positions, then rehire them as temporary or part-time employees without fringe benefits such as medical insurance and pension plans.

Further complicating matters for employees are continuously changing job descriptions. In this new work environment, employees are forced to learn new skills and upgrade present skills throughout their careers; lifelong education is the key to economic survival. And managers are not immune to these technologically induced changes. Experts predict that over the next ten years, there will be a dramatic increase in the use of electronic learning systems designed to assist in decision making, teaching job skills, and monitoring work performance. If this occurs, functions that managers have been performing for centuries will be handled by technology rather than humans.

Electronic learning systems can also make employees feel helpless, manipulated, and exploited. Because electronic learning systems record and evaluate employee job performance in minute detail, many workers feel that their managers are spying on them, constantly looking over their shoulders (Barner 1996). Recalling the discussion in Chapter 2 about the ethics of researchers using video cameras, you might anticipate your response if your actions at work were being monitored by a computer, creating a record of your behavior that could be replayed and minutely reexamined. In fact, there has actually been a dramatic rise in workplace surveillance. Over two-thirds of U.S companies now engage in electronic cybersnooping of employees, reviewing e-mail, examining computer files, and documenting Web sites visited each day (Naughton 1999).

Thus, digitally based technology offers many benefits, boosting productivity and creating many new employment opportunities. But like any technology that has wide-ranging effects on society, there are some undesirable consequences. Postindustrial societies are just beginning to deal with the dark side of a very bright technology.

Analyzing the Trends
Which theoretical perspective do you think underlies this research and speculation? Indicate specific features of the research to support your conclusion.

What are the major features of postindustrial society? Daniel Bell (1999) identifies five major features of postindustrial society:

1. *For the first time, the majority of the labor force is employed in services rather than in agriculture or manufacturing.* Today, the United States is the only country in the world in which more than half of all employment is in services—trade, finance, transportation, health, recreation, research, and government. In fact, about 75 percent of all employed workers in the United States are in service jobs.
2. *White-collar employment replaces much blue-collar work.* The shift to a service economy in postindustrial society—with an emphasis on office work—leads to a preponderance of white-collar workers. They outnumbered blue-collar workers in the United States for the first time in the mid-1950s, and the gap is still increasing. The most rapid growth among white-collar workers has been in professional and technical employment.

3. *Theoretical knowledge is the key organizing feature.* In postindustrial society, knowledge is used for the creation of innovations as well as for the formulation of government policy (computers permit economic forecasting so that various economic theories can be tested). As theoretical knowledge becomes more important, so do educational and research institutions.
4. *Through new means of technological forecasting, society can plan and control technological change.* In industrial society, technological change is uncontrolled. That is, the effects of a technological innovation are not assessed before its introduction. The pesticide DDT was introduced as a benefit to agriculture before its harm to living things had been determined. The internal combustion engine contributes to our affluence and economic growth, but contaminates the environment. Technology is more intensely scrutinized in postindustrial society. In the race to provide the automobile industry a lighter and cheaper metal, the aluminum industry claimed that a 1-ton increase of aluminum in place of steel would reduce carbon dioxide pollution by 20 tons over the life of an average car. This claim was successfully challenged by a scientific study showing that it would take thirty-two to thirty-eight years of driving aluminum-intensive vehicles to offset the pollution released by the production of aluminum required to build those vehicles (Chamberlain 1999). Technological assessment in postindustrial society will permit us to consider the effects—good and bad—of an innovation before it is introduced.
5. *Intellectual technology dominates human affairs.* A new intellectual technology dominates in postindustrial society, much as production technology has dominated industrial society for the past 150 years. With "intellectual technology," mathematical-type problem-solving rules can, in many cases, replace human judgment. With modern computers, it is possible to consider, simultaneously, a large number of interacting variables. This capability allows us to manage the large-scale organizations that prevail in postindustrial society. Intellectual technology enables complex organizations—including government at national, state, and local levels—to set rational goals and to identify the means for reaching them.

These are some of the most general features of postindustrial society. Considerable detail on the nature of politics and work in postindustrial society is presented in Chapter 14, "Political and Economic Institutions."

What effect has the postindustrial transformation had on the stability of society? According to Francis Fukuyama (1999a, 1999b), the United States and other developed nations have undergone a thirty-year economic transition equivalent in magnitude to the Industrial Revolution. This economic transition, he asserts, has brought with it social instability.

Fukuyama identifies several signposts of "the great disruption" in a number of deteriorating social conditions, starting in the mid-1960s:

> *Crime and social disorder began to rise, making inner-city areas of the wealthiest societies on earth almost uninhabitable. The decline of kinship as a social institution, which has been going on for more than 200 years, accelerated sharply in the second half of the twentieth century. Marriages and births declined and divorce soared; and one out of every three children in the United States and more than half of all children in Scandinavia were born out of wedlock. Finally, trust and confidence in institutions went into a forty-year decline.* (Fukuyama 1999a:55)

Does Fukuyama expect social instability to lessen or increase? Currently, Fukuyama sees indications of a return to social stability as the great disruption runs its course. The establishment of new social norms, he believes, is reflected in the slowing down of increases in divorce, crime, distrust, and illegitimacy. In the 1990s, Fukuyama notes, many developed societies have even seen a reversal of these rates; crime, divorce, illegitimacy, and distrust have actually declined:

> *This is particularly true in the United States, where levels of crime are down a good 15 percent from their peaks in the early 1990s. Divorce rates peaked in the early 1980s, and births to single mothers appear to have stopped increasing. Welfare rolls have diminished almost as dramatically as crime rates, in response both to the 1996 welfare-reform measures and to opportunities provided by a nearly full-employment economy in the 1990s. Levels of trust in both institutions and individuals have also recovered significantly since the early 1990s.* (Fukuyama 1999a:80)

Fukuyama finds the roots of a return to social order in two places. For one thing, human nature leads us away from social disorder toward social order. What's more, human reason permits us to create solutions to problems of social disorder. Thus, humans are geared by nature and by reason away from Durkheim's condition of anomie (normlessness): "The situation of normlessness . . . is intensely uncomfortable for us, and we will seek to create new rules to replace the old ones that have been undercut" (Fukuyama 1999a:76). Because culture can be changed, it can be used to create new social structures better adapted to changing social and economic circumstances.

FEEDBACK

1. A __________ is the largest and most self-sufficient group in existence.
2. The __________ society is the oldest known solution to the subsistence problem.
3. The __________ society solved the subsistence problem primarily through the domestication of plants.
4. The transition from horticultural to agricultural society was made possible largely through the invention of the __________.
5. __________ occurs when one social structure divides into two or more social structures that operate more successfully separately than the one alone would under the new circumstances.
6. A society based on __________ solidarity achieves social unity through a complex of highly specialized roles that makes members of a society dependent on one another.
7. Which of the following is *not* one of the major features of postindustrial society?
   a. emphasis on theoretical knowledge
   b. employment of the majority of the labor force in service industries
   c. reliance on intellectual technology
   d. increased dependence on skilled blue-collar workers
   e. shift toward the employment of white-collar workers

Answers: 1. society 2. hunting and gathering 3. horticultural 4. plow 5. Structural differentiation 6. organic 7. d

SUMMARY

1. Relationships among individuals are patterned. This underlying pattern of social relationships is called social structure. Social scientists have developed the concepts of status, role, role performance, and social interaction to explain the existence of social structure.
2. A status is a position that a person occupies within a social structure. Individuals occupying interrelated statuses usually behave toward one another in orderly and predictable ways. Some statuses are ascribed, or assigned, to people; other statuses are achieved, or earned. Still other statuses (master statuses) are so significant that they affect most other aspects of a person's life.
3. Roles—the glue that binds a network of social statuses—are the culturally defined rights and obligations attached to social statuses. Rights inform one person of the behavior that can be expected from another person; obligations inform individuals of the behavior that others expect from them. The rights of one status correspond to the obligations of another status.
4. Role performance occurs when roles are put into action. Role performance takes place through social interaction. Role conflict occurs when the performance of a role in one status clashes with the performance of a role in another status. Role strain occurs when the roles of a single position are inconsistent. Role conflict and role strain, which may lead to discomfort and confusion, can be reduced by setting priorities and segregating roles. Because of role conflict and role strain, an individual's conformity to all roles is impossible. This does not cause a problem so long as roles are carried out within certain tolerance limits.
5. A society is composed of a people living within defined territorial borders, sharing a common culture. The way a society solves the problem of subsistence heavily influences the society's culture and social structures. Historically, societies have become larger and more complex as the means for solving subsistence problems have improved. The major types of societies are hunting and gathering, horticultural, pastoral, agricultural, industrial, and postindustrial.
6. Although both the horticultural and agricultural revolutions brought significant change, the greatest transformation came with the Industrial Revolution. The leap from preindustrial to industrial society was so great that many scholars have been able to isolate the basic characteristics distinguishing these two types of societies.
7. Postindustrial society is on the rise. This type of society has a predominately white-collar labor force with the following features: concentration in service industries; base of technical knowledge; reliance on technical forecasting; and dependence on computer technology for organizing and making decisions.

REVIEW GUIDE

INFOTRAC® COLLEGE EDITION

http://www.infotrac-college.com/wadsworth

Another unique option available to you at the Student Resources section of the companion Web site is Infotrac College Edition, an online library with access to hundreds of scholarly and popular periodicals. Below are suggested search terms for this chapter. Results from these and other searches are found at the site.

Search keyword: **gemeinschaft**. Find two articles that use this term in a sociological perspective. Provide examples of how the authors used this term in their articles. Do you think the authors' portrayals are reasonable?

Search keywords: **postindustrial society**. Search Infotrac for three articles that discuss the social implications of postindustrial society. Do the authors foresee significant benefits to society, or potential problems?

Search keywords: **social structure**. Social structure is patterned, recurring social relationships. What are some of the elements of this structure, as reported in articles you find?

LEARNING OBJECTIVES REVIEW

- Explain what sociologists mean by social structure.
- Distinguish and illustrate the basic building blocks of social structure.
- Define the concept of society.
- Describe the means of subsistence practiced by the major types of preindustrial societies.
- Discuss the characteristics of industrial society.
- Compare and contrast preindustrial, industrial, and postindustrial societies.

CONCEPT REVIEW

Match the following concepts with the definitions listed below them.

____ a. society
____ b. urban society
____ c. formal organization
____ d. industrial society
____ e. organic solidarity
____ f. achieved status
____ g. rights
____ h. hunting and gathering society
____ i. social structure
____ j. role strain
____ k. master status
____ l. structural differentiation
____ m. gesellschaft
____ n. folk society
____ o. role performance

1. a slot within a social structure occupied because of an individual's efforts
2. a group deliberately created for the achievement of one or more specific goals
3. a society that solves the subsistence problem through hunting animals and gathering edible fruits and vegetables
4. roles informing individuals of the behavior that can be expected from others
5. when one social structure divides into two or more social structures that operate more successfully separately than the one alone would under the new circumstances
6. a society based on tradition, cultural and social consensus, family, personal ties, little division of labor, and an emphasis on the sacred
7. the actual conduct involved in putting roles into action
8. occurs when the roles of a single status are inconsistent with one another
9. patterned recurring social relationships among individuals and groups
10. people who live within defined territorial borders, sharing a common culture
11. a status that affects most other aspects of a person's life
12. Tönnies's term for the type of society characterized by weak family ties, competition, and impersonal social relationships
13. a society whose subsistence is based primarily on the application of science and technology to the production of goods and services
14. social unity based on a complex of highly specialized statuses that makes members of a society dependent on one another
15. a society in which social relationships are impersonal and contractual, kinship is deemphasized, cultural and social consensus are not complete, the division of labor is complex, and secular concerns outweigh sacred ones

CRITICAL-THINKING QUESTIONS

1. Suppose that a college friend of yours wants to know the meaning of the term *social structure*. Use the stage analogy to develop an understandable answer. Use examples.

2. Have you experienced role conflict or role strain lately? If so, describe the situation. If not, explain why you have been immune to role conflict and role strain, making clear the meaning of the concepts.

3. Discuss some of the basic distinguishing features of preindustrial and industrial societies. Illustrate your answer.

____

____

4. In what ways will your work life be different in a postindustrial society than in an industrial society?

____

____

MULTIPLE-CHOICE QUESTIONS

1. _________ is defined as patterned, recurring social relationships.
   a. Informal structure
   b. Social structure
   c. Constructive structure
   d. Anomic structure
   e. Symbolic interaction

2. The type of status that is neither earned nor chosen but is assigned to us is called
   a. achieved status.
   b. ascribed status.
   c. acquired status.
   d. assigned status.
   e. symbolic status.

3. The president of a corporation is an example of a/an
   a. achieved status.
   b. ascribed status.
   c. acquired status
   d. assigned status.
   e. symbolic status.

4. _________ are culturally defined rights and obligations attached to statuses, indicating the behavior expected of individuals holding them.
   a. Norms
   b. Values
   c. Mores
   d. Roles
   e. Beliefs

5. Rights inform individuals of behavior they can expect from others; _________ inform individuals of the behavior others can expect from them.
   a. exchange principles
   b. roles
   c. statuses
   d. cues
   e. obligations

6. _________ occurs when the performance of a role in one status clashes with the performance of a role in another status.
   a. Role conflict
   b. Role strain
   c. Role distancing
   d. Role conformity
   e. Role deformity

7. By ranking incompatible roles based on their importance to us, we can reduce role _________ and _________.
   a. conflict; strain
   b. conflict; behavior
   c. conflict; ambiguity
   d. strain; ambiguity
   e. ambiguity; distancing

8. A/An _________ is composed of a people living within defined territorial borders, sharing a common culture.
   a. society
   b. institution
   c. community
   d. social group
   e. formal organization

9. The _________ society solved the subsistence problem through the application of science and technology to the production of goods and services.
   a. hunting and gathering
   b. horticultural
   c. agricultural
   d. industrial
   e. postindustrial

10. Emile Durkheim used the term _________ to describe social unity that is achieved through a consensus of beliefs, values, and norms; strong social pressure for conformity; and a dependence on tradition and family.
   a. organic solidarity
   b. mechanical solidarity
   c. gesellschaft society
   d. gemeinschaft society
   e. traditional society

REVIEW GUIDE

FEEDBACK REVIEW

Fill in the Blank

1. __________ refers to the patterned relationships among individuals and groups.
2. __________ are culturally defined rights and obligations attached to statuses.
3. The rights of one social status correspond to the __________ of another one.
4. __________ occurs when the roles of a single status are inconsistent.
5. A __________ is the largest and most self-sufficient group in existence.
6. The transition from horticultural to agricultural society was made possible largely through the invention of the __________.
7. A society based on __________ solidarity achieves social unity through a complex of highly specialized roles that makes members of a society dependent on one another.

Multiple Choice

8. **Which of the following is *not* one of the differences between the stage and social life?**
   a. There is considerably more discrepancy between roles and role performance in social life than between a stage script and its performance.
   b. Unlike the stage, there are no cues and responses in real life.
   c. Role performance in real life is not the conscious process that actors go through on the stage.
   d. In social life, the cues and responses are not as programmed and predictable as on the stage.
9. **Which of the following is *not* one of the major features of postindustrial society?**
   a. emphasis on theoretical knowledge
   b. employment of the majority of the labor force in service industries
   c. reliance on intellectual technology
   d. increased dependence on skilled blue-collar workers
   e. shift toward the employment of white-collar workers

Matching

10. Match the type of status with the examples that best illustrate it.
    ____ a. ascribed status
    ____ b. achieved status
    ____ c. master status
    ____ d. status set
    ____ e. role
    ____ f. role performance
    ____ g. social interaction
    (1) basketball coach, mother, author, daughter, professor
    (2) occupation, gender, race, ethnicity
    (3) president of the United States
    (4) sex
    (5) a husband and wife discuss disciplining one of their children
    (6) a university president hands out diplomas at a graduation ceremony
    (7) a mother is expected to take care of her children

ANSWER KEY

Concept Review

a. 10
b. 15
c. 2
d. 13
e. 14
f. 1
g. 4
h. 3
i. 9
j. 8
k. 11
l. 5
m. 12
n. 6
o. 7

Multiple Choice

1. b
2. b
3. a
4. d
5. e
6. a
7. a
8. a
9. d
10. b

Feedback Review

1. Social structure
2. Roles
3. obligations
4. Role strain
5. society
6. plow
7. organic
8. b
9. d
10. a. 4
    b. 3
    c. 2
    d. 1
    e. 7
    f. 6
    g. 5

6

Groups and Organizations

© Jose Luis Pelaez, Inc./CORBIS

OUTLINE

LEARNING OBJECTIVES

- Define the concept of the group and differentiate it from social categories and social aggregates.
- Outline the basic characteristics of primary and secondary groups.
- Describe the five major types of social interaction.
- Discuss the nature of group processes: task accomplishment, leadership styles, and decision making.
- Define the concept of formal organization and identify the major characteristics of bureaucracy.
- Discuss the advantages and disadvantages of bureaucracy.
- Distinguish between formal and informal organization.
- Describe the iron law of oligarchy and demonstrate its importance with examples.
- Discuss prejudice and discrimination in organizations.
- Compare and contrast the major features of Western organizations and Japanese organizations.

USING THE SOCIOLOGICAL IMAGINATION

Most people assume that conflict should be avoided because it is disruptive and interferes with group effectiveness. Conflict does have some negative consequences. Not as well known, however, are some social benefits of conflict. Giving full consideration to the different opinions in a group, for example, can prevent a group from being a victim of the self-deceptive belief that the majority is right. In other words, openness to disagreement and conflict can prevent groupthink.

The space shuttle *Challenger* is an excellent example of a disastrous decision that was based on conformity to group ideas. The *Challenger* was launched from Kennedy Space Center on January 28, 1986. Just over a minute after the launch, the *Challenger* disintegrated, and in the instant of explosion, took the lives of all seven astronauts on board.

As is true for all space mission personnel, the *Challenger* team was composed of a number of specialists. The engineers on the team had recommended against takeoff that morning because crucial parts had never been tested at so low a temperature. As victims of groupthink, the NASA leaders screened out this opposition; they acted as if the engineers lacked the intelligence and courage to make the "right" decision. The leaders downgraded the opinions of the engineers and convinced most of the group that the decision, except for the engineers, was unanimous. By avoiding consideration of a dissenting view, the majority lost the shuttle's passengers and harmed NASA's long-term objectives (Moorhead, Ference, and Neck 1991).

Concept of the Group

Americans are socialized to think in individualistic terms. In fact, one of our country's values, as noted in Chapter 4 ("Socialization Over the Life Course"), is individuality. This individualistic orientation is often a stumbling block to those first introduced to the concept of the group. Perhaps it will ease the transition into thinking about groups (rather than individuals) if we begin by focusing on what is thought of as an individualistic activity—but in fact is a group endeavor. Art is a likely candidate for this purpose.

Think of painting and you are likely to have images of Rembrandt, Renoir, or Picasso. The art of ballet will call to mind Nureyev, Baryshnikov, or Kirkland. It may be Gershwin, Sondheim, or Webber who comes to the fore when the subject of musical achievement on Broadway is the topic. In short, when we think of art, we think of individual artists. Without denigrating the talent, hard work, and achievements of individual artists whose works are precious to us, Howard Becker wants to underscore the fact that even such an allegedly individualistic pursuit as art "involves the joint activity of a number, often a large number of people" (Becker 1982:1).

Becker begins his book *Art Worlds* with this story told by Anthony Trollope, the famous nineteenth-century English novelist:

> *It was my practice to be at my table every morning at 5:30 a.m.; and it was also my practice to allow myself no mercy. An old groom, whose business it was to call me, and to whom I paid £5 a year extra for the duty, allowed himself no mercy. During all those years at Waltham Cross he was never once late with the coffee which it was his duty to bring me. I do not know that I ought not to feel that I owe more to him than to any one else for the success I have had. By beginning at that hour I could complete my literary work before I dressed for breakfast.* (Becker 1982:1)

Trollope's groom is one member of the complex network of social relationships that permitted this artist to accomplish his work. Moreover, the activities of both Trollope and his groom had to mesh with those of many other people (printers, publishers, and editors, to mention only a few) if the literary works of this Victorian writer were ever to become available to the public. Some of these relationships were personal and face-to-face; others were not. This chapter elaborates on the variety of group relationships and the contexts in which they appear throughout social life.

A **group** is composed of people who are in contact with one another; share some ways of thinking, feeling, and behaving; take one another's behavior into account; and have one or more interests or goals in common. Because of their characteristics, groups play an important role in the lives of their participants as well as influencing the societies in which they exist.

Types of Groups

Categories, Aggregates, and Groups

A group should be distinguished from a **social category**—people who share a social characteristic. A taxpayer, a woman, and a college graduate each belong to a social category. A group is also sometimes confused with a **social aggregate**—people who happen to be at the same place at the same time, such as students waiting in line for concert tickets. Although neither categories nor aggregates are groups, some of their members may form groups. Victims of a disaster (an aggregate)—such as residents of a town devastated by a tornado—may work together to cope with the emergency. Citizens of a state (a social category) may band together in an organized tax revolt. These people may form a group if they begin to interact regularly; share ways of thinking, feeling, and behaving; take one another's behavior into account; and have some common goals.

Primary Groups

Primary and secondary groups are two principal types of groups. At the extremes, the characteristics of these two types of groups—and the relationships that occur within them—are opposites. But most groups sit at different points along a continuum from primary to secondary.

Why is a primary group described as tightly integrated? A **primary group** is composed of people who are emotionally close, know one another well, and seek one another's company. The members of a primary group have a "we" feeling and enjoy being together. These groups are characterized by relationships that are intimate, personal, caring, and fulfilling.

Charles Horton Cooley, one of the founders of symbolic interactionism, coined the term primary group. Primary groups are the most important setting for socialization. The family, neighborhood, and childhood play groups, Cooley observed, are primary because they are the first groups an infant experiences. People, of course, participate in primary groups throughout life. Close friends in high school and college, neighbors who keep an eye on one another's children, and friends who meet weekly for golf are examples of primary groups.

© Jose Luis Pelaez, Inc./CORBIS

Human accomplishments, even those attributed to individuals such as playwright Shakespeare or pianist Vladimir Horowitz, rest on the support of groups. The school newspaper these students are working on is produced by a complex web of social relationships.

What conditions promote the development of primary groups? A number of conditions favor the development of primary groups and primary relationships. Although primary relationships may occur in their absence, the likelihood of having a primary relationship is increased if these conditions exist.

- *Small group size.* It is difficult for the members of large groups to spend enough time together to

TABLE 6.1

Characteristics of Primary and Secondary Groups

You belong to both primary and secondary groups. Select one of each, and describe your experience with the functions it is supposed to perform.

	Nature or Relationship	Function	Examples
Primary Group	• Close and continuous social interaction	• Provides emotional support • Contributes to socialization process • Promotes conformity and social control	• Family • Soldiers in combat • Street gang
Secondary Group	• Impersonal social interaction	• Aids achievement of group goals	• Aerobics class • NFL football team • Telecommunications law firm

develop close emotional ties; the chances of knowing everyone fairly well are far greater in small groups. Thus, members of a bridge foursome are more likely to develop primary relationships than are members of a metropolitan chamber of commerce, even when both groups meet regularly.

- *Face-to-face contact.* People have maintained primary relationships despite separation by war, prison, or residential changes and have even established long-distance primary relationships through e-mail or telephone conversations, but primary relationships occur more easily when interaction is face-to-face. People who can see each other and who can experience such nonverbal communication as facial expressions, tone of voice, and touch are much more likely to develop close ties.
- *Continuous contact.* The probability of developing a primary relationship also increases with prolonged contact. Intimacy rarely develops in a short period of time. In spite of reported love at first sight, most of us require repeated social contact for development of a primary relationship.
- *Proper social environment.* The development of primary relationships is also affected by the social environment in which interaction occurs. If individuals are expected to relate to one another strictly on the basis of status or role, primary relationships are unlikely to develop. Total personalities are usually not considered, and personal concern for one another does not seem appropriate. Lawyers who see judges in court face-to-face, in the presence of a small number of people, and over a long period of time are nevertheless unlikely to develop primary relationships with them. Forming primary relationships when unequal statuses are involved is always difficult. This is the reason primary relationships do not usually develop between students and professors, bosses and employees, or judges and lawyers.

What are the functions of primary groups? First, primary groups provide emotional support through caring, personal, and intimate relationships. During World War II, the German army refused to disintegrate despite years of being outnumbered, undersupplied, and outfought. These conditions should have led to desertion and surrender, but they did not. According to Edward Shils and Morris Janowitz (1948), strong primary relationships within German combat units accounted for the Germans' stability and resilience against overwhelming odds. Although social cohesion developed among American soldiers during the Korean conflict, it was not the same as during World War II. Because of special combat conditions and a personnel rotation system, "the basic unit of cohesion was a two-man relationship rather than one that followed squad or platoon boundaries" (Janowitz 1974:109).

Second, primary groups contribute to the socialization process. For infants, the family is the primary group that provides the emotional support necessary for the development of both an integrated personality and a sense of self. The family also conveys information about culture so that children learn how to participate in social life. In addition, primary groups promote adult socialization as individuals adjust to new and changing

SOCIOLOGY AND THE NEWS MEDIA

Primary Relationships in Blended Families

A *blended family* is formed when a remarriage occurs involving at least one child from a previous marriage. Because a family is characterized by close, personal, and intimate relationships, it is a *primary* group. Members of a primary group usually like to be together and have a feeling of unity. As this CNN report demonstrates, conflict within blended families often prevents the development of such feelings.

1. Identify a problem that the blended family poses for the socialization of children.

____

____

2. Does a blended family that is not blending well actually constitute a secondary group? Why or why not?

____

____

An important function of primary groups is to provide emotional support for its members. Marital conflict, as in this blended family, can threaten the psychological well-being of its members.

social environments—as they enter college, take new jobs, change social classes, marry, and retire. Primary groups make these adjustments less painful because membership in them helps people fit into new social situations. Frequent interaction with people who demonstrate genuine concern for others aids adjustment to a new social setting.

Finally, primary groups promote conformity and contribute to social control. The stability and perpetuation of a society depend on the members' acceptance of the society's norms and values. Unless most of the members of a society support the norms and values, that society cannot continue to exist in its present form. Primary groups teach new members to accept norms and values and provide group pressure to promote conformity to them. William F. Whyte's (1993) study of an Italian slum gang illustrates social control within primary groups. Whyte found that rank within this gang determined how well the members performed in sports. One member of the gang, a former semiprofessional baseball player, was one of the gang's best players. Yet, after losing status in the gang, he could no longer play very well. According to Whyte, the man indicated that he would play baseball better when gang members were not involved. Whyte also reported that bowling scores corresponded with status in the gang—the higher the rank, the higher the score. If a lower-ranked member began to bowl better than those above him, the others used verbal remarks—"You're bowling over your head" or "How lucky can you get?"—to remind him that he was stepping out of line.

© Bob Daemmrich/Stock, Boston

People in primary groups are emotionally close, know each other well, enjoy each other's company, and share a feeling of "we-ness." The family is the most important primary group to which we belong.

Secondary Groups

Why is a secondary group not an end in itself? A **secondary group** exists to accomplish a specific purpose. Unlike a primary group, a secondary group is impersonal and goal oriented; it involves only a segment of its members' lives. Work groups, volunteer groups formed during disasters, and environmental organizations are all examples of secondary groups. The relationships of the group members are **secondary relationships**—impersonal interactions involving only limited parts of their personality. Interactions between salespersons and customers, employers and workers, and dentists and patients are secondary relationships.

Secondary relationships are not necessarily unpleasant. But the purpose of the group is to accomplish a task, not to enrich friendships. In fact, if friendship becomes more important than the task, a secondary group may become ineffective. If the members of a college basketball team become more interested in the emotional relationships among themselves or with their coach than in playing their best basketball, their play on the court could suffer.

Do secondary groups ever include primary relationships? Although primary relationships are more likely to occur in primary groups and secondary relationships in secondary groups, there are a number of exceptions. Many secondary groups are settings for primary relationships. Members of a work group may relate to one another in a manner that is personal, demonstrate genuine concern for one another as total personalities, and have relationships that are fulfilling in themselves. Similarly, members of a primary group occasionally engage in secondary interaction. One family member may, for example, lend money to another member of the family at a given interest rate with a specific repayment schedule.

Reference Groups

How do we use groups as a point of reference? We use certain groups to evaluate ourselves and to acquire attitudes, values, beliefs, and norms. Groups used in this way are called **reference groups**. Reference groups may include families, teachers, college classmates, student government leaders, college Greek organizations, rock groups, or professional football squads.

Reference groups influence self-esteem and behavior. Say, for instance, that because of the imagined reaction by reference groups (such as parents and teachers), a student is motivated to study. The resulting grade point average creates a sense of accomplishment and boosted self-esteem.

You may have a reference group without being a member; you may only aspire to be a member. For

SNAPSHOT OF AMERICA 6.1

Size of State Bureaucracies

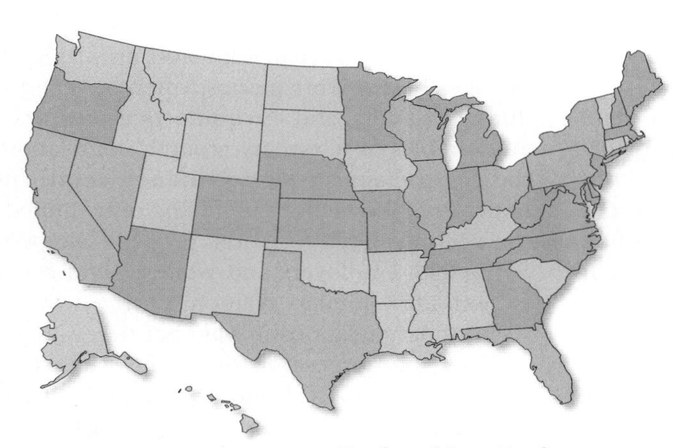

Source: U.S. Bureau of the Census, 1999.

Americans widely vilify the federal bureaucracy for being too large. In fact, it is smaller on a per capita basis than the smallest state bureaucracies. For every 100 Americans, there is 1 federal worker. Overall, the least densely populated states have the largest government bureaucracies on a per capita basis. Hawaii, for example, has 5 state workers per 100 residents.

1. Compare the size of your state government with the federal government. Are you surprised? How does this make you feel about your state government? Federal government?
2. Explain why the least populous states, particularly those that are larger geographically, have the highest ratio of state workers per capita.

example, you need not be a member of a ballet troupe to view dancers as a reference group. You need only assess yourself in terms of their standards and subscribe to their beliefs, values, and norms. Junior high school girls, for example, may imitate high school girls' leadership style or athletic interests. Junior high school boys may copy high school boys' taste in clothing and music.

Reference groups need not be positive. Observing the behavior of a group you dislike may reinforce a preference for contrary ways of acting, feeling, and behaving. For example, a violent gang may provide a blueprint for behavior to avoid.

In-Groups and Out-Groups

How is social life an interplay between in-groups and out-groups? In-groups and out-groups can exist only in relation to each other; one is the flip side of the other. An **in-group** encourages intense identification and loyalty. The level of identification and loyalty is sufficiently compelling that members tend to exclude others. An **out-group** is a group toward which in-group members feel opposition and competition. It is from membership in these groups that people divide into "we" and "they."

Where are these groups found? In-groups and out-groups develop in virtually every social arena—for example, college Greek organizations, athletic teams, political parties, racially or ethnically divided neighborhoods, and countries at war. Because of different tastes, perspectives, and values, everyone prefers some groups over others. And by virtue of being in some groups, you are out of others. As a college student, you can easily identify many of the in-groups and out-groups in your social landscape. For example, jocks and cheerleaders are in-groups for some, out-groups for others. Phi Beta Kappa is an in-group for its members and an out-group for students uninterested in academ-

Each of the participants in this massive office party is part of a broader social network. For each person, his or her relationships in this organization are merely one part of their large social network, which may include family members, friends, golf partners, and psychotherapy groups.

ics. At Columbine High School, students knew about a number of cliques beyond the publicized jocks and cheerleaders—preppies, stoners, gangbangers, skaters, nerds. Different schools have variations of these in- and out-groups. At Austin High School in Texas, "kickers" wear oversize belt buckles, large hats, and cowboy boots. California schools have surfer cliques (A. Cohen 1999). A *we* versus *they* mentality can even develop between generations. Baby boomers (Americans born between 1945 and 1964), for instance, are accused of bashing Generation X (America's twenty-somethings) as unambitious complainers who want the good life handed to them (Giles 1994).

The formation of in-groups and out-groups depends on the establishment and protection of group boundaries. If nothing distinguishes "us" from "them," then there can be no "ins" and "outs." Group boundaries remind in-group members that movement over the line places them on the outside. To the outsiders, group boundaries form an entrance barrier (Hogg and Abrams 2001).

How are group boundaries maintained? Maintaining something as important as group boundaries requires commitment from the group members. Unfortunately, this may involve clashes with outsiders. Skinheads (neo-Nazis) periodically attack Jews or destroy their property. Members of an urban gang may injure or kill a member of another gang who has entered their territory. Boundaries are also maintained through group symbols. Symbols may be as benign as a right-to-life bumper sticker or as intentionally threatening as having one's hair shaved. Groups also may adopt certain words and rituals (handshake, high five) to demarcate themselves. These are easily recognized at a social club initiation ceremony.

As this description of group boundaries suggests, in-group and out-group relationships can have serious consequences. Let's consider some of these repercussions.

What is the sociological importance of in-groups and out-groups? Membership in an in-group has both positive and negative aspects. Those on the inside may benefit from heightened self-esteem and a sense of social identity. However, in-group members may foster an inflated sense of their own worth while promoting an unrealistically negative view of others. This distortion of reality, which in-groups often deliberately create, is self-deluding. If such self-delusion becomes central to the identity of in-group members, psychological and social development is retarded. Moreover, some members may be personally damaged on realizing the false basis for their self-esteem.

Harm is not confined to in-group members. In fact, out-group members experience many more negative effects. The increase in self-esteem enjoyed by in-group members can come at the expense of outsiders. Even more serious is the physical damage that may be done by extremist groups such as neo-Nazis or certain religious factions willing to kill others in the name of their deity (O'Brien and Palmer 1993). Historically, racial or ethnic majorities have viewed minorities as outsiders—effectively denying them everything from a sense of worth to equal opportunity (Schaefer 2001). Males have been accused of using a similar pattern of dominance against females (Andersen 2002).

© Steven Rubin/The Image Works

In a 1999 conflict in Seattle, anti-WTO demonstrators intended to stop the World Trade Organization from implementing certain treaty provisions.

In-group–out-group relationships, of course, generally do not take such extreme forms. Spirited competition among sororities and friendly athletic rivalries are more typical expressions of these groups. The relative harmlessness of most cases, however, should not mask the potentially serious consequences of in-groups and out-groups, consequences related to conflict, power, and self-respect.

Social Networks

What are social networks? As individuals and as members of primary and secondary groups, we interact with many people. All of a person's social relationships constitute his or her **social network**—a web of social relationships that joins a person to other people and groups. The typical American knows between 500 and 2,500 people, and many of them know each other (Milgram 1967). This total set of social relationships—this social network—includes family members, work colleagues, classmates, physicians, church members, close friends, car mechanics, and store clerks. Social networks tie us to hundreds or thousands of people within our community, throughout the country, and even around the world (Doreian and Stokman 1997).

The Internet is expanding the amount of interaction and the flow of information within networks. Before the Internet, for example, environmental activists across the United States had to depend on slower, more cumbersome means of communication, such as the print media, the telephone, and letter writing. With the Internet, environmental organizations can supply almost unlimited information to anyone with access to the Web. For example, volunteers can now easily recruit others to join them in addressing the Chesapeake Bay's environmental problems. Protests in various regions of the country can be organized very quickly, and feedback among network members can be instantaneous.

This increased ease, speed, and frequency of social contact can promote a sense of membership in a particular network. Whereas opponents to gun control were once largely unaware of each other, they may now feel part of a nationwide social network.

Are social networks groups? Although a person's social network includes groups, it is not a group per se. A social network lacks the boundaries of a group, and it does not necessarily involve close or continuous interaction. Because many of the relationships are sporadic and indirect, not all participants experience a feeling of membership. Although a social network does not preclude close ties—like those of college alums who maintain contact via telephone, computers, and reunions—most ties in a social network are secondary in nature.

How strong are the ties in a social network? Social relationships within a network involve both strong and weak ties (Granovetter 1973; L. C. Freeman 1992). Strong ties exist in primary relationships, such as those with one's parents or spouse. These ties are emotionally close and intimate and are based on a genuine concern for the other person(s). Weak ties, most often found in secondary relationships, include salespersons, distant relatives, and most coworkers. These ties are more impersonal and goal oriented (Mizruchi and Stearns 2001).

The number and type of strong ties are heavily influenced by level of education. More highly educated people have a larger number of strong ties, a greater variety of strong ties, and more strong ties outside the family. Not unexpectedly, older people have fewer strong ties; and outside the home, urban dwellers have more strong ties than do rural residents. Although women have as many strong ties as men, more of these ties are within the family (C. S. Fischer 1982; Marsden 1987).

What are the functions of social networks? Social networks serve several important functions. First, networks

FEEDBACK

1. A _________ is composed of several people who are in contact with one another; share some ways of thinking, feeling, and behaving; take one another's behavior into account; and have one or more interests in common.
2. A _________ group is composed of people who are emotionally close, who know one another well, who seek one another's company because they enjoy being together, and who have a "we" feeling.
3. Listed below are some examples of primary and secondary relationships. Indicate which examples are most likely to be primary relationships (P) and which are most likely to be secondary relationships (S).
   ____ a. a marine recruit and his drill instructor at boot camp
   ____ b. a married couple
   ____ c. a manager and his professional baseball team
   ____ d. professors and students
   ____ e. car salespersons and their prospective customers
4. Which of the following is *not* a condition that promotes the development of primary groups?
   ____ a. small group size
   ____ b. face-to-face contact
   ____ c. continuous contact
   ____ d. interaction on the basis of status or role
5. Match the following terms and statements:
   ____ a. group (1) skinheads
   ____ b. in-group (2) spectators at a Fourth of July fireworks display
   ____ c. social aggregate (3) orchid growers
   ____ d. social category (4) teenage subculture
   ____ e. reference group (5) members of a regular Saturday night poker game
6. A _________ is a web of social relationships that joins a person directly to other people and groups and, through those individuals and groups, indirectly to additional parties.

Answers: 1. group 2. primary 3. a. (S) b. (P) c. (S) d. (S) e. (S) 4. d. 5. a. (5) b. (1) c. (2) d. (3) e. (4) 6. social network

provide a sense of belonging (Putnam 2000). Second, they furnish support in the form of help and advice. Finally, networks are a useful tool for those entering the labor market. Getting to know relevant people is often useful to one's career (Putnam 1996; Petersen, Saporta, and Seidel 2000).

The use of networks to locate new places of employment deserves special attention because of gender differences. Although the social networks of American females and males are equally extensive, men are more enmeshed in "good-old-boy networks" containing many higher-status contacts (Lin, Ensel, and Vaughn 1981). In the past, women tended to have more ties with family members and fewer higher-status ties. This situation is changing, however, as gender inequality diminishes in the United States and as more women are seeing networks as a pathway to higher status and more powerful positions.

Social Interaction in Groups

Social interaction is crucial to groups. Robert Nisbet (1970) describes five types of social interaction basic to group life: cooperation, conflict, social exchange, coercion, and conformity.

Cooperation

What is cooperation, and when are people most likely to cooperate? **Cooperation** is a form of interaction in which individuals or groups combine their efforts to reach some common goal. Cooperation usually occurs when reaching a goal depends on united resources and efforts. The survivors of a plane crash in a snow-covered mountain range must cooperate to survive. Victims of floods, mudslides, tornadoes, droughts, or famines must help one another to survive their crisis. This is not to say that cooperation exists only during emergencies. Children who agree to a set of rules for a game, couples who agree to share household duties, students who march in support of a community project, and groups who organize to prevent a major power-line installation are cooperating. Indeed, without some degree of cooperation, social life could not exist.

Conflict

What is conflict, and are all consequences negative? Individuals or groups who work together to obtain certain benefits are cooperating. Those who work against one another for a larger share of the rewards are in **conflict**. In conflict, defeating the opponent is considered essential. In fact, defeating the opponent may become more important than achieving the goal and may bring more joy than winning the prize.

As pointed out earlier in "Using the Sociological Imagination," conflict is usually considered a disruptive form of interaction—one to be minimized or eliminated. A cooperative, peaceful society is assumed to be better than one in conflict. Although conflict does have some undesirable consequences, it can also be socially beneficial.

According to Georg Simmel (1858–1918), a major benefit of conflict is the promotion of cooperation and unity *within* opposing groups. Although conflict may lead to the separation of groups, it can strengthen relationships within groups. Thus, the Revolutionary War drew American colonists together even though it brought them into conflict with the British. Similarly, a labor union becomes more integrated during the process of collective bargaining. And police officers who attempt to stop a husband from physically abusing his wife may find themselves under attack by both mates. According to historian Stephen Oates (1977), the "outside meddling" of Stephen Douglas and other Democrats united the Illinois Republican party and secured the U.S. Senate nomination for Abraham Lincoln. The threat of an impeachment trial of President Clinton united the Democratic party during the 1998 congressional elections. And the events of nine-eleven unified Americans with increased levels of patriotism.

Another positive consequence of conflict is the rejuvenation of norms and values. Equality of opportunity, freedom, and democracy have always been important values to Americans. But until the nonviolent civil rights demonstrations, the urban violence in large northern cities, and the abuse experienced by civil rights supporters, the poverty, inequality, and lack of opportunity among America's minorities were largely invisible to white society. Once these problems were forcibly called to the public's attention, actions were taken to reaffirm these basic values. Among these actions were Lyndon Johnson's War on Poverty, the Civil Rights Act of 1964, the Equal Employment Opportunity Act of 1972, and affirmative action programs (Fineman 1992).

Conflict may also be beneficial when it challenges norms, beliefs, and values. Domestic opposition to the Vietnam War in the 1960s—which tragically involved unparalleled conflict and violence on college and university campuses—ended the military draft, brought into question the legitimacy of political and military intervention in other countries, enlarged the American value of patriotism to include opposition to certain wars, and encouraged higher education to be more responsive to students' needs. These events did alter some norms, beliefs, and values—a lesson not lost on African Americans during the 1992 riots in Los Angeles.

A study by Douglas Maynard (1985) concludes that conflict can have benefits even for young children. In a study of first-grade children, Maynard observed a sense of social structure emerging from disputes and arguments. Authority and friendship patterns were also the product of social conflict.

Social Exchange

What is the nature of a social exchange relationship? In *The Nicomachean Ethics*, Aristotle observed:

> *All men, or most men, wish what is noble but choose what is profitable; and while it is noble to render a service not with an eye to receiving one in return, it is profitable to receive one. One ought, therefore, if one can, to return the equivalent of services received, and to do so willingly.*

In this passage, Aristotle touches on **social exchange**, a type of social interaction in which one person voluntarily does something for another, expecting a reward in return. If one person helps a friend paint her house, expecting that some equivalent help will be returned, the relationship is one of exchange. In an exchange relationship, it is the benefit derived rather than the relationship itself that is central. An individual who does something for someone else obligates that person. This obligation can be repaid only by a return favor. Thus, the basis of an exchange relationship is reciprocity, the idea that you should do for others as they have done for you (P. M. Blau 1986).

© Mark Richards/PhotoEdit

Conformity is the type of social interaction in which people behave as others expect them to behave. Without conformity, society could not exist. Imagine the chaos that would ensue if these people did not remain in line at the grocery checkout counter.

Exchange relationships are not always exploitative; many exchange relationships are based on trust, gratitude, and affection. You may help someone move out of a dormitory into an apartment because he has been friendly to you. You may expect some response—help when you move, or maybe just an expression of appreciation. We usually do anticipate some response even though we may be unaware of it.

What is the difference between cooperation and exchange? Although both cooperation and social exchange involve working together, there is a significant difference between these two types of interaction. In cooperation, individuals or groups work together to achieve a shared goal. Reaching this goal, however, may or may not benefit those who are cooperating. And although individuals or groups may profit from cooperating, that is not their main objective. For example, a group may work to build and maintain an adequate supply of blood for a local blood bank without thought of personal benefit. If, however, the primary objective of the group is to guarantee blood availability for its own members, then it has an exchange relationship with the blood bank. In cooperation, the question is, How can we reach our goal? In exchange relationships, the implied question is, What's in it for me?

Coercion

In what way is coercion an involuntary interaction? In **coercion**, an individual or group compels others to behave in certain ways. The central element is domination. In a sense, coercion is an unequal exchange: Because of superior power, someone can get something from someone else without repaying. Domination may occur through physical force, such as imprisonment, torture, or death. But coercion is generally expressed more subtly through social pressure—ridicule, rejection, ostracism, withdrawal of love, denial of recognition. Coercion is a part of many social relationships. Parents may coerce children with a curfew; employers may coerce employees with dismissal; professors may coerce students to study via the threat of a test. Obviously, conflict theory underlies this type of social interaction.

Conformity

Do most people conform to group pressures? **Conformity** is behavior matching group expectations. When we conform, we adapt our behavior to fit the behavior of those around us. While some people are more likely to conform than others are, most people conform to the expectations of some group most of the time. Social life—with all its uniformity, predictability,

TABLE 6.2

Illustrating Types of Social Interaction

A type of social interaction is illustrated from the viewpoint of a particular theoretical perspective. Each interaction can be viewed from either of the other two perspectives. Associate a type of social interaction with a different theoretical perspective, and make up your own example.

Theoretical Perspective	Type of Social Interaction	Example
Functionalism	Conformity	• Team integration is promoted when basketball players accept their roles on the floor.
Conflict theory	Coercion	• Conflict in prisons is kept in check by the superior power of the guards.
Symbolic interactionism	Social exchange	• Two college students may work together in preparing for an exam because passing a course carries the same symbolic value for each of them.

and orderliness—simply could not exist without this type of social interaction. Without conformity, there could be no churches, families, universities, or governments; without conformity, there could be no culture or social structures (see Doing Research).

The Business of Groups

Nineteenth-century social philosopher Georg Simmel was one of the first to recognize the importance of groups for society. But, he argued, group members face a dilemma. Belonging to a group requires that members submit to social control, losing some personal freedom in the process. If these people remain outside the group, however, they lose such possible benefits of membership as economic security and social acceptance (Simmel 1964; originally published in 1908). For this reason, most groups encounter both cooperation and conflict. Because these pressures can interfere with the smooth functioning of the groups, social scientists have devoted considerable attention to task accomplishment within groups, various leadership styles for groups, and group decision making.

What is the nature of task accomplishment within groups? Most groups are designed to accomplish certain tasks, which can range from planning leisure activities to manufacturing a product. At the same time, members of groups have self-images and desires that do not necessarily mesh with the goals of the group as a whole. Beginning with Robert Bales (1950), social scientists have tried to understand the ways in which task accomplishment within groups could expedite both the achievement of the group's goals and the needs of individual group members.

Bales developed a scheme, called interaction process analysis, to classify group interactions into several categories. He identified twelve separate categories, which can be understood in relation to two basic problems that, according to Bales, must be solved by any social system. First, *instrumental problems* must be solved. Instrumental problems are those that are directly related to achieving the goals of the group. A group trying to elect a political candidate must solve problems regarding fund-raising, advertising, arrangements for speeches and rallies, charges from the opposing candidate, and so on. Second, *social-emotional problems* must receive attention. These are the problems involving individual satisfaction, disagreements, and other related matters that inevitably arise whenever individuals try to coordinate their activities.

© Najlah Feanny/CORBIS

Eating at McDonald's seems like a harmless activity (fat content aside). However, many sociologists think the rationality and standardization undergirding fast-food restaurants promote dehumanization.

A number of laboratory studies conducted by Bales and his associates have further indicated that both instrumental and social-emotional processes are important, as Stephen Wilson notes:

> *. . . if the group is going to do its job, (1) task-oriented interaction must be high; (2) more answers must be given than questions asked; (3) there must be more interaction contributing to the solidarity of the group (positive social-emotional) than detracting from solidarity (negative social-emotional). In other words, there must be, as Bales's theory suggests, some degree of balance to keep the individuals together as a group.* (S. Wilson 1978:66–67)

What is the nature of leadership in groups? It was once believed that leaders are born and that people have to possess certain traits to be effective as leaders. It is now known that no leadership trait is absolutely necessary for effective leadership, but Ralph Stogdill has suggested that successful leaders are likely to have the following combination of traits:

> *The leader is characterized by a strong drive for responsibility and task completion, vigor and persistence in pursuit of goals, venturesomeness and originality in problem solving, drive to exercise initiative in social situations, self-confidence and sense of personal identity, willingness to accept consequences of decision and action, readiness to absorb interpersonal stress, willingness to tolerate frustration and delay, ability to influence other persons' behavior, and capacity to structure social interaction systems to the purpose at hand.* (Stogdill 1974:81)

Such traits are likely to increase the effectiveness of leaders only if they are translated into behavior, which is the reason appropriate leadership behavior has been the subject of considerable research. Studies by Rensis Likert (1961, 1967) led Likert to emphasize the importance of supportive behavior, in which leaders try to provide direction in a way that maintains others' sense of personal worth and importance. Similarly, Ralph White and Ronald Lippitt (1960) found that democratic leaders (who try to build consensus about decisions) are more effective than authoritarian ones (who give orders without considering the preferences of group members). Both, however, were found to be more effective than laissez-faire leaders (who make little or no attempt to organize group activities). Subsequent studies suggest that appropriate leadership behavior depends on the situation presented by the group to be led. Fred Fiedler (1967), for example, has argued that leaders should be more authoritarian if they face a situation in which they have very high or very low amounts of control. In intermediate situations, leaders should be more democratic and concerned with the quality of human relationships.

If Fiedler's theory is correct, the popular leader of an infantry platoon (high control) or the disliked supervisor of an ineffective work crew (low control) should be more authoritarian than the well-liked director of a research team facing an unstructured task (intermediate control). Fiedler's theory has been criticized on several grounds, but it is an indication of the increasing tendency for social scientists to understand leadership in its situational context (Ashour 1973; Schriesheim and Kerr 1977).

Can groups make better decisions than individuals? Groups are used in many settings for the solution of a variety of problems. In some situations, group decisions can be shown to be superior to the decisions that would be made by individuals acting alone. For example, researchers have found that groups are useful for solving intellectual tasks because they allow people to hear several points of view and provide a setting in which the merits of various ideas can be discussed (H. H. Kelley and Thibaut 1969).

However, group pressure can also harm the quality of decisions by leading to excessive conformity and loss of individuality. The Spanish writer José Rodó has written that "every society to which you remain bound robs you of a part of your essence and replaces it with a speck of the gigantic personality which is its own." Actually, the same thing could be said of each group to which an individual is strongly committed. For the sake of group membership, we may relinquish part of our individuality and independence.

The tendency to conform to group pressure has been dramatically illustrated in a classic experiment by Solomon Asch (1955) in which many participants publicly denied their own senses in order not to deviate from majority opinion. Asch assembled numerous small groups of male college students, ostensibly for experiments in visual judgment. He asked the groups to compare the length of lines printed on two cards (see Figure 6.1). One card showed a single vertical line, which was to be used as a standard for comparison with three lines of different lengths on a second card. Each participant was asked to identify the line on the second card that matched in length one of the lines on the first card. In each group, all but one of the subjects had been instructed by Asch to choose a line that obviously did not match. The naive subject—the only member of each group unaware of the real nature of the experiment—was forced either to select the line he actually thought matched the standard line or to yield to the unanimous opinion of the group. In earlier tests of individuals in isolation, Asch had found that the error rate in matching the lines was only 1 percent. Under group pressure, however, the naive subjects went along with the major-

FIGURE 6.1

Cards for Asch's Experiments

Which of the lines on card A matches the line on card B? You must be thinking, "What a no-brainer." You may be surprised to learn that in a group setting, many people associated the other two lines in card A with the line on card B. Read about Asch's experiment in the text.

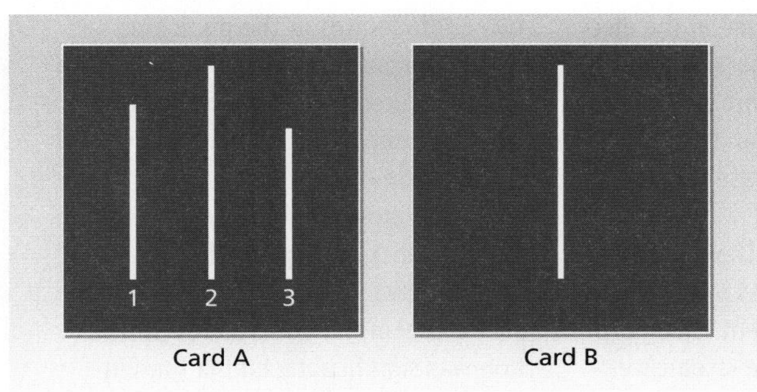

ity's wrong opinion over one-third of the time. If this large proportion of naive subjects yielded to group pressure in a group of strangers, it is not difficult to imagine the conformity in groups to which people are emotionally committed (Myers 2001).

How does "groupthink" discourage alternative points of view? Irving Janis (1982) has argued that many group decisions are likely to be the product of **groupthink**, a situation in which pressures toward uniformity discourage members of the group from expressing their reservations about group decisions. During the Kennedy administration, for example, a decision to launch the ill-fated Bay of Pigs invasion against Cuba was made only because several top advisers to the president failed to reveal their reservations about the plan. Another example of groupthink is found in the Reagan administration's plan to sell arms to Iran in exchange for the release of American hostages. The small number of designers of this plan—including Lieutenant Colonel Oliver North—minimized opposition by concealing information from officials in the administration who did not share their convictions. Such hesitancy to destroy the illusion of "assumed consensus" often leads to group decisions that have not been subjected to critical evaluation. Research indicates that groupthink can be avoided only when leaders or group members make a conscious effort to see that all group members participate actively and that points of disagreement and conflict are tolerated.

We have been viewing groups as relatively small social units. Now, we turn to larger social units known as formal organizations. It will become evident that groups are very much a part of any formal organization.

Formal Organizations

Even in the early part of this century, Charles Horton Cooley (1902) foresaw a shift from primary groups toward secondary groups. Until the 1930s, most Americans lived on farms or in small towns and villages. Nearly all of their daily lives were spent in primary groups—families, neighborhoods, and churches. As industrialization and urbanization advanced, however, Americans became more enmeshed in secondary groups. Born in hospitals, educated in large schools, employed by huge corporations, regulated by government agencies, cared for in nursing homes, and buried by funeral establishments, Americans—like members of other industrialized societies—find themselves within secondary groups known as *formal organizations*. This major change accompanies the transition from preindustrial to industrial society. The trend is continuing. Citizens of modern societies have become highly dependent on large organizations—both business and nonbusiness. Understanding the nature of organizations is essential to comprehending industrial and postindustrial society (Blau and Meyer 1987; Pfeffer 1997).

FEEDBACK

1. Match the following types of social interaction with the examples beside them:

____ a. cooperation	(1) paid blood donors
____ b. conflict	(2) students doing what a professor assigns
____ c. social exchange	(3) Saddam Hussein invading Kuwait
____ d. coercion	(4) flood victims helping each other
____ e. conformity	(5) employees forced to work overtime or be fired

2. According to Robert Bales, interaction within groups must be designed to solve both _________ problems and _________ problems.
3. Social scientists agree that leaders must have certain traits if they are to be effective. T or F?
4. Solomon Asch's experiment demonstrates the positive consequences of group pressure. T or F?
5. A situation in which pressures toward uniformity discourage members of a group from expressing their reservations about group decisions is called _________.

Answers: 1. a. (4) b. (3) c. (1) d. (5) e. (2) 2. instrumental; social-emotional 3. F 4. F 5. groupthink

Stanley Milgram—Group Pressure and Conformity

Can a group move a normally nonviolent person to physically punish a victim with increasing severity despite the victim's pleas for mercy? As incredible as this may sound, researcher Stanley Milgram (1964) revealed this as a possibility.

As noted in the text, Solomon Asch (1955) demonstrated that group pressure can influence people to such an extent that they will deny their true sense perceptions. Specifically, they can be pressured to match up uneven lines—lines they originally perceived as different lengths. Milgram wanted to know if group pressure could have the same effect on behavior. Can group pressure cause people to treat others in ways they otherwise would not?

To test this question, Milgram could have chosen a behavior relatively easy to induce, such as sharing food with a stranger or doing minor damage to someone else's property. Instead, Milgram asked research participants to administer increasingly higher electric shocks to people who appeared to be in pain. And these research participants were people just like you and me!

Milgram placed eighty males in an experimental situation in which a team of three individuals (one naïve subject who did not know the nature of the experiment and two "confederates" who did know the nature of the experiment) tested a fourth person (another confederate in on the experiment) on a learning task. Each mistake by the fourth party brought an apparent electric shock from the naïve subject. The two confederate "testers" (participants cooperating with Milgram) on each team suggested increasingly higher shock levels for successive mistakes made by the "learner" (the confederate in the electric chair). The researcher recorded the degree to which the team's naïve member resisted or went along with group pressure to increase the voltage levels.

After explaining to the naïve tester that the experiment was investigating the effects of punishment (electric shocks) on memory, the researcher strapped the learner into an electric-chair apparatus in full view of the other three team members. These three were seated in front of a large shock generator whose switches were labeled from 15 to 450 volts in 15-volt increments. The lowest-level group of switches was labeled "Slight Shock"; the highest-level group of switches read "Danger: Severe Shock."

The naïve subject controlled the maximum shock that could be administered. He could keep the shock level at 15 volts throughout or increase the shock level in line with the group's recommendation. It was his decision.

The learner in the electric chair not only answered questions; as the shocks became stronger, he grunted, protested, and finally demanded that the experiment stop. Each of these complaints was actually prerecorded and played in coordination with a given shock level. No discomfort was indicated until a 75-volt shock was administered. At 120 volts, the learner shouted that the shocks were becoming painful. Groans followed a 135-volt shock, and at 150 volts, the learner cried out to be released from the experiment, complaining of a heart condition. At 180 volts, the learner cried out, "I can't stand the pain," and 285 volts was followed by an agonized scream. From 300 to 450 volts, the victim continuously shouted desperately that he would no longer answer questions.

Milgram first needed to know how far the naïve subject would go in administering shocks *without* group influence. To accomplish this, an identical experiment was run minus the two confederates in the punishing group. Response to group pressure was measured by the difference in the naïve subject's behavior in the two situations.

Group pressure heavily affected the level of shock administered by the naïve subject. That is, the average shock level in the three-person situation was significantly higher than in the one-person situation. Perhaps most interesting are the results on *maximum* shock levels. When alone, only two naïve subjects went beyond the point where the learner first vehemently protested. Under group pressure, twenty-seven of the naïve subjects went beyond this point. Nineteen of the naïve subjects went above 255 volts ("Intense Shock"), and ten went into the group of voltages labeled "Danger: Severe Shock." Seven even reached 450 volts (the highest shock level possible).

The research by Milgram and Asch reveals the power of group pressure to create conformity in thought and behavior. Although conformity must occur to a degree for social structure and society to exist, many scholars worry about the extent to which social pressure can determine how humans think and act.

Thinking About the Research

1. Discuss the ethical implications of Milgram's experiment. (You may want to refer to Chapter 2 for pointers about ethics in social research.)
2. If the researcher had not been present as an authority figure during the experiment to approve the use of all shock levels, do you think group pressure would have been as effective? Explain.

Theoretical Perspectives and Formal Organizations

Although it is not always explicit, functionalism and conflict theory are intimately involved in the sociological study of formal organizations. Bureaucracy, for example, with its emphasis on structure, stability, rules, and hierarchy, is based squarely on functionalism. When sociologists examine the dynamics of organizations, they don't abandon functionalism, but conflict theory does play a predominant role.

The Structure of Formal Organizations

A **formal organization** is deliberately created to achieve one or more goals. Examples of formal organizations are high schools, colleges, corporations, government agencies, and hospitals.

How are formal organizations and bureaucracies related? Most formal organizations today are **bureaucracies**—formal organizations based on rationality and efficiency. Although bureaucracies are popularly thought of as monuments to inefficiency, they have proved to be an efficient form of organization for industrial society. Despite his deep concern about its negative consequences, Max Weber, who first analyzed the nature of bureaucracy, wrote of its efficiency.

What are the major characteristics of bureaucracy? From a study of history and an observation of events of his day, Max Weber sketched the basic organizational principles that were being developed to promote rational and efficient administration. Weber identified the following characteristics of a bureaucracy:

- *The organization has a division of labor based on the principle of specialization.* In a bureaucracy, each person is responsible for certain functions or tasks. (See Figure 6.2 for an organizational chart outlining the division of labor in a modern university.) A specialized division of labor improves organizational performance because an individual can become an expert in a limited area of organizational activity.
- *The organization has a hierarchy of authority.* To clarify the concept of authority, it is necessary to distinguish it from power. **Power** is to the ability to control the behavior of others, even against their will. **Authority** is the exercise of legitimate power—power that produces compliance because those subjected to it believe that obedience is the proper response. Bureaucratic organizations tend to have a pyramidal shape. The greatest amount of authority is concentrated in the few positions at the top, with decreasing amounts in the expanding number of lower positions (see Figure 6.2). Organizational effectiveness and efficiency are enhanced by a hierarchy that can coordinate the many statuses involved in a highly specialized organization.
- *Organizational affairs are based on a system of rules and procedures.* Rules and procedures guide the performance of work and provide a framework for decision making. They stabilize the organization because they coordinate activities and cover most situations.
- *Members of the organization maintain written records of their organizational activities.* These records of work and activities are kept in files. This organizational "memory" is essential to smooth functioning, stability, and continuity.
- *Statuses in the organization—especially managerial ones—are considered full-time jobs.* Bureaucratic statuses are not considered avocational. This increases the commitment of the organization's members and demands their full attention.
- *Relationships within the organization are impersonal, devoid of favoritism.* For example, personnel are selected on the basis of technical and professional qualifications and promoted on the basis of merit. Similarly, relationships with clients are conducted without regard to personal considerations. The norms prohibiting the influence of favoritism and personal relationships in organizational activities are intended to ensure equal treatment for those who work in the organization as well as for those who are served by it.
- *Employees of bureaucratic organizations do not own their positions.* Bureaucratic positions cannot be sold or inherited. This allows organizations to fill positions with the most qualified people. Those in bureaucratic positions are held responsible for their use of equipment and facilities.

For Weber, bureaucratic organization involved a set of strategies that would promote order in human relationships. Weber believed that the achievement of order would be increasingly difficult in societies undergoing industrialization. In preindustrial societies, activities in one aspect of life had been closely related to activities in every other aspect. As societies modernized, people's activities became less closely related, and traditional systems of norms and values began to weaken. As Kenneth Thompson indicates, Weber saw bureaucracy—with its hierarchy of authority, system of rules and procedures, and inherent safeguards—as a way to maintain control in an increasingly complex society:

> *The spread of the bureaucratic form of organization to all spheres was part of a general process of rationalization in modern society. The reason why it would inevitably spread was because its characteristics of precision, continuity, dis-*

FIGURE 6.2

University Organization Chart

Each organizational position and department within a bureaucracy has certain tasks associated with it. The connecting lines indicate who reports to whom and who has authority over whom.

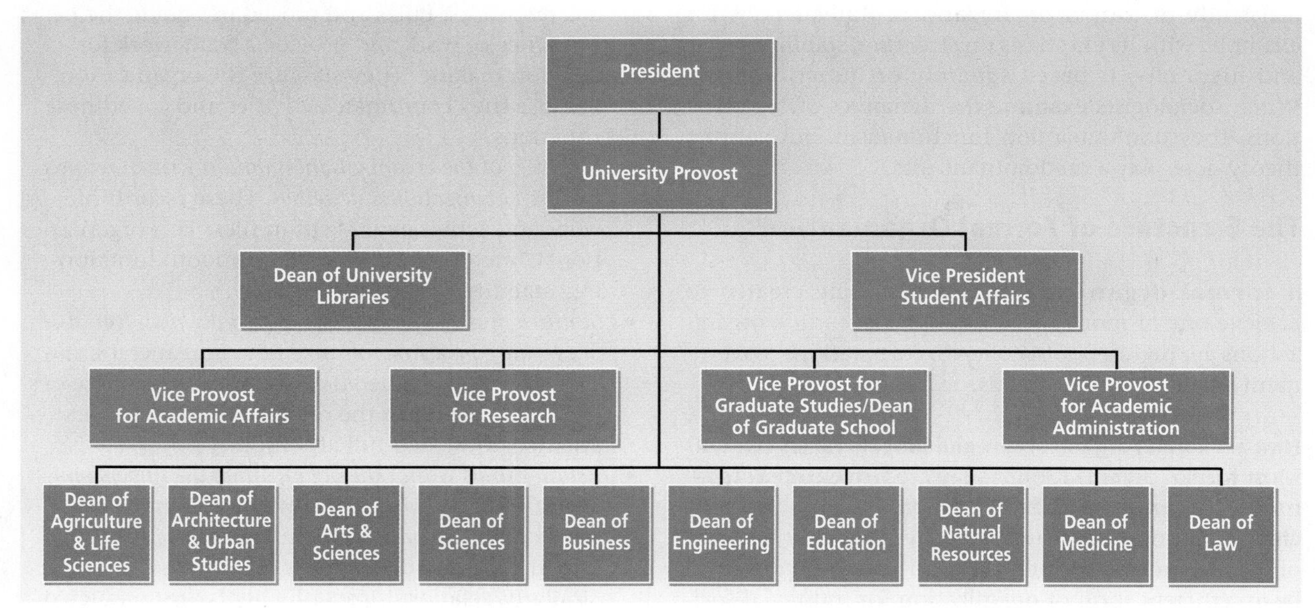

> *cipline, strictness and reliability make it technically the most satisfactory form of organization for those who sought to exercise organizational control.* (K. Thompson 1980:10–11)

Do modern organizations practice these characteristics of bureaucracy? In his formulation of bureaucratic characteristics, Weber used the **ideal-type method**. This method isolates, to the point of exaggeration, the most basic characteristics of some social entity. It involves the construction of a *pure* type. Because real life is seldom as pure as an abstract description, social entities do not have all the characteristics of an ideal type to the maximum degree. Universities, for example, are less bureaucratic than business firms, in part because universities employ large numbers of professionals whose quality of work is adversely affected by a bureaucratic environment (Litwak 1961).

Practically speaking, then, the characteristics of bureaucracy are best thought of as a series of continua; various organizations emphasize each characteristic of bureaucracy to different degrees. In fact, some units within an organization may be more bureaucratic than others (R. H. Hall 1962). In a university, for example, the president's office is much more bureaucratic than an academic department of philosophy or biology. The more closely an organization is based on the characteristics identified by Weber, the more bureaucratic it is.

Advantages of Bureaucracy

What are the advantages of bureaucracy? Citing the advantages of bureaucracy, Weber wrote:

> *The decisive reason for the advance of bureaucratic organization has always been its purely technical superiority over any other form of organization. The fully developed bureaucratic mechanism compares with other organizations exactly as does the machine with the nonmechanical modes of production.* (Gerth and Mills 1958:214)

Weber's model of bureaucracy was formulated in response to the inadequacies of earlier organization forms. In monarchies, for example, positions were appointed on the basis of favoritism and politics. Administrators had little or no training for the work they were doing. Because these officials were wealthy, administration was only an avocation. And because high positions were politically based, power could be lost overnight. Moreover, when a king or queen was replaced, the entire leadership structure changed drastically.

As Weber correctly saw, earlier types of organizational structure were out of step with the rise of the capitalist market economy, which required steadiness, precision, continuity, speed, efficiency, and minimum cost—advantages bureaucracy could offer. **Rationalism**—the solution of problems on the basis of logic, data, and planning rather than tradition and superstition—was on the rise, and the characteristics of bureaucracy were consistent with this sweeping trend.

As strange as it might sound, bureaucracy is designed to protect individuals. Without rules, procedures, and the norm of impersonal treatment (which everyone hates), decision making would be arbitrary. It might sound great, for example, to abolish final exams; but then grading might not be as objective. This, of course, is not to say that most professors in a nonbureaucratic setting would be abusive or that favoritism does not occur in bureaucratic organizations. Nevertheless, formally established rules and procedures and the norm of impersonality ensure an important degree of equal treatment.

A famous account of the benefits of specialization is found in Adam Smith's *The Wealth of Nations*. Published in 1776, this book attempts to answer the question: What is the source of wealth in a society? For centuries, the mercantilists had found wealth in the gold and silver in the royal treasury. Smith, expressing the new economic thought of the French physiocrats, gives a different answer: The wealth of nations resides in the productive power of labor, power that is best promoted by the division of labor. Smith illustrates with a description of a pin factory, citing the benefits that come when "one man draws out the wire, another straightens it, a third cuts it, a fourth points it, a fifth grinds it at the top for receiving the head." (A. Smith 1965:4/1776). By this process, Smith notes, ten men working in the factory can make 48,000 pins daily, whereas a man working alone could make only 20 or so pins. Although recognizing pin making as a "trifling" example, Smith emphasizes that the division of labor is readily applicable to any type of "art and manufacture."

Disadvantages of Bureaucracy

What are the disadvantages of bureaucracy? Although Weber praised bureaucracy as the most efficient type of organization, he was well aware of undesirable side effects. He expressed concern that the formal rationality of bureaucracy would lead to socially undesirable outcomes. For one thing, Weber noted that bureaucracy represented "the concentration of the material means of management in the hands of the master" (Gerth and Mills 1958:221). This leads to concerns about who controls bureaucratic organizations. In the words of Charles Perrow, one of the foremost contemporary scholars of organizations:

> *At present, without huge, disruptive, and perilous changes, we cannot survive without large organizations. Organizations mobilize social resources for ends that are often essential and even desirable. But organizations also concentrate those resources in the hands of a few who are prone to use them for ends we do not approve of, for ends we may not even be aware of, and, more frightening still, for ends we are led to accept because we are not in a position to conceive alternative ones. The investigation of these fearful possibilities has too long been left to writers, journalists, and radical political leaders. It is time that organizational theorists began to use their expertise to uncover the true nature of bureaucracy. This will require a better understanding both of the virtues of bureaucracy and its largely unexplored dangers.* (Perrow 1986:6)

Weber himself raised the specter of what he called the *iron cage of rationality*. He feared the likelihood of rationality spreading to all aspects of social life, creating a dehumanizing social environment and entrapping everyone. George Ritzer (1998, 2000b, 2002) refers to the "McDonaldization of society" as a metaphor to express Weber's iron cage of rationality. McDonald's, Ritzer argues, is the epitome of rationalism in organizations (see Doing Research, Chapter 1). On the dehumanizing effects of extreme rationality, Ritzer writes:

> *. . . fast-food restaurants are . . . often dehumanizing settings in which to eat or work. Customers lining up for a burger or waiting in the drive-through line, and workers preparing the food often feel as though they are part of an assembly line. Hardly amenable to eating, assembly lines have been shown to be inhuman settings in which to work.* (Ritzer 2000b:17)

It is the McDonaldization of everything from lube jobs to medicine that Ritzer believes threatens modern society with Weber's iron-cage nightmare. If the nightmare is still there when we awaken, then the rational has become irrational (Alfino, Caputo, and Wynyard 1998).

Although Weber was deeply concerned about the societal implications of bureaucracy, others have focused on bureaucracy's internal inefficiencies (Hummel 1987; Leibenstein 1987). The famous Parkinson's law, named after C. Northcote Parkinson (1993), states that work always expands to fill the time available for its completion. According to this principle, there is often little relationship between the amount of work to be done and the size of the staff assigned to do it. Thus, even when there is more staff than needed to do a given amount of work, those assigned to do the work will always find plenty to do, and their supervisors may even feel that additional help is needed. Parkinson's law is obviously intended to portray wastefulness in bureaucracies.

It is important to note that Parkinson's law applies to the bureaucratic form. The blame for wastefulness in bureaucracy lies not with lazy individuals but with the way this type of organization grows. According to Parkinson, the phenomenon he describes exists because

© Lee Snider/The Image Works

Informal groups develop spontaneously and help to meet the social needs of their members. The importance of informal groups within formal organizations has long been documented by research.

bureaucrats want to multiply the number of people under them in the hierarchy and because bureaucrats spend much of their time creating work for one another.

In addition, although promotion on the basis of merit is supposed to decrease favoritism and increase efficiency, it may also have some drawbacks. Those selected for promotion in an organization are, if the system is based on merit, those who are doing their present jobs well. As long as they continue to do well in their new jobs, these people are likely to continue to be promoted. The problem, according to Laurence Peter and Raymond Hull (1996), is that people are promoted until they find themselves in jobs that are beyond their capabilities. Thus, according to the Peter Principle, employees in a bureaucratic hierarchy tend to rise to their level of incompetence. Organization members who have reached their level of incompetence shift their concern from performing competent work to sustaining organization values. They also begin to evaluate subordinates on the basis of promptness, courtesy, and cooperativeness rather than competence.

In *The Organization Man*, William H. Whyte (1972; Whyte and Nocera 2002) maintains that organizations have become so influential that they shape our nature. Organizations, Whyte argues, reward people who are good team players and who can sell their personalities and promote the organization at the same time. "Organization men and women" sell themselves, adjust to different social environments, remain flexible, remain uncommitted to any set of values, and adjust their personalities to the people and groups around them.

These popular ideas about bureaucracy have gained wide acceptance, even though scientific evidence supporting them does not exist, and even though Parkinson's law and the Peter Principle were both satirically written. It is not that these ideas are necessarily wrong; we simply cannot evaluate their validity at this time because they defy scientific study. There are, however, other negative consequences of bureaucracy that are more widely recognized by social scientists. Among these are goal displacement and trained incapacity.

What is goal displacement? When organizational rules and regulations become more important than organizational goals, **goal displacement** occurs (Merton 1968). Professors who emphasize grades more than learning are practicing goal displacement. Social workers who become more concerned with eligibility requirements than with delivering services are engaging in goal displacement.

What is trained incapacity? According to Thorstein Veblen (1933), **trained incapacity** exists when previous training prevents someone from adapting to new situations. Trained incapacity involves tunnel vision, inflexibility, and inability to change. Let me offer an illustration based on personal experience. At one point during my graduate training in sociology, the federal government was giving me financial support. With a child on the way, I felt it was necessary to earn some additional money. The chairperson of the sociology department permitted me to teach one course, providing the university administration agreed that the rules would permit receiving money from the federal government and the university at the same time. I posed this question to the appropriate university official and watched a perplexed bureaucrat fumble through the code of regulations for a pat answer, which simply would not present itself. After minutes of obvious discomfort, he found what he was looking for. With the relief of a winner at Russian roulette, this man said excitedly, "Here it is!" Although the new regulation could have been interpreted otherwise, this administrator chose the interpretation that bolstered his original inclination to deny permission. I later learned that earlier policy had prohibited persons on grants from obtaining additional financial support from the university. This bureaucrat had become so accustomed to this way of doing things that he could not interpret the new ruling in any way other than through the old organizational viewpoint that had become so indelibly his own. By past training and experience, he had developed an incapacity to meet the new organizational realities.

More seriously, the crew of the USS *Vincennes*, on July 3, 1988, mistook an Iranian passenger airliner for an Iranian F-14 jet fighter heading their way. Instead of destroying a menacing military plane, the *Vincennes* crew killed all of the nearly 300 civilian passengers on a regularly scheduled flight over the Persian Gulf. This tragic accident occurred because the *Vincennes* was created to operate on the open sea, not on a small body of water with all sorts of air and water vehicles constantly coming into its radar sights. Instead of identifying the passenger airliner for what it was, the crew acted on the basis of its training and did what it thought was necessary for self-protection. The crew's training resulted in a tragedy, essentially preventing the crew from adapting to the new environment. The crew's incapacity stemmed from the nature of its training.

Do alternatives to bureaucracy exist? Weber developed his theory of bureaucracy for the organizations that were in the changing society in which he lived. During the 1960s, some sociologists began to see bureaucracy as an outmoded organizational form. Warren Bennis (1965), for example, contended that bureaucracy was no longer able to deal with the individual goals of employees and the increasing rates of social change. He therefore predicted that bureaucracies would be replaced by **organic-adaptive systems**—organizations based on rapid response to change rather than on the continuing implementation of established administrative principles:

> *The key word will be "temporary"; there will be adaptive, rapidly changing temporary systems. These will be "task forces" organized around problems-to-be-solved. The problems will be solved by groups of relative strangers who represent a set of diverse professional skills. The group . . . will evolve in response to the problem rather than to programmed role expectations. The "executive" thus becomes coordinator or "linking pin" between various task forces. . . . People will be differentiated not vertically, according to rank and role, but flexibly and functionally according to skill and professional training.* (Bennis 1965:34–35)

Bureaucracy, of course, has not disappeared, and Bennis (1979) has qualified some of his earlier predictions. Increasing numbers of organization researchers would agree, however, that bureaucracy is best suited for relatively stable, predictable situations. In situations characterized by rapid change and uncertainty, other forms of organization are likely to emerge (Mintzberg 1979). Such attempts to relate the structure of organizations to the situations they face have been labeled the *contingency* approach to organization theory (Burns and Stalker 1966; P. R. Lawrence and Lorsch 1967; Shepard and Hougland 1978). Although the contingency approach has been criticized on some grounds (Pennings 1975; Pfeffer 1978), it is considered a useful corrective to the traditional assumption that there exists "one best way" to organize.

A strong advocate of alternatives to the bureaucratic form of organization, Joyce Rothschild (1979, 1986) has studied cooperative organizations—organizations intended to be on the other end of the spectrum. Cooperative organizations, owned and managed by their employees, are characterized by full membership participation, minimization of rules, promotion of primary social relationships, hiring and promotion based on intimate ties and shared values, elimination of status differences, and a de-emphasis on job specialization.

Although some analysts contend that nonbureaucratic forms of organizations are best suited to new business technologies and the future environmental context (Drucker 1988; T. Peters 1988; Mazaar 1999), most scholars agree that bureaucracy will remain the dominant organizational form. Bureaucracies, it is contended, persist in part because they work best with the technologies and environments characteristic of the modern era. This does not mean, however, that some organizations cannot be nonbureaucratically structured. It is just that such organizations will be few in number. Nonbureaucratic organizational principles are most useful as a supplement to bureaucratic organization, as in research laboratories or in organizational work teams (S. P. Robbins 1990).

What are some of the differences among formal organizations? Sociologists, such as Amitai Etzioni (1975), have attempted to take differences in formal organizations into account by classifying organizations into categories. Some organizations, such as churches, political parties, universities, and service clubs, are joined because people want to join; they are joined because of personal interest or emotional commitment. Because members are free to join or not, these are voluntary organizations. Etzioni refers to them as *normative organizations*, because shared understandings provide an important basis for coordination of members' activities. Organizations that people are forced to join are *coercive organizations*. Examples include prisons, custodial mental hospitals, concentration camps, forced-labor camps, elementary schools, and military systems using the draft. Other organizations are *utilitarian organizations*; people join them because of the benefits they derive from membership. Work organizations, including factories, offices, and professional firms, are examples of utilitarian organizations. Utilitarian organizations fall somewhere between the first two types. People are not forced to work (it is not a coercive organization), and yet they must work to be self-supportive (it is not a voluntary organization). Thus, utilitarian organizations are neither entirely voluntary nor entirely coercive; they have some elements of both.

Whereas Etzioni classified organizations according to the motivation for membership, Peter Blau and Richard Scott (1982) have classified organizations according to the recipients receiving the most benefit. In *mutual-benefit organizations* (political parties, professional associations, labor unions, social clubs), the prime beneficiary is the membership. These organizations exist to promote the interests of those who belong to them. *Business organizations*, on the other hand, are expected to serve the interests of their owners. Because these organizations are profit oriented, they emphasize maximum gain at the least cost. The prime beneficiary of a *service organization* is the organization's clients. The goal of social work agencies, schools, and hospitals is to aid those people who qualify for their services. *Commonwealth organizations* (the military, the Department of State, police and fire departments, the Environmental Protection Agency) are intended to serve the general public.

Dynamics of Formal Organizations

The discussion of organizations thus far has described the formal framework on which organizations are based. But organizations also have dynamic processes. These processes unfold from the structure of the formal organization and lead to the modification of that structure (R. H. Hall 2001). This section is concerned with the informal structure within formal organizations as well as the tendency toward a concentration of power at the top.

Informal Organization

It is not necessary to observe any formal organization very long before realizing that it contains patterns of interaction and relationships that are not anticipated by the formal organization chart. Bureaucracies are designed to act as secondary groups. Relationships among organization members are supposed to be governed by the hierarchy of authority, by rules and procedures, and by impersonality. But, as anyone who has worked in a bureaucratic organization knows, there are primary relationships as well. Primary relationships emerge as part of the **informal organization**—a group (within a formal organization) in which personal relationships are guided by unofficial norms, rituals, and sentiments that are not part of the formal organization. Based on common interests and personal relationships, informal groups usually form spontaneously.

When was informal organization first recognized? The existence of informal organization within bureaucracies was first documented in the mid-1920s, when a group of Harvard researchers were studying the Hawthorne plant of the Western Electric Company in Chicago. In a study of fourteen male operators in the Bank Wiring Observation Room, F. J. Roethlisberger and William Dickson (1964; originally published in 1939) observed that work activities and job relationships were based on norms and social sanctions of that particular group of male operators. Group norms prohibited "rate busting" (doing too much work), "chiseling" (doing too little work), and "squealing" (telling group secrets to supervisors). Conformity to these norms was main-

FEEDBACK

1. A formal organization is ____
____.
2. The ability to control the behavior of others even against their will is _________.
3. Which of the following has not been seen as an advantage of bureaucracy?
   ____ a. its avoidance of the use of inappropriate criteria in hiring employees
   ____ b. its use of rules to provide definite guidelines for behavior within the organization
   ____ c. its success in hiding the true nature of authority relationships
   ____ d. its encouragement of administrative competence in managers
4. Social workers who become more concerned with eligibility requirements than with delivering services to their low-income clients are engaging in _________.

Answers: 1. a group that has been deliberately and consciously created for the achievement of one or more specific goals. 2. power 3. c 4. goal displacement

WORLDVIEW 6.1

Military Bureaucracy

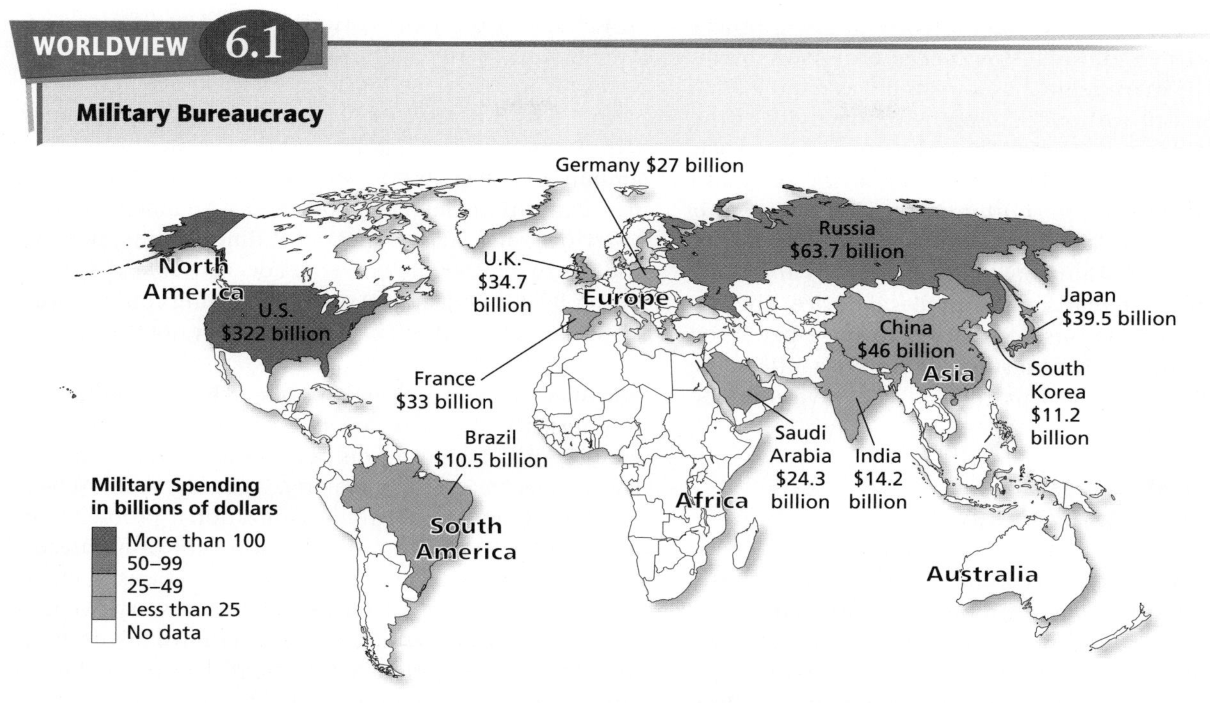

Source: *The Military Balance 2002–2003,* The International Institute for Strategic Studies, U.S. Department of Defense, 2002.

All societies have organized groups to defend themselves. In most preliterate societies, these groups are usually loosely organized and informal in nature. They are composed of group members who live normal lives except during defense emergencies. The ancient Roman army was a clear exception, of course. Still, widespread adoption of bureaucratic principles of organization for military purposes came only with industrialization. Today, as this map shows, virtually all societies have standing armies. This map shows the amount of money various countries spend annually to support their military bureaucracies.

1. Compare the amount the U.S. spends each year with the eleven next highest annual defense-spending nations (displayed on the map).
2. Discuss this ratio to demonstrate the distinction between the concepts of power and authority.

tained through ridicule, sarcasm, criticism, hostility, and "binging" (hitting deviants on the arm). The group also had its own unique stratification structure, with soldermen at the bottom, wiremen in the middle, and inspectors at the top.

The Need for Informal Organization

Why does informal organization exist? Informal groups spontaneously develop to meet needs that are ignored by the formal organization. Whereas modern organizations are designed to be impersonal, informal groups offer personal affection, support, and humor. In addition to humanizing the formal organization, informal organization is a means of protection. Informal organization encourages conformity, and the resulting solidarity protects group members from maltreatment by those outside the group.

How does informal organization affect the formal organization? Early studies, starting with the Bank Wiring Observation Room, concluded that informal organization resulted in behavior counterproductive to the organization's goals. In the Hawthorne study, for example, workers enforced the informal norm of producing an average of two units a day. Workers who turned out more than two units per day were forced by the group to stop exceeding the informal production quota.

Later research has revealed beneficial consequences as well. Among carpenters and bricklayers, Raymond Van Zelst (1952) found, for example, that establishing work groups on the basis of friendship patterns increased workers' job satisfaction, decreased friction and anxiety among work partners, and created a friendly, cooperative atmosphere—all of which contributed to lower turnover and higher productivity.

Joseph Bensman and Israel Gerver (1963) described the use of an illegal tool in an airplane factory. This tool, called a tap, is used to thread nuts so they will fit with wing bolts that are slightly out of alignment. Otherwise, the defect would have to be reported. The tap is outlawed—its use can lead to dismissal—because a tapped nut does not have proper holding power under flight conditions. However, use of the tap was tolerated by management as long as it was handled discreetly. In fact, without its use, the productivity of the plant would have suffered. Thus, the use of the tap was controlled informally by supervisors, inspectors, and workers themselves. In short, to avoid detection by Air Force inspectors, supervisors and workers evolved an informal organization complete with norms and sanctions for the systematic use of an illegal device. The formal organization considered itself to have benefited from this informal subterfuge. (Of course, the same could not be said for the passengers who traveled in the planes.)

The Role of Power in Formal Organizations

The word *organization* itself implies power. A hierarchy of authority is an expression of power distribution—in decreasing amounts from top to bottom. And organizations are political in more subtle ways. Every organization has people whose interests and goals conflict. Many preferences of top executives are not shared by wage workers; the outlooks of sales managers are often at odds with the views of production managers. Because a rational choice between all these conflicting views often requires more information than decision makers can assemble, decision making is usually political. Those with the most power usually get their way (Cyert and March 1963; H. A. Simon 1976; Pfeffer 1981).

Achievement of organizational goals requires the exercise of power. Although the employment of a power advantage is beneficial in organizations, power may be monopolized by individuals for their own limited purposes. This tendency is articulated in the *iron law of oligarchy* (Michels 1949; originally published in 1911).

How does the pyramidal shape of bureaucratic organizations promote the iron law of oligarchy? According to the **iron law of oligarchy** formulated by German sociologist Robert Michels, power tends to become increasingly concentrated in the hands of a few members of any organization. Michels observed that even in organizations intended to be democratic, leaders eventually gain control and other members become virtually powerless. In fact, he concluded that this increased concentration of power occurs because those in power want to remain in power:

> *It is organization which gives birth to the dominion of the elected over the electors, of the mandatories over the mandators, of the delegates over the delegators. Who says organization, says oligarchy.* (Michels 1949:401)

Politics in the United States is a prime example of Michels's principle. Voter apathy is reflected in low voter turnout for elections at all levels of government. Because so few are involved in the political process, those who are politically active gain an inordinate amount of power. As you will see, the same thing occurs in organizations such as labor unions. Membership passivity over an extended period of time permits those at the top to consolidate power. Leaders are able to stay in power by building a loyal staff, controlling communications, and marshalling resources (money, jobs, favors).

Why does organization lead to oligarchy? According to Michels, three organizational factors encourage oligarchy. First, organizations need a hierarchy of authority to delegate decision making. Agreeing with Weber, Michels contends that the organizational problems of any large group require the development of a bureaucracy. Even in highly democratic organizations, such as the socialist parties of Europe, it is necessary for the masses to delegate much of the decision making to a few leaders. Delegation means creating a hierarchy of authority.

Second, the advantages held by leaders at the top of the hierarchy allow them to consolidate their power. They are able to monopolize power because of their social and political skills, and once in power, they use these skills to strengthen their positions. Moreover, merely by doing their jobs, leaders of the organization gain special knowledge not available to the membership. They can create a staff that is loyal to them; they can control the channels of communication; they can use organizational resources to increase their power.

Third, oligarchy in organizations is also encouraged by the characteristics of the followers. The membership tends to defer to the greater skills possessed by their leaders. The nonelites believe that their leaders are more articulate and are doing a better job than they could do. In addition, the masses are passive, even indifferent; they do not want to spend a lot of time in organizational activities. The general membership may, in fact, feel grateful to those at the top for the leadership they are providing.

What are the consequences of oligarchy? Democratic organizations—the type of organization in which oligarchy presents the greatest problem—suffer from this tendency toward oligarchy, contended Michels, as those in power come to care more about perpetuating their power than about organizational goals. Oligarchic leaders become excessively careful and conservative as the maintenance of their positions becomes more important to them than considering the interests of the membership. As a result, the desires of the membership may be sacrificed.

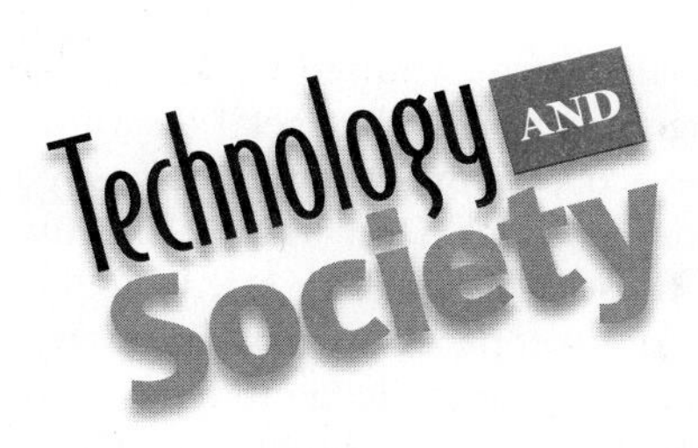

Privacy in the Workplace

In *The Unwanted Gaze*, Jeffrey Rosen (2000) attempts to document the destruction of privacy in the United States. He examines the legal, technological, and social changes eroding control over personal information. Because even deleted e-mail, for example, can be retrieved by others, a record of private self-disclosures exists permanently. Private preferences, tastes, and tendencies are revealed on every Web site we hit. During the impeachment of President Clinton, prosecutors legally gained access to Monica Lewinsky's bookstore purchases and unsent love letters on her own computer.

If Monica Lewinsky's loss of privacy is shocking to you, consider workplace vulnerability. Probably more than three-fourths of the Fortune 1,000 companies now have employee-monitoring software, and many of these firms keep their monitoring practices secret. More and more corporations are firing employees who violate prohibitions against the personal use or abuse of telecommunications technology. E-mailing a friend or co-worker a heated threat to file a discrimination suit against the company for failing to promote you could result in termination even if you later deleted it (Faltermayer 2000). Business ethicist Laura Hartman offers this sobering picture:

> *A multitude of basic and inexpensive computer monitoring products allows managers to track web use, to observe downloaded files, to filter sites, to restrict your access to certain sites, and to know how much time you have spent on various sites. These include products such as WebSense, Net Access Manager, WebTrack and Internet Watchdog.*
>
> *One particular firm, SpyShop.com, claims to service one-third of the Fortune 500 firms. This firm sells items such as a truth-telling device that links to a telephone. You are told that you can interview a job candidate on the phone and the device identifies those who lie. Another firm, Omnitracks, sells a satellite that fastens to the top or inside of a truck. The product allows trucking firms to locate trucks at all times. If a driver veers off the highway to get flowers for her or his partner on Valentine's Day, the firm will know what happened.*
>
> *SpyZone.com sells an executive investigator kit that includes the truth phone I mentioned earlier, as well as a pocket recording pen. Other outlets sell pinhole lens camera pens, microphones that fit in your pocket. The motto of one firm is "In God we trust. All others we monitor." That firm offers a beeper buster: a computer program that monitors calls placed to beepers within a certain vicinity. A screen on your computer will show you all the numbers so that you can determine whether the individual is being distracted during working hours.* (Hartman 2000:12)

Analyzing the Trends

To what extent do employees have a right to privacy in the workplace? Analyze workplace privacy from a functionalist, conflict, and symbolic interactionist perspective.

Must oligarchy always develop? Research done since Michels published his theory has tended to support the iron law of oligarchy, the operation of which has been documented in a variety of organizations, including political parties, legislatures, churches, the American Legion, the American Sociological Association, the American Medical Association, and labor unions.

There is, however, opposition to oligarchy in organizations; otherwise, all organizations would be oligarchic. According to Alvin Gouldner, if there is an iron law of oligarchy, there must also be an iron law of democracy: "If oligarchical waves repeatedly wash away the bridges of democracy, this eternal recurrence can happen only because men doggedly rebuilt them after each inundation" (Gouldner 1955:506). Sentiment from the masses is continuously bubbling up to depose oligarchic leaders in democratic organizations. The survival of democracy in the face of oligarchy can be seen in the democratic removal of entrenched labor leaders and politicians. Although the odds against turning out entrenched leaders are high, it happens often enough to remind us that oligarchy does not completely destroy democracy. Moreover, if they are to remain in power, oligarchic leaders have to pay attention to the desires of their followers. Although mere responsiveness to the membership may not be the ideal form of democracy, its presence indicates that even oligarchic leaders often do not have a totally free hand to do as they please (Sayles and Strauss 1967; Voss and Sherman 2000).

It is also possible for members of democratic organizations to erect barriers to the development of oligarchy. A classic example is the typesetters' International Typographical Union (ITU). According to Seymour Lipset, Martin Trow, and James Coleman (1956), the ITU has been able to avoid oligarchic leadership because of its internal two-party system. Every two years, each party offers a complete slate of candidates for all offices. Because the two parties are of equal strength, the turnover in office compares favorably with the turnover in national political elections.

Other research concludes that concentration of power does not inevitably lead to an abandonment of the organization's goals for the selfish benefit of the leaders. Research on the policies of churches toward civil rights and on other controversial issues has led James Wood to note that "many organizations exist precisely to foster values that are precarious when left to individuals" (J. R. Wood 1981:101). Wood found that such values—including those that would favor the enactment of controversial policies regarding civil rights—can most effectively be implemented when leaders have a strong organizational base of support. Contrary to Michels's predictions, leaders in such situations often enacted policies that resulted in controversy, a loss of membership, and a loss of professional positions within the organization. Given a choice between the leaders' personal benefits and the maintenance of the organization's values, the leaders often chose to maintain the organization's values.

Joyce Rothschild and J. Allen Whitt (1987) also question the alleged inevitable connection between organization and oligarchy. They challenge the iron law of oligarchy through an investigation of cooperatives—organizations owned and operated by the workers themselves. Although not all cooperatives prosper or even survive, Rothschild and Whitt maintain that the very existence and persistence of cooperatives contradict the assumption that organization leads to oligarchy.

Race, Ethnicity, Gender, and Social Class in Organizations

By the nature of organizations, power is concentrated in relatively few positions at the top. Mirroring the broader society, top positions in the hierarchy of organizations are held by white men from higher social classes. Consequently, minorities, women, and lower-class members are subject to the same patterns of prejudice and discrimination they experience in their nonwork lives (Miller 1999). The term *glass ceiling* has been used to reflect the fact that so few women and minorities get promoted to the more powerful (higher) positions.

How does the treatment of minorities mirror society? One of the most graphic reflections of inequality in organizations is found in the secretly taped meetings of Texaco's upper managers in 1996. The racial discrimination suit filed by Texaco's African American employees was settled in their favor for $115 million shortly after the content of these tapes hit the mass media. Texaco could hardly deny reparations for company racism when it became public knowledge that one top executive used the "N" word and another joked that "black jelly beans were stuck to the bottom of the bag" (Solomon 1996).

Much more systematic evidence of inequality exists. Even among employees with identical education, white men are disproportionately promoted over African Americans, Latinos, and Native Americans. It works the same way with dismissal: African American employees in the federal government (the environment most favorable to minorities) are discharged at a rate double that of whites (Zwerling and Silver 1992; DeWitt 1995). These and other studies provide convincing evidence of prejudice and discrimination against minorities in organizations (Feagin 1991; Cose 1995; see Chapter 9).

Is societal gender inequality reproduced in organizations? Yes it is. Like racial and ethnic minorities, women are less likely to be promoted than white men, even when they have the same level of education. For one thing, women are judged by higher criteria than men. Women, too, are more likely to be let go than men (White and Althauser 1984; Aldrich 1999; H. Smith 1999).

Sociologist Rosabeth Moss Kanter (1993) conducted a classic study of gender inequality in organizations. She found that women are usually treated as tokens, facing the charge that they owe their positions to just being women. The visibility of women due to their small number puts them under the glare of a spotlight. Stress is created by the fear of making mistakes easily seen by others in the organization. Women are more likely to be socially isolated, denied access to informal networks, and without powerful mentors. This isolation, in turn, reduces their chances for advancement in the hierarchy.

Because men overwhelmingly outnumber women in the upper ranks of organizations, Kanter found that even successful women find themselves in a minority status. Successful women in male peer groups become symbols of what-women-can-do. On the one hand, this makes them highly visible in an environment in which success depends on visibility. On the other hand, they sometimes feel the loneliness of being different from the other group members, having to sacrifice their privacy and anonymity. Because of their visibility, the few women in the upper ranks also feel extra pressure to avoid making mistakes. Even those women who have experienced little or no discrimination feel that they have to work harder than the average man to be successful.

Another indicator of gender inequality is **sexual harassment**—the use of one's superior power in making unwelcome sexual advances. Sexual harassment has been documented in nearly all types of organizations, including the military, the government, corporations, factories, and universities (Lott 1994). More will be said about this topic in Chapter 10, which more fully explores gender inequality in American society.

How is social class perpetuated in organizations? Organizations perpetuate social class levels through the placement of working-class people in low-paying, temporary, and dead-end jobs. Members of the upper and middle classes enter higher-level jobs with higher salaries and wages and greater opportunities for advancement. And this cannot be explained by social class differences in education. Persons from upper-class families tend to receive more favorable treatment than do lower-class persons with comparable educational credentials (Blau and Meyer 1987). Educational qualifications are often overshadowed by stereotypes associated with class differences in speech, dress, and behavior. Employees from lower-class backgrounds are more likely to be denied promotion, to receive lower pay raises, and to be dismissed than their higher-class coworkers who display proper "manners" and dress more stylishly.

Racial, ethnic, gender, and social class inequalities are mirrored in organizations. Not only do organizations reflect societal inequalities—they perpetuate them.

Organizations and Their Environments

The **organizational environment** consists of all the forces outside an organization that exert an actual or potential influence on the organization. An organization's environment includes the general technological, economic, political, cultural, and related conditions in which the organization operates (Hawley 1968; Meyer and Scott 1985; Hannan and Freeman 1989; W. R. Scott 1992; DiMaggio 2001). Interest in environments has increased in part because of the growing realization that organizations can survive only to the extent that they can generate enough support among environmental actors to continue receiving needed resources (in the form of money, new members, or whatever). It is equally true, however, that many organizations have sufficient resources to exert significant influence on their environments (Florida and Kenney 1991). Whether they are influencing outside actors or are the recipients of influence, organizations are frequently involved in interorganizational relationships.

What is the nature of interorganizational relationships? An **interorganizational relationship** is a pattern of interaction among authorized representatives of two or more formally independent organizations. Interorganizational relationships may be entered for many reasons, but most researchers would agree that a need for resources possessed by other organizations

FIGURE 6.3

Patterns of Cooperation and Conflict in Interorganizational Relationships Involving the Police

This figure illustrates the nature of interorganizational relationships. Even when the police need the help of other organizations to reach their goals, only three of their fourteen interorganizational relationships are cooperative.

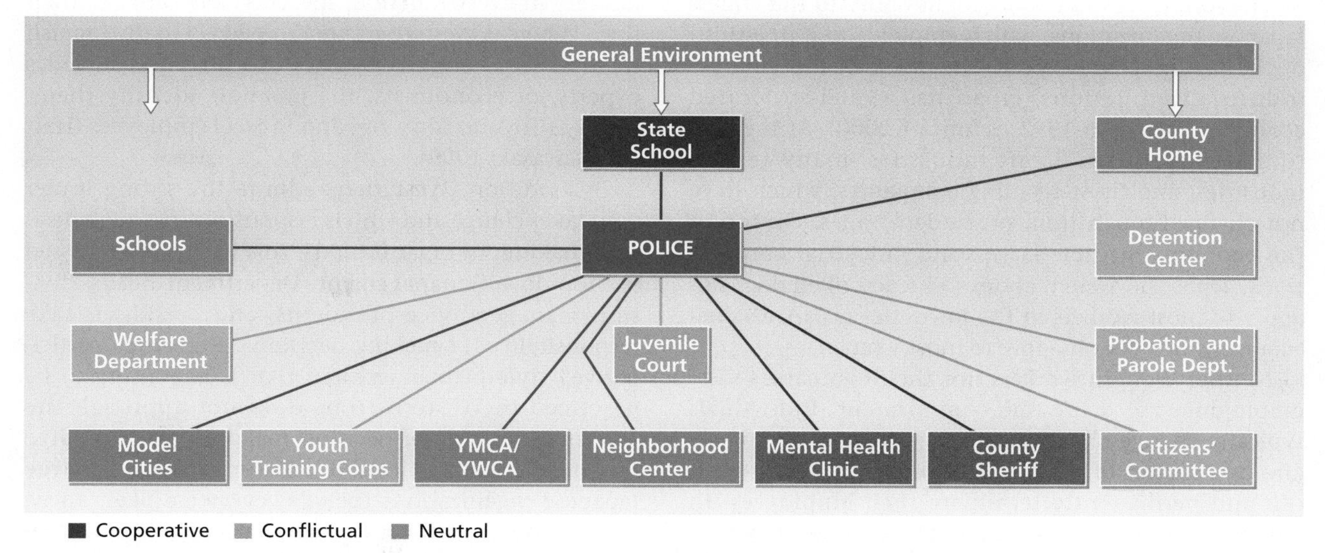

Source: Richard H. Hall, *Organizations*, 7th ed. (Englewood Cliffs, NJ: Prentice-Hall, 2001), 225. Reprinted by permission.

and an awareness of the availability of those resources in the other organizations contribute to the formation of interorganizational relationships (S. Levine and White 1961; Cook 1977; Hougland and Sutton 1978; Alter and Hage 1992).

Although organizations become involved in interorganizational relationships because outside resources are needed, all parties are concerned that the interests of their own organizations be protected. As Figure 6.3 shows, although the police must interact with fourteen other organizations in the area of problem youth, only three of these relationships are regularly cooperative. Six of the relationships, including those with important organizations such as the juvenile court and the detention center, are regularly marked by conflict (R. H. Hall 2001).

Formal Organizations in Japan

Our discussion of organizations thus far has assumed a broadly common cultural background—namely, that of the West. An examination of some of the unique features of formal organizations in Japan demonstrates that culture can be a powerful player in shaping the structure and dynamics of formal organizations (Lincoln and McBride 1987; McCormack and Sugimoto 1988; Lincoln and Kalleberg 1990).

Bureaucracies in both the East and the West share the fundamental characteristics identified by Weber. Culture, however, impinges to create some distinctive differences between Western and Japanese styles of bureaucracy (Pascale and Athos 1981; Hasegawa 1986; Ouchi 1993; Reischauer 1995).

The following discussion applies only to the largest Japanese organizations, which employ about one-third of the Japanese labor force. Like most of the younger industrialized nations, Japan has a dual-structured economy (Wokutch 1992; Schnitzer 2000). At the bottom are small handicraft industries, many service industries, and most retailing companies, which have not changed much from premodern times. On top of the economy sit new large-scale industrial corporations. Thus, the system about to be described does not apply to most workers in the lower tier of Japan's dual economy, nor does it apply to most women.

Neither Western workers nor their companies view employment as a lifetime commitment. Individuals typically change organizations to suit their advantage; and as a matter of policy, most organizations fire, lay off, and permanently terminate their employees. In contrast, a prime characteristic of the larger Japanese organizations is lifetime employment. Once hired, an employee is not expected to leave until reaching mandatory retirement age (55). In fact, those managers who are lured away by other companies are seen as deviants or as possessing serious character flaws. This policy of lifetime employment creates a strong sense of organizational loyalty and a workforce that is enthusiastic, proud of company products, and pleased to work overtime. Many employees do not take full vacation time, and workers are so conscientious that quality control inspectors are not needed. Although the concept of lifetime employment has been weakened somewhat by the economic decline Japan experienced in the late 1980s and 1990s, most employees still expect to work a lifetime for the same company (Wokutch and Shepard 1999).

Rapid employee turnover in Western firms requires frequent evaluation, financial reward, and promotion, partly to reduce the rate of departure. In large Japanese firms, employees who join an organization concurrently typically receive the same pay increases and promotions for about ten years. Only at the end of ten years is a formal evaluation of an employee made and differential promotions awarded. Consequently, much competition and friction are eliminated, both within and across age-groups. Individuals within the same age-group do not try to outshine their peers, and superiors in the organization do not face ambitious subordinates who want to replace them.

Specialization is very high in Western firms, both on the factory floor and in the managerial and professional ranks. This, on the part of higher management, is by design. At the same time, because American workers cannot expect to spend their lives in one firm, they prepare themselves to be marketable; they become competent in a set of specialized skills. In large Japanese organizations, training is much broader. At the peak of their careers, Japanese managers have been prepared for mastery in every function, specialty, and office of their firm. Whereas Westerners are more likely to distinguish themselves as engineers, sales executives, electronics experts, or economists, the Japanese identify themselves as Toyota, Sony, or Nippon Steel employees (Hart and Kawasaki 1999).

By tradition, Westerners admire the strong leader who takes charge and who is responsible for tough decision making. In organizations, this has been expressed in "top-down" management: Department heads, division managers, vice presidents, and presidents take responsibility for making decisions. Even the "participative" style of management that is beginning to be practiced in Western firms does not approach the "bottom-up" style of decision making typical of large Japanese organizations. Decision making in large Japanese organizations includes everyone likely to be affected. Sometimes this means that sixty or more people may be directly involved in making a decision. Although time consuming, this approach increases the employees' commitment and enables them to be more effective in implementing a resolution.

The relationship between employees and their employers in Western organizations tends to be narrowly limited to work-related matters. Large Japanese firms, in contrast, are very involved in their employees' nonwork lives. Concern for the total welfare of their workers extends to organizational provision of housing, loans, education, and recreational outlets. Imagine the intimate and holistic ties that begin to be forged during the induction of a group of bank trainees (Rohlen 1974). The bank president tells the employees that the firm will accept responsibility for their children's physical, moral, and intellectual development. The organizational "parent" representative for the ceremony rises to remind the new employees to display the same loyalty for their new business family that they have exhibited toward their biological families. Finally, a trainee pledges, for the entire class, a commitment to meet bank and parental expectations.

Origins of Japanese Organizational Structure

Obviously, differences between Japanese and Western organizations can be traced to environmental factors in one way or another. The process of linking environmental conditions to Japanese organizational characteristics, however, is not simple. As a broad generalization, the distinctive characteristics of large Japanese organizations reflect a culture that places the group above the individual: "The Japanese company is in effect a community of motivated people. In Japan a company sees itself as a cohesive group of people working toward a common goal" (Hasegawa 1986:1).

Although the nature of the Japanese organization is consistent with a collectivistically oriented culture, it does not mean that the organizational structure emerged totally formed from premodern times. For example, the practice of lifetime employment for nonexecutives, although consistent with a cultural emphasis on group solidarity, has not always existed in Japanese industry. Lifetime employment for all employees became a feature of large Japanese organizations only in response to a specific need—the need for skilled workers to perform more technically complicated work. To overcome the shortage of skilled workers that followed World War II, large Japanese corporations adopted the practice of lifetime employment for all employees. Much-needed skilled labor, the Japanese found, could be retained through offering employees a variety of special privileges, of which job security was the mainstay.

Environmental change continues in Japan. Although the Japanese system will probably remain intact for some time, Edwin Reischauer (1995) points to several signs of erosion. Among them is the disenchantment of some younger workers, who are not attracted to the idea of working hard in the present for rewards in the future. In addition, an increasing desire for more personal freedom makes company paternalism somewhat stifling for these young workers. To the shock of the older generation, more younger people, including executives, are showing less company loyalty as they display a greater willingness to accept new job offers. The reward system itself is evolving away from seniority toward skills and productivity.

FEEDBACK

1. A group in which personal relationships are guided by rules, rituals, sentiments, and traditions not provided for by the formal organization is known as a(n) __________ organization.
2. The effects of informal organization on the formal organization are almost always negative. T or F?
3. The theory known as the __________ emphasizes the danger of power becoming concentrated in the hands of a few members of any organized group.
4. Cooperative patterns tend to dominate in interorganizational analysis. T or F?
5. Japanese organizational structure has proved not to be the exception to Weber's analysis. T or F?

Answers: 1. informal 2. F 3. iron law of oligarchy 4. F 5. T

SUMMARY

1. Group members interact, share culture, consider one another's behavior, and have some common interests and goals. Groups are sometimes confused with social categories and social aggregates.
2. A primary group is composed of individuals who know one another well, interact informally, and like being together. A secondary group is goal oriented, and its members interact without personal or emotional involvement. Likewise, a primary relationship is intimate and personal, whereas a secondary relationship is impersonal.
3. Primary groups are more likely to develop when a relatively small number of people are involved, when

interaction is face-to-face, when people are in contact over a long period of time, and when the social environment is appropriate. Primary groups are formed in part because of the emotional support and sense of belonging they provide members. This type of group serves society by aiding in the socialization process and by helping to maintain social control.

4. We use reference groups to evaluate ourselves and to acquire attitudes, values, beliefs, and norms. Reference groups may be those we merely aspire to join, or those we use as negative role models.
5. Particularly intense feelings of identification and loyalty envelop in-groups. The feelings involve opposition, antagonism, and competition toward out-groups. Efforts must be made to maintain group boundaries, the consequences of which have significant social implications.
6. A social network (not a group per se) links a person with a wide variety of individuals and groups. The social relationships that are formed may be "strong" (parents and children) or "weak" (distant relatives, members of a college class). Social networks provide a sense of belonging and social support and help in the job market. More women are now using networking, once the sole province of men, for career advancement.
7. Five types of social interaction are basic to group life. In cooperation, individuals or groups combine their efforts to reach some common goal. When individuals or groups work against one another to obtain a larger share of the valuables, they are in conflict. In conflict, defeating one's adversary is thought to be necessary. Social exchange is a form of interaction in which one person voluntarily does something for another person with the expectation that a reward will follow. Coercion exists when one person or group is forced to behave according to another person's or group's demands. When a person behaves as others expect, conformity occurs.
8. Most groups experience both cooperation and conflict. This is partly because groups satisfy some personal needs, such as social acceptance, while thwarting others, such as personal freedom. Social scientists have been very interested in the ways in which interaction within groups can achieve group goals while fulfilling the needs of individual group members.
9. Moving beyond the trait theory of leadership, social scientists have focused on appropriate leadership behavior. More recently, studies of leadership suggest that appropriate leadership behavior depends on the situation a leader faces.
10. In some situations, group decision making has been shown to be superior to individual decision making. Intellectual tasks, for example, are better handled through groups. On the other hand, group pressure may damage the quality of decision making because of the excessive conformity of group members. Many group decisions are reached in situations in which group members are discouraged from expressing their reservations. At the extreme, this is called groupthink.
11. A formal organization—the epitome of a secondary group—is deliberately created to achieve some goal. There are many types of formal organizations in modern society, but most share one feature—they are bureaucratic. According to Max Weber, bureaucracy is the most efficient type of formal organization. Using the ideal-type method, Weber outlined the characteristics that give bureaucracy its technical superiority over other types of organizations.
12. Although Weber saw bureaucracy as an efficient organizational form, he was well aware of its negative side. When rules and regulations become more important than organizational goals (as they often do), those who are supposed to be served by the bureaucracy suffer. Adherence to rules and regulations may also make bureaucrats overly conservative; such people learn their jobs so well that they cannot adapt to changing times and circumstances. Such problems have led researchers to examine alternatives to bureaucracy. Although most social scientists agree that bureaucracy is appropriate for many situations, contingency theorists have suggested that other types of organizations may fit particular circumstances.
13. The existence of primary groups and primary relationships within formal organizations has received considerable attention. These informal groups can either help or hinder the achievement of formal organizational goals, depending on the circumstances.
14. The exercise of power in organizations is receiving increasing attention. Power can benefit organizations by facilitating change and adaptation to changing conditions, but power can also be used for the benefit of a small, unrepresentative group within the organization. The applicability of the iron law of oligarchy to modern organizations is a continuing source of debate among sociologists.
15. Minorities, women, and lower social classes suffer from power differentials in organizations. For this reason, organizations serve to perpetuate the inequality that exists in the broader society.
16. Organizations exist within an environment. They depend on factors within that environment for necessary resources, and they also attempt to exert influence on portions of the environment. Both of these situations encourage involvement in interorganizational relationships, which allow organizations to adapt to environmental pressures but involve conflict as well as cooperation.
17. An examination of the unique characteristics of organizations in Japan demonstrates how culture can shape the structure and dynamics of formal organizations. It will be interesting to see the effects of future environmental change on the nature of formal organizations in Japan.

INFOTRAC® COLLEGE EDITION

http://www.infotrac-college.com/wadsworth

Another unique option available to you at the Student Resources section of the companion Web site is Infotrac College Edition, an online library with access to hundreds of scholarly and popular periodicals. Below are suggested search terms for this chapter. Results from these and other searches are found at the site.

Search keyword: **bureaucracy**. Many articles on Infotrac use the term *bureaucracy*. Scan several of these works to determine the general attitude toward this organizational form. Based on your sociological understanding of the word, do you concur with this general attitude?

Search keyword: **groupthink**. In 1982 Irving Janis wrote a book entitled *Groupthink: Psychological Studies of Policy Decisions and Fiascoes*, in which he studied the flawed decision-making process exhibited by several cohesive groups. Find three articles that discuss groupthink. Based on your reading and your knowledge of sociology, do you think Janis's characterization of his study as psychological is accurate? Why or why not?

Search keywords: **social category**. British society is noted for its large number of social categories. Find an article that discusses this phenomenon. How do the British people view their social categories?

LEARNING OBJECTIVES REVIEW

- Define the concept of the group and differentiate it from social categories and social aggregates.
- Outline the basic characteristics of primary and secondary groups.
- Describe the five major types of social interaction.
- Discuss the nature of group processes: task accomplishment, leadership styles, and decision making.
- Define the concept of formal organization and identify the major characteristics of bureaucracy.
- Discuss the advantages and disadvantages of bureaucracy.
- Distinguish between formal and informal organization.
- Describe the iron law of oligarchy and demonstrate its importance with examples.
- Discuss prejudice and discrimination in organizations.
- Compare and contrast the major features of Western organizations and Japanese organizations.

CONCEPT REVIEW

Match the following concepts with the definitions listed below them.

____ a. social exchange
____ b. rationalism
____ c. ideal-type method
____ d. bureaucracy
____ e. out-group
____ f. iron law of oligarchy
____ g. organizational environment
____ h. group
____ i. conflict
____ j. secondary group
____ k. formal organization
____ l. primary relationship
____ m. social category
____ n. conformity
____ o. in-group

1. a type of formal organization based on rationality and efficiency
2. occurs when individuals or groups work against one another to obtain a larger share of the limited valuables in a society
3. a type of social interaction in which an individual behaves toward others in ways expected by the group
4. a social structure deliberately created for the achievement of one or more specific goals
5. several people who are in contact with one another; share some ways of thinking, feeling, and behaving; take one another's behavior into account; and have one or more interests or goals in common
6. a method that involves isolating the most basic characteristics of some social entity
7. a group toward which one feels intense identification and loyalty
8. the principle stating that power tends to become concentrated in the hands of a few members of any social structure
9. all the forces outside the organization that exert an actual or potential influence on the organization
10. a group toward which one feels opposition, antagonism, and competition
11. a relationship that is intimate, personal, based on a genuine concern for another's total personality, and fulfilling in itself
12. the solution of problems on the basis of logic, data, and planning rather than tradition and superstition
13. a group that is impersonal and task oriented and involves only a segment of the lives and personalities of its members
14. a number of persons who share a social characteristic
15. when one person voluntarily does something for another, expecting a reward in return

CRITICAL-THINKING QUESTIONS

1 Identify one of your primary groups. Discuss the three functions of primary groups using your own experience to illustrate.

2 Think of a club you belong to on campus or a club you joined as a high school student. Was this a primary or secondary group? Support your answer with the use of sociological concepts and actual experiences.

3 Recall an in-group of which you are (or were) a member. Discuss the consequences this group had for others. Defend the right of this group to exist on sociological grounds.

4 Think of an example of groupthink at your school. Analyze this situation in relation to positive or negative consequences.

5 Most Americans think of bureaucracies in a very negative way. Develop an argument you could use the next time someone endorses the prevailing view of bureaucracy in your presence.

6 Analyze your school as a bureaucracy. Give an example of the following characteristics of bureaucracy: (1) a system of rules and procedures; (2) impersonality and impartiality. Discuss a positive and negative consequence of each characteristic.

7 Do you think that the Japanese type of organizational structure would work in the United States? Defend your position using the characteristics of Japanese organizations and your knowledge of American culture.

MULTIPLE-CHOICE QUESTIONS

1. **Several people who are in contact with one another; share some ways of thinking, feeling, and behaving; take one another's behavior into account; and have one or more interests or goals in common are known as a**
   a. commune.
   b. clique.
   c. group.
   d. social category.
   e. social aggregate.
2. **Which of the following are considered the two principal types of groups?**
   a. primary and small groups
   b. primary and reference groups
   c. secondary and small groups
   d. secondary and corporate groups
   e. primary and secondary groups
3. **A group is sometimes confused with a__________, a number of people who happen to be in the same place at the same time.**
   a. social collectivity
   b. social category
   c. social class
   d. social aggregate
   e. mob
4. **A web of social relationships that joins a person directly to other people and groups and, through those individuals and groups, indirectly to additional parties is known as a**
   a. social network.
   b. relationship web.
   c. social structure.
   d. safety net.
   e. social complex.
5. **Robert Nisbet has described five types of social interaction basic to group life. Which of the following is *not* one of them?**
   a. cooperation
   b. conflict
   c. coercion
   d. consensus
   e. conformity
6. **Which of the following statements has been found to be most accurate concerning group decisions?**
   a. Individual decisions are always better than group decisions.
   b. Group pressure has little effect on group decisions.
   c. Group decisions can be superior to individual decisions.
   d. Mistakes made by groups are harder to correct.
   e. Group decisions are always better than individual decisions.
7. **The type of formal organization based on rationality and efficiency is known as**
   a. a commune.
   b. an ecclesia.
   c. a hierarchy.
   d. a profession.
   e. a bureaucracy.

8. **According to the text, _________ is the ability to control the behavior of others even against their will.**
   a. authority
   b. prestige
   c. privilege
   d. ostracism
   e. power
9. **The solution of problems on the basis of logic, data, and planning rather than on the basis of tradition and superstition is known as**
   a. groupthink.
   b. rationalism.
   c. mythology.
   d. existentialism.
   e. jurisprudence.
10. **A judge who is more concerned with following legal procedures than reaching just decisions is engaging in**
    a. contingency thinking.
    b. categorical thinking.
    c. goal displacement.
    d. coercion.
    e. trained incapacity.
11. **The part of a formal organization in which personal relationships are guided by rules, rituals, and sentiments not provided for by the formal organization itself is known as**
    a. the enforcement of mores zone.
    b. the application of legitimate authority network.
    c. informal organization.
    d. the Peter Principle unit.
    e. the contingency group.
12. **_________ states that power tends to become concentrated in the hands of a few members of any social structure.**
    a. The iron law of oligarchy
    b. The Peter Principle
    c. Parkinson's law
    d. The law of demographic transition
    e. The iron law of responsibility
13. **_________ consists of all the forces outside the organization that exert an actual or potential influence on the organization.**
    a. Organizational environment
    b. Bureaucratic environment
    c. Organic-adaptive environment
    d. Organizational context
    e. External organization
14. **Which of the following is *not* characteristic of formal organization in Japan?**
    a. collective decision making
    b. rapid evaluation and promotion
    c. lifetime employment
    d. broad organizational training
    e. blending of employees' work and nonwork lives

FEEDBACK REVIEW

True-False

1. Social scientists agree that leaders must have certain traits if they are to be effective. T or F?
2. The effects of informal organization on the formal organization are almost always negative. T or F?
3. Japanese organizational structure proved not to be the exception to Weber's analysis. T or F?

Fill in the Blank

4. A _________ is a web of social relationships that joins a person directly to other people and groups and, through those individuals and groups, indirectly to additional parties.
5. Social workers who become more concerned with eligibility requirements than with delivering services to their low-income clients are engaging in _________.

Multiple Choice

6. **Which of the following is *not* a condition that promotes the development of primary groups?**
   a. small group size
   b. face-to-face contact
   c. continuous contact
   d. interaction on the basis of status or role
7. **Which of the following has *not* been seen as an advantage of bureaucracy?**
   a. its avoidance of the use of inappropriate criteria in hiring employees
   b. its use of rules to provide definite guidelines for behavior within the organization
   c. its success in hiding the true nature of authority relationships
   d. its encouragement of administrative competence in managers

Matching

8. Listed below are some examples of primary and secondary relationships. Indicate which examples are most likely to be primary relationships (P) and which are most likely to be secondary relationships (S).
   ____ a. a marine recruit and his drill instructor at boot camp
   ____ b. a married couple
   ____ c. a manager and his professional baseball team
   ____ d. professors and students
   ____ e. car salespersons and their prospective customers

9. Match the following terms and statements.

____ a. group
____ b. in-group
____ c. social aggregate
____ d. social category
____ e. reference group

(1) skinheads
(2) spectators at a Fourth of July fireworks display
(3) orchid growers
(4) teenage subculture
(5) members of a regular Saturday night poker game

GRAPHIC REVIEW

Table 6.3 displays the sales, assets, and profits of corporations by size of the organization. Answering the following questions will check your understanding of this information as it relates to the material in this chapter.

1. Choose one characteristic of bureaucracy to show that this type of organization makes possible the development of extremely large corporations.

____
____
____

2. Relate the concentration of power in the top 100 companies to the iron law of oligarchy.

____
____
____

ANSWER KEY

Concept Review

a. 15
b. 12
c. 6
d. 1
e. 10
f. 8
g. 9
h. 5
i. 2
j. 13
k. 4
l. 11
m. 14
n. 3
o. 7

Multiple Choice

1. c
2. e
3. d
4. a
5. d
6. c
7. e
8. e
9. b
10. c
11. c
12. a
13. a
14. b

Feedback Review

1. F
2. F
3. T
4. social network
5. goal displacement
6. d
7. c
8. a. S
   b. P
   c. S
   d. S
   e. S
9. a. 5
   b. 1
   c. 2
   d. 3
   e. 4

Deviance and Social Control

7

© LICHTENSTEIN ANDREW/CORBIS SYGMA

OUTLINE

LEARNING OBJECTIVES

- Define deviance and explain its relative nature.
- Define social control and identify the major types of social control.
- Describe the biological and psychological explanations of deviance.
- Discuss the positive and negative consequences of deviance.
- Differentiate the major functional theories of deviance.
- Compare and contrast cultural transmission theory and labeling theory.
- Discuss the conflict theory view of deviance.
- Distinguish the four approaches to crime control.

USING THE SOCIOLOGICAL IMAGINATION

How often is rape committed by a stranger? Unfortunately, many young people today know someone who has been raped. When women find themselves outdoors and alone, they are often fearful of possible surprise attacks by dangerous strangers. This is a genuine concern, and women are smart to be cautious; yet, over 80 percent of reported rapes in the United States are committed by individuals who are known to their victims. Experts use the terms *date rape* and *acquaintance rape* to describe this reality.

As date or acquaintance rape illustrates, deviant behavior may involve perversion. Deviance, however, is not limited to "nuts, sluts, and perverts" (Liazos 1972). **Deviance** is *any* behavior that departs from societal or group norms. It can range from criminal behavior (recognized by almost all members of a society as deviant) to wearing heavy makeup (considered deviant by some religious groups).

Deviance and Social Control

The Nature of Deviance

Without conformity and predictability in human behavior, society could not exist. Order and stability are the cornerstones of social life. All the same, we know that social life includes people who breach norms. Some people violate norms by robbing banks, assaulting others, or committing murder. Some incidents of less severe deviance have been high profile because they involved prominent sports figures whose behavior was captured on national television. Dennis Rodman, the flamboyant NBA player, cross-dresses, has tattoos all over his body, and regularly changes the color of his hair (yellow, red, blue, green, purple and orange, multicolored, and so forth). During one season, Rodman head-butted a referee and deliberately kicked a photographer in the groin. Even more dramatic was the behavior of Mike Tyson, former heavyweight boxing champion, who twice bit the ear of an opponent in a world championship fight (Starr and Samuels 1997).

Most behavior tends to fall within a normal range of social expectations. **Deviance**, behavior outside this range, may be either *positive* (overconforming) or *negative* (underconforming).

What is the distinction between positive and negative deviance? **Negative deviance** involves behavior that underconforms to accepted norms. Negative deviants either reject the norms, misinterpret the norms, or are unaware of the norms. This is the kind of behavior popularly associated with the idea of deviance. There is, however, another type of deviance. **Positive deviance** encompasses behavior that overconforms to social expectations (Heckert 2003). Positive deviants conform to norms in an unbalanced way. Underlying this type of deviance is an idealization of group norms that leads people to extremes of perfectionism. In its own way, positive deviance can be as disruptive and hard to manage as negative deviance. Few people find a society characterized by slavish and unchallenged conformity any more desirable than a disorderly and unpredictable society.

This more complete picture of the nature of deviance is illustrated in Figure 7.1. Think about the norms related to personal appearance in American society. For example, the mass media are constantly telling young people that "lean is mean." Negative deviants will miss the mark on the obese side. Positive deviants may push themselves to the point of anorexia. Most young people will weigh somewhere between these two extremes.

Why is identifying deviance a challenge? Deviance is a matter of social definition. In a diverse society like the United States, it is often difficult to agree on what does and does not constitute deviance. Jerry Simmons asked a sample of respondents to list categories or types of persons they considered deviant:

© Rhoda Sidney/PhotoEdit

Sexual abuse is more common on college campuses than the public may think. Three-quarters of college women report being victims of sexual aggression; almost a quarter report being victims of date rape.

FIGURE 7.1

The Distribution of Deviance Relative to the Norm of Leanness

Most people conform to norms within a socially acceptable range. Only at the extremes of this normal range of behavior does deviance appear. We are accustomed to talking about negative deviance, behavior that involves underconformity. We are much less aware of positive deviance, or overconforming behavior. Leanness as the desirable standard for physical appearance is illustrated in this figure. Think of another example that exists among college students.

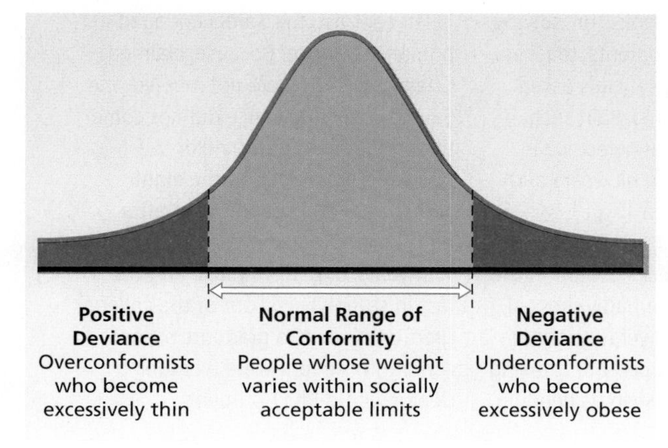

Source: Adapted from Jay J. Coakley, *Sport in Society*, 6th ed. (Boston, WBC/McGraw-Hill, 1998), p. 151.

> *The sheer range of responses predictably included homosexuals, prostitutes, drug addicts, radicals, and criminals. But it also included liars, career women, Democrats, reckless drivers, atheists, Christians, suburbanites, the retired, young folks, card players, bearded men, artists, pacifists, priests, prudes, hippies, straights, girls who wear makeup, the President of the United States, conservatives, integrationists, executives, divorcees, perverts, motorcycle gangs, smart-alec students, know-it-all professors, modern people, and Americans.* (Simmons 1969:3)

To this list, Leslie Lampert would add obese people. For a week she wore a "fat suit," adding 150 pounds to her normal body weight, to experience firsthand what it feels like to be an overweight woman in American society. According to Lampert, American "society not only hates fat people, it feels entitled to participate in a prejudice that at many levels parallels racism and religious bigotry" (Lampert 1993:154).

How do sociologists limit their study of deviance? Because minor instances of behavior that some might consider deviant occur frequently in modern societies, sociologists generally reserve the term *deviance* for violations of *significant* social norms. Significant norms are those that are highly important either to most members of a society or to the members with the most power. For sociologists, then, a **deviant** is a person who violates one or more of society's most highly valued norms.

Public reactions to deviants are usually negative and involve attempts to change or control the deviant behavior. Sociologists, however, do not make such value judgments. To them, deviant behavior is neither good nor bad; it is simply behavior that violates social norms and draws a negative response from others.

Conditions Affecting Definitions of Deviance

Identifying deviance is clearly a relative matter. Insight into the relativity of deviance began early in the history of sociology. According to William Graham Sumner, norms determine whether behavior is deviant or normal (Sumner 1906). Because norms vary from group to group, society to society, and time to time, the behavior considered to be deviant varies. Emile Durkheim contended that no behavior is deviant in itself, because deviance is a matter of social definition.

According to Sumner, Durkheim, and many other sociologists, the deviance label depends on three circumstances: the social status and power of the individuals involved, the social context in which the behavior occurs, and the historical period in which the behavior takes place.

How do social status and power influence the definition of deviance? Researcher William Chambliss (1973) demonstrated the importance of social status and power in defining deviance. As a participant observer, Chambliss watched two teenage gangs for two years at Hanibal High School. While at Hanibal High, Chambliss observed the school and nonschool activities of two gangs: the Saints, a gang of eight upper-middle-class white boys, and the Roughnecks, a gang of six lower-class white students. Although the Roughnecks and the Saints engaged in many of the same types of deviant behavior, the Roughnecks were judged by community members as worthless kids who were poor students, heavy drinkers, and violent individuals. Why, Chambliss asked, "did the community, the school, and the police react to the Saints as though they were good, upstanding, nondelinquent youths with bright futures but react to the Roughnecks as though they were tough, young criminals, who were headed for trouble?" (Chambliss 1973:28). The different treatment of these two gangs seemed to be traceable to their social class backgrounds. Lower-class people are more likely to be labeled deviants than are middle- or upper-class individuals, even when they behave in similar ways. (See Doing Research for more detail.)

Of course, high status does not always protect those who engage in deviant behavior. After discovery of

William Chambliss—Saints and Roughnecks

William Chambliss (1973), over a two-year period, observed the behavior of two teenage gangs at Hanibal High School. In addition to gang activity, Chambliss documented the responses of parents, teachers, and police to the delinquent behavior.

The Saints—eight children of white, stable, upper-middle-class parents—were pre-college, good students, and active in school affairs. The Roughnecks—six young men from lower-class white families—were poor students.

Both gangs engaged in similar amounts of delinquency. Yet, the Saints were slated for bright futures, while the Roughnecks were on track to dead-end lives. That was the picture held by parents, teachers, and law enforcement officials. And that was the way events actually turned out.

Why? The answer makes Chambliss's study sociologically fascinating. Deviance is relative to social class; it may be *primary* or *secondary*.

Hanibal townspeople simply never recognized the Saints' behavior as delinquent. When caught stealing or driving drunk, the Saints were seen as good boys "playing pranks" or "sowing wild oats." How else, parents, teachers, and law enforcement officials asked themselves, could you explain such behavior from well-mannered, well-dressed teenagers with nice cars and money in their pockets?

No such benefit of the doubt was extended to Roughnecks caught in the same acts. When these badly dressed, rude, not-so-rich boys were caught by the police stealing property or drinking excessively, they were seen as heading for trouble.

The Roughnecks engaged in secondary deviance—their delinquency evolved into a lifestyle and permeated their personal identities. Others responded to them as deviants, and they gradually incorporated this labeling into their self-definitions. The more they saw themselves as deviants, the more time they spent in deviant activities. Two of the boys dropped out of high school, one sentenced to thirty years for murder, the other sentenced for life on a murder conviction. Another Roughneck finished high school and become a gambler. There was scant information on a fourth Roughneck: he was "driving a truck up north." Two went to college on athletic scholarships and turned their lives around.

In contrast, the Saints engaged in primary deviance. Because, claimed Chambliss, they were not labeled, the Saints' acts of deviance did not come to dominate their lifestyle or self-image. In fact, seven of the eight Saints attended college right after high school. Five of these graduated in four years, one in six years, and the last in seven years. Four of the college graduates went to graduate school. Among these four were a lawyer, a doctor, and a Ph.D. student.

Thinking About the Research

1. From your understanding of Chambliss's study, is deviance socially created? Explain.
2. Which of the three major theoretical perspectives best explains Chambliss's findings? Support your choice.

some falsification of his resume in 2001, newly hired Notre Dame football coach, George O'Leary, had his job offer withdrawn (Kornheiser 2001). Noted historian Joseph Ellis was forced to take a year's leave of absence from his university teaching position following disclosure that, in his classes, he made false claims of personal military service. Although neither O'Leary nor Ellis committed prosecutable offenses, their deviant behavior was punished by official sanctions and personal embarrassment. Catholic priests accused of sexual child abuse have brought down the heavy hand of the church. If adopted, a new church policy would prohibit a molester from wearing a collar, celebrating public Mass, or publicly claiming to be a priest (Van Biema 2002). To address church officials who tolerate molesters under their supervision, an increasing number of state legislatures are passing laws requiring that allegations of sexual abuse of minors be reported to the authorities (Bayles 2002).

How do definitions of deviance vary with social context? Behavior can be considered conforming or deviant only within a particular social context; behavior considered deviant in one society may be acceptable and even encouraged in others. Suppose a group of people believe that gods live in rocks and trees and that these spirits are capable of granting them favors: power over their enemies, the ability to heal the sick, and the foresight to predict the future. Among the traditional Alaskan Eskimo, such people would be awarded high social status and deference. But in a highly industrialized society, members of such a group would be considered strange.

Deviance is also relative from group to group within the same society. Conformity to the norms of one group may earn intense disapproval from another. Teenagers, for example, are usually torn between conforming to the demands of their peers and meeting the expectations of their parents.

The possibility of being considered deviant from some group's point of view is likely in a society as complex as modern America. For instance, deviance can even vary within a geographical area. In Florida, it is illegal to play casino games on land. Three miles offshore in international waters, it is acceptable to play slot machines, cards, and other games of chance.

How do definitions of deviance vary over time? Societal approaches to cigarettes and drugs illustrate labeling variations over time. Gerald Markle and Ronald Troyer (1988), who studied the changing definition of cigarette smoking in American society, found that definitions of smoking have varied considerably. The definition of smoking was quite negative during the early 1900s, when smoking was banned by fourteen states. It became positive during the 1950s when advertising began to depict smoking as glamorous and sophisticated. Today, smoking is prohibited in places as diverse as restaurants, government buildings, corporate offices, airplanes, elevators, physicians' offices, and university buildings.

Like the United States, Europe has traditionally taken a strong stance against drug use. There is now a movement toward liberalization. While few in responsible positions favor legalizing hard drugs, some are less militantly opposed. As part of this shift from a war on drugs to an informal truce, some countries are tolerating marijuana use, and even the user of heroin or other hard drugs is less likely to be criminalized. Germany now authorizes local governments to permit citizen possession of drugs for "personal use." Other European countries are not ready for the legalization of drugs; their more tolerant view is, however, inviting less harsh ways to deal with drug use (Power 1999). Even though the 2002 Nevada ballot initiative to legalize possession of up to three ounces of marijuana failed, its very presence reflects growing support for the legalization of marijuana in the United States (Hawkins 2002; J. Stein 2002). Actually, one-third of Americans now favor legalizing marijuana, up from 12 percent in 1969 (Cauchon 2001).

Forms of Social Control

Human behavior is predictable most of the time. We generally know what to expect from others and what others expect of us. You know by now: this is no accident. Just as laws of physics help us predict the actions of the physical world around us, social and cultural mechanisms promote order, stability, and predictability in social life. We feel confident that drivers will not drive on the wrong side of the road, waiters will not pour wine over our heads, and strangers will not expect us to accept them as overnight guests. Without **social control**—means for promoting conformity to norms—social life would be capricious, even chaotic. There are two broad types of social control: internal control, which lies within the individual, and external control, which exists outside the individual.

What is the distinction between internal and external social control? Internal social control is self-imposed and is acquired during the socialization process. For example, most people most of the time refrain from stealing, not just because they fear arrest or lack the opportunity to steal, but because they consider theft to be wrong. The norm against stealing has become a part of the individual. This is known as the *internalization* of social norms.

Unfortunately for society, the internalization of norms is not complete in all societies or groups. If it were, we would have no deviance at all. Because the process of socialization does not ensure the conformity of all people all of the time, external social control must also be present.

External social control is based on sanctions—rewards and punishments—designed to encourage desired behavior. Positive sanctions, such as awards, salary increases, and smiles of approval, are used to encourage conformity. Negative sanctions, such as fines, imprisonment, and ridicule, are intended to stop socially unacceptable behavior. Sanctions may be formal or informal. Ridicule and gossip are examples of informal sanctions; imprisonment, capital punishment, exile, and honorary titles are formal sanctions.

1. __________ is behavior that violates norms.
2. A major problem that sociologists have in defining deviance is its __________.
3. __________ is the means for promoting conformity to a group's or society's rules.

Answers: 1. Deviance 2. relativity 3. Social control

Biological and Psychological Explanations of Deviance

As explained in Chapter 3, sociologists, despite the rise of sociobiology, remain skeptical of biological explanations of human behavior. The search for some biological foundation for deviance, however, has a long history and is still occurring.

Search for a Biological Basis for Deviance

What are some biological explanations of deviance? One of the earliest proponents of biological causes of deviance was Cesare Lombroso (1918), an Italian physician who saw a relationship between physical traits and crime. Lombroso believed that criminals were accidents of nature, throwbacks to an earlier stage of human evolutionary development. American psychologist William Sheldon (1949), a direct intellectual heir of Lombroso, identified three basic types of body shape: endomorphs (soft and round), mesomorphs (muscular and hard), and ectomorphs (lean and fragile). Sheldon concluded that criminals were more likely to be mesomorphs.

The theories of Lombroso and Sheldon are not generally considered valid today, but the search for a connection between physical characteristics and deviant behavior continues. Contemporary research in this area is much more sophisticated methodologically than the work of Lombroso and Sheldon (Fuller and Thompson 1978; Rowe 1983, 1986; Mednick, Gabrielli, and Hutchings 1984; Brunner et al. 1993). The most influential contemporary attempt to establish a link between biology and criminal behavior appears in a book entitled *Crime and Human Nature*, by James Q. Wilson and Richard Herrnstein (1985). The authors argue that Lombroso was thinking in the right direction. Although criminals are not born, the authors contend that many people are born with "constitutional factors"—personality, intelligence, anatomy—that predispose them to serious criminal behavior. Wilson and Herrnstein point to research on twins and adopted children. Criminality is said to be more common among identical twins than among fraternal twins. This is taken to be important because identical twins are the same genetically, whereas fraternal twins are only as genetically similar as brothers and sisters. Sons of chronic offenders raised in noncriminal homes are reported to have a three times greater probability of following a life of crime than sons of nonoffenders who have been adopted by law-abiding parents. Wilson and Herrnstein are careful to point out that they are referring only to genetic predispositions toward criminal behavior. People with these biological predispositions do not necessarily become criminals; these predispositions may be offset by socialization in an emotionally supportive environment, such as the family.

How do sociologists evaluate biological explanations of deviance? Sociologists generally have not placed much stock in biological explanations of deviance. There are five main reasons for this. First, biological theories ignore the fact that deviance is more widely distributed throughout society than are the hereditary and other physical abnormalities that are supposedly the causes of deviance. Second, biological theories almost totally discount the influence of social, economic, and cultural factors. These theories fail to explain how a person can have a genetic predisposition for something that is relative. Third, early biological theories of deviance were based on methodologically weak research. Fourth, there are ideological problems and controversial implications inherent in the biological approach. Forced sterilization of "defectives" and related eugenic strategies for reducing crime have created a storm of protests. Finally, biological factors are more often invoked to explain the deviance of armed robbers, murderers, and heroin addicts than, say, the crimes of corporate executives, government officials, and other high-status persons (Empey 1982).

Personality as the Source of Deviance

What common focus do psychological theories of deviance share? As varied as they are, psychological explanations of deviance all locate the origin of criminality in

© Jerry Koontz/The Picture Cube

Deviants are not just people who engage in obviously deviant acts such as prostitution. Any departure from social norms is considered deviant. This homeless woman is deviating from social expectations by rummaging through garbage. This makes most Americans feel uncomfortable.

© Mark Richards/PhotoEdit

Definitions of deviance change from time to time. Although same-sex marriage is viewed by some Americans as a form of deviance, prohibitions have loosened to the point that many gays are willing to publicly acknowledge their living relationship.

the individual personality (J. Hagan 1994). This is to be expected because psychology, as discussed in Chapter 1, concentrates on the development and functioning of mental-emotional processes in human beings. More specifically, psychological theories of deviance take for granted the existence of a "criminal personality," a pathological personality with measurable characteristics that distinguish criminals from noncriminals. Pathological personality traits may be either inherited (and therefore related to biological explanations of deviance) or socially acquired (Raine 1993).

Although Sigmund Freud did not attempt to explore the origins of deviance, his biologically based, instinctual approach to human personality lends itself to such exploration. Psychologists who have adapted Freud's theories to the study of deviance believe that deviance is due to unconscious personality conflicts created in infancy and early childhood. For example, criminality is said to be the result of biologically based desires beyond the control of a defective ego and superego. Other psychological explanations of crime are based on the identification of specific personality traits. Delinquents are said to be more aggressive, hostile, rebellious, or extroverted than nondelinquents (Glueck and Glueck 1950; Eysenck 1977). Some advocates of the psychological search for the criminal personality contend that criminals are born rather than made, an approach closely akin to biological explanations of deviance (Yochelson and Samenow 1994).

What is the sociological critique of psychological theories of deviance? First, psychological theories, like biological ones, often ignore social, economic, and cultural factors shown by sociological research to affect the likelihood, frequency, and types of deviance. Second, psychological theories focus on deviance such as murder, rape, and drug addiction with relatively little to say about such deviance as white-collar crime. Third, psychological theories tend to view deviance as the result of physical or psychiatric defects rather than as actions considered deviant by social and legal definitions. Fourth, psychological theories of deviance cannot explain why deviant behavior is engaged in by individuals not classifiable as pathological personalities. Finally, psychological theories, with their emphasis on pathology—deviants are "psychopaths" or "antisocial" personalities—suggest eugenic solutions to the crime problem that are unacceptable to some segments of society. For example, because more deviance occurs among certain racial and ethnic groups, birth control

FEEDBACK

1. Recent research has established a definite causal relationship between certain human biological characteristics and deviant behavior. T or F?
2. According to James Q. Wilson and Richard Herrnstein,
   a. studies of adopted children cannot help in understanding possible biological antecedents of criminal behavior.
   b. intelligence is not correlated with criminality.
   c. emotionally supportive families cannot help prevent criminal behavior.
   d. constitutional factors predispose individuals to criminality.
3. Which of the following is not one of the psychological approaches to deviance?
   a. extroversion
   b. race
   c. unconscious conflicts
   d. IQ
4. Psychological theories of deviance still have not developed the breadth to be able to explain white-collar crime. T or F?

Answers: 1. F 2. d 3. b 4. T

might be seen as a solution to the crime problem that is the most consistent with psychological theories.

This brings us to sociological explanations of deviance. The specific theories of deviance will now be discussed within the frameworks of functionalism, symbolic interactionism, and conflict theory.

Functionalism and Deviance

Functionalism sheds light on deviance, both in a general sense and via the strain theory of Robert Merton. We will examine both of these contributions as well as that of control theory.

Costs and Benefits of Deviance

Functionalism attempts to explain both the negative and the positive effects of social forces on society. It is easy to think of deviance as having negative social consequences. It is harder to imagine benefits of deviance, but functionalism tells us that there are some.

What are the negative effects of deviance on society? Widespread violation of norms threatens the foundation of social life. What would happen if bus drivers decided to create their own routes each morning, if local television stations aired what they preferred rather than what had been scheduled, if parents took care of their children's needs only when it was convenient? The result of such nonconformity is social disorder from which follows unpredictability, tension, and conflict.

Deviance erodes trust. If bus drivers do not follow the planned route, if television stations constantly violate their schedules, if parents support their children only sporadically, trust will be undermined. A society characterized by widespread suspicion and distrust cannot function smoothly. Take, for instance, the effects of increased rape in the United States. As pointed out in "Using the Sociological Imagination," rape overwhelmingly occurs between people who know each other. Almost one-fourth of college women report being victims of rape. Consequently, increasing female distrust often creates a barrier to normal dating relationships.

Unpunished deviance encourages others to be nonconformists. If bus drivers regularly pass people waiting for the bus, those people might bombard the buses with rocks. If television stations offer random programming, customers may damage their facilities in protest. If parents neglect their children, more teenagers may turn to delinquency. Deviance stimulates more deviance.

Deviant behavior is also expensive. For one thing, it diverts resources, both human and monetary. Police may have to spend their time dealing with wayward bus drivers, irate passengers, and mobs of annoyed television viewers rather than performing more serious duties. And if delinquency skyrockets, the criminal justice system will be even more jammed than it already is.

How does deviance benefit society? Despite its negative effects, deviance can sometimes benefit society. Emile Durkheim observed that deviance clarifies norms. By exercising social control, society defines, adjusts, and reaffirms norms; it defends its values. When parents are taken to court or lose their children because of neglect, for example, society's expectations are demonstrated to other parents and children.

Deviance can be a temporary safety valve. Teenagers listen to music, watch television programs, and wear clothes that adults may view as deviating from expected behavior. Involvement in this relatively minor deviant behavior may relieve some of the pressure teenagers feel from the social demands of the adult world (parents, teachers, clergy).

Deviance increases unity within a society or group. When deviance reminds people of something they value, it strengthens their commitment to that value. The exposure of government spies selling secrets to an enemy intensifies feelings of patriotism. Learning of parents who abuse their children may encourage other parents to rededicate themselves to proper child care.

Deviance promotes needed social change. Suffragettes who took to the streets in the early 1900s scandalized the nation but helped bring women the right to vote. Prison riots in the past have led to the reform of inhumane conditions. As deplorable as it sounds, the discovery of sexual molestation of young children may lead to an improvement of day-care facilities. Looting and other deviant behavior in inner cities may encourage government efforts to revitalize urban American.

Strain Theory

According to Durkheim (1964; originally published in 1893), **anomie** is a social condition in which norms are weak, conflicting, or absent. In the absence of shared norms, individuals are uncertain as to how they should think and act, and societies become disorganized. Robert Merton (1968) has adapted this concept of anomie to deviant behavior. He calls it strain theory.

According to Merton's **strain theory**, deviance is most likely to occur when there is a discrepancy between a culturally prescribed goal (economic success, for example) and a legitimate means of obtaining it (education). The resulting strain causes some people to engage in deviant behavior. Merton asserts that culture determines the things people should want (goals) and the legitimate ways (means) of obtaining these things. In American society, an important goal is success and the acquisition of the material possessions that usually accompany it.

© LICHTENSTEIN ANDREW/CORBIS SYGMA

Deviant groups, such as this Los Angeles Latino street gang, often engage in random, unpredictable acts that violate social norms. This creates a sense of mistrust among those outside the group.

Although everyone is taught to value material success, contends Merton, some are denied access to the legitimate means for achieving it (Passas and Agnew 1997).

How do people respond to strain? Merton suggests five general modes of adaptation. By way of illustration, Merton uses the American goal of success (see Table 7.1). One possible response is *conformity*—pursuing culturally approved goals through legitimate means. Here, people accept the goal and the means to achieve it. The very wealthy come to mind. But this category also includes poor people who continue to work hard in conventional jobs in the hope of improving life for themselves and their children. By definition, conformity is not deviant behavior. All of the other four responses are considered deviant, however.

In *innovation*, the individual accepts the goal of success but adopts illegitimate means for achieving it. Teenagers may steal an expensive automobile. Others may use prostitution, robbery, drug dealing, or other lucrative criminal behavior to be successful. Innovation is the most widespread and obvious type of deviant response.

In *ritualism*, the individual rejects the goal (success) but maintains the legitimate means. People go through the motions without really believing in the process. One example is the bureaucrat who continues to go about the daily routines of work while abandoning any idea of moving up the ladder. A college student may come to believe that college is without meaning, that the whole experience is a disappointment, that her goal of extreme economic success is not going to be realized. Yet, because of family expectations or lack of alternatives, she may continue to go through the motions needed to graduate.

Retreatism is a deviant response in which both the legitimate means and the approved goals are rejected. Skid-row alcoholics, drug addicts, and bag ladies have

TABLE 7.1

Conforming and Deviant Responses in Merton's Strain Theory

Describe the deviant behavior of some people you know using two of Merton's deviant responses.

Success as a Culturally Approved Goal	Hard Work as the Socially Accepted Way to Succeed	Conformity Response	Deviant Responses	Examples
Accepts goal of success	• Accepts hard work as the apppropriate way to succeed	**Conformity** —works hard to succeed		Business executive
Accepts goal of success	• Rejects hard work as the appropriate way to succeed		**Innovation**—finds illegal ways to succeed	Criminal
Rejects goal of success	• Accepts hard work as the appropriate way to succeed		**Ritualism**—acts as if he wants to succeed but does not exert much effort	Unmotivated teacher
Rejects goal of success	• Rejects hard work as the appropriate way to succeed		**Retreatism**—drops out of the race for success	Skid-row alcoholic
Rejects goal of success	• Rejects hard work as the appropriate way to succeed		**Rebellion**—substitutes new way to achieve new goal	Militia group member

Source: Adapted from Robert K. Merton, *Social Theory and Social Structure*, rev. ed. (New York: Free Press, 1968).

dropped out. Retreatists fail to be successful by either legitimate or illegitimate means; they do not even seek success.

In *rebellion*, people reject both success and the approved means for achieving it. At the same time, they substitute a new set of goals and means. Some militia group members in the United States illustrate this response. They may live in near isolation while pursuing their goal of changing society. They may use such means as creating their own currency, deliberately violating gun laws, and threatening (or engaging in) violent behavior against law enforcement officers.

How has strain theory been applied? Strain theory has been used most extensively in the study of juvenile delinquency. Albert Cohen (1977) uses the theory to explain the prevalence of gang delinquency among lower-class youth. According to Cohen, lower-class youth are poorly equipped to succeed in the middle-class world of school, which rewards verbal skills, academic success, neatness, and the ability to delay gratification. To protect their self-esteem, the youth reject these rules of success and self-worth and create their own status standards—standards they are able to achieve. Being tough, destructive, and daring is a vital part of this status system.

According to Richard Cloward and Lloyd Ohlin (1998), deviance does not always result from the strain created by a discrepancy between cultural goals and socially approved means for reaching these goals. Cloward and Ohlin refine strain theory, emphasizing that deviant behavior is not an automatic response; like any other type of behavior, deviance must be learned. And learning must take place through observation of others. Thus, deviance is more likely to occur as observation of others increases.

The Cloward and Ohlin concept helps to explain why deviance occurs more in some parts of a society than in others. In middle- and upper-class homes, young people are encouraged to value education—the most important socially approved means for success—and are exposed constantly to people who are, through legitimate means, successful. Poor young people, on the other hand, can more easily model their behavior after pimps, drug dealers, and other types of criminals who have illegitimately acquired material symbols of success—money, clothes, cars, and women. And when juvenile delinquents are sent to reform schools or prisons, they increase their opportunity to learn criminal behavior because they associate with well-practiced veterans. Cloward and Ohlin's use of strain theory overlaps with the symbolic interactionist approach to deviance, discussed in the next section.

How can strain theory be evaluated? Merton's strain theory has had great staying power, in part because of its applicability to juvenile delinquency and crime. Another plus is its emphasis on social structure rather than on individuals.

Strain theory has some shortcomings. Because it is in the functionalist tradition, it assumes a consensus in values; it assumes that everyone values success and defines success in economic terms. Also, strain theory does not seem to explain an individual's preference for one mode of adaptation over another. Why, for instance, do some people choose to conform rather than to rebel? Finally, although strain theory helps explain crime and delinquency, it offers no help in explaining other types of deviance, such as mental illness or drug abuse.

Control Theory

Travis Hirschi's (1972) control theory is also based on Durkheim's concept of deviance. According to **control theory**, conformity to social norms depends on a strong bond between individuals and society. If that bond is weak or is broken for some reason—if anomie is present—deviance occurs. In control theory, then, social bonds *control* the behavior of people; it is the social bond that prevents deviance from occurring. Individuals conform because they fear that deviance will harm the relationships they have with others. They don't want to "lose face" with family members, friends, or classmates.

What are the basic elements of social bonds? According to Hirschi, the social bond has four intertwined dimensions:

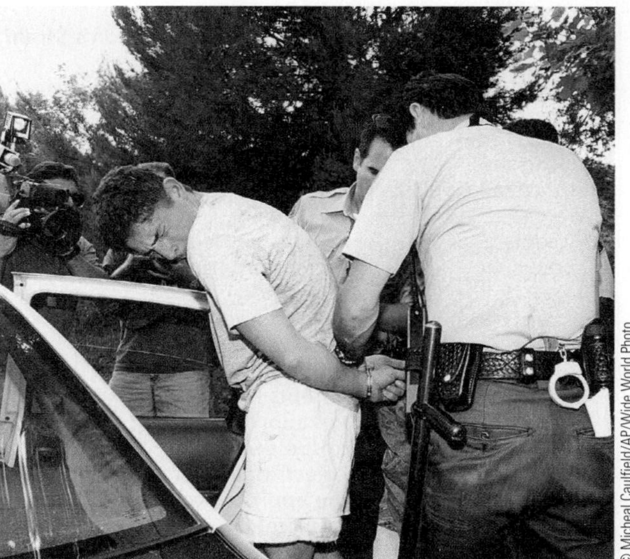

© Micheal Caulfield/AP/Wide World Photo

According to control theory, conformity to social norms relies on the presence of a strong bond between individuals and society. When social bonds are weak, the probability of deviance increases. When social bonds fail to control deviance, formal authorities such as the police supply the control.

1. *Attachment.* The likelihood of conformity varies with the strength of the ties one has with parents, friends, and institutions such as schools and churches. The stronger the attachment, the more likelihood of conformity.
2. *Commitment.* The greater one's commitment to legitimate social goals such as educational attainment and occupational success, the more likely one is to conform. The commitment of adults exceeds that of teenagers; and the commitment of individuals who believe their hard work will be rewarded is greater than the commitment of those who think they cannot compete within the system.
3. *Involvement.* Participation in legitimate social activities increases the probability of conformity, both because it positively focuses time and energy and because it encourages contact with others whose good opinion one values.
4. *Belief.* Subscription to norms and values of society promotes conformity. Respect for the rules of social life strengthens one against the temptations of deviance.

In short, when social bonds are weak, the chances for deviance increase. Individuals who lack attachment, commitment, involvement, and belief have little incentive to follow the rules of society.

How has control theory been received by other social scientists? Generally speaking, empirical support for Hirschi's control theory has been relatively strong. Several studies demonstrate the theory's power to explain both nondelinquent and deviant behavior (Bernard 1987; Rosenbaum 1987; Sampson and Laub 1990).

Scientific support weakened, though, when Hirschi attempted to broaden his control theory of delinquency into a more general theory encompassing all types of crime (Gottfredson and Hirschi 1990; Hirschi and Gottfredson 1990). Critics are skeptical that a specific theory of conformity and delinquency can explain crime in all its varieties, ranging from insider trading to murder. The existence of weak social bonds may satisfactorily account for pimps, drug dealers, and carjackers. But it is much more difficult for control theory to explain socially well-integrated individuals who embezzle corporate funds, use public monies for private purposes, or dump toxic waste into rivers (J. Hagan 1994; McCaghy and Capron 2002). In fairness, expansion of control theory needs further research.

Symbolic Interactionism and Deviance

According to theories based on symbolic interaction, deviance is learned through socialization. Concepts such as the looking-glass self, significant others, primary group, reference group, and the generalized other, discussed in Chapter 4, underlie both cultural transmission theory and labeling theory.

Cultural Transmission Theory

What is cultural transmission theory? Cultural transmission theory emerged from the "Chicago School" during the 1920s and 1930s. According to **cultural transmission theory**, deviance is part of a subculture; it is transmitted through socialization. Clifford Shaw and Henry McKay (1929) observed that delinquency rates stay high in certain neighborhoods, even though the ethnic composition changes over the years. According to Shaw and McKay, delinquency in

FEEDBACK

1. Which of the following is *not* one of the benefits of deviance for society?
   a. It decreases suspicion and mistrust.
   b. It promotes social change.
   c. It increases social unity among members of a society.
   d. It provides a safety valve.
   e. It promotes clarification of norms.
2. According to _________ theory, deviance is most likely to occur when there is a discrepancy between culturally prescribed goals and the socially approved means of obtaining them.
3. A college professor who simply goes through the motions of teaching classes without any thought of success is an example of which of the following responses in strain theory?
   a. rebellion   d. innovation   c. ritualism
   b. conformity   e. retreatism
4. The higher commitment to societal values and norms among adults than teenagers is consistent with the principles of _________ theory.

Answers: 1. a 2. strain 3. c 4. control

these neighborhoods persists because it is transmitted through play groups and gangs. As new ethnic groups enter the neighborhoods, they learn delinquent behavior from the neighborhood residents (Warr 2002).

Edwin Sutherland and Donald Cressey (1992) helped to extend cultural transmission theory to deviance that occurs from generation to generation and from one ethnic group to another. According to **differential association theory**, deviant behavior—like religious choice, political affiliation, or sport preference—is learned, principally, in primary groups. The more individuals are exposed to people who break the law, the more likely they are to become criminals.

At least three other factors impinge on the likelihood of deviance through differential association. First, an individual's vulnerability is affected by the ratio of exposure to deviants and nondeviants. Second, a person is more likely to become a deviant if his or her significant others are deviants. Finally, the earlier a person is exposed to deviants, the more likely that person will become a deviant.

How has cultural transmission theory been evaluated? Cultural transmission theory shows that conformity in one setting may be deviant in another. Beyond this, cultural transmission theory explains why individuals placed in prison may become even more committed to deviance; in prison one is exposed almost exclusively to criminals. **Recidivism**—a return to crime after prison—makes sense in the light of cultural transmission theory because ex-convicts normally return to the significant others, primary groups, and reference groups through which they originally learned to be deviants.

Cultural transmission theory leaves unanswered questions. It cannot account for the reason some individuals become deviant without exposure to great numbers of people who advocate breaking the law. Middle-class delinquents are an illustration. Conversely, cultural transmission theory cannot say why most people reared in crime-ridden environments do not become deviants. Cultural transmission theory is limited in scope because it deals only with crime. Finally, although cultural transmission theory explains how deviance is learned, it fails to explain why some behaviors are considered deviant whereas others are not. An explanation of this relativity of deviance is a major strength of labeling theory.

Labeling Theory

Is deviance defined by the act or the label? Although strain theory and cultural transmission theory help us understand why deviance occurs, neither explains the relativity of deviance. **Labeling theory** attempts to fill this gap. In this theory, no act is inherently deviant; deviance is a matter of social definition. According to labeling theory, deviance exists when some members of a group or society label others as deviant. Howard Becker, a pioneer of labeling theory, states the heart of this perspective:

> *Social groups create deviance by making the rules whose infraction constitutes deviance, and by applying these rules to particular people and labeling them as outsiders. From this point of view, deviance is not a quality of the act the person commits, but rather a consequence of the application by others of rules and sanctions to an "offender." The deviant is one to whom that label has successfully been applied; deviant behavior is behavior that people so label.*
> (Becker 1991:9)

Labeling theory explains, for example, why unmarried pregnant teen girls are more severely ostracized than the biological fathers. Even today, the males are seldom considered deviant because cultural ideas about their sexual responsibility are different. And, of course, it is easier to stigmatize women because advanced pregnancy is undeniably visible. Labeling theory also explains why a middle-class youth who steals a car may go unpunished for "borrowing" the vehicle, whereas a lower-class youth goes to court for stealing. Too often, lower-class youth are "expected" to be criminals while middle-class youth are not.

Does deviant behavior necessarily carry a label? Edwin Lemert's (2002) distinction between primary and secondary deviance helps clarify the labeling process. In cases of **primary deviance**, a person engages only in isolated acts of deviance. For example, when college students are asked to respond to a checklist of unlawful activities, most admit to having violated one or more norms. Yet, the vast majority of college students have never been arrested, convicted, or labeled as criminals. Certainly, those who break the law for the first time do not consider themselves criminals. If their deviance stops at this point, they have engaged in primary deviance; deviance is not a part of their lifestyle or self-concept. Juveniles, likewise, may commit a few delinquent acts without becoming committed to a delinquent career or regarding themselves as delinquents.

Secondary deviance, on the other hand, refers to deviance as a lifestyle and personal identity. A secondary deviant is a person whose life and identity are organized around deviance. In this case, the deviant status engulfs individuals and becomes a status that overshadows all other statuses. Individuals identify themselves primarily as deviants and organize their behavior largely in relation to deviant roles. Other people label them as deviant as well and respond to them accordingly. When this occurs, these individuals usually begin to spend most of their time committing acts of deviance; deviance becomes a way of life, a career (Kelly 1996). Secondary deviance is reflected in the words of Carolyn Hamilton-Ballard, known as Bubbles to her fellow gang members in Los Angeles:

> *Because of my size, I was automatically labeled a bully-type person. . . . I mean, people saw that Bloods jacket and since everybody thought I was crazy, I started acting crazy. At first it was an act, but then it became me. After being the target for drive-bys and going through different things, that became my life-style. I started retaliating back and I got more involved.* (Johnson and Johnson 1994:209)

What are the consequences of labeling? The labeling of people as deviants can cause pain and suffering. Erving Goffman revealed some of these negative effects of labeling when he wrote about **stigma**—an undesirable characteristic or label used by others to deny the deviant full social acceptance (Falk 2001). Ex-convicts, the unemployed, and the physically impaired are not accepted by many members of society. Why? Because stigmatic labels—such as jailbird, crazy, bum, and cripple—spoil an individual's entire social identity (*Mental Health* 1999; Heatherton, et al. 2000; Hudak and Kihn 2001; Link and Phelan 2001). One attribute—a missing leg, a stutter, a prison record—is used to form a general conception of an individual, and this stigma is then used to discredit the individual's entire worth. The words of a 43-year-old bricklayer, who was unemployed during the Depression, illustrate this point:

> *How hard and humiliating it is to bear the name of an unemployed man. When I go out, I cast down my eyes because I feel myself wholly inferior. When I go along the street, it seems to me that I can't be compared with an average citizen, that everybody is pointing at me with his finger. I instinctively avoid meeting anyone. Former acquaintances and friends of better times are no longer so cordial. They greet me indifferently when we meet. They no longer offer me a cigarette and their eyes seem to say, "You are not worth it, you don't work."* (Goffman 1963:17)

Does labeling cause deviance? According to labeling theory, labeling is all that is required for an act to be deviant. Deviance, however, cannot be defined solely based on the reactions of others. If an act is not deviant unless it is labeled, there can be no such thing as undetected deviance.

Deviance cannot be defined without some reference to norms, but this does not mean that labeling processes are unimportant. Future behavior may be affected by the way others respond to an individual's deviance. In this sense, labeling produces deviants. To assign a person the label of delinquent, drug addict, or prostitute is a form of social penalty that may lead to rejection and exclusion from conventional groups and consequently may evoke further deviant acts. Labeling leads to deviance when it produces secondary deviance.

Recall the study of the Roughnecks and the Saints, two teenage high school gangs. According to Chambliss (1973), labeling the Roughnecks as deviants promoted additional deviance. Labeling the Saints as upstanding young men promoted conformity to most conventional norms (see Doing Research).

What has been the evaluation of labeling theory? Labeling theory has contributed several insights. First, it has established that deviance is a matter of social definition, that deviance is relative. Second, labeling theory highlights, through concepts such as secondary deviance and stigma, the acquisition of a deviant self-concept that is likely to lead to a career of deviance. Third, labeling theory applies to a greater variety of deviance than the other theories. It can be applied to mental illness and prostitution as easily as to drug dealing and homosexuality.

Labeling theory also has its limitations. It does not explain habitual deviants who have never been detected and labeled, nor does it deal with isolated acts of deviance (primary deviance). It is also possible that labeling may halt rather than create deviance, as in the case of a man who frequents pornographic movie houses until he is suddenly discovered and labeled by his friends. Also, with its accent on relativity, labeling theory tends to overlook the fact that some acts, such as murder and incest, are banned in virtually all societies. Finally, labeling theory tends to generate sympathy for deviants, who are depicted as innocent victims of labeling.

Labeling theory implies that those who react against deviants are the culprits. Lawmakers do not cause people to commit criminal acts. It seems inappropriate to say that a police officer who arrests a person is somehow at fault. If the criminal law were abolished, acts that result in injuries would not disappear. In other words, a "mugging by any other name hurts just as much" (Nettler 1984).

Mental Illness and Labeling Theory

It is especially interesting to examine mental illness from the labeling perspective, because mental disorders, being a part of the medical field, are assumed to have an objective basis like cancer or arteriosclerosis. Labeling theory offers a unique way of viewing mental illness.

How does labeling theory approach mental illness? By virtue of its foundation in symbolic interactionism, labeling theory views mental illness as the result of social interaction in which others respond to us and we imagine what those responses mean. Mental illness is considered a matter of social definition (Collier 1993).

According to psychiatrist Thomas Szasz (1986), mental illness, as defined by the medical community, is a myth. Szasz sees the behaviors associated with mental disorder as adaptations to interaction-based stresses threatening to overwhelm an individual. In response to "problems of living," people (primarily unconsciously) impersonate a sick person to obtain help from others.

Thomas Scheff (1974, 1999) has formulated a theory of mental illness (based on the labeling perspective) in which most psychiatric symptoms of mental illness are seen as violations of social norms. Because of temporary pressures, contends Scheff, anyone might exhibit behaviors that violate norms—withdrawing socially, becoming overly suspicious, muttering to oneself. This primary deviance is usually considered transitory and is ignored by others. If this rule breaking is exaggerated and distorted by others, it can lead to a label—mentally ill. If the process of labeling goes far enough, mental illness may become a career path; it may reach the stage of secondary deviance in which it becomes part of a person's self-concept. Thus, a person may be successfully labeled mentally ill by a social process when, in fact, the original deviance may have been only temporary acts to handle temporary pressures.

Scheff's critics present data showing that the prime factor determining the negative reaction of others to mental patients is the patients' behavior, not the label such people have acquired (D. Phillips 1963, 1964; Link and Cullen 1983). More recent research concludes that labels do matter and that labeling theory cannot be dismissed as a framework for understanding responses to the mentally ill (Link 1987; Link et al. 1987).

Conflict Theory and Deviance

Deviance in Capitalist Society

Of the several conflict perspectives of deviance, Marxist researchers have forged an influential line of theory and research (Quinney 1980; Chambliss and Seidman 1982; Swaaningen 1997; Arrigo 1999). Focusing on class conflict, Marxist criminologists see deviance as a product of the exploitative nature of the ruling class. Deviance is behavior that the rich and powerful see as threatening to their interests. Consequently, the rich and powerful determine which acts are deviant and to what extent deviants should be punished.

According to the Marxian interpretation, the rich and powerful use the law to maintain their position. They can do this, argue the Marxists, because a society's beliefs, values, and norms (including, of course, laws) exist to protect the interests of the ruling class.

By what means do capitalism and deviance become linked? Steven Spitzer (1980) theorized some basic ways in which the culture of a capitalist society supports its economic system. First, critics of capitalism are considered deviants because their beliefs challenge the economic, political, and social basis of capitalism. Second, because capitalism requires a willing workforce, those who will not work are considered deviants. Third, those who threaten private property, especially that belonging to the rich, are prime targets for punishment. Fourth, because of society's need for respect of authority, people who show a lack of respect for authority—agitators on the job, people who stage nonviolent demonstrations against established practices—are treated as deviants. Fifth, certain activities and characteristics are encouraged or discouraged, depending on their congruence with the requirements of the economic system. Athletics are approved because they fos-

FEEDBACK

1. Which of the following did Edwin Sutherland mean by differential association?
   a. Crime is more likely to occur among individuals who have been exposed more to unfavorable attitudes toward the law than to favorable ones.
   b. People become criminals through association with criminals.
   c. Crime is not transmitted culturally.
   d. Crime comes from differential conflict.
2. _________ theory is the only sociological theory that takes into account the relativity of deviance.
3. _________ refers to deviance as a lifestyle and personal identity.
4. _________ is an undesirable characteristic used by others to deny the deviant full social acceptance.
5. According to labeling theory, mental illness is a matter of _________.

Answers: 1. a 2. Labeling 3. Secondary deviance 4. Stigma 5. social definition

ter competition, achievement, teamwork, and winning. Nonathletic males are wimps (Eder 1995; Adler and Adler 1998).

The Marxian perspective is not the only way to understand deviance from the viewpoint of conflict theory. The relationship between minorities and the judicial system is another way; white-collar crime is yet another (J. W. Coleman 1987, 1989; G. S. Green 1997; F. E. Hagan 2001).

Race, Ethnicity, and Crime

What is the relationship between race, ethnicity, and crime? Advocates of conflict theory point to the differential treatment minorities receive in the American criminal justice system. They cite statistics showing that, at all points in the criminal justice process, African Americans and Latinos are dealt with more harshly than whites—from arrest through indictment, conviction, sentencing, and parole (J. H. Skolnick 1998; Schaefer 2001). Even when the criminal offense is the same, African Americans and Latinos are more likely than whites to be convicted, and they serve more time in prison than whites (Huizinga and Elliot 1987; Bridges and Crutchfield 1988; Klein, Turner, and Petersilia 1988). Of those in prison, African Americans constitute a significantly larger proportion of those with death sentences. Although African Americans account for only 13 percent of the total population in the United States, more than 43 percent of inmates under the death penalty are African American.

Consider further illustrations of the differential experience of African Americans. During the four years following 1977, when the death penalty was legalized again in the United States, state attorneys were twice as likely to ask for the death penalty when an African American was found guilty of murdering a white person than when an African American murdered an African American or when a white killed a white (McManus 1985). Although nearly one-half of all homicide victims in the United States are African American, over 80 percent of prisoners on death row are there for murdering whites (Death Penalty Information Center 2002). Prosecutors are less likely to seek the death penalty when an African American has been killed, and juries and judges are less prone to impose the death penalty in cases involving African American victims. Finally, white murder defendants are more likely to work out plea bargains than African American defendants are (Pratt 1998).

As noted earlier, differential treatment of minorities and whites in the criminal justice system occurs in crimes less serious than murder. In the decade of the 1990s, the U.S. Department of Justice prosecuted only 252 police officers out of 17,000 formal complaints of police civil rights violations.

Why do minorities have a higher probability of being arrested, convicted, and severely punished? Conflict theory suggests several reasons for differences in the way minorities and whites are treated in the criminal justice system. For one thing, conflict theorists point to the fact that minorities generally do not have the economic

SOCIOLOGY AND THE NEWS MEDIA

Capital Punishment

This CNN video clip depicts the situation of Betty Lou Beets, a Texas woman convicted of the murder of her fifth husband. Scheduled for execution at the time, Beets appealed for clemency on the grounds of spousal abuse.

1. Using text materials, evaluate President Bush's belief that capital punishment saves lives.

   ____

   ____

2. If you had been the governor of Texas, would you have replaced her impending execution with a sentence of life imprisonment? Explain your decision on the basis of material in the text and information from the video clip.

   ____

   ____

About three-fourths of Americans would say that Betty Lou Beets's execution deters others from committing murder. What do you think? What do your parents think?

resources to buy good legal services. Thus, the outcomes of their trials are not likely to be as favorable to them.

Sociologists who follow the conflict perspective believe that crimes against whites tend to be punished more severely than crimes against minorities because society sees minority interests as less important than the interests of whites. **Victim discounting** reduces the seriousness of crimes directed against members of lower social classes (Gibbons 1985). According to the logic behind victim discounting, if the victim is less valuable, the crime is less serious, and the penalty is less severe.

White-Collar Crime

What is white-collar crime? According to Edwin Sutherland (1940, 1983), the originator of the term, **white-collar crime** is any crime committed by respectable and high-status people in the course of their occupations. Or, as one researcher put it, lower-status people commit crimes of the streets; higher-status people engage in "crimes of the suites" (Nader and Green 1972; D. R. Simon 2001). Officially, the term white-collar crime is reserved for economic crimes such as price fixing, insider trading, illegal rebates, embezzlement, bribery of a corporate customer, manufacture of hazardous products, toxic pollution, and tax evasion (Geis, Meier, and Salinger 1995; J. W. Coleman 1997; Calavita, Pontell, and Tillman 1999; Weisburd and Waring 2001).

U.S. citizens were dismayed in 2002 when it was discovered that many large corporations, including Enron and WorldCom, had illegally underreported their earnings in order to inflate their stock value. Moreover, we were shocked to learn, these corporations were aided in this deception by some accounting firms—such as Arthur Anderson—that were supposed to audit them. And all the while, brokers with national investment houses such as Merrill Lynch were encouraging investors to purchase stock they actually believed to be worthless in order to protect their firms' lucrative investment banking contracts with some of these same large corporations. Due to the subsequent collapse of stock values, thousands of workers lost their jobs (and their retirement nest eggs), and individual stockholders and pension funds lost billions.

How can white-collar crime be viewed from the conflict perspective? Advocates of conflict theory point out that white-collar crime is extremely harmful to society. Its perpetrators, however, are treated more leniently than other criminals because of their class position (Henry and Lanier 2001).

The costs of white-collar crime are higher than is generally thought. According to the U.S. Department of Justice, white-collar crime costs in excess of $200 billion annually—an amount eighteen times greater than the costs of street crime. Illegal working environments (for example, factories that expose workers to toxic chemicals) account for about one-third of all work-related deaths in the United States. Five times more Americans are killed each year from illegal job conditions than are murdered on the streets. None of these figures, of course, reflect the costs to society of a demoralized citizenry bombarded by news of criminal acts by corporate, political, and religious leaders (D. R. Simon 1998).

Despite this social harm, much more tolerance is shown white-collar criminals than criminals in the lower classes. Penalties are both tougher and more likely to be imposed for crimes that are committed by lower-class people than for those committed by people in higher social classes. Drug law violators (more likely to be from the lower class), for example, are treated more harshly than embezzlers. Although fewer than 400 embezzlers are in prison, thousands of drug law violators are incarcerated. In federal court, where most white-collar cases are tried, probation is granted to 40 percent of antitrust violators, 61 percent of fraud defendants, and 70 percent of embezzlers. In general, convicted white-collar criminals are less likely to be imprisoned; and if they are, these white-collar criminals receive shorter sentences. Moreover, when sentenced, white-collar criminals are more likely to be placed in a facility with extra amenities (for example, tennis courts or private rooms), which critics have dubbed "Club Fed" (Gest 1985; U.S. Department of Justice 1987; Reiman 2000).

In light of the conflict perspective, it will be interesting to examine the legal fate of top executives at corporations such as Enron, WorldCom, and Arthur Anderson. With so many corporations involved and with the magnitude of the personal and economic consequences, the legal response may be unusually harsh. These corporate officials (and their lax boards of directors) have already been rebuked by the public; and in 2002 the federal government enacted the stiffest legal penalties for financial fraud since the Great Depression in the 1930s (VandeHei and Hilzenrath 2002). Some criminal prosecutions by the federal government are likely—the process has already begun in a federal court with a guilty jury verdict for obstruction of justice against Arthur Anderson and in the "perp walks" (the public arrest and handcuffing of "perpetrators") forced on some executives at Adelphia Communications and WorldCom (Fineman and Isikoff 2002).

However, within a year after these scandals broke, penalties experienced by the perpetrators were relatively light. To avoid further investigations by New York State Attorney General Eliot Spitzer, major brokerage firms forged a financial settlement in 2002; but the amount of settlement was a fraction of either the money they made for themselves or the money they

TABLE 7.2

FOCUS ON THEORETICAL PERSPECTIVES: Illustrating Deviance

This table provides illustrations of deviance using concepts associated with a particular theoretical perspective. Construct some examples of your own.

Theoretical Perspective	Sociological Concept	Examples of Deviance
Functionalism	Anomie	• Delinquent gangs sell drugs because they do not agree with the norms and values against drug use advocated in the larger society.
Conflict theory	White-collar crime	• A convicted Wall Street stock manipulator (a more powerful member of society) may spend less time in prison than a factory worker (a less powerful member of society) found guilty of the same crime.
Symbolic interactionism	Labeling	• Some high school students reject dating because they have been consistently treated and described as "not cool."

lost for their trusting investors. While this earlier settlement with Spitzer did not prevent the Securities and Exchange Commission (SEC) from fining the 10 top investment firms in 2003, the $1.4 billion penalty pales beside the money the firms earned. Citigroup, for example, earned $43 billion between 2000 and 2002, but paid only a $400 million fine for its transgressions during the 1990s (Fogarty and Iwata 2003; Shell 2003). As of the beginning of 2003, only three individuals had been criminally indicted by the U.S. Department of Justice, and none of the investment firms were charged with fraud by the SEC.

What are the strengths and weaknesses of the conflict approach to deviance? Of course, not all definitions of deviance can be attributed to the exploitation of the weak by the powerful. Laws against murder and rape are meant to protect both the rich and the poor. Moreover, some laws—consumer and environmental laws—exist despite the opposition of big business. Most proponents of the conflict perspective would not conclude that deviance exists only because of cultural and social arrangements supportive of the ruling class. Nevertheless, the conflict approach to deviance, like labeling theory, underscores the relativity of deviance. Perhaps the greatest strength of conflict theory is its ability to link deviance to social inequality and power differentials.

Crime in the United States

Crime is the most important type of deviance in modern society. Most Americans think of **crime**—acts in violation of the law—as including a narrow range of behavior. On the contrary, in the United States, more than 2,800 acts are considered federal crimes, and many more acts violate state and local statutes.

If a society is to prevent and control crime, that society must be aware of just how much crime is being committed—and what type of crime. The major source of America's crime statistics is the FBI's *Uniform Crime Reports* (U.S. Department of Justice 2002). These official statistics are gathered from police departments across the country. Reports are submitted voluntarily by law enforcement agencies.

FEEDBACK

1. Marxist criminologists see deviance as a product of the exploitative nature of the ________.
2. Which of the following is *not* one of the basic ways the culture of a capitalist society supports the society's economic system?
   a. People whose beliefs clash with capitalism are labeled deviants.
   b. Capitalism requires a willing workforce.
   c. Innovation is rewarded.
   d. People who fail to show respect for authority are likely to be considered deviant.
3. The process of reducing the seriousness of crimes based on the lower social value placed on the victims is called ________.
4. ________ is any crime committed by respectable and high-status people in the course of their occupations.

Answers: 1. ruling class 2. c 3. victim discounting 4. White-collar crime

TABLE 7.3

Crimes in the United States

In this table you will find the rates for the major crime categories tracked annually by the FBI. Note the rate changes that occurred between 1992 and 2001. Pretending that you are a reporter for your school newspaper, use these data to summarize crime trends in the United States.

	2001		1992–2001	
Types of Crime	**Number of Crimes**	**Crime Rate (per 100,000 inhabitants)**	**Percent Change in Number of Crimes**	**Percent Change in Crime Rate**
Violent crime	1,436,611	504.4	–25.7	–33.4
Murder	15,980	5.6	–32.7	–39.8
Forcible rape	90,491	31.8	–17.0	–25.7
Robbery	422,921	148.5	–37.1	–43.7
Aggravated assault	911,706	318.5	–19.5	–27.9
Property crime	10,412,395	3,656.1	–16.7	–25.4
Burglary	2,109,767	740.8	–29.2	–36.6
Larceny-theft	7,076,171	2,484.6	–10.6	–19.9
Motor vehicle theft	1,226,457	430.6	–23.9	–31.8

Source: U.S. Department of Justice, Federal Bureau of Investigation, *Uniform Crime Reports*, 2001 (Washington, DC: U.S. Government Printing Office, 2002).

How is crime distributed in the United States? Data on arrests are collected for major categories of crimes (crime index offenses): murder, forcible rape, robbery, aggravated assault, burglary, larceny-theft, and motor vehicle theft (see Table 7.3). In 2001, the number of index crimes known to the police totaled 11,849,006, up 2.1 percent from 2000. Violent crime—murder, forcible rape, aggravated assault, and robbery—made up 12.1 percent of the known crimes, and property crime—burglary, larceny-theft, and motor vehicle theft—accounted for about 87.9 percent. Figure 7.2 displays the percent distribution of crime in America. Figure 7.3 on page 195 shows the *Uniform Crime Reports* estimate on the frequency at which crimes occur in the United States.

How much crime is there in the United States today? Crime increased sharply between the 1960s and the early 1990s. For example, the FBI index of violent crime has increased from a big-city offense rate per 100,000 of 860 in 1969 to over 1,100 in 2001. Violent crime rates are considerably higher than in most other industrialized countries:

> *Today the rate of homicide death for a young man is 23 times higher in the U.S. than in England. In 1995, handguns were used to kill 2 people in New Zealand, 15 in Japan, 30 in Great Britain, 106 in Canada, 213 in Germany, and 9,390 in the United States.* (To Establish Justice 1999:iv)

FIGURE 7.2

Types of Crimes Americans Commit: FBI Crime Index Offenses, 2001 (Percent Distribution)

This figure shows the contribution each major type of crime makes to the total of U.S. crime. What surprises you most about these figures?

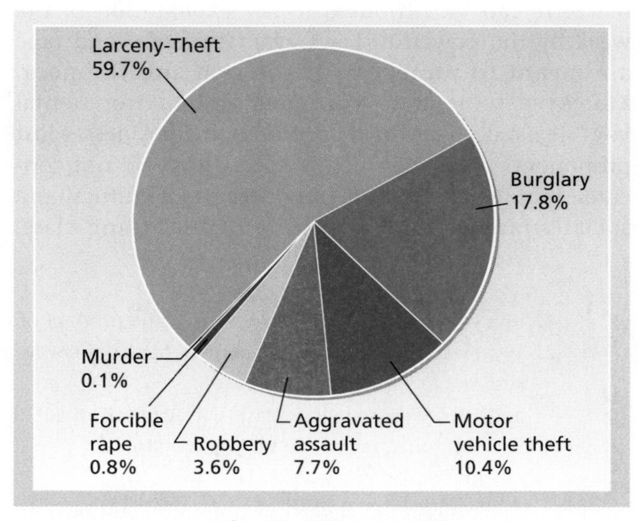

Source: U.S. Department of Justice, Federal Bureau of Investigation, *Uniform Crime Reports, 2001* (Washington DC: U.S. Government Printing Office, 2002).

FIGURE 7.3

How Often Do Americans Commit Crime? FBI Crime Clock: 2001

The numbers below, of course, do not mean that one crime index offense actually occurs every 2.7 seconds. It does mean that when *all* crime index offenses for 2001 are divided by the total number of seconds in a year, there are enough of them to be spaced out every 2.7 seconds. That is frightening enough.

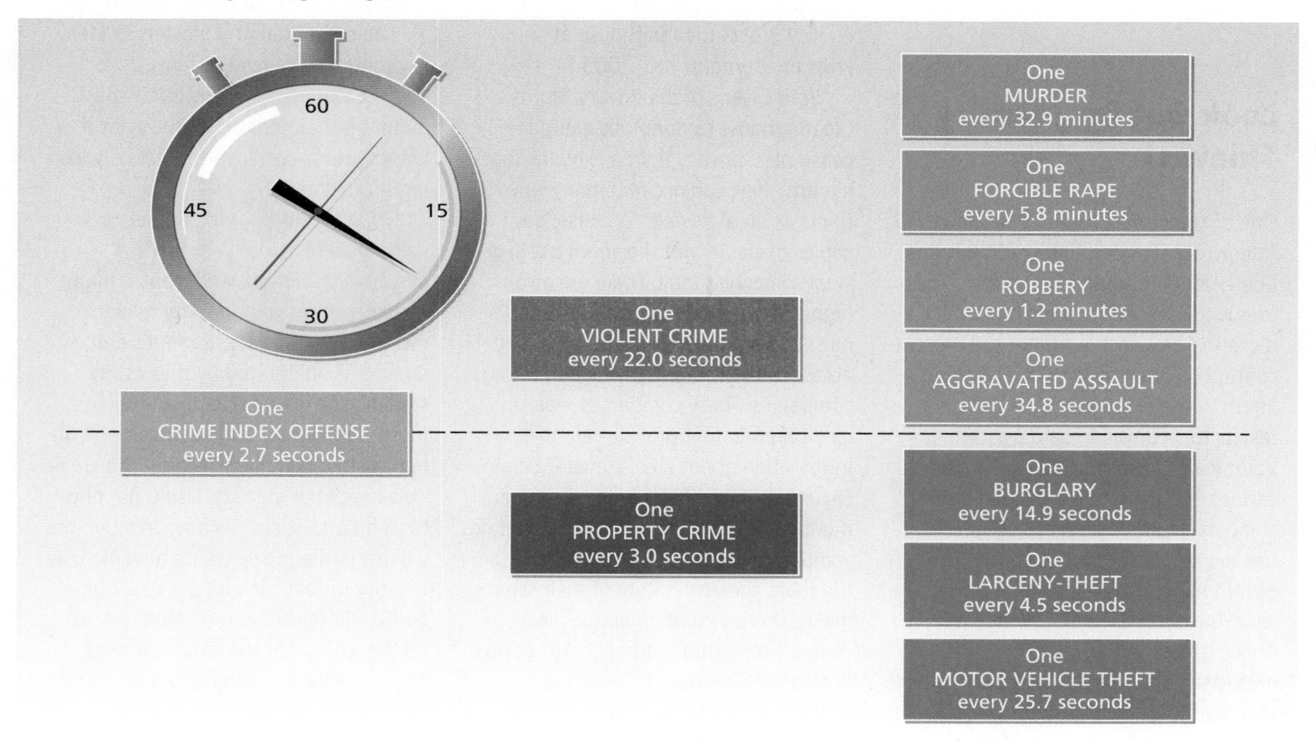

Source: U.S. Department of Justice, Federal Bureau of Investigation, *Uniform Crime Reports, 2001* (Washington DC: U.S. Government Printing Office, 2002).

INTERNET LINK The FBI publishes its *Uniform Crime Reports* annually. For additional information on the latest crime statistics for the United States, you can visit the FBI's home page at www.fbi.gov or you can access data from the U.S. Department of Justice at www/ojp.usdoj.gov

As Table 7.3 shows, however, both violent crime and property crime have declined since 1992. Because homicide receives the most publicity, it can be used to highlight this general, across-the-board reduction in crime. The murder rate in the United States has declined almost 40 percent since the early 1990s. This decline, however, has gained dramatic momentum only since the mid-1990s as the economy boomed and unemployment in the inner city diminished (Blumstein and Wallman 2000). Actually, the murder rate increased 1.3 percent between 2000 and 2001.

Measurement of Crime

How reliable are Uniform Crime Reports? Although the FBI statistics provide considerable information about crime, they have serious limitations. Some critics portray the FBI statistics as overrepresenting the lower classes and minorities but undercounting the middle and upper classes. Amateur thefts and minor assaults are not as likely to be reported to the police as murders and auto thefts. Prostitutes and intoxicated persons are subject to arrest in public places, but they are relatively safe in private settings, where the police cannot enter without a warrant.

Also, about two-thirds of U.S. crimes are not reported to the police. In addition, crime reporting varies from jurisdiction to jurisdiction and from crime to crime. For instance, government officials, corporate executives, and other white-collar offenders are seldom included in official crime statistics. Finally, in such areas as unfair advertising and job discrimination, the laws are enforced through special administrative boards rather than the criminal justice system (Box 1981; K. N. Wright 1987; Ermann and Lundman 1996; J. W. Coleman 1997).

In response to these criticisms, the National Crime Victimization Survey (NCVS) was launched in the

Look Out for Identity Thieves!

One of the newest forms of deviance is "identity theft"—stealing credit information belonging to another person. The frequency of identity theft began skyrocketing around the turn of the present century (O'Brien 2000; Grabosky and Smith 2001; O'Harrow 2001). The results to victims can be devastating. Veronica, a California college student, cannot get a student loan because someone in Michigan is using her identity to open credit card accounts and obtain telephone service. In another case, Meredith rented a room in her house to another woman. After discovering Meredith's Social Security number, this boarder used it to fill out several credit card applications in Meredith's name and to run up over $70,000 in charges. In yet another example, Graciela has been a victim of identity theft for more than ten years. A thief gained access to her Social Security number, birth certificate, and driver's license. With this information, the imposter has obtained credit cards, purchased furniture, bought cars, and obtained welfare. (All of these examples and more are available through the Privacy Rights Clearinghouse at www.privacyrights.org, 2003.)

Beth Givens of the Privacy Rights Clearinghouse (a nonprofit group for consumers' privacy rights) explains that identity theft can occur in many ways. A thief can steal a wallet or purse, get copies of credit card slips from trash, or steal someone's mail. There are also high-tech methods of identity theft. The most common method is through illegal access to the computers of credit-rating companies. These companies maintain credit reports that provide valuable information about a consumer—Social Security number, birth date, credit card numbers, address. Although credit-rating companies try to prevent high-tech identity theft, the very nature of their service makes this information accessible through computer terminals. This access is an open invitation to criminals.

The victims of identity theft obviously suffer great damage. Unless the thief is caught in the act, there seems to be little the police can do to stop this kind of crime. At the same time, many victims report that their victimization is compounded by abusive collection agencies. Victims also have to spend time and money cleaning up the mess imposters have created for them—sometimes up to ten years or more. Victims are often scarred emotionally, feeling violated, hopeless, and angry.

The main goal of imposters in identity theft is to purchase items at no cost. In these cases, the victims still maintain their identities. But what if identity theft also involved actually losing one's identity?

What would happen if a person's identity were actually "stolen"? A movie—*The Net*—showed what it might be like to have your identity taken away. In this movie, a woman's entire identity is obliterated by thieves. By stealing the documents that would prove her identity and destroying all of her existing computer records, the criminals hijack her identity. Using her photograph and Social Security number, the villains in the movie create a whole new identity for her, including a new name, bad credit report, and criminal record. As the woman in the movie laments, "They knew everything about me. It was all on the Internet!"

Analyzing the Trends
Which theoretical perspective would be most useful in analyzing identity theft? Explain your choice, and apply that perspective to identity theft.

early 1970s. This survey is conducted semiannually for the Bureau of Justice Statistics by the U.S. Census Bureau. The bureau uses state-of-the-art survey research techniques (U.S. Department of Justice 2001). At points, the NCVS and the *Uniform Crime Reports* contradict each other. For example, the NCVS indicates that the FBI statistics underestimate the total amount of crime victimization in the United States. Although recent FBI statistics reported just under 13 million crimes, the NCVS set criminal victimizations at about 26 million in 2000 (U.S. Department of Justice 2001).

Whom are we to believe? Both the *Uniform Crime Reports* and the National Crime Victimization Survey have their strengths. Because most people know relatively little about the technical definitions of crime, a major strength of the *Uniform Crime Reports* is its use of experienced police officers; they decide if an act has violated a statute. The NCVS has two attractive advantages. First, it helps to compensate for the massive official underreporting of crime. Second, the surveys are scientifically sound. At the very least, the NCVS is an increasingly important supplement to the FBI's official statistics. Together, these two sources provide a relatively complete account of the extent and nature of crime in the United States (K. N. Wright 1987; U.S. Department of Justice 2001, 2002).

Juvenile Crime

Juvenile crime refers to violations of the law committed by those less than eighteen years of age. Juvenile offenders are the third largest category of

FEEDBACK

1. According to the FBI's *Uniform Crime Reports*, crime in the United States has increased since 1989. T or F?
2. According to the National Crime Victimization Survey, does the *Uniform Crime Reports* underestimate or overestimate the amount of crime in America? ________

Answers: 1. F 2. underestimate

criminals in the United States. Teenage criminal activity includes theft, murder, rape, robbery, assault, and the sale of illegal substances. Juvenile delinquent behavior can extend beyond crime to deviance that only the young can commit (such as failing to attend school, fighting at school, and drinking and smoking under age).

What is the trend in juvenile crime? Violent juvenile crime reached its lowest level in a decade in 2001, a fall of about 30 percent since 1995 (U.S. Department of Justice 2002). Since 1993, the juvenile murder arrest rate has dropped by half, juvenile arrests for weapons violations have declined by a third, and the juvenile rape arrest rate has gone down by a fourth. There have also been fewer juvenile victims of murder—down from almost 3,000 to about 2,000. Juvenile crime, in short, returned to the rates typical of the years prior to the crack epidemic of the late 1980s.

Several factors are said to account for this decline in juvenile crime. First, there has been a decline in the demand for crack cocaine. Second, remaining crack gangs that provided guns to juveniles have reached truces. Third, police have clamped down on illegal guns, and repeat violent juvenile offenders have been given stiffer sentences.

Global Differences in Crime

Crime rates vary within any society. Sociological factors such as gender, age, race, ethnicity, social class, geographical region, and neighborhood each contribute to these variations. Crime rates also vary from society to society. One way to examine cross-cultural differences in crime is to compare crime rates in the United States with those of other countries.

Global Crime Comparisons

How do the levels of crime in other countries compare with crime in the United States? Data on crime rates in the 1990s reveal some striking differences between the United States and other countries (Kalish 1988; MacKellar and Yanagishita 1995). The following specific statements can be made:

- The U.S. homicide rate per 100,000 population is around 5.5; the rate of homicide in Europe is less than 2 per 100,000.
- The U.S. rape rate was 32 per 100,000, roughly six times higher than the average for Europe.
- U.S. rate for robbery was around 145 per 100,000 compared with the average European rate of less than 50 per 100,000.
- For crimes of theft and auto theft, the ratio of U.S. rates to average European rates was roughly 2 to 1.
- Burglary was the only crime examined for which the U.S. rate was less than double the average for European countries.
- U.S. crime rates were also higher than those of Canada, Australia, and New Zealand, but the differences were smaller than when compared with Europe. For burglary and auto theft, the rates were quite similar. (F. E. Hagan 2001)

Clearly, Americans watching the evening news do not have to fear being deceived by sensationalizing reporters telling them of the homicides, rapes, and robberies that seem to be occurring everywhere in the country. There *is* a lot of crime in the United States. Of the major industrialized countries, the United States has the highest murder, rape, and robbery rates and keeps pace, as well, in burglaries and auto thefts. Despite the rise in crime rates in Europe during the 1990s, crime remains a significantly greater problem in the United States than in any other country.

Are these cross-cultural data really to be trusted? You read earlier about the difficulties associated with gathering accurate U.S. crime data. So, it is not unreasonable to question the accuracy of data coming from many different cultures. There are problems associated with data recording, data collection, and varying definitions of crime from one country to another. Despite some notable dissenters (Balrig 1988; Gerber 1991), experts generally agree, though, that the differences in crime rates across societies are too great and too consistent to be denied credibility.

Crime and Modernization

According to one line of argument, increased crime is associated with economic development (Fenwick

FEEDBACK

1. In which of the following categories of crime is the United States *not* the leader in the world?
   a. murder
   b. rape
   c. robbery
   d. burglary
2. Crime in Japan refutes a predicted positive relationship between modernization and crime. T or F?

Answers: 1. d 2. T

1982). If this is the case, then the world's leading countries, economically speaking, will predictably have the highest levels of crime. This prediction, however, is contradicted by the relatively low rates for most types of crimes in such highly industrialized countries as Japan, Germany, and Switzerland.

Approaches to Crime Control

The **criminal justice system** comprises the institutions and processes responsible for enforcing criminal law. It includes police, courts, and a correctional system (Reichel 1998; Tonry 2000; Inciardi 2002; Senna and Siegel 2002). A criminal justice system may draw on four approaches to punishment—deterrence, retribution, incarceration, and rehabilitation—which may be used in varying combinations.

Deterrence

In 1994, Michael Fay, an American teenager convicted of car vandalism, was lashed four times with a 4-foot rattan cane in a Singapore prison (M. Elliott 1994). The experience of being "caned" is said to be sufficiently bad that future crime is effectively discouraged. Singapore is by no means alone in the use of corporal punishment. Practices such as caning are designed to protect society through deterrence. The **deterrence** approach emphasizes intimidation, using the threat of punishment to discourage criminal actions. There is considerable debate on the effectiveness of deterrence (DiIulio and Piehl 1991).

Does punishment deter crime? Until the mid-1960s, social scientists generally rejected the idea that punishment deters crime. This rejection was based partly on earlier studies, which indicated that the death penalty did not deter murder any more effectively than life imprisonment. More recent research has led to a reevaluation of the relationship between punishment and crime. Although the evidence to support this point is inconclusive, social scientists may have discounted the deterrence doctrine prematurely (Greenberg and Kessler 1982; Logan 1982; Hook 1989; Yunker 2001).

Investigation of the complicated relationship between punishment and crime is only in its infancy, but recent research indicates that the threat of punishment does deter crime if potential lawbreakers know two things: that they are likely to get caught and that the punishment will be severe. In the United States, however, punishment for crime is usually not certain, swift, or severe. Consequently, punishment does not have the deterrent effect that it could have (Pontell 1984).

Capital punishment is a special case. Murder is usually committed not after a rational consideration of the act or its consequences but during an outburst of emotion. Under such irrational circumstances, one would not expect the fear of capital punishment to be a deterrent, and research shows that it is not. If the death penalty were a deterrent to murder, then a decline in its use should be followed by an increase in the murder rate. But research in many countries, including the United States, indicates that the murder rate remains constant or even drops following a decline in the use of the death penalty. Other research indicates that the use of capital punishment as a sanction is neither swift nor certain (Acker, Bohm, and Lanier 1998; Sarat 1998; Schonebaum 1998; M. E. Williams 2000).

How do Americans feel about capital punishment? Actually, American attitudes toward capital punishment have shifted dramatically over the years. In colonial America, murderers, robbers, arsonists, counterfeiters, and many other types of criminals were regularly hung in the public square before large crowds. By the 1780s, the death penalty was beginning to be a sign of social backwardness by many. By the Civil War, three northern states had abolished it and the other northern states reserved it for murderers. Later in the nineteenth century, the death penalty lost even more public favor. In the late 1960s, the death penalty was abolished, only to gain acceptance less than a decade later. Since the mid-1970s, the frequency of executions has spiked (Banner 2003; Zimring and Zimring 2003) (see Snapshot 7.1).

Despite evidence to the contrary, about three-fourths of Americans believe that the death penalty is a deterrent to murder. But, of those Americans who favor the death penalty, over three-fourths say that they would continue to favor it even if confronted with con-

SNAPSHOT OF AMERICA 7.1

Number of Executions

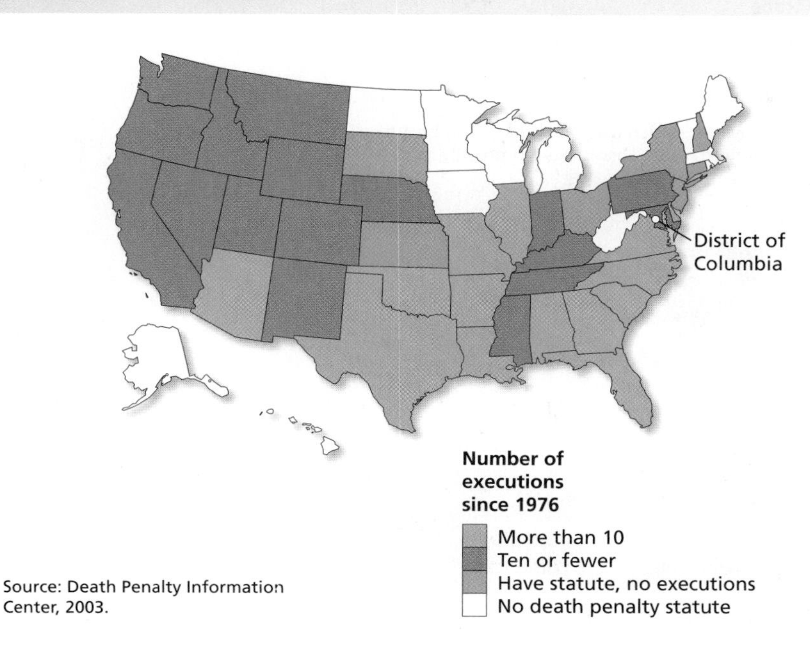

Source: Death Penalty Information Center, 2003.

Research shows that the existence of the death penalty does not appear to reduce the murder rate. Still, the death penalty is legal in most states. In fact, the United States is unique in leaving the decision on capital punishment to each state. This map shows the states that have no death penalty, and the degree of use in those states with capital punishment.

1. Does your state have the death penalty? When and why was the death penalty legalized or abolished in your state?
2. If the existence of the death penalty does not deter murder, why do you think it should still be legalized? Or should it? Explain.

clusive evidence that the death penalty does not act as a deterrent to murder and that it does not lower the murder rate. Rather than a need for deterrence, feelings of revenge and a desire for retribution, then, appear to contribute more to the support of capital punishment. (Zimring and Zimring 2003)

When asked to choose, a significantly higher proportion of the American population supports the death penalty for murder (71 percent) than opposes it (19 percent) (Longmire 1996). Attitude toward the death penalty in the United States varies according to race and ethnicity. As shown in Figure 7.4, 59 percent of whites favor the death penalty over life imprisonment with no chance of parole, compared with 10 percent of African Americans and 32 percent of other nonwhites. This racial and ethnic variation in attitude toward the death penalty is not surprising, considering that when compared to whites, African Americans are disproportionately sentenced to death row (Spohn 1995). Although African Americans make up only about 12 percent of the U.S. population, they make up 43 percent of death row inmates. And racial minorities constitute half of all inmates in U.S. prisons (Fins 2002; Herivel and Wright 2002).

In early 2003, there were new rumblings of the death penalty debate. Illinois Governor George Ryan, formerly a proponent of the death penalty, had emptied his state's death row. After a three-year review of Illinois death row cases, Ryan had taken some unprecedented steps. He began by pardoning four death row inmates, three of whom were released from prison immediately. And for the remaining 167 inmates, Ryan announced a commutation of the death sentences to life without parole. Citing injustice, arbitrariness, and capriciousness, Ryan indicated that he would no longer "tinker with the machinery of death" in a system "haunted by the demon of error" (Pierre and Lydersen 2003).

Supporters of the death penalty, including prosecutors and victims' groups, were extremely disturbed, charging that Ryan had taken justice into his own hands by subverting courts and legislatures. These critics point out that, though down slightly from 1990, Americans' support for the death penalty remains at about 70 percent.

Opponents of the death penalty had nothing but praise for Ryan's decision. They called attention to concern for mistakenly punishing the innocent. They attached great significance to the fact that since the death penalty was reinstated by the Supreme Court in the mid-1970s, one prisoner has been released from death row for every eight executed. Juries, it should be noted, issued the death penalty less frequently between 1998 (303) and 2001 (155), partly in recognition of the increasingly successful use of DNA testing and other forensic improvements in overturning convictions (Bower2003; Dirk Johnson 2003).

It remains to be seen if Ryan's actions spark increased opposition to the death penalty. His voice may or may not influence the ten or so states that have recently mandated studies on the fairness of the death penalty.

FIGURE 7.4

Attitudes Toward the Death Penalty and Life Imprisonment without Parole in the United States, by Race

This figure displays Americans' attitudes toward the death penalty and life imprisonment without parole. State the responses you would have made and explain why.

"If you could choose between the following two approaches, which do you think is the better penalty for murder—the death penalty or life imprisonment, with absolutely no possibility of parole?"

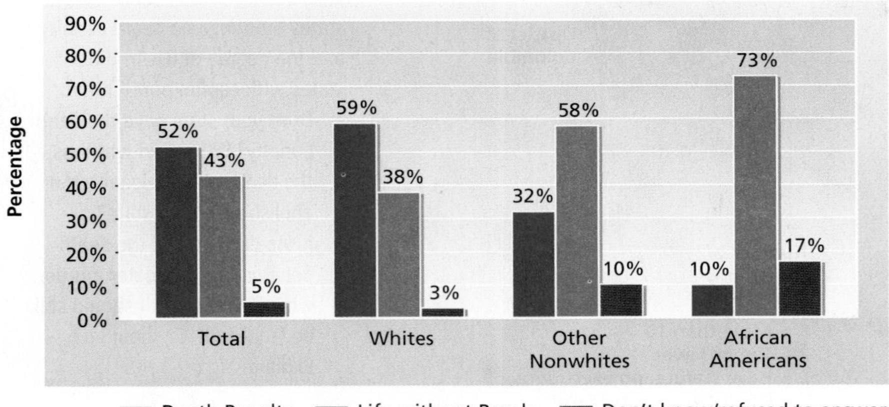

Source: Sourcebook of Criminal Justice Statistics Online, "Attitudes Toward the Penalty for Murder," 2002; available online at www.albany.edu/sourcebook/1995/pdf/t260.pdf.

Retribution

What is the philosophy behind the retributive approach to criminal punishment? The public may demand **retribution**—that criminals pay compensation equal to their offenses against society. When an eye is taken, an eye must be returned. Although the law allows designated officials to exact retribution, it does not tolerate acts of personal vengeance. The legal system removes responsibility for the punishment of a criminal from the hands of individuals and places it in the hands of officials. A convicted criminal repays the state rather than the harmed individual(s). Retribution, in the name of justice, is a social rather than a personal matter. If an individual whose husband is a victim of homicide exacts the price of wrongdoing by shooting her husband's killer, she is also responsible to society.

The retributive philosophy is concerned with past rather than future crime. Like deterrence, the incarceration and rehabilitation approaches to punishment, to which we now turn, focus on preventing further crime by already existing offenders.

Incarceration

What is the justification for incarceration? Criminals who are not on the street cannot commit crimes. The **incarceration** approach removes criminals from society. In repressive societies, such as Nationalist China and the former Soviet Union, people may spend their entire lives in prison camps for crimes ranging from political opposition to murder. The incarceration philosophy, of course, may take less extreme forms. The United States, for example, is less severe, even though recent decades have seen an increasing emphasis on incarceration (Grapes 2000; Mauer 2001). Compared to an historical average of the 1925–1970 period, the current rate of imprisonment in the United States has increased five times. Moreover, the current incarceration rate is six to eight times the rates of Western European countries (Garland 2001; Western and Pettit 2002).

What does the American public say? There are some signs of a decline in the "get-tough" approach. According to a recent study, American public opinion now favors crime prevention and treatment over severe punishment. In general, adults have shifted from several previously held convictions: deterrence through stricter sentencing, capital punishment for more crimes, and fewer paroles for convicted felons. By a 65 to 32 percent margin, they endorse prevention through job and vocational training, provision of family counseling, and more neighborhood activity centers for youth. In 1994, Americans were pretty evenly divided between addressing the underlying causes of crime or emphasizing deterrence through strict punishment. Almost two-thirds of Americans now agree that crime reduction is best achieved through rehabilitation *(Changing Public Attitudes Toward the Criminal Justice System 2002).*

Rehabilitation

Rehabilitation is an approach to crime control that attempts to resocialize criminals. Rehabilitative programs have traditionally been introduced within the prisons, with programs aimed at giving prisoners both social and work skills that will help them adjust to society after release.

Do prisons rehabilitate criminals? Unfortunately, because 30 to 60 percent of those released from penal institutions will be back in prison two to five years later, it is difficult to believe that rehabilitation programs in the United States are working (Elikann 1996; Zamble and Quinsey 1997). This high rate of recidivism—past offenders returning to prison—can be attributed partly to characteristics of the offenders and the problems they face while in the criminal justice system (such as the stigma of being an ex-convict). The tendency toward recidivism may also be due to the inadequacies of prisons. Prisons are schools for crime: ex-convicts often leave prison more committed as criminals and with a higher likelihood of continued criminal involvement (Abramsky 2001).

Social scientists also recognize that it is difficult to change attitudes and behavior within the prison subculture. Although prison guards and officials hold the power, the informal rules of the prison subculture have greater effect on prisoners' behavior. Conformity with the "inmate code" stresses loyalty among inmates as well as opposition to correctional authorities. This conformity is enforced verbally and physically. Consequently, involvement within prison criminal networks often increases in prisons, promoting a continued life of crime (J. Hagan 1993; Sherman 1993a).

There is yet another precipitating factor leading to continued criminal activity following imprisonment: the changing nature of employment in today's economy. Extended separation from a rapidly changing society leads to the obsolescence of employable job skills and to the deterioration of outside contacts who might be helpful in finding a job. Also, a released prisoner is likely to transfer the toughness reinforced in prison life to the workplace. This transfer of prison norms does not work, because most jobs in the service economy require interpersonal skills (J. Hagan 1994).

If prisons do not rehabilitate, what are some alternatives? Several alternatives are being tried. One is the combination of prison and probation. A mixed, or split, sentence, known as *shock probation*, is designed to shock offenders into recognizing the realities of prison life. Prisoners must serve part of their sentence in an institution. The remainder of the sentence is then suspended, and the prisoner is placed on probation.

Another approach relies on community-based programs designed to reintroduce criminals into society. By getting convicts out of prison for at least part of the day, community-based programs help break the inmate code. At the same time, prisoners have a chance to become part of society, participating in the community but under professional guidance and supervision.

A third strategy, *diversion*, is aimed at preventing, or greatly minimizing, the offender's penetration into the criminal justice system. Diversion involves a referral to a community-based treatment program rather than to a penal institution or a probationary program. Because offenders are handled outside the formal system of criminal law, the lawbreakers are not as likely to acquire stigmatizing labels and other liabilities (N. Morris and Tonry 1990; Lanier and Henry 1997).

Will any of these alternatives be successful? Most of the programs are relatively recent and have not yet been adequately evaluated. But as now being implemented, it appears that alternatives to prison offer little advantage over conventional probation programs. Still, defenders of alternative sanctions argue for their usefulness, if appropriately applied to those incarcerated who are a lower risk for recidivism (Tonry and Lynch 1996). And, given public concern about the skyrocketing costs of the criminal justice system, use of alternatives will probably increase with an eye to more effective application (Senna and Siegel 2002).

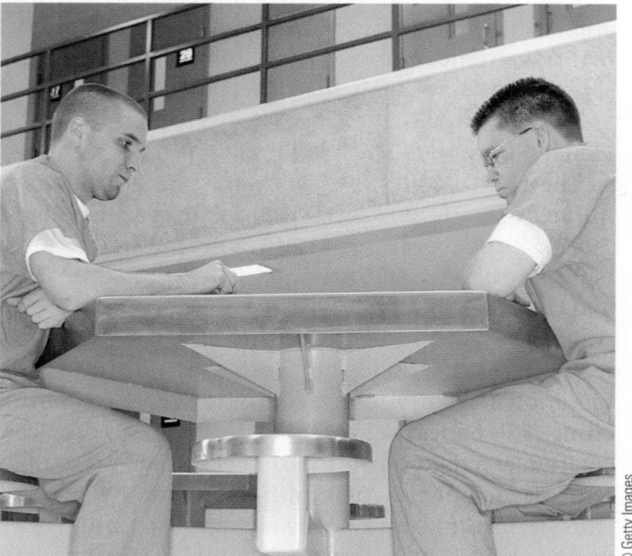

© Getty Images

Social scientists debate the effectiveness of imprisonment on the deterrence of crime. Even from model prisons like this one, between 30 and 60 percent of released inmates are reincarcerated within two to five years.

FEEDBACK

1. Match the approaches to punishment on the left side with the examples on the right.
   ____ a. rehabilitation — (1) imprisonment without parole
   ____ b. deterrence — (2) the practice of caning
   ____ c. retribution — (3) employment and educational programs in prison
   ____ d. incarceration — (4) death penalty for murder
2. Prisons currently do an adequate job rehabilitating inmates. T or F?
3. It appears that social scientists have prematurely discounted the deterrence doctrine. T or F?
4. Research supports capital punishment as a deterrent to murder. T or F?

Answers: 1. a. (3) b. (2) c. (4) d. (1) 2. F 3. T 4. F

Crime Prevention

The approaches just described target current criminals. A different sentiment for crime prevention concentrates on steering potential offenders away from criminal activity in the first place, by providing treatment for addicts and mentoring and counseling for poor children; creating living-wage jobs in poor neighborhoods; improving public education for poor children, and funding after-school and late-night recreation programs. Given Americans' punitive philosophy of corrections, it seems unlikely that we will invest heavily in "precrime" programs.

Crime Control: Domestic and Global

Crime Control in the United States

A society can adopt various combinations of the four approaches to crime control. Singapore, for example, takes deterrence, retribution, and incarceration very seriously. The upside is that Singapore has safe streets and clean subways. The downside is a reduction in individual liberty and increased brutality toward prisoners.

The United States does not have a consistent commitment to any one of the major approaches (Bouza 1993). Public opinion and public policy are split between the less punitive approach of rehabilitation and the other three approaches.

Public and political concern about violent crime has led to new penal policies that reflect a turn from rehabilitation toward a more punitive attempt to control crime. This endeavor involves more emphasis on incarceration as a method of deterring crime and on retribution as a method of compensation. By way of illustration, many Americans support California's "three strikes" statute, which doubles the standard penalty for second-time serious or violent crimes and requires twenty-five years to life for three-time felons (W. Booth 1994; Morin 1994; Shichor and Sechrest 1996). Due to this more severe approach, the prison population quadrupled during the 1980s and 1990s, standing at nearly 1.4 million in 1999. It is expected to exceed 2 million early in the twenty-first century.

Even recent rehabilitative initiatives have been harsher. Boot camps, for example, have been created for young nonviolent first offenders. The tough, military-style environment is intended to resocialize these young people by instilling self-discipline, commitment to work, and respect for the law (Katel 1994).

Although the latest national crime bill devoted some of its $30 billion to crime prevention programs (such as more police, youth programs, and the banning of nineteen semiautomatic assault-type weapons), it also has a strong punitive accent. This legislation vastly increases the number of federal crimes subject to the death penalty and imposes mandatory life sentences for repeat violent offenders. Funds are also available to states for new prisons.

A cross-cultural comparison in approaches to crime control will be informative. It reveals some striking differences between the United States and other countries (see Worldview 7.1).

Global Crime Control

All countries use incarceration. They vary significantly, however, in their emphasis on imprisonment (Christie 2000). Even Western European countries, which share many cultural traditions, have varying imprisonment rates. England and Wales are near the top in inhabitants per 100,000 in prison. Germany, France, and Spain are near the middle, and Greece and the Netherlands are near the bottom. As stated earlier, the imprisonment rate in the United States is six to eight times that of any Western European country.

Moreover, the United States is conspicuous when compared to the Eastern European countries. In Russia, for example, two out of every three crimes are not reported, and a relatively small proportion of reported crimes are solved ("Money Kills" 1996).

Not all industrialized countries have moved in the more punitive direction. In Sweden, emphasis remains

WORLDVIEW 7.1

National Death Penalty Policy

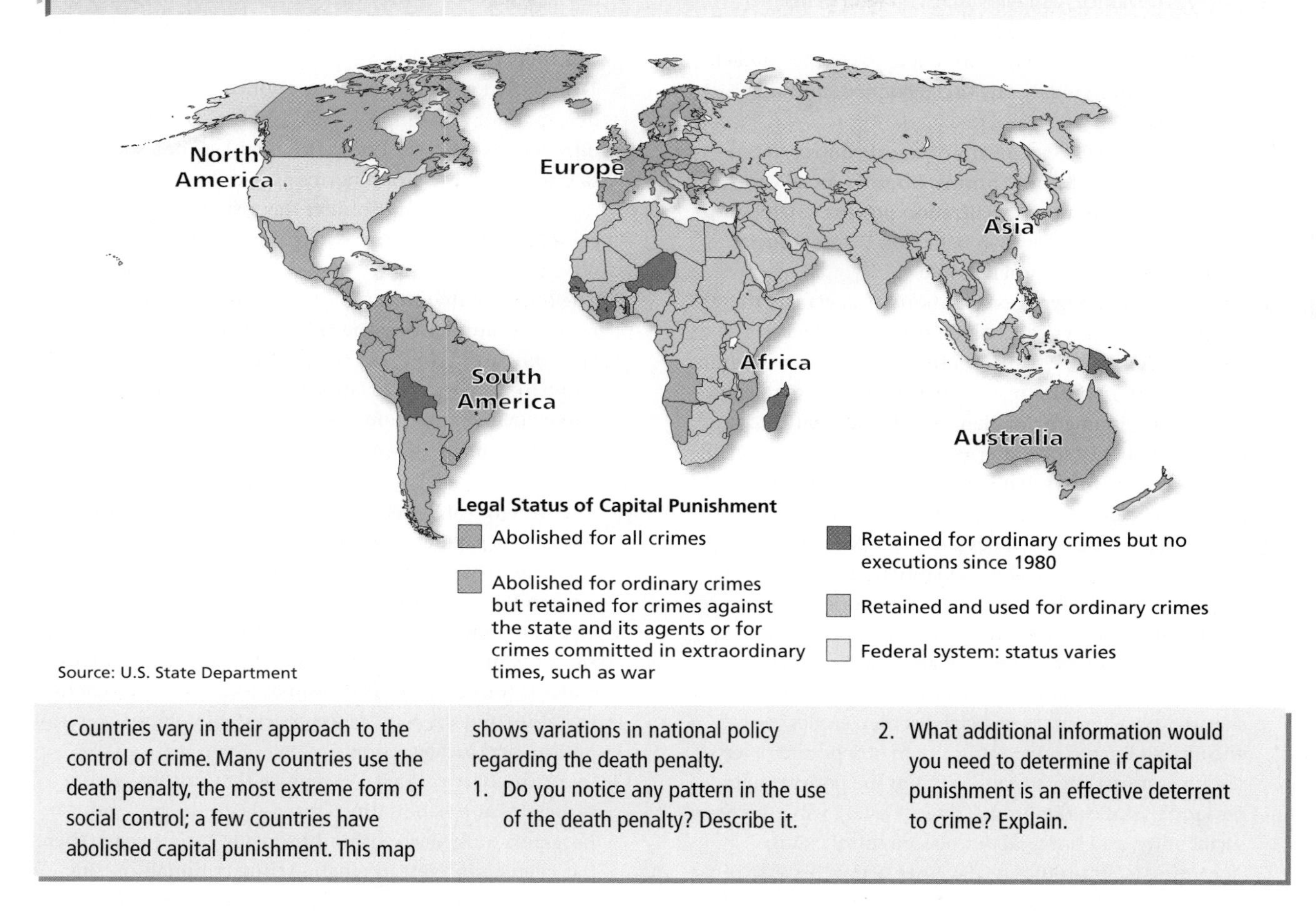

Countries vary in their approach to the control of crime. Many countries use the death penalty, the most extreme form of social control; a few countries have abolished capital punishment. This map shows variations in national policy regarding the death penalty.

1. Do you notice any pattern in the use of the death penalty? Describe it.
2. What additional information would you need to determine if capital punishment is an effective deterrent to crime? Explain.

on rehabilitation, treatment, and job training. Among the Japanese, criminals are resocialized to recognize their moral shortcomings; moral responsibility is placed on offenders (Feeley and Simon 1992). This stands in stark contrast to the current warehousing approach in the United States, where the prison population has increased sixfold since 1970 (Western and Pettit 2002).

FEEDBACK

1. Toward which approach to crime control has the United States been moving?
   a. rehabilitation c. incarceration
   b. deterrence d. retribution
2. Of these Western European countries, which has the highest rate of imprisonment?
   a. Spain c. Germany
   b. England and Wales d. Netherlands
3. The imprisonment rate in the United States is four times that of any Western European country. T or F?

Answers: 1. c 2. b 3. T

SUMMARY

1. Although social life requires conformity and order, it also involves deviance, behavior that violates norms. Deviance is difficult to define because it is a relative matter. What is considered deviant depends on the characteristics of the individuals, the social context, and the historical era.
2. All societies need some means of social control to promote conformity to their norms. Internal social control, promoted through the socialization process, enables individuals to control their own behavior. External social control is applied by other people.
3. According to biological explanations of deviance, there is a link between physical characteristics and deviant behavior. Early on, Cesare Lombroso believed that criminals were throwbacks to an earlier stage of human evolutionary development. William Sheldon attributed crime to body shape. Despite later research efforts, there is still no convincing proof that genetic characteristics cause people to be deviant.
4. Psychological theories of deviance attribute deviant behavior to pathological personality traits. Sociologists remain as skeptical of psychological theories of deviance as they are of those rooted in biology.
5. According to functionalists, deviance has both negative and positive consequences for society. On the negative side, deviance encourages social disorder, erodes trust, encourages further nonconformity in others, and diverts resources from other social needs. On the positive side, deviance helps clarify norms, offers a safety valve, increases social unity, and brings about needed social change.
6. According to strain theory, deviance occurs because of a discrepancy between cultural goals and socially acceptable means of achieving those goals. Control theory attempts to explain conformity rather than deviance. Conformity, which of course excludes deviance, is based on the existence of a strong bond between individuals and society.
7. The symbolic interactionist perspective yields two theories of deviance. Cultural transmission theory contends that deviance is learned, just like any other aspect of culture. According to labeling theory, an act is deviant only if other people respond to it as if it were deviant. Isolated norm violation may have no serious consequences for individuals. If individuals are labeled as habitual deviants, however, they may organize their lives and personal identities around deviance. Mental illness illustrates this theory.
8. The conflict perspective, when applied to the study of deviant behavior, emphasizes social inequality and power differentials. The most powerful members of a society are said to determine group norms and, consequently, the definition of deviant. Conflict theorists relate deviance to capitalism, pointing to the relationship between race, ethnicity, and crime.
9. The official crime statistics—the *Uniform Crime Reports* published annually by the FBI—are useful in providing an estimate of crime, but they underestimate national crime in America. The National Crime Victimization Survey (NCVS) was launched to overcome this limitation. There is some inconsistency in the findings of these two sources.
10. The United States has the greatest crime problem in the world. The gap between the United States and other countries remains substantial, despite the rising crime rates in Europe during the 1990s.
11. The four approaches to crime control are deterrence, retribution, incarceration, and rehabilitation. The United States is torn between the less punitive approach of rehabilitation and the harsher approaches of deterrence, retribution, and incarceration.
12. Increasingly, sociologists recognize that prisons are not successful at rehabilitating criminals. In the past, reform programs took place within prisons. In recent years, there has been an interest in rehabilitating criminals outside the prison system. During the 1970s, Americans began to take a harsher view of criminals. Even social scientists are beginning to believe that punishment can deter crime under certain circumstances, although there is no convincing evidence that capital punishment affects the murder rate.
13. Of late, the United States has increasingly moved toward incarceration as a solution to crime control. The imprisonment rate in the United States is currently six to eight times that of any Western European country and nearly double that of the former Soviet Union.

INFOTRAC® COLLEGE EDITION

http://www.infotrac-college.com/wadsworth

Another unique option available to you at the Student Resources section of the companion Web site is Infotrac College Edition, an online library with access to hundreds of scholarly and popular periodicals. Below are suggested search terms for this chapter. Results from these and other searches are found at the site.

Search keyword: **anomie**. Emile Durkheim introduced the concept of anomie to describe a social condition in which norms are weak, conflicting, or absent. Find articles that discuss anomie in current societies. Do you agree that anomic conditions exist in some societies today? Which societies suffer the most from anomie? What are some consequences of anomie in these societies?

Search keywords: **capital punishment**. The United States of America is one of the few wealthy industrialized countries that utilizes capital punishment. Read three articles that discuss either the relationship between capital punishment and crime rates or people's view of capital punishment. How did your opinion about capital punishment change, if at all?

Search keywords: **crime and retribution**. Use these two terms in a search to find out about one approach to punishment for crime. Does this approach appear to be effective in preventing crime?

Search keywords: **white-collar crime**. Find three articles that discuss the prevalence of white-collar crime and some typical punishments for these convictions. What issues are relevant to these topics?

LEARNING OBJECTIVES REVIEW

- Define deviance and explain its relative nature.
- Define social control and identify the major types of social control.
- Describe the biological and psychological explanations of deviance.
- Discuss the positive and negative consequences of deviance.
- Differentiate the major functional theories of deviance.
- Compare and contrast cultural transmission theory and labeling theory.
- Discuss the conflict theory view of deviance.
- Distinguish the four approaches to crime control.

CONCEPT REVIEW

Match the following concepts with the definitions listed below them.

____ a. labeling theory
____ b. differential association theory
____ c. retribution
____ d. control theory
____ e. social control
____ f. anomie
____ g. white-collar crime
____ h. cultural transmission theory
____ i. deviance
____ j. recidivism

1. behavior that violates norms
2. a social condition in which norms are weak, conflicting, or absent
3. the conception that conformity is based on the existence of a strong bond between individuals and society
4. the theory that deviance is part of a subculture transmitted through socialization
5. the public demand of compensation from criminals equal to their offense against society
6. the theory that crime and delinquency are more likely to occur among individuals who have been exposed to more unfavorable attitudes toward the law than to favorable ones
7. the theory that deviance exists when some members of a group or society label others as deviants
8. means for promoting conformity to a group's or society's rules
9. a crime committed by respectable and high-status people in the course of their occupations
10. a return to crime after imprisonment

CRITICAL-THINKING QUESTIONS

1. Identify a time in your life when you were considered a deviant by others. Discuss the attempts at social control you experienced. Be specific as to the types of social control and their concrete application to you.

____

____

2. Name some type of deviant behavior you would like yourself and/or others to engage in because you think society would benefit from it. State the sociological case for the social benefit of this type of deviance.

____

____

3. Describe someone you know who falls into one of the four deviant responses identified by strain theory. Use specific characteristics of this person to show that he or she fits the response you selected and not the other three responses.

____

____

4. Everyone has observed someone who has been labeled deviant by some members of society. Discuss the consequences of this labeling for the person identified as a deviant.

____

____

5. The text outlined several distinct approaches to crime control. Choose one approach and explain why you believe it has been successful or unsuccessful in relation to functionalism, conflict theory, or symbolic interactionism.

____

____

MULTIPLE-CHOICE QUESTIONS

1. **Behavior that violates norms is known as**
   a. deviance.
   b. anomie.
   c. stigma.
   d. recidivism.
   e. social control.
2. **Emile Durkheim contended that no behavior is deviant in itself, because deviance is a matter of social**
   a. pressure.
   b. biological determination.
   c. definition.
   d. control.
   e. indifference.
3. **The process of promoting conformity to a group's or society's rules is known as**
   a. coercion.
   b. influence.
   c. social cohesion.
   d. social control.
   e. groupthink.
4. **Imprisonment, capital punishment, medals, and honorary titles are examples of**
   a. informal sanctions.
   b. internal social control.
   c. formal sanctions.
   d. stigmas.
   e. differential rewards.
5. **According to Emile Durkheim, _________ is a social condition in which norms are weak, conflicting, or absent.**
   a. chaos
   b. ennui
   c. enmity
   d. melancholia
   e. anomie
6. **The strain theory has been used most extensively in the study of**
   a. white-collar crime.
   b. scientific fraud.
   c. alcoholism.
   d. prison riots.
   e. juvenile delinquency.
7. **Theories based on symbolic interaction say that deviance is the product of**
   a. behavior modification.
   b. cerebral mutations.
   c. socialization.
   d. trial and error.
   e. psychological adaptation.
8. **Clifford Shaw and Henry McKay used _________ to explain the continuation of high delinquency rates in certain neighborhoods even though the ethnic composition changed over the years.**
   a. cultural transmission theory
   b. differential association theory
   c. functionalism
   d. labeling theory
   e. deviant zone theory
9. **According to _________, deviance exists when some members of a group or society label others as deviants.**
   a. labeling theory
   b. functionalism
   c. strain theory
   d. cultural transmission theory
   e. symbolic interactionism
10. **Isolated acts of norm violations are known as**
    a. secondary deviance.
    b. noncriminal activity.
    c. learned deviance.
    d. primary deviance.
    e. independent deviance.
11. **Undesirable characteristics used to deny a deviant full social acceptance are known as**
    a. social shields.
    b. social brands.
    c. social stigmas.
    d. informal sanctions.
    e. formal sanctions.
12. **Marxist criminologists view deviance as a product of**
    a. socialization.
    b. language acquisition.
    c. differential association.
    d. the harmful effects of religion.
    e. class conflict.
13. **The official statistics from the FBI's *Uniform Crime Reports* are known to be biased mainly due to**
    a. undercounting of the lower class.
    b. overcounting of the middle and upper classes.
    c. overrepresentation of the lower classes and minorities.
    d. human error in compiling the statistics.
    e. the upper class not being included due to their high social status.

14. **The _________ is comprised of the police, the courts, and the correctional system.**
   a. justice complex
   b. social legal charter
   c. system of jurisprudence
   d. legal elite
   e. criminal justice system
15. **Which of the following is the most accurate statement regarding crime and punishment?**
   a. Crime and punishment are not related in industrial societies.
   b. Punishment deters crime if potential lawbreakers believe they are likely to be caught and severely punished.
   c. Punishment deters crime if potential lawbreakers believe that the criminal justice system is fair.
   d. Punishment deters crime only if children are taught that an eye for an eye is the best philosophy.
   e. Potential lawbreakers are less likely to commit crimes if they have friends or relatives who have been punished by the criminal justice system.

FEEDBACK REVIEW

True-False

1. Recent research has established a definite causal relationship between certain human biological characteristics and deviant behavior. T or F?
2. It appears that social scientists have prematurely discounted the deterrence doctrine. T or F?
3. The imprisonment rate in the United States is six to eight times that of any Western European country. T or F?

Fill in the Blank

4. A major problem that sociologists have in defining deviance is its _________.
5. According to labeling theory, mental illness is a matter of _________.
6. The process of reducing the seriousness of crimes based on the lower social value placed on the victims is called _________.

Multiple Choice

7. **Which of the following is *not* one of the benefits of deviance for society?**
   a. It decreases suspicion and mistrust among members of a society.
   b. It promotes social change.
   c. It increases social unity.
   d. It provides a safety valve.
   e. It promotes clarification of norms.
8. **Which of the following did Edwin Sutherland mean by differential association?**
   a. Crime is more likely to occur among individuals who have been exposed more to unfavorable attitudes toward the law than to favorable ones.
   b. People become criminals through association with criminals.
   c. Crime is not transmitted culturally.
   d. Crime comes from differential conflict.
9. **In which of the following categories of crime is the United States *not* the leader in the world?**
   a. murder
   b. rape
   c. robbery
   d. burglary

Matching

10. Match the approaches to punishment with the examples below.
   ____ a. rehabilitation
   ____ b. deterrence
   ____ c. retribution
   ____ d. incarceration
   (1) imprisonment without parole
   (2) the practice of caning
   (3) employment and educational programs in prison
   (4) death penalty for murder

GRAPHIC REVIEW

Attitudes toward the death penalty for murder vary among Americans. The results of a Gallup poll, contained in Figure 7.4, clearly reflect a difference between blacks and whites. Answer these questions to explore the meaning of these results.

1. Do you think that the strong support for capital punishment among white Americans is due more to its deterrent effects or to the desire for revenge? Explain.

____

____

____

2. Why do you think African Americans are less supportive of the death penalty than white Americans?

ANSWER KEY

Review

a. 7
b. 6
c. 5
d. 3
e. 8
f. 2
g. 9
h. 4
i. 1
j. 10

Multiple Choice

1. a
2. c
3. d
4. c
5. e
6. e
7. c
8. a
9. a
10. d
11. c
12. e
13. c
14. e
15. b

Feedback Review

1. F
2. T
3. T
4. relativity
5. social definition
6. victim discounting
7. a
8. a
9. d
10. a. 3
    b. 2
    c. 4
    d. 1

8

Social Stratification

© John Coletti/Stock, Boston

OUTLINE

LEARNING OBJECTIVES

- Explain the relationship between social stratification and social class.
- Compare and contrast the three dimensions of stratification.
- State the major differences between the functionalist, conflict, and symbolic interactionist approaches to social stratification.
- Identify the distinguishing characteristics of the major social classes in America.
- Discuss the extent of poverty and the perceptions of poverty in the United States.
- Evaluate U.S. commitment to poverty programs.
- Outline some of the consequences of social stratification.
- Describe upward and downward social mobility in the United States.
- Discuss the major features of global stratification.

USING THE SOCIOLOGICAL IMAGINATION

Is federal spending on social welfare programs excessive? A significant proportion of the U.S. population is unhappy with the amount of tax dollars being spent on the poor. Americans, however, seriously overestimate the proportion of the federal budget that is spent on welfare. Less money in real terms was spent on welfare recipients in 2002 than in 1997. Besides, these welfare expenditures account for less than 1 percent of total federal expenditures, a reduction of three-fourths since 1970, and less than one-fifth of 1 percent of the total gross national product (GNP) of the United States.

Negative attitudes toward welfare recipients are part of American culture, and these attitudes stigmatize the poor. Although all societies have social inequality, not all societies develop negative (and false) stereotypes of those at the bottom. We will return to this topic later in this chapter, but for now we turn to the dimensions on which social classes are created.

© Earl & Nazima Kowall/CORBIS

Social stratification creates layers of people—social classes—possessing unequal shares of scarce desirables such as financial resources, prestige, and power. These members of India's ancient caste system are relegated by birth to the very bottom of the stratification structure.

Dimensions of Stratification

Because of the human tendency to rank one another, inequality emerges. Soon some people are at the top of the hierarchy and others are at the bottom. This tendency is a recurring theme in literature. Dr. Seuss writes of the Sneetches, birds whose rank depends on whether or not they have a large star on their stomach. Star-bellied Sneetches have high status; plain-bellied Sneetches have low status. In the classic novel *Animal Farm*, George Orwell created a barnyard society where the pigs ultimately take over the previously classless animal society by altering one of their cardinal rules from "All animals are equal" to "All animals are equal—but some animals are more equal than others."

What is the difference between social stratification and social class? The fictional creatures mentioned in the previous paragraph are engaged in **social stratification**—the creation of layers (strata) of a population who possess unequal shares of scarce desirables, the most important of which are income, wealth, power, and prestige (R. Levine 1998). Each of the layers is a **social class**—a segment of the population whose members hold a relatively similar share of scarce desirables and who share values, norms, and an identifiable lifestyle. The number of social classes within a stratification structure varies. A stratification structure might include upper upper, middle upper, lower upper, upper middle, middle middle, lower middle, upper lower, middle lower, and lower lower classes. Or, like old coal-mining towns in America or like underdeveloped countries, there might be only an upper class and a lower class.

Karl Marx and Max Weber made the most significant early contributions to the study of social stratification. Marx demonstrated the importance of the economic foundations of social classes, while Weber accented the prestige and power aspects of stratification. Together, the work of these two great minds reveals to us that stratification is multidimensional (Gerth and Mills 1958; McCall 2001).

The Economic Dimension: Karl Marx

Although both Marx and Weber were concerned with inequality, their emphases were different. For Marx, the economic factor was an independent variable explaining the existence of social classes. Weber, on the other hand, viewed the economic dimension as a dependent variable. That is, Weber was more concerned with the economic consequences of stratification.

Marx identified several social classes in nineteenth-century industrial society—laborers, servants, factory workers, craftspeople, proprietors of small businesses, moneyed capitalists—but predicted that capitalist societies would ultimately be reduced to two social classes. Those who owned capital (the means of production)—the **bourgeoisie**—would be the rulers. Those who worked for wages—the **proletariat**—would be the ruled. The propertied capitalist class could exploit the labor of those without property. Even though the workers performed all the labor, they would be kept at a subsistence wage level, and the capitalists would enjoy the economic surplus created by that labor. In short, Marx predicted that because the capitalists owned the means of production (factories, land), they would both rule and exploit the working class. The working class would have nothing to sell but its labor.

Marx's analysis went deeper: Marx contended that all aspects of capitalist society—work, religion, government, law, morality—were economically conditioned. Consequently, the capitalists controlled all social institutions, which they used to their advantage. They could structure the legal system, educational system, and government to suit their own interests. For Marx, all of capitalist society was a superstructure resting on an economic foundation; the economy determined the nature of the society.

Whereas Marx foresaw only two social classes, Weber envisioned several social classes. Marx focused on the relationship to the means of production as the cause of social stratification; Weber examined the consequences of people's relationships to the economic institution. These consequences Weber termed **life chances**—the likelihood of securing the "good things in life," such as housing, education, good health, and food. The probability of possessing these desirables was directly related to economic resources, such as real estate, wages, inheritance, and investment profits.

By almost any measure—food, clothing, housing, or health—the standard of living has steadily improved in Europe and America. Still, there is considerable economic inequality, even within Western countries; and the United States is no exception. To understand the extent of economic inequality, it is necessary to distinguish income from wealth. **Income** is the amount of money received (within a given time period) by an individual or group. **Wealth** refers to all the economic resources possessed by an individual or group.

What is the extent of economic inequality in America? America was not supposed to have much economic inequality. As Richard Morin puts it:

> *It was the great story of the American Century: the fusing after World War II, of the broadest and most prosperous middle class the world had ever seen. In its ticky-tack ramblers and backyard barbecues and unbridled sense of confidence lay a historic reversal—the steady closing of physical and economic gaps between America's rich and poor.* (Morin 1994:14)

It has not turned out that way. Between 1980 and 1990, the share of America's wealth held by the top 1 percent of the population almost doubled, increasing from 22 percent to 39 percent. Since then the top 1 percent has accumulated over 70 percent of all earnings growth. In fact, the United States is now the most economically polarized and unequal of the major Western countries (K. Phillips 2002; Palast 2003).

In 2001, about 31 million Americans were living in poverty (U.S. Bureau of the Census 2002e).

At the other extreme, there were about 5 to 8 million millionaires in 2001 and some 200 billionaires (Collins and Yeskel 2000; Wolff and Leone 2002). Economist Paul Samuelson describes income inequality in America this way:

> *If we made an income pyramid out of building blocks, with each layer portraying $500 of income, the peak would be far higher than Mt. Everest, but most people would be within a few feet of the ground.* (Samuelson and Nordhaus 2001:386)

The truth in Samuelson's statement is supported by government figures on the distribution of income. The richest 20 percent of American families receive 50 percent of the nation's income; the poorest 20 percent control about 4 percent (Lichter and Crowley 2002; Wolff and Leone 2002).

Income inequality decreased slightly between 1950 and the late 1960s, but it has increased significantly since then (Doyle 1999; Goozner 1999; Mishel, Schmitt, and Bernstein 2001; Keister 2000). The 1980s

FIGURE 8.1

Percentage Change in After-Tax Income 1977–1999

What percentages do the labels "lowest fifth," "middle fifth," and "top fifth" refer to?

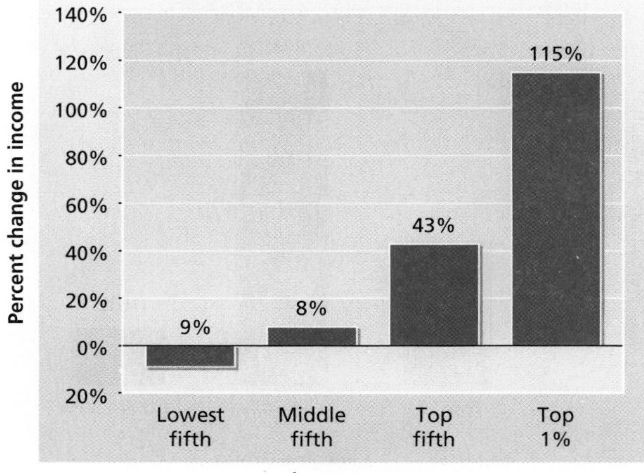

Source: Isaac Shapiro and Robert Greenstein, "The Widening Income Gulf" (Washington, DC: Center on Budget and Policy Priorities, September 4, 1999). Available online at www.cbpp.org/9-4-99tax-rep.htm, p. 2.

and 1990s saw the after-tax income of the poorest fifth of the nation decrease by 9 percent as the after-tax income of the middle fifth went up 8 percent and the richest fifth increased by 43 percent (see Figure 8.1). After-tax income among the top 1 percent rose by 115 percent between 1977 and 1999. To say it another way, the 2.7 million Americans with the highest incomes now have as much after-tax income as the 100 million Americans with the lowest incomes. Compare this with a 9 percent decline for the lowest fifth of the population. Clearly, income inequality between wealthy and other Americans has increased markedly. In fact, this gap is the largest since the U.S. Census Bureau first reported on the distribution of income, nearly thirty years ago (Shapiro and Greenstein 1999; K. Phillips 2002).

Although these income distribution figures reveal persistent economic inequality, they do not show the full nature of the concentration of wealth (what one owns) in America. Because the federal government regularly publishes data on national income, income comparisons are not difficult to calculate. Accurate data on wealth holdings are much harder to obtain (D. A. Gilbert 2003).

Recent research documents the concentration of wealth in the United States. On the low side, almost 25 percent of American households have a net worth (assets minus liabilities) below $5,000; almost one-third show a net worth below $10,000; and about 42 percent of American families have a net worth below $25,000.

FIGURE 8.2

Shares of Wealth

Is this picture of the distribution of wealth different from what you expected? Explain.

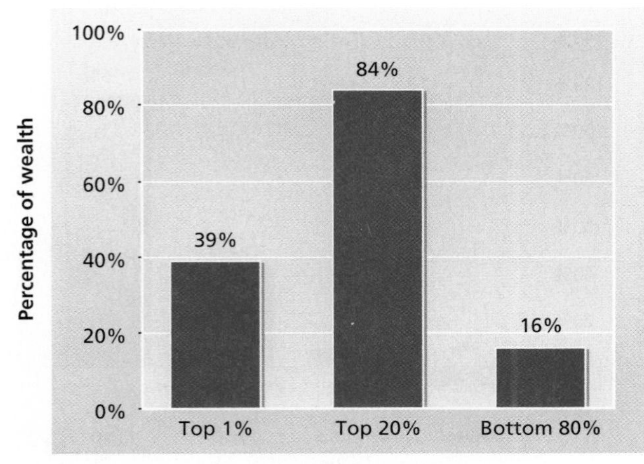

Source: Isaac Shapiro and Robert Greenstein, "The Widening Income Gulf" (Washington, DC: Center on Budget and Policy Priorities, September 4, 1999). Available online at www.cbpp.org/9-4-99tax-rep.htm, 1999, p. 11.

At the other extreme, 3.5 percent of American households have a net worth of $500,000 or more (U.S. Bureau of the Census 2001c). In 1995, the top 1 percent of American households held about 40 percent of the nation's wealth; the top 20 percent had almost 85 percent of the wealth; the bottom 80 percent of households owned less than 16 percent of the nation's wealth (see Figure 8.2). Prior to the fall in the stock market in the early 2000s, the concentration of wealth had become even more dramatic (Shapiro and Greenstein 1999; Collins and Yeskel 2000).

What are the consequences of growing income inequality? Inferior health care, loss of leisure time, restricted diets, low-quality housing, and poor police protection are just a few of the disadvantages brought about by the growing disparity between rich and poor in the United States. But, this is the widest rich-poor gap since the U.S. Bureau of the Census began keeping records in 1947, and it has other less visible negative consequences. The growing economic disparity between the rich and the poor is lowering relative educational levels because college attendance goes up with income level. This, in turn, decreases the proportion of the workforce with college backgrounds, thereby retarding the growth of the skilled workforce. The quality of the workforce is further damaged because the increase of low-income students leads to poorer overall academic performance at all grade levels. The rate of job growth is affected by the income gap, and the rich are even more empowered to influence government for their own interests (Bernstein 1994; Zepezauer and Naiman 1998; Mishel, Schmitt, and Bernstein 2001; Palast 2003).

The Power Dimension: Max Weber

You will recall that *power* is the ability to control the behavior of others, even against their will. Individuals or groups who possess power are able to use it to enhance their own interests, often—but not necessarily—at the expense of society.

Why is power a separate dimension of stratification? According to Marx, those people in a society who own and control capital have the power. Sociologists agree that economic success heightens the chances of increased power. The wealthy, for example, have the power to influence tax laws in their favor. However, many sociologists, beginning with Weber, have argued that economic success and power do not necessarily overlap. This argument is based on several points. First, the fact that money can be used to exert power does not mean that it will be used in this way. Money is a resource that can be used to enhance power, but a decision must be made to use it for that purpose. For example, there are other families as wealthy as the Kennedys who have not used their economic resources to gain political power.

Second, money and ownership of the means of production are not the only resources that can be used as a basis for power. Expert knowledge can be used to expand power, too. For example, many lawyers convert their expertise into substantial amounts of political power. Eloquence as a speaker is another possible resource. And fame was a factor in 1952, when Albert Einstein was offered the presidency of Israel. He refused by saying, "I know a little about nature, and hardly anything about men."

Power is also attached to the social positions we hold. In organizations, elected officers have more power than rank-and-file members do. People in top executive positions in the mass media are powerful, even if they themselves do not have great wealth.

Finally, we can overcome a scarcity of resources if we have large numbers of people on our side or if we are skillful at organizing our resources. Hitler, for example, was able to transform limited resources into a mass political movement. He gained absolute power by promising to deliver Germany from the economic destitution it suffered following World War I.

The Prestige Dimension: Max Weber

A third dimension of social stratification is **prestige**—recognition, respect, and admiration attached to social positions. Prestige is defined by one's culture and society (J. M. Berger et al. 1998; Webster and Hyson 1998). In the first place, favorable social evaluation is based on the norms and values within a group. Honor, admiration, respect, and deference are extended to dons within the Mafia, but despite America's fascination with the underworld, Mafia chiefs do not have high prestige outside their own circles. Second, prestige is voluntarily given, not claimed. Scientists cannot proclaim themselves Nobel Prize winners; journalists cannot award themselves Pulitzer Prizes; and corporate executives cannot grant themselves honorary doctorates. Recognition must come from others. Third, those accorded similar levels of prestige share identifiable lifestyles. The offspring of upper-class families are more likely to attend private universities and Episcopalian churches; children from lower-class homes are less likely to attend college at all and tend to belong to fundamentalist religious groups. In fact, some sociologists view social classes as subcultures.

How is conspicuous consumption related to the prestige dimension? People with sufficient economic resources may use their wealth to enhance their prestige; they may consume goods and services to display their wealth to others (Frank 2000). Thorstein Veblen (1995; originally published in 1899) called this phenomenon **conspicuous consumption**. Examples are numerous and often astonishing: guests at Cornelius Vanderbilt's seventy-room, $10 million "summer cottage" at Newport could choose between salt water and fresh water in their bathrooms; guests preferring milk at Baron Alfred de Rothschild's mansion had their choice of milk from Jersey, Hereford, or Shorthorn cows (Wallechinsky and Wallace 1981). There are many more contemporary illustrations of status symbols as well. While a top executive at the Ford Motor Company, Lee Iacocca and his colleagues regularly enjoyed lunches flown in from all over the world. The lunches were estimated at $104 a plate—and this was in 1960 dollars (Iacocca and Novak 1984). Millionaire Malcolm Forbes's son paid $156,450 for an almost 200-year-old bottle of wine believed to have belonged to Thomas Jefferson. High-income professional couples are installing "trophy kitchens" costing as much as $200,000 (Cavanaugh 1999). Wall Street broker Jack Grubman was investigated for temporarily upgrading his rating of AT&T stock in a successful effort to enroll his children in the prestigious 92nd Street Y preschool, an important status symbol in New York City (McGinn and Peterson 2002).

How is prestige distributed? Prestige distribution coincides with a society's value system. People in statuses that are considered the most important—that are valued the most highly—have the most prestige. Because Americans value the acquisition of wealth and power, they tend to assign higher prestige to persons in positions rewarded with wealth and power. The most stable

Homes in most societies connote prestige, whether high, middle, or low. Large, contemporary houses such as this one suggest that its inhabitants are reasonably high on the prestige hierarchy.

TABLE 8.1

Prestige Rankings of Selected Occupations in the United States

Occupation	Prestige Score
Surgeon	87
Lawyer	75
College professor	74
Engineer	71
School principal	69
Pharmacist	68
Registered nurse	66
Accountant	65
Professional athlete	65
Public grade school teacher	64
Banker	63
Druggist	63
Veterinarian	62
Police officer	61
Actor	60
Journalist	60
TV anchorman	60
Businessperson	60
Nursery school teacher	55
Social worker	52
Funeral director	49
Jazz musician	48
Mail carrier	47
Insurance agent	46
Secretary	46
Bank teller	43
Automobile dealer	43
Evangelist	41
Sales clerk	36
Bus driver	32
Waitress	29
Garbage collector	28
Janitor	22
Prostitute	14
Panhandler	11

Why do you think the highest-listed prestige score is 87? What occupations might rate a higher score?

Source: General Social Survey, National Opinion Research Center, 1996.

source of prestige in modern society is associated with occupations (Nakao, Hodge, and Treas 1990; Nakao and Treas 1990).

The prestige rankings of selected occupations in the United States are presented in Table 8.1. Obviously, white-collar occupations (doctors, ministers, schoolteachers) tend to have higher prestige than blue-collar occupations (carpenters, plumbers, mechanics). As expected, physicians, lawyers, and dentists have high occupational prestige, whereas janitors, cab drivers, and garbage collectors have low prestige. You may find it somewhat surprising, however, that college professors have more prestige than bankers—a demonstration that even though wealth and power usually determine prestige, there are exceptions.

What is the basis for occupational prestige? According to Robert Hodge, Paul Siegel, and Peter Rossi (1964), all societies rely on comparable factors when determining occupational prestige. In fact, Robert Hodge, Donald Treiman, and Peter Rossi (1966) have shown that the ranking of occupations based on prestige is similar in advanced as well as underdeveloped societies. Occupational prestige scores vary according to the following factors: compensation; education, skills, and ability required; power associated with the occupation; the importance of the occupation to the society; and the nature of the work (mental, or white-collar, work versus manual, or blue-collar, work). More recently, occupational prestige has been shown to be affected by gender.

Why do different cultures evaluate occupations similarly? Donald Treiman attempted to answer this question through an examination of nearly 100 studies of occupational prestige in sixty countries. From these studies, which included countries of Western and Eastern Europe, North America, Africa, and South America, Treiman concluded that all societies must accomplish certain objectives. For example, domestic peace must be maintained, and a variety of goods—including food, housing, clothing, and transportation—must be produced and distributed. Culturally dissimilar societies develop similar occupational positions because of the functions that each must perform. This is clearly a functionalist explanation.

Not only do societies around the world develop similar occupational positions, but these positions tend to be arranged in occupational prestige hierarchies that are substantially similar throughout the world:

> *In all societies, ranging from highly industrialized nations like the United States to peasant villages in up-country Thailand, the basic pattern of occupational evaluations is the same—professional and higher managerial positions are most highly regarded, lower white-collar and skilled blue-collar jobs fall in the middle of the hierarchy, and unskilled service and laboring jobs are the least respected.* (Treiman 1977:103)

This is the case, according to Treiman, because certain occupations in all societies have greater control over scarce resources than do other occupations.

Differential access to these resources—skill, authority, property—creates differential power. The power created by control over scarce resources is used to acquire special privilege. Power and privilege, in turn, are used to garner prestige. In this way, a single worldwide occupational prestige hierarchy is created. This is a conflict explanation.

Are the economic, power, and prestige dimensions related? Generally, there is a close relationship between these three dimensions of stratification. Those who are high on one dimension tend to be high on the other two. Power, particularly political power, is a decided economic advantage to its holders. For example, the wealthy pay for political favors either directly (as with money) or indirectly (as through inside information on lucrative business deals). Of the 535 members of the U.S. Congress, 170 have a net worth of at least 3.1 million. Ten members of President George W. Bush's first cabinet were millionaires (Edsall 2002; Keller 2002). And former President George Bush and his son George W. Bush are by no means the only Americans who have used their wealth as springboards to the presidency of the United States.

Prestige and wealth are also closely intertwined. High prestige is heavily dependent on economic resources. The lifestyle usually associated with high prestige cannot be maintained without the economic clout to purchase such necessities as a nice home in the proper neighborhood or to send one's children to the best schools. But those with high prestige normally have no cause for concern, for with prestige come special economic rewards and unique business opportunities for generating wealth. Most of the wealth amassed by Arnold Palmer and Jack Nicklaus did not come directly from prize money in professional golf tournaments. Because of their prestige, people want to buy Arnold Palmer sweaters and Jack Nicklaus golf clubs as much as corporations want their endorsements on products. In fact, Jack Nicklaus now owns the MacGregor Sporting Goods Company, the firm that first used his name on its golf equipment. In light of an initial $40 million contract with Nike, it is hard to imagine the wealth Tiger Woods will accumulate.

Prestige and power often mix quite well. Prestige is attached to social positions requiring important decision making. And prestige may be converted into power. Consider Ronald Reagan, the actor who became a California governor and then president of the United States, or Jack Kemp, the professional quarterback who became a member of the House of Representatives, a presidential candidate, a member of George Bush's cabinet, and a vice-presidential candidate.

Explanations of Stratification

Obviously, social class exerts a powerful influence on social life. The social class into which people are born and to which they rise or fall affects their existence as much as any other social factor. For this reason, if for no other, it is of benefit to understand why the process of stratification is a mainspring of society.

In most societies, inequality—in income, wealth, power, and prestige—seems to be a basic fact of life (Chris Tilly 1999). Even societies deliberately organized to eliminate inequality are not classless. Despite the former Soviet Union's attempt to create an unstratified society, great differences in rank and rewards developed. Leading members of the Communist Party, the dominant elite, were able to obtain privileges for themselves and their families. For example, Politburo members and national secretaries of the Communist Party drove black Zil limousines, hand-tooled cars worth over $100,000 each, while the majority of Russian people had no chance of ever owning any kind of car. The Chinese Communist Party has also apparently become an entrenched, self-interested ruling class (Wortzel 1987).

Why is stratification a universal feature of social life? There are several explanations.

The Functionalist Theory of Stratification

According to the functionalist theory of stratification, inequality renders a service. The most qualified people

FEEDBACK

1. _________ involves the creation of layers of people possessing unequal shares of scarce desirables.
2. Match the concepts with the examples.
   ____ a. life chances — (1) the respect accorded doctors
   ____ b. power — (2) a politician conforming to the interests of a lobby
   ____ c. prestige — (3) living in a neighborhood with good police protection
   ____ d. conspicuous consumption — (4) a lavish wedding
3. The top 1 percent of American households hold nearly 40 percent of the total wealth. T or F?
4. The most stable source of prestige in modern societies is associated with _________.

Answers: 1. Social stratification 2. a. (3) b. (2) c. (1) d. (4) 3. T 4. occupations

fill the most important positions and perform their tasks competently; they are then rewarded for their efforts. The functionalist theory recognizes that inequality exists not only because certain jobs are more important than others but also because these jobs often involve special talent and training.

To encourage people to make the sacrifices (years of education and so forth) necessary to fill these jobs, society attaches special monetary rewards and prestige to these positions. For example, doctors make more money and have more prestige than bus drivers because a high level of skill is required in the medical profession and our society's need for highly qualified doctors is great. Once qualified people assume positions, however, they must be motivated to do their jobs competently. It is not happenstance, then, that greater rewards go to those who do their work better than others do. Thus, the more competent doctors, secretaries, and salespersons usually receive more of the "just deserts" than those who perform their tasks less well.

The functionalist theory has dominated sociological thinking for a long time, but it has been a very difficult theory to test. Although some studies support at least some aspects of the functionalist theory (Grandjean 1975; Cullen and Novick 1979), several weaknesses should be noted.

What are the weaknesses of the functionalist theory? Although the functionalist theory is elegant, concise, and consistent with traditional American beliefs about inequality, it has several failings (Tumin 1953). First, there are many people who have power, prestige, and wealth whose contributions to society do not seem very important. Many top athletes, film stars, and rock singers make many times as much money as the president of the United States. Why should scientists make almost no money for groundbreaking books reporting their research whereas screenwriters regularly receive $500,000 plus per script?

© David Frazier/Photo Researchers

According to the functionalist theory of stratification, higher monetary rewards are distributed to the most qualified people in the most important positions. If this is so, we can feel more comfortable knowing that airline pilots earn high salaries.

Second, the functionalist theory ignores the barriers to competition faced by some members of society—the poor, women, the aged, African Americans, and various other ethnic groups. A wealth of talent exists among people in these categories, but much of it has never been tapped, because such people have not experienced equal opportunity to compete. They have not had the opportunity to become educated because they are too poor, the wrong color, or the wrong sex. And even if they are technically qualified, they may lose higher-level positions to less qualified individuals who happen to possess particular social characteristics.

Third, the functionalist theory overlooks the inheritance of social class level. In some societies, one's social class is determined at birth. Even in modern societies, in which the rate of mobility is supposed to be high, children tend to remain in or near the social class of their parents. This means that talented daughters and sons from lower-class homes cannot compete on an equal basis with the offspring of middle- and upper-class people. Consequently, many talented individuals—who might have preferred being surgeons, lawyers, or executives—are truck drivers, plumbers, and short-order cooks.

Finally, the functionalist theory of stratification has an ethnocentric bias. That is, it assumes that all people in all societies will be motivated to compete for a greater portion of the scarce desirables. In fact, in some societies, greater emphasis is placed on a more or less equitable distribution of wealth, power, and prestige than on individual competition for larger shares.

The Conflict Theory of Stratification

Melvin Tumin (1953) concluded that stratification exists not because it is necessary for the benefit of society but because it helps people holding the most power and economic resources to maintain the status quo. From his perspective, the functionalist theory is used as a justification for keeping some people at the top of the class structure and some at the bottom. This viewpoint is consistent with the conflict theory of stratification.

How does conflict theory account for inequality? According to the conflict theory of stratification, inequality exists because some people are willing to exploit others. The elite hold a monopoly over the desirables, and they use their monopoly to dominate others. Stratification, from this perspective, is based on

According to the conflict theory of stratification, inequality is a result of the elite exploiting the less powerful. The power of nineteenth-century industrialists is reflected in this caricature of J. P. Morgan.

force rather than on consent. Those with wealth, power, and prestige are able to maintain, perhaps even increase, their share of desirables in a society (Perrucci and Wysong 2002).

This conflict theory of stratification is based on Marxian ideas regarding class conflict. For Marx, all of history has been a class struggle between the powerful and the powerless, the exploiters and the exploited. Capitalist society is the final stage of the class struggle. Although, according to conflict theorists, the capitalists are outnumbered, they are able to control the workers by creating a belief system that legitimates the status quo. Those who own the means of production are able to disseminate their ideas, beliefs, and values through the schools, the media, the churches, and the government. In this way, the dominant ideas of the elite become the dominant ideas of the ruled. Marx used the term **false consciousness** to refer to working-class acceptance of the dominant ideology (even though the ideology worked against working-class interests).

Because Marx believed that class conflict occurs in every historical epoch, he predicted that the working class would eventually shed its false consciousness—and its chains—for a revolution based on a true class consciousness. That is, it would realize its exploitation and reject the dominant belief system. After overthrowing the capitalists, the working class would create a socialist society in which the means of production and all property would be owned equally by the people. This would be a transitional stage between capitalist society and the classless society of communism. Ultimately, socialism would be replaced by communism, which would put an end to human misery and exploitation.

Marx's work was somewhat neglected by American sociologists for a time, although it has indirectly influenced the field through the negative reactions of many sociologists. Why has Marx been neglected? For one thing, many of his ideas—exploitation, revolution, classless society, communism—do not fit well with the American emphasis on capitalism, achievement, and upward mobility. In addition, many of Marx's predictions appear to have been wrong. Socialist revolutions have occurred in noncapitalist societies; communism did not produce classless societies; the middle class rather than the working class has expanded in capitalist societies.

Marxists do not consider that any of the preceding factors disprove Marx's theory. First, the so-called socialist and communist revolutions in precapitalist societies were essentially the result of a misinterpretation of Marx, who insisted that capitalism was an essential precondition that must occur before a communist society is possible. Second, Marx's theory defines the middle class as part of the working class. This is because the middle class also does not own the means of production, and its gradual expansion fulfills Marx's prediction of the increasing "proletarianization" of labor (that is, more and more people, including small business owners, doctors, and other professionals, lose ownership of their labor power and come to work for someone else). Finally, according to Marxists, it is premature to assert that Marx was wrong simply because capitalism has not yet disappeared. Marx gave no time table for the socialist revolution, and Marxist theorists see evidence of increasing contradictions within capitalism (Therborn 1982, 1984; E. O. Wright 1985, 2000).

Some sociologists have begun to look past the apparent miscalculations in Marx's theories to appreciate some of the assumptions on which they are based. They have substantial appreciation for his insight into the ways the elite use society's institutions, including the state, to achieve their own ends.

German sociologist Ralf Dahrendorf (1959) has advocated shifting from the Marxist emphasis on class conflict to conflict between groups, such as unions and corporations. Dahrendorf has attributed stratification more to the possession of power than to ownership of property. America's legal system, for example, is used by the wealthy for their benefit; education helps the elite perpetuate their power; the political system is skewed toward the interests of the powerful. Classes, then, for the followers of the conflict perspective, are created through the struggle for scarce resources (Domhoff 1996; Giddens 1987, 2000).

TABLE 8.2

FOCUS ON THEORETICAL PERSPECTIVES: Social Stratification

This table summarizes what issues of social stratification might be of interest to each of the major perspectives and predictions that they would make. Why would the symbolic interactionists be more likely than the functionalists to look at issues of self-esteem?

Theoretical Perspective	Research Topic	Expected Result
Functionalism	• Relationship between job performance and pay	• Pay levels increase with job performance.
Conflict theory	• Relationship between social class level and the likelihood of punishment for a crime	• The chances for prosecution decrease as social class level increases.
Symbolic interactionism	• Link between social class level and self-esteem	• Self-esteem is higher among the upper class than the lower class.

Symbolic Interactionism and Stratification

Why was the inequality between former Soviet political leaders and average Russian people tolerated by the masses? Why does the poorest one-fifth of Americans (possessing less than 1 percent of the wealth) not revolt against the richest one-fifth of the population (holding 75 percent of the wealth)? Social stratification systems can persist over the long haul only if people believe in their legitimacy.

How is the legitimacy of a stratification structure incorporated into the minds of individuals? Social stratification persists only as long as its legitimacy is accepted. Symbolic interactionism helps us understand the legitimation process; it must occur through socialization. Symbols, in the form of language, explain to the young the existence of stratification structure and the reasons for people being located in particular strata. According to this perspective, American children are taught that one's place in the stratification structure is the product of talent and effort. Those "on top" have worked hard and have used their abilities, and those "on the bottom" have lacked the talent and motivation to succeed; therefore, it is not fair to challenge the system. In this way, people come to believe in the fairness of the existing system.

The legitimacy of a stratification structure penetrates even deeper. Legitimacy views are incorporated into an individual's self-concept.

FEEDBACK

1. Match the theories of stratification with the examples.
   ____ a. functionalist theory
   ____ b. conflict theory
   ____ c. symbolic interactionism
   (1) Corporate executives make more money because they determine who gets what in their organizations.
   (2) Engineers make more money than butlers because of the education they possess.
   (3) Ghetto children tend to have low self-esteem.
2. Which of the following is *not* a criticism of the functionalist theory of stratification?
   a. There are people at the top whose jobs do not seem very important.
   b. It has a bias toward cultural relativism.
   c. It overlooks the barriers to competition.
   d. It glosses over the inheritance of social class level.
3. The conflict theory of stratification is based on Marx's ideas on class _________.
4. According to the symbolic interactionists, the deepest penetration of the beliefs supporting the stratification structure comes as a result of
   a. the "I"
   b. evolution
   c. conflict
   d. the self-concept

Answers: 1. a. (2) b. (1) c. (3) 2. b 3. conflict 4. d

How do ideas about inequality become part of the self-concept? Certain self-images are implied in the legitimization of stratification. Those at the bottom of the stratification structure suffer from lower self-esteem. How could it be otherwise when messages from all sides tell them of their inferiority? After all, as the symbolic interactionists document, self-esteem is based on our perceptions of others' opinions. The looking-glass process is at work (see Table 8.2).

Moreover, a person's sense of self-worth affects the amount of rewards he or she feels entitled to receive. Thus, people at the bottom of the heap tend not to challenge the stratification structure because of the conviction that they are undeserving. Those at the top blame the victims; the victims blame themselves.

The reverse is true for the higher classes. Those profiting most from the stratification structure tend to have higher self-esteem. This, in turn, fuels their conviction that the present arrangement is just. In short, people's self-concepts help preserve the status quo.

Stratification in American Society

Class Consciousness

Americans have always been aware of inequality, but they have never developed a sharp sense of **class consciousness**—a sense of identification with the goals and interests of the members of one's own social class (Centers 1949; Hodge and Treiman 1968; Jackman and Jackman 1983). In part because the American public has historically shown relatively little interest in class differences, American scholars began to investigate inequality rather late. It was not until the 1920s that sociologists began to systematically identify social classes in American communities. Since that time, however, the material available on this subject has been voluminous. Early efforts to study stratification were through intensive case studies of specific communities (W. L. Warner and Lunt 1941; Hollingshead 1949; Hollingshead and Redlich 1958; W. L. Warner 1960). Only relatively recently have attempts been made to describe the stratification structure of America as a whole (Rossides 1997; Chris Tilly 1999; Perrucci and Wysong 2002; D. A. Gilbert 2003).

Identification of Social Classes

Any attempt to describe the social class structure of American society is hazardous. Social classes are fluid, replete with exceptions, and abstract. Nevertheless, sociologists have identified the major features of the American class structure. According to Daniel Rossides (1997), the upper class comprises about 1 percent of the population, the upper middle class about 14 percent, the middle middle class about 30 percent, the working class about 30 percent, and the lower class about 25 percent. Despite the hazards involved, a brief description of each of these social classes follows. (Figure 8.3 presents a similar picture of the American class structure.)

FIGURE 8.3

American Class Structure

What does the teardrop shape of the American class structure indicate about stratification in the United States?

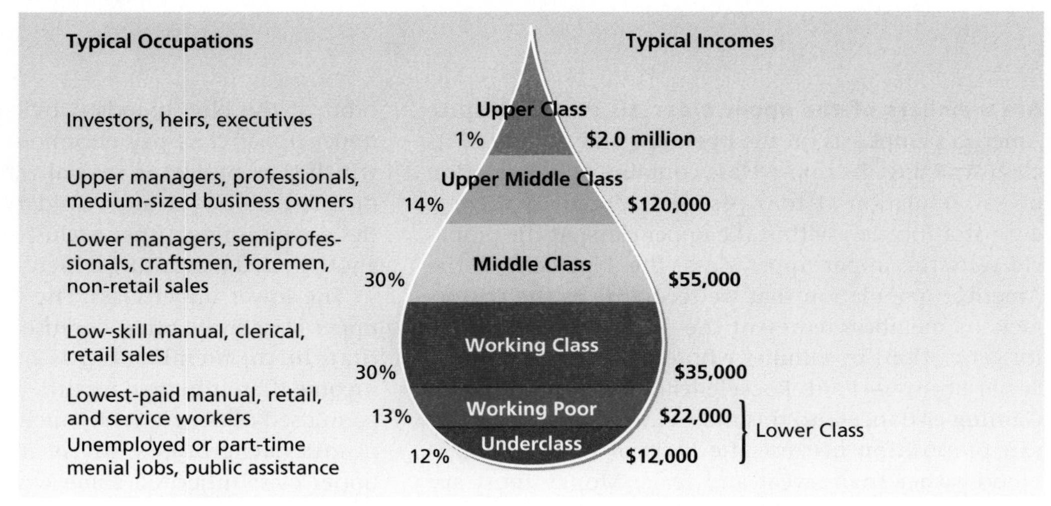

Source: Adapted from Dennis Gilbert, *The American Class Structure*, 6th ed. (Belmont, CA: Wadsworth, 2003), p. 17.

Donna Eder—Who's Popular, Who's Not?

Can you remember what it was like to be in middle school? Donna Eder (1995) formed a research team to answer this question. Focusing on the rituals and daily speech routines of twelve- to fourteen-year-olds, Eder paints a portrait of teenage school culture.

At Woodview Middle School, Eder and her research team observed lunchtime interaction and attended extracurricular activities—talent shows, athletic practices, athletic contests, cheerleading tryouts, cheerleading practice, and choir and band practices and performances. After several months of observation, the research team conducted informal interviews with individuals and groups. To capture interaction for closer study, the researchers obtained student and parental permission for audio and video recordings. Our discussion is confined to Eder's findings about social stratification in middle school.

In the sixth grade, there were no elite groups. Woodview seventh and eighth graders, however, did not see each other as equals. This central fact was evident to the researchers after only a few days of observation. Popular seventh graders were divided along gender lines, with popular boys in one group and popular girls in another. By the eighth grade, the two groups intermingled. In both grades, popularity was based not on who was best liked, but on who was most visible in the school. Being visible meant that many others knew who you were and wanted to talk with you.

Status differences could arise during the seventh and eighth grades because cheerleading and team sports existed as a vehicle to high visibility. Realizing the source of their prestige, male athletes took every opportunity to display symbols of their team affiliation. Team uniforms, jerseys, and athletic shoes were among the most important items of dress. Bandages, casts, and crutches were worn with pride.

Girls could not use sports to gain visibility because female athletics were not valued by faculty, administrators, or students. But since boys' athletic events were the most important and best-attended school activities, girls used cheerleading to make themselves widely known. Besides appearing at basketball and football games, cheerleaders appeared in front of the entire student body at pep rallies and other school events.

Boys made fun of this high-status female activity by mockingly imitating cheers. One male coach joined the mockery by telling football players to either practice harder or he would get them cheerleading skirts. He then pretended to cheer in a falsetto voice.

Girls, on the other hand, regarded cheerleaders highly. Popular girls in the seventh and eighth grades either were cheerleaders or were friends of cheerleaders. Flaunting their status (just as the male athletes did), cheerleaders put on their uniforms as far ahead of games as possible and wore their cheerleading skirts for extracurricular school activities. Some girls even had T-shirts emblazoned with "cheerleader" on the back.

Thinking About the Research

1. Although it happened a while ago, middle school is something you experienced firsthand. Which of the three major theoretical perspectives do you think best explains the stratification structure described in this feature? Give reasons for your choice.
2. Was the stratification structure at your school similar to Woodview's? Explain why or why not, giving examples to support your conclusion.

Are members of the upper class all alike? Despite America's emphasis on the openness and fluidity of its class structure, its upper class contains only a fraction of its population (1 to 3 percent). Moreover, there is even stratification within the upper class. At the pinnacle rests the upper upper class, the 1 percent of the American population that we recognize as the aristocracy. Its members represent the "old" money amassed for generations by families whose names appear in the Social Register—Ford, Rockefeller, Vanderbilt, du Pont. Gaining entrance into this most elite of clubs is a long-run proposition because the basis of membership is blood rather than sweat and tears. Money must age over generations to have any chance of entering the orbit of the blue bloods. Obviously, members of the upper upper class pay enormous attention to lineage, which they protect in several ways: sending their children to the best private secondary schools and universities, maintaining rather exclusive interaction with one another, and promoting a high rate of intramarriage.

The lower upper class, the "new rich," are in the upper class more often because of achievement rather than birth; membership is based more on earned income than inherited wealth. The lower upper class is composed of people with much more varied socioeconomic backgrounds. Many members of the lower upper class inherited some wealth, although usually not fortunes. Others may have amassed fortunes by being shrewd investors on the stock market, creating a chain of hardware stores, becoming a Metropolitan Opera star, or earning millions from athletic talent. Although they may actually be better off financially, people who have entered the lower upper class in their own lifetimes are generally barred from the exclusive social circles in which members of the upper upper class travel.

The upper class as a whole wields tremendous power over domestic and international decisions and over events affecting vast numbers of people (Dye 2001). If the United States has a power elite, this is it (see Chapter 14).

What is the composition of the middle class? The upper middle class (10 to 15 percent of the population) is composed of those who have been successful in business, the professions, politics, and the military. Basically, this class is made up of individuals and families who benefited from the tremendous corporate and professional expansion following World War II. Members of this class earn enough to live pretty well and to save some money. They are typically college educated and have high educational and occupational aspirations for their children. Although they do not exercise national or international power, their influence is felt in their communities, where they tend to be active in voluntary and political organizations.

The middle middle class is large (30 to 35 percent of the population) and heterogeneous. Its members include small business owners; small farmers; small, independent professionals (small-town doctors and lawyers); semiprofessionals (clergy, teachers, nurses, firefighters, social workers, and police officers); middle-level managers; and sales and clerical workers. Their income level, which is at about the national average ($22,199 in 2000) does not permit them to live as well as the upper middle class. Typically, these people have only a high school education, although many may have had some college or may have completed college.

Members of the middle class are interested in civic affairs. Their participation in political activities is less than in the classes above them but higher than in either the working class or the lower class. The 1980s witnessed a dramatic decrease in the proportion of Americans that can be classified as middle class. Since 1980, the proportion of middle-income Americans has declined by 20 percent (Morin 1991; Wallich and Corcoran 1992).

© John Coletti/Stock, Boston

The American working class is composed primarily of blue-collar workers and is one of the largest single segments of the population. These construction workers, enjoying a coffee break, are representative of the working class.

What is unique about the working class? The working class (often referred to as the lower middle class) comprises about one-third of the population. Its members typically include either blue-collar workers who work with their hands—roofers, delivery truck drivers, and machine operators—or lower-level salespeople and clerical workers (Rubin 1994). The economic resources of its members are, on the whole, fewer than those of the classes above, although some of these workers (plumbers, for example) may earn more than some middle-class people. Members of the working class experience considerable economic pressure because their incomes are below the national average, their employment is less stable, and they are generally without the hospital insurance and retirement benefits enjoyed by those in the middle class. The specter of unemployment or illness is both real and haunting. Outside of union activities, members of the working class have little opportunity to exercise power, and their participation in organizational life, including politics, is scant (Zweig 2000).

What groups of people fall into the lower class, and what is their most common shared characteristic? Lumped into the lower class (20 to 25 percent of the population) are the working poor, the frequently unemployed, the underpaid, and the permanently unemployed. The most common shared characteristic of the lower class is lack of value in the labor market. There are many routes into the lower class—birth, physical or mental disability, old age, loss of a marriage partner, lack of education or training, occupational failure, alcoholism. There are, however, very few paths out (W. J. Wilson 1987; W. J. Wilson et al. 1988; M. B. Katz 1993; D. A. Gilbert 2003).

The **working poor** (about 13 percent of the population) consists of people employed in low-skill jobs with the lowest pay. Its members are typically the lowest-level

FEEDBACK

1. _________ refers to a sense of identification with the goals and interests of the members of one's social class.
2. The upper upper class in the United States contains about _________ percent of the total population.
3. A major difference between the upper upper class and the lower upper class is that the latter has relied on _________ rather than birth.
4. The _________ is composed of people with an intergenerational history of unemployment.

Answers: 1. Class consciousness 2. 1 3. achievement 4. underclass

clerical workers, manual workers (laborers), and service workers (fast-food servers). Lacking steady employment, the working poor do not earn enough to rise above the poverty line. The working poor tend not to belong to organizations or to participate in the political process (Barrington 2000; Munger 2002).

The **underclass** (about 12 percent of the population) is composed of people who are usually unemployed and who come from families with a history of unemployment. They either work in part-time menial jobs (unloading trucks, picking up litter) or are on public assistance. In addition to a lack of education and skills, many of the members of the underclass have other problems. Physical or mental disabilities are common, and many are single mothers with little or no income. We will explore the underclass in the following chapter on racial and ethnic minorities. Poverty in the United States, another way to discuss the underclass, is the topic of the next section.

Poverty in America

Widespread poverty existed throughout the United States when Michael Harrington helped to make it a political issue in his book *The Other America* (1962). Forty years later, scholars continue to underscore the presence of poverty in affluent America (Newman 1999). In fact, 32.9 million Americans were below the government's poverty line of $18,104 (for a family of four) in 2001 (U.S. Bureau of the Census 2002a).

Measuring Poverty

On an absolute scale, poverty embodies a lack of housing, food, medical care, and other necessities for maintaining life. **Absolute poverty** is the absence of enough money to secure life's necessities. On a relative scale, though, it is possible to have those things

FIGURE 8.4

Number of Poor and Poverty Rate: 1959–2001

This graph shows two types of information: (1) the number of poor in the total population and (2) the poverty rate as a percentage of the total population. Why is it often helpful to have related information plotted on the same graph?

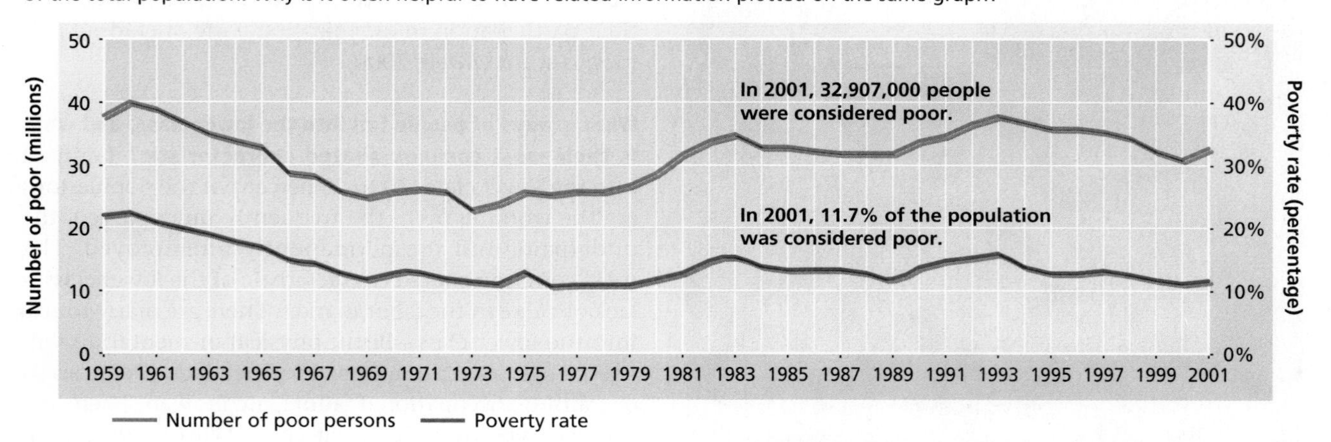

Number of poor persons — Poverty rate

Note: The data points represent the midpoints of the respective years. The latest recessionary period began in July 1990.
Sources: U.S. Bureau of the Census, *Poverty in the United States: 1995*, Current Population Reports, P60–185 (Washington, DC: U.S. Government Printing Office, 1996); U.S. Bureau of the Census, "Poverty 1997—Poverty Estimates by Selected Characteristics," *Current Population Survey, March 1998*, 1999a, at www.census.gov/hhes/poverty/poverty97/pv97est1.html; and U.S. Bureau of the Census, "Household Income at Record High. Poverty Declines in 1998, Census Bureau Reports." United States Department of Commerce News Release, September 30,1999; and U.S. Bureau of the Census, *Poverty in the United States: 2001*, Current Population Reports, P60-219 (Washington, DC: U.S. Government Printing Office, 2002).

FIGURE 8.5

The Distribution of Poverty in the United States

What are the most important conclusions you would reach from this figure?

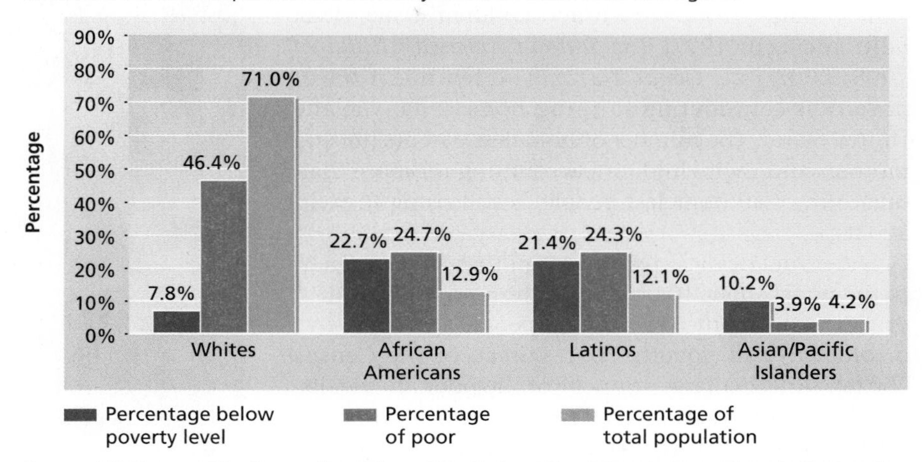

Sources: U.S. Bureau of the Census, "Projections of the Resident Populations by Race, Hispanic Origin, and Nativity: Middle Series, 2001 to 2005," at www.census.gov/population/projections/nation/summary/np-t5-b.pdf; and U.S. Bureau of the Census, *Poverty in the United States: 2001,* Current Population Reports, P60-219 (Washington, DC: U.S. Government Printing Office, 2002).

required to remain alive—and even to live in reasonably good physical health—and still be poor. **Relative poverty** is measured by comparing the economic condition of those at the bottom of a society with that of other strata. Because poverty is determined by the standards that exist within a society, a definition of poverty is not the same in India as in the United States.

How can relative poverty be measured? According to advocates of the relative measurement of poverty, the poverty threshold should be raised as the standard of living in a country rises. An increasing amount of spendable income does not make people feel less poor if the income levels of those around them are increasing at the same or a faster rate. A common measure of relative poverty is an income comparison between the lowest fifth of the population and the other four-fifths (refer to Figure 8.1). Such a measure means, of course, that a segment of the population will always be poor unless the total income distribution approaches equality.

How can absolute poverty be measured? Poverty in America has traditionally been measured in an absolute way. Absolute measurement of poverty involves drawing a "poverty line," an annual income level below which people are considered poor. In 2001, that figure was set at $18,104 for a family of four (see Figure 8.4). The poor, as measured by the federal government standard, comprised 11.7 percent of the American population in 2001 (U.S. Bureau of the Census 2002a). The following description of America's poor is based on this absolute measure of poverty.

Who are America's poor? Poverty is particularly prevalent in disadvantaged groups: minorities, female-headed households, children under eighteen years of age, the elderly, the disabled, and people who live alone or with nonrelatives. Although about 46 percent of the poor in America today are white, the poverty rate for African Americans and Latinos is higher than for whites (see Figure 8.5). In 2001, 7.8 percent of the white population fell below the poverty threshold, compared with 22.7 percent for African Americans and 21.4 percent for Latinos (Mishel, Bernstein, and Boushey 2003; U.S. Bureau of the Census 2002a).

Although African Americans and Latinos account for less than one-fourth of the total population, they make up over 49 percent of the poor population. The "foreignization of poverty" is now occurring. Over the past twenty years, the percentage of immigrant households in poverty rose from 16 to 22 percent (Camarota 1997).

There is a heavy concentration of the poor in households without husbands/fathers. Well over one-fourth of the poor are members of female-headed families, whereas almost 95 percent of the nonpoor are in families with a husband/father present. The poverty rate for female-headed families is 26.4 percent (compared with just 9.2 percent for all families). Of all those living in poverty, about 36 percent are children under eighteen years of age. Almost 12 million American children under age eighteen live in impoverished households.

The poverty rate for children under six years of age is about 18 percent, the highest poverty rate for any age-group in the United States (U.S. Bureau of the Census 2002a). This documents the **feminization of poverty**, the trend toward more of the poor in the United States being women and children (Sidel 1996; Edin and Lein 1997; *The State of America's Children* 1998; 2002 *Kids Count* 2002). The feminization of poverty is contributing to rising homelessness in the United States. The number of homeless parents (mostly mothers and their children), accounting for more than one-third of all homeless people, is still rising (J. Stein 2003).

Older Americans account for another large segment of the poor. About 10 percent of the poor are sixty-five or older, and about 10 percent of people aged sixty-five or older live in poverty (U.S. Bureau of the Census 2002e). Another large segment of the poor are the disabled—those who are blind, deaf, crippled, or otherwise disabled—who account for some 12 percent of America's poor. Finally, more than 29 percent of the poor—more than one out of every four—either live alone or live with nonrelatives.

© Nathan Benn/CORBIS

Poverty is greatest among American children under six years of age.

This description should not leave the impression that able-bodied poor in the United States do not work (Schwarz and Volgy 1992; O'Hare 2002). More than 22 million of the poor in America are over sixteen years of age. Of these, nearly 8.5 million worked either full-time or part-time in 2001. In other words, 38 percent of poor Americans of working age were working (U.S. Bureau of the Census 2002a). Fifty-five percent were ill, disabled, retired, at home or not working for family reasons, or looking for work. Over 11 percent were in school or not working for some unspecified reason. According to a recent study, at any particular point in time, about one-third of welfare mothers are working (K. M. Harris 1993). Nearly all of the poor in America are either underage children, working people, or persons not working for legitimate reasons. This evidence contradicts the prevailing assumption that the poor are able-bodied individuals

SNAPSHOT OF AMERICA 8.1

Percentage of the U.S. Population in Poverty

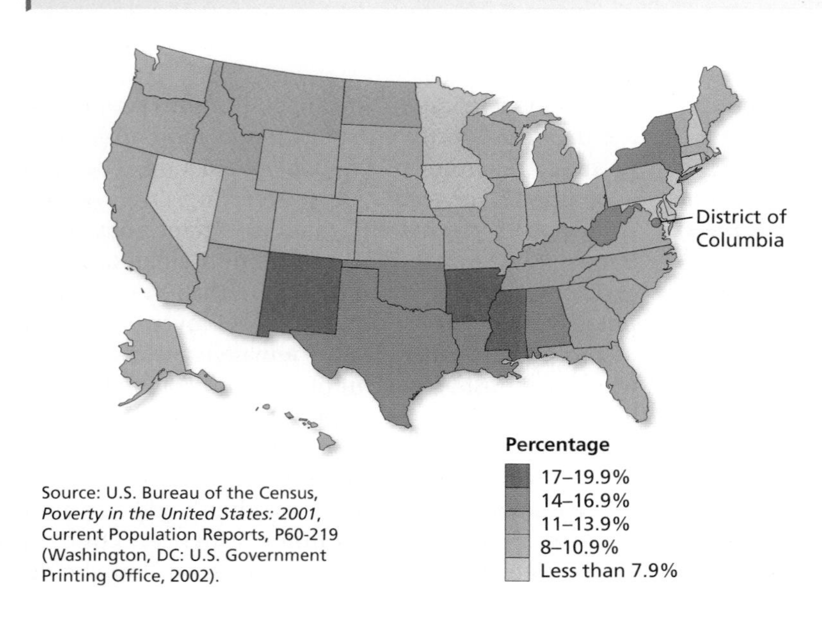

Source: U.S. Bureau of the Census, *Poverty in the United States: 2001*, Current Population Reports, P60-219 (Washington, DC: U.S. Government Printing Office, 2002).

Although the U.S. economy is booming, many people have not benefited from this prosperity. In fact, many still live in poverty. This map shows the percentage of the poor by state.

1. Can you make any generalization about poverty from this map?
2. If you were the governor of your state, what would your platform on poverty be? Be specific.

simply waiting for others to take care of them (Schwarz and Volgy 1992; O'Hare 1996; Swartz and Weigert 1996).

Is this an accurate picture of America's poor? At the very least, this absolute measure of poverty is a reasonable approximation. Reasonable, in part, because prior to 1963 the federal government had neither a definition nor a measure of poverty at all. Developed by Mollie Orshansky (1965) of the Social Security Administration, the federal government's absolute measure of poverty was based on an "economy" food plan: a plan designed by the U.S. Department of Agriculture as nutritionally adequate for use in emergencies or, temporarily, when family funds are low. The poverty cutoffs (based on factors such as family size, gender of the head of the household, and place of residence) were based essentially on the amount of income left after allowing for an adequate diet at minimum cost.

Critics point to some shortcomings of this approach. This definition reflects inflation, but it ignores social and economic changes such as the feminization of poverty and the higher costs of child care, transportation, and health care. The federal poverty level also considers the cost of living to be uniform across all states (except Hawaii and Alaska): a family living in Manhattan in New York is assumed to spend the same amount on food, shelter, and clothing as a family in Manhattan, Kansas (Bhargava and Kuriansky 2002). According to Mark Rank (2003), the number of people ever in poverty is a better indicator of the extent of poverty than the number of people in poverty in any particular year. By age 75, he contends, 59 percent of Americans will have spent at least a year below the poverty line during adulthood, and 68 percent will have experienced near poverty.

Perceptions of Poverty

How do Americans view the poor? The traditional American view of the poor has roots extending back to ancient Greece. In the seventh or eighth century B.C., the Greek poet Hesiod expressed for his contemporaries this view about work and the poor:

> *Work! Work, and then Hunger will not be your companion, while fair-wreathed and sublime Demeter will favor you and fill your barn with her blessings. Hunger and the idling man are bosom friends. Both gods and mortals resent the lazy man, a man no more ambitious than the stingless drones that feed on the bees' labor in wasteful sloth. Let there be order and measure in your own work until your barns are filled with the season's harvest. Riches and flocks of sheep go to those who work. If you work, you will be dearer to immortals and mortals; they both loathe the indolent.* (Hesiod 1983:74–75)

There is another example of this approach to poverty: it is the adaptation of Charles Darwin's theory of biological evolution to the social world. According to Darwin, only the fittest plants and animals survive in a particular environment. Those plants and animals that survive carry physical characteristics that give them a competitive edge for survival. The process of natural selection ensures that the fittest survive. When applied to human society, this view is called *social Darwinism*. It contends that the fittest individuals survive as a result of social selection and that the inferior are eliminated by the same process (Strickland and Shetty 1998).

Although social scientists have long rejected social Darwinism, this philosophy has contributed to that part of America's basic value system that is variously labeled the work ethic, the achievement and success ideology, and the ideology of individualism (Gilens 1999). Individualism in America involves several central beliefs:

1. Each individual should work hard and strive to succeed in competition with others.
2. Those who work hard should be rewarded with success (seen as wealth, property, prestige, and power).
3. Because of widespread and equal opportunity, those who work hard will, in fact, be rewarded with success.
4. Economic failure is an individual's own fault and reveals lack of effort and other character defects. (Feagin 1975:91–92)

According to the ideology of individualism, those at the bottom are where they belong because they lack the ability, energy, and motivation to survive in a competitive social world. From this vantage point, it is the poor who are to blame for their condition; those who fail deserve to do so (Ryan 1976; O'Toole 1998; N. P. Barry 1999; Zinn 2003). In its most extreme form, the ideology of individualism contends that legislation to protect the poor is harmful because protective legislation interferes with the process of natural selection, thus slowing social evolution (Lux 1990).

To test whether the ideology of individualism continues to affect the American public's view of the poor, James Kluegel and Eliot Smith (1990) conducted a nationwide survey. Respondents were asked to judge the reasons usually cited to explain poverty in America. They found that individualistic explanations were the most popular. Lack of thrift, lack of effort, lack of ability, and loose morals and drunkenness—all individualistic causes—were considered the most important reasons for poverty. Much less emphasis was placed on structural explanations that locate the causes of poverty in social and economic factors, factors beyond the control of the poor (see Table 8.3).

TABLE 8.3

Perceived Reasons for Poverty in the United States

Reasons for Poverty	Percentage Replying: Very Important	Somewhat Important	Not Important
Individualistic Explanations			
• Lack of thrift and proper money management by poor people	64%	30%	6%
• Lack of effort by the poor themselves	53	39	8
• Lack of ability and talent among poor people	53	35	11
• Loose morals and drunkenness	44	30	27
Structural Explanations			
• Low wages in some businesses and industries	40	47	14
• Failure of society to provide good schools for many Americans	46	29	26
• Prejudice and discrimination against blacks	31	44	25
• Failure of private industry to provide enough jobs	35	39	28
• Being taken advantage of by rich people	20	35	45

How would you have answered this survey? Explain your answers.

Source: Adapted with permission from James R. Kluegel and Eliot R. Smith, *Beliefs About Inequality: Americans' Views of What Ought to Be* (New York: Aldine de Gruyter, 1990), p. 79. Copyright © 1990 by James R. Kluegel and Eliot R. Smith.

Although structural explanations for poverty have grown in importance in recent years, the individualistic interpretation remains dominant, particularly among affluent, higher-status, white, older Americans, those furthest from the situation being studied (W. J. Wilson 1997; Goode and Maskovsky 2001; "National Survey on Poverty in America" 2001; Draut 2002; Eitzen and Smith 2003).

Does race influence attitudes toward the poor? In a complicated way, it does. Americans do not reject welfare in principle; the poor who are not lazy and who try to find jobs deserve help, most say. It is the "undeserving" poor—those perceived as shiftless—who are the target of public scorn (R. C. Lieberman 1998; M. K. Brown 1999; Neubeck and Cazenave 2001). This is where race is thought to enter the mix. Due, in part, to the media's long-standing negative portrayal of the African American poor, welfare has become enshrouded in racial stereotypes. While Americans do not dislike welfare because African Americans benefit from it, many do believe that welfare benefits undeserving African Americans who prefer a government handout to work (Gilens 1999; Blank and Haskins 2001).

Americans, of course, do not condemn just the African American poor. While race complicates the issue, many Americans are opposed to welfare for *all* "undeserving" poor, regardless of their race and ethnicity.

How accurate is the negative image of the poor? Several studies challenge the view that the poor, whatever their race or ethnicity, are lazy, opposed to work, and prefer to live off others. In a recent study of the working poor in the inner city (Harlem), Katherine Newman (1999) documents the many African Americans and Latinos who are trying to form families, work steadily, and support themselves. If, Newman argues, Americans realized that most minorities are members of the hard-working poor, Americans would be less ready to label them as undeserving.

For years, studies of welfare mothers have indicated that about three-fourths prefer employment to staying at home. Most welfare mothers say they are not presently working because of their children, housework, bad health, or lack of skills (Poddel 1968; Warren and Berkowitz 1969). A national survey indicates a strong work ethic among the poor. A representative sample of poor Americans was asked, "Which do you think is more important in life: working hard and doing what is expected of you or doing the things that give you personal satisfaction and pleasure?" Seventy-eight percent subscribed to the idea of hard work, which was higher than for the nonpoor in the sample (65 percent). In addition, the poor favored the idea of workfare—requirement by the government that physically able poor people work before they can receive poverty benefits—to a somewhat greater degree than the nonpoor did (59 percent versus 51 percent). This evidence suggests that the poor believe in the same work-related values and attitudes as other Americans (I. A. Lewis and Schneider 1985). More recent studies report no difference in desire to work between the poor and the nonpoor. Whether male or female, black or white, young or old, poor people appear to share the middle-class identification of self-esteem with work and self-support. The aspirations of the poor match those of the nonpoor; both groups desire an adequate education and a home in a good neighborhood (W. J. Wilson 1997; Lichter and Crowley 2002).

In summary, many Americans have tended to blame the poor for their poverty, despite evidence that the poor are not the lazy, shiftless, willingly dependent people they are popularly thought to be. Both the attitudes and the behavior of the poor contradict the negative label often applied to them (Blee and Billings 1999; O'Connor 2002).

Combating Poverty

Prior to the mid-1960s, combating poverty was not a prominent goal of the federal government (Noble 1997; Hamilton and Hamilton 1998; Trattner 1999; O'Connor 2002). Some programs, such as Social Security and Aid to Families with Dependent Children (AFDC, replaced in 1996 by Temporary Assistance for Needy Families or TANF) had been enacted during the Great Depression of the 1930s, but these measures did not reach the greatest levels of need. So in August 1964, President Lyndon Johnson marshaled the forces of the federal government, began the War on Poverty, and signed into law the Economic Opportunity Act (Levitan 1990; Yoo 1999). The Economic Opportunity Act of 1964 offered a baseline philosophy: help poor people help themselves (Patterson 1986). The intent was to enhance the income-generating capacity of the poor. If the chains of poverty were to be broken, it had to be through self-improvement rather than through temporary relief.

Has America's effort to help the poor been a failure? According to critics such as Charles Murray (1994) and Myron Magnet (2000), the war on poverty in America cannot be won through welfare programs. The reason poverty has not been reduced, they contend, lies in a welfare policy that breeds dependency on government aid. In particular, Murray asserts, between 1960 and 1970 there were alterations in the policies underlying the welfare system that encouraged recipients to abandon work and marriage and to have children out of wedlock to increase their benefits.

In response to Murray and other "warriors of welfare," Theodore Marmor, Jerry Mashaw, and Philip Harvey wrote *America's Misunderstood Welfare State* (1992). Although these authors concede that total federal social welfare spending increased greatly between 1960 and 1996 (when welfare reform was enacted by the U.S. Congress), less money in real terms was spent on welfare recipients in 1996 than in 1970. Programs of the American welfare state may not eliminate poverty, contend Marmor, Mashaw, and Harvey, but that was never their sole purpose in the first place. Social welfare programs were designed for several reasons: to encourage the poor to behave in more socially acceptable ways, such as working at any available jobs; to provide a "safety net" for the "deserving poor"; and to protect recipients from economic insecurities engendered by sickness, injury, retirement, involuntary unemployment, and widowhood (Jacoby 1997). Going further, Benjamin Page and James Simmons (2000) amass evidence on the effectiveness of government poverty programs. They challenge the charges that government poverty programs are ineffective, wasteful, and promoters of dependency.

This debate regarding the government's role in combating poverty is not over. The welfare critics' view, however, is clearly imprinted on the Personal Responsibility and Work Opportunity Reconciliation Act, the welfare reform legislation passed by Congress in 1996 (Handler and Hasenfeld 1997; R. M. Blank 1998; Chael 1999; Arrow, Durlauf, and Bowles 2000).

© David Grossman/Photo Researchers, Inc.

America's welfare efforts have been defended by its supporters and criticized by its detractors. Both groups, however, tend to agree that the Head Start program has been a success.

Plugged In at Street-Level

Many economically disadvantaged people fear computer technology. They see computers replacing people in the workplace or requiring job skills they do not have. In fact, the gap between the "haves"—those people with regular access to computers—and the "have nots" is widening (Bolt and Crawford 2000; Norris 2001). Even in environments where this gap should not exist—the public schools—research shows that children from disadvantaged families are likely to be enrolled in schools with less access to computers and other new technologies than children in wealthier school districts are. Even when they have access to some computers, students in poor schools have much less opportunity to work with them. These children also tend to have teachers who are less skilled on computers and who have negative attitudes toward the use of computers in the classroom (Wax 2000).

Because of this situation, educators are now designing special school-based programs to provide computers in low-income schools and to train teachers in those schools to use them. Community groups and businesses are also creating computer-training programs aimed at school age children and adults. Since 1985, well over 200 sites in the United States have established out-of-school computer centers. Many of these centers provide opportunity for the disadvantaged to gain computer skills, skills that are required to enter the job market in an information-based society.

Playing to Win is one of the first community computer centers designed to serve people in economically disadvantaged communities. Founded in New York City's Harlem, Playing to Win offers computer classes and workshops to nearly 400 people per week. It also assists other community groups that want to set up their own computer centers (George 1993).

Another successful program is Street-Level Youth Media in Chicago's inner city. Street-Level's mission is to educate disadvantaged young people about emerging technologies. Street-Level began its operations by offering inner-city youth the opportunity to make videos about their everyday lives on the streets of Chicago. When these videos were made available to the community, residents were able to see these young people as human beings trapped in desperate, life-threatening situations. Street-Level soon began using computers in its effort to reach the young, becoming one of the first groups in the country to offer new technologies to urban youth. This organization continues to work with youth who have been rejected by mainstream society. Street-Level helps the young people find solutions to their problems, strengthen their communities, and achieve economic success ("Street-Level Youth Media," 1999).

Another example of a computer community is Plugged In, located in East Palo Alto, California. This organization was created in 1992, just at the time the Internet was becoming popular. Plugged In teaches school age children to design Web pages, and it supports them in an information technology business that makes websites for paying customers. In addition to this teen-run business, Plugged In provides other technology and information services on a fee basis. It continues, though, its primary mission—delivering free educational and computer access services to the local community (Decrem 1998).

Analyzing the Trends

Social class is usually based on income, wealth, power, and prestige. Will the rise of computer technology lead to a new basis for social stratification? If you think so, what will it be? If not, why not?

Welfare Reform

When the term **welfare reform** is used in the United States, the reference is generally to Temporary Assistance for Needy Families (the old AFDC program; Dervarics 1998). As highlighted in "Using the Sociological Imagination," TANF welfare payments account for less than 1 percent of all federal government expenditures. In 2002, actual spending for education, training, employment, and social services (many more programs than "welfare") was $47 billion, or 2.5 percent of total expenditures (Executive Office of the President of the United States 2002). Still, even though the TANF welfare programs constitute a very small part of that figure, it remains a hot button of welfare reform.

What is the nature of welfare reform? During his 1992 presidential campaign, President Bill Clinton consistently promised "to end welfare as we know it." Although major differences existed between a reelected President Clinton in 1996 and the Republican majority in Congress, their major point of agreement was the need for workfare—welfare with limits on the amount of time the able-bodied can receive welfare payments.

The 1996 welfare reform legislation reflects this agreement (Handler and Hasenfeld 1997; R. M. Blank 1998). This workfare legislation places the federal government's relationship with the poor on a fundamentally different basis (O'Hare 1996; Solow, Himmelfarb, and Gutmann 1998; Chael 1999; Gilbert 2002). The bill has three major elements: it reduces welfare spending;

it increases state and local power to oversee welfare regulation; and it adds new restrictions on welfare eligibility. For example, benefits to children of unwed teenage mothers are denied unless the mother remains in school and lives with an adult; and cash aid to able-bodied adults is terminated if they fail to get a job after two years.

Is welfare reform working? This attempt at welfare reform is too novel, yet, for a definitive evaluation. A major study in 2000 documented that welfare rolls had decreased more dramatically than most predicted (O'Neill and Hill 2001). Just over 5 million U.S. families were now on the welfare rolls, down from over 12 million in 1996, the date of the latest welfare reform bill. About half of those leaving the welfare rolls reported finding jobs, and one-third of welfare recipients had jobs.

There is a darker side. Most of those leaving the rolls since 1996 hold entry-level jobs—in restaurants, cleaning services, retail stores—making less than $7 per hour. Most of those leaving public assistance are at the bottom of the economy with little hope of advancement. One-fourth work at night and over half report child-care problems. Most have jobs without health insurance, and a substantial minority report food shortage and difficulty paying rent. In short, many of those leaving welfare have shifted from welfare to poverty (Peck, Piven, and Cloward 2001; Fuller et al. 2002; Hofferth 2002; Nelson, 2003). Welfare reform actually promotes this trend. Because welfare recipients under TANF are significantly less likely to attend college than were AFDC recipients, welfare reform has erected an added barrier to upward mobility among the poor (Cox and Spriggs 2002).

© Phil Cantor/SuperStock

The 1996 welfare reform legislation defined time limits for able-bodied adults on welfare. These two adults are working as a result of that legislation.

To date, the downside of welfare reform leaves ample room for critics to question its success. The true test of welfare reform, the critics point out, is just now beginning—as the effects of the weakened economy are felt; as we get down to the harder cases; and as the last time limits take effect for the most difficult cases (Goode and Maskovsky 2001; Bernstein and Starr 2002; Kuttner 2002). In fact, a recent study of 31 cities found evidence of a reversal of the unprecedented rise in welfare recipient employment between 1996 and 2000 (M. Waller 2003). According to the mayors of 50 percent of these cities, welfare rolls increased in 2002,

FEEDBACK

1. _________ poverty is defined as the absence of enough money to secure life's necessities.
2. _________ poverty is measured by comparing the economic condition of those at the bottom of a society with that of other segments of the population.
3. Which of the following is *not* one of the major categories of the poor in the United States?
   a. children under eighteen
   b able-bodied men who refuse to work
   c. the elderly
   d. the disabled
   e. people who live alone or with nonrelatives
4. According to the ideology of _________, those at the bottom of the stratification structure are there because of their own inadequacies.
5. The philosophy behind the Economic Opportunity Act of 1964 was
   a. to provide temporary relief for the poor
   b. to make the poor dependent on the federal government
   c. to make it easy for the poor to receive help
   d. to help poor people help themselves
6. The problem with the poor is that they have lower economic and educational aspirations than the nonpoor. T or F?

Answers: 1. Absolute 2. Relative 3. b 4. individualism 5. d 6. F

and two-thirds of these cities reported an increase in TANF recipients facing substantial barriers to employment beyond their control.

Consequences of Stratification

The unequal distribution of income, wealth, power, and prestige has additional consequences for people. Two broad categories of class-related social consequences are *life chances* and *lifestyle* (Charles Tilly 1999).

Life Chances

Life chances refers to the likelihood of possessing the good things in life: health, happiness, education, wealth, legal protection, and even life itself. The probability of acquiring and maintaining the material and nonmaterial rewards of life is significantly affected by social class level. Power, prestige, and economic rewards increase with social class level. The same is true for education, the single most important gateway to these rewards. But additional more subtle life chances exist, and these, too, vary with social class (Putnam 1996; Pennar 1997). These less obvious life chances are in good part the product of inequality in the distribution of education, power, prestige, and economic rewards.

How does social class affect more subtle life chances? The probability of possessing life itself—the most fundamental, precious life chance—declines with social class level. Whether measured by the death rate or by life expectancy, the likelihood of a longer life is enhanced as people move up a stratification structure (U.S. Bureau of the Census 2001a). This disparity is generated by differences in several areas: value placed on medical attention, concern with proper nutrition, attention to personal hygiene, and ability to afford these things.

In light of differences in life expectancy, it is not surprising that physical health is affected by social class level. Those lower in a stratification structure are more likely to be sick or disabled and to receive poorer medical treatment once they are ill (Califano 1985; R. Wilkinson 1986; Conrad and Kern 1990). It is no different for mental health. Persons at lower class levels have a greater probability of becoming mentally disturbed and are less likely to receive therapeutic help, adequate or otherwise (Goodman et al. 1983; Link, Dohrenwend, and Skodol 1986; Miech and Hauser 1998).

Innumerable other life chance inequalities exist. For example, the poor often pay more for the same goods and services. They are more likely to get caught for committing a crime, and when they are apprehended, they stand a greater chance of being convicted and serving prison time. They are less likely to have connections to help them around the system, as in obtaining tickets to a football game or getting a favor from a political figure. And the public services they receive—garbage collection, police protection, street repair—are often inferior (Caplovitz 1963; R. J. Thomson and Zingraff 1981).

Lifestyle

How does social class affect lifestyle? The rich and the poor are separated by more than money. Social class differences in lifestyle can be observed in many areas of American life, including education, marital and family relations, child rearing, political attitudes and behavior, and religious affiliation.

People in higher social classes tend to marry later, experience lower divorce rates, and have better marital adjustment (Kitson and Raschke 1981; Fergusson, Horwood, and Shannon 1984). While working-class and middle-class parents are more alike in child-rearing practices than in the past, some differences remain. Compared to the middle class, lower-class parents tend to be less attentive to their children's social and emotional needs; in disciplining, they are more inclined to use physical punishment rather than reasoning. Middle-class parents are more interested in helping their children develop such traits as concern for others, self-control, and curiosity. Finally, working-class parents emphasize conformity, orderliness, and neatness, whereas middle-class parents stress self-direction, freedom, initiative, and creativity (Kohn 1963, 1977, 1990; Wright and Wright 1976; Williamson 1984).

The incidence of voting and political involvement increases with social class level. The middle and upper classes are more likely to be Republicans, whereas the working and lower classes tend to register as Democrats. Those in the lower class tend to be more liberal than middle- and upper-class Americans are on economic issues, but more conservative on social issues. Thus, lower-class people are more in favor of labor unions, government control of business, and social welfare programs but tend to exhibit less tolerance and sympathy than do higher social classes for social issues involving civil rights and international affairs (Wolfinger and Rosenstone 1980; Szymanski 1983).

The rate of church membership and attendance is lowest at the extremes of the stratification structure. Episcopalian, Congregational, and Presbyterian churches are significantly less populated by members of the lower class. Lower-class Americans lean more toward Baptist and fundamentalist churches.

FEEDBACK

Identify the following as examples of life chances (LC) or lifestyle (LS).

____ a. divorce rate
____ b. quality of neighborhood street repair
____ c. life expectancy
____ d. child-rearing practices
____ e. voting behavior
____ f. church attendance

Answers: a. lifestyle b. life chances c. life chances d. lifestyle e. lifestyle f. lifestyle

Methodist and Lutheran denominations fall in between (see Chapter 15).

Social Mobility

Social mobility refers to the movement of individuals or groups within a stratification structure. In the United States, the term *mobility* implies an elevation in social class level. Beyond this, it is possible either to move down in social class or to move with little or no change in social class.

Types of Social Mobility

Social mobility can be horizontal or vertical. Both types of mobility can be measured either within the career of an individual—**intragenerational mobility**—or from one generation to the next—**intergenerational mobility**.

How does horizontal mobility differ from vertical mobility? A change from one occupation to another at the same general status level is called **horizontal mobility**. Examples of intragenerational horizontal mobility are familiar: an army captain becomes a public school teacher, a minister becomes a psychologist, a restaurant server becomes a taxi driver. The daughter of an attorney who becomes an engineer illustrates intergenerational horizontal mobility. Because horizontal mobility involves no real change in occupational status or social class, sociologists are not generally interested in investigating it.

Vertical mobility, however, is investigated extensively. In **vertical mobility**, occupational status or social class moves upward or downward. Vertical mobility can also be intragenerational or intergenerational. The simplest way to measure intragenerational mobility is to compare an individual's present occupation with his or her first one. Someone who began as a dockworker and later became an insurance salesperson has experienced upward intragenerational mobility. Intergenerational mobility involves the comparison of a parent's (or grandparent's) occupation with the child's occupation. If a plumber's daughter becomes a physician, upward intergenerational mobility has occurred. If a lawyer's son becomes a carpenter, downward intergenerational mobility has occurred.

Caste and Open Class Stratification Systems

The extent of vertical mobility varies from society to society. Some societies have considerable mobility; others have little or none. This is the major difference between caste systems and open class systems.

How much mobility can we expect in a caste system? In a **caste system**, there is no social mobility because social status is inherited and cannot be changed. In a caste system, statuses (including occupations) are ascribed or assigned at birth; individuals cannot change their statuses through any efforts of their own. By reason of religious, biological, magical, or legal justification, those in one caste are allowed to marry only within their own caste and must limit relationships of all types with those below and above them in the stratification structure. Apartheid, as practiced in South Africa before the election of Nelson Mandela, was a caste system based on race.

The caste system in India is based on occupation and religion and is as complex as it is rigid. Hindus believe that there are four primary caste categories, ranked according to their degree of religious purity. The Brahmin, the top caste, is composed of priests and scholars. Next come the Kshatriyas—noble and warrior caste. Merchants form the third caste, called the Vaishyas. Finally, there is the Shudra caste, containing farmers, menial workers, and craftspeople. Actually, there is a fifth category called the "untouchables"—Indians thought to be so impure that any physical contact contaminates the religious purity of higher caste members.

How is the caste system kept intact? Traditional rules in India prevent movement into a higher caste. Not only are members of different castes not permitted to eat together, but higher-caste people will hardly accept anything to eat or drink from lower-caste persons. Untouchables, who must live apart from everyone else, cannot even drink water from the wells used by higher castes. Although the long-standing legal prohibition against dating or marrying someone in a higher caste no longer exists, such crossings are still extremely rare. Most important, the caste system is maintained due to the power of those in the higher castes, who use their political power, wealth, and prestige to prevent change.

How does an open class system differ from a caste system? In an **open class system**, an individual's social status is based on merit and individual effort. Individuals move up and down the stratification structure as their abilities, education, resources, and commitment to work permit. Although inequality exists in an open class system, it is supposed to be based on differences in monetary worth and personal accomplishment.

Just as some social mobility occurs in an actual caste system, the opportunity for upward mobility is sometimes denied individuals or groups within an open class stratification structure. Prominent illustrations in American society today—which is relatively, but not purely, an open class system—are African Americans, Native Americans, and Latinos (see Chapter 9).

Extent of Upward Mobility

How much upward mobility exists in America? In American society, economic and occupational success is assumed to be the product of talent and willingness to work. The classic Horatio Alger stories—in which a down-on-his-luck boy makes good through honesty, pluck, and diligence—embody the long-standing belief that America is the land of opportunity. The only thing standing between any American citizen and success, it is argued, is talent, willingness to work, and perseverance (Kluegel and Smith 1986). The careers of Abraham Lincoln, Cornelius Vanderbilt, John D. Rockefeller, and Henry Ford are used to support the idea of unlimited mobility in American society. These men, in reality, are exceptions to the rule. While few places in the world provide the opportunities available in the United States, countless Americans fail to be upwardly mobile, despite their talents and dedication to work.

The actual extent of upward mobility has never been as great as the rags-to-riches myth would have us believe (Schwarz 1997). Most American presidents, contrary to popular belief, were not born into the lower strata of society (Pessen 1984). This does not mean that upward mobility does not occur in American society. In fact, the considerable mobility that has occurred has been upward. It does mean, however, that great leaps in social class level have been rare. Studies have consistently shown that upward mobility, when it occurs, is usually slight (Sorokin 1927; E. F. Jackson and Crockett 1964; Blau and Duncan 1967; Featherman 1971; F. J. Davis 1982; D. A. Gilbert 2003).

Why isn't upward intergenerational mobility more dramatic? There is a reason that upward social mobility does not occur in large leaps, even in an open class system: one's occupational level is strongly influenced by the occupational level of one's parents. Sociologists have traditionally demonstrated this fact through data that show the influence of fathers' occupations on their

SOCIOLOGY AND THE NEWS MEDIA

Challenging the Caste System in India

In a caste system, social mobility is prohibited. All social statuses are determined at birth. Naturally, those at the bottom of the caste structure have historically been powerless. As this CNN report indicates, protest against the caste system is now occurring in India.

1. Discuss the protests in India within the context of power.

____

____

2. Do you think there is a caste system in American society? Has there ever been? Explain.

____

____

The caste system in India is finally being challenged.

sons' occupations. More recently, mobility studies have begun to include women (Baxter 1994; Kalmijn 1994; D. A. Gilbert 2003).

In the United States, there is considerable occupational succession between fathers and sons (sons at the same occupational level as their fathers). Despite a significant amount of occupational mobility up and down the occupational ladder, there is a barrier between white-collar and manual jobs. About two-thirds of sons born on either side of this manual-nonmanual line do not manage to permanently cross it. Moreover, most upward social mobility in the United States occurs in small increments. For example, the son of a janitor becomes a bus driver, or the son of a retail sales worker becomes a commercial artist.

The occupational level of females is also highly correlated with their fathers' occupational level. Just over half of the daughters of upper white-collar fathers are in similar positions; only one-fourth of the daughters of lower manual workers are in upper white-collar occupations. Similarly, nearly 40 percent of the daughters of fathers in lower manual jobs are in manual work, whereas a relatively small percentage of the daughters of white-collar men hold blue-collar jobs.

Why has upward mobility occurred in the United States? Although upward mobility in the United States has not typically involved dramatic leaps, it has occurred frequently. According to Peter Blau and Otis Dudley Duncan, there are three basic determinants of upward mobility. First, as less qualified immigrant or rural workers take lower-level jobs, the more qualified urban dwellers fill higher-level positions. Second, because the higher social classes have fewer children, members of the lower social classes are needed to fill higher-level positions. Third, as an economy advances, technology eliminates lower-level jobs and creates higher-level ones. Shrinking job opportunities at the lower end preclude the children of lower-level manual workers from following in their parents' occupational footsteps. They have to prepare themselves to move up to higher-level occupations. This type of social mobility, which occurs because of changes in the distribution of occupational opportunities, is called **structural mobility** (Hope 1982; DiPrete 1996; Sobel, Becker, and Minick 1998).

One example of structural mobility in the United States is the change in labor markets due to the new technology and the globalization of business. Emboldened by computer-driven production, improved means of communication, and better transportation, many U.S. companies are moving their manufacturing operations overseas; they are seeking lower costs and increased profits. And they are doing so often. As a result, high-paying U.S. jobs are being transferred to lower-paid foreign workers.

Without the higher-paying manufacturing jobs, and without the education needed to perform the new, more technologically sophisticated jobs, U.S. workers are being forced to take lower-paying jobs. Compared to their parents, more U.S. workers are now experiencing downward mobility (Newman 1999; Bernhardt et al. 2001).

Global Stratification

Identification of Economies

Thus far, the focus has been on stratification within societies. But scarce desirables—income, wealth, prestige, and power—are also differentially distributed among nations. The gross domestic product (GDP) of a country—the total value of the goods and services it produces in one year—is a reasonably good indicator for classifying countries into economic categories. GDP

FEEDBACK

1. __________ refers to the movement of individuals or groups within a stratification structure.
2. Match the major types of social mobility with the illustrations.
   ____ a. intergenerational mobility — (1) A restaurant waiter becomes a taxi driver
   ____ b. vertical mobility — (2) An auto worker becomes a manager
   ____ c. horizontal mobility — (3) The daughter of a hairdresser becomes a college professor
3. In a __________ system, social status is inherited at birth from one's parents and cannot be changed.
4. In an __________ system, rank based on the stratification structure is achieved.
5. Upward mobility in America usually involves only a small improvement in status. T or F?
6. The son of a steelworker finds that the steel industry has shrunk so much that a job in the mill is not available. Instead of following in his father's footsteps, he attends a community college and becomes a computer operator. This is an example of
   a. horizontal mobility.
   b. intragenerational mobility.
   c. caste mobility.
   d. structural mobility.

Answers: 1. Social mobility 2. a. (3) b. (2) c. (1) 3. caste 4. open class 5. T 6. d

Global Inequality

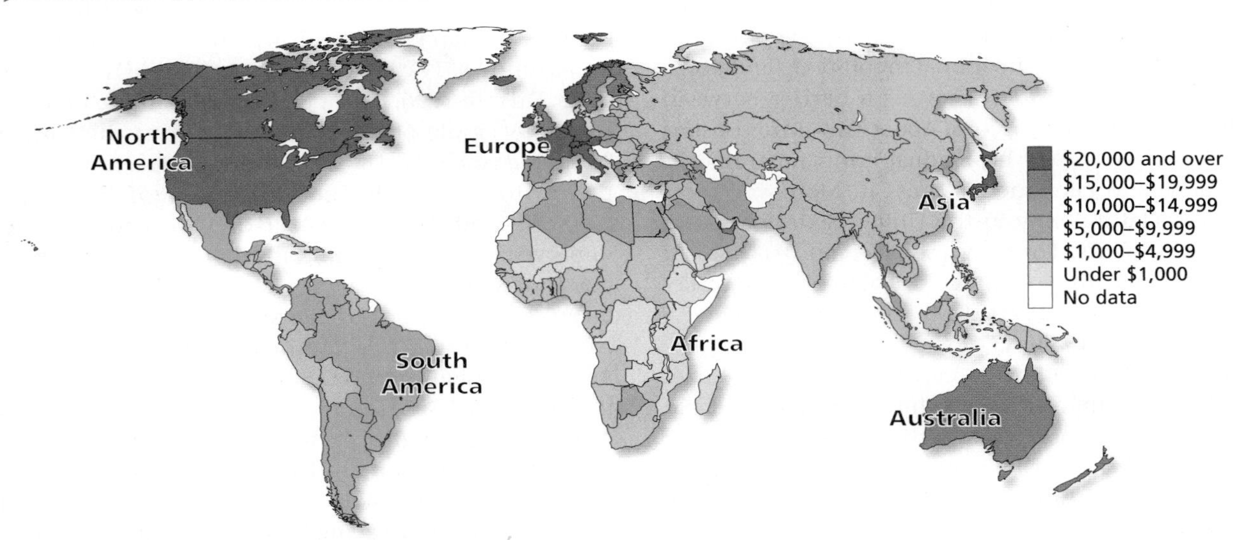

Source: *Human Development Report 1998*, United Nations Development Report (New York: Oxford University Press, 1998).

This map illustrates the extent of inequality among nations as measured by comparisons of real purchasing power per person across countries. The 20 percent of the world's population living in the wealthiest countries can afford to consume 16 times as much per person as the 20 percent residing in the poorest countries.

1. Write a paragraph describing the pattern of global inequality you see in this map.
2. Write another paragraph elaborating on your thoughts about the pattern (and degree) of global poverty you see.

is useful both because it is a single, measurable economic indicator and because a nation's economic rank is highly correlated with the extent of its prestige and power (World Bank Atlas 1999).

Where are the high-income economies, and how wealthy are they? The richest countries, nearly all capitalistic, include countries such as England, Germany, France, Norway, Finland, Denmark, Sweden, and Switzerland in Western Europe; the United States and Canada in North America; Australia and New Zealand in Oceania; and Japan in Asia. A few oil-rich nations, such as Kuwait and the United Arab Emirates, also have high-income economies. High-income economies, spanning approximately one-quarter of the Earth's land surface, have only 15 percent of the world's population (approximately 800 million people). Yet, these countries are in control of most of the wealth. Due to this concentration of wealth, the standard of living of nearly all the people living in these countries, even most of the poor people, is higher than that for the average person in low-income economies (Schnitzer 2000).

What is the nature of middle-income economies? Middle-income economies include, but are not limited to, nations historically founded on or influenced by socialist or communist economies. This encompasses members of the former Soviet Union, many of them renamed and now part of the new Commonwealth of Independent States that was created in 1991. Other countries that fell into the Soviet orbit but are not part of the Commonwealth of Independent States include Poland, Bulgaria, the Czech Republic, the Slovak Republic, Hungary, Romania, Estonia, Latvia, Lithuania, and Cuba.

From the end of World War II until 1989, all these countries, under the political, economic, and military control of the former Soviet Union, constituted the Eastern bloc in the Cold War against the West. These countries are no more densely populated than high-income economies—they occupy 15 percent of the Earth's land with only 10 percent of the world's population (about 500 million people). Although more prosperous than low-income countries, citizens of middle-income economies do not enjoy the standard of living characteristic of high-income economies. The extent of industrialization in these countries is variable.

Most inhabitants of these countries live in urban areas; but in comparison to countries with high-income economies, a greater percentage of their people live in rural areas and participate in agricultural economic activities.

In 1989, Eastern European countries (freed of Soviet domination, its socialist economy, and government-controlled bureaucracy) turned to capitalism and its market mechanism. A major source of current political and economic turmoil in the former Soviet Union revolves around the struggle between the socialist and capitalist roads to economic development (see Chapter 14).

The middle-income category contains other countries that have never been part of the old Soviet bloc. These countries are spread from Mexico, Latin America, and South America to parts of Africa to Thailand, Malaysia, and New Guinea in the Far East.

What is distinctive about low-income economies? The economic base of low-income economies is primarily agricultural. These countries span the globe from south of the United States border to Africa to China to Indonesia. Population density within these countries as a whole is very high. In fact, just over two-thirds of the world's population (over 5 billion people) live on 60 percent of the Earth's land (Livernash and Rodenburg 1998).

No single economic system is typical of these economies. Capitalism, socialism, and various combinations of each are practiced. If these countries share no economic ideology, they do share one important economic characteristic: they are unimaginably poor. Problems of the poor in high-income countries, which have already been explored in this chapter, are real and significant. But most people in low-income economies live annually on less than 7 percent of the official poverty line in the United States (U.S. Bureau of the Census 2001a). The plight of the poor in low-income countries is particularly bleak because they are trapped in a double bind. Their populations are exploding (due to declining infant mortality rates) while they maintain their traditionally high birth rate.

Global Poverty

Poverty persists despite a $400-trillion-plus global economy. Three recent international reports detail the extent and distribution of global poverty *(Human Development Report 1999; World Development Report 2001; Least Developed Countries Report 2002).*

What is the state of poverty at the end of the twentieth century? Despite progress, one-quarter of the Earth's population remains in extreme poverty. Almost one-fourth of the people in less developed countries—1.2 billion—now have incomes of less than $1 daily (see Table 8.4). Poverty by this standard is most extensive in South Asia—affecting about 500 million people. Southern and eastern Asia, Southeast Asia, and the Pacific have about 300 million of these 1.2 billion people. Sub-Saharan Africa now also has 300 million of its people living at this level, and the rate is increasing. In Latin America and the Caribbean, almost 80 million people have less than $1 a day on which to live (*World Development Report* 2001; Sheehan 2002).

In the more economically developed world, the greatest increase in poverty in the last ten years is found in Eastern European countries and among member nations of the former Soviet Union. Whereas poverty in these areas was formerly not widespread, about a third of these people (120 million) now have incomes below $4 daily. Even in industrial nations, over 100 million people subsist below the established poverty line.

According to the Food and Agriculture Organization of the United Nations, advancement in reducing world

TABLE 8.4

People Living on Less than $1 a Day, Selected Regions, 1987 and 1998

Many people in the world live on less than $1 per day. Describe any trends you see between 1987 and 1998. Any surprises?

	1987		1998	
Region	**Total (million)**	**Share of Population (percent)**	**Total (million)**	**Share of Population (percent)**
Sub-Saharan Africa	217.2	46.6	290.0	46.3
South Asia	474.4	44.9	522.0	40.0
Latin America & Caribbean	63.7	15.3	78.2	15.6
East Asia & Pacific	417.5	26.6	278.3	15.3
Eastern Europe & Central Asia	1.1	0.2	24.0	5.1
Middle East & North Africa	9.3	4.3	5.5	1.9
Total	**1,183.2**	**28.3**	**1,198.9**	**24.0**

Source: World Bank, *World Development Report 2000/2001* (New York: Oxford University Press, 2000), p. 23.

hunger has come to a virtual halt. Since 1990–1992, the number of undernourished people in the world has decreased by only 2.5 million per year. Even this slight reduction is due to significant progress in a few large nations. For example, if the progress in China is left out of the calculation, the number of undernourished people in the rest of the developing world has risen by over 50 million since 1990–1992 (*State of Food Insecurity in the World* 2002).

Has there been no progress in combating global poverty? Poverty has declined in many areas of the world (Bhalla 2002; *World Development Report* 2003). In fact, in all regions of the world, the past fifty years has seen poverty decrease more than in the preceding five centuries. In less than twenty years, China and fourteen other countries, whose populations exceed 1.6 billion, have cut in half the proportion of their people existing in poverty. An additional ten nations, who number almost 1 billion, have cut their poverty rate by one-fourth. Still, poverty is pervasive throughout the world and may increase as a result of slowed economic growth and persisting conflict within and between countries (*Least Developed Countries Report* 2002; Weisbrot 2002; Wermuth 2002; Renner 2003).

FEEDBACK

1. Classify the following examples as high-income (HI), middle-income (MI), or low-income (LI) economies.
   ____ a. Commonwealth of Independent States
   ____ b. agricultural economic base
   ____ c. movement toward capitalism
   ____ d. 800 million people
   ____ e. double population bind
   ____ f. one-fourth of the Earth's land surface
2. Do you think that classifying all the nations of the world into three broad categories leads to some distortion of reality? Explain.
3. What proportion of the Earth's population is considered poor today?
   a. one-tenth
   b. one-fifth
   c. one-fourth
   d. one-third
4. Over the past fifty years, global poverty has actually decreased. T or F?

Answers: 1. a. MI b. LI c. MI d. HI e. LI f. HI 3. c 4. T

REVIEW GUIDE

SUMMARY

1. There is a universal tendency for humans to create inequality. When individuals are ranked by wealth, prestige, and power, social stratification exists. Any stratification structure is composed of either social classes or segments of a population whose members hold a similar share of community resources. What's more, these class members share attitudes, values, norms, and an identifiable lifestyle.
2. A stratification structure is based on economic, power, and prestige dimensions. Income inequality has always existed in American society, but it has increased dramatically since 1980. Prestige, especially in developed societies, is based primarily on occupation. Occupations with the greatest prestige generally pay the most; require the greatest amount of training, skill, and ability; provide the most power; and are considered the most important.
3. Those individuals who are ranked high on one dimension of stratification are usually high on the other dimensions. Wealth, power, and prestige tend to go together.
4. Functionalists contend that stratification is necessary. It motivates people to prepare themselves for difficult and important jobs, and it motivates them to perform well once they are in those jobs. According to the conflict theory of stratification, inequality exists because some people have more power than others and are willing to use it to promote their self-interest.
5. The functionalist and conflict theories shed light on the reasons for stratification. Symbolic interactionism aids in understanding the manner in which stratification structures are perpetuated. This perspective emphasizes the legitimization of a stratification structure through symbols. These symbols, of course, are learned through socialization.
6. Although most sociologists believe that class lines in America are difficult to draw, they agree that the following social classes have been identified: the upper class, the middle class, the working class, and the lower class. Sociologists agree that the poor exist as an underclass.

7. Poverty can be measured in two ways—absolute or relative. Absolute poverty is determined by annual income. Anything below a determined (absolute) amount is categorized as poverty. Relative poverty is a comparative measure; it contrasts income groups at the bottom of the stratification structure with those above them. Because of the continuing influence of social Darwinism, Americans tend to blame the poor for their own situation, despite evidence to the contrary.
8. The federal government became serious about combating poverty during the mid-1960s. There have been objections, from some quarters, ever since. The debate on the role of government continues.
9. Americans do not wish to abolish all welfare benefits, but they do want a system that forces the able-bodied off the rolls. In 1996, the U.S. Congress passed a sweeping welfare reform act that dramatically changed the relationship between the poor and their government.
10. Social classes can be thought of as subcultures. For this reason, distinctive patterns of thinking, feeling, and behaving are associated with the various social classes. Both life chances and lifestyle characteristics are affected by one's social class level.
11. Social mobility, the movement of individuals or groups within the stratification structure, is usually measured by changes in occupational status. Sometimes movement within a stratification structure is horizontal. Sociologists, however, are much more interested in upward or downward (vertical) mobility, whether within an individual's lifetime or from one generation to the next.
12. Caste societies, such as India's, permit little vertical mobility; open class societies, such as those in industrialized countries, allow considerable upward mobility. Even in open class societies, however, an upward move tends to be a small one.
13. Nations are also stratified. Richer nations, primarily in the West, are classified as high-income economies. Other less developed industrialized nations, including member nations of the former Soviet Union, can be classified as middle-income economies. The rest of the world, poor and resting primarily on an agricultural base, is classified as low-income economies. In these societies, poverty is widespread.

INFOTRAC® COLLEGE EDITION

http://www.infotrac-college.com/wadsworth

Another unique option available to you at the Student Resources section of the companion Web site is Infotrac College Edition, an online library with access to hundreds of scholarly and popular periodicals. Below are suggested search terms for this chapter. Results from these and other searches are found at the site.

Search keywords: **executive compensation**. The level of executive compensation is a topic of significant concern in the United States. Find two articles that discuss this issue. What are the primary concerns about executive compensation in the United States? How does U.S. executive compensation compare to that in other countries?

Search keywords: **false consciousness**. Karl Marx introduced this term in his writings about communism. Search Infotrac to find what Marx meant by false consciousness. Based on your understanding of the term, do you agree that it exists? Why or why not?

Search keywords: **life chances**. The likelihood of obtaining "the good things in life" is affected by many things beyond an individual's control. Find and read three articles about life chances. What are some of the social factors that affect one's life chances?

Search keywords: **welfare reform**. The United States passed major welfare reform legislation in 1996. Find articles in Infotrac that discuss the outcomes of this law. How successful has the reform been?

LEARNING OBJECTIVES REVIEW

- Explain the relationship between social stratification and social class.
- Compare and contrast the three dimensions of stratification.
- State the major differences between the functionalist, conflict, and symbolic interactionist approaches to social stratification.
- Identify the distinguishing characteristics of the major social classes in America.
- Discuss the extent of poverty and perceptions of poverty in the United States.
- Evaluate the U.S. commitment to poverty programs.
- Outline some of the consequences of social stratification.
- Describe upward and downward social mobility in the United States.
- Discuss the major features of global stratification.

CONCEPT REVIEW

Match the following concepts with the definitions listed below them.

____ a. life chances
____ b. horizontal mobility
____ c. caste system
____ d. intergenerational mobility
____ e. relative poverty
____ f. power
____ g. social mobility
____ h. wealth
____ i. conspicuous consumption
____ j. structural mobility

1. all the economic resources possessed by an individual or group
2. the consumption of goods and services to display one's wealth to others
3. social mobility that takes place from one generation to the next
4. the ability to control the behavior of others, even against their will
5. movement of individuals or groups within a stratification structure
6. the type of stratification structure in which there is no social mobility because social status is inherited and cannot subsequently be changed
7. the likelihood of securing the "good things in life," such as housing, education, good health, and food
8. the measurement of poverty by a comparison of the economic condition of those at the bottom of a society with that of other segments of the population
9. a change from one occupation to another at the same general status level
10. mobility that occurs because of changes in the distribution of occupational opportunities

CRITICAL-THINKING QUESTIONS

1. Stratification can occur in any social setting. Think about a group with which you are familiar that is stratified. Discuss the "unequal shares of scarce desirables" used to rank members.

   ____

   ____

2. Consider the existence of a class at the bottom of the American social stratification structure. Compare the explanations for the enduring presence of this class that would be given by functionalism and conflict theory.

   ____

   ____

3. More and more of the poor in the United States are women and children. Relate the feminization of poverty to the prevailing view of the poor held by most Americans. Is there an inconsistency? Why or why not?

   ____

   ____

4. You are a member of a particular social class. Discuss your life chances and lifestyle in sociological terms.

   ____

   ____

5. Analyze the social mobility that has occurred in your family across as many generations as you can. Use specific sociological concepts in your analysis.

   ____

   ____

MULTIPLE-CHOICE QUESTIONS

1. **The creation of layers of people possessing unequal shares of scarce desirables is known as**
   a. social layering.
   b. social grading.
   c. social tracking.
   d. social stratification.
   e. resource accumulation.
2. **A _________ is a segment of a population whose members hold a relatively similar share of scarce desirables and who share attitudes, values, norms, and identifiable lifestyles.**
   a. social network
   b. social nexus
   c. social group
   d. social category
   e. social class
3. **Which of the following statements regarding Marx's analysis of class differences is false?**
   a. All aspects of capitalistic societies are economically based.
   b. The capitalists control all social institutions, which they use to their advantage.
   c. Capitalists will both rule and exploit the working class because they own the means of production.
   d. Prestige is the prime determinant of social stratification.
   e. Several social classes existed in nineteenth-century industrial society.
4. **In contrast to Karl Marx, Max Weber placed emphasis on**

a. the consequences of people's relationship to the economic institution.
b. the economic dimension as an independent variable.
c. the development of a single social class.
d. social stratification from an evolutionary perspective.
e. the "culture of poverty."

5. **Which is the true statement regarding economic inequality in the United States?**
a. There is surprisingly little economic inequality in the United States.
b. The majority of wealth in the United States is held by the richest fifth of the population.
c. The middle fifth of the American population holds approximately half of the wealth.
d. Economic inequality is declining in the United States today.
e. The majority of wealth in the United States is controlled by religious organizations.

6. **Max Weber argued that economic success and power do *not* always overlap completely. Which of the following statements does *not* reflect his line of argument?**
a. Money and ownership of the means of production are not the only types of resources that can be used as a basis of power.
b. The fact that money can be used to exert power does not mean that it will be used in this way.
c. An individual's power is a reflection of his or her relationship to the means of production.
d. Power is also attached to the social positions we hold.
e. Individuals can overcome a scarcity of resources if they have large numbers of people on their side.

7. **The most stable source of prestige in modern society is associated with**
a. occupation.
b. education.
c. social class.
d. physical beauty.
e. athletic performance.

8. **According to the text, which of the following statements best describes the relationship between the economic, power, and prestige dimensions of stratification?**
a. Individuals who are low on the prestige dimension tend to be high on the other two dimensions.
b. Individuals who are high on the economic dimension tend to be low on the other two dimensions.
c. Individuals who are high on one dimension tend to be high on the other two dimensions.
d. Individuals who are low on the power dimension tend to be high on the other two dimensions.
e. Sociologists now know that it is impossible to generalize about the relationships among the economic, power, and prestige dimensions of stratification.

9. **According to the functionalist theory of stratification, _________ is necessary to ensure that the most important positions are filled by the most qualified people.**
a. social equality
b. comparable worth
c. conflict
d. cooperation
e. social inequality

10. **_________ refers to a sense of identification with the goals and interests of the members of one's social class.**
a. Class consciousness
b. Class diversion
c. Goal placement
d. Class preference
e. Class knowledge

11. **The absence of enough money to secure life's necessities is known as**
a. relative poverty.
b. chronic poverty.
c. undeserved poverty.
d. lifestyle deprivation.
e. absolute poverty.

12. **_________ mobility is measured by an upward or downward change in occupational status over an individual's lifetime.**
a. Intergenerational
b. Horizontal
c. Lateral
d. Intragenerational
e. Caste

13. **The world poverty rate currently stands at _________ of the world's population.**
a. one-eighth
b. one-quarter
c. one-third
d. one-half
e. three-fourths

REVIEW GUIDE

FEEDBACK REVIEW

True-False

1. The top 1 percent of American households hold nearly 40 percent of the total wealth. T or F?
2. In America, working-class individuals who fail to experience upward mobility claim that it is the system rather than themselves. T or F?

Fill in the Blank

3. The most stable source of prestige in modern societies is associated with _________.
4. The conflict theory of stratification is based on Marx's ideas about class _________.
5. A major difference between the upper upper class and the lower upper class is that the latter has relied on _________ rather than birth.

Multiple Choice

6. **Which of the following is *not* a criticism of the functionalist theory of stratification?**
   a. There are people at the top whose jobs do not seem very important.
   b. It has a bias toward cultural relativism.
   c. It overlooks the barriers to competition.
   d. It glosses over the inheritance of social class level.
7. **The son of a steelworker finds that the steel industry has shrunk so much that a job in the mill is not available. Instead of following in his father's footsteps, he attends a community college and becomes a computer operator. This is an example of**
   a. horizontal mobility.
   b. intragenerational mobility.
   c. caste mobility.
   d. structural mobility.

Matching

8. Match the concepts with the examples.
   ____ a. life chances
   ____ b. power
   ____ c. prestige
   ____ d. conspicuous consumption
   (1) the respect accorded doctors
   (2) a politician conforming to the interests of a lobby
   (3) living in a neighborhood with good police protection
   (4) a lavish wedding
9. Match the theories of stratification with the examples.
   ____ a. functionalist theory
   ____ b. conflict theory
   ____ c. symbolic interactionism
   (1) Corporate executives make more money because they determine who gets what in their organizations.
   (2) Engineers make more money than butlers because of the education they possess.
   (3) Ghetto children tend to have low self-esteem.
10. Match the major types of social mobility with the illustrations.
   ____ a. intergenerational mobility
   ____ b. vertical mobility
   ____ c. horizontal mobility
   (1) A restaurant waiter becomes a taxi driver
   (2) An auto worker becomes a manager
   (3) The daughter of a hairdresser becomes a college professor

GRAPHIC REVIEW

Figure 8.3 graphically portrays the American class structure.

1. Write a paragraph using the information in Figure 8.3 to describe the American class structure.

   ____

   ____

   ____

2. Discuss the most surprising findings in Figure 8.3.

   ____

   ____

   ____

ANSWER KEY

Concept Review

a. 7
b. 9
c. 6
d. 3
e. 8
f. 4
g. 5
h. 1
i. 2
j. 10

Multiple Choice

1. d
2. e
3. d
4. a
5. b
6. b
7. a
8. c
9. e
10. a
11. e
12. d
13. b

Feedback Review

1. T
2. F
3. occupations
4. conflict
5. achievement
6. b
7. d
8. a. 3
   b. 2
   c. 1
   d. 4
9. a. 2
   b. 1
   c. 3
10. a. 3
    b. 2
    c. 1

Inequalities of Race and Ethnicity

© Jeff Greenberg/PhotoEdit

OUTLINE

LEARNING OBJECTIVES

- Distinguish between the concepts of minority, race, and ethnicity.
- Describe patterns of racial and ethnic relations.
- Differentiate prejudice from discrimination.
- Illustrate the different views of prejudice and discrimination taken by functionalists, conflict theorists, and symbolic interactionists.
- Describe, relative to the white majority, the condition of minorities in the United States.
- Describe the increasing racial and ethnic diversity in America.

USING THE SOCIOLOGICAL IMAGINATION

Have African Americans reached income parity with white Americans? Many people believe that as a result of the 1960s civil rights legislation, African American and white family incomes are now approaching comparable levels. In reality, evidence reveals a very different picture. The average income of African American households is considerably less than that of white households. Moreover, at each level of education (the gateway to jobs), African American males gain less income than do white males. On average, for example, white high school graduates can expect to earn annually nearly as much as African American college graduates.

Like all minorities, African Americans suffer the consequences of prejudice and discrimination. Latinos, Native Americans, Jewish Americans, white ethnics, and other minorities have felt the same sting of prejudice and unequal treatment. More recently, attention has been called to other categories of people who are subject to the fallout from prejudice and discrimination, particularly women and the elderly. Inequalities of age and gender are covered in subsequent chapters.

Racial and Ethnic Minorities

All Americans are either immigrants or the descendants of immigrants. Even Native Americans are thought to have migrated to this continent many centuries ago. In part because of this large, diverse migration to the United States, prejudice and discrimination have developed toward minority groups. The emergence of minorities is a recurring theme in the American experience, as immigrant and native groups alike have encountered barriers to full integration into American society. Due to the nature of prejudice and discrimination, much of the research on minorities is approached from the conflict perspective.

The Definition of Minority

Customarily, a minority refers to a relatively small number of people. A sociological definition of a minority, however, can apply to a people numerically larger than others in a society. Before the end of apartheid, for example, blacks in South Africa were a minority—even though they outnumbered whites. Women in the United States are referred to as a minority, and they outnumber men. Obviously, for sociologists, something more than size distinguishes a minority (M. E. Williams 1997).

If small numbers do not necessarily make a minority, what does? Louis Wirth identifies the most important characteristics of a **minority**:

> *We may define a minority as a group of people who, because of their physical or cultural characteristics, are singled out from the others in the society in which they live for differential and unequal treatment, and who therefore regard themselves as objects of collective discrimination. The existence of a minority in a society implies the existence of a corresponding dominant group with higher social status and greater privileges. Minority carries with it the exclusion from full participation in the life of the society.* (Wirth 1945:347)

This definition expresses several key ideas. First, a minority has distinctive physical or cultural characteristics. Such characteristics can be used as a reminder that the minority person is different. This is more evident for physical characteristics such as skin color, facial features, or disabilities. Minority cultural characteristics might include accent, religion, language, or parentage. Culturally, a minority member can pass as a member of the majority by a change in name, the loss of an accent, or the adoption of the majority's culture. Experience suggests, however, that where differences are not sufficiently visible to allow easy identification, other means of identification may be imposed. During the Nazi regime, for example, Jewish Germans were forced to wear yellow stars to separate them from non-Jewish Germans.

Second, minority status is reflected in a society's stratification structure. Almost any society has desired goods, services, and privileges. Largely because the majority is the dominating group, it holds an unequal share of the desired goods, services, and privileges. Further, minority members have less access to the desirable resources. Minorities, for example, often have difficulty getting good jobs because of inferior schooling or discriminatory hiring.

Third, the distinctive cultural or physical traits of a minority can be judged by the majority to be inferior to their own and can be used to justify unequal treatment. In other words, the alleged inferiority of a minority can become part of the majority's ideology—a set of ideas used to justify and defend the interests and actions of those in power (Eagleton 1994). A majority practices job discrimination more easily if, for example, its ideology depicts a minority as shiftless or lazy.

© Jeff Greenberg/PhotoEdit

A minority is composed of people who are defined by distinctive physical or cultural characteristics, are dominated by the majority, and are denied equal treatment. In this photograph, you can easily identify members of various American minorities.

Fourth, because members of a minority regard themselves as objects of discrimination, they have a sense of common identity. Within the minority there is a "consciousness of kind." It is from this sense of common identity that a "we" and "they" vocabulary is accepted within the minority. This vocabulary reflects a strong sense of solidarity and loyalty.

Finally, membership in a minority is ascribed. People do not make an effort to join a minority; they become members by birth. Because membership in a minority is an ascribed status, it is not easily shed. One does nothing to achieve this status, and there is little one can do to avoid it. Although some members of ethnic minorities may, through great effort and a name change, leave this ascribed status behind, it is nearly impossible for members of racial minorities to do so.

Throughout history, various peoples have been treated as minorities. For this reason, it is important to understand the nature of race and ethnicity (Smaje 2000; Parrillo 2002).

The Significance of Race

What is race? A **race** is a category of people who are alleged to share certain biologically inherited physical characteristics that are considered socially important within a society. Biologists have used such physical characteristics as skin color, hair color and texture, facial features, head form, eye color, and height to create broad racial classifications. The most commonly used system of racial classification has three major racial categories—Negroid, Mongoloid, and Caucasoid—along with some unclassified racial categories.

Are racial classifications valid? Sociologists now consider racial classifications, first developed by nineteenth-century biologists, arbitrary and misleading. Ashley Montagu refers to race as "man's most dangerous myth" (Montagu 1997). Although certain physical features have been associated with particular races, scientists have known for a long time that there is no such thing as a "pure" race. This long-standing belief has been reinforced recently through DNA studies. Geneticist Kenneth Kidd of Yale University concludes from the DNA samples he has investigated that there is

> *a virtual continuum of genetic variation around the world. There's no place where you can draw a line and say there's a major difference on one side of the line from what's on the other side. If one is talking about a distinct, discrete, identifiable population, . . . There's no such thing as race in* [modern] Homo sapiens. (Quoted in Marshall 1998:654)

Consequently and frequently, features—or markers—typical of one race show up in other races. For example, some people born into African American families are assumed to be white because of their facial features and light skin color (Appiah and Gutmann 1998).

Even defenders of racial classification acknowledge that their number of identifiable categories exceeds thirty. If racial classification categories are this numerous, scientists conclude, they lose any meaningful capacity to differentiate physical characteristics (Dobzhansky 1970; Hirschfeld 1998, Olson 2002, 2003; Wells 2003).

If arbitrary racial classifications are accepted, social definitions can then be imposed on rather superficial physical differences. For example, in the United States prior to the Civil War, a person born of African American and white parents was traditionally considered "black." If "white" blood were merely a matter of biology, there would have been no legal penalty attached to biracial offspring who claimed to be Caucasian. However, attempting to "pass for white" carried heavy legal sanctions. Some southern and border states defined a person as legally African if one of the great-grandparents was African (1/8) or if one of the great-great-grandparents was African (1/16). In practice, though, it took far less than one-sixteenth of the blood to be "black": anyone with any known African ancestry was usually considered black. This was known as the "one-drop" rule (Spickard 1992; Korgen, 1998). Absent the socially imposed restrictions attached to "black" blood, these individuals could have classified themselves as "white" if they had preferred. Fortunately, the legal system in the United States now recognizes the arbitrariness of these racial labels, and parents are able to decide for themselves into which census category to classify their children. For the first time, the 2000 census allowed residents to select more than one racial category.

What is the relationship between alleged racial physical characteristics and mental capacity? Racist thinkers, from Count de Gobineau in the mid-nineteenth century to Adolf Hitler in the twentieth century, have attempted to link physical differences among people to innate mental and physical superiority or inferiority. Physical characteristics, however, are superior only in the sense that they provide advantages for living in particular environments. For example, a narrow opening between eyelids protects against bright light or extreme cold such as found in Siberia or Alaska. A darker skin is better able to withstand a hot sun. But these physical differences are controlled by a very few genes. In fact, geneticists claim that there is more genetic difference between a tall person and a short person than between two people of different races who are the same height. Only about six genes in the human cell control skin color, while a person's height is affected by dozens of genes. Thus, a 6-foot white male is closer genetically to a black male of similar height than to a 5-foot white male.

The bulk of scientific evidence does not support **racism**: ideas that attempt to connect biological characteristics with innate racial superiority or inferiority (van den Berghe 1978; Wieviorka 1995; Hurley 1998). There are, for example, no biologically inherited differences in intelligence among the various races. Most social scientists endorse the following statement from a UNESCO Statement on Race, prepared by a distinguished group of social scientists:

> *According to present knowledge, there is no proof that the groups of mankind differ in their innate mental characteristics, whether in respect of intelligence or temperament. The scientific evidence indicates that the range of mental capacities in all ethnic* [or racial] *groups is much the same.* (Klineberg 1950:466)

A subsequent review of research supports UNESCO's earlier conclusion ("Statement on Race and Intelligence" 1969). Both of these reports attribute any differences in measured IQ among racial groups to differences in social environment, training, and education.

What do social scientists mean by "social" races? The consequences of racial discrimination based on alleged differences are real. Those people professing innate racial differences can make judgments regarding superiority/inferiority and then attempt to justify prejudice and discrimination. For this reason, many social scientists have become interested in "social" races (J. E. Farley 1995; Banton 1998; Reagin and McKinney 2002). African Americans, Native Americans, Latinos, and Asian Americans are examples of socially defined races that, in America, often continue to face barriers. It seems reasonable to conclude that race remains an influential part of the American experience (Omi and Winant 1994; Feagin 2001; Loury 2001).

The Significance of Ethnicity

A people can be considered a racial category at one time and not at another time. The Irish and Italians, for example, were once considered racial groups by Anglo-Saxon Americans. Later, as it became clear that they had no distinctive physical characteristics, the Irish and Italians became ethnic minorities.

What is an ethnic minority? The term *ethnicity* comes from the Greek word *ethnos*, originally meaning "people" or "nation." Thus, the original Greek word referred to cultural and national identity. Today an **ethnic minority** is socially identified by its unique characteristics related to culture or nationality. Just as physical characteristics define racial minorities, cultural differences define ethnic minorities.

Because of their differences from the host culture, ethnic minorities are subcultures. They have a way of life that is based on their own language, religion, values, beliefs, norms, and customs. Like any subculture, they are part of the larger culture—their members work in the majority (or host) economy, send their children through the dominant educational system, and are subject to the laws of the land—but they are also separate from the larger culture. This separation persists either because an ethnic minority wishes to maintain its cultural and national origins or because the majority erects barriers that prevent the ethnic group from blending in with the larger culture. Michael Novak (1996), himself a Slovak, has contended that members of white ethnic minorities from southern and eastern Europe—Poles, Slavs, Italians, Greeks—have not been able to blend completely into American society because they are more culturally different from white Anglo-Saxon Protestants (WASPs) than are immigrants from western and northern Europe. Western Europeans, for example, have an alphabet similar to English and have religions similar to Americans.

How are ethnic minorities received? Negative attitudes toward ethnic minorities exist in part because of ethnocentrism (the judgment of others based on one's own cultural standards). Many of the majority, out of loyalty to and preference for their own values, beliefs, and norms, may respond to other cultural ethnic views as

FIGURE 9.1

Attitudes of Americans Toward Immigrant Minorities

This figure displays attitudes toward immigrant groups in the United States. Do these attitudes coincide with your own attitudes? With those of your friends? Explain.

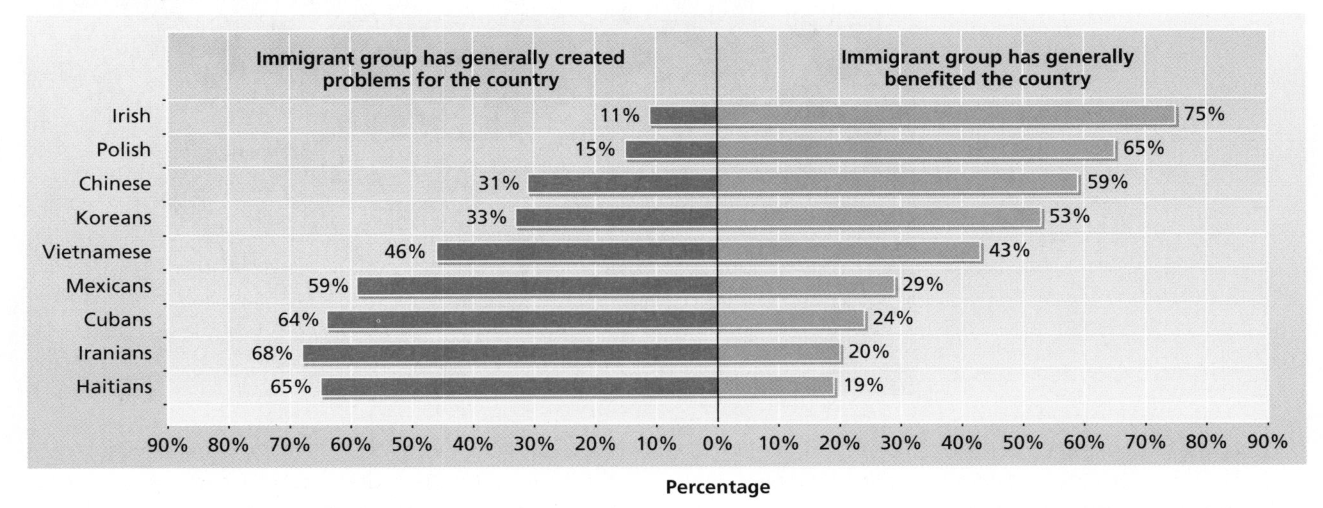

Source: George H. Gallup Jr., *The Gallup Poll: Public Opinion 1993* (Wilmington, DE: Scholarly Resources, Inc., 1994), pp. 250–251. Copyright © 1994 by Scholarly Resources, Inc. Reprinted by permission of Scholarly Resources, Inc.

inferior. Because members of ethnic minorities do not meet the majority's definition for appropriate ways of behaving, it may be assumed that something is wrong with the ethnic members. Ethnocentric judgments, of course, are often expressed via prejudice and discrimination. Jews, for example, have had to contend with prejudice and discrimination based on their religious beliefs.

A society's attitudes toward ethnic (and racial) minorities vary from minority to minority and from time to time (Gallup 1994). Figure 9.1 illustrates the opinions of many Caucasian Americans toward a variety of immigrant minorities. Clearly, older immigrant minorities (Irish and Polish) are viewed more favorably than later arrivals (Cubans, Haitians, Iranians). Whether you are Caucasian American or a member of an ethnic or racial minority, you may want to compare your attitudes with those depicted in Figure 9.1.

Patterns of Racial and Ethnic Relations

When people of various racial and ethnic backgrounds interact, a wide range of outcomes is possible (Wallace 1997). It is helpful, however, to divide these many outcomes into two major types: patterns of assimilation (minority groups are accepted) and patterns of conflict (minority groups are rejected).

Assimilation involves "those processes whereby groups with distinctive identities become culturally and socially fused together" (Vander Zanden 1990:274). It is the integration of a racial or ethnic minority into a society, where minority members are given full participation in all aspects of the society. The classic works in the study of assimilation are those of Milton Gordon, who has defined three basic assimilation patterns in American society: Anglo-conformity, melting pot, and cultural pluralism (M. M. Gordon 1964, 1978).

What is the most common pattern of assimilation? *Anglo* is a prefix indicating an American of English descent. So in *Anglo-conformity*, traditional American institutions are maintained. Basic to this pattern of assimilation is the acceptance of immigrants as long as they conform to the "accepted standards" of the host society. Anglo-conformity has been the most prevalent pattern of assimilation in America. It is the least egalitarian assimilation pattern because the immigrant minority is required to conform and, by implication, either give up or suppress its own values.

Is America more like a melting pot or a tossed salad? A second pattern of assimilation is the *melting pot*, in which all ethnic and racial minorities blend together. Israel Zangwill's drama *The Melting Pot* expresses the goal of

> *the completion of a vast American symphony which will express* [the immigrant's] *deeply felt conception of* [the] *adopted country as a divinely appointed crucible in which all the ethnic divisions of mankind will divest themselves of*

Racial classification categories are arbitrary and misleading. The futility of placing social definitions on physical differences was first seen in Jesse Owens's 1936 Olympic performance, in which he showed a black man to be superior to Hitler's blond, blue-eyed athletes.

their ancient animosities and differences and become fused into one group, signifying the brotherhood of man. (M. M. Gordon 1978:193)

Although the ethnic and racial pot did simmer, especially in American cities, there is a question as to how much melting of differences and resulting fusion have really taken place. Instead of a melting pot, many sociologists are now using the idea of a "tossed salad," where traditions and cultures exist side by side.

The third assimilation pattern is *cultural pluralism*. Unlike the ideal of a unified culture underlying the Anglo-conformity and melting-pot patterns, cultural pluralism recognizes the immigrants' desire to maintain at least a remnant of their "old" ways. Ethnic enclaves, or settlements, characterized American cities of the past two centuries, and many survive today. Such enclaves allowed immigrants to maintain many of their traditional ways while accommodating American values and norms.

Accommodation, an extreme form of cultural pluralism, occurs when a minority, despite its familiarity with the beliefs, norms, and values of the dominant culture, maintains its own culturally unique way of life. An accommodated minority learns to deal with the dominant culture when necessary, but remains independent in language and culture. Cubans in Miami are an example of a distinct community within a larger community (Wilson and Martin 1982). Cultural pluralism emphasizes the ethnic and racial diversity that still characterizes American society and is a more democratic and egalitarian pattern of assimilation than either Anglo-conformity or the melting pot. Like the other two patterns, however, cultural pluralism also overlooks the continuing inequities that characterize American racial and ethnic relations today. Each of these assimilation patterns stresses a kind of egalitarian outcome and thus omits patterns of conflict.

What are the patterns of conflict? Conflict patterns are associated with the dominance of a majority over racial and ethnic minorities. Three basic patterns of conflict are genocide, population transfer, and subjugation (Mason 1970).

Genocide, the most tragic expression of dominance, refers to politically motivated mass murder of most or all of a targeted population (Fein 1993; Chirot and Seligman 2001). One of the least-known examples occurred between 1492 and 1650 when Christopher Columbus totally eliminated 250,000 Arawak Indians of the Bahamas (Zinn 2001). One of the best-known examples is the Holocaust, Adolph Hitler's attempt to destroy all European Jews during the 1930s and 1940s. The Nazis succeeded in killing some 6 million Jews before World War II ended in 1945. Less known is the "Rape of Nanking," begun in 1937, during which the Japanese massacred an estimated 260,000 to 350,000 Chinese men, women, and children (Chang 1998). Tragically, genocide campaigns are more common in world history than might be supposed. The Serbian campaigns of "ethnic cleansing" against the Muslims in Bosnia and Kosovo and the ethnic slaughter of anywhere from 200,000 to 500,000 people by the Tutsi tribe of Rwanda are recent instances of genocide (Prunier 1997; Marks 1999; Naimark 2002; S. Power 2001, 2003).

Population transfer is another way in which a society dominates a minority. In this pattern of conflict, a minority is forced either to move to a remote location or to leave entirely the territory controlled by the

The infamous Trail of Tears symbolizes the destruction of Native American culture by a white majority that systematically and relentlessly confiscated valuable territory.

majority. Population transfer was the most common policy used against Native Americans, especially during the late 1800s. The Cherokees, for example, were once a single nation in the southeastern United States until 20,000 Cherokees were forcibly removed to Oklahoma reservations, where they became dependent on the U.S. government. Four thousand of these Cherokees died because of harsh conditions along the "Trail of Tears." Although Cherokee reservations can be found today in North Carolina and Oklahoma, many Native Americans have migrated to cities to find work, a move that in itself threatens the survival of some tribes. In another example, some believe that the 1999 Serbian aggression in Kosovo was actually closer to an attempt at population transfer than it was to genocide, since the Serbian intent was not to annihilate the Albanians but to move them (Wiesel 1999).

Subjugation is the most common pattern of conflict and most clearly reflects the characteristics of a minority relationship: the majority enjoys greater access to the culture and lifestyle of the larger society. Inequities appear in such areas as power, economics, and education as well as in other important indicators of the quality of life, such as health and longevity. In subjugation, the majority and the minorities may live in the same general area and may participate together in at least some aspects of social life, but members of the minority remain clearly subordinate. Subjugation may be based on the law—**de jure subjugation**—as in the segregation of African Americans following Reconstruction in the United States or as in the former apartheid system of law in South Africa. It may also be based in actual everyday practice—**de facto subjugation**—regardless of what the law is. In this case, subjugation is visible primarily in indicators like political underrepresentation and low occupational achievement. The continuing practice of housing discrimination in the United States is an illustration of subjugation. This practice has existed over many decades, even in the face of direct efforts, including open housing laws, designed to overturn it. According to a 1989 survey, African Americans and Latinos experience discrimination more than half the time they attempt to rent or purchase a home (M. A. Turner, Struyk, and Yinger 1991).

At any given time, the patterns of race and ethnic relations are many, diverse, and often coexisting. Also,

The nature of slavery in the United States limited the upward mobility of freed slaves both before and after the Civil War.

FEEDBACK

1. Which of the following is *not* always a characteristic of a minority?
   a. distinctive physical or cultural characteristics
   b. smaller in number than the majority
   c. dominated by the majority
   d. denied equal treatment
   e. a sense of collective identity
2. _________ is an ideology that links a people's physical characteristics with their alleged intellectual superiority or inferiority.
3. A(n) _________ is a people within a larger culture that is treated as a minority because of its distinctive cultural characteristics.
4. _________ refers to those processes whereby groups with distinctive identities become culturally and socially fused.
5. The pattern of conflict in which one people is forced to move to a remote location is called _________.

Answers: 1. b 2. Racism 3. ethnic group 4. Assimilation 5. population transfer

in the history of relations between two peoples, we can find wide variation through time in the ways the people have interacted. Many factors can influence the pattern of racial and ethnic relations adopted: the nature of the first contact, the reasons for contact and interaction, visibility of minority groups, views held by respective members, and general social conditions.

Theories of Prejudice and Discrimination

A variety of theoretical perspectives have been used to explain prejudice and discrimination. Before introducing these theories, however, the nature of prejudice and discrimination needs to be discussed.

The Nature of Prejudice and Discrimination

How is a sociological definition of prejudice different from the popular view of the term? Prejudice is often viewed as a general characteristic of human relationships. Students can be prejudiced against professors, professors against university administrators, Catholics against Protestants, and employees against employers. In sociology, however, the concept of **prejudice** has a more precise meaning. It refers to widely held preconceptions of a group (minority or majority) and its individual members. Because such attitudes are based on strong emotions and rooted in unchallenged ideas, they are difficult to change, even in the face of overwhelming invalidating evidence. New information that contradicts one's prejudices tends to be denied because of selective perception. It is easier to explain an individual who doesn't fit the stereotype as an exception rather than reexamine a whole set of established beliefs. Consequently, prejudiced attitudes generally are not altered either by new personal experiences or by favorable accounts from others. An Anglo student may prefer to believe that an Asian classmate who does poorly in math is the "exception to the rule" that Asians have a "gift" for mathematics.

Prejudice involves an either/or type of logic: a minority (or majority) is either good or bad, and it is assumed that every member possesses the characteristics attributed to the group. Prejudice, then, involves overgeneralization based on biased or insufficient information. Such overgeneralizations are usually defended by one of two responses: citing limited personal experiences with minority members or reciting stories told by others about their experiences.

How is discrimination different from prejudice? Whereas prejudice refers to biased attitudes, **discrimination** refers to unequal treatment. Prejudice does not always result in discrimination, but it often does. Discrimination can take various forms: avoiding social contact with members of minority groups, denying them positions that carry authority, or blocking their access to exclusive neighborhoods. It can also involve such extremes as physically attacking or killing targeted people (Allport 1979).

What is the relationship between prejudice and discrimination? Prejudice is usually considered the cause of discrimination. Although this is often the case, discrimination may also cause prejudice. For example, unskilled workers may believe that their jobs are in jeopardy because of the massive immigration of a new ethnic group. This fear of economic threat may lead to unequal treatment of the ethnic members. To justify this discrimination, threatened workers may attempt to show why a minority deserves unequal treatment. This occurs in part through the application of stereotypes. This is exactly what happened to Chinese Americans during the nineteenth and early twentieth centuries (Lyman 1974).

Spinning a Web of Hate

White supremacists, neo-Nazis, and other hate groups have discovered the Internet as an effective means of spreading hatred of Jews, African Americans, homosexuals, and fundamentalist Christians (Sandberg 1999). The Web boosts the recruiting efforts of these groups from leaflets and small meetings to a mass audience—more than 100 million worldwide. Reaching this audience is inexpensive and allows hate groups to present their message unedited and uncensored.

If you had browsed the World Wide Web in 1995, you would have found only one hate site, Stormfront. According to the Anti-Defamation League, there are now two thousand Web sites advocating racism, anti-Semitism, and violence. Aryan Nation identifies Jews as the natural enemy of whites; White Pride Network offers a racist joke center; Posse Comitatus defends alleged abortion clinic bomber Eric Robert Rudolph; World Church of the Creator is violently anti-Christian.

Because they use high technology to deliver their message to a mass audience, organized racists are no longer easily identifiable. While members of hate groups used to be recognized by their white hoods or neo-Nazi swastikas, they can now just as easily be wearing business suits instead of brown shirts. The Southern Poverty Law Center is especially concerned about the repackaging of hate-based ideologies to make them appear more respectable to mainstream America. To reach the young, hate Web sites offer such child-friendly attractions as crossword puzzles, jokes, cartoons, coloring books, contests, games, and interactive comic strips.

One of the most prominent racist groups is the Council of Conservative Citizens (CCC). This organization, founded in 1985, now has over 15,000 members in at least twenty-two states. Its members include thirty-four state legislators in Mississippi, and it has sought the political support of several state governors and U.S. senators and representatives. The CCC is an outgrowth of the White Citizens Councils from the 1950s and 1960s, whose primary purpose was to block the civil rights movement and school desegregation. While the CCC disavows racism, billing itself as "a conservative organization that tries to defend the Constitution," its Web site regularly publishes racist material. One of its featured columnists, H. Millard, made this prediction about the outcome of the interracial melting pot: "What will emerge will be just a slimy brown mass of glop. The genocide being carried out against white people hasn't come with marching armies. . . . Genocide via the bedroom chamber is just as long-lasting as genocide via the gas chamber."

Not all hate group activity comes from white supremacists who target African Americans. The Southern Poverty Law Center also tracks the activities of black separatists and documents several recent hate crimes committed by blacks against whites. In addition, the continued immigration of Asians and Central and South Americans is drawing the angry attention of hate groups of all types. More information on hate group activities can be found at the Southern Poverty Law Center Web site: www.splcenter.org.

Analyzing the Trends

When the economy is not performing well, membership in hate groups rises; membership declines when the economy is doing well. Relate this fluctuating membership pattern to scapegoating.

What is a stereotype? Like prejudice, stereotyping can appear throughout society. In the United States, for example, athletes are thought to be "all brawn and no brain," and politicians are assumed to be corrupt. As used in sociology, a **stereotype** is a set of ideas based on distortion, exaggeration, and oversimplification that is applied to all members of a social category (Stangor 2000; Loury 2001; Pickering 2001).

Stereotypes may be used as a justification for prejudiced attitudes and discrimination. For example, very early relationships between the colonists and Native Americans were relatively peaceful and cooperative, but as the population of the colonies grew, conflicts over land and resources became more frequent and intense. To justify expansion into Indian territory, the colonists nurtured a stereotype of Native Americans as lying, thieving, murdering heathen savages. This picture was frequently not held by the trappers and traders who moved individually among the Indians and, because of friendly contact, tended to judge the Indian differently. But the prejudice of the farmer-settler prevailed, leading to the continued seizure of Native American lands with a minimum of compensation and with the dramatic reduction of the Native American population from as many as 5 to 7 million to as few as a quarter million in 1890 (Snipp 1996).

Discrimination based on stereotypes is not merely past history. Take, for example, a recent field experiment designed to measure racial discrimination in the labor market. Marianne Bertrand and Sendhil Mullainathan (2002) sent resumes in response to help-wanted ads appearing in Boston and Chicago newspapers. To manipulate the perception of race, each resume was randomly assigned either a very white-sounding name or a very African American–sounding name. Resumes with white names drew about 50 percent more callbacks than those with African American names. In addition, the researchers introduced variations in the quality of credentials. Among whites, the higher quality resumes elicited 30 percent more callbacks than the lower quality ones. Among African Americans, in contrast, the higher-quality resumes did not elicit significantly more callbacks. In short, whites benefited from improving their credentials; African Americans did not.

What is a hate crime? In 1999, James Byrd Jr., an African American from Texas, was chained to a pickup truck and then dragged to death. That same year saw Matthew Shepard, a gay college student at the University of Wyoming, tied to a fence and beaten to death. Both murders were hate crimes. A **hate crime** is a criminal act motivated by prejudice (F. M. Lawrence 1999; Perry 2000; Jenness and Grattet 2001). Hate or "bias" crimes involve discrimination related to race, religion, sexual orientation, national origin, or ancestry (Levin and McDevitt 1993; Swain 2002). Victims include, but are not limited to, African Americans, Native Americans, Latinos, Asian Americans, Jews, Muslims, gay men and women, and people with disabilities (Levin and McDevitt 2002; F. M. Lawrence 2002; Perry 2000, 2003; Green, McFalls, and Smith 2001).

The term *hate crime* is relatively new; the behavior is not. The federal government has kept statistics since 1900. Hate crime, which still occurs in small numbers (almost 10,000 cases at last report), is increasing. For example, incidents targeting Muslims increased from 28 in 2000 to 481 in 2001 (C. Anderson 2002). By 2002, forty-five states had passed hate crime laws.

Obviously, prejudice and discrimination cannot be accounted for by any single factor. The causes are varied. Psychology, functionalism, conflict theory, and symbolic interactionism each contribute to the understanding of prejudice and discrimination.

The Psychological Perspective

Psychological explanations of prejudice and discrimination focus on the prejudiced person's personality, how it developed, and how it functions in the present. Questions like these have been asked about prejudiced persons: What was their relationship with their parents or with their significant others? What are their values, attitudes, and beliefs? How high is their self-esteem? Two prominent psychological explanations of prejudice and discrimination are the frustration-aggression explanation and the authoritarian personality.

How can frustration lead to prejudice and discrimination? According to the *frustration-aggression* explanation, prejudice and discrimination are the products of deep-seated hostility and aggression that stem from frustration. Aggression is most likely when built-up hostility cannot be directed at the actual source of frustration. Pent-up hostility and frustration may subsequently be

SOCIOLOGY AND THE NEWS MEDIA

Hate Crimes

Hate crimes are based on extreme prejudice. Traditionally, hate crime is associated with race, religion, sexual orientation, or ethnicity. The shooting of twelve students and one teacher at Columbine High School in 1999, however, may have broadened the application of the term *hate crime*. Some individuals in this CNN video believe that hate crimes can be committed against virtually anyone.

1. Do you think the shootings at Columbine High qualify as a hate crime? Why or why not?

   ____

   ____

2. Which of the three major theoretical perspectives do you think best explains hate crime?

   ____

   ____

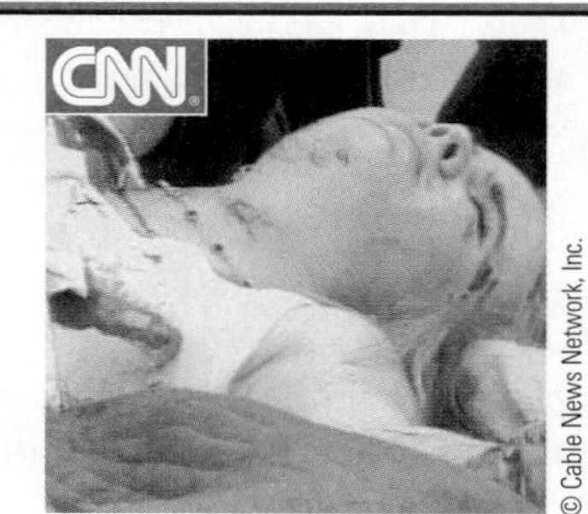

Many Americans have difficulty interpreting school shootings as hate crimes.

redirected toward some substitute object that is less threatening than the one causing the frustration. These substitute objects, known as **scapegoats**, serve as convenient and less feared targets on which to place the blame for one's own troubles, frustrations, failures, or sense of guilt.

John Dollard (Dollard et al. 1939), originator of the frustration-aggression explanation, contended that the frustrations experienced by the Germans after World War I help account for their acceptance of anti-Semitism. The loss of the war, the disappearance of international prestige, the forced acceptance of the Treaty of Versailles, and a ruined economy all contributed to tremendous frustration among the German people. Because direct aggression against the Allies was not an alternative, the Germans channeled their aggression toward other targets, one of which was the Jews. The best scapegoats are those who have already been singled out by the majority for unequal treatment and who therefore have the least chance to defend themselves.

What personality type may tend toward prejudiced attitudes? Another psychological explanation contends that there is a personality type—the **authoritarian personality**—that tends to be more prejudiced than other types. The authoritarian personality is characterized by excessive conformity; submissiveness to authority figures; inflexibility; repression of impulses, desires, and ideas; fearfulness; and arrogance toward persons or groups thought to be inferior. T. W. Adorno and his colleagues, the creators of this theory, summarized it this way:

> *The most crucial result of the* [study of the authoritarian personality] *is the demonstration of the close correspondence in the type of approach and outlook a subject is likely to have in a great variety of areas, ranging from the most intimate features of family and sex adjustment through relationships to other people in general, to religion, and to social and political philosophy. Thus a basically hierarchical, authoritarian, exploitative parent-child relationship is apt to carry over into a power-oriented, exploitatively dependent attitude toward one's sex partner and one's God and may well culminate in a political philosophy and social outlook which has no room for anything but a desperate clinging to what appears to be strong and a disdainful rejection of whatever is relegated to the bottom.* (Adorno et al. 1950:971)

The Functionalist Perspective

Functionalists have focused on the dysfunctions of prejudice and discrimination, and as the last half of this chapter will demonstrate, the negative consequences of prejudice and discrimination are wide and deep. The social, political, educational, and economic costs of the exploitation and oppression of minorities are extremely high. Furthermore, the safety and stability of the larger society are at risk because violence periodically erupts between the groups.

Do functionalists see any positive contributions of prejudice and discrimination? Because functionalists look for positive contributions that each aspect of a culture makes to that society's stability and continuity, they have identified some potential benefits. As noted in Chapter 7, Emile Durkheim identified the ways in which deviance contributes to the cohesiveness of society. The functionalist theory of stratification, as shown in Chapter 8, contends that social inequality helps society to channel the most qualified people into the most important positions and to ensure that people in these positions are motivated to perform their tasks competently. Another potential benefit is the social solidarity of the majority. This happens because the attempted exclusion of outside groups rests in part on ethnocentrism—the tendency to judge others based on one's own cultural standards. Once a majority is convinced of its superiority, it can unite around its own way of life. This strengthens the boundaries of the majority.

But what is the downside of ethnocentrism? As you know, what is functional for one part of society may be dysfunctional for other parts. The price of national unity via ethnocentrism is extremely high for its targets. Passage of the U.S. Constitution in 1789 permitted the retention of slavery. This consigned black slaves to nearly another 100 years of bondage. Documentation of the persistent damage of this extreme oppression to African Americans will be covered in this chapter. This early sense of national unity has also been expensive to the entire society. Consequences of the Civil War, which saw the country violently divided in the nineteenth century, still reverberate through regional factions.

The Conflict Perspective

According to conflict theory, a majority uses prejudice and discrimination as weapons of power in the domination of a minority. This theory, then, traces the existence of prejudice and discrimination to majority interests rather than to personality needs. Domination by the majority is often motivated by its desire to gain or increase its control over scarce goods and services. When those with power in a society are able to persuade the majority that a minority *should* be subjugated, the minority have been effectively eliminated or neutralized as competitors (W. J. Wilson 1973).

According to the conflict perspective, how do minority groups view each other, and what effect does this have for

the majority? Members of a majority tend to think of minorities as people unlike them but similar to one another. Despite being a common target of the majority's prejudice and discrimination, minorities tend to view one another as competitors, rather than as allies, in their struggle for scarce resources (Olzak and Nagel 1986). Conflict among minorities, particularly African Americans and Latinos, is increasing in the United States as whites leave cities and African Americans assume political power. On the one hand, many urban blacks believe that Latinos are benefiting from the civil rights movement waged by African Americans. On the other hand, many Latinos believe that African Americans are using their political clout to push an agenda that favors blacks. A Latino member of Chicago's board of education has charged the five African American members on the board with imposing "apartheid" on Latino children. The board members had voted to send more African American children to integrated schools while disregarding the overcrowding suffered by Latino students. It remains to be seen if urban African Americans and urban Latinos will become allies for their mutual welfare or if they will engage in fierce conflict over the scarce resources available to them (Salholz 1988).

From a Marxian viewpoint, the ruling class (capitalists) benefits from a continuation of conflict among minority groups. If various minorities fight with each other and attribute their job losses to each other, then the ruling elite can, with little or no resistance, downsize, move jobs offshore, and replace workers with technology. In other words, so long as the conflict is not too extreme, capitalists are the beneficiary of a divided working class.

Prejudice and discrimination cannot be accounted for solely by functionalist and conflict explanations. These theories do not explore the factors involved in learning prejudice and discrimination.

The Symbolic Interactionist Perspective

According to the symbolic interactionist perspective, prejudice and discrimination, just like other aspects of culture, are acquired through socialization. Members of a society learn to be prejudiced in much the same way that they learn to be patriotic (Van Ausdale and Feagin 2001). Gordon Allport (1979) described two stages in the learning of prejudice. In the first stage—*the pregeneralized learning period*—children may have been exposed to prejudice, but they have not yet learned to categorize people. The idea that Asian Americans form a distinct category is beyond their understanding. Later, when children enter the second stage of learning prejudice—*total rejection*—they are able to use physical clues to sort people into groups. At this point, if children hear their parents systematically denigrate a minority, they will reject all members of the group on all counts and in all situations. By this stage, the child has learned the name of the minority he or she is supposed to dislike and can identify individuals who belong to it. A nine-year-old Kosovar Albanian child who wants to "kill Serbs" when he grows up is in the stage of total rejection (Nordland 1999).

Language itself can provide a context supportive of prejudice and discrimination (J. M. Jones 1972). For example, in Anglo culture, although it is good to be "in the black" financially, many references using the term *black* are negative. Such terms as *blackball*, *blacklist*, and *black mark* illustrate the negative connotations associated with the term *black*. During the 1960s, militant African American groups coined the slogan "Black is beautiful" to combat this negative connotation. By comparison, it is difficult to think of similar instances in which *white* is used negatively. Such cultural referents, learned as a part of everyday culture, create a context in which some people may erroneously assume that the well-dressed African American male standing outside a fine restaurant is the doorman rather than a

TABLE 9.1

FOCUS ON THEORETICAL PERSPECTIVES: Prejudice and Discrimination

This table illustrates how a particular theoretical perspective views a central sociological concept. Switch the concepts around and illustrate how each theoretical perspective would view a different concept. For example, discuss some functions and dysfunctions of differential power or the self-fulfilling prophecy.

Theoretical Perspective	Concept	Example
Functionalism	Ethnocentrism	• White colonists used negative stereotypes as a justification for taking Native American land.
Conflict theory	Differential power	• African Americans accuse Latinos of using their political clout to win advantages for themselves.
Symbolic interactionism	Self-fulfilling prophecy	• Members of a minority fail because of low expectations they have for their own success.

FEEDBACK

1. ________ refers to the unequal treatment of individuals based on their membership in a minority.
2. A ________ is a set of ideas based on distortion, exaggeration, and oversimplification that is applied to all members of a social category.
3. The ________ is characterized by excessive conformity; submissiveness to authority figures; inflexibility; repression of impulses, desires, and ideas; fearfulness; and arrogance toward those thought to be inferior.
4. According to Gordon Allport, children learning prejudice move from a pregeneralized learning period to a stage of total ________.
5. A ________ occurs when an expectation leads to behavior that causes the expectation to become a reality.

Answers: 1. Discrimination 2. stereotype 3. authoritarian personality 4. rejection 5. self-fulfilling prophecy

customer. Such responses support the continuation of prejudice and discrimination.

Symbolic interactionists also underscore the labeling process. In 1996, Texaco Inc. agreed to pay more than $115 million as reparation for the economic effects of racism within the company. This settlement was fueled in part by a tape recording made by a high corporate official. On this tape, one executive appeared to impose racial labels (Solomon 1996).

What's more, symbolic interactionism underlies the concept of the **self-fulfilling prophecy**—when an expectation leads to behavior that then causes the expectation to become a reality (Merton 1968). For example, if two nations are convinced they are going to war, they may engage in hostile interaction that actually leads to war. Similarly, if members of a minority are constantly treated as if they are less intelligent than the majority, the minority members may eventually accept this limitation. This, in turn, may lead them to place less emphasis on education. Then, indeed, the minority members appear to themselves and others as less intelligent. Or members of a minority may be socialized to believe that they are not capable of holding important positions in a society and may begin to accept this definition. Given this negative image, and the lack of opportunity to develop their abilities, members of minorities may become locked in lower-level jobs.

Institutionalized Discrimination

Although it is popular to think of the United States as a society in which everyone has an equal chance to achieve their chosen goals, such freedom has always been limited. This is the "American dilemma" of which Gunnar Myrdal wrote with regard to African Americans (Myrdal 1944; Obie 1996). At various times in American history, such practices as slavery and the internment of Japanese Americans during World War II have reflected the open and legal practice of discrimination against members of various minorities. Virtually all minorities have encountered such practices to a greater or lesser degree (Feagin and Feagin 1986; Schaefer 2001; Luhman 2002).

With the passage during the 1960s of a series of civil rights laws, many Americans felt that racism was going to be eradicated. Although these statutes did stop many discriminatory practices, many people in this country still suffer from **institutionalized discrimination**. This type of discrimination is the result of unfair practices that are part of the structure of society and have grown out of traditionally accepted behaviors. The legacy of 300 years of discrimination is not easily erased from American life ("Black/White Relations in the United States" 1997; Pollard and O'Hare 1999; Anderson and Massey 2001; Feagin 2001; Feagin and McKinney 2002).

How does direct institutionalized discrimination differ from indirect institutionalized discrimination? To help clarify discriminatory practices, Joe and Clairece Feagin (1986) distinguish between direct and indirect institutionalized discrimination. **Direct institutionalized discrimination** refers to organizational or community actions intended to deprive a racial or ethnic minority of its rights. **Indirect institutionalized discrimination** refers to unintentional behavior that negatively affects a minority. Examples of direct institutionalized discrimination are laws that segregated African Americans or that denied Mexican American children the right to speak their native language in public schools. High school exit exams provide an example of indirect institutionalized discrimination. When required for graduation, these exit exams are disadvantageous to minority students, who are more likely to drop out because they are less likely than white and Asian American students to pass on their first attempt. (See Snapshot 9.1.) Seniority systems (promotion and pay increase with years of service), discriminate against the promotion of newly hired people—many of whom are minority workers. Because of past institutionalized discrimination, members of minorities are just now beginning to enter jobs with seniority systems. Because they have to wait their turn for advancement and

Katherine S. Newman—No Shame in My Game

Katherine Newman has created a rich portrait of minimum-wage workers employed in four fast-food restaurants in central Harlem. These people we label the "working poor"—they hold jobs and pay taxes, but they do not earn enough money to buy the basic necessities of life.

Her research involved 300 New Yorkers. Two hundred worked in four large fast-food restaurants in Harlem. The others could not find jobs but had applied at these four restaurants.

Newman's research team used a variety of research techniques. All respondents answered a survey, and half gave complete life histories in 3- to 4-hour interviews. After collecting this information, Newman's research team worked in the restaurants for four months. Twelve workers opened their lives to up-close observation. They were observed at home, holiday celebrations, movies, schools, and churches. The twelve also kept personal diaries for a year and agreed to let their friends be interviewed.

In the following passage, Newman argues that the working poor share the same basic values as white middle-class society. Therefore, there is shame in a cultural view that defines employment in fast-food jobs as degrading. The working poor do not prefer such work; it is one of the few employment opportunities open to them.

> *Swallowing ridicule would be a hardship for almost anyone in this culture, but it is particularly hard on minority youth in the inner city. They have already logged four or five years' worth of interracial and cross-class friction by the time they get behind a Burger Barn* [a fast-food restaurant in Harlem] *cash register. More likely than not, they have also learned from peers that self-respecting people don't allow themselves to be "dissed" without striking back. Yet this is precisely what they must do if they are going to survive in the workplace.*
>
> *This is one of the main reasons why these* [fast-food] *jobs carry such a powerful stigma in American popular culture: they fly in the face of a national attraction to autonomy, independence, and the individual's "right" to respond in kind when dignity is threatened. In ghetto communities, this stigma is even more powerful because—ironically—it is in these enclaves that this mainstream value of independence is most vigorously elaborated and embellished. Film characters, rap stars, and local idols base their claim to notoriety on standing above the crowd, going their own way, being free of the ties that bind ordinary mortals. There are white parallels, to be sure, but this is a powerful genre of icons in the black community, not because it is a disconnected subculture but because it is an intensified version of a perfectly recognizable American middle-class and working-class fixation.*
>
> *It is therefore noteworthy that thousands upon thousands of minority teens, young adults, and even middle-aged adults line up*

because they have fewer years of service than members of the majority, who have been in the system for years, members of minorities often find that their chances for quick promotion are slight, even though the seniority systems may not have been intentionally designed to obstruct minority progress.

Another example of institutionalized discrimination can be found in public education. Schools with large numbers of minority students are more common in large urban areas than in the wealthier suburbs. This is partly because of "white flight" to the suburbs and housing discrimination against minorities. As a result, minority children are more concentrated in school districts with a tax base too low to provide resources equal to those in the suburbs. This lack of funding means that teachers in minority schools receive fewer opportunities for training; textbooks, when students have them, are outdated; there is little, if any, money for new technology; and buildings are badly in need of repair. Moreover, parental and community support is generally not as strong in these school districts as in the suburbs.

Institutionalized discrimination in the United States is reflected in the experiences of minorities—African Americans, Latinos, Native Americans, Asian Americans, white ethnics, and Jewish Americans. For each minority, the social and economic costs of discrimination are enormous (Andrews 1996; Aguirre and Turner 1997; Spencer 1997; Pollard and O'Hare 1999; Smelser, Wilson, and Mitchell 2001).

African Americans

African Americans make up one of the two largest minority groups in the United States, numbering just over 35 million, or about 12 percent of the total population (see Figure 9.2). They are also one of the oldest minorities, first brought to America as indentured servants in 1619.

for jobs that will subject them, at least potentially, to a kind of character assassination. They do so not because they start the job-seeking process with a different set of values, one that can withstand society's contempt for fast food workers. They take these jobs because in so many inner-city communities, there is nothing better in the offing. In general, they have already tried to get better jobs and have failed, landing at the door of Burger Barn as a last resort. . . .

The stigma also stems from the low social status of the people who hold these jobs: minorities, teenagers, immigrants who often speak halting English, those with little education, and (increasingly in affluent communities afflicted with labor shortages) the elderly. To the extent that the prestige of a job refracts the social characteristic of its average incumbents, fast food jobs are hobbled by the perception that people with better choices would never purposely opt for a "McJob." . . . There is no quicker way to indicate that a person is barely deserving of notice than to point out he or she holds a "chump change" job at Kentucky Fried Chicken or Burger King.

Ghetto youth are particularly sensitive to the status degradation entailed in stigmatized employment. As Elijah Anderson . . . and others have pointed out, a high premium is placed on independence, autonomy, and respect among minority youth in inner-city communities—particularly by young men. No small amount of mayhem is committed every year in the name of injured pride. Hence jobs that routinely demand displays of deference force those who hold them to violate "macho" behavior codes that are central to the definition of teen culture. There are, therefore, considerable social risks involved in seeking a fast food job in the first place, one that the employees and job-seekers are keenly aware of from the very beginning of their search for employment.

It is hard to know the extent to which this stigma discourages young people in places like central Harlem from knocking on the door of a fast food restaurant. It is clear that the other choices aren't much better and that necessity drives thousands, if not millions, of teens and older job-seekers to ignore the stigma or learn to live with it. But no one enters the central Harlem job market without having to face this gauntlet.

Source: Katherine S. Newman, *No Shame in My Game: Working Poor in the Inner City* (New York: Alfred A. Knopf, 1999), pp. 93, 95. Reprinted with permission of Alfred A. Knopf, a division of Random House, Inc.

Thinking About the Research

1. Do you think that Newman is using the functionalist perspective or the conflict perspective? Explain.
2. Analyze this study from the viewpoint of symbolic interactionism.
3. Suppose you were a politician about to speak to a group strongly opposed to government support of poor people. How would you use Newman's research to counter their position?

What are the barriers to African American assimilation? Because African Americans are physically identifiable, and because of negative stereotypes based on physical characteristics, it has been very difficult for them to assimilate. A second reason for the continuing minority status of African Americans lies in the historical nature of slavery, which can be traced to the first shipment of slaves to Jamestown, Virginia, in 1619 (Zinn 2001). According to Frank Tannenbaum (1947), the attitudes and practices associated with *manumission*—the transition from slave to free person—are the most important aspects of any slave system. Despite an increase in the practice of manumission following the Revolutionary War, slavery in the United States, particularly in the South, was generally a permanent condition. Unlike African slavery, support for manumission was weak, and there were numerous barriers to it. Freed slaves constantly ran the risk of being returned to slavery. For instance, in Maryland in 1717, any freed slave who married a white person was returned to slavery for life. In Virginia, an emancipated slave could be sold into slavery again if he or she remained in the state after one year. Those slaves who did manage to gain and keep their freedom were barred from voting, holding public office, and pursuing most jobs (Johnson and Smith 1998; O. Patterson 1998).

Due to the unique nature of the American form of slavery, slaves freed before the Civil War were denied opportunities for upward social mobility. According to Richard Hofstadter:

> *The Anglo-Americans of the North American mainland quickly became committed to sharp race separation, took a forbidding view of manumission, defined mulattoes simply as Negroes, and made outcasts of free Negroes. Hence there was as little upward mobility from slavery as possible, especially in the Southern colonies and states, and even where masters chose to manumit slaves.* (Hofstadter 1973:114)

SNAPSHOT OF AMERICA 9.1

High School Exit Exams

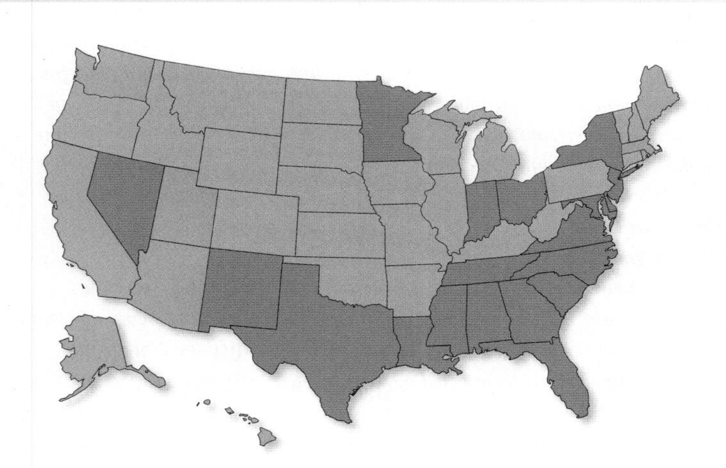

Source: Naomi Chudowski et al., *State High School Exit Exams: A Baseline Report;* at www.ctredpol.org/pubs/StateHighSchoolExitExams13Aug2002.pdf.

- States with exit exams in place
- States phasing in exit exams but not yet withholding diplomas
- States without exit exams

As this map indicates, eighteen states currently require passage of exit exams to receive a high school diploma. Exit exams may unintentionally discriminate against minority and poor children because their lower rate of passage increases their dropout rate.

1. Does your home state require passage of exit exams to graduate from high school? If so, did they appear to discriminate against minorities and the poor at your high school? If not, do you think exit exams would discriminate? Explain.
2. Describe the geographic distribution of states requiring an exit exam that you see in the map. Of these states, which are more likely to allocate state funds for instructional help to failed students? Explain.

FIGURE 9.2

Resident Minority Populations in the United States According to Race/Ethnicity, 1980–2002

This figure shows the percentage increase in the major minority populations in the United States since 1980. Are you surprised by the growth of any particular group?

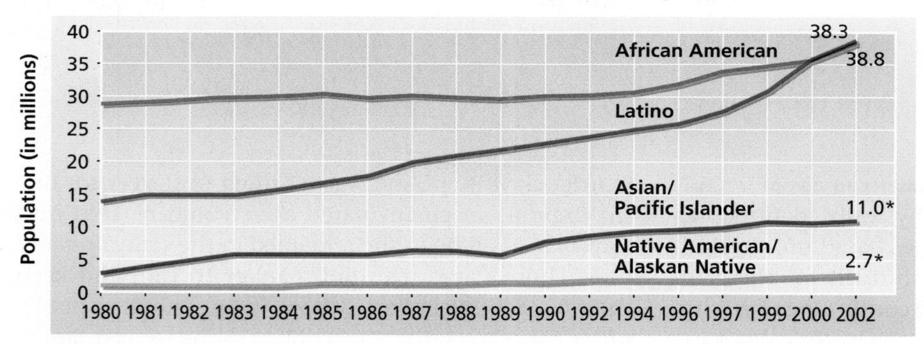

*July 2001 data.

Sources: National Center for Health Statistics, *Health, United States: 1990* (Hyattsville, MD: Public Health Service, 1991), p. 8; and U.S. Bureau of the Census, *Statistical Abstract of the United States: 1996* (Washington, DC: U.S. Government Printing Office, 1996), p. 19; U.S. Bureau of the Census, "Resident Population Estimates of the United States by Sex, Race, and Hispanic Origin: April 1, 1990 to April 1, 1999," *Population Estimates Program*, 1999f, at www.census.gov/population/estimates/nation/intfile3-1.txt; and U.S. Bureau of the Census, *The Population Profile of the United States: 2000* (Washington, DC: U.S. Government Printing Office, 2001).

The denial of opportunity for upward mobility did not end even though the Civil War was over and slavery was legally abolished by the Thirteenth Amendment (1865). By the late 1800s, de jure segregation (the separation of blacks and whites based on law) was institutionalized throughout the country, especially in the South. Such practices continued until the late 1960s, when they were made illegal by the passage of civil rights legislation and by court decisions. In a very real sense, then, African Americans have been

More African Americans are moving into the middle class. Still, their average income lags far behind that of whites.

legally free in much of the United States for little more than thirty years. The gap between African Americans and whites in education, income, and employment represents the legacy of centuries of prejudice and discrimination (Klinker and Smith 1999; Anderson and Massey 2001; Feagin 2001; Feagin, Vera, and Batur 2001).

What is the economic situation for African Americans today? The median income of African American households ($29,470) is substantially below that of white households ($46,305; see Figure 9.3). The poverty rate for African Americans (22.7 percent) is almost triple that of whites (see Table 9.2). Recent experience carries both good and bad news for African Americans. While African American workers have made significant economic gains since 1960, this advancement has not closed the gulf between black and white Americans (B. Anderson 2002). Whereas the median family income of whites increased by about 3.0 percent between 1989 and 2001, African American family income increased by 3.5 percent. Thus, African American income remains at just over two-thirds that of whites. For every $100 an average white family earns, an average African American family earns $63 (U.S. Bureau of the Census 2001a). Thus, as noted in "Using the Sociological Imagination," the average African American income in the United States is far from equal to the average income for whites.

Not surprisingly, there is also a vast gulf in wealth (home and car equity, net business assets, net liquid assets). The average African American family holds less than one-quarter of the wealth of the average white family (U.S. Bureau of the Census 2001b).

How do African Americans fare in the job market? According to recent "social audit" studies, African Americans experience employment discrimination despite stronger legal prohibitions. Audit studies involve sending white and minority researchers with comparable resumes to the same firms, applying for the same jobs. Research consistently shows that employers are more likely to interview or hire white applicants. This discrimination now takes more sophisticated and covert forms. For example, discriminatory employers interview or offer jobs to all white applicants before tapping their list of black applicants. Discriminatory firms also offer white applicants higher salaries and higher-status positions (Herring 2002).

TABLE 9.2

Socioeconomic Characteristics of Minorities

This table compares minorities on basic socioeconomic characteristics. Examine these data, looking for correlations among the characteristics. Describe your findings.

	Whites	African Americans	Latinos	Native Americans	Asian Americans
Poverty	7.8%	22.7%	21.4%	27%	10.2%
Median income	$46,305	$29,470	$53,635	$21,619	$53,635
High school graduation	88%	79%	57%	66%	86%
College	26%	16%	11%	9.4%	44%

Sources: Bernadette D. Proctor and Joseph Dalaker, U.S. Bureau of the Census, Current Population Reports, P60–219, *Poverty in the United States: 2001* (Washington, DC: U.S. Government Printing Office, 2002); U.S. Bureau of the Census, "Table PINC-01. Selected Characteristics of People 15 Years and Over, by Total Money Income in 2001, Work Experience in 2001, Race, Hispanic Origin, and Sex," at http://ferret.bls.census.gov/macro/032002/perinc/new01_000.htm: U.S. Bureau of the Census, *Statistical Abstract of the United States: 2001* (Washington, DC: U.S. Government Printing Office, 2001).

FIGURE 9.3

Median Household Income by Race and Latino Origin, 1989 and 2001

Explain why sociologists consider Asian Americans a minority group despite their relatively high annual income.

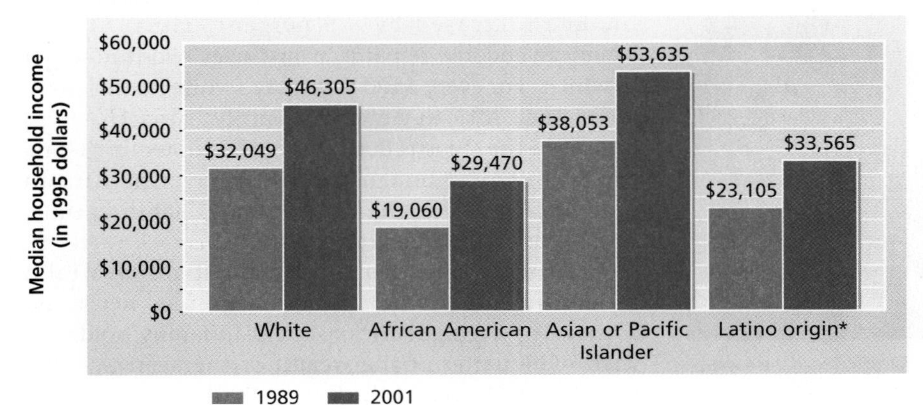

* Persons of Latino origin may be of any race.
Sources: U.S. Bureau of the Census, *Income and Poverty: 1990*, Economics and Statistics Administration (Washington, DC: U.S. Government Printing Office, 1991); Carmen DeNavas-Walt and Robert Cleveland, U.S. Bureau of the Census, Current Population Reports, P60-218, *Money Income in the United States: 2001* (Washington, DC: U.S. Government Printing Office, 2002).

Such employment practices perpetuate the long-standing minority overrepresentation in low-prestige, low-paying jobs. Thirty-five percent of African American men are employed in the highest occupational categories: professional, managerial, technical, and administrative, whereas 49 percent of white men have jobs in these categories. Similarly, about 61 percent of African American women are employed in these occupational categories, compared to 74 percent of white women. At the same time, African Americans are almost twice as likely as whites are to work in low-level service jobs (U.S. Department of Labor 2001a).

New long-term economic trends threaten to make matters worse. These trends include a shift from higher-paying manufacturing jobs to lower-paying service jobs; technological replacement of workers; and transfer of high-wage jobs to low-wage countries. Reduced job opportunities for African Americans and other minorities, especially in segregated neighborhoods, are an important negative consequence (Jacob 1994; W. J. Wilson 1997; "World Without Work" 1999; Cherry and Rodgers 2000).

Patterns of unemployment also affect the economic status of African Americans. Jobless rates among African Americans are more than double those of whites (U.S. Department of Labor 2001a). And these rates do not take account of all unemployed persons. Traditional unemployment rates are based on the number of unemployed people who are looking for jobs. They do not include so-called **hidden unemployment**—discouraged workers who have stopped looking or part-time workers who would prefer full-time jobs. When hidden unemployment is considered, the jobless rate for African Americans approaches one in four workers, the rate during the Great Depression of the 1930s (Swinton 1989; W. J. Wilson 1997).

It is among African American teenagers that the greatest unemployment problem exists. According to official statistics, about one out of every three African American teenagers is unsuccessfully looking for full-time work (U.S. Department of Labor 2001a). With hidden unemployment taken into account, it is estimated that over 40 percent of all African American teenagers are unemployed. Consequently, thousands of African American youths are becoming adults without the job experience vital to securing good employment (Jacob 1994; Pearlstein and Brown 1994; "World Without Work" 1999).

Are there gender differences among African Americans? Knowledge of the nationwide gap between men and women would lead you to answer yes. And you would be right. The long-standing advantage of black males over black females, however, appears to be eroding somewhat. At every educational level, black men still earn more than black women do. But African American women with college degrees earn substantially more than the median for all African American men. That is not true for white men and women. And African American college-educated women earn as much as white women with college degrees. Given that a college education opens more doors for African American women, the additional good news for them is their increasing educational level. While over a third of African American women enter college, only a fourth of black men do. Contrast this to 1970, when African American men and women had a roughly equal level of college enrollment (U.S. Bureau of the Census 2001; Cose 2003).

What are the educational opportunities for African Americans? Education is the traditional American path to economic gain and occupational prestige. The educational story for African Americans is mixed. As of 2001, 88 percent of whites had finished high school compared to 79 percent of African Americans. Similarly, whereas 26 percent of whites had completed college, only 16 percent of African Americans had done so (Daniels 2002). Moreover, higher educational attainment doesn't pay off for African Americans as it does for whites. Although income tends to rise with educational level for both African Americans and whites, it increases much less for African American men (and for women of both races) than for white men. At each level of schooling, African American men tend to gain less than their white peers do. White male high school dropouts have an average income almost equal to African American men with high school diplomas. White high school graduates, on the average, earn nearly as much each year as African American men with college degrees (refer to Figure 2.3).

What is the African American presence in politics? African Americans have enjoyed some political success since 1970. This progress is due in part to legal changes that occurred in the 1950s and 1960s, including the increase in African American voter registration. Also, African Americans constitute a growing political majority in many large cities. Consequently, more than 5,400 African Americans are serving as city and county officials, up from 715 in 1970. There are over 9,000 African American elected officials in the United States, a sixfold increase since 1970 (Bositis 2002).

Some see the slow emergence of a "biracial politics"—election of African Americans in predominantly white areas—as a hopeful sign (Kilson 2002). According to Richard Zweigenhaft and William Domhoff, some African Americans, though still vastly underrepresented, have entered the "power elite" of America:

> *Although the power elite is still composed primarily of Christian white men, there are now . . . blacks . . . on the boards of the country's largest corporations; presidential cabinets are far more diverse than was the case forty years ago; and the highest ranks of the military are no longer filled solely by white men.* (Zweigenhaft and Domhoff 1999:176)

It is in the statehouses (in this century, only one African American governor has been elected—L. Douglas Wilder of Virginia) and at the national level that African Americans have experienced the least political gain. Despite the visibility and influence generated by Jesse Jackson's two presidential bids and the naming in 1989 of an African American as chairman of the national Democratic Party, less than 1 percent of national political offices are held by African Americans. This is true even though there have been more African Americans in the U.S. House of Representatives in the last four Congresses (37 in 2003) than at any other time in American history. This is more than double the number of African American House members in 1981. Still, blacks comprise 12 percent of the U.S. population and only 8.5 percent of the House. In 2002, there were no African Americans in the U.S. Senate. Although African Americans are an emerging political force, they must accomplish much before they can claim proportionate political representation (Dentler 1995; Bositis 2001, 2002).

In fact, a 1995 U.S. Supreme Court decision appears to represent a roadblock to political power for African Americans. The Court ruled that race could no longer be the "predominant factor" in determining congressional districts (Fineman 1995). The Court created, at the same time, a legal test that makes it difficult, if not impossible, to keep in place government programs that give an advantage to minorities and women. In fact, "antipreference" forces are making significant progress toward repealing affirmative action laws. The Universities of California and Texas have abandoned all racially based admissions preferences; the state of California has repealed affirmative action laws; and Arizona, Ohio, New Jersey, and Washington are seriously considering similar legislation (Ponnuru 1997).

Voting patterns reflect another dimension of political inequality among African Americans. Because politicians respond most favorably and consistently to the desires of those who vote, it is in the interest of African Americans to make their presence known on election day. Yet, the African American voting rate is typically several percentage points lower than the white voting rate. A typical illustration is the 2000 presidential election, in which whites outvoted African Americans by six percentage points (64 percent vs. 58 percent). Adding significance to these data is the fact that African Americans are swamped, numerically, by whites.

African Americans vote at a lower rate than whites for at least two reasons. First, African Americans are disproportionately represented in the working class, among the working poor, and in the underclass—the socioeconomic categories least likely to vote. Second, because African Americans have much less confidence in the political system than whites, they are more likely to believe that their votes will count less. This distrust of politics is nurtured by voting experiences like the

2000 presidential election. Intense nationwide scrutiny was given to the undercounting of low-income minority voters in Florida (Parker and Eisler 2001). Unfortunately, Florida is not the exception. Undercounting of low-income and minority voters is a nationwide pattern. Voters in low-income, high minority congressional districts had greater than triple the likelihood of having their votes discarded than were voters in affluent, low-minority districts (Dooley 2001). Given this situation, it is easy to understand why many African Americans believe that they live in a "still-white" society (Roediger 2002).

What is meant by "two black Americas?" During the 1960s and 1970s, some occupational progress was made. The number of African Americans in professional and technical occupations—doctors, engineers, lawyers, teachers, writers—increased by 128 percent. The number of African American managers or officials is more than twice as high as it was in 1960. Due to the recent upward mobility of educated African Americans, some scholars predict the emergence of two black Americas—a black **underclass** composed of the permanently poor trapped in inner-city ghettos and a growing black middle class (W. J. Wilson 1984; Landry 1988; Kilson 1998; Pattillo-McCoy 1999). In Richard Freeman's (1977) terms, a black elite is said to be emerging in America. David Swinton offers this view of the mixed state of African Americans:

> *The empirical evidence does show that some individual blacks have made impressive economic gains. In fact, there has been an increase in the proportion of blacks who can be classified as upper middle class. This limited upward mobility for the few cannot offset the stagnation and decline experienced by the larger numbers of black Americans whose economic status has deteriorated. The central tendency for the group as a whole is revealed by the trends in the averages. And these trends tell a consistent story of stagnation or decline.* (Swinton 1989:130)

U.S. Census data confirm the existence of two black Americas—the rich and the poor. The number of African American households earning a minimum of $50,000 annually has more than tripled since 1970, rising from 11.5 percent to 27.8 percent (3.7 million). About 30 percent of African American households earn between $25,000 and $50,000 each year. The percentage of poor African Americans, however, continues to dwarf that of whites. Although less than 8 percent of white families have annual incomes under $10,000, about one out of six (17 percent) African Americans are at that low income level.

Is the significance of race declining for African Americans? The argument for the declining significance of race is applied to both successful African Americans and the black underclass. Let's examine each in turn.

Although African Americans remain behind whites on all dimensions of stratification, gains made since the 1960s have led some to conclude that race is of declining significance in America (Thernstrom and Thernstrom 1997; Cose 1999). According to some analysts, race is now less important than resources in determining the life chances of young, well-educated African Americans. Well-educated African Americans, it is argued, can now compete on equal ground with whites. African Americans with finances and education are said to be no longer shut out because of their color (W. J. Wilson 1987, 1993, 1997, 2002).

William Julius Wilson concurs that race is now less important than economic class for successful African Americans. Wilson goes further by saying that race is declining in significance even for the African American urban poor who are part of the underclass—those in poverty who are either continuously unemployed or underemployed due to the absence of job opportunities and/or required job skills. In some of his early work on this topic, Wilson stated it this way:

> *The recent mobility patterns of blacks lend strong support to the view that economic class is clearly more important than race in predetermining job placement and occupational mobility. In the economic realm, then, the black experience has moved historically from economic racial oppression experienced by virtually all blacks to economic subordination for the black underclass.* (W. J. Wilson 1980:152)

In his book *The Truly Disadvantaged*, Wilson (1987) elaborates on the thesis of the declining significance of race in the United States. It is true, Wilson acknowledges, that many African Americans (and other minority group members) are in the underclass as a result of historical patterns of discrimination. This, however, is in the past. Moreover, the argument continues, African Americans remaining in the underclass are there not because of prejudice and discrimination, but because of features of the American economy. Inner-city African Americans are adversely affected by the deindustrialization of the American economy. In the past, African Americans were drawn to Detroit, Chicago, and other large cities by the availability of high wages and stable employment in factories. These manufacturing jobs, which traditionally required little education, have gone to low-income countries where labor is cheap and labor unions are nonexistent. Consequently, today's inner-city poor are denied the type of jobs used for upward mobility. A second contributing economic factor is the internal movement of business and industry from central cities to the suburbs. A third economic factor contributing to the perpetuation of the underclass is the movement of upwardly mobile African Americans to the suburbs. Lacking the positive role models available in the past, the inner-city poor now see drug dealing, prostitution, and other illegal economic activities as the primary avenues to success (Vobejda 1991; O. Clayton 1996; Massey and Denton 1996; Small and Newman 2001).

If economic factors are perpetuating the underclass, Wilson argues, solutions are to be found in the economic sphere. Wilson calls for a federal economic policy designed to create a higher employment rate and better jobs for all Americans. Wilson looks to such things as higher-paying jobs, improved education, job training, relocation support to encourage members of the underclass to secure and keep decent jobs, publicly supported day care for the low-income employed, a markedly higher minimum wage, and medical insurance for the employed in low-paying jobs (W. J. Wilson 1993, 1997).

Use of the term *underclass* has been questioned (M. B. Katz 1993). Wilson, himself, suggests substituting the term *ghetto poor* because the concept of underclass seems to be a code word for inner-city African Americans and is often used by journalists to highlight unflattering behavior in the ghetto (W. J. Wilson 1991). Herbert Gans (1990) sees the term underclass as a pejorative, value-laden term now being used to describe the "undeserving" poor. Joining a growing number of social scientists, Gans believes that the term underclass is hopelessly mired in ideological connotations of undeservingness and blameworthiness. These social scientists, including Wilson, do not wish to obscure the harsh reality faced by the urban poor by continuing to use the term underclass if such use contributes to victim blaming. It remains to be seen whether the term underclass will lose its currency in sociological terminology. At any rate, some term, perhaps Wilson's *ghetto poor*, will be utilized to capture the problems associated with the inner-city poor.

There is some evidence that young African American college graduates are reaching an economic par with their white peers and that many African Americans in general have made significant gains in the past twenty years (R. Farley 1984; Payne 1998). It is not certain, however, that this situation is a permanent one. The strides since the 1960s may simply be the result of tremendous changes in the past two decades that will not be maintained in the future.

Many critics disagree with the idea that race is a declining force for African Americans, whether or not progress has been made (Thomas and Hughes 1986; Carnoy 1994; Hacker 1995; Cancio, Evans, and Maume 1996; Hughes and Thomas 1998; Pulera 2002). Doris Wilkinson (1995) believes that discrimination still determines the life chances of African Americans. The current racial dominance, she argues, is simply more sophisticated and subtle. Others point to inferior education in America's inner cities, a situation tied in part to racial discrimination (see Chapter 13). Still other critics point to housing discrimination and segregation as an enduring barrier to large-scale African American migration to the suburbs, where many jobs are located (Massey and Denton 1987; Massey 1990; Massey and Eggers 1990; Pattillo-McCoy 1999). According to the 2000 census, U.S. inner cities are more segregated today than they were before segregation was outlawed in the 1960s (Ware and Allen 2002). Others point to the high rate of incarceration among black males, who are eight times more likely to enter prison than white males (Western and Pettit 2002). Finally, some argue that the concept of a black underclass applies to rural areas. According to this viewpoint, the black underclass is more densely concentrated in rural parts of the South than in the inner cities of the North (O'Hare 1992a, 1992b; O'Hare and Curry-White 1992).

Latinos

How diverse is the Latino population? Latinos in the United States number almost 36 million (12 percent of the population). They are composed of many diverse ethnic minorities, including Mexican Americans, Puerto Ricans, Cubans, and increasing numbers of people from Central and South American countries (see Figure 9.4). High birth rates and immigration rates combine to make Latinos one of the most rapidly growing minorities in the United States. In fact, early in this century, Latinos will become America's largest ethnic

Latinos, including Mexican Americans, Puerto Ricans, Cuban Americans, and a variety of peoples from Central and South America, comprise a diverse minority in American society. Their socioeconomic conditions range from destitution to success.

FIGURE 9.4

Composition of the Latino Population in the United States

The Latino population will clearly become the largest minority in the United States during this century. This figure displays the diversity within that population itself. Do you see any reason why the composition of the Latino population will change over the next fifty years?

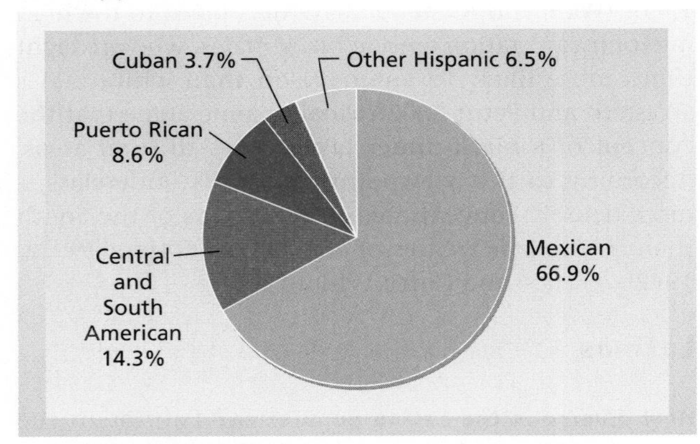

Source: Roberto R. Ramirez and G. Patricia de la Cruz, *The Hispanic Population in the United States: March 2002*, Current Population Reports, P20-545 (Washington, DC: US Census Bureau, 2002).

minority. By 2050, the Latino population is projected to reach nearly 100 million, constituting 24 percent of the U.S. population (del Pinal and Singer 1997; Pollard 1999; Suarez-Orozco and Paez 2002).

Just less than two-thirds of Latinos today are of Mexican descent; Puerto Ricans make up a little less than one-tenth the total Latino population. Most Puerto Ricans are concentrated in or near New York City, although greater geographical dispersion is occurring. Cubans make up the third most populous category of Latinos, with about 1 million people, most of whom are located in the Miami area *(2002 National Survey of Latinos 2002)*.

Latino peoples are diverse. Each group came to the United States under different circumstances and retains a sense of its own identity and separateness. In addition, there are significant internal differences within individual Latino minorities (Delgado and Stefancic 1998; Suarez-Orozco and Paez 2002). For example, the first large group of Cuban immigrants to enter the United States included successful middle- and upper-class people who fled from Cuba when Fidel Castro instituted a communist government there in the late 1950s. These Cuban Americans were substantially more educated than later Cuban immigrants were (Stavans 1996; Darder and Torres 1997; Stavans 2001).

What is the socioeconomic condition of Latinos? Latinos fall behind white Americans in formal education. The median educational level for the total American population is about 12.7 years, or high school graduation, but for all Latinos, it is 11.7 years. Although in 2001, 57 percent of Latinos age twenty-five or over had completed high school, 88 percent of non-Latino whites had done so (see Table 9.2). Mexican Americans have the lowest levels of educational attainment, Cubans the highest. This is due to the migration pattern and the history of these peoples. Mexican immigrants came to the United States with little education; they were unable to speak English and willing to accept the lowest-level jobs. Cubans, on the other hand, tended to be middle- and upper-middle class immigrants who had been educated and economically successful in an economy similar to that of the United States (Valdivieso and Davis 1988; Delgado and Stefancic 1998).

As shown in Figure 9.3, the average income for Latinos ($33,565) is now higher than that of African Americans, but significantly lower than that of non-Latino whites ($46,305). Cubans are the most affluent Latinos, but even their median income is only about 75 percent that of whites. The poorest among the large Latino groups are the Puerto Ricans, whose income is about 51 percent that of whites. Approximately 21 percent of Latino families were living below the poverty level in 2001, compared to about 7.8 percent of white non-Latinos (see Table 9.2). Puerto Rican families have the highest poverty rate, Cubans the lowest (U.S. Bureau of the Census 2001b).

As one might expect from these income figures, Latinos tend to be concentrated in lower-paying and lower-status jobs as semiskilled workers and unskilled laborers. Farmwork is still common, especially among Mexican Americans. Cuban men belong to the only Latino minority that approaches the occupational profile of Anglos (Moore and Pachon 1985; Trueba 1999).

The increase in Latino-owned businesses must be noted. In 1969, there were only 100,000 Latino-owned businesses, with less than $5 billion in revenue. By 2000, there were about 800,000 Latino-owned firms with revenues totaling nearly $50 billion.

No Latino minority is very close to the national average of 67 percent home ownership. Even the relatively well-off Cubans show a figure of only 52 percent. Moreover, Latino economic progress was curtailed by the recession beginning in 2001 (Lowell 2002).

How do Latinos stand politically? Politically, Latinos are becoming a more nationally visible force shaping American politics (Rodriguez 1999; de La Isla 2003). As of 2003, Latinos held twenty-two seats in the U.S. House of Representatives—a tripling from 1981. Of these members of Congress, the majority were Mexican Americans. In 2002, Cuban American Robert Menendez became the Democrat caucus chair, the highest post ever held in Congress by a Latino. On the state and local level, there were almost six thousand elected Latino public officials. Latinos, of course, still have a long way to go to achieve political parity. In 2003, Latinos represented only 5 percent of the U.S. House of Representatives and had no members in the U.S. Senate, despite constituting 12 percent of the population. Issues of education and immigration, as well as income and the quality of life, promise to keep Latinos politically active (Vazque and Torres 2002).

Native Americans

What is the composition of the Native American population? If a single word can be used to describe Native Americans, it is diversity: they are divided into 500, or so, tribes and bands. This diversity is generally unrecognized because of stereotypes perpetuated by old Hollywood films and cheap paperback adventures of the old West. Most Americans don't know that the Navajo and Sioux are entirely different nations with different cultures, social structures, and problems. Tribal groups are as different from one another as they are from the dominant culture, or as different as Anglo Americans are from Italians or Brazilians. Today, Native Americans number around 2.5 million, three-fourths of whom do not live on reservations (Snipp 1992, 1996; Pollard and O'Hare 1999; Lobo and Peters 2001).

What has been the U.S. policy toward Native Americans? Over the years, the federal government has vacillated between a policy of paternalism (that is, domination and care) and a policy of neglect. In the past, impoverishment, oppression, and deceit have typically, though not always, characterized relations between the government of the United States and various tribal governments (Marger 2003). Jonathan Turner sums up both past and present government policy toward Native Americans:

> *The legacy of economic exploitation, especially the great "land grabs" by whites in this century, has forced Indians into urban areas because they can no longer support themselves on their depleted reservations. Yet the legacy of isolation on the reservation prevents many Indians from making the cultural and psychological transition to urban, industrial life. And the burden of change has been placed on the individual Indian, for white institutions—from factories to labor unions and welfare agencies—display little flexibility in adjusting to Indian patterns. The contemporary Indian is therefore faced with impoverishment no matter what course of action he chooses: to stay on the reservations results in poverty, but to leave the reservation and encounter the hostile white economic system in urban areas also results in poverty.* (Turner 1976:185)

Both the paternalistic and neglectful approaches have left Native Americans outside the mainstream of social and economic opportunity (K. M. Dudley 1997; Johnson 2000).

What is the current situation of Native Americans? This neglectful approach is evident in the fact that the U.S. Census Bureau does not regularly report data for Native Americans. Other than a special study performed in the early 1990s, little data are available to describe their current status. Thus, though many of the descriptions that follow are dated, they are the most current available.

Historically the U.S. government's policy toward Native Americans has alternated between paternalism and neglect. The greatest impoverishment is experienced by those remaining on reservations.

Abject poverty remains a major fact of life among Native Americans, especially for many who remain on reservations. Approximately 27 percent of the Native American population live below the poverty line (see Table 9.2). The median income is $21,619 per year, and 14 percent have an income over $50,000 per year, as compared with 32 percent of the white population.

A gap in education also exists. Of Native Americans twenty-five years of age or older, only 66 percent were high school graduates, in contrast to 88 percent of the non-Latino white U.S. population. Only 9.4 percent had completed four years or more of college, compared to 26 percent of the white population.

Native Americans have made only scant penetration into the upper levels of the occupational structure. Although some gains were made during the 1960s and 1970s, only 20 percent of all employed Native American men and women currently hold professional, managerial, or administrative positions. One-third are in blue-collar jobs (craftsworkers, supervisors, operatives, and nonfarm laborers; U.S. Bureau of the Census 1993b). There is one Native American member of the U.S. Senate and two in the House of Representatives.

Are conditions better or worse on the reservations? For the approximately one-fourth of Native Americans living on reservations, the situation is considerably worse than for those living off the reservations. Fully 50 percent of those on reservations live below the poverty line (compared to just over 25 percent for the total Native American population). Reservation dwellers earn only $16,000 per year on average. The rate of college education for Native Americans living on reservations is only about half that of those living off reservations—5 percent versus 9.3 percent (U.S. Bureau of the Census 1993b, 1993d).

A relatively recent development on reservations is the introduction of casino-type gaming establishments. Gaming on and off reservations has grown unexpectedly into an enormous, rapidly expanding industry. In 1999, more than 184 tribes were operating over 300 gaming facilities. Gaming revenues had exceeded $5 billion, and the tribes had received almost $2 billion of this amount. Given the poor social and economic conditions on reservations, it is not surprising that many Native Americans embrace the gaming industry as a source of income. Because ten of the tribes received over half of this money, its long-term effects remain to be seen.

Is there any momentum for change? Recently, the relationship between the U.S. government and various tribal governments has been changing. According to Vine Deloria and Clifford Lytle, a new government policy grants tribes special political status. As a result, many tribes have proclaimed their nationhood, demanding some form of representation in the United Nations (Deloria and Lytle 1984; Wilkins 2001).

Deloria and Lytle cite several other possible directions for change. There is, first of all, the need for reform in tribal governments themselves to allow continuity with the past as well as movement into the future. There is also a need to establish a place for Native American cultures in the modern world. Another need is the creation of relationships with federal, state, and local governments based on mutual respect and parity. Change will not come easily, given the tradition of inequality and discrimination that has characterized Native American life.

From the time they first arrived in the United States, Japanese Americans experienced discriminatory practices. Prejudice and discrimination reached its peak in 1942 when President Roosevelt signed an executive order placing more than 100,000 of the 126,000 Japanese in American in concentration camps. Two-thirds of them were American citizens whose property, as well as their liberty, was taken from them. German Americans and Italian Americans did not suffer the same maltreatment despite Germany's and Italy's involvement in the war.

Asian Americans

Is the Asian American population increasing, decreasing, or stabilizing? Nearly 11 million Asians live in the United States, comprising 3.7 percent of the total population. Asian Americans, the fastest-growing minority in the United States, increased in population by 49 percent between 1990 and 2000. In 1984 alone, more Asian immigrants entered the United States than during the entire thirty years between 1930 and 1960. In 1996, approximately 220,000 Asians were permitted to immigrate to the United States. The Asian American population is predicted to double by 2025 and more than triple by 2050 (U.S. Bureau of the Census 2001a).

Like Latinos, Asians come from many different national and ethnic backgrounds. The largest groups are from China, the Philippines, Japan, India, Korea, and Vietnam. If a success story can be told for any minority groups in America, those groups are the Chinese and Japanese Americans, particularly the latter (S. M. Lee 1998). Even for Chinese and Japanese Americans, however, the road has not been smooth (Kitano and Daniels 1988; Espiritu 1996; Gatewood and Zhou 2000; Wu and Song 2000; Kibria 2002).

How have Chinese Americans fared over the years? Heavy Chinese immigration to California in 1848 caused a furor. Americans feared that the presence of another race would encourage the institution of slavery, cause internal discord, and introduce incurable diseases. Although Chinese labor was accepted, even exploited, in the mining and railroad industries in the West, there was fear that the Chinese immigrants would migrate across the country. Added to this fear was the belief that the Chinese would deprive white men of work in the declining mining and railroad industries. These fears, combined with long-standing prejudice—that existed even prior to their immigration—resulted in an anti-Chinese movement that lasted into the first part of the twentieth century. During this period of more than fifty years, there was much violence against the Chinese, including riots, lynchings, and massacres. In the late 1880s, when a group of Chinese in Tonapah, Nevada, asked for time to collect their belongings before being expelled from the town, they were robbed and beaten; one of them died as a result (Lyman 1974).

Finally, the "Chinese question" became a political matter of high priority. Many state laws were created to restrict Chinese from holding jobs that could deprive whites of employment. The Chinese Exclusion Act passed by the U.S. Congress in 1882 prohibited, for a ten-year period, the entry of any skilled or unskilled Chinese laborers or miners. Strict federal legislation against Chinese immigration continued to be passed until after 1940. During this period, Chinese Americans were driven into large urban ghettos known as Chinatowns, where they are still concentrated today.

Although Chinese Americans in many ways remain isolated from American life, their situation began to improve after 1940 (Loo 1998). American-born Chinese college graduates began to enter professional occupations, and Chinese American scholars and scientists began to make publicly recognized contributions to science and the arts. Their dedication to hard work and education and their contributions to American society have been widely recognized.

What has been the history of Japanese Americans? The earliest relations between Americans and Japanese were positive (unlike the Chinese experience). Early diplomatic relations were warm and cordial. Beginning in 1885, large numbers of Japanese men immigrated to the West Coast of the United States, but the timing for this massive immigration was wrong. The entry of the Japanese came on the heels of America's attempt to exclude Chinese immigrants. Although the Japanese suffered prejudice and discrimination during these early years, they moved from being laborers in certain industries (railroad, canning, logging, mining, meat packing) to being successful farmers (Kitano 1995).

When the Japanese began to compete with white farmers, however, anti-Japanese legislation was passed. The California Alien Land Bill of 1913, for example, permitted Japanese to lease farmland for a maximum of three years; it did not allow land they owned to be inherited by their families. In 1924, the U.S. Congress halted all Japanese immigration, and the 126,000 Japanese already in the United States became targets for prejudice, discrimination, stereotyping, and scapegoating. In 1942, Japan's attack on Pearl Harbor brought the United States into World War II. Wartime hysteria generated fear of a possible Japanese invasion. This led President Roosevelt to issue Executive Order 9066, which sent more than 110,000 of the 126,000 Japanese in America—two-thirds of whom were American citizens—into internment camps (euphemistically referred to as "relocation centers") away from the West Coast (Nagata 1993; Robinson 2001). It was argued that Japanese Americans posed a security threat during World War II. Even though the same argument could have been made about German and Italian immigrants—their countries were also at war with the United States—they were not relocated (U.S. Commission on Wartime Relocation and Internment of Civilians 1983). Eventually, in 1987, the Supreme Court ruled that the internment of Japanese Americans was "based upon racism rather than military necessity."

Despite their internment, Japanese Americans have not had to overcome the centuries of prejudice and discrimination endured by African Americans and Native Americans. Nevertheless, they have overcome great hardship and have become one of the most successful racial minorities in the United States. Through an emphasis on education and hard work, Japanese Americans are an amazing story of economic and occupational success (Montero 1981; Zweigenhaft and Domhoff 1999).

Why have Asian Americans been so successful? Asian Americans have been particularly successful at using the educational system for upward mobility (S. M. Lee 1998; Marger 2003). This fact is reflected in the academic achievement of school-age Asian Americans, whose average Scholastic Aptitude Test (SAT) scores are 45 points higher than that of the general high school population. Furthermore, over 44 percent of Asian Americans have completed four years of college, compared to 26 percent of whites and 11 percent of Latinos (U.S. Bureau of the Census 2001a).

Some claim that Asian Americans are academically successful because they are innately more intelligent than other Americans. Most scholars believe that Asian academic excellence is due to culture, socialization, and influence of the family (Gibson and Ogbu 1991; Caplan, Choy, and Whitmore 1992). Most Asians see education as the key to success. Moreover, in the Confucian ethic (which is so much a part of the Chinese, Japanese, Vietnamese, and Korean cultures), academic excellence is the only way of repaying the debt owed to parents. According to school principal Norman Silber, "Our Asian kids have terrific motivation. They feel it is a disgrace to themselves and their families if they don't succeed" (McGrath 1983:52). In addition, Asian immigrants are accustomed to a tougher academic regimen. For example, the average number of school days in America is 180, but it is 225 days in Japan. Whatever the reasons, Asian Americans have been successful in part because of their willingness to work hard inside and outside of school (Barrett 1990).

The struggle, of course, is not over. Although Asian Americans have been more successful than other American racial minorities, they still feel the effects of prejudice and discrimination. In fact, the relative success of Asian Americans has hurt them in some ways. The stereotype of success has led the public to conclude erroneously that Asian Americans are all well educated, are overrepresented in higher-level occupations, and are making as much money as white Americans—maybe even more. The facts speak otherwise. Vast socioeconomic differences exist among groups within Asian American communities. Popular emphasis on success has led to a disregard of those Asian Americans who have not done very well (Buckley 1991). Moreover, the accent on success camouflages the fact that even "successful" Asian Americans are worse off than are similarly educated and employed white Americans. Due to the stereotype of success, many experts believe that Asian Americans have been neglected and ignored by government agencies, educational institutions, private corporations, and other sectors of society (U.S. Commission on Civil Rights 1980).

In 1996, Gary Locke became the first Asian American governor on the U.S. mainland (Puente 1996). In 2002, there was a combined total of five Asian Americans holding seats in the House of Representatives and two in the Senate, including the first ever Korean American to serve in Congress.

White Ethnics

White ethnics are descendants from what parts of the world? White ethnics are the descendants of immigrants from eastern and southern European nations, particularly Italy and Poland, but they also include Greek, Irish, and Slavic peoples (Rubin 1994). Most of them are blue-collar workers living in small communities surrounding large cities in the eastern half of the United States (Palen 2001). During the 1960s, white ethnics gained the undeserved reputation of being conservative, racist, pro-war, and "hardhats."

What are white ethnics really like? This portrait of white ethnics is not an accurate one. In fact, the evidence is just the contrary. Surveys conducted during the 1960s showed white ethnics to be more against the Vietnam War than white Anglo-Saxon Protestants were. Catholic blue-collar workers were found to be more liberal than either Protestant blue-collar workers or the country as a whole: they were more likely to favor a guaranteed annual wage, more likely to vote for an African American presidential candidate, and more concerned about the environment. Finally, white ethnics, when compared to WASPs, were more likely to be sympathetic to government help for the poor and were more in favor of integration (Greeley 1974).

According to David Featherman (1971), white ethnics have not traditionally been victims of occupational or income discrimination. In fact, Andrew Greeley (1976) contended, white ethnics have been so successful that it is inaccurate to label them working class, as some have done. White ethnic success has not generally been recognized, according to Greeley, because America's elite are not willing to abandon the myth of the blue-collar ethnic.

Despite their relative success, many white ethnics have in recent years become notably conscious of their cultural and national origins (Schaeffer 2002) and have formed a white ethnic "roots" movement. The new trend toward white ethnic identity was influenced by the black power movement of the 1960s, when many African Americans expressed a desire to preserve their cultural and racial identities. Many white ethnics believe that "white ethnicity is beautiful" and that the price of abandoning one's cultural and national roots is simply too high.

Lillian Rubin (1994) links the continuing accent on white ethnicity to the economic decline of white ethnics over the last twenty years and the rising demands of minorities. White ethnics, she contends, are attempting to establish a public identity that enables them to take a seat at the multicultural table.

On the other hand, some sociologists contend that white ethnicity is fading. According to Richard Alba (1985, 1990), the remaining relatively small number of white ethnics may not survive long. Ethnic identity, he argues, cannot be maintained in the face of disappearing ethnic families, neighborhoods, and communities. Mary Waters (1990) agrees with Alba's description and sees this as the "twilight" of white ethnics. In researching white ethnics, she found ethnicity not very important to those she studied. Although her respondents felt that being Italian, Polish, or Irish might make one distinctive in some way, ethnicity had little effect on where they lived, whom they married, or what they did for a living.

Jewish Americans

The United States has the largest number of Jews in the world. America's 6 million Jews outnumber even Israel's 3.5 million. The majority of Jewish Americans are concentrated in several northeastern states, including New York, New Jersey, Massachusetts, and Connecticut, although a sizable number live in California, Florida, and Pennsylvania. The most recent immigrants—primarily well educated and highly skilled—come from Israel and the former Soviet Union. It is estimated that about 10 percent of all Jewish Americans have been in the United States less than ten years.

© Bill Aron/Getty Images

Despite prejudice and discrimination, Jewish Americans are one of the most successful ethnic groups ever to migrate to the United States.

The first Jewish immigrants landed in New Amsterdam in 1654. All of the colonies discouraged them from immigrating. The colonies, in fact, legally prohibited them from holding political office or voting. Still, by the 1840s, Jews began to arrive in large numbers. Anti-Semitism reached its peak in the 1920s and 1930s, particularly among America's upper and upper middle classes. New York was often called "Jew York," and Jewish Americans were subjected to occupational and social segregation (F. J. Davis 1978; Parrillo 2002).

How do Jewish Americans demonstrate success, and how do they respond socially? Throughout the first half of the twentieth century, Jewish Americans were almost totally excluded from top positions in most major industries, denied membership in social and recreational organizations, and subjected to quotas in colleges and universities. Although limited in their opportunities for movement into the top circles of the most economically powerful American corporate entities, Jewish Americans are one of the most successful ethnic groups ever to migrate to the United States. This is partly because Jewish men, particularly later immigrants, entered the country as skilled workers in a proportion far above the norm for immigrant groups. In addition to bringing skills necessary to fit into industrial society, they valued hard work, sacrifice, perseverance, family responsibility, and education. Although working at skilled blue-collar jobs, Jewish parents urged their children to pursue higher education, to become teachers, lawyers, and doctors. Consequently, Jewish Americans represent an above-average proportion of college graduates; they comprise 2 percent of the population, but account for 5 percent of college graduates (Gallup 1993).

Because of the exclusion they experienced, Jewish Americans in the past have tended to remain socially isolated from the rest of American society. This is undergoing some change. Whereas less than 6 percent of Jewish Americans married non-Jews in the 1950s, approximately 47 percent now do so (Goldman 1993). And in 2000 Joe Lieberman became the first Jewish American to run, as a major political party candidate, for the vice presidency of the United States. Although the social isolation of Jewish Americans has diminished somewhat, some separation will probably continue for two basic reasons. First, anti-Semitic sentiment appears to be increasing again. And second, many Jewish Americans do not want to abandon their cultural roots.

Beyond Direct Institutionalized Discrimination

The socioeconomic gap between many minorities in the United States and the general white population is largely due to past and present prejudice and discrimination. But the situation is now even more complex. Minorities are concentrated at the bottom of the stratification structure partly due to the nature of the economy, a factor that operates beyond the influence of direct, or consciously intended, institutional discrimination. This factor has been explored through the context of the **dual labor market**—the existence of a split between core and peripheral segments of the economy and the division of the labor force into preferred and marginalized workers (Hodson and Sullivan 1990). This perspective, which attributes persisting racial inequality to economic factors, originated in an effort to understand the continuing survival of large-scale racial (and gender) inequalities in the face of programs designed to overcome such disparities (Hodson and Kaufman 1982).

What is the dual labor market theory? Traditionally, it has been assumed that only one labor market existed in which all workers competed for jobs. It was also assumed that occupational advancement was based on hard work, education, and training. These assumptions have been successfully challenged by advocates of conflict theory. The existence of a dual labor market means that the rewards for hard work, education, and training vary in different segments of the labor market. Workers involved in the core sector of the labor market—durable manufacturing and petroleum industries, for example—enjoy high wages, good opportunities for advancement, and job security. Those involved in the peripheral sector, including such industries as textile manufacturing and retail trades, are employed in low-paying jobs with little hope for advancement. African Americans and other minority peoples tend to be trapped in these secondary labor markets and lack the resources to alter their situation. Thus, it is concluded, historical practices of racial and ethnic discrimination now interact with contemporary economic processes to lock a significant and growing number of minority peoples out of the core economy, thereby reducing the likelihood of any improvement in their life chances and lifestyles (Bonacich 1976; Szymanski 1976; Cummings 1980; Lord and Falk 1982).

How does indirect institutionalized discrimination operate in the dual labor market? Minority groups, who are overrepresented in the peripheral sector of the American economy and consequently tend to hold low-skill, low-paying jobs, are subjected to indirect institutionalized discrimination. A long-standing division of labor automatically reduces minority access to jobs in the core sector despite their education and training. In addition, minority members are disproportionately shut out of the more desirable jobs and out of the education and training needed to move into core sector jobs on the grounds that they actually prefer, and are better suited to, the marginal jobs they hold (Hodson and Sullivan 1990).

Increasing Racial and Ethnic Diversity in the United States

The United States is moving from a predominately white population of Western origin and culture to a society comprising larger numbers of racial and ethnic minorities. Almost one-half of all Americans will be members of a minority group by the middle of the twenty-first century (O'Hare 1992b; Pollard and O'Hare

FEEDBACK

1. _________ discrimination refers to unintentional organizational or community actions that negatively affect a racial or ethnic minority.
2. The attitudes and practices associated with _________ in the American slave system help account for the long-term prejudice and discrimination experienced by African Americans.
3. The evidence clearly shows that race is declining in importance in America. T or F?
4. Which of the following Latino ethnic minorities is in the best socioeconomic condition?
   a. Puerto Ricans b. Cubans c. Mexican Americans
5. The federal government's policy toward Native Americans has vacillated between _________ and almost total neglect.
6. Contrary to popular opinion, white ethnics in the United States are politically more liberal than Protestant blue-collar workers. T or F?
7. Anti-Semitism in America during the 1920s and 1930s was most prevalent among the _________ and upper middle classes.
8. Within the dual labor market, minorities tend to be disproportionately trapped in the _________ sector.

Answers: 1. Indirect institutionalized 2. manumission 3. F 4. b 5. paternalism 6. T 7. upper 8. secondary or peripheral

WORLDVIEW 9.1

Ethnic Diversity

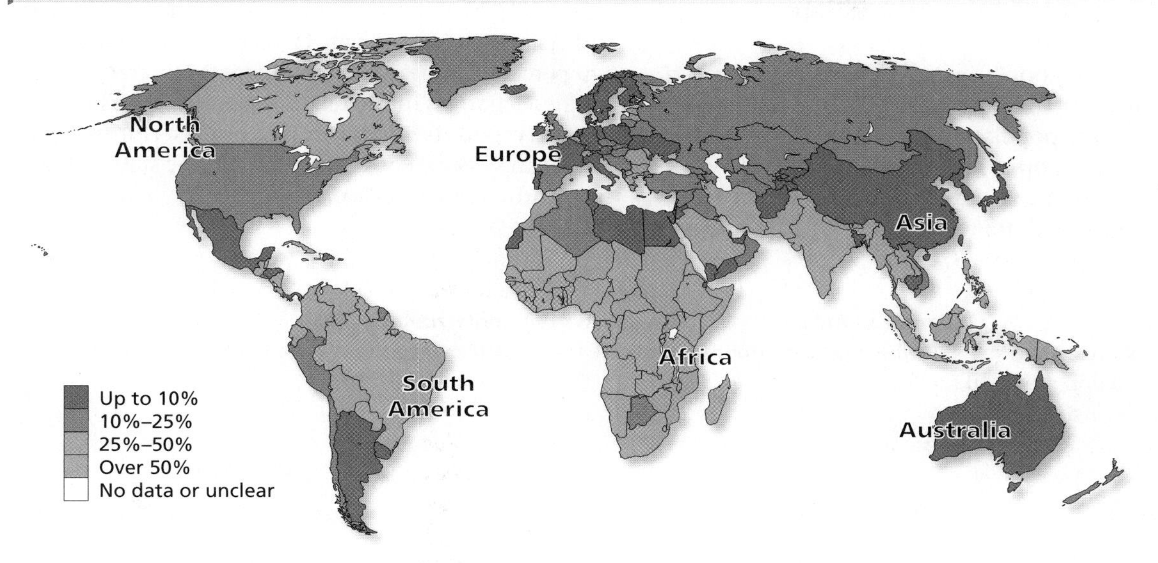

Source: Barry Turner (ed.), *The Statesman Yearbook* (New York: St. Martin's Press, 1996–97).

The degree of worldwide ethnic diversity is displayed in this map. The color code indicates the proportion of a national population represented by members of an ethnic, racial, or national minority.

1. Were you surprised by the comparative degree of ethnic diversity between any particular countries? Describe some differences between your expectations regarding national ethnic diversity and the actual degree of diversity.
2. Write a paragraph detailing your conclusions regarding the extent of ethnic diversity across the globe.
3. Do you believe that the extent of ethnic diversity will increase or decrease during the twenty-first century? Explain your position.

1999; Parrillo 2002). This growth and increasing diversity of the minority population is one of the most important developments in the United States.

Minorities' Share of the Total Population

At the end of the twentieth century, the combined number of African Americans, Latinos, Asian Americans, and Native Americans exceeded 84 million, up from just under 10 million in 1900 and 21 million in 1960. If these minority groups resided in a separate country, they would be the thirteenth most populous nation in the world. Great Britain, France, Italy, and Spain would all be smaller.

Figure 9.5 shows the share of the American population that minorities represented in 2000 and that they are expected to represent in 2050. For most of the twentieth century, America's minority population constituted a relatively constant percentage of the total population—the share of the population represented by minorities increased from 13.1 to only 14.9 percent between 1900 and 1960. Since then, however, the

FIGURE 9.5

The U.S. Population by Race and Ethnicity

The racial and ethnic composition is expected to look very different by 2050. Discuss some social consequences of this changing composition.

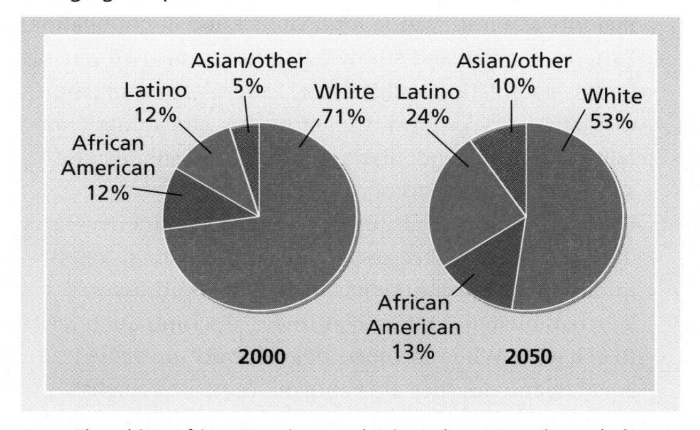

Note: The white, African American, and Asian/other categories exclude Latinos, who may be of any race. The Asian/other category includes American Indians, Eskimos, Aleuts, and Pacific Islanders.
Source: U.S. Bureau of the Census, *Statistical Abstract of the United States: 2001* (Washington, DC: U.S. Government Printing Office, 2001), p.17.

minority proportion of the population has increased to about 30 percent and, as noted earlier, is expected to constitute almost 50 percent by 2050.

Relative Growth of Minority Groups

Whereas the total population of the United States is expected to increase by 23 percent between 2000 and 2025, the minority population is projected to grow 64 percent. The predominant minority in the United States for most of the twentieth century was African American. A rather dramatic change in the relative growth rate of minorities is expected between 2000 and 2050. Although the African American population is expected to expand by about 60 percent, Latinos are projected to increase by almost 200 percent, Native Americans are expected to grow by almost 58 percent, and Asian Americans will have mushroomed by over 240 percent (Riche 2001; see Figure 9.5).

These changes in the relative size and diversity of the minority population promise to alter the nature of American society. Reconsideration of the status of racial and ethnic minorities seems probable.

FEEDBACK

1. Almost ________ percent of all Americans will be members of minority groups in 2050.
2. Rank order (1–4) the following minority groups to indicate their relative proportion of the total American minority population projected for 2050.
   ____ a. Asian Americans
   ____ b. Native Americans
   ____ c. Latinos
   ____ d. African Americans

Answers: 1. 50 2. a. (3) b. (4) c. (1) d. (2)

SUMMARY

REVIEW GUIDE

1. A minority consists of more than small numbers. A minority possesses some distinctive physical or cultural characteristics, is dominated by the majority, and is denied equal treatment. Minorities also have a sense of collective identity, tend to marry within their own kind, and inherit their minority status at birth.
2. A race is a category of people who share certain biologically inherited physical features. Racists use these physical characteristics as an index of a race's superiority or inferiority. Despite the lack of scientific support for this viewpoint, prejudice and discrimination are often justified by alleged differences in intelligence and ability.
3. Ethnic minorities have distinct subcultures. Unique cultural characteristics of ethnic minorities are used by the majority as a justification for prejudice and discrimination.
4. Patterns of racial and ethnic relations assume two general forms: assimilation and conflict. Patterns of assimilation include Anglo-conformity, melting pot, and cultural pluralism. Genocide, population transfer, and subjugation are the major patterns of conflict.
5. Prejudice—negative attitudes toward some minority—is difficult to change because prejudiced individuals reject information that contradicts their existing attitudes. Whereas prejudice refers to attitudes, discrimination refers to behavior. When members of a minority are denied equal treatment, they are being discriminated against. Although prejudice usually leads to discrimination, there are exceptions. In fact, in some instances, discrimination against a minority may lead to the creation of prejudiced attitudes, usually embodied in racial or ethnic stereotypes.
6. Many attempts have been made to explain the existence of prejudice and discrimination. Some psychologists attribute prejudice and discrimination to deep-seated frustration. Hostility and frustration are said to be released on scapegoats who had nothing to do with causing their development in the first place. Other psychologists point to the existence of a personality type—the authoritarian personality—that tends to be very prejudiced against those thought to be inferior. Although functionalism emphasizes the negative consequences of prejudice and discrimination, it also points to some benefits. According to conflict theory, prejudice and discrimination are used by the majority as weapons of power to dominate subordinate minorities. Symbolic interactionism concentrates on the learning of prejudice and discrimination through socialization.
7. Virtually all minorities have felt the consequences of institutionalized discrimination. Institutionalized discrimination may be direct, as in segregation laws; or it may be indirect, as in the form of seniority systems that prevent upward mobility of newly hired minorities.
8. Institutionalized discrimination has been harmful to America's minorities. As a result, African Americans, Native Americans, Latinos, some Asian Americans, and white ethnics lag behind the white majority occupationally, economically, and educationally. Successful cases do exist, but the gains of all minorities remain fragile.
9. Some scholars contend that the disadvantages that minorities are currently experiencing are due to economic as well as social factors. Because of the dual labor

market, minorities are trapped in the peripheral labor market, which offers low-paying, dead-end jobs.

10. Racial and ethnic diversity is increasing dramatically in the United States. By the middle of the twenty-first century, one-half of the entire American population will be minority group members. There are differential growth rates among minorities. African Americans, traditionally the predominant minority in the United States, will be surpassed in numbers by Latinos in the early part of the twenty-first century.

INFOTRAC® COLLEGE EDITION

http://www.infotrac-college.com/wadsworth

Another unique option available to you at the Student Resources section of the companion Web site is Infotrac College Edition, an online library with access to hundreds of scholarly and popular periodicals. Below are suggested search terms for this chapter. Results from these and other searches are found at the site.

Search keywords: **Affirmative Action**. Affirmative Action regulations were introduced in 1965 to require those companies doing business with the U.S. government to hire and to promote people from groups previously discriminated against. After nearly forty years, what progress has been made? How are Affirmative Action laws viewed today? Read three articles to answer these questions. Are the views that authors express based on evidence, or on some other basis?

Search keywords: **internment camps**. Find three articles that report about internment camps. What are they? Where have they existed? What were their conditions like? When have they been used? Did you find any common elements among the camps you read about?

Search keyword: **underclass**. Search Infotrac to find at least two articles about America's underclass. What groups comprise the underclass? What social conditions contribute to the existence of an underclass?

LEARNING OBJECTIVES REVIEW

- Distinguish between the concepts of minority, race, and ethnicity.
- Describe patterns of racial and ethnic relations.
- Differentiate prejudice from discrimination.
- Illustrate the different views of prejudice and discrimination taken by functionalists, conflict theorists, and symbolic interactionists.
- Describe, relative to the white majority, the condition of minorities in the United States.
- Describe the increasing racial and ethnic diversity in America.

CONCEPT REVIEW

Match the following concepts with the definitions listed below them.

____ a. ethnic minority
____ b. underclass
____ c. race
____ d. minority
____ e. de facto subjugation
____ f. assimilation
____ g. self-fulfilling prophecy
____ h. feminization of poverty
____ i. discrimination
____ j. ideology

1. the integration of a racial or ethnic minority into a society
2. subjugation based on common, everyday social practices
3. unequal treatment of individuals based on their minority membership
4. a people who are socially identified and set apart by others and themselves on the basis of their unique cultural or nationality characteristics
5. a set of ideas used to justify and defend the interests and actions of those in power in a society
6. a people who possess some distinctive physical or cultural characteristics, are dominated by the majority, and are denied equal treatment
7. a distinct category of people who are alleged to share certain biologically inherited physical characteristics
8. when an expectation leads to behavior that causes the expectation to become a reality
9. those in poverty who are continuously unemployed or underemployed because of the absence of job opportunities and/or required job skills
10. the trend toward more of the poor in the United States being women and children

REVIEW GUIDE

CRITICAL-THINKING QUESTIONS

1. Evaluate the definition of minority used by sociologists.

2. Do you think that both the pattern of assimilation and the pattern of conflict accurately describe the state of racial relations between blacks and whites in America today? State your position using terms and evidence from the text.

3. Compare and contrast the functionalist and conflict perspectives on prejudice and discrimination. Is one a better explanation than the other? Are they complementary? Defend your conclusion.

4. Taking into account the past and present situation of African Americans in the United States, construct a line of argument for or against affirmative action. Be specific.

MULTIPLE-CHOICE QUESTIONS

1. **A set of ideas used to justify and defend the majority's interests and actions is known as**
   a. an ideology.
   b. a doctrine.
   c. a dogma.
   d. a code of ethics.
   e. a philosophy.
2. **People who possess some distinctive physical or cultural characteristics, are dominated by the majority, and are denied equal treatment are known as**
   a. a minority.
   b. a race.
   c. an ethnic group.
   d. a subculture.
   e. a caste.
3. **Which of the following statements best represents contemporary social scientists' beliefs regarding differences in intelligence among various races?**
   a. The weight of current scientific evidence supports a connection between genetically determined physical characteristics and innate superiority or inferiority of certain races.
   b. There is no scientific proof that races differ in their innate mental characteristics.
   c. Recent reports attribute existing differences in measured IQ to racial differences rather than to the social environment.
   d. Social scientists are now becoming more interested in biological races than in social races.
   e. Social scientists have stopped doing research on what they consider a closed issue.
4. **_________ is a pattern of assimilation in which immigrants are allowed to maintain their traditional cultural ways while at the same time learning the values and norms of the host culture.**
   a. The melting pot
   b. Anglo-conformity
   c. Cultural pluralism
   d. Accommodation
   e. Population transfer
5. **The unequal treatment of individuals based on their membership in some minority is known as**
   a. stereotyping.
   b. prejudice.
   c. racism.
   d. the ideal-type method.
   e. discrimination.
6. **Believing that "all politicians are corrupt" is an example of**
   a. racism.
   b. stereotyping.
   c. discrimination.
   d. prejudice.
7. **The _________ contends that barring others from social, political, and economic opportunities strengthens the boundaries of the majority.**
   a. conflict perspective
   b. psychological perspective
   c. interactionist perspective
   d. functionalist perspective
8. **Two nations that are convinced they are going to war engage in hostile interaction that causes a war to occur. This is an example of**
   a. stereotyping.
   b. discrimination.
   c. a self-fulfilling prophecy.
   d. agitation.
9. **Forcing Native Americans onto reservations is an example of**
   a. Anglo-conformity.
   b. subjugation.
   c. demographic transition.
   d. partition.
   e. population transfer.
10. **According to the text, which of the following statements regarding the current condition of African Americans is false?**
    a. Their median income is substantially below that of white households.
    b. Their poverty rate is four times that of whites.

c. African American median family income decreased in the 1980s.
d. Less than one-third of African American men are employed in the highest occupational categories.
e. Young African American families hold less than one-fifth of the wealth of young white families.

11. **The United States policy toward the Native Americans over the years has**
a. been consistently paternalistic.
b. been consistently neglectful.
c. vacillated between paternalism and neglect.
d. attempted to unite the various tribes so they can help themselves.
e. been based on the concept of workfare.

12. **According to the text, Asian Americans have been especially successful because they**
a. possess superior intelligence.
b. profit from illegal activities such as gambling and prostitution.
c. refuse to work at jobs that do not pay well.
d. embrace education as the key to success.
e. place little value on human life.

13. **The success of Jewish Americans can be attributed to all of the following *except***
a. their willingness to work hard.
b. their emphasis on family responsibility.
c. the high value they place on education.
d. the skills many Jewish immigrants brought with them.
e. their easy assimilation into mainstream American society.

14. **According to dual labor market theory, minority workers tend to be trapped in the __________ labor market, with low-paying jobs and little hope for advancement.**
a. secondary or peripheral
b. core
c. primary
d. inferior
e. malfunctioning

FEEDBACK REVIEW

True-False

1. The evidence clearly shows that race is declining in importance in America. T or F?
2. Contrary to popular opinion, white ethnics in the United States are politically more liberal than Protestant blue-collar workers. T or F?

Fill in the Blank

3. According to Gordon Allport, children learning prejudice move from a pregeneralized learning period to a stage of total __________.
4. The attitudes and practices associated with __________ in the American slave system help account for the long-term prejudice and discrimination experienced by African Americans.
5. Almost __________ percent of all Americans will be members of minority groups in 2050.

Multiple Choice

6. **Which of the following is *not* always a characteristic of a minority?**
a. distinctive physical or cultural characteristics
b. smaller in number than the majority
c. dominated by the majority
d. denied equal treatment
e. a sense of collective identity

Matching

7. Rank order (1–4) the following minority groups to indicate their relative proportion of the total American minority population projected for 2050.
____ a. Asian Americans ____ c. Latinos
____ b. Native Americans ____ d. African Americans

GRAPHIC REVIEW

Some attitudes of Americans toward various immigrant minorities are summarized in Figure 9.1.

1. If you had to write a one-sentence summary of Figure 9.1 for the CNN evening news, what would you say?

____
____
____

2. What do the data in Figure 9.1 suggest regarding the future of ethnic relations in the United States?

____
____
____

ANSWER KEY

Concept Review

a. 4
b. 9
c. 7
d. 6
e. 2
f. 1
g. 8
h. 10
i. 3
j. 5

Multiple Choice

1. a
2. a
3. b
4. c
5. e
6. b
7. d
8. c
9. e
10. b
11. c
12. d
13. e
14. a

Feedback Review

1. F
2. T
3. rejection
4. manumission
5. 50
6. b
7. a. 3
   b. 4
   c. 1
   d. 2

Inequalities of Gender

10

© Michael Newman/PhotoEdit

OUTLINE

LEARNING OBJECTIVES

- Distinguish the concepts of sex, gender, and gender identity.
- Demonstrate the relative contributions of biology and culture to gender formation.
- Outline the perspectives of gender expressed by functionalists, conflict theorists, and symbolic interactionists.
- Describe the position of women in the United States with respect to work, law, politics, and sports.
- Discuss factors promoting change in gender roles as well as factors promoting resistance to change in traditional gender roles.
- Describe the future of gender roles.

USING THE SOCIOLOGICAL IMAGINATION

How well does the United States compare with other industrialized countries on gender income equality? Even if Americans recognize that women are not paid on a par with men, most believe that at the very least the United States is making more progress than other countries. On the contrary, among modern countries, America is surprisingly near the bottom in male-female income parity. Only Luxembourg and Japan have wider gender gaps than the United States. Whereas Swedish women in manufacturing jobs, for example, earn 87 percent of the wages paid men, females in the United States earn only 76 percent of the wages paid men—for the same work (U.S. Bureau of the Census 2002d).

Throughout history, men have predominantly performed the social, political, and economic functions outside the home, while women have assumed responsibility for child care and household tasks. Although this division of labor does not necessarily imply any status difference between the sexes, in practice it certainly has. The female domestic tasks are undervalued in industrial societies, where women—thought to be dependent, passive, and deferring—have usually been considered subordinate to independent, aggressive, and strong men. Historically, the division of labor based on sex has almost always led to gender inequality. And this chapter will demonstrate other disadvantages attached to the female gender role as well. But first let's delve into the cultural and social underpinnings of gender roles.

Antecedents of Gender

In the past, people believed that anatomy was destiny. Behavioral differences between men and women were popularly attributed to **sex**—the biological distinction between male and female. Males were assumed to be naturally more aggressive than women and to be built for providing and protecting. Thought of as being naturally more passive, females were believed to be designed for domestic work (Valian 1999). This way of thinking is called **biological determinism**—the attribution of behavioral differences to inherited physical characteristics. If this popular conception were true, all men and all women in all societies would behave uniformly in their unique biologically determined ways because of inborn biological forces beyond their control.

As will be shown, biological determinism lacks scientific validation. Although biology may create some behavioral tendencies in the sexes, such tendencies are so weak that they are easily overridden by cultural and social influences (Ridley 1996; Sapolsky 1997; Bearman and Bruchner 2002; Powell and Graves 2002). Therefore, **gender**—the expectations and behaviors associated with a sex category within a society—is acquired through socialization (Jackson and Scott 2001).

From the moment of birth, males and females are treated differently. Few parents in American society point with pride to the muscular legs and broad shoulders of their baby girl or to the long eyelashes, rosebud mouth, and delicate curly hair of their baby boy. Rather, parents stress the characteristics and behavior that fit the society's image of the ideal male or female, including modes of dress, ways of walking and talking, play activities, and life aspirations. In response, girls and boys gradually conform to these definitions and learn to behave as expected. From this process comes **gender identity**—an awareness of being masculine or feminine, based on culture. Margaret Andersen succinctly captures the difference between sex and gender:

> *The terms sex and gender have particular definitions in sociological work. Sex refers to the biological identity of the person and is meant to signify the fact that one is either male or female. . . . Gender refers to the socially learned behaviors and expectations that are associated with the two sexes. Thus, whereas "maleness" and "femaleness" are biological facts, becoming a woman or becoming a man is a cultural process. Like race and class, gender is a social category that establishes, in large measure, our life chances and directs our social relations with others. Sociologists distinguish sex and gender to emphasize that gender is a cultural, not a biological, phenomenon.* (Andersen 2002:31)

These two young children at play are engaging in the stereotypical behavior associated with their respective genders. The small girl, with a bow in her hair, is taking care of her "baby." The small boy, with a baseball cap on his head, is playing with a truck.

Controversy over explanations for gender differences persists; the nature (biology) versus nurture (socialization) distinction remains at the heart of the debate (J. Q. Wilson 1993; Canary et al. 1998). All the same, a notable difference exists today: research is beginning to replace bias and opinion. Explanations for masculinity and femininity are now based on research rather than on tradition and common knowledge. This research investigates the biological and cultural antecedents of gender (Moir 1991; Hrdy 1999; Andersen 2002).

Biological Evidence

In what ways are men and women biologically different? The obvious biological differences between the sexes include distinctive muscle and bone structure and fatty tissue composition. The differences in reproductive organs, however, are much more important, because they result in certain facts of life: only men can impregnate; only women are able to produce an ovum, carry and nurture the developing fetus, and give birth; and only women can secrete milk for nursing infants. In addition, the genetic composition of the body cells of men and women is different. All human beings have twenty-three pairs of chromosomes (the components of cells that determine heredity), but one of those pairs is sex distinctive. Males have an X and a Y chromosome in the pair that determines sex, and females have two X chromosomes. At about the eighth week of prenatal development, the information coded into the chromosomes determines whether the embryo—which is still sexually neutral—will begin to develop testicles or ovaries. These reproductive glands discharge the hormones that are characteristic of either females or males: estrogen and progesterone in females, testosterone and androgens in males. These hormone combinations influence development in both males and females throughout life (Geary 1998).

More recently, some research indicates that the brains of men and women are slightly different in structure (Gur et al. 1995; Fisher 1999). For example, men show more activity in a region of the brain thought to be more tied to adaptive, evolutionary responses such as fighting. Women have more activity in a newer region of the brain; this region is more highly developed and thought to be linked to emotional expression. The female brain is less specialized than the male brain. Women tend to use both sides of the brain simultaneously when performing a task. For example, whereas men tend to process verbal tasks on the left side of the brain, women are more likely to use both sides. Women tend to use both ears when listening, and men tend to use the right ear. And while males are more muscular and stronger as a group than females, females have greater physical endurance and generally outlive males.

Do biological differences necessarily lead to differences in social behavior? Biological determinists point to research indicating that men and women in dozens of different cultures (at varying stages of economic development) are associated with some distinctly different ways of behaving. For example, men and women differ in what they look for in romantic and sexual partners. Men value physical appearance more than women do. Women place more emphasis on social class and income. Men tend to prefer slightly younger mates, while women favor slightly older ones. In addition, males in general tend more toward physical aggressiveness in conflict situations (Buss 1994a, 1994b, 1995; Maccoby 1997; Wodak 1997; Buss, Malamuth, and Winstead 1998).

On the other hand, researchers investigating behavioral differences between the sexes have been unable to consistently establish biological differences as an independent variable (Fausto-Sterling 1987; Stockard and Johnson 1992; Kimmel 2000; Andersen 2002). One researcher's findings tend to contradict another's. Furthermore, many studies seek to find differences between males and females but ignore the similarities. To compound the problem, researchers often fail to document the variation in characteristics that exists *within* sex categories. Some men, for example, tend to be submissive, weak, and noncompetitive, and some women are aggressive, strong, and competitive.

Cultural Evidence

Is there cross-cultural evidence that gender-related behavior is not solely the product of biology? In her classic study of three primitive New Guinean peoples, Margaret Mead (1950) demonstrated the influence of culture and socialization on gender role behavior. Among the Arapesh, Mead found that both males and females were conditioned to be cooperative, unaggressive, and empathetic. Both men and women in this tribe behaved in a way that is consistent with the more traditional concept of the female gender role. Among the Mundugumor, both men and women were trained to be "masculine"—they were aggressive, ruthless, and unresponsive to the needs of others. In the Tchambuli tribe, the gender roles were the opposite of those found in Western society. Women were dominant, impersonal, and aggressive, and men were dependent and submissive. On the basis of this evidence, Mead concluded that human nature is sufficiently malleable to rule out biological determination of gender roles. Cross-cultural research since Mead's landmark work has clearly substantiated her findings that gender roles are not fixed at birth (Janssen-Jurreit 1982; Montagu 2000).

For example, look at the gender behavior of the five-nation League of the Iroquois, which was in territory now part of Pennsylvania and upper New York. In a matrilineal society (family line traced through females) the Iroquois women enjoyed power and respect. Senior women chose the male members of the ruling council, and if the men failed to follow the women's wishes, the males were removed from office. Male dominancy and female subordination were absent in a culture where women took charge of village affairs, including military activities (Zinn 2001).

Studies have also been conducted on infants whose parents intentionally treated their children as if they belonged to the opposite gender. Apparently, individuals can fairly easily be socialized into the gender of the opposite sex. What's more, after a few years, these children resist switching back. These studies indicate that biological tendencies can be greatly influenced by culture and society (Schwartz 1987; L. Shapiro 1990; Ridley 1996; Sapolsky 1997).

What are the most common male and female roles? Societies such as the Arapesh, Mundugumor, and Tchambuli are exceptions to the rule. The general pattern in cross-cultural studies of gender roles in preliterate societies is one of male dominance and female nurturance (Zelditch 1955; Barry, Bacon, and Child 1957; Collins 1971; Reiss 1976; G. R. Lee 1977; Sinclair 1979). Although both men and women perform economic tasks that provide for family welfare, women's tasks usually involve domestic chores, child rearing, and emotional harmony. Men, on the other hand, are more likely to provide financially for the family and to represent the family in activities outside the home (Tilly and Scott 1978). This is probably because, beginning with early hunting societies, men have had the advantage of greater size and physical strength, and women have had the capability of childbearing and breast-feeding. Larger physical size has led to social, economic, and political advantages for men. Rarely, if ever, have these advantages been shared equally with women.

Do gender definitions change? Cultural definitions of gender roles can and do change. In his book *American Manhood*, E. Anthony Rotundo (1993) identifies three distinct culturally created conceptions of manhood. These definitions and redefinitions of gender roles developed in the North American middle class prior to the twentieth century. The first phase, *communal manhood*, was developed in the socially integrated society of colonial New England. During this period, the definition of manhood was embedded in one's obligations to community. A man's social class of birth gave him his place in the community. He fulfilled himself through public usefulness more than through economic success. As head of the household, a man expressed his value to his community and bestowed a social identity on his wife and children. During this era, before 1800, men were considered superior to women. In particular, men were held to be more virtuous. Thought to possess greater powers of reason, men were considered better able than women were to control passions such as ambition and envy.

The second phase, *self-made manhood*, became the dominant male gender definition in the first decades of the nineteenth century. Social status of birth was replaced with a social identity based on personal achievement. A man's work role became his source of male identity. "Male" passions were allowed more fullness of expression. Because a man was supposed to prove his superiority, the drive for dominance was viewed as virtuous. As males turned from the tradition of public usefulness to the pursuit of self-interest, the nature of women was redefined. Women were attributed a stronger moral sense than men and were assigned the role of protecting the common good in an age of rampant male individualism. Despite her moral superiority, a woman's social identity still depended on her husband's place in society, and she was denied participation in the pursuit of individualism. Her primary purposes were to control men's passions, to make others happy, and to take care of her husband and children.

Late in the nineteenth century there came another shift. The third phase, *passionate manhood*, emphasized not just achievement, but ambition, combativeness, and aggression. Male toughness was admired, tenderness scorned. The gender definition of men and women changed over these periods, but there was a constant: the exclusion of women from economic, political, and social power. In these arenas, women were consistently relegated to a subservient position.

FEEDBACK

1. _________ is the biological distinction between male and female.
2. _________ refers to the expectations and behaviors associated with a sex category within a society.
3. Gender is acquired through the process of _________.
4. _________ is an awareness of being masculine or feminine.
5. Societies that do not follow traditional gender definitions are exceptions to the rule. T or F?

Answers: 1. Sex 2. Gender 3. socialization 4. Gender identity 5. T

Where does the nature/nurture debate stand? The nature/nurture controversy has unfortunately been posed as an either/or situation by both the mass media and some researchers. In general, researchers investigating behavioral differences between the sexes have not been able to prove that any particular behavior has a biological cause (Fausto-Sterling 1987; Stockard and Johnson 1992; Andersen 2002). In light of this, it is reasonable to credit *both* biology and culture for gender differences. Biological characteristics exist, but they can be modified through social influences (Parker and Easton 1998). In other words, men and women can learn to be submissive or aggressive by mirroring the behaviors of influential role models, such as parents or siblings. Also, this is a good time to remind ourselves that human behavior is the result of multiple causes.

Theoretical Perspectives on Gender

The functionalist perspective focuses on the origins of gender differences, whereas conflict theory concentrates on the reasons gender differences continue to exist. Symbolic interactionism attempts to explain the ways in which gender differences are acquired.

Functionalism and Gender

If some pattern of behavior is not beneficial to society, functionalists argue, it will ultimately cease to be important. According to functionalism, the division of responsibilities between males and females survived because it was beneficial for human living. Early humans found sex-based division of labor efficient. In part because of their size and muscular strength, men performed hunting and defense tasks. In addition, men were assigned these tasks because men were more expendable than women. Because these duties were more dangerous and because one male was enough to ensure sufficient procreation to sustain the survival of a group (the same is not true of a single woman with several men), it hurt the group's chances of survival less to lose a man. Women were assigned domestic and childcare tasks. Because this sexual division of labor promoted the survival of the species, assert the functionalists, it was retained.

Although giving credit to the functionalist perspective for its insights into the original emergence of gender roles, some critics have contended that it is too conservative and inadequate to explain recent changes in gender (Peplau 1983). Gerald Maxwell (1975), like most sociologists, concedes that the traditional division of labor may have been appropriate at one time. Today, however, argues Maxwell, rapid social change has undermined the value of functionalist theory in understanding gender. Maxwell believes that it is to the advantage of modern society to treat people based on their ability rather than on an outdated conception of instrumental and expressive roles. According to Maxwell and other critics, then, the contribution of the functionalist view of gender is restricted to an explanation of the development of gender.

In response to Maxwell's criticism, functionalists now recognize that the traditional division of labor has dysfunctions, especially for modern society. Consequently, they point to harmful gender definitions resulting from this division of labor. These dysfunctions are examined later in the discussion on gender inequality.

Conflict Theory and Gender

Conflict advocates are concerned not only with the origins of the traditional conception of gender but also with gender's persistence. The focus here will be on the latter. According to conflict theory, men and women have differential access to the necessary resources for outside-the-home success. It is to the advantage of men (higher-status members of the society) and the disadvantage of women (lower-status members of the society) that women not gain access to political, economic, and social resources. By keeping the traditional division of labor intact, men can maintain the status quo and preserve the privileges they enjoy.

Perhaps the most recent example of maintaining the gender status quo was found in Afghanistan, when the then ruling Taliban militia practiced "gender apartheid." This gender war trapped women in a way of life unknown elsewhere in the modern world (O'Dwyer 1999; Rashid 2001). The Taliban prohibited girls from attending school, banned women from all work outside the home, punished women who left home without a male relative, enforced the wearing of black sacks over the body with only a wire mesh for vision, required windows of houses to be painted black, forced women to wear silent shoes, and required women to remain mute in public.

Conflict theorists see traditional gender roles as outdated. Although these conventional gender roles may have been appropriate in hunting and gathering, horticultural, and agricultural societies, they are inappropriate for the industrial and postindustrial era. Male physical strength may have been important when hunting was the major means of subsistence, but work in modern society does not position men with that advantage. In addition, demographic changes make women more available for outside work. Women are marrying later, are having fewer children, are younger when their last child leaves home, are remaining single in greater numbers, and are increasingly choosing to be single parents. According to conflict theorists, women who prefer careers in fields formerly reserved for men have every right to make that choice, whether or not it is "functional" for society. For example, any potential harm to the family must now be weighed against the damage done to women who are not permitted to develop their occupational potential. According to conflict advocates, research suggests that there are benefits to the family associated with working wives and mothers (Hoffman and Nye 1975; Moen 1978; Kate 1998; Zuo and Tang 2000; Hochschild 2001). Also, they say, the barriers blocking women from entry into higher-level jobs waste talents and skills that modern societies need.

Marxist and socialist feminists see the position of women in capitalist society as the result of two interrelated influences: patriarchal (male-dominated) institutions and the historical development of industrial capitalism. Historically, women have been relegated to subordinate roles because their unpaid labor in the home is needed to produce the future labor force. Home work provides for the care and feeding of workers, the creation of new workers and consumers (through childbearing and child rearing), and the creation of people who can devote their time to the consumption activities necessary to keep factories running and workers working. In addition, women provide a reserve army of cheap labor for times of crisis (for example, war) and for expansion of the labor force when more workers are needed. Thus, the subordination of women is seen as a key component in the maintenance of the political and economic institutions in capitalist society (Sokoloff 1980; Chafetz 1984, 1990; Jaggar and Rothenberg 1993; Nierenberg 2002).

How is sexual harassment related to conflict theory? The conflict perspective is built into the concept of **sexual harassment**—the use of one's superior power in making unwelcome sexual advances. Sexual harassment may be experienced as raw superior physical power. This was clearly the case in the 1992 Tailhook scandal in which many women Navy personnel, including officers, were sexually harassed by naval aviators at a Las Vegas convention. The subsequent investigation, however, revealed a pattern of sexual harassment in the Navy that went beyond the physical power of an individual or the collective physical strength of a group.

Sexual harassment in all branches of the military has subsequently been revealed to be quite high (Thomas and Vistica 1997). According to a 1995 army survey, 4 percent of all female soldiers reported that they had been the victim of an actual or attempted rape or sexual assault within the previous year. That is almost ten times greater than the rate of rape and sexual assault in the general population. The highly publicized court cases during the 1990s involving the sexual abuse of military recruits by their superiors are apparently only the tip of an ugly and frightening iceberg. In fact, evidence of the prevalence of sexual harassment in the military surfaced again in early 2003 at the Air Force Academy, when the country learned of over fifty reported cases of male cadet rape or sexual assault of female cadets. As the number of cases increased, the Air Force Secretary James G. Roche admitted that the actual number of rapes is probably three times the reported number ("Air Force Reports 54 Rapes, Assault" 2003; Lorch 2003). Many observers consider this assessment as far too conservative.

In addition to physical superiority, sexual harassment can be a matter of social and economic power. Two prominent cases illustrate the difficulties involved in legally establishing sexual harassment. In 1991 Anita Hill, a law professor, brought the issue of sexual harassment to public attention when she testified in the televised Senate Judiciary Committee confirmation hearings of now Supreme Court Justice Clarence Thomas. Professor Hill accused Justice Thomas of sexually harassing her during the time she worked on his staff. The Senate committee, then totally composed of men, initially discounted the sexual harassment charge and subsequently confirmed Justice Thomas. The truth was not determined. The second case in 1993 involved Senator Robert Packwood of Oregon. Several former

© Michael Newman/PhotoEdit

Work-related sexual harassment involves the use of superior power in making unwelcome sexual advances. Although sexual harassment is usually committed by men, men may be harassed by women as well.

employees of Senator Packwood accused him of sexual harassment. A legal struggle involving the senator's diaries, the women involved, the Senate Ethics Committee, and the Justice Department followed. Packwood was eventually forced to resign.

Symbolic Interactionism and Gender

Symbolic interactionists focus on **gender socialization**—the social process in which boys learn to act the way boys are supposed to act and girls learn to act as girls are expected to act. This process occurs in large part through interaction with parents, teachers, and peers as well as through the mass media. This socialization process is very powerful, with gender being incorporated into the self-concept through role taking and the looking-glass self.

How do parents contribute to gender socialization? Parents are vital in gender socialization because they transfer values and attitudes regarding how boys and girls should behave. The learning of gender begins at birth and is well established by the time the child is two and a half years old (Mussen 1969; Frieze et al., 1978; Davies, 1990). Actually, parents begin laying the groundwork for gender socialization before the child is born. Parents often painstakingly select a name to represent the gender-related characteristics they expect their child to possess. Immediately after the baby's birth, friends and relatives give gifts "appropriate" to the child's sex—blue or pink blankets, the baseball playsuits or frilly dresses, balls and trucks or dolls. Numerous studies report that gender definitions lead to differential treatment of boys and girls. Studies of infant care have found that girls are cuddled more, talked to more, and handled more gently. Parents expect boys to be more assertive, and they discourage them from clinging.

Parental differential treatment of the young does not stop at infancy. For example, the bedrooms of preschoolers reflect masculine and feminine themes. Girls' rooms contain dolls and household items, whereas the rooms of boys are more likely to contain guns, footballs, and trucks. Gender is even taught and reinforced in the assignment of family chores. In an investigation of almost 700 children between the ages of two and seventeen, Lynn White and David Brinkerhoff (1981) found that boys were often given "masculine" jobs, such as cutting grass and shoveling snow, whereas girls were more often assigned "feminine" chores, such as washing dishes and cleaning up the house.

As parents respond to their children's behavior, they usually intentionally and unconsciously continue to transfer their gender-related values. Parents often even evaluate children according to their level of conformity to gender definition.

In what ways do schools reinforce gender socialization? Although the most critical period of gender socialization occurs within the family during early childhood, gender socialization occurs through interaction in schools as well (K. A. Martin 1998). Teachers may not recognize that they encourage different behaviors from boys and girls, but observation of preschool teachers reveals a clear sex-based difference in many teacher-student relationships. This pattern continues into the elementary school years. Myra and David Sadker, in an extensive study of fourth-, sixth-, and eighth-grade students, found boys to be more assertive in class: boys were eight times more likely than girls to call out answers, whereas girls sat patiently with their hands raised. Sadker and Sadker linked this classroom behavior to the differential treatment given boys and girls by teachers. Teachers, report these researchers, are more likely to accept the answers given by boys who call out answers. Girls who call out in class are given such messages as, "In this class we don't shout out answers, we raise our hands." According to Sadker and Sadker, the message is subtle and powerful: "Boys should be academically assertive and grab teacher attention; girls should act like ladies and keep quiet" (Sadker and Sadker 1985:56).

In their later book *Failing at Fairness*, the Sadkers (1995) examine sexism from elementary school through college. The conclusion is consistent with their earlier findings: through differential treatment, America's schools often shortchange females; gender bias results in an inferior education for girls. Academically, girls typically outperform boys in the early years of school. Through the transmission of gender role values, well-intentioned teachers often dampen female competitiveness. Girls, the study concludes, are systematically taught passivity, a dislike of math and science, and a deference to the alleged superior abilities of boys. Females carry this impairment into adult life—into the working world, where its disadvantageous effects take full toll.

How do peers contribute to gender socialization? Even children in day-care centers and nursery schools are aware of and encourage gender definitions. When three-year-olds behave according to their gender definition, their peers encourage, praise, and imitate them. Children who imitate the opposite sex meet considerable opposition within the group. This opposition includes criticism as well as efforts to modify the behavior (Lamb, Easterbrooks, and Holden 1980).

Interaction among adolescent peers also affects gender conceptions because adolescence is the time when

© Mark Lennihan/AP/Wide World Photos

The teen peer group reinforces the American conception of beauty. Because slimness is a cornerstone of this conception, teenage girls tend to compare themselves with fashion models.

an individual's identity is being firmly established, and identity is closely linked with definitions of masculinity and femininity (Erikson 1973, 1982; Adler and Adler 1998). Because adolescents want approval from their peers, acceptance or rejection by one's friends can significantly influence a teenager's self-concept. Teens who most closely mirror traditional gender definitions—such as football players and cheerleaders—are generally given the greatest respect, whereas "feminine" boys and "masculine" girls are often accorded low status. This peer group pressure encourages teenage conformity to society's idealized role models. To do otherwise is to risk rejection and a significant loss of self-esteem.

What is the role of the media in gender socialization? All members of society are subject to the continual bombardment of images from television, books, magazines, radio, movies, and advertising. For example, in the video game "Grand Theft Auto: Vice City," players can have sex with a female prostitute and earn additional points by beating and murdering her. Many studies document the distorted view of women's roles prevalent in the media and the ways in which this distorted view affects people. In general, the media present the most stereotypical version of gender definitions, thus reflecting and reinforcing the limits on the options available to both sexes (Craig 1992; Kilbourne 2001; Rutledge Shields, and Heinecken 2001; Dines and Humez 2002; Gauntlett 2002).

In the 1970s, illustrated children's books presented a conservative, traditional image of the sexes, with women underrepresented in titles and as major characters. One study examining over 2,700 children's stories found 5 stories about boys for every 2 stories about girls, 119 biographies of men compared with 27 biographies of women, and 65 stories degrading girls compared with 2 degrading boys (American Association of School Administrators 1974). In a study of a library display of children's books, Deane Gersoni-Stavn (1974) found less than half included pictures of women. Of these books, 80 percent included a picture of a woman wearing an apron. In the few books that pictured women working outside the home, the jobs were traditionally feminine occupations, such as teaching and

TABLE 10.1

FOCUS ON THEORETICAL PERSPECTIVES: Gender Inequality

Each of the major theoretical perspectives can focus on gender inequality in its own unique way. Explain why the examples given fit each theoretical perspective. How would each of the other theories approach the same social arrangement differently?

Theoretical Perspective	Social Arrangement	Example
Functionalism	• Gender-based division of labor	• Women are expected to perform household tasks for the benefit of society.
Conflict theory	• Patriarchy (male domination)	• Women are denied high-status occupations for the benefit of men.
Symbolic interactionism	• Favoring males over females in the classroom	• Few females believe they can become scientists.

nursing, or unusual female roles, such as queen or trapeze artist. In a content analysis of several prizewinning children's books—the Caldecott Medal winners for the years 1967–1971—Leonore Weitzman and Deborah Eifler (1972) found a distinct gender bias. Human male characters outnumbered human female characters by 11 to 1; the ratio of male to female nonhuman animals was 95 to 1. Males were the central characters who dominated the action, whereas females were cast in passive roles.

But haven't things changed dramatically? A later study of Caldecott winners published between 1972 and 1979 showed some improvement: human males appeared more often than females by a ratio of 1.8 to 1; nonhuman male animals outweighed females by 2.66 to 1 (Kolbe and LaVoie 1981). A gender bias reminiscent of the Weitzman and Eifler study was still apparent in these books, however. The characters in seventeen of the nineteen books portrayed the traditional gender definitions, and not one female character was shown to be employed outside the home.

When researchers later replicated Weitzman and Eifler's study, they also found only slight improvement. Boys were still depicted as competitive, active, and creative; girls were cast as submissive, dependent, and passive (Williams et al. 1987; McDonald 1989; Bigler and Liber 1990). Although male and female characters were about the same numerically and one-third of the principal characters were female, only one female, a waitress, was shown working for pay. Despite the increased appearance of females, they still were unequally represented in active roles. The power of children's literature to influence young minds is now being recognized. Although this heightened sensitivity is currently reflected in the publication of more books with tempered depictions of the sexes, not all traces of gender bias have disappeared (Purcell and Stewar 1990). When Myra and David Sadker (1995) examined the content of grade-school textbooks, they found that the ratio of boys/men to girls/women in the language arts textbooks ranged from a ratio of 2 to 1 to a ratio of 3 to 1 in favor of males. And out of a 631-page textbook on world history, only seven pages related to women.

Harlequin romance books are another example of the staying power of gender role stereotypes in books. In excess of 160 million of these books are published in over 20 languages. In fact, Harlequin romance book sales account for almost one-third of trade paperbacks sold in the United States and Canada annually. In Harlequin books, women are consistently depicted as more sexually passive; men are the sexual aggressors (Grescoe 1997).

Television continues to reflect traditional gender roles. Despite a reduction in gender bias, most of those women on television who are portrayed as working outside the home are in female-dominated jobs such as nursing and secretarial work (Kalisch and Kalisch 1984; Signorielli and Bacue 1999). Male characters are either heroes or villains, whereas females tend to be adulterers or victims (Goff, Goff, and Lehrer 1980). In general, television consistently depicts men as aggressive, independent, and in charge of events. Women are portrayed as dependent and passive. Characters such as the females in television's *Judging Amy* exist only as dramatic exceptions.

Nowhere in the mass media are gender role stereotypes more prevalent than in advertising. Gender is used to promote everything from automobiles to underwear. Despite an increasing number of gender role reversal ads, men usually appear in advertising as symbols of dominance and aggression and women are portrayed as submissive (Thomas and Treiber 2000). This depiction actually starts targeting audiences quite early in life. Advertising designed for children uses the

FEEDBACK

1. According to the ________ perspective, the division of labor based on sex has survived because it is beneficial and efficient for human living.
2. According to the ________ perspective, traditional gender definitions exist because they provide greater rewards and privileges to men than to women.
3. ________ is the theoretical perspective that attempts to explain how gender is acquired.
4. Which of the following is *not* a true statement?
   a. Boys are less assertive in class than girls.
   b. Illustrated children's books still tend to present very traditional gender definitions.
   c. Parents' perceptions of the physical characteristics of their children are heavily influenced by their children's gender.
   d. Nursery school children who fail to conform to their "appropriate" gender are met with resistance from their peers.

Answers: 1. functionalist 2. conflict 3. Symbolic interactionism 4. a

same stereotypes without sexual overtones—girls are more often cast as frivolous and helpless, and boys are more likely to play dominant and aggressive roles (Browne 1998).

Terry Fruch and Paul McGhee (1975) found that children who viewed television in excess of 25 hours per week had more traditional gender conceptions than children viewing television 10 or fewer hours a week. Similar findings have been reported among preschoolers and young adults (Gross and Jeffries-Fox 1978; Eisenstock 1984). Thus, the gender images seen on television still appear to restrict the gender definitions of viewers (Garst and Bodenhausen 1997; MacKay and Covell 1997; Eagley and Wood 1999).

Sex Stereotypes and Gender Roles

Sex Stereotypes

In Chapter 9, **stereotype** was defined as a set of ideas based on distortion, exaggeration, and oversimplification that is applied to all members of a social category. A stereotype does not take into account individual differences, so every member of the group is assumed to have the same traits, strengths, and weaknesses. The stereotype is a standard used to judge individuals. Men in American society are expected to be virile, brave, sexually aggressive, unemotional, logical, rational, mechanical, practical, dominating, independent, aggressive, confident, and competitive. Women are expected to be the opposite: weak, fearful, sexually passive, emotional, insecure, sentimental, "arty," dependent, submissive, modest, shy, and noncompetitive (Williams and Best 1990). In American society, these stereotypes reflect a belief in the innate superiority of men and the inferiority of women.

Stereotypes do not allow for a range of alternatives. One label is applied to all members of the group. Moreover, stereotypes tend to be exaggerations or extremes infrequently found in real life. For example, consider the stereotype of men offered by Marc Fasteau:

> *The male machine is a special kind of being, different from women, children, and men who don't measure up. He is functional, designed mainly for work. He is programmed to tackle jobs, override obstacles, attack problems, overcome difficulties, and always seize the offensive. He will take on any task that can be presented to him in a competitive framework. His most positive reinforcement is victory.*
>
> *He has armor plating that is virtually impregnable. His circuits are never scrambled or overrun by irrelevant personal signals. He dominates and outperforms his fellows, although without excessive flashing of lights or clashing of gears. His relationship with other male machines is one of respect but not intimacy; it is difficult for him to connect his internal circuits to those of others. In fact, his internal circuitry is something of a mystery to him and is maintained primarily by humans of the opposite sex.* (Fasteau 1975:60)

Gender Roles

Gender roles involve culturally based expectations associated with each sex. They represent an ideal, because in reality people do not always behave as expected. Although mothers are expected to be patient and loving with their children, many are not. Consequently, individuals can decide to emphasize different aspects of their gender according to their preferences and abilities. Thus, as noted earlier, there is a wide range of actual feminine and masculine behavior. Not all men try to be strong, fearless, and aggressive; not all women attempt to be sweet, submissive, and deferring.

Role Conflict and Role Strain

What female role conflicts and strains are created by gender roles? Women suffer from conflicts and strains built right into the female role. And, a changing society is creating even greater conflicts. Because married women are increasingly working outside the home (see Figure 10.1), they must balance the requirements of work with the demands of housework and child care (Burros 1993; Hochschild 1997). Married women must juggle, first,

Gender socialization promotes a sense of masculinity or femininity. Males in American society are made to feel more masculine when engaged in competitive activities.

FIGURE 10.1

Marital Status of Women in the U.S. Civilian Labor Force

These data show the relationship between marital status and employment. Write a paragraph describing the important trends you can see.

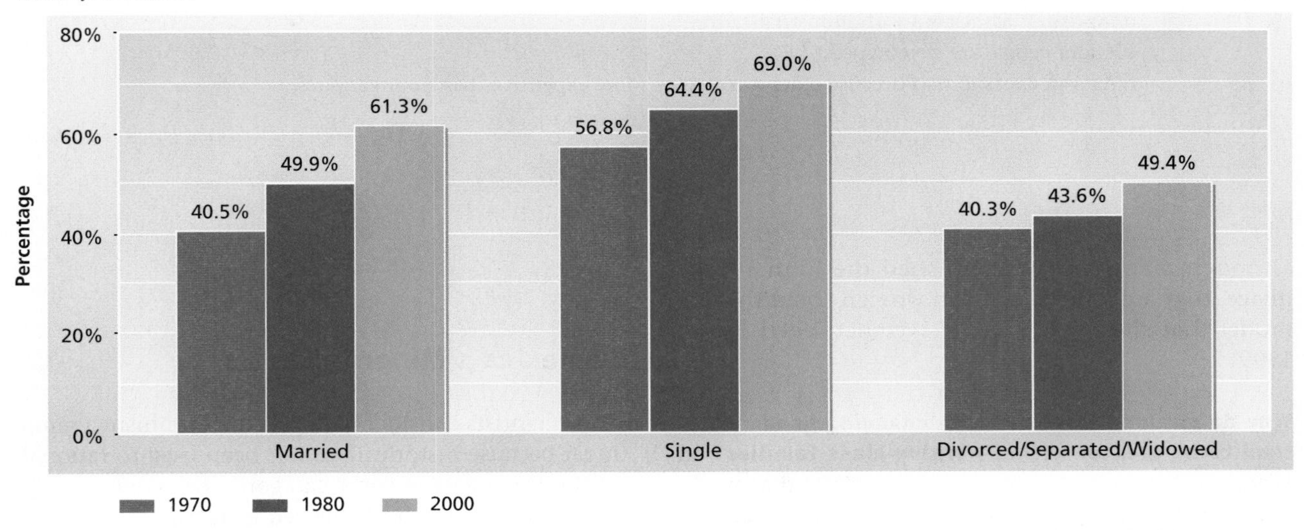

Source: U.S. Bureau of the Census, *Statistical Abstract of the United States, 2001* (Washington, DC: U.S. Government Printing Office, 2001), table 576.

the expectation that they will move from one place to another as they or their husbands climb occupationally and, second, their concern for the effects of rootlessness on their children.

While most married women now share responsibility for the breadwinner role, men, by and large, have been slow to share domestic responsibilities. Furthermore, no other institution in society—whether public or private—has stepped in to help working women. As a result, women have shouldered the responsibilities of work and family largely on their own (Reskin and Padavic 2002).

According to past research, housewives report higher rates of illness than women in the labor force (Nathanson 1980). This apparent positive health effect of women working outside the home, however, may now be offset by the increasing stress that accompanies the combined demands of work and family (Rosenfield 1989). In addition, greater stress and depression occur among women whose husbands do not support their working than among women with supportive husbands (Ulbrich 1988). Poorer health is also more likely among minority working women with husbands opposed to their employment (Andersen 2002).

What male role conflicts and strains are created by gender roles and sex stereotypes? Like women, men experience conflicts and strains within the traditional masculine role. Many working-class men, whose wives must work just to help support their families, are bothered by a feeling of having failed at their task of provider:

> *A guy should be able to support his wife and kids. But that's not the way it is these days, is it? I don't know anybody who can support a family anymore, do you? . . . Well, I guess those rich guys can, but not some ordinary Joe like me. (Rubin 1994:78)*

Men are expected to be occupationally and economically successful while simultaneously devoting time to their wives and children. Men are encouraged to admire other women sexually, but they are expected to remain faithful to their wives. The "macho man" ideal conflicts with women's preferences for warm, gentle, romantic lovers. Men also have to handle the inconsistency of being cool and unemotional at work but loving at home (Doyle 1995).

Role conflict and strain for men are intensifying in today's environment of changing gender roles. Although many women want to be liberated from male domination, they sometimes still expect men to be protective of them. This is partly because appropriate gender role behavior is currently conditioned by the situation. For example, some women want to be treated as equal to men while at work but treated in more traditional ways off the job. Other women resent any attempt by men to treat them differently from the way they would treat other men. Consequently, men are often confused about what women expect. Moreover,

FEEDBACK

1. Following are several statements about the elderly. Identify each statement with one major theoretical perspective: functionalism (F), conflict theory (C), or symbolic interactionism (S).
   ____ a. Ageism results in part from an oversupply of labor.
   ____ b. The young may wish to distance themselves from the elderly to avoid facing their own mortality.
   ____ c. The stigma attached to aging promotes a low self-concept among the elderly.
   ____ d. Ageism is associated with industrialization.
   ____ e. Older people are stereotyped.
   ____ f. Ageism exists in part because older workers are more expensive than younger ones.

Answers: 1. a. (C) b. (F) c. (S) d. (F) e. (S) f. (C)

although many men wish to shed the John Wayne image, they are less certain than women about the role model that should replace it (Wetcher 1991; Faludi 1999).

Why do strains associated with changing gender roles tend to be greater within working-class families? In *Worlds of Pain*, Lillian Rubin (1992) reports that working-class Americans advocate more traditional gender roles; yet, for economic reasons, the women are more likely to work outside the home. She also reports that working-class husbands with wives in the labor force do not participate very much in domestic tasks. This assertion has since been bolstered by other research reporting that less educated, lower-income men are less likely to share household labor when their wives work outside the home (Miller and Garrison 1982; Sweet, Bumpass, and Call 1988). In addition, Rubin concludes, working-class husbands of working women see their wives' employment as a challenge to their self-esteem and masculinity. The contradiction between traditional gender role attitudes and the sharing of the provider, household, and child-care roles among working-class couples, Rubin asserts, leads to personal strain. Rubin wrote that, among other things, this contradiction led working-class men to demand more submission and obedience from their wives and children in order to bolster their own egos and sense of masculinity.

In *Families on the Faultline*, Rubin (1994) reports that although more contemporary working-class men now believe in a woman's right to full independence, most still want women with sufficient dependency and need to make them feel manly. Nearly all the working-class men Rubin studied did some work inside the family—washing dishes, vacuuming, grocery shopping—but only 16 percent shared family work equally with their wives. Almost all of these men, however, either worked different shifts than their wives or were unemployed. The high price paid for the retention of traditional sex stereotypes and gender roles remains much higher for working-class women.

Gender Inequality

Women as a Minority Group

Most scientists consider biological determinism a moral threat because historically it has been used to rationalize the treatment of some categories of people as inferior. This view, in short, has led to racism and sexism. Minorities suffer the effects of racism; women are hurt by **sexism**: a set of beliefs, norms, and values used to justify sexual inequality. Sexist ideology—the belief that men are naturally superior to women—has been used and is still being used to justify men's leadership and power positions in the economic, social, and political spheres of society. It explains women's dependence on men as well (Clayton 1996; Benokraitis 1997; Swann, Langlois, and Gilbert 1998).

Although there is no factual basis to support the maintenance of sexism, its influence remains. Women are easily identified, biologically and culturally, and are subject to many forms of discrimination (Blau and Ehrenberg 1997; McCall 2001; *The World's Women* 2000).

But isn't sex discrimination disappearing? The answer is yes and no. Some segments of American society have more positive attitudes about women (R. M. Jackson 1998). And a few women hold positions traditionally reserved for men. In 1999, for example, Carleton Fiorina became the first female CEO of one of the thirty companies making up the Dow Jones industrial average. In that same year, Eileen Collins became the first female NASA shuttle commander. Women now head a number of top universities, including Princeton, Brown, Duke, Chicago, and the University of Michigan. The percentage of women in the Army went from 9.4 percent in 1981 to 15.5 percent in 2001. Women also comprise 18.3 percent of the Air Force and 13.3 percent of the Navy. And approximately 130 women are Army combat pilots, some of whom fought in the 2003 Iraq war (Gibbs 2003). Still, a careful examination reveals

Harriet Bradley—Men's Work, Women's Work

In a 1989 study, Harriet Bradley focused on occupational sex segregation and job sex-typing in Britain after the Industrial Revolution, around 1750. She presented case studies of various industries and occupations—food and raw material production, agriculture, fishing, mining, manufacturing, pottery, hosiery, shoemaking, shopwork, medicine, teaching. In these studies, Bradley documented the development of the division of labor along gender lines. Her database consisted of a combination of secondary sources in the form of previously published research, along with new material she collected from primary sources. This study is an excellent example of the integration of sociology and history.

According to Bradley, pervasive sex-typing of jobs—the allocation of certain work-related activities to men and others to women—results in men and women rarely being allocated the same types of jobs. Even when men and women are working side by side, Bradley points out, they are doing different things. This allocation of certain work to men and certain work to women, she argues, is based on a simple principle: men occupy jobs that place them in control, and women do work that requires obedience. In the fields, men cut and women gather; in the factory, men stamp out parts and women sew them together; in the office, men handle accounts and women do the typing and filing.

The social definition of what is "women's work" and what is "men's work" may vary from time to time and place to place. Jobs thought of as the province of women may have historically been assigned to men, or vice versa. Prior to the industrialization of the cotton industry, for example, men were weavers, while women did the spinning. These roles were flip-flopped when power-driven machinery replaced hand labor. What does not vary, Bradley asserts, is the assignment of jobs based on the principle of men as superordinate and women as subordinate. Weavers had higher status prior to industrialization and lower status afterward.

Although occupational sex segregation and job sex-typing existed in preindustrial society, Bradley contends that capitalism dramatically increased both. From the case studies, she concludes that the tremendous expansion of capitalist production in the 1880s and 1890s created the foundation for the current patterns of gender job segregation and job sex-typing.

During the capitalist expansion, according to Bradley, hiring women in jobs labeled as "women's work" served the interests of capitalists as well as male workers. Capitalists could increase their profits by hiring women for lower wages because of the low social standing of female labor. In addition, capitalists used women as a threat in their battle to shatter the control male workers held (as a result of their expertise in preindustrial craft production techniques). As men fought back by demanding that certain jobs remain "men's work," occupational sex segregation deepened; and women were forced into low-paying "female" jobs. Males and females, Bradley concludes, continued to be steered into socially appropriate lines of work. This played into their desire to assert unique sexual identities. Gender-based ideologies emphasize sex differences. She quotes from C. Cockburn:

> *What a man "is," what a woman "is," what is right and proper, what is possible and impossible, what should be hoped and what should be feared. The hegemonic ideology of masculinism involves a definition of men and women as different, contrasted, complementary and unequal. It is powerful and it deforms both men and women.*
> *(C. Cockburn 1986:85)*

Thinking About the Research

1. Which theoretical framework do you think Bradley used? Explain your choice.
2. Do you think that the current changes in gender relations in the United States will alter the state of occupational sex segregation described by Bradley? Support your conclusion.

many gaps in social rights, privileges, and rewards for women in the United States (Valian 1999). These gaps, although they have closed somewhat in recent years, are reflected in the continuing occupational, economic, legal, and political inequality experienced by American women (Bianchi and Spain 1996; Riley 1997). For example, Southern Baptists recently voted to ban female ministers (Rosin 2000).

Occupational and Economic Inequality

By far the most important U.S. labor market development over the last thirty years has been the dramatic increase in the number and proportion of women in the workforce (U.S. Bureau of the Census 2002b). From 1972 through 1989, more than a million additional women entered the labor force each year. In 1960, 23 million women worked outside the home for pay; by 2000, the figure had reached nearly 66 million. In 2000, 60 percent of women worked outside the home, compared to 75 percent of men (see Figure 10.2). That same year, women represented approximately 47 percent of the U.S. labor force (see Figure 10.3).

The increase in labor force participation among women twenty-five to fifty-four years of age has been particularly steep. For example, the labor force participation rate of married women with children under six years of age has skyrocketed from 19 percent in 1960 to 37 percent in 1975 and reached about 64 percent in 2000 (U.S. Bureau of the Census 2002b).

Has increased labor force participation eliminated occupational inequality? Despite their increased participation in the labor force, women are still concentrated in different occupations than men. This is known as **occupational sex segregation** (Reskin 1993; Farley 1995; Yamagata et al. 1997; Perlmann and Margo 2001; Tomaskovic-Devey and Skaggs 2002). Women are underrepresented in high-status occupations and overrepresented in low-status occupations. To begin with, females account for only about 10 percent of engineer positions and about 30 percent of attorney jobs. By contrast, women occupy nearly all of the pink-collar jobs—secretaries, clerks, stenographers—whose purpose is to support those higher up the occupational ladder (U.S. Bureau of the Census 2002b). Moreover, when women are in high-status occupational groups, they are concentrated in lower-prestige, lower-paid jobs. Even within female-dominated occupations, men fill a disproportionate share of higher positions. Occupational sex segregation, then, produces a dual labor market—a split of the economy into a core segment of preferred workers and a peripheral segment of the marginalized workers.

Have women and men reached financial equality? Earnings inequality has followed a pattern since the 1960s. After increasing in the late 1960s, earnings inequality stabilized for most of the 1970s, decreased

WORLDVIEW 10.1

Women in the Workplace

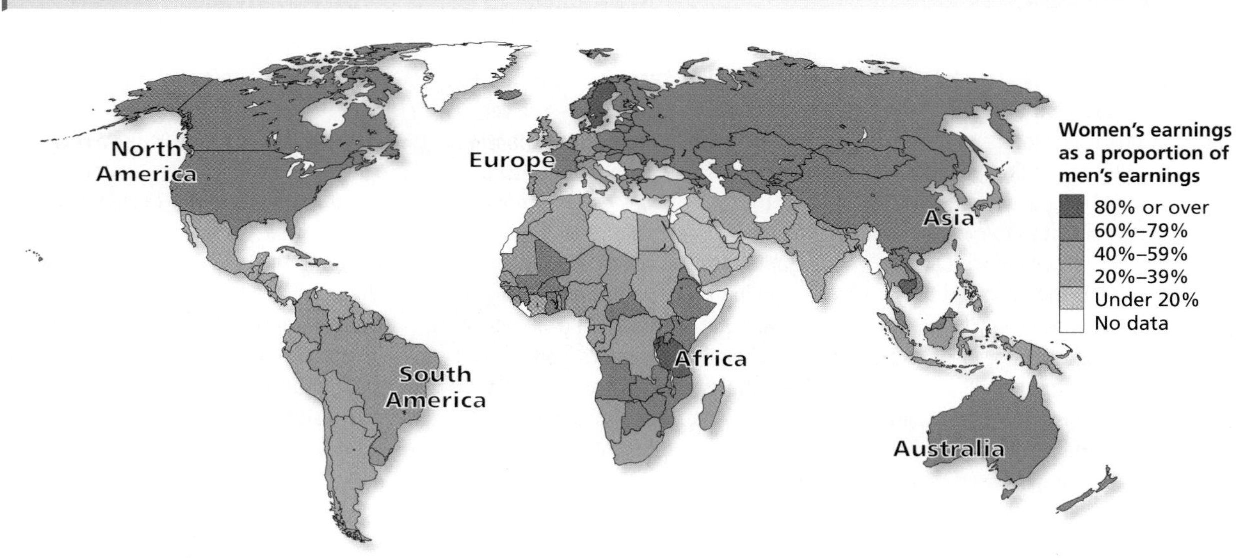

Source: *Human Development Report 1998*, United Nations Development Report (New York: Oxford University Press, 1998).

In nearly all countries of the world, the higher-paying jobs and better occupational opportunities still go overwhelmingly to men. This map shows women's earnings as a proportion of men's earnings.

1. Countries with the highest gender income equality include highly developed countries like the United States and France as well as relatively underdeveloped countries like Angola and Kenya. Can you think of some reasons for this?
2. Do you see any other patterns?

FIGURE 10.2

U.S. Labor Force Participation Rates by Sex: 1890–2000

This figure tracks changes in the percentage of women and men in the U.S. labor force since 1890. The participation gap between the sexes has changed radically. Describe this shift using specific percentages.

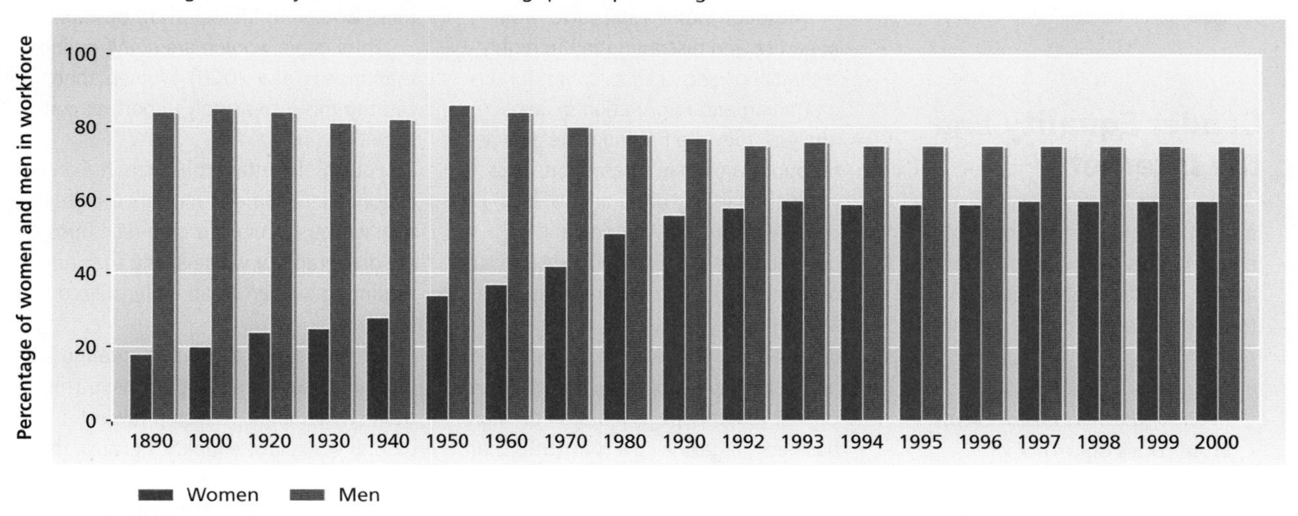

Source: U.S. Department of Labor; Bureau of Labor Statistics, *Employment in Perspective: Women in the Labor Force*, No. 865 (Washington DC: U.S. Government Printing Office, 1993), p. 1; U.S. Department of Labor, Bureau of Labor Statistics, *Employment and Earnings*, (Washington, DC: U.S. Government Printing Office, 1997), p. 6; and U.S. Bureau of the Census, *Statistical Abstract of the United States, 2001* (Washington, DC: U.S. Government Printing Office, 2001), p. 367.

FIGURE 10.3

Composition of the U.S. Labor Force by Sex, 1870–2008

As this figure demonstrates, the male-female composition of the U.S. labor force has steadily moved toward parity. The female percentage of the labor force has moved from less than 15 percent in 1870 to just over 46 percent in 2000. What do you think is the most important social consequence of this change?

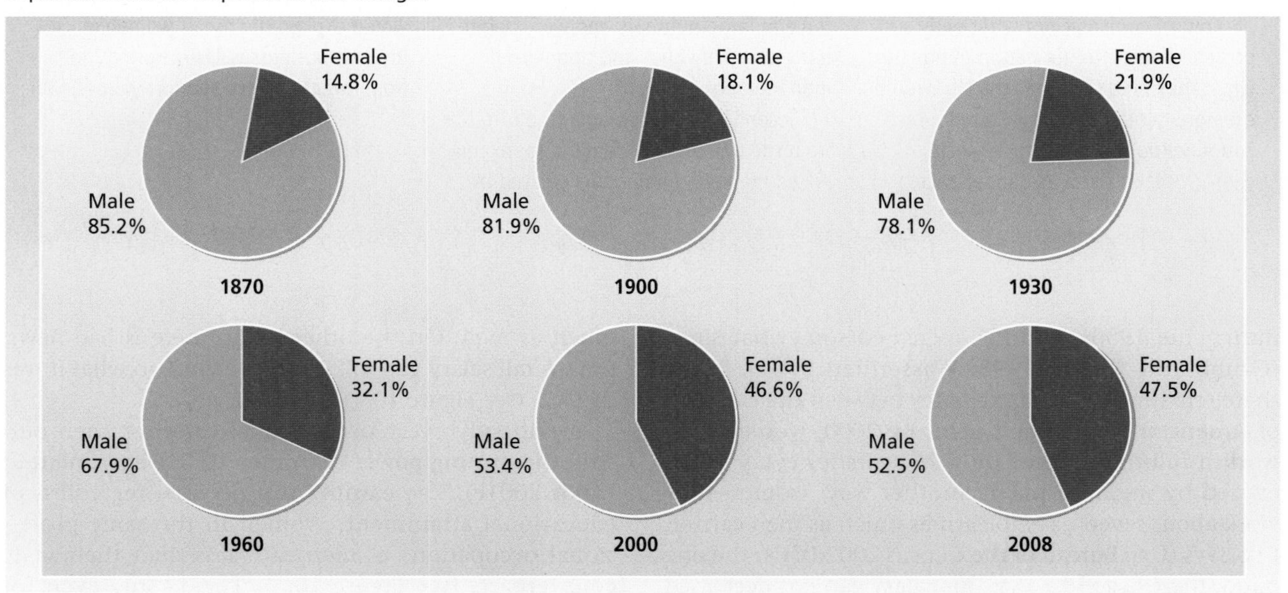

Source: U.S. Bureau of the Census, *Statistical Abstract of the United States: 2001* (Washington, DC: U.S. Government Printing Office, 2001), pp. 367–369.

Gender Equality and the Internet

Internet communication does not require full disclosure of social characteristics. This has a downside. Take, for example, a marriage made on the Internet:

> *In one of the more bizarre events in America's experience with cyberspace, a Virginia woman met a man through the Internet, and after several dates and visits decided to get married. Only later did the Virginia woman discover that she had actually married another woman who through various ruses had tricked her into believing that she was a he. Consequently she is suing the former "husband" for a variety of harms.* (J. E. Katz and Aspden 1997:81)

This, of course, is not what early supporters of the Internet had in mind when they predicted that the Net would promote gender equality. Communication without knowledge of social characteristics (such as gender, race, ethnicity, and age), they asserted, could reduce bias. While this evaluation of ideas on merit alone may contribute to less discrimination, gender equality has been advanced more by unanticipated social forces.

Although women still suffer from prejudice and discrimination in highly technical careers such as computer programming and information systems analysis, they are finding more and better opportunities in Internet business fields than in any other area of business. Women have founded and become chief executive officers (CEOs) of many high-tech companies, such as Marimba, Oxygen Media, iVillage, DoubleClick, Women.com, togglethis, and The Mining Company. Kim Polese, CEO of Marimba, was featured on more business magazine covers in 1998 than Bill Gates, founder and CEO of Microsoft (K. E. Hoffman 1998).

These unanticipated opportunities for women are due to several factors. First, the Internet has gained economic importance so rapidly that an "old boys'" network does not exist. Whereas most American industries developed when women were expected to stay at home, the system that would become the Internet was started only thirty years ago and did not become popular until the late 1990s. By that time, women had already entered the workforce in large numbers and were beginning to occupy mid- and upper-level management positions.

Second, women are able to profit from the tremendous demand for experienced marketing managers created by the Internet. Because women are responsible for some 85 percent of purchasing decisions in non-Internet businesses, they have the experience to move into marketing management positions. Internet companies have turned to these women to fill important positions.

Third, more women are going online than men (Weise 2000). Women, then, can tap this increasingly important consumer group.

Fourth, the Internet has created an astronomical demand for skilled high-tech workers. American high-tech firms are desperate for workers, and they are turning to women as an underutilized resource.

Of course, not all women entering Internet businesses escape sexism. This trend, nonetheless, is a step toward greater workplace equality. Because the Internet has rapidly become such a large part of the U.S. economy and it will only continue to grow, the information age holds considerable promise for gender equality (K. E. Hoffman 1998).

Analyzing the Trends

According to the functionalist perspective, a sex-based division of labor involved some benefits to society. The conflict perspective stresses unequal access to the resources that are needed to succeed in the workplace. Choose one of these perspectives and discuss how it would explain the rise of women in Internet businesses. Use information from this chapter to support your answer.

during the 1980s, and has increased somewhat since (Grubb and Wilson 1992; Wasserman 2001). Thus, there remains a wide discrepancy between the earnings of American women and men. By 2000, women who worked full-time earned only 76 cents for every dollar earned by men. To put it another way, women now work about seven days to earn as much as men earn in five days (U.S. Bureau of the Census 2002d). On the one hand, there is good news: this salary gap has decreased since 1980, when women were earning 60 percent as much as men. On the other hand, there is bad news: the overall salary gap in 2000 is too close to what it was in 1955 (see Figure 10.4).

In virtually every occupational category, men outstrip the earning power of women (U.S. Department of Labor 2001c). The earning gap persists, regardless of educational attainment. Women in the same professional occupations as men earn less than their male counterparts (see Figure 10.5). This is true even for women who have pursued careers on a full-time basis

FIGURE 10.4

Women's Median Annual Earnings as a Percentage of Men's Among Full-Time Workers: 1955–2000

This figure traces the ratio of women's to men's earnings since 1955. Discuss the two most important conclusions you can make from these data.

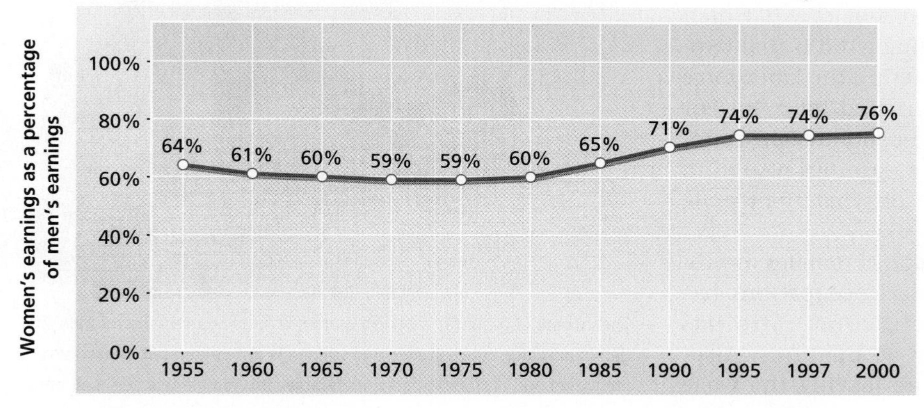

Sources: *The Wage Gap: Women's and Men's Earnings*, Institute for Women's Policy Research (Washington, DC: U.S. Government Printing Office, 1993), p.1; U.S. Department of Labor, Bureau of Labor Statistics, *Employment in Perspective: Women in the Labor Force*, No. 865 (Washington, DC: U.S. Government Printing Office, 1993), p.1; U.S. Bureau of the Census, *Money Income in the United States: 1995* (Washington, DC: U.S. Government Printing Office, 1996), pp. 28–29; and U.S. Bureau of the Census, *Statistical Abstract of the United States: 1998* (Washington, DC: U.S. Government Printing Office, 1998), p. 436; and U.S. Department of Labor, Bureau of Labor Statistics, *Highlights of Women's Earnings in 2000*, No. 952 (Washington DC: U.S. Government Printing Office, 2001), p. 3.

INTERNET LINK For additional information on employment and earnings of women and men in the United States, you can visit the Web site of the Department of Labor, Bureau of Labor Statistics, at http://stats.bls.gov/

FIGURE 10.5

Female-to-Male Earnings: 2001

On average, women in the United States earn about 76 cents for every dollar a man earns. In what way do the data in this figure support the contention that gender inequality is real?

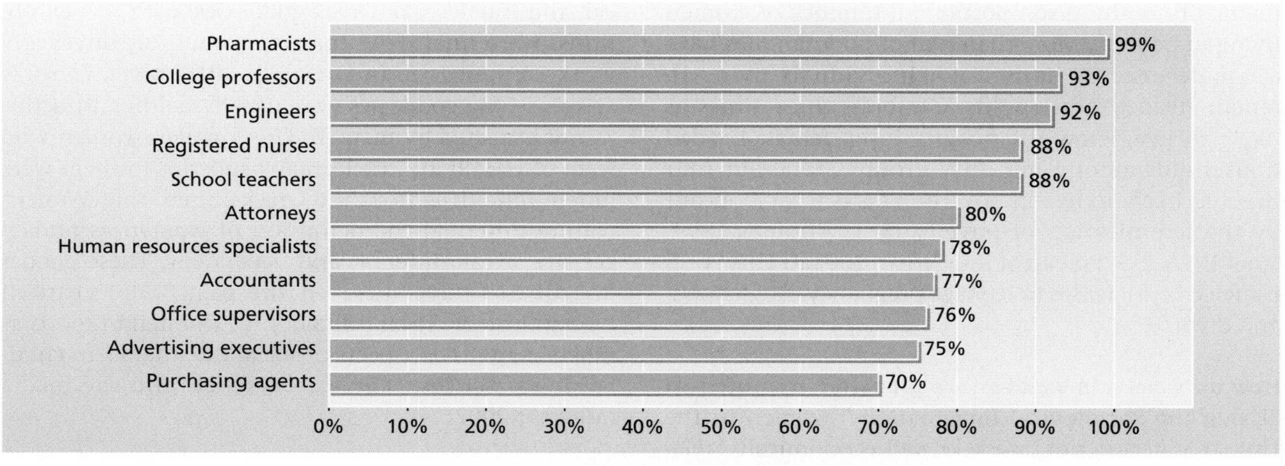

Women's Earnings as a Percentage of Men's Earnings

Source: "The 2001 Salary Survey," *Working Woman* (July/August 2001), pp. 44–47.

for all of their adult lives. Furthermore, males in female-dominated occupations typically earn more than women (Stoltenberg and McCrum 1986; C. L. Williams 1989; Stewart 1999).

Is there hope of closing the salary gap? Although the salary gap will remain indefinitely, there are a few signs of hope. Women now earn 76 percent of what men earn. Probably the most encouraging word is that two-thirds of the women currently entering the labor force are taking jobs historically dominated by men. For instance, as more women became computer programmers over the past few years, their earnings rose from 70 percent to almost 85 percent of what their male counterparts earned (Stewart 1999).

More generally, as the percentage of females in professional, managerial, and technical occupations has increased, the female-to-male income ratio has improved in a number of these occupations. Also, increasing numbers of women are leaving the wage system behind as they become entrepreneurs. Excluding large corporations, women now own about one-fourth of all businesses. However, females are still more likely than males to hold jobs that pay relatively low wages; and as you have seen, the differential in female-to-male earnings persists within occupational categories.

Why does the salary gap persist? Research on wage differences has shown that even taking into account all possible factors that might reasonably account for wage differences between women and men—for example, age, experience, amount of absenteeism, and educational background—a large difference remains. With the elimination of all other explanations, the conclusion reached from these studies is that sex discrimination is the major source of economic differences.

One of the most serious indicators of economic inequality is the disproportionate number of women living in poverty. More than half of poor families have a female head of household. Individuals living in female-headed households are about three times as likely to live below the poverty level; female-headed households among minority groups are about four times as likely to live in poverty. This is what is meant by the feminization of poverty (S. L. Thomas 1994; Sidel 1996; U.S. Bureau of the Census 2001a). This trend is fueled by increases in teenage pregnancy, singlehood, and divorce.

How do American women fare globally? As noted in "Using the Sociological Imagination," women in the United States do not fare very well economically, relative to women in other developed countries. Here, of course, we are talking about relative earning power, not absolute dollar amounts. American women in nonagricultural employment earn only 76 percent of that earned by men, compared to 80 percent in Germany and 88 percent in Sweden (see Figure 10.6). (Van Der Lippe and Van Der Lippe 2001; *The World's Women* 2000).

© Bob Daemmrich/Stock, Boston

The income of American working women is three-quarters that of men. The gap in earnings between men and women is not obliterated even when education and occupation are comparable. This female teacher in all probability is paid less than her male counterparts.

Sexism in Sports

Women experience sexism in athletics just as they do in other aspects of their lives. The source of this sexism has roots in an ancient Greek culture, whose gods had masculine characteristics. Greek gods were depicted as athletic, strong, powerful, competitive, rational, physical, and intellectual. Greek goddesses, with few exceptions, were unathletic, passive, beautiful, physically weak, supportive, and sexually attractive. The few active, strong goddesses were usually neither attractive to nor attracted by men. To Greek males, women who were physically or intellectually superior to them were unfeminine. It is from the Greeks, then, that Western culture inherited the definition of women as nonaggressive, weak, inferior, and dependent. These gender definitions have survived the past 2,500 years of human history. Their influence is exhibited in sports as much as in other aspects of social life, thanks to child-rearing practices, schools, churches, and the media (Messner 2002).

What are some consequences of sexism for females in sports? One aspect of sexism exists in the negative

FIGURE 10.6

Women's Wages Compared with Men's Wages in Nonagricultural Employment

This figure reveals where the United States stands worldwide in wage parity. Explain this fact based on information in this chapter.

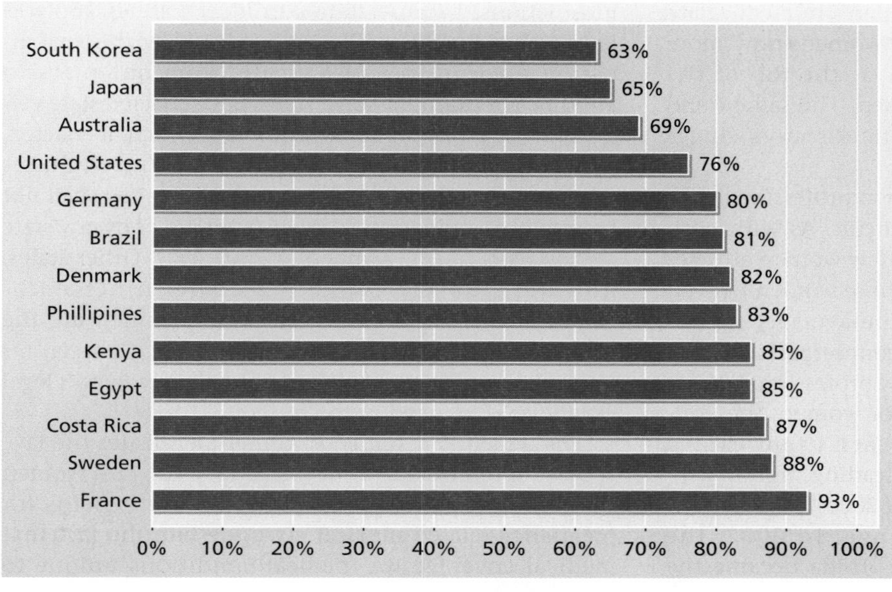

Source: International Labour Office, Bureau of Statistics, http://laborsta.ilo.org/ (accessed 18 October 2003).

stereotypes of female athletes. The idea that athletics makes females masculine has discouraged many females from participating in sports and has tyrannized many of those who have participated. Another deterrent to female sports activity is the long-standing argument that such activity will harm a woman's health, particularly as it relates to a woman's reproductive system.

Sexism also denies females equal participation in organized sports. It was not until the mid-1970s that the national Little League organization, under legal threat, rescinded its males-only policy. Resistance to females at the local level continues. Only in 1972 did the Educational Amendment Act require schools and colleges (those that receive federal funds) to provide both sexes equal access to sports. As late as 1984, the U.S. Supreme Court ruled that Title IX—the provision for equal access included as part of the 1964 Civil Rights Act—applies only to "specific" programs or departments receiving federal funds. This is a departure from the original interpretation that an institution receiving federal funds has to provide equal opportunity to females in "all" of its sports programs, and it opens the door to decisions that could undermine the progress

The Western cultural tradition defines women as weak, dependent, and nonaggressive. Tell that to tennis superstar Venus Williams.

made since 1972. Supporters and critics of Title IX now vigorously debate its consequences for men's sports (Orecklin 2003).

Women are still denied equal access to the power structure of sports. In fact, although Title IX increased equality for female athletes, it led to a decrease in the number of coaching and administrative positions held by women. In the early 1970s, women's intercollegiate teams were led almost entirely by women; now more than one-half of women's teams at the top of the NCAA's structure are coached by men. The same trend is appearing among female athletic directors (Suggs 2000).

Nowadays, women have their own professional soccer league, golf tour, and tennis circuit. As with both white and minority men, relatively few women athletes make it to the professional ranks. Those women who do become professionals make, on the average, significantly less money than their male counterparts. In golf, for example, which is one of the few professional sports offering significant opportunities for women, the leading money winner on the 2003 men's tour earned almost four times as much as the leading money winner on the women's tour (*Golfworld* 2003).

There are some signs of change. In 2002, the University of New Mexico's Katie Hnida became the first female to play in a Division 1-A football game. In 2003, Annika Sorenstam became the first woman to play on the men's golf tour in fifty-eight years. In addition to her Nike commercials, U.S. soccer star Mia Hamm has a lucrative deal with Gatorade. Chamique Holdsclaw, an extremely talented professional basketball player for the Washington Mystics, obtained an unheard of (for women athletes) five-year contract with Nike, plus her own signature Holdsclaw shoe (Hammel and Mulrine 1999). In 2000, tennis champion Vensus Williams obtained a $40 million contract with Nike.

Periodically, blatant reminders of the pervasiveness and persistence of sexism in sports are made public, as in the popular press coverage of Augusta National Golf Club's 2002 announcement to keep its membership all male. A strong commitment to the club's traditional gender segregation is reflected in its willingness to forfeit corporate sponsors of its annual Masters golf tournament at the cost of some $7 million in foregone revenues (McCarthy and Brady 2002; Saporito 2002).

So, despite significant gains, gender equality in sports is not a reality (Nakamura 1997). Huge gaps persist between men's and women's athletic programs (Brady 2002; see Figure 10.7).

Legal and Political Inequality

Sexism is similarly built into America's legal and political institutions. National, state, and local legal codes reflect sexual bias; and important differences exist between women and men in positions of political power.

Is sexism built into the law? According to the U.S. Civil Rights Commission, more than 800 sections of the U.S. legal code are sexually biased. Legal inequities and complexities are compounded by overlapping state and local jurisdictions. Because state and local statutes apply as long as they do not conflict with federal law, the legal situation for women varies greatly from one place to another. For example, some states have enacted state versions of the Equal Rights Amendment, which, if enacted, would enter this language into the United States Constitution: "Equality of rights under the law shall not be denied or abridged by the United States or by any state on the basis of sex" (Andersen 2002:292). Other states, however, have retained archaic nineteenth-century laws based on the idea that women are the property of either a husband or a father. Georgia, for example, has an 1863 law that "defines a woman's legal existence as 'merged in the husband'."

The dependent status of women permeates the law. A U.S. Supreme Court decision refused to grant women the legal guarantee of health insurance benefits for pregnancy-related medical costs, despite the fact that medical coverage for medical conditions unique to men—such as prostate problems and vasectomies—was routinely provided. At the state level, some states have traditionally refused women the right to keep their own surnames after marriage. Other states have had "protective legislation" restricting women's rights. Supporters of these laws viewed them as safeguards against the abuse and exploitation of women. However, the end result was that women were denied certain jobs, many of which were better paid than more traditional occupations for women (Andersen 2002). Such protective legislation limited the number of hours women could work, the conditions under which they could work (such as barring women from toxic areas because of potential birth defects in their children), and the kinds of work they could do (such as setting a 30-pound limit on the amount of weight a woman could lift).

Due to the passage of Title VII of the Civil Rights Act of 1964, such laws have been preempted; but despite the prohibition of such laws, the practices still linger. Moreover, the Family and Medical Leave Act of 1993, which requires that employees be given up to twelve weeks without pay for childbirth, adoption, serious illness, and so forth, inadvertently affects women negatively. Because women are more likely than men to take parental leave, this legislation gives employers another reason to hire men rather than women.

There are gender differences in criminal law as well. Certain crimes are typically associated with one gender or the other. For example, laws against prostitution are

FIGURE 10.7

Persistent Gender Inequality in College Sports

Examine these data on college sports gender equality in relation to recruiting expenses, scholarships, and coaches' salaries. Is the requirement of gender equality in Title IX of the 1965 Civil Rights Bill being fulfilled? Use the data to support your conclusion.

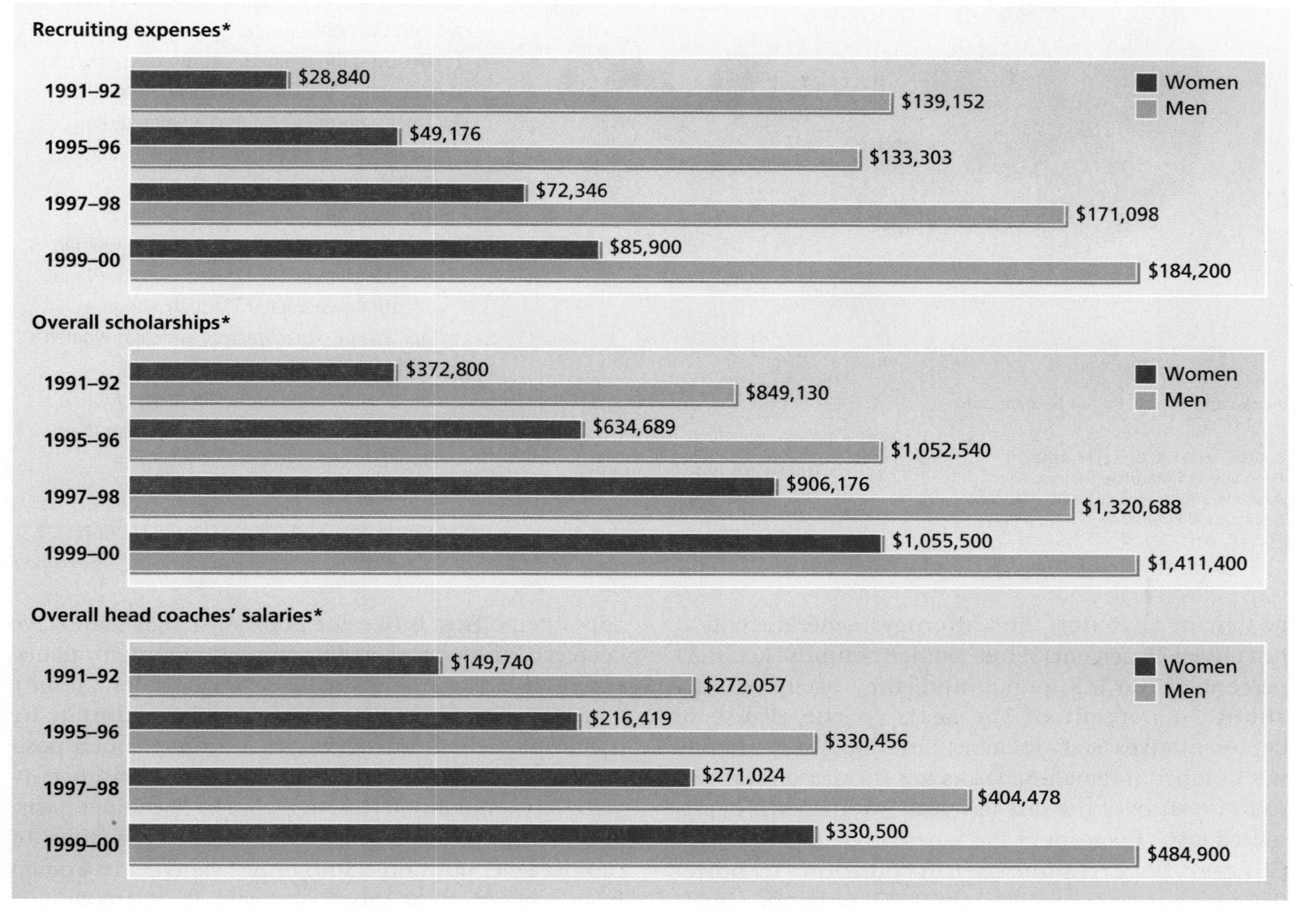

* Average per institution
Source: National Collegiate Athletic Association.

generally enforced only against women, while male customers go free. On the other hand, sexual relations with a minor female, even though she consents, may result in prosecution of the man for statutory rape.

Rape has been one of the most controversial legal areas. Nowhere is sexism more explicitly built into the language and application of the law. For example, no other crime besides rape requires independent corroboration that a crime has been committed. There are two assumptions of many rape laws: women will lie about the occurrence of rape; and unless evidence of rape is produced, no prosecution can occur. For women raped without witnesses (presumably the majority of cases), or for women who submit to rape due to the threat of violence, there is often no legal protection. Furthermore, when cases actually do come to trial, the victim is often treated more harshly than the defendant. Many observers have described rape trials as trials of the victims, with evidence of the victims' past sexual activity used to undermine their testimony and acquit the accused. Even convicted rapists have been allowed to go free on the grounds of provocation. A notorious case involved a Wisconsin judge who refused to send a convicted rapist to prison on the grounds that the way women dress today causes men to lose control.

How does sexism relate to political power? One possible explanation for legal bias is the underrepresentation of women in the lawmaking institutions. In cities of over 30,000 people, only 20 percent had female mayors in 2002. In the same year, less than a fourth of state legislators were women. The number of female governors,

SNAPSHOT OF AMERICA 10.1

Women's Right to Vote Before 1920

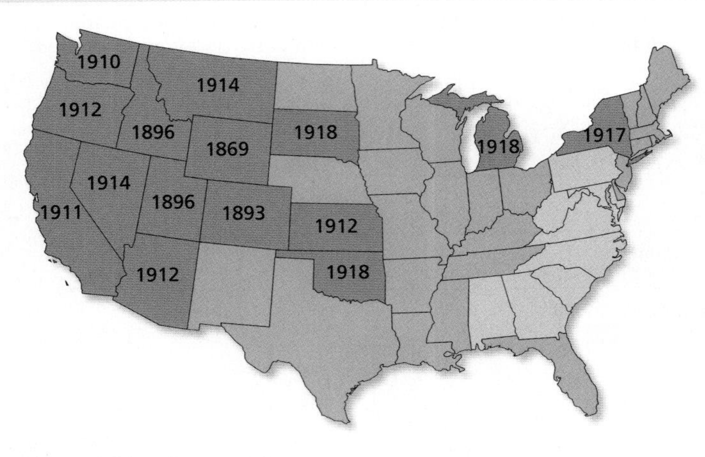

Alaska and Hawaii did not become states until 1959.

Source: James R. Giese, Matthew T. Downey, and Mauricio Mazon, *The American Century* (Cincinnati, OH: West Educational Publishing, 1999), p. 377.

Full suffrage and year
Partial suffrage (right to vote varied by locality within a state)
No suffrage

This map shows a state-by-state picture of female suffrage prior to 1920. In 1920, the states ratified the Nineteenth Amendment to the U.S. Constitution, which granted all U.S. women the right to vote.

1. Describe the regional pattern of voting for women in the United States before 1920.
2. Does it surprise you that women did not have the full legal right to vote in so many states prior to 1920? Explain this situation sociologically.
3. What was the voting situation for women in your state prior to passage of the Nineteenth Amendment? Through research, analyze the social factors affecting women's suffrage in your state before 1920.
4. Research the general response in your state to the passage of the Nineteenth Amendment. Analyze this response sociologically.

lieutenant governors, and attorneys general stood at just above 25 percent. While women comprise about 51 percent of the U.S. population, they occupied only about 14 percent of the seats in the House of Representatives (sixty members) in 2003. And although the number of female Senators has increased from two to fourteen over the last ten years, women still represented just 14 percent of the Senate in 2003. Women in Congress have seldom risen to positions of power (Foerstel and Foerstel 1996). In 2002, only one female was chairing a House or Senate standing committees, and none were ranking members of these committees (Center for the American Woman and Politics 2002).

Over the last two decades, a few women have reached the national political stature previously held exclusively by men. In 1988, Geraldine Ferraro became the first female vice-presidential candidate in the history of the United States; Madeleine Albright was named the first female secretary of state in 1996; Elizabeth Dole ran for her party's presidential nomination in 2000. In 2002, Democrat Nancy Pelosi became the first female House minority whip and was subsequently elected House minority leader, the highest-ranking position a female has ever held in Congress. At the same time, as the newly named Chair of the House Republican Conference, Deborah Pryce became the highest-ranking Republican woman ever to serve in Congress.

Female exclusion from political power is not restricted to elective offices. The record for women in appointed offices is likewise poor. Although there have been recent increases in the number of appointments, the total is extremely small. When President Jimmy Carter appointed four women to his cabinet, he quadrupled the number ever to have held such positions at one time. President Bill Clinton, almost twenty years later, appointed thirteen women to cabinet posts, and George W. Bush has appointed five women to cabinet-level positions. Still, only twenty-nine women have ever served as cabinet officers. President Ronald Reagan appointed the first woman Supreme Court Justice, Sandra Day O'Connor, in 1981; and President Clinton elevated Ruth Bader Ginsburg to the high court in 1993. Only a small percentage of federal judges are women. Even though women comprise over 40 percent of the federal workforce, they have only recently begun to gain access to higher pay grades (S. Barr 1999). About 40 percent of women are employed in midlevel grades. Women hold less than one-third of the supervisor and manager positions and account for only one-fifth of senior executives.

The political position of women in the United States is among the worst in the Western world. With some notable exceptions, Western European nations have much greater female political participation. In the Scandinavian countries, for example, up to 20 percent of parliament members are women. The 1979 election of Margaret Thatcher as England's prime minister marked a significant breakdown in that country's barrier to women for the country's highest public office. In

FEEDBACK

1. __________ is a set of beliefs, norms, and values used to justify sexual inequality.
2. On the average, women earn about __________ of what men earn.
   a. three-fourths
   b. one-half
   c. two-thirds
   d. 95 percent
3. Despite Title IX, females are still not represented more equally in the collegiate sports power structure. T or F?
4. Although women are discriminated against in civil law, they are not at a disadvantage in criminal law. T or F?
5. The exclusion of women from political power applies mainly to elective offices. T or F?

Answers: 1. Sexism 2. a 3. T 4. F 5. F

the United States, on the other hand, the election of a woman president has yet to become a reality.

On the positive side, 91 percent of men and 93 percent of women say they would vote for a woman as their political party's nominee. When Gallup first asked this question in 1937, only 33 percent of Americans said they would vote for a female presidential candidate. On the downside, almost half of the American public believes that a man would be a better president than would a woman (W. W. Simmons 2001).

Richard Zweigenhaft and William Domhoff (1999) do point out that women are now part of the power elite. The power elite is no longer the exclusively male enclave it used to be. Still, women are seriously underrepresented. And most of those women who do make it come predominantly from upper-class backgrounds.

Changing Gender Roles

Gender Roles and Social Change

What are the possibilities for change in gender roles? What are the consequences of change? One way to begin answering these questions is to look at the factors discouraging and encouraging gender role changes.

Why is there resistance to change? First, although men may not be aware that their suppression of women is to their own advantage, like all dominant groups, they can see little reason to change the way things have always been. Second, even if men were willing to relinquish their dominant position, it would be costly to them to do so. Third, some women resist giving up some advantages they see for themselves in the traditional gender role arrangement. Some fear losing the option to be full-time homemakers and do not want to relinquish their husbands' legal obligation to provide for them and their children. Others do not wish to lose conventional male courtesies and protection. Finally, an ideology still exists to support members of both sexes who wish to maintain the traditional gender roles.

Resistance to gender role change can hardly be surprising to anyone who understands the durability of deeply entrenched culture (Risman 1998). Traditional gender roles are embedded in the long-standing cultural myth of male superiority. Again, the idea of male superiority, revealed as a myth today, actually began as a myth in the first writings of the ancient Greeks sometime in the seventh or eighth century B.C. In his account of the birth of the Greek gods and the social structure of daily life, the poet Hesiod, who was writing before the Greeks used prose, chronicled the creation of the myth of male superiority (Hesiod 700 B.C./1983). Zeus, the king of gods, decided to act harshly after another god, his son Prometheus, went too far in his constant deviousness by giving fire to man. In addition to requiring hard work for subsistence living, the myth goes, Zeus created woman as the carrier of all the evils that she eventually unleashed on men. Even in rural Greece today, peasants only reluctantly acknowledge that they have a female child, and it is not uncommon for a man to say that he has three children (sons) and one girl (Hesiod 1983:97–98). Considered one of the more moderate Islamic nations, Jordan remains a patriarchy. Daughters are killed or abused by their fathers if they "dishonor" the family. Jordanian wives can renew their passports only with their husbands' permission (Khouri 2003).

It was, of course, from this myth, as well as similar ideas from many early cultures around the world (O'Brien and Palmer 1993), that male supremacy was founded. This supremacy underlies the establishment of **patriarchy**—a hierarchical social structure in which women are dominated by men (see Chapter 12).

What are some of the factors promoting change? There are a number of agents promoting change in gender roles. Demographic trends—longer life expectancy, smaller families, higher education, increased numbers of jobs requiring skilled labor—have had a profound

SOCIOLOGY AND THE NEWS MEDIA

The Evils of Dowry Payment

This CNN report provides a glimpse into an ugly manifestation of gender inequality around the world. Until 1980, it was legal in Bangladesh to charge the fathers of brides a dowry as part of arranged marriages. Though now illegal, the practice continues, with often devastating consequences for young women and their children.

1. Identify some of the consequences of the dowry system for Bangladeshi women.

   ____

   ____

2. Discuss the dowry system within the context of sexism.

   ____

   ____

Imagine the emotions of this woman. Her parents arranged her marriage and her father was charged for the privilege.

impact on everyone's lives. For women, these trends have led to their increased participation in activities outside the home—in school, work, politics, and voluntary organizations. In 2000, 62 percent of all married women held jobs outside the home. Another source of change in gender roles is the feminist movement. Although the goals and objectives of the feminist movement appear radical to many people, its central ideas are not new.

The Feminist Movement

What is feminism? **Feminism** is a social movement aimed at the achievement of sexual equality—socially, legally, politically, and economically. Recent feminism is best viewed as the reemergence of a long historical struggle that began as far back as 1792 with the publication in England of *A Vindication of the Rights of Women*, by Mary Wollstonecraft. The movement picked up tempo in the United States in 1848, when a group of women abolitionists drew up the *Declaration of Sentiments*, which demanded equal rights for women. By the end of the nineteenth century, feminism was a mass movement (Kraditor 1971; S. M. Evans 1989; West and Blumberg 1990; Tobias 1997; Andersen 2002).

The women's movement almost perished in 1920 when, with the passage of the Nineteenth Amendment, women gained the right to vote. However, many of the problems that had prompted feminism in the first place—the economic dependence of women and the restriction of their full participation in work and politics—remained. These remaining problems, coupled with other developments, such as effective birth control technology and the growth of certain sectors of the labor market (clerical, sales, technical), kept the women's movement alive (V. Taylor 1989).

During the 1960s, what appeared to be a new social movement, the women's movement, gained national attention through its efforts for gender role and societal changes. Beginning with a few isolated voices, such as the publication of Betty Friedan's (1963) enormously influential book *The Feminine Mystique*, the women's movement gained momentum. By the 1970s, it was a mass movement with widespread support (S. M. Evans 1980).

Advocates of feminism offer several basic rationales for the movement. One, male-oriented social, legal, political, and economic structures have inhibited women in their development as complete human beings. Two, the results of early socialization, as well as some blatant forms of discrimination, have restricted women's options and personal growth. (More recently, it has been recognized that men are also the victims of limited options based on gender roles, although with different outcomes.) Three, problems created by sexism are not created by individuals and cannot be solved by individuals. Like racism, the problem of sexism is embedded in the entire structure of society, and therefore its eradication requires changes in social structures as well as in individuals. To obtain such change, group action is required (Ferree and Hess 1994; Tobias 1997).

Gender Roles in the Future

Can we predict the future of gender roles? Making predictions is risky. For one thing, an economic recession

© Ilene Perlman/Stock, Boston

These Arab women live in a patriarchal society, a hierarchical social structure in which women are dominated by men.

or depression could seriously curtail the occupational progress of women, who could easily fall victim once again to the "last hired, first fired" policy. For another, people's attitudes toward gender roles are inconsistent and unstable. For example, although polls show that most Americans favor equal pay for equal work, there is far less agreement on whether or not sex bias actually exists in American society. Consequently, there is disagreement as to whether special steps should be taken—such as affirmative action programs and quotas. Because some Americans support traditional gender roles and others favor a "unisexual" society, it is difficult to know in which direction the society will go.

The ups and downs of the Equal Rights Amendment (ERA) are a good indicator of the uncertainties involved. First introduced in the U.S. Congress in 1923, it was almost fifty years before the ERA passed both the Senate and the House of Representatives and was sent to the individual states for ratification. As of 1979, thirty-five of the necessary thirty-eight states had ratified it. Opponents from a wide variety of right-wing political and religious groups mobilized opposition to the ERA on grounds ranging from fear of women being drafted to the specter of unisex toilets. The strong opposition of President Reagan and many of the conservative politicians elected in 1980 signaled the defeat of the amendment in 1982. Efforts were immediately under way, however, to reintroduce it.

The 1980s and early 1990s were a time of both right-wing backlash against the aims of the feminist movement and increased mobilization of supporters (Faludi 1999). Efforts to ban abortion, to restrict access to contraception, and to weaken Title IX regulations (which ensure equality in education) are all attempts at reasserting traditional gender roles. At the same time, feminists have mobilized, with increasing militancy, around such issues as spouse abuse, rape, sexual harassment, and pension discrimination (Richardson and Taylor 1983; A. L. Taylor 1983; Andersen 2002). Such inconsistencies illustrate the problem of forecasting the future.

Notwithstanding these problems, some speculation can be made. The changes that have taken place provide a foundation for discussing the future of gender roles and the consequences of gender role changes.

What could gender roles be like in the future? Women are living longer, bearing fewer children, becoming better educated, and entering the labor force in greater numbers than at any other time in history (Reskin and Padavic 2000). If these trends continue—and there is every reason to believe that they will—certain gender role changes should occur. For example, more and more women can be expected to plan for and pursue careers. No longer will female gender roles define work and career as unfeminine pursuits, nor will certain occupations be considered off-limits to women. The increasing enrollment of women in medical, law, and business schools indicates that this development is already under way.

Women may be placing less emphasis on children and marriage as standards for femininity. Men may become less preoccupied with work and success. Many observers have speculated on the nature of gender roles in future generations. Robert Hefner and Rebecca Meda (1979) have summarized the various alternatives in this way:

- *Emergent pluralism.* There would be no restrictions on the options available to either sex. Individuals following the traditional gender roles would exist alongside those who are pursuing a different course.
- *Conservative pluralism.* Men's and women's roles would be different but valued equally. In other words, men might still have major economic responsibility for the family, and women would perform housework and child care. Or men might be doctors and lawyers, and women would work as nurses and secretaries. However, all roles, whether male or female, would be considered equally important and rewarded accordingly.
- *Melting pot.* There would be no important differences in gender roles for men and women. Male and female roles would be combined to create androgynous individuals who combined the

strengths of the two traditional roles, but without their weaknesses. Both men and women would be strong and self-sufficient and yet capable of expressing emotion and sensitivity.

- *Assimilation to the male model of success.* Women would be encouraged to follow the traditional masculine gender role if they wanted to participate in traditionally male activities. Women could become doctors, lawyers, and businesspersons as long as they behaved just like men. This alternative would assume that feminine traits are only a hindrance and can make no contribution to the performance of societal roles.
- *Female exclusion.* There would be a continuation of traditional gender roles. On the assumption that sex differences are biological, men and women would be excluded from exchanging tasks. In particular, women would be discouraged from assuming male roles, and their own tasks would still be under-rewarded.

It is difficult to predict which of these gender role patterns will prevail in the future, because unforeseen events could result in the reversal of present trends. However, the changes that have already occurred suggest a future in which more alternatives will be available to both men and women. If so, the most restrictive pattern (female exclusion) should continue to break down, and a variety of factors and efforts will probably be directed toward creating an atmosphere that permits individuals to choose from a number of options (emergent pluralism).

FEEDBACK

1. The women's movement can be traced as far back as the ________ century.
2. ________ is a social movement aimed at the achievement of sexual equality.
3. A central assumption of the feminist movement is that the problems created by sexism can be solved by individuals. T or F?
4. Match the gender role alternatives for the future with the descriptions beside them.

____ a. emergent pluralism	(1) Male and female gender roles are basically identical.
____ b. conservative pluralism	(2) Traditional gender roles exist along with alternatives.
____ c. melting pot	(3) Traditional gender roles continue.
____ d. assimilation to the male model of success	(4) Women are socialized into the masculine role.
____ e. female exclusion	(5) Men's and women's roles are different but equally valued.

Answers: 1. eighteenth 2. Feminism 3. F 4. a. (2) b. (5) c. (1) d. (4) e. (3)

REVIEW GUIDE

SUMMARY

1. All societies have definitions of gender—expectations and behaviors associated with a sex category. Through socialization, members of a society acquire an awareness of themselves as masculine or feminine.
2. Aside from obvious biological differences, such as genital characteristics and muscular structure, there are no established, genetically based behavior differences between the sexes. Behavioral differences between men and women are culturally and socially created.
3. Anthropological research reveals cultural diversity in gender characteristics. All in all, though, the traditional gender roles remain the predominant pattern.
4. Functionalists and conflict theorists have different explanations for traditional gender definitions. According to functionalists, a division of labor based on sex was a convenient and obvious solution for preindustrial societies. Conflict theorists, who are more concerned with the continued existence of traditional roles than with their origins, contend that men want to maintain the status quo; they want to protect their dominance of women and society. According to conflict advocates, traditional gender definitions are not appropriate in the modern world.
5. According to symbolic interactionism, gender definitions are imparted through the socialization process. They are learned and reinforced through interaction with parents, teachers, peers, and the media. This interaction is especially important because gender becomes intertwined with the self-concepts of those affected by it. Gender socialization, which begins before birth and continues throughout life, occurs through elements of the mass media such as books, television, and advertising.
6. Gender roles permit some deviation from the ideal, but sex stereotypes are labels applied to all members of each sex. Sex stereotypes are employed to encourage all men to be masculine and all women to be feminine and allow for no variation from the generalized picture.
7. Sex stereotypes and gender roles are damaging to both men and women. Among other things, they intensify role conflict and strain for men and women.
8. Although both men and women are damaged by sexism, it is women who feel its full effect. Women, like other minorities, are subject to prejudice and discrimination occupationally, economically, athletically, politically, legally, and socially.

9. There are several trends promoting gender role changes, but there are also barriers. Consequently, it is risky to make concrete predictions. Most likely, the exclusion of women from traditionally male domains will continue to diminish, and both sexes will increasingly be able to choose from a variety of gender role alternatives.

INFOTRAC® COLLEGE EDITION

http://www.infotrac-college.com/wadsworth

Another unique option available to you at the Student Resources section of the companion Web site is Infotrac College Edition, an online library with access to hundreds of scholarly and popular periodicals. Below are suggested search terms for this chapter. Results from these and other searches are found at the site.

Search keywords: **sex stereotypes and gender roles**. Find articles that use these terms in combination. How are the two concepts related, if at all? Do you view the different sexes in typical stereotypes? If so, how did you learn or develop these stereotypes? If not, why not?

Search keywords: **sexual harassment**. Find articles that attempt to define sexual harassment. Based on these definitions, do you have a clear concept of what constitutes sexual harassment? How can an understanding of sociology help clarify this concept?

Search keywords: **work-life balance**. Although the topic of work-life balance is becoming more of a concern for all workers, it is particularly relevant to women. Read two articles that discuss this topic. What is meant by work-life balance? Why is it of special concern to women?

LEARNING OBJECTIVES REVIEW

- Distinguish the concepts of sex, gender, and gender identity.
- Demonstrate the relative contributions of biology and culture to gender formation.
- Outline the perspectives of gender expressed by functionalists, conflict theorists, and symbolic interactionists.
- Describe the position of women in the United States with respect to work, law, politics, and sports.
- Discuss factors promoting change in gender roles as well as factors promoting resistance to change in traditional gender roles.
- Describe the future of gender roles.

CONCEPT REVIEW

Match the following concepts with the definitions listed below them.

____ a. sex
____ b. gender
____ c. patriarchal society
____ d. feminism
____ e. occupational sex segregation
____ f. sexual harassment
____ g. dual labor market

1. a split between core and peripheral segments of the economy and the division of the labor force into preferred and marginalized workers
2. the type of society that is controlled by men who use their power, prestige, and economic advantage to dominate women
3. a social movement aimed at the achievement of sexual equality—socially, legally, politically, and economically
4. the expectations and behaviors associated with a sex category within a society
5. the concentration of women in different occupations than men
6. the biological distinction between male and female
7. the use of one's superior power in making unwanted sexual advances

CRITICAL-THINKING QUESTIONS

1. Suppose that one of your high school teachers invites you back to speak to her class on the nature versus nurture gender debate. Write a brief essay you could use to summarize the scientific knowledge and to initiate class discussion.

____

____

2. Describe the major influences on the development of your gender identity. Use the categories of influence mentioned in the chapter in laying out the specifics of your experiences.

____

____

3. Some sociologists draw a parallel between racism and sexism. Discuss the consequences of sex stereotypes to substantiate or refute this assertion.

____

____

4. Draw on the information in this chapter to discuss the need for affirmative action programs. Substantiate a case either for or against continuation of these programs.

____

____

5. Use information throughout this chapter to predict the future of gender roles. Choose one of the five alternatives for the future of gender roles as the framework for your analysis.

____

____

MULTIPLE-CHOICE QUESTIONS

1. **Whereas _________ is a biological term, _________ is a psychological and cultural term.**
   a. gender; sex
   b. identity; role
   c. gender; gender identity
   d. sex; gender
   e. phenotype; genotype
2. **The debate over explanations for sex differences persists, with the _________ distinction still at the heart of the controversy.**
   a. nature versus nurture
   b. culture versus socialization
   c. innate forces versus genetic forces
   d. gender roles versus gender identity
   e. testosterone versus estrogen
3. **Although given credit for its insights into the original emergence of gender roles, functionalism is often criticized on the grounds that**
   a. it tends to support a male bias.
   b. it is too conservative and inadequate to explain recent changes in gender definitions.
   c. its proposition that gender roles are functional is detrimental to women.
   d. it does not address the issue of gender-based inequality.
   e. it fails to take the biological basis of gender roles into account.
4. **According to conflict theory, gender roles persist because**
   a. men and women have differential access to the resources needed to succeed outside the home.
   b. they are beneficial and efficient for human living.
   c. gender roles are incorporated into the self-concept through role taking and the looking-glass self.
   d. they are essential to human survival.
   e. women are naturally better suited to perform certain tasks.
5. **Symbolic interactionists believe that traditional gender roles persist because**
   a. men and women have differential access to the resources needed to succeed outside the home.
   b. they are beneficial and efficient for human living.
   c. gender is incorporated into the self-concept through role taking and the looking-glass self.
   d. they are essential for human survival.
   e. of innate biological differences.
6. **Which is the true statement regarding the contribution of peers to gender socialization?**
   a. Nursery schools now attempt to ignore gender roles and do not encourage conformity to them.
   b. Teenagers who most closely fulfill gender role expectations are given the greatest respect.
   c. Acceptance or rejection by peers is having less and less influence on the adolescent's self-concept.
   d. Boys are more concerned than girls with gender role socialization.
   e. In schools today, children are often encouraged to act out the roles appropriate to the opposite sex.
7. **Women are hurt by the set of beliefs, norms, and values used to justify sexual inequality, known as**
   a. sexism.
   b. prejudice.
   c. gender identity.
   d. discrimination.
   e. gender roles.
8. **On the average, women in the United States are now earning _________ cents for every dollar earned by men.**
   a. 45
   b. 52
   c. 60
   d. 76
   e. 83
9. **According to the text, the political position of women in the United States is**
   a. among the worst in the Western world.
   b. more favorable than that of women in Scandinavian countries.
   c. harmed by "women's lib."
   d. actually superior to that of men because women vote in greater numbers.
   e. measured by the dependency ratio.
10. **The women's movement was an identifiable mass movement by the end of the _________ century.**

a. sixteenth
b. seventeenth
c. eighteenth
d. nineteenth
e. twentieth

11. **The social movement aimed at the achievement of sexual equality is generally known as**
    a. sexism.
    b. feminism.
    c. genderism.
    d. communism.
    e. comparable worth.
12. **According to the text, the most likely form of gender roles in the future is**
    a. conservative pluralism.
    b. the melting pot.
    c. assimilation to the male model of success.
    d. female inclusion.
    e. emergent pluralism.

FEEDBACK REVIEW

True-False

1. Societies that do not follow traditional gender roles are exceptions to the rule. T or F?
2. Sex stereotypes and gender roles are essentially the same thing. T or F?

Fill in the Blank

3. Gender is acquired through the process of _________.
4. According to the _________ perspective, the division of labor based on sex has survived because it is beneficial and efficient for human living.
5. According to the _________ perspective, traditional gender definitions exist because they provide greater rewards and privileges to men than to women.
6. _________ is the theoretical perspective that attempts to explain how gender is acquired.

Multiple Choice

7. **Which of the following is not a true statement?**
    a. Boys are less assertive in class than girls.
    b. Illustrated children's books present very traditional gender definitions.
    c. Parents' perceptions of the physical characteristics of their children are influenced very little by their children's gender.
    d. Nursery school children who fail to conform to their "appropriate" gender are met with resistance from their peers.
8. **The idea that victory is the most positive reinforcement for all men is an example of**
    a. sex.
    b. a sex stereotype.
    c. a gender role.
    d. a value uniquely held by men.

Matching

Match the gender role alternatives for the future with the descriptions beside them.

____ a. emergent pluralism
____ b. conservative pluralism
____ c. melting pot
____ d. assimilation to the male model of success
____ e. female exclusion

(1) Male and female gender roles are basically identical.
(2) Traditional gender roles exist along with alternatives.
(3) Traditional gender roles continue.
(4) Women are socialized into the masculine role.
(5) Men's and women's roles are different but equally valued.

GRAPHIC REVIEW

Figure 10.4 displays American women's median annual earnings as a percentage of men's from 1955 to 2000.

1. How would you explain this trend to your sixteen-year-old daughter (if you had one)?

____
____
____

2. Assuming that the upward trend since 1980 continues, which of the five gender role patterns outlined in the text would be most descriptive? Why?

____
____
____

ANSWER KEY

Concept Review

a. 6
b. 4
c. 2
d. 3
e. 5
f. 7
g. 1

Multiple Choice

1. d
a
b
a
c
b
a
d
a
10. d
11. b
12. e

Feedback Review

1. T
2. F
3. socialization
4. functionalist
5. conflict
6. Symbolic interactionism
7. a
8. b
9. a. 2
   b. 5
   c. 1
   d. 4
   e. 3

Inequalities of Age

11

© Bachmann/Photo Researchers, Inc.

OUTLINE

LEARNING OBJECTIVES

- Describe the aging of the world's population.
- Discuss the graying of America.
- Explain the importance of the dependency ratio.
- Distinguish between age stratification and ageism.
- Contrast the theoretical perspectives of ageism.
- Outline the inequality experienced by older Americans.

USING THE SOCIOLOGICAL IMAGINATION

Why does health begin to decline as people retire? Actually, deteriorating health upon retirement is only a stereotype. Research fails to show any adverse health consequences soon after retirement (Ekordt 1987). On the contrary, physical and mental well-being tends to improve at the onset of retirement (Gall, Evans, and Howard 1997).

Thus far in our exploration of social inequality, we have examined race, ethnicity, and gender as major factors in the differential distribution of the scarce desirables (power, wealth, privileges) in a society. Since chronological age is another basis for social ranking, sociologists are interested in **age stratification**—the age-based unequal distribution of scarce desirables in a society. Although age can operate to the advantage or disadvantage of any age-group in a society, sociologists are especially focused on the inequality existing among older Americans (Falk and Falk 1997; Pampel 1998; Schaie and Schooler 1998) because as the median age of the U.S. population grows older, this form of stratification affects more and more people.

Age stratification, like most aspects of a society, must exist within a social and cultural context: it must make sense to those involved, and it must be socially justified. The justification for negative age stratification among the elderly in a society comes in the form of **ageism**—a set of beliefs, attitudes, norms, and values used to justify age-based prejudice and discrimination. An understanding of ageism—in this context, directed at the elderly—will be further enhanced through the study of this chapter. Let's begin with aging and stratification in preindustrial societies.

Aging and Stratification

Both Tiriki and Irigwe, African societies with economies based on hoe agriculture, illustrate a type of stratification structure that places the elderly at the top (Sangree 1989). A combination of land control and religiously based wisdom constituted the foundation of status and power for the elders in agricultural societies. As we shall see, the bases of stratification, in either preagricultural or postagricultural (industrial) societies, do not necessarily follow this favorable pattern for the elderly (Kertzer 1989).

As described in Chapter 5 ("Social Structure and Society"), hunting and gathering societies have no social classes. Their division of labor is based on sex and age. All members of this type of society are poor, living on the edge of survival as they move seasonally to take advantage of an available food supply. The absence of social classes, however, does not preclude the existence of social inequality. Obtaining an adequate food supply depends on having the necessary number of individuals with the physical ability to hunt or gather. Consequently, it is not surprising that the loss of strength and stamina that comes with age (actually, this often occurred in early adulthood) often turns the elderly from an economic asset to an economic drain. They feel the effects of social inequality as their ability to contribute radically diminishes. **Geronticide**—the socially accepted killing of the elderly within a society—has been documented in several hunting and gathering societies (Hoebel 1983; Glascock and Braden 1981; Guemple 1983). However, not all hunting and gathering societies living on a thin margin of economic safety have cultural support for eliminating their elderly. For example, among the !Kung of the Kalahari Desert—an extremely egalitarian society—the elderly maintain their place in society by controlling waterholes (Halpern 1997). Still, according to the evidence, the prevailing pattern in hunting and gathering societies disfavors the elderly.

To some degree in horticultural society, but especially in agricultural society, technological advances permit the creation of a genuine, dependable economic surplus. With the existence of desirables above the subsistence level, social stratification, including age stratification, becomes more prominent. Particularly in agricultural society, wealth and power rest on the ownership of land by ruling elites. At this level of economic development, the elderly benefit. These societies may be modeled on a **gerontocracy**, the type of society in which the elderly hold the greatest share of scarce desirables. Abkhasia, an autonomous republic that is administratively part of the Republic of Georgia in the former Soviet Union, constitutes an enlightening illustration of positive age discrimination for elderly males (Benet 1971).

How does Abkhasian culture favor the elderly? Among the Abkhasians, who draw their living primarily from agriculture, the elderly benefit from age stratification. Respect and prestige for males increases with age because of the shared belief that wisdom increases with advancing age. Consequently, the elders are expected to pass the benefit of their experience to the younger and to take center stage at ceremonial functions.

Two aspects of Abkhasian culture contribute to the central place of the elderly: attitudes toward work and social integration. They have no word in their vocabulary for retirement. Abkhasians are guided by the maxim, "It is better to move without purpose, than to sit still." For them, then, retirement from work would be the same as retirement from life. A full workload for men—including heavy lifting and plowing—may be reduced at age eighty or ninety. Women at this age may contribute less to household management and work less in the fields. Nevertheless, a four-hour workday is common for the average Abkhasian centenarian.

Because elder Abkhasians remain at the center of social life, they do not view themselves as a burden or as marginalized. Their continued integration into the full round of social life is reflected in their continuous engagement in work, participation in games with the young, and maintenance of the same diet as everyone else.

Recall from Chapter 5 the lines of demarcation between preindustrial and industrial societies. A defining characteristic of industrial society, you'll remember, is the loss of social standing among the elderly.

Aging, Stratification, and Modernization

With some exceptions (Rhodes 1984), modernization in most societies appears to be associated with a loss of power, prestige, privileges, and economic resources among the elderly (Cowgill and Holmes 1972; Shelton 1972; Goldstein and Beall 1982; Goldstein et al. 1983; Gilleard and Gurkan 1987).

Why is there an association between modernization and age stratification? In agricultural societies, the elderly have high status and power in part because others are dependent on the knowledge they have gained from years of experience. These societies depend on tradition and ceremonies. Because the elderly know the most about the societies, their opinions and advice are considered to be a valuable resource. This situation changes with modernization. Formal education replaces personal experience as a source of knowledge, so industrial societies depend less on the older generation. Ownership of land is also a source of power and status for the older generation in agricultural societies, but industrialization pushes the elderly from the land and into competition with the young for positions in the nonagricultural labor market. In most agricultural societies, elderly men and women remain in their lifelong occupational roles until they can no longer physically perform the tasks involved. Thus, they retain status for most of their lives. In modern societies, the older generation loses status and power because they are forced to retire at an arbitrary age. Agricultural societies are based on the extended family structure in which the older generation retains a strong position. The rise of the nuclear family in industrial societies diminishes the status of the elderly because they are usually forced to live in separate households or in institutions for the aged (Nolan and Lenski 1999).

In industrial societies, death is associated with old age because low death rates result in people's living longer. In preindustrial societies, high death rates mean that people die at all ages; thus people in these societies do not associate death with old age. Because of the connection between death and old age in modern societies, the young may wish to put distance between themselves and the elderly to avoid facing their own mortality. One way to do this is to stigmatize the aged.

One of the most prominent characteristics of modern societies is change. Nowhere is change more apparent than in the area of occupational knowledge. Since formal education (including basic job preparation) occurs early in life, the passage of years leads to knowledge obsolescence. Individuals whose knowledge base and skills have become outdated are treated as though they had never been educated and are incapable of undergoing necessary retraining. At best, they are regarded as incompetent; at worst, they are deprived of their jobs (Atchley 2000).

Does negative age stratification among the elderly occur in all modern societies? If there is one society that most Westerners would cite as an exception to the reduction in social economic standing associated with modernization, it is Japan (Palmore 1978). Elderly Japanese are commonly thought to live with respect and honor. Confucianism teaches devotion to parents. Japanese children are expected to provide economic support for their parents. Also, the extended family, in which the elderly have an important place, predominates.

Changes within Japan since World War II, however, are beginning to undermine the accuracy of long-standing images of the high social standing of the Japanese elderly (Plath 1980, 1983; Tobin 1987). A brief examination of a few of these changes will provide some further insight into the link between industrialization and the loss of social standing among the elderly (Hillier and Barrow 1999).

Like all highly industrialized countries (much more will be said about the United States shortly), the Japanese are experiencing population pressures as the elderly population explodes and the birth rate declines. Since the mid-1940s, life expectancy at birth for the Japanese has increased from 50 to 78. In 1991, Japanese males had an average life expectancy of 76 years and females 82.5 years. At the same time, the birthrate has dropped to 1.8 children per family. If population projections hold, almost one-quarter of the Japanese population will be 65 and over by 2025; and half of that group will be over 75 years of age (Dentzer 1991). Not only is Japan the fastest-aging country in the world, but its elderly population is increasing at a rate double that of other industrialized societies (Rosario 1990).

These population trends obviously mean that more elderly have fewer younger people to care for them. This is compounded by the increasing employment of women outside the home and the increasing number of children moving from home for employment reasons. Consequently, more of the sick elderly are now being cared for by nonfamily members, and more elderly are living in public housing and private retirement homes (Berger and Gaffney 1987). Prior to 1948, the oldest son inherited the father's wealth—an incentive for these sons to care for their fathers and mothers. Since all brothers and sisters now have equal inheritance rights, the traditional old-age system of care for the elderly has been altered. According to a study comparing Japanese over sixty years of age and Japanese under sixty years of age, a higher percentage of the elderly believe the eldest son has an obligation to take care of his parents (Martin 1989). In addition, larger numbers of young adults are leaving rural areas for employment in cities. As a result, greater numbers of the elderly are attempting to live off their land without the aid of their children. Moreover, those elderly who do move from rural areas to the city with their children often must adjust to new, disadvantageous social arrangements.

A study of historical records and anthropological field work suggests an interesting new hypothesis on the relationship between the level of economic development and the status of the elderly. According to Harold Cox (1990), the elderly do, in fact, experience negative age stratification in hunting and gathering societies, have high status in agricultural societies, and again experience a loss of social standing in industrializing countries. This generalization confirms the analysis presented in this section. Cox, however, goes on to suggest that the elderly are accorded higher status in postindustrial societies. The basis for this shift, according to Cox, is the de-emphasis on the importance of work and the elevation of leisure activity that occurs in the most developed countries, whose high productivity based on advanced technology reduces the need for labor. Evidence sufficient to inspire high confidence in this conclusion is not yet available.

The Graying of America

America is not alone in its growing proportion of older persons. The situation in the United States, therefore, should be seen in perspective.

Aging of the World's Population

Graying is occurring in all highly industrialized countries. Economic development carries with it a train of factors, not the least of which is improved medical care and knowledge of birth-control methods (see Chapter 17, "Population and Urbanization"). These factors increase life expectancy while reducing the birth rate. For example, the proportion of the world's population sixty-five and older is increasing at a rate of about 6 percent annually, whereas the world's population as a whole has a growth rate under 2 percent. Consequently, the world in the twenty-first century will be an older world. The percent increase in the population aged sixty-five and over in China, for example, is expected to increase by over 200 percent by 2025 (see Table 11.1). If this occurs, the proportion of China's elderly population will increase to nearly 20 percent by the end of the first quarter of this century. By that same time, over one-half of the world's elderly will inhabit East and South Asia. In Africa, where the most dramatic increases in the growth of the aged population are expected, the population aged sixty-five and over will have gone from 23 million persons in 1980 to over 100 million in 2025 (Curran and Renzetti 2000).

Aging in the United States

An **age cohort** consists of persons born during the same time period in a particular population. The baby

FEEDBACK

1. ____________ exists when the unequal distribution of scarce desirables in a society is based on chronological age.
2. In which of the following types of society is geronticide most likely to be practiced?
   a. hunting and gathering
   b. agricultural
   c. industrial
   d. gerontocratic
3. A ____________ is a stratification structure in which the greatest share of scarce desirables is held by the elderly.
4. The general relationship between industrialization and social status for the elderly appears to be
   a. J-shaped
   b. negative
   c. positive
   d. nonexistent

Answers: 1. Age stratification 2. a 3. gerontocracy 4. b

TABLE 11.1

Percent Increase in Population Age 65 and Older, 1985–2025, for Selected Countries

This table shows the percentage increase in the elderly population for certain countries. What generalization(s) can you make from these data?

Country	Percent Increase	Country	Percent Increase	Country	Percent Increase
Guatamala	357	Australia	125	Belgium	49
Singapore	348	Poland	122	Greece	48
Mexico	324	Japan	121	Uruguay	47
Philippines	310	Israel	116	Denmark	36
Indonesia	301	United States	105	Federal Republic	
Brazil	292	New Zealand	99	of Germany	36
India	264	France	67	Austria	35
China	238	Bulgaria	65	Norway	34
Hong Kong	219	Luxembourg	56	United Kingdom	23
Bangladesh	201	Hungary	51	Sweden	21
Canada	135	Italy	51		

Source: U.S. Department of Commerce.

boomers, for example, constitute an age cohort born between roughly 1946 and 1964. The important thing about an age cohort is that its members pass through life together. Consequently, an age cohort not only exerts a certain influence on the rest of society, but experiences in common some effects of the particular nature of its society.

The age cohort of persons sixty-five years and over—35 million strong—comprises 13 percent of the total U.S. population. By 2040 there will be about 82 million elderly Americans representing 20 percent of the nation's population (see Figure 11.1). The median age of the population will be forty-three years by 2080; it was about sixteen in 1800 and thirty-six in 2000 (see

FIGURE 11.1

Percentage of the U.S. Population 65 and Over

This graph charts the increase in elderly Americans from 1900 to 2040. Why has this trend occurred?

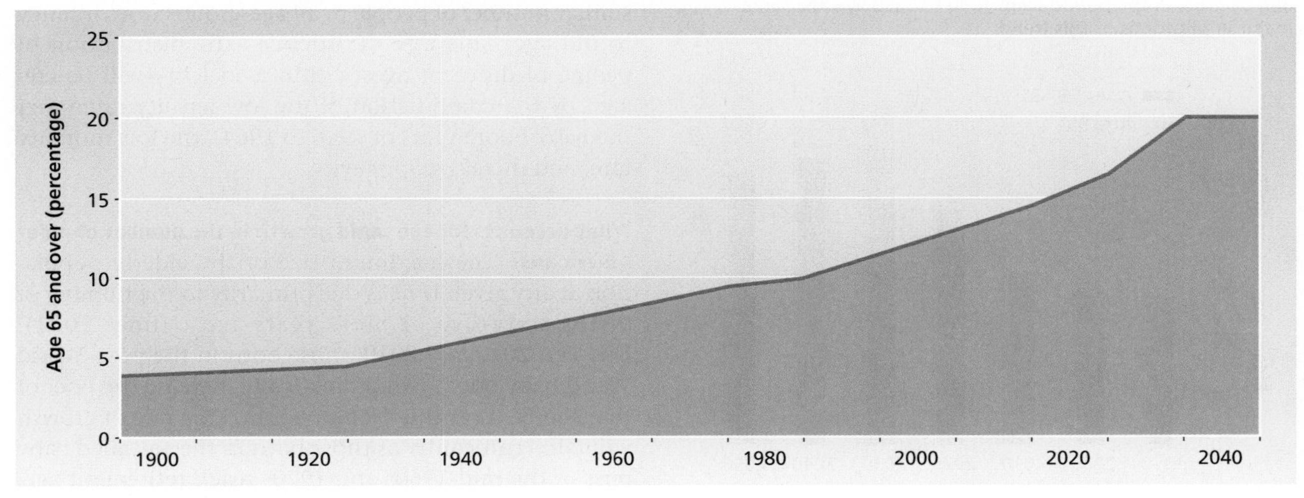

Source: U.S. Bureau of the Census.

FIGURE 11.2

Median Age of the U.S. Population: 1800–2080

As this graph shows, the median age of Americans is increasing. Discuss this trend in relation to the dependency ratio.

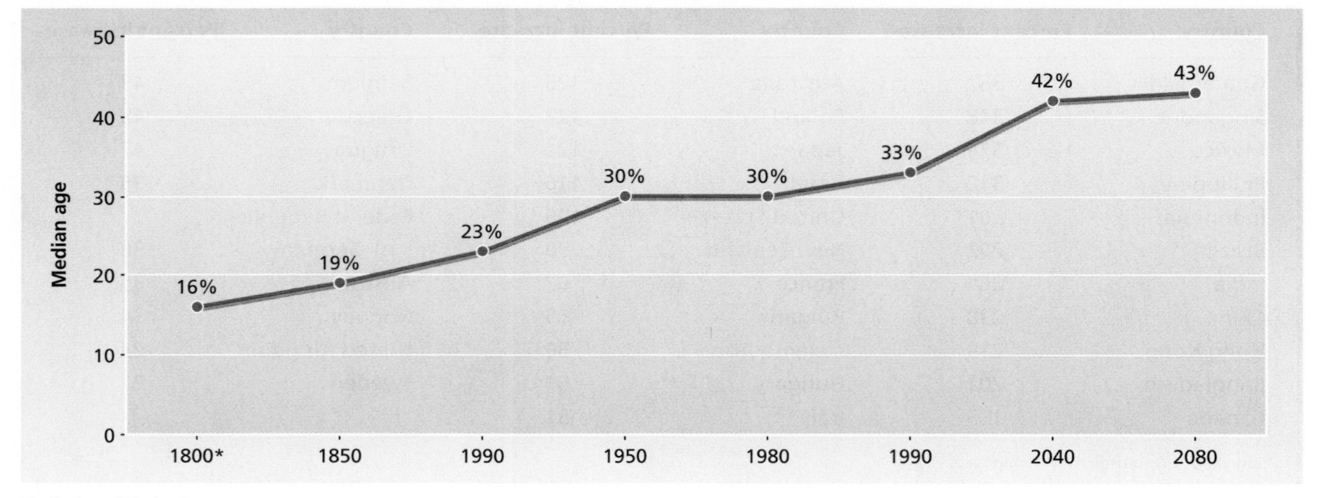

*Includes whites only.
Source: U.S. Bureau of the Census.

Figure 11.2). Due to the baby-boom generation, there will be an increasing proportion of elderly in the advanced age cohort. The number of Americans sixty-five to eighty-four will increase from about 30 million in 1990 to over 50 million in the middle of the twenty-first century (Haupt and Kane 1997). Moreover, the number of Americans over age eighty-five will increase from about 4 million in 2000 to about 8 million in 2010 to over 17 million by 2050 (see Figure 11.3).

FIGURE 11.3

U.S. Elderly Population 1990 –2050

This bar chart documents the number of Americans ages 65 –85, and over 85, in the period from 1990 to 2050. Discuss some major implications of this trend.

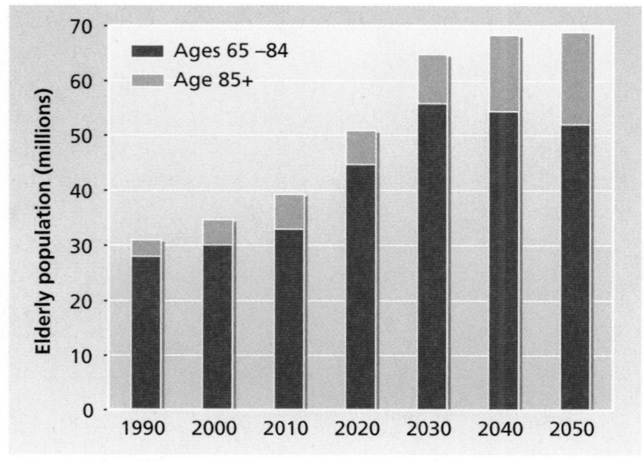

Source: U.S. Bure au of the Census.

Population pyramids are graphic representations illustrating the age and sex distribution of a population. Their use enhances understanding of an aging society. The pyramid on the left in Figure 11.4 shows the number of American males and females by age-group in 1945. This is the classic pyramid shape, representing a population that has many younger people at the bottom, fewer adults in the middle, and a still smaller number of older people at the top. The pyramid on the right in Figure 11.4 portrays the projected age structure of the United States in 2050. Obviously, this is more of a rectangle or square than a geometric pyramid, with a similar number of people in all age-groups from infancy to old age. This **age structure**—the distribution of people of different ages within a society—will be created by the combination of the low fertility rate (after the baby-boom years of 1946 to 1964), the low morality rate, and increased longevity.

What accounts for the rapid growth in the number of older Americans? The maximum size of the elderly population at any given time is due primarily to the number of births sixty-five or more years ago (Himes 2001). Between 2010 and 2030, those born in the post–World War II baby boom will dramatically increase the pool of the elderly. After this "senior boom," the rate of growth will fall significantly as those born in the so-called baby bust of the mid-1960s and 1970s reach retirement age. Around 2045, another dramatic increase in the elderly

FIGURE 11.4

Age and Sex Structure of the United States: 1945 and 2050

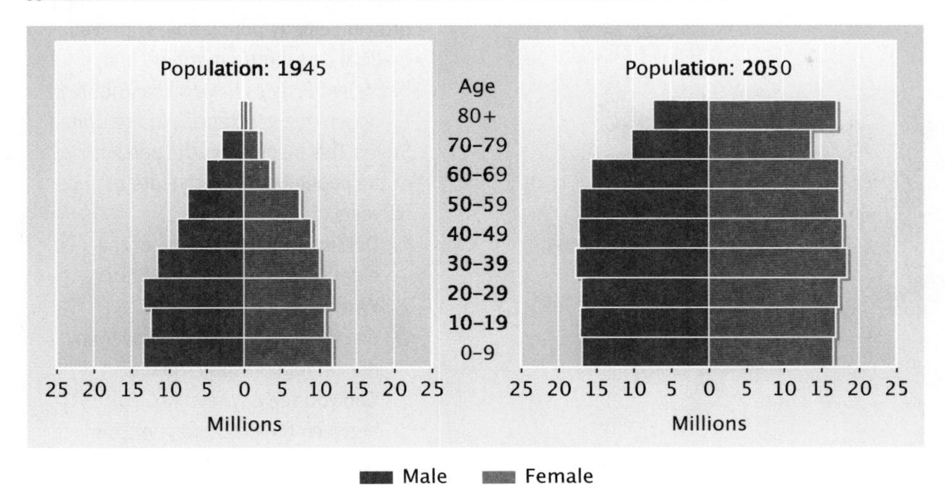

Source: U.S. Bureau of the Census.

will occur as the children of the baby boomers reach age sixty-five. This uneven growth rate will see the number of the elderly increase from 26 million in 1980 to over 53.7 million in 2020 and then to almost 82 million in 2050.

The second most significant factor in the growth of the elderly population is the decline in the death rate. Thanks to medical advancements such as improved sanitation, antibiotics, and vaccines, the death rate among infants and children dropped precipitously between 1900 and 1954. Although only 40 percent of those Americans born at the turn of the century had a life expectancy of sixty-five, today 80 percent are expected to reach this age. The life expectancy figures are, of course, even better than this. The average male life span, now 74.2 years, is expected to increase to 75.6 by 2010. Among females, the life span is projected to increase from the current 79.9 years to 81.4 years in 2010 (U.S. Bureau of the Census 2001a).

Immigration has had its effect through the massive pre–World War I immigration into the United States. Most of those immigrants were young adults with a very high fertility rate. Their children, born in the United States, are now among the elderly.

The **dependency ratio** is the proportion of persons in the dependent ages (under fifteen and over sixty-four) in relation to those in the economically active ages (fifteen to sixty-four). A higher dependency ratio indicates that each worker has to support more dependent people; a lower ratio indicates that each worker has fewer people to support. Suppose a population of 200 persons was composed of 45 juveniles, 15 seniors, and 100 people of economically active ages. This would yield a dependency ratio of 60/100 or .60, denoting that each worker supports just over half a dependent person; or to put it another way, there would be just over one dependent for every two working-age people. There are two subtypes of dependency: *youth dependency* and *old-age dependency*. The fictitious population just described would have a youth dependency ratio of .45 (45/100) and an old-age dependency ratio of .15 (15/100). That is, for every youthful dependent, there would be two-and-a-fraction workers; for every elderly person, there would be over six-and-a-half workers. Less developed nations have much higher youth dependency than more developed nations. More developed nations have significantly higher old-age dependency.

Why is the dependency ratio important? For less developed countries such as Mexico, a high youth dependency means that national income must be diverted from savings (and the capital needed for economic development) to take care of its large population of children (food, housing, education). In more developed countries such as the United States, rising old-age dependency creates other challenges. There are fewer working-age people in the labor force to support the growing number of seniors. For example, in the United States in 1985, there were about 19 persons age sixty-five and over for every 100 Americans fifteen to sixty-four, an old-age dependency ratio of .19, or about five times as many in the economically active ages as there were elderly. By 1995, there were 25 dependent elderly Americans for every 100 economically active ones (ratio of .25), or just about four times as many Americans 15–64 years of age as there were Americans 65 and older. In 2030, there will be 50 dependent

SNAPSHOT OF AMERICA 11.1

Percentage of Population 65 and over

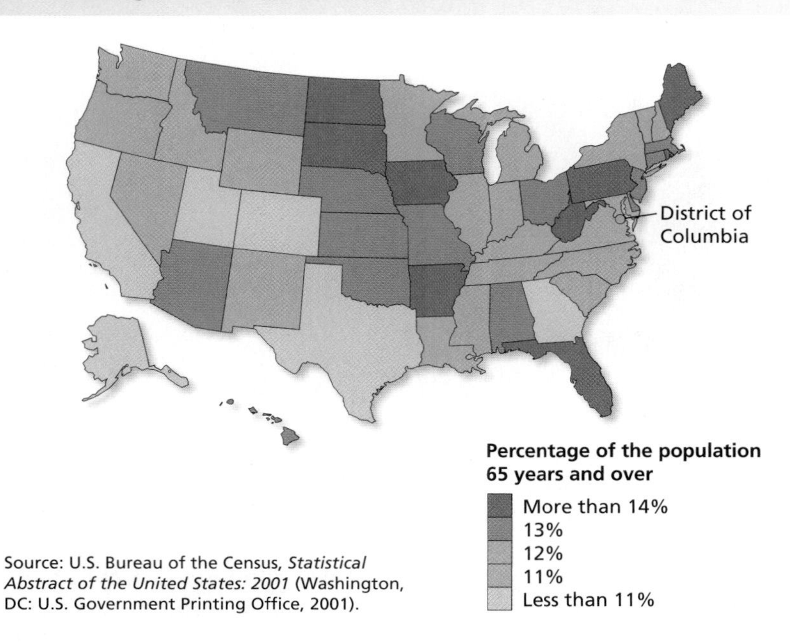

Source: U.S. Bureau of the Census, *Statistical Abstract of the United States: 2001* (Washington, DC: U.S. Government Printing Office, 2001).

The "graying of America" refers to the growing elderly population. Improved medical care, better nutrition, and healthier lifestyles have all contributed to longer life expectancies in the United States. This map shows the percentage of the population in each state that is 65 years old and over.

1. Describe the distribution of the elderly across the United States.
2. Where is your state in this distribution? Can you discover any reason for you state's place?
3. Can you see any relationship between the percentage of your state's population that is elderly and their political power?

FEEDBACK

1. The aging of the population is confined to only the most highly industrialized societies. T or F?
2. By 2050, the size of the elderly population in the United States will constitute ____________ percent of the population.
   a. 13
   b. 17
   c. 20
   d. 33
3. By 2050, the population pyramid of the United States is expected to resemble a
   a. triangle
   b. a classic pyramid
   c. a rectangle
   d. an inverted pyramid
4. By 2030, the old-age dependency ratio in the United States is expected to
   a. more than double its 1985 level
   b. just about triple its 1985 level
   c. remain relatively stable
   d. be about half its 1985 level

Answers: 1. F 2. c 3. c 4. a

elderly individuals for every 100 Americans of working age (ratio of .50). In other words, every two working-age Americans will have to support one elderly person. This shift will dramatically increase the burden on the economically active to pay for Social Security and Medicare (Kent and Mather 2002). This economic burden will be even heavier because of escalating health-care costs, the need for increasing health-care services, prescription drugs, and institutional arrangements for the long-term care of an aging population.

Theoretical Perspectives on Ageism

Functionalism and Ageism

How do functionalists view ageism? Functionalists concentrate on the contributions of ageism—contributions to the order and stability of society. According to functionalists, elderly people are treated differently in different societies because they serve different functions in each.

As noted earlier in this chapter, aging is not stigmatized in all societies. In agricultural society, elderly males usually—though not always—play important roles, such as priests or elders. Donald Cowgill and Lowell Holmes point out why, in such societies, the aged are usually respected and highly valued:

> *In all of the African societies, growing old is equated with rising status and increased respect. Among the Igbo, the older person is assumed to be wise: this not only brings him respect, since he is consulted for his wisdom, it also provides him with a valued role in his society. The Bantu elder is "the Father of His People" and revered as such. In Samoa, too, old age is "the best time of life" and older persons are accorded great respect. Likewise, in Thailand, older persons are honored and deferred to and Adams reports respect and affection for older people in rural Mexico.* (Cowgill and Holmes 1972:310)

In early colonial America, no stigma was attached to age. In fact, to be elderly brought respect with the opportunity to fill the most prestigious positions in the community. It was believed that God looked with favor on those who reached old age: the longer one lived, the more likely he or she was to have been chosen to go to heaven. The Bible linked age with living a moral life: "Keep my commandments, for length of days and long life shall they add unto thee." During the 1600s and 1700s, Americans even tried to appear older than they actually were. Men and women wore clothing that made them appear older. Men covered their heads with white powdered wigs. During the 1700s, people often inflated their age when reporting to census takers (D. H. Fischer 1977).

Attitudes about aging shifted as industrialization modified the nature of work. In a technical society, an adult's value can lessen when he or she no longer contributes fully to the common good. And because modern societies change rapidly, younger workers are more likely to possess the current skills needed in the workplace. As individuals get older, their skills are more likely to be out of date in the workplace; as a result, they lack the "wisdom" that is most highly valued. Thus, aging tends to lead to lower status.

This loss of status with older age might help explain the increase in the suicide rate among elderly men. Men may have greater difficulty in the declining years than women do, because they have been socialized in a culture that encourages young men to identify strongly with work but denies them a sense of value after retirement.

© Adam Crowley/Getty Images

Societies vary in their approaches to the elderly. In traditional Eskimo society, it was socially acceptable to kill the elderly. Though the practice of gerontocide has, in fact, been documented in several hunting and gathering societies, the elderly in America are encouraged to take steps (such as surfing) to prolong their lives.

Conflict Theory and Ageism

What is the relationship between competition and ageism? Competition over scarce resources lies at the heart of the conflict perspective. According to this perspective, the elderly compete with other age-groups for economic resources, power, and prestige. In agricultural societies, older people are often accorded a relatively large share of the scarce resources. The young and middle-aged are more likely to defer to the senior citizens. Younger people in industrial society, in contrast, do not permit the elderly to compete as freely for scarce resources (Palmore 1981).

Why are the elderly not permitted to compete? According to the conflict perspective, the elderly are treated relatively well in preindustrial society because their labor is needed by the other age-groups. Because work in preindustrial society is labor intensive, all available hands must be utilized.

Industrial society, in contrast, usually has more workers than it needs, so the elderly are removed from the competition for jobs. In addition, industrial societies save scarce resources by replacing high-priced older workers with less costly younger ones. Forced retirement is one way the more powerful age-groups remove elderly competitors.

According to the theory of differential power, prejudice and discrimination are used by the majority as

TABLE 11.2

FOCUS ON THEORETICAL PERSPECTIVES: Age Inequality

This table shows how each major theoretical perspective depicts the elderly in modern societies. Do you think the depictions are accurate? Why or why not? Create a social arrangement and example for yourself for each theoretical perspective.

Theoretical Perspective	Social Arrangement	Example
Functionalism	• In modern societies, aging is associated with lower economic value.	• Companies prefer to hire and retain younger employees.
Conflict theory	• As Americans age, their ability to compete economically is undermined by age-related prejudice and discrimination.	• The elderly are labeled as duller and less productive than younger workers.
Symbolic interactionism	• A stigma is attached to the elderly in modern societies.	• Older people tend to suffer lower self-esteem.

weapons in the domination of subordinate groups. In this way, the minority cannot effectively compete with members of the majority for scarce goods and services. If the elderly can be labeled as intellectually dull, closed-minded, inflexible, and unproductive, forcing their retirement from the labor market becomes relatively easy. This leaves more jobs available for younger workers.

Symbolic Interactionism and Ageism

What is the role of stereotypes in ageism? Prejudice and discrimination, as noted earlier, are justified in part through the creation and use of stereotypes. As with racism, ageism involves negative stereotyping. According to symbolic interactionists, children learn negative images of older people just as they learn other aspects of culture. Through the process of socialization, stereotypes of the elderly are often firmly implanted into a child's view of the world. In fact, negative images of older people have been observed in children as young as three years old (Seefeldt et al. 1977; Hillier and Barrow 1999).

When stereotypes are used to justify prejudice and discrimination, they harm the elderly. They also harm older people when the stigma attached to aging promotes the development of a negative self-concept (R. B. Ward 1977; Friedan 1993).

What are the stereotypes of the elderly? Robert Butler has succinctly summarized the many negative stereotypes of the elderly:

> *An older person thinks and moves slowly. He does not think as he used to or as creatively. He is bound to himself and can no longer change or grow. He can learn neither well nor swiftly and, even if he could, he would not wish to. Tied to his personal traditions and growing conservatism, he dislikes innovations and is not disposed to new ideas. Not only can he not move forward, he often moves backward. He enters a second childhood, caught up in increasing egocentricity and demanding more from his environment than he is willing to give to it. Sometimes he becomes an intensification of himself, a caricature of a life-long personality. He becomes irritable and cantankerous, yet shallow and enfeebled. He lives in his past; he is behind times. He is aimless and wandering of mind, reminiscing and garrulous. Indeed, he is a study in decline, the picture of mental and physical failure. He has lost and cannot replace friends, spouse, job, status, power, influence, income. He is often stricken by diseases which, in turn, restrict his movement, his enjoyment*

FEEDBACK

1. Following are several statements about the elderly. Identify each statement with one major theoretical perspective: functionalism (F), conflict theory (C), or symbolic interactionism (S).
   ____ a. Ageism results in part from an oversupply of labor.
   ____ b. The young may wish to distance themselves from the elderly to avoid facing their own mortality.
   ____ c. The stigma attached to aging promotes a low self-concept among the elderly.
   ____ d. Ageism is associated with industrialization.
   ____ e. Older people are stereotyped.
   ____ f. Ageism exists in part because older workers are more expensive than younger ones.

Answers: 1. a. (C) b. (F) c. (S) d. (F) e. (S) f. (C)

Aging in the Mass Media

Few question the powerful influence of the mass media on the attitudes and behavior of Americans (and the world). Scholars are continually documenting these influences in research studies. For example, consider these recent studies of the effects of advertising on gender-related attitudes. Garst and Bodenhausen (1997) report that advertising images of men influence their gender role attitudes soon after the men are exposed to the images. According to MacKay and Covell (1997), there is a link between the exposure to images of women in advertisements and attitudes toward feminism and the push for gender equality. The elderly have not been spared from the effects of the mass media on American attitudes regarding aging.

If CNN conducted a poll on beliefs about the elderly, most respondents would agree that older people tend to be irritable, inefficient, uncreative, mentally slow, unfit for work, and unadaptable. Why do these negative stereotypes of the elderly persist despite research contradicting each of them? An important factor is ageism, focusing as it does on those elderly who are in the worst shape. And the mass media, which historically tends to rely on negative stereotypes, perpetuates this ageism.

Television is the most popular vehicle in the mass media (Brunner 2001). On average, children watch television about 3 hours per day. Men aged 25–54 watch television about 5 hours each day; women in that age bracket view television slightly more. Television occupies more leisure time than any other single activity among Americans fifty-five and over (men—over 5 hours daily; women—almost 6 hours daily). As would be expected from such constant and considerable exposure, television greatly influences the attitudes of the young toward the elderly as well as the elderly's view of themselves (Hillier and Barrow 2001).

Unfortunately, the influence of television on attitudes toward the elderly has been primarily unfavorable. Aronoff (1974) reported that older people are more frequently depicted as evil, inept, and unhappy. According to Jeffreys-Fox (1977), actors in elderly roles tend to be portrayed as less physically appealing, intelligent, and friendly than actors in younger roles are. While there is some evidence that television's portrayal of older people is becoming more accurate and favorable (Dail 1988), negative stereotyping of them remains prominent in this most pervasive medium. Older people are usually cast in minor roles (Robinson and Skill 1995). It is even worse for women, who have an even lower probability of appearing in major roles; and when they do, they are more likely to be in stereotypical roles (Vasil and Wass 1993; Robinson and Skill 1995; Hilt and Lipschultz 1996). Even when positive, stereotyping of older people on television comedies comes in the perverse form of "reversed stereotyping" —presenting an older character as supercompetent in ways that defy more common negative stereotypes. An older character with sexual prowess, super intelligence, or physical vigor readily provokes laughter because it is generally believed that such performances are simply beyond the capabilities of older people. By showing older people performing activities thought to be antithetical to them, the audience receives reinforcement for the negative views in which the reversed (and positive) stereotypes are grounded. Laughter provoked in this way is a putdown.

Television shares the blame for negatively stereotyping the elderly with other mass media. Since most movie watchers are between the ages of sixteen and twenty-four, commercial films are marketed for teenagers and young adults. Like television, movies tend to emphasize the virtues of youth and the downsides of age. Older women are portrayed even more negatively than older men are. As in American culture in general, movie producers and directors assume that men age more attractively than women do. Consequently, in Hollywood there are no female parallels to Paul Newman, Sean Connery, or Michael Douglas. While these older male actors are still cast in romantic lead parts, their love interests are almost invariably much younger.

Analyzing the Trends

Do you think that the aging of the baby boomers will affect the mass media's portrayal of the elderly? Why or why not? Describe any changes in attitudes toward the elderly that you think will be brought about by an increase in their numbers.

of food, the pleasures of well-being. He has lost his desire and capacity for sex. His body shrinks, and so too does the flow of blood to his brain. His mind does not utilize oxygen and sugar at the same rate as formerly. Feeble, uninteresting, he awaits his death, a burden to society, to his family, and to himself. (Butler 1975:6–7)

Are these stereotypes accurate? By definition stereotypes are usually inaccurate because they do not apply to all members of a group. Stereotypes of older people are no exception, as much research has shown. Most elderly people are not senile. Old age is not a sexless period for the majority of those over sixty-five. There are few age differences in response to job-related challenges; most elderly people are able to learn new things and can enthusiastically adapt to change (Palmore 1980, 1981; Hendricks and Hendricks 1987; Atchley 2000).

In summary, there is enough evidence to challenge the popular stereotypes about elderly people. It is, of course, true that some older people fit one or more of these stereotypes (as do some younger people) and that more individuals are likely to fit one or more of them as they reach age seventy. This, however, does not justify applying these stereotypes to all senior people at any age or for mindlessly applying them to individuals in their fifties and sixties.

Age and Inequality

Elderly People as a Minority Group

Are America's elderly a minority group? Jack Levin and William Levin (1980) contend that the field of **social gerontology**—the scientific study of the social dimensions of aging—has in the past blamed the elderly for their situation in much the same way that Americans in general have blamed the poor for their plight. Because early researchers tended to study older people in institutions, their studies focused on people with diminished mental and physical capacities. This perspective coincides with the American public's negative view of the elderly. Many scholars, however, join Levin and Levin in their quest to avoid blaming the elderly for their situation and view them instead as a minority (M. L. Barron 1953; Comfort 1976; Palmore 1978; Hillier and Barrow 1999).

Richard Posner (1996) suggests that it is not necessary to consider the elderly (or women) as a minority to recognize them as victims of prejudice and discrimination. The existence of ageism alone substantiates the persistence of age-based inequality in America.

Economics and the Elderly

What is the economic condition of the elderly? Because the economic situation among America's elderly has improved since 1960, the public thinks of the elderly as not only well-off, but as taking economic resources from the young (Duff 1995). It is true that the poverty rate for Americans over sixty-five years of age has declined since 1960 (see Figure 11.5). The conclusion that the elderly are well-off economically, however, fails to capture the complexity of their economic condition. Among the factors masking the true economic situation of the elderly are distorted measures of poverty and factors of race, ethnicity, and gender.

What is being left out of this rosy economic picture? The measurement of poverty among older people distorts the reality: the poverty line is drawn at a higher dollar amount for the elderly than for younger Americans. Although the elderly spend proportionately more on health care and housing than younger people do, the federal government assumes a lower cost of living for the

FIGURE 11.5

Poverty Rates Among Americans 65 and Over, 1960–2000

This figure documents the changing poverty rate among Americans 65 and over since 1960. Explain why it would be misleading to cite the current poverty rate as evidence that older people are economically well off.

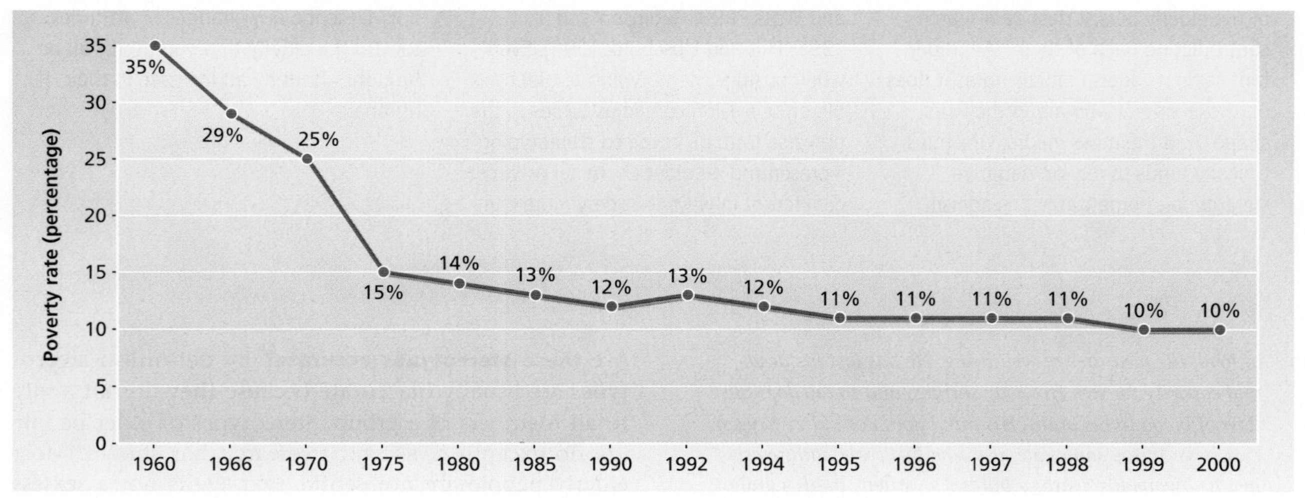

Source: U.S. Bureau of the Census, *Poverty in the United States: 2000* (Washington, DC: U.S. Government Printing Office, 2001).

elderly. If the standard used for younger age categories were applied to elderly people, their poverty rate would increase from 10 to 15 percent. Gross poverty rates also fail to take into account the just over 7 percent of the elderly who are officially considered "near poor." Counting these at-risk elderly people, about one in six of those over age sixty-five is poor (U.S. Bureau of the Census 2001b). Nor do official statistics include the "hidden poor" among the elderly population, who live either in institutions or with relatives because they cannot afford to live independently. Inclusion of these people would substantially raise the poverty rate for America's elderly.

Diversity is another factor complicating an evaluation of the economic situation of the elderly (Lee and Haaga 2002). There is a wide financial gap among the elderly. Some older people have moderate to high incomes based on dividends from assets, cash savings, and private retirement programs (Davis and Rowland 1991). Most elderly Americans, however, do not have sources of income beyond Social Security benefits. The existence of a small percentage of high-income older people gives the false impression that most elderly people are economically well-off (Himes 2001).

How do race and ethnicity compound age inequality? The elderly who are members of racial or ethnic minority groups are generally in worse condition than older white Americans (see Table 11.3). Whereas the median income of whites over sixty-five was approximately $14,961 in 2001, the median income for elderly African Americans was $10,631; for elderly Latinos, it was about $9,133. The poverty rate among the African American elderly is nearly triple that of whites. For older Latinos, the poverty rate is two-and-one-half times that of white Americans. Unquestionably, this disparity is intimately linked to the lifetime of prejudice and discrimination experienced by these racial and ethnic groups. The problem just becomes magnified in old age (Calisanti and Slevin 2001; Olson 2001).

TABLE 11.3

Median Income of Americans Age 65 and Older by Race, Ethnicity, and Gender, 2001

Race/Ethnicity	Total 65+	Male 65+	Female 65+
All races	$14,152	$19,688	$11,313
White	14,961	20,856	11,929
African American	10,631	13,776	9,051
Latino	9,133	12,338	7,585

Source: U.S. Bureau of the Census, *Historical Income Tables—People;* at www.census.gov/hhes/income/histinc/incperdet.html

© Bachmann/Photo Researchers, Inc.

Not all of America's elderly are at the bottom of the stratification structure. Consider this affluent retired couple with their grandchildren.

How does gender heighten age inequality? Because women the world over live longer than males, there is a global trend toward the feminization of poverty among the elderly. Since the highest poverty rates exist among single women, the longer life expectancy of women means that, if poor, they will live longer than men will in that condition. Elderly women constitute one of the poorest segments of American society (see Table 11.3). Women over age sixty-five are 63 percent more likely to live in poverty than are their male counterparts (U.S. Bureau of the Census 2001b). Elderly women most likely to be poor are single women who have never married or who are divorced, separated, or widowed. About 44 percent of women over sixty-five have annual incomes below $10,000, compared to 18 percent among older males. Although the median income of elderly women is $11,313, it is $19,688 for males over sixty-five. Thus, the average income of elderly women is almost 60 percent that of elderly males, lower than the amount that American women under sixty-five earn relative to working males (U.S. Bureau of the Census 2001a). This is not surprising, because the roots of poverty among older women lie in the work-related experiences women have had throughout their lives (Sidel 1996). The consequences of gender inequality are compounded by the inequalities of age.

Political Power and Older Americans

Clearly, any power older Americans have cannot be based on wealth or occupational prestige. Their major

WORLDVIEW 11.1

Gender Differences in Life Expectancy Around the Globe

Source: *2001 World Population Data Sheet* (Washington, DC: Population Reference Bureau, 2001).

This map shows graphically that women tend to outlive men the world over. In more developed countries, the average life expectancy at birth is 79 years for women and 72 years for men. While life expectancy for both genders is lower in less developed societies, women still live longer than men on average (67 versus 63 years).

1. Identify the major factors that account for the persistent gender difference in longevity worldwide.
2. Do you think this gender gap in longevity can be eliminated? Reduced? Explain.
3. Discuss the implications of the data on this map for poverty among women.

© Alain Keler/CORBIS SYGMA

Senior citizens of minority racial or ethnic groups experience the highest rates of poverty. This African American man living on a farm is existing close to the level of subsistence.

source of influence must be through the political process: political interest groups and the voting booth.

Do the elderly use interest group participation to their advantage? Seniors do comprise a large segment of the American population, but because large numbers do not necessarily translate potential political power into reality, the elderly also attempt to exercise their power through political interest groups. The assault on ageism is being waged by millions of Americans who belong to such "gray lobby" organizations as AARP (American Association for Retired Persons), the Gray Panthers, the National Council of Senior Citizens (NCSC), the National Association of Retired Federal Employees (NARFE), and the National Caucus and Center on Black Aged (NCCBA).

What is the voting turnout among the elderly? Voting turnout in the United States increases with age. Since the mid-1980s, Americans sixty-five and over have been the most active voters in presidential and congressional elections (see Figure 11.6). In 2000, for example, 68 percent of the elderly voted in the presidential elec-

FIGURE 11.6

Percentage of Americans Voting, by Age

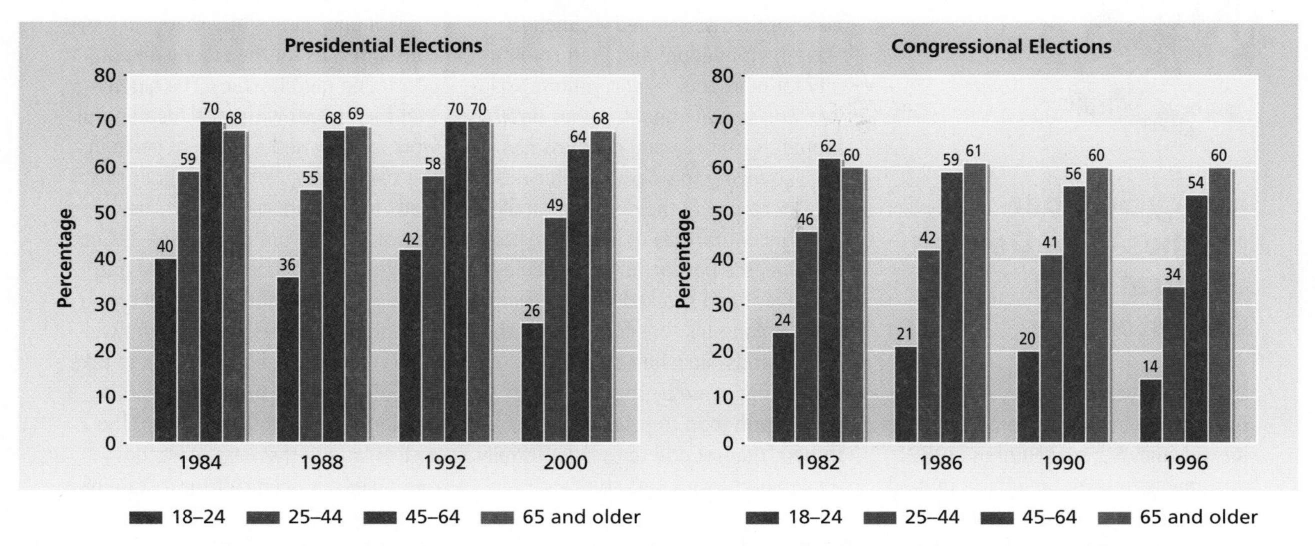

Source: U.S. Bureau of the Census, *Statistical Abstract of the United States: 2001* (Washington, DC: U.S. Government Printing Office, 2001), p. 251.

tion, compared to 26 percent of the eighteen- to twenty-four-year-olds and just over 49 percent of the twenty-five- to forty-four-year-olds (U.S. Bureau of the Census 2002c). Given current trends, by 2022 there will be four times as many voters sixty-five and older as there are voters twenty-nine and under (Goldstein and Morin 2002).

Do the elderly actually exercise significant political power in the United States? The potential political power of the elderly is not fully realized because of the diversity among the older population. Because the elderly cut across many important divisions in American society—social class, ethnicity, race, geographical area, religion—they do not speak with a unified political voice. In fact, the elderly do not vote as a bloc on any political question, even on issues related directly to their interests. Consequently, there is a debate on whether the elderly are an effective political force in the United States. Some analysts see the elderly as a strong political force influencing both the political elections and the decisions of public officials. They point out, for example, that just over one-third of the U.S. Senate is sixty-five years or older. Some of these analysts even fear excessive gray power. Others believe that the elderly are not a significant political force. The elderly, this line of argument goes, have been successful only in a defensive mode. Granted, they have successfully shielded their benefits programs from budget cuts and harmful legal changes. Yet, the elderly have not been very effective either in changing existing laws that have been harmful or in creating new policies that might be helpful (Jacobs 1990). In addition, some researchers assert, the elderly are too heterogeneous to truly be politically effective.

Like most debates on social issues, the truth seems to lie between these two extremes. Although the actual political clout of seniors today may be overestimated, their voting power and national political organizations make them a political force that cannot be ignored.

Will the political power of the elderly increase as America grays? Because of the stigma associated with aging, many older Americans identify with the middle-aged. If the elderly are to become a truly strong political force, they must identify with their own age-group. Such a step, leading to greater political power, would be a significant move against the negative effects of age stratification and ageism experienced by elderly Americans (Powell, Williamson, and Branco 1996).

Prospects for the Future

This chapter has documented some problems the elderly experience as a result of prejudice and discrimination in the United States. Closer examination of the aging population is now suggesting a brighter outlook for seniors. The baby boom generation—the 76 million Americans born between 1946 and 1964—are followed immediately by the much smaller Generation X, the 41 million born between 1965 and 1976. The arithmetic is easy for corporate America to understand: there is a looming shortage of new workers in line to replace the older workers leaving. And population relief is not

Nancy Scheper-Hughes—The Demise of Rural Irish Gerontocracy

Based on a year of fieldwork in Ballybran, a mountain parish of west Kerry, Nancy Scheper-Hughes (1983) documents the eroding effects of modernization on the elderly of western Ireland. This area of Ireland, rural and resting on an agricultural base, has traditionally been a gerontocracy in which the power resided in the hands of the "old ones," particularly elderly males.

Scheper-Hughes identifies the basic demographic shift that has led to the death of the rural Irish gerontocracy, describes the negative consequences of this change for the elderly, and discusses several areas in which the loss of social standing among the elderly is reflected.

Like many rural Irish communities, Ballybran's population has declined dramatically over the last century and a half. The origin of this demographic bust lies in the Great Famine of 1845–1849, which halved the parish from 2,772 to 1,500 people. The population now is below 500. The rural youth, seeing little future in the farming economy, migrate to the cities, leaving behind both a reduced pool of marriageable people and a male-female ratio imbalance. After age eighteen, males outnumber females by about three to one. Thus, in 1983, although Ballybran had more than thirty males between twenty-one and thirty-five who wished to marry, there were only five young unmarried women in this age category. Moreover, none of the five seemed inclined to leave their freedom and their community for marriage. This migration pattern and sexual imbalance obviously promotes celibacy and childlessness.

Underlying this demographic picture, of course, is the devaluation of the agricultural way of life. At an earlier time, the patriarchal father delayed retirement and sparked intense competition among his sons for rights to the family lands. Now heir selection is determined more by the process of elimination than the choice of the father: "the last one to escape (usually the youngest son) gets stuck by default with an unproductive farm and saddled with a life of celibacy and greatly resented service to the 'old people'" (Scheper-Hughes 1983:134).

The result for the aged parents is fairly clear; they no longer have the economic power base to control the younger generation and to maintain their superior status in the family and community. Because young people prefer to be "liberated" from the land, the "old ones" control little that the youth want. The awe and respect for the elderly that once characterized the community has, in many cases, been replaced not only by pity but by contempt. The demise of the traditional family farming-based culture leaves the elderly father, in Scheper-Hughes's words, a "broken figure." Toleration from his adult children is the most he can expect; open ridicule the worst. With the erosion of their economic power, the elderly have also lost their cherished role as preservers of the ancient Celtic traditions—the myths, stories, songs, prayers, and proverbs. In fact, the young tend to reject these traditions. Worse, most of the high school students resent having to study the Irish language—a "dead" language that they believe will be of no use to them in the commercial and professional world outside the rural community.

This ageism has had many negative consequences for the elderly inheritors of a faded gerontocracy. The elderly find themselves without a meaningful work identity and a sense of place in the community. Not surprisingly, this leads to rampant alcoholism, loss of sexual interest, diminished self-esteem, and considerable depression among those over fifty, many of whom are bachelors, spinsters, widows, and widowers without family or friends to take care of them. The author offers this succinct, poignant summation: "The Irish village of the west coast today embodies a broken culture: a state of affairs most detrimental to the aged who are unable to flee or to accept new values, and who, consequently, are left to contemplate the wreckage" (Scheper-Hughes 1983:145).

Thinking About the Research

1. Would you feel any greater confidence in the results of this study if survey research techniques had been used? Why or why not? Be specific in your analysis and evaluation.
2. Discuss the application of functionalism, conflict theory, and symbolic interactionism to this study. Which theoretical perspective, or combination of perspectives, contributes most to our understanding of the demise of the rural gerontocracy in western Ireland? Defend your position.
3. Does the government have an active role to play in any society in which modernization is destroying a long-standing gerontocracy? Explain and defend your position to a small group of Ballybran's elderly. Discuss how they would probably respond to your viewpoint.

expected until after 2010, when most of Generation Y, the boomers' children, enters the workforce (Sherrid 2000). Consequently, many employers are now beginning to view older employees as a resource rather than a liability. They are just beginning to underscore experience, job skills, and flexibility of older workers rather than their alleged lack of productivity and creativity. At the same time, older Americans are increasingly remaining in the labor force. The number of sixty-four- to sixty-nine-year-olds still working has risen 25 percent over the last ten years (Kadlec 2002). If some of the negative stereotypes about the aged weaken in the process, increased longevity may result (Levy et al. 2002).

FEEDBACK

1. The scientific study of the social dimensions of aging is known as __________.
2. Of the following, which is an accurate statement?
   a. Since 1960, the economic situation for the elderly in the United States has deteriorated.
   b. The poverty rate for Americans over sixty-five is lower than the official count indicates.
   c. The elderly who are members of racial or ethnic minorities are in much worse condition than other older people are.
   d. Only about one-fourth of American families in poverty have female heads of household.
3. America's elderly for the first time now speak with a unified political voice. T or F?

Answers: 1. social gerontology 2. c 3. F

SOCIOLOGY AND THE NEWS MEDIA

Baby Boomer Marketing

Baby boomers have always made their presence known. For example, this post–World War II population juggernaut caused the expansion of public schools and universities as they grew up. Now, as they grow older, their numbers are affecting advertisers who have always recognized them as a mass market.

© Cable News Network, Inc.

Personal computers didn't exist in the marketplace when this grandfather was growing up. Computer marketing is now clearly targeting him to buy the latest laptop.

1. Describe a few recent television advertisements that reflect baby boomer marketing. Include some cultural symbols that advertisers are using to attract baby boomers.

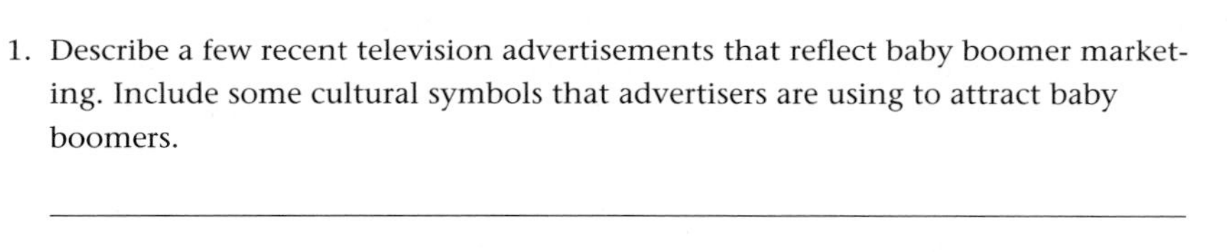

2. Discuss two major changes in society that will occur as baby boomers begin to retire.

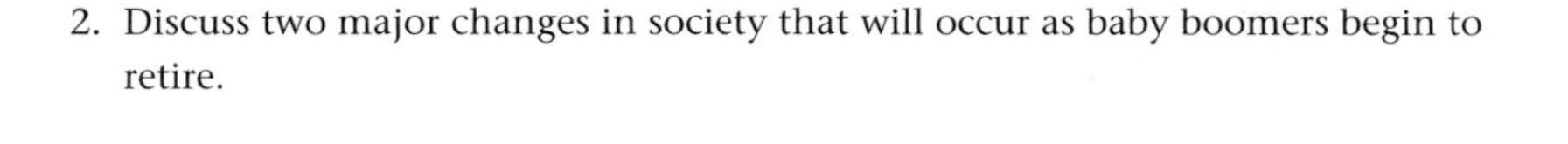

SUMMARY

1. Age stratification—the distribution of scarce desirables based on chronological age—can affect any age-group. The social inequality reflected in age stratification is always justified socially and culturally. The justification for the relatively low social standing among the elderly in a society is based on ageism—a set of beliefs, attitudes, norms, and values that permit age-based prejudice and discrimination.
2. In hunting and gathering societies, age usually works against the elderly. In fact, geronticide—the socially accepted killing of the elderly—has been documented in many hunting and gathering societies. The situation tends to change in agricultural societies, which are often gerontocracies based on the type of stratification structure in which the greatest share of scarce desirables are held by the elderly.

3. Social inequality for the elderly tends to reemerge with modernization. Even Japan, an industrial society that has traditionally revered the elderly, appears to be beginning to conform to this pattern.
4. The aging of the population is occurring in all highly industrialized societies. It is, therefore, not surprising that the graying of America is one of the most significant changes in the American population. Whereas persons over sixty-five years of age comprise 13 percent of the population today, they will constitute 20 percent in 2040.
5. One of the important consequences of the graying of America is the rising dependency ratio. Thus, as the population ages, there will be proportionately fewer employed people to support the growing retired population.
6. Each of the three major theoretical perspectives has a unique slant on ageism. According to functionalism, ageism increases as industrialization proceeds because the contributions to society by the elderly diminish. Conflict theorists attribute ageism to a battle for scarce resources. Whereas the elderly are needed to work in preindustrial society, they become competitors with younger workers in industrial society, where a surplus of labor exists. Symbolic interactionism traces ageism to the process of socialization, in which stereotypes of the elderly are learned. Although erroneous, these stereotypes are used to justify prejudice and discrimination that harm the elderly.
7. Although the economic situation of America's elderly has improved over the last several decades, their poverty rate still stands at about 10 percent. Moreover, the gross poverty rate masks the true economic situation of most elderly Americans. The elderly who are members of racial or ethnic minorities or who are female are in the worst economic condition.
8. Power for the elderly is not and cannot be financially or occupationally based. It is expressed through the political process—through their high voting rate and their support of relevant interest groups. The political influence of older people, however, is diluted significantly because of their racial, ethnic, social class, religious, and geographical diversity.
9. The future prospects for the elderly appear somewhat brighter. The looming shortage of labor as more older Americans leave the workforce make the elderly more valuable in the job market.

INFOTRAC® COLLEGE EDITION

http://www.infotrac-college.com/wadsworth

Another unique option available to you at the Student Resources section of the companion Web site is Infotrac College Edition, an online library with access to hundreds of scholarly and popular periodicals. Below are suggested search terms for this chapter. Results from these and other searches are found at the site.

Search keywords: **baby boomers, Generation X, Generation Y**. Find articles that discuss these three generations of Americans. What age cohorts make up each of these groups? How are they similar to each other? How are they different? To what extent do they influence American society? Which generation do you belong to? Do you agree with the description of your generation?

Search keywords: **Gray Panthers**. Scan the articles you find on Infotrac relating to the Gray Panthers. Who are they? Based on the article titles you find, what seems to be the group's primary focus?

Search keywords: **social security**. Read a few articles related to America's social security system. What is its primary purpose? What difficulties are currently being encountered in this system? Use the information you learned in Chapter 11 to develop a logical explanation for the current problems.

LEARNING OBJECTIVES REVIEW

- Describe the aging of the world's population.
- Discuss the graying of America.
- Explain the importance of the dependency ratio.
- Distinguish between age stratification and ageism.
- Contrast the theoretical perspectives of ageism.
- Outline the inequality experienced by older Americans.

CONCEPT REVIEW

Match the following concepts with the definitions listed below them.

____ a. ageism
____ b. age stratification
____ c. age cohort
____ d. social gerontology
____ e. geronticide
____ f. population pyramids
____ g. dependency ratio
____ h. gerontocracy
____ i. age structure

1. a set of beliefs, attitudes, norms, and values used to justify age-based prejudice and discrimination
2. the unequal distribution of scarce desirables in a society based on age
3. a type of society in which the greatest share of scarce desirables are held by the elderly
4. persons born during the same time period in a particular population
5. the socially accepted killing of the elderly within a population
6. the distribution of people of different ages within a society
7. the proportion of persons in the dependent ages relative to those in the economically active ages
8. graphic representations illustrating the age and sex distribution of a population
9. the scientific study of the social dimensions of aging

CRITICAL-THINKING QUESTIONS

1. Considering the economic situation of America's elderly, do you think that ageism is at work? Provide the detailed information you need to convince your best friend, who disagrees with you.

____

____

2. Discuss factors that led to the graying of America.

____

____

3. Based on information in Chapter 11, would you expect that the dependency ratio has increased, decreased, or remained stable over the past fifty years? What other factors influence the dependency ratio? How will America's dependency ratio affect you as you enter the labor force?

____

____

4. Compare and contrast the functionalist and conflict theory perspectives of aging.

____

____

5. Examine the stereotypes commonly associated with the elderly. Based on your personal experience with older people, do you think those stereotypes are accurate? Why or why not?

____

____

MULTIPLE-CHOICE QUESTIONS

1. **Of the following, which is an accurate statement?**
   a. Since 1960, the economic situation for the elderly in the United States has deteriorated.
   b. The poverty rate for Americans over sixty-five is lower than the official count indicates.
   c. The elderly who are members of racial or ethnic minorities are in much worse condition than other older people.
   d. Only about one-fourth of American families in poverty have female heads of household.
2. **Age stratification is justified by**
   a. scientific research.
   b. legal precedent.
   c. society.
   d. economic reality.
3. **Geronticide is most prevalent in ___________ societies.**
   a. hunting and gathering
   b. horticultural
   c. industrial
   d. postindustrial
4. **Modernization in most societies is associated with**
   a. geronticide.
   b. gerontocracy.
   c. lower dependency ratios.
   d. age stratification.
5. **Graying (the aging of a population) is most common in**
   a. developing countries.
   b. highly industrialized countries.
   c. preindustrial societies.
   d. African and Asian countries.
6. **The most famous age cohort in the United States is**
   a. the Depression population.
   b. baby boomers.
   c. the Gray Panthers.
   d. the Vietnam protestors.
7. **Which of the following factors does *not* affect the age structure of a population?**
   a. fertility rate
   b. age stratification
   c. mortality rate
   d. life expectancy
8. **The maximum size of an elderly population at a given time is primarily reliant on**
   a. migration rates.
   b. mortality rates.
   c. the number of births sixty-five or more years ago.
   d. morbidity rates.
9. **The proportion of persons under age fifteen and over sixty-four to those between fifteen and sixty-four is known as the**

a. dependency ratio.
b. age cohort.
c. population pyramid.
d. ageism.

10. **Conflict theory explains ageism as a response to**
a. reduced contributions by the elderly.
b. stereotypes.
c. competition for scarce resources.
d. prejudice.

11. **Including the "near poor" in the number of at-risk elderly means that about one in ______ of those age sixty-five and older is poor.**
a. ten
b. twenty
c. two
d. six

12. **The primary reason that the elderly have a degree of political power in the United States is**
a. their economic status.
b. the esteem in which they are held by society.
c. their high voting rate.
d. their accumulated wisdom.

FEEDBACK REVIEW

True-False

1. America's elderly for the first time now speak in a unified political voice. T or F?
2. The aging of the population is confined to only the most highly industrialized societies. T or F?

Fill in the Blank

3. ________ refers to a set of beliefs, attitudes, norms and values used to justify age-based prejudice and discrimination.
4. A ____________ is a stratification structure in which the elderly hold the greatest share of scarce desirables.
5. The scientific study of the social dimensions of aging is known as ______________.

Multiple Choice

6. **Of the following, which is an accurate statement?**
a. Since 1960, the economic situation for the elderly in the United States has deteriorated.
b. The poverty rate for Americans over sixty-five is lower than the official count indicates.
c. The elderly who are members of racial or ethnic minorities are in much worse condition than other older people are.
d. Only about one-fourth of American families in poverty have female heads of household.

7. **By 2050, the population pyramid of the United States is expected to resemble**
a. a triangle.
b. a classic pyramid.
c. a rectangle.
d. an inverted pyramid.

Matching

8. Following are several statements about the elderly. Identify each statement with one major theoretical perspective: functionalism (F), conflict theory (C), or symbolic interactionism (S).
____ a. Ageism results in part from an oversupply of labor.
____ b. The young may wish to distance themselves from the elderly to avoid facing their own mortality.
____ c. The stigma attached to aging promotes a low self-concept among the elderly.
____ d. Ageism is associated with industrialization.
____ e. Older people are stereotyped.
____ f. Ageism exists in part because older workers are more expensive than younger ones.

GRAPHIC REVIEW

Figure 11.4 depicts the age and sex structure of the United States as of 1945 and its projected age and sex structure in 2050. Answer the following questions to check your understanding of this information as it relates to the material in the chapter.

1. What was the largest age cohort in 1945?

____

____

2. What factors contribute to the change in the shapes of the population pyramids in 1945 and 2050?

____

____

3. Why is the 70–79 age cohort markedly smaller than nearly all of the other age cohorts in the 2050 population pyramid? Check your answer against what you know about the baby boomers, Generation X, and Generation Y.

____

____

ANSWER KEY

Concept Review

a. 2
1
4
9
5
8
7
3
6

Multiple Choice

1. c
2. c
3. a
4. d
5. b
6. b
7. b
8. c
9. a
10. c
11. d
12. c

Feedback Review

1. F
2. F
3. Ageism
4. gerontocracy
5. social gerontology
6. c
7. c
8. a. C
   b. F
   c. S
   d. F
   e. S
   f. C

12 Family

OUTLINE

© Mark Richards/PhotoEdit

LEARNING OBJECTIVES

- Describe the types of family structure, dimensions of family structure, societal norms for mate selection, and types of marital arrangements.
- Compare and contrast views of the family proposed by functionalists, conflict theorists, and symbolic interactionists.
- Outline the extent and cause of divorce in America.
- Give an overview of family violence in the United States.
- Describe contemporary alternatives to the traditional nuclear family structure.
- Discuss the future of the family in the United States.

USING THE SOCIOLOGICAL IMAGINATION

U.S. divorce rates are soaring! Family life is on the ropes. Yes? No? Films, television, music, and print media suggest that America's family life is in jeopardy. Actually, data on divorce provide grounds for some optimism. Although a dramatic rise in the divorce rate did begin in the early 1960s, the divorce rate began a decline in 1985.

The facts regarding divorce are not as clear-cut as commonly believed, and the family is not as simple an institution as it appears. It is a complex social unit with many dimensions. In addition, there is great diversity in the family structure—across various societies and even within one society (United Nations 1996; Sussman, Steinmetz, and Peterson 1998).

Marriage and Family Defined

In the simplest terms, a **marriage** is a legal union based on mutual rights and obligations (Eshleman 2002), and a **family** is a group of people related by marriage, blood, or adoption. The **family of orientation** is the family a person is born into, or the family of birth. It provides children with a name, an identity, and a heritage. In other words, it gives the child an ascribed status in the community. The family of birth "orients" children to their neighborhood, community, and society. The forms you fill out to register for college typically ask, What is your father's name? What is your mother's name? Where do you live? Where do your parents work? The answers to these questions indicate who you are and to whom you belong. The family of orientation locates you in the world. The **family of procreation** is established upon marriage. The marriage ceremony signifies that it is legal (officially sanctioned) for a couple to have offspring and to give the children a family name. The family of procreation, then, becomes the family of orientation for the children created from that marriage.

All known societies have families and marriage; the permutations, however, are staggering. Exposure to these variations requires something of an anthropological excursion.

Cross-Cultural Analysis of Family and Marriage

The structure of family and marriage involves many variations and many cultures. This cross-cultural analysis examines the predominant family and marriage forms in modern society as well as an appreciation for alternative lifestyles.

Types of Family Structure

The **nuclear family**, the smallest group of individuals that can legitimately be called a family, is generally composed of a mother, a father, and any children. Because the nuclear family is usually based on marriage,

The dominance of the nuclear family is a fairly recent phenomenon. At the beginning of the twentieth century, the extended family—in which several adult generations live together—was the dominant family type in the United States.

it is sometimes called the *conjugal family*. However, the two are not always synonymous. A nuclear family can be composed of a single parent and children or of a brother and sister. The **extended family** consists of two or more adult generations of the same family whose members share economic resources and live in a common household. Extended families may contain grandparents, children, grandchildren, aunts, uncles, and so forth. Because it is based on blood ties, the extended family is often identified as the *consanguine family*.

In general, extended families are most characteristic of preindustrial societies and rural parts of industrial societies. This is not to say, however, that nuclear families were not common in preliterate societies or in preindustrial Europe (Nydeggar 1985). Moreover, it would be inaccurate to conclude that the nuclear family—most characteristic of modern society—exists in isolation from a larger kinship network. Both of these issues—the association of the nuclear family with industrialization and the question of the isolation of the modern nuclear family—require elaboration.

How are industrialization and the nuclear family related? As Figure 12.1 indicates, there is a curvilinear relationship between types of family structure and industrialization. Before humans began to domesticate animals and cultivate crops, most economies were based on hunting and gathering. Small bands of nuclear families followed herds of animals and changing seasons, moving around constantly, never staying long in any one place. When humans learned to domesticate animals to help with tilling the soil and cultivating crops (about

FIGURE 12.1

Economy and Family Structure

The nature of family structure varies with the type of economy. Agricultural society promotes the extended family because of the need for labor from a large number of people. The nuclear family is more compatible with hunting and gathering societies and modern societies.

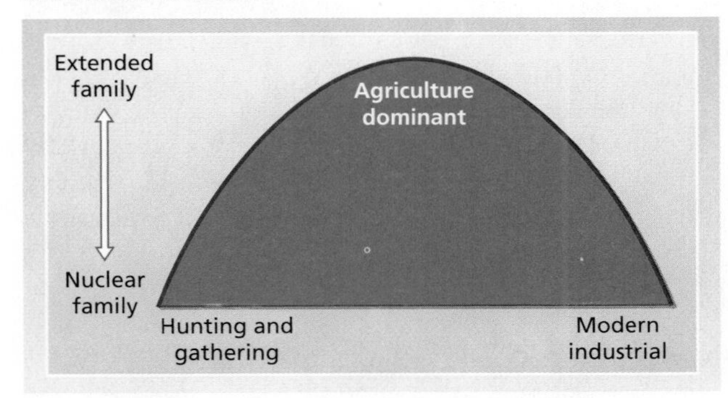

Source: Richard R. Clayton, *The Family, Marriage, and Social Change*, 2d ed. (Lexington, MA: D.C. Heath, 1979), p. 95. Reprinted by permission of the publisher.

9,000 years ago), they no longer needed to be mobile to maintain a food supply. Families began to farm, settle down, and establish roots. Large families were then needed to plow, harvest, and fulfill the many other tasks needed to live. Thus, the emergence of family farms and permanent residences encouraged the development of the extended family. When agriculture became the dominant economic base, the extended family became the major type of family. In general, when societies move from agricultural economies to industrialized ones, the extended family is replaced by the nuclear family because large families are no longer needed to supply the necessities of life (W. J. Goode 1970).

What accounts for the relationship between industrialization and nuclear family structure? One factor is increasing geographical mobility. People in rural areas must leave their relatives to secure industrial employment in cities. Furthermore, for occupational reasons, those already living in urban areas come to accept frequent changes in residence. Either way, geographical mobility makes it difficult to maintain close ties with large numbers of relatives; usually, only the most immediate kin can be taken along. Also, before industrialization, family members were expected to take care of the sick and the elderly. (Actually, they scarcely had a choice, because there were few outside resources to rely on.) With the growth of special-purpose organizations and government programs, however, the sick can be cared for in hospitals—which insurance companies often pay for—and the elderly can live in assisted living homes, paid for in part by their Social Security payments. In short, because outsiders have taken over many of the services formerly performed by the family, relatives do not have to depend on one another as much as they once did. Thus, certain conditions associated with industrialization are more compatible with the nuclear family than with the extended family.

How isolated is the modern nuclear family? Although families in industrialized societies tend to have a nuclear structure and maintain separate residences, there is evidence that they are not as isolated from their relatives as was once believed. Early research depicting the American nuclear family as isolated was conducted prior to some recent changes—changes that are now occurring in other industrialized societies as well. For one thing, modern transportation and communication make it easier for relatives to maintain family ties even though separated geographically. Second, because of metropolitan growth, children no longer need to leave their parents and other relatives to go to college or to pursue their occupations.

Early on, Marvin Sussman (1953) observed considerable cooperative effort and mutual aid between adults in urban nuclear families and their parents. According to

Sussman, families of orientation and procreation are linked together through a kin network. He found, for example, parents offering a variety of services to their married children, including financial aid, home repair and maintenance, baby-sitting, and various kinds of help during illnesses. Subsequent researchers have also found this network of kin relationships (Litwak 1960; Adams 1968, 1970; Troll 1971). A study of Muncie, Indiana, by Theodore Caplow and his colleagues (1982) showed that a high proportion of the close kin of the residents surveyed lived within 50 miles of the city. Most of the residents with parents still living saw their parents at least once a month, and nearly half saw them either weekly or more often. Apparently, the overwhelming majority of Americans maintain frequent contact with a substantial number of relatives. At the very least, it must be concluded that the modern family has either a modified nuclear or extended family structure that falls somewhere between the isolated nuclear family and the classic extended family (De Luca 2001).

Whatever the basic family type—nuclear or extended—several dimensions of family structure must be recognized. These patterns take into consideration three circumstances: descent and inheritance, authority within the family, and place of residence.

Dimensions of Family Structure

Who owns what? Different arrangements exist for determining descent (who becomes head of the family) and inheritance (who owns family property). If the arrangement is patrilineal, descent and inheritance are passed from the father to his male descendants. Iran, Iraq, and Tikopia in the Western Pacific are examples of **patrilineal** societies. Should descent and inheritance be transmitted from the mother to her female descendants, the arrangement is a **matrilineal** one. Some Native American tribes, such as the five belonging to the League of the Iroquois in Pennsylvania and upper New York, were **matrilineal** (Zinn 2001). In other societies, descent and inheritance are bilateral, passed equally through both parents. Thus, both the father's and mother's relatives are accepted equally as part of the kinship structure. For most American families, this is the situation today.

Who has authority? With **patriarchal control**, the oldest man living in a household has authority over the rest of the family members—as in Jordan, Iraq, and China (Khouri 2003). In the purest form of patriarchy, the father is the absolute ruler. Usually, patriarchal families are extended families structured along bloodlines. So rare is **matriarchal control**, in which the oldest woman living in a household holds the authority, that controversy exists over whether any society has ever had a genuinely matriarchal family structure. With **democratic control**, authority is split evenly between husband and wife. Many families in the Scandinavian countries and in the United States follow the democratic model.

Who lives where? In a nuclear family, a married couple establishes a new residence of its own. Such a residential arrangement is a **neolocal residence**. Extended families, of course, have different residence norms. A **patrilocal residence**, as in premodern China, calls for living with or near the husband's parents. Residing with or near the wife's parents is expected under a **matrilocal residence**, such as the Nayar caste of Kerala in southern India.

Mate Selection

In the United States, most people assume they will have the freedom to choose their mate. It is true that American parents generally no longer arrange marriages for their children. American women no longer fill hope chests with quilts, china, silverware, and other household goods in anticipation of marriage. An American man is not expected to negotiate, with a future father-in-law, a price for the woman he wishes to marry. Therefore, we assume that freedom of choice prevails in the selection of a marriage partner. Actually, this is a false assumption. All societies, including the United States, have norms regarding who may marry whom.

What does society tell us about whom to marry? **Exogamy** refers to mate selection norms requiring individuals to marry someone outside their kind. Exogamous norms are usually referred to as incest taboos. In the classic Chinese culture, for example, two people with identical family names could not marry unless their family lines were known to have diverged at least five generations previously (Queen, Habenstein, and Quadagno 1985). In the United States, you are not legally permitted to marry a son or daughter, a brother or sister, a mother or father, a niece or nephew, or an aunt or uncle. In twenty-nine states, marriage to a first cousin is prohibited, and it is illegal to marry a former mother-in-law or father-in-law. Incest is almost universally prohibited. In fact, incest has seldom existed as an established pattern of mate selection. Royalty in ancient Europe, Hawaii, Egypt, and Peru were prominent exceptions. Even in these instances, most members of the royal families chose partners to whom they were not blood related.

Endogamy involves mate selection norms that require (or at least encourage) individuals to marry within their own kind. In the United States, an endogamous norm requires that marriage partners be of the same race, ethnicity, religion, and social class. The degree to which these endogamous norms are enforced is reflected in the small percentage of interracial and interfaith marriages (2 percent).

What is the difference between endogamy and homogamy? While endogamy refers to cultural pressure to marry someone similar, **homogamy** refers to the tendency to marry someone similar to oneself based on personal preference. Such choices are made because of the greater likelihood of being attracted to and becoming intimate with someone like ourselves. Some factors leading to homogamous choice are age, educational background, degree of physical attractiveness, shared interests, and extent of religious commitment. Of course, endogamy and homogamy can converge when norms encouraging mate selection with one's race, religion, or social class are so completely internalized that they determine personal preference. But people do tend to marry someone based on one or more shared characteristics. Most marriages in the United States occur between individuals who are about the same age. Most people who have never before married become married to someone who also has always been single. Divorced people tend to marry those who have been previously married. Finally, people tend to choose marriage partners from their own communities or neighborhoods (Knox and Schacht 2002).

Although still the exception in the United States, **heterogamy** is rising. In heterogamous marriages, partners are dissimilar with respect to some important social characteristics. Presently, in America, more marriages are crossing traditional barriers of age, race, social class, and ethnicity. This trend is due to several factors. America has become a bit more racially and ethnically integrated. Although they represent only 2 percent of all marriages in the United States, mixed marriages have doubled since 1980 (Suro 1999). (Figure 12.2 shows the racial and ethnic breakdown of intergroup marriages in 1998.) Also, class lines are crossed with greater frequency because more Americans of all social classes are attending college together. Finally, norms separating age-groups have weakened.

As you can see, who may marry whom in a society is not completely a matter of choice. The field of eligible marital partners is limited by norms of mate selection (Kephart and Jedlicka 1997; Kennedy 2003; Romano, 2003).

FIGURE 12.2

Intergroup Married Couples in the United States: 1998

In the United States, 2 percent of marriages are mixed. Use that fact and the information in this figure to generalize about heterogamy in the United States.

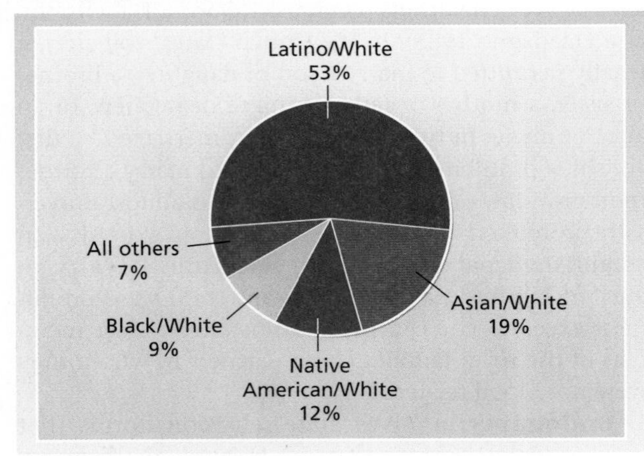

Note: White, black, Asian and Pacific Islander, and Native American categories do not include Latinos.
Sources: Roberto Suro, "Mixed Doubles," *American Demographics* (November 1999):61. Population Reference Bureau. U.S. Bureau of the Census, *Statistical Abstract of the United States: 1998* (Washington, DC: U.S. Government Printing Office, 1998), p. 60.

Types of Marriages

Mention a wedding, and most Americans think of a bride dressed in a long, white gown walking down an aisle; the husband-to-be waiting for his bride at the end of the aisle; the repetition of ancient vows to love and honor; an exchange of rings; and a kiss. In addition, though rarely, some modern couples get married while standing on mountaintops, while diving out of planes, and while swimming underwater. In other cultures, the wedding ceremony looks even more different. Here is a description of part of the ceremony among the Reindeer Tungus of Siberia:

> *After the groom's gifts have been presented, the bride's dowry is loaded onto reindeer and carried to the groom's lodge. There the climax of the ceremony takes place. The bride takes the wife's place—that is, at the right side of the entrance of the lodge—and members of both families sit around in a circle. The groom enters and follows the bride around the circle, greeting each guest, while the guests, in their turn, kiss the bride on the mouth and hands. Finally, the go-betweens spit three times on the bride's hands, and the couple are formally husband and wife. More feasting and revelry bring the day to a close.* (Ember and Ember 1999:310–311)

Whatever form it takes, the marriage ceremony is an important ritual announcing to the world that a man and woman have become husband and wife, that a new family has been formed, and that any children born to the couple can legitimately inherit the family name and property. The traditional marital arrangement in the United States is merely one of several possibilities.

What are the possible types of marriage? Monogamy—the marriage of one man to only one woman at a

WORLDVIEW 12.1

Types of Marriages

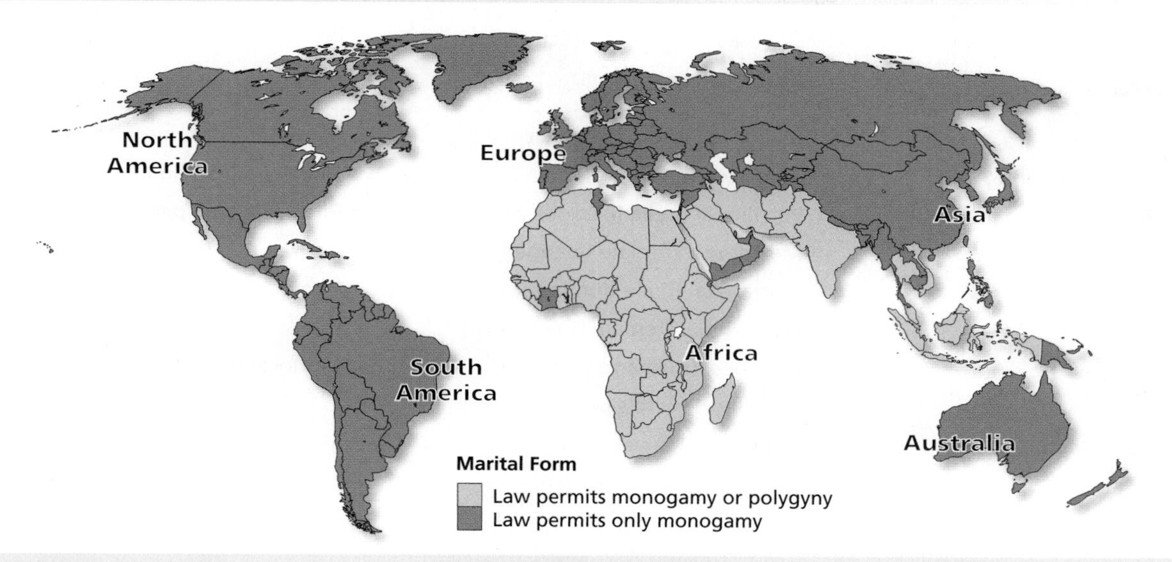

Monogamy—the marriage of one man to only one woman at a time—is the only legal form of marriage in all industrial and postindustrial societies. It is also the only legal arrangement in the Western Hemisphere. However, in many African and southern Asian nations, where Islam is the predominant religion, polygyny—marriage of one man to two or more women at one time—is legal. This map shows the countries where monogamy and polygyny are legal forms of marriage.

1. Is Islam related to polygyny? Explain.
2. Other than the influence of Islam, can you think of any reasons why polygyny would be legal?
3. Can you think of any reasons why monogamy is the only legal form of marriage in the West?

time—is the most widely practiced form of marriage in the world today. In fact, it is the only form of marriage that is legally acceptable in the United States and most other societies. Polygamy, on the other hand, which is the marriage of a male or female to multiple partners of the other sex, takes two forms: **polygyny** and **polyandry**. *Polygyny* is the marriage of one man to two or more women at the same time. The most obvious example of *polygyny* can be found in the Old Testament, in which King Solomon is reported to have had 700 wives and 300 concubines. George Murdock (1957) found that polygyny was practiced in 75 percent of the 575 preliterate societies he studied. However, polygyny is not legal in any Western society (Paddock 1999). As stated in its 1948 statement on human rights, the United Nations explicitly frowns on the practice of polygyny. **Polyandry**—the marriage of one woman to two or more men at the same time—is an even rarer form of marriage, found exclusively in the South Asian countries of Tibet, Nepal, and India (Stockard 2002). It is known to have been common in only two societies: in Tibet and among the Todas of India (Queen, Habenstein, and Quadagno 1985). Where polyandry has existed, it usually has consisted

The marriage of one man to one woman at a time (monogamy) is currently the only legal form of marriage in the United States. Whatever form the marriage ceremony takes—the traditional form, such as chosen by this couple, or a very unusual form chosen by a couple taking the vows while skydiving—the ceremony ritualistically announces to others that a new family has been created.

TABLE 12.1

Major Characteristics of Extended and Nuclear Families

Examine the major characteristics of extended and nuclear family structures. Do you live in a nuclear or extended family? Explain by describing your family based on each characteristic.

Characteristics	Extended Family	Nuclear Family
Family structure	Extended (also some nuclear)	Nuclear (also some extended)
Basis of family bond	Blood (consanguine)	Marriage (conjugal)
Line of descent and inheritance	Patrilineal (male lineage)	Bilateral (dual parental lineage)
	Matrilineal (female lineage)	
Locus of control	Patriarchal (male dominance)	Democratic (sexes share power)
	Matriarchal (female dominance)	
Place of residence	Patrilocal (husband's parents)	Neolocal (independent)
	Matrilocal (wife's parents)	
Marriage structure	Monogamy (one spouse)	Monogamy (one spouse)
	Polygyny (several wives)	
	Polyandry (several husbands)	
	Group (several husbands and wives)	

of several brothers sharing a wife (fraternal polyandry). Instances of a father sharing a wife with his sons have been reported, but not with any degree of regularity. A **group marriage** consists of two or more men married to two or more women at the same time. This form of marriage is also rare. In fact, researchers cannot identify even one society in which group marriage has been the normal form. There have, however, been communes that have practiced group marriage (Berger, Hackett, and Millar 1972; Kanter 1972; Ferrar 1977).

This analysis of the family and marriage is complicated because of the large number of concepts. Table 12.1 summarizes these concepts within the context of the traditional (extended) and modern (nuclear) family structures.

Theoretical Perspectives and the Family

The family is as central to society as it is complex. It is not surprising that each of the major theoretical perspectives sheds light on this institution.

FEEDBACK

1. A _________ is a group of people related by marriage, blood, or adoption.
2. A family of _________ is the family into which a person is born.
3. A family of _________ is the family established upon marriage.
4. The _________ family consists of two or more generations of the same family sharing a common household and economic resources.
5. There is a _________ relationship between types of family structure and industrialization.
6. Industrialization promotes the shift from the extended family to the _________ family.
7. In a _________ family, the oldest man living in the household controls the rest of the family members.
8. Indicate which type of mate selection norm is reflected in each of the following situations.
   ____ a. Jews are supposed to marry Jews. (1) exogamy
   ____ b. A father is not permitted to marry his daughter. (2) endogamy
   ____ c. Members of the same social class marry. (3) homogamy
9. _________ is the marriage of a woman to two or more men at the same time.

Answers: 1. family 2. orientation 3. procreation 4. extended 5. curvilinear 6. nuclear 7. patriarchal 8. a. (2) b. (1) c. (3) 9. Polyandry

Functionalism

You were born into a family with whom you have lived for almost twenty years. You are likely to marry, and despite the high rate of divorce, the overwhelming majority of your adult life will be spent within a family.

The family performs many essential functions. First, it provides the initial learning experiences that make people human. Second, it generally provides a supportive and loving atmosphere that fulfills basic human social and emotional needs. Third, it is the only legitimate source of reproduction for a society. Fourth, it regulates sexual activity. Fifth, it places people in a social class at birth. Finally, the family serves an important economic function. Because of these vital functions, the family is often referred to as the basic institution of society. Each function deserves elaboration.

How does the family contribute to the socialization process? A major function of the family begins immediately after a baby has been born. In addition to caring for an infant's physical needs, parents begin the vital process of teaching the child necessities for participating in society. During the first year, the infant begins to mimic words and, later, sentences. During the second and third years, parents begin to teach the child customary ways of behaving. By providing role models, training, and education, the process of family socialization continues as the child enters new stages of development.

What is the socioemotional function of the family? Another major function of the family is "socioemotional maintenance" (Nimkoff 1965). This begins in the first year with intimate holding, touching, stroking, and talking to the baby. Generally, the family provides the one supportive environment in society where an individual is unconditionally accepted and loved. Family members accept one another as they are; every member is special and unique. Without this care and affection provided by parents (or other adults), children will not develop normally. They may have low self-esteem, fear rejection, feel insecure, and eventually find it difficult to adjust to marriage or to express affection to their own children. In fact, some young children who do not receive love and attention become retarded or die (see Chapter 4). Even individuals who are well integrated into society require support when adjusting to changing norms and when developing and continuing healthy relationships. Here again, the family can provide socioeconomic maintenance.

What is the reproductive function of the family? Society cannot survive without new members. The family's reproductive function provides an orderly mechanism for producing generation after generation. So important is this function that for many cultures and religions, it is the primary purpose for sexual relations. In many societies in developing nations, the failure of a wife to bear children can lead to divorce. Residents of places like the Punjab region of North India, for example, view children as an economic necessity. The significance of having children is also seen in the hundreds of rituals, customs, and traditions associated with pregnancy and childbirth in virtually all cultures around the world. It is customary in the United States, for instance, to honor the mother-to-be with a baby shower. After birth, babies are displayed in hospital nurseries so that friends and relatives can marvel at them. The parents send out birth announcements that report their baby's weight, height, and name. Among the Sirono Indians of eastern Bolivia, there are three days of rituals celebrating the birth of a child. These rituals are designed to protect the life of the infant and to ensure the child's good health (Holmberg 1969).

In what ways does the family regulate sexual activity? In no known society are people given total sexual freedom. Even in sexually permissive societies, like that of the Hopi Indians, there are rules about mating and marrying (Queen, Habenstein, and Quadagno 1985). The function of sexual regulation is usually assigned to the family.

Norms regarding sexual activities vary from place to place, but whatever the norms, it is almost always up to the family to enforce them. To illustrate, in the Trobriand Islands off the coast of Papua New Guinea, all children are encouraged to engage in premarital sex (Malinowski 1929). Thus, much of childhood is spent going from partner to partner, as a bee goes from flower to flower. By contrast, other societies, like those in Iran or Afganistan, go to great lengths to prevent any contact between unrelated single males and females. In Moslem countries, for instance, the bride's family must guarantee that she is a virgin; the honor of the family rests on that guarantee (E. L. Peters 1965). In the United States, the norm has traditionally fallen between these two extremes.

Sexual norms also change within a society. In colonial America, sex outside of marriage was considered a sin. But after Alfred Kinsey's studies of sexual behavior hit the American public like a bombshell (Kinsey, Pomeroy, and Martin 1948; Kinsey and Gebhard 1953), most experts agreed that a sexual revolution had occurred.

How does the family transmit social status? As noted in Chapter 8, an individual's place in the stratification structure is largely assigned at birth. If people move up or down the stratification structure, it is usually only a very slight move. Transmission of social class occurs primarily through the family. This is partly because families provide economic resources that open and close occupational doors. The sons and daughters of high-income professionals, for example, are more likely to attend college and graduate school than are the children of blue-collar workers. Consequently, the children of professionals are more likely as adults to enter professional occupations. The family also transmits values that later affect social status. The children of professionals, to cite only one illustration, tend to feel a greater need to pursue a college degree than do their counterparts from blue-collar families. In these and many other ways, the family affects the placement of children in the stratification structure.

What is the economic function of the family? At one time, families were self-sufficient economic units whose members all contributed to the production of needed goods—growing food, making cloth, and taking care of livestock. Malls, corporations, and supermarkets were not part of the scene. In contrast, the modern American family is a unit of consumption rather than production. Adult members, increasingly including working mothers, work outside the home and pool their resources. But the end result is the same: the family provides the economic necessities of life.

Conflict Theory

Functionalism emphasizes the ways in which the family attempts to contribute to the order and stability of society. The conflict perspective draws attention to social concerns such as sexual inequality.

How was the conflict perspective first applied to the family? The interest of conflict theorists in family gender inequality can be traced to Friedrich Engels, Karl Marx's collaborator. Engels saw women's oppression as the result, first, of women's loss of a productive role, and second, of men's interest—with the development of private property and surplus wealth—in utilizing monogamy to confirm paternity (Engels 2001; originally published in 1884). To Engels, then, the family has historically been used to maintain male domination of females.

How do the ideas of early feminist writers fit with Engels's approach? Some feminists share with conflict theorists the belief that families reflect social inequality and experience conflict over resources (Benokraitis 2001). Because these feminists focus on patriarchy (control by men over women), they point out that the domination by men over women predates private ownership of property and capitalism (P. Mann, 1994). Conflict in the family, according to these feminists, is primarily the result of women's attempt to gain more power within the family structure.

Do conflict theories still endorse this viewpoint? This theme, which is still being advanced by conflict theorists, bears on at least three issues: the domination of women by men, family rules of power and inheritance, and the male-dominated economic division of labor. Because most family structure throughout history has been patriarchal, women have traditionally been considered the property of men—first of fathers, then of husbands. Only since 1920 have American women been permitted to vote in federal elections (after a tremendous struggle), and only since 1974 have women gained the right to make contracts and obtain credit independently of husbands or fathers. Even so, the connotation of women as the property of men often continues, despite the trend toward gender equality in advanced industrial societies (Collins 1971). Moreover, because most arrangements have been patrilineal as well, male control of family members and male ownership of property typically have been transmitted through male bloodlines. This has created built-in gender inequality in most family systems because male dominance is considered "natural" and "legitimate."

The predominant patrilineal and patriarchal family patterns are also related to the traditional economic division of labor, which again perpetuates gender inequality. In the traditional division of labor, males work outside the home for finances to support the family, while women remain at home to prepare meals, keep house, and care for the children. Women are unpaid laborers who make it possible for men to earn wages. This economic power enables males to maintain dominance over the economically dependent wife and mother.

These traditional patterns have been altered by industrialization. But despite some changes, the persistent social, economic, occupational, political, and legal inequality of women continues, as we saw in Chapter 10.

Symbolic Interactionism

What is the relevance of symbolic interactionism to the study of the family? Symbolic interactionism is frequently used in the study of marriage and the family (Waller and Hill 1951; Leslie and Korman 1989). According to this perspective, a key to understanding behavior within the family lies both in the interactions among family members and in the meanings that members assign to these interactions. Within the family, all the major concepts of symbolic interactionism can be observed—socialization, looking-glass self, role taking, primary group, reference group, significant others, symbolic interpretation of events, and symbolic communication (Waller and Hill 1951; Leslie and Korman 1989).

Technology and the Family

The effects of technology are evident in the workplace. Computers are on almost everyone's desk, fax machines and cell phones encourage instant communication, and more people are working at home instead of at the office. But what, if any, effects does the new technology have on the family? According to many experts, the influence of technology is just as far-reaching in the home as it is in the office. As this passage from a newspaper article notes, activities in the home are changing dramatically due to recent technological innovations:

> *Notes on the refrigerator still have a place in Gene Kunitomi's house, but the traditional family message center is slowly being replaced by a nondescript box on a table in his home office. It is the home server, the hub of an internal data network that connects the family's seven personal computers—one in each of his three children's bedrooms and four in the home office. Strung together with high-speed Ethernet cables snaking behind the walls, the equipment makes up a kind of domestic cyber-space, over which family members exchange e-mail reminders about chores and the children post school reports and hone their home-page design skills on an internal Web site.* (Wingfield 1998:R18)

Because more American families are living farther apart from relatives, more are using the Internet to stay in touch with each other. Birth announcements, reunion plans, gift registries for weddings, and funeral arrangements are now being shared with families and friends online (Bulkeley 1997). Although somewhat impersonal, these social connections may reduce family social isolation and fragmentation.

Many, however, see a darker side to new technology for the family. For example, one technology critic offers this concern:

> *If we wish to raise our children as androids who respond to Internet packets rather than parental guidance, I can't think of a better way to do that than to put computer networks in homes.* (Wingfield 1998:R23)

Another critic contends that high-tech electronics like cable television, the Internet, and video games increasingly rule the lives of American families. Because children spend such an enormous amount of time alone with these technological wonders, they are deprived of frequent and intense social contact with other children, their parents, and other adults in the neighborhood. Consequently, the current generation of children could very well be the first to grow up with highly deficient social skills. Offering indirect support for this conclusion is the fact that almost three-fourths of Americans say they do not know their neighbors, and the number of Americans who admit they have spent no time with the people living next to them has doubled in the last twenty years (Quintanilla 1996).

Technology can also separate, socially, those family members who use the new technology from those who do not. For example, some couples who depend on Web pages to inform their relatives of family news have found that some family members cannot share in this information. Older members of the family who do not have access to the Internet often feel cut off from the rest of the family (Bulkeley 1997). Quentin Hardy writes of another "great divide":

> *In one corner is the technology lover, the early adapter who just has to have the most spectacular new gadget. In the other corner is the technology avoider, who would much prefer a good book to, say, a Powerbook. And as more exotic electronic gear moves into our homes, the split between technophiles and technophobes is only getting wider.* (Hardy 1997:R10)

In fact, some women whose husbands are addicted to their computers have become "computer widows." Some individuals become so attached to their computers that they often head straight for their machines after work, leaving little or no time for talking with their mates or their children.

Analyzing the Trends

A dark picture of the Internet has been emphasized. Think of some positive social consequences of this technology for the family. Discuss two of them.

Socialization begins within the family. As family members share meanings and feelings, children develop self-concepts and learn to put themselves mentally in the place of others. Interaction with adults permits children to acquire the personality and social characteristics associated with human beings. With the repertoire of personality and social capabilities learned within the family, children are able to develop further as they interact with people outside the home.

FEEDBACK

1. Which of the following functions of the family is not shared with any other institution?
   a. socialization
   b. reproduction
   c. socioemotional maintenance
   d. sexual regulation
2. Match the following examples with the major theoretical perspectives:
   ____ a. a father "giving away" the bride
   ____ b. sexual regulation
   ____ c. development of self-concept
   ____ d. a newly married couple adjusting to each other
   ____ e. child abuse
   ____ f. social class passed from one generation to another

   (1) functionalism
   (2) conflict theory
   (3) symbolic interactionism

Answers: 1. b. 2. a. (2) b. (1) c. (3) d. (3) e. (2) f. (1)

According to symbolic interactionists, relationships within the family are constantly being redefined. A newly married couple will spend many months (perhaps years) testing their new relationship. As time passes, the initial relationship changes, along with some aspects of the partners' personalities—including self-concepts. These changes occur as the partners grapple with such problem issues as distribution of responsibilities, personality clashes, and in-law distractions. With the arrival of children comes a new set of adjustments. Parental views may differ on number of children desired, child-rearing practices, and education for the children. The situation is made even more complex by the new member(s) of the family, who must also become part of the interaction patterns.

Family and Marriage in the United States

The Nature of the American Family

The **marriage rate**—the number of marriages per year for every one thousand members of a population—has fluctuated in the United States since 1940. As shown in Figure 12.3, the marriage rate peaked at over 16 immediately following World War II. Since then, the marriage rate, with some ups and downs, has been cut in half.

Some claim that there is no "typical American family." After all, the United States is a large, diverse society in which various groups—reflecting vastly different cultural heritages—have blended together. The early white settlers came primarily from Holland and England. Not long after they arrived, many Africans were brought in as slaves. Subsequently came waves of immigrants from northern Europe, Ireland, Italy, the Slavic and Baltic countries, China, Japan, Southeast Asia, and, more recently, countries in Latin America.

Still, there are more similarities than differences among American families. As the various ethnic groups assimilate into American life, their families tend to follow the American pattern: containing only the parents and children in the same household (nuclear); tracing lineage and passing inheritance equally through both parents (bilateral); dividing family decision making

FIGURE 12.3

Marriage and Divorce Rates: 1940–2001

Can you apply what you learned in history to interpret this chart? What happened in the early 1940s that caused the dramatic rise in the marriage rate during this period? Why do you think the marriage rate dropped so low in the 1950s? What are some possibilities that caused the divorce rate to peak in 1980?

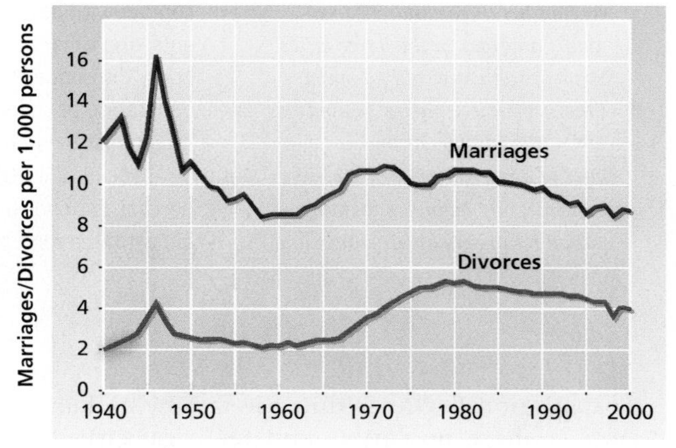

Sources: Centers for Disease Control and Prevention, *Monthly Vital Statistics Report* vol. 43, no. 9(s) (March 22, 1995), vol. 43, no. 12(s) (July 14, 1995), vol. 44, no. 12, (July 24,1996); and "Birth, Marriages, Divorces, and Deaths: Provisional Data for 2001," *National Vital Statistics Report* 50, no. 11 (June 26, 2002).

© Janeart/The Image Bank

This young mother represents a significant departure from the past. According to conflict theory, male dominance has been maintained in part by controlling the economic division of labor. As industrialization advances, however, women increasingly work outside the home, and male dominance is challenged.

equally (democratic); establishing a separate residence (neolocal); and following the norm of one woman married to one man at any particular time (monogamous).

A large family was once the norm in American society. In 1971, almost three-fourths of American adults thought that three or more children would be the most desirable number (Roper Organization 1985). Currently, the average number of children born to American women is about two (World Population Data Sheet 2002).

Why have large families declined in popularity? Many factors contribute to the smaller American family. Both men and women are marrying later and staying in school longer. Increasing numbers of women are in the labor force, for financial rewards as well as personal fulfillment (McLaughlin and Associates 1988). Legalized contraception and abortion have also reduced family size. Two additional factors promoting a lower fertility rate are the high cost of children and the removal of the stigma on childless marriages. (More is said about childless marriages later in the chapter.)

Americans are increasingly choosing alternatives to the family pattern of mother (as homemaker), father (as breadwinner), and children: slightly less than one-fourth of American households consist of a married couple with children under age eighteen. These new family and marriage arrangements include single-parent families, childless marriages, dual-employed marriages, single life, cohabitation, and gay and lesbian couples, all to be discussed later in the chapter.

Romantic Love as the Basis for Marriage

Most Americans agree with these lyrics from an old song: "Love and marriage go together like a horse and carriage." Actually, love is one of the most widely abused and ill-defined words in the English language. It is used to describe feelings of affection for dogs, cats, horses, homes, motorcycles, cars, parents, children, wives, and mistresses. The word love can be molded, modified, and stretched to mean just about anything we want. Yet, when someone sighs, "I'm in love!" we all empathize and say, "Isn't it great!"

In a recent poll of the American public, 83 percent of both men and women rated "being in love" as the most vital reason to marry. The relationship between love and marriage, though, has not always been viewed in this way. Among the British feudal aristocracy, romantic love was a game of pursuit played outside of marriage. Marriage was not thought to be compatible with deeply romantic feelings. In ancient Japan, love was considered a barrier to the arrangement of marriages by parents. Among Hindus in India today, parents or other relatives are expected to find suitable mates for the young. Criteria for mate selec-

TABLE 12.2

FOCUS ON THEORETICAL PERSPECTIVES: Perspectives on the Family

Both functionalism and conflict theory are more concerned with the ways social norms affect the nature of the family. Symbolic interactionism tends to examine the relationship of the self to the family. If functionalism and conflict theory were used to focus on the self, what examples would you give?

Theory	Topic	Example
Functionalism	Sex norms	Children are taught that sexual activity should be reserved for married couples.
Conflict theory	Male dominance	Husbands use their economic power to control the ways money is spent.
Symbolic interactionism	Developing self-esteem	A child abused by her parents learns to dislike herself.

SNAPSHOT OF AMERICA 12.1

U.S. Marriage Rates

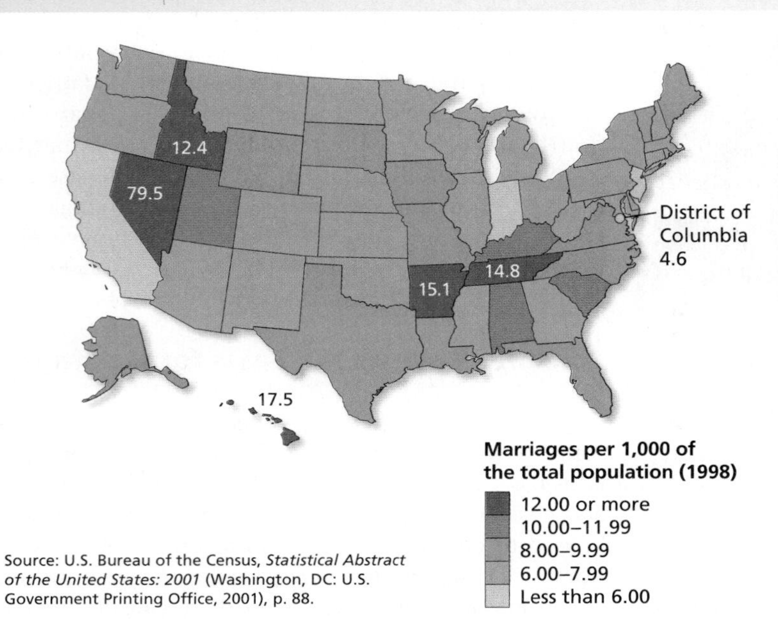

Source: U.S. Bureau of the Census, *Statistical Abstract of the United States: 2001* (Washington, DC: U.S. Government Printing Office, 2001), p. 88.

As noted in the text, the U.S. marriage rate overall has declined dramatically since 1940. Variation in the marriage rate among individual states is also interesting. The lowest marriage rate occurs in Washington, DC. Nevada has far and away the highest marriage rate.

1. Compare the marriage rate in your state with other states, keeping in mind that the national average is just over 8. Describe your reaction to your state's position in the marriage rate ranking.
2. Why is the Nevada marriage rate so high?
3. Does the distribution of marriage rates across the country suggest any pattern to you? Explain.
4. Would you expect the divorce rates of states to be correlated with their marriage rates? Make a prediction before looking up the divorce rates for comparison. Report your findings to the class.

SOCIOLOGY AND THE NEWS MEDIA

Cyber Sperm

Many childless women around the world want to become mothers. This list includes infertile married couples, infertile cohabiting adults, single women, and same-sex partners. This CNN clip describes an Internet service that provides the medical and socioeconomic background of sperm donors. These companies are attracting many customers who wish to determine the characteristics their children will have.

1. Discuss potential effects of this practice on the American family.

   ____

   ____

2. Would you use this service? Should all other adults be permitted to use it? Explain.

   ____

   ____

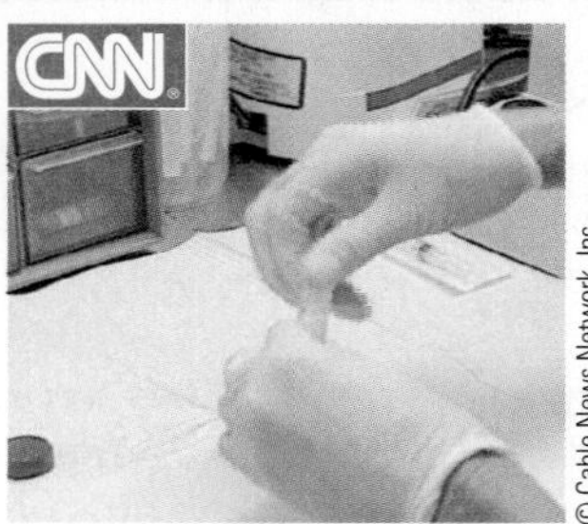

There seems to be no limit to what can be obtained via the Internet. But, just when you think you've seen it all, along comes the capability of sperm selection.

tion include caste, wealth, family reputation, and appearance. Love is not absent in Hindu marriages, but love follows marriage rather than the other way around (F. D. Cox 2002).

Are modern marriages based on romantic love alone? In modern societies, romantic love is almost always stated as a condition for marriage, but it is seldom the only one. People marry for many reasons, and love may or may not be one of them. They may marry to obtain regular sex partners, to legitimate living together, to enter a rich and powerful family, or to advance a career. One of the strongest motivations for marriage is conformity. Throughout our formative years, we are told what to do: "Feed yourself," "Join the Girl Scouts," "Learn to drive," "Go to college," "Get a job." Many parents also expect their children to marry after a certain age and worry about them—perhaps even pressure them—if they remain single very long. Peers are another source of pressure. Because well over 90 percent of all adults in the United States do marry, conformity must certainly be a motivating factor. In fact, most Americans are married by their mid-twenties.

Although Americans typically believe that a marriage that is not based on romantic love cannot last, it may be more accurate to say that a marriage based only on romantic love is almost sure to fail. According to research, love may be a good start, but it is only the beginning. For a marriage to last, a couple must build a relationship that goes beyond romantic love (J. F. Crosby 1985).

Divorce

What is the current divorce situation? The **divorce rate** is the number of divorces annually for every one thousand members of the population. Except for a peak and decline after World War II, the divorce rate in the United States increased slowly between 1860 and the early 1960s. A dramatic increase in the divorce rate occurred over the next twenty years. It more than doubled from 2.2 in 1960 to 5.3 in 1981. Since then, as indicated in "Using the Sociological Imagination," the rate has leveled off, declining slightly since 1985. The divorce rate, which peaked at 5.3 in 1981, stood at 4.1 in 2001 (see Figure 12.3).

In 1940, the marriage rate was 12.1 per 1,000 people, whereas the divorce rate was only 2 per 1,000 persons. Now the rate of divorce (4.1) is half the rate of marriage (see Figure 12.3). Based on these data, it is correct to say only that the divorce rate in the United States is now half its marriage rate. It is wrong to conclude that half of all marriages end in divorce, because the divorce and marriage rates are snapshots of a single year and do not take into account the total number of divorces and intact marriages from all years past.

This is why sociologists also employ the **divorce ratio**—the number of divorced persons in the population divided by the number of persons who are married and living with their spouses. In the United States in 2000, there were 19.8 million divorced persons and 120.1 million persons married and living with their spouses. Dividing the number of divorced persons by the number of married persons yields a divorce ratio of 165. This divorce ratio of 165 in 2000 is almost five times the 1960 ratio of 35. Women generally have a higher divorce ratio than men (187 versus 143 in 2000), reflecting the greater tendency of men to get remarried after divorce and to do so more quickly than women (U.S. Bureau of the Census 2001a).

These figures provide some basis for optimism for the future of the American family, just as the divorce rates of recent years do. Between 1960 and 1970, the divorce ratio in the United States increased by an annual average of 1.2 (from 35 in 1960 to 47 in 1970). During the 1970s, the average annual increase rose sharply to 5.2 (from 47 in 1970 to 100 in 1980). Since 1980, the average annual increase has actually declined, to an average annual increase of 3.25 (from 100 in 1980 to 165 in 2000). In other words, although the divorce ratio is still rising each year, its rate of increase has declined since 1980 (Cherlin 2002). Still, the U.S. divorce rate is high among industrialized countries (Ahlburg and DeVita 1992; F. D. Cox 2002).

Why the high U.S. divorce rate? There are no easy answers to this question; like family and marriage, divorce is a complex matter. Those who study marriage and divorce have sought answers at both individual and societal levels.

At the individual level, the following factors seem to be associated with divorce (Knox and Schacht 2002). First, the earlier one marries, the greater the likelihood of divorce. Second, the longer a couple has been married, the lower the probability that their marriage will end in divorce. The average length of first marriages ending in divorce in the United States is about six years. However, the largest number of first marriages ending in divorce occurs between the second and third anniversaries. This suggests that poor decisions in mate selection may be as important as marital conflict. Third, as might be expected, divorce is related to the nature and quality of the marital relationship. The more respect and flexibility taking place between partners, the lower the chance of divorce.

At the societal level, several factors seem to affect the divorce rate. First, divorce rates increase during economic prosperity and decrease during economic recession or depression. This is probably because economic prosperity permits people to concentrate on issues beyond survival and to consider options other than marriage for their personal happiness. Second, the rise in the divorce rate in the United States in the 1960s reflects the passage of the baby boom generation into the marriageable ages. Not only are there greater numbers of them, but baby boomers as a group are much more forgiving of divorce and remarriage than are earlier generations. Third, because women are more economically independent and because they have a wider availability of child care, they are less hesitant to dissolve a bad marriage. Finally, American values and attitudes about marriage and divorce have changed. The stigma once associated with divorce is much weaker today (R. Farley 1998).

What about future divorce rates? The recent decline in the U.S. divorce rate and divorce ratio may continue for several reasons. First, we know that the later people

FIGURE 12.4

Median Age at First Marriage

This figure displays changes in the median age at first marriage in the United States since 1900. The marrying age for both men and women has been on the increase since the 1970s. Discuss the implications of this trend for the future divorce rate.

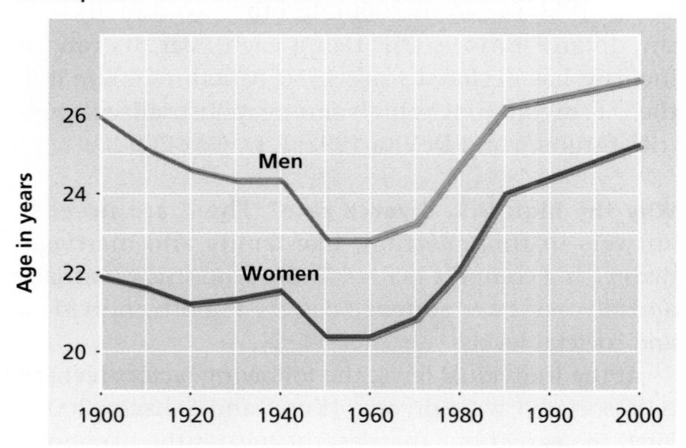

Source: U.S. Bureau of the Census, *Current Population Reports*, Series P20-537, "America's Families and Living Arrangements: 2000," and earlier reports; at www.census.gov/prod/2001pubs/p20-537.pdf

marry, the less likely they are to divorce. (This happens in part because more mature individuals have more realistic expectations about their relationships and because they have fewer economic and career problems that create emotional strain.) The average age at first marriage is, in fact, increasing in the United States (see Figure 12.4). In 1970, the average age at first marriage was 23.2 for men and 20.8 for women. By 2000, the average age had increased to 26.8 for men and 25.1 for women. This trend is likely to continue well into the twenty-first century. Second, the average age of the population of the United States is increasing as baby boomers grow older. It was this exceptionally large generation that set records for divorce in the late 1960s and 1970s. Baby boomers now range in age from late thirties to mid fifties, removing them from the age bracket that produces the highest divorce rates. Third, American couples are having fewer children and the children are spaced farther apart. This reduces pressure.

The recent decline in the U.S. divorce rate is encouraging. At the very least, it is the first break in a long-term increase. Still, many experts are cautious about predicting the future, pointing to the fact that the extent of the divorce rate is cyclical and that the recent decline may be only temporary.

Is the rising divorce rate unique to the United States? The mass media often leave the impression that the United States is experiencing a rising divorce rate due to social factors peculiar to this country. In fact, since 1960 the divorce rate has risen more than fivefold in Canada and the United Kingdom, almost tripled in France, and doubled in Germany and Sweden. The divorce rate has also gone up in Japan to a lesser degree (see Figure 12.5). On the other hand, countries in Asia, Central America, and South America have significantly lower divorce rates.

To understand this worldwide variation in divorce rates, it is necessary to identify the major socioeconomic factors influencing these rates. Increasing prosperity, to cite one important factor, is associated with higher divorce rates. Consider China, whose divorce rate has risen with its recent economic development. China's divorce rate is now twice as high as it was in 1990. Because of social and economic development, more women are initiating divorce as their economic opportunities and rights increase.

The correlation between economic development and divorce rate is far from perfect. For example, the divorce rate is higher in Belarus despite its lower economic development, while the divorce rate is lower in a highly prosperous country such as Japan. This is because the divorce rate is affected by additional social factors.

Religion is a second factor affecting divorce rates, particularly in less developed countries. The predominance of Roman Catholicism in Central and South America, for example, diminishes the frequency of divorce. Religion also affects divorce independently of level of socioeconomic development, as seen in Roman Catholic–dominated Italy and Ireland.

A third factor retarding the divorce rate is a patriarchal power structure. The more patriarchal a society, the lower its divorce rate. In many Middle Eastern countries, for example, a husband can dissolve his marriage by saying "I divorce thee" three times in the presence of witnesses. A woman wishing a divorce, however, must convince a court of her husband's economic neglect and his morally corrupting influence on the family. And only recently have Egyptian women even been allowed to file for divorce in the absence of extraordinary maltreatment. Unless they give considerable wealth to a willing husband, Taiwanese women are discouraged from seeking divorces because their children will automatically be in the sole custody of their husbands (Chang 1998).

Family Violence

Americans, like the rest of the world, have traditionally denied the existence of widespread family violence (*The World's Women* 2000). Its occurrence was mistakenly associated most often with lower-class families, with maybe a few exceptional cases in the middle and upper classes.

There is a good reason for this erroneous picture. Early research was done using law enforcement and public medical records. Because law enforcement offi-

FIGURE 12.5

Divorce Rates by Country

This figure compares divorce rates by country. Using the two extremes (Belgium and Turkey), describe to a friend the relative condition of marriage in these two countries. Also describe a few instances where the divorce rates surprised you.

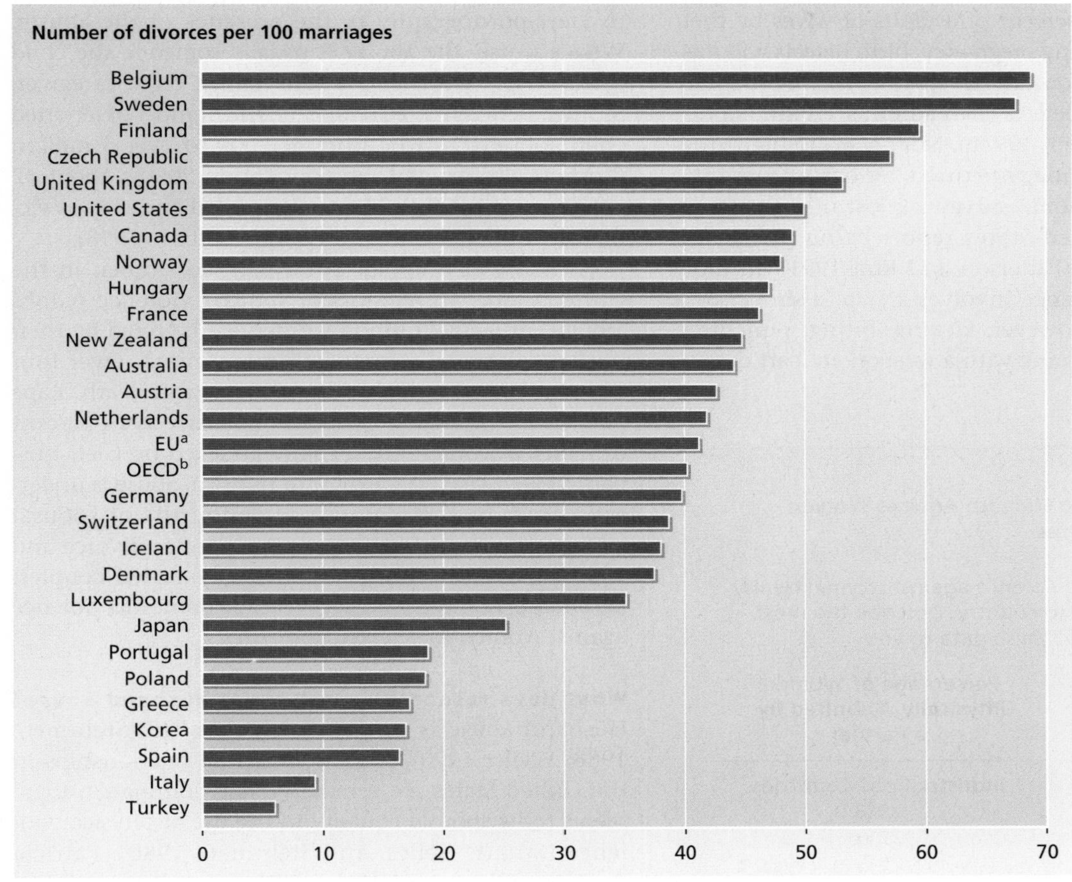

Sources: Eurostat and national sources (see the annex on the Internet). *Society at a Glance: OECD Social Indicators* (Paris, France: Organisation for Economic Co-operation and Development, 2001), p. 33.
[a]Eu = European Union
[b]OECD = Organisation for Economic Co-operation and Development; members include Australia, Austria, Belgium, Canada, the Czech Republic, Denmark, Finland, France, Germany, Greece, Hungary, Iceland, Ireland, Italy, Japan, Korea, Luxembourg, Mexico, the Netherlands, New Zealand, Norway, Poland, Portugal, the Slovak Republic, Spain, Sweden, Switzerland, Turkey, the United Kingdom, the United States.

cials and public hospitals were more likely to be aware of lower-class family violence, the statistics were skewed toward these families. Middle- and upper-class families disproportionately conceal their domestic violence (S. Weitzman 2000).

However, the 1990s brought heightened public attention to domestic violence (Ingrassia and Beck 1994; Smolowe, 1994; Jasinski, 2000). The mass media made every effort to give Americans the details of several dramatic legal cases. Erik and Lyle Menendez admitted planning and carrying out the murder of parents they accused of abusing them. In a televised trial, O. J. Simpson was tried for the wrongful death of his wife, Nicole Brown Simpson, and her friend. Evidence indicated that Simpson had abused his wife when they were married. And claiming past abuse, humiliation, and forced sex, Lorena Bobbitt severed her husband's penis while he slept (Kaplan 1994).

How widespread is family violence? Although the family does, as the functionalists contend, usually provide a safe and warm emotional haven, it can also be a hostile environment. Assault and murder are more likely to be the acts of someone living in the home. Over one-fifth of all reported cases of aggravated assault involve domestic violence. The situation is actually much worse because most episodes of domestic violence go unreported. Perhaps as little as 10 percent of information about such cases reaches the authorities (Gelles and Loseke 1993; Hampton et al. 1993; Gelles 1997). Domestic violence is not reserved for any particular members of the family. It involves children, spouses, siblings, and older people (Barnett, Miller-Perrin, and Perrin 1997; "Violence Against Women" 2002).

Parental abuse of children may begin in the womb. Approximately 40 percent of assaults of wives by their husbands occur during pregnancy; birth defects and miscarriages are common results. In fact, such abuse may lead to more birth defects than all other childhood diseases combined (Gibbs 1993b). Nor do all children grow up in a loving home, nurtured by caring parents. According to a national survey, almost one-quarter of adults in the United States report being physically abused as children (Patterson and Kim 1991). In most cases, physical violence involves a slap, a shove, or a severe spanking. However, kicking, biting, punching, beating, and threatening with a weapon are part of abusive violence on the part of many parents. Furthermore, according to estimates, one of every four girls and one in ten boys are victims of sexual aggression either within the home or outside (Pryor 1999). Reported child sexual abuse in the United States has skyrocketed in recent years. Child sexual abuse goes beyond physical contact. Some children are forced into pornography or are made to view pornography in the presence of the abuser. What's worse, the abuser is usually someone the child trusts—a parent, friend of the family, child caregiver, brother. Between 1976 and 1999, the number of reported child abuse cases rose from 662,000 to over 3 million (National Exchange Club Foundation 2000). About 47 out of every 1,000 children are reported annually as victims of child maltreatment (Wang and Daro 1998).

As many as one-half of all married women in the United States are victims of spousal violence (Gibbs 1993b). At least 4 million women are battered by their husbands annually, probably many more. Over four thousand women each year are beaten to death. Rape and sexual assault are common. As many as 14 percent of married women are sexually attacked by their husbands every year. The extent of physical abuse is underestimated, in part because three-fourths of spousal violence occurs during separation or after divorce and most research is conducted among married couples. (See Table 12.3 for data on levels of domestic violence against women in selected countries.)

TABLE 12.3

Levels of Domestic Violence Against Women in Selected Countries

Levels of domestic violence against women clearly vary from country to country. Describe the most surprising aspect of these data to you.

Country	Percentage of Women Physically Assaulted by a Partner
	Industrialized Countries
Canada	29%
United States	28%
United Kingdom	25%
Switzerland	21%
Norway	18%
	Asia and Western Pacific
Korea	38%
Thailand	20%
India	19%
	Eastern Mediterranean
Egypt	34%
Israel	32%
	Africa
Kenya	42%
Uganda	41%
South Africa	13%
	Latin America and the Caribbean
Mexico	30%
Nicaragua	28%
Chile	26%
Paraguay	10%

Sources: World Health Organization, www.who.int/, July 1997; and *World Report on Violence and Health* (Geneva, Switzerland: World Health Organization, 2002), pp. 90–91.

What does research reveal about husband abuse? Husband abuse is frequently overlooked (Steinmetz 1988; Wiehe 1990). Although marital relationships in the United States are generally male dominated, there seems to be spousal equality in the use of physical violence (Straus, Gelles, and Steinmetz 1980). Patricia Pearson and Michael Finley (1997) found that almost one-third of the husbands in their survey had acted violently against their wives and that wives were almost as likely to have used physical violence against their husbands. Other studies also show that husbands and wives assault each other at about the same rate (Hotaling et al. 1988; Flynn 1990). Much of the violence on the part of women, however, involves self-protection or retaliation. Whoever initiates the violence, women are more likely to suffer greater injury because the average man is bigger, stronger, and more physically aggressive (Gelles 1997).

Is abuse always physical? Domestic violence, of course, is not limited to physical abuse. Verbal and psychological abuse are also a part of many families (M. D. Schwartz 1997; Jacobson and Gottman 1998; Knox and Schacht 2002). Furthermore, psychologists report that the feelings of self-hate and worthlessness that are often the effects of such abuse can be as damaging as physical wounds. In addition, neglect, a condition of

© Bill Aron/PhotoEdit

Although the family can provide a safe and warm emotional haven, it does not always do so. This woman has been the victim of a hostile family environment, as the expression on her face and the bruises on her face attest.

being ignored rather than abused, occurs at double the rate of physical abuse, involving as many as 9 million children each year (Helfer, Kempe, and Krugman 1999).

What other forms of family violence are of concern? Probably the most frequent and most tolerated violence in the family occurs between children. Much violence occurs between children who are too young to channel their frustrations in nonphysical channels. Abuse among siblings may be based on rivalry, jealousy, disagreements over personal possessions, or incest. Although violence among siblings declines somewhat with age, it does not disappear. Sibling violence appears to be prevalent and on the rise. According to one study, violence in the family is most likely to come from children (Hotaling et al. 1988).

Little is known about the abuse of elderly people because less research has been done in this area. Abuse of older people usually takes the form of physical violence, psychological mistreatment, economic manipulation, or neglect. Normally, the abuser is a family member (Pillemer and Finkelhor 1988). Estimates of elder abuse range from 500,000 to 2.5 million cases annually (Gelles 1997). There is fear of an increase in abuse of older people as the population bulge, represented by the baby boomers, enters old age. For one thing, there will be fewer working adults to support a growing, aging segment of the population (Kart 1990). The most likely to be abused are white Protestant women over seventy-five years of age who are mentally, physically, or financially disadvantaged (Garbarino 1989). Older women represent a higher proportion of the abused, in part because they live longer than men. There are simply more elderly women with greater dependency needs. Abuse of the elderly is partly due to the psychological, physical, and economic strain of caring for elderly persons. These strains of caretaking may lead to hostility and depression, followed by violence. Elder abuse may involve a dependent adult child who has moved back into his or her parents' home as a result of divorce or financial problems (Gelles 1997), or it may occur in nursing homes.

The high incidence of divorce, violence, and abuse has motivated sociologists to examine the strength and durability of the American family. One area of research focuses on what is being called *family resiliency*.

Family Resiliency

What is family resiliency? Synonyms for the word resilient include flexible, supple, and buoyant. Thus, resiliency implies an ability to cope successfully with shock. **Family resiliency** refers to the family's capacity to emerge from crises as stronger and more resourceful. A resilient family is able to flourish despite distress.

What factors promote family resiliency? Sociologists have identified four sets of factors promoting resiliency. First, family resiliency is enhanced by *individual* characteristics such as self-esteem, autonomy, sense of humor, and problem-solving skills. Second, resiliency is engendered by *family* characteristics such as emotional support, commitment, warmth, affection, and cohesion. Third, resiliency is nurtured by *community* characteristics such as ample opportunities for participation in community life, emphasis on helping others, avenues for communication with friends and adults, and availability of youth activities. Fourth, resiliency is increased via family-friendly *public policy*. Because it is the most abstract factor, public policy requires some elaboration.

What role does public policy play in promoting family resiliency? **Public policy** is a broad course of governmental action expressed in specific laws, programs, and initiatives; it is created and changed in democratic societies through the interplay of government, lobbyists, and interest groups (Bardes, Shelly, and Schmidt 2002; Parenti 2002; Patterson, 2002). The making of public policy involves trade-offs, compromises, winners, and losers. The 1993 Family and Medical Leave Act granted American workers the right to twelve weeks of unpaid leave for family emergencies without the threat of job loss. But the United States remains the only major industrialized country without paid maternity leave. Germany, for example, provides new parents with fourteen weeks of fully paid leave (Knox and Schacht 2002).

Murray A. Straus, David B. Sugarman, and Jean Giles-Sims—Spanking and Antisocial Behavior

Many children in the United States experience spanking and other legal forms of physical punishment from their parents. In the 1980s over 90 percent of parents used corporal punishment on young children, and more than half persisted in its use during the early teen years. Though high, this extent of corporal punishment was less than in the 1950s (99 percent) and in the mid-1970s (97 percent). Despite a further decline since 1985, nearly all American children still experience some form of corporal punishment.

The use of corporal punishment to correct or control the behavior of children is widely accepted in American culture. "Spare the rod and spoil the child" is a warning deep in our national consciousness. Murray Straus, David Sugarman, and Jean Giles-Sims (1997), however, present evidence contradicting the supposed salutary effects of corporal punishment on children's behavior.

These researchers interviewed over 800 mothers of children, aged six to nine years, in a national longitudinal (involving before and after measurement) study. This research compared parents' use of corporal punishment with antisocial behavior in children. The study defined corporal punishment as "the use of physical force with the intention of causing a child to experience pain, but not injury, for the purpose of correction or control of the child's behavior" (Straus, Sugarman, and Giles-Sims 1997:761). Slapping a child's hand or buttocks and squeezing a child's arm are examples. A measure of antisocial behavior was based on the mothers' reports of their children's behavior: "cheats or tells lies," "bullies or is cruel or mean to others," "does not feel sorry after misbehaving," "breaks things deliberately," "is disobedient at school," and "has trouble getting along with teachers."

Since this was a longitudinal study, data on the frequency of parents' use of corporal punishment was collected *prior* to reports on subsequent antisocial behavior. Contrary to common expectation, Straus found that the higher the use of corporal punishment, the higher the level of antisocial behavior two years later.

Straus and his colleagues suggest that the reduction or elimination of corporal punishment could have significant benefits for lowering antisocial behavior in children. In addition, given the research indicating a relationship between childhood antisocial behavior and violence on the one hand and other crime in adulthood on the other, society at large could benefit from abandoning the use of corporal punishment in child rearing.

Thinking About the Research

1. Explain why a longitudinal study inspires greater confidence that a relationship is causal.
2. Suppose that you are on a panel reporting on child rearing to the president of the United States. Describe the study you would conduct on a possible relationship between childhood corporal punishment and adult crime.
3. How do you anticipate corporally punished children will discipline their own children later in life?
4. Describe what you think would be more effective means of discipline.
5. How were you disciplined? How will you discipline your children? Why?

FEEDBACK

1. Which half of each of the following pairs best describes the typical American family?
   a. nuclear/extended
   b. patrilineal/bilateral
   c. authoritarian/democratic
   d. neolocal/matrilocal
   e. polygynous/monogamous
2. The idea of love as a crucial factor in the mate selection process did not become common practice until which century?
   a. seventeenth
   b. eighteenth
   c. nineteenth
   d. twentieth
3. The _________ is the number of divorces in a particular year per 1,000 members of the total population.
4. Which of the following is not stated in the text as one of the factors associated with divorce?
   a. decline of religious influence
   b. age at first marriage
   c. length of marriage
   d. economic conditions
5. Contrary to what one would expect from the traditional male dominance in American family life, women are spousal abusers almost as frequently as men are. T or F?
6. _________ refers to the family's capacity to emerge from crises as stronger and more resourceful.

Answers: 1. a. nuclear b. bilateral c. democratic d. neolocal e. monogamous 2. d 3. divorce rate 4. a 5. T 6. Family resiliency

In the United States, this lack of a universal public policy places the American family at a disadvantage in coping with the unique pressures of modern life.

For example, the lack of a national health insurance program has resulted in some 40 million uninsured Americans with millions more underinsured. This adds to the stress of low-income, poor, and unemployed families already disproportionately characterized by high rates of divorce, violence, and abuse. The United States also reduces family resiliency through the absence of paid parental leaves surrounding childbirth. Longer paid family leaves promote such vital conditions as improved maternal health, lower infant mortality, enhanced infant and child development, and greater likelihood of women resuming employment after childbirth. These conditions, in turn, encourage family resiliency.

Lifestyle Variations

American marital and familial arrangements are undergoing dramatic change. Although the future of the American family is difficult to predict, an examination of current alternatives to the traditional nuclear family structure will provide some hints. Social approval of most of these alternative lifestyles is a continuing long-term trend, although certain divergences continue to be spurned by some Americans (Thornton and Young-DeMarco 2001; Skolnick and Skolnick 2002).

Blended Families

The relatively high divorce rate in the United States has led to the creation of the **blended family**—a family formed when at least one of the partners in the marriage has been married before and has one or more children from a previous marriage. This type of family can become extremely complicated (Ganong and Coleman 1994; Herbert 1999). Consider the number and complexity of relationships in the following blended family:

> *Former husband (with two children in the custody of their natural mother) marries new wife with two children in her custody. They have two children. Former wife also remarries man with two children, one in his custody and one in the custody of his former wife, who has also remarried and had a child with her second husband who also has custody of one child from his previous marriage. The former husband's parents are also divorced and both have remarried. Thus, when he remarries, his children have two complete sets of grandparents on his side, plus one set on the mother's side, plus perhaps two sets on the stepfather's side.* (F. D. Cox 2002:530)

Part of this complexity arises because more marriages are ending through divorce rather than death. For children, a parent is being added, not replaced. Along with the third parent come his or her relatives. Blended families, then, create a new type of extended family, a family that is not based strictly on blood relationships. As the preceding excerpt shows, it is possible for a child in a blended family to have eight grandparents, if each of the child's biological parents remarries. Although not all blended families are this complicated, it is significant that about 39 percent of households in the United States contain biologically unrelated individuals (Barnes et al. 1998).

What are the reasons for greater instability in blended families? Although many blended families are harmonious, especially if they make adjustments during the first few years, children from previous marriages are one factor in the higher divorce rate among second marriages (Baca Zinn and Eitzen 2000). Sociologists point to three major problems facing blended families: financial difficulties, stepchildren's antagonism, and unclear roles (Lamanna, Riedmann, and Riedmann 2000).

- *Financial difficulties.* Incomes are lower in stepfamilies. This is compounded by the financial demands from both the former and present families. Remarried husbands are often legally obligated to support children from their previous marriage. Second wives may resent the income denied their own children to support children from that previous marriage.
- *Stepchildren's antagonism.* Hoping for a reunion of their original parents, stepchildren may attempt to undermine the new marriage. Even five years after divorce, about a third of stepchildren continue to strongly disapprove of their original parents' divorce. This is especially true for teenagers, who can be very critical of their stepparent's values and personality.
- *Unclear roles.* The roles of stepparents are often vague and ambiguous. Stepchildren often don't consider their parent's new spouse as a "real" father or mother. It is also uncertain to stepparents or stepchildren how much power the new spouse really has. As a result, questions of control and discipline can become very contentious, especially when teenagers are involved.

Single-Parent Families

How widespread are single-parent families? Single-parent families, on the rise since 1970, now account for slightly more than one out of four American families. The percentage of families with children headed by a single parent has more than doubled in the last thirty years, going from 12 percent in 1970 to 26 percent in

FIGURE 12.6

Percentage of Single-Parent Families: 1970–2000

This figure compares the percentage of African American, Latino, and white families that have one parent. Pick three trends from these data, and explain their significance for the American family.

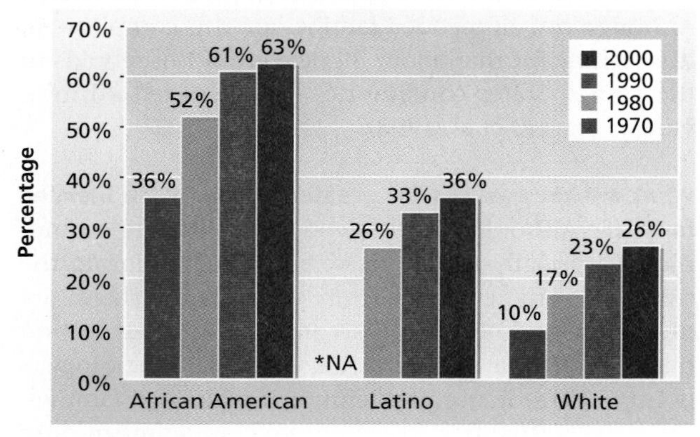

*NA = not available
Source: U.S. Bureau of the Census, "Family Household by Type, Age of Own Children, Age of Family Members, and Age, Race and Hispanic Origin of Householder: March 2000" (June 29, 2001), at www/census.gov/population/socdemo/hh-fam/p20-537/2000/tabF1.pdf

2001. The greatest proportion of children in single-parent families live with their mother (24 percent). Only 4 percent of children living with one parent are in a male-headed household (*Child Trends DataBank* 2002).

In 2000, 63 percent of all African American families were headed by a single parent, compared with 36 percent among Latinos and 26 percent among whites. African American and Latino children are more likely than white children to live alone with their mothers because of high divorce and out-of-wedlock birth rates, combined with lower rates of marriage and remarriage (U.S. Bureau of the Census 2001a). While the proportion of one-parent families has increased for all races since 1970, the gap between them has remained relatively constant (see Figure 12.6).

Is social class an ingredient in the vast majority of single-parent households headed by women? Although courts today are more sensitive to fathers' claims, women in all social classes are still more likely to win custody of their children in cases of separation and divorce. Frequently, though, there are factors unique to poor single-parent women. Poor single-parent households are generally created by unwed mothers or by women abandoned by their husbands and/or the fathers of their children. And poor women are able to marry (or remarry) at a very low rate.

Though significantly fewer, there is an increasing number of well-educated, professional women who head single-parent households. With the stigma of unwed motherhood declining, more affluent unmarried women are choosing to have children and to care for them alone. In addition, these women have the economic resources to support an independent family. Finally, well-educated women are adopting higher standards for selecting husbands (Seligmann 1999).

What are the effects of single-parent families? Females heading single-parent households face considerable time and economic constraints (Kissman and Allen 1993). Single working parents can seldom provide the time, attention, and guidance to their children that two parents can give, although research has been inconsistent on the effects of this situation (F. D. Cox 2002). One study concluded that children in female-headed families are as socially, intellectually, and psychologically well adjusted as children from two-parent families (Cashion 1982). In contrast, most of the research indicates an increase in negative effects on children. Sheppard Kellam, Margaret Ensminger, and R. Jay Turner (1977) found that children from one-parent families consistently score lower on tests of psychological well-being and are less able to adapt in social settings, such as school, than their counterparts from two-parent or two-adult families. They concluded that the problem lies not so much in the absence of the father, but in the aloneness of the mother. Children living in two-adult families fared almost as well as children from two-parent families. Researchers now generally agree that growing up in a fatherless house places children at risk (Mclanahan and Schwartz 2002).

Adolescents who live with one parent or with a stepparent have much higher rates of deviant behavior than adolescents living with both natural parents (Dornbush et al. 1985; Popenoe 1999). David Popenoe's study (1999) of a national sample of twelve- to seventeen-year-olds indicates that arrests, school discipline, truancy, runaways, and smoking occur more often in single-parent and stepparent families, regardless of income, race, or ethnic background. Daniel Mueller and Philip Cooper (1986), in a study of nineteen- to thirty-four-year-olds, found that compared to individuals from two-parent families, persons raised by single parents (especially single mothers) tend to have lower educational, occupational, and economic success. Adults who grew up in single-parent homes are also more likely to be divorced, have an illegitimate child, be receiving psychological counseling, be school truants, and use alcohol and drugs (Bianchi 1990; P. Taylor 1991; McLanahan and Sandefur 1996).

Are the negative effects for these children actually due to divorce and living in single-parent homes? Some researchers contend that the achievement and behavioral problems noted here may not be attributable to single-parent families alone. Such factors as poor parent-child relationships, failure to communicate, lack of economic resources, and family disharmony may be more important than the structure of the family (Demo and Acock 1991). In a series of longitudinal studies of children (one of the studies compared children in the United States and Great Britain), some researchers conclude that the adjustment problems we associate with divorce and single-parent families may actually be reflections of conditions existing prior to separation or divorce (Block, Block, and Gjerde 1986; Cherlin et al. 1991; Furstenberg and Cherlin 1991).

This line of research will be carefully followed by other sociologists. Only later will we have more definite data showing, in general, the effects of single-parent families and, in particular, the causal relationship between divorce and the subsequent adjustment of the children.

On one point all researchers agree. Female single parents and their children suffer from insufficient economic resources. Although single mothers comprise only about one-fifth of all family households, they account for about half of households with minor children with income below the poverty line. Because only about 16 percent of the single-parent families headed by men are below the poverty line, experts are now writing about the feminization of poverty (see Chapters 8 and 10). Earning a living is a problem for these single-parent mothers, not only due to the difficulty of arranging for child care (which all working mothers must face) but also because most of these women do not have the educational credentials to do well in the job market. Many were housewives or part-time workers before a divorce; others are young, uneducated women who never married. Thus, if they do not get an education or marry a solidly employed male, these mothers are unlikely to work their way out of poverty. Their children, of course, suffer the consequences along with them.

Childless Marriages

In the past, there was a stigma attached to marriages without children. Due in part to the assumption that all women have a maternal instinct, married women without children were seen as failing to fulfill their biological destiny. In addition, as we learned in Chapter 5, children of that time were needed for farm labor. As the traditional negative societal response to childless marriages diminishes, more American couples are choosing not to have children (Gallup and Newport 1990).

To what extent are married couples choosing to be childless? Around 22 percent of ever-married American women are childless. This is up from about 15 percent in 1970 (U.S. Bureau of the Census 2001a). Experts predict that the number of married couples choosing never to have children is likely to increase (Benokraitis 2001).

Why are some married women choosing not to have children? The reasons married women give for choosing not to have children are diverse. An important factor, of course, is the disappearing social stigma attached to childless marriages. Also, some women are so involved in their careers that they see children either as an impediment or as too important to shortchange if brought into the world. Sometimes women want their independence in order to pursue personal goals. According to others, today's world is a questionable place into which to bring a child. Some women simply do not enjoy the presence of children. For some couples, having children has been delayed so long that it becomes difficult to make the necessary adjustments for raising a family. Finally, some childless marriages are not a result of choice but are due to physical or mental limitations.

Aren't childless couples missing an important part of life? Because of the strains placed on a marriage, the presence of children tends to reduce marital happiness and satisfaction. Marital satisfaction is higher at the start of marriage, lowest during children's teenage years, and higher again when the children leave home. It is true, however, that among childless couples who want children, marital happiness is generally lower than for married couples with children (Singh and Williams 1981). This is not the case among women who prefer not to have children. Most voluntarily childless couples appear to be happier and more satisfied with their marriages and lives than

Despite the happy expressions on the faces of this dual-employed couple, they face special strains. If the husband does not assume a significant role in household duties, the burden falls heavily on his spouse.

couples with children (Polonko, Scanzoni, and Teachman 1982; Callan 1983; F. D. Cox 2002).

Dual-Employed Marriages

In a dual-employed marriage, both husband and wife are in the labor market. This is quite different from the marriage in which the wife works at a job merely to supplement the family income, a pattern long observed in the lower and working classes (Bielby and Bielby 1989). Although a relatively new trend, the dual-employed marriage is now the norm, as suggested in Chapter 10 (Hertz and Marshall 2001; Moen 2003).

Do dual-employed marriages create special strains? Women in dual-employed marriages are apparently expected to handle the bulk of household tasks and child-care responsibilities in addition to the pursuit of full-time jobs (Goldscheider and Waite 1991; Thompson and Walker 1991; Spain and Bianchi 1996). Combining employment with child care and domestic tasks, married working women work about 15 hours more a week than men do. This additional month of 24-hour days a year Arlie Hochschild (1997) calls the "second shift." Although men now spend an average of 16 hours per week (up from 12 hours in 1965) in household and child-care duties, women clearly bear the larger burden (Williams, Swisher, and Stalcup 1997; Juster, Ono, and Stafford 2002).

In addition to this greater workload, women in dual-employed marriages often must cope with role conflict. They are torn between the time requirements of their jobs and their desire to spend more time with their children and husbands (Rubenstein 1991).

Despite their general unwillingness to assume household responsibilities equal to their wives, men in dual-employed marriages also may feel the negative effects of conflicting roles and excessive demands on their time (Moen 1992). In addition, having an employed wife, particularly if she earns more, may not fit with a man's image of himself as provider (Menaghan and Parcel 1991; G. Spitze 1991).

The special strains experienced by dual-employed couples will probably not greatly decrease the number of people adopting this alternative. A national survey of high school seniors in 1991 found that only 2 percent of the females questioned indicated that they expected to be homemakers at age thirty, down 10 percentage points since 1976 (Morin 1991).

The challenge in the future will be to find ways of reducing the strains placed on members of dual-employed families (L. A. Gilbert 1993). Some pressure on working women could be relieved if men became more participative in domestic duties. Many experts believe that the combined pressures of job and family will lead to accommodative changes in the workplace. Alterations in company policies and practices will likely involve more flexible working schedules, family leaves for both males and females, work-at-home situations, and on-site child-care centers (Deutschman 1991; Labich 1991).

Some evidence suggests that men are assuming a more active role in child care. By 1991, 20 percent of all children under age five were being cared for by their fathers, up from 15 percent only three years earlier (O'Connell 1993). This mirrors the long-term trend toward approval of gender equality in families (Thornton and Young-DeMarco 2001). It could signal even greater movement toward willingness on the part of working fathers to offer more help at home.

How do women, children, and men feel about their dual employment? Thus far, emphasis has been placed on the strains experienced by dual-employed couples. There can be a positive side. On balance, most recent research reveals beneficial effects for the psychological well-being of women (Moen 1992; Crosby 1993; F. D. Cox 2002). Working outside the home can provide a wider set of social relationships and greater feelings of control, independence, and heightened self-esteem. Employment also appears to provide a socioemotional cushion for women when their children leave home. Compared with women who do not work outside the home, employed women tend to have more alternative channels for self-expression (Adelmann et al. 1989; Wolfe 1997).

As economic pressures are eased, more discretionary money is available for purchases that improve the quality of life of all family members. In addition, sons and daughters of working mothers can benefit in noneconomic ways. Daughters of working mothers may see themselves as working adults, as capable of being economically independent, and as benefiting from further education. Sons may choose wives with similar attitudes toward education and employment.

For men, benefits of a dual-employed marriage may include freedom from sole economic responsibility, increased opportunity for job changes, and opportunity for continuing an education (Spitze 1991; Crosby 1993). Men with employed wives can often share the triumphs and defeats of the day with someone who is in a similar situation. If their wives are happier working outside the home, husbands also enjoy a better marital relationship. Those husbands who take advantage of the opportunity can form a closer relationship with their children by being more active parents (Booth and Crouter 1998).

Whatever happens between partners, there seems to be a role here for government and business (Kamarck 1992; *Starting Points* 1994). As Table 12.4 shows, when compared to the United States, other industrialized countries have been more diligent in requiring employers to participate in relieving the strains associated with

TABLE 12.4

Parental and Maternity Leave Policies, United States vs. Selected Countries

This table compares the paternal and maternity leave policies of selected countries. Does the rank of the United States among these industrialized countries surprise you? Why?

Country	Leave Duration (Weeks)	Percentage of Pay/Weeks	Recipient
Canada	17–41	60%/15 weeks	Mother
Italy	22–48	80%/22 weeks	Mother
West Germany	52	100%/14–18 weeks	Mother or father
Sweden	12–52	90%/38 weeks	Mother or father
Austria	16–52	100%/20 weeks	Mother
Chile	18	100%/18 weeks	Mother
United States	12	0	Mother or father

Source: John J. Sweeney and Karen Nussbaum, *Solutions for the New Work Force* (Cabin John, MD: Seven Locks Press, 1989), p. 108. Reprinted by permission.

dual employment. Sweden has been the most advanced. Following the birth of an employee's child, Swedish employers are required to provide thirty-eight weeks of work furlough at 90 percent of pay for either the father or mother and up to one year at a further-reduced compensation rate.

In 1993, under the Clinton administration, the Family and Medical Leave Act was passed. Prior to this federal law, no employers had been required to offer family leave for childbirth or medical emergency. This act requires that employers with fifty or more employees provide up to twelve weeks of unpaid leave for either the father or mother of a newborn child, the adoption of a child, the placement of a foster child, or for some other family member's medical emergency. A serious health condition is broadly defined to include illness, injury, and mental problems. The employer must permit the employee to return to the same or an equivalent job and must continue group health insurance for the employee during the leave period (Snarr 1993; Stranger et al. 1993).

There are several shortcomings to this step, which is relatively modest in comparison to government policies in other industrialized societies. First, because most American workers are employed in organizations with fewer than fifty employees, this benefit is unavailable to them. Second, less than 40 percent of dual-employed couples can afford to take an unpaid six-week leave. Third, employees must have been employed by their company for at least twelve months to be eligible (Shaller and Qualiana 1993; Reskin and Padavic 2000; Wisensale 2001).

Reversing a twenty-five year trend, there is a recent decline in employment among women with infants under one year of age, from 60 percent in 1998 to 54 percent in 2000. It would be premature to predict a continuing decrease in working mothers with infants. Moreover, there has been no decrease in employment among mothers of older children (Bachu and O'Connell 2001).

Single Life

To what extent are Americans remaining single? The increased age at first marriage for both sexes and the high divorce rate have combined to create an increase in the percentage of single adults. In 1980, about 17 percent of women and 24 percent of men over the age of seventeen had never been married. By 2000, these percentages had increased to just over 21 percent for women and almost 27 percent for men. More than 26 million Americans over the age of fifteen now live alone, an increase of 46 percent since 1980. Although many of these people will eventually marry, an increasing percentage will remain single all their lives (U.S. Bureau of the Census 2001a).

Why are more Americans choosing to live alone? Remaining single has always been an alternative, but it has carried a stigma. During colonial days, bachelors were taxed more heavily than married men; spinsters were viewed as millstones around the necks of their families. Adults were expected to marry and to begin having children as soon as morally and physically possible. Failure to marry was seen as a form of inadequacy and deviance.

The stigma attached to remaining single has faded since colonial times, especially in the past two decades, and more Americans are either choosing this alternative or at least marrying later than previous generations. In addition to the lifting of the stigma, there are other factors contributing to the popularity of singlehood. More single Americans are choosing to remain childless, to obtain sexual gratification outside of marriage, to pursue careers, to rear children from a former marriage, to adopt children, to have strictly homosexual relationships, or to rear out-of-wedlock children.

Will the current trend continue? It is too early to predict confidently whether the increase in singlehood among the young will eventually lead to a decline in marriage at all ages. It is safe to say that singlehood is an increasingly popular alternative to traditional marriage. This does not necessarily mean a rejection of marriage, but it does imply a desire to expand the period of "freedom" after leaving home and an unwillingness to rush into the responsibilities of early marriage and parenthood.

Cohabitation

What is cohabitation, and how widespread is it? **Cohabitation**—living with someone in a marriage-like arrangement without the legal obligations and responsibilities of formal marriage—has emerged as an alternative to traditional monogamy. It is still not known if cohabitation has increased because more Americans are delaying marriage or because cohabitation is being substituted for marriage.

It is presently impossible to predict accurately whether cohabitation will ever become a commonly accepted practice in the mate selection process, something experienced by virtually everyone. Certainly, as the average age at first marriage goes up, increased toleration for this lifestyle is a possible outcome; and given the high divorce rate, it may be that cohabitation will become a more popular practice in the selection of a second marital partner as well. In fact, the number of American adults cohabiting increased from 439,000 to nearly 5 million between 1960 and 2000, accounting for 48 percent of unmarried householders (Cohn 2001; Kantrowitz and Wingert 2001). One-third of American women age fifteen to forty-four report that they have cohabited at some time in their lives. For women age twenty-five to twenty-nine, the figure is 45 percent (U.S. Bureau of the Census 2001a).

Cohabitation has risen among people of all ages and all marital statuses, but particularly among the young and the divorced. By 1999, about 75 percent of all unmarried-couple households were maintained by someone under forty-five years of age, and about one-third involved at least one child under fifteen (Peterson 2000; U.S. Bureau of the Census 2001a).

Is cohabitation a workable alternative to marriage? Research reports on cohabitation are not encouraging. Only about 25 percent of cohabiting couples stay together more than four years, reflecting a lower level of certainty about commitment than is found in marriage. This lack of commitment is probably an important reason for the lower satisfaction among cohabiting couples than among married couples (Nock 1995). Another factor is the higher rate of abuse among cohabiting women than among married, divorced, or separated women.

Cohabitation, some assumed, would provide good experience for future marriage, but it has not fulfilled the promise (F. D. Cox 2002; Manning and Smock 2002). Cohabitation does not appear to improve the quality of later marriage. Couples who cohabited have shown lower marital adjustment than that of couples who had not lived together. In addition, premarital cohabitation is associated with a higher risk of divorce (Brown and Booth 1996).

Same-Sex Domestic Partners

How prevalent is homosexual cohabitation? Because of the stigma that surrounds homosexuality, it is impossible to know what proportion of the American population is homosexual. The Institute of Sex Research, founded by Alfred Kinsey, estimates that homosexuals constitute about 10 percent of the U.S. population (13 percent of the males, 5 percent of the females). The number of homosexuals cohabiting is increasing even

© Deborah Davis/PhotoEdit

In the United States, the number of homosexual families is small compared with heterosexual marriages, but it is on the rise. Whether by divorce, adoption, or artificial insemination, these lesbian women have become parents.

though homosexual "marriages" are recognized only by Massachusetts (Van Drehl 2003).

Homosexual families—same-sex partners living together with children—are also increasing in number, although their number is small compared with heterosexual marriages. Homosexual families are created in several ways. A divorced person may take his or her children into a new homosexual relationship. In fact, about 15 percent of lesbian mothers are currently given custody of their children, and homosexual fathers are beginning to seek custody as well. Homosexual couples are also adopting children, a movement encouraged by a 1979 New York case in which the court gave permanent custody of a thirteen-year-old boy to an openly homosexual minister who had been caring for the boy (Maddox 1982). Even before this case, welfare agencies in large cities were placing orphaned children with homosexual couples. Although it is as yet uncommon, lesbian couples can purchase frozen semen from sperm banks and have their families by artificial insemination, and homosexual male couples can hire surrogate mothers who are inseminated with the couple's mixed semen.

Several relatively recent legal and religious decisions reflect change in the social landscape for same-sex domestic partners. In 1996, the U.S. Supreme Court ruled that gays have equal rights under the Constitution (Thomas 2003).

Gay rights advocates were heartened by a 1999 Vermont Supreme Court ruling that gay and lesbian couples are due the same benefits and protections as heterosexual married couples. While this ruling was the first significant legal advance for gays, the justices stopped short of legalizing gay marriage (Rosin 1999). In 2000, the Vermont legislature also failed to legalize gay marriage. Vermont did, however, become the only state to legalize "civil unions" between two gays. Those who enter this type of arrangement will have the same rights as married couples with regard to such matters as taxes, health benefits, and inheritance (Rosenberg 2000).

The U.S. Supreme Court dropped a legal bombshell in a 2003 decision striking down the nation's anti-sodomy laws. Although the decision did not specify the implications of this law for gay marriage, gay adoption, and related disputes, the glee experienced by the gay rights community was matched by the gloom among social conservatives (Cooperman 2003; Lane 2003; Von Drehle 2003). An openly gay Episcopal bishop, the Rev. Gene Robinson, was elected Bishop of New Hampshire in 2003. The next day Episcopal Church leaders adopted a resolution making official blessing of same-sex unions an option available to the clergy. Finally, in 2003 the Ontario province of Canada joined Belgium and the Netherlands in recognizing the legality of same-sex marriage. Because the United States accepts Canadian marriage licenses, confusion is likely to follow the expected border crossings by gays wishing to be married.

Debate may become even more complicated if recent laboratory research on animals proves applicable to humans. In 2003, scientists reported turning ordinary mouse embryo cells into egg cells in laboratory dishes. If the same can be done with human cells, then gay male couples could produce children. This would make it possible for male gay partners whose ordinary cells are transformed into eggs to become biological mothers even though gestation would take place in females. Moreover, if it ever becomes possible to make sperm from stem cells, then a lesbian couple could reproduce a child in complete independence from males (R. Weiss 2003).

Adult Children Returning Home

In nineteenth-century America, the transition from childhood to adulthood (for males) varied in length and was very loose in its boundaries. As historian E. Anthony Rotundo writes:

> *During this transitional phase, young males lived in settings that ranged from boardinghouses to college dormitories to their own family homes, and they often shuttled back and forth between these settings. A leading historian has aptly called this a period of "semidependence," since a youth's relation to his family was ambiguous and—in some cases—frequently shifting.* (Rotundo 1993:56)

For different reasons, young American adults (eighteen- to thirty-four-year-olds) have a much higher probability of living in their parents' home than they did thirty years ago. For example, the percentage of adults aged eighteen to twenty-four living at home increased from 43 percent in 1960 to 57 percent in 2000 (Bianchi and Casper 2000). Moreover, 26 percent of adults eighteen to thirty-four years old now live with their parents (U.S. Bureau of the Census 2001a). These offspring have been labeled boomerang kids because they are returning to their point of origin (Quinn 1993; Goldscheider and Goldscheider 1994).

Why are more adult children returning home? Several factors combine to produce this higher proportion of young adults living with their parents (Ward, Logan, and Spitze 1992; F. D. Cox 2002). Because young adults are marrying later, more stay at home longer. In addition, more of them are continuing their education and find living at home the best solution to the challenges of self-support and school expenses. Many young adults return home (or remain home) even after completing their education because the high cost of living outstrips their earning capacity. The high divorce rate also increases the proportion of young adults living at home because parents tend to accept their children in the home after a failed marriage.

What are some possible consequences of this boomerang effect? An added financial burden can create significant strain for older parents whose adult children live at home—costs associated with education, day-to-day living, and perhaps even a grandchild or two. Many parents complain that their adult children do not share in expenses, fail to help around the house, rob them of their privacy, and prevent them from developing relationships with spouses and friends. It is not surprising that higher marital dissatisfaction among middle-aged parents is associated with adult children living at home (Glick and Lin 1986).

Adult children who find themselves in this situation could suffer as well. Adult children who have returned home normally do so from necessity rather than choice. They are likely to be having difficulties associated with balancing school and work, making their way economically, forming a family, or surviving the aftermath of a divorce. They know the burden they represent. In addition, returning home usually means forfeiting some freedom and being subject to unwanted parental control.

Despite these strains, most families appear to adjust well with the return of older children (Mitchell and Gee 1996). This is especially true when the returning older child contributes financially and helps with household duties.

The Sandwich Generation

The current middle-aged generation faces problems associated with new family forms. Due to prolonged life expectancy, refusal to place parents in nursing homes, and fewer siblings to share the burden, more middle-aged adults are finding mothers or fathers (their own or their spouse's) living with them (Knox and Schacht 2002). *Sandwich generation* is the term applied to those adults caught between caring for their parents and caring for the family they formed after leaving home. Sandwiching, of course, can occur whether or not the elderly parent(s) live with the younger couple.

What are the repercussions? On the positive side, elderly parents unable to take care of themselves usually receive better care from those who love them and feel responsible for their well-being. Older children may enjoy taking care of those who reared them. The caregiving is usually not one-way. Aging parents can offer emotional support and financial resources (Walker, Pratt, and Oppy 1992).

There are also negative repercussions. Taking care of an elderly parent is often not easy. A parent with severe arthritis or Alzheimer's disease demands close and constant attention. Younger adults in this situation may resent the social and emotional intrusion in their own family life. Guilt feelings, then, may arise because of this resentment, and the younger individuals may suffer a loss of self-esteem because they perceive themselves to be selfish. The older parents, too, may feel guilt and anger about the burden they are placing on those they love. They also experience stress and depression from the problems they see their adult children, their spouses, and their grandchildren undergoing (Pillemer and Suiter 1991).

One additional negative consequence deserves special mention. The burden of caring for an aging parent falls much more heavily on women. According to one study, in 90 percent of the cases the primary caregiver is a female (Dychtwald 1990). It is typically daughters, rather than sons, who take charge of the care for an aging parent (Brubaker 1990; Spitze and Logan 1990). The average woman in the United States is likely to devote more years caring for her aging parents than she did caring for her own children (Clabes 1989). According to one study, over one-fourth of nonworking women were out of the labor force because of their parents' needs (Hull 1985).

Looking Forward

In early 2000, Darva Conger and Rick Rockwell, a couple who had never met, married as part of a television contest called "Who Wants to Marry a Millionaire?" The marriage was annulled less than two months later. Most Americans shook their heads, wondering if these events marked the final stages of deterioration for the American family. While this was truly a bizarre event, it is not representative of the state of the American family (C. Wilson 2000).

If the frequency of marriage and remarriage is any indication, the nuclear family is not disappearing. Over 90 percent of men and women in the United States marry sometime during their lives, and about three-fourths of men and women between the ages of twenty-five and sixty-five are married at any given time (Goldstein and Kenney 2001). It is estimated that two-thirds of all divorced persons remarry. In fact, 45 percent of all weddings each year are remarriages. Sixty percent of second marriages end in divorce, but three-fourths of people twice divorced marry a third or even a fourth time. This is known as *serial monogamy*, an alternative that is on the rise.

Despite all the current experimentation with alternatives, the nuclear family remains the most popular choice among Americans. In 2000, 53 percent of all households were composed of married couples. Of all families, 77 percent were married couples. Of the remaining families, 5.6 percent were families headed by males

FIGURE 12.7

Families in the Labor Force: 1940–2000

The nature of families in the United States has changed since 1940. What do you believe will be the most significant trend by 2025? Explain.

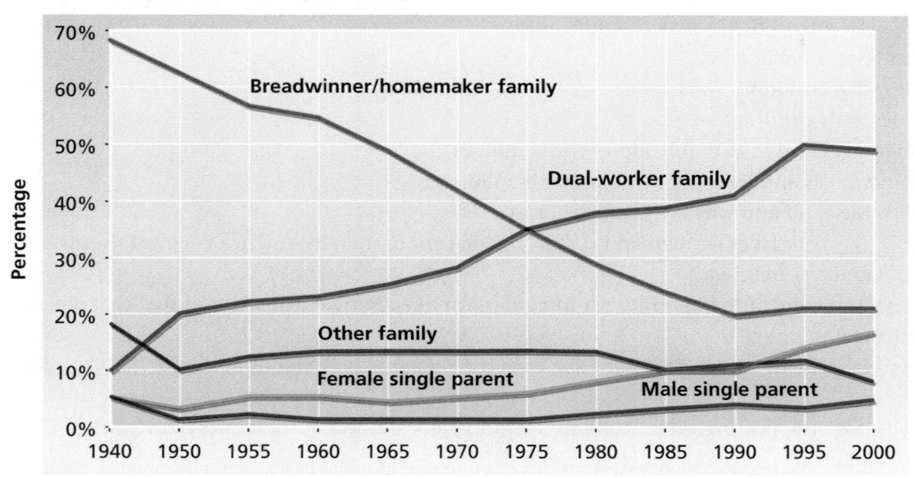

Sources: Howard V. Hayghe, "Family Members in the Work Force," *Monthly Labor Review* 113 (March 1990):16; and U.S. Bureau of the Census, *Statistical Abstract of the United States: 2001* (Washington, DC: U.S. Government Printing Office, 2001), p. 374.

INTERNET LINK Considerable information on the marital status and living arrangements in the United States can be accessed from the Web site of the U.S. Bureau of the Census at www.census.gov/population/www/socdemo/ms-la.html

and 17.6 percent were families headed by females (U.S. Bureau of the Census 2001a). Moreover, after decades of a decrease in percentage of two-parent families, 69 percent of American children now live with two parents.

The nuclear family, then, is not being abandoned. Contrary to a long-standing fear, most Americans are not avoiding marriage permanently; they are simply postponing it or sampling it often (Popenoe, Elshtain, and Blankenhorn 1996).

This is not to say that change in the American family is not occurring. The proportion of all households with the traditional husband wage earner, wife homemaker, and two children is expected to continue to account for less than one-fourth of all American households. Continued increases are expected for other family lifestyles such as the dual-employed family and the single-parent family (see Figure 12.7). As sociologist Judith Stacey notes, there is no longer "a single culturally dominant family pattern to which the majority of Americans conform or aspire" (1990:17).

But, change is not the equivalent of obliteration. Contrary to Arlie Hochschild's (1997) highly influential thesis, a shift toward women's view of work as more satisfying than family life does not appear to have occurred. In the first longitudinal study on this question, K. Jill Kiecolt (2003) reports an increase in satisfaction with home over work among women since 1973. And a relatively stable proportion of younger Americans continue to value marriage and family life, complete with children (Thorton and Young-DeMarco 2001).

Although variations will become more prevalent and socially acceptable, the family system is not about to disappear from American society. In the United States, as well as in other modern societies, the nuclear family will remain the bedrock of the family system, despite the growing diversity in family form (Wolfe 1999; Bianchi and Casper 2000; Knox and Schacht 2002).

FEEDBACK

1. A family composed of at least one remarried man or woman with at least one child from a previous marriage is a __________ family.
2. Although the proportion of single-parent families has increased for both blacks and whites since 1970, the gap between them has
   a. decreased somewhat.
   b. increased significantly.
   c. remained the same.
   d. decreased significantly.
3. Voluntarily child-free couples are generally
   a. less educated than couples with children.
   b. less emotional than couples with children.
   c. less satisfied with their marriages and lives than couples with children.
   d. more satisfied with their marriages and lives than couples with children.
4. In a __________ marriage, both husband and wife are in the labor market.
5. In 1997, approximately __________ percent of women and __________ percent of men in America between the ages of twenty-five and twenty-nine had never married.
6. __________ involves living with someone of the opposite sex in a marriage-like arrangement without the legal obligations and responsibilities of formal marriage.
7. The percentage of adults aged eighteen to twenty-four who live with their parents now stands at over __________ percent.
8. Although the nuclear family is not disappearing, Americans are spending less and less of their lives married. T or F?

Answers: 1. blended 2. c 3. d 4. dual-employed 5. 35, 51 6. Cohabitation 7. 53 8. T

SUMMARY

1. A family is a group of people related by marriage, blood, or adoption. Each of us may belong to two families—the family into which we were born and the family we create upon marrying.
2. The extended family and the nuclear family are two basic types of family structure. A curvilinear relationship exists between family structure and industrialization. In hunting and gathering societies, the nuclear family was the most prevalent type. With the rise of agriculture as the means of subsistence, the extended family came to prevail. The nuclear family regained popularity in modern industrial society.
3. Whether the family is nuclear or extended, there are several important dimensions of family structure. These dimensions pertain to descent and inheritance, family authority, and residence pattern.
4. Mate selection is never completely a matter of individual choice. Exogamous, endogamous, and homogamous mate selection norms exist in all societies.
5. There are four basic types of marriages: one man and one woman may marry; one man and several women may marry; one woman and several men may marry; or two or more men may marry two or more women.
6. In all societies, the family has been the most important institution. It is the institution that produces new generations, socializes the young, provides care and affection, regulates sexual behavior, transmits social status, and provides economic support.
7. Although functionalism emphasizes the benefits of the family for society, the conflict perspective depicts the traditional family structure as the instrument of male domination over women. Evidence of this domination, they assert, is reflected in the traditional ownership of women by men, family rules of power and inheritance, and the male-dominated economic division of labor.
8. Symbolic interactionism is used frequently in the study of the family. It is within the family that socialization of children begins and children develop a self-concept. Most of the interactions within families can be analyzed within this theoretical perspective.
9. Modern marriages are primarily based on love. This is a relatively new development in the formation of marriages.
10. The divorce rate in the United States rose dramatically in recent times. Factors promoting divorce include the quality of the marital relationship, the increasing economic independence of women, and the social acceptability of divorce. The dissolution of marriage, however, is not expected to continue to escalate. In fact, the divorce rate has declined slightly.
11. Although the American family provides social and emotional support for its members, violence in this intimate setting is not uncommon. Spouses use violence against each other, parents abuse children, siblings are physically violent with each other, and the elderly are abused.
12. Problems such as the high divorce rate and frequent family violence and abuse have raised the question of the strength and durability of the American family. Some researchers approach this area of research via the concept of family resiliency—the family's capacity to emerge from crises as stronger and more resourceful.

13. The nuclear family is not going to be replaced on any broad scale. Still, there is considerable experimentation with new patterns of marriage and family: blended families, single-parent families, child-free marriages, people remaining single, cohabitation, gay and lesbian couples, adult children returning home, and the sandwich generation. In addition, the American family is being affected by the rise of the dual-employed marital arrangement.

INFOTRAC® COLLEGE EDITION

http://www.infotrac-college.com/wadsworth

Another unique option available to you at the Student Resources section of the companion Web site is Infotrac College Edition, an online library with access to hundreds of scholarly and popular periodicals. Below are suggested search terms for this chapter. Results from these and other searches are found at the site.

Search keywords: **blended family**. Because of the relatively high divorce rate in the United States, blended families are becoming much more common. What are some of the special problems encountered in blended families? Search Infotrac to learn more about blended families and the unique issues they face.

Search keyword: **cohabitation**. Historically, in the United States, cohabitation was considered deviant behavior. Read at least two articles to find out how cohabitation is viewed today. Is it becoming more or less popular? What are some of the contributing factors to the change you noted?

Search keywords: **family planning**. Read three articles about family planning. What is meant by this term? Do each of the authors define it in the same way? What are some of the problems, if any, encountered by advocates of family planning? From a sociological perspective, how do you view family planning?

Search keywords: **gay rights**. Some people have likened the current gay rights movement to the civil rights movement of the 1960s. Read two articles about gay rights in the United States, and compare that information with what you know about the civil rights movement. What similarities are there? What differences?

LEARNING OBJECTIVES REVIEW

- Describe the types of family structure, dimensions of family structure, societal norms for mate selection, and types of marital arrangements.
- Compare and contrast views of the family proposed by functionalists, conflict theorists, and symbolic interactionists.
- Outline the extent and cause of divorce in America.
- Give an overview of family violence in the United States.
- Describe contemporary alternatives to the traditional nuclear family structure.
- Discuss the future of the family in the United States.

CONCEPT REVIEW

Match the following concepts with the definitions listed below them.

____ a. matrilinear descent
____ b. exogamy
____ c. nuclear family
____ d. family
____ e. polyandry
____ f. blended family
____ g. marriage
____ h. democratic control
____ i. patrilineal descent
____ j. family of procreation
____ k. divorce ratio
____ l. monogamy
____ m. family resiliency

1. composed of a remarried man or woman and at least one child from a previous marriage
2. the form of control in which authority is split evenly between husband and wife
3. the number of divorced persons per 1,000 persons who are married and living with their spouses
4. mate selection norms requiring individuals to marry someone outside their kind
5. a group of people related by marriage, blood, or adoption
6. the family group established upon marriage
7. a heterosexual union in which public approval is given to sexual activity, having children, and assuming mutual rights and obligations
8. the familial arrangement in which descent and inheritance are passed from the mother to her female descendants
9. the form of marriage in which one man is married to only one woman at a time
10. the smallest group of individuals (mother, father, and children) that can be called a family

11. the familial arrangement in which descent and inheritance are passed from the father to his male descendants
12. the form of marriage in which one woman is married to two or more men at the same time
13. the family's capacity to survive crises

CRITICAL-THINKING QUESTIONS

1. The text outlines six functions of the family. Describe how you have experienced these functions within your family of orientation. Be specific.
2. Does conflict theory explain any of the social relationships you have seen in your family or the families of your friends? Explain.
3. Focusing on the looking-glass self, show how symbolic interactionism can be used to explain your experiences in your family of orientation. Provide examples that apply to you or other members of your family.
4. What do you think the divorce rate trend line in the United States will be in the twenty-first century? Develop your answer within the context of information in this chapter.
5. Given all the current marital and familial lifestyle variations in the United States today, many observers argue that the nuclear family is going to become a choice of the minority of Americans. Do you agree or disagree? Use data and social trends to support your position.

MULTIPLE-CHOICE QUESTIONS

1. **The family that provides one with a name, an identity, and a heritage is known as the**
   a. matrilineal family.
   b. family of orientation.
   c. patriarchal family.
   d. nuclear family.
   e. neolocality family.
2. **The _________ family consists of two or more adult generations of the same family sharing a common household and economic resources.**
   a. nuclear
   b. extended
   c. adult-child
   d. patriarchal
   e. matriarchal
3. **As societies move from an agricultural to an industrial base, the _________ family tends to be replaced by the _________ family.**
   a. nuclear; extended
   b. matriarchal; patriarchal
   c. matrilocal; patrilocal
   d. extended; nuclear
   e. democratic; patriarchal
4. **The family arrangement in which descent and inheritance are transmitted from the mother to her female descendants is called a**
   a. female-centric family.
   b. matrilineal family.
   c. neolineal family.
   d. matrilocal family.
   e. bilateral family.
5. **A residential arrangement in which a married couple establishes a new residence of its own is known as a/an _________ arrangement.**
   a. bilateral
   b. democratic
   c. matrilocal
   d. neolocal
   e. independent
6. **A mate selection arrangement in which norms encourage individuals to marry within their own kind is known as**
   a. endogamy.
   b. exogamy.
   c. monogamy.
   d. polyandry.
   e. incest.
7. **Which of the following is the most widely practiced form of marriage around the world today?**
   a. polygyny
   b. polygamy
   c. polyandry
   d. monogamy
   e. group marriage
8. **Which of the following is not considered a vital function of the family?**
   a. the provision of initial learning experiences that make people human

b. the provision of a warm and loving atmosphere that fulfills basic human social and emotional needs
c. the legitimization of reproduction in a society
d. the regulation of sexual activity
e. the promotion of social mobility

9. **According to conflict theorists,**
   a. the family has historically been an important means of maintaining male dominance over females.
   b. the family in modern society is helping to shatter male dominance.
   c. relationships within the family are failing to undergo the proper kinds of changes.
   d. male dominance over females is due to the dynamics of intimate and personal relationships within the family.
   e. the family remains the one place in society where an individual is unconditionally accepted.
10. **The number of divorced persons, per 1,000 persons in the population, who are married and living with their spouses is known as the**
    a. divorce rate.
    b. divorce ratio.
    c. marital failure rate.
    d. marriage-divorce ratio.
    e. marital success ratio.
11. **According to research cited in the text, voluntarily child-free couples appear to be**
    a. pathologically self-centered.
    b. more committed to religion than couples with children.
    c. declining in number compared to couples with children.
    d. lacking in self-esteem.
    e. more satisfied with their marriages and their lives than couples with children.
12. **_________ involves living with someone in a marriagelike arrangement without the legal obligations and responsibilities of formal marriage.**
    a. Dual-consent marriage
    b. Blended marriage
    c. Duolocal residence
    d. Homogamy
    e. Cohabitation
13. **It is estimated that about _________ percent of the U.S. population is gay or lesbian.**
    a. 2
    b. 5
    c. 10
    d. 12
    e. 20
14. **Approximately _________ percent of American adults between the ages of eighteen and thirty-four live at home with their parents.**
    a. 12
    b. 19
    c. 26
    d. 31
    e. 35
15. **Which of the following is the most accurate statement regarding the future of the American family?**
    a. The nuclear family is no longer the most popular choice.
    b. Childless couples are now so prevalent that the growth of the population is being seriously affected.
    c. Most Americans are not avoiding marriage permanently.
    d. Homosexual marriage is undermining the institution of the family.
    e. Americans are spending more and more of their lives married.

FEEDBACK REVIEW

True-False

1. Contrary to what one would expect from the traditional male dominance in American family life, women are spousal abusers almost as frequently as men are. T or F?
2. Although the nuclear family is not disappearing, Americans are spending less and less of their lives married. T or F?

Fill in the Blank

3. Industrialization promotes the shift from the extended family to the _________ family.
4. In a _________ family, the oldest man living in the household controls the rest of the family members.
5. In 1997, approximately _________ percent of women and _________ percent of men in America between the ages of twenty-five and twenty-nine had never married.

Multiple Choice

6. **Which of the following functions of the family is not shared with any other institution?**
   a. socialization
   b. reproduction
   c. socioemotional maintenance
   d. sexual regulation
7. **Which of the following is not stated in the text as one of the factors associated with divorce?**
   a. decline of religious influence
   b. age at first marriage
   c. length of marriage
   d. economic conditions
8. **Although the proportion of single-parent families has increased for both blacks and whites since 1970, the gap between them has**
   a. decreased somewhat.
   b. increased significantly.

c. remained the same.
d. decreased significantly.

Matching

9. Indicate which type of mate selection norm is reflected in each of the following situations.
____ a. Jews are supposed to marry Jews.
____ b. A father is not permitted to marry his daughter.
____ c. Members of the same social class marry.
(1) exogamy
(2) endogamy
(3) homogamy

10. Match the following examples with the major theoretical perspectives.
____ a. a father "giving away" the bride
____ b. sexual regulation
____ c. development of self-concept
____ d. a newly married couple adjusting to each other
____ e. child abuse
____ f. social class passed from one generation to another
(1) functionalism
(2) conflict theory
(3) symbolic interactionism

GRAPHIC REVIEW

U.S. trends in age at first marriage for males and females are graphically depicted in Figure 12.4.

1. If this upward trend in age at first marriage for both sexes continues, what do you predict will happen to the divorce rate?

____

____

____

2. Explain the reasons for the relationship between age at first marriage and the divorce rate you assumed in answering the first question.

____

____

____

ANSWER KEY

Concept Review
a. 8
b. 4
c. 10
d. 5
e. 12
f. 1
g. 7
h. 2
i. 11
j. 6
k. 3
l. 9
m. 13

Multiple Choice
1. b
2. b
3. d
4. b
5. d
6. a
7. d
8. e
9. a
10. b
11. e
12. e
13. c
14. c
15. c

Feedback Review
1. T
2. T
3. nuclear
4. patriarchal
5. 35, 51
6. b
7. a
8. c
9. a. 2
b. 1
c. 3
10. a. 2
b. 1
c. 3
d. 3
e. 2
f. 1

13 Education

© Tony Freeman/PhotoEdit

OUTLINE

LEARNING OBJECTIVES

- Describe the relationship between industrialization and education.
- Discuss schools as bureaucracies and the attempts to debureaucratize education in the United States.
- Outline the basic functions (manifest and latent) of the institution of education.
- Evaluate the meritocratic model of public education.
- Discuss educational inequality.
- Describe the ways in which schools socialize students.
- Identify and describe the dominant issues in higher education.

USING THE SOCIOLOGICAL IMAGINATION

Is federally funded Head Start another example of government waste? Americans tend to believe that the federal "bureaucracy," including its efforts in Head Start, is doomed to failure. This is government "interference" in our lives. Research, to the contrary, has consistently shown the Head Start program to produce positive benefits for underprivileged children: improvements in intelligence tests, general ability tests, and learning readiness.

Education serves many functions in modern society, including the promotion of a common identity among all members of society. Due to educational inequality among minority groups, the achievement of this integrative function has been retarded in the United States. Although Head Start has helped in this regard, barriers in the educational system remain. Before turning to the issue of educational inequality, however, some other topics require coverage, beginning with the development and structure of the American educational institution.

The Development and Structure of Education

The basic purpose of education is the transmission of knowledge. Education was originally a family responsibility, but industrialization changed that, dramatically.

Industrialization and Education

Why did schools emerge as the primary educational agent? Before industrialization, the family, the rural community, and the church were the major socializing groups in a child's life. The family taught children the values, norms, and farming skills needed to survive in an agricultural society; but as societies industrialized, this knowledge increasingly lost its relevance. And as knowledge increased in volume and complexity, family members gradually lost the ability to adequately educate their children.

As a consequence, the educational institution developed. Many leaders of the American Revolution created blueprints for publicly financed school systems. These plans for state-sponsored education for all citizens finally bore fruit in the common school movement of the early nineteenth century (G. S. Wood 2002).

What was the objective of early schools? The founding fathers considered education an investment in democracy because it would create a literate, active, and informed public. Schools, they believed, would create some minimal "democracy of knowledge," thus preventing the privileged few from having a monopoly on knowledge. The teaching of democratic values and marketable skills became especially popular in urban areas inundated by the many ethnic groups who immigrated to the United States between 1850 and 1910. Labor leaders, fearing the loss of adult jobs to child laborers, urged city governments to keep children in school. Many new questions about education were raised. Should children attend school with other children like themselves? Should neighborhoods operate schools? Should "professional" teachers be paid for their work? Should education be viewed as charity for poor children? On the whole, there was no consensus.

Eventually, some agreement about the mission of education was reached. Most late-nineteenth-century proponents of public education believed the objective of the early school to be

> *not so much . . . intellectual culture, as the regulation of the feelings and dispositions, the extirpation of vicious propensities, the preoccupation of the wilderness of the young heart with seeds and germs of moral beauty, and the formation of a lovely and virtuous character by the habitual practice of cleanliness, delicacy, refinement, good temper, gentleness, kindness, justice, and truth.* (M. B. Katz 1975:31)

Thus, public schooling in nineteenth-century America was not designed primarily to teach sophisticated intellectual skills; it was a mechanism of social control. Otherwise, it was feared, children would grow up "vicious" or the "wilderness of the young heart" would lack "seeds of moral beauty."

How did education develop after the turn of the twentieth century? Just as farming slowly gave way to manual factory labor in the latter part of the nineteenth century, relatively simple manual skills began to be less and less important after the beginning of the twentieth century. New middle-management and clerical jobs were seen as highly desirable by many Americans seeking to better their own lives and, more important, their children's future. Business leaders and parents began to demand more advanced training in schools. School administrators responded by developing secondary schools that emphasized advanced learning. Figure 13.1 indicates the percentage of Americans (including major racial

FIGURE 13.1

High School Graduates by Race: 1970 and 2000

This figure indicates the percentages of persons twenty-five years and older who have completed high school, by racial and ethnic category. In today's context, it is important to see that the proportion of high school graduates in each group has increased sharply between 1970 and 2000. As a result, each of these groups is placing more pressure on public schools to accomodate them.

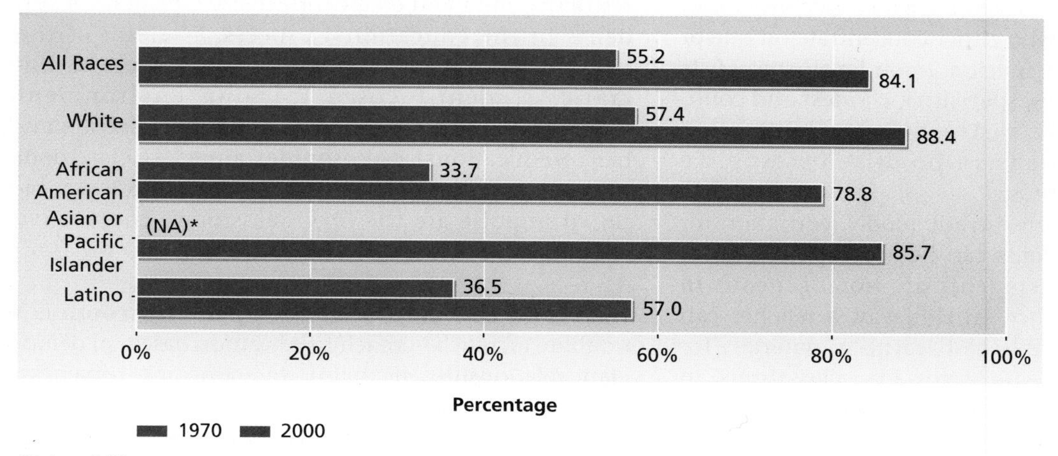

*Not available.
Source: U.S. Bureau of the Census, "Table 1: Educational Attainment of the Population 15 Years and Over, by Age, Sex, Race, and Hispanic Origin: March 2000," 2000; at www.census.gov/population/socdemo/education/p20-536/tab01.pdf

and ethnic groups) over twenty-five years of age who have at least completed high school.

Bureaucracy in Education

Organizationally, early school administration was based on the factory model of education. Children were to be educated in much the same way as cars were mass-produced:

> *Schooling came to be seen as work or the preparation for work; schools were pictured as factories, educators as industrial managers, and students as the raw materials to be inducted into the production process. The ideology of school management was recast in the mold of the business corporation, and the character of education was shaped after the image of industrial production.* (Cohen and Lazerson 1972:47)

© Mark Richards/PhotoEdit

Government spending on the poor is under attack, an attack based in part on the claim that these government programs do not work. Head Start, a federally funded program for preschoolers from disadvantaged homes, has proven to be a success.

Although this philosophy softened considerably, the concept of bureaucratized public education remains prevalent, with its specialization, rules and procedures, and impersonality.

In what ways are schools bureaucratic? In keeping with the bureaucratic model, the process of educating children is divided into specializations: administrators who run schools; teachers who either concentrate on one or two topics or teach a specific grade level; specialists who purchase materials; specialists who test and counsel students; and specialists who manage libraries. There are even specialists who drive buses, serve lunches, and type memos.

According to the bureaucratic model, education for large numbers of students can be accomplished most efficiently when the students are homogeneous in development and ability. In this way, a teacher can focus attention on one set of learning materials for large numbers of students. Age-graded classrooms, in which all students receive the same instruction, reflect the impersonal, bureaucratic nature of schools.

It is also thought more efficient (the ultimate goal of a bureaucracy) to have all teachers of a given subject teaching the same, or at least similar, material. Materials can be approved and purchased in bulk, and testing can be standardized. In addition, if parents need to know what subjects are being taught at any given time, school administrators can readily advise them. This practice also allows students to transfer from one school to another without losing continuity of curriculum. Rules and procedures exist to ensure that this happens.

Schools are also part of a much larger bureaucratic system. This system begins with the federal government and progresses layer by layer through state and local governments.

Critics of the bureaucratic model claim that children are not inorganic materials to be processed on an assembly line. Children are human beings who come into school with previous knowledge and who interact socially and emotionally with other students. Moreover, it is hard to determine whether children have been "processed" correctly. How can we know what takes place inside a child's head? According to critics of formal schooling, the school's bureaucratic nature is unable to respond to the expressive, creative, and emotional needs of individual children.

Reforms in the Classroom

Are there alternatives to the bureaucratic model? American educational reform occurs in cycles. The U.S. progressive education movement of the 1920s and 1930s was a reaction to the Victorian authoritarianism of early-nineteenth-century schools. Educational philosopher John Dewey (1859–1952) led the progressive education movement, with child-centered focus and an emphasis on work-related knowledge (Dewey 2001). This movement almost disappeared in the 1950s but had a resurgence in the 1960s and 1970s, in the humanistic education movement. The humanistic movement advocated such steps as the elimination of restrictive rules and codes and the involvement of students in the educational process. The aim of the humanistic movement was to create a more democratic, student-focused learning environment (Ballantine 2001). This movement has spawned a number of educational philosophies aimed at classroom reform. Three formulations of the humanistic educational impulse are the open classroom, cooperative learning, and the integrative curriculum.

What is the open classroom? The **open classroom** is a nonbureaucratic approach to education based on democratic relationships, flexibility, and noncompetitiveness:

> *In the 1960s, advocates of the open classroom proposed alternatives to the bureaucratically organized school. They agreed on an underlying philosophy: Children are naturally curious and motivated to learn by their own interests and desires. The most important condition for nurturing this natural interest is freedom, supported by adults who enrich the environment and offer help. In contrast, coercion and regimentation only inhibit emotional and intellectual development.* (Graubard 1972:352)

The open classroom philosophy advocates elimination of the following customs: the sharp authoritarian line between teachers and students, the predetermined curriculum for all children of a given age, the constant comparison of students' performance, the use of competition as a motivator, and the grouping of children according to performance and ability. In short, schools adhering to the open classroom approach wish to avoid the bureaucratic features of traditional public schools (Silberman 1973; Kozol 1985; Kohl 1988; Holt 1995).

The open classroom approach was not fulfilled in a clear-cut way. Success or failure was affected by the availability of money, student-teacher ratio, and teacher endorsement. However, approaches to school structure continue to be debated (Hallinan 1995). Some argue that schools must operate bureaucratically because limited money requires the less expensive mass instruction. Others claim that concern for children should supersede administrative needs, and individualized programs developing the child's whole personality should be encouraged. According to another viewpoint, the bureaucratic nature of schools prepares children for the bureaucratic society to which they must ultimately adapt.

Though further experimentation continues and appears to be warranted, it remains to be seen whether Americans are in the mood for it. In fact, the "back-to-basics" movement, which emerged in the mid-1980s and 1990s, conflicts directly with the open classroom philosophy. The back-to-basics movement represents an attempt to reinstate a traditional curriculum ("reading, writing, and arithmetic") based on more bureaucratic methods.

What is the back-to-basics movement? In 1983, America received an "educational wakeup call" when the National Commission on Excellence in Education issued a report dramatically entitled *A Nation at Risk.* Catching the attention of politicians and the general public, the report warned of a "rising tide of mediocrity" in America's schools. Because of deficiencies in its educational system, the report claimed, America is at risk of being overtaken by some of its economic world competitors (D. P. Gardner 1983). As evidence, the report pointed to several indicators:

- Twenty-three million American adults, 40 percent of minority young people and 13 percent of all seventeen-year-olds, are functionally illiterate.
- Scholastic Aptitude Test (SAT) scores dropped dramatically between the early 1960s and 1980.
- Only one-third of all seventeen-year-olds can solve a mathematics problem requiring several steps.
- American children are being outperformed by children of other nations on achievement tests.

Unlike the recommendations of the progressive and humanistic reform movements, most of the solutions offered by the commission tended toward the bureaucratic. The report urged a return to basic skills such as reading and mathematics. High school graduation requirements should be strengthened to include four years of English, three years of mathematics, three years of science, three years of social studies, and a half year of computer science. School days, the school year, or both should be lengthened. Standardized achievement tests should be administered as students move from one level of schooling to another. High school students should be given significantly more homework than they are assigned at present. Discipline should be tightened through the development and enforcement of codes for student conduct.

The commission's report stirred intellectual debate (Adelson 1984; Bunzel 1985; Gross and Gross 1985) and aroused the public. Americans generally favor the commission's recommendations, especially its call for more homework and greater discipline (Barrett 1990).

In a series of books, English professor E. D. Hirsch contends that American schoolchildren are not being taught the knowledge they need in today's world (Hirsch 1987, 1996). Hirsch has labeled American schoolchildren "culturally illiterate." Surprising numbers of schoolchildren don't know when the American Revolution occurred, who George Washington Carver was, or where Washington, D.C., is located. Although concerned for all Americans, Hirsch is particularly concerned that the failure of schools, and the accompanying cultural illiteracy, is dooming children from poor and illiterate homes to perpetual poverty. Keeping poor and illiterate children in the same condition as their parents is for Hirsch an unacceptable failure of American schools.

Late in 1987, U.S. Secretary of Education William Bennett, under President Ronald Reagan, proposed a back-to-basics curriculum for American high schools. Emphasis was placed on a minimum core curriculum that all students must take (Bennett 1988). In 1991, the first President Bush's secretary of education, Lamar Alexander, formulated a national educational reform plan with six goals (*America 2000* 1991):

1. Every child must begin school ready to learn.
2. The national high school graduation rate must be 90 percent.
3. Competence in core subjects must be shown after grades 4, 8, and 12.
4. American students should be the best educated in the world in math and science.
5. All adults must be literate and possess the skills necessary for citizenship and competition in a global economy.
6. Schools should be free of drugs and violence.

Implementation of the plan included a longer school year, national standards and state exams, and merit pay for teachers.

The latest manifestation of the back-to-basics philosophy is the "high stakes" testing movement. As part of this movement, President George W. Bush signed in January 2002 a bill he considers the cornerstone of his administration's domestic policy. Known as "No Child Left Behind," this federal law requires that each student racial and demographic subgroup improve each year (grades 3–8) on mandatory standardized tests. Schools failing to accomplish this improvement for two years running must help students transfer to better schools and use public money for tutoring students. A school that continues to fall short of the federal standard must either replace its principal and teachers or reopen as a charter school (Fletcher 2003; *From the Capital to the Classroom* 2003).

To date, there is no evidence that the back-to-basics movement, on its own, can live up to its advance billing (Kantrowitz 1993a; Hancock and Wingert 1995; Wingert 1996). (Some successes have been seen when parents and community members become part of the

School's Out—Forever?

In a recent book, *The Age of Spiritual Machines*, author Ray Kurzweil (1999) forecasts modern life in the twenty-first century. He claims that by the end of the century, humans will no longer be the most intelligent or capable beings on the planet—computers will be. To support his view, he makes specific predictions about various aspects of life. The following is his idea of education in 2009:

© Kevin Radford/SuperStock

The American education system was founded on the industrial factory model. Although this mass production, bureaucratic conception of education is still prevalent, classrooms are more informal than in the past.

In the twentieth century, computers in schools were mostly on the trailing edge, with most effective learning from computers taking place in the home. Now in 2009, while schools are still not on the cutting edge, the profound importance of the computer as a knowledge tool is widely recognized. Computers play a central role in all facets of education, as they do in other spheres of life.

The majority of reading is done on displays, although the "installed base" of paper documents is still formidable. The generation of paper documents is dwindling, however, as the books and other papers of largely twentieth-century vintage are being rapidly scanned and stored. Documents circa 2009 routinely include embedded moving images and sounds.

Students of all ages typically have a computer of their own, which is a thin tabletlike device weighing under a pound with a very high resolution display suitable for reading. Students interact with their computers primarily by voice and by pointing with a device that looks like a pencil. Keyboards still exist, but most textual language is created by speaking. Learning materials are accessed through wireless communication.

Intelligent courseware has emerged as a common means of learning. Recent controversial studies have shown that students can learn basic skills such as reading and math just as readily with interactive learning software as with human teachers, particularly when the ratio of students to human teachers is more than one to one. Although the studies have come under attack, most students and their parents have accepted this notion for years. The traditional mode of a human teacher instructing a group of children is still prevalent, but schools are increasingly relying on software approaches, leaving human teachers to attend primarily to issues of motivation, psychological well-being, and socialization. Many children learn on their own using their personal computers before entering grade school.

Preschool and elementary school children routinely read at their intellectual level using print-to-speech reading software until their reading skill level catches up. These print-to-speech reading systems display the full image of documents, and can read the print aloud while highlighting what is being read. Synthetic voices sound fully human. Although some educators expressed concern in the early '00 years that students would rely unduly on reading software, such systems have been readily accepted by children and their parents. Studies have shown that students improve their reading skills by being exposed to synchronized visual and auditory presentations of text.

Learning at a distance (for example, lectures and seminars in which the participants are geographically scattered) is commonplace. (Kurzweil 1999:191–192)

Analyzing the Trends

If Kurzweil's predictions come true, how will education's role in the socialization of students change? Will social stratification become more of a factor in education than it is now—or less? Use information from the chapter to support your answers.

overall approach.) Evidence indicates that implementation of some of the solutions actually inflicted greater harm. The dropout rate increased as students' time increased in school, on testing, and on homework assignments. The low self-esteem of many low-performing students was exacerbated. Stricter grading with increased academic requirements discouraged low-income and minority students even more. Bureaucratically oriented solutions appear to fail as the cure-all they seemed to be. Despite the efforts, it appears that the nation remains at risk (Wagner 2001).

How does cooperative learning contribute to a less bureaucratic classroom? Evidence indicates that a less bureaucratic approach to education should not be dismissed (R. A. Horwitz 1979; Hurn 1993). In fact, the open classroom philosophy that became prominent in the 1960s and 1970s is currently being implemented again (emerging alongside the back-to-basics movement). The philosophy this time is in the form of cooperative learning. **Cooperative learning** takes place in a nonbureaucratic classroom structure in which students study in groups, with teachers acting as guides rather than as the controlling agents:

> *Cooperative learning approaches are based in part on the assumption that students learn best when actively working with others in small heterogeneous groups. . . . It occurs when teachers have students work together in small groups on a task toward a group goal—a single product (a set of answers, a project, a report, a piece or collection of creative writing, etc.) or achieving as high a group average as possible on a test—then reward the entire group on the basis of the quality or quantity of its product according to a set standard of success.* (Oakes 1985:208, 210)

According to the cooperative learning method, students learn more if they are actively involved with others in the classroom (Sizer 1996). The traditional teacher-centered approach rewards students for being passive recipients of information and requires them to compete with others for grades and teacher recognition. Cooperative learning, with its accent on teamwork rather than individual performance, is designed to concentrate on results rather than on performance comparisons among students. Cooperation replaces competition. Students typically work in small groups on a specific task. Credit for the task is given only if all group members do their part.

Some benefits of the cooperative learning approach have been documented. Uncooperativeness and stress among students is reduced; academic performance is elevated; students have more positive attitudes toward school; racial and ethnic antagonism decreases; and self-esteem increases (Oakes 1985; Aronson and Gonzalez 1988; Children's Defense Fund 1991).

Although Japanese schools have been using group learning in the early grades for years, resistance remains strong among many American teachers (Kantrowitz 1993b). They fear loss of control with younger children and disciplinary problems among older students, especially teenagers. Many American teachers and principals believe that mapping a student's progress is more difficult in a cooperative learning setting. This approach also requires more preparatory work for teachers, who must plan lessons allowing for less frequent teacher intrusion. Grading is also more complex when blending an individual and group evaluation.

What is the integrative curriculum? The **integrative curriculum**, another extension of the open classroom, is an approach to education based on student-teacher collaboration. In the traditional classroom, the curriculum is predetermined for students; in the integrative curriculum, the curriculum is created by students and teachers together. Because students are asked to participate in curriculum design and content, the integrative curriculum is democratic in nature. Giving students such power obviously deviates from the traditional subject-centered curriculum (Barr 1995). Rather, subject matter is selected and organized around certain real-world themes or concepts, as exemplified in a Washington State sixth-grade unit of study on water quality:

> *The unit became a part of an actual water quality project that originated in the Great Lakes region of the United States but now spans the globe. Lessons were organized around the actual work of determining water quality in Puget Sound. These lessons culminated in students' reporting to community groups about the quality of the water. In this way learning was relevant to a real-world problem that the students contributed to solving.* (Simmons and El-Hindi 1998:33)

Instruction in this unit was organized around hands-on experience and a "multiple intelligence" approach, recognizing that students in a classroom are not all identical. Students bring to any unit of study a variety of learning styles, interests, and abilities, and different units of study will engage students in varying ways.

Competitors to the Traditional Public School

While the debate over the most effective classroom methods continues, educators and politicians are looking for answers beyond the classroom—to the way schools are organized, funded, and administered. Meanwhile a school choice movement promotes the idea that the best way to improve schools is by using the free enterprise model—creating competition for the public school system. According to school choice advocates, parents and students should be able to select the school that best fits their needs and provides the greatest educational benefits. There are several approaches to the goal of school choice: the voucher system, charter schools, for-profit schools, and magnet schools. Results are mixed (Fuller and Elmore 1996).

What is a voucher system? Proponents of **vouchers** contend that the government should make available to families with school-age children a sum of money that they can use in private or parochial schools. Families who chose a public school would pay nothing, just as in the current system. Parents who chose a religious or other private school would have the tuition paid up to the amount of the government voucher and would have to make up the difference (if any) between the amount of the voucher and the tuition amount. A voucher plan in Cleveland, for example, provides publicly funded scholarships of just over $2,000 annually to almost four thousand city children. Most parents have chosen to spend the money at private schools rather than to keep their children in public schools. In Wisconsin, adoption of a voucher plan ($5,300 per voucher for students whose parents earn less than $30,000 for a family of four) has increased from less than ten schools to 103 (M. A. Fletcher 2001).

Proponents point to the freedom vouchers give parents, the increased quality that would presumably come from schools competing for students, and the ability to place students in educational environments best suited to their particular needs. Critics fear that free choice of school would make urban schools even more of a dumping ground for the poor and minorities, because poor and minority parents in the inner cities cannot make up the difference between the amount of the voucher and the cost of sending their children to higher-quality schools. Critics contend that the use of vouchers will erode national commitment to public education, leaving the public school system in even worse shape (Kozol 1992; Henig 1994; Martinez et al. 1995; Schrag 1999).

Public reaction to the voucher approach has been mixed, with public vouchers affecting only about 0.1 percent of American schoolchildren. In 1999, Florida initiated the first statewide public voucher program, but large-scale public programs exist only in three cities—Cleveland, Milwaukee, and Washington, D.C. Low-income African American and Latino parents tend to prefer a voucher system, permitting them to remove their children from public schools that they believe have not met their children's needs. Because most whites seem to be satisfied with the public schools, they have not embraced the voucher system in large numbers (Thomas and Clemetson 1999).

Claiming that vouchers violate separation of church and state, courts traditionally ruled voucher systems unconstitutional. However, a 1998 U.S. Supreme Court ruling validated a Wisconsin law allowing state money for low-income students to attend either private or parochial schools (Perry and McGraw 1999). And in a 2001 decision, the Supreme Court declared the Cleveland vouchers to be neutral with respect to religion because parents still retained choice of where and how to school their children (Morse 2002).

Evidence on the effectiveness of the voucher system is inconsistent. Although some voucher programs have improved student test scores (compared to public schools), other programs produced no improvement (Toch and Cohen 1998; Moe 2001; Witte 2000; *School Vouchers* 2001; Blum and Mathews 2003). In the most comprehensive study yet, Rand Corporation researchers conclude that many of the key questions about vouchers still await answers. With regard to academic achievement, these researchers found that small-scale experimental privately funded voucher programs aimed at low-income students may have a modest academic benefit for African American students after one to two years in these programs. While even this effect must be tentatively reported, no conclusion at all can be drawn regarding children of other racial or ethnic groups (Gill et al. 2001).

The harshest critics draw several negative conclusions: public schools lose money to private voucher schools, making public education less able to be successful; voucher programs do not necessarily lead to higher academic performance; voucher schools are not accountable to taxpayers (Miner 2000). American public opinion appears to swing toward the critics. As the approval rating of public schools rises higher than it has ever been, support of vouchers is diminishing, from 44 percent in 1997 to 34 percent in 2001. More generally, the momentum of the school choice movement is declining. Whereas 51 percent of the public favored government financial backing for parents selecting nonpublic schools in 1999, 45 percent gave their endorsement in 2000 (T. Henry 2001).

What are charter schools? **Charter schools** are publicly funded schools operated like private schools by public schoolteachers and administrators. Freed of answering to local school boards, charter schools have the latitude to use nontraditional or traditional teaching methods and to shape their own curricula.

The Mosaica Academy in Bensalem, Pennsylvania, which opened in 1998, is deliberately not organized along public school lines. The school day is nearly 2 hours longer than at public school, and the school year is twenty days longer. This school created its own curriculum with the goal of immersing students in a study of civilizations that developed over a period of four thousand years (Symonds 2000).

The number of charter schools exceeded 2,100 in 2002, with more than 500,000 students in thirty-seven states and Washington, D.C. After ten years of charter school operations, the American Federation of Teachers (AFT) has concluded that such schools have not fulfilled their promise. Among other problems, the AFT reported that charter schools spend more on administration and less on instruction than public schools do; student performance is generally no better, and often worse; and charter schools are more homogeneous in race and class than their public counterparts (Wingert 2002).

Like vouchers, for charter schools the jury is still out. Some charter schools have improved education, but others have not (Schorr 2000). The Rand researchers conclude that it is too soon to know if the fears of opponents or the hopes of supporters will be confirmed (Gill et al. 2001).

What is the nature of for-profit schools? Some reformers do not believe that local or federal government is capable of improving the educational system. Government, they say, is too wasteful and ineffective. Why not look to business and market forces to solve the problems facing schools today? **For-profit schools** would be supported by government funds but run by private companies. By borrowing from modern business practices, the argument goes, these schools could be efficient, productive, and cost-effective. Marketplace forces are alleged to ensure that the "best" schools would survive.

The most comprehensive for-profit organization is Edison, which launched a $40 million, three-year campaign in 1992 to develop its program. Edison schools feature challenging curricula, along with a schedule that has children in school almost a third longer than in the average public school. Beginning in the third grade, students are equipped with a computer and modem that they can take home to access the Edison's intranet system (Symonds 2000). In 2002, Philadelphia privatized 42 schools, 20 of which were turned over to Edison (Steinberg 2002; Winters 2002).

Critics of this approach are bothered by the idea of mixing profit and public service. What happens to the students when their needs are weighed against the profit margin? Will for-profit schools skimp on equipment, services, and training? Another problem involves oversight: with a for-profit system, voters lose the power to influence officials and educational policy (Saltman 2001). In one of the few studies of for-profit schools, researchers found no increase, over a three-year period, in student achievement outcomes in the first ten Edison schools opened in 1995 and 1996 (Miron and Applegate 2000).

What is unique about magnet schools? Magnet schools are public schools that attempt to achieve excellence by specializing in a particular area. One school may emphasize the arts, another may stress science. If their public school district includes magnet schools, families can decide which particular school best suits their children's needs. Magnet schools are designed to enhance

SOCIOLOGY AND THE NEWS MEDIA

School Test Trouble

According to many national and state politicians, an emphasis on standardized testing will improve public schooling. As this CNN clip states, 29 states already have laws that will require passage of statewide proficiency exams in order to graduate from high school. Troubling many Americans is the especially high failure rate on state-mandated high school proficiency exams in certain types of schools.

1. Discuss the most important factors in performance on standardized tests.

____

____

2. Referring to material in the text, discuss standardized proficiency testing in light of America as a meritocracy.

____

____

Does your state require statewide proficiency exams? Describe the results so far.

© AP/Wide World Photos

Charter schools are part of the back-to-basics movement in education. Legal in only a few states, these publicly supported schools are established under a charter specifying certain performance requirements. This charter school in Michigan features mandatory Spanish and French, daily homework, and uniforms.

school quality and promote school desegregation. They have emerged as a significant factor in urban education (Blank, LeVine, and Steel 1996).

The Functionalist Perspective

Functions of Education

According to the functionalists, social institutions develop because they meet one or more of society's basic needs. The educational institution performs several vital functions in modern society: cultural transmission; social integration; selection and screening of talent; promotion of personal growth and development; the dissemination, preservation, and creation of knowledge.

How does the educational institution transmit culture? Cultural transmission must occur if a society is to endure. Indeed, the educational institution perpetuates a society's cultural heritage from one generation to the next. When we see evidence of the teaching of Marxist-Leninist doctrine in communist societies, we call it indoctrination. We are less aware of the process of indoctrination when, through the study of

FEEDBACK

1. The _________ institution developed to fulfill society's need to transmit knowledge to the young.
2. Contemporary public education is important primarily because of
   a. the quality of teachers in modern schools.
   b. efficient operation of school bureaucracies.
   c. historical changes in the economic and social structures of industrial societies.
   d. lack of qualified individual tutors.
3. Early public schooling in America was viewed as a mechanism of _________.
4. Schools in America today are based on the _________ model of organization.
5. _________ attempt to avoid most of the bureaucratic features of traditional public schools.
6. _________ is a nonbureaucratic classroom structure in which students study in groups with teachers acting as guides rather than as the controlling agents determining all activities.
7. A school that concentrates on science in its curriculum is known as a/an _________ school.
   a. magnet
   b. for-profit
   c. integrative
   d. charter school

Answers: 1. educational 2. c 3. social control 4. bureaucratic 5. Open classrooms 6. Cooperative learning 7. a

Washington, Jefferson, and Lincoln, we Americans learn about the merits of democracy. Thus, although schools teach basic academic skills such as reading, writing, and mathematics, they also transmit culture by instilling in students the basic values, norms, beliefs, and attitudes of the society. The value of competition, for example, is taught through an emphasis on grades, sports, and school spirit. (More is said about this function later.)

How does education contribute to social integration? Although television is now a strong competitor, the educational system remains the major agent of social integration. Formal education transforms a diverse population into a community with a common identity. Learning an official language, sharing in national history and patriotic themes, and being exposed to similar informational sequences facilitate a shared identity. The result is a society with relatively homogeneous values, norms, beliefs, and attitudes. Even newly arrived immigrant children without the ability to speak and write English attend local schools and soon learn to participate in the American way of life.

The current debate on bilingual education touches on this function. People who emphasize the need for recognizing and honoring cultural diversity (*multiculturalists*) usually support instruction in the student's own language, at least for some period of time. Opponents of bilingual education argue that it hinders the development of a common American cultural identity and that it has not been proven to help students succeed academically. A conservative political backlash to the bilingual approach has led twenty-three states to adopt English as their official language. The creation of a similar law for the nation is being discussed in Congress.

How do schools select and screen students? Scores on intelligence and achievement tests are used for grouping students. The stated purpose of testing (administered for over fifty years now) is to identify an individual's talents and aptitudes. Test scores are also used for **tracking**—placing students in curricula consistent with the school's expectations for the students' eventual occupations (Oakes 1985; Oakes and Lipton 1996; Taylor et al. 1997). Counselors use test scores and early performance records to predict careers for which individuals may be best suited. Counselors usually feel a moral obligation to ensure that students pursue careers consistent with their abilities. (The concept of tracking plays an important role in the conflict perspective of education, discussed shortly.)

How do schools promote personal growth and development? Schools expose students to a wide variety of perspectives and experiences that encourage them to develop creativity, verbal skills, artistic expression, intellectual accomplishment, and cultural tolerance. In this way, education provides an environment in which individuals can improve the quality of their lives. In addition, schools promote personal growth and development by preparing students for their occupational pursuits.

How does education encourage the dissemination, preservation, and creation of knowledge? Educational organizations disseminate knowledge not only via the classroom but through contributions such as written or audiovisual materials that can reach hundreds of thousands of people. Preservation of knowledge is perpetuated by the transmission of knowledge from generation to generation. The task of preserving knowledge is also achieved by such activities as deciphering ancient manuscripts, protecting artifacts, and recording knowledge in written, video, or audio form. Innovation—the creation or discovery of new knowledge through research or creative thinking—can take place at any level within the educational system, but it traditionally receives more emphasis at the university level.

Latent Functions of Education

The functions of schools discussed to this point are recognized and intended—**manifest functions**. The educational institution has other functions as well that are unrecognized and unintended—**latent functions**.

What are the latent functions of education? We do not usually count among the functions of schools the provision of day-care facilities for dual-employed couples or single parents. Nor do we think that our tax money goes to schools so that people can find marriage partners. Also, schools are not consciously designed to prevent delinquency by holding juveniles indoors during the daytime. Nor are schools intended as training grounds for athletes. Finally, schools are not generally recognized for inculcating the discipline needed to follow orders in a bureaucratized society (DeYoung 1989; also see Doing Research in Chapter 4).

Each of the latent functions just mentioned, although unrecognized and unintended, is considered a positive contribution to society. But some consequences, of course, are negative, or dysfunctional. Determining what is dysfunctional depends on the reference point. Tracking, for example, has just been presented as a positive function of the educational institution. Most would agree with this assessment. Most see this function as a manifest, positive function of schools, because it seems to fulfill a societal need. According to the conflict perspective, however, tracking serves a latent dysfunction: it is a means for those

Illiteracy Rates

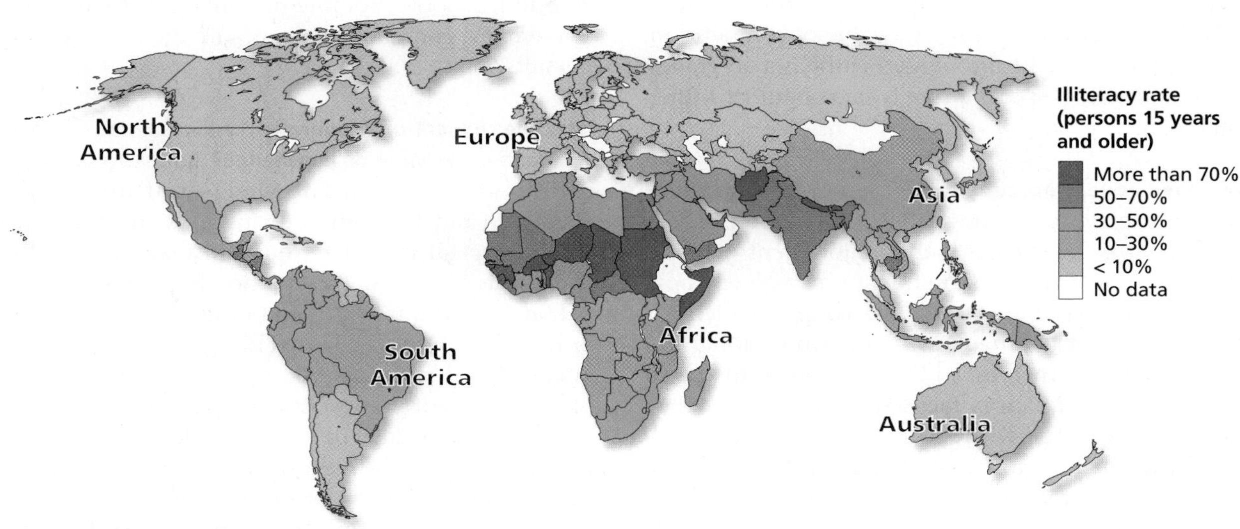

Source: Central Intelligence Agency.

One function of education is to promote literacy, the key to continued learning, problem solving, and information analysis. This map shows rates of illiteracy among those fifteen years and older in different countries of the world. Due to neglect and discrimination, more women than men are illiterate.

1. Do you see a pattern in the rates of illiteracy? Explain.
2. How does the United States stack up?

benefiting from the stratification structure to continue to benefit. In other words, tracking is one mechanism for perpetuating the social class structure. Moreover, evidence suggests that tracking is harmful to those placed on "slower" tracks (Hurn 1993; Oakes and Lipton 1996). The conflict perspective of education is presented in the following discussions on meritocracy, educational inequality, and cognitive testing. These topics are central to the relationship between education and social inequality, a research question in the sociology of education since World War II (Hallinan 1988).

The Conflict Perspective

Meritocracy

In a **meritocracy**, social status is based on ability and achievement rather than social class or parental status (M. Young 1967; Bell 1976). According to the meritocratic model, all individuals have an equal chance to succeed and to develop their abilities for the benefit of themselves and their society. A meritocratic society is free of barriers that might prevent individuals from developing their talents. Meritocracy, then, by definition, involves competition. For this reason, sports are seen as the ultimate meritocracy. Although some sports have glaring shortcomings in this regard, they do fit very closely the definition of **competition** as "a social process that occurs when rewards are given to people on the basis of how their performances compare with the performances of others doing the same task or participating in the same event" (Coakley 1998:78).

Is America a meritocracy? Many Americans believe that the United States is a meritocratic society. However, a number of recurring problems suggest that the model does not always work the way we think it does (Lieberman 1993; Lemann 1999).

Although America purports to be a meritocratic society, researchers recognize the existence of barriers to true merit-based achievement. For example, critics argue that public education primarily serves the economic elite. Because schools are not equal for children of all social classes, students do not all have an equal starting point (Bowles and Gintis 1976; Giroux 1983; Gittell 1998).

Randall Collins (2000) contends that America is not a meritocratic society, but is composed of various status groups competing with one another for wealth, power, and prestige. **Status groups** are made up of individuals who share a sense of status equality based on their participation in a common culture, with similar manners, hobbies, opinions, and values. In their struggle for advantage in society, individuals frequently turn to others like themselves as a resource in their struggle with other status groups. In America, for example, white Anglo-Saxon Protestant males dominate most of the positions of wealth, power, and prestige. Collins argues that this elite has structured public schools to reinforce the shared culture of their group. Entry into higher-status positions requires acceptance of their values. People from higher-status homes who have been educated in "better" schools, for example, have an easier time getting into the higher-level jobs. Those who are from lower social classes, who are of a different color or sex, or who do not possess the appropriate values, manners, language, and dress find it more difficult to enter better colleges and universities and to obtain higher-level jobs.

In an earlier book, Collins (1979) takes the conflict interpretation even further with the concept of **credentialism**—the idea that credentials (educational degrees) are unnecessarily required for many jobs. Why, he asks, must American physicians be trained almost into middle age when communist countries can produce qualified doctors in a relatively short time? Credentials are required, according to Collins, so that those who can afford them can maintain a strong hold on the elite occupations. Otherwise, why would people spend so much time in school learning things that are useless to them on the job (Labaree 1997)?

Racial and ethnic minorities face related barriers to achievement. An important example is the lower performance on college entrance examinations.

How do minorities perform on college entrance exams, and why? African Americans, Latinos, and Native Americans have lower average scores on the Scholastic Assessment Test (SAT) than whites (see Table 13.1). Sociologists attribute this fact, in part, to the differences in school quality noted earlier. And both school quality and SAT performance are related to social class levels of parents. Children from upper-class and upper-middle-class families attend more affluent schools. These children also have higher SAT scores. Social class clearly affects SAT performance.

How does SAT performance influence occupational and economic achievement? The SAT, originally created in 1926, was subsequently used to enable talented youth, regardless of social class background, to attend premier colleges and universities (Lemann 1999). Ironically, as we have just seen, social class is a major factor in SAT performance. Consequently, social class (through SAT performance) still influences who will attend the elite institutions, which are the gateway to America's higher social classes.

TABLE 13.1

SAT Scores by Race and Ethnicity

An examination of this table reveals the gap in average SAT scores for white and Asian Americans versus African American, Latinos, and Native Americans. Interpret these data as a typical conflict theorist would in the context of the United States as a meritocracy.

Racial/Ethnic Category	SAT Verbal Mean Scores	SAT Match Mean Scores	Totals
Native American or Alaskan Native	482	481	963
Asian, Asian American, or Pacific Islander	499	565	1064
African American	434	426	860
Latino Background			
Mexican or Mexican American	453	460	913
Puerto Rican	456	451	907
Latin American, South American, Central American, of other Hispanic or Latino	461	467	928
White	528	530	1058

Source: The College Board, New York: 2002.

But don't the rewards accompanying high SAT performance indicate that America is a meritocracy? On the surface, it does seem that merit is being rewarded. After all, it is those who do better academically who enjoy higher levels of success.

But there are two problems. First, the advantage some people have because of their parents' social class can create an unlevel playing field; talent in the lower social classes often does not get recognized and developed. Second, the assumption that SAT performance measures academic ability and the likelihood of success in both college and life requires examination. For example, African American students who attend the most prestigious schools—including students with SAT scores below 1000—complete college at a higher rate than black students attending less rigorous institutions. They are also more likely to go on to graduate or professional schools (Bowen and Bok 2000).

At the least, these findings raise doubts. Recognizing this, an official at the Educational Testing Service (ETS)—developer and marketer of the SAT—announced in 1999 that ETS was creating a "Strivers" score. The idea was to adjust a student's SAT score to factor in social class as well as racial and ethnic characteristics thought to place him or her at a competitive disadvantage. Any student whose original score exceeded by 200 points the score predicted for his or her social class or racial or ethnic category would be considered a striver. The striver's score would be made available for colleges and universities to use, if they desired, in their admissions decisions (Glazer 1999; Wildavsky 1999). The proposal was quickly withdrawn after a firestorm of criticism. But, concern did not die. By 2005 the SAT is supposed to be transformed from an IQ-like test to a measure of academic preparedness (J. E. Barnes 2002).

Of course, the basic problem is not an SAT score deficiency (Darragh Johnson 2002). The larger issue is educational inequality.

Educational Inequality

How has the definition of educational inequality changed? From the beginning, American public education stressed educational equality (Demaine 2001). That originally meant that all children must be provided a free public education; that all children, regardless of social and economic background, must be exposed to the same curriculum; that all children within a given locality must attend the same school; and that all schools within a locality should be equally financed by local tax dollars (Coleman 1969; E. W. Gordon 1972; Brint and Brint 1998).

Despite this ideal, argue conflict theorists, not all children have been considered deserving of a formal education. Native Americans and African Americans were specifically excluded. During the 1850s, it was believed that slaves could not and should not be educated. It was not until slavery was abolished and former slaves began to assume industrial jobs that African Americans eventually were included among those Americans thought to be educable. Even then, African Americans were not given equal educational treatment. In the *Plessy v. Ferguson* decision of 1896, the Supreme Court made "separate but equal" part of America's legal framework. This doctrine held that equal treatment was being given when African Americans and whites had substantially equal facilities, even when these facilities were separate.

It was not until 1954 that the separate-but-equal doctrine was seriously challenged by the judicial system. In the *Brown v. Board of Education* decision of 1954, Chief Justice Earl Warren wrote, as part of the Court's opinion:

> *Does segregation of children in public schools solely on the basis of race, even though the physical facilities and other "tangible" factors may be equal, deprive the children of the minority group of equal educational opportunities? We believe that it does.*

Warren went on to say that the separate-but-equal doctrine had no place in America's public educational structure and that separate educational facilities are inherently unequal. Still, it took the civil rights movement of the 1950s and 1960s to make the long-existing educational inequality for African Americans and other minorities a socially recognized problem for America. Thus the Civil Rights Act of 1964 prohibits discrimination on the grounds of race, color, religion, or national origin in any educational program or activity receiving financial support from the federal government.

What is meant today by educational equality? The most recent notion of **educational equality** addresses equality in relation to "effects" of schooling. Educational equality exists when schooling produces the same results (achievement and attitudes) for lower-class and minority children as it does for other children. Results, not resources, are the test of educational equality (Coleman et al. 1966; Hallinan 2000).

In 1964, James Coleman was commissioned by Congress to conduct a survey on discrimination and educational opportunities. From information on approximately 600,000 students and 60,000 teachers from some 4,000 schools, Coleman concluded that some consistent differences in black and white schools did exist. On the average, African American students had fewer school facilities: laboratories for physics, chemistry, and languages were less plentiful; textbooks were in shorter supply; and fewer books per pupil were available in school libraries. Coleman reported that minorities also suffered in relation to the majority when

SNAPSHOT OF AMERICA 13.1

Public School Expenditures per Pupil

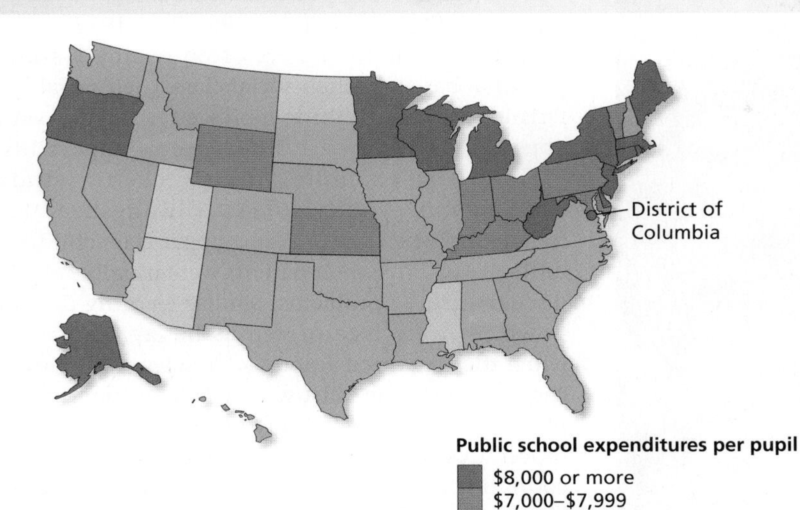

Source: U.S. Bureau of the Census, *Statistical Abstract of the United States: 2001* (Washington, DC: U.S. Government Printing Office), 2001.

Everybody says, "you get what you pay for." Because of this idea, many people use the amount of money spent on public schools as a measure of the quality of education. This map shows that some states, such as New York and New Jersery, spend more than twice as much on their students as other states, such as Utah.

1. How does your state stack up with other states? Why?
2. Do you see a regional pattern in public school funding? Explain.
3. What other factors contribute to educational quality? Be specific.

SNAPSHOT OF AMERICA 13.2

High School Dropouts

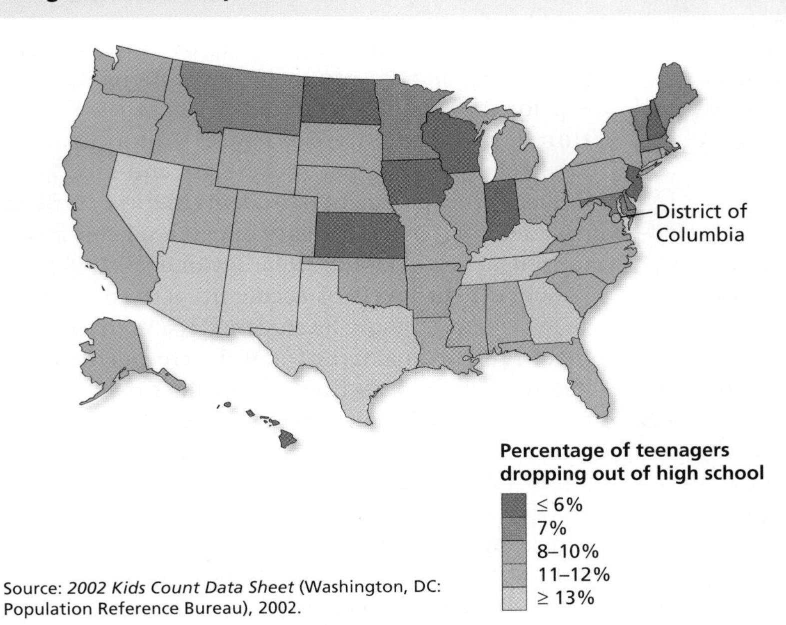

Source: *2002 Kids Count Data Sheet* (Washington, DC: Population Reference Bureau), 2002.

For many jobs, a high school diploma is a minimum requirement. People who do not complete high school earn only about three-fourths as much as high school graduates. This map shows the percentage of teenagers (ages sixteen through nineteen) in each state who are high school dropouts.

1. Compare this map with the one in Snapshot 13.1. Do you see any relationship between the money spent per student and the percentage of high school dropouts?
2. How does the percentage of high school dropouts in your state compare with the percentage in other states?
3. Compare this map with Snapshot 8.1. Do you see any relationship between the percentage of high school dropouts and the percentage living below the poverty line?

it came to curricular and extracurricular matters: African Americans in secondary schools were less likely to attend accredited schools; college-preparatory and accelerated curricula were less accessible to minorities; and whites generally enjoyed better access to such academically related extracurricular activities as debate teams and student newspapers. An important source of these differences was the inequality of school financing. Using the survey data collected by Coleman, Christopher Jencks (1981) reported that about 15 to 20 percent more funds were spent for the average white schoolchild per year than for the average African American student.

© Getty Images

Some evidence suggests that ethnic and racial integration in schools improves attendance and academic achievement. Despite these potential advantages, over two-thirds of African American and Latino students are in segregated classrooms.

It was not surprising that Coleman found some differences in resources allocated to majority and minority schools. It was surprising, however, that the differences in resources accounted for only a small part of the differences in majority and minority student performance on standardized achievement tests. According to Coleman, if educational equality is to exist, schools must overcome the educational inadequacies children bring with them from home. It is not enough, therefore, to provide minority children with equal school resources (Coleman 1968).

Such logic supported the mandatory busing programs instituted in urban areas from the late 1960s to the present. As interpreted by the courts, public schools must provide not only equal resources such as texts, teachers, and buildings but also equal "social climates," which are the result of students' personalities and motivational factors. Because students influence one another's learning capabilities, school districts must provide similar types of students within each school based on racial, social class, and ethnic characteristics (Coleman 1990).

How do teacher evaluation and tracking affect educational equality? According to Ray Rist (1977), even the best-intentioned teachers often evaluate students' potential based on their social class, their racial characteristics, and their ethnic heritage rather than on their objective abilities. There is evidence to support this allegation. For example, in his classic study of "Elmstown's" youth, August Hollingshead (1949) found that when middle- and upper-class children had difficulty in class, their parents were usually called in to discuss learning problems. Similar learning difficulties, however, were interpreted as behavioral problems when they occurred among lower-class children. Aaron Cicourel and John Kitsuse (1963) later found the same phenomenon in their study of high school counseling procedures. When test scores were consistent with the counselor's expectations, there was no problem. The counselor, however, was much more willing to find excuses for poor performance among high-status students than among low-status ones. Rist (1977) reported a similar pattern among classroom teachers. When asked questions in class, the students perceived as more capable by teachers were given more time to answer than "less able" students were given. Also, lower-status students were more likely to be considered by teachers as less capable.

Most high schools have a series of curricula (tracks): college preparatory, commercial, vocational, and general for those who are undecided about their occupational futures (Kilgore 1991; Gamoran 1992). Researchers report that social class and race heavily influence the placement of students in these curricula—regardless of their intelligence or past academic achievement (Oakes and Lipton 1996; Taylor et al. 1997).

Subsequent levels of academic performance have been influenced more by students' tracks than by their intelligence or current scholastic performance. Regardless of earlier school performance or intelligence, the overall academic performance of those bound for college increases, whereas the performance of those on a noncollege track decreases. Through tracking, then, schools may deny educational equality to many students, especially minority children, who often come from disadvantaged families (Oakes 1985; Gibson and Ogbu 1991). In part because tracking has been linked to educational inequality, American high schools have been retreating from the practice. Some research, however, casts doubt on the idea that detracking promotes greater educational equality (Lucas 1999). Rather than assigned courses being based on students' course-specific abilities, mathematical achievement has become a gateway to college-prep courses.

Testing Cognitive Ability

The technical term for intelligence, today, is **cognitive ability**—the capacity for thinking abstractly. Dating back to the turn of the twentieth century, there has been a tradition in schools to attempt to measure cognitive ability. Because cognitive ability testing is an important element in sorting and tracking students, it contributes to educational inequality. Whenever cognitive ability tests are discussed, the question of inherited intelligence always arises.

Is intelligence inherited? In the past, some people assumed that individual and group differences in measured intellectual ability were due to genetic differences. This assumption, of course, underlies social Darwinism.

A few researchers still take this viewpoint. More than thirty years ago Arthur Jensen (1969), an educational psychologist, suggested that the lower average intelligence score among African American children may be due to heredity. A recent book by Richard Herrnstein and Charles Murray (1996), entitled *The Bell Curve*, is also in the tradition of linking intelligence to heredity. According to these authors, humans inherit 60 to 70 percent of their intelligence level. Herrnstein and Murray further contend that the fact of inherited intelligence makes largely futile the efforts to help the disadvantaged through programs such as Head Start and affirmative action.

Social scientists have generally criticized these advocates of genetic differences in intelligence for failing to consider adequately the effects of the social, psychological, and economic climate experienced by children of various racial and ethnic minorities. Even those social scientists who grant genetics an important role in intelligence criticize both the interpretations of the evidence and the public policy solutions contained in *The Bell Curve*. They point to the body of research that runs counter to Herrnstein and Murray's thesis. More specifically, they see intelligence not as an issue of nature versus nurture, but as a matter of genetics and environment (Morganthau 1994; L. Wright 1995; Fischer et al. 1996). We know, for example, that city dwellers usually score higher on intelligence tests than do people in rural areas, that African Americans with higher socioeconomic status score higher than lower-status African Americans, and that middle-class African American children score about as high as middle-class white children. We also have discovered that as people get older, they usually score higher on intelligence tests. These findings, and others like them, have led researchers to conclude that environmental factors affect test performance at least as much as genetic factors (Samuda 1975; Schiff and Lewontin 1987; Jencks and Phillips 1998; Arrow, Durlauf, and Bowles 2000). One of these environmental factors is a *cultural bias* in the measurement of cognitive ability.

How can intelligence tests be culturally biased? Robert Williams (1974) is one of many early social scientists who argued that intelligence tests have a **cultural bias**—an unfair measure of cognitive abilities for people in certain social categories. Specifically, intelligence tests are said to be culturally biased because they are designed for middle-class children and because the results measure learning and environment as much as intellectual ability. Consider this intelligence test item cited by Daniel Levine and Rayna Levine (1996:67):

> *A symphony is to a composer as a book is to what?*
>
> *( ) paper ( ) sculptor ( ) author ( ) musician ( ) man*

According to critics, higher-income children find this question easier to answer correctly than lower-income children because higher social class members are more likely to have been exposed to information about classical music. (The same charge was made by critics of a recent SAT question that used a Bentley automobile as its illustration.) Several studies indicate that because most intelligence tests assume fluency in English, minorities cannot respond as well. Some researchers suggest that many urban African American students are superior to their white classmates on several dimensions of verbal capacity, but this ability is not recognized, because intelligence tests do not measure these areas (S. J. Gould 1981; Goleman 1988; Hurn 1993).

The testing situation itself can affect performance. Low-income and minority students, for example, score higher on intelligence tests when tested by adult members of their own race or income group. Apparently, they feel threatened when tested in a strange environment by someone dissimilar to them. Middle-class children are frequently eager to take the tests because they have been taught the importance of tests and academic competitiveness. Because low-income children do not recognize the importance of tests and have not been taught to be academically competitive, they ignore some of the questions or look for something more interesting to do. Nutrition also seems to play a role in test performance. Low-income children with poor diets may do less than their best when they are hungry or when they lack particular types of food over long periods of time.

Promoting Equality in Education

Although it is difficult to overcome the barriers of economics and social class, policy makers and educators are exploring avenues for educational equality. Four of the avenues—school desegregation, compensatory education, community control, and private schooling—are discussed here.

Does school desegregation improve academic achievement of minority children? The Coleman Report supports the idea that it does. Specifically, James Coleman et al. (1966) reported that African American children in classrooms that are more than one-half white scored higher on achievement tests. The finding that minority children perform better academically when placed in desegregated, middle-class schools has been verified by several other studies. In a broad review of research on school desegregation, Meyer Weinberg (1970) concluded that African American children consistently experience higher academic achievement in desegregated schools. It does not seem coincidental that the period of higher school desegregation witnessed major academic progress for African American students, and that the resegregation of the 1990s is associated with a growing gap in racial academic achievement (Kaufman 2000, 2001).

On the other hand, several studies have shown that racially balanced classrooms can have either positive or negative effects on the academic achievement of minority children. Mere physical desegregation without adequate support may have detrimental effects on children. However, desegregated classrooms with an atmosphere of respect and acceptance promote improved academic performance (U.S. Commission on Civil Rights 1967; I. Katz 1968; Longshore 1982a, 1982b; Patchen 1982; Orfield et al. 1992). For example, a study of seniors at a racially and ethnically diverse high school in Cambridge, Massachusetts, revealed a positive effect of diversity on higher-education aspirations ("The Impact of Racial and Ethnic Diversity on Educational Outcomes" 2002).

Some postschooling benefits of desegregation include higher occupational positions and higher incomes for minority students who attended desegregated public schools than for minority students who attended segregated schools. These differences cannot be attributed solely to differences in educational attainment. Rather, increased social contact with middle-class students also contributes to college attendance and better employment opportunities, because middle-class students provide norms of behavior, dress, and language often required by employers in the middle-class hiring world (Crain 1970; Crain and Weisman 1972; W. D. Hawley 1985). Today, more than ever, the social and economic benefits of school equality are an important consideration. Good jobs require postsecondary education. Those well-paying factory jobs that require minimal education are a shrinking part of the economy (Orfield 2001).

Some evidence suggests that school desegregation may also lead to better racial and ethnic relations because students are exposed to people of different backgrounds (W. D. Hawley 1985; Hawley and Smylie 1988; Kurlaender and Yun, 2000; Eaton, 2001). Support for this finding also appeared in the aforementioned study of seniors in a racially and ethnically diverse high school. These students felt well prepared for life as adults in a diverse community, indicated increased understanding of different points of view, and expressed enhanced appreciation of the background of other groups ("The Impact of Racial and Ethnic Diversity on Educational Outcomes" 2002).

On such evidence rests the promise of **multiculturalism**: an approach to education in which the curriculum accents the viewpoints, experiences, and contributions of minorities—including women as well as ethnic and racial minorities (Katsiaficas and Kiros 1998; Watson 2000). Multiculturalism attempts to dispel stereotypes and to make the traditions of minorities valuable assets for the broader culture. According to an earlier study, school attendance and academic performance have increased in some schools with multicultural curricula (Marriott 1991).

As noted earlier, public schools in the United States are drifting back toward segregation. In a reversal of the trend toward integration from the mid-1960s until 1988, African American and Latino youngsters are becoming racially and ethnically segregated. Since 1986, school districts have tended to show lower levels of inter-racial exposure (Frankenberg and Lee 2002). This intensification of school resegregation throughout the 1990s is hardly surprising given the three major U.S. Supreme Court decisions permitting a shift back toward segregated neighborhood schools. Around three-fourths of African American and Latino students are now in classrooms with a predominantly minority enrollment (Orfield and Eaton 1997; Orfield 2001).

Does compensatory education increase school achievement for disadvantaged children? It appears so (Entwisle and Alexander 1989, 1993; Ramey and Campbell 1991; Zigler and Styfco 1993; Campbell and Ramey 1994). The best-known attempt at remedial education is the Head Start program, a federally supported program for preschoolers from disadvantaged homes. Head Start originated as part of the Office of Economic Opportunity in 1965. It was conceived, in part, as a means of preparing disadvantaged preschoolers for entering the public school system. Its goal is to provide disadvantaged children an equal opportunity to develop their potential (Waldman 1990–1991). Initial studies on the progress of Head Start report positive long-term results. Low-income youngsters between the ages of nine and nineteen who had been in preschool compensatory programs performed better in school, had higher achievement test scores, and were more motivated academically than low-income youth who had not been in compensatory education programs (Bruner 1982; Etzioni 1982).

Follow-up research also supports the benefits of Head Start (Schweinhart 1992; Zigler and Muenchow 1992; K. Mills 1998). For example, in one study children entering a Head Start program scored lower on intelligence than a group of their peers, but follow-up research revealed benefits for those Head Start students: better school attendance, completion of high school at a higher rate, and entrance into the workforce in greater proportion (Passell 1994).

As noted in "Using the Sociological Imagination," children's scores on IQ tests, general ability tests, and learning readiness assessments significantly improve after exposure to Head Start programs. Studies report that the scores of Head Start participants eventually reach the national average on tests of general ability and learning readiness. Attitudinal, motivational, and social progress have been observed as well.

Politicians, both Republicans and Democrats, currently recognize the benefits of Head Start. After overwhelming support in both the House and Senate, new Head Start authorization legislation was signed by President Clinton on May 19, 1994. Under this legislation, funding was significantly increased, and coverage was extended to approximately 800,000 children (Dewar and Vobejda 1994).

What is the relationship between community control and educational equality? The authority to finance and administer elementary and high schools lies within the states rather than with the federal government. The states in turn delegate most of this authority to local school systems. Ultimately, then, the power rests with school administrators who carry out policies established by the local school boards.

The local school board is thought to represent the interests of the community residents while, it is assumed, administrators represent the professional needs of teachers and the educational needs of students. These assumptions have been challenged. It is argued that the current school governance arrangement has become so centralized and bureaucratic that accountability to parents and neighborhood is rare. It is also argued that political decentralization in urban areas, which would place direct control of each school in the hands of neighborhood residents, is desirable because the present remote, bureaucratic structure fails to meet the needs of children, needs that vary from one neighborhood to another. This is held to be a particular disadvantage by minorities, who charge that schools are oriented toward white students' needs and are not responsive to the problems of minority students.

Although many school administrators have agreed to share major educational decisions with parents, it remains to be seen whether they are willing to share their power as well. There is also the question of whether neighborhood schools can raise enough money to ensure quality education. Finally, there is the possibility that community control will increase school segregation.

Do private schools offer a better teaching environment for the disadvantaged? Amid the controversy surrounding educational opportunity in public schools has come some interesting research by James Coleman and his colleagues (Coleman, Hoffer, and Kilgore 1981; Coleman, Kilgore, and Hoffer 1982; Coleman et al. 1982a, 1982b). Coleman and his colleagues attempted to investigate whether actual achievement among students attending private schools was better than that among public school students. Much to the dismay of many public school educators, the private schools studied produced higher student achievement levels in mathematics and vocabulary, even when controlled for background characteristics of students. Of particular interest to the researchers was the observation that minority children in these private schools, particularly Catholic schools, performed better than their counterparts in public schools. The authors argued that higher academic standards, stricter discipline, more homework, and the accompanying student perceptions of academic emphasis accounted for the differences in achievement levels. These findings were used by Coleman and his associates to argue for changes in public school policies that could enhance achievement levels for all students and to suggest that the growing private school movement might have beneficial consequences for many of America's children.

The results of the Coleman study have been criticized. Arthur Goldberger and Glen Cain (1981, 1982) accuse the researchers of being biased toward private schools. Others contend that Coleman has overstated his conclusions. The government agency sponsoring the Coleman research, for example, contends that the performance comparisons are inaccurate because a greater percentage of private school students (70 percent) pursue college-bound academic programs than do public school pupils (about one-third). After a reanalysis of his data, Coleman agreed that precollegiate public school seniors perform as well on standardized tests as Catholic school seniors.

These criticisms should not be allowed to overshadow the results of the Coleman study. Diane Ravitch places the debate into proper focus:

> *What it [the Coleman study] does show is that good schools have an orderly climate, disciplinary policies that*

FEEDBACK

1. _________ refers to a social system in which social status is a function of ability and achievement rather than background or parental status.
2. The idea that some educational requirements are necessary for many jobs today is called
   a. status grouping.
   b. credentialism.
   c. meritocracy.
   d. conflict.
3. A significantly larger proportion of higher social class children attend college than lower-class children. This is largely due to differences in ability among social classes. T or F?
4. Educational equality exists when schooling produces the same _________ for lower-class children and minorities as it does for other children.
5. Most sociologists agree that IQ test scores do not measure inherited learning ability. T or F?
6. _________ is an approach to education in which the curriculum accents the viewpoints, experiences, and contributions of women and diverse ethnic and racial minorities.
7. Compensatory education appears to benefit deprived children. T or F?

Answers: 1. Meritocracy 2. b 3. F 4. results 5. T 6. Multiculturalism 7. T

students and administrators believe to be fair and effective, high enrollment in academic courses, regular assignment of homework, and lower incidence of student absenteeism, class-cutting, and other misbehavior. These conditions are found more often in private than in public schools (though they are not present in all private schools, and they are not absent in all public schools). (Ravitch 1982:10)

Symbolic Interactionism

Schools are in the business of transmitting culture, and it is in this socialization that symbolic interactionists are interested. Socialization takes place primarily through three sources: the hidden curriculum, textbooks, and teachers.

The Hidden Curriculum

How does the hidden curriculum contribute to socialization? Modern society places considerable emphasis on the verbal, mathematical, and writing skills an adult needs to obtain a job, read a newspaper, balance a checkbook, and compute income taxes. However, schools teach much more than these basic academic skills. They also, through a **hidden curriculum**, transmit to children a variety of nonacademic values, norms, beliefs, and attitudes (Sadker and Sadker 1995; Ballantine and Spade 2001). The hidden curriculum teaches children skills such as discipline, order, cooperativeness, and conformity. These skills are thought to be necessary for success in modern bureaucratic society, whether one becomes a doctor, a college president, a computer programmer, or an assembly-line worker. Over the years, schools, for example, socialize children for the transition from their closely knit, cooperative families to the loosely knit, competitive adult occupational world. The school provides systematic practice for children to operate independently in the pursuit of personal and academic achievement. The values of conformity and achievement are emphasized through individual testing and grading. Because teachers evaluate young people as students—not as relatives, friends, or equals—students participate in a model for future secondary relationships (for example, employer-employee, salesperson-customer, or lawyer-client).

Textbooks and Teachers

How do textbooks socialize children? A by-product of the hidden curriculum is the development of patriotism and a sense of civic duty. Partly for this reason, courses such as history and government are generally slanted in favor of a particular view of history. Accounts of the American Revolution, for example, are not the same in British and American textbooks. Few societies are willing to admit imperfections, and this reluctance is often reinforced in school materials. Most U.S. history textbooks, for example, have not accurately portrayed the government's treatment of Native American peoples. And the resistance to critical accounts of history is especially apparent when a teacher attempts to introduce a controversial book, such as Alice Walker's *The Color Purple*.

Textbooks implicitly convey values and beliefs. Sociologists interested in gender stereotypes point out that older primary textbooks tended to present men in challenging and aggressive activities while portraying women as homemakers, mothers, nurses, and secretaries. Women were not only placed in traditional roles, they appeared far less frequently than men. Similarly,

George Ritzer—The McDonaldization of Higher Education

Max Weber, you remember, was concerned that rationalization would go too far. He feared that rationalization would spread to all aspects of society, resulting in an "iron cage of rationality." This dehumanizing cage of rationality, Weber asserted, would eventually deprive everyone of freedom, autonomy, and individuality.

George Ritzer uses the metaphor of the "McDonaldization of society" to illustrate Weber's image of the iron cage. McDonaldization, Ritzer states, is the process of increasing efficiency, calculability, predictability, and control by replacing human technology with nonhuman technology. *Efficiency* refers to the relationship between effort and result. An organization is most efficient when the maximum results are achieved with minimum effort. For example, fast-food restaurants are efficient in part because they transfer work usually done by employees to customers. *Calculability* involves estimation based on probabilities. High calculability exists when the output, cost, and effort associated with products can be predicted. A McDonald's manager trains employees to make each Big Mac within a rigid time limit. *Predictability* pertains to consistency of results. Predictability exists when products turn out as planned. Big Macs are the same everywhere. *Control* is increased by replacing human activity with technology. McDonald's drink machines stop after a cup has been filled to its prescribed limit. McDonald's restaurants, Ritzer argues, ideally reflect the dehumanizing effects of extreme rationality.

> *The fast-food restaurant is often a dehumanizing setting in which to eat or work. People lining up for a burger, or waiting in the drive-through line, often feel as if they are dining on an assembly line, and those who prepare the burgers often appear to be working on a burger assembly line. Assembly lines are hardly human settings in which to eat, and they have shown to be inhuman settings in which to work.* (Ritzer 2000b:12).

Ritzer uses a wide variety of materials to document the McDonaldization of many important aspects of American society (1998). His sources of information include newspapers, books, magazines, and industry publications. As a college student, you will find Ritzer's conclusions on the "McUniversity" of immediate interest. Higher education, Ritzer concludes, is well on the way to becoming another consumer item. Increasingly, students and parents view a college degree as a necessary credential to compete successfully in the job market.

This consumer orientation, Ritzer asserts, can be observed on most college campuses in the United States. For example, students want education to be conveniently located and available as long as possible each day. They seek inexpensive parking, efficient service, and short waiting lines. Students want a high-quality service at the lowest cost. A "best buy" label in national academic rankings catches the attention of parents and students.

Public colleges and universities, Ritzer contends, are responding to this consumer orientation. They are doing so in part because government funding for higher education is becoming scarcer. To meet reduced funding, colleges and universities are cutting costs and paying more attention to "customers." For example, Ritzer points to student unions, many of which are being transformed into mini malls with fast-food restaurants, video games, and ATMs.

Ritzer predicts that a far-reaching customer-oriented tactic will be to McDonaldize through new technology. The McUniversity will still have a central campus, but it will also have convenient satellite locations in community colleges, high schools, businesses, and malls:

> *Students will "drop by," for a course or two. Parking lots will be adjacent to McUniversity's satellites (as they are to fast-food restaurants) to make access easy.* (Ritzer 1998:156)

McDonaldization, Ritzer contends, will dehumanize the process of education. Most instructors at satellites will be part-timers hired to teach one or more courses. They will come and go quickly, depriving students of the opportunity to form relationships with permanent, tenured faculty. To make the courses alike from satellite to satellite, course content, requirements, and materials will be highly standardized, losing the flavor individual professors bring to their classes. Students will not be able to choose a particular instructor for a course because there will be only one per satellite. In fact, there may be no teacher physically present at all. More courses will be delivered by professors televised from distant places.

It is not, Ritzer states, that colleges and universities will become a shopping mall or a chain of fast-food restaurants. Institutions of higher education will retain many traditional aspects. But there will undoubtedly be a significant degree of McDonaldization.

Thinking About the Research

1. Ritzer's study is based on one of the three theoretical perspectives. Explain which one you think it is and why.
2. Could you use another perspective instead? Why or why not?
3. Analyze the functions and dysfunctions of the McDonaldization of higher education.

the little white house with the picket fence may be part of the worldview of middle-class Americans, but parents of low-income or inner-city children complained that such pictures of middle-class life harmed their children. Underprivileged children who compared their homes with middle-class homes felt out of place (Trimble 1988; Gibson and Ogbu 1991). Today, active parent groups, minority special interest groups, and state boards of education are working with textbook authors and publishers to ensure that a more balanced picture of society is presented to students.

How do teachers socialize children? Teachers, usually a child's first authority figures outside the family, intentionally socialize students by having them perform academic tasks in predetermined ways. But teachers unintentionally affect children as well, as can be seen in a study by Robert Rosenthal and Lenore Jacobson (1989). In their book *Pygmalion in the Classroom*, Rosenthal and Jacobson demonstrate the **self-fulfilling prophecy**—when an expectation leads to behavior that causes the expectation to become a reality (Merton 1968). (Self-fulfilling prophecy was discussed in Chapter 9.) In their study, elementary schoolteachers were given a list of children in their classrooms who, according to the researchers, were soon to blossom intellectually. Actually, these children were picked at random from the school roster and were no different from the student body in general. At the end of the year, the authors found that this randomly selected group of children had scored significantly higher on the same intelligence test they had taken at the beginning of the year. Their classmates as a group did not. According to Rosenthal and Jacobson, the teachers expected these "late bloomers" to spurt academically. Consequently, the teachers treated these students as if they were special. This behavior by the teachers encouraged the students to become higher academic achievers (see Doing Research). Eleanor Leacock (1969) also found the self-fulfilling prophecy at work in a study of second and fifth graders in black and white low- and middle-income schools. Both studies demonstrated the transmission of negative self-impressions as well as positive ones. Subsequent research generally supports dramatic effects of teacher (and parent) expectations (Gamoran 1986). To quote the Children's Defense Fund: "Educational research shows that children adjust their performance and self-image to meet adult expectations, and children who are placed in lower ability classes soon come to believe that they cannot excel in school" (*Children's Defense Fund* 1991:80).

How do teachers foster sexism? Following a long line of researchers, Myra Sadker and David Sadker (1995) contend that well-meaning teachers unconsciously transmit sexist assumptions, basing these assumptions on stereotypes for appropriate gender behavior. America's teachers, therefore, are often unfair to girls, crippling them through differential classroom treatment. Girls, for example, learn to talk softly, avoid certain subjects (especially math and science), defer to the alleged intellectual superiority of boys, and emphasize appearance over intelligence.

Moreover, in a coeducational setting, girls, compared with boys, are

- five times less likely to receive the most attention from teachers
- three times less likely to be praised
- eight times less likely to call out in class
- three times less talkative in class
- half as likely to demand help or attention
- half as likely to be called on in class

Throughout their book, Sadker and Sadker emphasize that boys are more talkative in class, raise their hands more often, move around more, argue with teachers more, and get more of the teachers' attention than do girls. All of this tends to make young women mature as second-class citizens, these researchers conclude, and deprives the world of important contributions women can make.

FEEDBACK

1. According to symbolic interactionists, schools are very much in the business of transmitting ________.
2. The ________ teaches children such things as discipline, order, cooperativeness, and conformity—skills thought to be necessary for success in a modern bureaucratic society.
3. ________ are the first authority figures children encounter on a daily basis outside the family.
4. The ________ occurs when an expectation leads to behavior that causes the expectation to become a reality.
5. Match the following research topics with the three theoretical perspectives.

____ a. functionalism	(1) requirement of unnecessary credentials by schools
____ b. conflict theory	(2) benefit of schools for society
____ c. symbolic interactionism	(3) effects of teacher reactions on the self-esteem of students

Answers: 1. culture 2. hidden curriculum 3. Teachers 4. self-fulfilling prophecy 5. a. (2) b. (1) c. (3)

But what about the upbeat media reports? Writers who paint an optimistic picture fail to produce convincing evidence (Deak 1998). Inequalities remain. There is objective evidence that girls are guided in school toward traditional female jobs and away from high-paying, powerful, and prestigious jobs in science, technology, and engineering (Millicent 1992). True, significantly more high school girls want to go into engineering today than in the past. But five times more men than women receive bachelor's degrees in engineering.

Significantly, these gender-based discrepancies cannot be explained by ability differences. Girls perform almost as well as boys on math and science tests (O'Sullivan, Reese, and Mazzeo 1997). Girls score higher than boys at reading and writing at all grade levels and are more likely to attend college (Greenwald et al. 1999). Moreover, females fare better in single-gender schools and single-gender classes in coeducational schools:

> *Girls in these situations, in general, get better grades, report that they learn more and are more positive about the learning situation, have higher self-esteem, and more often move on to advanced courses than do girls in regular coeducational situations.* (Deak 1998:19–20)

Higher Education

College Attendance

About 15.5 million students are enrolled in over 3,600 American colleges and universities. Immediately following World War II, as a result of the G.I. Bill and the influx of veterans, enrollment in institutions of higher education began to grow; it has not stopped (see Figure 13.2). Since 1980, the percentage of high school graduates attending college has increased from 49 percent to 63 percent (U.S. Bureau of the Census 2001a). This increase has occurred despite predictions of a downturn in college enrollment—a prediction based on a decline in the number of eighteen- to twenty-four-year-olds, the traditional college-age category.

Why didn't college enrollment drop as anticipated? Several trends account for the increase. First, despite a decline in the eighteen- to twenty-four-year-old population, a substantially larger proportion of this age category is attending college. In addition, the number of students over age twenty-four attending college, particularly women, has continued to grow, and the number of students remaining in college after their first year has increased.

FIGURE 13.2

College Enrollment in the United States: 1965–2010

This figure traces the increase in college enrollment in the United States since 1965. Explain why college enrollment continues to increase.

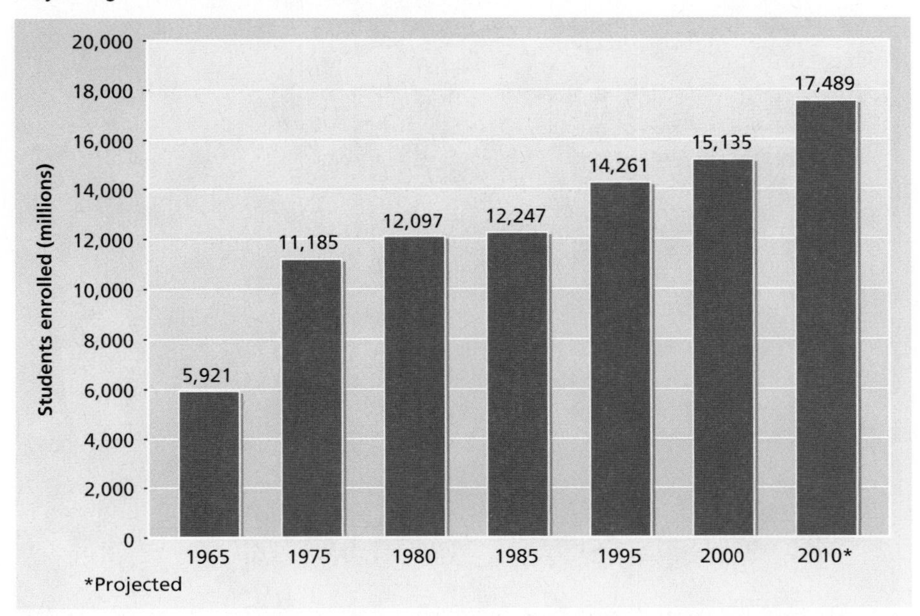

Source: U.S. Bureau of the Census, *Statistical Abstract of the United States: 2001* (Washington, DC: U.S. Government Printing Office, 2001), table 205.

At least three factors account for these trends. Fearful of declining enrollments, colleges and universities mounted strong student recruitment campaigns. This stimulated student demand and enlarged university applicant pools. Also, low-paying service jobs increasingly replace high-paying manufacturing jobs, resulting in fewer attractive noneducational options for those with only high school diplomas. With the loss of high-wage unionized jobs in blue-collar industries such as automobiles and steel, the incentive to attend college increases. Finally, the increase in the number of community and branch colleges offers an opportunity for many who could not otherwise continue their education.

What can we expect in the future? A new baby boom is about to hit colleges and universities. The leading edge of the children and grandchildren of baby boomers—the so-called "baby boomlet"—is beginning to work its way into colleges. This is only the beginning. The year 2009 is predicted to have the largest high school graduating class in the history of the United States—some 3.2 million students. The college enrollment of 12.8 million in 1987 will reach 17.7 million in 2012, a 28 percent increase.

At the same time that college enrollments are accelerating—across the country—state budget shortfalls are spawning deep cuts in higher education. The impending train wreck between student demand and limited college financial resources will last at least until 2012. Colleges will scramble to revamp their enrollment policies, to expand services, to create adequate housing and class space, and to generate new sources of funding. College students and their parents will meet head-on with rising tuition and fees as well as more intense competition at nationally reputed schools (DeBarros 2003).

Increased tuition can burden community college students even further. A tuition increase accounts for a larger share of available personal financial resources among students who are employed and older than 21. And most community college students are beyond the typical college age, supporting themselves. Independent from their families and denied financial aid because they have incomes, many community college students find this additional obstacle the last straw; they drop out. Others will reduce the number of hours they take each term and/or work more hours. Either of these alternatives further retards their often already slow progress caused by work and/or family responsibilities.

Education, Income, and Occupation

Because of advanced industrialization, education is a ticket to better jobs and higher incomes. A college degree has a significant payoff, financially and occupationally (Guinzburg 1983; Katchadourian and Boli 1994).

What is the economic payoff? As first noted in Chapter 2, college education results in higher annual income, regardless of race or gender. Among African American and white males who have a college education or more,

Community colleges, such as this one in Boston, perform an important function in the sorting and channeling of college students. Community colleges, designed for career education, train students for jobs in such areas as business, computers, and health care.

FIGURE 13.3

Median Annual Income by Gender, Race, and Education

Clearly, this figure documents the income advantage that white males in the United States have over white females and African Americans of both sexes. Explain how this situation can be used to challenge the existence of a true meritocracy.

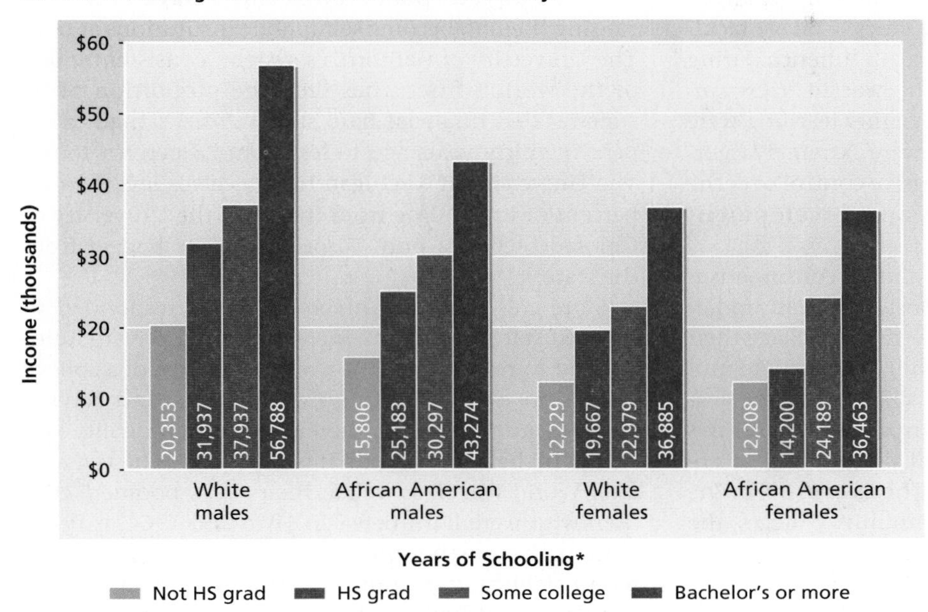

*Based on highest grade completed.

Note: These figures include the total money income of full-time and part-time workers, ages 25 and over, as of March 2001.

Source: U.S. Bureau of the Census, "PINC-03, Educational Attainment—People 25 Years Old and Over, by Total Money Earnings in 2001, Work Experience in 2001, Age, Race, Hispanic Origin, and Sex"; available at http://ferret.bls.census.gov/macro/032002/perinc/new03_000.htm.

INTERNET LINK To learn more about income in the United States, visit the U.S. Bureau of the Census Web site at www.census.gov. On its home page, select "Subjects A–Z," then select "I" ("Income") and then your choice of topics.

annual average income is more than double that of high school dropouts. White male high school graduates earn annually about 59 percent of that earned by white college graduates. African American males with high school diplomas earn about 61 percent of that earned by African American male college graduates. The results are even more dramatic among females. White college graduates earn about three times the amount earned by white high school female dropouts; for African American females, the increase is also threefold. White female high school graduates earn each year only about 52 percent of that earned by white college graduates. African American females with high school diplomas earn only 49 percent of the earnings of African American female college graduates.

The downside of the data in Figure 13.3 is the comparison between African Americans and whites and between males and females. Although African Americans and women benefit from a college education, they clearly suffer in comparison to white males with identical educational credentials. (See Chapters 9 and 10 for more detailed analyses of the socioeconomic situation of African Americans and women in the United States.)

How strong is the relationship between occupational status and education? If anything, the benefits of college are even more apparent occupationally. For African Americans, whites, males, and females, occupational status (measured by blue-collar versus white-collar status) increases dramatically with higher education. The percentage for all race and gender combinations employed in white-collar occupations rises dramatically with educational level, whereas the percentages in blue-collar jobs decline sharply. Although African Americans and females are more concentrated in lower-level jobs within the higher-status occupations, the significant relationship between occupational status and higher education cannot be dismissed (U.S. Department of Labor 2001).

Changes in Higher Education

How would you describe university environments: static or fluid? College studies and college life have changed dramatically during the past forty years. College faculties and programs reached their peak in America during the fifteen-year period before the war in Vietnam. During this period, institutions of higher learning were very selective, academic programs were extremely rigorous, and college faculties had extensive control over the direction of student learning and development (Riesman 1980).

A shift from academic merit to student consumerism began in the mid-1960s, when most American undergraduate and graduate programs began to package their curricula more attractively. As a result, colleges and universities are losing some of their ability to shape academic and intellectual development. Students increasingly demand courses and programs that are more beneficial for employment. This is evident in the rise of vocational programs in community colleges, the shift of students from liberal arts into business and economics, and grade inflation.

Stanley Aronowitz and Henry Giroux (1994) are also concerned about the influence of labor market considerations. They point to the rise of business enrollments and the concomitant decline in the liberal arts majors. They see social science and humanity departments becoming service departments for nonmajors. They witness the diminution of graduate programs and faculty in liberal arts. As a result, they contend, a generation of liberal arts scholars has been lost.

An explosion was set off in 1987 when Allan Bloom, a University of Chicago professor, published *The Closing of the American Mind*. This best-selling book became the touchstone for those who believe that higher education is on its way to intellectual chaos. According to these critics of higher education, academia is now under the control of administrators and faculty who were part of the radical left in the 1960s. Universities, they believe, are using a radical egalitarian philosophy to elevate works of popular culture to the level of Platonic studies. The most frequently cited illustration is a Stanford University faculty decision, in 1988, to drop from a required Western culture course some traditionally classic works by white male authors in favor of some women and minority authors. A basic fear of these academic critics is that the gender, racial, and ethnic background of authors is supplementing merit as a basis for designing programs, courses, and reading lists. This is a debate that promises to continue for some time.

Another problem in higher education, declining public financial support, threatens the traditional high quality and low cost of public higher education. The Association of Governing Boards of Universities and Colleges describes the situation as "the new depression in higher education." Since World War II, the public, through its tax dollars, has supported higher education. That commitment is currently diminishing. Even the most prestigious public universities have been cut, causing them to become semipublic institutions at best. The University of California's system, consistently one of the world's finest, has seen the proportion of its state-funded financial base shrink from a high of 70 percent thirty years ago to less than 25 percent today. The University of Michigan now receives less than 20 percent of its funding from the state; the University of Colorado receives only 10 percent of its budget from the state.

State colleges and universities have responded to reduced state funding by increasing both private funding and tuition charges. As a result, the cost of a college education is skyrocketing (Morganthau and Nayyar 1996; Symonds, 2003). The low-cost, high-quality system of higher education that was provided the veterans after World War II, as well as their "baby boomer" children who were born between 1946 and 1964, is being denied the postboomer generation. In response to recent public outcry, tuition increases have leveled off. If the costs of higher education increase at 8 percent annually, students entering kindergarten when President Clinton was inaugurated in January 1993 can expect to pay as much as $75,000 for a degree at a public college or university or about $300,000 for a degree from premier institutions such as Yale and Harvard (The College Board 2001).

Parents and the general public are in shock. Between 1981 and 2001, tuition increased 112 percent at private colleges and universities and 106 percent in public institutions, whereas family income rose only 27 percent (The College Board 2001). With college tuition rising at three times the rate of inflation, and with the current no-new-taxes mood of the country, the call is for universities to restructure. If companies such as IBM and Xerox can downsize and do more with less, the argument goes, so can public colleges and universities. "Time to Prune the Ivy" was the clever title of an article in *Business Week* (Farrell 1993). Higher education and the public are obviously in conflict. The taxpayers are generally opposed to tax increases, and university administrators and professors fear a decline in quality due to restructuring and downsizing. Is the future at risk for a country that has built its prosperity on higher education?

An important casualty of reduced public funding may be the democratization of higher education achieved during the twentieth century. Prior to World War II, colleges were attended by only a small elite group. Through extensive federal and state investment after 1945, the United States became the first country to support mass higher education. As government financial backing declines, the costs of higher education are increasingly shifted onto students and their parents. While most students feel the pinch, the most damaged are those from low-income, immigrant, and minority families. Reduced access jeopardizes the promise of higher education as a means to upward social mobility (Symonds 2003).

One final issue: colleges and universities need to redesign themselves. The technologically driven information age is transforming colleges and universities from a geographically bound campus to a virtual one. Although the traditional residence-base will continue, higher education will increasingly be characterized by geographically dispersed faculty and students. Distance learning will be necessary both because technology makes it possible and because higher education is being redefined from a set number of semesters to a lifelong process of learning. Indeed, this trend is being reinforced by a rapidly changing global economy (Brody 1997; Ritzer 1998).

FEEDBACK

1. As expected, college enrollment in the United States declined in the 1980s because of a decrease in the number of eighteen- to twenty-four-year-olds. T or F?
2. Although a college education pays off financially for white males, it does not benefit African American men and women. T or F?
3. Occupational status gains produced by a college education are less apparent than the relationship between income and higher education. T or F?
4. According to David Riesman, most colleges and universities have shifted from academic merit to ________.
5. Public commitment to higher education is as strong for the postboomer generation as it was for their parents. T or F?

Answers: 1. F 2. F 3. F 4. student consumerism 5. F

SUMMARY

1. With industrialization and urbanization, the ability of the family to prepare its young for the future declined. As a result, formal schooling became necessary. The early emphasis in American schools was on "civilizing" the young; but after the turn of the twentieth century, the emphasis shifted to education for jobs.
2. Early schools were modeled after businesses and have increasingly become more bureaucratic. Advocates of open classrooms and cooperative learning contend that bureaucratically run schools fail to take into account the emotional and creative needs of individual children. The evidence suggests that some aspects of these alternatives provide benefits that should not be overlooked. The back-to-basics movement, however, seems to clash with the open classroom philosophy.
3. The debate over the best classroom methods has led to the research for alternatives. Major competitors to the traditional public school are vouchers, charter schools, for-profit schools, and magnet schools.
4. Functionalists see the emergence of the educational institution as a response to society's needs. The manifest functions of education include the transmission of culture; social integration; selection and screening of talent; promotion of personal growth and development; and the dissemination, preservation, and creation of knowledge. Schools also serve latent functions.
5. America is supposed to be a meritocracy in which social status is achieved rather than determined by family background. Advocates of the conflict perspective point to flaws in the meritocratic model, contending that the elite use the educational institution to preserve their privileged position.
6. Many of the flaws in the meritocratic model appear within the context of educational inequality, which exists when schooling produces different results for lower-class and minority children than it does for other children. Schools themselves promote educational inequality by reacting to students' social class backgrounds and racial or ethnic characteristics rather than to their abilities and potential. Schools use intelligence testing—which ignores the environmental deficiencies of lower-class students—to channel students into noncollege tracks.
7. Although there is disagreement, some evidence suggests that school desegregation reduces educational inequality. Three other approaches to promoting educational equality are compensatory education, increased community control of schools, and private schooling.
8. Symbolic interactionists emphasize the nonacademic socialization that occurs within schools. Through the hidden curriculum, children are taught a variety of values, norms, beliefs, and attitudes thought to be desirable by a society. This nonacademic socialization is accomplished in large part through textbooks and teachers. Much of this socialization is designed to help young people make the transition from home to the larger society.
9. Contrary to expectation, college enrollment has increased. This occurred partly because of active college recruiting and partly because of the decline of high-paying union-

ized manufacturing jobs. College education still pays off handsomely in the United States. Both income and occupational level increase with higher education.

10. Critics are disturbed about changes in higher education that began in the mid-1960s. Some see a shift from academic merit to student consumerism. The financial crisis in higher education looms as a genuine threat to the long-standing reliance of American society on an affordable, high-quality system of public higher education. Some observers contend that colleges and universities will have to redesign themselves to adapt to the information age.

INFOTRAC® COLLEGE EDITION

http://www.infotrac-college.com/wadsworth

Another unique option available to you at the Student Resources section of the companion Web site is Infotrac College Edition, an online library with access to hundreds of scholarly and popular periodicals. Below are suggested search terms for this chapter. Results from these and other searches are found at the site.

Search keywords: **Affirmative Action**. In Chapter 9, you were asked to use Infotrac to learn more about Affirmative Action in business situations. These federal regulations have also influenced colleges and universities. Search Infotrac to find articles about current events in higher education that are related to Affirmative Action. What are some of these events? How can the sociological perspective aid your understanding of these events?

Search keywords: **grading and education**. As a college student, you are keenly aware of the importance that grades play in America's educational system. However, not everyone supports the emphasis on grades. Search for articles that discuss different views on grades and the grading process. Based on your reading, discuss the impact that grades have on individual students. What impact does the grading process have on society?

Search keywords: **hidden curriculum**. The "hidden curriculum" implies that schools do more than teach students how to read, write, and do arithmetic. Read two articles that discuss hidden curricula. What are some of the elements contained in the hidden curriculum? How do these elements serve society? (As a quick test, which theoretical perspective does the preceding question come from?)

LEARNING OBJECTIVES REVIEW

- Describe the relationship between industrialization and education.
- Discuss schools as bureaucracies and the attempts to debureaucratize education in the United States.
- Outline the basic functions (manifest and latent) of the institution of education.
- Evaluate the meritocratic model of public education.
- Discuss educational inequality.
- Describe the ways in which schools socialize students.
- Identify and describe the dominant issues in higher education.

CONCEPT REVIEW

Match the following concepts with the definitions listed below them.

____ a. multiculturalism
____ b. hidden curriculum
____ c. vouchers
____ d. cooperative learning
____ e. self-fulfilling prophecy
____ f. dysfunction
____ g. manifest function

1. a negative consequence of some element of a society
2. the nonacademic agenda in schools that teaches children skills thought to be necessary for success in modern bureaucratic society
3. an intended and recognized consequence of some element of a society
4. an approach to education in which the curriculum accents the viewpoints, experiences, and contributions of women and diverse ethnic and racial minorities
5. when an expectation leads to behavior that causes the expectation to become a reality
6. a nonbureaucratic classroom structure in which students study in groups with teachers acting as guides rather than as the controlling agents determining all activities
7. tax allowances granted to cover part of the costs for an academic year at the school selected

CRITICAL-THINKING QUESTIONS

1. The open classroom philosophy represents an alternative to the bureaucratic model of education. Use the symbolic interactionist perspective to convince a local school board of the desirability of adopting the open classroom approach.

2. Assume for the moment that schools cannot provide educational equality. Analyze the consequences of the persistence of educational inequality within the context of the functions of education discussed in the text.

3. With the hidden curriculum in mind, think back to your early school years. Provide specific examples from your own experience to support the existence of the hidden curriculum.

4. Identify what you believe are the two biggest issues facing higher education in the twenty-first century. Develop recommendations for solving these problems.

MULTIPLE-CHOICE QUESTIONS

1. **Schools emerged as the primary educational agent in industrializing societies because**
   a. the skills needed for survival in an agricultural society became increasingly important.
   b. knowledge increased in volume and complexity and family members gradually lost the ability to educate their children adequately.
   c. too many parents were qualified to teach their children, and not enough parents were willing to work in agriculture.
   d. parents no longer received the income they needed to educate their children at home.
   e. the birth rate increased dramatically in the early 1800s.
2. **According to the text, critics of the bureaucratic model of education claim that**
   a. bureaucracies are notoriously inefficient.
   b. the United States school system should have more "vouchers."
   c. teachers should treat all students the same.
   d. administrators are useless and should therefore be eliminated.
   e. it is hard to determine whether children have been "processed" correctly.
3. **Education contributes to social integration by**
   a. allowing individuals to learn several languages.
   b. exposing individuals to similar educational traditions.
   c. exposing individuals to different ethnic and cultural traditions.
   d. presenting various religious doctrines.
   e. promoting personal growth and individual development.
4. **Schools promote personal growth and development by**
   a. preparing students to assume a place in the occupational world.
   b. discouraging emotional experiences.
   c. tracking lower-class students into vocational programs.
   d. discouraging group activities.
   e. teaching students the importance of following orders in a bureaucratic society.
5. **The school's efforts at the dissemination of knowledge most commonly take the form of**
   a. field trips outside the classroom.
   b. using television as a teaching tool.
   c. classroom teaching.
   d. personal experience.
   e. homework assignments.
6. **According to Randall Collins, credentials are required for many jobs because**
   a. they allow those who can afford to get them a strong hold on elite occupations.
   b. they perpetuate a meritocratic society.
   c. they fulfill a vital service for society by verifying educational attainment.
   d. they prevent the best talent from being lost.
   e. they provide worthwhile goals for those who lack advanced degrees.
7. **A significantly larger proportion of upper-class children attend college than do lower-class children. According to the text, this is due in part to differences in**
   a. academic ability.
   b. economic resources.
   c. desire for intellectual growth.
   d. motivation to learn.
   e. study habits.
8. **The most recent concept of educational equality involves**
   a. equality of resources.
   b. equality of opportunity.

c. equality in busing.
d. equality in effects of schooling.
e. equal education for equal ability.

9. Which of the following is a false statement?
a. City dwellers usually score higher on intelligence tests than people in rural areas.
b. Performance on IQ tests increases with age.
c. Middle-class African American children score about as high on intelligence tests as middle-class white children.
d. Intelligence tests are culturally biased.
e. Recent evidence has convinced most sociologists that about half of human intelligence is genetically based.

10. Central to _________ is the idea that special preschool programs can help deprived children to overcome an inferior intellectual environment.
a. forced busing
b. tracking based on standardized testing
c. expanding the scope of the Special Olympics
d. postsecondary school remedial reading
e. compensatory education

11. In his study on the value of a private school education, James Coleman found that
a. public schools produce higher student achievement levels.
b. private schools produce higher student achievement levels.
c. students in public and private schools produce similar levels of achievement.
d. minority children in public schools perform better than minority children in private schools.
e. minority students perform equally well in private and public schools.

12. When girls are successfully taught in school to see their self-worth in relation to their physical appearance rather than in relation to their abilities to contribute to society, the _________ has been at work.
a. functionalist perspective
b. self-fulfilling prophecy
c. "blooming" effect
d. multicultural perspective
e. policy of tracking

13. Which of the following is a false statement?
a. A college education results in higher annual income regardless of race and gender.
b. College-educated African American males and white males make more than double the annual income of high school dropouts.
c. African American females earn annually three times as much as white high school dropouts.
d. African American male high school graduates earn annually over one-half of that earned by African American college graduates.
e. A college education pays off economically for white Americans, but not for African Americans.

14. In his analysis of higher education, David Riesman contends that
a. university administrators are not paying enough attention to keeping educational organizations in business.
b. universities should devote more attention to intercollegiate athletics in order to retain alumni interest and financial contributions.
c. universities have increasingly lost their ability to shape development along academic and intellectual avenues.
d. universities are now under the control of administrators who were part of the radical left of the 1960s.
e. universities should do more to get students involved in activist causes and social movements.

FEEDBACK REVIEW

True-False

1. Most sociologists agree that IQ test scores do not measure inherited learning ability. T or F?
2. Public commitment to higher education is as strong for the postboomer generation as it was for their parents. T or F?

Fill in the Blank

3. Early public schooling in America was viewed as a mechanism of _________.
4. Educational equality exists when schooling produces the same _________ for lower-class children and minorities as it does for other children.
5. According to symbolic interactionists, schools are very much in the business of transmitting _________.
6. According to David Riesman, most colleges and universities have shifted from academic merit to ________.

Multiple Choice

7. Contemporary public education is important primarily because of
a. the quality of teachers in modern schools.
b. efficient operation of school bureaucracies.
c. historical changes in the economic and social structures of industrial societies.
d. lack of qualified individual tutors.

8. The idea that some educational requirements are unnecessary for many jobs today is called
a. status grouping.
b. credentialism.
c. meritocracy.
d. conflict.

Matching

9. Match the following research topics with the three theoretical perspectives:
   ____ a. functionalism
   ____ b. conflict theory
   ____ c. symbolic

(1) requirement of unnecessary credentials by schools
(2) benefit of schools for society
(3) effects of teacher reactions on the self-esteem of students

GRAPHIC REVIEW

The trend in educational level among whites and major minority groups since 1970 is displayed in Figure 13.1. Answer these questions, referring to the data in this figure.

1. Relate the educational levels in the various categories to the process of industrialization.

____

____

____

2. Would functionalists and conflict theorists interpret the gap between the educational level of whites and the educational attainment of African Americans and Latinos differently, or in the same way? Explain.

____

____

____

ANSWER KEY

Concept Review

a. 4
b. 2
c. 7
d. 6
e. 5
f. 1
g. 3

Multiple Choice

1. b
2. e
3. b
4. a
5. c
6. a
7. b
8. d
9. e
10. e
11. b
12. b
13. e
14. c

Feedback Review

1. T
2. F
3. social control
4. results
5. culture
6. student consumerism
7. c
8. b
9. a. 2
   b. 1
   c. 3

14 Political and Economic Institutions

OUTLINE

LEARNING OBJECTIVES

- Distinguish among power, coercion, and authority and identify three basic forms of authority.
- Discuss the nation-state, comparing and contrasting the functionalist and conflict approaches to this form of political authority.
- Identify the major differences between democracy, totalitarianism, and authoritarianism.
- Differentiate the views of functionalism and conflict theory on the distribution of power in America.
- Describe the major characteristics of capitalism, socialism, and mixed economic systems.
- Discuss the effects of the modern corporation on American society.
- Describe America's changing workforce composition and occupational structure.
- Discuss corporate downsizing and its consequences.

USING THE SOCIOLOGICAL IMAGINATION

Do Japanese or Americans work more hours per week? Less than a decade ago, Americans looked at workers in Japan with awe. Stories of men and women slaving away for 10 or more hours per day, six days per week, made the rounds of corporate America and surprised many American workers. The reality today is different. The average American works longer than workers in any other industrialized country, including Japan. Between 1977 and 2000, the average workweek among salaried American workers lengthened from 43 to 48 hours. In that same period, the number of workers putting in more than 50 hours per week went from 24 percent to 37 percent. In fact, Americans work an equivalent of eight weeks longer every year than Western Europeans. In this chapter on politics and the economy, we examine where, why, and how Americans work and vote.

Power and Authority

All societies must deal with the challenges of economics and politics. Economic issues exist because of the need to produce goods and services. The institution that carries out the production and distribution of goods and services is the **economy**. Because economic decisions affect organizations and the general public, severe conflicts often arise. Every society must develop some means of handling these conflicts, and such decisions are usually made by those responsible for the general welfare of the society. This responsibility lies in the institution through which power is obtained and exercised—the **political institution**.

© Paul Conklin/PhotoEdit

The most stable form of power is viewed as legitimate by those subjected to it. This photograph reflects familiar symbols of authority as two lawyers converse beside the massive columns of a Washington, D.C., federal building.

The 1997 strike by the Teamsters Union against United Parcel Service (UPS) illustrates the interaction between the economy and the political institution. This labor confrontation was based on a strong objection to the UPS corporate policy of downsizing—changing permanent jobs into temporary ones. UPS, by asking President Clinton to intervene early in the dispute, demonstrated the intimate interdependence between business and government in modern society. The interplay of these two institutions will become apparent in this chapter.

Defining Power and Authority

Max Weber (1946), a founder of political sociology, defined **power** as the ability to control the behavior of others, even against their will. Power can exist in one's personal appeal or magnetism. For example, in his classic study of power and decision making, Floyd Hunter (1953) wrote about James Treat, a man whose extensive power was not evident from his simple, nonthreatening appearance. Treat's power was based not on his visible personal characteristics but on his ability to influence the actions of others—including Georgia's governor. His power existed in his relationships with others.

Weber recognized another form of power, power through force, or **coercion**. A blackmailer might extort money from a politician. A government might confiscate, without compensation, the property of one of its citizens. In neither case do the victims recognize this use of power as valid. Moreover, because victims of coercive power see it as illegitimate, they normally are resentful and wish to fight back. Weber recognized that a political system based on coercive power is inherently unstable. Because force produces compliance only in the short run and only when those in power can exercise surveillance, Weber believed that a political institution must rest on a more stable form of power. This more stable form of power is **authority**—power accepted as legitimate by those subjected to it. For example, students take exams and accept the evaluations they receive because they believe that their teachers have the right to determine grades. Citizens pay taxes because they believe that their government has the right to collect the money. Because authority is based on others' acceptance, it is the most stable form of power. Its existence reduces the likelihood that coercion will be needed to maintain order.

Forms of Authority

Weber identified three forms of authority—*charismatic*, *traditional*, and *rational-legal*. Those invested with these forms of authority can expect compliance from others, who recognize authority figures as holders of legitimate power.

What is charismatic authority? **Charismatic authority** arises from a leader's personal characteristics. Charismatic leaders can lead because of their magnetic personalities or the feelings of trust they inspire. Martin Luther King Jr., Nelson Mandela, and César Chávez are examples of charismatic leaders. The power of a charismatic figure may be unlimited while it lasts; but for modern nation-states, charismatic authority alone is too unstable to provide a permanent basis of power. It is linked to an individual and is therefore difficult to transfer to another. When charismatic leaders die, they relinquish their authority, leaving no traditional norms for the continuation of power. Adolf Hitler, himself a charismatic leader, made a futile attempt at the end of World War II to name his successor. But as historian John Toland has noted:

> *Hitler's death brought an abrupt, absolute end to National Socialism. Without its only true leader, it burst like a bubble. . . . What had appeared to be the most powerful and fearsome political force of the twentieth century vanished overnight. No other leader's death since Napoleon had so completely obliterated a regime.* (Toland 1976:892)

So, even governments controlled by charismatic leaders must eventually come to rely on other types of authority. The two alternatives identified by Weber are traditional authority and rational-legal authority.

What is traditional authority? In the past, most states relied on **traditional authority**, authority in which the legitimacy of a leader is rooted in custom. Early kings, for example, often claimed rule by divine right. Orderly perpetuation was a key benefit. This peaceful transfer of power was possible because only a few individuals were eligible to become the next ruler. Until recently, custom was sufficiently strong to legitimate one's inherited right to rule. Because, for example, the kings in eighteenth-century Europe could depend on the custom of loyalty from their subjects, their traditional authority provided a stable political foundation—more stability than charismatic authority could have provided.

What is rational-legal authority? Most modern governments are based on a system of **rational-legal authority**; the power of government officials is based on their offices. Those who hold government office are expected to operate on the basis of specific rules and procedures that define and limit their rights and responsibilities. Because rational-legal authority is invested in positions rather than in individuals, persons lose their authority when they leave their formal positions of power. Citizens are expected to follow the directives of the political leaders, and leaders are expected to stay within the boundaries of their legal authority. Even presidents (Richard Nixon, for example) can lose their power if excessive abuse is made public. Legal authority thus limits the power of government officials.

How are these forms of authority ideal-typical? In the real world, people may hold more than one form of authority. Weber knew this, but he intentionally distinguished these three forms of authority through use of the ideal-type method defined in Chapter 6. As you may recall, this method involves isolating the most basic characteristics of some social entity. The purpose here is to crystallize the meaning of some social entity by removing that entity from the context in which it is embedded with other entities. Stripped of the surrounding complexity, the essential characteristics of a social entity can be identified.

FEEDBACK

1. The institution within which power is obtained and exercised is the __________ institution.
2. __________ is the ability to control the behavior of others, even against their will.
3. Power that is considered legitimate by those subjected to it is called __________.
4. On which of the following types of authority are kingships based?
   a. charismatic
   b. traditional
   c. rational-legal
5. Which of the following types of authority places the strongest limits on government officials' freedom of action?
   a. charismatic
   b. traditional
   c. rational-legal

Answers: 1. political 2. Power 3. authority 4. b 5. c

Although the power of any American president is fundamentally grounded in rational-legal authority, authority forms are somewhat mixed. The concept of the "imperial presidency," for example, suggests that some element of traditional authority is attached to the office. Particular presidents (Theodore Roosevelt, John Kennedy, Ronald Reagan) may also hold charismatic authority, whereas other presidents (Gerald Ford, Jimmy Carter) may not.

No matter the form, because authority is based on others' acceptance, it is the most stable expression of power. This is particularly true of modern nation-states.

The Nation-State

Distinguishing Between Government and Nation-State

By combining several independent political units, a nation-state consolidates power over a large geographical area. The United States, for instance, became a nation through the Constitutional Convention of 1787, whereby particular powers of individual states were transferred to the federal government.

Historically, the nation-state is a recent development. For 99 percent of human history, people lived in small hunting and gathering bands, and there was little need for large legal systems of control. Even the Roman Empire, with a few exceptions, consisted of self-governing cities (A. H. M. Jones 1966; Service 1975; Bottomore 1993; P. James 1996). Nation-states began to emerge in Europe during the fifteenth century, but as recently as 100 years ago, there were still politically unclaimed territories in the world. Many people lived without formal political rule. No more. Presently, the Earth is divided into over 100 nation-states, and nearly everyone is a member of one of them. One is born, lives, and dies only by the official recognition of a nation-state (Alford and Friedland 1985; Skocpol 1985; Skocpol and Amenta 1986; P. James 1996).

What is a nation-state? Although the nature of nation-states may vary, the **nation-state** is always the political authority over a specified territory (Charles Tilly 1990). One key characteristic of a nation-state is the absolute sovereignty it has over its citizens. Citizens can appeal no higher than the laws of the state, and the state has supreme authority over its citizens. Only state representatives have a legal right to act on behalf of the state. Only they, for example, can impose taxes, imprison criminals, or declare war. A second key characteristic of a nation-state is its devotion to **nationalism**—a people's commitment to a common destiny based on a recognition of a common past and a vision of a shared future (Stoessinger 1981).

It is important to distinguish between a state, which has ultimate authority over a territory, and a **government**—the political structure that rules a nation. All societies have government; arrangements for settling disputes are always necessary. A state exists as an entity, however, only if governmental functions are separated from other institutions, such as the family and religion. As societies modernize, states become increasingly common; governmental functions are handled by groups of officials whose roles in society are distinctly governmental.

The emergence of the state is associated with advances in agriculture and industry. In hunting and gathering societies, most decisions could be reached by consensus. Agricultural and industrial developments, however, led to surpluses that could be converted into wealth and power for certain segments of society. When surpluses lead to conflict over who will control them, a clearly defined state is needed to maintain order (Nolan and Lenski 1999). Both functionalists and conflict theorists have explanations for the emergence of nation-states.

A Functionalist View of the Nation-State

According to functionalism, nation-states are formed because they are needed by society. Unless a strong state exists, argued English philosopher Thomas Hobbes (1958; originally published in 1651), people will act only according to their own selfish interests. Without a state to control these selfish impulses, there will be a "war of every one against every one" and life will be "solitary, poor, nasty, brutish, and short." To escape such chaos, people agree to create nation-states on the basis of a "social contract" of cooperation.

A similar argument was developed by Emile Durkheim (1964; originally published in 1893). He feared the rapid and extensive social change experienced within modern societies. Unless sufficient controls are developed, admonished Durkheim, modern societies will experience disruptive internal conflicts. Although attempting to act in their self-interest, people will actually create a situation in which disruption prevents the achievement of personal goals. For Durkheim, the role of the state is to provide centralized regulation of economic life. Like Hobbes, Durkheim believed that the absence of a state invites chaos.

Talcott Parsons (1959) agrees with Hobbes; humans must have external constraints. Through social constraints, a government can perform necessary functions. For example, a system of laws can be developed and enforced. Government officials can mobilize resources to plan, coordinate, and achieve commonly shared goals. They can formulate policy decisions reflecting society's highest values. State representatives can oversee relationships with other nation-states—whether political, economic, or military.

A Conflict Perspective of the Nation-State

According to the conflict perspective, nation-states exist primarily to serve the interests of a society's elite. One of Hobbes's earliest critics was the French philosopher Jean-Jacques Rousseau. In the original state of nature, Rousseau argued, people were noble savages living in harmony with their environment. They were probably socially indifferent rather than at war with one another. There was no reason to start fighting until the creation of private property led to the rise of social inequality. At this point, Rousseau agreed, a social contract may indeed restore order by creating a state. This social contract, however, carries advantages primarily for the privileged. He described the contract in this fashion:

> *You need me, for I am rich and you are poor. Let us therefore make a contract with one another. I will do you the honor to permit you to serve me under the condition that you give me what little you have left for the trouble I shall take in commanding you.* (quoted in Cassirer 1951:260)

Rousseau argued for a new social contract in which obedience to the privileged class is replaced by obedience to the "common will," a will aimed at the welfare of all, rich and poor alike.

Karl Marx offers the most influential expression of the conflict perspective. The nature of a society, Marx contended, is determined by the relations between those who own property and the propertyless. Industrial societies, in Marx's view, are dominated by the relatively small number of people who own factories, machines, and other productive resources. As noted in Chapter 1, those who own productive property are the ruling class, and those who do not own property are the workers. The **bourgeoisie** (the few in the ruling class) and the **proletariat** (the mass of workers) form the major social classes in capitalist societies.

According to Marx, political life is a reflection of relations between these two social classes. Because the ruling class controls the resources, the government serves its needs: "The executive of the modern State is but a committee for managing the affairs of the whole bourgeoisie. . . ." (Tucker 1972:337; originally published by Marx in 1867). The state, through the power of force, provides a means for controlling workers. In most cases, however, the use of force is unnecessary, because the state draws on common values and experiences to promote an ideology that encourages general acceptance of the system. Marx used the term **false consciousness** to describe acceptance of a system that works against one's own interests.

According to Marx, then, those in charge of the state in capitalist societies—the government, military, police, bureaucracy—make decisions to benefit the ruling class. It is only through the revolution of the proletariat that the state will serve the interests of everyone. A short-term "dictatorship of the proletariat" had to exist during the transition from the capitalist state to communism, the final stage of historical development. Because Marx spent his time analyzing the problems of capitalist society, he did not refine a conception of the communist society. It does seem, however, that the state under communism had to be nonauthoritarian and nonbureaucratic and to exist for purely administrative reasons. Ownership of the means of production, including land, had to be in public hands, not in the hands of a few. To ensure that state functionaries did not form a new elite class, their pay had to be kept commensurate with those outside government.

Marx failed to foresee the difficulties in reducing the power of state bureaucracy and in persuading government officials to relinquish their positions and power (Orum 1983). Modern conflict theorists broaden Marx's concept of class conflict to encompass social conflict of all types, whether between classes or between political interest groups.

FEEDBACK

1. A _________ is the political entity that holds authority over a specified territory.
2. A _________ is the political structure that rules a nation.
3. According to Thomas Hobbes and Emile Durkheim, nation-states exist
   a. to serve the interests of powerful, privileged social classes.
   b. because of custom.
   c. to control the selfish impulses that would otherwise make social order impossible.
4. According to Jean-Jacques Rousseau, it is primarily the privileged classes who benefit from the existence of the state. T or F?
5. According to Karl Marx, those in charge of the state in capitalist societies make decisions that serve the interests of the owners of the means of production. He used the term _________ to describe acceptance of a system that works against one's own interests.

Answers: 1. nation-state 2. government 3. c 4. T 5. false consciousness

Political Systems

Nation-states embrace tw po types of political systems—*democracy* and *to arianism.* Because few societies are pure enough t aintain all characteristics of each polar type, the world contains variations between the two (Orun 3).

Democracy

There are two c ns of democracy. One concep-tion, inspired cient Greeks, assumes a form of self-governn involves the political participa-tion of *all* is classical conception of democ-racy app ns in the fifth century B.C. and to old Ne wn meetings. A second, more famil-iar c ed on realities of modern times, con-cei **cy** as a system of elected officials r ulfilling majority wishes. The United est representative democracy (republic) llis 2000).

What assumptions are made in a representative democracy? Representative democracy is based on two assumptions. The first assumption recognizes that not everyone in modern society can be actively involved in all political decision making. Citizens are not expected to be involved deeply in politics; they merely need to vote. Given the minute division of labor in contemporary society, it would be hopelessly inefficient for all qualified citizens to be preoccupied with political activities. Who would manage companies, teach school, design buildings, prepare taxes, rear children? The classical practice of democracy in ancient Greece could exist only because the masses (women and slaves), excluded from the political process, were available to perform nonpolitical functions. In the second assumption of representative democracy, politicians who fail to satisfy the wishes of the majority will not be elected (or reelected).

In the two-party system of the United States, we have a "winner take all" form of representative democracy. Here, the party with the most votes wins the election. In many European countries, where third-party

w 14.1

cal Freedom

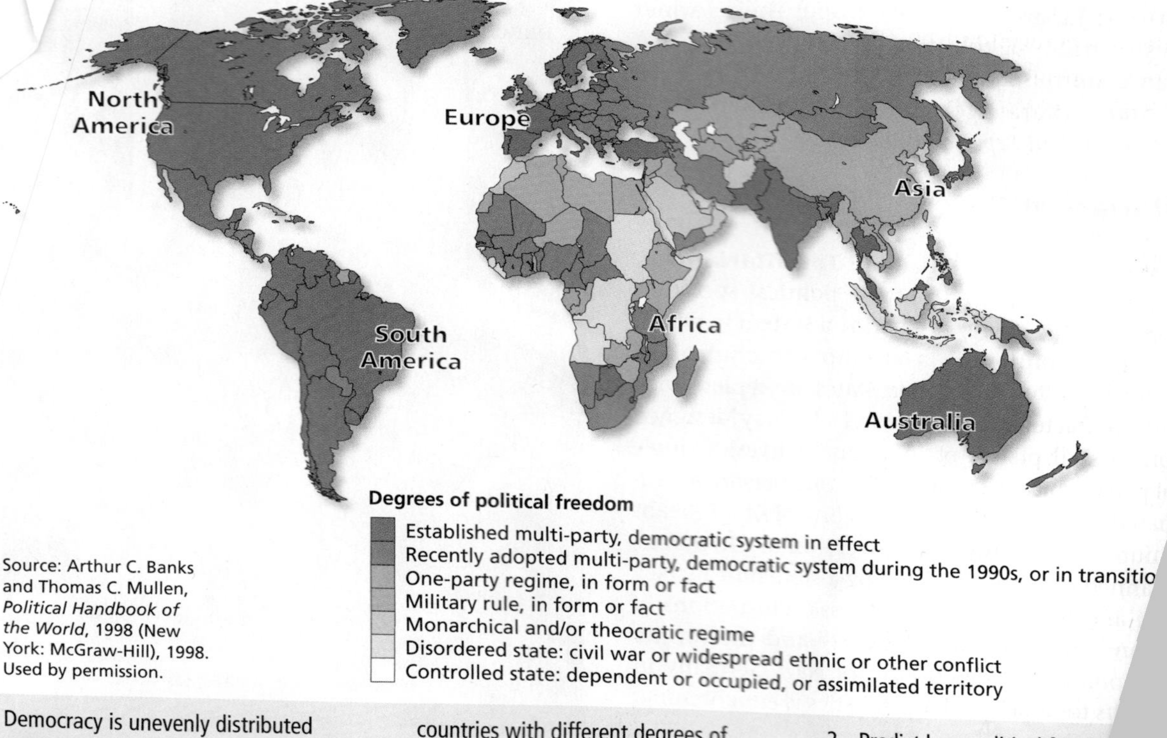

Source: Arthur C. Banks and Thomas C. Mullen, *Political Handbook of the World*, 1998 (New York: McGraw-Hill), 1998. Used by permission.

Democracy is unevenly distributed worldwide. There is now a trend toward a more democratic form of government. The accompanying map distinguishes countries with different degrees of political freedom.

1. Do you see any pattern in the degree of political freedom around the world?
2. Predict how political freedo the world will change in th years.

systems are common, parties participate in the government to the extent that they win representation in general elections. For example, one party might win 40 percent of the vote and control 40 percent of the legislature. Three other parties might take 20 percent each and still control a combined 60 percent of the legislature. This proportional representation system seems to be more democratic; it tends to encourage compromises and cooperation. Governments formed under this system can be fragile, however, and shifting political alliances may force new elections, even after relatively short periods of time.

Is democracy increasing throughout the world? The collapse of Soviet communism and the end of the Cold War created opportunities for more democratic societies. Over the last 20 years, 81 countries advanced toward democracy with civilian governments replacing 33 military dictatorships. And 140 of the world's nearly 200 countries have multiparty elections, the highest number in history (*Human Development Report* 2002). Still, there is little evidence of sustained worldwide progress (see Worldview 14.1). Only about half of the 81 new democracies are fully democratic: some did not continue their transition to democracy, others slipped back into authoritarianism or war. For the most part, "free" political systems are associated with advanced ...nomic development and are found primarily in a ... nations: Japan; Western Europe; their former ...nies, notably Canada, Australia, some Latin ...an countries, several African nations, and the ... States (Karatnycky 1995; Vanhanen 1997; ...evelopment Report 2002).

...anism

...ure of totalitarianism? Totalitarianism, ...he other end of the political spectrum ... is the type of political system in which ...olute power attempts to control all ...fe. Totalitarian states are replete with ...ristics: a detailed ideology designed ...hases of individuals' lives; a single ...ally controlled by one person; a well- ...of terror; total control of all means ...monopoly over military resources; ...y directed by a state bureaucracy ... 1965). Classic illustrations of ...zi Germany and the former ...ription of Nazi Germany in ...racteristics is enlightening.

o one

around
next fift...

...i government? Hitler's ...ment, which came to ... 1930s, offers a good ...ystem works. Hitler,

in his autobiog phy, *Mein Kampf,* laid the ideological
foundation for t National Socialist Party. At the center of this ideolog is the myth of Aryan racial superiority and its role saving the world, in part by
destroying Jews. Desp presenting a facade of democracy, Hitler and the Na Socialist Party held the
power. In fact, Hitler help solidify his own power
by setting up identical office state bureaucracy
and the party. The ensuing co allowed Hitler
to divide and conquer any oppo Hitler's absolute
control was also bolstered by the secret police)
and SS troops, whose terrorist pr elched all
dissenters and eliminated state e ith few
exceptions, all news media were ei the
party or eliminated. Hitler dominated
often devising his own military plans o
tion of his generals. Nazi Germany's forces,
nomic plans included strategies fo
production, organization of factories, and fo

Authoritarianism

Because authoritarianism is a residual (leftover) ...gory between democracy and totalitarianism, it is m... difficult to define than either polar type. As a type ... political system, it is, of course, closer to totalitarianis...

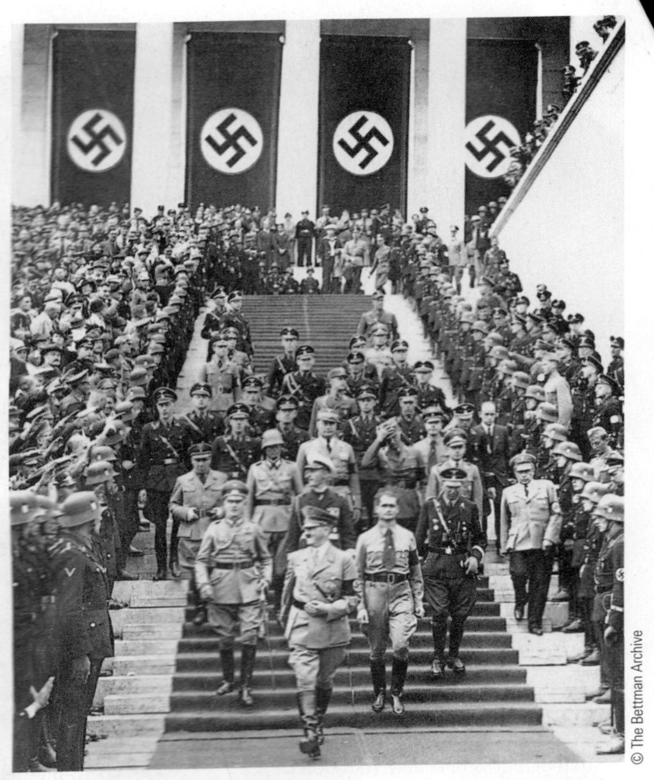

© The Bettman Archive

At the opposite end of the political spectrum from democracy is totalitarianism, the type of political system in which an absolute ruler controls all aspects of society. This photo chillingly illustrates a totalitarian state—Adolf Hitler and his staff touring Nuremberg at the zenith of Nazi power.

Technology and Protection Against Terrorism

Do your parents have a disaster supply kit? Does it include plastic sheeting and duct tape? If so, they have this in common with millions of other Americans. And they were only heeding the advice of Tom Ridge, Secretary of the Department of Homeland Security. Who can blame them? Americans feel uneasy. For the first time in our history we fear an attack on our shores from an enemy willing to use weapons of terrorism or mass destruction.

Personal safety isn't ensured by simply avoiding obvious symbols of our society such as the U.S. Capitol, Chicago's Sears Tower, or an Exxon oil refinery. Americans feel vulnerable in the many "soft targets" they enter daily—offices, shopping malls, apartment buildings.

The only consensus Americans have regarding self-protection is that duct tape is not enough. We seek security like all animals, but what price are we willing to pay for it? In an article entitled "Fortress America," Matthew Brzezinski (2003) explores several aspects of these costs. Two of these costs are relevant here—the use of electronic frisking and government surveillance.

Technologies available for electronic frisking come in several forms. National identity cards could contain a microchip that transfers your identity to, say, a digital subway fare card. Once the digital fare card is swiped into the turnstile reader, a computer at the mass transit authority would record your presence. While national ID cards aren't here yet, the Transportation Security Administration (TSA) is experimenting with one for transportation workers, from airport luggage handlers to dockworkers to truckers. This identity card, planned for nationwide distribution in 2004, will have at least one type of biometric information, most likely fingerprints (V. Novak 2003).

It would also be possible to install radiation scanners into turnstiles. Then someone you know could be strip-searched—like a cancer patient in a New York subway station whose residue from radiation treatments tripped a police handheld detector.

With regard to government surveillance, video cameras could be installed to monitor a commuter from his entrance into a subway station to the escalator ride out. A "suspicious" individual's identity could be confirmed by transmitting an image of his face through fiber-optic cables to a police command center. A computer there could compare this facial image to a databank of criminal mug shots, national identity card photos, and a government watch list. Facial recognition systems could also be installed at places like border crossings, DMV offices, and ATMs.

Software programs, now in existence, scan for suspicious words in e-mail messages. Corporations use them to block employee e-mails. The National Security Agency already processes the content of up to 2 million calls and e-mail messages per hour around the globe, thanks to spy satellites and supercomputers.

Government surveillance technology exists for still other purposes. How many places do you go where video cameras record your movements? Except for inside your home, it is legal for the government to track you via video cameras. While a warrant is required to use thermal imaging that could penetrate to your house's basement, there are no restraints on locations outside the home.

Washington, D.C., already has an experimental facility to access a network of video cameras that are located all over the city. According to the Washington, D.C., chief of police, it is possible to interface the system with schools, subways, highways, and the streets. After asking for a demonstration, Brzezinski was shown the air-conditioning unit on his roof, his garden furniture, and a newly planted cypress hedge. He writes:

> *The fact that government officials can, from a remote location, snoop into the backyards of most Washingtonians opens up a whole new level of information they can find out about us almost effortlessly. They could keep track of when you come and go from your house, discovering in the process that you work a second job or that you are carrying on an extramarital affair. Under normal circumstances there's not much they could do with this information. And for the time being, that is the way most Americans want it. But this is the kind of issue that will come up over the next few years. How many extra tools will we be willing to grant to the police and federal authorities? How much will we allow our notions of privacy to narrow?* (Brzezinski 2003:43)

An American security state, Bzrezinski concludes, would resemble Israel's. The former security chief of Israel's national airline, El Al, makes this statement: "Security is a balancing act. And there are always tradeoffs. . . . The question is, What are you willing to pay or put up with to stay safe?" (Brzezinksi 2003:76).

Analyzing the Trends

Brzezinski's article, of course, calls to mind George Orwell's *1984*. But, considering the trade-off of privacy for security, should the American government embrace these technologies? Discuss the social implications of such a move.

FEEDBACK

1. _________ is the type of political system in which elected officials are held responsible, through the mechanism of voting, for fulfilling the wishes of the majority of the electorate.
2. The United States is characterized as a
   a. classical democracy   c. representative democracy
   b. lineage democracy   d. democracy of hegemony
3. The type of political system in which a ruler with absolute power attempts to control all phases of social life is called _________.
4. _________ refers to the type of political system controlled by nonelected rulers who generally permit some degree of individual freedom.

Answers: 1. Democracy 2. c 3. totalitarianism 4. Authoritarianism

than to democracy. **Authoritarianism** refers to a political system controlled by nonelected rulers who generally permit some degree of individual freedom (Linz 1964). Countless regimes have leaned toward totalitarianism but have fallen short of all its defining characteristics. These authoritarian governments include monarchies (King Fahd of Saudi Arabia), military seizures of power (Pakistan's General Musharraf Parvez), and dictatorships (Saddam Hussein of Iraq).

Political Power in American Society

Political elitism is inevitable, even in democratic societies, because the public, which cannot govern directly, must delegate political decisions and governing responsibilities to representatives. By this act of delegation, an elite group is created (Dye 2001). Still, citizens in democratic societies are supposed to be able to influence the actions of their representatives, through voting (usually via political parties) and through joining interest groups. Voting is covered next, followed by a discussion of interest groups within the context of the pluralist model.

The Vote

Like all contemporary democracies, the United States emphasizes political participation through voting. Voting is supposed to be an important source of power for citizens, and it does in fact enable people to remove incompetent, corrupt, or insensitive officials from office and to influence issues at the local, state, and national levels. Unfortunately, American political parties are more concerned with winning elections than with maintaining a clear-cut ideological position, contributing to a politics of "blandness."

How much actual power do Americans exercise through the ballot box? In current U.S. practice, voting has severe limitations as a means of exercising power (see Snapshot 14.1). In the first place, the range of candidates from which to choose is restricted, because of the power of political parties—organizations designed to gain control of the government through the election of candidates to public office (Bardes, Shelley, and Schmidt 2002). Because the United States has only two major political parties, and because only candidates endorsed by those parties have any chance of winning a major political office, the political views of candidates for any particular office often resemble each other more than they differ. Consequently, many groups are not effectively represented (S. Hill 2002). Second, the limited choice of political candidates is reinforced by the high cost of political campaigns. Only people who are wealthy themselves or who are able to attract large contributions from others can mount effective campaigns. For example, by summer 2000, George W. Bush had raised almost $100 million for his presidential campaign, and Al Gore had collected almost $50 million (Federal Election Commission 2002). Not only are such people often not representative of the public, but they must face powerful contributors who expect favors once their candidates are in office (VandeHei and Hamburger 2002).

A third limitation lies in the way power is exercised by those at the top of the American political structure. According to Howard Zinn (2003), the ballot box is a poor tool for coping with such matters as racial discrimination, economic justice, and the formation of foreign policy. Singling out foreign policy as the most clear-cut example, he documents the powerlessness of voters with respect to waging war. The Constitution gives Congress the power to declare war, but Zinn cites examples throughout U.S. history of presidents waging war on the advice of a few *appointed* officials, even after campaigning as peacemakers (Dreyfuss and Vest, 2004). And once war is declared, presidents are rarely seriously challenged by elected representatives or the people.

How fully do Americans take advantage of the voting privilege? American interest in voting is low, partly

Cybernews

Cable News Network (CNN) was forced to retract a deceitful story about the use of nerve gas during the Vietnam War. The *Cincinnati Enquirer* apologized for its news-gathering techniques in a report about Chiquita Brands International; it paid $10 million in damages. Two other newspapers admitted that their reporters had recently fabricated stories without any basis in fact (Lissit 1998).

Marvin Kalb, a thirty-year veteran of CBS and NBC news, blames the decline in journalism ethics on the profit motive. News has become a big, big business. Media giants are sending profits into the stratosphere. In the process, "NBC manufactures news in much the same way and with much the same motivation as GE [which owns NBC] manufactures light bulbs" (Kalb 1998).

There is reason to worry that as the Internet's economic potential is tapped by media megacorporations, cybernews will magnify this ethical problem. There is already evidence of greater inaccuracy associated with Internet reporting.

Central to the changes is the fact that anyone with access to the Internet is free to "report" the news. Internet journalist Matt Drudge says that now, "any citizen can be a reporter" (Trigaboff 1998). Drudge portrays the Internet as a democratizing institution for the media, eliminating differences between reporters and readers.

Many reporters, however, note the risks the Internet carries. Internet news suffers from a news cycle that runs 24 hours a day, seven days a week. Instant reporting on the Internet means that sources for stories often go unchecked as reporters sacrifice accuracy for speed (Lissit 1998; Rust and Danitz 1998). Reporters on the Internet generally do not have the traditional overseers: editors to review their stories, in-house attorneys to be concerned about lawsuits, or publishers to make judgments about the appropriate nature of news stories (S. Levy 1998). Joseph C. Goulden, director of media analysis for Accuracy in Media, a nonprofit, grassroots citizens watchdog of the news media, describes the reporting style on the Internet as "ready, fire, aim" (Rust and Danitz 1998:23).

In the United States, one of the justifications for the freedom of the press is its delivery of accurate information to voters. If Internet reporting represents a trend toward greater inaccuracy, this traditional contribution of a free press to American democracy could be weakened. What if voters begin to distrust, even more, the information they receive and thus become increasingly cynical about the political process? At this time, only one thing can be said with certainty about Internet journalism. It will have a profound impact on the way news is reported (Kinsley 1998).

Analyzing the Trends

The Internet will affect our democracy. Identify some of its positive and negative consequences.

SNAPSHOT OF AMERICA 14.1

Distribution of Electoral College Votes

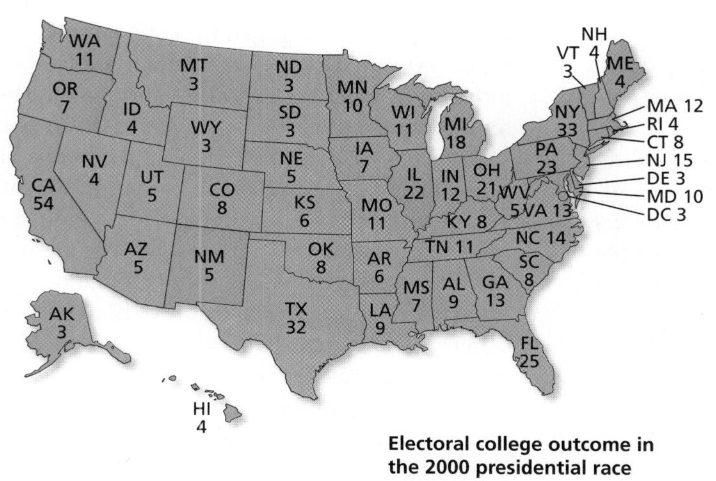

Source: Federal Election Commission, www.fec.gov, November 2000.

Electoral college outcome in the 2000 presidential race

Gore
Bush

As the American population learned during the 2000 presidential election, it is the electoral college rather than the popular vote that determines the winner. This map shows the distribution of electoral votes by state within the context of whether Bush or Gore won the state.

1. What does the distribution of electoral votes across the country (without respect to the candidates) say about political power in the United States? Explain.
2. Look now at the states Bush and Gore won. What does the pattern of states won by each candidate indicate about differences in the characteristics of Republican and Democrat supporters? Explain.

TABLE 14.1

Top Fifteen Nations in Voter Participation: 1945–2001

Listed here is a rank ordering of the fifteen nations with the highest rates of voter participation. Compare these rates with U.S. elections in both presidential and off-year elections.

Country	Elections (number)	Average Share of Voters Participating in National Elections (percent)
Australia	22	94.5
Singapore	8	93.5
Uzbekistan	3	93.5
Liechtenstein	17	92.8
Belgium	18	92.5
Nauru	5	92.4
Bahamas	6	91.9
Indonesia	7	91.5
Burundi	1	91.4
Austria	17	91.3
Angola	1	91.2
Mongolia	4	91.1
New Zealand	19	90.8
Cambodia	2	90.3
Italy	15	89.8

Sources: Molly O. Sheehan, "Voter Participation Declines," in *Vital Signs 2002: The Trends That Are Shaping Our Future* (New York: Norton, 2002), p.159.

because of the public's relatively low confidence in political leaders (see Figure 14.1). In 2000, 51.3 percent of eligible U.S. voters exercised their right, up slightly from 49 percent in 1996. This means that about a fourth of the American people voted in the 2000 presidential election, a proportion comparable to the 20 percent who elected President Reagan in 1980 (C. Lewis 2000). U.S. voter participation is even lower in off-year (nonpresidential) elections. Less than 40 percent of eligible voters went to the polls in 2002. In fact, the United States has one of the lowest voter turnout rates in the industrialized world (Center for Voting and Democracy 2002; J. T. Patterson 2002a). Note the place of the United States in world voter participation rates in Table 14.1.

In general, the lower social classes and the working class in the United States tend to vote in smaller proportions than the middle and upper classes. Members of minorities, people with less education, and people with less income are less likely to vote in both congressional and presidential elections (Walsh 2000; U.S. Bureau of the Census 2002c).

On what do Americans base their vote? Most attitudes and beliefs that are expressed as political opinions are gained through a learning process called political socialization. This can be formal, as in government class, or informal, as through the family, the media, economic status, and educational level. Studies have shown that most political socialization is informal:

- *The family*. Children learn political attitudes the same way they learn values and norms, by listening to everyday conversations and by watching the actions of other family members. In one study, more high school students could identify their parents' political party affiliation than any other of their parents' attitudes or beliefs.
- *Education*. Level of education influences political knowledge and participation. For example, more highly educated men and women tend to show more knowledge about politics and policy. They also tend to vote and participate more often in politics.
- *Mass media*. Television is the leading source of political and public affairs information for most of us. Television and other mass media determine which issues, events, and personalities are in the public eye. By publicizing some issues and ignoring others and by giving some stories high priority and others low priority, the media decide the relative importance of issues. While it is clear that the media play an important role in shaping public opinion, the extent of that role is unclear. Studies indicate that the mass media have the greatest effect on those people who have not yet formed an opinion about the issue being discussed.

FIGURE 14.1

Level of Public Confidence in the Executive and Legislative Branches of Government: 1966–1996

Varying levels of public confidence in the president and Congress are displayed in this figure. The downward trend began in the 1960s due to the turmoil surrounding the Vietnam War. The decline for Congress continued after President Nixon's Watergate hearings in 1973. Confidence began to go up for both branches of government with the election of Jimmy Carter in 1976 and continued an upward trend during the first four years of President Reagan's years. The decline resumed after Reagan's reelection in 1984, particularly for the president, and continued during the first President George Bush's years. The lowest point was reached during Clinton's impeachment in 1999.

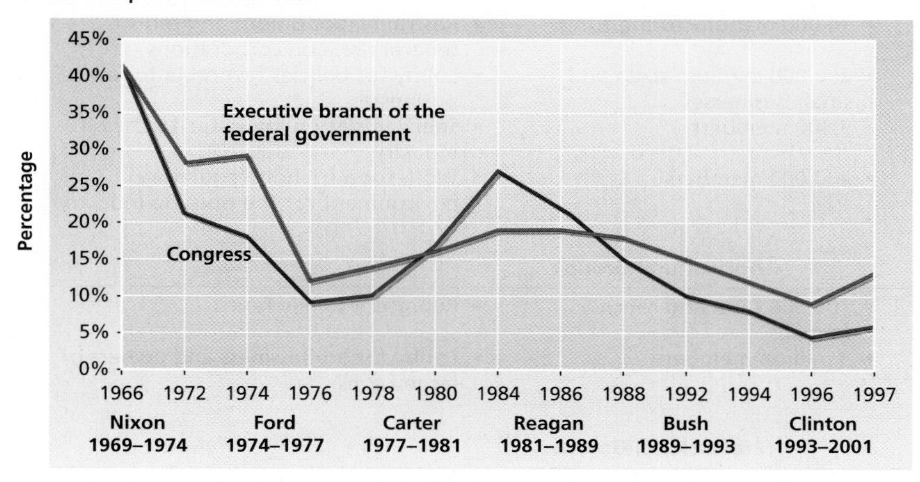

Sources: Humphrey Taylor, "Confidence in Leaders of Institutions Falls to Lowest Level Ever," The Harris Poll (March 7, 1994): 3–4; Washington Post/Kaiser Family Foundation/Harvard University Survey Project, *Why Don't Americans Trust the Government?* (Menlo Park, CA: The Henry J. Kaiser Family Foundation, 1996); Lydia Saad, "Americans' Faith in Government Shaken But Not Shattered by Watergate," Gallup News Service (June 19, 1997), www.gallup.com

- *Economic status and occupation.* Economic status clearly influences political views. Disadvantaged people are more likely to favor government assistance programs than wealthier people, for example. Similarly, where you work affects how you vote. Corporate managers are more likely than hourly factory workers to favor tax shelters and aid to businesses.
- *Age and gender.* On most issues, young adults are a bit more liberal than older Americans. Young adults, for example, tend to be more progressive than older persons on racial and gender equality. Women tend to be more liberal on abortion rights, women's rights, health care, and government-supported child care.

Pluralism: The Functionalist Approach

In a democratic society, two major models of political power are evident—*pluralism* and *elitism*. According to **pluralism**, decision making is the result of competition, bargaining, and compromise among diverse special interest groups. In this view, no one group holds the majority of power; rather, power is widely distributed throughout a society or community. The second major model of political power, elitism, is in direct contrast to pluralism. According to **elitism**, a community or society is controlled from the top by a few individuals or organizations. Whereas the elitist model is based on the conflict perspective, the pluralist model is consistent with the functionalist perspective.

How do pluralists view the distribution of power? According to David Riesman (2001), major political decisions in the United States are made via competition among special interest groups, each of which has its own stake in the issue. In addition to attaining their own ends, interest groups try to protect themselves from opposing interest groups.

What is an interest group? An **interest group** is a group organized to influence political decision making. Group members share one or more goals—either of their own members (as in the case of the National Rifle Association) or of a larger segment of society (as

TABLE 14.2

Types of Interest Groups

The U.S. government is influenced by a wide variety of interest groups. This table illustrates the most important types of interest groups. Do you believe that the influence of all these interest groups promotes or hinders democracy? Explain your answer using conflict theory or functionalism.

Organization	Membership	Objectives
	ECONOMIC GROUPS	
	Business Groups	
National Association of Manufacturers (NAM)	• 14,000 manufacturing firms	• Restrain labor unions and reduce federal taxes on corporations
U.S. Chamber of Commerce	• 200,000 medium-sized and small businesses	• Lobby for policies favorable for all businesses
American Petroleum Institute	• 9,400 members	• Seek legislation favorable to the oil industry
National Association of Home Builders	• 200,000 members	• Work for a friendly regulatory environment for the housing industry
	Agricultural Groups	
National Farmers Union	• 300,000 farm and ranch families	• Represent family farms
American Farm Bureau Federation	• 3 million members	• Lobby for agribusiness and owners of large farms
	Professional Groups	
American Medical Association (AMA)	• 271,000 members	• Oppose government involvement in medical practices
American Bar Association	• 360,000 members	• Protection of the civil and political rights of all U.S. citizens
American Association of University Professors	• 44,000 members	• Dedicated to the defense of academic freedom and tenure and to the promotion of sound academic standards and due process

in the case of ecology-oriented groups such as the Sierra Club).

Interest groups are not new to American politics. In the nineteenth century, they were active in extending women's rights and promoting the abolition of slavery. Furthermore, a history of twentieth-century America could not be written without describing a wide range of interest groups, such as the Women's Christian Temperance Union and the early labor union movement. The 1960s—with controversies surrounding civil rights, the Vietnam War, the environment, the women's movement, and corporate power—strengthened many interest groups and led to the creation of new ones (Clemens 1997).

New interest groups are born all the time. The environmental lobby is one example. There were relatively few environmental interest groups prior to the passage of major environment legislation (such as the Clean Water Act) in the 1960s. The success of the then existing environmental groups that lobbied for this legislation spawned additional groups, now numbering three times the total number of groups before the legislation. This added clout produced additional environmental legislation (for example, the 1990 Clean Air Act Amendments), which subsequently led to the creation of still other interest groups (Bardes, Shelley, and Schmidt 2002). The health-care industry is another illustration. In response to the managed health-care industry (representing its stockholders), the American Medical Association in 1999 endorsed unionization to represent physician interests. Predictably, consumers will organize to represent their interests (Lore 1999).

Gary Wasserman illustrates interest group bargaining and compromise:

When Ralph Nader, the consumer advocate, proposes that automobile makers be required to install more safety devices such as air bags in their cars, numerous interests get involved in turning that proposal into law. The car manufacturers worry about the increased costs resulting in fewer sales

TABLE 14.2

Types of Interest Groups (continued)

Organization	Membership	Objectives
	Labor Groups	
AFL-CIO	• Over 100 affiliated-unions (14 million members)	• Lobby to protect member unions from business influence in government
United Mine Workers	• 130,000 members	• Advance the goals of mine workers
National Air Traffic Controllers Association (NATCA)	• 14,000 members	• Shape policy decisions affecting air traffic controllers
	CITIZENS' GROUPS	
	Public Interest Groups	
League of Women's Voters	• 1,100 local leagues, 130,000 members and supporters	• Work on the drive to simplify voter registration and related issues
Common Cause	• 265,000 members	• Advocate political reform such as more restrictive campaign financial laws
Public Citizen	• 100,000 members	• Focus on consumer issues such as auto safety and pollution control
	Single-Issue Groups	
Sierra Club	• 550,000 members	• Advocate the protection of scenic areas
National Audubon Society	• 550,000 members	• Dedicated to the conservation and restoration of natural ecosystems
Greenpeace USA	• 2,100,000 members	• Pursue environmental protection
	Ideological Groups	
Americans for Democratic Action (ADA)	• 65,000 members	• Support liberal positions on social, economic, and foreign policy issues
Christian Coalition	• 2,000,000 members	• Promote the restoration of "Christian values"
National Organization for Women (NOW)	• 250,000 members	• Protect the rights of women

Sources: Adapted from Thomas E. Patterson, *The American Democracy*, 4th ed. (Boston: McGraw-Hill, 1999), pp. 262–269.

and reduced profits, and they may try to limit the safety proposals. Labor unions may want to make sure that the higher costs do not result in lower wages. Insurance companies may be interested in how greater safety will affect the claims they have to pay out. Oil companies may worry about the effect on gasoline consumption if fewer cars are sold because of their increased cost. Citizens' groups like Common Cause may try to influence the legislation so that it provides the greatest protection for the consumer. The appropriate committees of the House and Senate and the relevant parts of the bureaucracy will weigh the competing arguments and pressures as they consider bills covering automobile safety. The resulting legislation will reflect the relative power of the competing groups as well as the compromises they have reached among themselves. (Wasserman 2001:254)

In all of this, responsibility falls to government leaders for balancing the public welfare with the desires of special interest groups. (See Table 14.2 for a sample of interest groups and their primary issues.)

How prevalent are interest groups? Alexis de Tocqueville depicted Americans as a nation of joiners:

> *As soon as several Americans have conceived a sentiment or an idea that they want to produce before the world, they seek each other out, and when found, they unite. Thenceforth they are no longer isolated individuals, but a power conspicuous from the distance whose actions serve as an example; when it speaks, men listen.* (Tocqueville 1955:488; originally published in 1835)

The presence of over 100,000 associations in the United States today, however, might have surprised even Tocqueville. About two-thirds of all Americans belong to some voluntary formal association. Most of these, of course, are not interest groups attempting to affect government policy. The Washington, D.C., telephone directory contains listings for over 2,500 organizations whose names begin with "National Association of." These organizations, along with many other interest

SOCIOLOGY AND THE NEWS MEDIA

The Politics of Smoking

The American tobacco industry is an interest group organized to influence political decision making regarding cigarette smoking. As noted in this CNN report, the tobacco industry is charged with buying favors from state lawmakers across the country. During 1995–1996, three of the top ten contributors to Republican national party committees were tobacco companies. Philip Morris gave over twice as much as the second highest contributor ($2.5 million). Philip Morris also donated nearly half a million dollars to Democrats during this period. According to critics, this interest group pushes for watered-down laws that appear to be effective but that do not discourage smoking among children and teenagers.

1. State the view of this situation through the eyes of a functionalist.

____

____

2. How would a conflict theorist interpret this situation?

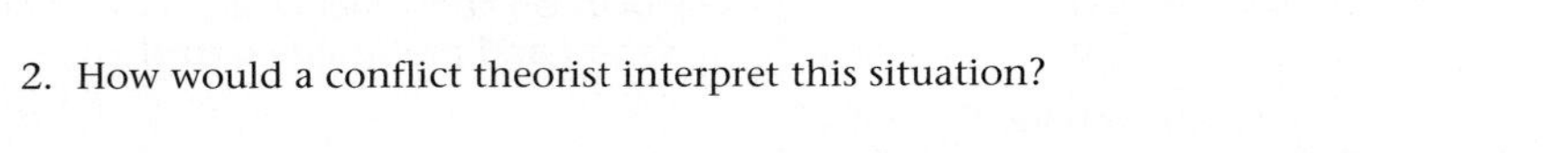

____

____

The American tobacco industry is currently experiencing intense scrutiny from medical, legal, and political quarters. According to one charge, cigarette makers are targeting minorities with advertising campaigns designed to promote smoking.

groups, energetically push the economic interests of their members. In Washington, the number of lobbyists attempting to influence Congress is estimated at about 14,000 and rising (C. Lewis 2000).

From a study of New Haven, Connecticut, Robert Dahl (1989) made an interesting observation: power in New Haven was not concentrated in the hands of one elite group. The groups, for example, trying to influence political decisions on public education were not the same groups competing, bargaining, and compromising on the issue of highway construction. Power was sufficiently dispersed that few segments of the community were without power. This system of checks and balances worked on the local level just the way James Madison, in the Federalist Papers, predicted it would in a large republic (Ellis 2000).

Pluralists point to the beneficiaries of the 1997 tax-cut bill. Tax breaks came not only to the wealthy, such as Microsoft shareholders, but also to groups with more modest resources, such as churches and people with mental disabilities (P. Hill 1997). We can further examine the pluralist model by focusing on political action committees.

Political Action Committees

American political parties were influential in the past because they controlled the nominations for major offices, because they oversaw campaign financing, and because many citizens tended to vote the straight party ticket. These bases of political party power appear, however, to be eroding. Less than one-fourth of Americans indicate that party affiliation makes a great deal of difference in their decision to vote for a candidate for Congress, whereas over one-third say that party affiliation would have little or no influence on their vote. Over one-third of Americans consider themselves "independents" (U.S. Bureau of the Census 2001a). With the decline of citizen participation in political parties, political interest groups are currently most often represented by political action committees that in turn lobby the political parties (Greider 1992; Birnbaum 2000; M. Green 2002).

What are political action committees? **Political action committees (PACs)** are organizations established by interest groups to raise and distribute funds to selected political candidates. Although PACs are prohibited from contributing directly to political candidates, they can channel up to $10,000 into any candidate's election campaign. Moreover, there is no legal restriction on the number of PACs that may support a candidate.

Since PACs were authorized in the mid-1970s, their number has mushroomed. Whereas only about 600 political action committees existed in 1974, there are now nearly 4,500 (see Figure 14.2). Moreover, contributions from PACs are beginning to dwarf the role of political parties in financing political campaigns. PACs

FIGURE 14.2

Growth in the Number of PACs: 1974–2000

This figure graphically depicts the sharp increase in the number of PACs in the United States. Evaluate the presence of PACs within the context of pluralism and elitism.

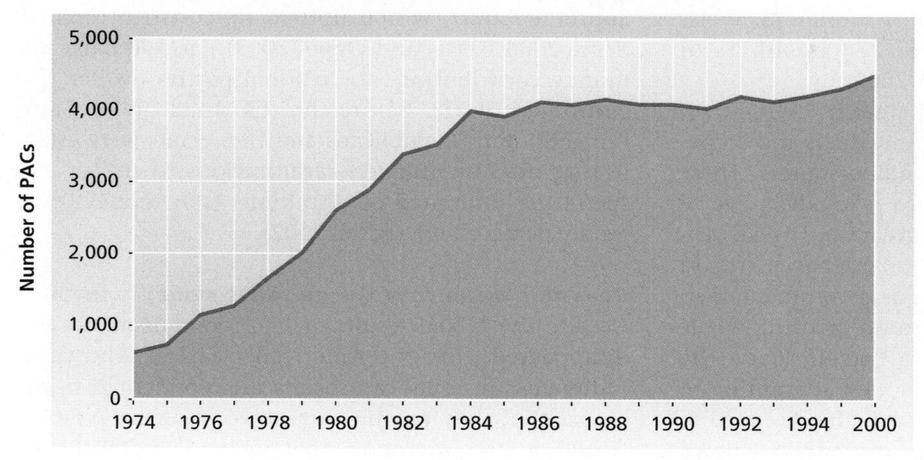

Source: Federal Election Commission, February 11, 1994; U.S. Bureau of the Census, *Statistical Abstract of the United States: 1996* (Washington, DC: U.S. Government Printing Office, 1996a), p. 290; Federal Election Commission,1998, and Federal Election Commission, 2002.

now contribute more than nine times as much support to congressional candidates as do political parties. Dollar amounts of PAC contributions to federal candidates (so-called "hard" money, limited by legal restrictions) increased from just under $12 million in 1974 to almost $260 million in 1999–2000 (Federal Election Commission 2002).

For a complete report of campaign contributions, one has to add the $495 million in "soft" money (large unregulated contributions) given to political parties during the 1999–2000 election cycle. This is nearly twice the amount contributed during the previous national election cycle (Federal Election Commission 2002). Records were set in the spring of 2000 when the Republican party raised over $21 million from donors in one fund-raising dinner—only to be topped by the Democratic Party, which held a $26.5 million fund-raising affair. The price tag for the 2000 federal elections is estimated at over $3 billion, nearly 50 percent more than in 1996 (G. Gordon 2000).

Political contributors are active in state legislatures, as well. In 2003, the Virginia Senate Finance Committee voted overwhelmingly against a proposed cigarette tax. Each of the twelve state senators voting against the tax had received, in the most recent election cycle, campaign contributions from Philip Morris ranging from $500 to $3,500. Among the three senators voting in favor of a tax increase on cigarettes, only one had received campaign money ($500) from the giant cigarette maker (Blackwell 2003; Copsey and McNeill 2003).

Do PACs contribute to pluralism? PACs were originally designed to provide a mechanism for small contributors to become involved in politics, to allow candidates without personal wealth to run for office, and to encourage independence from the viewpoints of political parties. Critics are concerned, however, that PACs may be undermining pluralism, because most PACs are connected with business-related interests (Eismeier and Pollock 1988; Watzman, Youngclaus, and Shecter 1997; Bardes, Shelley, and Schmidt 2002).

Charles Lewis (1998) makes a strong case for the influence of business-related PACs on members of Congress. His book *The Buying of Congress* is replete with purported links between business-related PAC money and congressional votes. As one illustration, consider the possible political influence by business-related PACs associated with the Tax Reform Act of 1986. As the congressional tax-writing panels were considering tax reform, PACs contributed $6.7 million—$119,643 per member—to fifty-six members of the House Ways and Means and Senate Finance Committees. This was two and a half times more money than the members of those same committees received from PACs two years earlier.

Finances heavily influence who can run for office and who wins elections in the United States (M. Green 2002; K. Phillips 2002; Palast 2003). According to the Federal Election Commission (FEC), 90 percent of the primary races in the 2002 congressional primary elections were won by candidates who raised the most money. In fact, victorious candidates raised four times as much as their opponents, an average of $464,000 versus $99,000. And most of this money was given by relatively few large donors, many of whom were out-of-state residents (Loiz and Cassady 2002). In the actual 2002 midterm elections, just over 95 percent of U.S. House races and 75 percent of Senate races were won by the candidate who spent the most money (S. Weiss 2002). At that same time, over 170 members of Congress were millionaires (Keller 2002).

Worse yet were the so-called stealth PACs—shell groups that raised millions from individuals and corporations without revealing their funding sources. These organizations emerged as a way to take advantage of loopholes in campaign and tax legislation. Unlike conventional PACs, these new groups did not have to report any of their activities to the public or government oversight organizations. A business group favoring whale hunting, for example, could call itself "Save the Whales" while spending lots of money attempting to block, repeal, or soften legislation actually designed to protect whales. The most widely known, issue-advocacy PAC was Republicans for Clean Air (formed by two supporters of George W. Bush), which spent $2.5 million against Bush's rival John McCain during the 2000 presidential primaries (Isikoff 2000). In the first alteration of the campaign finance laws in twenty years, Congress in 2000 passed legislation forcing stealth PACs to disclose their political donors and spending. After Pesident Clinton signed the bill, these PACs could no longer work in the shadows for partisan purposes (Dewar 2000; Trister 2000). Further reforms were passed in March 2002, when the Bipartisan Campaign Reform Act (more popularly known as the McCain-Feingold bill) was approved. By ruling the McCain-Feingold ban on unlimited soft money contributions to political parties unconstitutional in early 2003, a federal district court set this latest campaign reform act on a path to the Supreme Court, which upheld its constitutionality (Lane 2003; Lewis and Oppel 2003).This bill bans soft money contributions to national parties (Miller and Penniman 2002). Before the ink dried on McCain-Feingold, both Republicans and Democrats were establishing new, unaffiliated organizations to collect and spend the unlimited contributions (soft money) now legally denied them (Edsall 2002).

John McCain ran for the presidency of the United States in 2000. His candidacy highlighted the need for campaign finance reform.

Does this mean that the pluralist model is invalid? Despite the mixed results of PACs, pluralism has not disappeared from the American political landscape. Individuals in American society are represented to the extent that they belong to groups—whether political parties or interest groups—that are capable of influencing political decision making. Therefore, interest groups are one of the most important avenues for political representation. Groups representing civil rights, the environment, and consumer concerns have won some concessions that otherwise would not have been obtained. They have accomplished this through lobbying, making contributions to political campaigns, gathering information, generating publicity to express their viewpoint, and filing lawsuits. The National Rifle Association, for example, launches major letter-writing campaigns whenever gun control legislation is introduced. Many environmental organizations have taken legal action against companies that violate pollution laws and have forced some government regulatory agencies to investigate illegal environmental practices.

Doesn't the public have any influence over its government? Political pressure is routinely channeled through voting, despite the limitations of the ballot box. Two recent cases, moreover, demonstrate that an aroused public can have a political voice that is heard above the din of special interest groups. The popular will was credited with the passage of the Brady handgun bill in 1993 and the ban on the sale of assault weapons in 1994, despite strong opposition by the National Rifle Association (Lacayo 1993; J. Alter 1994). The tobacco lobby is also on the wrong end of public opinion. For example, 96 percent of the American public believes that smoking should either be banned from workplaces or restricted to special areas (C. J. Farley 1994). In part as a response to public opinion, steps are being taken to control cigarette smoking. Consider the national level alone. In 1994, President Clinton signed a bill prohibiting smoking in all public and some private schools. The U.S. Department of Defense, in that same year, banned smoking in all military work spaces, whether they are on military bases or inside battlefield tanks. The Smoke-Free Environment Act bans smoking in all buildings entered by ten or more people each day. All structures—bars, restaurants, workplaces—except homes would be smoke-free zones. Violators would face fines of up to $5,000 a day. In 1994, a panel of the Food and Drug Administration concluded that nicotine is an addictive drug, setting the stage for possible regulation of nicotine products (Infante 1994). Regulation of nicotine has been publicly supported by the American Medical Association. State and local governments are also becoming more aggressive, as are many private companies such as McDonald's and many owners of professional baseball franchises (C. J. Farley 1994; Solomon 1994). During 1997, a group of state attorneys general negotiated a settlement with the tobacco companies that called both for the payment of $368 billion and for measures designed to reduce smoking by children. For their part, the tobacco companies would be granted immunity from future civil lawsuits claiming damages (J. Hall 1997). The U.S. Congress failed to approve the settlement, but in 1999 state attorneys general negotiated civil settlement for all states; the settlement was worth $206 billion.

These two illustrations, of course, are not meant to pronounce dead the gun and tobacco lobbies. According to functionalism and the pluralist model, interest groups will continue their influence. They are part of the fabric of the American political system. At the same time, claim the functionalists, don't assume that the American public is excluded from all important political decisions, because the public can express its will through a strong collective voice. There is, however, a competing viewpoint (J. Q. Wilson 1995).

Elitism: The Conflict Approach

According to **elitism**, power is concentrated in the hands of a relatively few individuals and organizations. This approach, of course, has its roots in the conflict perspective.

What is the power elite? The pluralist model has been dominant in American social science since the mid-twentieth century. But C. Wright Mills (1956) and others (F. Hunter 1953, 1980) developed an influential alternative view. According to Mills, the United States no longer has separate economic, political, and military leaders; rather, the prominent people in each sphere overlap to form a unified elite. Whereas most pluralists saw two levels of power—special interest groups and the public—Mills outlined three levels. Overshadowing special interest groups and the public is the **power elite**—a unified coalition of top military, corporate, and government leaders (the executive branch in particular). Mills placed U.S. senators and representatives at the middle levels of power rather than at the higher levels. It is at the higher levels where decisions are made regarding national policy, war, and domestic affairs.

According to Mills, members of the power elite share common interests and similar social and economic backgrounds. Members of the elite tend to be educated in select boarding schools, military academies, and Ivy League schools; belong to the Episcopalian and Presbyterian churches; and come from upper-class families. Members of the power elite have known each other for a long time, have mutual acquaintances of long standing, share many values and attitudes, and

TABLE 14.3

FOCUS ON THEORETICAL PERSPECTIVES: Characteristics of Two Models of Political Power

This table illustrates the way the functionalist and conflict perspectives view political power. Several key features of the political system are compared. Which theory do you think best describes power in the United States? Explain.

Characteristics	FUNCTIONALIST PERSPECTIVE Pluralist Model	CONFLICT PERSPECTIVE Power Elite Model
Who exercises power?	Bargaining and compromising interest groups	National political, economic, and military leaders
What is the source of power?	Resources of interest groups	Leadership positions in major institutions
Where is power located?	Spread widely among interest groups	Concentrated in hands of elites
How much influence do nonelites have?	Considerable influence on public policy	Very little influence on public policy
What is the basis for public policy decisions?	Goals and values are shared by the general public	Preferences of the elites

intermarry. All this makes it easier for them to coordinate their actions and obtain what they want.

Although advocates of elitism do not always use Marxian analysis, their explanations have often been influenced by Marx's thinking. Like Marx, elitist theorists tend to see political leaders as captives of business and corporate interests (Wolfe 1999). Pluralists, who are more likely to have been influenced by classical theorists of democracy than by Marx, view political leaders as referees among competing interest groups.

In a democratic society, in which power is supposed to rest with the people, the powerful prefer that their activities not be widely advertised. Power, according to conflict theory, is exercised behind the scenes through informal channels. Although the invisibility of power is an obstacle, it does not stop social scientists from investigating it (Zweigenhaft and Domhoff 1999; Parenti 2002).

Is there a national power elite in America? G. William Domhoff (1990, 1996) has spent much of his career investigating a ruling class, as described by Mills. Using such measures as membership in private schools, prestigious urban clubs, and the Social Register, Domhoff concluded that the very top of the upper class actually consists of less than 1 percent of the population (perhaps as little as 0.5 percent). Thomas Dye corroborates this conclusion. He reviewed studies of eminent leaders in political, military, and corporate positions. These institutional leaders are not typical Americans: "They are recruited from the well-educated, prestigiously employed, older, affluent, urban, white, Anglo-Saxon, upper- and upper-middle-class male populations of the nation" (Dye 2001:189). For example, about half of the elite group graduates from only twelve prestigious private universities, only 5 percent are female, and only one-fourth graduates from public universities.

The extensive socioeconomic homogeneity among military, political, and corporate institutional leaders does not *prove* the existence of the power elite. To make a convincing case for the existence of a power elite, it is necessary to document the presence of an elite group that is unified and that cooperates for the promotion of its shared interests.

Although it is impossible to definitively identify a national power elite, conflict supporters offer evidence for its existence. Thomas Dye (2001) identified 7,314 first-rate national positions in thirteen different institutions: education, industrial corporations, banking, investments, insurance, mass media, law, government, utilities, foundations, the military, civic/cultural organizations, transportation, and communications. The "elite" are those individuals who occupy the top-drawer positions in these institutional sectors. They control one-half of America's industrial and financial assets, almost one-half of the assets of private foundations, and over two-thirds of the assets of private universities. In addition, they control the mass media, the most prestigious civic and cultural organizations, and all three branches of the federal government.

At the time of Dye's study, these 7,314 top positions were occupied by 5,778 individuals. The existence of fewer occupants than positions means that some individuals held more than one top position. Approximately 15 percent of the elite held more than one top position at a time. What's more, almost one-third of all top positions were interrelated because some individuals held three or more positions. Some even held six or more top positions at the same time. The members of this inner group of multiple position holders are situated to allow communication with one another and to coordinate the activities of a wide variety of powerful institutions. Still, it must be remembered that 85 percent of the top position holders were not interrelated. While this leaves considerable room for disunity and conflict among the top institutional leaders at any given time, this latitude was offset by the fact that some of these individuals held top leadership positions in a number of institutional areas over their careers. A study by Kevin Hallock found 9,804 director seats in 602 large American corporations occupied by 7,519 persons (Hallock 1997). In

FEEDBACK

1. The model of political power known as ________ depicts political decision making as the outcome of competition, bargaining, and compromise among diverse special interest groups.
2. The model of political power known as ________ contends that political control is held by a unified and enduring few.
3. The majority of political action committees (PACs) are now associated with
   a. labor unions. c. political parties.
   b. citizen groups. d. corporations and other businesses.
4. Interest groups are an important means by which the nonelite can influence political decision making. T or F?
5. The operation of PACs in America today invalidates the pluralist model. T or F?
6. According to C. Wright Mills, which of the following is *not* part of the power elite?
   a. military organizations c. large corporations
   b. U.S. senators and representatives d. executive branch of the government

Answers: 1. pluralism 2. elitism 3. d 4. T 5. F 6. b

addition, we know that prominent people move from business to government, from government to universities, from universities to foundations, and so on (Useem 1984; B. Mintz and Schwartz 1985; Zweigenhaft and Domhoff 1999).

Economic Systems

The earlier discussion of interest groups established the fact that many groups are tied to corporate and other business interests. Discussions of elitist and pluralist models revealed that business interests play an important role in the distribution of power. These two points illustrate the close link between the polity and the economy. Societies differ, however, regarding the kinds of political and economic systems they develop. Variations in economies play particularly important roles in relationships between nations because fundamental differences in worldviews are likely to underlie differing solutions to problems associated with the production and distribution of goods and services. Because preindustrial societies were covered in Chapter 5, industrial societies are featured in this chapter.

What is the basic distinction between capitalism and socialism? **Capitalism**, as an economic system, is founded on two assumptions: the sanctity of private property and the right of individuals to profit from their labor. Successful individuals, according to capitalists, deserve to own and control land, factories, raw materials, and the tools of production. In addition, economic success is most likely promoted by free competition with no government interference. Other economic systems assume that property belongs to the society. These economies are carefully planned by a centralized government. This latter type of economy, in which production is owned by the people and the economy is controlled by the government, is called **socialism** (Gregory and Stuart 1989; Schnitzer 2000).

When we speak of the political/economic systems of capitalism and socialism, we are referring to ideals rather than actual operating systems. As we saw in Chapter 6, much can be learned by considering ideal types that involve an isolation of the most basic characteristics of some social entity. No society is likely to embody all the principles described in the ideal type, but a contrast of ideal types can help us understand how societies become very different by placing different degrees of emphasis on particular sets of ideals (Schnitzer 2000).

Capitalism

Capitalist societies are founded on the belief that private individuals and organizations have the right to pursue their own private gain and that society will benefit from their activities. Although economic activity under capitalism is motivated by the pursuit of profit, the individuals and organizations that are pursuing profit also run the risk of losing money:

> *In a typical year* [in the United States] *about four out of ten corporations report net losses. Of ten business firms started in an average year, five close down within two years, and eight within ten years—lack of success being the main reason.* (Ebenstein, Ebenstein, and Fogelman 2000:132)

This combination of profit seeking and risk taking is possible only under certain economic ownership arrangements.

Risk taking and profit seeking are at the center of capitalism. Nowhere are these defining characteristics more apparent than on the floor of a stock market such as the Chicago Mercantile Exchange.

© Joseph Sohm; ChromoSohm Inc./CORBIS

Who owns the property in a capitalist system? Under capitalism, most property belongs to private individuals and organizations rather than to the state or the community. Private ownership is based on a conviction that people have an inalienable right to hold and to control their own property. While advocates of capitalism recognize that private ownership leads to concentrations of power, they believe that distributing power among a large number of owners trumps concentrating power in the state. Moreover, because of private ownership, the pursuit of profit is possible. And, defenders of capitalism argue, private ownership benefits society.

How is capitalism thought to benefit society? According to Adam Smith, eighteenth-century Scottish social philosopher discussed in Chapter 1, a combination of both private ownership of property and the pursuit of profit brings advantages to society. Because of competition from other capitalists, Smith stated, individual capitalists will always be motivated to provide the goods and services desired by the public at prices the public is willing and able to pay. Capitalists who produce inferior goods or who charge too much will soon be out of business because the public will turn to their competitors. The public, Smith reasoned, will benefit through the "invisible hand" of market forces. Not only will the public receive high-quality goods and services at reasonable prices, but capitalists will always be searching for new products and new technologies to reduce their costs. As a result, capitalist societies will use resources efficiently. Efforts by government to control the economy will distort the economy's ability to regulate itself for the benefit of everyone.

Do contemporary capitalist economies work this perfectly? We cannot say how perfectly or imperfectly a purely capitalist economy operates, because none exists. At the same time, the United States is generally considered the purest existing example of a capitalist system. Most Americans view their own affluence as evidence that capitalism works.

What are some deficiencies of the capitalist model? In practice, there are important deviations from the ideal capitalist model, one of which involves the tendency to form a **monopoly**, a single company controlling a particular market, or an **oligopoly**, a combination of companies controlling a market. This muzzling of competition is reflected in several ways. When capitalist organizations experience success, they tend to grow until they become giants within their particular industries. When this happens, new organizations and entrepreneurs find it difficult to enter the industry. There is little hope of competing on an equal basis. Worse yet, the creation of monopolies and oligopolies permits price fixing, leaving consumers with the choice of either buying or not buying a good or service (Ebenstein, Ebenstein, and Fogelman 2000).

A recent example of monopolistic practices in the U.S. economy involves the Microsoft Corporation. Microsoft manufactures, among other products, the Windows operating system—by far the most popular operating system for personal computers. Computer manufacturers typically include Windows on the machines they sell. In the 1990s, Microsoft began to insist that these manufacturers include its Internet browser, Explorer, on their computers as well. Not only that, but the manufacturers were not to install another browser in addition to Explorer. If they refused, Microsoft would withhold their license to sell Windows on their computers. Because Microsoft had so much power over computer manufacturers, other makers of Internet browsers, such as Netscape, were essentially excluded from the market (Chandbasekaran 1999). Eventually, the federal government took Microsoft to court, where it was decided that Microsoft was engaging in monopolistic practices. Microsoft reached a settlement with the federal government and nine state governments in November 2001. However, nine other state governments are contesting this settlement, and terms of the remedies required of Microsoft are still being negotiated (*The Economist* 2002).

Also, the rise of widespread stock ownership can affect the profit motive. When a firm is owned and operated by one or a few individuals, the motivation to make a profit is usually very high. When a company has thousands of shareholders and the managers are not the owners, the profit motive may have to compete with other motives, such as company growth or self-aggrandizement by the managers.

How does the government contribute to the U.S. economy? Adam Smith is often misinterpreted as saying that government should take a hands-off approach where the economy is concerned. Although Smith did strongly oppose overregulation by government, he reserved a place for it. In fact, the U.S. government has always been involved in the workings of the economy.

The Constitution expressly provided a role for the national government in the promotion of a sound economy. Government functions included the regulation of commerce, development of a strong currency, creation of uniform standards for commerce, and the provision of a stable system of credit. In 1789, Congress supported our shipping industry through a tariff on goods imported by foreign ships, and since this initial move into the economy, the federal government has continued to augment business, labor, and agriculture. For example, the federal government aids private industry through loan guarantees, as in the 1980s government guarantee (up to $1.5 billion) that bailed out the Chrysler Corporation. U.S. labor is strengthened by the government through minimum wages, maximum working hours, health and safety conditions, and unemployment support. The agricultural industry feels the influence of government through price controls, price supports, and grain embargoes against other countries. Small farmers and agribusinesses receive financial assistance that runs into the billions of dollars each year (Katznelson, Kesselman, and Draper 2001; T. E. Patterson 2002b).

Consider other examples of government economic and regulatory assistance:

- Public utilities are often owned and operated by state or local governments.
- Antitrust legislation exists to control the growth of corporations.
- The federal government is heavily involved in the defense industry.
- Businesses rely on publicly financed roadways, airports, and waterways.
- Publicly funded public schools, colleges, and universities supply business with its skilled workforce and provide basic research for product development.
- The U.S. military protects American international business interests.
- Government supports business through tax breaks.
- Legislation requires labor and business to follow labor laws.

Socialism

The quest for private property and the pursuit of profit, according to many critics of capitalism, leads to serious inequities because workers are paid less than the value of their products. In keeping with Karl Marx, contradictions within the capitalist system would lead to its overthrow by the proletariat, who would replace it first with socialism and later with communism. Marx, however, was not the first to propose a socialist system. Plato's *Republic* and some portions of the Bible also contain elements of socialist thought.

Who owns property in a socialist system? According to socialist thinkers, the means of production must be removed from the hands of wealthy capitalists. The state, as the people's representative, should own and control property. The state is expected to ensure all members of society a share in the monetary benefits.

How is socialism thought to benefit society? Capitalist thinkers, of course, see a role for the state in a capitalist economy—military protection, a system of criminal justice, and a limited number of public works and institutions (highways and public schools, for example). Socialist thinkers envision a far more ambitious role for the state. They see state control of the economy as a means for the people to maintain control over the production and distribution of goods.

Socialist theory also points to important benefits for workers. Whereas workers under capitalism receive wages below the value that their labor produces and experience little control over their work, workers under socialism should profit because both the state and the workplace exist for their benefit (J. E. Elliott 1985).

Do contemporary socialist economies work this perfectly? Cases of pure socialism have been as rare as cases of pure capitalism. In the socialist economy of the former Soviet Union, for example, some agricultural and professional work was performed privately by individuals who worked for a profit, and significant portions of housing were privately owned as well. Managers received salaries that were considerably higher than those received by workers, and managers were eligible for bonuses such as automobiles and housing.

Private enterprise existed in Poland under Russian communist rule. Service businesses, such as restaurants and hotels, had a significant degree of private ownership. Hotels, in fact, were typically built and managed by multinational chains. Because Poles could travel abroad, they formed business relationships, learned about capitalist methods, imported goods to fill demand, and brought back hard currency. They then used the hard currency earned abroad to create private businesses (Schnitzer 2000).

Some socialist societies even introduced economic incentives for higher productivity—a step that would seem to encourage the workers' desire for profit. Socialist systems have not been successful in eliminating income inequalities, nor have they been able to develop overall economic plans that guarantee sustained economic growth.

Mixed Economic Systems

Because capitalist and socialist economic systems are based on different assumptions, their relationships are often strained. The historic tension between the United States and Russia illustrates the point. This conflict is often predicated on seeing the other as a pure case. In reality, however, every nation violates some of the assumptions of its economic system, and most nations fall between the extremes of capitalism and socialism. Countries in Western Europe, for example, have developed capitalist economic systems in which both public and private ownership play important roles. In these nations, highly strategic industries (banks, transportation, communications, and some others) are owned and operated by the state, while other industries are privately owned but are more closely regulated than in the United States (Esping-Andersen 1996; C. D. Harris 1997; Ollman 1998; Amsden 2001).

Is there movement toward mixed systems in socialist societies? As the former Soviet Union lost control over its republics and Eastern Europe, many of the formally socialist countries began to move toward a market system (Elster, Offe, and Preuss 1998; Stark and Bruszt 1998). Czechoslovakia, in several ways, has shifted from public to private ownership of businesses. Private property nationalized after the Russians took over in 1948 has been returned to the original owners or their heirs. These assets that moved from the public to the private sector were valued at about $5 billion. Many small shops and businesses have been sold in public auctions. In 1992, Czechoslovakia sold over one thousand of its bigger state enterprises to its citizens. During 1992–1993 alone, 25 percent of the nation's assets were privatized. In Hungary, state-owned enterprises have been allowed to become privately owned companies. Over 1 million Hungarians have been given the right to buy land, businesses, buildings, or other property taken over by the Russians in 1949. Nearly all of the state-owned small businesses are now in the hands of private owners. Agricultural cooperatives have also been privatized (Schnitzer 2000). In 1991, Cuba's Communist Party adopted some degree of capitalism by permitting handymen, plumbers, carpenters, and other tradespeople to work for profit.

Economic change has not been limited to Soviet-bloc countries. In 2002, the Communist nation of North Korea announced the creation of a free capitalist zone near its Chinese border. Freed of central government control for fifty years, this zone will solicit private capital from its neighbors and the West (French 2002). After the death of Mao Zedong, the longtime communist leader of the People's Republic of China, the Chinese began market-related economic reforms. The changes initiated in 1979—restoration of financial incentives, free-market prices, education, and economic relations with the West—accelerated in the early 1980s. Private enterprise was legitimated in the Chinese Constitution in 1999. Market forces increasingly replace central economic planning as farmers are permitted to work the land freely (rather than pay the government a predetermined amount of produce), markets and local entrepreneurship are allowed to develop, foreign businesses are encouraged to open facilities on the mainland, and capitalists are permitted to join the Community Party (Nee and Stark 1989; Ollman 1998; Muldavin 1999; Pomfret 2001).

High inflation and public demand for greater political freedoms (to match the newly granted economic liberation in the late 1980s) led to political turmoil. Communist hardliner Li Peng gained control after a popular revolt in Tiananmen Square in June 1989 was suppressed. Mainland China, nevertheless, did not revert to central economic planning. It continues its rapid move toward what its now deceased leader, Deng Xiaoping, called the creation of a "socialist market economy" (Nelan 1993a, 1993b; Schell 1997). When ownership of the port city of Hong Kong reverted to China in 1997, the Chinese government pledged not to interfere with the city's capitalistic economy and so far has lived up to that promise (Kraar 1997). Most recently, China's Community Party began a path to "state capitalism." In this system, the government promotes the success of selected companies via supports such as tax breaks, contracts, and land deals. The payback from these companies comes in the form of party and government loyalty, taxes, and kickbacks (Pomfret 2002). In 2002, China's party revised its constitution to permit membership of private entrepreneurs, formerly considered capitalist "exploiters" (Liu 2002).

Mainland China is attempting to develop a mixed economy, or what has been termed a "socialist market economy." One test of mainland China's commitment to capitalism lies in its keeping the promise it made after the 1997 takeover: a promise not to interfere with Hong Kong's capitalistic economy.

According to some scholars, the demise of communism ensures the triumph of capitalism. The twenty-first century, it is argued, will see a global economy dominated by capitalism (Fukuyama 1991; Thurow 1994; Reich 1993, 1997, 2001). Whether or not socialism passes from the economic and political scene, capitalism is clearly spreading.

The Global Cultures of Capitalism

What are the cultures of capitalism? Although several variants of capitalism can be identified (Hampden-Turner and Trompenaars 1995), they can be collapsed into two basic types. **Individualistic capitalism** is founded on the principles of self-interest, the free market, profit maximization, and the highest return possible on stockholder investment. This variant of capitalism dominated the world in the nineteenth century (the United Kingdom) and twentieth century (the United States). European and Japanese capitalism of the late twentieth century is known as **communitarian capitalism**—the type of capitalism that emphasizes the interests of employees, customers, and society. As will be seen, these two types of capitalism deserve the label "cultures of capitalism" because each rests on a distinctive value base.

What is culturally distinctive about individualistic capitalism? Individualistic capitalism is based on the ideas of John Locke and Adam Smith. Locke, a seventeenth-century philosopher who strongly influenced the leaders of the American war for independence and the nature of the U.S. Constitution, placed individual rights, including the right to own property, at the heart of his *Second Treatise of Government* (1980; originally published in 1690). Government is needed to protect the property of individuals from others, Locke argued, but it must have very limited powers. Although Smith has been widely misinterpreted (Shepard, Shepard, and Wokutch 1991; Werhane 1991; Muller 1993), some of his ideas were used to shape individualistic capitalism and continue to be used to maintain it. The concept of the "invisible hand," mentioned only once in his *Wealth of Nations* (2000; originally published in 1776), has been particularly influential in individualistic capitalism. According to the popular interpretation, if individuals are released to pursue their own self-interest, their separate activities will, through the free-market mechanism, automatically benefit everyone else by increasing the supply of consumable goods and services.

These cultural underpinnings are easy to observe in individualistic capitalism (Thurow 1994). Both individuals and firms have their own economic strategies for success. The individual wants to prosper, and firms wish to satisfy their stockholders. Firms seek to maximize profits because shareholders invest to maximize profits. Customers and employees are the firm's means to reach the goal of higher shareholder return. Because higher wages cut into profits, firms try to keep them as low as possible and lay off employees whenever necessary. Short-term thinking in the name of profit maximization is the norm. Employees are expected to leave one job for another if they can get higher wages. Employees and employers have no mutual obligations outside the legal employment contract.

Within individualistic capitalism, government is not to interfere with the workings of the free market. Government is supposed to make minimal rules and act as an umpire to settle disputes (Friedman 1982). There is no place for government in investment funding or economic planning. So long as government protects private property rights, the pursuit of profit maximization will ensure the greatest prosperity.

What is the value base of communitarian capitalism? Communitarianism, not to be confused with communism, is also rooted in the ideas of a philosopher. Jean-Jacques Rousseau, writing in the eighteenth century, was, like Locke, a social contract theorist. Both believed that individuals forfeit some of their freedom to obtain protection from a central government. Locke saw the social contract as an individualistic device, but Rousseau thought it a communitarian one. Locke's notion of individualism appeared in the seventeenth century and influenced the American Revolution. Rousseau's idea of the "General Will" emerged in the eighteenth century and helped to fuel the French Revolution. The General Will always puts the community first: "The French state thus became an instrument of intervention in the interest of the General Will, to plan and guide the nation's institutions" (Lodge 1975:13).

Japan shares a communitarian viewpoint with France and many other Western European countries. They each emphasize cooperation, interpersonal harmony, and the subordination of the individual to the community. Roots of Japanese communitarianism, however, lie in its own philosophical and religious ideas and in its unique demographic and geographical situation.

In communitarian capitalism, individuals and firms also have strategies (Thurow 1994). But given the cultural foundations just outlined, those strategies are very different from strategies in individualistic capitalism. Individuals in communitarian capitalism attempt to select the best firm and then work hard to be part of the company team; personal success or failure is identified with the fate of the company. Job switching is not widespread. In fact, in many Japanese firms, the person who voluntarily leaves a company is a "traitor" ("Graduates Take Rites of Passage into Japanese Corporate Life" 1991). At the same time, both Japanese and German workers are less often laid off.

Communitarian firms place the interests of shareholders behind those of their employees and customers. Because employees are of first importance, high wages are a top priority. Maintenance of high wages and job security supersede profits. Long-range planning is emphasized over profit maximization, and shareholders, in fact, earn relatively low returns.

In both Europe and Japan, the government is expected to play a significant role in economic funding, planning, and growth. There is cooperation between business and government as well as between business and labor. Labor officials sit on the boards of directors of German firms, and German banks are major stockholders in German companies. Job training is the responsibility of both business and government. In short, there is no line of demarcation between private and public interests.

The contrast between the two cultures is conceptually clear, though in actual practice, it is more complicated. Individualistic capitalists in Britain, Holland, and the United States believe that by concentrating on individual self-interest, they will automatically better serve customers and society. Communitarian capitalists, on the other hand, assume that by concentrating on serving customers and society, they will automatically serve their own interests.

Which culture of capitalism is superior? Some scholars believe that in the long run, communitarian capitalism outperforms individualistic capitalism (Lodge and Vogel 1987; Albert 1993; Thurow 1994). In fact, according to George Lodge, the American economy has already moved, albeit kicking and screaming, toward communitarian capitalism (Lodge 1986). However, only future unfolding events will answer our question.

The Corporation

Within capitalist and mixed economic systems, corporations are extremely important actors. In their provocative book, *America Inc.*, Morton Mintz and Jerry Cohen contend that large corporations now own and operate the United States. Government power is a derivative of corporate power:

FEEDBACK

1. Which of the following is *not* characteristic of capitalist thought?
   a. Private individuals have a right to pursue private gain.
   b. Society will benefit from attempts to make a profit.
   c. The state must make an active attempt to control the economy.
   d. Individuals pursuing a profit must be willing to risk losing money.
2. Under ________, most property belongs to private individuals and organizations rather than to the state or the community as a whole.
3. As capitalist systems mature,
   a. the economy is increasingly dominated by small businesses.
   b. the willingness to take risks increases.
   c. consumer goods become very scarce.
   d. many industries are dominated by a few large firms.
4. According to economic theorists, which system should have the greatest degree of control by the state?
   a. capitalism　　c. democratic socialism
   b. socialism　　d. welfare capitalism
5. Proponents of socialism contend that it will
   a. prevent workers from exerting significant control over social policy.
   b. lead to increased profits for owners.
   c. allow workers to exert significant control over work organizations.
   d. result in very inactive governments.
6. Currently, there seems to be a movement toward mixed economic systems in socialist societies. T or F?
7. Several characteristics of capitalist economic systems are listed here. Indicate in the space beside each characteristic whether it best describes individualistic capitalism (IC) or communitarian capitalism (CC).
   ____ a. government and business cooperation
   ____ b. high job mobility
   ____ c. profit maximization
   ____ d. more value placed on the interests of employees and consumers
   ____ e. self-interest paramount
   ____ f. long-range economic planning

Answers: 1. c 2. capitalism 3. d 4. b 5. c 6. T 7. a. (CC) b. (IC) c. (IC) d. (CC) e. (IC) f. (CC)

The danger that this superconcentration poses to our economic, political and social structure cannot be overestimated. Concentration of this magnitude is likely to eliminate existing and future competition. It increases the possibility for reciprocity and other forms of unfair buyer-seller leverage. It creates nation-wide marketing, managerial and financing structures whose enormous physical and psychological resources pose substantial barriers to smaller firms wishing to participate in a competitive market. . . . [This] superconcentration creates a "community of interest" which discourages competition among large firms and establishes a tone in the marketplace for more and more mergers. This leaves us with the unacceptable probability that the nation's manufacturing and financial assets will continue to be concentrated in the hands of fewer and fewer people. (Mintz and Cohen 1972:39)

The political and economic effects of modern corporations are seen at the local, national, and international levels, a point discussed later. For now, let's examine the nature of modern corporations.

The Nature of Modern Corporations

The economy of the United States has experienced substantial and continual growth. The gross domestic product (GDP)—the domestic output of goods and services—has steadily increased since 1933. This increase in the GDP was accompanied by an increasing domination of the economy by large business corporations. While the emergence of giant business organizations contributed to economic growth, it also created problems. These problems will be better understood if we first examine the nature and extent of modern corporations.

What is a corporation? A **corporation** is an organization owned by shareholders who have limited liability and limited control over organizational affairs. A key feature is the separation of ownership from control (Berle and Means 1968; originally published in 1932). Although shareholders receive reports of gains and losses from the corporation's transactions, they are not legally liable for them. And although shareholders are formally entitled to elect a board of directors, in reality they routinely approve the candidates recommended by the existing board. The actual control of a corporation rests with the board of directors and the corporate managers. Some critics contend, however, that this widely assumed separation between ownership and management is a myth.

How important are corporations today? At the beginning of the twentieth century, only a few American industries, such as railroads, shipping, steel, oil, and mining, were organized as corporations (Roy 1999; Perrow 2002). Today, U.S. corporations dominate the economy, providing everything from diapers to retirement communities.

Total corporate assets are concentrated in the hands of a relatively few giant corporations. The top 100 corporations, which account for less than 0.1 percent of all corporations in the United States, control over 9 percent of total corporate assets; the top 200 corporations control nearly 11 percent. This has changed little since 1960. There is also a very high concentration of assets *within* the top corporations. The top 100 corporations have three-fourths of the assets held by the top 500 corporations, and the top 200 corporations possess 88 percent of the assets held by the top 500 corporations. Although only 0.2 percent of American corporations have assets of $250 million or more, these corporations control over 81 percent of the corporate assets in the United States. In sharp contrast, the 98.5 percent of corporations with assets of less than $10 million control only about 7 percent of the nation's corporate assets.

The Effects of Modern Corporations

Corporations represent massive concentrations of economic resources. Because of their economic muscle, corporations like Microsoft, AT&T, and General Motors make their voices heard by government decision makers. Many government policies, such as those regarding consumer safety, tax laws, and relationships with other nations, reflect corporate influence.

How do corporations affect political decision making? The tremendous political influence of corporate officials exists for several reasons. First, corporate leaders develop influential political and social ties. James Treat, the businessman whom Floyd Hunter found to be an influential adviser to Georgia's politicians, described his status in these terms:

I am asked occasionally to appear before state legislative committees, but I am not registered as a lobbyist. I just happen to know a lot of people, and I get called in on a lot of things. Sometimes, I think, on too many! (F. Hunter 1953:164)

Mr. Treat, of course, knew a multitude of people—and exerted influence—primarily because of his business position and the business activities of his relatives. On the federal level, many influential, wealthy individuals alternate between high corporate and political positions. For example, Robert McNamara was president of the Ford Motor Company before becoming secretary of defense in the Kennedy administration; Michael Blumenthal was chairman of Bendix Corporation before becoming Jimmy Carter's secretary of the treasury; and Donald Regan was chairman of the board and chief executive officer of Merrill Lynch before entering Ronald Reagan's cabinet as secretary of the treasury. In addition to being a CEO himself, George W. Bush picked a former CEO as his running mate and appointed three former CEO's to his cabinet and one as his White House Chief of Staff (M. Moore 2001; Page 2002). The movement between government and corporate positions occurs on lower levels as well, as indicated by the National Association of Home Builders' tendency to hire former Federal Housing Authority (FHA) officials.

Second, corporate officials—because of their personal wealth and organizational connections—are able to reward or punish elected government officials through investment decisions. It is unlikely, for example, that Detroit's $500 million Renaissance Center would have been constructed without the personal and financial support of Henry Ford II. On a more ominous note, financially troubled International Harvester extracted considerable concessions from both Springfield, Ohio, and Fort Wayne, Indiana, by announcing that it would close its plant in one of the two cities. The decision to allow the Springfield plant to remain open was reached only after the city found $30 million in local and state funds to buy the plant and lease it back to the company.

Third, in addition to affecting political processes through investment decisions, wealthy members of society, including corporate officials, affect a political candidate's chance of being elected in the first place. State and national politicians hesitate to jeopardize their chances for election (or reelection) by offending corporate officials or wealthy potential contributors, who are frequently the same people (Keen 2002).

Fourth, the political clout of large corporations is multiplied because of **interlocking directorates**—members of corporations sitting on one another's boards of directors. Although interlocking directorates are illegal if the corporations are competitors, this does not prevent noncompeting corporations with common interests from forming interlocking directorates. For example, members of the General Motors board of directors are also on the boards of many other corporations, including Eastman Kodak, Bristol-Myers Squibb, and Merck and Company. Moreover, these interlocks are only the beginning. The corporations having interlocking directors with General Motors have interlocks with hundreds of other corporations. In addition, there are indirect interlocks that permit two board members of competing firms to be members of a third corporation. It is not difficult to imagine the political power emanating from this spider's web of interlocks among America's most powerful corporations (Mintz and Schwartz 1981a, 1981b, 1985; Useem 1984; Dye 2001).

Fifth, the political power of corporations is enhanced through conglomerates. A **conglomerate** is a network of unrelated businesses operating under a single corporate umbrella. The parent corporation does not actually produce a product or provide a service; it simply has controlling interest in a set of diverse enterprises. RJR Nabisco, Inc., for example, owns companies in such different areas as tobacco, pet foods, candy, cigarettes, food products, bubble gum, research, and technology. A list of the company's North American subsidiaries covers almost an entire page in *Who Owns Whom* (1998; *Lexis-Nexis Corporate Affiliations* 2002).

The informal contacts and global power, then, give corporations important advantages in the political decision-making process. And although corporate officials may disagree with one another and may not share all of the same goals, they have the ability, when they see their interests threatened, to effectively use their organizational resources. This influence is reflected in what critics call **corporate welfare**—economic benefits government regularly gives to corporations. Corporate welfare comes in many forms, including tax breaks, subsidies, and lax regulation and law enforcement for white-collar crimes. Mark Zepezauer and Arthur Naiman (1996) estimate that the government hands out about half a trillion dollars annually to big business and wealthy individuals.

Finally, the political influence of corporations is not confined to their countries of origin. The world is increasingly influenced by **multinational corporations**, firms in highly industrialized societies with operating facilities throughout the world. Multinationals are concentrated in a few industries, each controlled by a small number of giant companies. The oil and automobile industries alone account for about half of all multinational activity. Although multinational corporations existed for centuries, they became more powerful after World War II. Improvements in communication and transportation technology allowed them to exert wide control over their worldwide operations.

How economically powerful are multinational corporations? Of the world's hundred largest political and economic entities, fifty-one are multinational corporations rather than nation-states. Seven corporations based in the United States—ExxonMobil, IBM, General Motors, Ford Motors, AT&T, Wal-Mart Stores, and General Electric—have sales volumes on a par with the gross national products of the world's forty most productive nations. Nations ranking below both General Motors and Ford Motors include Argentina, South Africa, Venezuela, Greece, Portugal, Israel, and Ireland (Steedley 1994). If the 300 largest multinational corporations were to band together, their combined economic resources would be exceeded only by the combined wealth of the United States and Japan (*World Development Report* 1994). Table 14.4 compares some multinational corporations with selected nations.

TABLE 14.4

Total Revenue of Selected Multinational Corporations versus Gross Domestic Product of Selected Nations

This table compares the revenue of selected multinational corporation in comparison to the gross domestic product (value of all goods produced and consumed domestically) of selected nations. State the most important conclusion that can be drawn from these data. Were you surprised by any of the information?

Wal-Mart Stores, Inc. $119.3 billion	vs.	Greece $119.1 billion
Volkswagen AG $65.3 billion	vs.	New Zealand $65 billion
IBM International Business Machines Corp. $78.5 billion	vs.	Egypt $75.5 billion
Mitsubishi Corporation $128.9 billion	vs.	South Africa $129.1 billion
Sony Corporation $55 billion	vs.	Czech Republic $54.9 billion
General Electric Company $90.8 billion	vs.	Israel $92 Billion

What are the negative effects of multinational corporations? Defenders of multinationals argue that the corporations provide developing countries access to technology, capital, foreign markets, and products that would otherwise be unavailable. Critics claim that multinationals actually harm economies of host nations—exploiting their natural resources, disrupting local economies, introducing inappropriate technologies and products, and increasing income inequality. Multinationals, these critics note, often rely on inexpensive labor or abundant raw materials in the developing nations while returning their profits to corporate headquarters and shareholders in rich nations. Multinationals' domination of their industries make it difficult for the economically developing nations to establish new companies that can effectively compete with the multinationals. As a result, multinationals may retard rather than promote the economic development of some regions of the world.

Although the evidence is not always clear, these firms do possess huge concentrations of power whose abuses cannot always be checked, even by national governments. According to critics, evidence suggests that executives of International Telephone and Telegraph Corporation (ITT) worked with elements of the U.S. government in an effort to overthrow the government of Chile, headed by Salvador Allende (J. Anderson 1974; Wise 1976). Although such actions may be atypical of multinationals, they illustrate the difficulties faced by a government wishing to control the multinationals within its borders.

The available evidence suggests that some problems introduced by multinationals are not effectively controlled. For example, Haitian sugarcane cutters employed by Gulf & Western in the Dominican Republic are said by some observers to

> *labor under the most inhuman conditions known in Latin America in the twentieth century. They are transported to the Gulf & Western plantation in huge trucks, receiving only cane or a mixture of cane juice and brown sugar for nourishment. Their drinking water is the same water used to wash animals, and only 57 percent of the shacks have latrines.* (Kowalewski 1982:144)

Environmental impacts are similarly problematic. For example, in Jamaica such companies as Alcoa and Revere are polluting the lakes with "red mud," a by-product of aluminum-producing activities. Union Carbide and Sun Oil refineries operating in the Caribbean are linked to increased incidences of respiratory disease. Products marketed by multinationals can also create problems when they are incompatible with local conditions. For example, an infant formula marketed by Nestlé led to significant illness and death in several low-economy nations when the dry formula was mixed with contaminated local water supply.

Finally, multinationals often move operations to less developed countries that offer cheap labor and weak regulatory controls. This could severely affect the U.S. economy, as well as economies of other developed nations.

FEEDBACK

1. A __________ is an organization owned by shareholders who have limited liability and limited control over organizational affairs.
2. Which of the following is *not* one of the reasons behind corporate political influence?
   a. Corporate managers and directors have informal and organizational ties that lead them to be consulted about matters of political policy.
   b. Many corporate officials have reward and coercive power over elected government officials.
   c. Many elected officials are also top corporate officials.
   d. Interlocking directorates exist.
   e. Conglomerates are present.
3. Indirect interlocks permit two board members of different corporations to sit on each other's boards of directors. T or F?
4. Firms with operating facilities in several different countries are called __________.
5. Multinational corporations are concentrated in a few industries, each dominated by a small number of giant companies. T or F?
6. Multinational corporations
   a. are easily manipulated by the governments of low-economy nations.
   b. sometimes create problems involving unsafe working conditions and inappropriate products.
   c. are known to have overthrown large numbers of low-economy governments.
   d. carefully avoid environmental damage in their areas of operation.

Answers: 1. corporation 2. c 3. F 4. multinational corporations 5. T 6. b

Work in the Contemporary Economy

The United States economy is dynamic. This is reflected in the radical alterations projected for the workforce composition. Also, current change in the occupational structure is expected to accelerate, with some serious consequences (Rifkin 1995; Milkman 1997; Mazaar 1999).

The Changing Workforce

To understand work in modern society, it is necessary to be familiar with the three basic economic sectors: primary, secondary, and tertiary.

How do the economic sectors differ? The **primary sector** of an economy depends on the natural environment to produce economic goods. The types of jobs in this sector vary widely—farmer, miner, fishery worker, timber worker, cattle rancher. In the **secondary sector**, manufactured products are made from raw materials. Occupations in this sector include blue-collar workers of all types, from those who produce computers to those who turn out Pokémon cards, as well as associated white-collar employees in management and office positions. Employees in the **tertiary sector** provide services. If today you went to class, filled your car with gas, stopped by the bank, and visited your doctor, you spent most of your time and money in the tertiary (service) sector. Other service industries include insurance, real estate, retail sales, and entertainment. We generally refer to people in these industries as white-collar workers since they are the predominant type of employee in the tertiary sector.

Historically, how have the three sectors evolved? Obviously, the primary sector dominated the preindustrial economy. At that stage of economic development, most citizens were involved in the battle for economic survival. Very few people could be spared for religious, medical, or educational purposes. And physical goods were made by hand.

This balance began to change with the mechanization of farming in the agricultural economy. Mechanical innovations (cotton gin, plow, tractor), along with the application of science (seed production, fertilization, and crop rotation) reduced the proportion of the labor force needed in the primary sector. During the 1800s, the average farmer could feed five workers or so. Today, the figure is eighty. Not surprisingly, the proportion of workers in the farming sector has declined from almost 40 percent in 1900 to about 2 percent today (see Figure 14.3).

Another change began occurring along with the mechanization of farming. With other technological advancements (power looms, motors of all types, electrical power) came the shift of agricultural workers from farms to factories, ushering in the secondary sector. As Figure 14.3 indicates, the percentage of the U.S. labor force engaged in blue-collar jobs reached almost 40 percent in 1900.

FIGURE 14.3

Changes in U.S. Labor Force by Occupational Category

This figure tracks changes in the U.S. labor force from 1900 to 2000. Describe the most significant finding in these trends, based on *speed* of change.

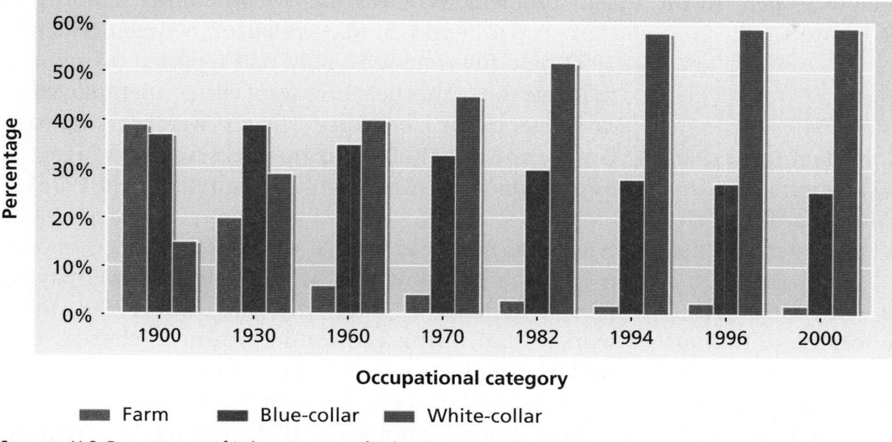

Sources: U.S. Department of Labor, Bureau of Labor Statistics, "News" (Washington, DC: U.S. Government Printing Office, May 6, 1994); and U.S. Bureau of the Census, *Statistical Abstract of the United States: 2001* (Washington, DC: US Government Printing Office, 2001).

INTERNET LINK The U.S. Department of Labor collects a great deal of information on the American labor force. Go to its Web site at www.dol.gov/

Just as in agriculture, technological developments in the secondary sector permitted greater production with fewer workers. After World War II, the fastest-growing occupations were the white-collar jobs—managers, professionals, sales workers, clerical workers. In 1956, white-collar workers for the first time accounted for a larger proportion of the U.S. labor force than did blue-collar workers. In manufacturing industries, white-collar workers are now triple the number of blue-collar workers.

Technological progress did not stop with the secondary sector. As relative growth of workers in goods-producing jobs was decreasing, the demand for labor in the tertiary section was increasing. Fueled by computer technology, the U.S. economy moved from a

The American economy is shifting from assembly-line production with blue-collar workers to a service economy with white-collar employees.

manufacturing base to a "knowledge" base (see Chapter 5). Demand for people who could manage information and deliver services became the new workforce hallmark. This development, along with the growth of white-collar employment in the secondary sector, increased the proportion of white-collar workers in the United States from just below 30 percent in 1930 to about 60 percent today (see Figure 14.3).

Occupational Structure

We read in Chapter 9 about America's dual labor market. The emphasis there was on institutionalized discrimination suffered by minorities. Actually, the implications of the dual labor market are much broader.

What is the nature of the two-tiered occupational structure? One tier—the **core tier**—contains dominant positions in large firms. Computer technology, pharmaceutical, and aerospace firms are prime examples. Between 30 and 40 percent of U.S. workers are in the core. The **peripheral tier** is composed of jobs in smaller firms that either are competing for business left over from core firms or are engaged in less profitable industries such as agriculture, textiles, and small-scale retail trade. Most U.S. workers—60 to 70 percent—are employed in the peripheral tier.

Historically, jobs in the core paid more, offered better benefits, and provided longer-term employment. This is not surprising, since the firms involved are large and highly profitable. Peripheral jobs are characterized by low pay, little or no benefits, and short-term employment (Weakliem 1990; Hudson 2000). These features follow from the weaker competitive position and the smaller size of the employing firms.

How are the core and peripheral tiers changing? During the last twenty years the core industries scaled back, laying off experienced workers. As early as 1983, for example, a steel mill in Hibbing, Minnesota, that once employed 4,400 people had a payroll of only 650 ("Left Out" 1983). Since 1983, the Weirton Steel Company cut its production capacity by 30 percent and laid off more than half of its workforce. In fact, since 1979, more than 43 million jobs were eliminated in the United States. Over 570,000 job cuts were announced in the United States in 1998, more than half of which occurred in manufacturing plants (McNamee and Muller 1998; Riederer 1999). American corporations created more than 21,000 mass layoffs in 2001, affecting nearly 2.5 million workers (Marchese 2002). Of course, as these top-tier jobs have been disappearing, peripheral jobs have become a larger share of the total jobs.

Newer industries, such as those based on microchip technology, are beginning to replace manufacturing jobs in some areas. These industries, however, provide few jobs suited to the skills and backgrounds of laid-off manufacturing workers. Moreover, most jobs in high-tech industries pay minimal wages and offer few chances for promotion. Responsible positions with high pay are held by a very small proportion of high-tech employees.

Thus, reemployment of laid-off workers is a significant problem. Whereas the overwhelming majority of the over 5 million U.S. workers laid off between 1979 and 1992 held full-time jobs, only half reported taking new full-time jobs. Another third were either unemployed or no longer in the labor force. The rest were working part-time, running their own businesses, or engaged as unpaid family workers (Uchitelle and Kleinfield 1996).

What does this mean for U.S. workers? The U.S. economy is losing higher-paying jobs and gaining lower-paying jobs. This helps explain why, since the 1970s, most workers are losing economic ground. Whereas one American worker could support a family thirty years ago, the dual-employed married couple is the norm today.

This process, known as "downwaging," is expected to continue, propelled, in part, by the movement offshore of higher wage blue-collar jobs and higher income white-collar jobs as a result of globalization policies such as NAFTA (North American Free Trade Agreement; Thurow, 2003). Of the ten occupations projected to expand between 1994 and 2005, five pay below the poverty level. Only two of the top ten shrinking jobs fail to meet the poverty threshold (U.S. Bureau of the Census 2001a). Many sociologists believe that these trends of job loss and downwaging will threaten the American Dream (Newman 1993; Barlett and Steele 1996; see Doing Research in Chapter 18).

How useful is the dual labor market perspective? The dual labor market perspective has been criticized for at least two reasons: because some of its assumptions have not been clarified and because it does not always provide accurate predictors of labor market experiences. Studies within each sector reveal considerable variation in experiences of individuals. Despite a need for refinement, however, dual labor market theory alerts us to shifts in industries and firms that will be offering jobs in the future (Hodson and Kaufman 1982; M. R. Kelley 1990; Sakamoto and Chen 1991; Hudson 2000).

Downsizing and Contingent Employment

Clearly, the American labor landscape is changing, due in part to corporate management decisions. Two interrelated steps by management, downsizing and contingent employment, deserve special attention (Mazaar 1999).

Why do corporations downsize? Downsizing refers to the reduction in a corporation's workforce designed to cut costs, increase profits, and enhance stock values. To justify downsizing, business leaders point to several factors: rising health-care costs; the need to become leaner and less hierarchically structured in an increasingly competitive environment; and the necessity to eliminate levels of middle management that can be replaced by computers and information technology (Gleckman et al. 1993; Huber and Korn 1997).

Top management points also to lower profits caused by increasing global competition. And it is true that about 20 percent of all U.S. workers are directly exposed to international competition (McNamee and Muller 1998). Many companies respond by moving operations overseas or by replacing domestic full-time employees with part-time workers.

To what extent is downsizing occurring? Downsizing became popular during the 1990s (Boroughs 1996; Sloan 1996; Belton 1999; Burke and Cooper 2000; Cohen 2001). Since 1985, it is estimated that over 8 million employees were downsized, half of whom held white-collar jobs. In 2001, corporate layoffs increased nearly 40 percent over the previous year (Marchese 2002). For a multitude of reasons, the momentum for downsizing and layoffs continues: the collapse of the stock market, the economic downturn, the exposure of CEO scandals, and the 9/11 terrorist attack (Altman 2002; Foss 2002; Iwata 2002; Noguchi 2002).

How are contingent employees a factor in the downsizing process? Part of the motivation for downsizing is top management's belief that their companies employ a surplus of people and that work can be done by fewer employees without reduction in efficiency and effectiveness. The desire to shrink the size of the workforce, however, is not the only incentive. Part of the impetus for downsizing is the belief that work done by permanent, full-time employees can be done in other, less expensive ways. Profitable companies are increasingly using the services of **contingent employees**—individuals hired on a part-time, short-term, or contract basis. These contingent employees are hired to replace full-time employees ("The New World of Work" 1994; T. S. Moore 1996; Cappelli et al. 1997). The contingent label is appropriate because their employment is contingent on daily, weekly, or seasonal company need. They may also be referred to as "marginal" workers because they are out of the traditional system of job and economic security. Some even refer to them as "disposable" workers (Kirkpatrick 1988; Castro 1993).

Contingent employees now make up over one-third of the U.S. labor force (over 40 million people). Temporary employment has increased over 250 percent since 1982. Temporary employees earn about 60 percent of the amount earned by the average full-time worker and are not entitled to health insurance or retirement benefits. The contingent part of the labor force has traditionally included groups from the bottom of the occupational structure—minorities, women, youth, the elderly. Most contingent workers have been in low-level, low-paying clerical, retail, and service jobs. Increasingly, however, downsizing is affecting higher-level white-collar people—engineers, accountants, nurses, managers, physicians (Heckscher 1995; Rifkin 1995; Altman 2002).

What will be the effects of downsizing and contingent employment? These new trends may portend a fundamental change in the U.S. labor market (Harrison 1994; Heckscher 1995; Osterman et al. 2001). For some it marks the dawn of a brave new workplace of individual entrepreneurialism (Kiechel 1993; Sherman 1993b). Others insist that it constitutes a fundamental change in the implied social contract between businesses and employees (K. M. Dudley 1997; Wallulis 1998; Fraser 2002). As evidence of a firm's need to be lean, adaptable, and flexible, Daniel Bell (1993), representative of the entrepreneurial view, cites the decline of the American steel industry. On the other hand, Robert Kuttner (1993a, 1993b) points out, these corporate actions are fracturing job security, with its underlying social contract. Similarly, Robert Reich (1993, 1997, 2001), former secretary of the U.S. Department of Labor under President Clinton, expresses concern that these actions will create greater polarization between those who control capital and those who do not. As such critics believe, the disposable workforce is the most important trend in business today; it is fundamentally changing the relationship between Americans and their employers. Downsizing and the increased use of contingent workers, they contend, will threaten worker faith in a valued future relationship with a firm (Wolman and Colamosca 1997; Armour 2002; Greider 2002).

For corporations as well, the disadvantages of extensive downsizing may outweigh the advantages. Many companies that downsized did not increase their earnings and are now less competitive (Baumohl 1993). Stress among managers who make and carry out downsizing decisions is an increasing problem. A *Fortune* magazine article entitled "Burned-Out Bosses" concludes with this warning:

> *Constant restructuring has become a fact of business life in this era of change. Well and good, but companies that don't acknowledge the stress that survivors undergo and support those who are in danger of burning out may find that their glistening, reengineered enterprises end up being run by charred wrecks.* (L. Smith 1994:52)

According to a survey of 2,500 employees across the United States, there is a growing split in employees' attitudes toward their work and their employers (*Towers Perrin Workplace Index* 1997). Although employees express higher job satisfaction, there are indicators of

Kathryn Marie Dudley—The End of the Line

Because she grew up near Chrysler's auto plant in Kenosha, Wisconsin, researcher Kathryn Marie Dudley had an interest in studying the cultural fallout from the plant's closing in 1988. She offers Kenosha as a representative example of the cultural consequences of the current restructuring of the U.S. economy. As Dudley indicates, the plant relocations, downsizings, and job eliminations undertaken by large corporations over the past few decades are seen as part of the shift from industrial to postindustrial society:

> *What was once a fundament al segment of the American economic structure—heavy industry and durable goods manufacturing—has now become a marginal part of the national portfolio. As this sector of the economy gives way to the new "knowledge industries," workers in this sector are being superseded as well. In America's new image of itself as a postindustrial society, individuals still employed in basic manufacturing industries look like global benchwarmers in the competitive markets of the modern world.* (Dudley 1997:161)

Dudley's research is a case study of a large plant in a one-industry community experiencing the ongoing shrinking of core jobs in postindustrial society.

During the year following the shutdown, Dudley conducted in-depth interviews with autoworkers and with a wide variety of professionals in the Kenosha area. Interview questions were open-ended to give informants freedom to roam where their thoughts and feelings took them. Dudley's only restriction was that the interviews be geared to the cultural meaning of repercussions suffered by the community because of its declining employment base.

For Dudley, the demolition of the auto plant was a metaphor for the dismantled lifestyle of U.S. blue-collar workers in core manufacturing industries. These increasingly displaced blue-collar workers, contends Dudley, find themselves caught between two interpretations of success. On the one hand, middle-class professionals justify their place in society by reference to their educational credentials and "thinking" jobs. Blue-collar workers, on the other hand, legitimize their place in society by the high market value placed on their hard labor. One ex-autoworker, whom Dudley calls Al Tirpak, captures the idea beautifully:

> *We're worth fifteen dollars an hour because we're producing a product that can be sold on the market that'll produce that fifteen dollars an hour. . . . I don't know if you want to [base a person's value] strictly on education. You can send someone to school for twelve years and they can still be doing something that's socially undesirable and not very worthwhile for society. I don't know if they should get paid just because they had an education. In my mind, yuppie means young unproductive parasite. We're gonna have an awful lot of yuppies here in Kenosha that say they are doing something worthwhile when, really, they aren't.* (Dudley 1997:169)

Due to the massive loss of high-paying factory jobs, Dudley contends that the blue-collar vision of success is coming to "the end of the line." They have lost their cultural niche in a postindustrial world where work is based on education and the application of knowledge.

Dudley documents the blue-collar workers' view of this new reality. From her extensive interviews, she constructs a portrait of their struggle to preserve their cultural traditions in a social world that is removing the type of employment on which these traditions were built. The penalty for not creating new cultural supports for a sense of social worth, Dudley concludes, will be life in a state of confusion with a sense of failure.

Thinking About the Research

1. What does Dudley mean when she says that the blue-collar workers caught in the wave of corporation restructuring are experiencing the shift from industrial to postindustrial society?
2. Do you think that Dudley's research methods are strong enough to support her conclusion?
3. Do you believe that Dudley can be objective in this study of her hometown? Explain.

eroding trust in management. Workers seem to be losing faith in management's commitment to them. For instance, only 41 percent of those surveyed believe that their company considers employees' interests. A sense of injustice is heightened by the growing compensation gap between top corporate executives and employees. While median CEO annual compensation tripled to $6 million between 1993 and 2002, pay for all workers increased only a third (R. J. Samuelson 2003). Ronald J. Carty, then CEO of American Airlines, was forced in early 2003 to resign following disclosure of company plans to pay retention bonuses to executives and shield their pensions in the event of bankruptcy while asking employees to absorb up to 23 percent pay cuts (Goo

FEEDBACK

1. Which of the following is *not* one of the major labor-force trends discussed in the text?
   a. decrease in primary sector employment
   b. decline in agricultural employment
   c. increased employment of service workers
   d. increase in blue-collar jobs
   e. increase in white-collar proportion of the labor force
2. Most of the jobs that were added to the American economy over the past twenty years were in which of the following sectors?
   a. service and white-collar
   b. manufacturing
   c. mining
   d. skilled blue-collar
3. The _________ industrial sector contains large, profitable firms with dominant positions within major industries.
4. The _________ industrial sector contains small, less profitable firms operating in more competitive industries.
5. _________ is the decision on the part of top management to reduce its workforce to cut costs, increase profits, and enhance stock values.

Answers: 1. d 2. a 3. core 4. peripheral 5. Downsizing

2003). Trust and loyalty are difficult to maintain when employees do not believe that their companies' policies treat them fairly (S. Greenhouse 2002).

Correspondingly, organized labor is reacting negatively to contingent hiring. As noted in "Using the Sociological Imagination," the 1997 Teamsters Union strike against United Parcel Service, the first in the company's ninety-year history, was based squarely on union opposition to the UPS policy of increasing contingency employment (60 percent of its workers are part-time). John Sweeny, president of the AFL-CIO, publicly called the policy of contingency work "corporate greed" and cites it as an issue for the labor movement to pursue. The strike did not last very long, in part because the American public generally sided with the Teamsters (L. Greenhouse 1997).

Because research in this area is just beginning, we do not know the full consequences that downsizing and contingent employment will have on either companies or employees. What will happen to employee loyalty, performance, and product quality? Will trust deteriorate between managers and employees? Will the survivors of downsizing suffer more illnesses due to greater workloads and higher stress? Will there be a deluge of lawsuits based on charges of wrongful discharge? Will employees, blue- and white-collar alike, turn more to unions (Watson and Shepard 1997)? Characterizing survivors of downsizing as "victims," William Snizek, in a comprehensive review of existing research, concludes:

> *A growing body of evidence is beginning to accumulate which shows that the deleterious effects of corporate layoffs are as great, if not greater, on those who survive, as they are on those initially laid off.* (Snizek 1994:3)

One conclusion seems clear: if expected economic benefits are to be realized, management will at the very least have to acknowledge the problems of survivors and initiate appropriate remedial responses (Sloan 1996; Cappelli et al. 1997). The needs are basic: maintaining honest and constant communication between top management and survivors of layoffs, setting goals with survivors, further empowering survivors on the job, exercising accessible and sensitive leadership, and creating a positive organizational culture for survivors (O'Neill and Lenn 1995; Snizek and Kestel 1999).

SUMMARY

REVIEW GUIDE

1. The economic and political institutions are closely related. The economy is the institution designed for the production and distribution of goods and services, and it is in the political institution where power is obtained and exercised.
2. Power is the ability to impose one's will on others whether or not they wish to comply. Authority is power accepted as legitimate by those subjected to it. Max Weber identified three types of authority: charismatic authority, which is too unstable; traditional authority, which is too outdated to serve as the primary basis of power for modern nation-states; and rational-legal authority, which is the authority of most nation-states today. The power of most modern government officials, then, is limited by legal rules and regulations.
3. A nation-state is the political entity that holds authority over a specified territory; a government is the political

structure that rules a nation. The nation-state is relatively new in human history, emerging with the rise of agriculture and industrial production. According to functionalism, the nation-state developed to maintain social order. From the conflict perspective, nation-states exist primarily to promote the interests of the societal elite.

4. Democracy and totalitarianism are the two polar types of political systems. In modern societies, democracies are representative with minimal citizen involvement in political affairs. Totalitarian political systems have absolute rulers who control all aspects of social life. Between these two polar types lies authoritarianism. This third type of political system involves nonelected rulers who possess absolute control but frequently permit some individual freedom.
5. Voting seems to have limitations as an effective means for the nonelite to influence political decision making. This is partly because America does not have a high voter participation rate and partly because the most disadvantaged tend to be nonvoters.
6. The two major models of power distribution are pluralism and elitism. Pluralists, whose view is associated with functionalism, depict power in the United States as widely distributed among diverse special interest groups. Interest groups appear to be on the rise as the public becomes more politically sophisticated and as the influence of political parties declines. The growing importance of political action committees (PACs) reflects the same social conditions. Some powerful interest groups, notably those PACs associated with business interests, thwart public opinion by using their power for the attainment of narrow, selfish goals. Still, interest groups provide one of the few mechanisms for the representation of nonelite segments of a society. Some of these interest groups have successfully challenged the political and economic establishment.
7. Advocates of the conflict perspective contend that American society is controlled by a unified and enduring elite. This is a "power elite" comprised of a unified coalition of top military, corporate, and government leaders. Although some evidence documents the existence of a national power elite, its presence cannot be definitively asserted.
8. Economies differ in their organization and underlying assumptions. Capitalist economies are based on private property and the pursuit of profit without government interference. In socialist economies, the means of production are owned by the people, and government has an active role in planning and controlling the economy. There is movement toward capitalism in socialist countries. Capitalist and socialist ideologies appear to be becoming less of a barrier in the relationships among nations.
9. Individualistic capitalism and communitarian capitalism are the two basic variants of capitalism. Self-interest, the free market, profit maximization, and highest return possible on stockholder investment are foundational principles for individualistic capitalism. In communitarian capitalism, the emphasis is on the interests of employees, customers, and society.
10. The rise of the corporate economy has intensified the link between political and economic institutions. Some critics contend that the political institution in the United States is now controlled by vast corporations whose economic assets are owned by a relatively small segment of the population.
11. Corporations are rapidly increasing in size, number, and power. Their influence is felt domestically and internationally. Corporations affect domestic political decision making and also influence the political and economic institutions of countries around the world.
12. Workers in the contemporary economy face a changing labor force composition. The occupational structure is changing as traditional industrial jobs are replaced with white-collar and service jobs. More American corporations are downsizing and replacing full-time employees with contingent, or temporary, workers. Granted, the full implications of these changes for workers and for the future of American society are not yet known, but evidence is beginning to establish some negative consequences.

INFOTRAC® COLLEGE EDITION

http://www.infotrac-college.com/wadsworth

Another unique option available to you at the Student Resources section of the companion Web site is Infotrac College Edition, an online library with access to hundreds of scholarly and popular periodicals. Below are suggested search terms for this chapter. Results from these and other searches are found at the site.

Search keyword: **downsizing**. Read three articles to find out what effects downsizing is having on American society. What does the preponderance of evidence say about the effects of downsizing? What factors have led to the popularity of downsizing in the 1990s and 2000s?

Search keywords: **hydrogen economy**. Jeremy Rifkin has recently written a book entitled *The Hydrogen Economy*. Search Infotrac for information about this. What is meant by this term? Why might it be important to the study of sociology?

Search keywords: **power elite**. C. Wright Mills, an important U.S. sociologist, coined the term *power elite*. Use Infotrac to explore current thinking on this concept. If a power elite does exist, what implications does its existence have for society? Which theoretical perspective does this concept come from?

LEARNING OBJECTIVES REVIEW

- Distinguish among power, coercion, and authority and identify three basic forms of authority.
- Discuss the nation-state, comparing and contrasting the functionalist and conflict approaches to this form of political authority.
- Identify the major differences between democracy, totalitarianism, and authoritarianism.
- Differentiate the views of functionalism and conflict theory on the distribution of power in America.
- Describe the major characteristics of capitalism, socialism, and mixed economic systems.
- Discuss the effects of the modern corporation on American society.
- Describe America's changing workforce composition and occupational structure.
- Discuss corporate downsizing and its consequences.

CONCEPT REVIEW

Match the following concepts with the definitions listed below them.

____ a. individualistic capitalism
____ b. political action committees
____ c. oligopoly
____ d. government
____ e. nation-state
____ f. democracy
____ g. monopoly
____ h. downsizing
____ i. communitarian capitalism
____ j. authoritarianism
____ k. traditional authority
____ l. power elite
____ m. rational-legal authority
____ n. political party
____ o. bourgeoisie
____ p. elitism
____ q. interest group
____ r. corporation

1. the type of political system controlled by nonelected rulers who generally permit some degree of individual freedom
2. members of a society who own productive property
3. the type of political system in which elected officials are held responsible for fulfilling the goals of the majority of the electorate
4. the theory of power distribution that sees society in the control of a few individuals and organizations
5. a group organized to achieve some specific, shared goals by influencing political decision making
6. a situation in which a single company controls a market
7. the political entity that holds authority over a specified territory
8. organizations established by interest groups for the purpose of raising and distributing funds to selected political candidates
9. an organization designed to gain control of government through the election of candidates to public office
10. a unified coalition of top military, corporate, and government leaders
11. authority based on rules and procedures associated with political offices
12. the political structure that rules a nation
13. legitimate power rooted in custom
14. the type of capitalism that emphasizes the interests of employees, customers, and society
15. the decision of top management to reduce its workforce to cut costs, increase profits, and enhance stock values
16. the type of capitalism founded on the principles of self-interest, the free market, profit maximization, and the highest return possible on stockholder investment
17. a situation in which a combination of companies controls a market
18. an organization owned by shareholders who have limited liability and limited control over organizational affairs

CRITICAL-THINKING QUESTIONS

1. Like all organizations, universities are based on some form of authority. Discuss each of the three types of authority and show which type is most basic to the university structure. Be careful to show why each type of authority is applicable or inapplicable to the university's organizational structure.

____

____

2. Functionalism and conflict theory each have a unique perspective on the nature and purpose of the nation-state. Do you think that one of these perspectives more accurately describes American society than the other? Why or why not?

____

____

3. Define the concept of the power elite. Is American society best characterized as a pluralist society or a society controlled by a power elite? Defend your position.

____

____

4. Distinguish between the two global cultures of capitalism. Do you envision a convergence of these two approaches to capitalism within the rapidly developing global marketplace? Support your viewpoint.

____

____

5. American corporations have in recent years turned to the practices of downsizing and contingent employment. State the reasons for this managerial strategy, and discuss whether or not it will promote the organizational goals it is designed to reach.

____

____

MULTIPLE-CHOICE QUESTIONS

1. **Max Weber defined _________ as the ability to control the behavior of others, even against their will.**
   a. authority
   b. charisma
   c. power
   d. elitism
   e. influence
2. **_________ is power accepted as legitimate by those subjected to it.**
   a. Pluralism
   b. Authority
   c. Elitism
   d. Influence
   e. Coercion
3. **The political structure that rules people in a specific territory is known as a**
   a. Leviathan.
   b. nation-state.
   c. democracy.
   d. political party.
   e. government.
4. **According to the conflict perspective, nation-states exist primarily to**
   a. serve the interests of a society's elite.
   b. eliminate tensions between social classes.
   c. solve the Hobbesian problem of order.
   d. perpetuate competition for new economic markets.
   e. establish the legitimacy of charismatic authority.
5. **What two types of political systems exist in nation-states?**
   a. democracy and authoritarianism
   b. totalitarianism and authoritarianism
   c. Republican and Democratic
   d. communism and Republican
   e. democracy and totalitarianism
6. **According to _________, a community or society is controlled from the top by a few individuals or organizations.**
   a. symbolic interactionism
   b. elitism
   c. communism
   d. functionalism
   e. pluralism
7. **Decision making resulting from competition, bargaining, and compromise among diverse special interest groups is based on the _________ model.**
   a. elitist
   b. laissez faire
   c. conflict
   d. pluralist
   e. functional
8. **According to C. Wright Mills, power in the United States is controlled by**
   a. the people through a democratic process.
   b. labor unions that are able to exert covert influence on management and the general public.
   c. top military, corporate, and government leaders.
   d. Fortune 100 companies.
   e. competing interest groups.
9. **A system founded on the belief that the means of production should be owned by the people as a whole and that government should have an active role in planning and controlling the economy is known as**
   a. socialism.
   b. capitalism.
   c. communism.
   d. industrialism.
   e. pluralism.
10. **As capitalist systems mature,**
    a. consumer goods come under the influence of the law of diminishing returns.
    b. the willingness to take economic risks increases.
    c. the desire to turn to a mixed economic system increases.
    d. many industries are dominated by a few large firms.
    e. many small businesses are able to compete with larger firms.
11. **An organizational form in which shareholders have limited liability and limited control over organizational affairs is known as a/an**
    a. corporation.
    b. cooperative.
    c. unilateral organization.
    d. voluntary organization.
    e. employees organization.
12. **Which of the following statements is false?**
    a. Communist countries have not been slow to move to some sort of market system after the fall of the former Soviet Union.
    b. Economic change in recent years has not been limited to Soviet-bloc economies.
    c. Cuba has now adopted some degree of capitalism.

d. Capitalism now threatens to triumph over communism.
e. China, with the repossession of Hong Kong, is reverting to a planned economy.

13. **According to the text, the workforce hallmark today is**
a. a decline in knowledge-based workers.
b. a reduction in white-collar workers.
c. an increase in blue-collar workers.
d. the management and delivery of services.
e. the resurgence of agricultural employment due to corporate farming.

14. **Which of the following generalizations regarding corporate downsizing in the United States now appears to be true?**
a. Extensive downsizing is a fading corporate strategy.
b. Contingent employment has passed its peak.
c. The disadvantages of extensive downsizing may outweigh the advantages.
d. Management no longer has to be concerned about problems associated with downsizing and contingent employment because employees have become accustomed to these practices.
e. Since 1990, job loss has been receding from the pinnacle of employee fears.

FEEDBACK REVIEW

True-False

1. According to Jean-Jacques Rousseau, it is primarily the privileged classes who benefit from the existence of the state. T or F?
2. The operation of PACs in America today invalidates the pluralist model. T or F?

Fill in the Blank

3. According to Karl Marx, those in charge of the state in capitalist societies make decisions that serve the interests of the owners of the means of production. He used the term _________ to describe acceptance of a system that works against one's own interests.
4. _________ refers to the type of political system controlled by nonelected rulers who generally permit some degree of individual freedom.
5. Firms with operating facilities in several different countries are called _________.
6. The _________ industrial sector contains large, profitable firms with dominant positions within major industries.

Multiple Choice

7. **Most political action committees (PACs) are now associated with**
a. labor unions.
b. citizen groups.
c. political parties.
d. corporations and other businesses.

8. **Which of the following is *not* one of the reasons behind corporate political influence?**
a. Corporate managers and directors have informal and organizational ties that lead them to be consulted about matters of political policy.
b. Many corporate officials have reward and coercive power over elected government officials.
c. Many elected officials are also top corporate officials.
d. Interlocking directorates exist.
e. Conglomerates are present.

9. **Which of the following is *not* one of the major labor force trends discussed in the text?**
a. decrease in primary sector employment
b. decline in agricultural employment
c. increased employment of service workers
d. increase in blue-collar jobs
e. increase in white-collar proportion of the labor force

Matching

10. Several characteristics of capitalist economic systems are listed here. Indicate in the space beside each characteristic whether it best describes individualistic capitalism (IC) or communitarian capitalism (CC).
____ a. government and business cooperation
____ b. high job mobility
____ c. profit maximization
____ d. more value placed on the interests of employees and consumers
____ e. self-interest paramount
____ f. long-range economic planning

REVIEW GUIDE

GRAPHIC REVIEW

Figure 14.3 displays changes in the composition of the U.S. labor force since 1900.

1. Which occupational category had the steepest decline? Describe this change between 1900 and 2000.

____
____
____

2. In which occupational category has growth been the most rapid? Describe this change.

3. Discuss the most important implications regarding employment in the United States that can be suggested from these data.

ANSWER KEY

Concept Review

a. 16
b. 8
c. 17
d. 12
e. 7
f. 3
g. 6
h. 15
i. 14
j. 1
k. 13
l. 10
m. 11
n. 9
o. 2
p. 4
q. 5
r. 18

Multiple Choice

1. c
2. b
3. e
4. a
5. e
6. b
7. d
8. c
9. a
10. d
11. a
12. e
13. d
14. c

Feedback Review

1. T
2. F
3. false consciousness
4. Authoritarianism
5. multinational corporations
6. core
7. d
8. c
9. d
10. a. CC
    b. IC
    c. IC
    d. CC
    e. IC
    f. CC

15 Religion

OUTLINE

LEARNING OBJECTIVES

- Explain the sociological meaning of religion.
- Demonstrate the different views of religion taken by functionalists and conflict theorists.
- Distinguish among the basic types of religious organization.
- Discuss the meaning and nature of religiosity.
- Define secularization and describe its relationship to religiosity in the United States.
- Differentiate between civil and invisible religion in America.
- Describe the current resurgence of religious fundamentalism in the United States.
- Identify a wide variety of new religious movements in America.
- Outline the relationship of baby boomers to religion.

USING THE SOCIOLOGICAL IMAGINATION

Is secularization destroying religion in the United States? Many self-proclaimed prognosticators fear the decline of religion in the United States. On the contrary, evidence reveals that America, compared with other industrialized nations, remains a fairly religious nation.

Religion in America, then, remains a viable topic to explore. Before doing so, we will view religion within the context of sociology, define religion as an institution, examine religion from the functionalist and conflict perspectives, explore religious organization, and identify the ways people express their religious beliefs.

Religion, Science, and Sociology

Religion and Science

Can religion and science coexist? Because religion involves matters beyond human observation and because science is all about observation, these two institutions can potentially conflict. It is not that scientists cannot also be religious. Rather, scientists separate their professional work (based on reason and observation) from their religious life (based on faith and the unobservable).

Science and religion do conflict, however, when people try to mix the two. Some American religious fundamentalists do just that when they argue that *creationism* (explanation of the natural world through the Bible) is as scientifically based as evolutionary biology. The evolution versus creation debate has intensified in several states (Nelson 2002).

Will science replace religion? Science need not, and doubtless will not, eradicate religion. Because all societies have some form of religion, we can assume that humans seek answers to some fundamental questions about existence: Is there a higher purpose to life? Why does life unfold for us in particular ways? Is there an afterlife? Answers to such questions are simply beyond science. We can therefore expect religion to endure as long as human life exists. Furthermore, it is likely that science and religion will continue on their historical path of separation and coexistence.

Religion and Sociology

The study of religion attracted the attention of sociologists in the nineteenth century. These early sociologists were interested in the place religion holds in social life. Auguste Comte, the man who coined the term sociology, saw sociology as a new religion (a religion of science) and sociologists as high priests. Other pioneering sociologists, such as Emile Durkheim, Max Weber, and Karl Marx, viewed the practice of religion as a key to some of the mysteries of social life. Religion, they believed, could help explain the presence of social order and the cultural variation of societies. Despite this early scholarly interest, the sociological study of religion disappeared during the first half of the twentieth century. Only after World War II did social scientists again become interested in the scientific study of religion (Sharot 2001).

Sociologists of religion neither explain nor endorse particular religions but rather are interested in all sociological aspects of religion. This interest must begin with a sociological definition of religion.

What is the sociological meaning of religion? **Religion** is a unified system of beliefs and practices relating to sacred things. This definition comes from Emile Durkheim, whose work was based on studies of the Australian aborigines in the late nineteenth century. According to Durkheim, every society distinguishes between the **sacred**—entities that are set apart and

© International Stock Photography

Why does religion exist in all societies? According to Emile Durkheim's functionalist explanation, religion is ubiquitous because sacred symbols (such as the cross on top of this church) serve as mirrors through which members of society are able to see their common unity.

Is Human Cloning Ethical?

Aldous Huxley's 1932 novel *Brave New World* described a society in which humans were able to reproduce asexually. Another novel—*The Boys from Brazil*, written by Ira Levin and published in 1972—features a story about German Nazis cloning Adolf Hitler. Both of these books play on our fears about the effects and ethics of human cloning (a nonsexual creation of a genetically identical copy). Although no human has yet been cloned, the reproduction of a sheep called Dolly in February 1997 along with several subsequent clonings of mice, sheep, and pigs have made the question much more pertinent today than it was a few years ago.

Even though the technology is not yet available to clone humans, companies and scientists are already offering their services to interested individuals. Dr. Richard Seed, an American physicist, announced in 1998 that he plans to clone humans, using himself as the first subject. He also plans to open a for-profit clinic to assist childless couples in cloning themselves. A company called Valiant Venture, Ltd., has been formed to offer cloning services to humans for as "little" as $200,000. Valiant Ventures, Ltd., is owned by the Raelian Movement, an international religious organization whose members claim that life on earth was created in laboratories by extraterrestrials.

More traditional religious groups have expressed serious concerns about cloning. According to the general argument of Judaism and Christianity, human cloning allows the sacred generation of life to enter the profane realm. A group of scientists sponsored by the Church of Scotland reached the following conclusions:

- If humans are cloned, people will be placing themselves in a position only God has occupied.
- If humans are cloned, the basic dignity and uniqueness of each individual will be violated.
- Political power could influence the creation of clones.
- Cloning will be limited to those who can afford it.

But, might it not be beneficial to clone a Bill Gates, Billy Graham, or Michael Jordan? What about the potential contributions from a new Christiaan Barnard, the South African physician who did the first heart transplant in 1967? And can we prevent an organization from providing a service to those who can afford it?

Human cloning is the latest in a long line of medical technologies that affect the length and quality of life. Society will have to decide if cloning is so different from other scientific advances that it should be legally prohibited (Reaves 2001; R. Weiss 2001).

Analyzing the Trends
What role should religion play in the debate over human cloning? Include some information from this chapter to support your answer.

given a special meaning that transcends immediate human existence—and the **profane**—nonsacred aspects of life. (Profane in this context does not mean unholy or defiled, but simply commonplace and not involving the supernatural.)

Sacred things take on a public character that makes them appear important in themselves; the profane do not. Temple Mount in Jerusalem is a particularly illuminating example because its sacred meaning is not the same for the three religions that revere it. Jews hold sacred the Temple Mount "Wailing Wall" because it is the remnant of the First and Second Temples destroyed by the Romans in A.D. 70. Christians consider it holy because it was a site of Jesus' preaching. For Muslims, it is sacred as a place from which Mohammed ascended to heaven. Because Hindus do not share any of these religious beliefs, Temple Mount is part of their nonsacred, or profane, world.

Interestingly, some nonreligious aspects of culture can assume something of a sacred character. Babe Ruth's bat illustrates the difference between the sacred and the profane:

> *When Babe Ruth was a living idol to baseball fans, the bat he used to slug his home runs was definitely a profane object. It was Ruth's personal instrument and had little social value in itself. Today, however, one of Ruth's bats is enshrined in the Baseball Hall of Fame. It is no longer used by anyone. It stands, rather, as an object which in itself represents the values, sentiments, power, and beliefs of all members of the baseball community. What was formerly a profane object is now in the process of gaining some of the qualities of a sacred object.* (Cuzzort and King 1976:27)

Babe Ruth's bat illustrates two points about the sociological study of religion. First, a profane object can become sacred, and vice versa. Second, sociologists can deal with religion without becoming involved in theological issues. By focusing on the cultural and social

FIGURE 15.1

Division of World Population by Religions

This figure shows the percentage of the world's population belonging to various religions (also including the nonreligious and atheists). Compare these data against Worldview 15.1. Write down the parts of the world in which each religious faith predominates.

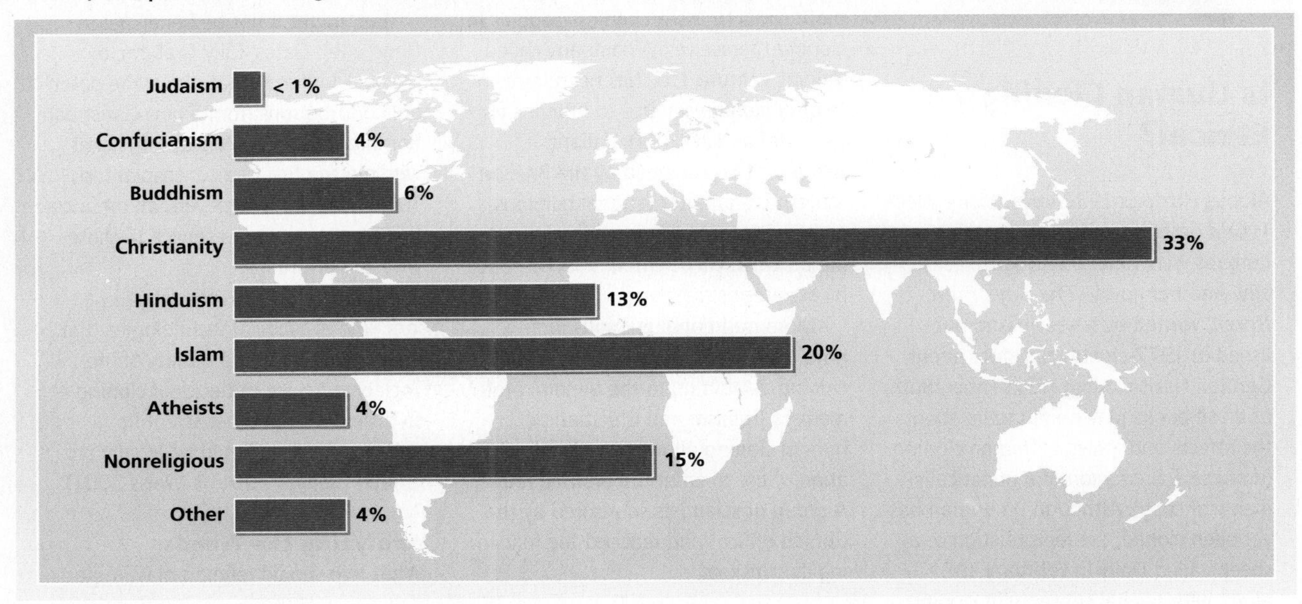

Source: Religious Tolerance.org. "Religions of the World: Numbers of Adherents; Rates of Growth," 2002, at www.religioustolerance.org/worldrel.htm

TABLE 15.1

Major World Religions

This figure outlines selected characteristics of the major world religions. Identify any similarities and differences among these religions. Explain anything that surprises you about this figure.

Religion	Date of Beginning	Founder	Main Beliefs	Main Areas	Number of Followers
Confucianism	200 B.C.	Confucius	• Moral conduct and right human relationships	China	5,086,000
Judaism	1200 B.C.	Abraham	• God established a covenant with the people of Israel, who are called to lives of justice and mercy.	USA, Israel	13,866,000
Buddhism	500 B.C.	Siddhartha Gautama	• By adherence to the Eightfold Path (correct thought and behavior), one can escape from desire and suffering and achieve nirvana.	Far East and Southeast Asia	325,275,000
Hinduism	2000 B.C.	No specific founder	• Brahman is the principle and source of the universe. Life is determined by the law of karma.	India	793,076.000
Islam	A.D. 600	Muhammad	• Muhammad received the Koran from God. Believers go to an eternal Garden of Eden.	Africa, Middle East, Southeast Asia	1,126,325,000
Christianity	A.D. 0	Jesus Christ	• Jesus is the son of God. Through God's grace and profession of faith, people achieve eternal life with God.	Europe, North America, South America	1,955,229,000

aspects of religion, sociologists avoid questions about the ultimate validity of any particular religion. This point needs elaboration.

How do sociologists avoid theological issues? Religion involves a *transcendent reality*—a set of meanings attached to a world beyond human observation. Because sociologists must deal with this dimension of unobservability, their task is particularly difficult. For example, in this realm, sociologists face unique problems: Can an empirical science actually study the unobservable? Can scientific objectivity be maintained while investigating something as value laden as religion? Can sociologists retain religious faith after scientifically probing religion? Each of these questions deserves a brief answer.

Obviously, sociologists cannot study the unobservable. As part of an empirical science, sociologists who investigate religion must formulate theories and hypotheses on aspects of religion that can be measured and observed. Consequently, sociologists avoid the strictly spiritual side of religion and focus on social aspects of religion that can be measured and observed.

Objectivity, as we saw in Chapter 2, is a goal for which scientists strive. Because there are so many emotional overtones in the study of religion, researchers must be especially vigilant of their own biases. Thus, sociologists investigating religion do not attempt to determine the truth or falseness of any religion. Because definitions of good or bad, truth or falsity are avoided, theological issues lie beyond the sociology of religion. Sociologists can point out that conceptions of God, religious beliefs, and religious practices vary from culture to culture without attempting to make value judgments of any of the beliefs and practices involved. Because the validity of any religion is founded in faith, sociologists restrict themselves to the social manifestations of religion.

Sociologists, then, are not in the business of determining which religious faith people ought to follow. For this reason, it is possible to retain a personal religious faith while investigating the social dimensions of religion. In fact, sociologists, like other people, vary in their religious orientations. They have varying degrees of religious commitment, exhibit a myriad of religious behaviors, and belong to quite different religious organizations (see Figure 15.1 and Table 15.1).

Theoretical Perspectives

The functionalist and conflict perspectives have different views of the sociological dimensions of religion. We begin with functionalism, which examines the role of religion in relation to the functions it serves for society.

Functionalism and Religion

We know from archaeological discoveries and anthropological observations that religion has existed in some form in virtually all societies (Nielsen et al. 1993). The earliest evidence of religion has been traced as far back as 50,000 B.C., when humans had already begun to bury their dead—a practice that suggests belief in some existence after death. In early Rome, there were specific gods for objects and events—a god of trees, a god of money, a goddess of fever. According to the ancient Hebrew religious texts (both the Book of Genesis and the Book of Leviticus), pigs were unclean animals whose pollution would spread to all who touched or tasted them. To the tribes of New Guinea and the Pacific Melanesian Islands, on the other hand, pigs are holy creatures worthy of ancestral sacrifice (Harris 1974).

Emile Durkheim, the first sociologist to examine religion scientifically, wondered why it is that all societies have some form of religion. In one of his books, *The Elementary Forms of Religious Life* (1995; originally published in 1915), Durkheim offered a sociological explanation for the ubiquity of religion, an explanation rooted in the function religion performs for society. The essential function of religion, he believed, was to provide through sacred symbols a mirror for members of society to see themselves. Through religious rituals, people worship their society and thereby remind themselves of their shared past and future existence. Following Durkheim's lead, sociologists identify the following social functions of religion: (1) to legitimate social arrangements, (2) to encourage a sense of social unity, (3) to provide a sense of meaning, and (4) to promote a sense of belonging.

How does religion legitimate social arrangements? *Legitimation* justifies and explains the status quo. It explains why a society is—and should be—the way it is.

FEEDBACK

1. A _________ is a unified system of beliefs and practices relative to sacred things.
2. _________ things are set apart and given special meaning that transcends immediate existence.
3. Sociologists who study religion are normally forced to abandon their religious beliefs and convictions. T or F?

Answers: 1. religion 2. Sacred 3. F

Religions of the World

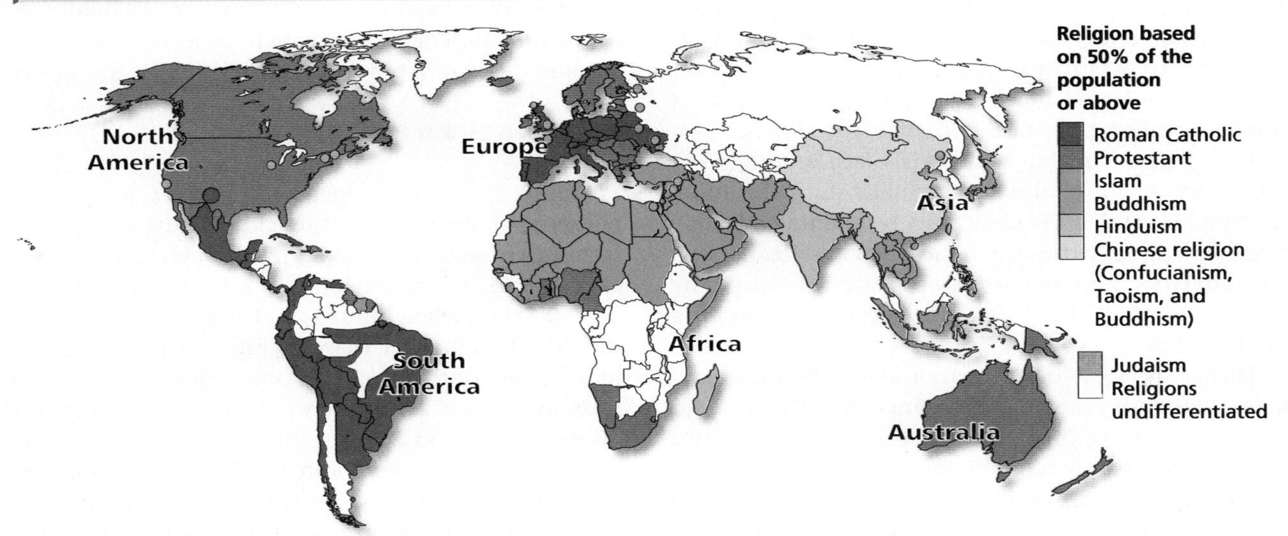

Source: After a map compiled by the Office of the Geographer, Department of State.

This map displays the worldwide distribution of all religions. Emile Durkheim showed that suicide rates vary according to group characteristics. One of these characteristics was religious background. For example, Durkheim showed that the suicide rate is lower among Catholics than among Protestants.

1. Based on the information shown in this map, where would you expect to find lower and higher rates of suicide?
2. What information on the map did you use in your analysis?

Legitimations tell us why some people have power and others do not, and why some are rich and others poor. Many such legitimations are based on religion. According to Durkheim, legitimation is the central function of religion. For him, ideas of the sacred within a society represent the society itself. When people think of their souls, Durkheim believed, they are really thinking of the social element within themselves—an element that is superior to them and will live on after them.

If, as Durkheim believed, religion is a representation of society, its power to provide legitimation for society is considerable. Through its system of beliefs, a religion explains the nature of social life, the existence of evil, and the imperfect nature of the world. Religion has been used to justify slavery in America (many ministers in the nineteenth-century South preached that African Americans were descendants of Ham, a son of Noah whose ancestors were condemned to slavery) and male superiority (Eve was created from one of Adam's ribs, and it was her encouragement that led Adam to eat the forbidden apple). One of the most important and widely agreed-on social functions of religion, then, is the legitimation or justification of social arrangements.

How does religion promote social unity? Religion, according to Durkheim, is a glue that holds society together. Without religion, society would be chaotic. As Cuzzort and King have stated, Durkheim "provided the greatest justification for religious doctrine ever granted by a social scientist when he claimed that all societies must have religious commitments. Without religious dedication there is no social order" (Cuzzort and King 1976:43). In some cases, however—Northern Ireland being one example—religion provokes societies to fragment, even to the point of civil war. Thus, whereas it is accurate to say that religion is usually a source of social unity, it can also divide a society.

How does religion provide a sense of meaning? Religion not only legitimates existing social arrangements and encourages social unity but also provides individuals meaning beyond day-to-day life. People mark important events in life—birth, sexual maturity, marriage, death—with religious ceremonies and explain such events in religious terms. Religion gives believers cosmic significance and gives eternal significance to a short and uncertain earthly existence.

SOCIOLOGY AND THE NEWS MEDIA

The Sacred Ganges

This CNN report explores the importance of the Ganges River in the religious life of the people of India. Hindus are taught that the Ganges is sacred. According to the Hindu faith, the Ganges is so spiritually pure that it cannot possibly be polluted. Consequently, despite its polluted state, over a million Hindus enter its waters every day. To die and be cremated on the banks of the Ganges is believed to ensure salvation among devout Hindus.

© Cable News Network, Inc.

For most Americans, there is nothing sacred about river water. Hindus in India, on the other hand, attribute to the Ganges River a special meaning that rises above immediate existence.

1. Using India's religious view of the Ganges, distinguish between the sacred and the profane.

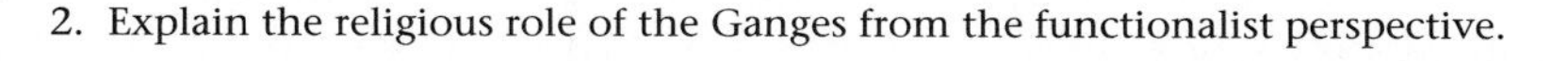

2. Explain the religious role of the Ganges from the functionalist perspective.

How does religion promote a sense of belonging? Religious organizations provide opportunities for people to share significant commonalties—ideas, a way of life, an ethnic background. Religion supplies a kind of group identity. People usually join religious organizations freely and feel a degree of influence within these organizations. For many people in modern society, membership in a religious organization provides a sense of community. This counteracts depersonalization, powerlessness, and rootlessness.

Of course, religious organizations are not unique in providing a sense of belonging. Fraternal groups, clubs, and other voluntary associations do the same thing. Because of religion's other functions, however, the sense of belonging to a religious organization can be particularly appealing.

Conflict Theory and Religion

As noted, Durkheim believed that religion supports societies by providing legitimation for social arrangements. Durkheim's ideas are still being debated, however, and other sociological theorists have identified quite different relationships between religion and society.

Georg Friedrich Hegel, an eighteenth-century German philosopher, believed that ideas and beliefs, including religious ones, determine the nature of social life. Ideas and beliefs, Hegel thought, cause social life to be structured in certain ways. Hegel's position was very popular within the religious and intellectual establishment of his day, but it was not without critics. By the early nineteenth century, a group of critics known as the young Hegelians had begun to develop a radical makeover of Hegel's position. They claimed to set Hegel "on his head." Social structure determines ideas and beliefs, not the other way around. More specifically, the economic structure of a society determines the nature of the society's religion. Karl Marx, deciding the group had gone too far the other way, developed a view encompassing both ideas and structure.

How did Marx view religion? Marx believed that once people create a unified system of sacred beliefs and practices, they act as if it were something beyond their control. They become "alienated" from the religious system they set up. Although humans have the power to change (or, better yet, in Marx's mind, to abandon) the religion they have created, they don't do so because they see it as a binding force to which they must conform.

Religion, Marx wrote, is used by the ruling class to justify its economic, political, and social advantage over the oppressed. The elite attribute poverty, degradation, and misery to God's will. So to eliminate inequalities and injustices is to tamper with God's plan. Change is sacrilegious to the elite. Many illustrations of ties between the elite and religion exist; in fact, a close relationship between the dominant classes and the church has been the rule in human history. The Catholicism of medieval Europe, for example, was so closely identified with the upper classes that the lower and lower middle classes welcomed the Protestant Reformation. In his classic study *Millhands and Preachers*, Liston Pope (1942) documented a case of religious control by the managers of a mill in a North Carolina community (Gastonia):

> *The churches were inextricably bound to mill management by their finances if not by their ideology. Organized religion was a key instrument in creating a manipulable labor force and in resisting the union's effort to break with the exploitative status quo. Ministers not only justified management practices; they were mute in reaction to the violent recriminations exerted by the community against the union organizers.* (Pope 1942:xx)

In a restudy of Gastonia, churches were still providing latent support for long-standing economic relations in the community.

In addition, Marx contended, the oppressed employ religion to explain and justify their deplorable existence and to provide hope for a better life after death. In this way, they can find self-fulfillment in their oppression. Marx thought of religion as a narcotic for the oppressed:

> *Religious suffering is at the same time an expression of real suffering and a protest against real suffering. Religion is the sign of the oppressed creature, the sentiment of a heartless world, and the soul of soulless condition. It is the opium of the people.* (quoted in Bottomore 1963:43–44; originally published by Karl Marx in 1844)

If people abandoned the religion of their oppressors, Marx alleged, they would turn from the comforts offered by an afterlife and instead concentrate on the present. This would lead them to reject their oppressed economic condition and decide to do something about it. Nonetheless, Marx recognized that it would be difficult for the oppressed to view their religion as a manipulation of the elite. A contemporary scholar of ideology highlights the latter point:

> *The most efficient oppressor is the one who persuades his underlings to love, desire and identify with his power, and any practice of political emancipation thus involves that most difficult of all forms of liberation, freeing ourselves from ourselves.* (Eagleton 1994:xiii–xiv)

Capitalism and the Protestant Ethic

How did Weber view the place of religion in social life? Whereas Marx contended that religion retards social change because the elite wish to maintain the status quo, Max Weber (1958; originally published in 1904) supposed that religion sometimes encourages social change. Weber looked for a case that would demonstrate his point and found it in the relationship between the ideas of early Protestantism and the rise of capitalism.

Weber began with the fact that capitalism was widespread only in northwestern Europe and America. Why had this happened? Why had capitalism not emerged as the basic economic system in other parts of the world? Through an extensive study of various religions of the world, Weber (1951, 1952, 1958, 1964) found a possible answer in the basic compatibility between what he termed the spirit of capitalism and the Protestant ethic.

What is the relationship between the spirit of capitalism and the Protestant ethic? Capitalism, Weber noted, involved a radical redefinition of work. With capitalism, work became a moral obligation rather than a mere necessity. Capitalism also required the reinvestment of profits, if businesses were to grow. Because capitalism involved the reinvestment of capital, investment for the future was more important than immediate consumption. All of this Weber called the **spirit of capitalism**.

Although most major religions did not define hard work as an obligation or demand the reinvestment of capital for further profits (rather than for immediate enjoyment), some Protestant sects did. Here, then, embodied in Protestantism, was a cluster of values, norms, beliefs, and attitudes that favored the emergence of modern capitalism. Weber referred to this cluster of values and attitudes stressing hard work, thrift, and discipline as the **Protestant ethic**.

What is the nature of the Protestant ethic? Weber found the roots of the Protestant ethic in the seventeenth-century Puritan theology of Calvinism. Of particular importance was the Calvinist belief that every person's eternal fate was unalterably predetermined by God, at birth—predestination. Society, the Calvinists believed, was divided into the few who were elected to salvation and the many who were predestined to spend eternity in hell. Individuals could do nothing to alter their fate.

Calvinism taught its followers to look for signs of God's blessing because, according to Calvin, God identifies his elect by rewarding them in this world. This was translated into the belief that the more successful people were in this life, the more sure they were of being a member of God's select few. In addition, consumption beyond necessity was considered sinful; those who engaged in self-pleasure were agents of the devil. Finally, Calvinists believed that there was an underlying purpose of life: glorification of God on earth through one's occupational calling. Because everyone's material rewards were actually God's and the purpose of life was to glorify God, profits should be multiplied (through reinvestment) rather than used in the pursuit of personal pleasures.

Other Protestant religious beliefs may have contributed to the development of capitalism. For example, Protestants endorsed the idea that salvation came from unselfish good works here on earth. Protestants were encouraged to do more and more work while receiving less and less of the material rewards.

Does the Protestant ethic still exist? The debate over Weber's Protestant ethic continues (MacKinnon 1988a, 1988b; Zaret 1992; Lehrmann and Roth 1993; R. F. Hamilton 1996; J. Cohen 2002). However, any link between Protestantism and economic behavior that may have existed 200 years ago appears to be rather weak today. Surveys show that Protestants do not value success and hard work any more than Catholics do. Moreover, Jews have higher incomes and are more successful occupationally than Protestants. If the Protestant ethic still exists, it does not appear to be the exclusive property of Protestants. As a matter of fact, it may not be the property of any religious group.

Did Protestantism precipitate the development of capitalism? Weber did not contend that capitalism developed *because* of Protestantism. (Capitalism had actually flourished among Catholics in some parts of the world before the Protestant Reformation.) He did, however, suggest that early Protestantism seems to have provided a social environment *conducive* to the emergence of this economic system. Weber's emphasis on the influence of ideas on social structure remains important as a counterpoint to Marx's accent on the effect of social structure (economy) on ideas. He recognized that capitalism could exist apart from Protestant dogma. Weber himself pointed out that capitalism helped destroy many of the foundations of Protestantism. He was well aware that from the seventeenth century on, religion gradually lost its hold over capitalist society. Jere Cohen (2002) concludes that Weber's exaggeration of the effects of Protestantism on the development of modern capitalism should not cause us to overlook the real economic impact of Protestantism.

Gender and Religion

Conflict theorists view religion as a primary source of division and strife within societies and between societies. They point out that most wars are grounded in religious conflict. The Arab–Israeli–United States conflict is only one of the most current illustrations. War, however, is only one manifestation of religious-based conflict. Less visible forms of conflict are generated by the efforts of religious elites to maintain control over an entire society or subgroups within a society. An important instance is male domination of women that exists in most of the world's major religions (McGuire 2001).

For starters, feminists see the assumed maleness of the Supreme Being, in all major religions, as blatant sexism. Female inferiority and subordination, they argue, is prominent in all sacred texts of the major religions. Orthodox Jewish males thank God daily that they were not born female. In the Muslim Koran, the superiority of men over women comes from God-given gender differences. The account of creation in the Christian Bible blames Eve for the eating of forbidden fruit and announces an eternal punishment for women: "In sorrow thou shalt bring forth children; and thy desire shall be to thy husband, and he shall rule over you" (Genesis 3:16).

Despite the plethora of positive statements about women attributed to Jesus in the New Testament, the negative view in Genesis has historically dominated the Christian view of gender. And even after the many significant contributions women made to American religious denominations over the centuries, gender inequalities in modern religious practices and organization continue to reflect the maintenance of male patriarchy (Benowitz and Vowels 1998).

Since the early 1980s, feminists have worked against male patriarchy in Christian religious organizations. Their results are meager. This is partly because many major religious organizations, including Roman Catholics, Eastern Orthodox churches, conservative and Orthodox Jews, and Missouri Synod Lutherans, continue to bar female ordination. And women are at an extreme disadvantage even within religious organizations that accept their ordination. Although the increasing number of female seminarians and ordained ministers is one important consequence of feminist efforts to eradicate sexism in churches, their number remains small. Even after ordination, women face a glass ceiling, salaries below those of male ministers, and assignment to small congregations. These conditions frustrate female ministers as well as discourage women from preparing for ordination.

An Evaluation

Interpretations of the relationship between religion and social change, as the preceding paragraphs show, are varied. Marx, writing from the conflict perspective, believed that religion retards social change in part because of its ties to an economic elite. Unlike Marx, Weber argued that religion could encourage social change by supporting new social arrangements.

Although the two views appear contradictory, each may be viable under particular conditions. Meredith McGuire (2001) suggested that the relationship between religion and social change depends on several situations: quality of religious beliefs and practices existing in a society, dominant ways of thinking in a culture, social location of religion within a larger society, and internal structure of religious organizations and movements. For example, McGuire has argued that religion is likely to promote change if it provides a critical standard against which the established social system can be measured. When religion provides such a standard, it allows the social critic to say, "This is what we say is the right way to act, but look how far our group's actions are from these standards!" For example, during the American civil rights movement religious leaders pointed to American values of equality and opportunity and their incompatibility with racial discrimination.

The degree of a society's network for change is another factor. In many Latin American countries, for example, the Roman Catholic Church plays an important role in social change and protest because it is the sole dissenting grassroots organization with an established network and sympathetic leanings. When other channels for action exist—such as interest groups—religion may play a minor role in social change.

The relationship of religion to social change, then, is quite variable. Although the interpretations developed by Marx and Weber call attention to specific aspects of that relationship, no single interpretation appears applicable to all situations.

Symbolic Interactionism and Religion

Sociologist Peter Berger (1990) captured the relationship between religion and **symbolic interactionism** in his book *The Sacred Canopy*. In this book, Berger explores the idea that humans create from their religious traditions a canopy, or cover, of symbolic meanings to lay over the secular world. These meanings are used to illuminate the difference between the sacred and the profane and to guide everyday social interaction. A canopy of religious beliefs, rituals, and ideas provides stability and security in a changing and uncertain existence.

Symbolic interactionism, for example, helps us understand the expression "There are no atheists in foxholes." Insecurity and uncertainty, of course, are at a peak in the life-and-death situation of war, and the desire to regain security and certainty is a natural human response. Religious meanings, especially those related to a possible afterlife, can offer some relief. Like the Japanese kamikaze pilots in World War II, Middle Eastern terrorists can infuse into suicidal behavior the meaning of a reward beyond life. Less dramatically, people enduring troubled marriages can be strengthened by their commitment to uphold their holy vows of matrimony spoken in a place of worship.

So far, the interplay between religion and society has been portrayed as one-way. Religion does bestow meanings on social life, but the reverse is also true; social change within a society also influences religious beliefs, rituals, and ideas. As scientific knowledge began to increase with the Enlightenment, religious beliefs in the West also began to change. Gradually, the belief that devils inhabited witches weakened. An understanding of the universe slowly emptied the prisons of scientists who disputed the theological truism of an earth-centered solar system. In our own time, most Protestants now commit the former heresy of practicing birth control, without negative church sanction. Given the widespread practice of birth control among the Catholic laity, one can imagine the eventual acceptance of birth control methods by the Church itself. And the general rejection of female ministers within Protestant faiths seems certain to be eroded by the tide of gender equality (Eck 2002).

Each of the three major theoretical perspectives aids in the scientific study of religion. Table 15.2 shows the unique light each perspective sheds.

TABLE 15.2

FOCUS ON THEORETICAL PERSPECTIVES: Religion

This table shows that in examining religion, the three major perspectives focus on different aspects. Discuss the conclusion on any one of the theories in light of your experience with the institution of religion.

Theoretical Perspective	Concept	Example
Functionalism	• Religion makes specific contributions to society.	• Legitimates social arrangements • Promotes social unity • Provides a sense of understanding • Encourages a sense of belonging
Conflict theory	• Elites use religion to manipulate the masses.	• Religion is used by the most powerful to justify their economic, political, and social advantage.
Symbolic interactionism	• People create symbolic meanings from their religious beliefs, rituals, and ideas.	• People use their socially created symbolic meanings to guide everyday social interaction.

FEEDBACK

1. According to Karl Marx, religion is a force working for
   a. the good of all people.
   b. the good of the proletariat.
   c. maintenance of the status quo.
   d. the creation of conflict between the classes.
2. According to Marx, religion is created by the _________ to justify its economic, political, and social advantages over the oppressed.
3. The _________ is the belief that work is an obligation and that capital should be reinvested for further profits rather than used for immediate enjoyment.
4. According to Max Weber,
   a. early Protestantism provided a social environment conducive to capitalism.
   b. capitalism developed because of Protestantism.
   c. capitalism strengthened Protestantism.
   d. religion will always be linked with economy.
5. Religion
   a. has been found to promote social change under almost all conditions.
   b. is unrelated to social change.
   c. promotes social change only under certain conditions.
   d. always supports the interests of underprivileged groups in society.

Answers: 1. c 2. elite 3. Protestant ethic 4. a 5. c

Religion: Structure and Practice

Religious Organization

Major religious faiths in the world include Christianity, Judaism, Islam, Hinduism, and Buddhism. Because most people practice religion through some organizational structure, religious organization is an important component of the sociological study of religion. Early scholars identified four basic types of religious organization: church, denomination, sect, and cult (Troeltsch 1931; Niebuhr 1968; originally published in 1929). A religious faith may be practiced through any one of these types of religious organizations.

How do sociologists distinguish among the basic types of religious organization? In this context, **church** is a life-encompassing religious organization to which all members of a society belong. This type of religious organization exists when religion and the state are closely intertwined. King Henry VIII's divorce troubles stemmed from the Roman Catholic Church's hold on sixteenth-century England. In Elizabethan England, Anglican Archbishop Richard Hooker of the Church of England wrote that "there is not any man of the Church of England but the same man is also a member of the commonwealth; nor any man a member of the commonwealth which is not also of the Church of England."

A **denomination** is one of several religious organizations that most members of a society accept as legitimate. Because denominations are not tied to the state, membership in them is voluntary, and competition among them for members is socially acceptable. Being one religious organization among many, a denomination generally accepts the values and norms of the secular society and the state, although it may at times be in opposition to them. Most American "churches"—Methodist, Episcopalian, Presbyterian, Baptist, Roman Catholic, Reformed Jewish—are actually denominations.

A **sect** is a religious organization formed when members of an existing religious organization break away in an attempt to reform the "parent" group. Generally, sect members believe that some valuable beliefs or traditions were lost by the parent organization, and they form their own group to save these features. Thus, they see themselves not as establishing a new religious faith but as redeeming an existing one. The withdrawal of a sect from the parent group is usually psychological, but some sects go farther and form communal groups apart from the larger society. The Pilgrims, who landed in Jamestown in 1609, wished to reform the Church of England from which they had separated. Another example is the Amish, a sect formed in 1693 when a Swiss bishop named Jacob Amman broke from the Mennonite Church in Europe (Kraybill and Olshan 1994). Less extreme sects in the United States today include the Seventh-Day Adventists, the Quakers, and the Assemblies of God.

A **cult**, by contrast, is a religious organization whose characteristics are not drawn from existing religious traditions within a society (Karson 2000; Kaplan and Loow 2002). Whether imported from outside the society or created within the society, cults bring something new to the larger religious environment. We may think of cults as engaging in extreme behavior. The world was shocked in 1997, for example, when reports came of

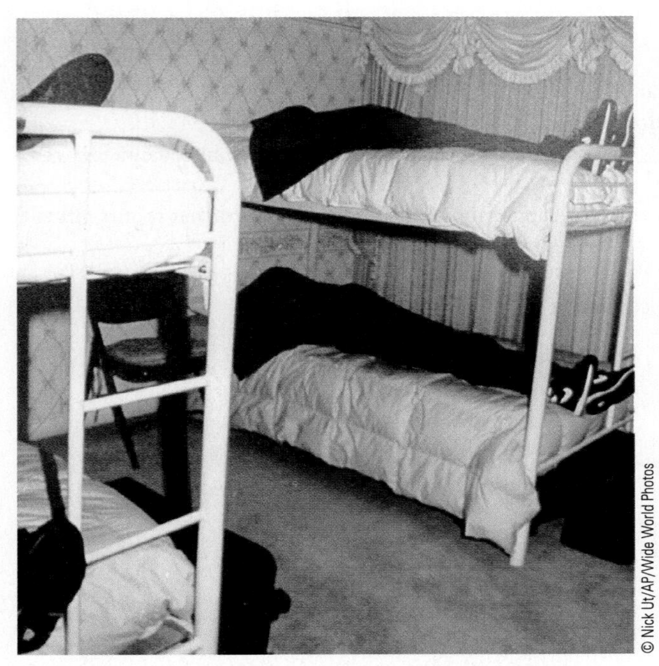
© Nick Ut/AP/Wide World Photos

The Heaven's Gate cult in California was an extremist religious group. This is a photograph of a few of the thirty-nine members who participated in a mass suicide in 1997.

the ritualistic suicides of thirty-nine members of the Heaven's Gate cult in California (E. Thomas 1997). Dwarfing this incident was the mass killing of approximately one thousand members of the Ugandan cult called the Movement for the Restoration of the Ten Commandments of God ("Cult Killings Exceed Jonestown Toll" 2000). Cults do not usually appear in such an extreme and bizarre form, however. More conventional examples of cults are the Unification Church, the Divine Light Mission, and the Church of Scientology (C. S. Clark 1993).

Although the term cult covers a wide range of groups and organizations, the religious cults in the United States today share several characteristics. At their center are an authoritarian structure, rejection of the secular world's laws and ways, strict discipline of adherents, rigidity in thinking, conviction of sole possession of truth and wisdom, belief in the group's moral superiority, and discouragement of individualism (Appel 1983; J. J. Collins 1991). More is said about cults later.

Religiosity

For many years, Charles Glock and Rodney Stark (1965, 1968) have explored religious experiences in relation to **religiosity**—the ways in which people express their religious interests and convictions. These authors' work focuses on the types of religious attitudes and behavior people display in their everyday lives.

How do people display religiosity? Glock and Stark distinguish five dimensions of religiosity: belief, ritual, intellectual, experience, consequences. *Belief* refers to some idea about what a person considers to be true. People may, for example, believe that Christ died on the cross for everyone, or that there is no god but Allah. A *ritual* is a religious practice that members of a religion are expected to perform. A ritual may be private (personal prayer) or public (attending mass). The *intellectual* dimension of religiosity may involve knowledge of the

SNAPSHOT OF AMERICA 15.1

Religious Believers

District of Columbia

Religious adherents as a percentage of state population

≥ 70%
60–69%
50–59%
40–49%
Less than 40%

Source: U.S. Bureau of the Census, *Statistical Abstract of the United States: 2001* (Washington, DC: U.S. Government Printing Office, 2001).

Religion is common to all societies. Although most Americans are Christian, many other faiths are represented in the United States. This map shows U.S. religious believers as a percentage of the population of each state.

1. Compare your state with other states in your region based on number of religious believers.
2. What does this map tell you about the state of religion in the United States? Explain.

Exhibiting religious practices (rituals) is an important dimension of religiosity. These Baptists are engaging in a traditional practice of baptism.

Bible or an interest in such religious aspects of human existence as evil, suffering, and death. Religious persons are expected to be informed about their faith. *Experience* encompasses certain feelings attached to religious expression. This dimension is the hardest to measure, but James Davidson (1975) contended that it can be ascertained by asking people whether or not they think specific religious experiences are desirable and whether or not they have had specific religious experiences. *Consequences* are the decisions and commitments people make as a result of religious beliefs, rituals, knowledge, or experiences. Consequences may be social (opposing or supporting capital punishment or abortion) or personal (restricting one's sex life to marriage or telling the truth regardless of the cost).

Why use a dimensional approach to religiosity? Although the dimensions of religiosity may be related to one another (Clayton and Gladden 1974), they may also operate independently at times. People who attend a place of worship (ritual dimension), for example, may know little about religious doctrines (intellectual dimension). And some people who worship regularly may very well believe that people who do not go to church will spend eternity in hell (belief dimension).

Religion in the United States

The nature rites of South Sea Islanders and the well-ordered church of the Middle Ages were relatively simple and stable forms of religion. By comparison, advanced industrial society presents a challenge to the sociological study of religion.

The Development of Religion in America

Although the search for religious freedom was only one of the reasons the Puritans came to America, there was a genuine religious element in the American colonization and revolution (Brauer 1976; Marty 1985). Robert Bellah has described the American religious connection this way:

> *In the beginning, and to some extent ever since, Americans have interpreted their history as having religious meaning. They saw themselves as being a "people" in the classical and biblical sense of the word. They hoped they were a people of God.* (Bellah et al. 1991:2)

FEEDBACK

1. Match the following descriptions with the types of religious organizations.
   ____ a. church ____ c. denomination
   ____ b. sect ____ d. cult
   (1) a religious organization whose characteristics are not drawn from existing religious traditions within a society
   (2) an exclusive and cohesive religious organization based on the desire to reform the beliefs and practices of another religious organization
   (3) a life-encompassing religious organization to which all members of a society belong
   (4) one of several religious organizations within a society that are considered legitimate
2. ________ refers to the ways people express their religious interests and convictions.
3. Match the dimensions of religiosity with the examples beside them.
   ____ a. belief (1) prayer in church
   ____ b. ritual (2) knowing the content of the Koran
   ____ c. intellectual (3) having seen an angel
   ____ d. experience (4) conviction that the Bible is divinely inspired
   ____ e. consequences (5) marching in support of prayer in schools

Answers: 1. a. (3) b. (2) c. (4) d. (1) 2. Religiosity 3. a. (4) b. (1) c. (2) d. (3) e. (5)

Religion in the colonies was varied, but beginning in the second half of the eighteenth century, the approach to religion was increasingly influenced by the leaders of the Enlightenment—Bentham, Voltaire, Diderot, Hume, and many others—in Europe and the British Isles. The same ideas that inspired the questioning of arbitrary colonial government and the writing of the Declaration of Independence also led to a critical examination of religion and its relationship to the state. Although the framers of the U.S. Constitution seldom raised arguments against religious faith, they were sharply critical of any entanglement between religion and the state. Indeed, the ideas of separation of church and state and freedom of religious expression are cornerstones of American life. Despite this tradition, people in the United States have experienced incidents of religious persecution, including some directed at immigrant groups.

Secularization: Real or Apparent?

Religion has been an important influence during most of human history, including the history of the United States. The industrialization (and accompanying secularization) of many nations, however, diluted the influence of religion. Our examination of religion in the present-day United States will therefore begin with the process of secularization (Swatos and Olson 2000).

What is secularization? **Secularization** is a profane process through which the sacred loses influence over society. Through this process, other social institutions are emptied of religious content and freed from religious control (B. Wilson 1982). Religion itself becomes a specialized, isolated institution. This is in sharp contrast to preindustrial society, where religion and social life were inseparable. Hester Prynne, for example, the central character of Nathaniel Hawthorne's novel *The Scarlet Letter*, was forced by religious leaders and peers in her Puritan community to live forever with shame and ostracism for her act of adultery. Religion and social life were so intertwined in seventeenth-century America that the discovery of Hester's adultery led to her loss of a normal social existence.

Evidence is mixed concerning the relative importance of religion in the United States today. There are some indications of a diminution of the importance of religion in the United States (Swatos and Olson 2000; Hout and Fischer 2002). The percentage of Americans claiming that religion is very important in their lives declined from 75 percent in 1952 to 59 percent in 2001, but has increased a bit since hitting a low of 52 percent in 1978 (see Figure 15.2). The Princeton Religion Index, comprised of six to eight leading indicators, shows a decline since the 1940s. Finally, 14 percent of the public in 1957 indicated that religion was losing influence on American life; in 2001, 58 percent of the public saw a lessening of religion's influence on American life (Gallup 2002).

FIGURE 15.2

Percentage of Americans Saying Religion Is Very Important in Their Lives: 1952–2001

This figure tracks changes in the percentage of Americans who say that religion is very important in their lives. Do you think the percentage will continue to rise, or will it decline again? Explain.

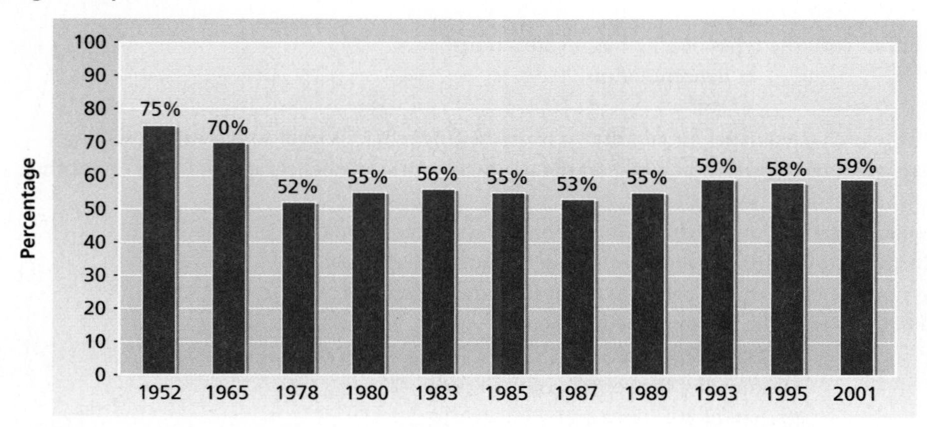

Source: George H. Gallup Jr., *Religion in America 2002* (Princeton, NJ: Princeton Religion Research Center, 2002), p. 18.

INTERNET LINK The Gallup Organization compiles a tremendous amount of data on the American public's views. You can go to its Web site at www.gallup.com/

But is secularization destroying religiosity? Researchers characterize America as being unusually religious among industrialized nations (Caplow 1985; Lipset 1996; Inglehart and Baker 2000; Kohut 2000; Gallup 2002). Another recent study shows that poorer nations tend to be very religious and wealthy nations tend to be less religious, except for the United States (Kohut 2002).Theodore Caplow, Howard Bahr, and Bruce Chadwick (1983) contend that residents of Middletown (which they take to be representative of the United States) are as committed to religion as preceding generations. Whether they measured by the number of houses of worship per capita, the proportion of regular churchgoers, or financial support of the churches, Caplow and his colleagues observed a trend toward greater involvement in religious affairs (Caplow 1998; De Luca 2001).

As suggested in "Using the Sociological Imagination," America still appears to be a religious nation when compared with other industrialized countries (Wolfe 1999). Only 7 percent of the American population is without some religious preference. About 83 percent identify themselves as Protestants, Catholics, Jews, or Mormons. There are now over 300 recognized denominations and sects and thousands of independent congregations in the United States (Linder 2000). About seven out of ten Americans have some religious affiliation, and over half of these claim to be active in their congregations. About one-third of Americans claim they attend a church or synagogue at least once a week. (In England, the average weekly church attendance is 14 percent.) Since 1939, weekly church or synagogue attendance in the United States fell from 41 percent to 32 percent in 2001. The proportion of Americans belonging to a church or synagogue declined somewhat from a high of 76 percent in 1947 to 68 percent in 2001. American teenagers tend to support traditional religious beliefs on a par with adults. Ninety-eight percent of American teenagers believe in God or a universal spirit, 67 percent believe in life after death, 91 percent believe in heaven, and 76 percent believe in hell. Seventy-six percent believe in the existence of angels (Gallup 2002).

Why is there a continuing interest in religion? Caplow and his colleagues suggest that religion is sought as a buffer against the insatiable demands of the state. According to Gallup, Americans are searching for spiritual moorings for three basic reasons: the threat of nuclear war; loneliness; and disenchantment with what is perceived as a society without rules. Several sociologists contend that religion remains important in helping many Americans deal with problems of grief, misery, fear of the unknown, hopelessness, hunger, death, and the search for meaning beyond life in this world (Wind and Lewis 1994; Greeley 1996; Stark and Finke 2000).

Does this mean that secularization is not an important force in American society today? No, it does not. As Robert Wuthnow (1985) emphasizes, when Weber, Durkheim, and Marx wrote about secularization, they were not thinking of short-term, individual religious attitudes and behavior such as belief in God or synagogue attendance. They were observing major historical trends in European history since the Middle Ages, such as the separation of church and state. From this long-term, large-scale perspective, secularization accompanied modernization throughout the industrial world, including the United States (Bruce 2002; Chaves and Hagaman 2002).

In addition to reducing the social influence of religion, secularization promotes the mixing of the sacred and the profane (Fenn 1978). American places of worship are becoming more secular, and the American clergy are increasingly involved in community affairs. Religious leaders were heavily involved in the civil rights movement and the Vietnam War protests of the 1960s (Hadden and Rymph 1973) and are still concerned with such political issues as world peace and poverty. Although the clergy's participation in public affairs has receded in favor of parishioners' spiritual needs, the clergy remain involved in solving such social problems as drug abuse, juvenile delinquency, hunger, malnutrition, homelessness, and school violence. Secular society is entering religion in other ways as well. Worship services, for example, use "worldly" musical instruments such as saxophones and electric guitars, as well as jazz and rock liturgies. The televangelists of the "electronic church" use many marketing and show-business techniques to attract viewers and bring in vast amounts of money (Hadden and Swann 1981; Hoover 1988).

Despite the declining influence of religion in American life, the United States is relatively religious compared with other industrialized nations. Whereas 40 percent of Americans indicate they normally attend a church or synagogue weekly, only 4 percent of those in England report doing so.

According to C. Kirk Hadaway, Penny Long Marler, and Mark Chaves (1993), characterizations of religious commitment in America typically rely on poll data. Self-reports of church attendance, they contend, inflate estimates of church attendance. Comparing church attendance rates based on actual counts of those who attend church with self-reported rates of church attendance, these researchers conclude that religious service attendance rates for Protestants and Catholics in the United States are actually about one-half the generally cited rates.

Robert Wuthnow (1976, 1978, 1990) argues that secularization, like increased religious fervor, should be seen as a process that fluctuates in strength. He claims that the religious revival of the 1950s waned because of opposition to it within the countercultural generation of the 1960s. Wuthnow thinks that a return to religious commitment may occur as successive age-groups, unaffected by the counterculture, mature.

Civil and Invisible Religion

There is yet other evidence of the enduring religious influence on American society. This can be seen in the existence of civil religion and invisible religion.

What is civil religion? Every American president's inaugural address (except Washington's second one) has made reference to God. Similar religious references exist throughout American life: Thanksgiving Day is a day of public thanksgiving and prayer; our currency proclaims "In God We Trust"; the pledge of allegiance to our flag contains the phrase "under God"; many formal public occasions open with prayer. Even voluntary organizations such as the Rotary Club and the Boy Scouts carry out this religious theme.

None of these public religious allusions, however, are supposed to involve any specific religion. References are not made to Jesus Christ, Moses, or the Virgin Mary but are made to the general concept of God, a concept compatible with nearly all religions. **Civil religion**, then, is a public religion that expresses a strong tie between a deity and a culture; it is broad enough to encompass almost the entire nation. Robert Bellah (1967, 1992; Bellah and Hammond, 1980), the sociologist most closely identified with the concept of civil religion, has summarized it this way:

> *The separation of church and state [in America] has not denied the political realm a religious dimension. Although matters of personal religious beliefs, worship, and association are considered to be strictly private affairs, there are, at the same time, certain common elements of religious orientation that the great majority of Americans share. These have played a crucial role in the development of American institutions and still provide a religious dimension for the whole fabric of American life, including the political sphere. This public religious dimension is expressed in a set of beliefs, symbols, and rituals that I am calling the American civil religion.* (Bellah and Hammond 1980:171)

The concept of civil religion continues to generate debate (Gehrig 1981; Demerath and Williams 1985; Audi and Wolterstorff 1996; Horsley 2003), but many studies support Bellah's claim for the existence of a civil religion in the United States (Wimberley et al. 1976; Christenson and Wimberley 1978; Wimberley and Christenson 1980; Kearl and Rinaldi 1983; Warner 1994; Heclo and McClay 2002). Martin Marty (1985) points out that civil religion is alive and well today, thanks in part to its adoption by the conservative evangelical churches, which in the 1980s linked God and nationalism. The public outcry over a federal court's ruling that bans the phrase "under God" from the Pledge of Allegiance provides strong evidence of civil religion in the United States.

What is invisible religion? Those focusing only on traditional religious beliefs and practices underestimate the religiosity of Americans (Luckmann 1967; Yinger 1969, 1970). Many Americans who do not belong to churches or sects nevertheless practice a type of religion. These individuals, according to Luckmann, practice an **invisible religion**—a private religion that is substituted for formal religious organizations, practices, and beliefs.

In a study of 208 households in Baton Rouge, Louisiana, Richard Machalek and Michael Martin (1976) found support for the existence of an invisible religion, even in a region of the United States that is defined as deeply religious in the conventional sense. The authors asked participants in their survey to list the most important concerns of life, beyond immediate daily problems, and to indicate the means they used to cope with them. They found that 18 percent of the respondents cited such traditional religious concerns as life after death, salvation, and the existence of heaven. More significantly, 8 percent listed a variety of less religiously conventional concerns, such as consideration and love for others, the need for less selfishness in the world, national corruption, lack of political leadership, poverty, the need for social equality, and the meaning of life. One-third of the sample relied on conventional religious practices—prayer, Bible study, formalized church activities, and informal religious study groups—to help them handle their daily concerns. For two-thirds of the sample, coping strategies were not related to institutionalized religion. They relied on such humanistic strategies as reflection and meditation, participation in civic organizations, and informal discussion groups composed of family and friends.

Even people who ordinarily remain outside formal religion tend to observe religious ceremony during such rites of passage as birth, marriage, and death. For example, Gorer (1965) found that although 25 percent of his sample did not believe in life after death, 98 percent arranged religious rites when they buried their relatives.

Whether or not the United States is as religious as in the past, it is clear that religious effects abound throughout the American way of life. Americans join churches, enter synagogues, and attend services in comparatively large numbers, and civil religion and private religion influence their lives daily. Moreover, there has recently been a revival of religious fundamentalism (Riesebrodt 1998).

© Mark Richards/PhotoEdit

Contemporary religious fundamentalism in the United States has its roots in the nineteenth century. True to their nineteenth-century antecedents, American religious fundamentalists today interpret the Bible literally, believe in Satan as an active force for evil, and wait for the eventual destruction of the world prior to the Messiah's second coming.

The Resurgence of Fundamentalism

Since the late 1960s, most mainline American Protestant denominations—Methodists, Lutherans, Presbyterians, Episcopalians—have either declined in membership or fought to hold their own. In contrast, contemporary fundamentalist denominations, the conservatives of Protestantism, have been growing (see Figure 15.3). Fundamentalists (also referred to as born-again or evangelical Christians) exist in all Protestant organizations, but they are predominantly found in such religious bodies as the Assemblies of God, the Seventh-Day Adventists, the Southern Baptists, and the Jehovah's Witnesses (Finke and Stark 1992; Aldridge 2000; Antoun 2001). More than half of Protestants (54 percent) identify themselves as born-again or evangelical Christians (Gallup 2002).

It is, of course, inaccurate to limit fundamentalism to Protestants alone. Fundamentalism is found in all religions, including the Roman Catholic (21 percent), Jewish (7 percent), and Mormon (24 percent) faiths. In this section, however, we will concentrate exclusively on Protestant fundamentalism.

FIGURE 15.3

American Church Membership Trends: 1990–1999*

This figure displays trends in church membership in the United States. Do you believe that this pattern will continue in the twenty-first century? Explain your conclusion using text materials.

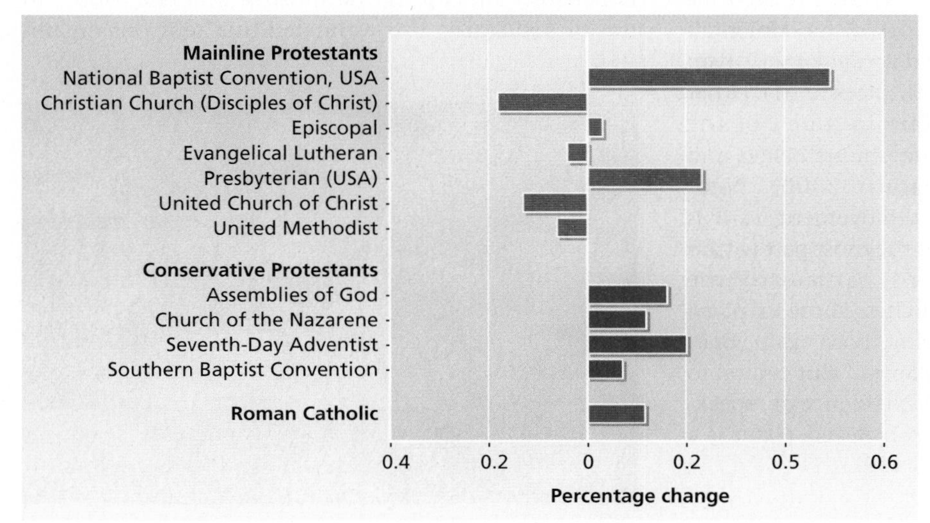

Sources: Constant H. Jacquet Jr., (ed.), *Yearbook of American and Canadian Churches, 1990.* Copyright © 1990 by the National Council of the Churches of Christ in the U.S.A.; and Eileen W. Linder, (ed.), *Yearbook of American and Canadian Churches, 1999.* Copyright © 1999 by the National Council of the Churches of Christ in the U.S.A.

What is the nature of fundamentalism? Fundamentalism is based on the rejection of secularization and the corresponding close adherence to traditional beliefs, rituals, and doctrines. It is not surprising that most religious fundamentalists are politically conservative, given that the roots of contemporary religious fundamentalism can be found in the latter part of the nineteenth century. Two issues disturbed the early fundamentalists. First, fundamentalists were concerned about the spread of secularism. Science was challenging the Bible as a source of truth, Marxism was portraying religion as an opiate for the masses, Darwinism was challenging the biblical interpretation of creation, and religion was losing its traditional influence on social institutions. Second, fundamentalists rejected the movement away from the traditional message of Christianity toward an accent on social service (Johnstone 2000).

The theological agenda of today's fundamentalists is very close to that of their nineteenth-century forebears. Today's Protestant fundamentalists believe in the literal truth of the Scriptures, being "born again" through acceptance of Jesus Christ as the Son of God, the responsibility of all believers to give witness for God, the presence of Satan as an active force for evil, and the destruction of the world prior to the Messiah's return to establish his kingdom on earth (H. Cox 1996; Hunt, Hamilton, and Walter 1998; C. Smith and Emerson 1998).

Are all fundamentalists alike? Variations exist, of course, among fundamentalists. The term *evangelical,* for example, is often used as a synonym for fundamentalism. However, when compared to evangelicals, most fundamentalists are more theologically rigid and less accommodating in their opposition to the rest of the world. Most evangelicals are on the political right or in the center, but some are on the ideological left. Evangelicals share much of the traditional theology of fundamentalism, but often reject its political and social conservatism (Quebedeaux 1978; J. D. Hunter 1987; Roof 2001). Other religious organizations that share in much of the fundamentalist theology have some unique beliefs and practices of their own (C. Smith 2000). Neo-Pentecostalism—or the charismatic movement, as it is sometimes called—has occurred for the most part within traditional religious organizations, particularly the Roman Catholic and Episcopal churches. Those involved in this movement often speak of being "born again" or of receiving "the baptism of the Holy Spirit." But central to most neo-Pentecostal groups is the experience of "speaking in tongues" (glossolalia), which believers claim is a direct gift of the Holy Spirit (H. Cox 1992).

Why has fundamentalism reappeared? There are several reasons for the revival of fundamentalism. First, many Americans feel their world is out of control. The social order of the 1950s has been shattered by a string of traumatic events beginning with the civil rights movement and progressing through campus violence, political assassinations, the Vietnam War, Watergate, and destruction of the World Trade Center towers. Fundamental religion, with its absolute answers and promise of eternal life, provides a strong anchor in a confusing, bewildering world. Second, by placing emphasis on warmth, love, and caring, fundamentalist churches provide solace to people who are witnessing and experiencing the weakening of family and community ties. Mainline churches are more formal and impersonal. Third, fundamentalist churches offer what they consider a more purely sacred environment, in contrast to mainline denominations that fundamentalists see as accommodating to secular society. Finally, the electronic church has articulated the appropriate messages to a receptive nationwide audience. As part of the mass media, the electronic church has been an important contributing factor in the growth of religious fundamentalism (Hadden and Swann 1981). The place of the electronic church deserves some elaboration.

What has been the place of the electronic church in the resurgence of fundamentalism? In the past, religious evangelists typically traveled from community to community and preached to the faithful in tents, open fields, or rented meeting halls. Modern technology has changed all that. Television, radio, and direct-mail campaigns via computers have made it possible for religious broadcasters to reach millions of people easily, rapidly, and repetitively. Ministers, particularly from the fundamentalist flank of Protestantism, have created vast and profitable television ministries. Prime-time preachers (Jerry Falwell, Jimmy Swaggart, Oral Roberts, and Pat Robertson) are among the most prominent people in the United States. It is estimated that religious organi-

Robert Tilton is part of the "electronic church." Controversy exists regarding the effects of televangelists. Although some attribute significant influence to the electronic church, others think its effects are limited.

Fenggang Yang and Helen Rose Ebaugh—The Secularization of Religion?

Yang and Ebaugh (2001) challenge the long-standing view that religious pluralism accompanying modernization secularizes religions. They present evidence for a new view: instead of promoting the decline of religions, religious pluralism leads to their revitalization. Organizational and theological transformations provoked by exposure to a wide variety of religions are said to energize religions.

Their research is based on the study of immigrants who came to the United States after 1965. These "new immigrants" arrive in much larger numbers from Asia and Latin America than from Europe, the predominant home of "old immigrants." Whereas earlier European immigrants left mainly Judeo-Christian countries, many of the new immigrants from Asia are Muslim, Hindu, Buddhist, or adherents to other religions (Eck 2002). Even immigrants from South and Central America grew up with forms of Catholicism and Protestantism distinct from U.S. versions.

Data came from thirteen immigrant churches, ranging from Buddhist temples to Protestant churches. In 1997–1998, intensive interviews were conducted with clerical and lay leaders, new immigrants, and more established residents.

According to Yang and Ebaugh, these churches have prospered by adopting new organizational and theological forms. New immigrant religions are becoming congregational, theologically pure, and open to outsiders.

Adopting Congregation Forms

New immigrants often abandon the religious institutional forms brought from their home countries in favor of U.S. Protestant-style congregations, which are communities that gather voluntarily. Rather than remaining for life in their religions of birth, new immigrants feel freer to join or leave a religious group. Also, these new immigrants had been taught reliance on periodic visits to formal religious leaders at their often isolated temples and monasteries. These immigrants now emphasize lay leadership within their local congregations, and they are providing social services and recreational centers.

Returning to Theological Roots

New immigrants come from societies dominated by one or two religions. In such societies, religion is intertwined with the culture. Consequently, being a Muslim in one country may be different in certain ways from practicing Islam in another. For example, in the United States Pakistani Muslim men, accustomed to praying with their caps on, find themselves worshipping with Arab Muslims who pray bareheaded. Which custom does one's pluralistic congregation adopt? Rather than attempting to make a host of such decisions, American Muslim congregations return to the original teachings of Prophet Muhammad, which they share worldwide. Various culturally based adaptations of Islam are dropped as congregations seek commonalities. Reference to original theological foundations provides culturally diverse Muslim congregations with a unified rationale for determining what aspects of their faith to keep and to discard.

Openness to Outsiders

Diverse congregations forge a distilled theological base. By reducing religious idiosyncrasies, new immigrant congregations can incorporate members from their own faith as well as from other religious traditions. Not only can Pakistani and Arab Muslims worship together, they can attract native-born American converts. This is appealing because it speeds up the process of Americanization.

Thinking About the Research

1. Why would these organizational and theological processes of change work to prevent a weakening of the new immigrants' religions?
2. Explain the interrelationships among these three processes of change.
3. Discuss the link between these processes of change and secularization.

zations own over 1,400 radio stations and 60 television stations. According to the National Religious Broadcasters Association, there are about 350 television ministries. From the thousands of weekly hours of religious programming come well over 500 million tax-free dollars each year (Johnstone 2000).

Sexual and financial scandals surrounding Jim Bakker, Jimmy Swaggart, and the Catholic Church have damaged the religious institution. The proportion of Americans indicating that they have "a great deal" or "quite a lot" of confidence in religion as an institution decreased from 66 percent in 1985 to 45 percent in 2002, a drop of 21 percentage points (Herlinger 2002). It remains to be seen whether this drop in public confidence weakens the electronic church and the fundamentalist religions themselves.

New Religious Movements

Many new religious movements have emerged alongside the resurgence of fundamentalism (Heelas 1996; Bainbridge 1997; J. R. Stone 1997; Bromley 1998; Zellner and Petrowsky 1998; Group 2002). Speaking broadly, these new religious movements embrace such varied organizations as the Jesus People (on the fringe of the Christian establishment), the Reverend Moon's Unification Church, Scientology, the Guru Maharaj Ji, the Hare Krishnas, the Messianic Jews, and politically oriented movements such as the Christian Coalition.

Religious cults are an interesting part of the new religious movements. Three examples, varying in their closeness to Christianity, are the Raelians, the Unification Church, and Scientology.

Who are the Raelians? In January 2003, a religious cult known as the Raelians proclaimed to the world their successful attempt to clone the first human being (named "Eve"). The group's promise to verify this claim with DNA evidence from the baby and the egg donor, however, was never kept. Claiming 55,000 adherents, this cult was founded by Claude Vorilhon (a.k.a. Raël), a French-born Canadian whose thirty-year-old messianic vision was sparked by being aboard an alien spaceship for a trip to a distant planet whose inhabitants created life on earth. Raël's mission is to replace the "myth of God" with a scientific vision of creation. As part of this mission, the Raelians see human cloning as a path to achieving external life. The Raelians also are preparing a nucleus of attractive females ("Order of the Angels") to welcome the return of their alien creators sometime prior to 2035 (Adler 2003).

What is the Unification Church? One of the most controversial of the new religious movements is the Reverend Sun Myung Moon's Unification Church, one of the new religions whose adherents are generally (and negatively) referred to as Moonies. Mr. Moon is a Korean industrialist turned prophet–minister–messiah–newspaper publisher who proclaims a religion that is a combination of Protestantism and anticommunism (Barker 1984).

Several interesting sociological observations can be made about the Unification Church. First—as in some of the other new mystical movements, such as the Hare Krishnas—the Unification Church expects total conformity of its members. Second, despite parental fears, its morality is generally conservative, sometimes rigidly so. More often than not, it endorses such basic American values as a powerful God, a strong sense of family, and a belief in individualism. Third, the more the Unification Church grows, the more it resembles a denomination. For example, it now has its own seminary with faculty members who belong to respected national professional societies.

Although the Unification Church is small, it has attracted considerable attention. It forced society to deal with the legal rights of converts whose family members believe their kin were brainwashed (Richardson 1982; Shepherd 1982; Bromley and Shupe 1984). More recently, the Unification Church was in the public spotlight because of its economic activities. The Reverend Moon's movement now controls more than $10 billion in business assets around the world. These assets include an automobile plant in Asia, the University of Bridgeport (Connecticut), the *Washington Times* newspaper, Atlantic Video, and the Nostalgia television network (Byron 1993).

What is Scientology? Scientology is a spiritual therapy inspired by the late science fiction writer L. Ron Hubbard. Hubbard claimed to have discovered the therapy and technique of Dianetics, a new approach to mental health that was received with faddish seriousness in the 1950s. Later, Hubbard developed an elaborate system of Scientology that included far more metaphysical speculation, particularly the doctrine of reincarnation. Hubbard and his followers claim the ability to perform mental and physical healing. Its practitioners charge for their services; they have very little formal religious ritual; and, like the Unification Church and some other newer religious movements, they demand absolute conformity to stated doctrines. In recent years, however, the requirement that followers sever all non-Scientological connections was relaxed (Hubbard 1998).

Are these new religious movements unique? These new religious movements resemble older religions in many ways. The New Testament, for example, demands that Christians place the gospel ahead of family and friends. And throughout the history of Western Christendom, movements have periodically arisen that demand absolute conformity, self-denial, and separation from family and friends who could not be converted (Beckford 1985). Similarly, the political activities of religious fundamentalists such as Jerry Falwell and Pat Robertson have many parallels in American history, and political involvement by fundamentalist religious groups also occurs in other nations (Marty 1980; Yinger and Cutler 1982). Nor are exotic cults new to our nation. Consider this example from the early 1880s:

> *Some frontier religions were mildly eccentric, while others were extremely bizarre. Possibly the most flamboyant messiah of all was Isaac Bullard. Clad in nothing but a bearskin loincloth, he led his troop of adherents from Vermont through New York and Ohio and finally south to Missouri, proclaiming a religion compounded of free love, communism and dirt—Bullard often boasting that he hadn't washed in seven years.* (Wise 1976:20)

In late November 1978, news began to arrive in the United States, first slowly and then with a tragic rapidity, that a semireligious, socialistic colony in Guyana, South America, headed by the Reverend Jim Jones—founder of the California-based People's Temple—had been the scene of one of the most shocking suicide-murder rites in recent times. The bizarre accounts appearing in the media prompted many people to ask how anyone could have become involved in the cult.

Some tried to answer the question by dismissing the participants as ignorant or mentally unbalanced individuals. But as more facts were uncovered, people learned that many of the members were fairly well educated young people and that Jones was trusted and respected by many in the California political establishment. They also learned that such events, although rare, had occurred before.

Moreover, it would happen again. Like Jim Jones, David Koresh was a charismatic cult leader. In 1993, under FBI assault, he and many of his Branch Davidians (some sixty people, including seventeen children) went up in flames in their Waco, Texas, compound. Koresh saw himself as Jesus Christ in sinful form, and his devoted followers probably believed they were on their way to join Jesus in heaven after the Apocalypse (Edmonds and Potok 1993; Gibbs 1993a).

Why are people willing to go to extremes? Sociology cannot answer this question, but it can help us understand why people may be disposed to join religious cults.

Why have these new religious movements emerged? Reasons for movements closely related to fundamentalism were covered in the discussion of the revival of fundamentalism. However, such movements should not be identified with the likes of the People's Temple of Jim Jones or David Koresh's Branch Davidians. In this section we look exclusively at some explanations for the current rise of cults and why people join them.

Thomas Robbins, Dick Anthony, and James Richardson (1978) contend that cults are the product of a value crisis or normative breakdown in modern industrial society. In this view, some people attempt to escape moral ambiguity—a sense of personal helplessness, doubt, and uncertainty—by grasping onto a new "truth."

According to Charles Glock and Robert Bellah, interest in cults is a reaction to the excessive emphasis on self-interest in American Protestantism. The exaggeration of the parts of the Bible emphasizing self-interest has resulted in a failure to provide either social benefits or a sense of personal worth in complex urban society. The new religious yearning seen in cults is geared toward socialism and mysticism, which are incompatible with American denominations that emerged from an agricultural or frontier environment:

> *The demand for immediate, powerful, and deep religious experience, which was part of the turn away from future-oriented instrumentalism toward present meaning and fulfillment, could on the whole not be met by the [traditional] religious bodies. . . . In many ways Asian spirituality provided a more thorough contrast to the rejected utilitarian individualism than did biblical religion. To external achievement it posed inner experience; to the exploitation of nature, harmony with nature; to the impersonal organization, intense relation to a guru.* (Glock and Bellah 1976:340–341)

Harvey Cox (1977) identifies four seductive features of religious cults. First, most of the cult converts are looking for friendship, companionship, acceptance, warmth, and recognition. The cult provides a supportive community that helps overcome past loneliness and isolation. The Eastern religious cults provide emotional ties that converts cannot find at home, school, church, or work. Many of them even use kinship terms to give recruits new identities—sister, brother, Hare Krishna. Often a convert is renamed. Entertainer Steve Allen's son Brian joined the Church of Jesus Christ at Armageddon in Seattle and was "reborn" under the name Logic Israel.

Second, most of the Eastern religious cults emphasize immediate experience and emotional gratification rather than deliberation and rational argument. Converts report "experiencing" religion rather than merely thinking about it. Whether by meditation, speaking in tongues, or singing hymns, adherents have the frequent and intense emotional experiences they could not find elsewhere.

Third, Eastern religious cults emphasize authority. By having a firm authority structure and a clear, simple set of beliefs and rules, they offer converts something in which to believe. Converts profess to exchange uncertainty, doubt, and confusion for trust and assurance.

Finally, these cults purport to offer authenticity and naturalness in an otherwise artificial world. By emphasizing natural foods, communal living apart from "civilization," a uniform dress code, and sometimes nudity, these groups attempt to show that they are not part of the "plastic society."

Social Correlates of Religion

Social class and politics are two important social correlates of religious preference. Before discussing them, however, it will be helpful to glance at a picture of religious affiliation in the United States.

What are the religious preferences of Americans? As noted earlier, 93 percent of Americans profess some religious preference. Although there are over 300 denominations and sects in the United States, Americans are largely Protestant (56 percent) and belong to a few major denominations—Baptist (20 percent), Methodist (10 percent), Lutheran (6 percent), Presbyterian (4 percent), and Episcopalian (4 percent). Fourteen percent prefer other Protestant denominations. Catholics constitute a relatively large proportion of the American population (27 percent) and Jews a small proportion (2 percent) (U.S. Bureau of the Census 2001a).

What is the relationship between social class and religious characteristics? There are marked differences in social class (as measured by education and income) among the adherents of various religions in the United States. Generally speaking, Presbyterians, Episcopalians, and Jews are on the top of the stratification structure. Below them are Lutherans, Catholics, and Methodists. When measured by education and income, Baptists, on the average, come out the lowest. Because these are average figures, there are, of course, many individual exceptions to these rankings.

The social class differences are due partly to self-selection: people tend to prefer churches with members who have socioeconomic characteristics similar to their own. The socioeconomic differences occur partly because religious organizations place varying emphasis on worldly success versus otherworldly rewards (Niebuhr 1968). Finally, a historical dimension exists. With the exception of the Jewish groups, the higher-status denominations have experienced this social benefit for a long time. Thus, members of these denominations have had tremendous socioeconomic advantages. Lutherans and Roman Catholics, in contrast, tend to be later immigrants who have not had as many opportunities to achieve higher status (Johnstone 2000).

Differences in religiosity exist between the upper and lower classes. Reflecting Marx's view of religion as an opiate of the masses, many early sociologists saw religion primarily as a source of solace for the deprived poor. This entrenched view was challenged by research in the 1940s showing higher church affiliation among the wealthy. Further research in the 1950s and 1960s revealed religion as important at both extremes of the stratification structure. It is just that the upper and lower classes express their religious beliefs in different ways. The upper classes display their religiosity through church membership, church attendance, and observance of ritual, whereas lower-class people more often pray privately and have emotional religious experiences (Stark and Bainbridge 1985).

What is the relationship between religion and politics? Because American religious groups vary widely in socioeconomic characteristics, we must be careful in interpreting empirical correlations between religion and politics (Audi and Wolterstorff 1996). Some of the apparent influence of religion on political attitudes and behavior is in part due to social class.

Followers of the Jewish faith are particularly aligned with the Democratic Party; they are followed in strength of support by Catholics and Protestants. This is predictable, because Protestants generally are more politically conservative than Catholics or Jews, and the Democratic Party is generally not associated with political conservatism. Of the major Protestant denominations, the greatest support for the Republican Party is among Episcopalians and Presbyterians, which is hardly surprising, because the upper classes are more likely to be identified with the Republican Party.

There are some contradictions in this general pattern. Despite their affiliation with the more conservative Republican Party, Episcopalians and Presbyterians are less socially conservative than Baptists, who have been the strongest supporters of the Democratic Party of all Protestant denominations, especially in the South.

At the presidential level, white Protestants, significantly more than all Catholics or Jews, are likely to vote for Republican candidates (Green 1997). Since 1952, Catholics voted for Democratic candidates, except in 1972, when they voted for Richard Nixon, and in 1980 and 1984, when they voted for Ronald Reagan. Protestants broke their string of Republican presidential votes only in 1964, when 55 percent voted for Lyndon Johnson.

Religion and the Baby Boomers

What is unique about baby boomers? Two points need to be established about America's baby boomers, the generation born between 1946 and 1964 (whose oldest members are now in their fifties). First, there are a lot of them. Their number—72 million—represents about 36 percent of the adult population and about 26 percent of the entire population of the United States. When baby boomers say "Jump," American society generally replies "How high?" America's schools, political system, mass media, and housing industry have adapted to accommodate their needs and take advantage of their consumption potential at every stage of the life cycle. Second, having grown up during a period of startling change, baby boomers are not the homogeneous generation that their collective label implies. Their ranks include all social classes, ethnic groups, and races. They are also separated by gender. Despite this diversity, baby boomers have a shared identity sufficient to view them as a socially meaningful generation.

What are the baby boomer effects on religion? The religious institution first felt the effects of the baby boomers when their parents swelled membership rolls

FIGURE 15.4

Age Cohorts and Attitudes Toward Organized Religion

As this figure reveals, religious attitudes in the United States are changing. How do you think your generation would compare to pre–baby boomers and baby boomers?

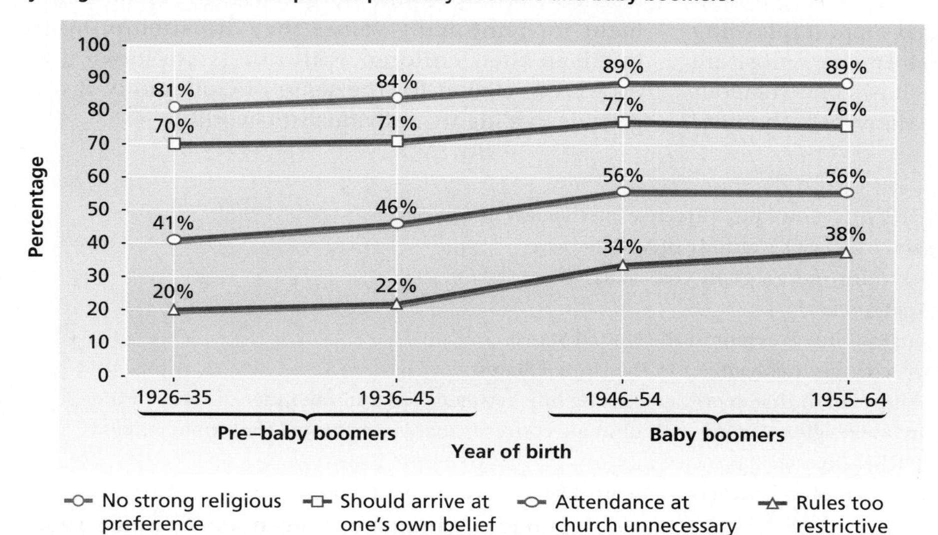

Source: "Age Cohorts and Attitudes Toward Organized Religion," *A Generation of Seekers,* by Wade Clark Roof. Copyright © 1993 by Wade Clark Roof. Reprinted by permission of HarperCollins Publishers, Inc.

by faithfully bringing them to houses of worship. The second shock came when the baby boomers, after leaving home, dropped out of their religious organizations, in record numbers (Bouvier and De Vita 1991). As revealed in Figure 15.4, the two **age cohorts**—persons born during the same time period in a particular population—born prior to and during World War II (1926–1935 and 1936–1945) have a much stronger orientation toward organized religion than the baby boomers' birth cohorts (1946–1954 and 1955–1964). These differences in religious attitudes help explain why the bulk of baby boomers left their religious organizations after leaving home and why a substantial proportion still do not return.

According to some researchers, large numbers of baby boomers are reverting to the spiritual roots they abandoned. If that is true, American religion is once again about to feel the shift made by the baby boom generation. Organized religion may be in for another landscaping.

About one-third of the baby boomers never left their synagogue or denomination, and over 40 percent remain dropouts from organized religion. It is the returning one-fourth that is altering the nature of formal religion in America and the religious practices of millions of its inhabitants. These are some of the observations made by Wade Clark Roof, who conducted an extensive survey on religion among the baby boomers (Roof 1994, 2001).

© Peter Southwick/Stock, Boston

Some evidence suggests that many baby boomers, who left their churches in large numbers in their early twenties, returned to their religious roots. This trend, however, did not produce the expected shock wave that baby boomers typically cause when they touch an institution.

Why are many baby boomers returning to church? Baby boomers are at the age when preceding generations of Americans have typically become more active in organized religion. According to a Gallup poll, many baby boomers are following this pattern (Gallup 2002).

Although some baby boomers are returning to religious organizations for needed social support following divorces or severe career disappointments, most are doing what parents normally do—show concern about the well-being of their children. Many baby boomer parents believe it their duty to expose their children to religious tenets. Later, the children can, as a matter of their own choice, embrace or reject the religious influence (Aldridge 2000). Others see their children without exposure to moral values except within a religious context. For some, the church or synagogue is an instrument for reinforcing values they are attempting to instill in their children. Still others are looking to organized religion for the sense of community it can provide in a highly individualistic world.

FEEDBACK

1. _________ is the process whereby religion loses influence over society as a whole.
2. According to Caplow, Bahr, and Chadwick's restudy of Middletown,
   a. American ministers are becoming more socially conscious.
   b. secularization has proceeded too far.
   c. the trend is toward less religious involvement in the United States.
   d. the trend is toward greater religious involvement in the United States.
3. A _________ religion is a public religion that expresses a strong link between God and the American way of life.
4. An _________ religion is a private religion dealing with ultimate concerns that is substituted for formal religious organizations, practices, and beliefs in a secular society.
5. _________ is based on the rejection of the effects of modern society on traditional religious beliefs, rituals, and doctrines.
6. Which of the following is *not* one of the reasons presented for the resurgence of fundamentalism in the United States?
   a. fear that the world is out of control
   b. provision of solace by fundamentalist organizations
   c. the electronic church
   d. the rise of invisible religion
7. The origins of new religious movements appear to lie in
   a. social conditions in modern industrial society.
   b. desecularization.
   c. the rising influence of the Middle East.
   d. the decline of civil religion.
8. Which of these groupings of religious organizations is at the top of America's stratification structure?
   a. Jewish, Presbyterian, Episcopalian
   b. Lutheran, Jewish, Catholic
   c. Catholic, Episcopalian, Methodist
   d. Methodist, Baptist, Presbyterian
9. Which of the following is the major reason many baby boomers are turning to churches?
   a. concern for the ethical and religious training of their children
   b. need for comfort following divorces
   c. guilt over letting their own religiously oriented parents down
   d. fear of dying without practicing a religious faith

Answers: 1. Secularization 2. d 3. civil 4. invisible 5. Fundamentalism 6. d 7. a 8. a 9. a

REVIEW GUIDE

SUMMARY

1. Religion is a unified system of beliefs and practices about sacred things. From his studies of Australian aborigines, Emile Durkheim concluded that every religion distinguishes between the sacred and the profane.
2. Because religion involves a set of meanings attached to a world beyond human observation, sociologists who study it scientifically face some unique problems. They do not evaluate the validity of various religions, leaving theological issues to theologians and philosophers.
3. Religion is found in nearly all societies. Its importance is due in part to the functions it serves. Religion legitimates the structure of society, promotes social unity, and provides a sense of meaning and belonging.
4. Writing from the conflict perspective, Karl Marx contended that religion is used by the elite to justify and maintain their power, and it is used by the oppressed to soothe their misery and maintain hope for improvement. Religion, in his view, is the opiate of the masses.
5. According to the conflict interpretation, religion retards social change because those in power wish to maintain the status quo. Max Weber, on the other hand, believed that religion could promote social change. He attempted to support this idea by connecting the Protestant ethic with the rise of capitalism. The Protestant ethic emphasizes hard work and reinvestment for profit, characteristics that were highly compatible with the needs of emerging capitalism.

6. Most of the world's major religions suppress women. Starting in the early 1980s, feminists have challenged the male patriarchy characteristic of Christian religious organizations.
7. Two of the most important components of religion are religious organization and religiosity. The major forms of religious organization are churches, denominations, sects, and cults. The ways people express their religious interests and convictions (religiosity) can be analyzed in relation to five dimensions: belief, ritual, intellectual, experience, and consequences.
8. Through secularization, the sacred and the profane are increasingly intermixed in modern society. The existence of secularization does not mean, however, that a society is not religious. Although many observers have noted a decline in the role of religion in the United States, others contend that religion may be stronger than supposed. For one thing, Americans practice both a civil (public) religion and an invisible (private) religion.
9. Moreover, there has been a revival of religious fundamentalism in the United States. This fundamentalism, which is occurring largely outside the mainline religious organizations, shares a theological conservatism with nineteenth-century American fundamentalism. In addition, contemporary fundamentalism has its own conservative social and political agenda.
10. The resurgence of religious fundamentalism is due to several factors, not the least of which is the electronic church. Television, radio, and computers (for direct-mail campaigns) have enabled fundamentalist organizations to spread their messages to vast numbers of people on a continuous basis. It remains to be seen whether recent scandals among some televangelists will damage the electronic church and/or fundamentalism itself.
11. The rise of new religious movements since the 1960s may also be an indication of a religious revival. The Raelians, the Unification Church, Scientology, and other new religious movements may be a response to contemporary feelings of confusion, doubt, helplessness, and uncertainty.
12. Two important social correlates of religion are social class and politics. The major religious faiths can be ranked based on educational level, income, and occupational prestige. Also, the upper and lower classes express their religiosity in quite different ways.
13. The baby boomers, a generation that has influenced all aspects of American social life since they began being born in 1946, are having a new effect on organized religion. Having dropped out of churches early in their lives, many are returning, primarily out of concern for the ethical and religious training of their children.

REVIEW GUIDE

INFOTRAC® COLLEGE EDITION

http://www.infotrac-college.com/wadsworth

Another unique option available to you at the Student Resources section of the companion Web site is Infotrac College Edition, an online library with access to hundreds of scholarly and popular periodicals. Below are suggested search terms for this chapter. Results from these and other searches are found at the site.

Search keywords: **civil religion**. Find three articles that discuss civil religion in the United States. What evidence do the authors present to support the existence of a civil religion? Do any of the authors claim that the United States has no civil religion? If so, why? How does the constitutionally guaranteed separation of church and state affect the existence of a civil religion, if at all?

Search keywords: **cults and Raelians**. The Technology and Society feature in this chapter presents information about human cloning. The Raelians, who claim to have cloned two humans already, are considering building a hotel to house people who want to be cloned. Search Infotrac to find out more about the Raelians. Do a separate search on cults. Based on these articles, define what constitutes a cult. Do the Raelians fit the definition of a cult? Why or why not?

Search keywords: **religious righ**t. Discover the leaders of the religious right in the United States. Do you recognize any of the names? If so, from where? What impact has the religious right had on American society?

LEARNING OBJECTIVES REVIEW

- Explain the sociological meaning of religion.
- Demonstrate the different views of religion taken by functionalists and conflict theorists.
- Distinguish among the basic types of religious organization.
- Discuss the meaning and nature of religiosity.
- Define secularization and describe its relationship to religiosity in the United States.
- Differentiate between civil and invisible religion in America.
- Describe the current resurgence of religious fundamentalism in the United States.
- Identify a wide variety of new religious movements in America.
- Outline the relationship of baby boomers to religion.

CONCEPT REVIEW

Match the following concepts with the definitions listed below them.

____ a. secularization
____ b. invisible religion
____ c. sacred
____ d. age cohort
____ e. profane
____ f. civil religion
____ g. denomination

1. persons born during the same time period in a particular population
2. a public religion that expresses a strong tie between a deity and a society's way of life
3. the process whereby religion loses its influence over society as a whole
4. one of several religious organizations that members of a society accept as legitimate
5. a private religion that is substituted for formal religious organizations, practices, and beliefs in a secular society
6. nonsacred aspects of our lives
7. things that are set apart and given a special meaning that transcends immediate human existence

CRITICAL-THINKING QUESTIONS

1. Does the distinction between the sacred and the profane still have validity in America today? Explain why or why not, using examples.

____

____

2. Consider the four functions of religion identified by sociologists. Which of these functions are being performed by religion in the contemporary United States? Illustrate each.

____

____

3. Describe Weber's Protestant ethic. Explain why Marx could not have endorsed the relationship between the Protestant ethic and the spirit of capitalism proposed by Weber.

____

____

4. What do you think is the relationship, at the present time, between secularization and religiosity in the United States? Develop your answer with material from the text.

____

____

5. Evaluate the place of civil religion and invisible religion in America today.

____

____

6. Identify what you think are the three most important effects of fundamentalism on religion in contemporary America. Support your case with specifics.

____

____

____

MULTIPLE-CHOICE QUESTIONS

1. **Which of the following statements is false?**
   a. A profane object can become a sacred object.
   b. Profane objects do not take on a public or social character that makes them appear important in themselves.
   c. Religion is a unified system of beliefs and practices relative to sacred things.
   d. In their research, sociologists must avoid the strictly spiritual side of religion.
   e. Sociologists cannot successfully avoid questions about the ultimate validity of any particular religion.
2. **Aspects of life that are everyday, ordinary, and taken for granted are labeled**
   a. sacred.
   b. transcendental.
   c. profane.
   d. legitimate.
   e. material.
3. **Archaeological discoveries and anthropological observations have demonstrated that religion**
   a. began around the eleventh century.
   b. tends to dissolve into a limited number of basic denominations.
   c. has been more successful at governing societies than political parties.
   d. always contains an element of salvation and afterlife.
   e. has existed in some form in nearly all known societies.
4. **Emile Durkheim contended that _________ was the central function of religion.**
   a. sacredness
   b. supernaturalism
   c. legitimation
   d. transcendentalism
   e. salvation
5. **Religious ceremonies marking important events in life, such as birth, sexual maturity, marriage, and death, illustrate the contribution of religion to**
   a. a sense of unity.
   b. a sense of anomie.

c. a sense of time.
d. a sense of meaning.
e. the legitimation of social arrangements.

6. Karl Marx contended that religion _________ social change.
a. promotes
b. retards
c. refracts
d. serves no function in
e. precludes

7. The Calvinist belief that every person's eternal fate is unalterably predetermined by God at birth is known as
a. sacred legitimation.
b. transcendentalism.
c. predestination.
d. deism.
e. immaculate salvation.

8. According to Max Weber,
a. early Protestantism provided a social environment conducive to capitalism.
b. religion tends to promote the economy in most societies.
c. capitalism developed as a direct result of Protestantism.
d. Protestantism developed as a direct result of capitalism.
e. religion serves as an economic opiate for the masses.

9. The ways in which people express their religious interests and convictions is called _________ by sociologists.
a. piety
b. religiosity
c. worship
d. secularization
e. ritualism

10. _________ is the process through which religion loses influence over society as a whole.
a. Dereligiosity
b. Transcendentalism
c. Segmentation
d. Secularization
e. Pragmatism

11. According to studies of religiosity in the United States,
a. secularization is destroying religion in the United States.
b. only half of all Americans now believe in God or a universal spirit.
c. only 20 percent of all Americans now believe in life after death.
d. the proportion of regular churchgoers has declined sharply.
e. compared with other industrialized countries, America still appears to be a fairly religious nation.

12. Researchers cited in the text attribute Americans' continuing interest in religion to three basic reasons: the threat of nuclear war, loneliness, and
a. the need to impress others with their piety.
b. disenchantment with what they perceive as an anomic society.
c. basic spiritual and psychological needs.
d. the absence of competing institutions to fulfill their needs.
e. guilt and fear of divine punishment after death.

13. The involvement of contemporary clergy in community affairs demonstrates
a. the mixing of the sacred and the profane.
b. their resistance to secularization.
c. the strong tie between a deity and a society's way of life.
d. the demise of civil religion.
e. the increasing significance of religion.

14. Many of the new religious movements in the United States have emerged alongside the resurgence of
a. fundamentalism.
b. escapism.
c. desecularization.
d. invisible religion.
e. civil religion.

15. Which of the following statements about baby boomers is not incorrect?
a. About one-third of baby boomers never left their churches.
b. Baby boomers are religiously heterogeneous.
c. Older baby boomers with children are more involved in church than younger members of their generation.
d. Baby boomers do not appear to be returning to the spiritual roots they had abandoned.
e. The bulk of baby boomers abandoned their churches after leaving home.

FEEDBACK REVIEW

True-False

1. Sociologists who study religion are normally forced to abandon their religious beliefs and convictions. T or F?

Fill in the Blank

2. According to Marx, religion is created by the _________ to justify its economic, political, and social advantages over the oppressed.
3. _________ is based on the rejection of the effects of modern society on traditional religious beliefs, rituals, and doctrines.

Multiple Choice

4. According to Karl Marx, religion is a force working for
a. the good of all people.
b. the good of the proletariat.
c. maintenance of the status quo.
d. the creation of conflict between the classes.

5. **According to Max Weber,**
   a. early Protestantism provided a social environment conducive to capitalism.
   b. capitalism developed because of Protestantism.
   c. capitalism strengthened Protestantism.
   d. religion will always be linked with economy.
6. **Which of the following is *not* one of the reasons presented for the resurgence of fundamentalism in the United States?**
   a. fear that the world is out of control
   b. provision of solace by fundamentalist organizations
   c. the electronic church
   d. the rise of invisible religion
7. **The origins of new religious movements appear to lie in**
   a. social conditions in modern industrial society.
   b. desecularization.
   c. the rising influence of the Middle East.
   d. the decline of civil religion.
8. **Which of these groupings of religious organizations is at the top of America's stratification structure?**
   a. Jewish, Presbyterian, Episcopalian
   b. Lutheran, Jewish, Catholic
   c. Catholic, Episcopalian, Methodist
   d. Methodist, Baptist, Presbyterian
9. **Which of the following is the major reason many baby boomers are turning to churches?**
   a. concern for the ethical and religious training of their children
   b. need for comfort following divorces
   c. guilt over letting their own religiously oriented parents down
   d. fear of dying without practicing a religious faith

Matching

10. Match the dimensions of religiosity with the examples beside them.

____ a. belief	(1) prayer in church
____ b. ritual	(2) knowing the content of the Koran
____ c. intellectual	(3) having seen an angel
____ d. experience	(4) conviction that the Bible is divinely inspired
____ e. consequences	(5) marching in support of prayer in school

REVIEW GUIDE

GRAPHIC REVIEW

Data in Figure 15.4 compares attitudes of pre–baby boomers and baby boomers toward organized religion.

1. Did the change in attitudes toward organized religion begin with the baby boomers? Explain your answer by using the statistics presented.

____

____

____

2. The typical analysis contrasts the attitudes and behavior of pre–baby boomers with baby boomers. Use the data in Figure 15.4 to discuss differences in attitudes toward organized religion *within* each generation.

____

____

____

ANSWER KEY

Concept Review

a. 3
b. 5
c. 7
d. 1
e. 6
f. 2
g. 4

Multiple Choice

1. e
2. c
3. e
4. c
5. d
6. b
7. c
8. a
9. b
10. d
11. e
12. b
13. a
14. a
15. d

Feedback Review

1. F
2. elite
3. Fundamentalism
4. c
5. a
6. d
7. a
8. a
9. a
10. a. 4
    b. 1
    c. 2
    d. 3
    e. 5

16

Health and Health Care

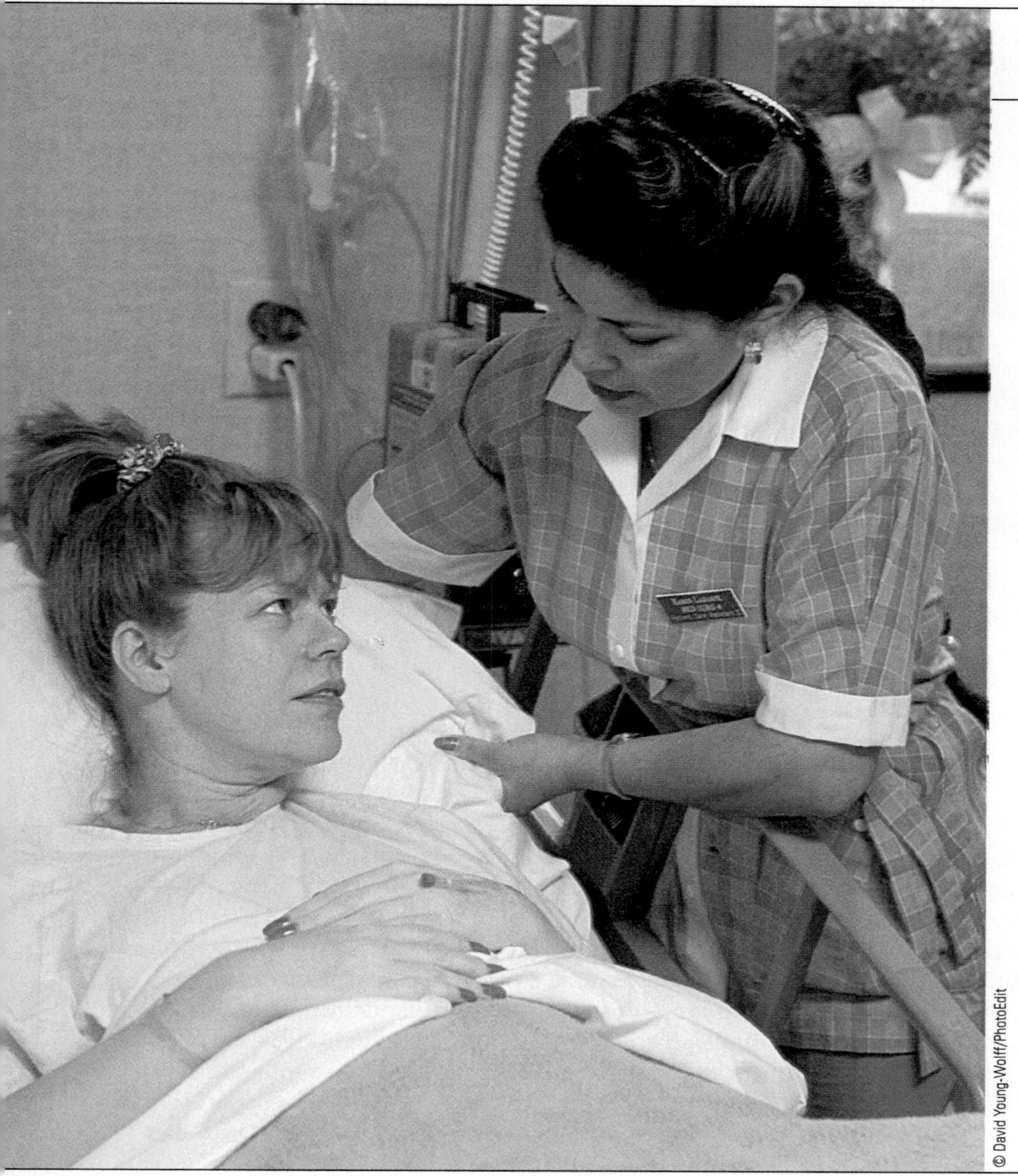

© David Young-Wolff/PhotoEdit

OUTLINE

LEARNING OBJECTIVES

- Define the concept of a health-care system and identify its major components.
- Apply functionalism, conflict theory, and symbolic interactionism to the health-care system in the United States.
- Discuss the distribution of diseases in the United States according to age, gender, race, ethnicity, and social class.
- Describe what is meant by viewing the health-care system as a medical-industrial complex.
- Describe the major health-care delivery innovations in America today.
- Discuss current health-care reform in the United States.

USING THE SOCIOLOGICAL IMAGINATION

Is Canada our only model for a national health-care system? Canada is not the only industrialized country with a health-care system organized and operated by the federal government. In fact, the United States is the only highly developed nation that does *not* provide a federally centralized, free (or inexpensive) health insurance program, a service for all citizens on an equal access basis.

To this point, we have explored the most thoroughly developed institutions in modern society—family, education, polity, economy, and religion. Neither modern society nor sociology is static, however. As society changes, sociology expands to encompass the study of new social phenomena. Several additional institutions, such as health-care and sports, are beginning to assume increasing relevance in contemporary society. In this chapter we explore one of these emerging institutions—the health-care system.

Health Care as a Social Institution

Health Care and Society

Health is an issue in every society. Whether they rely on shamans, barber-surgeons, magic, rituals, herbs, or hospitals, people develop cultural patterns to cope with sickness and death. In the 1830s, Alexis de Tocqueville noted the concern Americans tend to express for their well-being. But even he might be overwhelmed by the current mania over physical health among Americans. Burgeoning industries supply the needs of joggers, weight lifters, health food devotees, tennis players, and aerobic dancers.

This American propensity for personal health is embedded in culture. Americans tend to respond aggressively to their health-care needs. This aggressiveness, contends Lynn Payer (1996), is reflected in U.S. medical practice as compared with other industrialized countries. Compared with their European colleagues, for example, American physicians practice extremely aggressive medicine, and their patients expect it. Thus, American physicians are much more likely than their European counterparts are to perform surgery, run tests, and prolong life, even for individuals facing imminent death.

The Nature of the Health-Care System

The **health-care system** embraces the professional services, organizations, training academies, and technological resources committed to the treatment, management, and prevention of disease. The American health-care system constitutes a very large sector of the total national economy. Its growing claim on national resources reveals it to be an institution of increasing import. As shown in Figure 16.1, Americans spent over $1.2 trillion on health care in 1999, up from approximately $75 billion in 1970. Since 1980, total health expenditures increased nearly 400 percent (U.S. Bureau of the Census 2001a). When viewed as a proportion of the gross domestic product (GDP)—the monetary value of all the goods and services produced by the economy during a given year—these numbers reveal even more clearly the size of the health-care system. Health-care expenditures now account for about 14 percent of the GDP; the comparable figure in 1970 was about 7 percent (Centers for Disease Control 2002; S. H. Altman et al. 2003; Pear 2003; see Figure 16.2).

What are the components of the health-care system? By way of introduction, we will briefly discuss four major components of the health-care system: physicians, nurses, hospitals, and patients (L. D. Weiss 1997). Each of these components is covered later in greater depth.

Although physicians constitute only about 10 percent of health-care workers in the United States, they establish the working framework for everyone else. Only physicians are authorized by most states to diagnose illness, prescribe medicine, and certify such events as birth and death. Hospitalized patients may be attended by many professionals and other employees, but decisions about their diagnosis and treatment are made by physicians. American physicians' responsibilities are matched by high levels of social prestige and monetary rewards.

Nursing became a recognized profession only in the late nineteenth century, and its identity continues to undergo change. When Florence Nightingale was sent (at her insistence) with nurses and supplies to help British army doctors in Turkey in 1854, she encountered substantial resistance. One colonel wrote: "The ladies seem to be on a new scheme, bless their hearts. . . . I do not wish to see, neither do I approve of, ladies doing the drudgery of nursing" (quoted in Mumford 1983:289). After finally winning acceptance, Nightingale influenced the establishment of nursing schools in England and elsewhere. Although this represented a major advance in medical care, Nightingale's vision of the nurse's role was narrower than most nurses would now accept:

> *She insisted that nurses should be "clean, chaste, quiet, and religious." She saw the nurse as providing wifely support,*

FIGURE 16.1

Rising Health-Care Expenditures in the United States: 1970–1999

This figure shows the steep increase in health-care expenditures in the United States since 1970. What do you think accounts for this trend?

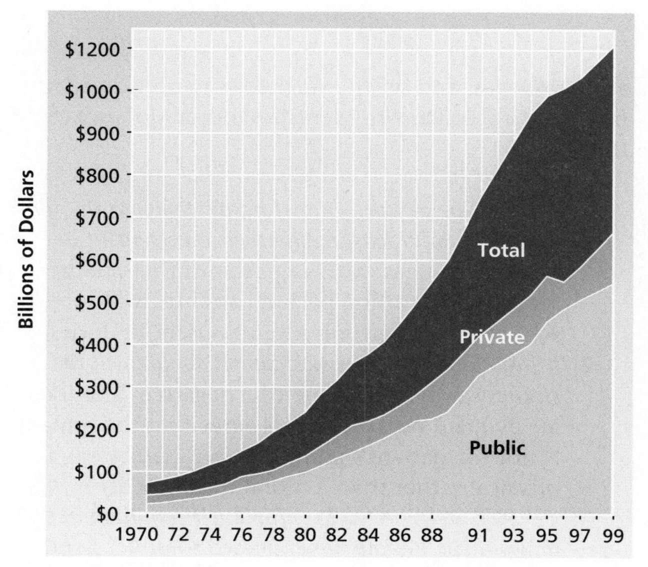

Sources: U.S. Department of Health and Human Services, *Health United States: 1992*, Public Health Service, Centers for Disease Control and Prevention, National Center for Health Statistics (Washington, DC: U.S. Government Printing Office, 1993), p. 160; U.S. Bureau of the Census, *Statistical Abstract of the United States 1998* (Washington, DC: U.S. Government Printing Office, 1998); and Centers for Disease Control, National Center for Health Statistics, www/cdc.gov/nchs/fastats/hexpense.htm (updated 8/02).

FIGURE 16.2

United States Health-Care Expenditures as Percentage of GDP: 1970–1999

This figure displays the increase in U.S. health-care expenditures as a percentage of GDP. Do you think anything should be done about this development? Why or why not?

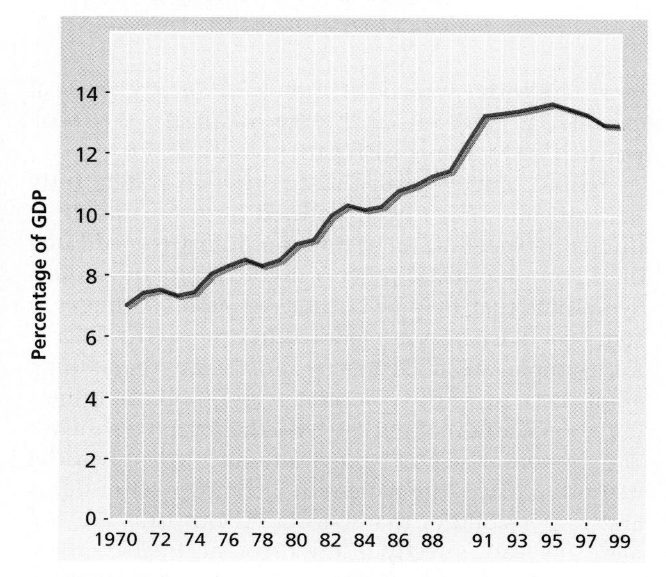

Sources: U.S. Department of Health and Human Services, *Health United States: 1992*, Public Health Service, Centers for Disease Control and Prevention, National Center for Health Statistics (Washington, DC: U.S. Government Printing Office, 1993), p. 160; U.S. Bureau of the Census, *Statistical Abstract of the United States 1998* (Washington, DC: U.S. Government Printing Office, 1998); and Centers for Disease Control, National Center for Health Statistics, www/cdc.gov/nchs/fastats/hexpense.htm

> *keeping the hospital going so the physician could do his important work.* (Mumford 1983:290)

Possibly as a result of this initial orientation, nursing has experienced frequent controversy regarding education, professional roles, and compensation.

Training programs for nurses shifted from three-year diploma programs in hospitals to associate, baccalaureate, and masters programs in colleges and universities. Despite a need for increasingly sophisticated knowledge, college and university nursing programs were slow to develop because hospital administrators enjoyed the plentiful supply of cheap labor that was generated within their own hospital training programs. The shift of nurses' training to colleges and universities is now well established, but the fact that it took an interminable half a century symbolizes the professional and economic problems of nurses (Lynaugh and Brush 1996).

Hospitals provide specialized medical services to a variety of inpatients and outpatients. Hospitals range from small facilities specializing in short-term, uncomplicated care to large medical centers providing long-

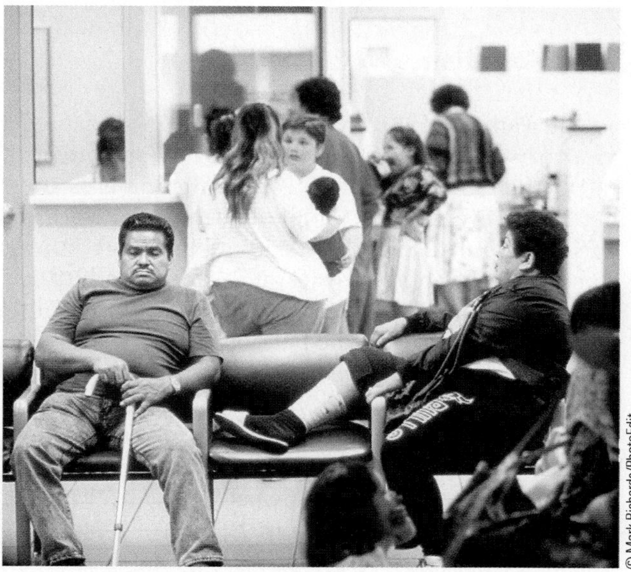

Americans have always been health conscious, a point made by Alexis de Tocqueville in the 1830s. It is not surprising, therefore, that the health-care system constitutes a very large proportion of the total productive apparatus of the U.S. economy.

FEEDBACK

1. The __________ embraces the professional services, organizations, training academies, and technological resources that are committed to the treatment, management, and prevention of disease.
2. Health-care expenditures now account for about __________ percent of the American GDP.
   a. 5
   b. 10
   c. 14
   d. 21

Answers: 1. health-care system 2. c

term care with complex technology. Typically, medical centers, as well as some large hospitals, combine research and training with patient care.

People usually enter the health-care system only because they have been defined by others as ill or injured. The definition of illness or injury is predicated on a complex social process involving many people. Symptoms that may be considered proof of illness in some circumstances are ignored in others. The elderly, for example, are labeled ill far more easily than young adults because of the common assumption that illness is a normal aspect of old age. Whether or not a patient's designation as sick is biologically justified, this social definition places the patient in a complex set of rights and responsibilities that Talcott Parsons calls the *sick role*. This leads directly to the theoretical perspectives, because the analysis of patients in relation to the sick role is part of the functionalist approach to the health-care system.

Theoretical Perspectives and the Health-Care System

Functionalism

Talcott Parsons (1951, 1964a, 1964b, 1975) proposed a view of sickness that was distinctively sociological, rather than merely medical. For the first time, sickness became relevant to sociological theory and research. Parsons assumed, first, that health problems are a threat to society. If people are sick and cannot fulfill their roles, society will not function smoothly. Either functions will not be performed at all, or they will be performed in an inferior manner by others who substitute for the sick. According to Parsons, society responds in two ways. As a deterrent to fake illness, society defines sham sickness as a form of deviant behavior. But society also institutionalizes legitimate patterns of behavior for a sick role (Weiss and Lonnquist 1996).

What is the sick role? The **sick role** is a confluence of appropriate behavior patterns for people who are ill. It serves to remove the sick from active involvement in daily routines, give these individuals special protection and privileges, and set the stage for their return to normal social roles. Parsons identifies four major aspects of the sick role:

1. *The sick are permitted to withdraw temporarily from other roles or at least reduce their involvement in them.* As long as others agree with the sick prescription, the ailing person can miss work or school or not perform usual household duties.
2. *It is assumed that the sick cannot simply will the sickness away.* Consequently, negative sanctions are avoided because others agree that the malady is not the patient's fault. Illness is defined as a physical rather than a moral matter.
3. *The sick are expected to define their condition as undesirable.* People are expected to prefer good health again. They are expected to neither resign themselves to illness nor take unfair advantage of an ailment's benefits (relief from normal activities, displays of sympathy, service from others). Thus, individuals who feign an illness have no valid claim to the sick role.
4. *The sick are expected to seek and to follow the advice of competent health-care providers.* Although not blamed for the onset of their illness, the sick are responsible for their own improvement. If the sick fail to cooperate in efforts to cure them, they will be tagged as deviants.

If any of these elements are missing, the privileges of the sick role may be withdrawn. For example, the quality of care given to those who are "uncooperative" may be decreased. Some illnesses, including those linked to obesity and anorexia, may even be denied the usual privileges of the sick role if people judge the disorders to be self-imposed. The existence of the sick role and the relationships surrounding it demonstrate the social nature of illness.

What are the criticisms of the concept of the sick role? Like any important theoretical formulation, Parsons's concept of the sick role has been closely scrutinized. Several criticisms have been leveled. One criticism depicts the sick role as too narrow in scope: it speaks only to Western society; it does not capture all aspects of illness; and it does not seem to apply to such conditions

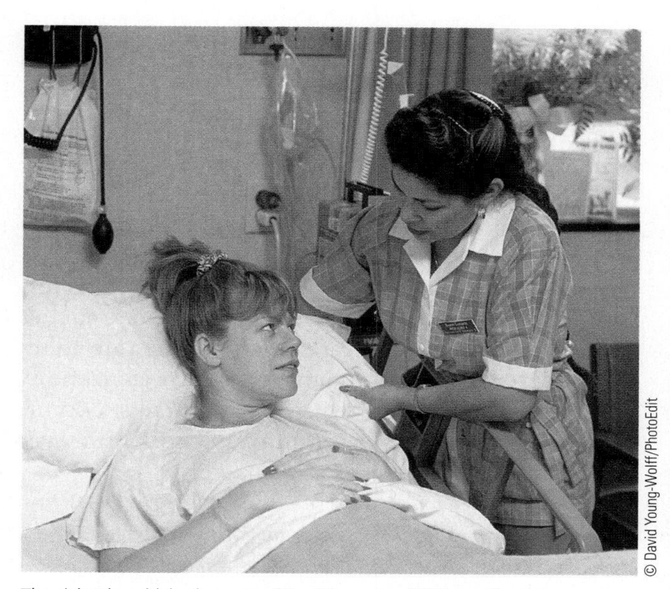

© David Young-Wolff/PhotoEdit

The sick role, which places considerable responsibility on the sick person to return to normal activities, can be applied only to curable illnesses. This sick role does not cover individuals with incurable diseases, such as terminal cancer or AIDS. This woman will be expected to abandon the sick role as soon as she can return to her life as it was before her accident.

as pregnancy. According to another objection, the sick role applies primarily to acute illnesses, such as viral infections, which are curable. In the case of an incurable disease, such as inoperable cancer, the patient is not expected to leave the sick role. Also, charge some critics, the sick role is not applied to stigmatized illnesses such as AIDS and mental illness. Some have also contended that the sick role is related to the professional health-care system and ignores other means of coping with sickness, such as self-treatment or the use of religious faith. Although analyzing sickness sociologically requires more than the concept of the sick role, Parsons's effort was pioneering and is still quite useful (Twaddle 1981; Twaddle and Hessler 1987; Aronowitz 1998).

Why has the American medical profession risen to its present heights? A second relevant issue to the functionalist perspective is the medical profession's rise to preeminence in American society. In his Pulitzer Prize–winning book, *The Social Transformation of American Medicine*, Paul Starr (1983) offers a functionalist explanation for the advancement of American physicians. Using a pluralist orientation, Starr contends that physicians gained the recognition of medicine as a profession through the efforts of the American Medical Association (AMA), especially after its reorganization around the turn of the twentieth century. As an interest group, the AMA was able to wield power vis-à-vis other competing interest groups such as patent medicine producers. The patent medicine companies were direct competitors with the doctors because early physicians, in addition to treating patients, often prepared their own drugs, and the patent medicine producers offered therapy and advice in addition to selling medicine.

According to Starr, the AMA persuaded Americans that they "needed" physicians to handle their health problems. The combination of a convincing lobby and a responsive American public elevated physicians in the eyes of the American public. In fact, today the American public accords greater confidence to those people controlling medicine than to organized religion or the U.S. Congress (PollingReport.com 2002).

Conflict Theory

The conflict perspective challenges many health-care practices. Their concerns include an alternative explanation of the medical profession's prestige, an explanation of the high income and status of physicians, the professional and economic problems of nurses, and the inequality of health care (Freund and McGuire 1998; Annandale 1998; Pringle 1998; Waitzkin 2000; Scambler 2002).

How do conflict theorists explain the prestige of the medical profession? According to Vicente Navarro (1976, 1986), the evolution of American medicine reflects the values and beliefs of American society. Navarro credits Starr with recognizing the role of the AMA in persuading American people of the need for physicians and the resulting preeminence of medicine. Nevertheless, Navarro offers a very different interpretation.

He begins with American power differentials based on class, race, and gender. The success of the medical profession, Navarro claims, is due to the power it possesses because of its alliances with the dominant capitalist class, the white race, and the male gender.

The same political and economic forces that determine the nature of capitalism determine the nature of the medical institution. Thus it was not merely persuasion that promoted the interests of the physicians but, more important, the coercive and repressive power physicians possess because of the capitalist power structure. Navarro would point, for example, to physician attempts at monopolization through repression of alternative (competing) approaches to health care—spiritual healing, acupuncture, osteopathy (based on the idea that disease results from loss of structural integrity), and homeopathy (treatment of disease by administering minute doses of a remedy that would in healthy persons produce symptoms of the disease being treated). In fact, a U.S. Supreme Court decision ruled that the American Medical Association was guilty of unfair restraint of trade in its efforts to keep osteopaths from practicing. It is these continuous conflicts and struggles, claims Navarro, that determine alterations in American society and in medicine.

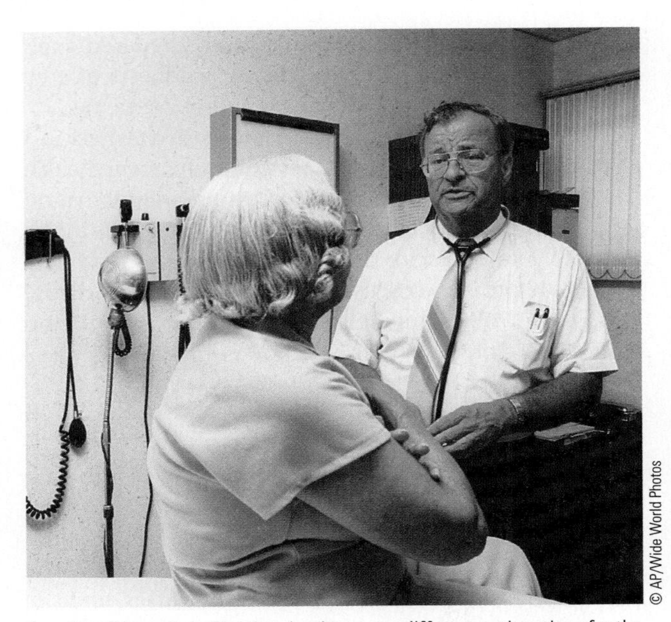
© AP/Wide World Photos

Functionalists and conflict theorists have very different explanations for the high rewards enjoyed by physician specialists in the United States. While functionalists attribute the handsome income of physicians to their talent and training, advocates of conflict theory credit the medical establishment with limiting the number of available physicians.

Why do physicians receive a hefty reward? Conflict theorists reject the functionalist explanation for the high financial and prestige rewards of physicians. According to functionalist theory, the high rewards accorded any occupation, including physicians, are necessary for attracting talented people into demanding professions that require extensive training. But, according to the conflict perspective, occupational groups use various mechanisms (licensing, credential requirements) to limit competition. Thus, physicians' rewards are high in part because the medical establishment keeps medical school enrollments artificially low, thereby limiting price competition among physicians (Weeden 2002).

Whatever the reason, not all physicians have enjoyed the privileges accorded those in the United States today. In eighteenth-century England and early-twentieth-century France, physicians were poorly paid and were socially marginal. In the former Soviet Union, physicians received only about 75 percent as much income as the average industrial worker. Contemporary Great Britain has a few specialists who command very high incomes, but most general practitioners are only moderately well paid. Although physicians in most contemporary societies have higher status than was the case in ancient Rome (where physicians were primarily slaves, freedmen, and foreigners and medicine was considered a very low occupation), the situation in the United States is unusual (P. E. Starr 1983).

Can physicians maintain their rewards? Whether physicians in the United States will maintain their privileged and powerful position is not known. For one thing, the number of physicians per 10,000 population in the United States has risen sharply since 1950. Also, because of declines in population growth, increasing costs of malpractice insurance, and the movement from general practice to specialized medicine, physicians are becoming less likely to work as independent professionals. Increasing numbers of physicians are taking salaried positions with organizations, or they are going into group practice and managed care organizations where expenses are shared. As these changes occur, individual physicians are beginning to lose control over their work:

> *Perhaps the most subtle loss of autonomy for the profession will take place because of the increasing corporate influence over the rules and standards of medical work. Corporate management is already thinking about the different techniques for modifying the behavior of physicians, getting them to accept management's outlook and integrate it into their everyday work.* (Starr 1983:447–448)

Medicine retains enough mystery that physicians are unlikely to lose all of their social status. As long as physicians use professional jargon, write illegible prescriptions, maintain control over patients' records, and force patients to wait in a waiting room "because the doctor's time is so valuable" (Mumford 1983:334), they will be held in awe by many members of society. It appears, however, that their ability to dictate the nature of the medical system is eroding.

One reflection of a growing sense of powerlessness is the 2003 strike of more than two dozen surgeons in West Virginia. They were protesting the rise in malpractice insurance premiums. A similar walkout in Pennsylvania was averted only because the state's governor proposed a $220 million fund to partially offset insurance premiums. In early 2003, protests, work slowdowns, and work stoppages occurred in New Jersey, Florida, and Mississippi. Other doctors are protesting rising malpractice costs by refusing to perform high-risk procedures, taking early retirement, or moving from states with the highest malpractice insurance premiums (L. Parker 2003; R. Stein 2003).

What are some of the professional and economic problems nurses face? Role expectations and authority relationships are problematic for nurses. Although nurses tend to patients far more frequently than do physicians, the nurses lack the authority to make any but the most routine decisions. In many situations, when nurses convey recommendations to physicians, they must appear to be passive. Leonard Stein illustrates this problem of domination in a hypothetical telephone conversation between a nurse and a physician:

> *This is Dr. Jones.*
>
> *(An open and direct communication.)*
>
> *Dr. Jones, this is Miss Smith of 2W—Mrs. Brown, who learned today of her father's death, is unable to fall asleep.*
>
> *(This message has two levels. Openly, it describes a set of circumstances, a woman who is unable to sleep and who that morning received word of her father's death. Less openly, but just as directly, it is a diagnostic and recommendation statement; i.e., Mrs. Brown is unable to sleep because of her grief, and she should be given a sedative. Dr. Jones, accepting the diagnostic statement and replying to the recommendation statement, answers.)*
>
> *What sleeping medication has been helpful to Mrs. Brown in the past?*
>
> *(Dr. Jones, not knowing the patient, is asking for a recommendation from the nurse, who does know the patient, about what sleeping medication should be prescribed. Note, however, his question does not appear to be asking her for a recommendation. Miss Smith replies.)*
>
> *Phenobarbital mg 100 was quite effective night before last.*
>
> *(A disguised recommendation statement. Dr. Jones replies with a note of authority in his voice.)*
>
> *Phenobarbital mg 100 before bedtime as needed for sleep, got it?*
>
> *(Miss Smith ends the conversation with the note of a grateful supplicant.)*
>
> *Yes I have and thank you very much doctor.* (L. Stein 1967:700)

Despite the difficult tasks and inconvenient hours associated with nurses' work, nurses' pay ranks below that of most professionals. As noted in Chapter 10, women receive substantially lower pay for their work than men do. This is certainly the case for nurses compared with physicians. The low pay received by nurses may relate to the fact that about 93 percent of registered nurses are female. In contrast, only about 28 percent of physicians are female.

Past efforts by nurses to improve their salaries were resisted by hospitals, but the nature of contemporary medical work could lead to an improvement. Medical work is increasingly defined as more than prescribing medicine and performing surgery. As demonstrated by Anselm Strauss and his colleagues (1985), successful treatment requires attention to safety, comfort, emotion, and the coordination of a variety of activities. Nurses, they note, play critical roles in these tasks. Nursing itself is becoming increasingly specialized, and the need for coordination with other professionals is accelerating (Kovner 1990).

How do conflict advocates view inequality in health care in the United States? The concern of conflict advocates for health-care inequality begins with **epidemiology**—the study of the distribution of diseases within a population. Social epidemiologists study the relationship of illness to the social and physical environment (Rockett 1994, 1999). Conflict theorists are interested in the reasons minorities and the poor have shorter life expectancies and higher incidence of certain diseases than the general American population. (They are aware that certain illnesses are associated with some minorities more than others. For example, African Americans are more prone to heart disease, and alcoholism is more prevalent among Irish Americans.) Conflict proponents trace the proclivity for ill health among minorities and the poor to the nature of capitalist society. Specifically, they point to the role of the health-care establishment in maintaining unequal access to medical care (Navarro 1976, 1986; McKinlay 1985; Waitzkin 1986; Weitz 2000). In addition, underprivileged people more often live in areas with heavy industrial pollution.

How does the medical establishment promote unequal access to health care? The market philosophy promotes unequal access to medical care, making medical care a market commodity. Those without the money to purchase health care are likely to be ill more often, have more serious diseases, and die younger. Unable to purchase good health care, they suffer the consequences. A suggested solution to this problem is the institution of a national health-care program. In fact, the topic of health-care reform in the United States will close this chapter.

Why is affordable health care not provided to the poor in the United States? According to conflict theorists, medical care is a capitalistic industry, an industry that brings to its participants—hospitals, nursing homes, nurses, physicians, pharmaceutical companies, medical schools, medical suppliers, pharmacies—over $1.2 trillion dollars annually. According to conflict theorists, these combined vested interests ensure the perpetuation of health-care inequality.

Symbolic Interactionism

As we have seen repeatedly, symbolic interactionism is grounded in the concepts of socialization and symbols.

Howard S. Becker—Socialization of Students in White

Using participant observation, sociologist Howard Becker (1961) examined the socialization of medical students. His was a symbolic interactionist approach. His picture of student culture in medical school is not glamorous. Medical students grapple with a sense of eroding idealism, diminished concern for patient well-being, and bewilderment over the shortcomings of medicine.

Bright college graduates who enter medical school face a rude awakening. The gratifying thoughts of helping others through the practice of a high-status, honored profession are, in the first year in medical school, supplanted by the recognition of unforeseen barriers. Before becoming a physician, it is first necessary to learn to be a medical student. The medical student preoccupation blots out any images of how to perform as a doctor in the future.

Much of Becker's book concentrates on the ways students recognize and solve problems posed by their teachers. Medical students turn their eyes from the glorious distant mountain to the swamp that threatens to overwhelm them. Becker, in short, focuses on the social interaction between medical students and their instructors as they undergo a prolonged rite of passage from student to doctor.

Becker began his research without firm, testable hypotheses and without preset data-gathering research instruments. Because he wished to "discover" how the medical school as an organization influenced the basic perspectives of physicians in the making—something Becker did not know beforehand—he could not design formal research instruments in advance. This is not to say, however, that he lacked any research design. He states that his commitment to symbolic interactionism led him to use participant observation as the major research technique. Although he and his colleagues did not pose as medical students, they did "participate" in the life of medical students. The researchers attended classes and laboratories with students, listened to student conversations outside of class, accompanied students on rounds with attending physicians, ate with students, and took night calls with students.

In the end, Becker challenged the view (on the part of both the public and the medical school faculty) that medical students shed their idealism for cynicism as they progress through their training. Often cited evidence for this alleged shift from idealism to cynicism is the detached way medical students come to view their patients. Death and disease become depersonalized for medical students. Because medical students view death and disease unemotionally, as a medical responsibility rather than a tragedy, they are viewed by laypeople as cruel and heartless.

Becker does not deny that medical students appear to become cynical and detached in medical matters. He attributes this appearance not to the loss of an idealistic long-range perspective but to the need to survive the overwhelming demands of the training experience. A cadaver, for example, is not viewed as somebody's daughter or mother but as a necessary means for gathering information for a forthcoming examination. Successful medical students ask not what they can do for a cadaver but what a cadaver can do for them. It is, in short, the manifestations of the urgent and immediate need to survive medical training that give the appearance of cynicism to outsiders.

As graduation approaches, another change occurs. As medical students lose concern for the immediate situational demands of school, they begin to display the concern for service to others they had felt upon entering medical school. This new idealism, however, is more informed than the original.

The erroneous conclusions regarding cynicism are understandable within the framework of symbolic interactionism. The cynicism fails to acknowledge that individuals can hold conflicting values and rely on the relevant one in a particular situation. As the situation varies, so does an individual's definition of the situation. Thus, as the students' circumstances revert near graduation, the dormant values of idealism replace the expressions of cynicism. Corresponding alterations in behavior follow.

Thinking About the Research

1. Do you think the method of investigation is appropriate? Why or why not? What other methods could have been used?
2. Does the symbolic interactionist interpretation of medical school socialization ring true? Explain.
3. If you think they would be appropriate, explain how functionalism and conflict theory could be brought to bear on the study of the socialization of medical students. If you don't think one or both of these two theoretical perspectives are applicable, explain why.

This section examines the socialization of physicians, the personal attributes and socialization of nurses, and the influence of labeling.

The socialization process of most physicians begins early. Those who decide to become doctors tend to do so early in life, many by the time they are sixteen years old. This is not accidental. Early decision occurs in part because many of those who enter medical school have been influenced by relatives who are themselves physicians. This, of course, also helps explain why physicians tend to come from the higher social classes. This presocialization, however important in the decision to become a physician, does not prepare medical students for the socialization they experience in medical school.

What are the most common socialization experiences of medical students? That medical students acquire technical knowledge and skills and learn to diagnose and treat illness goes without saying. In addition, medical school socializes students to accept the beliefs, norms, values, and attitudes associated with the medical profession. They learn, for example, to be dispassionate and unemotional about illness, suffering, and death (see Doing Research); to never criticize another physician publicly; and to avoid revealing a lack of knowledge. They also learn gradually to view themselves as doctors and to display a confident, professional appearance (Colombotos and Kirchner 1986; Twaddle and Hessler 1987; Kronenfeld 1998). Medical school also transforms students from eager neophytes who wish to absorb all medical knowledge to more experienced individuals who realize that their human limitations require a selective learning strategy. They must determine what is considered safe not to learn by assessing what they think the faculty wants them to learn, a social process involving meanings, symbols, and subjective interpretations (Becker et al. 1961).

How are nurses socialized? Although nurses are both male and female and come from all social classes, they tend to be white, middle-class females. Compared to the general female college student population, nursing students are more altruistic, benevolent, and generous and are less interested in power, control, and self-advancement. They prefer to make their own decisions and they want to be treated empathetically. These characteristics are supported during the four years of undergraduate nursing education:

> *During their four years of nursing education, students are socialized to a nurse role that values individualized, direct patient care, rational knowledge, and innovation, and they are taught to think of themselves as autonomous, professional persons.* (Twaddle and Hessler 1987:215)

Upon graduation, however, nurses are faced with a rigid, bureaucratized work environment. Their work lives are spent in a setting that contradicts the values they brought to their education and that their school socialization reinforced. This creates the potential for many conflicts and difficulties because nurses are constantly attempting to interpret their occupational world with symbols and meanings that do not fit their work situation.

How is the labeling perspective applied to illness? As discussed in Chapter 7, the labels and stigmas applied to people affect the way others behave toward them, which in turn affects the behavior of the labeled people. "Sickness" and "illness" are social labels that can be used to stigmatize people.

The nature of health-related labels is based on culture. Because culture varies, definitions associated with health and illness are also diverse. Mark Zborowski (1952, 1969) found Jews and Italians more likely to talk openly about their pain, whereas "old Americans" and the Irish, with objectively the same levels of pain, tended not to complain. Similarly, Irving Zola (1966) reported that although the Irish did not see any relationship between their illnesses and their social relationships, the Italians presented relationship problems as a significant part of their illnesses.

Given this background, the existence of labels and stigma associated with certain illnesses is not surprising. The elderly are erroneously labeled "disabled" (for example, forgetful) far more often than young adults because of the common assumption that illness is a normal aspect of aging. People with mental health problems are often stigmatized, which may lead to the adoption of a personal identity based on the properties of the stigma.

There is no more timely illustration of labeling and stigmatizing than the current view of AIDS in the United States. Because AIDS was initially discovered among homosexual men, who are already stigmatized in American society, people who acquire AIDS from such sources as blood transfusions or heterosexual partners are also labeled and stigmatized (Dennis Altman 1986). The taint of deviance and immorality is often attached to AIDS patients. This stigma is reflected in many responses. A substantial number of Americans believe that AIDS is a punishment for the decline in moral standards. An AIDS-related stigma persists even though the disease cannot be acquired through casual social interaction.

The three theoretical perspectives provide very different insights into the health-care system; that is to be expected. In fact, in the absence of a single unifying sociological theory, sociology is fortunate to have three powerful theoretical perspectives. Together, these theoretical approaches underscore the social nature of sickness and health care (see Table 16.1).

TABLE 16.1

FOCUS ON THEORETICAL PERSPECTIVES: Health Care in the United States

Each theoretical perspective opens unique avenues for research on health care. Can you think of other research questions for each theory?

Theoretical Perspective	Research Topic	Hypothesis
Functionalism	Sick role	• Society requires (needs) a timely exit from the state of illness.
Conflict theory	Power of physicians	• Physicians repress competing approaches to health care.
Symbolic interactionism	Stigmatization of illness	• AIDS victims are labeled as immoral and deviant.

FEEDBACK

1. The __________ serves to remove people from active involvement in everyday routines, give them special protection and privileges, and set the stage for their return to normal social roles.
2. If a sick person in American society does not attempt to recover, he or she will be labeled a __________.
3. According to the __________ perspective, physicians rose to dominance in the American health-care system by taking such actions as attempting to create a monopoly on medical services.
4. AIDS appears most frequently among homosexuals and intravenous drug users. This is an example of __________.
5. According to __________ theory, affordable health care is not provided to the poor in the United States because of the desire for profit.
6. Nurses are caught in a conflict between their values and their work environment. Which of the following perspectives best helps us understand this situation?
   a. functionalism c. symbolic interactionism
   b. conflict theory d. exchange theory
7. America's negative reaction to AIDS victims illustrates the effects of __________.

Answers: 1. sick role 2. deviant 3. conflict 4. epidemiology 5. conflict 6. c 7. labeling

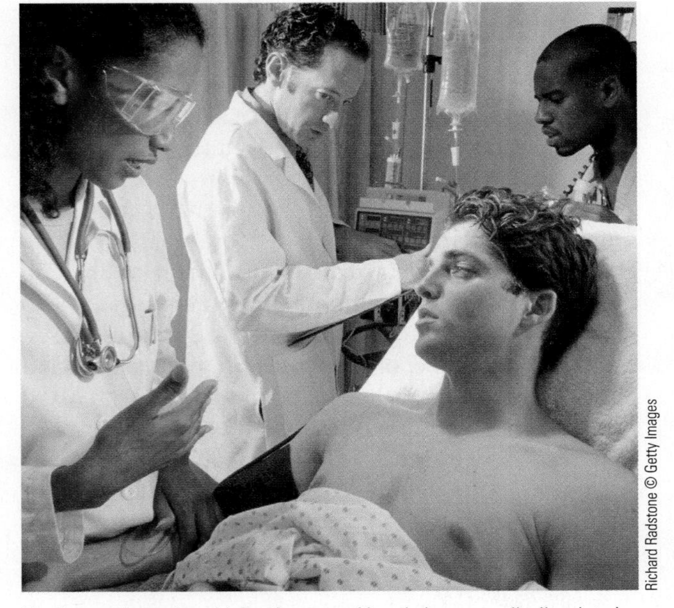

Nursing students place high value on making their own medically related decisions and on their being treated with empathy. These expectations usually are not met in the inflexible organizational setting in which they work after their training is completed.

Health in the United States

Epidemiologists, as noted earlier, study distribution patterns of diseases. Health and sickness, social epidemiologists have demonstrated, are strongly influenced by social class and social factors, as reflected in several demographic characteristics. Prominent among these demographic characteristics are age, gender, race, and ethnicity (Rockett 1999).

Age and Health

What is the relationship between illness and age? Causes of death vary with age. Infectious diseases, such as pneumonia and influenza, are the biggest killers of children under six years of age. Adolescents and young adults die more from suicide, murder, and accidents (notably automobile related). Chronic illnesses, such as heart disease and strokes, are most likely to take the lives of the aged.

What is life expectancy now? Thanks to improvements in medical treatment, working conditions, diet, living

WORLDVIEW 16.1

Life Expectancy at Birth

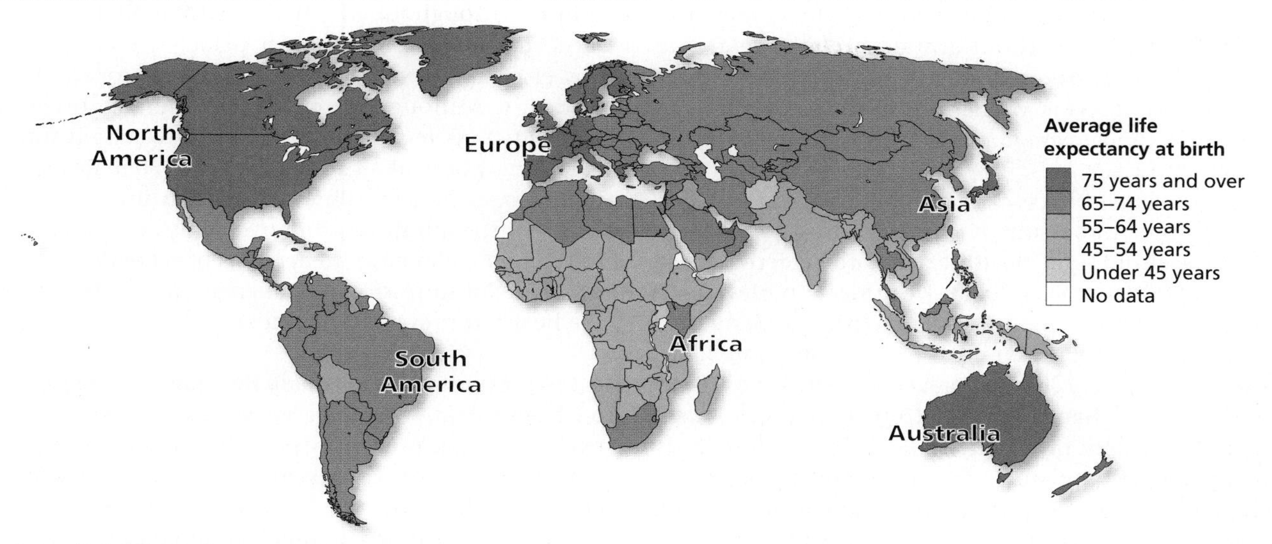

Source: United Nations Development Program.

Although the life expectancy picture in industrialized nations is rosy, it is not so for much of the world. Life expectancy in the industrialized world currently stands at 75 years, compared to 64 years in less developed countries. Japan has the highest life expectancy (81 years), Malawi in East Africa has the lowest (36 years).

The variation in life expectancy at birth around the globe is apparent in this map. Seventeen of the twenty nations in which less than half the population has access to modern medicine are in Africa.

1. Is your view on the need for health-care reform in the United States affected by the data presented in this map? Explain your perspective.
2. Should America be involved in bringing modern health care to deprived nations? Frame your answer within the context of America values.

conditions, and sanitation, Americans are living longer. The average life expectancy—average age at death—for a newborn baby in the United States today is about seventy-seven years, an increase of almost 50 percent since the beginning of the twentieth century, when life expectancy was forty-seven years (*World Population Data Sheet* 2002). Still, among the more developed countries, the United States has a relatively high infant mortality rate.

Life expectancy may soon reach eighty-five in industrialized countries (see Worldview 16.1). If so, more American citizens will be fairly healthy until near the end of their lives. Whether or not life expectancy in industrialized countries reaches eighty-five, the health of the aged is going to be better than it was for previous generations.

What are some of the biggest health-care problems of America's elderly? The lack of prescription drug coverage is one major health-care problem for the elderly. Although advances in medical treatment do not come cheap, Medicare offers, to older Americans, a partial shield from the relentlessly rising health-care costs. But because this federally administered health insurance plan has basically no prescription drug benefit, more older Americans are being forced to spend a higher percentage of their incomes (often low and fixed) on increasingly expensive prescription drugs, or do without them (Noonan 2000). Not surprisingly, the creation of a Medicare prescription drug benefit is an object of heated political debate. In July 2002, the U.S. Senate failed on a 49–50 vote (passage required 60 votes) to provide a prescription drug benefit through Medicare. In his 2003 State of the Union speech, President Bush proposed a Medicare prescription drug benefit. He restricted the benefit to seniors willing to leave their traditional fee-for-service plans for managed care programs. Congressional Republicans and Democrats both feared a backlash to any plan leaving out seniors who prefer their present arrangement of choosing their own doctors. Although Bush's plan was given little chance of enactment, it was passed by Congress in 2003 (Dickerson and Tumulty 2003; Pear 2003).

Until this legislation is activated in 2006, the 25 percent of Medicare beneficiaries with no prescription drug coverage, and many others with partial coverage, will have to pay for their prescription medications (outside a hospital) from their own pockets. One-half of older Americans without any drug coverage are on annual incomes below $15,000. These people will have to continue making wrenching financial decisions to purchase their average of over $500 in prescription drugs each year.

The negative effects of ageism on health-care quality are another important issue. According to a recent study, ageism harms health care for the elderly in several ways (McCracken 2003). Health-care professionals have insufficient training in geriatrics to properly treat many older patients. Younger patients are more likely than older ones to receive preventive care. The elderly are less likely than the young to be tested for diseases and other health problems. The elderly are more likely to receive incomplete or erroneous treatment due to the failure of health-care professionals to use proven medical interventions. Senior citizens almost never are included in clinical trials, although they are the biggest consumers of approved drugs.

How does increasing life expectancy affect the baby boomers? The combination of a longer life span and skyrocketing health-care costs places a substantial financial burden on baby boomers in their sixties. An estimated 22 million American households already provide aid to aging parents, a number heading upward. Organized social responses to cope with the increased number and rising health-care needs of seniors are just beginning.

How can we care for the expanding elderly? Novel solutions are coming from community-based organizations, governmental agencies, and commercial firms (Koss-Feder 2003). Residents 50 and over in Boston's Beacon Hill Village participate in a nonprofit community program providing significant discounts on senior health-care services for an annual fee of $500–$600. New Mexico has created a statewide program that pays family members and friends of Medicaid recipients $9 an hour for caring for them. Almost 60 percent of those receiving care in this governmental program are sixty-five or over. U.S. corporations suffer an $11 billion annual loss from the absenteeism, turnover, and diminished productivity of employees caring for aging parents. Recognizing the situation, many corporations now address elder care with employee-assistance programs.

Gender and Health

How do health differences vary by gender? Across the world, females have a significantly greater life expectancy than males. In the United States, for example, the average life expectancy for females is eighty years, compared to seventy-four years for males (*World Population Data Sheet* 2002). Starting at birth and continuing throughout life, males exhibit higher death rates than females. Except for diabetes, men die at a higher rate than women do at all ages and for all causes of death. Although women have the advantage of longevity over men, they are at a relative disadvantage regarding physical sickness. American females suffer higher rates of acute illnesses, such as infectious diseases and respiratory ailments. Only in the case of injuries do males have higher rates of acute illness than females. Not surprisingly, women are more involved in the health-care system than men.

Is there a consensus explaining this irony: men enjoy better health than women do, but die earlier? No. Several explanations, however, are plausible. It may be that the female gender role creates greater stress than the masculine gender role. And because they have a greater interest in health, women may be more alert to illness (Verbrugge 1986). The nature of the male gender role may also contribute to this situation. First, men do not seem to handle stress as well as women. The "macho" role may influence men to avoid physicians and ignore medical advice. Males are also not as likely as females to relieve stress through supportive, intimate relationships (Nathanson 1989; Verbrugge 1989). As a result, stress may be more likely to lead to death for men than for women. Moreover, the male gender role encourages behavior injurious to health. Society encourages the males to engage in dangerous activities such as drinking, smoking, fighting, and fast driving. In fact, young males are killed in automobile accidents at a rate three times higher than the rate for females, and males are twice as likely as females to die of cigarette- and alcohol-related diseases during middle age.

Race, Ethnicity, and Health

What health patterns are associated with race and ethnicity? Except for Asian Americans, racial and ethnic minorities in the United States have poorer physical and mental health than the general white population (Weisse 1998; Keppel, Pearcy, and Wagener 2002). The infant mortality rate for African American infants, for example, is nearly 2.5 times that of white infants (see Figure 16.3). And although African Americans have experienced an increase in life expectancy since the beginning of the twentieth century, white males typically outlive African American males by about five years. White females have a life expectancy over five years longer than African American females. African Americans are more likely than whites are to die from thirteen of the fifteen major causes of death (O'Hare et al. 1990; Riche 2001).

FIGURE 16.3

Infant Mortality Rates in the United States According to Race/Ethnicity of Mother

As this figure reveals, infant mortality rates in the United States vary a good bit. If you were the secretary of Health and Human Services, would this lead you to make any recommendations to the sitting president? Explain.

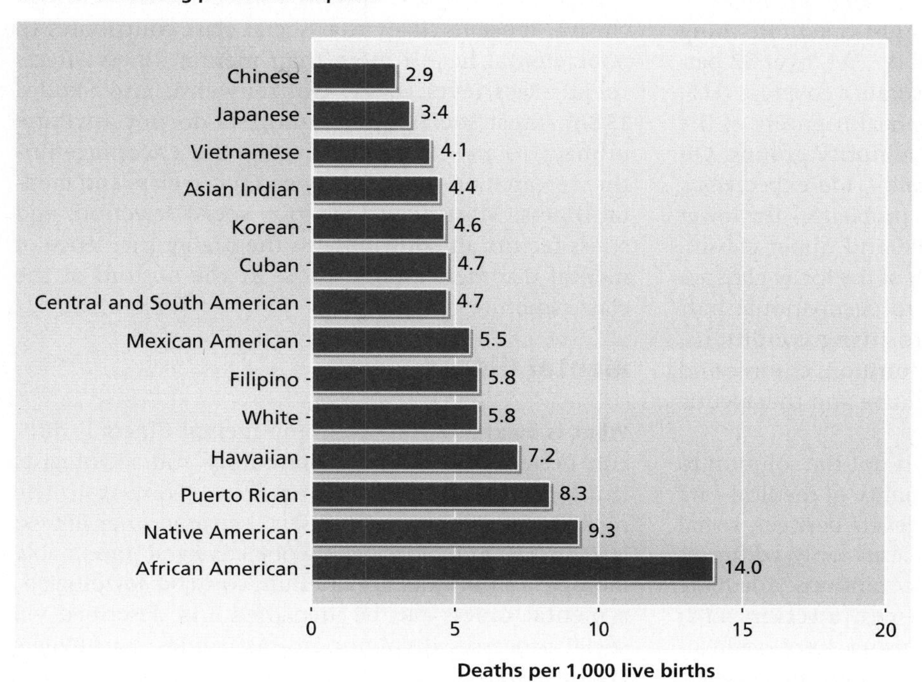

Deaths per 1,000 live births

Source: National Center for Health Statistics, *National Vital Statistics Reports* 50, no. 4 (Jan 30, 2002): 3–4.

INTERNET LINK Additional information is available on infant mortality rates in the United States. Go to the Web site for the U.S. Bureau of the Census at www.census.gov/

Although Latino Americans have lower mortality rates from heart disease and cancer than either African Americans or non-Latino whites, Latino Americans have a shorter life expectancy and higher death rates from diabetes, pneumonia, tuberculosis, and influenza than non-Latino white Americans. Despite improvements in health since 1950, Native Americans still suffer disproportionately from diabetes, venereal disease, hepatitis, alcoholism, tuberculosis, malnutrition, and suicide.

Although some diseases such as sickle-cell anemia, hypertension, and diabetes seem to be biologically based, most minority group illnesses are traced to the disadvantages associated with poverty. Minorities in the United States die younger and suffer more illness than the white majority, not primarily because of their racial and ethnic characteristics but because of low income. People in poverty get infectious diseases from living in unsanitary, overcrowded conditions; are exposed to extreme and frequent violence; cannot afford to maintain a nutritious diet; and do not receive proper medical treatment, either preventive or when ill (Smedley, Stith, and Nelson 2002). Consequently, minorities tend to die more from infectious diseases and malnutrition rather than from long-term illnesses such as heart disease and cancer.

The link between poverty and illness can also be seen when we compare Asian Americans with other minority Americans. Asian Americans, who are the most prosperous minority group in the United States, have the lowest age-adjusted mortality rate of all minority groups in the country. Asian Americans also have the lowest infant mortality rate and have significantly lower suicide and homicide rates.

Debilitating health among minorities may be due to additional factors beyond low income. Even within the same income groups, African Americans have a higher death rate than whites do. In fact, lower-income whites have a significantly lower death rate from hypertension than that of the highest-income African Americans. One prominent explanation is the physical and psychological stress experienced by minority group members, who remain targets of prejudice and discrimination despite social and economic success (see Doing Research in Chapter 9).

Social Class and Health

What are the effects of social class on the incidence of illness? For the poor, good health is just one more thing they have less of than the affluent do. Yet, over 32 percent of the poor have no health insurance coverage (U.S. Bureau of the Census 2001a). The health profile of the poor pretty well mirrors that of minority groups. On nearly all measures—infant mortality, life expectancy, infectious diseases, chronic health problems—the lower class is worse off than the middle and upper classes. Contributing factors to poor health of the lower class are quite familiar by now: exposure to occupational hazards, unsanitary and substandard living conditions, exposure to violence, inadequate nutrition, chronic and acute stress, and inadequate preventive and therapeutic medical care (Lantz et al. 1998).

In itself, access to medical care is not the solution to poor health in the lower class. Equality of medical care alone would not equalize health levels between social classes. Providing quality medical care only addresses the results of poor environmental conditions. Adequate medical care will help only in conjunction with lifestyle changes such as better housing, improved eating habits, and reduction of stress (Marmot, Shipley, and Rose 1984; Blane, Davey-Smith, and Bartley 1998; Cockerham 2000).

Similarly, social class level is related to mental illness. A series of highly respected studies on social class and mental health establish the existence of a higher incidence of mental illness within the lower class (Faris and Dunham 1939; Hollingshead and Redlich 1958; Link and Dohrenwend 1989; Kessler et al. 1994; Srole et al. 1962, 1978; "Mental Health, Culture, Race, and Ethnicity" 2001). Although the lower class has a higher incidence of mental problems, including personality disorders and schizophrenia, the upper classes tend to have more anxiety and mood problems (Cockerham 1985, 1999; Link, Dohrenwend, and Skodol 1986).

Why do members of the lower class have a higher incidence of mental illness? Three basic explanations have been offered for the higher incidence of mental illness among the poor (Cockerham 1995). According to the *genetic* explanation, there is a biological tendency to mental illness within the lower social class. But studies of twins indicate that genetics by itself is an insufficient explanation. Social factors also contribute. According to the *social stress* explanation, the lower class is more susceptible to mental disorders because of the extra stress of trying to cope with deprivation. Evidence on this explanation does not speak with a single voice. Other scholars, supporters of the *social selection* explanation, account for the higher rates of mental disorders in the lower class by a mobility process: individuals with mental disorders in other social classes tend to end up in the lower class, and the mentally healthy born into the lower class move up to higher social classes. Although the evidence is inconclusive, it seems likely that social class contributes to poor mental health more than mental illness affects social class level (Link, Dohrenwend, and Skodol 1986). Most social epidemiologists do not attribute primacy to any single explanation but encourage further research on the link between social class and mental illness. Most likely, genetics, social selection, and stress factors all contribute to the higher incidence of mental disorders among those at the bottom of the class structure.

Mental Illness

What is mental illness? Defining mental illness is difficult because the medical community and sociologists think differently about its nature. According to the medical model, a mental illness, like any other illness, is identified and described by objective and measurable biological indicators. In addition, contend sociologists, a mental illness can be identified and described via social processes involving norms, values, and beliefs (Weitz 2001).

Evidence of the social construction of mental illness exists even in the medical community's attempt to categorize them. For example, until 1974 psychiatrists listed homosexuality as a separate mental disorder in its bible, the *Diagnostic and Statistical Manual of Mental Disorders (DSM)*. Following a long period of disagreement about the nature of homosexuality within psychiatry itself and considerable lobbying on the part of gay rights activists, the American Psychiatric Association voted in 1974 to drop homosexuality from *DSM-III* (Spitzer, Williams, and Skodol 1980; Kirk 1992).

How prevalent is mental illness in the United States? Despite changes in the psychiatric definitions of various mental illnesses, there is sufficient stability in the categories of mental disorders to measure its extent. According to the best estimate available, almost one-third of adults 18–64 suffered a diagnosable mental disorder in the previous year. Over a lifetime, almost half indicated an episode of mental illness at some time in their lives. These episodes ranged from the most widely experienced illness (depression and alcohol-related difficulties) to much rarer forms of mental illness such as manic depression and psychotic symptoms (Weitz 2001).

FEEDBACK

1. Although American women have a longer life expectancy than men, women are more involved in the health-care system as patients. T or F?
2. With the exception of __________, racial and ethnic minorities in the United States have poorer health than the general white population.
3. Although the upper class has a higher incidence of serious mental problems, the lower class tends to have more anxiety and mood problems. T or F?
4. Higher rates of mental illness among the lower class are pretty clearly due primarily to the genetic factor. T or F?

Answers: 1. T 2. Asian Americans 3. F 4. F

Who tends to experience mental illness? Similar to other types of diseases, mental illness appears disproportionately by social class, gender, and race and ethnicity. As just noted, sociologists have consistently documented a higher rate of mental illness within the lower class.

Gender does not appear to influence in consistent ways the incidence of several mental disorders, including schizophrenia. It is the case, however, that women experience a greater incidence of minor depression and anxiety disorders than men do. Men are more likely to experience problems associated with substance abuse and personality disorder such as compulsive gambling.

The most notable variations in mental illness among racial and ethnic groups appear in rates of depression and manic episodes. In descending order of prevalence, differences in the rate of these mental illnesses appear among African Americans, whites, and Latinos. Although not strictly considered a mental illness, psychological distress is markedly higher among African Americans than among whites, regardless of income level. According to the most prevalent interpretation, stresses created by racism lie at the root of these differences (Kessler and Neighbors 1986).

The Changing Health-Care System

The health-care system, like other American institutions, is undergoing rapid change, as seen in the medicalization of society, increased reliance on hospitals, the rise of managed care, and the development of hospice organizations.

The Medicalization of Society

How is the United States a medicalized society? Within contemporary society, physicians possess the formal power to define people as ill. On the basis of physicians' diagnoses, students may (or may not) be excused from school and employees may (or may not) be allowed to miss work. Social critic Ivan Illich (1982) contends that such power is potentially dangerous to society. We are, he believes, experiencing the process of **medicalization**, in which problems that were once considered matters of individual responsibility and choice are now considered a part of the domain of medicine and physicians:

> *In a medicalized society the influence of physicians extends not only to the purse and the medicine chest but also to the categories to which people are assigned. Medical bureaucrats subdivide people into those who may drive a car, those who may stay away from work, those who must be locked up, those who may become soldiers, those who may cross borders, cook, or practice prostitution, those who may not run for the vice-presidency of the United States, those who are dead, those who are competent to commit a crime, and those who are liable to commit one.* (Illich 1982:76–77)

The effects of medicalization may or may not be as pervasive as Illich contends. However, many aspects of life—mental illness, drug abuse, homosexuality, behavior disorders in children, and certain criminal behaviors, to name a few—that once were considered evidence of character or spiritual flaws are now interpreted in medical terms. Such interpretations can lead to more humane treatment, but the nature of the treatment may also serve to profit certain groups. Pharmaceutical companies, for example, gain opportunities to sell drugs for concerns that individuals may once have handled on their own. Similarly, some categories of physicians may profit from opportunities to diagnose and treat people who would not otherwise be considered ill. The question of whether the treatments that serve the interests of corporations and physicians are necessarily beneficial for those who are now defined as ill is a matter of continuing concern (L. D. Weiss 1997; Spragins 1999; Cowley and Turque 1999).

Hospitals

Hospitals are organizations that provide specialized medical services to a variety of patients, including some who continue to live in their homes and others who will spend a prolonged period in the facility. As mentioned earlier, hospitals range from small facilities specializing in patients who need short-term, uncomplicated care to large medical centers that provide long-term care with complex technology. Medical centers and some large hospitals combine patient care with medical research and training.

What are the benefits and handicaps of hospitals? The United States has become highly dependent on hospitals. Trends in childbirth illustrate this. In 1940, over one-half of American babies were born at home; today, despite some resurgence in interest in home delivery, almost all babies are born in hospitals (Hoff and Schneidermann 1985).

The increased reliance of people on hospitals and their advanced technology results in a number of tangible benefits. Childbirth is safer for mothers and infants who experience complications. The ability to treat cancer and heart disease has increased dramatically. Diseases such as tuberculosis have declined, measurably.

Problematic consequences are also experienced. Many hospitals, with their highly specialized professionals and technical equipment, are contributing to an increasingly specialized, incomplete view of health. Medical historian Stanley Joel Reiser acknowledges that health care has advanced considerably since physicians had to make diagnoses with the benefit of little information besides what patients told them, but he sees dangers in the increasing reliance on sophisticated technology:

> *Machines inexorably direct the attention of both doctor and patient to the measurable aspects of illness, but away from the "human" factors that are at least equally important. Insofar as technological evidence occupies the time and commands the chief allegiance of both doctor and patient, it diminishes the possibility that a close personal relationship will develop between the two. So, without realizing what has happened, the physician in the last two centuries has gradually relinquished . . . unsatisfactory attachment to subjective evidence—what the patient says—only to substitute a devotion to technological evidence—what the machine says.* (Reiser 1978:229–230)

The physician's view of the patient, then, is becoming increasingly incomplete. Physicians, according to Reiser, are likely to become estranged from their patients and their own judgment as a result of their increasing reliance on technological devices.

James James (1979) demonstrated the lack of a clear-cut, positive relationship between the health of community citizens and the medical practice in their hospitals. He found that during a 1976 work slowdown by physicians in Los Angeles, death rates actually fell below the usual rate. James speculates that the reduced number of elective operations may have lowered the death rate during the period of the slowdown.

How are hospitals contributing to rising health-care costs? Health-care costs increased dramatically in recent years, and hospital expenses account for a large proportion of the increase. When patients check into a hospital, they are paying for more than a room and a bed. A large staff of professional and nonprofessional employees and increasingly expensive equipment are underwritten by patients and their insurance companies. Patients are not in a position, as an organized group, to fight rising costs; yet, individual patients may insist on the very best facilities. Moreover, physicians generally lack expertise in the efficient planning and use of health-care resources (Jonas 1978). As a result, efforts to contain hospital medical costs are poorly organized.

Managed Care

Paul Starr (1982, 1988) writes of the rise of the **medical-industrial complex**—the network of physicians, medical schools, hospitals, medical equipment suppliers, pharmaceutical companies, and health insurance carriers. This coalition, with its shared interests, was instrumental in the rise to power of the American medical profession. Entry of the for-profit corporation brought another major transformation: *managed care*.

Managed care is any health-care system based on cost containment, control of medical decisions, and regulation of treatments. The adoption of managed care—sometimes called "managed competition"—has been spurred by large organizations looking for ways to lower employee health costs in the face of escalating medical expenses. Given this motivation, it is not surprising that all managed care systems share these characteristics:

- *Delivery of a comprehensive set of health services for a prepaid premium;*
- *Utilization and quality controls that providers agree to accept (for example, the insurer may not sanction or pay for certain procedures, may require prior authorization for others, require the submission of specified outcome data, etc.);*
- *Financial incentives for patients to use the provider's facilities or designated physicians only;*
- *The assumption of some financial risk by doctors to motivate them to balance patients' needs against the need for cost control.* (Ranade 1998:7)

Traditionally, Americans have relied on indemnity insurance (full payment to patients for costs) and fee-for-service medicine (payment to doctors for each item of service). According to large employers, this system increases their insurance premium costs because neither patients nor providers have any incentive to control costs. Their argument has been so successful that three-fourths of working Americans are now covered by managed care (Bransten 1997; W. R. Scott et al. 2000).

With the emergence of large health-care corporations comes the increasing employment of medical providers in group or large corporate settings. Both the American Medical Association and physicians are abandoning their traditional commitment to the private (solo) practice of medicine (Twaddle and Hessler 1987; Salmon et al. 1990a, 1990b; Ranade 1998). This move is not without consequences.

Corporate employment involves, for doctors, a significant loss of autonomy and independence. Office routines and pace of work are more regulated. Choice of retirement time is no longer the sole prerogative of the physicians. Doctors are increasingly measured against standardized performance criteria. Mistakes are monitored carefully to reduce the risk of malpractice losses (Robinson 1999; Ludmerer 2000).

Will the medical profession lose power due to the corporatization of medicine? Without question, the new health-care corporate giants wield considerable power, forcing the medical profession to relinquish some of its own traditionally held autonomy. In that sense, the profession of medicine is no longer the sole dominating factor in determining and controlling the nature of the health-care system. Some doctors now believe that these health-care organizations are practicing medicine without a license, and that they are practicing insurance rather than medicine (Gorman 1999). But as medical sociologist David Mechanic writes:

> *It would be naive to assume that the enormous power of the professional, painstakingly and skillfully cultivated over this century, will be quickly or substantially reversed despite changing conditions. . . . Physicians will continue to affect substantially the patterns of medical care, appropriate standards, and the costs associated with varying episodes of illness.* (Mechanic 1986: vii)

Patients are also affected by the managed care system. Most Americans now complain that they are paying more but getting less. The advent of managed care has, in fact, not stopped spiraling medical costs (Weitz 2000). And patients have some grounds for complaining about a deteriorating quality of medical care:

> *When organizations provide care as their raison d'être—whether they are for-profit firms, nonprofit agencies, or private or public financing programs—they must be concerned with things other than the people getting and giving the care. They must pay attention to productivity, costs, waste, fraud, and solvency. They must worry about pleasing their shareholders, boards of directors, bill payers, insurance companies, Medicare and Medicaid offices, auditors, accrediting agencies, and congressional oversight committees. All of a sudden, the goals of staying in business, balancing the books, and lowering costs displace the goals of making patients feel cared for and improving their well-being.*
> (D. Stone 1999:63)

Actually, criticism of managed care is coming from all directions: patients lament the loss of freedom to choose their own doctors and hospitals; the American Medical Association endorses the unionization of doctors to regain some of their lost prerogatives; and the federal government worries about the loss of patient medical "rights" (King 1996; Neal and Daniel 1998; Tumulty 1998; Lore 1999; Welch 1999).

The managed health-care system involves several organizational forms. Prominent among these are for-profit hospitals, health maintenance organizations, preferred provider organizations, and physician-hospital organizations.

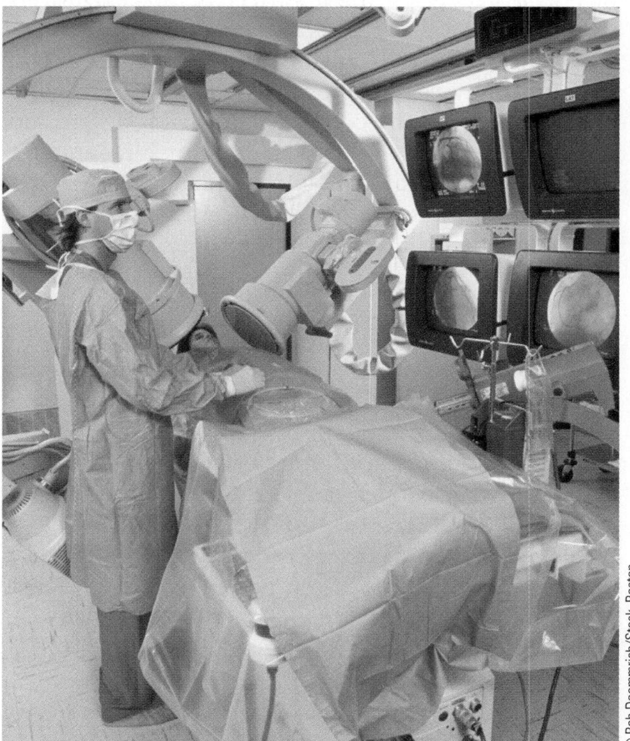

© Bob Daemmrich/Stock, Boston

Much of the skyrocketing cost of medical care in the United States is due to the changing nature of hospitals. This operating room scene reflects the sophisticated technology that most hospital administrators and physicians appear to desire. Once obtained, this costly technology is underwritten by patients, insurance companies, and the federal government.

Are for-profit hospitals a solution to rising costs? There is a trend toward for-profit hospitals. On the assumption that corporate efficiency will aid in medical cost containment, more and more local hospitals are being absorbed by huge, privately owned hospital chains, such as the Hospital Corporation of America and Humana. Research, however, runs counter to the early optimism for cost savings through privatization. An extensive study of investor-owned and nonprofit community hospitals revealed that costs per hospital visit were nearly 20 percent higher in the for-profit hospitals and that overhead expenses per admission were 4 percent lower in community hospitals (Pattison and Katz 1983). Two additional studies also indicated significantly higher costs for corporate hospitals than for nonprofit hospitals (State of Florida 1981–82, 1982–83). It appears, then, that chain hospitals are less cost-effective than community not-for-profit hospitals. (On the other hand, the for-profit hospitals may be a good investment for themselves because of higher costs to patients.)

What are health maintenance organizations? **Health maintenance organizations (HMOs)** are health plans in which members pay set fees on a regular basis and in return receive all necessary health care at little or no additional cost. Most HMOs are group practices with physicians representing a variety of specializations. Originally, HMOs were supposed to provide physicians with access to a full range of expertise while saving individual physicians from worry about the mechanics of a business. On the consumer side, HMOs were supposed to "improve the market" by providing an alternative to health plans in which benefits were received only after the onset of an illness or accident. At least originally, the major orientation of HMOs was to curb disease through early detection and preventive medicine (Beck 1993).

What are the pluses and minuses of HMOs? Because the plan is based on prepayment, members do not avoid necessary health care simply because they will be unable to afford it. Some critics believed that the availability of extensive treatment at little or no additional cost would lead to the HMOs' downfall; members would use the services excessively. Despite some important exceptions, this has not been the case. As Emily Mumford notes:

> *Many people do not enjoy going to the doctor; some even very rich people avoid doctors. The record of savings in HMOs suggests that most people do not use health services excessively when they are free.* (Mumford 1983:384)

If HMOs reduce chronic illness and unhealthy personal habits, such health plans can reduce the overall cost of medicine. Whether or not HMOs can do so is a matter of serious debate. Aligned on one side are those who see health care as a professional relationship between doctor and patient (Anders 1996; Relman 1997). On the other side are supporters of HMOs who see health care as a service industry in which "consumers" rather than "patients" select products on the basis of market-driven criteria such as quality, convenience, and cost (Herzlinger 1997). Evidence is not sufficient to support either side.

There are negatives. First, coverage is restricted: patients have to leave their regular physicians if those doctors are not members of the patients' HMOs. Restriction also occurs when physicians turn patients away because of the quota for subscribers. And physicians are more likely to limit enrollment because HMO reimbursements are below other forms of insurance (Curran and Renzetti 2000). Second, HMOs use copayment requirements; this discourages subscribers from visiting their doctors. Third, HMOs control referrals to more costly specialists: the HMO permits the primary care doctors to divide among themselves the referral money left over at the end of the year. Moreover, primary care doctors who fail to limit referrals are more likely to be released by HMOs. Fourth, more HMOs are encouraging doctors to keep costs low: they pay the physicians a predetermined amount annually for each patient (subscriber) they have. Because doctors can keep the difference between their total allocation and the actual costs of patient care, they have an incentive to undertreat (Bodenheimer 1999; Weitz 2001).

How prevalent are HMOs? Originated during the 1920s, HMOs have operated on a large enough scale for their effects on the quality and cost of health care to begin to be understood. Since 1976, HMO enrollment increased tenfold (Spragins 1999). This growth, however, appears to be slowing. In 2000, HMOs had an enrollment of 81 million members, down from 103 million in 1995. This decline may be due partly to growing public disenchantment with HMOs (Julie Appleby 1999; Cowley and Turque 1999). Most Americans were generally positive about managed care in 1995, but HMO support has waned considerably since then. Fifty-nine percent of Americans surveyed in 1995 believed that HMOs and other managed care plans would help contain costs. Only 34 percent felt that way in 2002. The percentage agreeing that HMOs improve health-care quality declined from 48 percent in 1995 to 27 percent in 2002. Whereas 59 percent thought the trend toward managed care to be a good thing in the middle of the last decade, only 36 percent agree today (Taylor and Leitman 2002).

What are preferred provider organizations, and how are they formed? **Preferred provider organizations (PPOs)** are small groups of physicians providing a limited set of medical services to pre-enrolled patients at reduced prices. There are purported advantages on both sides: doctors have a dependable supply of patients, and patients save money. PPOs are flexible in that patients can visit physicians outside the group. Patients must, however, pay extra if the alternative fees are higher. Physicians can accept patients not in the group plan. PPOs may be formed by hospitals, employers, managed insurance companies, or physicians.

What services do physician-hospital organizations provide? **Physician-hospital organizations (PHOs)** are joint ventures between hospitals and physicians to reach predetermined business objectives. Typically, activity is centered on managed care contracting. PHOs provide the same services as most managed care organizations: claims processing, quality assurance, receipt and distribution of fees, and recruitment and credentialing of health-care personnel. PHOs are formed mostly in areas where community hospitals do not have access to other organized medical groups (Friend 1996).

Hospices

How do hospices contribute to patient care? Although HMOs speak to the need for preventive medicine and controlled expenses, the medical system also addresses the needs of the terminally ill. Advances in medical procedures and technology may prolong life, but these advances do not necessarily improve the situation faced by patients who are experiencing physical pain, psychological trauma, and the knowledge that their disease is terminal. **Hospices**—organizations designed to provide support for the dying and their families—are a recent response to this problem.

The hospice movement has its roots in England, but during the past two decades it has gained importance in the United States. Some hospices are located in hospitals, some are in separate buildings, and some provide services to patients in their homes. Regardless of their location, however, they make the services of physicians, nurses, counseling psychologists, and social workers available to patients and their families.

Whereas hospitals are oriented toward curing diseases, hospices are oriented toward making patients with apparently incurable diseases as comfortable as possible. These differing orientations result in very different strategies. For example, painkillers may be administered much more freely in a hospice setting. Hospices can do nothing to reverse an unfavorable medical diagnosis. They can, however, assist patients in maintaining their dignity during their final months of life (Cloud 2000).

Euthanasia

Euthanasia, the deliberate termination of life, is rendered to a person suffering from painful and incurable illness. Because the technology exists for sustaining life—even when patients are comatose and when

SOCIOLOGY AND THE NEWS MEDIA

Do Americans Love Their HMOs?

A revolution in health-care delivery has occurred in the United States. Many Americans, who have traditionally relied on private doctors, are now enrolled in HMOs. This CNN report offers a view of American attitudes toward their HMOs.

1. Describe the prevailing attitude of Americans toward HMOs. Do you agree with this attitude? Why?

   ____

   ____

2. Relate HMOs to the new medical-industrial complex. Which major theoretical perspective best describes this relationship?

   ____

   ____

This young woman is being treated in an HMO. Is she a "patient" or a "consumer"?

improvement is unlikely—serious questions exist as to when or whether treatment should be withheld. A contemporary advocate of euthanasia raises the issue:

> *I know of a woman who tried, unsuccessfully, to kill herself. She was brought to the hospital in a coma, with a bullet lodged in her spine. Using heroic measures the surgeon kept her alive, and he considers her case a success: she lives, but she is totally paralyzed; he no longer has to worry about her ever attempting suicide again.* (Illich 1982:198)

Many states are dealing with difficult questions: When is a patient legally dead? Do patients, their families, and their physicians have the right to withhold or discontinue treatment when there is no apparent prospect of recovery? Only Oregon has legalized physician-assisted suicide—a law that so far has weathered an attempt to overturn it, in 2002, by Attorney General John Ashcroft's Department of Justice (W. Booth 2002).

How and when is euthanasia administered? Opponents of euthanasia, of course, believe that there is a duty to maintain human life, under all circumstances. Even supporters of euthanasia do not agree on the form it should take. Advocates of so-called "passive euthanasia" wish to limit the practice to the cessation of heroic measures to prolong a life. Other supporters would go further. Dr. Christiaan Barnard (the pioneer of heart transplant operations) recommends that physicians practice "active euthanasia," ending the lives of patients rather than allowing them to continue enduring pointless suffering. According to Barnard (quoted in the press), unplugging machines and ceasing treatment is insufficient, for some patients, when suffering is severe and recovery impossible (Mumford 1983). Dr. Jack Kevorkian, who accepted credit for helping more than 130 individuals commit "assisted suicide," was the first to announce publicly that he was helping terminally ill patients who seek an end to their lives. For a long while, U.S. courts upheld Dr. Kevorkian's right to this practice, but in 1999 he was convicted of second-degree murder in the death of a fifty-two-year-old man (Belluck 1999). Debate continues (Dowbiggin and Dowbiggin 2003).

Health-Care Reform in the United States

As mentioned in "Using the Sociological Imagination," most modern societies provide free or inexpensive health care through their governments. The United Kingdom, although it has a system of private medical care, promotes equality of medical care through its National Health Service, which provides the vast majority of its services without charge. Canada also has a public health-care delivery system committed to the provision of equal access. As in the United Kingdom, health care in Canada is financed through the government.

Of the highly developed countries of the world, the United States is the only one without national health insurance for all its citizens. Moreover, except for the Veterans Administration hospital system, the military, and the Indian Health Service, the United States is without any direct mechanism for the delivery of health-care services (Waitzkin 2001).

Except for President George Bush's successful alteration of Medicare in 2003, efforts at health-care reform, have been defeated (Kronenfeld 1997). Consequently, health insurance in the United States is a privilege largely enjoyed, with the exception of the elderly, by those who have full-time jobs with well-established firms. Those with the power—including business interests and the medical establishment—have thus far successfully frustrated all efforts to reform the health-care system. By invoking threats of socialized medicine and profit reduction, contend conflict proponents, the elite in the United States have used capitalist ideology to prevent equality of access to medical care. In fact, say some conflict theorists, some form of socialized medicine will be required if equality of health care is to become a reality in the United States.

The Need for Health-Care Reform

Why is there a need for health-care reform? One motivation for health-care reform is economic. America annually devotes a larger share of its gross domestic product

FEEDBACK

1. The term used to describe the process through which problems that were once considered matters of individual responsibility and choice become a part of the domain of medicine and physicians is called _________.
2. A "new" medical-industrial complex has been created because of the entry of the _________ into the health-care system.
3. Investor-owned hospitals are less efficient than not-for-profit community hospitals. T or F?
4. Health plans in which workers pay set fees and receive all necessary health care at no additional cost are called _________.
5. Organizations providing support for the dying and their families are called _________.

Answers: 1. medicalization 2. corporation 3. T 4. health maintenance organizations 5. hospices

(GDP) to health care than any other nation. In 1999, health care accounted for 13 percent of America's (GDP), or $1.2 trillion. This amounts to over $4,300 annually for every man, woman, and child. If current trends continue, health-care spending in the United States is expected to reach 18 percent of GDP early in the twenty-first century (*World Development Report 1993*, 1993). And as you saw at the beginning of this chapter, the costs of health care are skyrocketing. The escalating costs of medical care also negatively affect the overall health of the economy. According to critics, economic resources of government—in excess of $800 billion annually—desperately needed in other areas of public life are being siphoned off by a health-care system out of control (Marmor and Barr 1992).

A second reason for reform is the substantial proportion of Americans who, except on an emergency basis, do not have access to medical care (see Figure 16.4). A large segment of the 42 million uninsured citizens are the most disadvantaged, those on the bottom of the stratification structure. Even with Medicaid, about half of America's poor are without medical coverage. Poor children are particularly vulnerable. This vulnerability extends beyond the poor. In 2001, 60 percent of the uninsured were employed. Two-thirds earned $25,000–$50,000, and one-third earned more than $50,000 (Connolly and Goldstein 2003). Increasingly, either employers are requiring employees to pay both high premiums and deductibles for the coverage of their children and spouses, or they are providing health plans that simply do not cover dependents. Nearly one-half of uninsured children live with parents who are insured. Nearly 10 percent of children in the United States (8.5 million) were without health insurance in 2001 (Goldstein 2002).

Quality of life is a third reason for health reform. One obvious aspect of a lowered quality of life is the lack of health care or inferior health care. Perhaps less

FIGURE 16.4

Percentage of Persons Not Covered by Health Insurance in the United States by Age: 2001

This figure displays the percentage of various age categories in the United States without health insurance. If you were running for the presidency of the United States, what would you say about this situation?

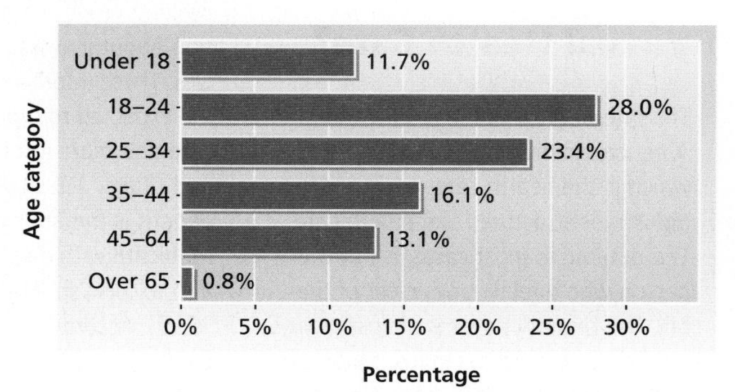

Source: U.S. Bureau of the Census, *Current Population Survey*, 2001 and 2002 Annual Demographic Supplements.

SNAPSHOT OF AMERICA 16.1

Americans without Health Insurance, 2001

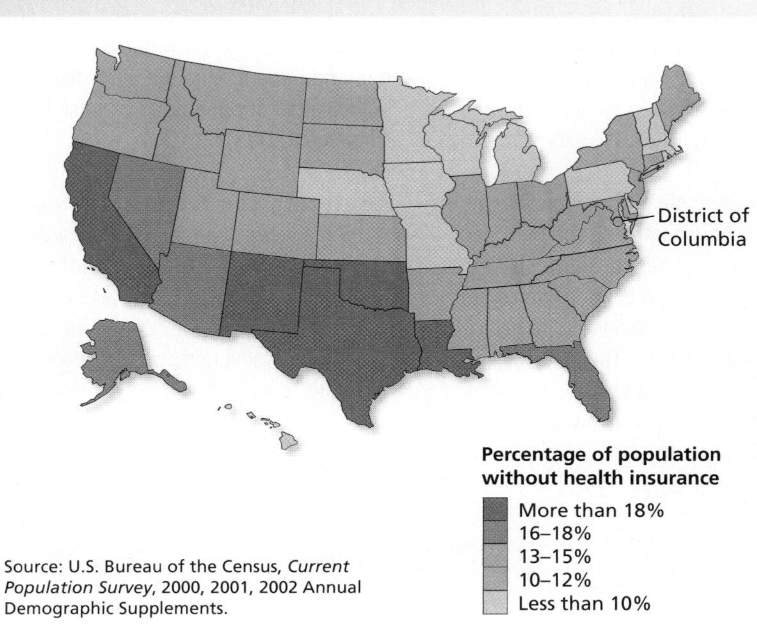

Source: U.S. Bureau of the Census, *Current Population Survey*, 2000, 2001, 2002 Annual Demographic Supplements.

Those favoring health-care reform point to the over 40 million Americans without health insurance coverage. Critics of the existing health-care system also point to inequities in insurance coverage among states. This Snapshot documents how the percentage of the population without health-care insurance varies significantly among states.

1. How does your state compare with other states? Can you explain its relative place?
2. Is guaranteed health insurance for all U.S. citizens consistent or inconsistent with American values? Explain your answer by drawing on the discussion of values in Chapter 3.

Virtual Health Care?

The Internet is profoundly affecting America's institutions, not the least of which is the health-care industry. No industry is structured like health care. The patient, in most cases, is not the person who directly pays most of the bill. The primary payer is almost always a third party, such as an insurance company or government agency. Knowledge that the patient is not paying most of the bill has been a disincentive to cost cutting. Therefore, doctors have provided as much care as possible at the highest cost. Given this setup, medical providers have had little reason to invest in information technology such as Internet services. It was not until the advent of managed care in the mid-1990s that concern over rising health insurance premiums became an issue in the health-care industry (Francis 1999).

But at the start of the twenty-first century, the Internet is beginning to make a significant impact. First, the Internet is selling medical "stuff" such as prescription drugs, medical supplies, and durable medical equipment. Insurance claims are also being filed, investigated, and paid online (Francis 1999; McTighe 1999). These mundane business tasks, however, do not directly affect the delivery of medical care.

Advances in the care of patients through the Internet have been longer in coming. Because the health-care industry is fragmented (thousands of individual physicians treat patients and gather information without a central means of communication), patients and doctors have not been able to access information easily. The Internet, with its standardized language and communication formats, is overcoming this obstacle and providing vast stores of easily accessed information. This central store of data makes analysis of medical research much easier and allows patients to obtain much more information about illnesses, diseases, and doctors. These improvements in health care are expected to produce better-quality care and more informed patients (Nobel and Cherry 1999). Another potential benefit is the ability of patients to communicate with each other in chat rooms and to form support groups (McTighe 1999). According to one expert in the field:

> *The Web has the power to enable patients who have not had access to vast sources of health and wellness information to make them better health consumers. The Internet equalizes the playing field for physicians who, with inefficient communications and information access tools, have been at the mercy of large payer and provider service organizations seeing them, the docs, as the endless source of cost savings. The Web empowers all participants in the health care market to better and more cost efficiently communicate with one another. The result: the potential for tremendous cost and quality efficiencies that have heretofore eluded the health care market.* (Francis 1999:17)

One example of the improvement in quality and efficiency is the treatment of chronic illnesses such as diabetes, asthma, and heart disease, which are becoming more prevalent due to increased longevity. This field of medicine, known as disease management, seeks to keep patients away from expensive health care as much as possible while delivering high-quality service. The Internet permits this because patients can take medical readings from equipment provided by physicians and transmit them to health-care providers. The data, transmitted over the Internet, is then analyzed by computer, and patients requiring special attention are made known to the physician (Nobel and Cherry 1999).

Although e-mail communication between doctors and patients is the exception, the American Medical Association states that about one-quarter of physicians use e-mail with some of their patients (Kritz 2003). Obvious advantages are less phone tag for doctors and fewer office visits and more rapid feedback for patients.

There are concerns about the use of the Internet in health care. The most notable of these worries is the possible breach of patients' privacy and information security. As you can imagine, the information about patients' health is very sensitive and must be well protected (McTighe 1999). Another concern is the relative lack of software programs to gather, compile, sort, analyze, and access medical data. This latter problem is expected to be solved in the near future, as software developers turn their attention to the health-care field (Francis 1999). Physicians also worry about reimbursement and increased liability. With respect to increased liability, many doctors fear harming patients through a misdiagnosis by e-mail or patient malpractice lawsuits (Kritz 2003).

Analyzing the Trends

If the predicted changes in the health-care industry are realized, what effects will they have on the social status of physicians? Use one or more of the three sociological perspectives in your analysis.

obvious is the fear and desperation that the uninsured feel. They know that an illness or accident experienced by themselves or a member of their family could either go untreated or absorb whatever economic resources they have. Moreover, even the insured live with the prospect of losing coverage should they lose their jobs. As you saw in Chapter 14, more Americans, both blue- and white-collar, are experiencing job loss or, even if employed, lower compensation. In addition, companies are requiring employees to pay for more of their health-care coverage. In response, 17,000 General Electric workers went on strike in 2003 to protest the estimated $200 hike in their health-care costs (S. Greenhouse 2003).

Health-Care Reform Options

What are the health-care reform options? There are a few basic approaches to delivering health-care services (Samuelson and Nordhaus 2001). The "modified competitiveness" option is based on market principles such as consumer cost sharing. This approach, however, depends on universal health coverage as a precondition to health-care reform.

A second option, managed competition, is related to the first option, but is a combination of free-market forces and government regulation. Health care would be structured around several large plans modeled after HMOs. To increase competitive edge, employers would form large purchasing networks, with health-care providers bidding on their business. Individuals would be free to select a plan, and the costs would be paid either through an income or a payroll tax. The idea here is to create competition and to reward those health-care providers that have superior performance in cost, quality, and patient satisfaction (*World Development Report* 1993).

A third option is the single payer approach, like the Canadian model. The government finances medical services. Canadians choose their doctors and hospitals and bill the government according to a standardized fee structure. A public opinion comparison of the U.S. and Canadian health-care systems favors the Canadian system. The Canadian approach engenders higher levels of consumer satisfaction with patient services, and it is regarded much more favorably than the U.S. system (H. Taylor 1994; Budrys 2001).

A fourth option—based primarily on the German model—uses a play or pay mechanism. Here, universal coverage is provided by employers who either offer employees health coverage (play) or pay into a public fund for covering the uninsured. Access to medical care in Germany, considered among the best in the world, is guaranteed for life (R. Atkinson 1994; Budrys 2001).

A significant proportion of the population in the United States is medically uninsured. Of the over 40 million uninsured, the poorest Americans make up a large share. Where can the uninsured in poverty obtain the cash required for health care?

FEEDBACK

1. Approximately ________ million Americans are without health insurance.
2. Which of the following highly industrialized countries does not have national health-care coverage?
   a. Germany c. United States
   b. France d. Canada
3. The major reason the United States does not have a national health-care program is because its citizens want different results from health-care programs than citizens of other developed societies. T or F?
4. Of those Americans without health-care protection, the poor and children comprise the majority. T or F?
5. The approach to health-care delivery under which medical services are financed governmentally is called
   a. managed competition. c. all payer.
   b. single payer. d. play or pay.

Answers: 1. 44 2. c 3. F 4. T 5. b

REVIEW GUIDE

SUMMARY

1. All societies must develop ways of coping with the issues of health care and dying. Modern societies have evolved large and complex health-care systems. In the United States, for example, 14 percent of the gross domestic product is accounted for by health-care expenditures alone.
2. The major components of the health-care system are physicians, nurses, patients, and hospitals. Physicians control the health-care system in the United States and are highly rewarded in terms of power, prestige, and money. Partly because nursing began as a low-status, female-dominated occupation, it suffers from low power and pay.
3. Functionalists conceptualize a sick role, which carries with it prescriptions for the treatment and behavior of patients. Patients are expected to withdraw from or reduce their involvement in other roles, seek the advice of physicians, and attempt to recover. Failure to cooperate in recovery efforts may result in loss of accommodating privileges.
4. Functionalists and conflict theorists offer opposing views of the U.S. medical profession. The medical profession, functionalists contend, has become preeminent because Americans perceive it as essential. Conflict theorists attribute the dominance of the medical profession, the prestige of physicians, the professional and economic problems of nurses, and the inequality of health care to the power of the medical establishment.
5. Symbolic interactionism highlights the socialization of physicians and nurses. This perspective also deals with the labeling process in illness. It explains, for example, the attachment of a stigma to certain illnesses.
6. The distribution of diseases within a population is related to a variety of social factors. Because of increasing life expectancy, America's elderly have serious health-care problems. Women in the United States live longer than men on the average, but are more involved in the health-care system as patients. Generally, health is poorest among members of racial and ethnic minorities. Both physical and mental health improve with social class level.
7. The health-care system in the United States is undergoing significant change. Some critics fear a double downside to the medicalization of society: it excessively benefits corporations and physicians; and health-related matters, once thought to be the responsibility of individuals, are becoming the province of medicine and physicians. Others point to the entry of the corporation into the health-care system, a movement that threatens the autonomy of physicians and the power of the medical establishment.
8. Americans have become very dependent on hospitals. This dependence has both positive and negative consequences. Childbirth has become safer, and many diseases are more easily treated and controlled. On the negative side, hospitals, through the promotion of specialization, contribute to an incomplete view of health. Hospitals are also contributing to rising health-care costs. Investor-owned hospitals do not appear to be a solution to this problem. Although profitable to investors, for-profit hospitals are more costly to run and deliver more expensive health care.
9. The emergence of managed care is another important change in the U.S. health-care system. Intended to curb spiraling health-care costs, managed care is receiving criticism from the medical establishment, the public, and the federal government.
10. Managed care involves a shift from patient care by private physicians to organizationally based health-care delivery. Among these organizational forms are for-profit hospitals, health maintenance organizations, preferred provider organizations, and physician-hospital organizations.
11. The public and politicians agree that the United States, the only major industrial power in the world without national health insurance for everyone, must undergo major health-care reform. There are several distinct options for such reform, but the surrounding political, social, and economic struggles cloud the path.

INFOTRAC® COLLEGE EDITION

http://www.infotrac-college.com/wadsworth

Another unique option available to you at the Student Resources section of the companion Web site is Infotrac College Edition, an online library with access to hundreds of scholarly and popular periodicals. Below are suggested search terms for this chapter. Results from these and other searches are found at the site.

Search keyword: **euthanasia**. Find three articles that describe different societies' views of euthanasia. What are the ranges of opinions about the subject? Is euthanasia legal in any society? What are the primary reasons given to support euthanasia? What are the main objections to it?

Search keywords: **national health-care systems**. Attempt to find articles that compare America's health-care system with those of other countries. Based on your reading, how would you compare America's system to those of other countries? What advantages does the U.S. system have? What disadvantages? If you had the power to design a health-care system for America, what features would it have?

Search keywords: **stem cell research**. This subject has been hotly debated since President Bush took office in 2001. Read two articles to find out what the main issues are. How could the study of sociology help resolve this debate?

LEARNING OBJECTIVES REVIEW

- Define the concept of a health-care system and identify its major components.
- Apply functionalism, conflict theory, and symbolic interactionism to the health-care system in the United States.
- Discuss the distribution of diseases in the United States according to age, gender, race, ethnicity, and social class.
- Describe what is meant by viewing the health-care system as a medical-industrial complex.
- Describe the major health-care delivery innovations in America today.
- Discuss current health-care reform in the United States.

CONCEPT REVIEW

Match the following concepts with the definitions listed below them.

____ a. medicalization
____ b. hospitals
____ c. health maintenance organizations (HMOs)
____ d. epidemiology
____ e. sick role

1. health plans in which members pay set fees on a regular basis and in return receive all necessary health care at little or no additional cost
2. organizations that provide specialized medical services to a variety of patients, including some who continue to live in their homes and others who will spend a prolonged period in the facility
3. a process in which problems that were once considered matters of individual responsibility and choice are now considered a part of the domain of medicine and physicians
4. a social definition serving to remove people from active involvement in everyday routines, give them special protection and privileges, and set the stage for their return to their normal social roles
5. the study of disease distribution patterns within a population

CRITICAL-THINKING QUESTIONS

1. Evaluate the idea that under certain circumstances sickness is seen as a form of deviant behavior. Draw on the concept of the sick role in forming your answer.

____

____

2. The medical profession in the United States enjoys a lofty position on the stratification structure. Compare the functionalist and conflict explanations for this fact.

____

____

3. Members of the lower class in American society have a higher incidence of mental illness than is documented in other social classes. Evaluate the various explanations for this phenomenon.

____

____

4. Do you think the health-care system in the United States will be reformed in the next ten to fifteen years? Why or why not? If it is reformed, describe the basic direction you think it will take.

____

____

MULTIPLE-CHOICE QUESTIONS

1. **The size of the health-care system in America is**
   a. decreasing slightly.
   b. leveling off.
   c. steady, as it has been for the last two decades.
   d. increasing.
   e. decreasing substantially.
2. **Although _________ constitute only about 10 percent of health-care workers in the United States, they establish the framework within which everyone else works.**
   a. hospice personnel
   b. physicians
   c. registered nurses
   d. practical nurses
   e. medical consultants
3. **Talcott Parsons uses the term _________ to describe the complex set of rights and responsibilities associated with biological illness.**
   a. illness syndrome
   b. sick role
   c. doctor-patient nexus
   d. instrumental role nexus
   e. mutual contact

4. **According to Marxist scholar Vicente Navarro, the success of the medical profession in the United States is due to**
   a. clever lobbying by the American Medical Association.
   b. the high quality of American medical schools.
   c. the dedication of nurses and volunteers.
   d. its connection with dominant societal power blocs.
   e. its roots in religion.
5. **The concern of conflict advocates for health-care inequality begins with _________, the study of patterns of disease distribution within a population.**
   a. demography
   b. epidemiology
   c. immunology
   d. medical geography
   e. structural medicine
6. **Which of the following statements would conflict theorists be most likely to endorse?**
   a. Medical schools socialize students to accept the beliefs, norms, values, and attitudes associated with the medical profession.
   b. Those political and economic forces that determine the nature of capitalism also determine the nature of the health-care system.
   c. "Sickness" and "illness" are now seldom used social labels to stigmatize people in the United States.
   d. The experience of pain is the one area of illness that is not affected by cultural differences.
   e. Because it is now known that AIDS cannot be acquired through casual social interaction, this disease is on the way to losing its negative symbolic baggage.
7. **Epidemiologists have demonstrated that health and illness are strongly influenced by _________ factors.**
   a. social psychological
   b. demographic and social class
   c. anthropological
   d. genetic
   e. spiritual
8. **Which of the following statements is *not* correct?**
   a. The average life expectancy of Americans has increased almost 50 percent since 1900.
   b. Around the world, females have a significantly greater life expectancy than males.
   c. Asian Americans have the lowest age-adjusted mortality rate of all minorities in the United States.
   d. Social class level has been consistently found to be related to mental illness.
   e. Among highly industrialized countries, the United States has a relatively low infant mortality rate.
9. **With the exception of _________, racial and ethnic minorities in the United States have poorer health than the general white population.**
   a. Native Americans
   b. Dominicans
   c. Asian Americans
   d. Aleuts
   e. Puerto Ricans
10. **Minority groups in the United States die younger and suffer more illness than the white majority. Epidemiologists attribute this to**
    a. low income.
    b. genetic diseases.
    c. racial and ethnic characteristics.
    d. poor health habits.
    e. lack of knowledge about preventive health and adequate nutrition.
11. **The entry of large health-care organizations into America's health-care system has led to the creation of**
    a. an atmosphere inhospitable to health-care reform.
    b. medicalization.
    c. a medical-industrial complex.
    d. the professionalization of nurses.
    e. physicians' assistants.
12. **The current major threat to the power of the medical profession lies in**
    a. the new medical-industrial complex.
    b. medicalization.
    c. demedicalization.
    d. hospices.
    e. the failure of the U.S. Congress to pass a health-care reform bill.
13. **_________ are health plans in which members pay set fees on a regular basis and in return receive all necessary health care at little or no cost.**
    a. Cafeteria plans
    b. Payroll deduction plans
    c. Fee-for-service plans
    d. Hospices
    e. Health maintenance organizations
14. **Which of the following statements regarding health-care reform in the United States is accurate?**
    a. Few modern societies provide health care as a free or inexpensive service through their governments.
    b. The United States is one of the few highly developed countries with national health insurance.
    c. Only the poorest Americans are without access to medical care on a nonemergency basis.
    d. Only a little over 1 million children in the United States are not covered by health insurance.
    e. A major reason for health-care reform in the United States is the substantial proportion of Americans who do not have access to medical care, except on an emergency basis.
15. **Which of the following was not discussed as a basic health-care reform option for the United States?**
    a. modified competitiveness
    b. managed competition
    c. all payer
    d. single payer
    e. play or pay

FEEDBACK REVIEW

True-False

1. Although American women have a longer life expectancy than men, women are more involved in the health-care system as patients. T or F?
2. Investor-owned hospitals are less efficient than not-for-profit community hospitals. T or F?
3. The major reason the United States does not have a national health-care program is because its citizens want different results from health-care programs than citizens of other developed societies. T or F?

Fill in the Blank

4. The _________ embraces the professional services, organizations, training academies, and technological resources that are committed to the treatment, management, and prevention of disease.
5. If a sick person in American society does not attempt to recover, he or she will be labeled a _________.
6. According to _________ theory, affordable health care is not provided to the poor in the United States because of the desire for profit.
7. A new medical-industrial complex has been created because of the entry of the _________ into the health-care system.

Multiple Choice

8. **Health-care expenditures now account for about _________ percent of the American GDP.**
   a. 5
   b. 10
   c. 14
   d. 21
9. **Nurses are caught in a conflict between their values and their work environment. Which of the following perspectives best helps us understand this situation?**
   a. functionalism
   b. conflict theory
   c. symbolic interactionism
   d. exchange theory
10. **The approach to health-care delivery under which medical services are financed governmentally is called**
    a. managed competition
    b. single payer
    c. all payer
    d. play or pay

GRAPHIC REVIEW

Figure 16.3 displays U.S. infant mortality rates by race and ethnicity of mothers. Answer these questions to test your understanding of the data in this figure.

1. Discuss two findings that you weren't aware of previously. Can you explain them?

____

____

____

2. Use the Internet to compare U.S. infant mortality rates with other countries. Compare rates in the United States with some other more developed countries, as well as with some less developed countries. How does the United States stack up?

____

____

____

ANSWER KEY

Concept Review

a. 3
b. 2
c. 1
d. 5
e. 4

Multiple Choice

1. d
2. b
3. b
4. d
5. b
6. a
7. b
8. e
9. c
10. a
11. c
12. a
13. e
14. e
15. c

Feedback Review

1. T
2. T
3. F
4. health-care system
5. deviant
6. conflict
7. corporation
8. c
9. c
10. b

17 Population and Urbanization

© Spencer Grant?Stock, Boston

OUTLINE

LEARNING OBJECTIVES

- Identify and distinguish among the three population processes.
- Discuss major dimensions of the world population growth problem.
- Relate the ideas of Thomas Malthus to the demographic transition.
- Predict world population trends.
- Discuss demographic trends of older Americans.
- Differentiate basic concepts of urbanization.
- Trace the historical development of preindustrial and modern cities.
- Describe some of the consequences of suburbanization.
- Discuss world urbanization.
- Compare and contrast four theories of city growth.
- Compare and contrast traditional and contemporary views on the quality of urban life.

USING THE SOCIOLOGICAL IMAGINATION

Do you perceive urbanites as being part of the "lonely crowd," living amid a sea of strangers? If so, you can find support in the pioneering studies of urbanism. Subsequent research, however, presents another depiction: urban residents engage in personal relationships and these social ties contribute to a positive emotional state of mind. There are those without such social relationships, but they tend to be recently relocated residents who have not yet had time to form close human connections.

The quality of urban life is an important consideration. There are, of course, many other issues associated with population growth and urbanization. As a prelude to these issues, the chapter begins with the concept of demography.

The Dynamics of Demography

The Nature of Demography

Why is demography essential in the discussion of population change? **Demography**, the scientific study of population, encompasses all measures of population: size, distribution, composition, age structure, and change. Although basically a social science, demography draws from many disciplines, including biology, geography, mathematics, economics, sociology, and political science.

Demographic research can be divided into two subareas. **Formal demography** deals with gathering, collating, analyzing, and presenting population data. For example, formal demography looks at the data demonstrating changes in the entire American population. Another example of formal demography is the documentation of the dramatic increase in the Latino population of California and Texas. **Social demography**, on the other hand, is the study of population patterns within a social context. For example, social demography examines the relationship between population growth and congressional districting. Historically, social demographers have found that the projected growth of minorities in the United States benefited Democrats more than Republicans (Tilove 1999). Today, though, with respect to Latinos, that particular demographic trend is changing. Now the largest minority in the United States, Latinos are not firmly aligned with either political party. Regardless of political affiliation, the growth of minorities affects the way congressional districts are drawn and is one reason census taking can be a controversial topic. For a second example, to help plan for hospitals and long-term nursing facilities, a social demographer might study trends in the population shifts of aging baby boomers.

Why study population trends? There is a demographer in all of us, observing and recording important events related to population change—births, deaths, the relocation of friends and family. The professional demographer considers these events in two ways: first, by gathering, organizing, and analyzing the patterns of population size, structure, composition, and distribution; and second, by attempting to identify and understand relationships between demographic and social processes. In other words, the demographer studies the link between the quantity of people and the quality of their lives.

Knowledge of population trends has never been more important. The study of population provides information about population growth, the characteristics of population, the location of population, and the probable long- and short-run effects of demographic trends. Demography plays a major role in policy formation, planning, and decision making in both public and private sectors of modern economies. Population information is used, for example, to plan for the health, education, transportation, and recreation needs of virtually every community in the United States. Demographers assist government policy makers and decision makers in meeting the needs of various socioeconomic groups, including the young, the poor, the unemployed, and the elderly. Demography generates information for identifying variable demands for products and services.

Three population processes—fertility, mortality, and migration—are responsible for population growth and decline. Any alterations in population characteristics originate from these three processes (Namboodiri 1996; McFalls 1998). Sociologists use these processes to chart population shifts, past and future.

Fertility

How is fertility related to population growth? **Fertility** measures the number of children born to a woman or to a population of women. Whereas fertility refers to the "actual" number of children women produce, **fecundity** is the maximum rate at which women can potentially produce children. The estimate for the upper limit of a society's average fecundity is fifteen births per woman. The record fertility rate for a group probably is held by the Hutterites, who migrated from Switzerland to North and South Dakota and Canada in

SNAPSHOT OF AMERICA 17.1

Population by Region

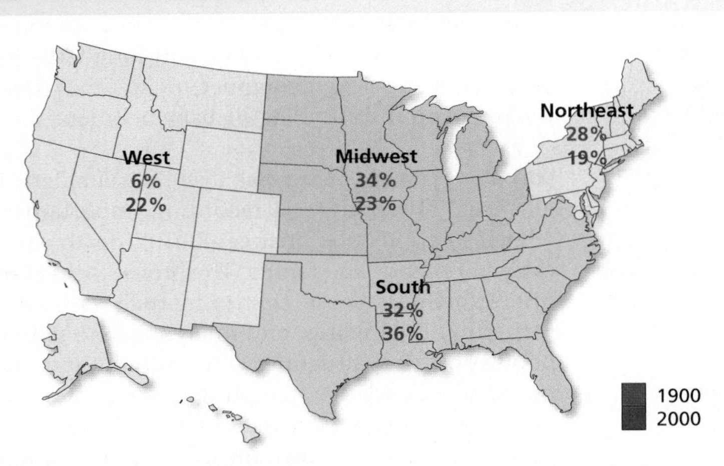

Source: U.S. Census Bureau, 1900, accessed online at www.census.gov/population/censusdata/table-16.pdf on January 20, 2000; 1999 accessed online at www.census.gov/population/estimates/state/st-99-2.txt, on January 20, 2000.

This map shows the percentage of the population in the four regions of the United States at the beginning of the twentieth century and the beginning of the twenty-first century.

1. Discuss this map within the context of migration patterns.
2. Identify and discuss some implications of these population shifts for American politics.

the late nineteenth century. Hutterite women in the 1930s produced an average of more than twelve children each (Westoff and Westoff 1971). The Hutterites give us a good estimate of fecundity, because they are the best example of *natural fertility*—the number of children born to women in the absence of conscious birth control. No other known society has exhibited the fertility level of the Hutterites (Weeks 2002).

How is fertility measured? The **crude birth rate** is the annual number of live births per one thousand members of a population. Birth rate varies considerably from one country to another. The birth rate for the United States is 15 per 1,000. Kenya, in East Africa, experiences a very high birth rate of 34 per 1,000; and Germany, a very low rate of 9 per 1,000 (*World Population Data Sheet* 2002).

To calculate birth rate, one simply needs to know the number of births in a year and the size of the population. One then divides the annual number of live births by the total population and multiplies that number by one thousand. As a formula, it would appear as

$$\text{Crude birth rate} = \frac{\text{Live births} \times 1{,}000}{\text{Total population}}$$

This formula is "crude," however, because, being based on the entire population, it fails to identify those women in the population most likely to give birth; and it ignores the age structure of the population—two factors that affect the number of live births in any given year. Consequently, in addition to the crude birth rate, demographers use the **fertility rate**—the annual

© Robert Brenner/PhotoEdit

The birth rate of a society is affected by social factors, including societal attitudes toward reproduction. Due to changing attitudes regarding the desirable number of children, fewer women are having children, and more women are having fewer children. This couple, like more and more of their contemporaries, could well decide to have no more than two children.

number of live births per one thousand women ages fifteen to forty-four. **Age-specific fertility** is the number of live births per one thousand women in a specific age-group, such as 20 to 24 or 35 to 39. The rate easiest to comprehend is the **total fertility rate**—the average number of children born to a woman during her lifetime. Currently, total fertility rates in the world range from 5.2 in Africa to 1.4 in Europe. In the United States, the total fertility rate stands at 2.1.

The birth rate of a population is influenced by both biological and social factors. A major biological influence is health. For example, widespread disease (especially rubella, or German measles) causes the birth rate to decline because many pregnancies end in miscarriage. Another biological factor, fecundity, influences fertility data: it still takes an average of nine months to give birth to a baby. Primary social factors affecting the birth rate include the average age at marriage, the level of economic development, the availability and use of contraceptives and abortion, the number of women in the labor force, the educational status of women, and social attitudes toward reproduction.

The U.S. birth rate has steadily declined over the last thirty years. It is slightly below the average of two children per woman of child-bearing age, which is required to keep the population size stable. An obvious precipitating factor here is the dramatic change in family planning. More Americans are using contraceptives and having abortions. A greater number of couples believe that two children—or even one child—is a desirable size for a family. And work patterns affect the birth rate as more American women are postponing having children until their late twenties and early thirties. As a result, fewer women are having children, and more women are having fewer children. Even though these factors remain in place, 1998 saw the first increase (1 to 2 percent) in the total fertility rate since 1990 (J. Martin et al. 2002). The significance—if any—of the rising birth rate for women in their twenties and thirties is not known yet.

Mortality

What are the dimensions of mortality? Mortality refers to deaths within a population. The important dimensions of mortality are life span and life expectancy. **Life span**—the most advanced age to which humans can survive—can be inferred only from the person with the greatest authenticated age, a Japanese man who lived nearly 121 years (Russell 1987). Although this reveals that people are biologically capable of living to be 120 years old, few even approach this age. **Life expectancy** is the average number of years that persons in a given population born at a particular time can expect to live. World life expectancy is sixty-seven years (*World Population Data Sheet* 2002).

Morbidity refers to rates of disease and illness in a population. For example, in early twentieth century America, influenza, pneumonia, and tuberculosis accounted for nearly 25 percent of all deaths; but their share of all deaths has declined to less than 3 percent. As more people survive to older ages, chronic degenerative diseases such as heart disease, stroke, and cancer have come to represent an increasing share of deaths (nearly seven out of ten). These diseases represent the consequences of an industrialized lifestyle, including stress, alcohol consumption, cigarette use, and pollution.

How is mortality measured? The **crude death rate** is the annual number of deaths per one thousand members of a population. Similar to the crude birth rate, the death rate is figured by dividing the annual number of deaths by the total population and multiplying by one thousand. Like the crude birth rate, the crude death rate varies widely throughout the world. The worldwide average death rate is 9 per 1,000 persons. Looking at specific regions of the world, the death rate varies from a low of 5 per 1,000 in Central America to a high of 16 per 1,000 in parts of Africa. The death rate in the United States is about 9 per 1,000 (*World Population Data Sheet* 2002).

Demographers are also interested in the variations in birth rates and death rates for specific groups. They have devised **age-specific death rates** to measure the number of deaths per one thousand persons in a specific age-group, such as 15 to 19 or 60 to 64. This measure allows one to compare the risk of death to members of different groups. Although death eventually comes to everyone, the rate at which it occurs depends on many factors, including age, sex, race, occupation, social class, standard of living, and health care.

The **infant mortality rate**—the number of deaths among infants under one year of age per one thousand live births—is considered a good indicator of the health status of any group, because infants are extremely susceptible to variations in food consumption, availability of medical care, and public sanitation. Infants in less developed countries are over seven times more likely to die before their first birthday than infants in the more developed nations are; this fact illustrates the wide variation in living standards among these nations.

Migration

In what two ways can we describe migration? **Migration** refers to the movement of people from one geographical area to another for the purpose of establishing a new residence. We can describe migration in relation to a move either within a country or from one country to another. A current example of international migration is the resettlement of Asian refugees from Vietnam and Cambodia to the United States and other countries around the world. Many of the refugees who settle in

© David Turnley/CORBIS

The infant mortality rate among African children is the highest in the world. These children have beaten the odds. Among African children under one year of age, 89 per 1,000 perish. This compares with 7 per 1,000 in the United States.

the United States in one particular city or region later move to another region, thus becoming internal migrants, as when people move from New York State to Arizona (Weeks 2002).

How is migration measured? The **gross migration rate** is the number of persons per one thousand members of a population who, in a given year, enter (immigrants) or leave (emigrants) a geographical area. Net migration is the combined effect of immigration and emigration on the size of a population. Thus, the **net migration rate** is the annual increase or decrease per one thousand members of a population resulting from movement into and out of the population. The United States, for example, has a net migration rate of about 3.5 per 1,000 population. That is, 3.5 more persons per 1,000 population enter the country than leave the country (Central Intelligence Agency 2002).

When the U.S. Census Bureau reports migration rates, it refers only to the number of legal immigrants. Many people violate immigration quotas in the United States. In the 1970s, illegal entry into the United States—primarily from Latin American and Caribbean countries—became a major concern. It continues to be controversial. There are no precise statistics on either the illegal immigration rate or the total number of illegal immigrants living in the United States, but the current number of illegal immigrants is over 8 million (Martin and Widgren 2002).

As of 2002, 33.1 million legal and illegal immigrants were living in the United States, more than double the number (13.5 million) reached in 1910 during the last massive influx of immigrants. Thus, immigration has become the determinate factor in U.S. population growth, accounting for almost 90 percent of the U.S. population increase since the 2000 census. Immigrants now comprise 11.5 percent of the total population—the largest proportion in seventy years. Given current trends, the end of this decade will see the immigrant portion of the total population exceed the previous high of 14.8 percent, documented in 1890 (Camarota 2002).

Why do people migrate? The most general explanation for migration is the so-called push-pull theory. That is, people move either because they are attracted elsewhere or because they feel impelled to leave their present location. Often "push and pull" factors exist simultaneously. The United States was settled by people who were fleeing economic, religious, and social circumstances and who were attracted by the promise of a better future. People, of course, may not migrate despite significant pushing and pulling factors. Certain barriers, such as the cost of moving or bad health, may outweigh the push-pull factors.

In cases of voluntary movement, economic reasons are the most compelling. Voluntary migration is also associated with stages in the life cycle. The greatest movement is found among young people leaving home to attend college, to take a job, to marry, or to join the military. Retired people may move from a metropolitan area to a rural community or from a colder climate to a warmer one.

Unfortunately, not all migration decisions are freely made. People may be forced to leave an area or prevented from leaving a locale, both of which occurred among Jews in Hitler's Germany and in Bosnia and Yugoslavia. An environmental crisis or an inhospitable climate producing famine or drought may give people no choice but to abandon their place of residence.

© Brown Brothers

America is, of course, a land of immigrants. People came to the United States in massive numbers either because circumstances forced them to leave their homeland or because they envisioned the attainment of a better life. The latter was the case for this immigrant family at New York City's Ellis Island in the early twentieth century.

FEEDBACK

Match the following terms and definitions.

____ a. fertility	____ e. fecundity	____ i. migration
____ b. crude death rate	____ f. age-specific fertility	____ j. gross migration rate
____ c. crude birth rate	____ g. age-specific death rate	____ k. net migration rate
____ d. demography	____ h. infant mortality rate	____ l. morbidity

(1) the number of deaths per 1,000 members of a population per year
(2) the movement of people from one geographical area to another for the purpose of establishing a new residence
(3) the increase or decrease per 1,000 members of a population per year as a result of people leaving and entering the population
(4) the number of live births per 1,000 women aged 15 to 44 years
(5) the number of deaths per 1,000 persons in a specific age-group
(6) the study of the growth, distribution, composition, and change of a population
(7) the number of live births per 1,000 women in a specific age-group
(8) the number of persons per 1,000 members of a population who enter or leave the population each year
(9) the number of deaths to infants under one year of age per 1,000 live births
(10) rates of disease and illness in a population
(11) the maximum rate at which women can physically produce children
(12) the number of live births per 1,000 members of a population per year

Answers: a. 4 b. 1 c. 12 d. 6 e. 11 f. 7 g. 5 h. 9 i. 2 j. 8 k. 3 l. 10

Population Growth

World Population Growth

On October 12, 1999, the United Nations officially declared that the world's population had reached 6 billion. How big is 6 billion? If you counted 100 numbers every minute for 8 hours a day, five days a week, it would take you 500 years to reach 6 billion!

According to Population Connection, the world's population is growing at a rate of 80 million people per year. If asked about the cause of rapid world population growth, what would you say? Like most people, you would probably refer to the high birth rate in poorer countries. You could point out that 135 million new infants are born each year, enough people to more than duplicate Japan's population. You could note that every time you watch a 30-minute television program, 4,860 infants are born. This explanation, however, is only half the story. It leaves out the death-rate side of the equation. Developing countries have passed through the stage when both birth rates and death rates were high. Now these countries are growing rapidly because their birth rates remain high while their death rates have dropped sharply, thanks to modern medicine, improved sanitation, and better hygiene. The growth and the distribution of the world's population—over 6 billion—vary greatly among the nations of the world. Not only that, but population has grown at markedly different rates throughout history (Livi-Bacci 1997; Haupt and Kane 1998 McFalls, 2003).

How fast is the world's population growing? When describing the growth of human population, it is impossible to avoid a certain amount of conjecture, because there has never been a complete worldwide numeration of the population. Although most countries now take a census, there are still national populations that are not calculated. Furthermore, the quality of census data varies a great deal from country to country. Nevertheless, it is possible to indicate world population growth patterns by using historical information and recently collected data (Gelbard, Haub, and Kent 1999).

Rapid world population growth—now standing at 1.3 percent or 79 million people per year—is a relatively recent phenomenon: "Those of us born before 1950 have seen more population growth during our lifetimes than occurred during the preceding four million years, since our early ancestors first stood upright" (L. R. Brown 1996:3). It is estimated that only about 250 million people were on the earth in A.D. 1. It was not until 1650 that the world's population doubled, to half a billion (see Figure 17.1). Subsequent doublings have taken less and less time. The second doubling occurred in 1850, bringing the world population to 1 billion. By 1930, only eighty years later, another doubling had taken place. Only forty-five years later, in 1976, a fourth doubling raised the world's population to 4 billion. At the current growth rate, the world's population is expected to double again in approximately fifty-one years, and it will exceed 9 billion persons by the year 2050 (*World Population Data Sheet* 2002). Obviously, the time between each doubling of the population is getting shorter and shorter (see Table 17.1).

FIGURE 17.1

World Population Growth

This figure shows estimated world population growth from 8000 B.C. to 2150. Population growth since 1850 has been phenomenal. The first doubling of the world's population occurred between 8000 B.C. and 1650. The fourth doubling had occurred by 1976, and world population is on schedule to double for the fifth time in 2025.

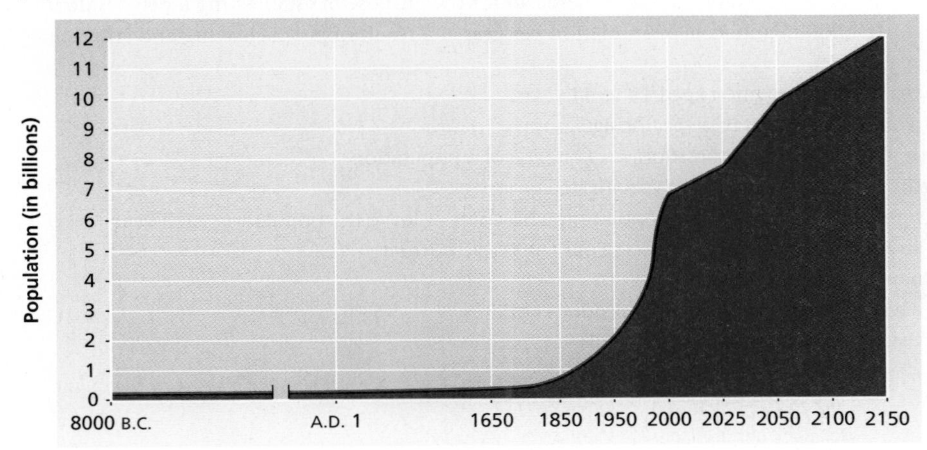

Sources: Data from 8000 B.C. to 2000 comes from Population Reference Bureau, Washington, DC. Projections from 2000 are based on Carl Haub, "New UN Projections Depict a Variety of Demographic Futures" (Washington, DC: Population Reference Bureau, 1997), p. 3.

Why is the world growing so rapidly? The population is increasing dramatically, in part because of the mathematical principle by which population increases. We are accustomed to thinking in terms of linear growth—arithmetical increases of a constant amount within a given time period (as in the progression 1, 2, 3, 4, 5 . . .). If you save $100 a year for ten years, you will end up with $1,000, accumulated in ten equal amounts. Population, however, does not grow linearly. It increases geometrically (as in 2, 4, 8, 16, 32 . . .), following the principle of **exponential growth**—the absolute growth that occurs within a given time period becomes part of the base for the growth rate in the next time period. This means that if the growth rate of a population remains the same for two successive years, the absolute growth will be larger in the second year.

Suppose that a city of 100,000 has an annual growth rate of 5 percent for two years. At the end of the first year, its population will increase by 5,000 (100,000 × 0.05). During the second year, the population will increase by an additional 5,250 (105,000 × 0.05). Consider a concrete case. The world population grew at about 2 percent per year between 1960 and 1970, which produced an increase of 650 million people. At the same rate, by 1980, the world population would increase by 800 million. This was nearly the case. In 1970, the world population was 3,261 million; in 1980, it stood at 4,414 million, an actual increase of 1,153 million.

A classic story offers an example of exponential growth. The story tells of a clever courtier who presented a beautiful chess set to his king and in return asked only that the king give one grain of rice for the first square on the chess board, two grains, or double the amount, for the second square, four (doubling again) for the third, and so forth. The king, being mathematically naive, agreed and ordered the rice brought forth. The eighth square required 128 grains, and the twelfth took more than a pound of rice. Long before reaching the sixty-fourth square, the king's coffers were depleted. Even today, the world's richest king could not produce enough rice to fill the final square. It would require more than 200 billion tons, or the equivalent of the world's current total production of rice for the next 653 years.

A term often used in analyzing population growth is **doubling time**—the number of years needed to double the original population size (given its current rate of growth). If a population is growing at 1 percent per year, it takes only seventy years to double. The number of people added each year becomes part of the total population, which then increases by another 1 percent in the following year.

Whereas the current world population growth rate of 1.3 percent may not seem high, it represents an increase of nearly 220,000 people every day. At this rate, the world's population grows every seven days as much as the size of the entire population of Utah (*World Population Data Sheet* 2002). To put it another

TABLE 17.1

Population Projections by World Regions

This table displays projections by regions of the world from 2002 to 2050. Note the dramatic difference in population doubling time between less economically developed areas and more economically developed areas.

Region	Population 2002 (in millions)	Population Projection for 2025 (in millions)	Population Projection for 2050 (in millions)	Projected Population Change Between 2002 and 2025	and 2050	Doubling Time (in years)
North America	316	382	450	+21%	+42%	139
Asia	3,720	4,714	5,262	+27%	+41%	50
Oceania	31	40	46	+29%	+48%	63
Latin America	525	697	815	+33%	+55%	41
Africa	818	1,268	1,800	+55%	+120%	29
Europe	727	717	662	−1%	−9%	*
Less economically developed areas	4,944	6,570	7,794	+33%	+58%	43
More economically developed areas	1,193	1,248	1,242	+5%	+4%	693
World	6,137	7,818	9,036	+27%	+47%	53

*Doubling not projected to occur.
Source: Adapted from *World Population Data Sheet* (Washington, DC: Population Reference Bureau, 2002).

way, the human population is growing at the rate of 152 per minute, nearly 220,000 per day, almost 80 million per year. The pace of growth is so rapid that additions to the world's population could equal the population of the United States in just over three years. At this rate, more people will be born in the year 2050 than were born in the 1,500 years after the birth of Christ.

On the encouraging side, the rate of world population growth has been declining since the 1970s. Data from fifty-six countries with an estimated population of 10 million or more indicate that the average annual rate of increase dropped from 2.1 percent in 1970 to 1.7 percent in the 1980s, to 1.3 percent in 2002 (*World Population Data Sheet* 2002).

The top part of Table 17.1 shows the population in various regions of the world, along with the projected populations for 2025. Consider the projected percentage of population increase in the less developed regions: Asia, 27 percent; Africa, 55 percent; Latin America, 33 percent; Oceania, 29 percent. Among the more developed regions, the highest percentage of population increase is 21 percent (in the United States and Canada). Doubling times for less developed regions vary from a low of twenty-nine years (Africa) to a high of sixty-three years (Oceania). In contrast, doubling times for more developed regions range from 139 years in the United States and Canada to no expectation of doubling in Europe.

The concentration of the world's population in less developed regions is also apparent in the bottom half of Table 17.1. About 81 percent of all people lived in the less developed regions of the world in 2002; 84 percent are projected to do so by 2025. Thus, the majority of people live in nations least able to provide for their needs. When the less developed and more developed regions are categorized separately, as they are in the bottom half of Table 17.1, two things become clear: not only is the population projected to increase by 33 percent in less developed regions between 2002 and 2025, but the rate of increase is almost seven times that of more developed regions.

By 2050, the more developed countries are projected to increase from 1.197 billion to 1.249 billion, an increase of only 52 million people. Over the same period, the population of less developed countries will rise from 5.018 billion to 7.873 billion, an increase of over 2.8 billion. To put it another way, nearly 99 percent of the world's population growth will take place in less developed countries, while deaths exceed births annually in Europe (*World Population Data Sheet* 2002).

What difference does one child make? The importance of limiting family size, even by one child, can be illustrated by population projections for the United States (see Figure 17.2). Even though the United States is not going to a three-child average in the future, the hypothetical American case can help us understand the importance of population control. Figure 17.2 contrasts the projected population of the United States in the year 2070 for an average family size of two children and an average family size of three children. When small decreases in the crude death rate and a stable net migration are assumed, an average two-child family size would result in a population of 300 million in 2015. Taking the hypothetical average family size of three children, the U.S. population would grow to 400 million by 2013. As time passed, the difference of only one extra child per family would assume added significance. By 2070, the two-child family would produce a population of 350 million, but the three-child family would push the population to close to 1 billion! To say it another way, with an average family of two children, the U.S. population would not quite double itself between 1970 and 2070. But should the three-child family be the average, the population would double itself twice during this same period.

The importance of limiting population in developing regions becomes clear when the effect of even one child added to the average number of children in a family is recognized. Moreover, the addition of one child per family has a greater effect as the population base gets larger; not only is one extra person added, but theoretically that one person will be involved with the reproduction of yet another three; and on it goes. The largest populations are found in developing countries, which also have the largest average number of children per family.

The Malthusian Perspective

Although concern about population growth is strong, it is not new. In 1798, Thomas Robert Malthus, an English minister and economist, published *An Essay on the Principle of Population*, one of the most important works ever written on population growth and its repercussions. Thanks to the pessimism of Malthus's thesis, economics earned the label "dismal science."

In his essay, Malthus proposed a set of relationships between population growth and economic development:

1. Population, if left unchecked, will tend to exceed the available food supply. This is because population increases exponentially; the food supply does not.
2. Checks on population can be positive—those factors that increase mortality, such as famines, disease, wars—or preventive—those factors that decrease fertility, such as later marriages or abstaining from sexual relations in marriages. (At the time, there was no reliable birth control. For this conservative minister, sexual abstinence was the only morally acceptable—and practical—way to reduce the number of births.)
3. For the poor, any improvement in income is lost to additional births; this leads to reduced food consumption and lower standards of living and eventually to the operation of positive checks.

FIGURE 17.2

Projected Population of the United States, Comparing a Two-Child Average per Family (Replacement Level) to a Three-Child Average

This figure illustrates the importance of reaching the population replacement level (two children per family). Are you surprised with the difference in U.S. population growth caused by an average three children per family versus two children? Relate this difference to the principle of exponential growth.

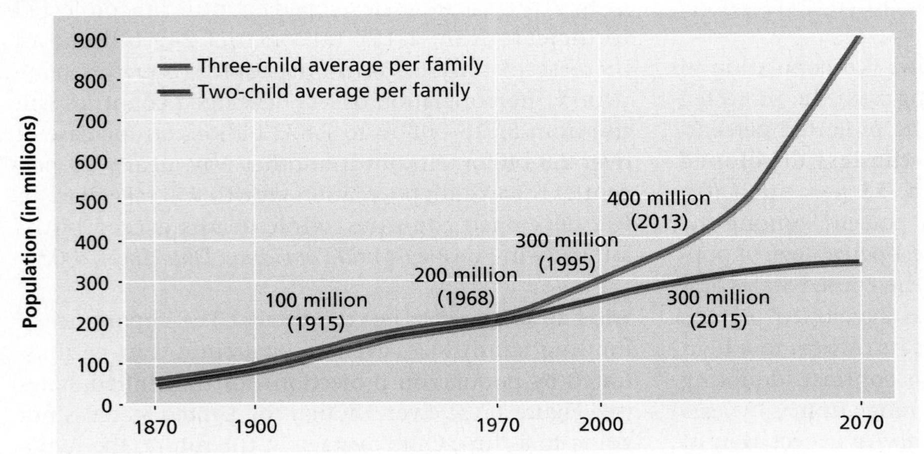

Source: *World Population Data Sheet*, Washington, DC: Population Reference Bureau, 2000.

4. The wealthy and better educated already exercise preventive checks.

This last point leads to some of Malthus's suggestions for solving the "population problem." Because the wealthy and better educated had achieved controlled fertility, the method of extending this to the poor would be through universal education, which would offer all members of the population the chance of improving their situation: "The desire of bettering our condition, and the fear of making it worse, has been constantly in action and has been constantly directing people into the right road" (Malthus 1798:477). The beneficial effects of universal education on population control could be enhanced, Malthus suggested, by raising people's aspirations for a higher standard of living. This would be accomplished by raising wages above the minimum required for subsistence, thus providing the poor an opportunity to choose between more children at a minimal standard of living or smaller families with a higher quality of life.

Has the Malthus perspective had any effect? Malthus's writings have profoundly influenced the study of population change. Events during the late nineteenth and early twentieth centuries brought his predictions into question. The last half of the twentieth century, however, has again raised the specter of a Malthusian crisis: population growth outstripping both food resources and the ability of different societies to provide even minimal quantities of basic necessities (housing, health care, education, employment). Some social scientists—the neo-Malthusians—modified Malthus's propositions in an attempt to explain the current world situation and to predict possible futures. First, neo-Malthusians note that the development of reliable contraceptives has not distorted marital relations as Malthus feared that it might. Second, historical developments since Malthus's time indicate that values promoting, and norms supporting, smaller families are positively related to certain kinds of social and economic changes. In other words, values on family size and norms for fertility regulation are adapted to socioeconomic conditions, enabling society to avoid Malthusian checks through self-regulation. Third, neo-Malthusians argue that many nations have a rate of population growth that overloads this self-regulating process because population growth is excessive to the extent that resources are diverted from socioeconomic change to population maintenance (Humphrey and Buttel 1986).

A debate is taking place between those who believe that the world's population is exploding beyond control and those who are convinced that the population explosion will be defused. Those predicting a slackening of the current high population growth rate often base their optimism on the increasing worldwide acceptance of birth control. This debate is more easily understood when viewed within the context of the demographic transition.

The Demographic Transition

How does the demographic transition develop? To the extent that developed nations have defied Malthus's theory, they have done so through the **demographic transition**—the process by which a population, as a result of economic development, gradually moves from high birth rates and death rates to low birth rates and death rates (see Figure 17.3). The number of stages of population growth in the demographic transition ranges from three to five, depending on how finely the process is broken down. The description that follows is a four-stage model with examples:

Stage 1. Both the birth rate and the death rate are high, and population growth is modest. This is characteristic of preindustrial societies.

Stage 2. The birth rate remains high, but the death rate begins to drop sharply because of modernization—sanitation, increased food production, medical advances. The rate of population growth is very high. Most sub-Saharan African countries are presently at this stage.

Stage 3. The birth rate declines sharply; but because the death rate continues to drop, population growth is still rapid.

Stage 4. Both the birth rate and the death rate are low, and the population grows slowly if at all. North America, Europe, and Japan are at this stage today. Europe, in fact, is in what is being called the "second demographic transition."

What is the second demographic transition? According to Table 17.1, the expected population decrease for Europe between 2000 and 2050 is 1 percent. This reflects the "second demographic transition," which Europe alone is experiencing. Europe's first demographic transition began with a gradual death-rate decline in the early nineteenth century, followed by a birth-rate decline beginning around 1880. The transition to both low birth rate and low death rate was completed by the 1930s. Europe's second demographic transition began in the mid-1960s.

The primary demographic characteristic of the second demographic transition is the decline in fertility from slightly above **replacement level**—the birth rate at which a couple replaces itself in the population—to a fertility level of 1.7. Should Europe maintain this lower fertility rate, and if immigration

FIGURE 17.3

Stages of the Demographic Transition

This figure illustrates the demographic transition. Stage 1 begins with smaller population growth due to a balance between birth rates and death rates (both at high levels). In Stage 2, population grows dramatically because the death rate decreases so much faster than the birth rate. Population growth begins to slow in Stage 3, when the birth rate belatedly drops sharply. Stage 4 is again a condition of smaller population growth because birth rates and death rates come into balance (both at low levels). Given this description, what would you call the gap between the two graphs representing birth and death rates?

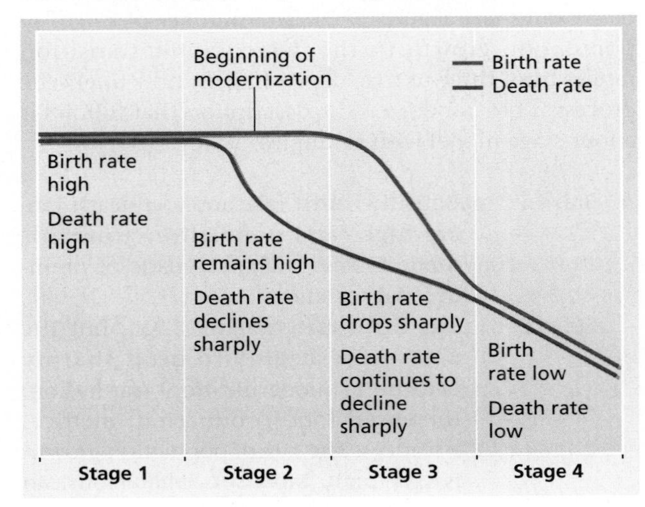

does not interfere, Europe's population will decline, something that has already happened in Italy, Sweden, Germany, Hungary, and Russia. The original formulation of the demographic transition, as noted earlier, ended with stable or nongrowing population.

Does the demographic transition occur in developing societies? Traditionally, demographers assumed that developing societies would proceed through the same stages as the developed world. This assumption is being criticized for its ethnocentric bias. Why? Because developing societies did not begin the transition at the same point as Western societies. For example, less developed countries started the transition with birth rates and death rates higher than existed in the now-developed societies when they began the transition. And mortality declined in the West because of domestic industrialization rather than from the importation of sophisticated techniques of disease control.

Population Control

Because death rates in both developing and developed nations have already dropped dramatically, efforts to curb world population growth must concentrate on lowering birth rates. Thus, **population control** is the conscious attempt to regulate population size through national birth control programs.

What is the history of birth rates? Historically, many societies were more concerned with increasing their population size than with overpopulation. High birth rates were needed to offset the high death rates from disease and poor hygiene. For some societies, larger populations enhanced security because of the ability to maintain larger armies. Agricultural societies needed large numbers of people to work the land. Aging parents wanted to be more secure in old age. High birth rates were also encouraged in countries with religious prohibitions against birth control.

Since the mid-twentieth century, however, more governments have come to view high birth rates as a threat to their national well-being. By 1990, most countries had in place formal programs to reduce birth rates—either voluntary or compulsory.

What is voluntary population control, and how successful has it been? Voluntary population control is generally known as **family planning**—making it technically possible for women to choose the number of children they will have. Beyond making contraceptive devices available, family planning programs provide birth prevention information and other services. Though women in the poorest countries still have the highest fertility, their overall fertility has been reduced by 50 percent since 1969 (Kluger and Dorfman 2002).

Even when effective, family planning programs merely enable families to achieve their *desired* family size. Unfortunately for effective population control, in many nations the desired family size is quite high. The World Fertility Survey reported that the average preferred family size of women in African nations is 7.1; in Middle Eastern nations, 5.1; in Latin American nations, 4.3; and in Asian Pacific nations, 4.0. In European countries, the average preferred family size ranges from 2.1 to 2.8.

Family planning has worked in Taiwan, where by the turn of the twenty-first century, the birth rate declined below replacement level. Taiwan's family planning efforts were launched under very favorable conditions. When the Japanese withdrew from Taiwan after World War II, they left behind a labor force trained for industrial work. Consequently, the Taiwanese were able to use this advantage to build an expanding economy. With economic development came a decline in both death rates and birth rates. In short, they went through the demographic transition fairly rapidly.

India was a different story. Family planning there got off to a very slow start, and the country has been unable to reduce the rate of population growth through voluntary means. Family planning efforts failed because government officials and family planners underestimated the barriers to birth control—they did not take the broader social context into account. For one thing, India did not have Taiwan's advantage of relatively rapid economic development. In addition, the Indian officials and planners made insufficient efforts to overcome cultural and religious opposition to birth control. Nor did they address the lack of channels to effectively distribute birth control information and technology. Finally, the implementation of the national birth control policy was left in the hands of individual state governments.

Efforts at population control began to succeed in India only after the government turned to a sterilization program in 1976. Although the government did not use the force of law, a system of disincentives had the effect of compulsion. Those who could not produce official proof of a sterilization were denied such things as business permits, gun licenses, and ration cards for the purchase of basic goods (Weeks 2002). Still, India currently has a total fertility rate of 3.2 children per woman (the United States has a total fertility rate of about 2.1 children per woman). India will have the largest population in the world by about 2040.

Have any countries successfully enforced compulsory population control methods? Both China and Singapore have required population control policies that seem to achieve their goals. In an attempt to reach zero population growth, the Chinese government first tried to discourage a third child in a family. The second step was to push for the one-child family. China was successful in reducing its total fertility rate from 7.5 in 1963 to 1.8 today through coercion and a system of rewards and punishments. In cities, couples with only one child can obtain a one-child certificate. This certificate ensures them a monthly allowance for the costs of children up to age fourteen. One-child families receive a larger retirement pension and enjoy preference in housing, school admission for their children, and employment. Families with more than one child are subject to an escalating tax on each child, and they get no financial aid from the government for the medical and education costs of their extra children.

Singapore, an island city-state on the Malaysian peninsula, began formally discouraging large families in 1969, installing some economic penalties to encourage replacement-level fertility. The government included the following measures: denial of a paid eight-week maternity leave, loss of an income tax allowance, diminished access to public housing, increased maternity costs for each additional child, and lower likelihood of children entering the better schools.

These policies worked so well that the total fertility rate in Singapore dropped from 4.5 children per woman to 1.4 between 1966 and 1985. Worried over the reduction in population size, the government in 1987 switched to the guideline of three or more children for people able to afford them (Yap 1995). Despite this effort, Singapore's total birth rate is now 1.6, still below replacement level.

Future World Population Growth

World population growth has reached a watershed. After more than 200 years of acceleration, the annual

SOCIOLOGY AND THE NEWS MEDIA

Family Planning in Egypt

Because of an acute land shortage, the Egyptian government is concerned about overpopulation. Thanks to government subsidies, citizens pay a nominal price for contraceptive devices. Despite a tradition of large families, Egypt's family planning program appears to be working. This CNN report offers some details about population control in this North African nation.

1. In what stage of the demographic transition do you think Egypt finds itself? Explain your choice.

2. Discuss the concept of zero population growth within the context of the situation in Egypt.

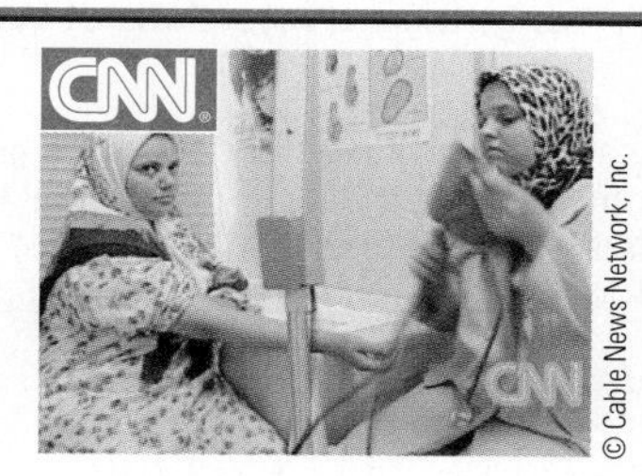

These Egyptian women are involved in a government-supported birth control facility.

population growth rate is declining. From 1950 to 1975, world population rose 64 percent; since 1975 it has increased 48 percent (Kluger and Dorfman 2002). The current growth rate of 1.3 percent compares favorably with the peak of 2.04 percent in the late 1960s. Moreover, the rate is projected to drop to zero by the end of the twenty-first century (United Nations Population Division, 2000). The birth rate is already low in developed countries and, as noted earlier, is now declining in many developing countries. Similarly, the population growth rate is declining somewhat in both developed and developing areas of the world (Gelbard, Haub, and Kent 1999; Singer 1999).

Demographers are unsure of future world population growth in part because they do not know for sure how many children today's youth will have. Nor do they know what will happen to change life expectancy, particularly in developing countries.

Some analysts are now foreseeing a potential shift from concern about a "population explosion" to worry about a "population implosion" or decline due to lower fertility rates (Eberstadt 2001). Other demographers, however, point out that there are two very different worlds of population growth. In developed countries, fertility rates are at or below two children per couple (Europe, the United States, Canada, Japan, and some rapidly developing countries such as China, South Korea, and Thailand). In developing societies, women average four children each (most of Africa, Asia, and Latin America). If the accelerating population growth that the world has known since the 1950s is to stop, there must be a dramatic decrease in fertility in developing countries (Gelbard and Haub 1998; Ashford 2001).

What is the future of world population growth? The United Nations offers three possible world population growth scenarios (see Figure 17.4). These three futures vary, depending on their assumptions regarding the average number of children women will bear. For the medium scenario to come true, women will have to average two children. If so, world population will rise to 9.4 billion by 2050 and continue to 11 billion in 2150 before leveling off. The medium scenario depicts **zero population growth**—when deaths are balanced by births so that the population does not grow. To make the high scenario a reality, women worldwide will have to average about 2.5 children. If this happens, world population will rise to 11.2 billion in 2050 and continue to grow indefinitely, exceeding 27 billion 100 years later. In the low scenario, fertility will have fallen to 1.6 children per woman or less. World population, under this condition, will peak at 7.7 billion in 2050 and drop to 3.6 billion by the middle of the twenty-first century (Ashford 2001; see Figure 17.4).

But as we have seen, despite the reduction in the annual growth rate and birth rate, the world's population will continue to increase. Nearly 7 billion people are expected to inhabit the globe by 2010. Throughout the first half of the twenty-first century, the annual growth rate is expected to decline until world population stabilizes at about 12 billion people. At this point, the world will have reached zero population growth,

FIGURE 17.4

Long-Range Projections of World Population: 2000–2150

As this figure illustrates, the future growth of world population is uncertain. Which of these scenarios do you think is most likely to occur? Defend your answer with information from this chapter or from additional research.

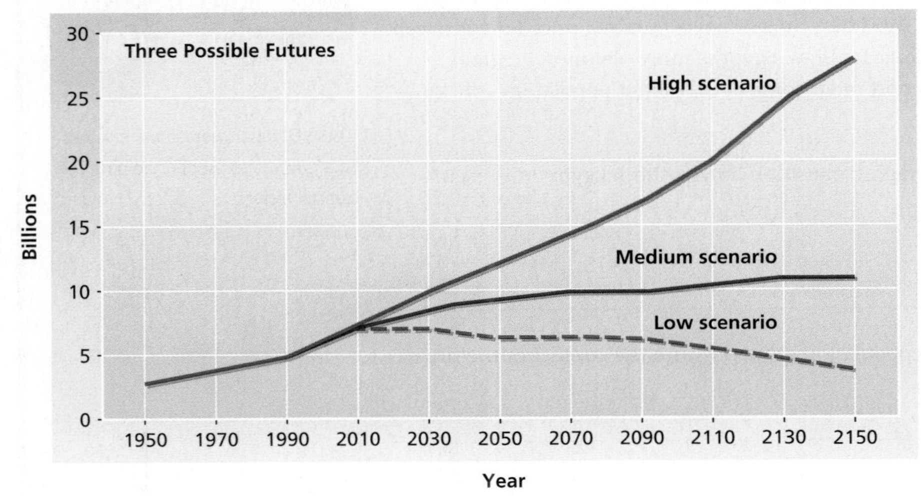

Source: *World Population Data Sheet* (Washington, DC: Population Reference Bureau, 2002).

when deaths are balanced by births so that the population does not increase.

Contrary to popular opinion, limiting the average family size to two children does not immediately produce zero population growth. This is because there is a time lag of sixty to seventy years, due to the high proportion of young women of childbearing age. Even if each of these women had only two children, the world population would grow because there are so many young women. Zero population growth will be reached only when this disproportionate number of young mothers disappears.

The time lag is what demographers call **population momentum**—a population continues to grow, regardless of a recent drop in the birth rate, because of the existing population base created by past growth. The increase in the world's population, like a huge boulder careening down a mountain, cannot be suddenly stopped. But the sooner the momentum of current population growth is halted, the better. The sooner the world fertility rate reaches the replacement level, the sooner zero population growth will be reached. The ultimate size of the world's population, when it does stop growing, depends greatly on the timing of the replacement level. To state it another way, each decade it takes to reach replacement level, fertility will increase the world's population by 15 percent.

Population Growth in the United States

The population of the United States has increased dramatically since the first census in 1790, growing steadily from less than 4 million in 1790 to 288 million in 2002. At the same time, the population growth rate has, with some deviations, tended to decline since 1790. Between 1990 and 2000, however, the American population surged by nearly 33 million, after growing more slowly between 1970 and 1990. The population of the United States is expected to continue growing despite the average American family reproducing at the replacement level of 2.1 children per family. In part, this is because of the baby-boom bulge in the age structure that produced the large number of women whose children and grandchildren are now part of the population (the "baby boomlet"). In addition to fertility, American population growth will be affected by its rates of mortality and migration (Kent and Matter 2002). If the current predictions of fertility, mortality, and immigration rates are accurate, the total U.S. population will reach 420 million by 2050. This would represent an almost 50 percent increase, considerably higher population growth than is expected in other more developed countries. In fact, as shown in Figure 17.5, U.S. population growth is projected to increase between now and 2050, while population growth in other more developed countries is expected to decline. Thus, the United States will remain the largest of the more developed countries.

FIGURE 17.5

Population Growth in the United States and Other More Developed Countries: 1950–2050

This figure compares the curves of population growth in the United States with the rest of the more developed world. Describe any differences you see between 1950 and 1990. Describe the current differences in population growth between the two areas. Explain.

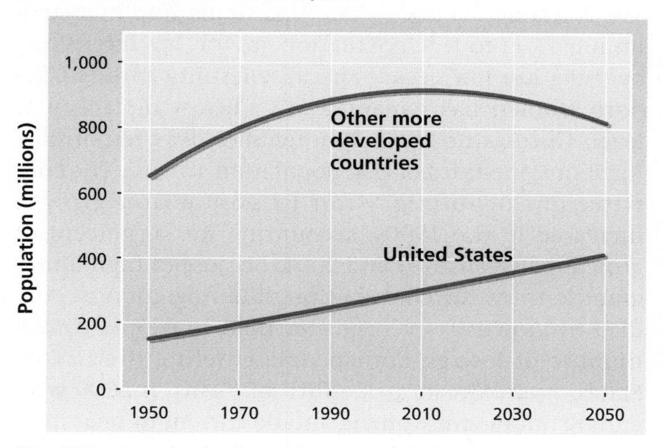

Note: "Other more developed countries" includes all European countries, Australia, Canada, Japan, and New Zealand.
Source: U.S. Census Bureau, International Data Base, October, 10, 2002, release; www.census.gov/ipc/www/idbagg.html (accessed October, 12, 2002).

Why will the U.S. population grow faster than other more developed countries? Generally, fertility is the primary factor in population change. The United States is no exception. Its population experienced **natural increase** (excess of births over deaths) each decade of the twentieth century. This trend continued between 2000 and 2001, when 4 million births were offset by only 2.4 million deaths. This natural increase of 1.6 million accounted for just over 60 percent of the total population growth that year (2.7 million).

Demographers often use the total fertility rate (the average number of children born to a woman during her lifetime) to predict population change. The total fertility rate ebbed and flowed during the twentieth century. In the first few decades of that century, native-born white women had an average of 3.5 children over their lifetimes. In part because of a lower total fertility rate, natural increase hit a low during the Great Depression of the 1930s. This declining natural increase was reversed after World War II when the baby boom occurred. Another total fertility rate low occurred in the early 1970s (1.7 children per woman), followed by an upswing beginning in the late 1970s. This small natural increase in the late 1970s and 1980s did not approach that of the baby boom of the 1950s, during which women were once again averaging 3.5 children.

A positive net migration rate has always, except during the Great Depression in the 1930s, played a larger role in the population growth in the United States than in most other nations. Among more developed countries, the United States has the largest foreign population, by a considerable margin ("The Longest Journey" 2002). In 2000, immigrants accounted for 11 percent of America's population, the highest share since the early 1900s (Camarota 2001). The traditional importance of immigration to U.S. population growth has intensified over the last few decades because fertility among U.S.-born women has remained at or below replacement level. During the 1980s, immigration was responsible for about one-third of U.S. population growth. The contribution of immigration to population growth increased in the 1990s, accounting for 40 percent of growth between 2000 and 2001. Because of high immigrant fertility, the contribution of immigration to population growth is even greater than suggested by the number of foreign-born people entering the United States. Approximately one-fifth of births in 2000 were among immigrant women. Should current immigration and fertility trends continue for 100 years, four in ten Americans will be post-1980 immigrants or their descendants. Currently, approximately one-fifth of U.S. residents is either foreign born or born of immigrant parents (Schmidley 2001).

© Chuck Savage/CORBIS

Americans over sixty-five are more active today, whether participating in leisure group activities or remaining in the workforce.

During the 1950s, almost half of all immigrants came from Northern and Western Europe, whereas only two in ten came from less developed nations. Since 1960, the sources of immigration have shifted from Europe to Latin America, the Caribbean, and Asia. Less than one-sixth of the legal immigrants now come from Europe, whereas more than three-fourths arrive from Latin American, the Caribbean, and Asia. For example, the number of Asians in the United States doubled to 3.5 million during the 1970s and doubled again during the 1980s. Eleven million Asian Americans now comprise almost 4 percent of the population. When countries are ranked according to the number of immigrants sent to the United States, the top four are Latin American (U.S. Bureau of the Census 2001a).

Although mortality rates can influence population growth rates, they have little effect among more devel-

FEEDBACK

1. According to the principle of __________, the absolute growth that occurs within a given time period becomes part of the base on which the growth rate is applied for the next time period.
2. Thomas Malthus saw famines as a __________ check on population growth.
3. The process by which a population gradually moves from high birth and death rates to low birth and death rates as a result of economic development is called the __________.
4. Europe's second demographic transition involves what kind of population growth?
   a. stable b. increasing c. declining
5. Which of the following figures is the world's population most likely to reach before it stops growing?
   a. 4 billion b. 8 billion c. 12 billion d. 25 billion
6. ________ exists when deaths are balanced by births so that the population does not increase.
7. Between now and 2050, the U.S. population is expected to
   a. remain stable
   b. increase by almost 50 percent
   c. decrease by 33 percent
   d. dip in 2036 and start growing again.
8. A population with an excess of births over deaths will experience ________.

Answers: 1. exponential growth 2. positive 3. demographic transition 4. c 5. c 6. Zero population growth 7. b 8. natural increase

oped countries, where mortality is already low (Kent and Mather 2002). At nine deaths per one thousand persons, the death rate in the United States is relatively low, but because of its aging of the population, the number of deaths has been increasing slightly since the early 1980s. Life expectancy, of course, has increased as the death rate has declined. On the average, an American born now can expect to live 77 years, up from 71 years at the beginning of the 1970s. Life expectancy remains lower for males (74 years) than for females (80 years).

Actually, mortality rates could affect future U.S. population growth because they are higher at most ages than in other more developed countries, outside Eastern Europe. U.S. mortality rates could decline further if preventable deaths due to such factors as poor medical care and poverty among younger Americans were lowered, particularly among African Americans. Death rates could also be reduced at later ages due to improved medical care and the adoption of healthier diets and lifestyles (Kent and Mather 2002).

Urbanization

For centuries, people have lived in large population settlements. The nature of those settlements has undergone important changes, and as the settlements have mutated, the language to discuss them has also been altered. As an aid to understanding the volatile conditions in urban areas, we will first examine the terminology used to describe such areas.

Basic Concepts

What defines a community? Long before the rise of cities, people depended on communities for their common needs. In fact, a **community** is a concentration of people whose major social and economic needs are satisfied primarily within the area where they live. It is easy to identify the network of social relationships and organizations that fulfill basic social and economic needs in a small community of several hundred people, but the situation is far more complex when several million people live within a given geographical area. Some sociologists even doubt that such large concentrations of people can be described as communities. Consequently, social scientists are now concentrating on cities and metropolitan areas rather than on communities.

When does a village become a city? The census bureaus of most nations define cities as population aggregates exceeding a certain size. In Denmark and Sweden, an area with 200 inhabitants officially qualifies as a city. Japan uses a much higher cutoff point—30,000 inhabitants. The cutoff point used by the U.S. Census Bureau to define a city is a population of 2,500. These relatively low definition points were set at a time when urbanization had just begun and population concentrations were small; they are clearly low for modern times. Besides, using mere numbers of inhabitants to define a city overlooks other important aspects of cities. In addition to containing a reasonably large number of people, cities are characterized by permanence, density, heterogeneity, and occupational specialization of inhabitants. In brief, a city is a dense and permanent concentration of people who live in a limited geographical area and who earn their living primarily through nonagricultural activities.

Are cities the largest urban units? A city may grow to a point where its official political boundaries fail to keep pace with its population and economic activities. But extending its boundaries may be unfeasible because outlying areas may resist being incorporated into a city (because of higher taxes or loss of political control). What's more, expansion may be impeded because of state and county lines or nearby cities. Consequently, urban areas often extend far beyond the boundaries of the original city.

New terms identify such enlarged urban areas (Palen 2001). One such term is **metropolitan statistical area (MSA)**—a grouping of counties that contains at least one city of 50,000 inhabitants or more or an "urbanized area" of at least 50,000 inhabitants and a total metropolitan population of at least 100,000. The establishment of criteria for designating MSAs allows the Census Bureau to identify large areas whose populations have interdependent relationships with a central city. This is important because the influence of any city extends far beyond its official boundaries. There are 255 MSAs in the United States (U.S. Bureau of the Census 1997b). Yet even the MSA is too limited to identify some urban areas. Many MSAs do not exist in isolation. For example, although Newark, New Jersey, qualifies as an MSA, it must be considered in relation to surrounding MSAs, including New York City. Because of such situations, the Census Bureau now identifies a **primary metropolitan statistical area (PMSA)**—adjacent MSAs that are closely integrated economically and socially. The United States now has seventy-three PMSAs.

Urban growth has led to yet a third category. A **consolidated metropolitan statistical area (CMSA)** consists of two or more sets of neighboring PMSAs (also known as a megalopolis). There are eighteen CMSAs in the United States. At the present time, 13 percent of the American population resides in the top two CMSAs, 21 percent in the top five, 31 percent in the top ten, and 38 percent in the entire eighteen CMSAs.

The lower Pacific Coast of California has developed into a megalopolis. This night shot of the L.A. Hollywood Freeway hints at the mammoth size of a megalopolis.

The Urban Transition

Urbanization is the process by which an increasingly larger portion of the world's population lives in urban areas (see Worldview 17.1). Urbanization has been such a prevalent trend that it is now taken for granted in many parts of the world. As you have just seen, this process has resulted in cities so large and interdependent that it is now necessary to think of strips of interlocking cities. At the beginning of the twenty-first century, almost as many people live in urban areas as in rural areas (Pacione 2001; Scripps-Howard 2002). This is a fairly recent development in human history.

What was the population of the early cities? The size of the first cities—founded only five thousand years ago—was quite small by contemporary standards. Ur, located at the point where the Tigris and Euphrates Rivers meet (in modern-day Iraq), was one of the world's first major cities; but at its peak, it held only about 24,000 people. Later, during the time of the Roman Empire, it is unlikely that many cities had populations larger than

WORLDVIEW 17.1

World's Largest Cities

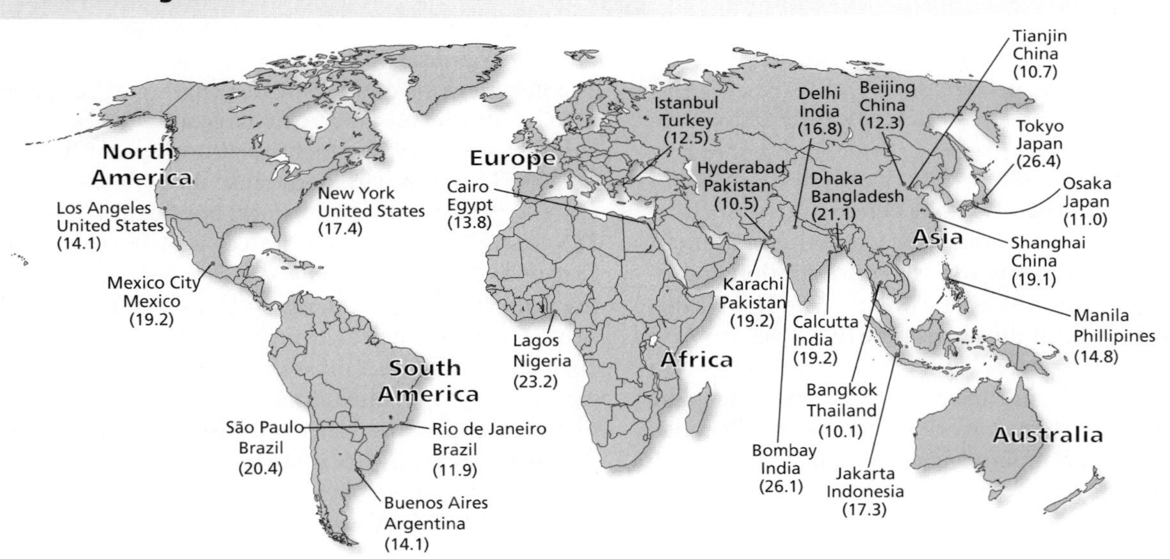

Source: U.N. Population Division, March 2000.

Within five years, half of the world's population will reside in metropolitan areas. This means that the populations of less developed countries are also coming to cities. Whereas 18 percent of the people in less developed countries lived in urban areas in 1950, 40 percent do today, and 56 percent will by 2030. This map locates cities in the world that will have 10 to 30 million residents by 2015.

1. What are major implications of this trend toward global urbanization for the demographic transition and world population growth?
2. Discuss the size and the distribution of these megacities over the globe in relation to overurbanization.
3. Specify some of the environmental effects of urbanization of this magnitude.

33,000. The population of Rome itself was probably under 350,000 (A. H. Hawley 1971).

In addition to their small size, the cities of ancient and medieval periods contained only a small portion of the world's population. As recently as 1800, less than 3 percent of the world's population lived in cities of 20,000 or more (K. Davis 1955). By contrast, 45 percent of the world's population now lives in urban areas. In North America, 80 percent of the population lives in cities (*World Population Data Sheet* 2002). Only about 5 percent of the American population lived in urban areas in 1790 (U.S. Bureau of the Census 1993c).

What accounts for this increase in urbanization? To find the answer to this question, we must take a closer look at urbanization before and after the Industrial Revolution. Although rapid urbanization is a fairly recent occurrence, its groundwork has been developing for several centuries.

Preindustrial Cities

Because people in cities hold primarily nonagricultural jobs, they require an agricultural surplus. Before 3500 B.C., few areas of the world were capable of producing a dependable supply of extra food for nonagriculturally productive people. For this reason, cities were initially slow to develop (Robert Adams 1965).

What enabled the first cities to develop? According to anthropological research, the first urban settlements were established around 3500 B.C. in Mesopotamia (in southwestern Asia between the Tigris and Euphrates Rivers). The Mesopotamian region is among the world's most fertile areas, and at that time, its people were relatively advanced in the development of plant cultivation. So this area provided the necessary surplus of food for people in cities.

The existence of a surplus food supply clarifies the location of cities, but it does not explain people's attraction to them. Cities tended to attract four basic types of people: the elite, functionaries, craftspeople, and the poor and destitute. For the elite, who comprised but a small segment of the population, the city provided a setting for consolidating political, military, or religious power. The jewelry and other luxury items found in the tombs of the elite acknowledge the benefits these people gained from their consolidation of power and control. The functionaries were the political or religious officials who carried out the plans of the elite and whose lives were undoubtedly easier than those of the peasant-farmers in the countryside. Craftspeople, still lower on the stratification structure, came to the city to work and sell their products to the elite and functionaries. The poor and destitute, who were lured to the city for economic relief, were seldom able to improve their condition (Gist and Fava 1974).

Do preindustrial types of cities still exist? Many parts of Africa, Asia, and Latin America are not industrialized and therefore still have some preindustrial types of cities. This is particularly true of capital cities, which attract migrants from rural areas because of the symbolic importance of these cities and the promise of a better life. Unfortunately, those who migrate to many of these cities are disappointed, because the expected employment opportunities do not exist. Also, many cities in developing nations often lack the housing, sanitation, transportation, and communication facilities needed to accommodate their residents. As a result, many city dwellers in developing nations are forced to live in makeshift housing with few or none of the facilities that most people in modern societies consider necessities. Although cities in developing nations are unable to fulfill the social and economic needs of their residents, many people are driven to them because their life chances in the overpopulated rural areas and small towns are even worse.

Rise of the Modern City

As merchants gained power and wealth during the late medieval period, they often established small industries that would give them a stable supply of goods. As these industries gradually increased in size, the stage was set for rapid city growth.

How did the Industrial Revolution encourage the growth of cities? The Industrial Revolution, beginning in the eighteenth century, created major changes in transportation, agriculture, commerce, and industry. Technological developments during this period allowed for better food production and more efficient transportation systems with less human labor. As food production and transportation became more dependable, it became possible for more people with nonagricultural jobs to live in cities. The major reason for the growth of cities following the Industrial Revolution, however, was the increased use of factories.

Factories were not established to encourage the growth of cities, but they almost immediately had that effect; one factory tends to attract others to the same location. While taking advantage of natural features such as water power and river transport, factories located together can share raw materials and transportation costs. And at a time when transportation and communication were slow and undependable, it was desirable for producers of machinery and equipment to be near the factories they were supplying. The concentration of industry led to dense population settlements of factory workers, which in turn attracted retailers, innkeepers, entertainers, and a wide range of people offering services to city dwellers. Of course, the greater availability of urban services often enticed still more factories, keeping the cycle going. Although subsequent technological and social developments affected the growth of cities in different ways, the overall trend toward urbanization was set. The industrial world was becoming urbanized.

How does modern urban living differ from life during preindustrial times? Life in preindustrial cities was directed toward the city's center. The walls, moats, and other protective devices found in these cities symbolized their role in defending the cities' inhabitants. It was safer to live in cities than outside them. The importance of the center of the city in preindustrial times is also reflected in residential patterns. Members of the upper class lived near the center of preindustrial cities, which gave them access to the most important government and religious buildings and, as a rule, the main market. The outskirts (including what we would now consider the suburban areas) were populated by the poor and those with low-status craft occupations, such as tanning and butchering. These people lived on the outskirts partly because more privileged residents wanted to have minimal contact with them. More important, though, the outskirts were the least desirable areas due to their lack of accessibility to the central city (Sjoberg 1966).

Modern cities, like preindustrial cities, located their major economic and government activities in their central core. Cities in industrialized nations, however, are connected by sophisticated communication and transportation networks, which have had the opposite effect of the walls and other protective devices of preindustrial cities. Thus, industrial cities tend to have an outward orientation. Most of their development has occurred away from the central city in the surrounding suburban areas. In industrialized nations, population movement from rural to urban areas has been accompanied by suburbanization. In fact, many central cities are now losing population to the areas surrounding them (Brockerhoff 2000).

Suburbanization

Suburbanization—which occurs when central cities lose population to the area surrounding them—is a hallmark of contemporary American urban growth (Baldassare 1992; Gober 1993). By 1980, seventeen of the twenty-six central cities with a population of half a million or more in 1970 registered losses in population. Still, the population outside central cities is expanding several times as fast as the central cities themselves. According to the U.S. Census Bureau, the population of the central cities of the 255 MSAs increased by only 3.9 percent between 1990 and 1998, whereas the population of MSA areas outside the central cities (almost entirely suburbs) increased by 12.5 percent during the same period (U.S. Bureau of the Census 2000). During 1991–1992, the percentage of Americans moving from central cities to suburbs was twice the rate of movement from suburbs to central cities (U.S. Bureau of the Census 1993c). This is a continuation of a trend that more than doubled the proportion of the population living in suburbs since 1950. Thus, the United States is now predominantly suburban.

What makes suburbanization possible? Suburbanization has become an important trend partly as a concomitant of technological developments. Improvements in communication—such as telephones, radios, television, and (later) computers, fax machines, and the Internet—allow people to live away from the central city without losing touch with its activities. Developments in transportation (especially highways, automobiles, trucks, and metro systems) make it possible both for people to commute to work and for many businesses to leave the central city for suburban locations (Baxandall and Ewen 2000).

Technology is not the only cause of suburbanization, however. Both cultural and economic pressures have encouraged the development of suburbs. Partly because of America's frontier heritage, American culture has always had a bias against urban living. Some Americans prefer urban life, but most report that they would rather live in a rural setting. Even those who, on balance, choose to live in the city believe they are giving up some advantages. Suburbs, with their low-density housing, allow many people to escape the problems of urban living without completely leaving the urban areas. Suburbs are attractive because of decreased crowding and traffic congestion, lower taxes, better schools, less crime, and reduced pollution.

Also, the scarcity and high cost of land in the central city encourage suburbanization. Developers of new

In part because of their frontier heritage, many Americans continue to hold a bias against urban living. Suburbs, such as this one in Maryland, permit people to remain near the advantages of urban centers while living in areas with lower-density housing.

housing, retail, and industrial projects often find suburban locations to be far less expensive than those near the central city.

Finally, government policy has often increased the impact of economic forces. Federal Housing Administration regulations, for example, favor the financing of new houses (which can be built most cheaply in suburban locations) rather than the refurbishing of older houses in central cities (Steinnes 1982).

As people move to suburban areas, increasing numbers of businesses and jobs follow. In fact, suburbanization is leading to the development of edge cities, and "Boomburgs" that are changing the face of urban America.

What are edge cities and boomburgs? Originally, most suburbs were merely places where people lived. Suburbanites commuted daily from their "bedroom" communities to their jobs in the central city. Over the last decade, the nature of suburbanization has changed to include the movement of jobs and businesses to the suburbs as well. This change in suburbanization has created the **edge city**—a suburban unit specializing in a particular economic activity (Garreau 1991; Bingham and Bowen 1997). Employment in one edge city may focus on computer technology, another on financial services or health care. An edge city, of course, will have many other types of economic activities—industrial tracts, office parks, distribution and warehousing clusters, and home offices of national corporations. These are actually little cities in themselves with a full range of services, including schools, retail sales, restaurants, malls, medical facilities, hotels, motels, recreational complexes, and entertainment centers.

Edge cities do not have legal and physical boundaries separating them from the larger urban area in which they are located. This does not prevent names from being attached to several of them: Tyson's Corner is located in northern Virginia near Washington, D.C.; Las Colinas is close to the Dallas–Fort Worth airport; and King of Prussia is northwest of Philadelphia. Some edge cities bear the name of highways, such as Route 128 outside of Boston.

Boomburgs are places of over 100,000 residents that are not the largest cities in their metropolitan areas and have experienced double-digit population growth in recent decades (Lang and Simmons 2001). According to the U.S. Census Bureau, the United States has 53 boomburgs, 41 of which exceed 100,000 people. Accounting for 51 percent of the 1990s growth are cities between 100,000 and 400,000 residents, boomburgs now contain one-fourth of all Americans living in cities of this size.

Despite developing into large cities, most boomburgs retain their suburban character. They are not without traditional urban problems, however, such as sprawl, overused government services, and traffic congestion.

Consequences of Suburbanization

When suburbanization first became noticeable in the 1930s, only the upper and middle classes could bear the cost. It was not until the 1950s that the working class joined them. But until the 1970s, the suburbs have remained largely white, despite federal legislation prohibiting housing discrimination (Feagin 2002).

Central-city minorities are beginning to move to the suburbs in greater numbers (Frey 1984, 1985; Massey and Denton 1987; R. Farley 1998). Although America's suburban counties remain predominantly white, minorities have poured into the suburbs over the past ten years. There are now 3 percent more Latinos in suburban counties than there were in 1990, 5 percent more Asians, and 5 percent more African Americans (Nasser 2001). A greater number of Latinos and Asian Americans, in fact, now reside in suburban counties than live in central cities. The percentage of African Americans living in suburbs increased almost 15 percent since 1980 (now 39 percent).

Despite this decline in the degree of residential segregation, African Americans remain the most isolated ethnic or racial minority in the United States (O'Hare et al. 1990; Farley and Frey 1994; South and Crowder 1997; Palen 2001). Although the percentage of African Americans living in suburbs increased from 16 percent to 26 percent between 1970 and 1990, the percentage of African Americans living in central cities declined only slightly. The situation is better for Latinos and Asians, as already noted. Not only are Latinos and Asians more suburbanized—and thus less concentrated in the central cities—than African Americans, they are also less segregated in suburban areas (R. Farley 1997; Palen 2001). Even this progress has a downside. As the more successful minority members move to the suburbs, the "underclass" accounts for a larger proportion of the remaining central-city population (Frey and Farley 1996; Massey and Denton 1996; Orfield 2001; see Chapter 9).

What is the central-city dilemma? Suburbanization created a major problem in the United States. The problem is not merely that minorities remain trapped in inner cities. Businesses have now followed the more affluent people to the suburbs, where they can find lower tax rates, less expensive land, less congestion, and their customers who have already left the city. Accompanying the exodus of the middle class, the manufacturers, and the retailers is the shrinking of the central-city tax base (Mouw 2000). As a result, the central city has become increasingly populated by the poor, the unskilled, and the uneducated. The result is a

John Hagedorn—Gang Violence

Gangs were a constant feature of the American urban landscape during most of the twentieth century. John Hagedorn (1998), however, contends that the extent and nature of gang violence has substantially changed with the emergence of postindustrial society in the 1960s. Hagedorn's conclusions are based on a combination of three methods: a review of the research of others, secondary analysis of precollected data (data collected by other researchers), and original data he gathered.

Gangs (mostly male) in the industrial period were tied to specific neighborhoods and new immigrant groups. These gangs emerged among unsupervised teens whose fathers worked in factories. Gang violence primarily centered on "turf" battles among neighborhood peer groups. Pride in violence came from defending the gang's space. Violence provided excitement and a sense of place in a group. Nevertheless, these working- and lower-class boys would eventually move on to hold decent jobs, have families, and live in better neighborhoods.

Gangs today still tend to form around racial and ethnic groups and neighborhoods. Currently, gangs tend to be African American, Latino, or Asian, just as earlier gangs were mostly European immigrants, such as those from Ireland, Italy, or Eastern Europe. Gangs form in the same poor urban areas, and members continue to come from dysfunctional families who do not supervise them. Juvenile gangs continue to engage in small-time crime, substance abuse, and periodic violence.

According to Hagedorn, however, the extent and nature of violence in postindustrial gangs underwent a major change. First, gang violence significantly increased. Second, gang-related homicides rose dramatically.

The question is, why? In answering this question, Hagedorn pinpoints important changes in postindustrial gangs, particularly gang functions. Gang violence, he notes, skyrocketed at the same time that American corporations were moving well-paying jobs from the central city. As legitimate work disappeared in inner cities, gangs turned from their earlier territorial emphasis to participation in the illegitimate drug market. The common outlook of gang members today is expressed by this gang member:

> *I got out of high school and I didn't have a diploma, wasn't no jobs, wasn't no source of income, no nothing. That's basically the easy way for a . . . young man to be—selling some dope—you can get yourself some money real quick, you really don't have nothing to worry about, nothing but the feds. You know everybody in your neighborhood. Yeah, that's pretty safe just as long as you don't start smoking it yourself.* (Hagedorn 1998:390)

Significantly, this gang member was not a teenager. More urban gang members today remain members into adulthood. Rather than getting a job, marrying, and settling down, more inner-city young adults hang out on the street, making a living on the drug trade.

Although drug selling was a small part of the gang scene before the 1980s, it was not a business. By contrast, Hagedorn found more young African American males in drug trafficking than any other area of employment.

There is still movement out of urban gangs, however. A minority of gang members remain committed to the drug economy, but most seek "legit" jobs as they approach their thirties.

Thinking About the Research

1. Relate Hagedorn's findings to Merton's strain theory in Chapter 7.
2. What do you think is the most important step in solving this problem? Justify your answer.

concentration of economic problems as well as a social dependency in cities that are financially unprepared to deal with them. This situation has been captured in what is known as the **central-city dilemma**—the concentration of a large population in need of public services (schools, transportation, health care) but without the money to provide for them (Wolfe 1999).

What are the consequences of the central-city dilemma? Poor inner-city African American children are handicapped by inferior educational facilities and teachers, and the dropout rate among them is considerably higher than among whites (Kozol 1992). During prosperous times, their unemployment rate is at least twice as high as it is for urban whites, and it skyrockets when the economy falters. Those who are employed generally must take dead-end, low-paying jobs because the more desirable jobs go to the better educated. A catalog of social ills flows from this socioeconomic condition in the inner city. African Americans in large-city slums exist in a world of poverty, congestion, prostitution, drug addiction, broken homes, and brutality. Crime rates, for example, appear to increase in central cities that have been affected by suburbanization, both because suburbanization creates stratification along city-suburban lines and because it contributes to an absolute decline in the socioeconomic status, the economic and tax base, and the physical structure of the central city (Farley and Hansel 1981; Ganz 1985; Yinger 1995).

Can the central-city dilemma be solved? Some countertrends exist. There are city governments now requiring certain public employees to live in the city. Some parts of inner cities are being restored through **gentrification**—the development of low-income areas by middle-class home buyers, landlords, and professional developers ("Race, Place, and Segregation" 2002). Finally, there is a fairly significant movement of whites back to the central city. This movement is particularly evident among baby boomers who are remaining single or establishing childless or two-income families. Because these people are not as heavily involved in child-rearing, they prefer central-city living more than the previous generation did (Palen 2001). The importance of these countertrends for easing the central-city dilemma remains to be seen. They certainly have not been sufficiently important to stop the emergence of edge cities.

World Urbanization

Urbanization is a worldwide movement. From 1800 to the mid-1980s, the number of urban dwellers increased almost a hundredfold, whereas the population increased only about fivefold. More than 2.8 billion people—46 percent of the world's population—now live in urban areas. In more developed countries, 76 percent of the population lives in urban areas compared to 40 percent in less developed countries (*World Population Data Sheet* 2002; Brockerhoff 2000).

The pattern of urbanization is different in more developed countries than it is in less developed countries. One distinction is the rapidity of the urban growth rate. In less developed countries, most of the urban growth before the turn of the twentieth century occurred through colonial expansion. Western countries, which were involved in colonial expansion after the late fifteenth century, held half the world under colonial rule by the latter part of the nineteenth century. Only since World War II have many of these colonial countries become independent nations (Bardo and Hartman 1982). Since gaining independence, these former colonies have experienced rapid urbanization. In fact, urbanization in these areas is now proceeding faster than it did in the West, even during that area's biggest urban expansion period from 1850 to 1950 (Gelbard, Haub, and Kent 1999).

What are some additional differences in the pattern of urbanization in less developed and more developed countries? Other differences are worth noting. In the first place, industrial development in less developed countries has not kept pace with urbanization. In the cities of less developed countries, the supply of labor from the countryside exceeds the demand for labor. A high rate of unemployment is the obvious result. The term **overurbanization** has been created to describe a situation in which a city is unable to supply adequate jobs and housing for its inhabitants. In the Western experience, on the other hand, industrial development kept pace with urbanization, and cities of North America and Europe had jobs for most migrants from rural areas.

Another difference between urbanization in more developed and less developed countries is the number

The pattern of urbanization is different for high-income and low-income economies. In high-income countries there are a few very large cities, many medium-sized cities, and a much larger number of small cities. In many low-income countries, there is one large city that dominates a myriad of villages.

TABLE 17.2

World's Ten Largest Urban Areas: 1000, 1800, 1900, and 2001

Here is a comparison of the world's ten largest urban areas from the year 1000 to now. Discuss any patterns of change or stability in the listings between 1000 and 2001. Explain the relationship between economic development and relative size of urban areas.

1000		1800		1900		2001	
			(million population)				
Cordova[1]	0.45	Peking[3]	1.10	London	6.5	Tokyo	26.5
Kaifeng	0.40	London	0.86	New York	4.2	São Paulo	18.3
Constantinople[2]	0.30	Canton[4]	0.80	Paris	3.3	Mexico City	18.3
Angkor	0.20	Edo[5]	0.69	Berlin	2.7	New York	16.8
Kyoto	0.18	Constantinople	0.57	Chicago	1.7	Mumbai[7]	16.5
Cairo	0.14	Paris	0.55	Vienna	1.7	Los Angeles	13.3
Baghdad	0.13	Naples	0.43	Tokyo	1.5	Calcutta	13.3
Nishapur	0.13	Hangchow[6]	0.39	St. Petersburg	1.4	Dhaka	13.2
Hasa	0.11	Osaka	0.38	Manchester	1.4	Delhi	13.0
Anhilvada	0.10	Kyoto	0.38	Philadelphia	1.4	Shanghai	12.8

[1]Cordoba today. [2]Istanbul today. [3]Bejing today. [4]Guangzhou today. [5]Tokyo today. [6]Hangzhou today. [7]Formerly Bombay.
Source: 1000–1900 From Tertius Chandler, *Four Thousand Years of Urban Growth: An Historical Census* (Lewiston, NY: Edwin Mellen Press, 1987); 2001 from U.N. Population Division, *World Urbanization Prospects: The 2001 Revision* (New York: 2002).

and size of cities. When grouped by size, cities in developed countries form a pyramid: a few large cities at the top, many medium-sized cities in the middle, and a large base of small cities (Light 1983). In the less developed world, many countries have only one big city, a tremendously large city that dwarfs a large number of villages. Calcutta, India, and Mexico City are examples. In 1950, of the world's ten largest cities, only two—Shanghai and Calcutta—were in less developed areas. By 2000, seven of the top ten largest urban areas were in less developed countries. By the turn of the twenty-first century, there were more than twenty-one "megacities" with populations of 10 million or more. Of these, eighteen were in developing countries, including the most impoverished societies in the world. Similarly, seven of the world's largest urban areas are in less developed countries (see Table 17.2).

Why do people in developing countries move to large cities where they will find inadequate jobs and housing? Urban sociologists point to the operation of push-pull factors in the rural-to-urban migration in less developed countries. Peasants are pushed out of their villages because expanding rural populations cannot be supported by the existing agricultural economy. Peasants are also attracted to cities by perceived opportunities for better education, employment, social welfare support, and availability of good medical care, even though they are likely to be disappointed (Firebaugh 1979; Sassen 2000).

Theories of City Growth

Every city is unique, but the patterns of all cities can be interpreted based on **urban ecology**—the study of the relationships between humans and their urban environments.

In the 1920s and 1930s, sociologists at the University of Chicago studied the effects of the city environment on urban residents. It turns out that the

FEEDBACK

1. A _________ is a concentration of people whose major social and economic needs are satisfied within the area in which they live.
2. _________ is the process by which an increasingly large proportion of the world's population lives in urban areas.
3. Thanks to worldwide urbanization, preindustrial cities no longer exist. T or F?
4. The _________ exists because diminishing public revenues are not sufficient to cope with the influx of the poor, unskilled, and uneducated.
5. _________ occurs when a city is unable to supply adequate jobs and housing for its inhabitants.

Answers: 1. community 2. Urbanization 3. F 4. central-city dilemma 5. Overurbanization

FIGURE 17.6

Theories of City Growth

This figure displays in graphic form the four major theories of city growth. Discuss one important advancement these theories make to understanding urban growth.

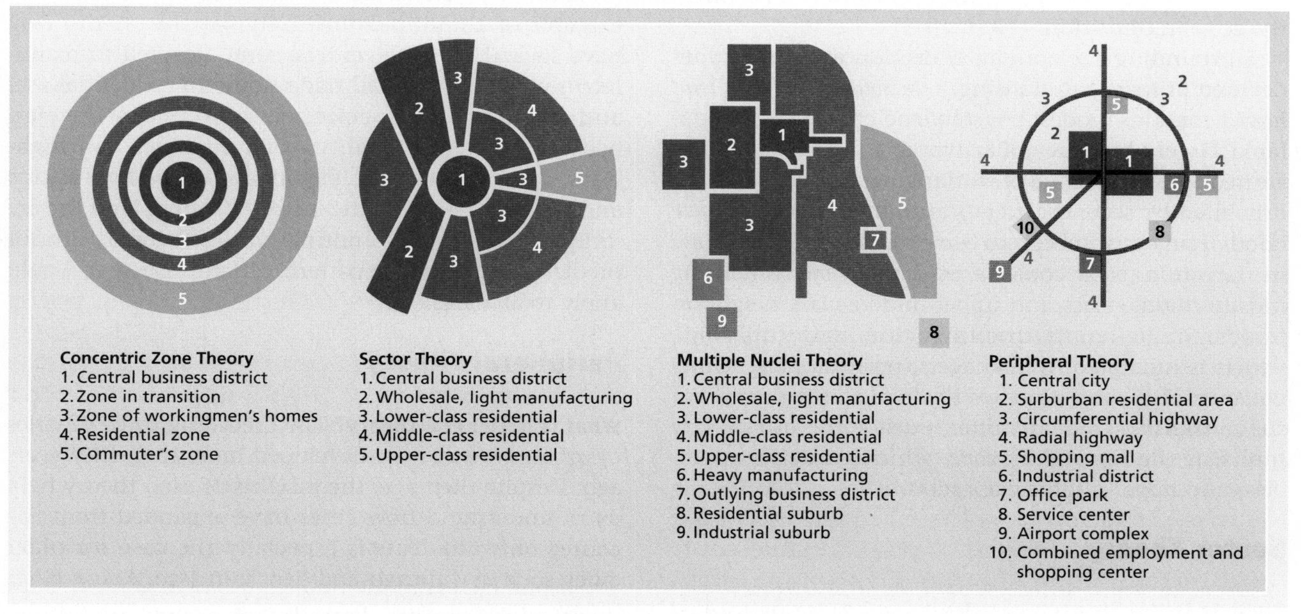

Source: Chauncy D. Harris and Edward L. Ullman, "The Nature of Cities," *The Annals* (1945):13. Copyright © 1945 by The American Academy of Political and Social Science; and Chauncy D. Harris, "'The Nature of Cities' and Urban Geography in the Last Half Century," *Urban Geography* 18(1997):17. Used by permission of V.H. Winston & Son, Inc.

effects varied in different parts of the city. For example, Harvey Zorbaugh (1929) found significant differences between Chicago's wealthy Gold Coast and its slum areas. The realization that such differences existed led to additional questions: Why do differences exist between areas of a city? How do different areas affect one another? What are the processes through which areas are changed? To answer these and other questions, the University of Chicago sociologists developed theories of urban ecology, including theories of city growth (Micklin and Poston 1998; Flanagan 2001; Kleniewski 2001).

There are four major theories of city growth. **Concentric zone theory** describes urban growth in relation to circular areas that grow from the central city outward. **Sector theory** emphasizes the importance of transportation routes in the process of urban growth. **Multiple-nuclei theory** focuses on specific geographical or historical influences. These three theories, developed over fifty years ago, have been joined by a more contemporary model, the **peripheral theory**, which accents the growth of suburbs around the central city. These four theories lead to quite different images of urban space (see Figure 17.6). Indeed, all four tell us more when considered together than they tell us separately. To understand why this is so, we must first examine each theory.

Concentric Zone Theory

What is concentric zone theory? Ernest Burgess (1925), like other early sociologists at the University of Chicago, was interested in the causes and consequences of Chicago's growth. His work led to the concentric zone theory, which describes city growth in relation to distinctive zones that develop from the central city outward, in a circular pattern. As illustrated in Figure 17.6, the innermost circle is the *central business district*, the heart of the city, containing major government and private office buildings, banks, retail and wholesale stores, and entertainment and cultural facilities.

The central business district exerts an especially strong influence on other parts of a city. This influence is especially clear in the zone immediately surrounding it. Burgess called this the *zone in transition* because it is in the process of change. As new businesses and activities enter the central business district, the district expands by invading the next zone. This area may have been a residential area inhabited by middle- or upper-class families, who left because of the invasion of business activities. Most of the property in this zone is bought by those with little interest in the area. Rather than investing money in building maintenance, landowners simply extract rent from the property or sell it at a profit after the area is more commercialized.

Until the zone in transition is completely absorbed into the central business district (which may never occur), it is used for slum housing, warehouses, and marginal businesses that are unable to compete economically for space in the central business district itself. In short, the invasion of business activities creates deterioration in the zone in transition.

Surrounding the zone in transition are three zones devoted primarily to housing. The *zone of workingmen's homes* contains modest but stable neighborhoods populated largely by blue-collar workers. In the northern United States, the zone of workingmen's homes is often inhabited by second-generation immigrants who have enough financial success to leave the deteriorating zone in transition. Next comes a *residential zone* containing mostly middle-class and upper-middle-class neighborhoods. Single-family dwellings dominate this zone, which is inhabited by managers, professionals, white-collar workers, and some well-paid factory workers. On the outskirts of the city, often outside the official city limits, is the *commuter's zone*, which contains upper-class and upper-middle-class suburbs.

Sector Theory

What is sector theory? Based on a study he performed for the Federal Housing Administration, Homer Hoyt (1939) developed sector theory. Hoyt's study indicated that patterns of land use do not necessarily change according to distance from the central business district. Instead, land is more strongly affected by major transportation routes. As Figure 17.6 shows, sectors tend to be pie-shaped wedges radiating from the central business district to the city's outskirts. Each sector is organized around a major transportation route. Once a given type of activity is organized around a transportation route, its nature tends to be set. Thus, some sectors will be predominantly industrial, others will contain stores and professional offices, others will be "neon strips" with motels and fast-food restaurants, and still others will be residential sectors, each with its own social class and ethnic composition.

As in concentric zone theory, sector theory depicts cities as generally circular in shape. But due to the importance of transportation routes extending from the central business district, the boundaries of many cities form a star-like pattern, rather than a uniformly circular shape. The exact shape of a city, however, is not a major issue in sector theory. Emphasis here is on land patterns around transportation routes.

Multiple-Nuclei Theory

What is multiple-nuclei theory? Many cities have areas that cannot be explained by either concentric zone or sector theory. For example, the Beacon Hill area of Boston defies concentric zone theory; it remains an expensive, prestigious neighborhood despite its location near the central business district. Because of such cases, Chauncy Harris and Edward Ullman (1945) have suggested that cities do not always follow a pattern contingent on a central district. (Figure 17.6 shows one example of a multiple-nuclei city pattern.) A city may have several separate centers, some devoted to manufacturing, some to retail trade, some to residential use, and so on. These specialized centers can develop because of the availability of automobiles and highways. They reflect such factors as geography, history, and tradition. Because the factors affecting land use are unique to each city, multiple-nuclei theorists do not predict any particular pattern of land use that would apply to all cities.

Peripheral Theory

What is peripheral theory? The preceding three theories of urban growth were developed more than fifty years ago. Despite their age, the insights of each theory help us to understand how cities have expanded from the center outward. This is especially the case for older cities such as Chicago and San Francisco. Many cities today, however, no longer have a central-city core to which other parts of the metropolitan area are oriented.

Dependence on shipping, railroads, and heavy manufacturing has been replaced by more flexible means of transportation, such as cars and trucks. And large urban areas are now encircled by highways. New technologies (fax machines, cell phones, computers, the Internet) are also loosening the ties of most parts of the city to the central-city core. As a result, many cities are now oriented *away* from the older urban core.

As mentioned earlier, many Americans have moved from the city to the suburbs. They have done so in part because many businesses—offices, factories, schools, retail stores, restaurants, health centers—are also in the suburbs. To encompass changes in urban areas today, urban geographer Chauncy Harris (1997) has formulated the peripheral theory. The dominant feature of this model, obviously, is the growth of suburbs (and edge cities) around and away from the central cities (see Figure 17.6). Peripheral theory brings urban growth research up to date.

Of what value are these diverse theories of city growth? Although each of these theories calls attention to the dynamics of city growth, none of them applies to all cities. Each theory, however, does emphasize the importance of certain factors that cannot be overlooked by anyone interested in city growth. Concentric zone theory underscores the fact that growth in any one area of a city is part of the processes of segregation, invasion, and succession. This theory also reveals the importance of economic power in city growth: the distribution of space in cities is heavily influenced by those with the money to buy the land they want for the purposes they have in mind. Concentric zone theory is in need of modification, however, because it is based on the assumption that cities will continue to grow and that change will occur as one area encroaches on another. Many central cities, however, have ceased to grow. In those cases, the "push-from-the-center dynamic no longer applies" (Choldin and Hanson 1981:561).

Sector theorists also contributed to an understanding of urban dynamics. As they noted, transportation routes have a strong influence on cities. Decisions about the placement of railroad lines had important effects on the growth of cities in the nineteenth and early twentieth centuries. Highways and major streets have an even larger impact now.

Although multiple-nuclei theory is vague in its predictions, the types of geographical and historical factors it emphasizes are also important for understanding any specific city. Moreover, the apparent decline of many central business districts may increase the importance of specific local factors in the future growth of cities. Finally, peripheral theory brought urban growth research up to date by emphasizing the development of suburbs around the central city.

No single ecological theory describes the growth pattern of all cities; each is valuable in itself. In recognition of this, some researchers explain patterns of land use by combining the insights of these theories. Several researchers, for example, have reported that people distribute themselves in cities in a way that can be predicted only by combining concentric zone theory and sector theory (Berry 1965; Timms 1971; Berry and Kasarda 1977).

For another example, different sectors attract people from different social classes. The geography and history associated with some transportation routes out of a city encourage people in higher social classes to settle there. Other sectors, which are less appealing, are left to lower-income people. Within any particular sector, however, family status and lifestyle seem to vary as concentric zone theory predicts. The newer single-family homes at the edge of the city attract different types of people than the apartment complexes and older homes that are likely to be found near the center of the city. Exact patterns of land use in any urban area, however, are likely to depend on the cultural values affecting that particular area (Abu-Lughod 1980).

Regardless of the ultimate fate of the concentric zone, sector, multiple-nuclei, and peripheral theories, these theories have performed an important service. By calling attention to the factors influencing land use in urban areas, these theories remind us that cities do not grow in a random manner.

The Quality of Urban Life

As just noted, sociologists from the University of Chicago contributed to our understanding of urban land use. Their interests were not, however, confined simply to the ecological characteristics of cities. These sociologists were also concerned with the effects of cities on people. Their basic conclusion: people in cities live a unique way of life. The researchers designated this distinctive way of life—**urbanism**.

Cities have always been thought to produce a wide variety of lifestyles. During medieval times, cities were viewed as places to escape the domination of powerful landowners, and it was said that "city air makes one

FEEDBACK

1. Match the following zones of the concentric zone theory with the appropriate descriptions.
   ____ a. central business district
   ____ b. commuter's zone
   ____ c. residential zone
   ____ d. zone in transition
   ____ e. zone of workingmen's homes
   (1) zone that is the next zone out from the zone in transition
   (2) site of major stores and offices
   (3) zone with physical deterioration stemming from few repairs
   (4) zone the wealthiest people live in
   (5) zone that is largely middle class in composition
2. The theory that emphasizes the importance of transportation routes is called ________ theory.
3. The ________ theory contends that the growth of cities does not necessarily follow any specific pattern.
4. ________ theory of city growth brings urban growth research up to date.
5. One of the ultimate contributions of the four major theories of city growth is their demonstration that cities do not grow in a random fashion. T or F?

Answers: 1. a. (2) b. (4) c. (5) d. (3) e. (1) 2. sector 3. multiple-nuclei 4. Peripheral 5. T

free." Many of today's cities are also considered sources of intellectual and cultural diversity, sophistication, and freedom. Louis Wirth (1938), a University of Chicago sociologist and the first to develop the idea of urbanism as a way of life, agreed in part with the intellectual tradition that associated the city with freedom and tolerance. He believed that because urban residents are confronted with a variety of lifestyles, they are more tolerant of diversity and less ethnocentric than residents of small towns or rural areas (T. C. Wilson 1985, 1991). However, Wirth also drew on an intellectual tradition that saw impersonality and social disorganization in urban areas, a tradition represented in the work of German sociologist Ferdinand Tönnies. Later sociologists, as noted in "Using the Sociological Imagination," have a more positive view of the quality of urban social life.

The Traditional Perspective on Urbanism: Loss of Community

What was Tönnies's viewpoint? Although he did not use the term urbanism, Tönnies was the first sociologist to contrast the culture in rural and urban settings. As noted in Chapter 5, Tönnies distinguished between *gemeinschaft* (community) and *gesellschaft* (society). In the **gemeinschaft**—the type of society existing before the Industrial Revolution—social relationships were intimate because people knew each other. Daily life revolved around family and was based on tradition. There was a sense of a shared way of life and an interest in the common good of the community. Urbanization changes all that, according to Tönnies. In the **gesellschaft**, social relationships are impersonal because most people with whom one interacts are strangers. Although the immediate family is still important, the wider kinship network disappears. With the fading of tradition goes the sharing of common values. Interest shifts from the community's welfare to self-interest.

How did Wirth conceptualize urbanism? In Wirth's view, the urban conditions that promote sophistication, freedom, and diversity also involve a high cost. Urbanism, Wirth said, is marked by a depersonalization of human relationships. Rather than involving their total personalities in interaction, urban residents interact on the basis of rigidly defined roles. For example, a customer in a rural store may combine shopping with a lengthy conversation, but interactions in most urban stores are confined to the exchange of money and products. This emphasis on rigid role prescriptions, Wirth concluded, makes it difficult for urban residents to maintain a sense of their own individuality; it makes them indifferent and blasé. Heavy reliance on roles occurs partly because of sheer number: it is impossible to know everyone. Urbanites may actually know more people than do rural residents, but they know fewer people well enough to interact with them on a personal basis.

According to Wirth, urban residents behave as they do because of three demographic characteristics of cities: large size, high density (a large number of residents per square mile), and much diversity within the population. Wirth contended that as cities become larger, more densely settled, and more socially diverse, residents become more sophisticated and more reliant on impersonal relationships.

The Contemporary Perspective on Urbanism: Community Sustained

Wirth's theory of urbanism is criticized on several grounds, the most frequent being that it is an overgeneralization. Some critics charge that Wirth's view of urbanism mistakenly assumes that all urban residents

All urban residents do not share the same way of life. Many of the people pictured here are "cosmopolites"—they are well educated, have high incomes, and are drawn to the inner city because of its cultural advantages.

TABLE 17.3

FOCUS ON THEORETICAL PERSPECTIVES: Urban Society

This table illustrates how functionalism, conflict theory, and symbolic interactionisms might approach the study of urban society. State a research topic for the study of population for which any one of the theories would be appropriate.

Theoretical Perspective	Concept	Sample Research Topic
Functionalism	Urbanization	• Study of the relationship between population density and the suicide rate
Conflict theory	Overurbanization	• Investigation of the relationship between the distribution of scarce resources and social class
Symbolic interactionism	Urbanism	• Research on the effect of the degree of urbanization and the extent to which social interaction is based on shared meanings

share a common lifestyle (I. M. Young 1990; Macionis and Parillo 2004). A leading spokesperson for this line of thought is Herbert Gans (1962b, 1968).

Do all urban residents share the same way of life? Gans described five basic population groups found within inner-city areas, the first three of which live in the inner city by choice and are protected from the extreme negative consequences depicted by Wirth:

- *Cosmopolites*. People in this category are well educated and have high incomes. They live in inner-city areas because they are attracted to the area's cultural advantages, such as art museums, theaters, and symphony orchestras.
- *The unmarried and childless*. People in this category also live in the inner city by choice. They may not feel a strong commitment to the inner city, but living there allows them to meet other people, and the area is convenient to jobs and entertainment facilities.
- *Ethnic villagers*. People in this category live in inner-city neighborhoods with strong ethnic identities. Residents of such neighborhoods are suspicious of the city as a whole and participate little in its affairs. The neighborhood itself, however, is an important source of intimate, enduring relationships.
- *The trapped*. Some people—such as the elderly living on pensions or public assistance—cannot afford to leave the inner city. They may identify with their own neighborhood, but this does not prevent them from becoming annoyed over changes in the central-city area that they consider undesirable.
- *The deprived*. Poor minorities, the psychologically and physically challenged, some divorced mothers, and other deprived people see the city as a source of increased opportunities for employment or welfare assistance. The poverty, however, prevents such individuals from choosing where they will live in the city. These people are therefore exposed to the most extreme effects of population size, density, and diversity.

According to Gans, then, no blanket statement (such as that formulated by Wirth) applies to every urban resident. Although the trapped and the deprived are unable to escape many of the adverse effects described by Wirth, members of the other three categories have chosen to live in the inner city. They would not have made this choice had they not seen many advantages.

Are city dwellers isolated, with no personal relationships? According to Wirth's theory, urban residents have relatively few opportunities for participation in informal relationships. Subsequent research shows, however, that urban residents do participate in informal relationships, and such relationships play an important role in maintaining a positive emotional state (Kadushin 1983; Creekmore 1985; C. S. Fischer 1982, 1984, 1995). In an intentional test of Wirth's theory, John Kasarda and Morris Janowitz (1974) found that population size and density do not necessarily produce a lack of involvement in local friendship, kinship, or other social relationships. Those who are not highly involved in such relationships in a large city tend to be those who have not lived there long enough to become a part of it. John Esposito and John Fiorillo (1974) found essentially the same thing in a study of working-class neighborhoods in New York City.

Other evidence indicates that families continue to provide urban residents with emotional support and mutual aid (see Chapter 12). Studies based on observations of actual behavior in areas as diverse as lower-class slums and middle-class suburbs are showing the importance of personal relationships among family members and friends. For example, Gans (1962a) observed the

Technology and Society

Virtual Communities

Before the United States was a country, when it was just a loosely connected group of English colonies, the ideals of community were easily achieved. Local groups of citizens held town meetings in which all citizens came together in the same building and hashed things out until they reached an agreement. Now that the population has grown and has both concentrated in urban areas and dispersed across the entire continent, this type of interaction is impossible. Most Americans count this loss of community as a serious problem. In response, there is a growing effort to use the Internet to rebuild social relationships through community networks (Etzioni 2001).

These "virtual communities" use electronic communications technology to link people who live in the same area, city, or neighborhood (Virnoche and Marx 1997). Organizers of these community networks share the goals of local participation, community building, and democracy (Schuler 1996). As with the New England colonies' town meetings, the ideal of the new community networks is to include everyone. Supporters of the new technology claim that electronic communications will allow people to reestablish the types of interpersonal relationships that Tönnies described in his concept of gemeinschaft and that Wirth believed were lost in the urban way of life.

Even the most ardent supporters of community networks admit to the problem of "electronic stratification"—unequal access to technological advances. Those who lack the economic resources to participate in electronic community networks will be excluded. The most obvious obstacle is the initial costs involved. Low-income individuals and families cannot afford computers or Internet access, and public agencies are not ready to supply sufficient funding. Furthermore, as computers become more sophisticated, people who are not already computer literate (especially lower-income people) will have an increasingly difficult time catching up. The (technologically) poor will become (technologically) poorer. Finally, those with access to the Internet only through publicly supported computers will often be passive receivers of one-way communication. People with private access will be able to engage in interactive communication (Virnoche 1998).

The Boulder (Colorado) Community Network (BCN), established in the mid-1990s, experienced many of these problems. The founders of BCN trained many different groups within the city to use the technology of community networks. They found that acceptance and use of the new tools varied widely among groups. For example, residents at a local senior citizens home became avid users of the community computers placed in their facility. In contrast, a group of low-income single parents virtually ignored the computers and the Internet, even after extensive training (Virnoche 1998).

If community networks are firmly established, critics warn that the "human factor" will still be lacking. When people meet through the Internet, they have no social clues, such as body language and facial expressions, with which to learn about their new acquaintances. No matter how much you learn about another person online, critics say, you have not met someone for real until you meet in person (Herbert 1999).

Analyzing the Trends

How do you think the development of "virtual communities" will affect the urban way of life? Explain your answer using material in the chapter.

FEEDBACK

1. The distinctive way of life allegedly developed in urban areas is referred to as _________.
2. According to Ferdinand Tönnies, urbanism corresponds to the _________ (gemeinschaft/gesellschaft).
3. Louis Wirth relied on three characteristics to explain the development of urbanism as a way of life. What were they?
   a. age of city, population density, political structure
   b. population size, economic base, political structure
   c. population size, population density, population diversity
   d. economic base, political structure, population structure
4. According to Herbert Gans, which of the following groups have relatively little choice within the inner city?
   a. cosmopolites, the unmarried and childless
   b. ethnic villagers, the deprived
   c. the unmarried and childless, ethnic villagers
   d. the trapped, the deprived
5. Research shows that urban residents are largely isolated from family and friends. T or F?

Answers: 1. urbanism 2. gesellschaft 3. c 4. d 5. F

West End of Boston, where many aspects of Italian rural life continue to be honored. Personal relationships are encouraged there, some have argued, because the area is dominated by one ethnic group. However, neighborhoods in general appear to be important sources of interaction and support. Take, for example, research of a Chicago slum (Addams) by Gerald Suttles (1968): even in a low-income area populated by four different ethnic groups, a social order based on informal and personal relationships developed.

SUMMARY

1. Demography is the scientific study of population. Population data are very important today, in part because of their use by government and industry.
2. Demographers use three population processes to account for population change: fertility, mortality, and migration.
3. Because population grows exponentially, it increases more each year in absolute numbers, even if the rate of growth remains stable. Most developed nations have undergone what is known as the demographic transition. This is the process accompanying economic development in which a population gradually moves from high birth and death rates to low birth and death rates. Europe seems to be experiencing a second demographic transition, which involves an eventual decline in population size.
4. In more developed societies, economic growth has led to a lower fertility rate for a number of reasons. But this type of economic development, along with continued inequalities, has not led to a reduction in fertility in less developed countries. These latter countries may or may not pass through the demographic transition.
5. Although the world's population growth rate is declining after 200 years of acceleration, the world's population will continue to grow. It is projected that world population growth will stop at just over 12 billion by the end of the twenty-first century.
6. The American population is expected to grow more rapidly than do those in other developed countries, due to a combination of the U.S. total fertility rate, low mortality rates, and high net migration rate.
7. As urban areas increased in size, sociologists gradually shifted their attention from communities to larger units, such as metropolitan statistical areas (MSAs) and consolidated metropolitan statistical areas (CSMAs). However, the basic unit of study is the city, which is characterized by a large number of inhabitants, permanence, density, and occupational specialization of inhabitants.
8. Although the process of urbanization is now commonplace throughout the world, the first cities developed only 5,000 to 6,000 years ago. The first preindustrial cities developed in fertile areas where plant cultivation was sufficiently advanced to support a nonagricultural population. A surplus food supply permitted relatively large numbers of people to live in cities and perform nonagricultural work. Preindustrial types of cities still exist in underdeveloped parts of the world.
9. The Industrial Revolution, with its technological innovations and alterations in social structure, caused a major increase in the rate of urbanization. The development of factories was an especially important influence on the location of cities. Although cities and metropolitan areas are important in industrialized societies, they are strongly affected by suburbanization and a recent tendency for medium-sized metropolitan and nonmetropolitan areas to grow more rapidly than large metropolitan areas.
10. Suburbanization has created severe problems for central cities and their residents. The problems are particularly serious for minorities, the poor, and the elderly in the inner city.
11. Less developed countries are having a different experience with urbanization than did the West. In less developed countries, the urban growth rate has been faster, industrialization has been slower, and a base of small and middle-sized cities has not developed. Still, peasants in less developed countries migrate to cities both because they are forced off the land and because cities appear to provide unique opportunities.
12. Urban ecologists study the adjustment processes of human populations to their environments. Urban ecologists developed four major models of spatial development in cities: concentric zone theory, sector theory, multiple-nuclei theory, and peripheral theory. Each of these theories receive some empirical support, but it appears that combining insights from all of them is the most useful.
13. According to the traditional view of urbanism, cities provide a free and sophisticated environment but are characterized by impersonality and social disorganization. According to more recent empirical research, this description is not applicable to all urban residents; many people in urban areas are involved in personal and social relationships.

INFOTRAC® COLLEGE EDITION

http://www.infotrac-college.com/wadsworth

Another unique option available to you at the Student Resources section of the companion Web site is Infotrac College Edition, an online library with access to hundreds of scholarly and popular periodicals. Below are suggested search terms for this chapter. Results from these and other searches are found at the site.

Search keyword: **demography**. As you learned in this chapter of the textbook, demography is the scientific study of population. Although it may not sound like it, there are many interesting topics being explored in demography today. Scan several articles related to demography and determine what some of these current issues are.

Search keywords: **illegal immigration**. Much of America's population growth is the result of immigration to this country. However, not all of that immigration is welcomed, or even legal. Read two articles about illegal immigration and determine why it is occurring. What areas of the United States are most affected by illegal immigration? What are some of the problems caused by illegal immigration? What benefits, if any, are provided by illegal immigration?

Search keywords: **white flight**. This is a term that was widely used in the 1970s and 1980s. Although the term is not as popular as it once was, it still describes a common occurrence in cities. Read an article to find out what white flight is. What effects does it have on cities? Why does it take place?

LEARNING OBJECTIVES REVIEW

- Identify and distinguish among the three population processes.
- Discuss major dimensions of the world population growth problem.
- Relate the ideas of Thomas Malthus to the demographic transition.
- Predict world population trends.
- Discuss demographic trends of older Americans.
- Differentiate basic concepts of urbanization.
- Trace the historical development of preindustrial and modern cities.
- Describe some of the consequences of suburbanization.
- Discuss world urbanization.
- Compare and contrast four theories of city growth.
- Compare and contrast traditional and contemporary views on the quality of urban life.

CONCEPT REVIEW

Match the following concepts with the definitions listed below them.

____ a. gemeinschaft
____ b. multiple-nuclei theory
____ c. crude death rate
____ d. urban ecology
____ e. fecundity
____ f. replacement level
____ g. sector theory
____ h. age-specific fertility
____ i. social demography
____ j. doubling time
____ k. urbanization
____ l. community
____ m. gross migration rate
____ n. Industrial Revolution
____ o. consolidated metropolitan statistical area (CMSA)
____ p. morbidity
____ q. demography
____ r. net migration rate
____ s. metropolitan statistical area (MSA)
____ t. fertility rate
____ u. life span

1. the birth-rate level at which married couples replace themselves in the population
2. rates of disease and illness in a population
3. the maximum rate at which women can physically produce children
4. a description of the process of urban growth emphasizing the importance of transportation routes
5. the branch of demography that studies population in relation to its social setting
6. a grouping of counties that contains at least one city of 50,000 inhabitants or more or an "urbanized area" of at least 50,000 inhabitants and a total metropolitan population of at least 100,000
7. Tönnies's term for the type of society based on tradition, kinship, and intimate social relationships
8. a social group whose members live within a limited geographical area and whose social relationships fulfill major social and economic needs
9. the annual number of live births per 1,000 women ages 15 to 44
10. the annual number of deaths per 1,000 members of a population
11. the annual increase or decrease per 1,000 members of a population resulting from people leaving and entering the population
12. the scientific study of population
13. a description of the process of urban growth that emphasizes specific and historical influences on areas within cities
14. the most advanced age to which humans can survive

15. the combination of technological developments beginning in the eighteenth century that created major changes in transportation, agriculture, commerce, and industry
16. the number of live births per 1,000 women in a specific age-group
17. the study of the relationships between humans and their environments within cities
18. an urban area composed of two or more sets of neighboring primary metropolitan statistical areas (PMSAs)
19. the number of years needed to double the size of a population
20. the process by which an increasingly larger proportion of the world's population lives in urban areas
21. the annual number of persons per 1,000 members of a population who enter or leave a geographical area

CRITICAL-THINKING QUESTIONS

1. Suppose that the average number of children per family in the United States increased from 1.8 to 3. Discuss the short- and long-term effects this would have on population growth.

2. Should the slow-growing nations, such as the United States, have immigration policies that help relieve the population pressures on countries with high-growing populations? Defend your position using sociological concepts.

3. Explain and evaluate the usefulness of the following theories of city growth: concentric zone theory, sector theory, multiple-nuclei theory, and peripheral theory.

4. According to Louis Wirth, what are the major characteristics of urbanism as a way of life? Evaluate the accuracy of Wirth's explanation in view of more recent research findings.

MULTIPLE-CHOICE QUESTIONS

1. **The three population processes that are responsible for population growth and decline are**
   a. fecundity, morality, and migration.
   b. fecundity, morbidity, and migration.
   c. fertility, morbidity, and migration.
   d. fertility, mortality, and migration.
   e. fecundity, mortality, and morbidity.
2. **Which of the following best describes contemporary Americans' behavior regarding reproduction?**
   a. Fewer American women are having abortions.
   b. Fewer Americans are using contraceptives.
   c. More American women are having children in their late teens and early twenties.
   d. More American women are now beginning to have larger families.
   e. Fewer American women are having children, and more American women are having fewer children.
3. **The demographic term for deaths within a population is**
   a. morbidity.
   b. mortality.
   c. terminality.
   d. finality.
   e. turnover.
4. **Rates of disease and illness in a population are known as**
   a. life expectancy.
   b. mortality.
   c. morbidity.
   d. life span.
   e. social pathology.
5. **The movement of people from one geographical area to another for the purpose of establishing a new residence is known as**
   a. fecundity.
   b. exponential transition.
   c. migration.
   d. morbidity.
   e. the demographic transition.
6. **_________ is the principle by which a population increases.**
   a. Linear growth
   b. Natural logarithms
   c. Multiplication
   d. Exponential growth
   e. Arithmetic explosion
7. **For Thomas Malthus, the solution to the population problem was**
   a. early marriage.
   b. improved contraception.
   c. universal education.
   d. federally financed abortions.
   e. infanticide.

REVIEW GUIDE

8. ________ is the process by which a population gradually moves from high birth and death rates to low birth and death rates as a result of economic development.
   a. Exponential growth
   b. The demographic transition
   c. The dependency ratio
   d. The fertility-morbidity ratio
   e. The net migration rate
9. According to the text, three areas of the world are presently in the fourth stage of the demographic transition. These areas are
   a. China, Japan, and North America.
   b. Australia, Iran, and Europe.
   c. South America, Europe, and North America.
   d. Iran, North America, and Europe.
   e. Europe, North America, and Japan.
10. The primary characteristic of the second demographic transition is
   a. a decline in fertility from slightly above replacement level to slightly below.
   b. a reduction in the net migration rate.
   c. an increase in mortality.
   d. the progress achieved regarding adult literacy.
   e. the shift from gesellschaft to gemeinschaft patterns of social organization.
11. ________ has been reached when deaths are balanced by births so that a population does not increase.
   a. Zero population growth
   b. Exponential growth
   c. Equilibrium ratio
   d. The fecundity level
   e. Demographic transition
12. The primary resource needed by a preindustrial city was
   a. a strong trading center.
   b. a surplus food supply.
   c. adequate employment opportunities.
   d. a good system of waterways.
   e. a completed demographic transition.
13. Metropolitan dispersal refers to
   a. the spreading out of industry throughout a city.
   b. the dispersal of retail outlets outward due to inner-city deterioration.
   c. middle- and upper-class whites buying up downtown property after urban renewal.
   d. cities losing population to the areas surrounding them.
   e. cities gaining population over surrounding areas.
14. Which of the following theories of city growth explicitly deals with the zone in transition?
   a. concentric zone theory
   b. peripheral theory
   c. sector theory
   d. ecological invasion theory
   e. multiple-nuclei theory
15. Which of the following statements best describes the personal relationships of city dwellers?
   a. Urban residents conduct their social relationships on the basis of rigidly defined role prescriptions.
   b. Urban dwellers rely on impersonal relationships to protect themselves from the social claims of others.
   c. Kinship networks have disappeared from large cities.
   d. Urban residents know few people well enough to interact on a personal basis.
   e. Large population size does not necessarily produce a lack of involvement in local friendship, kinship, and other social relationships.

FEEDBACK REVIEW

True-False

1. Thanks to worldwide urbanization, preindustrial cities no longer exist. T or F?
2. One of the ultimate contributions of the four major theories of city growth is their demonstration that cities do not grow in a random fashion. T or F?

Fill in the Blank

3. Thomas Malthus sees famine as a ________ check on population growth.
4. ________ is the process by which an increasingly large proportion of the world's population lives in urban areas.
5. A population with an excess of births over deaths will experience ________.
6. ________ occurs when a city is unable to supply adequate jobs and housing for its inhabitants.
7. The distinctive way of life allegedly developed in urban areas is referred to as ________.

Multiple Choice

8. Which of the following figures is the world's population most likely to reach before it stops growing?
   a. 4 billion
   b. 8 billion
   c. 12 billion
   d. 25 billion
9. Europe's second demographic transition involves what kind of population growth?
   a. stable
   b. increasing
   c. declining
10. Which of the following characteristics is not included in the definition of a city?
   a. permanence
   b. dense settlement
   c. advanced culture
   d. occupational specialization in nonagricultural activities
11. According to Herbert Gans, which of the following groups have relatively little choice within the inner city?

a. cosmopolites, the unmarried and childless
b. ethnic villagers, the deprived
c. ethnic villagers, the unmarried and childless
d. the trapped, the deprived

Matching

12. Match the following zones of the concentric zone theory with the adjacent descriptions.
____ a. central business district
____ b. commuter's zone
____ c. residential zone
____ d. zone in transition
____ e. zone of workingmen's homes
(1) zone that is the next zone out from the zone in transition
(2) site of major stores and offices
(3) zone with physical deterioration stemming from few repairs
(4) zone the wealthiest people live in
(5) zone that is largely middle class in composition

GRAPHIC REVIEW

Population projections by world regions appear in Table 17.1. Answer these questions to test your grasp of world population growth between 2002 and 2050.

1. Compare the projected population growth by world region between 2002 and 2050. Explain any surprises. If you were not surprised, explain why projections went just as you expected.
____
____
____

2. Compare the population doubling time in less and more economically developed areas. Explain the difference.
____
____
____

ANSWER KEY

Concept Review

a. 7
b. 13
c. 10
d. 17
e. 3
f. 1
g. 4
h. 16
i. 5
j. 19
k. 20
l. 8
m. 21
n. 15
o. 18
p. 2
q. 12
r. 11
s. 6
t. 9
u. 14

Multiple Choice

1. d
2. e
3. b
4. c
5. c
6. d
7. c
8. b
9. e
10. a
11. a
12. b
13. d
14. a
15. e

Feedback Review

1. F
2. T
3. positive
4. Urbanization
5. natural increase
6. Overurbanization
7. urbanism
8. c
9. c
10. c
11. d
12. a. 2
b. 4
c. 5
d. 3
e. 1

REVIEW GUIDE

18 Social Change and Collective Behavior

© David Turnley/CORBIS

LEARNING OBJECTIVES

- Illustrate three social processes contributing to social change.
- Discuss, as sources of social change, the role of technology, population, the natural environment, conflict, and ideas.
- Compare and contrast the cyclical and evolutionary perspectives.
- Explain the why and how of social change within the context of the functionalist and conflict perspectives.
- Define modernization and explain some of its major consequences.
- Define world-system theory and summarize some of the consequences flowing from its description of the world economy.
- Describe social activities engaged in by dispersed collectivities.
- Describe the nature of a crowd and identify the basic types of crowds.
- Contrast contagion theory, emergent norm theory, and convergence theory.
- Define the concept of social movement and identify the primary types of social movements.
- Apply the concepts of relative deprivation and unfulfilled rising expectations to the emergence of social movements.
- Compare and contrast relative deprivation theory with value-added theory and resource mobilization theory.

USING THE SOCIOLOGICAL IMAGINATION

Do successful revolutions bring radical social change? It seems so, almost by definition. There are dramatic examples of truly revolutionary changes in a society: France in the eighteenth century, China in the twentieth century, and Russia in the early twentieth century. Generally, however, victory celebrations are not followed by a new social order. Change after the American Revolution was not radical; it was gradual and long-term. It is also a fact that the revolution was merely *one* of the precipitative factors creating change.

Revolution is only one factor contributing to the alteration of social structure. After defining social change, we turn immediately to several of its sources.

Social Change

Social Change a Constant

Imagine for a moment the history of the earth as a 365-day period, with midnight January 1 marking the starting point and December 31 being today. Within this time span, each "day" represents 12 million years. The first form of life, a simple bacterium, appeared in February. More complex life, such as fish, appeared about November 20. On December 10, the dinosaurs appeared; by Christmas they were extinct. The first recognizable human beings did not appear until the afternoon of December 31. *Homo sapiens* emerged shortly before midnight that last day of the year. All of recorded history occurred in the last 60 seconds of the year (Ornstein and Ehrlich 1991).

All societies change. The speed of social change may vary from the glacial to the mercurial. Nonetheless, social change is a constant. For sociologists, **social change** involves societal alterations with long-term and relatively important consequences (Abu-Lughod 1999).

Can social change be predicted? Predicting precise change is difficult, partly because a society absorbs change in ways consistent with its own culture. For instance, two societies that both adopt a democratic form of government may develop in very different ways. Both Britain and the United States are democracies. But their histories prior to becoming democracies (Britain had a royal tradition) led to divergent forms of government. In addition, people in a society can modify their behavior and consciously decide for themselves the ways in which change will occur. They can, for example, deliberately avoid a predicted state of affairs (Caplow 1991). These facts should not, however, discourage an attempt to understand alterations in society. But in doing so, one must base the accuracy of predictions on sound assumptions, just as Frenchman Alexis de Tocqueville did during the early 1800s in his remarkably penetrating study of American society (see Table 18.1).

Why do some societies change faster than others? This is a complex question with no easy answer. Technology, discussed shortly, is a prime force in social change. Technologically complex societies change much faster than technologically simple societies (Nolan and Lenski 1999). This is because technology increases

TABLE 18.1

Key Assumptions in Predicting Social Change in America

The most accurate prognosticator of trends in American society has been the Frenchman Alexis de Tocqueville. Tocqueville's *Democracy in America*, which was published in the 1830s, displayed an amazing grasp of American society. Tocqueville's success has been attributed to several key assumptions he made.

- *Major social institutions would continue to exist.* Unlike many of his contemporaries—and many of ours—Tocqueville did not expect the family, religion, or the state to be greatly changed.
- *Human nature would remain the same.* Tocqueville did not expect men and women to become much better or worse or different from what history had shown them to be.
- *Equality and the trend toward centralized government would continue.*
- *The availability of material resources (such as land, minerals, and rich soils) limits social change.*
- *Change is affected by the past, but history does not strictly dictate the future.*
- *There are no social forces aside from human actions.* Historical events are not foreordained by factors beyond human control.

Sources: Adapted from Theodore Caplow, *American Social Trends* (New York: Harcourt Brace Jovanovich, 1991), p. 216.

A moderate view of postmodernism has envisioned the electronic information superhighway. This bank of television monitors suggests how far along this road we currently are.

human control over the physical world. And as control over the environment increases, nontechnological changes occur in other parts of society—family, economy, education, medicine. For example, new technologies associated with the computer and information processing are radically changing health care. There are patients today who are diagnosed by physicians they will never meet in person.

Technology is, however, not the only reason some societies change faster than others. Sociologists have identified three important social processes and several more specific factors that influence the pace of social change.

Processes for Change

Three interrelated social processes lead to social change. Although related, the processes of discovery, invention, and diffusion are distinct.

How does each of these three social processes promote social change? In the **discovery** process, something is either learned or reinterpreted. Early ocean navigators, with skill and courage, demonstrated by experience the rounded shape of the earth. With this new interpretation, then, many worldwide changes followed, including new patterns of migration, intercontinental commerce, and colonization. Salt, another example, was first used to flavor food. Because it was so highly valued, it was subsequently used as money by Africans and as a religious offering among early Greeks and Romans. More significantly, fire, which prehistoric peoples used for warmth and cooking, was later used (in about 7000 B.C.) by the first agriculturists to clear fields and to create ash for fertilizer. Fire, of course, was central to the ensuing transition to agricultural society.

Invention is the creation of a new element by combining two or more already existing elements, with a spawning of new rules for its use. Examples of inventions come easily to mind. It was not so much the materials Orville and Wilbur Wright used to make an airplane (most of the parts were available), but the way they manipulated and combined the elements, that led to their successful flight at Kitty Hawk.

The pace of social change through invention is closely tied to the complexity of the cultural and social base. As the base becomes more complex and varied, the number of elements, as well as the number of ways the elements can be combined into inventions, increases dramatically. Consequently, the more complex and varied a society, the more rapidly it will change. This helps to explain why even though several million years passed between the evolution of the human species and the invention of the airplane, we reached the moon in less than seventy years after the Wright brothers' first flight. Social change, of course, occurs faster when an invention can be combined with many other aspects of a society. NASA was able to reach the moon relatively quickly because the United States was already advanced in physics, aerodynamics, and the manufacturing of specialized materials.

When one group borrows something from another group—norms, values, roles, styles of architecture—change occurs through the process of **diffusion**. Whereas the extent and rate of invention depend on the complexity of a society, the extent and rate of diffusion depend on the degree of social contact. Borrowing from others may involve entire societies, as in the American importation of cotton growing that was first developed in India. Or diffusion may take place between groups within the same society, as in the social and cultural spin-offs of the jazz subculture, which was created by African American musicians.

As noted earlier, a borrowed element must harmonize with the group culture. Despite the upsurge of unisex fashions in America today, wearing a Scottish kilt on the job could get a construction worker laughed off the top of a skyscraper. Wearing kilts still clashes with the American definition of manhood. If skirts are ever to become as acceptable for American men as pants are for women, either their form will be altered or the cultural concept of masculinity will be modified.

Integrating a borrowed element may involve merely a selection of certain aspects. The Japanese, for example, accept capitalism but resist the American form of democratic government, the American style of conducting business, and the American family structure. Diffusion, then, almost always involves selectivity and modification.

In modern society, most aspects of culture are borrowed rather than created. The processes of discovery and invention are important, but usually far more elements enter a society through cultural diffusion.

Sources of Social Change

Discovery, invention, and diffusion are social processes through which social change occurs. A multitude of factors affect society through these processes. Most important among the factors encouraging social change are technology, population, the natural environment, revolution and war, and ideas.

Technology

Time magazine selected Albert Einstein as the man of the century, a choice reflecting the significance of science and technology in the twentieth century (Golden 1999). **Technology** is the combination of knowledge and hardware that is used to achieve practical goals. It has historically been viewed as a precursor to social change (Teich 1993; MacKenzie and Wajcman 1999; Volti 2000). For instance, Karl Marx placed the importance of technology in sharp perspective: "The windmill gives you society with the feudal lord; the steammill, society with the industrial capitalist" (Marx 1920:119). Although the relationship between technology and social change is not this simple and direct—and Marx knew it—their interconnectedness is firmly established.

Do technological advances accelerate social change? Consider the social effects of even minor technological changes. The invention and diffusion of the stirrup

WORLDVIEW 18.1

Computers Connected to the Internet (per 100,000 people)

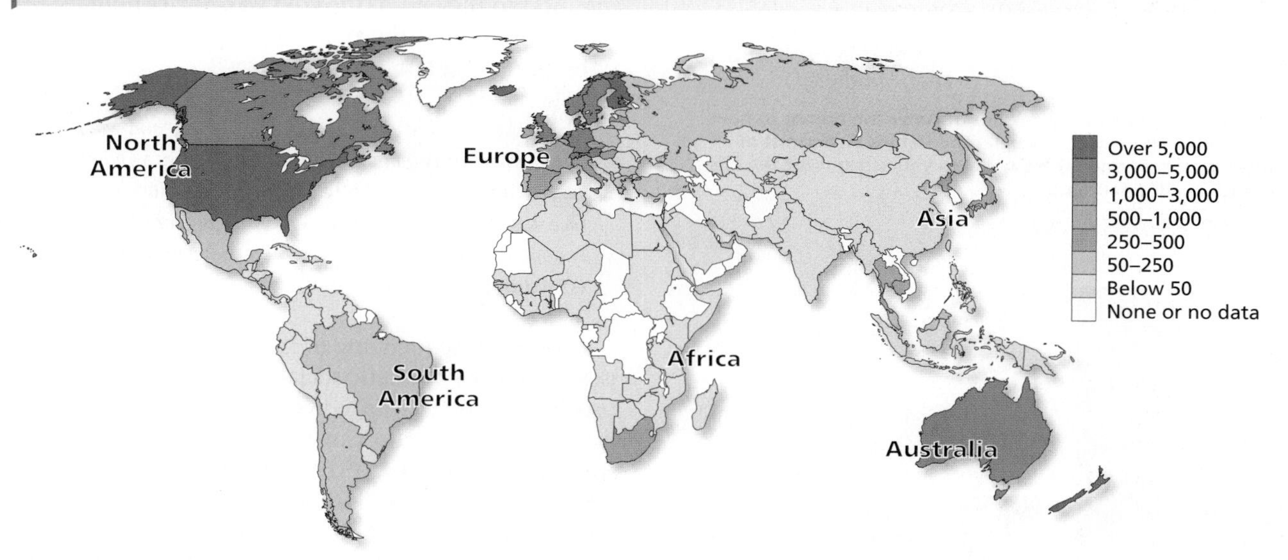

Source: Ian Pearson, *Atlas of the Future* (New York: Macmillan, 1998), pp. 58–59.

The number of Internet connections is exploding, revolutionizing how people communicate and live their lives. But, as this map shows, the number of people connected to the Internet varies widely from country to country. As of the late 1990s, nearly 60 percent of Internet connections were on the North American continent.

1. Do you see a pattern in the numbers of connections to the Internet? Explain.
2. What implications might this distribution have for future social change?

© Bettmann/CORBIS

The relationship between technology and social change is firmly established. Contrast modern family activities in homes with television sets to the activities of this 1920s family in a home with only a radio.

altered medieval warfare. Soldiers on foot were easily defeated by knights on horseback because the stirrup permitted knights to continuously hold their lances, rather than throw them. This new military technology—a combination of the knight, the horse, and the lance—contributed to the rise of the feudal system, with the knightly class at the top (Lynn White 1972). Although

FIGURE 18.1

Becoming Wired: Time It Took for 30 Percent of Americans to Acquire Selected Technologies

It took forty years for 30 percent of American homes to have telephones. How much time elapsed before 30 percent of Americans were online? How would you account for the difference in adoption time?

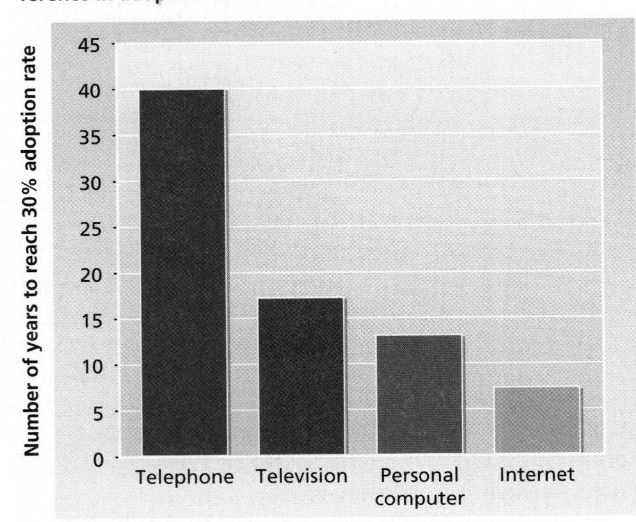

Source: David R. Francis, *The Internet in Healthcare* (Nashville, TN: J. C. Bradford, August 1999), p. 11.

originally thought to be a mere means of transportation, the automobile has had countless effects on modern society. It changed dating and courtship patterns, contributed to a new sexual morality, created new industries, promoted industrial expansion, produced suburbanization, and contributed immeasurably to environmental pollution. The creation of the silicon chip introduced technological change at an unprecedented rate. It took more than a century for telephone possession in the United States to reach 94 percent; but in less than a decade (from 1992 to 2000), the Internet reached about 30 percent of Americans; in just over twenty years, the personal computer market reached its present 40 percent household penetration; and in less than fifteen years, a cell phone was in almost 25 percent of American households (see Figure 18.1).

The societal transformations wrought by the microprocessor are many and far-reaching. The workplace, by way of illustration, is undergoing a transformation, with effects as far-reaching as those of the Industrial Revolution (Fukuyama 1999a, 1999b). Telecommunications technology, for example, will allow many to work from their homes, but it will result in far less human interaction (McGinn and Raymond 1997–98). In the field of medicine, the microprocessor radically changed many surgical techniques, from microsurgeries to radio wave therapy (Cowley and Underwood 1997–98). On the Autobahn in Germany, drivers using cell phones or electronic consoles in their cars can get real-time computer-generated information on traffic problems. Perhaps the most profound effect of the microprocessor was its contribution to the end of the cold war. Soviet military personnel recognized the potential of computer guidance systems for long-range munitions:

> *The realization that the Soviet Union did not, and would not, possess the computer or sensor capabilities to produce such weapons—and so, in the long run, must inevitably face defeat at Western hands—was a significant factor behind Mikhail Gorbachev's decision to sue for peace.* (Watson and Barry 1997–98)

The recent sequencing of the human genome promises even more dramatic social change. Now that scientists have cloned sheep, cows, pigs, domestic cats, and a human embryo, who can know where this capability will lead ("Hello, Dolly" 1997; R. Weiss 2000; J. Fischer 2001; Fukuyama 2002; Kluger 2002)? In fact, Nobelist and Cal Tech president David Baltimore sees biology as a driving force in the transformation of society: "Instead of guessing about how we differ from one from another, we will understand and be able to tailor our life experiences to our inheritance" (Brave 2001:18). The 2001 national debate on stem-cell research policy is only the beginning of public debate; where will human genetic engineering take us?

Why should we be careful of technological determinism? Technological determinism assumes that technology is the primary determinate of the nature of a society. This is a seductive theory. By use of dramatic examples, it is easy to conclude that the nature of a society can be traced directly to the society's technological base. Actually, the relationship between technology and social change is more complicated (Lauer 1991). First, social change can occur without technological developments. Ancient Greek commercial development was not based on improved production methods. Second, the introduction of technology does not necessarily lead to social change. A South Indian village did not change its community social structures following a switch from dry to wet cultivation of crops (Epstein 1962). Third, as just implied in the discussion of diffusion, the particular effects of technology will vary from society to society because the adoption and use of technology are always filtered through the nature of a given society. Although Russia, for example, has computer technology, it places much tighter controls on the use of the technology by individual citizens.

Although it seems inadvisable to see technology as a necessary and sufficient cause of social change, the contribution of technology to change should not be underestimated. It is best to view technology as an extremely important factor in a network of causes. Nor should we ignore the fact, as noted earlier, that technological innovation can occur very rapidly. David Freeman (1974) also makes this point by compressing the total lifetime of the earth (some 5 billion years) into an eighty-day period and calculating when some significant technological changes occurred: the Stone Age began 6 minutes ago, the agricultural revolution took place 15 seconds ago, metals were first used 10 seconds ago, and the Industrial Revolution started three-tenths of a second ago.

Population

Alterations in population size and composition have interesting effects. By way of illustration, population pressures in China led to a conservation of resources. This in turn produced a state policy replacing the more expensive, traditional burial rites with cremation ceremonies. Yet another consequence of population stress is found in the conflict and strain within Chinese families. Moreover, China instigated a national population control program to create the one-child family. This is already affecting Chinese social structure, as in the dishonesty engendered by some anxious local officials who underreport birth statistics to meet the government quotas. Less developed countries offer another illustration; they are undergoing many dramatic changes because their declining mortality is not being matched by lowered fertility.

America's baby boom following the return of American soldiers at the end of World War II is illustrative as well. The increased birth rate of Americans born between 1946 and 1964 required expanded medical personnel and facilities for child health care and then, in the 1950s and 1960s, created the need for more teachers and schools. On the other hand, the generation following the baby boomers, now in their thirties and in the labor market, is experiencing increased competition for jobs and fewer opportunities to move up the career ladder. As the baby boomers retire, problems of health care and Social Security loom large. Longer working hours, retraining programs, and reeducation for senior people will probably become prevailing patterns. As America's population continues to age, more attention is being paid to our senior citizens. Already, there are more extended-care homes for the aged, an increase in geriatric emphasis, and more television advertising and programming targeting the aging population.

The Natural Environment

Interaction with the natural environment, from the earliest times, has transformed American life. The vast territory west of the thirteen colonies permitted expansion, ultimately to the Pacific Ocean. This western movement helped shape our cultural identity and wrought untold changes, most tragically the destruction of many Native American cultures. Then there was the Great Depression of the 1930s, due in part to a long drought that hit the agricultural Midwestern plains states. Overplanting and plowing had upset the fragile ecosystem, turning the prairies into a giant "dust bowl." In the early 1970s, when the Organization of the Petroleum Exporting Countries (OPEC) launched an embargo, refusing to sell its oil to other countries, oil products became scarce and expensive. The embargo sent shock waves throughout the world, contributing to American economic inflation in the 1970s and early 1980s. As a result, Americans began driving smaller, more fuel-efficient automobiles.

Revolution and War

From the Greek historian Polybius to Karl Marx, many early thinkers viewed conflict as the central form of human interaction. Marx and more contemporary conflict theorists, who see social conflict as the prime impetus for social change, are covered in the next major section. For now, let's focus on revolution and war as sources of social change that involve conflict.

What is a revolution? A **revolution** is a social movement that involves the violent toppling of a political regime. According to Charles Tilly, a revolution results in the replacement of one set of power holders by another (Tilly 1978, 1997). Most revolutionaries, of course, expect more fundamental changes than this. Marx, for example, expected workers' revolutions to eliminate class-based inequality and therefore to profoundly affect the social and economic structures of the societies in which they occurred.

Are revolutions normally followed by radical changes? As suggested in "Using the Sociological Imagination," revolutions are normally not followed by radical social change. In Crane Brinton's (1990) view, the changes introduced by revolutionary leaders are followed eventually by restoration of a social order similar to that existing in prerevolutionary days. In other words, even successful revolutions fail to tear down the old order and bring in an entirely new one. Radical changes are rarely permanent; after victory celebrations are over, a society operates with a great deal of continuity with the past—a compromise between the new and the old.

What sorts of changes do follow revolutions? There is a middle ground between Marx's position that revolutions produce major changes and Brinton's position that revolutions have little lasting effect. Tilly argues that revolutions do have lasting effects if there is a genuine transfer of power. Most revolutions, Tilly writes, do not have lasting effects, because revolutionaries seize control of a government too weak to effect change. Eighteenth-century French revolutionaries, in contrast, took control of a powerful, centralized state, allowing them to seize and redistribute the property of aristocrats and churches. Despite a backlash, the French Revolution had lasting effects. Similarly, the Chinese Communists created a strong central state of their own, allowing them to reorganize village structure in a way that permanently altered Chinese society.

In the 1940s, a cornerstone of the communist revolution in China was sexual equality; liberation from sexism was a revolutionary plank. The situation for Chinese women has improved, but complete sexual equality is still a far-distant dream ("Closing the Gap" 1995). Revolutions, then, can be a powerful mechanism for social change, whether or not they create the total change promised by those who lead them and expected by those who participate in them.

What is the relationship between war and social change? Social change, as already seen, can be brought about by revolutionary conflict within societies. Change is also created by conflict in the form of **war**—armed conflict that occurs within a society or between nations. War and change are closely intertwined. The Golden Age of Athens (fifth century B.C.) started with the Persian War and ended with the Peloponnesian War. Rome's flowering in the first and second centuries cannot be separated from wars in Europe and Asia. War was a backdrop

This is a scene of "nuclear winter" after the bombing of Japan at the end of World War II. As a result of American intervention, tremendous social and economic change followed.

to the Age of Enlightenment. The founding of the United States is no exception; it required a war. The Civil War was instrumental in America's transition from an agricultural to an industrial society. Momentous technological, economic, and social changes also followed World War I, World War II, and the Vietnam War.

How does war promote social change? Robert Nisbet (1989) writes eloquently on the effects of war. Change is created through diffusion as war breaks down insulating barriers between societies. The crossbreeding and intermingling of people and cultures during and following a war leave participating societies different than they were before the conflict. Wars also promote invention and discovery. During the First World War (1914–1918), the U.S. government, due to the pressure of war, promoted and financed the development of such technologies as the airplane, automobile, nylon, and radio, each of which contributed to the social and cultural revolution that followed the war. And America's culture, both during and after World War I, was exported to societies all over the world.

Ideas

Indeed, ideas can lead to social change. Max Weber, for example, believed that the ideas inherent in Calvinism produced behavior conducive to the development of industrial capitalism.

Alfred North Whitehead wrote, "A general idea is always a danger to the existing order" (Whitehead 1933:22). We have seen the validity of this observation in the earlier discussion on revolution. In the next section, we will see that two major theories of social movements also emphasize the crucial influence of ideas.

Do ideas always lead to social change? Sometimes ideas retard social change. During the Middle Ages, Christian religious ideas forestalled change (Gay 1977). In fact, the Catholic Church based its strength on total resistance to change (Manchester 1993). For example, economic development was slowed in part because usury (charging interest on loans) was considered a sin. For many years, slavery in the United States was successfully defended on the basis of fallacious ideas that depicted African Americans as inferior. Similar unreasoned ideas have been used to deny women equal opportunity in many societies, including the United States.

How are these sources of change interrelated? Several, or all, of these factors can be operating in combination. Consider abortion. Abortions are possible because of medical inventions that have been widely diffused. Underlying the ability to perform abortions is medical technology. This technology affects the population by lowering the birth rate, which in turn affects family size and a number of other social conditions (child-free marriages, dual-employed couples, sexual freedom). A lowered birth rate also reduces demand on limited natural resources. Conflict between pro-life and pro-choice forces in the United States has existed ever since the Supreme Court legalized abortion in 1973 in the *Roe v. Wade* decision. This conflict, of course, has been based on disparate ideas regarding the moment at which a fetus constitutes a human being.

Theoretical Perspectives

It is through theory that we go beyond description of social change toward explanation. Before our theoretical discussion of functionalism, conflict theory, and symbolic interactionism, let's look, briefly, at some early theories specific to social change. This will make the three contemporary theoretical approaches more meaningful.

The Cyclical Perspective

Scholars were baffled. Why would the most developed nations in the world spend $337 billion between 1914 and 1918 (World War I) to kill over 8 million soldiers? This was very poor evidence indeed for a belief in the natural progression of civilization. Scholars began to ask whether change was necessarily progress. Was Western civilization on the road to improvement or on the way down?

How did advocates of the cyclical perspective describe the effects of social change? The dramatic upheaval during the early part of the twentieth century led scholars to consider the possibility that civilizations rise and fall rather than develop in a straight line. The rise and fall of Rome, after all, had been documented in six large volumes by British historian Edward Gibbon; his work was published between 1776 and 1788. Although many scholars disagreed with Gibbon's conclusions, they did not miss the point that social change may very well follow a pattern of growth and decay. Three of the most prominent advocates of the cyclical perspective were historians Oswald Spengler (1880–1936), Arnold Toynbee (1889–1975), and sociologist Pitirum Sorokin (1889–1968).

Spengler (1926–28) published *The Decline of the West*, in which he compared societies to living things, as did social Darwinist Herbert Spencer (1820–1903), but with a vital difference. Whereas Spencer saw advanced societies as superior survivors of the struggle of the fittest, Spengler believed that civilizations are born, ripen, decay, and perish. Specifically, he predicted that Western civilization—which had traded zest for materialism—was going to be replaced by a civilization from Asia. Spengler's work is now considered more literature than scientific explanation.

FEEDBACK

1. __________ refers to alterations in social structures that have long-term and relatively important consequences.
2. The more complex a society, the more rapidly it will change through invention. T or F?
3. The process of __________ usually involves modification and selectivity.
4. The theory of social change that assumes technology to be the primary cause of the nature of social structure is called __________.
5. The baby boom in the United States is a good example of the effects of __________ on an entire society.
6. __________ is a type of social movement that involves the toppling of a political regime through violent means.
7. Contrary to conventional wisdom, war retards social change. T or F?
8. Was it Karl Marx or Max Weber who emphasized the role of ideas in creating social change? __________

Answers: 1. Social change 2. T 3. diffusion 4. technological determinism 5. population 6. Revolution 7. F 8. Weber

Almost twenty years later, British historian Arnold Toynbee (1946) published *A Study of History*, in which he presented his challenge-response theory of social change. According to Toynbee, civilizations rise only when a proper challenge in the environment is presented to a people who can respond successfully to it. If a challenge from the environment is not issued or if the society fails to meet a challenge, that society will either fail to rise in the first place or it will decline.

Like Spengler and Toynbee, Sorokin did not see in history any linear pattern of change. To him, sociocultural history involves a cyclical alternation among three reasonably homogeneous cultural types. *Ideational* cultures are based on the conviction that truth and value come from God. Emphasis is on the spiritual and the next world. In *sensate* cultures, truth is discovered through empirical observation by the senses. It is experience in the current world—gained through touching, smelling, hearing, seeing, and tasting—that reveals truth and reality. Sensate cultures are not absolutist but allow for reinterpretation of reality. People in ideational cultures focus on spiritual needs; members of sensate cultures place priority on physical needs and gratification of the senses. Whereas ideational cultures tend to be spiritual, sensate cultures are characterized by pleasure seeking and materialism. Sorokin's third cultural type, the *idealistic* culture, is a blend of the ideational and the sensate, with somewhat more accent on the ideational.

For Sorokin, then, history is best described as a continuous movement from the predominance of one type of culture to another. According to Sorokin, sensate culture dominated in Christ's time. Christian medieval Europe typified the shift from the sensate to the ideational. Idealistic culture emerged in thirteenth- and fourteenth-century Europe. Modern Western society is again grounded in sensate culture.

Regarding the cyclical approach to social change, what kinds of questions are left unanswered? Because history tells us that societies do, in fact, rise and fall, cyclical perspectives are especially appealing. The proof is there. Cyclical perspectives have a major problem, however. They are not so much explanations for change as descriptions of change. Spengler does not indicate why societies mature and decay; he just tries to show that they do. Toynbee does not state why a formerly successful society fails to meet a new challenge; he just writes that societies failing to respond to a challenge will decline. Sorokin provides no explanation for the cyclical pattern he describes; he just attempts to substantiate its existence with an amazing amount of historical information.

Is the cyclical perspective even worth considering? The cyclical perspective is worthy of consideration for several reasons. First, it represents an early, historically important attempt to understand social change. Second, it helps to legitimate the scientific study of social change. Third, it is useful as a backdrop for distinguishing between description and explanation.

The Evolutionary Perspective

What was the nature of early evolutionary theory? "Every day in every way, things get better and better." In simple terms, this idea of progress, of societies constantly moving toward improvement, was at the heart of the evolutionary perspective that dominated nineteenth-century social thought. In its nineteenth-century form, the evolutionary perspective held that societies must pass through a set series of stages, each of which results in a more complex and advanced stage of development.

The evolutionary perspective became popular during the nineteenth century in part because of the times. During the middle and late nineteenth century, western Europeans prided themselves on their superiority over other peoples. Progress and territorial expansion were watchwords. During colonial expansion, contact with preliterate people only offered ethnocentric Europeans further proof that their societies were the most highly civilized ever to exist. "Primitives" simply represented early stages of social evolution.

Another important influence of the evolutionary perspective was the work of Charles Darwin (1809–1882), which provided an intellectual framework for interpreting the place of primitive and advanced societies in the world. Darwin's idea that all living things tend to improve and become more complex as they develop seemed to fit the fact that some societies were more advanced than others.

In the hands of the English sociologist Herbert Spencer, Darwin's biological ideas were transformed into social Darwinism. Spencer drew a parallel between living organisms and societies and coined the expression "survival of the fittest" to explain why some peoples had become "civilized" whereas others remained "savages" and "barbarians." Social Darwinism was so widely accepted in the Western world that governments in western Europe and America believed that they had the right to dominate, protect, and tutor less developed societies. Such was thought to be the destiny and burden of nature's hardiest people.

Social Darwinism and classical evolutionary perspectives survived into the first part of the twentieth century, but they are no longer considered valid. Social scientists no longer believe that societies necessarily improve as they change, or that developing societies are destined to follow the Western model. Nevertheless, more recent perspectives on social change have overcome some of the faults of the classical evolutionary perspective (Sanderson 1990; Chirot 1994).

What is the viewpoint of modern evolutionary theorists? The contemporary evolutionary perspective rejects the unilinear, or single-direction, assumption of earlier perspectives. Given the social and cultural diversity around the world, modern evolutionary theorists ask, how can we believe that all societies develop in a single direction? Because we know that societies can develop in many ways—some even toward greater deterioration and unhappiness—evolution must be multilinear, or multidirectional. Deliberately missing from the current evolutionary perspective, then, are the ideas of definite and orderly stages of development and of change inevitably producing progress and greater happiness. There are many paths and directions to evolutionary change (Steward 1979).

The most recent sociocultural evolutionary theory comes from Patrick Nolan and Gerhard Lenski (1999), whose initial assumption is that both change and stability must be considered if human societies are to be understood. Despite the dramatic changes that have taken place in the world over the past 10,000 years, they argue, most individual societies have successfully resisted change. This, of course, is a paradox; how can rapid social change occur when individual societies successfully resist change? To resolve this apparent paradox, Nolan and Lenski offer a proposition: a total system can change despite resistance to change in most of its parts, providing the parts that fail to change do not survive. This, they assert, is precisely what has happened in the "world system of societies." Of those societies that did not change or changed only a little, almost none are around today. Conversely, nearly all surviving societies are those that did alter their social structures. Thus, say Nolan and Lenski, in the world system of societies, a process of natural selection has operated that favors innovative societies over those that resist change.

For Nolan and Lenski, the key to societal survival is the accumulation of information, particularly information relevant to subsistence. In fact, improvements in subsistence technology are said to be a "necessary" precondition for a society to grow in complexity, size, wealth, or power. Subsistence technology is also important because it stimulates improvement in other types of technology, as in production, transportation, communication, and war. These further technological advances promote additional advances in subsistence technology. Nolan and Lenski are not technological determinists. For example, they think that a society's beliefs and values are crucial because, in part, they affect the extent to which a people are open to innovation and change.

What does modern evolutionary theory explain? Sociocultural evolutionary theory helps us understand how societal changes are made. It shows that as subsistence technologies improve, societies have the means for gradually progressing from a hunting and gathering to a postindustrial form. As we shall see, modern evolutionary theory is informed, in part, by both functionalist and conflict theory.

The Functionalist Perspective

Because functionalism emphasizes social stability and continuity, it may seem contradictory to refer to a functionalist theory of social change. However, two functionalist theories of social change—proposed by William Ogburn and Talcott Parsons—are especially interesting. Both of these theories are based on the concept of *equilibrium* discussed in Chapter 1.

What is the connection between functionalism and the concept of equilibrium? The word equilibrium implies balance and consistency. Concerned with equilibrium, the tightrope walker on a narrow rope inches his way across a deep chasm, continually shifting his body and using his long balancing pole to counterbalance the effects of the wind as well as the effects of his own motions. When applied to social life, equilibrium connotes attempts to reestablish stability after some disturbance. According to functionalists, societies act as inherently stable wholes that react to changes by making adjustments and eventually assimilating change into a new state of equilibrium. A society in change, then, moves from stability to temporary instability and back to stability. Sociologists refer to this as a "dynamic," or "moving," equilibrium. For example, in 1972, a broken dam led to the destruction of the community of Buffalo Creek, West Virginia. The physical destruction of the community was accompanied by death and the rending of the old fabric of life. Despite the ensuing chaos, residents of the community pulled their lives together again. Although things were not the same as before, a new equilibrium was built out of the physical, social, and human wreckage (K. T. Erikson 1976). The 1960s were a time when the norms of sexual behavior changed radically. But, after skyrocketing during the sixties, teenage pregnancy is now declining. Although Americans do not follow the norms of the 1950s, a retreat from extremes is occurring as new norms of sexual behavior are being established.

Functionalism is based on the concept of equilibrium. Kingsley Davis writes:

> *It is only in terms of equilibrium that most sociological concepts make sense. Either tacitly or explicitly, anyone who thinks about society tends to use the notion. The functional-structural approach to sociological analysis is basically an equilibrium theory.* (Davis 1949:634)

Although not all sociologists would agree that most sociological concepts make sense only in relation to equilibrium, most would concur that functionalism involves the concept of equilibrium.

Before turning to two specific functionalist theories of social change, we should consider the similarity and difference between the evolutionary and functionalist approaches, because both of the functionalist theories of change we will examine draw on evolutionary theory. Both the evolutionary and functionalist perspectives view society as composed of many highly differentiated parts, all contributing to the maintenance of a smooth-running, stable society. However, whereas the traditional evolutionary perspective emphasizes a constant forward direction spurred by change, the functionalist perspective places emphasis on society's ability to recover its balance if equilibrium is upset by change. In functionalism, the various parts of society are seen as highly integrated, so that if a change occurs in one part, other parts are affected. The chain reaction is supposed to eventually restore balance because change is absorbed and distributed among a society's elements until equilibrium is reestablished. The emphasis in the functionalist approach is always on a return to stability and order after some restructuring has taken place.

What is Ogburn's theory of social change? If a new equilibrium is achieved, the parts of society, asserted William Ogburn (1964; originally published in 1922), do not all reach the new balance at the same time. Some parts lag in time behind others. Ogburn applied the term **cultural lag** to any situation in which disequilibrium is caused by one segment of a society failing to change at the same rate as an interrelated segment. More specifically, Ogburn believed that changes in the nonmaterial dimensions of culture (norms, values, beliefs) lag behind alterations in the material culture (technology, inventions). Significant social change occurs when the nonmaterial culture is forced to change because of a prior change in the material culture. Ogburn's best known example is the lag between the technological ability to cut down entire forests and the subsequent emergence of the conservationist movement. More recently, we can point to the sexual norms, values, and beliefs now in the process of attempting to catch up with the widespread distribution of birth control technology that occurred years ago. Cultural lag still exists. The Roman Catholic Church, officially, continues to oppose birth control, while many members of the church are waging personal struggles over contraception. The continuing conflict (sometimes violent) between pro-choice and pro-life advocates over the issue of abortion is a prime illustration of cultural lag. Finally, in a related area, consider the currently unresolved legal, ethical, and social dilemmas flowing from the relatively new technology permitting surrogate motherhood.

How does Parsons approach social change? The concept of equilibrium is at the heart of Talcott Parsons's theory of social change. In his early work, Parsons (1937, 1951) did not emphasize social change. He depicted societies as systems attempting to resist change in order to maintain their current state of equilibrium. It was only later that Parsons (1966, 1971, 1977) began to depict change as a contribution to a new equilibrium with novel characteristics.

Consistent with his roots in evolutionary theory, Parsons was interested in the processes by which societies become more complex. In the first process, *differentiation*, aspects of a society are broken into separate parts. In simple societies, the family was responsible for nearly all functions—economic, educational, medical, emotional, and recreational. As societies became more complex, these functions began to differentiate from the family. Jobs were found in factories, education occurred in schools, medical needs were cared for by doctors and hospitals, emotional support could be found outside the home, and amusement was supplied by a variety of nonfamilial sources.

Differentiation brought on the need for the second process—*integration*. All the newly evolving social units formed workable links for a new equilibrium different from the previous one. Ways were developed for schools and families to mesh, for parents to accept their children's leaving for work in the factories, and for parents to cope with their offspring's independent lives outside their local communities. The process of integration not only leads to a new state of equilibrium but also, in conjunction with the process of differentiation, helps to produce a much more complex type of society.

What are the contributions and criticisms of the functionalist perspective on social change? A major contribution of the functionalist perspective is its attempt to explain both stability and change. Ogburn tries to show that a society's ways of thinking, feeling, and behaving are constantly attempting to catch up with prior technological change. Technology, for him, is an independent variable leading to further social change. Parsons identifies differentiation and integration as processes integral to the maintenance of a moving equilibrium. In addition, he emphasizes that change does not imply total change; continuity as well as change exists.

Functionalists have been criticized for having a conservative bias. The concept of equilibrium, argue some critics, assumes internal societal resistance to change. Also, the equilibrium model of functionalism is charged with depicting change as external to societies: societies are changing not because of their own internal dynamics, but as a result of outside forces. By failing to explore internal sources of change, it is argued, functionalists ignore many forces for change that influence a society. Finally, functionalists have been faulted for

This closed factory represents a significant loss in jobs for individuals as well as in tax revenue for the local community and state. According to conflict theorists, the ability and willingness of a corporation to move away despite these negative consequences lies in its superior control over resources.

© Gail Oskin/AP/Wide World Photos

focusing on gradual change and failing to consider radical change (R. Collins 1975; Giddens 1979, 1987; Giddens and Duneier 2000). The conflict perspective addresses these criticisms.

The Conflict Perspective

Whereas the functionalist perspective assumes society to be inherently stable, conflict theorists depict society as unstable. Whereas functionalism views society as an integrated whole whose various parts work harmoniously to achieve balance, the conflict perspective emphasizes the separate parts of society and the conflict that occurs between them. According to the conflict perspective, social change is the result of struggles between groups for scarce resources. Social change is created as these conflicts are resolved.

What are the origins of the conflict perspective? Many of the basic assumptions of the conflict perspective emerge from the writings of Karl Marx, who wrote that "without conflict, no progress: this is the law which civilization has followed to the present day" (Marx; in Feuer 1959:7). Ralf Dahrendorf, a modern advocate of the conflict perspective, summarizes Marx's view of society in this succinct passage:

> *For Marx, society is not primarily a smoothly functioning order of the form of a social organism, a social system, or a static social fabric. Its dominant characteristic is, rather, the continuous change of not only its elements, but its very structural form. This change in turn bears witness to the presence of conflicts as an essential feature of every society. Conflicts are not random; they are a systematic product of the structure of society itself. According to this image, there is no order except in the regularity of change.* (Dahrendorf 1959:28)

More specifically, according to Marx, the struggle for scarce economic resources, particularly property, is the primary stimulus for change. By way of review, Marx claims that change in capitalist society is authored by class struggle, and as a capitalist society develops, it begins to divide into two classes: those who possess property and own the means for production (the bourgeoisie) and the exploited workers who sell their labor merely to survive (the proletariat). These two classes polarize as the rich get richer and the poor get poorer. At some point, a revolution permits the oppressed proletariat to seize power. With the proletarian revolution comes the development of a classless society. The absence of classes eliminates the source of conflict.

What are the contributions and criticisms of Marx's approach to social change? Defenders of Marx contend that contrary to the charge, Marx avoided economic determinism. Although Marx contended that the economic base of any society is the ultimate determinant of the society's nature and course, he also allowed for noneconomic influences on this economic base. Also, Marx is credited with effectively showing that how people think, feel, and behave reflects the basic underlying economic foundation of their society. In addition, Marx refined the view that social conflict is built into all stratification structures and that the distribution of power is rooted in the economic system (Lauer 1991).

According to critics, Marx placed too much emphasis on economic factors as the determinants of social change while downplaying relevant social and cultural forces. Others criticized Marx for failing to see that conflict can integrate societies as well as tear them apart and for not recognizing the prevalence of cooperation over conflict within societies. Marx's views on revolution were also challenged. Contrary to Marx's prediction, the critics assert, most revolutions of this

Katherine S. Newman—The Withering of the American Dream?

After the economic recovery from the Great Depression, each subsequent generation of Americans expected a standard of living higher than that of their parents. A boost to this expectation was given by the unprecedented economic expansion following World War II. Upwardly spiraling affluence was assumed to be America's destiny.

Instead, according to Katherine Newman (1993), social and economic change are placing the American dream in jeopardy. The downscaling of jobs and pay that occurred during the 1980s and 1990s has replaced earlier optimism with anger, doubt, and fear. The traditional middle-class way of life is threatened by declining work compensation and a staggering national debt.

Newman analyzes this new reality among middle-class Americans as she tries to understand the effects on family life, political attitudes, and personal identities as well as the "rage, disappointment, and . . . sense of drift in communities across the land" (Newman 1993:x).

Bypassing the "bloodless" statistics, Newman talked with the Americans whose lives lie behind the graphs and dire popular media accounts. She spent two years conducting personal interviews with 150 Americans living in "Pleasanton," a representative suburban community. The population of Pleasanton corresponds to the hallmark of suburbia: a mix of skilled blue-collar workers and white-collar professionals from a variety of ethnic and religious origins. Her respondents were schoolteachers, guidance counselors, and sixty families whose children were then grown.

The promise of America had taken an unexpected wrong turn, and the residents of Pleasanton were trying to make sense of it. Newman attempted to understand the residents' view of this downward mobility. She looked for hidden meanings of culture as much as the more apparent behavior and norms. The stresses associated with changing economic conditions, she believed, would bring cultural expectations, disappointments, and conflicts close enough to the surface for a trained social scientist to see. In fact, she saw intergenerational conflict, marital discord, and racial and ethnic division. The following statement reveals a baby boomer's shattered confidence in the American dream:

> *I'll never have what my parents had. I can't even dream of that. I'm living a lifestyle that's way lower than it was when I was growing up and it's depressing. You know it's a rude awakening when you're out in the world on your own. . . . I took what was given to me and tried to use it the best way I could. Even if you are a hard worker and you never skipped a beat, you followed all the rules, did everything they told you you were supposed to do, it's still horrendous. They lied to me.*

century have come from the middle class rather than the working class. Also, the critics continue, communist revolutions did not occur in highly industrialized Western societies such as the United States, Great Britain, or Germany but in the agrarian societies of China, Russia, and Cuba. Finally, non-Marxists note the lack of a polarization of capitalist societies into ruling classes and working classes. They point to the emergence of a large middle class—composed of neither workers nor owners—in modern capitalist societies (Vago 1998).

Were these changes in the reemergence of the conflict perspective? Although the conflict perspective had little influence in sociology during the first half of the twentieth century, it has been revitalized. This reemergence was the work of sociologists, like Ralf Dahrendorf, who follow some of Marx's ideas while rejecting others (Dahrendorf 1958a, 1958b, 1959). Although Dahrendorf attributes change to the struggle over resources, the resources at stake are more than economic. Dahrendorf credits the quest for power—conflict over who controls whom—as the cause of social change. Whereas Marx envisioned two opposing social classes, Dahrendorf sees conflict between groups at all levels of society. Rather than change emerging from a single grand conflict, it comes from a multitude of competing interest groups. By rejecting Marx's contention that history is created by class conflict alone, Dahrendorf recognizes conflict between all types of interest groups—political, economic, religious, racial, ethnic, gender based. Thus, society changes as power relationships between interest groups change.

Historical events seem to support Dahrendorf's interpretation. Class conflict has not occurred in any capitalist society; social classes have not polarized into major warring factions. Rather, capitalist societies are composed of countless competing groups. In America, for example, racial groups struggle over the issue of equal economic opportunity; and environmentalists and industrialists debate the proper balance between environmental protection and economic development.

> *You don't get where you were supposed to wind up. At the end of the road it isn't there. I worked all those years and then I didn't get to candy land. The prize wasn't there, damn it.* (Newman 1993:3)

After a detailed and often personal exploration of what Newman calls the "withering American Dream," she turns to the larger social and political implications for society. She explores the transition from a society of upward mobility based on effort and merit to a society in which social classes of birth increasingly dictate future social and economic positions.

According to Newman, the soul of America is at stake. She raises these questions: Will Americans turn to exclusive self-interest, or will they care for others as well as themselves? Will suburbanites turn a blind eye to the rapidly deteriorating inner cities? Will the generational, racial, and ethnic groups turn inward, or will they attempt to bridge the divides that threaten to separate them further? A partial answer to these questions is reflected in public opinion about federal, state, and local tax revenues. If the residents of Pleasanton are any guide, Americans do not wish to invest in the common good. Public schools, colleges, universities, and inner cities, for example, are receiving a rapidly declining share of public economic support. In conclusion, Newman states:

> *This does not augur well for the soul of the country in the twenty-first century. Every great nation draws its strength from a social contract, an unspoken agreement to provide for one another, to reach across the narrow self-interests of generations, ethnic groups, races, classes, and genders toward some vision of the common good. Taxes and budgets—the mundane preoccupations of city hall—express this commitment, or lack of it, in the bluntest fashion. Through these mechanistic devices, we are forced to confront some of the most searching philosophical questions that face any country: What do we owe one another as members of a society? Can we sustain a collective sense of purpose in the face of the declining fortunes that are tearing us apart, leaving those who are able to scramble for advantage and those who are not to suffer out of sight?* (Newman 1993:221)

Thinking About the Research

1. Think about your experiences at home and in other social institutions (schools, churches). State the conception of the American dream these experiences provided. Critically analyze the ways in which society shaped your conception.
2. Newman's research was done in the early 1990s. Do you believe that she is right about the fate of the American dream? Explain.
3. If the American dream is withering, many social changes are in store. Describe the major changes you foresee.
4. Suppose Katherine Newman had decided to place her study in the context of sociological theory. Write a conclusion to her book from the theoretical perspective—functionalism, conflict theory, or symbolic interactionism—that you think is most appropriate.

These protesting South Africans see a link between world capitalism, poverty in noncapitalist societies, and insufficient money for preventing and treating AIDS. They illustrate the conflict perspective and its emphasis on competing interest groups as a source of social change.

Some sociologists have concluded that no single perspective on social change is clearly superior to others in all respects. They assert that social and cultural change is too complex to be captured by any single theory, given our present state of knowledge. They believe that because each perspective has its strengths and weaknesses, the best way to explain social change is through the combination of various aspects of each theory. These efforts have been called functionalist conflict theory (R. Collins 1994).

Reconciling the Functionalist and Conflict Perspectives

What attempts have been made to accommodate the differences in these two perspectives? There has been no fully successful synthesis of the functionalist and the conflict perspectives addressing social change, but some interesting attempts have been made. Although he does not reconcile all the differences between these two perspectives, Pierre van den Berghe (1978) points to some

TABLE 18.2

FOCUS ON THEORETICAL PERSPECTIVES: Social Change

This table gives examples of how the functionalist, conflict, and symbolic interactionist perspectives view social change. Describe how a functionalist would look at an interest group and how a conflict theorist would view equilibrium.

Theoretical Perspective	Concept	Example
Functionalism	Equilibrium	• The nature of the presidency has continuity, despite scandals in the Nixon and Clinton administrations.
Conflict theory	Interest group	• Civil rights laws were enacted in the 1960s as a result of the struggle over racial equality.
Symbolic interactionism	Urbanism	• A smaller proportion of social interaction in a large city is based on shared meanings.

commonalities. First, both perspectives observe society as systems with interrelated parts. Second, both perspectives are capable of viewing conflict as a contributor to social integration and integration as a producer of conflict. Third, both theoretical perspectives assume an evolutionary view of social change. Finally, van den Berghe sees both perspectives as equilibrium models.

Like Marx, Lewis Coser (1998) sees the pressure for social change coming from conflict: competing groups and classes rise and fall in the struggle for power and for protection of their interests. Unlike Marx, however, Coser does not believe that cataclysmic social change necessarily follows social conflict, or that change produced by conflict must necessarily come about abruptly. In fact, Coser distinguishes between change *within* a system and change *of* a system. A society flexible enough to adjust to a new environment will experience change while maintaining its basic structure. Examples of change within a society would be the civil rights and women's movements in the United States. If a society is not flexible enough, a change of the system may occur as a result of conflict. The Chinese Communist Revolution illustrates change of a society.

According to Coser, the behavior of a society's elite strongly influences the outcome of conflict, affecting whether conflict will produce a readjustment within the society or a breakdown and formation of a new society. If the dominant groups are flexible enough to allow free expression of complaints and make appropriate adjustments, change within the society is more likely to occur. Should those with the power choose to protect their interests by resisting change and stifling grievances, they run the risk of intensifying conflict and producing a change of the society.

Radical change of a society may occur suddenly, but it does not necessarily have to, according to Coser. Although basic institutions, values, and social relationships did change immediately following the Chinese Communist Revolution, change following the American War of Independence was gradual. The United States is a fundamentally different society than it was before 1776, but this restructuring cannot be linked to any specific event. In other words, for Coser, the fundamental restructuring of a society can take place as a result of gradual, cumulative changes within the society.

Nolan and Lenski (1999) also attempt to combine the functionalist and conflict perspectives. In the first place, they give equal weight to stability and change. Second, they explicitly incorporate conflict into their theory. For example, Nolan and Lenski contend that when societies have come into conflict with one another for territorial and other vital resources, those that have been technologically more advanced have usually been dominant. They make ample provision for conflict between societies, pointing out the need for societies to defend themselves and their territories from invasion by outsiders. Nor do they ignore conflict within societies, highlighting, for example, the struggle in industrializing agrarian societies between the elite in control of their society's resources and the masses who supply the labor for subsistence-level returns.

Regarding social change in particular, do these attempts at reconciliation mean that the functionalist and conflict perspectives can be abandoned? Although these attempts at reconciliation may be interesting, they do not eliminate the need for separate functionalist and conflict perspectives. Continued work by functionalists will tell us more about the processes involved in the maintenance of a dynamic equilibrium—how societies maintain stability and order while undergoing change. Conflict theorists continue to provide insights into how change occurs as a result of the struggles between those with differential power and opposing interests.

FEEDBACK

1. According to the cyclical perspective on social change, societies rise and fall rather than move continuously toward improvement. T or F?
2. Cyclical theories explain the reason social change occurs. T or F?
3. Classical evolutionary theory assumed that change leads to improvement. T or F?
4. According to modern evolutionary theory, is social change unilinear or multilinear? __________
5. According to the __________ perspective, society is inherently stable, and any change that occurs is eventually assimilated so that a new state of equilibrium is achieved.
6. __________ refers to any situation in which disequilibrium is caused by one aspect of a society failing to change at the same rate as an interrelated aspect.
7. According to the __________ perspective, social and cultural change occur as a result of the struggles between groups representing different segments of a society.

Answers: 1. T 2. F 3. T 4. multilinear 5. functionalist 6. Cultural lag 7. conflict

Modernization

The Nature of Modernization

It seems as if the entire world is either already modernized or attempting to move in that direction. Modernization refers to the updated social changes—there are a host of them—that accompany economic development. The trend toward modernization is one of the most significant trends in human history (Haferkamp and Smelser 1992; Chase-Dunn 1999; Fukuyama 1999a, 1999b; Giddens 2000).

What are some of the major consequences of modernization? Some significant demographic changes are associated with modernization. Population growth occurs as the death rate declines and life expectancy increases. Between the thirteenth and seventeenth centuries, for example, Europe had a life expectancy as low as twenty years; by 1930, life expectancy had increased to over sixty years and now stands at seventy-four years (*World Population Data Sheet* 2002). The population moves from rural to urban areas. Whereas most members of traditional societies work the land, most members of modern societies live and work in towns and cities, where industry and jobs are located.

Stratification structures are altered by modernization. Traditional society is characterized by a bipolar stratification structure—the wealthy at one end, the poor masses at the other. Modernization brings an expansion of the middle and upper classes. As more emphasis is

SOCIOLOGY AND THE NEWS MEDIA

Modernization and the American Family

This CNN report links modernization with the nature of contemporary family life. Modern transportation permits high rates of mobility: the average American moves eleven times during a lifetime. One possible result is the replacement of the extended family with the nuclear family.

This family is enjoying a large suburban home made possible by modernization. Social costs accompany such affluence.

1. Discuss in more detail this alleged connection between modernization and family life. Do you agree or disagree with the thesis of this CNN report? Explain.

   ____

   ____

2. Considering the nature of the contemporary American family, argue for or against the idea of convergence in modernized societies.

   ____

   ____

placed on personal effort and achievement, social mobility increases and inequality generally declines.

The political institution is altered by modernization. The role of the state expands as it becomes more centralized and more involved in social and economic affairs. At the same time, modernization promotes political democracy, even though the extent of democratization varies considerably from society to society. Although political power is not equalized as a result of modernization—there is always a powerful elite—power is more widely dispersed.

Modernization transfers education from the family to a formal system of education. Education is no longer designed just for the privileged few but for the entire society; primary education, in fact, is intended for all members of a modernizing society. A literate population is absolutely necessary to create a workforce suitable for an industrialized economy.

Modernization also affects family life. The nuclear family replaces the extended family. Because the economy ceases to be based on a familial division of labor, people must move to cities to work. Extended family ties can be obliterated; they simply become much harder to maintain, and much of their intimacy can be lost. With modernization, functions formerly of the family's domain are adopted by other institutions. For example, government assumes more responsibility for the elderly, schools take care of children's educational needs, and the mass media provide entertainment.

How comparable are the social and cultural characteristics of modernizing societies? This question is debatable (Waters 2001). Advocates of the pattern of convergence foresee the development of social and cultural similarity among modernizing nations. They contend that as societies adopt the newest technological arrangements, they tend to develop social and cultural commonalities. This is said to occur because a particular technology carries in its wake a particular occupational structure, which in turn exercises similar wide-ranging effects on various social structures and social relationships. The necessity for an industrializing society to develop an educated populace is another major unifying force. Other pressures for uniformity include the enlarged role of government, the influence of multinational corporations within modernizing countries, and the accelerated growth of large-scale organizations (Inkeles 1998).

Movement toward globalization, for example, was reflected in the television coverage of millennium celebrations. Never before had the world shared in witnessing New Year's celebrations from Australia west to Hawaii. This international exposure—fueled by the almost universal adoption of computer technology and broadcasted by commonly shared telecommunications technology—may foretell increased cross-cultural influence in the twenty-first century.

Supporters of the pattern of divergence, on the other hand, do not see social and cultural homogeneity as an inevitable result of modernization. Other sociologists emphasize the effects of idiosyncratic social and cultural forces to explain the emergence of new social patterns in modernizing societies. Advocates of divergence offer as evidence less developed countries that are now undergoing modernization along lines somewhat different from the path taken by Western societies. Many less developed countries who chose the Soviet economic model from the 1950s onward are, along with former Communist countries in Eastern Europe, adopting the model of market economies (*World Development Report* 1991; Schnitzer 2000). Could this signal greater convergence in the future?

As in most complex debates, there is a middle ground that is probably closer to the truth. Social and cultural elements of traditional societies do not necessarily vanish in the process of modernization; traditional elements may blend with the new (Mazaar 1999; Guillen 2001). Changes in the family during modernization, for example, depend in part on the type of traditional family a society has. Some traditional family structures are more compatible with modernization than others are. To cite another example, democratization usually accompanies modernization, but its extent and form may vary due to the specific social and cultural traditions of a society. Although modern Japan is more democratic than it was in premodern times, it is less democratic than most Western societies. And although the Soviet Communist Party leaders voted overwhelmingly in July 1991 to endorse the principle of private property, freedom of religion, and a pluralistic political system, serious political opposition has since arisen. It remains to be seen how such legislative changes will be reflected in democratic forms.

World-System Theory

To this point, we have examined the process of modernization and its consequences *within* countries. When we consider modernization from an international perspective, quite different theories emerge. One of these is *world-system theory*, which also depicts social and cultural diversity among modernizing nations. Its approach is unique (Chase-Dunn and Hall 1997; Chase-Dunn 1999; Kardulias 1999; Guilllen 2001).

What is world-system theory? According to **world-system theory**, the pattern of a nation's development hinges on its place in the world economy. The world economy is divided into segments: core and peripheral nations. Core nations, such as the United States, dominate the world economy by providing managerial expertise and technological innovation. This domination leads to the control and exploitation of the rest of the world; to benefit themselves, core nations use the cheap labor and natural resources of peripheral nations (less developed nations) (Wallerstein 1979, 1989, 2001).

What are the consequences of this arrangement? As a result of this world-system arrangement, the standard of living is much higher in core nations, whose skilled workers operate in a free labor market. Workers in peripheral nations provide unskilled, coerced labor and suffer a low standard of living. Peripheral nations do not develop as much as they could if they were not dominated and exploited by the core nations. In fact, world-system advocates contend, this deterrent to peripheral national development is the prime perpetuating impetus for the existing arrangement. Consequently, the economic, social, and cultural gap between core and peripheral nations is increasing (Rossides 1997; Wallerstein 2000; Robbins, 2001; Stiglitz 2002).

According to many advocates of globalization, free markets promote democracy, prosperity, and peace (Mandelbaum 2002). Free markets are said to promote enfranchisement of the masses, increase economic equality, and reduce conflict within and between societies. Looking at Russia and developing countries in Latin America, Africa, and Asia, Amy Chua (2002) questions this happy mix assumed by globalists. Conceding that free markets promote democracy, the results, Chua argues, do not include increased economic equality and reduced conflict. Instead, the introduction of the free market in a developing country creates a new class of extremely rich elites who tend to belong to the same minority group. This "market-dominant minority" is subsequently subjected to ethnic hatred and violence from the recently enfranchised majority. The majority government uses its newly acquired political freedom and power to wreak revenge on the politically weaker minority elites and confiscate or destroy their property. In Indonesia, to cite one of Chua's cases, the suppressed Muslim majority unleashed a vicious attack against the economically dominant ethnic Chinese, looting their businesses, breaking their store windows, and raping their women. Because Chua is going against the more conventional view, she is certain to be challenged by globalists.

Collective Behavior

Another area of sociological study—collective behavior—often focuses on short-term behavior. Some forms of collective behavior are thought to be inconsequential; other forms have far-reaching consequences. We will examine both.

Collective behavior is the relatively spontaneous and unstructured social behavior of people who are responding to similar stimuli. It is *collective* because it usually occurs among a large number of people. The phrase "responding to similar stimuli" emphasizes collective behavior as a reaction by people to some person or event outside themselves, such as the attacks on the World Trade Center on September 11, 2001. Collective behavior involves social interaction in which loosely connected participants influence one another's behavior.

The study of collective behavior poses some problems. In the first place, sociologists are accustomed to studying structured behavior. Second, how are researchers going to investigate a social phenomenon that occurs spontaneously? Despite these difficulties, sociologists have conducted fascinating research and formulated serviceable theories of collective behavior. It turns out that there is more structure and rationality to collective behavior than appears on the surface.

Initiating our discussion are the more disorganized, unplanned, and short-term forms of collective behavior—rumors, mass hysteria, panics, fads, and fashions. Crowds are covered next. Social movements, the most highly structured, enduring, and rational form of collective behavior, closes the chapter.

Dispersed Collectivities

In the more structured forms of collective behavior, such as crowds and social movements, people are in

FEEDBACK

1. __________ is a process involving all those social and cultural changes accompanying economic development.
2. Which of the following is *not* one of the major consequences of modernization discussed in the text?
   a. the social and cultural convergence of modernizing countries
   b. increased urbanization
   c. greater equality
   d. more political democracy
   e. widespread development of the nuclear family
3. According to __________ theory, the pattern of a nation's development largely depends on the nation's location in the world economy.
4. World-system theory would predict a (convergence/divergence) between core and peripheral nations.

Answers: 1. Modernization 2. a 3. world-system 4. divergence

physical contact. Other forms of collective behavior occur among dispersed members of a mass society. These **dispersed collectivities** engage in the less structured forms of collective behavior—rumors, mass hysteria, panics, fads, and fashions. Behavior among members of dispersed collectivities is not highly individualized. Members of dispersed collectivities are not physically connected but uniformly respond to some common object of attention; they are aware of being a member of a collectivity:

> *When people are scattered about, they can communicate with one another in small clusters of people; all of the members of a public need not hear or see what every other member is saying or doing. And they can communicate in a variety of ways—by telephone, letter, Fax machine, computer linkup, as well as through second-, or third-, or fourth-hand talk in a gossip or rumor network.* (Goode 1992:255)

Rumors

In the *Aeneid*, Virgil wrote these lines:

> *Rumor! What evil can surpass her speed? In movement she grows mighty, and achieves strength and dominion as she swifter flies.*

Rumor, as Virgil's words underscore, has a negative connotation. Rumors may be benign, as in the case of continual Elvis Presley sightings, or they can do considerable damage. They are often communicated as the truth when in fact they may be false. At best, rumors are usually inaccurate and misleading. In any event, the likelihood of an individual spreading a rumor depends in part on the degree of anxiety a person feels, the extent of uncertainty about events that the person is experiencing, the credibility of the person passing the rumor, and the relevance of the rumor to that particular person (Rosnow 1988, 1991).

Why is rumor considered an unstable form of collective behavior? A **rumor** is a widely circulating story of questionable truth. Rumors usually focus on people or events that are of great interest to others. Segments of the mass media exploit the public's fascination with rumors: entertainment magazines devoted exclusively to rock idols and movie stars; tabloid newspapers loaded with titillating guesswork, half-truths, and innuendos; even mainstream news publications catering to accounts of the rich, famous, and offbeat. As these examples suggest, rumors and gossip are closely related.

Three days after a nuclear accident at Three Mile Island near Harrisburg, Pennsylvania, a rumor flashed throughout the surrounding area: the plant would explode, devastate everything nearby, and spread radioactive fallout for miles. According to another rumor, a fast-food restaurant chain was increasing the protein content of their hamburgers with ground worms. Then there was warning about the combination of a soft drink and a popular candy; the combination would supposedly cause the stomach to explode. A rumor circulated after Saddam Hussein invaded Kuwait in the summer of 1990: Iraqi troops had crossed the Kuwaiti border into Saudi Arabia. The rumor drove up the price of the dollar on the world foreign exchange markets for several hours (Frankel 1990). Rumors swirled around Michael Dukakis, then governor of Massachusetts, during his 1988 presidential campaign against George H. W. Bush. One widely dispersed rumor had Dukakis twice receiving psychiatric care for depression. Within a matter of days, Dukakis dropped 8 points in the national public opinion polls (H. Johnson 1991). The crash of TWA Flight 800 was attributed, by some, to the accidental firing of a surface-to-air missile from an American warship (Hosenball 1996). None of these rumors proved true; but they were spread and believed, in part because they touched on people's insecurities, uncertainties, and anxieties (Rosnow 1991; Fine and Turner 2001). And did rumors about Y2K prove true? Did power grids fail, elevators stop working, and the stock market crash as the year 2000 began (Hillis 1999)? They did not.

The effects of rumors are not always frivolous. The potential damage caused by rumors is revealed in the following statement from the Report of the National Advisory Commission on Civil Disorders. This report focused on the urban riots and violence in large American cities during the summer of 1967:

> *Rumors significantly aggravated tension and disorder in more than 65 percent of the disorders studied by the Commission. Sometimes, as in Tampa and New Haven, rumor served as the spark which turned an incident into a civil disorder. Elsewhere, notably Detroit and Newark, even where they were not precipitating or motivating factors, inflaming rumors made the job of police and community leaders far more difficult.* (Report of the National Advisory Commission on Civil Disorders 1968:326)

Potentially harmful rumors persisted into the winter of 1967–1968 following the "long, hot summer" of 1967. One of these rumors was the Detroit castration rumor:

> *A mother and her young son are shopping at a large department store. At one point the boy goes to the lavatory. He is a long time returning, and the mother asks the floor supervisor to get him. The man discovers the boy lying unconscious on the floor. He has been castrated. Nearby salesclerks recall that several teenage boys were seen entering the lavatory just after the young boy and leaving shortly before he was discovered.* (Rosenthal 1971:36)

Smart Mobs

Have you ever joined a "smart mob?" A "smart mob" is an acting collectivity organized through the mobile Internet. Smart mob is the label Howard Rheingold (2003) gives to the new technology created by the combination of mobile communications (including your cell phone), computing, and the Internet.

Rheingold, one of the world's recognized authorities on the social implications of technology, wrote earlier books foreseeing the PC revolution and the Internet era. Over the next decade, he predicts, mobile communications and pervasive computing technologies will "change the way people meet, mate, work, fight, buy, sell, govern, and create" (Rheingold 2003:xiii). These new technologies, he writes, permit people to join together in collective action not possible before.

No Pollyanna, Rheingold envisions both positive and negative social consequences of the new technologies. Negative repercussions include the potential loss of privacy in a surveillance state, the increased capacity for terrorists to operate undercover, and the creation of new opportunities for criminal behavior. On the positive side, smart mobs can overthrow dictators and organize mass protests.

Some foreshocks of the coming revolutionary application of the new technologies are already visible. Examples include the following.

The first smart mobs Rheingold noticed were teenagers in Tokyo and Helsinki who used text messages on cell phones to stalk celebrity targets. A rock musician might emerge from a subway stop to be greeted by crowds of teenage fans tipped off to his or her whereabouts.

On November 30, 1999, internetworked groups of demonstrators protested the meeting of the World Trade Organization in Seattle. Those who participated represented a wide range of interest groups who opposed the policies of the World Trade Organization. The Direct Action Network permitted autonomous groups to select from a menu of actions—provoking mass arrests, offering nonviolent opposition, and performing acts of civil disobedience.

In September 2000, thousands of British citizens used various means to protest a sudden spike in gasoline prices. Using mobile phones, e-mail from laptop PCs, and CB radios in taxicabs, demonstrators were organized to block fuel delivery at certain service stations.

On January 20, 2001, President Joseph Estrada of the Philippines became the first head of state to be overthrown by a smart mob utilizing text messaging. Tens of thousands of Filipinos converged in a mass demonstration on Edsa Avenue within an hour of this text message: "Go 2/EDSA, wear blck." Four days later, more than a million citizens came to Edsa, mostly wearing black. Estrada was deposed without a single shot fired.

In 2003, *Moveon.org*, an antiwar group with four paid staff members and no office, can communicate instantly with 750,000 potential protestors. Using only e-mail and instant messaging, this group joined the millions demonstrating, around the world, against the coming war in Iraq (Taylor 2003).

In 2003, 400,000 antiwar activists protested without hitting the streets. They jammed White House and congressional switchboards with phone calls, faxes, and e-mails. This was said to be the first national virtual demonstration (Taylor 2003).

In 2003, a Brazilian drug lord coordinated riots, bombs, and the burning of vehicles using a smuggled cell phone in his prison cell (Taylor 2003).

Analyzing the Trends

In the past sociologists have not considered crowds and social movements as dispersed collectivities. Do you think that smart mobs are dispersed collectivities? Discuss why and why not.

In the white version, the castrated youth was white and his assailants black; in the black version, the colors were reversed. In a city as volatile as Detroit at that time, this rumor may have seemed quite believable and could have led to further rioting and violence.

How is an urban legend dissimilar from a rumor? Akin to rumors are what Jan Harold Brunvand calls **urban legends** (Brunvand 1981, 1998, 2000). Although urban legends may incorporate current rumors, they tend to have a longer life and wider acceptance. They are moralistic tales passed along by people who swear the stories happened to someone they know, or to an acquaintance of a friend or family member (Heath, Bell, and Sternberg 2001). The tales often focus on current places, concerns, and fears such as AIDS and inner-city gangs. Consider the case of "The Baby in the Oven," in which a babysitter gets high on drugs while babysitting and does a tragic thing:

> *This couple with a teenage son and a little baby left the baby with this hippie-type girl who was a friend of the son's. They went to a dinner party or something, and the mother called in the middle of the evening to see if everything was all right. "Sure," the girl says. "Everything's fine. I just stuffed the turkey and put it in the oven." Well, the lady couldn't remember having a turkey, so she figured something was wrong. She and her husband went home*

> *and they found the girl had stuffed the baby and put it in the oven. Now the son used a lot of drugs; she was his friend, so I guess they figure she took them, too.* (Brunvand 1980:55)

Another typical story tells about a man who wakes up in a hotel room missing a kidney. Another describes alligators roaming the sewer systems of big cities. As cautionary tales, urban legends warn us against engaging in risky behaviors by pointing out what has supposedly happened to others who did what we might be tempted to try. Like rumors, urban legends permit us to play out some of our hidden fears and guilt feelings by being shocked and horrified at others' misfortune. "The Baby in the Oven" story, for example, may represent guilt feelings for parents who sometimes leave their children with strangers or harbor a deep fear of outsiders entering their homes.

Mass Hysteria and Panics

What is mass hysteria? Mass hysteria is a collective anxiety created by the acceptance of one or more false beliefs. Orson Welles's famous "Men from Mars" radio broadcast in 1938, though based entirely on H. G. Welles's novel *The War of the Worlds*, caused nationwide hysteria. About 1 million listeners became frightened or disturbed and thousands of Americans hit the road to avoid the invading Martians. Telephone lines were jammed as people shared rumors, anxieties, fears, and escape plans (Houseman 1948; Cantril 1982; J. Barron 1988).

Another historical example of mass hysteria is found in the responses to imagined witches in seventeenth-century Salem, Massachusetts. In Salem, twenty-two people who had been labeled witches died—twenty by execution—before the false testimony of several young girls began to be questioned. The mass hysteria dissipated only after the false beliefs were discredited.

A third example followed the death from AIDS of actor Rock Hudson in the mid-1980s. A 1987 Gallup poll showed that a substantial proportion of Americans held false beliefs regarding the spread of AIDS: 30 percent believed that insect bites could spread the disease; 26 percent related the spread to food handling or preparation; 26 percent thought AIDS could be transmitted via drinking glasses; 25 percent saw a risk in being coughed or sneezed on; and 18 percent believed that AIDS could be contracted from toilet seats (Gallup 1988). These mistaken ideas persisted on a widespread basis despite the medical community's conclusion that AIDS is spread through sexual contact, by sharing hypodermic needles, and by transfusion of infected blood. By the late 1990s, knowledge, toleration, compassion, and understanding of AIDS had increased substantially enough that the frequency of these rumors had dissipated.

What is the difference between mass hysteria and a panic? A **panic** occurs when people react to a real threat in fearful, anxious, and often self-damaging ways. Panics usually occur in response to such unexpected events as fires, invasions, and ship sinkings. For example, 117 people died in two night clubs in Chicago and Rhode Island in 2003, when panic reactions to a fire caused a jamming of escape routes (D. Johnson 2003).

Interestingly enough, people often do not continue to panic after the initial chaos triggered by natural disasters such as earthquakes and floods. Although panic may occur at the outset, major natural catastrophes usually involve highly structured behavior (K. T. Erikson 1976, 1995; Dynes and Tierney 1994; Quarantelli 2001). An example of both initial panic and subsequent structured behavior occurred during the terrorist attacks of September 11, 2001. Some occupants of the buildings panicked, jumping from windows, whereas firefighters exhibited highly structured behaviors during their rescue efforts. Agreeing that rational behavior usually accompanies a catastrophe, Lee Clark writes:

> *When the World Trade Center started to burn, the standards of civility that people carried with them everyday did not suddenly dissipate. The rules of behavior in extreme situations are not much different from rules of ordinary life. . . . When danger arises, the rule—as in normal situations—is for people to help those next to them before they help themselves.* (Clark 2002:24)

Panics and episodes of mass hysteria are based on repulsion. Other forms of collective behavior, such as fads, crazes, and fashions, are rooted in attraction.

Fads, Crazes, and Fashions

Do fads have staying power? No. **Fads** are unusual behavior patterns that are spread rapidly, embraced zealously by a particular segment of society, and then disappear after a short time. The widespread popularity of a fad rests largely on its novelty. The "streaking" fad (running naked across college grounds or through occupied classrooms) delighted students in the early 1970s (Aguirre, Quarantelli, and Mendoza 1988). A reminder of that fad occurred in the summer of 1991, when a sixteen-year-old female, wearing only her eye shadow, ran down the first fairway at the British Open golf tournament in Southport, England. Ironically, on that same day, two male fans celebrated the Atlanta Braves' five-run sixth inning performance by baring it all on the field. One of these streakers impressed spectators and players alike with a head-first slide into home plate. Some fads in more recent years include body tattoos and piercing, text messaging, camera phones, and Ugg boots. Wearing pajamas as outerwear may be a fad in the making (C. J. Farley 1999; J. Gleick 1999; Edwards 2000; Croal 2002; Krupp 2003). By now, of course, you

A fad is a form of collective behavior that is unusual, spreads quickly, is zealously adopted, and is short-lived. Is the Harry Potter rage a fad?

will be able to identify the latest fads. Fads are not limited to hobbies, clothing, and entertainment; they also emerge regarding serious matters. Ron Insana (2002) has documented past investment fads, including the bull market of the 1990s.

What is the difference between a fad and a craze? A **craze** is a type of fad that can have serious consequences for its adopters. The fads previously mentioned are primarily recreational and harmless in nature. But consider the potential downside to fads such as losing weight through extreme diets, getting "high" from cold medication, using the date rape drug called "roofies," and injecting Botox (a toxin causing botulism) to diminish facial wrinkles.

Do fashions reflect a society's culture? Yes. **Fashions** are behavior patterns that evolve over time and are widely approved but expected to change periodically (Crane 2001). As societies modernize, fashion becomes more salient and may change more rapidly. Within contemporary American society, the "in" fashions for clothing are introduced seasonally. Women are "on top" of the defined dress or skirt length, and men are questioning if pant cuffs are fashionable or passé. High school students wishing to be fashionable wear the labels of Tommy Hilfiger, the Gap, and Nike.

Although the most widely recognized examples of fashion are related to appearance (clothing, jewelry, hairstyles), fashions also come and go in such diverse arenas as automobile design, home decorating, architecture, and politics. Slang, another example, goes in and out of favor (J. Lofland 1993). What self-respecting teenager today would be caught calling something "neat"? Fleeting though it may be, awesome is the current word describing something impressive. *Cool, the cat's pajamas, groovy, tubular, tough, fine, rad, bad, phat,* and *sick* were all slang terms of approval among young people at some point in time.

Why is fashion particularly prevalent in modern society? In the first place, modern societies are based on growing economies fueled by mass consumption. Styles in clothing, automobiles, eyewear, and sporting equipment must change if profits are to be made and people employed. Second, without traditions to supply the brakes, people in modern societies are eager to respond to the tempting new fashions that are constantly being created by entrepreneurs and established corporations. Finally, the relative affluence in modern societies provides enough disposable income for people to indulge their desire for fashion novelty and change.

Crowds

Distinguishing Characteristics of Crowds

The most dramatic form of collective behavior is the crowd, because it involves intense emotions. Look at the content of local and national television news. If this is any indication, people are alternately fascinated and horrified by passions that are unleashed in crowds. We are engrossed by reports of crowds, mobs, and rioters, whether they are celebrating a joyous event or lamenting a sad one.

How different is a crowd from an aggregate of people? A **crowd** is a temporary collection of people who share an immediate common interest. The temporary resi-

FEEDBACK

1. A __________ is a widely circulating story whose truth is questionable.
2. __________ exists when collective anxiety is created by acceptance of one or more false beliefs.
3. A __________ occurs when people react to a genuine threat in fearful, anxious, and often self-damaging ways.
4. Unusual patterns of behavior that spread rapidly, are embraced zealously, and disappear after a short time are called __________.
5. __________ are patterns of behavior that are widely approved but expected to change periodically.

Answers: 1. rumor 2. Mass hysteria 3. panic 4. fads 5. Fashions

© William H. Edwards/Getty Images

A "conventional" crowd gathers for a particular purpose and follows some predetermined norms defining appropriate behavior. It is expected that the Armenian and American fans at this soccer match would prominently display their respective national flags.

dents of a large campground, each occupied with his or her own activities, constitute an *aggregate*. But if some event, such as the landing of a hot-air balloon or the appearance of a bear, serves as a common stimulus to draw the campers together, the aggregate becomes a crowd. What will happen next is highly unpredictable. The specific subject of mutual interest—the common stimulus—is not important. What is significant is the collection of individuals into a situation that has the potential to stimulate an emotional reaction.

A crowd situation involves ambiguity and uncertainty; participants have no predefined ideas about the way they should behave toward one another or toward some target on which their attention is converging. Members of a crowd, however, are certain about one thing: they share the urgent feeling that something either is about to happen or should be made to happen (McPhail 1991; Marx and McAdam 1994).

How do crowds differ from one another? Although it is accurate to make generalizations about crowds, not all crowds are alike. Herbert Blumer (1969a) has distinguished four basic types of crowds. A *casual crowd* is the least organized, least emotional, and most temporary type of crowd. Although members of a casual crowd share some point of interest, it is minor and fades quickly. Members of a casual crowd may gather with others to observe the aftermath of an accident, to watch someone threatening to jump from a building, or to listen to a street-rap group.

A *conventional crowd* has a specific purpose and follows accepted guidelines for appropriate behavior. People watching a film, flying on a chartered flight to a university ballgame, or observing a tennis match are in conventional crowds. As in casual crowds, there is little interaction among members of conventional crowds. The fact that the activity of the conventional crowd follows some established procedures distinguishes this type of crowd from a casual crowd.

Expressive crowds have no significant or long-term purpose beyond unleashing emotion. Their members are collectively caught up in a dominating, all-encompassing mood of the moment. Free expression of emotion—yelling, crying, laughing, jumping—is the defining characteristic of this type of crowd. Hysterical fans at a rock concert, the multitude gathered at Times Square on New Year's Eve, and some 250,000 Americans at the Woodstock 1999 music festival are all examples of expressive crowds.

Finally, a crowd that takes some action toward a target is an *acting crowd*. This type of crowd concentrates intensely on some objective and engages in aggressive behavior to achieve it. The some 5 million demonstrators who congregated across the United States and the world in protest to the 2003 impending war on Iraq were acting crowds (Campo-Flores 2003; Tyler 2003). A conventional crowd may become an acting crowd, as when a crowd of European soccer fans attacked the officials. Similarly, an expressive crowd may become an acting one, as in the case of celebrating sports fans who destroy property following their team's victory. In June 2000, a peaceful celebration of the Los Angeles Lakers' NBA championship deteriorated as hundreds of fans destroyed two police vehicles and two television vans by fire, looted businesses, and damaged more than seventy cars. One person was killed and at least twelve others were injured (Gildea 2000). Another example occurred in New York City in the summer of 2000 when a number of males at the annual National Puerto Rican Day Parade turned from participating in normal parade festivities to sexually attacking several females in Central Park (Campo-Flores and Rosenberg 2000).

What is unique about mobs? A **mob** is an emotionally stimulated, disorderly crowd that is ready to use destructiveness and violence to achieve a specific purpose. Through group understanding, a mob knows what it wants to do and considers all other things as distractions. In fact, individuals who are tempted to deviate from the mob's purpose are pressured to conform. If a mob is storming a seat of government, for example, it is inappropriate for participants to waste their time raping or stealing, although that may come later. Concentration on the main event is maintained by strong leadership.

Mobs have a long and violent history. The French Revolution is a classic example of mob action. With the cry "To the Bastille! To the Bastille!" a Parisian mob stormed this symbol of oppression on July 14, 1789:

> *The population of Paris was aroused, the unruly element of the city was in the streets, their wrath directed against the prison-fortress, the bulwark of feudalism, the stronghold of oppression, the infamous keeper of the dark secrets of the kings of France. The people had always feared, always hated it, and now against its sullen walls was directed the torrent of their wrath.* (C. Morris 1893:269)

The overthrow of royalty and the desire for political freedom were the larger objectives of the French Revolution, but this mob's more specific and limited goal was to destroy the symbol of tyranny, oppression, and fear. As in all revolutions, there were other mob actions (Charles Tilly 1992). In October 1789, a largely female mob forcibly entered the royal palace at Versailles, and peasant mobs throughout France overthrew their feudal masters. Ultimately, King Louis XVI and Queen Marie Antoinette were escorted to Paris for execution by yet another mob.

The formation of mobs is not limited to revolutions. During the mid-eighteenth century, American colonists mobbed tax collectors as well as other political officials appointed by the British. During the Civil War, more than a thousand people were killed or injured as armed mobs protested against the Union Army's draft. Beginning in the late nineteenth century, some mobs in the American South acted as judges, juries, and executioners in the lynching of African Americans (as well as some whites). Approximately five thousand lynchings are estimated to have occurred between 1882 and 1968; no such illegal hangings have been recorded since 1968. Early labor-management relations in the United States were also marked with mob violence. And the 1960s saw mass violence in urban ghettos and on college campuses. In 1989, mob violence occurred in response to Iran's Ayatollah Khomeini's death threat to Salman Rushdie, author of *The Satanic Verses*. As a result, at least thirty-five people, including fifteen policemen, were injured in rampaging mob action in New Delhi, India. Later six anti-Rushdie demonstrators were killed by police in Pakistan, and ten protestors were killed in clashes with police in Bombay, India. A huge rally opposing Serbian President Slobodan Milosevic's continuation in office, after his 2000 election loss, turned into a popular uprising that forced his resignation (Hammer and Cirjakovic 2000).

How do riots differ from mobs? Some acting crowds, although engaging in deliberate destructiveness and violence, do not display a mob's sense of common purpose. These episodes of crowd destructiveness and violence are **riots**. Riots involve a much wider range of activities than mob action. Whereas a mob surges to burn a particular building, to lynch an individual, or to throw bombs at a government official's car, riots may include actions of several different crowds, with rioters often directing their violence and destructiveness at targets simply because they are convenient. People who participate in riots typically lack power and engage in destructive behavior as a way to express their frustrations. A riot, usually triggered by a single event, is best understood within the context of long-standing tensions. The 1967 summer ghetto riots tore through many large American cities, occurring against a background of massive unemployment, uncaring slum landlords, poverty, discrimination, and charges of police brutality. The rioters in the twenty-odd cities involved did not, however, lash out at the underlying cause of the riots—the enduring gap between black and white America (*Report of the National Advisory Commission on Civil Disorders* 1968). Although the rioters were protesting against discrimination and deprivation and although white-owned businesses were damaged more frequently than those owned by African Americans, the fact remains that African Americans did more damage to themselves (the overwhelming majority of the dead were African American) and their neighborhoods than they did to the establishment. As further evidence of random behavior in riots, many participants saw the riots as a way to loot stores or to have some destructive, violent fun.

Riots, of course, did not end with the 1960s. A vicious race riot occurred in Miami in 1980 (Porter and Dunn 1984). Several rioting incidents occurred in Europe in the 1980s and 1990s (Moody 1985; Bierman 1990). For example, in West Germany, thousands of youths rampaged in sixteen cities; five days of violence (smashed windows, looted shops, burned cars) resulted in dozens of injured people and about 500 persons detained by police. In 1990, thousands of angry citizens stormed the secret police headquarters in East Berlin. Although no one was killed or injured, the protest aroused widespread fear that the country was about to drop into anarchy (Bierman 1990). Violence also occurred in south London in 1985 in an area that had been hard hit by race riots in 1981. After two nights of arson and looting, 74 people were injured and over 200 arrested. Later in that same week, rioting broke out in Liverpool, England, an area that had also seen racial disturbances in 1981. In 1992, Los Angeles experienced America's deadliest riots in twenty-five years. Two days of rioting followed the acquittals of the police officers who had been charged in the brutal beating of Rodney King. The aftermath found the City of Angels with at least 53 dead, 2,300 injured, over 16,000 arrested, and an estimated $800 million in damages from looting and burning (Duke and Escobar 1992; Mathews 1992). Muslim riots in 2002 surrounding the Miss World contest resulted in the deaths of more than 100 Nigerians, and radical South Korean students joined anti-U.S. protests by firebombing American military bases (Masland 2002; Wehrfritz 2003).

Theories of Crowd Behavior

Theories have been developed to explain crowd behavior. The three most important are contagion theory, emergent norm theory, and convergence theory.

What is the focus for contagion theory? **Contagion theory** emphasizes the irrationality of crowds that is created by participants stimulating one another to higher and higher levels of emotional intensity. As emotional intensity in the crowd increases, people temporarily lose their individuality to the "will" of the crowd. This makes it possible for a charismatic or manipulative leader to direct behavior, at least initially.

This theory has its roots in the classic work of Gustave Le Bon (1960; originally published in 1895), a French aristocrat who disdained crowds composed of the masses. People in crowds, Le Bon thought, were reduced to a nearly subhuman level:

> *By the mere fact that he forms part of an organized crowd, a man descends several rungs in the ladder of civilization. Isolated, he may be a cultivated individual; in a crowd, he is a barbarian—that is, a creature acting by instinct. He possesses the spontaneity, the violence, the ferocity, and also the enthusiasm and heroism of primitive beings.* (Le Bon 1960:32)

Herbert Blumer (1969a) offers another version of contagion theory, but he avoids Le Bon's elitist bias. Blumer's theory is more refined but still implies that crowds are irrational and out of control. For Blumer, the basic process in crowds is a "circular reaction"—people mutually stimulating one another. There are three stages to this process. In *milling*, the first stage, people move around in an aimless and random fashion, much like excited herds of cattle or sheep. Through milling, people become increasingly aware of and sensitive to one another.

The second stage, *collective excitement*, is a more intense form of milling. At this stage, crowd members become impulsive and highly responsive to the actions and suggestions of others. Individuals begin to lose their personal identities and take on the identity of the crowd.

The last stage, *social contagion*, is an extension of the other stages. Excitement begins to spread. Behavior in this stage lacks caution and judgment and is a nonrational transmission of mood, impulse, or behavior. For example, fans at soccer games in Europe have launched attacks on referees to such proportion that games have been interrupted and people killed or injured. Taking a less extreme case, people who observe an auction can find themselves buying white elephants because they have become immersed in the contagious excitement of bidding.

Blumer's theory is more refined than Le Bon's, but it still implies that people in crowds are irrational and out of control. Sociologists today, though, realize that much of crowd behavior, even within an acting crowd such as a mob, is actually very rational (McPhail 1991). Emergent norm theory illustrates the structure and rationality of crowd behavior.

Where is the emphasis for emergent norm theory? **Emergent norm theory** stresses the similarity between daily social behavior and crowd behavior. In both situations, norms guide behavior (R. H. Turner 1964; Turner and Killian 1987). So, even within crowds, rules develop. These rules, of course, are *emergent* norms, because the crowd participants are not aware of the rules until they find themselves in a particular situation. These norms develop on the spot as crowd participants pick up cues for expected behavior. Emergent norm theory contends, in short, that crowd behavior is no different from noncrowd behavior, except that crowds do not have ready-made norms.

Also, whereas contagion theory proposes a collective mind that motivates crowd members to action, emergent norm theory views people in a crowd as present for a variety of reasons; they do not all behave in the same way (McPhail and Wohlstein 1983; Zucher and Snow 1990). Conformity may be active (some people in a riot may take home as many watches and rings as they can carry) or passive (others may simply not interfere with the looters, although they take nothing for themselves). In Nazi Germany, for instance, some people destroyed the stores of Jewish merchants while others watched silently, either in support or afraid to disagree for fear that others would ridicule or hurt them.

What is convergence theory? Both the contagion and emergent norm theories of crowd behavior assume that individuals are merely responding to those around them. It may be a more emotional response (as in contagion theory) or a more rational response (as in emergent

FEEDBACK

1. A _________ is a temporary group of people who are reacting to the same event or individual.
2. An _________ crowd has no purpose or direction beyond the unleashing of emotions.
3. Mob and riot are simply two terms for the same type of crowd. T or F?
4. Much crowd behavior is structured and rational. T or F?
5. Some individuals at a lynching do not participate or give verbal support, but do not attempt to stop it. Which of the following theories of crowd behavior best explains this response?
   a. contagion theory
   b. crowd decision theory
   c. emergent norm theory
   d. casual crowd theory

Answers: 1. crowd 2. expressive 3. F 4. T 5. c

norm theory). In other words, the independent variable in crowd behavior is the crowd itself. In contrast, in **convergence theory**, crowds are formed by people who deliberately congregate with others whom they know to be like-minded. According to convergence theory, the independent variable in crowd behavior is the desire of people with a common interest to come together.

There have been many instances of crowds gathering in front of clinics to discourage abortions. This behavior, say convergence theorists, does not simply occur because people happened to be at the same place and are influenced by others. Such a crowd is motivated to form because of shared values, beliefs, and attitudes (Berk 1974).

Contemporary sociologists view crowd behavior as a more structured and rational phenomenon than is apparent on the surface. Sociologists also agree that a social movement—discussed in the next section—is an even more highly structured, rational, and enduring form of collective behavior (Goode 1992; Marx and McAdam 1994).

Social Movements

The Nature of Social Movements

The next time you observe a **social movement**, you will be aware of its four defining elements: a large number of people, a common goal to promote or prevent social change, some degree of leadership and organization, and activity sustained over a relatively long period of time. It is the form of collective behavior that has the most structure, lasts the longest, and is the most likely to create social change (Lofland 1996). Most social movements are mounted to stimulate change. This was as much the case for Nazism and the American Revolution as it is for the U.S. militia movement (Cozic 1997). The pro-life and pro-choice movements and the green (environmental) movement are more contemporary examples (Powell, Williamson, and Branco 1996; Costain and McFarland 1998; A. Goldstein 1998; Diani and della Porta 1999).

Despite commonalities, various social movements have unique characteristics. It is difficult to compare the civil rights movement with the environmental movement. This has led sociologists to study differences among social movements (Crossley 2002; Goodwin and Jasper 2003).

What are the primary types of social movements? David Aberle (1991) has identified four basic types of social movements. A **revolutionary movement** attempts to change a society totally. An example is the revolutionary movement led by Mao Zedong in China; it entrenched a communist form of government. A **reformative movement** aims to effect only partial change in a society; it can either advocate change or resist change. Many reformative movements seek to

© David Turnley/CORBIS

A social movement involves a larger number of people acting together with some degree of leadership and organization to promote or to prevent social change. Most social movements, such as the movement to end apartheid in South Africa, seek social change. The presence of black South Africans at the ballot box is one important indicator of change.

SNAPSHOT OF AMERICA 18.1

Percentage of Females in the Workplace

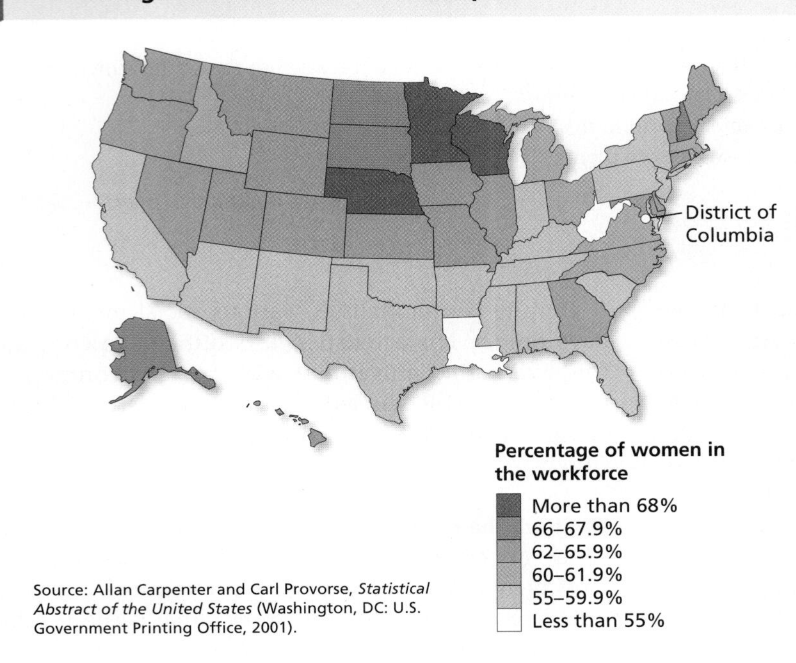

Source: Allan Carpenter and Carl Provorse, *Statistical Abstract of the United States* (Washington, DC: U.S. Government Printing Office, 2001).

Variation in female labor force participation in the United States is related to social change and social movements. The number of women in the U.S. workforce shot up during World War II. Once the soldiers returned home, however, a large percentage of those working women quit work to raise families. Owing in part to the women's movement, the United States has witnessed a peacetime resurgence of women entering the workforce. This map shows the percentage of women in each state who are active in the labor force.

1. Relate relative deprivation, the women's movement, and increased female labor force participation.
2. How does your state compare with other states in relation to female employment? Describe.

promote change. The women's liberation movement illustrates this type of social movement (Baxandall and Gordon 2000). Other reformative social movements oppose change. Conservative political and fundamentalist religious organizations have mounted a concerted effort to oppose abortion (Chesler 2001). The movement against corporate-led globalization among liberal youth is another example (Ferguson 2002). A **redemptive movement** focuses on changing individuals. The religious cult of David Koresh (the Branch Dividians) was a redemptive movement. Finally, an **alternative movement** seeks only limited changes in individuals. Population Connection (formerly known as Zero Population Growth) illustrates such a movement. It attempts to persuade people to limit the size of their families, but it does not advocate sweeping lifestyle changes, nor does it advocate legal penalties for large families.

Social movements are the most highly structured form of collective behavior. Three illuminating theories of social movements are relative deprivation theory, value-added theory, and resource mobilization theory.

Relative Deprivation Theory

It is the frustrated and discontent who want change and who are the most willing to fight for it. Discontent with present conditions, in short, is necessary for collective action. People must see existing conditions as unfair and unjust (Rose 1982). Thus, American colonists protested taxation without representation; Fidel Castro pointed to the vast gap between rich and poor Cubans; and Iranian revolutionaries in 1979 felt strongly that the Shah had gone too far with the processes of modernization, Westernization, and secularization. Discontent is more likely to lead to a social movement if it is linked with relative deprivation and unfulfilled rising expectations.

When does one feel relative deprivation? **Relative deprivation** is felt when people compare themselves with others and believe that they should have as much as those others have. Women's liberationists compare the situation of women to men, and gays in the United States underscore the penalties they suffer when they reveal their sexual preference. Government statistics indicate that African Americans receive less income than whites of comparable educational background. African Americans who are aware of this fact are likely to experience relative deprivation. Because a comparison is made between one's own situation and the situation of others, deprivation of this type is purely relative. There is no absolute standard for comparison—only the conviction among certain people that they, wrongfully, have less than some specific others have.

What are unfulfilled rising expectations? **Unfulfilled rising expectations** occur when newly raised hopes for a better life either are not satisfied at all or are not satisfied as rapidly as people had expected. For exam-

FIGURE 18.2

The J-Curve Theory of Revolution

According to the J-curve theory, a revolutionary social movement is most likely to occur when rising expectations cause expected need satisfaction to exceed actual need satisfaction by an unacceptable amount. Is this expected to be true even when total need satisfaction has actually increased? Explain.

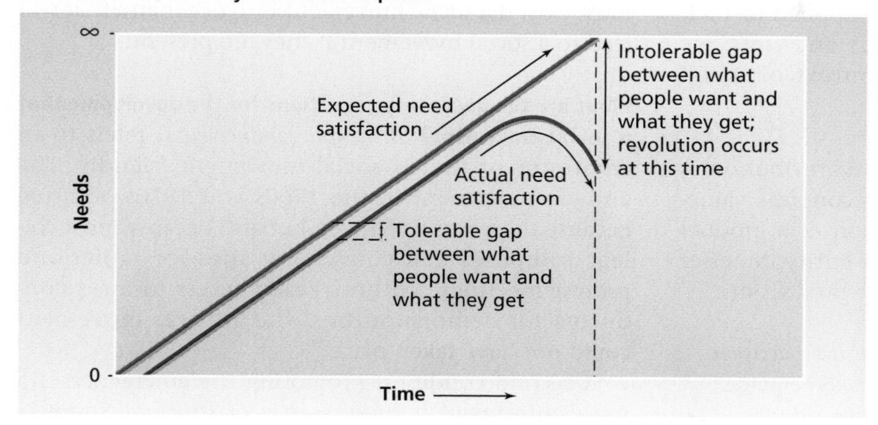

Source: James Chowning Davies, "The J-Curve of Rising and Declining Satisfactions as a Cause of Revolution and Rebellion," in Hugh Davis Graham and Ted Robert Gurr (eds.), *Violence in America*, rev. ed. (Beverly Hills, CA: Sage, 1979), p. 416. Reprinted by permission of J. C. Davies and the publisher, Sage Publications, Inc. (Beverly Hills/London).

ple, newly industrializing countries are likely to experience some revolutionary discontent when people who have been poor all their lives are suddenly promised a better life. They revolt not just because of their poverty but because their expectations about their material well-being have changed more rapidly than their actual material condition. The phenomenon of unfulfilled rising expectations helps explain why many revolutionary situations arise only after people have experienced some economic and social improvement. Alexis de Tocqueville (1955; originally published in 1835) observed improvement in the French peasant's economic situation:

> *It is a singular fact that this steadily increasing prosperity, far from tranquilizing the population, everywhere promoted a spirit of unrest. The general public became more and more hostile to every ancient institution, more and more discontented; indeed, it was increasingly obvious that the nation was heading for a revolution.*
>
> *Moreover, those parts of France in which the improvement in the standard of living was most pronounced were the chief centers of the revolutionary movement. Such records of the Ile-de-France region as have survived prove clearly that it was in the districts in the vicinity of Paris that the old order was soonest and most drastically superseded. In these parts the freedom and wealth of the peasant had long been better assured than in any other pays d'élection.*
> (Tocqueville 1955:175)

It will be interesting to observe events in the new Russian Federation and in China subsequent to the changes occurring in those societies.

James Davies (1979) has linked unfulfilled rising expectations to revolutionary social movements. According to Davies's J-curve theory, a revolutionary movement is most probable when a period of rising expectations accompanied by actual economic improvement is followed by a decline in the fortunes of the masses (see Figure 18.2). According to this model, once expectations begin to rise, they continue to do so. As long as expected need satisfaction and actual need satisfaction are reasonably close, people will tolerate the gap between what they want and what they are getting. It is when actual need satisfaction falls off sharply (note the upside-down J formed by the curve in Figure 18.2) that the gap between what people want and what they have becomes intolerable. At this point, a revolutionary social movement is most likely to occur.

What is the major shortcoming of relative deprivation theory? The theory cannot explain why some social discontent can exist without a subsequent social movement. Although African Americans had long been discontented and suffered from relative deprivation, a social movement was not mounted until the 1950s. We must conclude, then, that although discontent and deprivation are necessary conditions of social movements, they are not sufficient ones. Discontent and deprivation must precede a movement, but they cannot alone produce one. The other two theories of social movements point to other factors.

Value-Added Theory

One of the strengths of Neil Smelser's **value-added theory** is its applicability to many forms of collective behavior. Although the theory can be applied to such types of collective behavior as panics and riots, we describe it here only within the context of social movements.

What is value-added theory? Smelser's contribution is based on an economic theory. In the economic value-added process, each step in the creation of a product contributes (adds value) to the final entity. Smelser gives an example involving automobile production:

> *An example of* [the value-added process] *is the conversion of iron ore into finished automobiles by a number of stages of processing. Relevant stages would be mining, smelting, tempering, shaping, and combining the steel with other parts, painting, delivering to retailer, and selling. Each stage "adds its value" to the final cost of the finished product. The key element in this example is that the earlier stages must combine according to a certain pattern before the next stage can contribute its particular value to the finished product, an automobile. Painting, in order to be effective as a "determinant" in shaping the product, has to "wait" for the completion of the earlier processes. Every stage in the value-added process, therefore, is a necessary condition for the appropriate and effective condition of value in the next stage. The sufficient condition for final production, moreover, is the combination of every necessary condition, according to a definite pattern.* (Smelser 1971:13–14)

According to Smelser's theory, six conditions are necessary and sufficient for the development of a social movement. That is, Smelser specifies conditions that must exist if a social movement is to occur and that will lead to a social movement if they are present.

What are the necessary conditions for the development of a social movement? *Structural conduciveness* refers to an environment that is social movement friendly. The antiwar movement in the 1960s and 1970s occurred because there was a war, yes, but also because most college campuses had convenient sites for rallies and protest meetings. Without ready access to areas conducive for demonstrations, the antiwar movement could not have taken place.

A second condition promoting the emergence of a social movement is the presence of *structural strains*—conflicts, ambiguities, and discrepancies within a society. Without some form of strain, there is no impetus for change. Undoubtedly, conflict contributed to the 1960s antiwar movement: conflict between American credence in political self-determination, on the one hand, and the U.S. government's desire to save Southeast Asia from communism, on the other. Probably a key discrepancy in this case was the government's continued stance that there was no war (no legal war had been declared), despite the vast resources devoted to battle and the obvious combat casualties

TABLE 18.3

Hot Buttons for College Activists

Protest movements are based on an issue of burning importance. For college activists, these issues change with generations. Do you agree that the issues identified for the new millenium will ignite college activists over the next few years?

1960s	1970s	1980s	1990s	The New Millennium
Hot Topics	**Hot Topics**	**Hot Topics**	**Hot Topics**	**Hot Topics**
Vietnam war	Clean air and clean water	International human rights	Gay rights	Globalization and corporate dominance
Civil rights	Female empowerment	Endangered species	Sweatshop labor	Immigration
		Sexual assault	Legalization of marijuana	

Sources: Based on *The National College Magazine* (February 2000), p. 16.

suffered. (Table 18.3 identifies major structural strains that have mobilized college students since the 1960s.)

The combination of structural conduciveness and structural strains increases the probability that a social movement will occur. When a third condition—*generalized beliefs*—is also present, a social movement is even more likely. Generalized beliefs include a general recognition not only that there is a problem, but that something should be done about it. Two shared beliefs were crucial to the antiwar movement. One was the belief that the Johnson and Nixon administrations were not telling the truth about the war. It was also believed that the Vietnam War was so morally wrong that it had to be stopped.

Even when structural conduciveness, structural strains, and generalized beliefs exist, a social movement might not occur. One or more *precipitating factors* must occur to galvanize people into action. On April 30, 1970, President Nixon ordered the invasion of the neutral country of Cambodia. This event was a show of force to the North Vietnamese government, with whom the U.S. government was negotiating to end the war. President Nixon's announcement precipitated student antiwar protests and a chain of events that led to a confrontation between students and the Ohio National Guard, mobilized by the governor of Ohio. The encounter left many injured and four Kent State University students dead (G. H. Lewis 1972).

Once these first four conditions exist, the only remaining step to the emergence of a social movement is the *mobilization of participants for action*. It is at this point that leaders become very important. In the 1960s, radical student leaders, such as Mario Savio and Jack Weinberg, led efforts to mount an antiwar protest movement that began at the University of California, Berkeley. And the invasion of Cambodia fanned the flames of a student-led protest in California as well as across the country. Massive demonstrations were part of the political furor, with more than 100,000 opponents of the Vietnam War marching on Washington, D.C. Hundreds of colleges were forced to close as a result of strikes by 1.5 million students. Martin Luther King Jr., another historical example, moved into a leadership role when he seized an opportunity provided by Rosa Parks, the African American woman who in 1955 refused to move to the back of the bus in Montgomery, Alabama. King then mobilized sympathizers for a massive boycott that led to the desegregation of Montgomery's bus system.

The sixth determinant of a social movement is ineffective *social control*—efforts on the part of society (media, police, courts, community leaders, political officials) to prevent, minimize, or interrupt the momentum for a social movement. If appropriate techniques of social control are applied, a potential social movement may be prevented, even though the first five determinants are present. Sometimes, social controls can be applied after a panic, riot, or mob action has already started. At that point, control efforts may block the social movement, minimize its effects, or make matters worse. Social control efforts in the antiwar movement were ineffective, however, in part because heavy-handedness on the part of politicians and law enforcement officials only stimulated further protest that hastened the ending of the war. In addition to the students killed on the Kent State University campus, two African American students were killed during an antiwar protest at Jackson State University in Mississippi. Repressive tactics (beatings, jailings) were also inflicted on antiwar protestors at the 1968 Democratic National Convention.

What are the strengths and weaknesses of value-added theory? According to value-added theory, more than discontent and deprivation are necessary for a social movement to emerge. Beyond structural strains there must be structural conduciveness, generalized beliefs, precipitating factors, and the mobilization of participants for action. Critics reject the traditional view of social movements as irrational, spontaneous, and initially unstructured (Oberschall 1973; Tilly 1978; Opp 1989; Gamson 1990; Opp and Roehl 1990). According to other critics, the value-added approach works best for crowd behavior, but even in that type of collective behavior, all six factors are not always present (Goode 1992). Value-added theory has also been criticized for failing to consider the importance of resources—funds, people, abilities—for the emergence and success of social movements (Olzak and West 1991).

Resource Mobilization Theory

What is involved in resource mobilization theory? According to resource mobilization theory, a key to galvanizing people for collective action is the mobilization of resources. **Resource mobilization** is the process through which members of a social movement secure and use the resources needed to advance their cause. Resources encompass human skills such as leadership, organizational ability, and labor power as well as material goods such as money, property, and equipment (Cress and Snow 1996; McCarthy and Wolfson 1996; Freeman and Johnson 1999). The civil rights movement of the 1960s succeeded because of the commitment of African Americans and because people of other races, and college students in particular, contributed money, energy, and skills necessary to stage repeated protests. In contrast, the gay movement in the United States experienced difficulty partially because of a relative shortage of money, foot soldiers, and affluent supporters.

Martin Luther King Jr. is leading the 1963 Washington march. Which necessary condition for a social movement is illustrated here?

John Lofland (1979) writes of the 1970s "white-hot" mobilization efforts of the Unification Church ("Moonies"). Upon taking up residence in the United States in 1971, the Reverend Sun Myung Moon was shocked by the scant resource mobilization work that had been done. He set out to change the situation, establishing a set of long- and short-term goals for the movement, including new-member and fund-raising quotas. Elaborate publicity campaigns and events were staged, including speaking tours and rallies. An organizational structure was established and a national training center created. The trainees were subsequently sent across the country in evangelistic teams. The United States was divided into ten regions, each headed by a regional director. Each regional director supervised several state directors who were in charge of local center directors. These efforts required about $15 million to be generated each year, some of it coming from Korean and Japanese branches of the movement. (Allegations were made that some of the money came from the South Koreans and American CIA.) Many local groups made money from their own businesses, such as housecleaning services, gas stations, and restaurants. Between 1971 and 1974, membership increased from 500 to 2,000. (For more detail on the current resources of the Unification Church, see Chapter 15.)

The case of the Unification Church makes a point central to resource mobilization theory: the collective action is organizationally based and led by rational people who calculate the likelihood of achieving their ends (A. D. Morris 1981, 1984). Resource mobilization advo-

TABLE 18.4

Major Forms of Collective Behavior

You can test your understanding of these forms of collective behavior by suggesting an additional example for each one.

Forms of Collective Behavior	Definition	Example
Rumor	• A widely circulating piece of information that is not verified as being true or false	• Continuously repeated prediction that airplanes would crash on a massive scale on January 1, 2000
Urban legend	• A moralistic tale that the teller swears happened to someone he or she knows or to an acquaintance of a friend or family member	• Fierce alligators in New York City's sewer system
Fad	• An unusual behavior pattern that spreads rapidly and disappears quickly	• Swing dancing
Fashion	• A widely accepted behavior pattern that changes periodically	• Wearing Nike shoes
Crowd	• A temporary collection of people who share an immediate interest	• New Year's celebrants at Times Square in New York City
Mob	• An emotional crowd ready to use violence for a specific purpose	• Lynch mob
Riot	• An episode of largely random destruction and violence carried out by a crowd	• Destructive behavior following the acquittal of police officers who were filmed using extreme force against Rodney King
Social movement	• Movement whose goal is to promote or prevent social change	• Civil rights movement

cates contend that although other theories of social movements depict organization as something that emerges as the movement develops, resource mobilization theory sees preexisting organizational structure (and associated resources) as central to launching a social movement.

Several scholars have emphasized the role of "outsiders" in the creation of modern social movements (McCarthy and Zald 1977; Zald and McCarthy 1987; Burstein 1991). Outsiders may be volunteers who care about the plight of a category of people, or they may be professionals knowledgeable about the organization and management of social movements. For example, white students and adults from the North brought commitment and organizational skills to the civil rights movement in the 1960s.

How has resource mobilization theory been criticized? Although critics praise resource mobilization theory for shedding light on the importance of resources in mounting and sustaining social movements, they fault it for deemphasizing the necessary social discontent and strain needed for a social movement to emerge. Resources are of little use if people are not sufficiently dissatisfied with present conditions (Jenkins 1983; Klandermans 1984). Although recognizing the importance of organization and planning, some social movement theorists foresee a danger in dismissing the important role of spontaneity in the emergence and development of social movements. Social movement theorists, critics write, must not lose sight of the emotional factors that lead people to join a movement whose likelihood of success is small, and they must also consider the ways in which organization leaders are affected by the unpredictable, spontaneous actions of the people involved (Killian 1984; Opp 1988; Rule 1989; A. Scott 1995).

The Future Direction of Social Movement Theory

The three theories of social movements just discussed are more complementary than they are mutually exclusive. Relative deprivation theory emphasizes discontent from a social-psychological, or micro, viewpoint. Value-added theory implies discontent (within the concept of structural strain) but focuses on the operation of several factors at the macro level of analysis. Value-added theory, also at the macro level, implies the need for resources through its mobilization factor, but resource mobilization theory is needed to spell out this contributing factor to the rise of social movements.

Social movement theory in the future will likely encompass both preexisting structure and spontaneity, both rationality and irrationality. It seems inevitable that the viewpoints of resource mobilization theory and other social movement theories will be considered by those on both sides of the debate.

A distinction is being made between "old" and "new" social movements (Eyerman 1992; Haferkamp and Smelser 1992; Buechler 2000). Old social movements, such as the American civil rights movements and the labor movement, are based more on the interests of the poor and the working class. They are centered on the struggle for power and control over economic conditions within the context of industrial capitalism. New social movements, such as the women's movement, the ecology movement, the peace movement, the gay movement, and the animal rights movement, are not so embedded in economics (although women and gays seek economic equality) and opposition to capitalism (although environmentalists seek to moderate capitalism). Rather than expressing conflicts of industrial society and industrialization,

FEEDBACK

1. A _________ is the form of collective behavior that has the most structure, lasts the longest, and is the most likely to create social change.
2. Which of the following is an example of a reformative social movement?
   a. the French Revolution
   b. Zero Population Growth
   c. the Jesus People
   d. Women's Christian Temperance Union
3. _________ is felt when people compare themselves with others and believe that they should have as much as those others have.
4. _________ occur when newly raised hopes for a better life either are not satisfied at all or are not satisfied as rapidly as people expect them to be.
5. Once widespread discontent exists within a society, a social movement is bound to occur. T or F?
6. According to the _________ theory, several conditions are necessary and sufficient for the emergence of a social movement.
7. _________ is the process through which members of a social movement secure and use the resources needed to press for social change.

Answers: 1. social movement 2. d 3. Relative deprivation 4. Unfulfilled rising expectations 5. F 6. value-added 7. Resource mobilization

new social movements rest more on conflicts unique to postindustrial society. The new social movements are fueled less by economic than by cultural conflicts, aimed at redefinitions of norms and values rather than at questions of economics and who gets what. New social movements are more global in focus, tend to center on quality-of-life issues, and are fueled by the interests of the middle and upper-middle class (Melucci 1980; McAdam, McCarthy, and Zald 1988; Kriesi 1989; A. Scott 1995).

SUMMARY

1. Social change refers to social structure alterations that have long-term and relatively important consequences. For most of the world, social change is accelerating at a dramatic rate.
2. Although predicting the precise nature of social change within a society is hazardous, several sources of social change are pretty well understood. Discovery, invention, and diffusion are the major social processes through which social change occurs. Important sources of social change are technology, population, the natural environment, conflict, and ideas. There are many interrelationships among these sources of social change.
3. Sociologists go beyond identifying sources of social change to developing theoretical perspectives of social change. The evolutionary, functionalist, and conflict perspectives view social change in very different ways. The cyclical perspective, short on explanatory power, envisions societies as changing through the process of growth and decay. Contemporary evolutionary theory emphasizes multilinear change (that is, various societies may evolve in many diverse directions). The functionalist perspective depicts societies as relatively stable. Following a major change, these integrated systems seek a new equilibrium. According to the conflict perspective, societies are unstable systems that are constantly undergoing change. Attempts to combine the insights of the functionalist and conflict perspectives, although interesting, do not eliminate the need to retain separate perspectives.
4. With economic development comes modernization. This process involves numerous social and cultural changes. Consequences of modernization include rapid population growth, expanding urbanization, increased equality, greater political democracy, push for universal education, and a trend toward the nuclear family.
5. Even though modernization is a worldwide phenomenon, all modernizing societies are not on the road to social and cultural convergence. This is partly because modernization is filtered through existing social and cultural arrangements.
6. According to world-system theory, a nation's development hinges on its place in the world economy. So long as core nations control and exploit peripheral nations, the development gap between the former and the latter will increase.
7. Most areas of sociological study assume that social life is predictable, orderly, and recurrent. Collective behavior can be an important exception; much of this behavior is spontaneous, short-term, and relatively unstructured. Some forms of collective behavior, though, are planned, structured, and enduring.
8. On the more unstructured end of the collective behavior continuum are dispersed collectivities—rumors, fads, crazes, panics, mass hysteria, and fashions. Even these forms of collective behavior are structured to some degree.
9. Rumors have a negative reputation, usually deserved, because they are composed of inaccurate, distorted, or false information. Rumors can be playful and harmless, or they can be damaging and hurtful.
10. Mass hysteria occurs when people accept false beliefs, then become anxious. Panics take place when there is a collective reaction to a real threat.
11. Unusual patterns of behavior that are adopted quickly, accepted enthusiastically, and disappear soon are called fads. Fashions are patterns of behavior that are widely approved but expected to change periodically. Fads and fashions are more central to modern than premodern societies.
12. Crowd behavior is fascinating to most people because it usually involves intense feelings and sometimes outrageous behavior. There are casual, conventional, expressive, and acting crowds.
13. Three quite different theories attempt to explain crowd behavior. Contagion theory stresses the irrationality of crowds and the buildup of intense emotion—a result of social interaction within a large collection of people. Emergent norm theory depicts crowd behavior as being more rational. According to emergent norm theory, crowd behavior is guided by norms that arise spontaneously. In convergence theory, crowds are formed by people with a common interest.
14. Social movements are closer to conventional social behavior than is crowd behavior because social movements are more permanent and more organized. Even so, they are not as permanent and structured as most aspects of social life.
15. There are three major explanations of social movements. According to relative deprivation theory, discontent and unfulfilled rising expectations create a breeding ground for social movements. Value-added theory outlines six necessary and sufficient conditions for the development of a social movement: structural conduciveness, structural strains, generalized beliefs, precipitating factors, mobilization of participants for action, and social con-

trol. According to the resource mobilization theory, resources are a crucial ingredient in the mounting of a social movement. Resources include human skills and economic assets, both of which often come from outside the movement.

INFOTRAC® COLLEGE EDITION

http://www.infotrac-college.com/wadsworth

Another unique option available to you at the Student Resources section of the companion Web site is Infotrac College Edition, an online library with access to hundreds of scholarly and popular periodicals. Below are suggested search terms for this chapter. Results from these and other searches are found at the site.

Search keywords: **fads, fashions, urban legends**. These terms are examples of the less structured forms of collective behavior. Use Infotrac to discover what some of the current and former fads, fashions, and urban legends are. Which of these have affected you? What was the strangest one you found?

Search keywords: **technological determinism**. Search the database to find out more about technological determinism. How widely believed is this theory? What evidence supports it? What evidence discredits this theory? If the theory is true, what implications does that have for society?

Search keywords: **women's movement**. Read three articles about the women's movement in the United States. What is its history? What are some of the major acts accomplished by the movement? What is its current status? What are some relatively current political decisions/policies that help or hinder this movement?

LEARNING OBJECTIVES REVIEW

- Illustrate three social processes contributing to social change.
- Discuss, as sources of social change, the role of technology, population, the natural environment, conflict, and ideas.
- Compare and contrast the cyclical and evolutionary perspectives.
- Explain the why and how of social change within the context of the functionalist and conflict perspectives.
- Define modernization and explain some of its major consequences.
- Define world-system theory and summarize some of the consequences flowing from its description of the world economy.
- Describe social activities of dispersed collectivities.
- Describe the nature of a crowd and identify the basic types of crowds.
- Contrast contagion theory, emergent norm theory, and convergence theory.
- Define the concept of social movement and identify the primary types of social movements.
- Apply the concepts of relative deprivation and unfulfilled rising expectations to the emergence of social movements.
- Compare and contrast relative deprivation theory with value-added theory and resource mobilization theory.

CONCEPT REVIEW

Match the following concepts with the definitions listed below them.

____ a. diffusion
____ b. technology
____ c. invention
____ d. social change
____ e. collective behavior
____ f. mass hysteria
____ g. emergent norm theory
____ h. relative deprivation
____ i. panic
____ j. rumor
____ k. crowd

1. the social process that occurs when members of one group borrow social and cultural elements from another group
2. the creation of a new element by combining two or more already existing elements and creating new rules for their use as a unique combination
3. alterations in social structures that have long-term and relatively important consequences
4. the part of culture, including ideas and hardware, that is used to reach practical goals
5. a condition that exists when people compare themselves with others and believe that they should have as much as those others have
6. the relatively spontaneous and unstructured social behavior of people who are responding to similar stimuli
7. the form of collective behavior in which general social anxiety is created by acceptance of one or more false beliefs
8. a widely circulating story whose truth is questionable
9. the theory of crowd behavior that stresses the similarity between typical, everyday social behavior and crowd behavior

10. collective behavior that occurs when people react to a genuine threat in fearful, anxious, and often self-damaging ways

11. a temporary collection of people who share a common point of interest

CRITICAL-THINKING QUESTIONS

1. What do you think is the proper perspective on technological determinism? In the process of answering this question, be sure to incorporate the perspective of Karl Marx.

2. Identify a major social change that has occurred in your lifetime. What do you think are the major sources of this change? Be careful to relate each source of change to the nature of the change itself.

3. Is evolutionary theory compatible or incompatible with functionalism and conflict theory? Make clear the nature of all three perspectives on social change.

4. Are you convinced that modernizing societies are becoming socially and culturally similar? Why or why not?

5. Select a rumor, fad, craze, or fashion to demonstrate your understanding of a dispersed collectivity.

6. Sociologists take collective behavior to be a legitimate area of study on the grounds that it is more structured than it appears at first glance. Develop three examples of collective behavior of which you have some knowledge to support this claim by sociologists.

7. You have participated in a crowd at some time. Think of an instance and identify it as one of the four types of crowds described in the text. Using your personal experience, provide examples of behavior within the crowd (yours or someone else's) that illustrate why you think it was a particular type of crowd.

8. Which theory of social movements do you think best explains the women's movement? Link the theory to the nature of the women's movement.

MULTIPLE-CHOICE QUESTIONS

1. **Alterations in social structures that have long-term and relatively important consequences are known as**
   a. modernization.
   b. demographic change.
   c. cultural lag.
   d. social change.
   e. social movements.
2. **Which of the following statements best describes the relationship between technology and social change?**
   a. Social change cannot occur without technological developments.
   b. Technology does not necessarily lead to social change.
   c. The effects of technology are similar in all societies.
   d. Technology can be viewed as a single cause of social change.
   e. Technology has little effect on social change.
3. **Max Weber, partially in response to Marx, emphasized the role of _________ in social change.**
   a. technology
   b. revolution
   c. ideas
   d. inventions
   e. territorial discoveries
4. **A major criticism of the cyclical perspective is that it**
   a. assumes that society is inherently stable.
   b. assumes that society must pass through a set of progressive stages.
   c. fails to take into account the history of societies.
   d. describes rather than explains change.
   e. emphasizes inventions rather than discoveries.
5. **According to the conflict theory, _________ is the major source of social instability in societies.**
   a. technology
   b. revolution
   c. ideas
   d. politics
   e. scarcity of desired resources
6. **_________ is the process involving all those social and cultural changes accompanying economic development.**
   a. Modernization
   b. Postindustrialization
   c. Evolution
   d. Convergence
   e. Technological determinism

7. **The development of social and cultural similarity among modernizing nations is known as**
   a. convergence.
   b. world-system theory.
   c. congruence.
   d. consistency.
   e. modernization.
8. **Which of the following statements best fits with world-system theory?**
   a. The standard of living is much lower in core nations.
   b. Workers in peripheral nations experience a high standard of living.
   c. The economic gap between core and peripheral nations is increasing.
   d. Peripheral nations develop rapidly due to the aid given by core nations.
   e. The process of integration comes before the process of differentiation.
9. **The study of collective behavior poses some particular problems for sociology because**
   a. sociologists are accustomed to studying social interaction.
   b. collective behavior is difficult to define in sociological terms.
   c. sociologists are accustomed to studying structured behavior.
   d. sociological theories do not account for collective behavior.
   e. sociological research methods are inappropriate for the study of collective behavior.
10. **When parents in a local school rebel against an AIDS victim attending school with their children, they are said to be exhibiting**
    a. faddish behavior.
    b. mass hysteria.
    c. victimization.
    d. a public interest reaction.
    e. collective hypocrisy.
11. **When a crowd of soccer fans abandons the guidelines for watching the match in order to attack the officials, a/an _________ crowd has become a/an _________ crowd.**
    a. casual; acting
    b. conventional; acting
    c. expressive; acting
    d. conventional; expressive
    e. casual; expressive
12. **An emotionally stimulated, disorderly crowd that is ready to use destructiveness and violence to achieve a specific purpose is known as a/an**
    a. mob.
    b. expressive crowd.
    c. unconventional crowd.
    d. revolution.
    e. interest group.
13. **In his reformation of contagion theory, Herbert Blumer defined _________ as the stage in which people become increasingly aware and sensitive to one another and enter something akin to a hypnotic trance.**
    a. rising expectations
    b. social contagion
    c. milling
    d. collective excitement
    e. crowd decision
14. **According to _________ theory, rules develop within crowds to tell people how they are expected to behave.**
    a. emergent norm
    b. crowd decision
    c. value-added
    d. value-clarification
    e. contagion
15. **The role of "outsiders" in the creation of modern social movements is most prominent in _________ theory.**
    a. value-added
    b. relative deprivation
    c. emergent norm
    d. contagion
    e. resource mobilization

REVIEW GUIDE

FEEDBACK REVIEW

True-False

1. According to the cyclical perspective on social change, societies rise and fall rather than move continuously toward improvement. T or F?
2. Classical evolutionary theory assumed that change leads to improvement. T or F?
3. Mob and riot are simply two terms for the same type of crowd. T or F?
4. Much crowd behavior is structured and rational. T or F?

Fill in the Blank

5. The theory of social change that assumes technology to be the primary cause of the nature of social structure is called _________.
6. Was it Karl Marx or Max Weber who most emphasized the role of ideas in creating social change? _________
7. According to the _________ perspective, social and cultural change occurs as a result of the struggles between groups representing different segments of a society.
8. Unusual patterns of behavior that spread rapidly, are embraced zealously, and disappear after a short time are called _________.
9. An _________ crowd has no purpose or direction beyond the unleashing of emotions.

10. A __________ is the form of collective behavior that has the most structure, lasts the longest, and is the most likely to create social change.

Multiple Choice

11. Which of the following is *not* one of the major consequences of modernization discussed in the text?
   a. the social and cultural convergence of modernizing countries
   b. increased urbanization
   c. greater equality
   d. more political democracy
   e. widespread development of the nuclear family

12. Some individuals at a lynching do not participate or give verbal support, but do not attempt to stop it. Which of the following theories of crowd behavior best explains this response?
   a. contagion theory
   b. crowd decision theory
   c. emergent norm theory
   d. casual crowd theory

GRAPHIC REVIEW

Figure 18.1 shows the varying speeds with which Americans have adopted certain important technologies. Answer these questions to make sure you understand the information presented.

1. How would you explain the difference in adoption speed for the telephone and the Internet?

____
____
____

2. Discuss some social implications of the increasing use of the Internet.

____
____
____

ANSWER KEY

Concept Review

a. 1
b. 4
c. 2
d. 3
e. 6
f. 7
g. 9
h. 5
i. 10
j. 8
k. 11

Multiple Choice

1. d
2. b
3. c
4. d
5. e
6. a
7. a
8. c
9. c
10. b
11. b
12. a
13. c
14. a
15. e

Feedback Review

1. T
2. T
3. F
4. T
5. technological determinism
6. Max Weber
7. conflict
8. fads
9. expressive
10. social movement
11. a
12. c

Glossary

Absolute poverty The absence of enough money to secure life's necessities.

Achieved status A status within a social structure occupied because of an individual's efforts.

Age cohorts Persons born during the same time period in a particular population.

Ageism A set of beliefs, norms, and values used to justify age-based prejudice and discrimination.

Age-specific death rate The number of deaths per one thousand persons in a specific age-group.

Age-specific fertility The number of live births per one thousand women in a specific age-group.

Age stratification The unequal distribution of scarce desirables based on chronological age.

Age structure The distribution of people of different ages within a society.

Agricultural society A society whose subsistence relies primarily on the cultivation of crops with plows drawn by animals.

Alternative movement The type of social movement that seeks only limited changes in individuals.

Anomie A social condition in which norms are weak, conflicting, or absent.

Anticipatory socialization The process of preparing oneself for learning new norms, values, attitudes, and behaviors.

Ascribed status A status within a social structure that is not earned or chosen, but is assigned.

Assimilation The integration of a social or ethnic minority into a society.

Authoritarianism The type of political system controlled by nonelected rulers who generally permit some degree of individual freedom.

Authoritarian personality A personality characterized by excessive conformity; submissiveness to authority figures; inflexibility; repression of impulses, desires, and ideas; fearfulness; and arrogance toward persons or groups thought to be inferior.

Authority Power accepted as legitimate by those subjected to it.

Beliefs Ideas concerning the nature of reality.

Bilateral descent The familial arrangement in which inheritance and descent are passed equally through both parents.

Biological determinism The attribution of behavioral differences to inherited physical characteristics.

Blended family A family formed when at least one of the marriage partners has been married before and has one or more children from a previous marriage.

Boomburgs Places of over 100,000 residents that are not the largest cities in their metropolitan areas and have experienced double-digit population growth in recent decades.

Bourgeoisie Members of a society who own the means for producing wealth.

Bureaucracies Formal organizations based on rationality and efficiency.

Capitalism An economic system founded on the sanctity of private property and the right of individuals to profit from their labor.

Case study A thorough, recorded investigation of a small group, incident, or community.

Caste system The type of stratification structure in which there is no social mobility because social status is inherited and cannot subsequently be changed.

Causation The idea that events occur in a predictable, nonrandom way and that one event leads to another.

Central-city dilemma The concentration of a large population in need of public services, but without the tax-generated money to provide them.

Charismatic authority Legitimate power based on an individual's personal characteristics.

Charter schools Publicly funded schools that are operated like private schools by public schoolteachers and administrators.

Church A life-encompassing religious organization to which all members of a society belong.

City A dense and permanent concentration of people living in a limited geographical area who gain their living primarily through nonagricultural activities.

Civil religion A public religion that expresses a strong tie between a deity and a culture.

Class conflict The conflict between those controlling the means for producing wealth and those laboring for them.

Class consciousness A sense of identification with the goals and interests of the members of one's own social class.

Closed-ended questions Questions a person must answer by choosing from a limited, predetermined set of responses.

Coercion Social interaction in which individuals or groups are forced to give in to the will of other individuals or groups.

Cognition The process of thinking, knowing, or mentally processing information.

Cognitive ability The capacity for thinking abstractly.

Cohabitation A marriage-like living arrangement without the legal obligations and responsibilities of formal marriage.

Collective behavior The relatively spontaneous and unstructured social behavior of people who are responding to similar stimuli.

Communitarian capitalism The type of capitalism that emphasizes the interests of employees, customers, and society.

Community A concentration of people who live within a limited geographical area and whose social relationships fulfill major social and economic needs.

Competition A social process that provides rewards to people based on how their performance compares with that of others.

Concentric zone theory A description of the process of urban growth emphasizing circular areas that develop from the central city outward.

Conflict A form of social interaction in which individuals or groups work against one another to obtain a larger share of the valuables.

Conflict theory The theoretical perspective that emphasizes conflict, competition, change, and constraint within a society.

Conformity Behavior matching group expectations.

Conglomerate A network of unrelated businesses operating under a single corporate umbrella.

Consolidated metropolitan statistical area (CMSA) An urban area composed of two or more sets of neighboring PMSAs.

Conspicuous consumption The consumption of goods and services to display one's wealth to others.

Contagion theory The theory of crowd behavior that emphasizes the irrationality of crowds that is created by participants stimulating one another to higher and higher levels of emotional intensity.

Contingent employees Individuals hired on a part-time, short-term, or contract basis.

Continuity theory Presumes that most aging people maintain consistency with their past lives and use their life experiences to intentionally continue to develop in self-determined channels.

Control group The group in an experiment that is not exposed to the experimental variable.

Control theory The idea that conformity to social norms depends on the presence of a strong bond between individuals and society.

Convergence theory The theory that crowds are formed by people who deliberately congregate with others whom they know to be like-minded.

Cooperation A form of social interaction in which individuals or groups combine their efforts to reach some common goal.

Cooperative learning A nonbureaucratic classroom structure in which students study in groups, with teachers acting as guides rather than as the controlling agents determining all activities.

Core tier The tier of the occupational structure composed of large firms dominating their industries.

Corporate welfare Economic benefits government regularly gives to corporations.

Corporation An organization owned by shareholders who have limited liability and limited control over organizational affairs.

Correlation A statistical measure in which a change in one variable is associated with a change in another variable.

Counterculture A subculture that is deliberately and consciously opposed to certain central aspects of the dominant culture.

Craze A type of fad that can have serious consequences for its adopters.

Credentialism The idea that credentials (educational degrees) are unnecessarily required for many jobs.

Crime Acts in violation of the law.

Criminal justice system A system for controlling crime; comprised of police, courts, and corrections.

Crowd A temporary collection of people who share an immediate common interest.

Crude birth rate The annual number of live births per one thousand members of a population.

Crude death rate The annual number of deaths per one thousand members of a population.

Cult A religious organization whose characteristics are not drawn from existing religious traditions within a society.

Cultural bias In regard to intelligence tests, the unfair measurement of the cognitive abilities of people in some social categories.

Cultural lag Any situation in which disequilibrium is caused by one segment of a society's failing to change at the same rate as an interrelated segment.

Cultural particulars The widely varying, often distinctive ways societies demonstrate cultural universals.

Cultural relativism The idea that any given aspect of a particular culture should be evaluated in relation to its place within the larger cultural context of which it is a part rather than according to some alleged universal standard that is applied across all cultures.

Cultural transmission theory The theory that deviance is part of a subculture transmitted through socialization.

Cultural universals General cultural traits thought to exist in all known cultures.

Culture A people's way of life, consisting of material objects as well as the patterns of thinking, feeling, and behaving, that is passed from generation to generation among members of a society.

De facto subjugation Subjugation based on common, everyday social practices.

De jure subjugation Subjugation based on the law.

Democracy The type of political system in which elected officials are held responsible for fulfilling the goals of the majority of the electorate.

Democratic control The form of control in which authority is split evenly between husband and wife.

Demographic transition The process by which a population gradually moves from high birth rates and death rates to low birth rates and death rates as a result of economic development.

Demography The scientific study of population.

Denomination One of several religious organizations that members of a society accept as legitimate.

Dependency ratio The proportion of persons in the dependent ages relative to those who are economically active.

Dependent variable A variable in which a change (or effect) can be observed.

Desocialization The process of relinquishing old norms, values, attitudes, and behaviors.

Deterrence Intimidation of members of society into compliance with requirements of the legal system.

Deviance Any behavior that departs from societal or group norms.

Deviant A person who violates one or more of society's most highly valued norms.

Differential association theory The theory that deviant behavior is learned—that the more individuals are exposed to people who break the law, the more likely they are to become criminals.

Diffusion The process by which members of one group borrow social and cultural elements from another group, resulting in change.

Direct institutionalized discrimination Organizational or community actions intended to deprive a racial or ethnic minority of its rights.

Discovery The social process of learning something or reinterpreting something.

Discrimination Unequal treatment of individuals based on their minority membership.

Dispersed collectivities Participants in the less-structured forms of collective behavior such as rumors, panics, and fads.

Divorce rate The number of divorces in a particular year for every one thousand members of the total population.

Divorce ratio The number of divorced persons per one thousand persons in the population divided by the number of persons who are married and living with their spouses.

Doubling time The number of years needed to double the size of a population.

Downsizing The process by which companies reduce the size of their full-time workforce.

Dramaturgy The symbolic interactionist approach that depicts social life as theater.

Drives Impulses to do something to reduce discomfort.

Dual labor market The existence of a split between core and peripheral segments of the economy and the division of the labor force into preferred and marginalized workers.

Dynamic equilibrium The assumption by functionalists that a society both changes and maintains most of its original structure over time.

Dysfunction A negative consequence of some element of a society.

Economic determinism The idea that the nature of a society is based on the society's economy.

Economy The institution designed for the production and distribution of goods and services.

Edge city A suburban unit specializing in a particular economic activity.

Educational equality When schooling produces the same results (achievement and attitudes) for lower-class and minority children as it does for other children.

Elitism The theory of power distribution that sees society in the control of a relatively few individuals and organizations.

Emergent norm theory The theory of crowd behavior that stresses the similarity between typical, everyday social behavior and crowd behavior.

Endogamy Marriage within one's own group as required by social norms.

Epidemiology The study of the distribution of diseases within a population.

Ethnic minority A group of people who are socially identified by their unique characteristics related to culture or nationality.

Ethnocentrism The tendency to judge others in relation to one's own cultural standards.

Ethnomethodology The study of the processes people develop and use in understanding the routine behaviors expected of themselves and others in everyday life.

Euthanasia Deliberate termination of the life of a person suffering from a painful and incurable illness.

Exogamy Mate selection norms requiring individuals to marry someone outside their kind.

Experiment A laboratory procedure that attempts to eliminate all possible contaminating influences on the variables being studied.

Experimental group The group in an experiment exposed to the experimental variable.

Exponential growth When the absolute growth of a population that occurs within a given time period becomes part of the base for the growth rate in the next time period.

Extended family A family consisting of two or more adult generations that share a common household and economic resources.

Fads Unusual patterns of behavior that spread rapidly, are embraced zealously, and disappear after a short time.

False consciousness Acceptance of a system that works against one's own interests.

Family A group of people related by marriage, blood, or adoption.

Family of orientation The family into which an individual is born.

Family of procreation The family group established upon marriage.

Family planning The voluntary use of population control methods.

Family resiliency The capacity of the family to emerge from crises as stronger and more resourceful.

Fashions Behavior patterns that evolve over time and are widely approved, but are expected to change periodically.

Fecundity The maximum rate at which women can potentially produce children.

Feminism A social movement aimed at the achievement of sexual equality—socially, legally, politically, and economically.

Feminist theory A theoretical perspective that links the lives of women and men to the structure of gender relationships within society.

Feminization of poverty The trend toward more and more of the poor in the United States being women and children.

Fertility A measurement of the number of children born to a woman or to a population of women.

Fertility rate The annual number of live births per one thousand women ages fifteen to forty-four.

Field research A research approach for studying aspects of social life that cannot be measured quantitatively and that are best understood within a natural setting.

Folk society A society that rests on tradition, cultural and social consensus, family, personal ties, little division of labor, and an emphasis on the sacred.

Folkways Norms without moral overtones.

Formal demography The branch of demography that deals with gathering, collating, analyzing, and presenting population data.

Formal organization A social structure deliberately created for the achievement of one or more goals.

Formal sanctions Rewards and punishments that may be given only by officially designated persons.

For-profit schools Schools run by private companies on government funds.

Functionalism The theoretical perspective that emphasizes the contribution (functions) made by each part of a society.

Fundamentalism The rejection of secularization and the adherence to traditional religious beliefs, rituals, and doctrines.

Game stage According to Mead, the stage of development during which children learn to consider the roles of several people at the same time.

Gemeinschaft Tönnies's term for the type of society based on tradition, kinship, and intimate social relationships.

Gender The expectations and behaviors associated with a sex category within a society.

Gender identity An awareness of being masculine or feminine, based on culture.

Gender roles Culturally based expectations associated with each sex.

Gender socialization The social process in which boys learn to act the way boys are supposed to act and girls learn to act as girls are expected to act.

Generalized other An integrated conception of the norms, values, and beliefs of one's community or society.

Genocide Politically motivated mass murder of most or all of a targeted population.

Gentrification The development of low-income areas by middle-class home buyers, landlords, and professional developers.

Geronticide The socially accepted killing of the elderly within a society.

Gerontocracy The type of society in which the elderly hold the greatest share of scarce desirables.

Gesellschaft Tönnies's term for the type of society characterized by weak family ties, competition, and impersonal social relationships.

Gesture Facial expression, body movement, or posture carrying culturally defined and shared symbolic meanings.

Goal displacement The situation that occurs when organizational rules and regulations become more important than organizational goals.

Government The political structure that rules a nation.

Gross migration rate The annual number of persons per one thousand members of a population who enter or leave a geographical area.

Group A number of people who are in contact with one another; share some ways of thinking, feeling, and behaving; take one another's behavior into account; and have one or more interests or goals in common.

Group marriage The form of marriage in which two or more men are married to two or more women at the same time.

Groupthink A situation in which pressures toward uniformity discourage members of a group from expressing their reservations about group decisions.

Hate crime A criminal act motivated by prejudice.

Health-care system The combination of professional services, organizations, training academies, and technological resources that are committed to the treatment, management, and prevention of disease.

Health maintenance organizations (HMOs) Health plans in which members pay set fees on a regular basis and in return receive all necessary health care at little or no additional cost.

Heterogamy Marriage between people with differing social characteristics.

Hidden curriculum The educational curriculum that transmits to children a variety of nonacademic values, norms, beliefs, and attitudes.

Hidden unemployment Unemployment that is not measured because of discouraged workers who have stopped looking for work or part-time workers who would prefer full-time jobs.

Homogamy The tendency to marry someone similar to oneself based on personal preference.

Horizontal mobility A change from one occupation to another at the same general status level.

Horticultural society A society that solves the subsistence problem primarily through the domestication of plants.

Hospices Organizations designed to provide support for the dying and their families.

Hospitals Organizations providing specialized medical services to a variety of patients, including some who continue to live in their homes and others who will spend a prolonged period in the facility.

Humanist sociology The theoretical perspective that places human needs and goals at the center of sociology.

Hunting and gathering society A society that solves the subsistence problem through hunting animals and gathering edible fruits and vegetables.

Hypothesis of linguistic relativity The idea that one's perception of reality is based on language.

Hypothesis A tentative, testable statement of a relationship between particular variables.

"I" The spontaneous and unpredictable part of the self.

Ideal culture Cultural guidelines publicly embraced by members of a society.

Ideal-type method A method that involves isolating the most basic characteristics of some social entity.

Ideology A set of ideas used to justify and defend the interests and actions of those in power in a society.

Imitation stage The developmental stage during which, according to Mead, a child imitates the physical and verbal behavior of a significant other without comprehending the meaning of what is being imitated.

Incarceration The removal of criminals from society.

Income The amount of money received (within a given time period) by an individual or group.

Independent variable A variable that causes something to happen.

Indirect institutionalized discrimination Unintentional behavior that negatively affects a racial or ethnic minority.

Individualistic capitalism The type of capitalism founded on the principles of self-interest, the free market, profit maximization, and the highest return possible on stockholder investment.

Industrial society A society whose subsistence is based primarily on the application of science and technology to the production of goods and services.

Infant mortality rate The number of deaths among infants under one year of age per one thousand live births.

Informal organization A group in which personal relationships are guided by rules, rituals, and sentiments not provided for by the formal organization.

Informal sanctions Rewards and punishments that may be applied by most members of a group.

In-group A group toward which one feels intense identification and loyalty.

Instincts Genetically inherited, complex patterns of behavior that always appear among members of a particular species under appropriate environmental conditions.

Institutionalized discrimination Discrimination that results from unfair practices that are part of the structure of society and have grown out of traditionally accepted behaviors.

Integrative curriculum An approach to education based on student-teacher collaboration.

Interest group A group organized to achieve some specific shared goal by influencing political decision making.

Intergenerational mobility Social mobility that occurs from one generation to the next.

Interlocking directorates Members of corporations sitting on one another's boards of directors.

Interorganizational relationship A pattern of interaction among authorized representatives of two or more formally independent organizations.

Intervening variable A variable that influences the relationship between an independent variable and a dependent variable.

Interview A set of questions asked by a trained interviewer.

Intragenerational mobility Social mobility that occurs from one generation to the next.

Invention The creation of a new element by combining two or more already existing elements and creating new rules for their use as a unique combination.

Invisible religion A private religion that is substituted for formal religious organizations, practices, and beliefs in a secular society.

Iron law of oligarchy The principle that power tends to become concentrated in the hands of a few members of any organized group.

Juvenile crime Violations of the law committed by those under eighteen years of age.

Labeling theory The theory that deviance exists when some members of a group or society label others as deviants.

Latent function An unintended and unrecognized consequence of some element of a society.

Laws Norms that are formally defined and enforced by officials.

Liberation sociology The theoretical approach to sociology that seeks to replace human oppression with greater democracy and social justice.

Life chances The likelihood of securing the "good things in life," such as housing, education, good health, and food.

Life expectancy The average number of years that persons in a given population born at a particular time can be expected to live.

Life span The most advanced age to which humans can survive.

Looking-glass self One's self-concept based on perceptions of others' judgments.

Macrosociology The level of analysis that focuses on relationships among social structures without reference to the interaction of the people involved.

Magnet schools Public schools that attempt to achieve excellence by specializing in a certain area.

Managed care Any health-care system based on cost containment, control of medical decisions, and regulation of treatments.

Manifest function An intended and recognized consequence of some element of a society.

Marriage A legal union based on mutual rights and obligations.

Marriage rate The number of marriages per year for every one thousand members of a population.

Mass hysteria The form of collective behavior in which general social anxiety is created by acceptance of one or more false beliefs.

Mass media Those means of communication that reach large heterogeneous audiences without any personal interaction between the senders and the receivers of messages.

Master statuses Statuses that affect most other aspects of a person's life.

Matching Process in which participants in an experiment are matched in pairs according to all factors thought to affect the relationship being investigated.

Material culture The concrete, tangible objects within a culture.

Matriarchal control The form of control in which the oldest female living in a household has the authority.

Matrilineal descent The familial arrangement in which descent and inheritance are passed from the mother to her female descendants.

Matrilocal residence When a married couple is expected to live with the wife's parents.

"Me" The socialized part of the self.

Mechanical solidarity Social unity based on a consensus of values and norms, strong social pressure for conformity, and dependence on tradition and family.

Medical-industrial complex The network of physicians, medical schools, hospitals, medical equipment suppliers, pharmaceutical companies, and health insurance carriers.

Medicalization A process in which problems that were once considered matters of individual responsibility and choice are now considered a part of the domain of medicine and physicians.

Meritocracy A society in which social status is based on ability and achievement rather than background or parental status.

Metropolitan statistical area (MSA) A grouping of counties that contains at least one city of 50,000 inhabitants or more or an "urbanized area" of at least 50,000 inhabitants and a total metropolitan population of at least 100,000.

Microsociology The level of analysis concerned with the study of people as they interact in daily life.

Migration Movement of people from one geographical area to another for the purpose of establishing a new residence.

Minority A people who possess some distinctive physical or cultural characteristics, are dominated by the majority, and are denied equal treatment.

Mob An emotionally stimulated, disorderly crowd that is ready to use destructiveness and violence to achieve a specific purpose.

Modernization The great number of social changes accompanying economic development.

Monogamy The form of marriage in which one man is married to only one woman at a time.

Monopoly A single company controlling a particular market.

Morbidity Rates of disease and illness in a population.

Mores Norms that have great moral significance and that should be followed by the members of a society.

Mortality Deaths within a population.

Multiculturalism An approach to education in which the curriculum accents the viewpoints, experiences, and contributions of women and diverse ethnic and racial minorities.

Multinational corporations Firms in highly industrialized societies with operating facilities throughout the world.

Multiple causation The idea that an event occurs as a result of several factors operating in combination.

Multiple-nuclei theory A description of the process of urban growth that emphasizes specific geographical and historical influences on areas within cities.

Nationalism In a nation-state, a people's commitment to a common destiny based on recognition of a common past and a vision of a shared future.

Nation-state The political entity that holds authority over a specified territory.

Natural increase Population growth due to an excess of births over deaths.

Negative correlation A statistical measure in which the independent and dependent variables change in opposite directions.

Negative deviance Behavior that underconforms to accepted norms.

Neolocal residence When a newly married couple establishes a residence separate from either of their parents.

Net migration rate The annual increase or decrease per one thousand members of a population resulting from people leaving and entering the population.

Norms Rules defining appropriate and inappropriate behaviors.

Nuclear family The smallest group of individuals (mother, father, and children) that can be called a family.

Objectivity The principle stating that scientists are expected to prevent their personal biases from influencing the interpretation of their results.

Obligations Roles informing individuals of the behavior others expect from them.

Occupational sex segregation The concentration of men and women in different occupations.

Oligopoly A combination of companies controlling a market.

Open classroom A nonbureaucratic approach to education based on democracy, flexibility, and noncompetitiveness.

Open class system The type of stratification structure in which an individual's social status is based on merit and individual effort.

Open-ended questions Questions a person must answer in his or her own words.

Operational definition A definition of an abstract concept in terms of simpler, observable procedures.

Organic-adaptive systems Organizations based on rapid response to change rather than on the continuing implementation of established administrative principles.

Organic solidarity Social unity based on a complex of highly specialized roles that makes members of a society dependent on one another.

Out-group A group toward which one feels opposition, antagonism, or competition.

Overurbanization A situation in which a city cannot supply adequate jobs and housing for its inhabitants.

Panic Collective behavior that occurs when people react to a genuine threat in fearful, anxious, and often self-damaging ways.

Participant observation The type of field research technique in which a researcher becomes a member of the group being studied.

Pastoral societies Populations that depend primarily on raising and caring for domesticated animals for food.

Patriarchy The pattern in which the oldest man living in a household has authority over the rest of the family members. Also called *patriarchal control*.

Patrilineal descent The familial arrangement in which descent and inheritance are passed from the father to the male descendants.

Patrilocal residence When a married couple is expected to live with the husband's parents.

Peer group A group composed of individuals of roughly the same age and interests.

Peripheral theory Theory of city growth that emphasizes the growth of suburbs and edge cities around and away from central cities.

Peripheral tier The tier of the occupational structure composed of smaller, less profitable firms.

Personality The relatively organized complex of attitudes, beliefs, values, and behaviors associated with an individual.

Physician-hospital organizations (PHOs) Joint ventures between hospitals and organized groups of physicians to reach predetermined business objectives.

Play stage According to Mead, the stage of development during which children take on the roles of individuals, one at a time.

Pluralism The theory of power distribution that sees decision making as the result of competition, bargaining, and compromise among diverse special interest groups.

Political action committees (PACs) Organizations established by interest groups for the purpose of raising and distributing funds to selected political candidates.

Political institution The institution within which political power is supposed to be lodged and exercised.

Polyandry The form of marriage in which one woman is married to two or more men at the same time.

Polygamy Marriage of a male or female to multiple partners of the other sex.

Polygyny The form of marriage in which one man is married to two or more women at the same time.

Population All those people with the characteristics a researcher wants to study within the context of a particular research question.

Population control The conscious attempt to regulate population size through national birth control programs.

Population momentum When a population continues to grow, regardless of a recent drop in the birth rate to zero, because of the existing population base.

Population pyramids Graphic representations of the age and sex compositions of populations.

Positive correlation A statistical measure in which the independent and dependent variables change in the same direction.

Positive deviance Behavior that overconforms to social expectations.

Positivism The use of observation, experimentation, and other methods of the physical sciences in the study of social life.

Power The ability to control the behavior of others, even against their will.

Power elite A unified coalition of top military, corporate, and government leaders.

Preferred provider organizations (PPOs) Small groups of physicians providing a limited set of medical services to pre-enrolled patients at reduced prices.

Prejudice Widely held preconceptions of a group (minority or majority) and its individual members.

Presentation of self The ways that we, in a variety of social situations, attempt to create a favorable evaluation of ourselves in the minds of others.

Prestige Social recognition, respect, and admiration that a society attaches to a particular status.

Primary deviance Isolated acts of norm violation.

Primary group People who are emotionally close, know one another well, seek one another's company because they enjoy being together, and have a "we" feeling.

Primary metropolitan statistical area (PMSA) An urban area consisting of adjacent metropolitan areas that are closely integrated economically and socially.

Primary sector That part of the economy producing goods from the natural environment.

Profane Nonsacred aspects of our lives.

Proletariat Members of a society who labor for the bourgeoisie at subsistence wages.

Protestant ethic A set of values, norms, beliefs, and attitudes stressing hard work, thrift, and self-discipline.

Public policy A broad course of governmental action expressed in specific laws, programs, and initiatives.

Qualitative variable A variable that consists of variation in kind rather than in number.

Quantitative variable A variable that can be measured and given a numerical value.

Questionnaire A written set of questions participants fill out by themselves.

Race A category of people who are alleged to share certain biologically inherited physical characteristics that are considered socially important within a society.

Racism Ideas that attempt to connect biological characteristics with innate racial superiority or inferiority.

Randomization Process by which subjects are assigned to the experimental or control group on a random or chance basis.

Random sample A sample selected on the basis of chance so that each member of a population has an equal opportunity of being selected.

Rationalism The solution of problems on the basis of logic, data, and planning rather than tradition and superstition.

Rationalization The tendency to use knowledge and impersonality in social relationships to gain increased control over society.

Rational-legal authority Authority based on rules and procedures associated with political offices.

Real culture Actual (subterranean) patterns of thinking, feeling, and behaving.

Recidivism A return to crime after prison.

Redemptive movement The type of social movement that focuses on changing individuals rather than society.

Reference group A group one uses to evaluate oneself, and from which one acquires attitudes, beliefs, values, and norms.

Reflexes Simple, biologically inherited, automatic reactions to physical stimuli.

Reformative movement The type of social movement that aims to bring partial change to a society.

Rehabilitation The resocialization of criminals into conformity with the legal code of society.

Relative deprivation A condition that exists when people compare themselves with others and believe that they should have as much as those others have.

Relative poverty The designation of poverty resulting from a comparison of the economic condition of those at the bottom of a society with that of other segments of the population.

Reliability The ability of a measurement technique to yield consistent results.

Religion A unified system of beliefs and practices relating to sacred things.

Religiosity The ways in which people express their religious interests and convictions.

Replacement level The birth rate at which a married couple replaces itself in the population.

Resocialization The process of learning to adopt new norms, values, attitudes, and behaviors.

Resource mobilization The process through which members of a social movement secure and utilize the resources needed to press for social change.

Retribution The public demand that criminals pay compensation equal to their offenses against society.

Revolution The type of social movement that involves the toppling of a political regime through violent means.

Revolutionary movement The type of social movement that attempts to change society totally.

Rights Roles informing individuals of the behavior that can be expected from others.

Riots Crowds that engage in episodes of destructiveness and violence without a mob's sense of common purpose.

Role conflict Conflict between the performance of a role in one status with the performance of a role in another status.

Role performance The actual conduct involved in putting a role into action.

Roles Culturally defined rights and obligations attached to social statuses indicating the behavior expected of individuals holding them.

Role strain Conflicting roles within a single status.

Role taking The process of mentally assuming the viewpoint of another individual and then responding to oneself from that imagined viewpoint.

Rumor A widely circulating story of questionable truth.

Sacred Entities that are set apart and given a special meaning that transcends immediate human existence.

Sample A limited number of cases drawn from a population.

Sanctions Rewards and punishments used to encourage conformity to norms.

Scapegoat A substitute object that serves as a convenient and less feared target on which to place the blame for one's own troubles, frustrations, failures, or sense of guilt.

Secondary analysis The use of information already collected by someone else for another purpose.

Secondary deviance Deviance as a lifestyle and personal identity.

Secondary group A group that is impersonal and task oriented and involves only a segment of the lives and personalities of its members.

Secondary relationships Relationships that are impersonal, that involve only limited aspects of others' personality, and that exist to accomplish a specific purpose beyond the relationship itself.

Secondary sector That part of the economy engaged in manufacturing goods from raw materials.

Sect A religious organization formed when members of an existing religious organization break away in an attempt to reform the "parent" group.

Sector theory A description of the process of urban growth emphasizing the importance of transportation routes.

Secularization The process whereby religion loses its influence over society as a whole.

Self-concept An image of oneself as an entity separate from other people.

Self-fulfilling prophecy The situation in which an expectation leads to behavior that causes the expectation to become a reality.

Sex The biological distinction between male and female.

Sexism A set of beliefs, norms, and values used to justify sexual inequality.

Sexual harassment The use of one's superior power in making unwelcome sexual advances.

Sick role A confluence of appropriate behavior patterns for people who are ill, serving to remove them from active involvement in everyday routines, give them special protection and privileges, and set the stage for their return to their normal social roles.

Significant others Those persons whose judgments are the most important to an individual's self-concept.

Social aggregate A number of people who happen to be physically located together.

Social category A group of persons who share a social characteristic.

Social change Alterations in social structures that have long-term and relatively important consequences.

Social class A segment of a population whose members have a relatively similar share of the desirable things and who share attitudes, values, norms, and an identifiable lifestyle.

Social control Means for promoting conformity to norms.

Social demography The branch of demography that studies population patterns within a social context.

Social dynamics The study of social change.

Social exchange A form of social interaction in which one person voluntarily does something for another, expecting a reward in return.

Social gerontology The scientific study of the social dimensions of aging.

Social interaction The process by which people influence one another's behavior as they relate.

Social mobility The movement of individuals or groups within a stratification structure.

Social movement A large number of people acting together with some degree of leadership and organization over a relatively long period with the goal of promoting or preventing social change.

Social network A web of social relationships that joins a person to other people and groups.

Social statics The study of stability and order in society.

Social stratification The creation of layers (strata) of people possessing unequal shares of scarce desirables.

Social structure Patterned, recurring social relationships.

Socialism An economic system founded on the belief that the means of production should be controlled by the people as a whole and that government should plan and control the economy.

Socialization The process of learning to participate in group life through the acquisition of culture.

Society People who live within defined territorial borders and participate in a common culture.

Sociobiology The study of the biological basis of human behavior.

Sociological imagination The set of mind that allows individuals to see the relationship between events in their personal lives and events in their society.

Sociology The scientific study of social structure.

Spirit of capitalism The preference to accumulate capital as a moral obligation (not as a necessity) to reinvest rather than to use for consumption.

Spurious correlation An apparent relationship between two variables that is actually produced by a third variable that affects both of the original two variables.

Status The position that a person occupies within a social structure.

Status groups Groups composed of individuals who share a sense of status equality.

Status set All the statuses that an individual occupies at any particular time.

Stereotypes Ideas based on distortion, exaggeration, and oversimplification that are applied to all members of a social category.

Stigma An undesirable characteristic or label used by others to deny the deviant full social acceptance.

Strain theory The theory that deviance is most likely to occur when there is a discrepancy between culturally prescribed goals and legitimate (socially approved) means of obtaining them.

Stratified random sample A sample drawn from a population that has been divided into categories such as sex, race, or age. (The sample is selected at random from each category.)

Structural differentiation The division of one social structure into two or more social structures that operate more successfully separately than the one alone would under the new circumstances.

Structural mobility Mobility that occurs because of changes in the distribution of occupational opportunities.

Subculture A group that is part of the dominant culture but differs from it in some important respects.

Subjective approach A research method in which the aim is to understand some aspect of social reality through the study of the subjective interpretations of the participants themselves.

Suburbanization The process by which central cities lose population to the areas surrounding them.

Survey A research method in which people are asked to answer a series of questions.

Symbol Something that stands for, or represents, something else.

Symbolic interactionism The theoretical perspective that focuses on interaction among people based on mutually understood symbols.

Taboo A norm so important that its violation is considered repugnant.

Technology The combination of knowledge and hardware that is used to achieve practical goals.

Tertiary sector That part of the economy providing services.

Total fertility rate In a given population, the average number of children born to a woman during her lifetime.

Total institutions Places in which residents are separated from the rest of society and are controlled and manipulated by those in charge.

Totalitarianism The type of political system in which a ruler with absolute power attempts to control all phases of social life.

Tracking A process used in schools to place students in curricula consistent with the school's expectations of the student's eventual occupation.

Traditional authority Legitimate power rooted in custom.

Trained incapacity The situation that occurs when previous training prevents someone from adapting to new situations.

Underclass People typically unemployed who come from families with a history of unemployment for generations.

Unfulfilled rising expectations When newly raised hopes for a better life are either not satisfied at all or not satisfied as rapidly as people had expected.

Urban ecology The study of the relationships between humans and their urban environments.

Urbanism The idea that urbanization involves a distinctive way of life.

Urban legends Moralistic tales passed along by people who swear the stories happened to someone they know, or to an acquaintance of a friend or family member.

Urban society A society in which social relationships are impersonal and contractual, kinship is de-emphasized, cultural and social consensus are not complete, the division of labor is complex, and secular concerns outweigh sacred ones.

Urbanization The process by which an increasingly large proportion of the world's population lives in urban areas.

Validity The ability of a measurement technique to actually measure what it is designed to measure.

Value-added theory A theory outlining six conditions that are necessary and sufficient for the emergence of a social movement.

Value-free research Research in which personal biases are not allowed to affect the research process and its outcomes.

Values Broad cultural principles that most people in a society consider desirable.

Variable Something that occurs in different degrees among individuals, groups, objects, and events.

Verifiability A principle of science by which any given piece of research can be duplicated (replicated) by other scientists.

Verstehen The method of understanding social behavior by putting oneself in the place of others.

Vertical mobility An upward or downward change in status level.

Victim discounting The act of seeing a crime as less serious if the victim is a member of a minority or a lower social class.

Vouchers Government money made available to families with school-age children for use toward public, private, or parochial schools.

War Armed conflict that occurs within a society or between nations.

Wealth All the economic resources possessed by an individual or group.

White-collar crime A crime committed by respectable and high-status people in the course of their occupations.

Working poor People employed in low-skill jobs with the lowest pay who do not earn enough to rise out of poverty.

World-system theory A theory of modernization that sees the pattern of a nation's development as largely dependent on that nation's location in the world economy.

Zero population growth When deaths are balanced by births so that a population does not increase.

References

Abbott, Andrew. *Department and Discipline: Chicago Sociology at One Hundred.* Chicago: University of Chicago Press, 1999.

Aberle, David. *The Peyote Religion Among the Navaho.* Norman: University of Oklahoma Press, 1991.

Abramsky, Sasha. "Hard-Time Kids." *The American Prospect* (August 27, 2001):16–21.

Abu-Lughod, Janet L. *Rabat.* Princeton, NJ: Princeton University Press, 1980.

———. *Sociology of the Twenty-First Century: Continuities and Cutting Edges.* Chicago: University of Chicago Press, 1999.

Acker, James R., Robert M. Bohm, and Charles S. Lanier (eds.). *America's Experiment with Capital Punishment.* Durham, NC: Carolina Academic Press, 1998.

Adams, Bert N. *Kinship in an Urban Setting.* Chicago: Markham, 1968.

———. "Isolation, Function, and Beyond: American Kinship in the 1960s." *Journal of Marriage and the Family* 32 (November 1970):575–597.

Adams, Rebecca G. "The Sociology of Jerry Garcia." *Garcia: Reflections* (1995):16–18.

Adams, Robert. *The Evolution of Urban Society.* Chicago: Aldine, 1965.

Addams, Jane. *Twenty Years at Hull House.* New York: New American Library, 1981.

Adelmann, P. K., T. C. Antonucci, S. E. Crohan, and L. M. Coleman. "Empty Nest, Cohort, and Employment in the Well-Being of Midlife Women." *Sex Roles* (1989):173–189.

Adelson, Joseph. "Why the Schools May Not Improve." *Commentary* 78 (October 1984):40–45.

Adler, Emily Stier, and Roger Clark. *How It's Done: An Invitation to Social Research.* 2nd ed. Belmont, CA: Wadsworth, 2003.

Adler, Jerry. "Spaced Out." *Newsweek* (January 13, 2003):48–51.

Adler, Patricia A., and Peter Adler. *Peer Power.* New Brunswick, NJ: Rutgers University Press, 1998.

Adorno, T. W., Else Frenkel-Brunswik, Daniel J. Levinson, and R. Nevitt Sanford. *The Authoritarian Personality.* New York: Harper, 1950.

Aguirre, Adalberto, and Jonathan H. Turner. *American Ethnicity.* 2nd ed. New York: McGraw-Hill, 1997.

Aguirre, B. E., E. L. Quarantelli, and Jorge L. Mendoza. "The Collective Behavior of Fads: The Characteristics, Effects, and Career of Streaking." *American Sociological Review* 53 (1988):569–584.

Ahlburg, Dennis A., and Carol J. De Vita. "New Realities of the American Family." *Population Bulletin* 47 (August 1992):1–44.

Ahrens, Frank. "FCC Eases Media Ownership Rules." *Washington Post* (June 3, 2003):A1, A6, A7.

"Air Force Reports 54 Rapes, Assault." Associated Press, March 6, 2003.

Alba, Richard D. "The Twilight of Americans of European Ancestry: The Case of Italians." *Ethnic and Racial Studies* 8 (January 1985):134–158.

———. *The Transformation of White America.* New Haven, CT: Yale University Press, 1990.

Albert, Michel. *Capitalism vs. Capitalism.* New York: Four Walls Eight Windows, 1993.

Aldrich, Howard. *Organizations Evolving.* Thousand Oaks, CA: Sage, 1999.

Aldridge, Alan. *Religion in the Contemporary World.* Malden, MA: Blackwell, 2000.

Alexander, Jeffrey C., Bernard Giesen, Richard Munch, and Neil J. Smelser (eds.). *The Micro-Macro Link.* Berkeley: University of California Press, 1987.

Alfino, Mark, John S. Caputo, and Robin Wynyard. *McDonaldization Revisited.* Westport, CT: Praeger, 1998.

Alford, Robert R., and Roger Friedland. *Power of Theory.* Cambridge, UK: Cambridge University Press, 1985.

Allport, Gordon. *The Nature of Prejudice.* Reading, MA: Addison-Wesley, 1979.

Alter, Catherine, and Jerald Hage. *Organizations Working Together.* Newbury Park, CA: Sage, 1992.

Alter, Jonathan. "How America's Meanest Lobby Ran Out of Ammo." *Newsweek* (May 16, 1994):24–25.

Alterman, Eric. *What Liberal Media? The Truth About Bias and the News.* New York: Basic Books, 2003.

Altman, Daniel. "We Saved Your Job, But Gave You More Work." *New York Times* (October 29, 2002):G4.

Altman, Dennis. *AIDS in the Mind of America.* Garden City, NY: Anchor Books, 1986.

Altman, Stuart H., et al. "Escalating Health Care Spending: Is It Desirable or Inevitable?" *Health Affairs* (January 8, 2003). Available online at www.healthaffairs.org.

America 2000: An Educational Strategy. Washington, DC: U.S. Department of Education, 1991.

American Association of School Administrators. *Sex Equality in Educational Materials.* Executive Handbook Series, Vol. IV. Washington, DC: American Association of School Administrators, 1974.

American Sociological Association. "Code of Ethics." Washington, DC, 1997.

Amsden, Alice H. *The Rise of "the Rest": Challenges to the West from Late-Industrializing Economies.* Oxford, UK: Oxford University Press, 2001.

Anders, George. *Health Against Wealth.* Boston: Houghton Mifflin, 1996.

Andersen, Margaret L. *Thinking About Women.* 6th ed. Boston: Allyn & Bacon, 2002.

Anderson, Bernard E. "The Black Worker: Continuing Quest for Economic Parity." In Lee A. Daniels (ed.), *The State of Black America 2002.* New York: National Urban League, 2002, pp. 51–67.

Anderson, Craig A., and Brad J. Bushman. "The Effects of Media Violence on Society." *Science* 295 (29 March 2002):2377–2378.

Anderson, Curt. "Anti-Muslim Hate Crimes Soared Last Year, FBI Says." *The Philadelphia Inquirer* (November 26, 2002).

Anderson, Elijah, and Douglas S. Massey (eds.). *Problem of the Century: Racial Stratification in the United States.* New York: Russell Sage, 2001.

Anderson, Jack. *The Anderson Papers.* New York: Ballantine Books, 1974.

Anderson, T. B. "Widowhood as a Life Transition: Its Impact on Kinship Ties." *Journal of Marriage and the Family* 46 (February 1984):105–114.

Andrews, Lori B. *Black Power, White Blood.* New York: Pantheon Books, 1996.

Annandale, Ellen. *The Sociology of Health and Medicine: A Critical Introduction.* Malden, MA: Blackwell, 1998.

Antoun, Richard T. *Understanding Fundamentalism: Christian, Islamic, and Jewish Movements.* Thousand Oaks, CA: AltaMira Press, 2001.

Appel, Willa. *Cults in America.* New York: Holt, Rinehart and Winston, 1983.

Appiah, Kwame Anthony, and Amy Gutmann. *Color Conscious: The Political Morality of Race.* Princeton, NJ: Princeton University Press, 1998.

Appleby, Joyce. *Inheriting the Revolution: The First Generation of Americans.* Cambridge, MA: Harvard University Press, 2000.

Appleby, Julie. "Rethinking Managed Care." *USA Today* (October 7, 1999):1A, 2A.

Armour, Stephanie. "Faced with Less Time Off, Workers Take More." *USA Today* (October 29, 2002):1A.

Aronoff, Craig. "Old Age in Prime Time." *Journal of Communication* 24 (1974):86–87.

Aronowitz, Robert A. *Making Sense of Illness*. New York: Cambridge University Press, 1998.

Aronowitz, Stanley, and Henry A. Giroux. *Education Still under Siege*. 2nd ed. Toronto: OISE Press, 1994.

Aronson, E., and A. Gonzalez. "Desegregation, Jigsaw, and the Mexican American Experience." In P. A. Katz and D. A. Taylor (eds.), *Eliminating Racism*. New York: Plenum Press, 1988, pp. 301–314.

Arrigo, Bruce A. *Social Justice/Criminal Justice: The Maturation of Critical Theory in Law, Crime, and Deviance*. Belmont, CA: Wadsworth, 1999.

Arrow, Kenneth Joseph, Steven N. Durlauf, and Samuel Bowles (eds.). *Meritocracy and Economic Inequality*. Princeton, NJ: Princeton University Press, 2000.

Asch, Solomon E. "Opinions and Social Pressure." *Scientific American* 193 (November 1955):31–35.

Ashford, Lori S. "New Population Policies: Advancing Women's Health & Rights." *Population Bulletin* 56 (March 2001):1–44.

Ashour, Ahmen Sakr. "The Contingency Model of Leadership Effectiveness: An Evaluation." *Organizational Behavior and Human Performance* 9 (June 1973):339–355.

Atchley, Robert C. *Social Forces and Aging*. 9th ed. Belmont, CA: Wadsworth, 2000.

———. *Continuity and Adaptation in Aging: Creating Positive Experiences*. Baltimore, MD: Johns Hopkins University Press, 2001.

Atchley, Robert C., and S. Miller. "Types of Elderly Couples." In T. H. Brubaker (ed.), *Family Relationships in Later Life*. Beverly Hills, CA: Sage, 1983, pp. 77–90.

Atkinson, Paul. "Ethnomethodology: A Critical Review." In W. Richard Scott and Judith Blake (eds.), *Annual Review of Sociology* 14 (1988):441–465.

Atkinson, Rick. "Germany's Health-Care System Such a Success It Needs Reform." *Washington Post* (June 22, 1994):A1.

Audi, Robert, and Nicholas Wolterstorff. *Religion in the Public Square: Debating Church and State*. Lanham, MD: Rowman & Littlefield, 1996.

Axtell, Roger E. *Gestures: The Do's and Taboos of Body Language Around the World*. New York: Wiley, 1998.

Babbie, Earl R. *The Basics of Social Research*. 2nd ed. Belmont, CA: Wadsworth, 2003.

Baca Zinn, Maxine, and D. Stanley Eitzen. "Economic Restructuring and Systems Inequality." In Margaret L. Andersen and Patricia Hill (eds.), *Race, Class, and Gender: An Anthology*. 4th ed. Belmont, CA: Wadsworth, 2000, pp. 229–233.

Bachu, Amara, and Martin O'Connell. *Fertility of American Women: June 2000*. Current Population Reports. Series P20–543. Washington, DC: U.S. Government Printing Office, 2001.

Bagdikian, Ben. *The Media Monopoly*. 6th ed. Boston: Beacon Press, 2000.

Bainbridge, William Sims. *The Sociology of Religious Movements*. New York: Routledge, 1997.

Baldassare, Mark. "Suburban Communities." *Annual Review of Sociology* 18 (1992):475–494.

Bales, Robert F. *Interaction Process Analysis*. Reading, MA: Addison-Wesley, 1950.

Ballantine, Jeanne H. *The Sociology of Education*. 5th ed. Englewood Cliffs, NJ: Prentice Hall, 2001.

Ballantine, Jeanne H., and Joan Z. Spade (eds.). *Schools and Society: A Sociological Approach to Education*. Belmont, CA: Wadsworth/Thomson, 2001.

Balrig, Flemming. *The Snow-White Image*. Oslo: Norwegian University Press, 1988.

Banner, Stuart. *The Death Penalty: An American History*. Cambridge, MA: Harvard University Press, 2003.

Banton, Michael. *Racial Theories*. New York: Cambridge University Press, 1998.

Bardes, Barbara A., Mack C. Shelley, and Steffen W. Schmidt. *American Government and Politics Today: The Essentials*. Belmont, CA: Wadsworth/Thomson Learning, 2002.

Bardo, John W., and John J. Hartman. *Urban Sociology*. Itasca, IL: Peacock, 1982.

Barker, Eileen. *The Making of a Moonie*. London: Basil Blackwell, 1984.

Barlett, Donald L., and James B. Steele. *America: Who Stole the Dream?* Kansas City, KA: Andrews and McMeel, 1996.

———. *The Great American Tax Dodge: How Spiraling Fraud and Avoidance are Killing Fairness, Destroying the Income Tax, and Costing You*. Boston: Little, Brown and Company, 2000.

Barner, Robert. "The New Millennium Workplace: Seven Changes That Will Challenge Managers—and Workers." *The Futurist* 30 (1996):14–19.

Barnes, Gill Gorell, et al. *Growing Up in Stepfamilies*. New York: Oxford University Press, 1998.

Barnes, Julian E. "The SAT Revolution." *U.S. News and World Report* (November 11, 2002).

Barnett, Ola W., Cindy L. Miller-Perrin, and Robin D. Perrin. *Family Violence Across the Lifespan*. Newbury Park, CA: Sage, 1997.

Barr, C. "Pushing the Envelope: What Curriculum Integration Can Be." In E. Brazee and J. Capelluti (eds.), *Dissolving Boundaries: Toward an Integrative Curriculum*. Columbus, OH: National Middle School Association, 1995.

Barr, Stephen. "Women Make Gains in Parity." *Washington Post* (July 21, 1999):A19.

Barrett, Michael J. "The Case for More School Days." *The Atlantic Monthly* (November 1990):78–106.

Barrington, Linda. "Does a Rising Tide Lift All Boats? America's Full-Time Working Poor Reap Limited Gains in the New Economy." New York: The Conference Board, Inc., 2000.

Barron, James. "Fondly Recalling a Martian Invasion." *New York Times* (September 16, 1988):B1, B3.

Barron, Milton L. "Minority Group Characteristics of the Aged in American Society." *Journal of Gerontology* 8 (October 1953):477–482.

Barrow, Georgia M. *Aging, the Individual, and Society*. 7th ed. St. Paul, MN: West, 1999.

Barry, Herbert III, Margaret K. Bacon, and Irvin L. Child. "A Cross-Cultural Survey of Some Sex Differences in Socialization." *Journal of Abnormal and Social Psychology* 55 (November 1957):327–332.

Barry, Norman P. *Welfare*. 2nd ed. Minneapolis: University of Minnesota Press, 1999.

Baruch, Grace, and Jeanne Brooks-Gunn (eds.). *Women in Midlife*. New York: Plenum, 1984.

Baumohl, Bernard. "When Downsizing Becomes Dumbsizing." *Time* (March 15, 1993):55.

Baumol, William J., and Alan S. Blinder. *Economics*. 8th ed. New York: Harcourt, 2001.

Baumol, William J., and Julian N. May. *Downsizing*. New York: Russell Sage Foundation, 2003.

Baxandall, Rosalyn, and Elizabeth Ewen. *Picture Windows: How the Suburbs Happened*. New York: Basic Books, 2000.

Baxandall, Rosalyn, and Linda Gordon. "Second-Wave Soundings." *The Nation* (July 3, 2000):28–32.

Baxter, Janeen. "Is Husband's Class Enough? Class Location and Class Identity in the United States, Sweden, Norway, and Australia." *American Sociological Review* 59 (April 1994):220–235.

Bayles, Fred. "States Add Clergy to Sex-Abuse Laws." *USA Today* (July 5–7, 2002):1A.

Beals, Ralph L., Harry Hoijer, and Alan R. Beals. *An Introduction to Anthropology*. 5th ed. New York: Macmillan, 1977.

Bearman, Peter S., and Hannah Bruchner. "Opposite Sex Twins and Adolescent Same-Sex Attraction." *American Journal of Sociology* 107 (March 2002):1179–1205.

Beck, Melinda. "Doctors Under the Knife." *Newsweek* (April 5, 1993):28–33.

Becker, Howard S. *Art Worlds*. Berkeley: University of California Press, 1982.

———. *Outsiders*. New York: Free Press, 1991.

Becker, Howard S., et al. *Boys in White*. Chicago: University of Chicago Press, 1961.

Beckford, James A. *Cult Controversies*. London: Tavistock, 1985.

Begley, Sharon. "Not Just a Pretty Face." *Newsweek* (November 1, 1993):63–67.

———. "The 3–Million-Year-Old Man." *Newsweek* (April 11, 1994):84.

———. "The Science Wars." *Newsweek* (April 21, 1997):54–57.

———. "Getting Inside a Teen Brain." *Newsweek* (February 28, 2000):58–59.

Behr, Peter. "Enron Skirted Taxes Via Executive Pay Plan." *Washington Post* (February 14, 2003):E1, E3.

Bell, Daniel. "Equality and Equity." *The Seminar on Technology and Culture at MIT.* Boston: Massachusetts Institute of Technology, 1976.

———. "Downfall of the Business Giants." *Dissent* (Summer 1993):316–323.

———. *The Coming of Post-Industrial Society.* New York: Basic Books, 1999.

Bellah, Robert N. "Civil Religion in America." *Daedalus* 96 (Winter 1967):1–21.

———. *The Broken Covenant.* 2nd ed. Chicago: University of Chicago Press, 1992.

Bellah, Robert N., and Phillip E. Hammond. *Varieties of Civil Religion.* New York: Harper and Row, 1980.

Bellah, Robert N., et al. *The Good Society.* New York: Knopf, 1991.

Belluck, Pam. "Dr. Kevorkian Is a Murderer, The Jury Finds." *New York Times* (March 27, 1999).

Belton, Beth. "Despite Humming Economy, Workers Sweat Job Security." *USA Today* (March 2, 1999):A1.

Bendix, Reinhard. *Max Weber.* Garden City, NY: Doubleday, 1962.

Benet, Sula. "Why They Live to Be 100, Or Even Older, in Abkhasia." *The New York Times Magazine* (December 26, 1971):3ff.

Bennett, William J. "American Education: Making It Work." *Chronicle of Higher Education* 34 (May 4, 1988):29–41.

Bennis, Warren G. "Beyond Bureaucracy." *Transaction* 2 (July/August 1965):31–35.

———. "Response to Shariff: Beyond Bureaucracy Baiting." *Social Science Quarterly* 60 (June 1979):20–24.

Benokraitis, Nijole V. *Subtle Sexism.* Thousand Oaks, CA: Sage, 1997.

———. *Marriages and Families: Changes, Choices and Constraints.* 4th ed. Upper Saddle River, NJ: Prentice Hall, 2001.

Benowitz, June Melby, and Yvonne M. Vowels. *Encyclopedia of American Women and Religion.* Santa Barbara, CA: ABC-CLIO, Incorporated, 1998.

Bensman, Joseph, and Israel Gerver. "Crime and Punishment in the Factory: The Function of Deviancy in Maintaining the Social System." *American Sociological Review* 28 (August 1963):588–598.

Berger, Bennett M., Bruce Hacket, and R. Mervyn Millar. "The Communal Family." *The Family Coordinator* 21 (October 1972):419–427.

Berger, Joseph, M. Hamit Fisek, Cecilia L. Ridgeway, and Robert Z. Norman. "The Legitimization and Delegitimization of Power and Prestige Orders." *American Sociological Review* 63 (June 1998):379–405.

Berger, Michael, and Charles Gaffney. "The Rush Is On: Mine the 'Silver Generation'." *Business Week* (June 15, 1987):52–53.

Berger, Peter L. *The Sacred Canopy: Elements of a Sociological Theory of Religion.* New York: Doubleday, 1990.

Berk, Richard A. *Collective Behavior.* Dubuque, IA: William C. Brown, 1974.

Berle, Adolph A. Jr., and Gardiner C. Means. *The Modern Corporation and Private Property.* New York: Harcourt Brace Jovanovich, 1968.

Berlin, Brent, and Paul Kay. *Basic Color Terms.* Berkeley: University of California Press, 1969.

Bernard, Thomas J. "Structure and Control: Reconsidering Hirschi's Concept of Commitment." *Justice Quarterly* 4 (1987):409–424.

Bernhardt, Annette D., et al. *Divergent Paths: Economic Mobility in the New American Labor Market.* New York: Russell Sage Foundation, 2001.

Bernstein, Aaron. "Inequality: How the Gap Between the Rich and Poor Hurts the Economy." *Business Week* (August 15, 1994):78–83.

Bernstein, Aaron, and Alexandra Starr. "Welfare Reform, Round 2." *Business Week* (June 24, 2002):60–62.

Berry, Brian J. L. "Internal Structure of the City." *Law and Contemporary Problems* 30 (Winter 1965):111–119.

Berry, Brian J. L., and John D. Kasadra. *Contemporary Urban Ecology.* New York: Macmillan, 1977.

Bertrand, Marianne, and Sendhil Mullainathan. "Are Emily and Brendan More Employable than Lakisha and Jamal? A Field Experiment on Labor Market Discrimination." Chicago: University of Chicago Graduate School of Business, Working Paper (November 18, 2002):3–28.

Best, Joel. *Damned Lies and Statistics: Untangling Numbers from the Media, Politicians, and Activists.* Berkeley, CA: University of California Press, 2001.

Bhalla, Surjit. *Imagine There's No Country: Poverty, Inequality, and Growth in the Era of Globalization.* Washington, DC: Institute for International Economics, 2002.

Bhargava, Deepak, and Joan Kuriansky. "Really Out of Line." *Washington Post* (September 15, 2002):B2.

Bianchi, Suzanne M., "America's Children: Mixed Prospects." *Population Bulletin* 45 (June 1990):3–41.

Bianchi, Suzanne M., and Lynne M. Casper. "American Families." *Population Bulletin* 55 (December 2000):1–44.

Bianchi, Suzanne M., and Daphne Spain. *Women, Work, and Family in America.* Washington, DC: Population Reference Bureau, 1996.

Bickman, Leonard, and Debra J. Rog (eds.). *Handbook of Applied Social Research Methods.* Thousand Oaks, CA: Sage, 1997.

Bielby, William T., and Denise D. Bielby. "Family Ties: Balancing Commitments to Work and Family in Dual Earner Households." *American Sociological Review* 54 (October 1989):776–789.

Bierman, John. "Frustration and Fury." *Maclean's* (January 29, 1990):29.

Bierstedt, Robert. "Sociology and General Education." In Charles H. Page (ed.), *Sociology and Contemporary Education.* New York: Random House, 1963, pp. 40–55.

Bigler, Rebecca, and Lynn Liber. "The Role of Attitudes and Interventions in Gender-Schematic Processing." *Child Development* 61 (1990):1440–1452.

Billson, Janet Mancini, and Bettina Huber. "Embarking Upon a Career with an Undergraduate Degree in Sociology." Washington, DC: American Sociological Association, 1993.

Bingham, Richard D., and William M. Bowen (eds.). *Beyond Edge Cities.* New York: Garland Publishing, 1997.

Birnbaum, Jeffrey. *The Money Men: The Real Story of Political Power in the USA.* New York: Crown, 2000.

Birnbaum, Jeffrey H., and Alan S. Murray. *Showdown at Gucci Gulch.* New York: Vintage, 1988.

Black, Clifford. "Clinical Sociology and Criminal Justice Professions." *Free Inquiry in Creative Sociology* 12 (1984):117–120.

Black, Thomas R. *Doing Quantitative Research in the Social Sciences.* Thousand Oaks, CA: Sage, 2001.

Blackwell, John Reid. "Chance of Higher Cigarette Tax Dims." *Richmond Times Dispatch* (January 23, 2003).

"Black/White Relations in the United States." *The Gallup Poll Social Audit.* Princeton, NJ: The Gallup Organization, 1997.

Blane, David, George Davey-Smith, and Mel Bartley (eds.). *The Sociology of Health Inequalities.* Malden, MA: Blackwell, 1998.

Blank, Rebecca M. *It Takes a Nation: A New Agenda for Fighting Poverty.* Princeton, NJ: Princeton University Press, 1998.

Blank, Rebecca M., and Ron Haskins. *The New World of Welfare.* Washington, DC: Brookings Institution Press, 2001.

Blank, Rolf K., Roger E. LeVine, and Lauri Steel. "After 15 Years: Magnet Schools in Urban Education." In Bruce Fuller and Richard F. Elmore (eds.), *Who Wins, Who Loses?* New York: Teachers College Press, 1996, pp. 154–172.

Blau, Francine D., and Ronald G. Ehrenberg (eds.). *Gender and Family Issues in the Workplace.* New York: Russell Sage Foundation, 1997.

Blau, Peter M. *Exchange and Power in Social Life.* New Brunswick: Transaction Books, 1986.

Blau, Peter M., and Otis Dudley Duncan. *The American Occupational Structure.* New York: Wiley, 1967.

Blau, Peter M., and Marshall W. Meyer. *Bureaucracy in Modern Society.* 3rd ed. New York: Random House, 1987.

Blau, Peter M., and W. Richard Scott. *Formal Organizations*. London: Routledge and Kegan Paul, 1982.

Blee, Kathleen M., and Dwight B. Billings. *The Road to Poverty: The Making of Wealth and Hardship in Appalachia*. Cambridge, UK: Cambridge University Press, 1999.

Block, Jeanne H., Jack Block, and Per F. Gjerde. "The Personality of Children Prior to Divorce." *Child Development* 57(1986):827–840.

Bloom, Allan. *The Closing of the American Mind*. New York: Simon & Schuster, 1987.

Blum, Deborah. *Love at Goon Park: Harry Harlow and the Science of Affection*. Cambridge, MA: Perseus Publishing, 2002.

Blum, Justin, and Jay Mathews. "Quality Uneven, Despite Popularity." *The Washington Post* (June 19, 2003):A1,A12,A13.

Blumer, Herbert. "Collective Behavior." In Alfred M. Lee (ed.), *Principles of Sociology*. 3rd ed. New York: Barnes and Noble, 1969a, pp. 65–121.

———. *Symbolic Interactionism*. Englewood Cliffs, NJ: Prentice Hall, 1969b.

Blumstein, Alfred, and Joel Wallman (eds.). *The Crime Drop in America*. New York: Cambridge University Press, 2000.

Bodenheimer, Thomas. "The American Health Care System—Physicians and the Changing Medical Marketplace." *New England Journal of Medicine* 340 (1999):584–588.

Bolt, David, and Ray Crawford. *Digital Divide: Computers and Our Children's Future*. New York: TV Books, 2000.

Bonacich, Edna. "Advanced Capitalism and Black/White Race Relations in the United States: A Split Labor Market Interpretation." *American Sociological Review* 41 (February 1976):34–51.

Booth, Alan, and Ann C. Crouter (eds.). *Men in Families: When Do They Get Involved? What Difference Does It Make?* Mahwah, NJ: Erlbaum, 1998.

Booth, William. "1990's Social Reformer Now Toes the Line." *Washington Post* (July 12, 1994):A1, A4.

———. "Oregon's Assisted Suicides Upheld." *Washington Post* (April 18, 2002):A1, A7.

Boroughs, Don L. "Winter of Discontent." *U.S. News & World Report* (January 22, 1996):47–54.

Bositis, David A. *Black Elected Officials 99*. Lanham, MD: University Press of America, 2001.

———. *Black Elected Officials: A Statistical Summary 2000*. Washington, DC: Joint Center for Political and Economic Studies. 2002.

Bottomore, Tom B. *Karl Marx*. New York: McGraw-Hill, 1963.

———. *Political Sociology*. 2nd ed. Minneapolis: University of Minnesota Press, 1993.

Bouvier, Leon F., and Carol J. De Vita. "The Baby Boom—Entering Midlife." *Population Bulletin* 46 (November 1991):1–35.

Bouza, Anthony. *How to Stop Crime*. New York: Plenum, 1993.

Bowen, William G., and Derek Bok. *The Shape of the River: Long-Term Consequences of Considering Race in College and University Admissions*. Princeton, NJ: Princeton University Press, 2000.

Bower, Amanda. "Dead Men Walking." *Time* (January 20, 2003):40.

Bowles, Samuel, and Herbert Gintis. *Schooling in Capitalist America*. New York: Basic Books, 1976.

Box, Steven. *Deviance, Reality and Society*. 2nd ed. New York: Holt, Rinehart and Winston, 1981.

Brady, Erik. "Time Fails to Lessen Title IX Furor." *USA Today* (June 19, 2002):1C, 2C.

Bradley, Harriet. *Men's Work, Women's Work: A Sociological History of the Sexual Division of Labor in Employment*. Minneapolis: University of Minnesota Press, 1989.

Brajuha, Mario, and Lyle Hallowell. "Legal Intrusion and the Politics of Fieldwork: The Impact of the Brajuha Case." *Urban Life* 14 (January 1986):454–478.

Branscum, Deborah. "Go On, Break the Chain." *Newsweek* (March 1, 1999):58.

Bransten, L. "The Americas: U.S. Health Care Costs." *Financial Times* 25 (March 1997).

Brauer, Jerald C. (ed.). *Religion and the American Revolution*. Philadelphia: Fortress Press, 1976.

Brave, Ralph. "Governing the Genome." *The Nation* (December 10, 2001):18–24.

Bridges, George S., and Robert D. Crutchfield. "Law, Social Standing, and Imprisonment." *Social Forces* 66 (March 1988):699–724.

Brint, Steven, and Steve Brint. *Schools and Societies*. Thousand Oaks, CA: Pine Forge Press, 1998.

Brinton, Crane. *Ideas and Men*. 2d ed. Englewood Cliffs, NJ: Prentice-Hall, 1963.

———. *The Anatomy of Revolution*. New York: Knopf, 1990.

Brockerhoff, Martin P. "An Urbanizing World." *Population Bulletin* 55 (September 2000):1–44.

Brody, William. "Building a New Kind of Academy." *Planning for Higher Education* 25 (Summer 1997):72–74.

Bromley, David G. (ed.). *The Politics of Religious Apostasy: The Role of Apostates in the Transformation of Religious Movements*. Westport, CT: Praeger, 1998.

Bromley, David G., and Anson D. Shupe Jr. *New Christian Politics*. Macon, GA: Mercer University Press, 1984.

Bronfenbrenner, Urie. *Two Worlds of Childhood*. New York: Russell Sage Foundation, 1970.

Brouilette, John R. "The Liberalizing Effect of Taking Introductory Sociology." *Teaching Sociology* 12 (January 1985):131–148.

Brown, Lester R. *State of the World 1996*. New York: Norton, 1996.

Brown, Michael K. *Race, Money, and the American Welfare State*. Ithaca, NY: Cornell University Press, 1999.

Brown, Susan L., and Alan Booth. "Cohabitation versus Marriage: A Comparison of Relationship Quality." *Journal of Marriage and the Family* 58 (August 1996):668–678.

Browne, Beverly A. "Gender Stereotypes in Advertising on Children's Television in the 1990s: A Cross-National Analysis." *Journal of Advertising* 27 (Spring 1998):83–96.

Brubaker, T. H. "Family in Later Life: A Burgeoning Research Area." *Journal of Marriage and the Family* 52 (November 1990):959–981.

———. "Families in Later Life: A Burgeoning Research Area." In A. Booth (ed.), *Contemporary Families*. Minneapolis: National Council on Family Relations, 1991, pp. 226–248.

Bruce, Steve. *God is Dead: Secularization in the West*. Malden, MA: Blackwell, 2002.

Bruner, Jerome. "Schooling Children in a Nasty Climate." *Psychology Today* (January 1982):57–63.

Brunner, Borgna (ed.). *Time Almanac 2001*. Boston: Family Education Company, 2000.

Brunner, H. G., M. Nelsen, X. O. Breakefield, H. H. Ropers, and B. A. van Oost. "Abnormal Behavior Associated with a Point Mutation in the Structural Gene for Monoamine Oxidase A." *Science* 262 (October 22, 1993):578–580.

Brunvand, Jan Harold. "Urban Legends: Folklore for Today." *Psychology Today* 14 (June 1980):50 ff.

———. *The Vanishing Hitchhiker*. New York: Norton, 1981.

——— (ed.). *American Folklore: An Encyclopedia*. New York: Garland Publishing, 1998.

———. *Too Good to Be True: The Colossal Book of Urban Legends*. New York: Norton, 2000.

Brzezinski, Matthew. "Fortress America." *New York Times Magazine* (February 23, 2003):38–45, 54, 70, 75–76.

Buckley, Stephen. "Shrugging Off the Burden of a Brainy Image." *Washington Post* (June 17, 1991):D1, D6.

Budrys, Grace. *Our Unsystematic Health Care System*. Lanham, MD: Rowman & Littlefield, 2001.

Buechler, Steven M. *Social Movements in Advanced Capitalism*. New York: Oxford University Press, 2000.

Bulkeley, William M. "Family Portrait: Web Pages Allow Families to Keep in Touch." *Wall Street Journal* (June 16, 1997):R24.

Bunzel, John H. (ed.). *Challenge to American Schools*. New York: Oxford University Press, 1985.

Burgess, Ernest W. "The Growth of the City." In Robert E. Park, Ernest W. Burgess, and Robert D. McKenzie (eds.), *The City*. Chicago: University of Chicago Press, 1925, pp. 47–62.

Burgess, J. K. "Widowers." In C. S. Chilman, E. W. Nunnally, and F. M. Cox (eds.), *Variant Family Forms*. Beverly Hills, CA: Sage, 1988, pp. 150–164.

Burke, Ronald, and Cary Cooper (eds.). *The Organization in Crisis: Downsizing, Restructuring and Privatization*. Cambridge, MA: Blackwell, 2000.

Burns, T., and G. M. Stalker. *The Management of Innovation*. 2nd ed. London: Tavistock, 1966.

Burros, Marian. "Even Women at the Top Still Have Floors to Do." *New York Times* (May 31, 1993):A1, A11.

Burstein, Paul. "Legal Mobilization as a Social Movement Tactic: The Struggle for Equal Employment Opportunity." *American Journal of Sociology* 96 (March 1991):1201–1225.

Buss, David M. "The Strategies of Human Mating: People Worldwide Are Attracted to the Same Qualities in the Opposite Sex." *American Scientist* 82 (May 1994a):238–249.

———. *The Evolution of Desire: Strategies of Human Mating*. New York: Basic Books, 1994b.

———. "Psychological Sex Differences: Origins Through Sexual Selection." *The American Psychologist* 50 (March 1995):164.

Buss, David M., Neil M. Malamuth, and Barbara A. Winstead. "Sex, Power, Conflict: Evolutionary and Feminist Perspectives." *Contemporary Psychology* 43 (April 1998).

Butler, Joseph. *Five Sermons*. Indianapolis: Hackett, 1983.

Butler, Robert N. *Why Survive?* New York: Harper and Row, 1975.

Byron, Christopher. "Seems Like Old Times." *New York* (September 27, 1993):22–23.

Calasanti, Toni M. "Participation in a Dual Economy and Adjustment to Retirement." *International Journal of Aging and Human Development* 26 (1988):13–27.

Calasanti, Toni M., and Kathleen F. Slevin. *Gender, Social Inequalities, and Aging*. Walnut Creek, CA: AltaMira Press, 2001.

Calavita, Kitty, Henry N. Pontell, and Robert H. Tillman. *Big Money Crime*. Berkeley: University of California Press, 1999.

Califano, Joseph A. Jr. *America's Health Care Revolution*. New York: Random House, 1985.

Callan, Victor J. "The Voluntarily Childless and Their Perceptions of Parenthood and Childlessness." *Journal of Comparative Family Studies* 14 (Spring 1983):87–96.

Camarota, Steven A. *The New Americans: Economic, Demographic, and Fiscal Effects of Immigration*. Washington, DC: National Research Council, 1999.

———. "Immigrants in the United States—2000: A Snapshot of America's Foreign-Born Population." *Backgrounder*. Washington, DC: Center for Immigration Studies, 2001.

———. "Immigrants in the United States—2002." Washington, DC: Center for Immigration Studies, 2002.

Camic, Charles. *Reclaiming the Sociological Classics*. Malden, MA: Blackwell, 1998.

Campbell, Frances A., and Craig T. Ramey. "Effects of Early Intervention on Intellectual and Academic Achievement: A Follow-up Study of Children from Low-Income Families." *Child Development* 65 (1994):684–698.

Campo-Flores, Adrian. "Giving Protest a Chance." *Newsweek* (February 3, 2003):28.

Campo-Flores, Adrian, and Yuval Rosenberg. "A Return to Wilding." *Newsweek* (June 26, 2000):28.

Canary, Daniel J., et al. "Diversity—Sex and Gender Differences in Personal Relationships." *Contemporary Psychology* 43 (1998):764.

Cancio, A. Silvia, T. David Evans, and David J. Maume Jr. "Reconsidering the Declining Significance of Race: Racial Differences in Early Career Wages." *American Sociological Review* 61 (August 1996):541–556.

Cantril, Hadley. *The Invasion from Mars*. Princeton, NJ: Princeton University Press, 1982.

Caplan, Nathan, Marcella H. Choy, and John K. Whitmore. "Indochinese Refugee Families and Academic Achievement." *Scientific American* 266 (February 1992):36–42.

Caplovitz, David. *The Poor Pay More*. New York: Free Press, 1963.

Caplow, Theodore. "Contrasting Trends in European and American Religion." *Sociological Analysis* 46 (1985):101–108.

———. *American Social Trends*. New York: Harcourt Brace Jovanovich, 1991.

———. "The Case of the Phantom Episcopalians." *American Sociological Review* 63 (February 1998):137–145.

Caplow, Theodore, Howard M. Bahr, and Bruce A. Chadwick. *All Faithful People*. Minneapolis: University of Minnesota Press, 1983.

Caplow, Theodore, Howard M. Bahr, Bruce A. Chadwick, Rueben Hill, and Margaret Holmes Williamson. *Middletown Families*. Minneapolis: University of Minnesota Press, 1982.

Cappelli, Peter, et al. *Change at Work*. New York: Oxford University Press, 1997.

Carnoy, Martin. *The Politics and Economics of Race in America*. Cambridge, UK: Cambridge University Press, 1994.

Carter, Cynthia, and Kay C. Weaver. *Violence and the Media*. Philadelphia: Taylor and Francis, 2003.

Casbergue, Renee M., and Judith Kieff. "Marbles Anyone? Traditional Games in the Classroom." *Childhood Education* (Spring 1998):143–147.

Cashion, Barbara G. "Female-Headed Families: Effects on Children and Clinical Implications." *Journal of Marital and Family Therapy* (April 1982):77–85.

Casler, Lawrence. "The Effects of Extra Tactile Stimulation on a Group of Institutionalized Infants." *Genetic Psychology Monographs* 71 (1965):137–175.

Casper, Lynne M., and Suzanne M. Bianchi. *Continuity and Change in the American Family: Anchoring the Future*. Thousand Oaks: Sage, 2002.

Cassirer, Ernst. *The Philosophy of the Enlightenment*. Princeton, NJ: Princeton University Press, 1951.

Castro, Janice. "Disposable Workers." *Time* (March 15, 1993):47.

Cauchon, Dennis. "Marijuana Attains Record Support." *USA Today* (August 24–26, 2001):1A.

Cavanaugh, John C., and Fredda Blanchard-Fields. *Adult Development and Aging*. 4th ed. Pacific Grove, CA: Duxbury, 2001.

Cavanaugh, Stephanie. "The Trophy Kitchen." *Washington Post* (June 10, 1999):T14–T15.

Cavender, Gary, and Mark Fishman (eds.). *Entertaining Crime*. Hawthorne, NY: Aldine de Gruyter, 1998.

Center for the American Woman and Politics. "Fact Sheet." June 2002.

Center for Voting and Democracy. "International Voter Turnout, 1991–2000." 2002. Available online at www.fairvote.org/turnout/intturnout.htm

Centers for Disease Control, National Center for Health Statistics. 2002. Available online at www.cdc.gov/nchs/fastats/hexpense.htm.

Centers, Richard. *The Psychology of Social Classes*. New Haven, CT: Yale University Press, 1949.

Central Intelligence Agency. *Factbook: United States*. 2002. Available online at www.cia.gov/cia/publications/factbook/geos/us.html.

Chael, David. *New Poverty*. Westport, CT: Greenwood Publishing Company, 1999.

Chafetz, Janet Saltzman. *Sex and Advantage*. Totowa, NJ: Rowman & Allanheld, 1984.

———. *Gender Equity*. Newbury Park, CA: Sage, 1990.

Chagnon, Napoleon A. *Yanomamö*. 5th ed. Fort Worth, TX: Harcourt Brace College Publishers, 1997.

Chamberlain, Gary. "Metal Wars." *Design News* (October 4, 1999):115.

Chambliss, William J. "The Saints and the Roughnecks." *Society* 11 (November/December 1973):24–31.

Chambliss, William J., and Robert Seidman. *Law, Order, and Power*. 2nd ed. Reading, MA: Vantage, 1982.

Chandbasekaran, Rajiv. "Microsoft Trial Ends After 8 Months, " *Washington Post* (June 25, 1999):A1, A20.

Chang, Iris. *The Rape of Nanking: The Forgotten Holocaust of World War II*. New York: Viking Penguin, 1998.

Chang, Kenneth. "Panel Says Bell Labs Scientist Faked Discoveries in Physics." *New York Times* (September 26, 2002):A1, A20.

Changing Public Attitudes toward the Criminal Justice System. Open Society Institute. Criminal Justice Initiative, 2002.

Chase-Dunn, Christopher. *Global Formation: Structures of the World Economy.* Lanham, MD: Rowman & Littlefield, 1999.

Chase-Dunn, Christopher, and Thomas D. Hall. *Rise and Demise: Comparing World Systems.* Boulder, CO: Westview Press, 1997.

Chaves, Mark. "Abiding Faith." *Contexts* 1 (Summer 2002):19–26.

Cherlin, Andrew J., et al. "Longitudinal Studies of Effects of Divorce on Children in Great Britain and the United States." *Science* 7 (June 1991):1386–1389.

Cherlin, Andrew J. "A 'Quieting' of Change." *Contexts* (Spring 2002):67–68.

Cherry, Robert, and William M. Rodgers III (eds.). *Prosperity for All? The Economic Boom and African Americans.* New York: Russell Sage Foundation, 2000.

Chesler, Ellen. "New Options, New Politics." *The American Prospect* (Fall 2001):12–14.

Child Trends DataBank. Washington, DC: Population Reference Bureau, 2002. Available online at www.childtrendsdatabank.org.

Children's Defense Fund. *The State of America's Children – 1991.* Washington, DC: Children's Defense Fund, 1991.

Chirot, Daniel. *How Societies Change.* Thousand Oaks, CA: Pine Forge Press, 1994.

Chirot, Daniel, and Martin E. P. Seligman (eds.). Ethnopolitical Warfare: Causes, Consequences, and Political Solutions. Washington, DC: American Psychological Association, 2001.

Choi, N. G. "Correlates of the Economic Status of Widowed and Divorced Elderly Men." *Journal of Family Issues* 12 (March 1992):38–54.

Choldin, Harvey M., and Claudine Hanson. "Subcommunity and Change in a Changing Metropolis." *Sociological Quarterly* 22 (Autumn 1981):549–564.

Christenson, James A., and Ronald C. Wimberley. "Who Is Civil Religious?" *Sociological Analysis* 39 (Spring 1978):77–83.

Christie, Nils. *Crime Control as Industry.* 3rd ed. London: Routledge, 2000.

Chriswick, Barry R. "The Skills and Economic Status of American Jewry: Trends over the Last Half-Century." *Journal of Labor Economics* 11(1993):229–242.

Chua, Amy. *World on Fire: How Exporting Free Market Democracy Breeds Ethnic Hatred and Global Instability.* New York: Doubleday Publishing, 2002.

Cicourel, Aaron, and John Kitsuse. *The Educational Decision Makers.* New York: Bobbs-Merrill, 1963.

Citizens for Tax Justice. "Year-by-Year Analysis of the Bush Tax Cuts Shows Growing Tilt to the Very Rich." 2002. Available online at www.ctj.org/html/gwb0602.htm.

Clabes, J. "Caught in the Middle: Sandwich Generation." *Santa Barbara News Press* (May 1, 1989):B-10.

Clark, Charles S. "Cults in America." *CQ Researcher* 3 (May 7, 1993):385–408.

Clark, Lee. "Panic: Myth or Reality." *Contexts* 1 (Fall 2002):21–26.

Clausen, John A. *The Life Course.* Englewood Cliffs, NJ: Prentice Hall, 1986.

Clayton, Obie Jr. (ed.). *An American Dilemma Revisited.* New York: Russell Sage Foundation, 1996.

Clayton, Richard R., and James W. Gladden. "The Five Dimensions of Religiosity: Toward Demythologizing a Sacred Artifact." *Journal for the Scientific Study of Religion* 13 (June 1974):135–143.

Clemens, Elisabeth S. *The People's Lobby.* Chicago: University of Chicago Press, 1997.

"Closing the Gap." *The Economist* (September 2, 1995):74.

Cloud, John. "A Kinder, Gentler Death." *Time* (September 18, 2000):60–67.

Cloward, Richard A., and Lloyd E. Ohlin. *Delinquency and Opportunity.* London: Routledge, 1998.

Coakley, Jay J. "The Sociological Perspective: Alternate Causations of Violence in Sport." In D. Stanley Eitzen (ed.), *Sport in Contemporary Society.* 4th ed. New York: St. Martin's Press, 1993, pp. 98–111.

——— (ed.). *Sport in Society.* 6th ed. Boston: Irwin/McGraw-Hill, 1998.

Cockburn, Alexander. "Beat the Devil." *The Nation* (January 17, 1994):42–43.

Cockburn, C. "The Relations of Technology." In R. Crompton, and M. Mann (eds.), *Gender and Stratification.* Cambridge, UK: Polity, 1986.

Cockerham, William C. "Sociology and Psychiatry." In H. Kaplan and B. Sadeck (eds.), *Comprehensive Textbook of Psychiatry.* Vol. 1, 4th ed. Baltimore: Williams and Wilkins, 1985, pp. 265–273.

———. *Medical Sociology.* 6th ed. Englewood Cliffs, NJ: Prentice Hall, 1995.

———. *Sociology of Mental Disorder.* 5th ed. Upper Saddle River, NJ: Prentice Hall, 1999.

———. *Medical Sociology.* 8th ed. Upper Saddle River, NJ: Prentice Hall, 2000.

Cohen, Adam. "A Curse of Cliques." *Time* (May 3, 1999):44.

Cohen, Adam, and Cathy Booth Thomas. "Inside a Layoff." *Time* (April 16, 2001):38–40.

Cohen, Albert K. *Delinquent Boys.* New York: Free Press, 1977.

Cohen, David, and Marvin Lazerson. "Education and the Corporate Order." *Socialist Revolution* 2 (March/April 1972):47–72.

Cohen, Jere. *Protestantism and Capitalism: The Mechanisms of Influence.* Hawthorne, NY: Aldine de Gruyter, 2002.

Cohn, D'Vera. "Married-With-Children Still Fading." *Washington Post* (May 15, 2001):A1, A9.

———. "Hispanics Declared Largest Minority." *The Washington Post* (June 19, 2003):A1,A6.

Coleman, James S. "The Concept of Equality of Educational Opportunity." *Harvard Educational Review* 38 (Winter 1968):7–22.

———. *Equal Educational Opportunity.* Cambridge, MA: Harvard University Press, 1969.

———. *Equality and Achievement in Education.* Boulder, CO: Westview Press, 1990.

Coleman, James S., Thomas Hoffer, and Sally B. Kilgore. "Questions and Answers: Our Response." *Harvard Educational Review* 51 (November 1981):526–545.

Coleman, James S., Sally B. Kilgore, and Thomas Hoffer. "Public and Private Schools." *Society* 19 (January/February 1982):4–9.

Coleman, James S., et al. *Equality of Educational Opportunity.* Washington, DC: U.S. Government Printing Office, 1966.

———. "Cognitive Outcomes in Public and Private Schools." *Sociology of Education* 55 (April/July 1982a):65–76.

———. "Achievement and Segregation in Secondary Schools: A Further Look at Public and Private School Differences." *Sociology of Education* 55 (April/July 1982b):162–182.

Coleman, James W. "Toward an Integrated Theory of White-Collar Crime." *American Journal of Sociology* 93 (September 1987):406–439.

———. *The Criminal Elite: The Sociology of White Collar Crime.* New York: St. Martin's Press, 1989.

———. *The Criminal Elite.* 4th ed. New York: St. Martin's Press, 1997.

Coleman, L. M., T. C. Antoncucci, P. K. Adelmann, and S. E. Crohan. "Social Roles in the Lives of Middle-Aged and Older Black Women." *Journal of Marriage and the Family* 49 (November 1987):761–771.

The College Board. *Trends in College Pricing 1999.* New York: College Entrance Examination Board, 1999.

The College Board. *Trends in College Pricing 2001.* New York: College Entrance Examination Board, 2001.

Collier, Theresa J. "The Stigma of Mental Illness." *Newsweek* (April 26, 1993):16.

Collins, Chuck, and Felice Yeskel. *Economic Apartheid in America: A Primer on Economic Inequality and Insecurity.* New York: New Press, 2000.

Collins, John J. *The Cult Experience.* New York: Charles C. Thomas, 1991.

Collins, Randall. "Functional and Conflict Theories of Educational Stratification." *American Sociological Review* 36 (December 1971):1002–1019.

———. *Conflict Sociology.* New York: Academic Press, 1975.

———. *The Credential Society: An Historical Sociology of Education.* New York: Academic Press, 1979.

———. *Four Sociological Traditions.* New York: Oxford University Press, 1994.

———. "Conflict Theory of Educational Stratification." In Jeanne H. Ballantine, and Joan Z. Spade (eds.), *Schools and Society: A Sociological Approach to Education.* Belmont, CA: Wadsworth/Thomson, 2001.

Colombotos, John, and Corinne Kirchner. *Physicians and Social Change.* New York: Oxford University Press, 1986.

Comfort, Alex. "Age Prejudice in America." *Social Policy* 17 (November/December 1976):3–8.

Comte, Auguste. *The Positive Philosophy.* Translated and edited by Harriet Martineau. London: Bell, 1915.

Connolly, Ceci, and Amy Goldstein. "Health Insurance Back as Key Issue." *Washington Post* (March 16, 2003):A5.

Conrad, Peter, and Rochelle Kern (eds.). *Sociology of Health and Illness.* 3rd ed. New York: St. Martin's Press, 1990.

Cook, Karen S. "Exchange and Power in Networks of Interorganizational Relations." *Sociological Quarterly* 18 (Winter 1977):62–82.

Cooley, Charles Horton. *Human Nature and the Social Order.* New York: Scribner's, 1902.

Cooley, Charles Horton and Hans-Joachim Schubert. *On Self and Social Organization.* Chicago: University of Chicago Press, 1998.

Coon, Dennis. *Psychology: A Journey.* Belmont, CA: Wadsworth/Thomson Learning, 2002.

Cooperman, Alan. "Episcopal Vote Allows Blessings of Gay Unions." *The Washington Post* (August 7, 2003):A1,A9.

Copsey, William, and Brian McNeill. "'Sin Tax' Vote Raises Suspicions." *Collegiate Times* (January 31, 2003):1–2.

Cose, Ellis. *The Rage of the Privileged Class.* New York: HarperCollins, 1995.

———. "The Good News About Black America." *Newsweek* (June 7, 1999):29–40.

———. "The Black Gender Gap." *Newsweek* (March 3, 2003):48–51.

Coser, Lewis A. *The Functions of Social Conflict.* London: Routledge, 1998.

Costain, Anne N., and Andrew S. McFarland (eds.). *Social Movements and American Political Institutions.* Lanham, MD: Rowman & Littlefield, 1998.

Cowgill, Donald O., and Lowell D. Holmes (eds.). *Aging and Modernization.* New York: Appleton-Century-Crofts, 1972.

Cowley, Geoffrey, and Bill Turque. "Critical Condition." *Newsweek* (November 8, 1999):59–61.

Cowley, Geoffrey, and Anne Underwood. "Surgeon, Drop That Scalpel." *Newsweek* (February 1998, Winter 1997–98 Special Edition):77–78.

Cox, Frank D. *Human Intimacy.* 9th ed. Belmont, CA: Wadsworth, 2002.

Cox, Harvey. *Turning East: The Promise and Peril of the New Orientalism.* New York: Simon & Schuster, 1977.

———. *Religion in the Secular City.* New York: Simon & Schuster, 1992.

———. *Fire from Heaven.* London: Cassell, 1996.

Cox, Howard G. "Roles for Aged Individuals in Post-Industrial Societies." *International Journal of Aging and Human Development* 3 (1990):55–62.

Cox, Kenya L. C., and William E. Spriggs. "Special Research Report: Negative Effects of TANF on College Enrollment." In Lee A. Daniels (ed.), *The State of Black America 2002.* New York: National Urban League, 2002, pp. 223–251.

Cozic, Charles P. (ed.). *The Militia Movement.* San Diego, CA: Greenhaven Press, 1997.

Craig, Steve. *Men, Masculinity and the Media.* Thousand Oaks: Sage, 1992.

Crain, Robert L. "School Integration and Occupational Achievement of Negroes." *American Journal of Sociology* 75 (January 1970):593–606.

Crain, Robert L., and Carol S. Weisman. *Discrimination, Personality and Achievements.* New York: Seminar Press, 1972.

Crane, Diana. *Fashion and Its Social Agendas: Class, Gender, and Identity in Clothing.* Chicago: University of Chicago Press, 2001.

Creekmore, C. R. "Cities Won't Drive You Crazy." *Psychology Today* (January 1985):46–53.

Cress, Daniel M., and David A. Snow. "Mobilization at the Margins: Resources, Benefactors, and the Viability of Homeless Social Movement Organizations." *American Sociological Review* 61 (December 1996):1089–1109.

Crewdson, John. "Fraud in Breast Cancer Study." *Chicago Tribune* (March 13, 1994):A1.

Croal, N'Gai. "Sims Family Values." *Newsweek* (November 25, 2002):46–49.

Crosby, F. E. *Juggling.* New York: Free Press, 1993.

Crosby, John F. *Reply to Myth: Perspectives on Intimacy.* New York: Wiley, 1985.

Crossley, Nick. *Making Sense of Social Movements.* Buckingham: Open University Press, 2002.

Crow, Barbara A. (ed.). *Radical Feminism: A Documentary Reader.* New York: New York University Press, 1998.

Cullen, John B., and Shelley M. Novick. "The Davis-Moore Theory of Stratification: A Further Examination and Extension." *American Journal of Sociology* 84 (May 1979):1424–1437.

"Cult Killings Exceed Jonestown Toll." *Los Angeles Times* (April 1, 2000):A14.

Cummings, Scott. "White Ethnics, Racial Prejudice, and Labor Market Segmentation." *American Journal of Sociology* 95 (January 1980):938–950.

Curley, Tom. "Hawaii Gay Marriage Case Has National Implications." *USA Today* (December 5, 1996):4A.

Curran, Daniel J., and Claire M. Renzetti. *Social Problems: Society in Crisis.* 6th ed. Upper Saddle River, NJ: Prentice Hall, 2000.

Curtiss, Susan. *Genie.* New York: Academic Press, 1977.

Cushman, Thomas. *Notes from Underground: Rock Music Counterculture in Russia.* New York: St. Martin's Press, 1995.

Cuzzort, R. P., and E. W. King. *Humanity and Modern Social Thought.* 2nd ed. Hinsdale, IL: Dryden Press, 1976.

Cyert, Richard M., and James G. March. *A Behavioral Theory of the Firm.* Englewood Cliffs, NJ: Prentice Hall, 1963.

Dahl, Robert. *Who Governs?* New Haven, CT: Yale University Press, 1989.

Dahrendorf, Ralf. "Out of Utopia: Toward a Reorientation of Sociological Analysis." *American Journal of Sociology* 64 (September 1958a):115–127.

———. "Toward a Theory of Social Conflict." *Journal of Conflict Resolution* 2 (June 1958b):170–183.

———. *Class and Class Conflict in Industrial Society.* Stanford, CA: Stanford University Press, 1959.

Dail, Paul W. "Prime-Time Television Portrayals of Older Adults in the Context of Family Life." *Gerontologist* 28 (1988):700–706.

Daly, Martin, and Margo Wilson. "Crime and Conflict: Homicide in Evolutionary Psychological Perspective." *Crime and Justice* 22 (1997).

Daniels, Lee A. (ed.). *The State of Black America 2002.* New York: National Urban League, 2002.

Darder, Antonia, and Rodolfo D. Torres. *The Latino Reader.* Malden, MA: Blackwell,1997.

Davidson, James D. "Glock's Model of Religious Commitment: Assessing Some Different Approaches and Results." *Review of Religious Research* 16 (Winter 1975):83–93.

Davies, Bronwyn. *Frogs and Snails and Feminist Tales.* New York: Pandora Press, 1990.

Davies, Christie. "Sexual Taboos and Social Boundaries." *American Journal of Sociology* 87 (March 1982):1032–1063.

Davies, James Chowning. "The J-Curve of Rising and Declining Satisfactions as a Cause of Revolution and Rebellion." In Hugh Davis Graham and Ted Robert Gurr (eds.), *Violence in America.* Rev. ed. Beverly Hills, CA: Sage, 1979.

Davis, F. James. *Minority-Dominant Relations.* Arlington Heights, IL: AHM Publishing Corporation, 1978.

———. "Up and Down Opportunity's Ladder." *Public Opinion* 5 (June/July 1982):11 ff.

Davis, Kingsley. "Final Note on a Case of Extreme Isolation." *American Journal of Sociology* 52 (March 1947):232–247.

———. *Human Society*. New York: Macmillan, 1949.

———. "The Origin and Growth of Urbanization in the World." *American Journal of Sociology* 60 (March 1955):429–437.

Davis, Kingsley, and D. Rowland. "Old and Poor: Policy Challenges in the 1990s." 2 (1991):37–59.

Dawkins, Richard. *The Selfish Gene*. 2nd ed. New York: Oxford University Press, 1990.

Dawson, Lorne L. *Cults and New Religious Movements: A Reader*. Malden, MA: Blackwell Publishers, 2003.

Deak, JoAnn. *How Girls Thrive*. Washington, DC: National Association of Independent Schools, 1998.

Death Penalty Information Center. "Race of Defendants Executed Since 1976." 2002. Available online at www.deathpenaltyinfo.org/dpicrace.html.

DeBarros, Anthony. "New Baby Boom Swamps Colleges." *USA Today* (January 2, 2003):1A, 2A.

Decrem, Bart. "Plugged In." 1998. Available online at www.pluggedin.org.

Deegan, Mary J. *Women in Sociology: A Bio-Bibliographical Source Book*. Westport, CT: Greenwood, 1991.

———. *Jane Addams and the Men of the Chicago School, 1892–1918*. New Brunswick, NJ: Transaction Publishers, 2000.

Deegan, Mary Jo, and Michael Stein. "American Drama and Ritual: Nebraska Football." *International Review of Sport Sociology* 3 (1978):31–44.

DeFleur, Melvin L., and Sandra Ball-Rokeach. *Theories of Mass Communication*. 4th ed. New York: Longman, 1982.

de La Isla, Jose. *The Rise of Hispanic Political Power*. Santa Maria, CA: Archer Books, 2003.

De Luca, Rita Caccamo. *Back to Middletown: Three Generations of Sociological Reflections*. Stanford, CA: Stanford University Press, 2001.

del Pinal, Jorge, and Audrey Singer. "Generations of Diversity: Latinos in the United States." *Population Bulletin* 52 (October 1997):1–48.

Delgado, Richard, and Jean Stefancic (eds.). *The Latino Condition: A Critical Reader*. New York: New York University Press, 1998.

Deloria, Vine Jr., and Clifford Lytle. *The Nations Within*. New York: Pantheon Books, 1984.

Demaine, Jack (ed.). *Sociology of Education Today*. New York: Palgrave, 2001.

Demerath, N. J. III, and Rhys H. Williams. "Civil Religion in an Uncivil Society." *The Annals* 480 (July 1985):154–166.

Demo, D. H., and A. C. Acock. "The Impact of Divorce on Children." In A. Booth (ed.), *Contemporary Families*. Minneapolis: National Council on Family Relations, 1991, pp. 162–191.

Dentler, Robert A. "The Political Sociology of the African American Situation: Gunnar Myrdal's Era and Today." *Daedalus* 124 (Winter 1995):15–36.

Denzer, Susan. "The Graying of Japan." *U.S. News and World Report* (September 30, 1991):65–73.

Dervarics, Charles. "Is Welfare Reform Reforming Welfare?" *Population Today* 26 (October 1998):1–8.

DeSpelder, L. A., and D. L. Strickland. *The Last Dance*. Mountain View, CA: Mayfield Press, 1991.

Deutschman, Alan. "Pioneers of the New Balance." *Fortune* (May 20, 1991):60–68.

de Waal, Frans B. M. "Cultural Primatology Comes of Age." *Nature* 399 (June 1999):635–636.

Dewar, Helen. "Senate Approves Disclosure Measure." *Washington Post* (June 30, 2000):A1, A4.

Dewar, Helen, and Barbara Vobejda. "Clinton Signs Head Start Expansion." *Washington Post* (May 19, 1994):A1.

Dewey, John. *The School and Society: And the Child and the Curriculum*. Mineola, NY: Dover, 2001.

DeWitt, Karen. "Battle Is Looming on U.S. College Aid to Poor Students." *New York Times* (May 28, 1991):1, 9.

———. "Blacks Prone to Job Dismissal in Organizations." *New York Times* (April 20, 1995):A19.

DeYoung, Alan J. *Economics and American Education*. White Plains, NY: Longman, 1989.

Diamond, Jared. *The Third Chimpanzee: The Evolution and Future of the Human Animal*. New York: HarperPerennial, 1993.

———. *Guns, Germs, and Steel: The Fates of Human Societies*. New York: Norton, 1999.

Diani, Mario, and Donatella della Porta. *Social Movements: An Introduction*. Malden, MA: Blackwell Publishers, 1999.

Dickerson, John F., and Karen Tumulty. "Has Bush's Medicare Plan Got a Chance?" *Time* (February 10, 2003).

DiIulio, John J. Jr., and Anne Morrison Piehl. "Does Prison Pay?: The Stormy National Debate Over the Cost-Effectiveness of Imprisonment." *The Brookings Review* (Fall 1991):28–35.

DiMaggio, Paul (ed.). *The Twenty-First-Century Firm: Changing Economic Organization in International Perspective*. Princeton, NJ: Princeton University Press, 2001.

Dines, Gail, and Jean M. Humez (eds.). *Gender, Race, and Class in Media: A Text-Reader*. 2nd ed. Thousand Oaks, CA: Sage, 2002.

DiPrete, Thomas A. *Collectivist vs. Individualist Mobility Regimes?* Stockholm, Sweden: Stockholm University, 1996.

Directory of Corporate Affiliations. New Providence, NJ: LexisNexis Group, 2001.

Dobzhansky, Theodosius. *Mankind Evolving*. Toronto: Bantam Books, 1970.

Dollard, John et al. *Frustration and Aggression*. New Haven, CT: Yale University Press, 1939.

Domhoff, G. William. *The Power Elite and the State*. Hawthorne, NY: Aldine de Gruyter, 1990.

———. *State Autonomy or Class Dominance*. New York: Aldine de Gruyter, 1996.

———. *Who Rules America Now?* New York: Touchstone Books, 1997.

———. *Changing the Powers That Be: How the Left Can Stop Losing and Win*. Lanham, MD: Rowman and Littlefield, 2003.

Dooley, Calvin M. *Income and Racial Disparities in the Undercount in the 2000 Presidential Election*. Washington, DC: U.S. House of Representatives Committee on Government Reform, 2001.

Doreian, Patrick, and Frans N. Stokman. *Evolution of Social Networks*. Amsterdam: Gordon and Breach Publishers, 1997.

Dornbush, Sanford M., et al. "Single Parents, Extended Households, and the Control of Adolescents." *Child Development* 56 (1985):326–341.

Douglas, Mary. "Accounting for Taste." *Psychology Today* 13 (July 1979):44 ff.

Dowbiggin, Ian Robert, and Ian Dowbiggin. *Merciful End: The Euthanasia Movement in Modern America*. New York: Oxford University Press, 2003.

Doyle, James A. *The Male Experience*. 3rd ed. Madison, WI: Brown and Benchmark, 1995.

Doyle, Rodger. "Income Inequality in the U.S." *Scientific American* (June 1999):26–27.

Draut, Tammy. "New Opportunities?: Public Opinion on Poverty, Income Inequality and Public Policy: 1996–2002." New York: D_mos, 2002.

Dreyfuss, Robert, and Jason Vest. "The Lie Factory." *Mother Jones* 29 (January/February 2004):34–41.

Drucker, Peter F. "The Coming of the New Organization." *Harvard Business Review* 66 (January/February 1988):45–53.

Du Bois, W. E. B. *The Philadelphia Negro: A Social Study*. New York: Schocken, 1967.

Du Bois, W. E. B., and David Levering Lewis (ed.). *W. E. B. Du Bois: A Reader*. New York: Henry Holt, 1994.

Du Bois, William, and R. Wright. *Applying Sociology: Making a Better World*. Boston: Allyn & Bacon, 2000.

Dudley, Kathryn Marie. *The End of the Line: Lost Jobs, New Lives in Postindustrial America*. Chicago: University of Chicago Press, 1997.

Dudley, William. *Media Violence: Opposing Viewpoints*. San Diego, CA: Greenhaven Press, 1999.

Duff, Christina. "Profiling the Aged: Fat Cats or Hungry Victims." *Wall Street Journal* (September 28, 1995):B1, B16.

Duggan, Paul. "Texas Executes Gary Graham for '81 Slaying." *Washington Post* (June 23, 2000):A1, A14.

Duke, Lynne, and Gabriel Escobar. "A Looting Binge Born of Necessity, Opportunity." *Washington Post* (May 10, 1992):A1, A23.

Durant, Thomas J., and J. David Knottnerus (eds.). *Plantation Society and Race Relations: The Origins of Inequality.* Westport, CT: Greenwood Publishing Group, 1999.

Durkheim, Emile. *Suicide.* Translated by John A. Spaulding and George Simpson. Edited by George Simpson. New York: Free Press, 1964.

———. *The Rules of Sociological Method.* Translated by Sarah A. Solovay and John H. Mueller. Edited by George E. G. Catlin. New York: Free Press, 1966.

———. *The Elementary Forms of Religious Life.* New York: Free Press, 1995.

———. *The Division of Labor in Society.* New York: Free Press, 1997.

Dychtwald, Ken. *Age Wave.* New York: Bantam Books, 1990.

Dye, Thomas R. *Who's Running America?* 7th ed. Englewood Cliffs, NJ: Prentice Hall, 2001.

Dynes, Russell R., and Kathleen J. Tierney (eds.). *Disasters, Collective Behavior, and Social Organization.* Newark: University of Delaware Press, 1994.

Eade, John, and Christopher Mele (eds.). *Understanding the City: Contemporary and Future Perspectives.* Malden, MA: Blackwell Publishers, 2002.

Eagleton, Terry. *Ideology.* London: Longman, 1994.

Eagley, Alice H., and Wendy Wood. "The Origins of Sex Differences in Human Behavior: Evolved Dispositions versus Social Roles." *American Psychologist* 54 (1999):408–423.

Eaton, Susan E. *The Other Boston Busing Story: What's Won and Lost Across the Boundary Line.* New Haven, CT: Yale University Press, 2001.

Ebenstein, Alan O., William Ebenstein, and Edwin Fogelman. *Today's Isms.* 11th ed. Upper Saddle River, NJ: Prentice Hall, 2000.

Eberstadt, Nicholas. "The Population Implosion." *Foreign Policy* (March/April 2001):42–53.

Eccles, Jacquelynne, et al. "Development during Adolescence: The Impact of Stage-Environment Fit on Young Adolescents' Experiences in Schools and in Families." *American Psychologist* 48 (February, 1993):90–101.

Eck, Diana L. *A New Religious America: How a "Christian Country" Has Become the World's Most Religiously Diverse Nation.* New York: HarperCollins Publishers, 2002.

The Economist. "On the Right Track?" (March 2, 2002):59–60.

Eder, Donna. *School Talk.* New Brunswick, NJ: Rutgers University Press, 1995.

Edin, Kathryn, and Laura Lein. *Making Ends Meet: How Single Mothers Survive Welfare and Low-Wage Work.* New York: Russell Sage Foundation, 1997.

Edmonds, Patricia, and Mark Potok. "Followers 'See Themselves on Side of Jesus.'" *USA Today* (April 20, 1993):A1, A2.

Edsall, Thomas B. "New Ways to Harness Soft Money in Works." *Washington Post* (August 25, 2002):A1, A6, A7.

———. "Bush Has A Cabinet Full of Wealth." *Washington Post* (September 18, 2002):A27.

Edwards, Ellen, "Kids on Scooters Are Swarming the Streets." *Kids Post* (August 30, 2000):C15.

Ehrenreich, Barbara. "Helping the Rich Stay That Way." *Time* (April 18, 1994):86.

Eisenstock, Barbara. "Sex-Role Differences in Children's Identification with Counterstereotypic Televised Portrayals." *Sex Roles* 10 (1984):417–430.

Eismeier, Theodore J., and Philip H. Pollock III. *Business, Money, and the Rise of Corporate PACs in American Elections.* Westport, CT: Greenwood Press, 1988.

Eitzen, D. Stanley. *Sport in Contemporary Society: An Anthology.* 6th ed. New York: Worth Publishers, 2000.

———. *Fair and Foul: Beyond the Myths and Paradoxes of Sport.* 2nd ed. Lanham, MD: Rowman & Littlefield, 2003.

Eitzen, D. Stanley, and Kelly Eitzen Smith. *Experiencing Poverty.* Belmont, CA: Wadsworth, 2003.

Ekordt, D. J. "Why the Nation Persists that Retirement Harms Health." *Gerontologist* 27 (1987):454–457.

Elikann, Peter T. *The Tough-On-Crime Myth: Real Solutions to Cut Crime.* New York: Plenum, 1996.

Elkin, Frederick, and Gerald Handel. *The Child and Society.* 6th ed. New York: McGraw-Hill, 2001.

Elliott, John E. *Comparative Economic Systems.* 2nd ed. Belmont, CA: Wadsworth, 1985.

Elliott, Michael. "Crime and Punishment." *Newsweek* (April 18, 1994):18–22.

Ellis, Joseph J. *Founding Brothers: The Revolutionary Generation.* New York: Knopf, 2000.

El Nasser, Haya. "Judges Say 'Scarlet Letter' Angle Works." *USA Today* (June 25, 1996):1A–2A.

Elster, Jon, Clause Offe, and Ulrich Preuss. *Institutional Design in Post-Communist Societies.* Cambridge, UK: Cambridge University Press, 1998.

Ember, Carol R., and Melvin Ember. *Anthropology.* 9th ed. Upper Saddle River, NJ: Prentice Hall, 1999.

Empey, LaMar T. *American Delinquency.* 2nd ed. Homewood, IL: Dorsey, 1982.

Encyclopedia of American Religions. 7th ed. Farmington Hills, MI: Gale Group, 2002.

Engels, Friedrich. *The Origin of the Family, Private Property, and the State.* New York: International Publishing, 2001.

Entwisle, Doris R., and Karl L. Alexander. "Early Schooling as a 'Critical Period' Phenomenon." In K. Namboodiri and R.G. Corwin (eds.), *Sociology of Education and Socialization.* Greenwich, CT: JAI Press, 1989, pp. 27–55.

———. "Entry into School: The Beginning School Transition and Educational Stratification in the United States." *Annual Review of Sociology* 19 (1993):401–423.

Epstein, T. Scarlett. *Economic Development and Social Change in South India.* New York: Humanities Press, 1962.

Erickson, Rosemary J., and Rita J. Simon. *The Use of Social Science Data in Supreme Court Decisions.* Urbana, IL: University of Illinois Press, 1998.

Erikson, Erik H. *Childhood and Society.* Harmondsworth, UK: Penguin Books, 1973.

———. *The Life Cycle Completed.* New York: Norton, 1982.

Erikson, Kai T. "A Comment on Disguised Observation in Sociology." *Social Problems* 14 (Spring 1967):366–373.

———. *Everything in Its Path.* New York: Simon & Schuster, 1976.

———. *A New Species of Trouble: Explorations in Disaster, Trauma, and Community.* New York: Norton, 1995.

———. *Sociological Visions.* Blue Ridge Summit, PA: Rowman & Littlefield, 1997.

Ermann, M. David, and Richard J. Lundman. *Corporate and Governmental Deviance.* 5th ed. New York: Oxford University Press, 1996.

Eshleman, J. Ross. *The Family.* 10th ed. Boston: Allyn & Bacon, 2002.

Esping-Andersen, Gosta. *Welfare States in Transition.* Thousand Oaks, CA: Sage, 1996.

Espiritu, Yen Le. *Asian American Men and Women.* Thousand Oaks, CA: Sage, 1996.

Esposito, John, and John Fiorillo. "Who's Left on the Block? New York City's Working Class Neighborhoods." In Hans Spiegel (ed.), *Citizen Participation in Urban Development.* Fairfax, VA: Learning Resources, 1974, pp. 307–334.

Etzioni, Amitai. *A Comparative Analysis of Complex Organizations.* Rev. ed. New York: Free Press, 1975.

———. "Education for Mutuality and Civility." *Futurist* 16 (October 1982):4–7.

———. *The Spirit of Community.* New York: Crown Publishers, 1993.

———. *Rights and the Common Good.* New York: St. Martin's Press, 1995.

———. *The Monochrome Society.* Princeton, NJ: Princeton University Press, 2001.

Evans, Jean. *Three Men*. New York: Knopf, 1954.

Evans, Sara M. *Personal Politics*. New York: Vintage, 1980.

———. *Born for Liberty: A History of Women in America*. New York: Free Press, 1989.

Executive Office of the President of the United States. "A Citizen's Guide to the Federal Budget." 2002. Available online at w3.access.gpo.gov/usbudget/fy2002/guidetoc.html.

Eyerman, Ron. *Studying Collective Action*. Newbury Park, CA: Sage, 1992.

Eysenck, Hans. *Crime and Personality*. 3rd ed. London: Routledge and Kegan Paul, 1977.

Fagot, Joël, Edward A. Wasserman, and Michael E. Young. "Discriminating the Relation Between Relations: The Role of Entropy in Abstract Conceptualization by Baboons (*Papio papio*) and Humans (*Homo sapiens*)." *Journal of Experimental Psychology: Animal Behavior Processes* 27 (October 2001):316–328.

Falk, Gerhard. *Stigma: How We Treat Outsiders*. Amherst, NY: Prometheus Books, 2001.

Falk, Ursula A., and Gerhard Falk. *Ageism, the Aged, and Aging in America*. Springfield, IL: C.C. Thomas, 1997.

Faltermayer, Charlotte. "Cyberveillance." *Time* (August 14, 2000):B22–B25.

Faludi, Susan. *The Betrayal of the American Man*. New York: William Morrow, 1999.

Farhi, Paul. "TV Ratings Agreement Reached." *Washington Post* (July 10, 1997):A1, A16.

Faris, Robert E., and H. Warren Dunham. *Mental Disorders in Urban Areas*. Chicago: University of Chicago Press, 1939.

Farley, Christopher John. "The Butt Stops Here." *Time* (April 18, 1994):58–64.

———. "Hip-Hop Nation." *Time* (February 8, 1999):54–63.

Farley, John E. *Majority-Minority Relations*. 3rd ed. Englewood Cliffs, NJ: Prentice Hall, 1995.

Farley, John E., and Mark Hansel. "The Ecological Context of Urban Crime: A Further Exploration." *Urban Affairs Quarterly* 17 (September 1981):37–54.

Farley, Reynolds. *Blacks and Whites*. Cambridge, MA: Harvard University Press, 1984.

———. "Modest Declines in U.S. Residential Segregation Observed." *Population Today* 25 (February 1997):1–2.

———. *The New American Reality: Who We Are, How We Got There, Where We Are Going*. New York: Russell Sage Foundation, 1998.

Farley, Reynolds, and William H. Frey. "Changes in the Segregation of Whites from Blacks During the 1980's: Small Steps Toward a More Integrated Society." *American Sociological Review* 59 (February 1994):23–45.

Farrell, Christopher. "Time to Prune the Ivy." *Business Week* (May 24, 1993):112–118.

Fasteau, Marc. "The Male Machine: The High Price of Macho." *Psychology Today* 8 (September 1975):60.

Fausto-Sterling, Anne. *Myths of Gender*. New York: Basic Books, 1987.

Feagin, Joe R. *Subordinating the Poor*. Englewood Cliffs, NJ: Prentice Hall, 1975.

———. "The Continuing Significance of Race: Antiblack Discrimination in Public Places." *American Sociological Review* 56 (February 1991):101–115.

———. *Racist America: Roots, Current Realities and Future Reparations*. New York: Routledge, 2001.

———. *Racial and Ethnic Relations*. 7th ed. Englewood Cliffs, NJ: Prentice Hall, 2002.

Feagin, Joe R., and Clairece Booker Feagin. *Discrimination American Style*. Malabar, FL: Krieger, 1986.

Feagin, Joe R., and Karyn D. McKinney. *The Many Costs of Racism*. Lanham, MD: Rowman & Littlefield, 2002.

Feagin, Joe R., and Hernán Vera. *Liberation Sociology*. Boulder, CO: Westview, 2001.

Feagin, Joe R., Hernán Vera, and Pinar Batur. *White Racism*. 2nd ed. New York: Routledge, 2001.

Featherman, David L. "The Socio-Economic Achievement of White Religio-Ethnic Sub-groups." *American Sociological Review* 36 (April 1971):207–222.

Fedarko, Kevin. "Bodies of Evidence." *Time* (December 6, 1993):70.

Federal Election Commission. "Campaign Finance Reports and Data." 2002. Available online at www.fec.gov/finance_reports.html.

Feeley, Malcolm, and Jonathon Simon. "The New Penology: Notes on the Emerging Strategy of Corrections and Implications." *Criminology* 30 (1992):449–474.

Fein, Helen. *Genocide*. Newbury Park, CA: Sage, 1993.

Fenn, Richard. *Toward a Theory of Secularization*. Storrs, CT: Society for the Scientific Study of Religion, 1978.

Fenwick, Charles R. "Crime and Justice in Japan: Implications for the United States." *International Journal of Comparative and Applied Criminal Justice* 6 (1982):62–71.

Ferber, Abby L. *White Man Failing: Race, Gender, and White Supremacy*. Lanham, MD: Rowman & Littlefield, 1998.

Ferguson, Sarah. "Testing Protest in New York." *Mother Jones* (January 24, 2002). Available online at www.motherjones.com/web_exclusives/features/news/wef.html.

Fergusson, D. M., L. J. Horwood, and F. T. Shannon. "A Proportional Hazards Model of Family Breakdown." *Journal of Marriage and Family* 46 (August 1984):539–549.

Ferrar, Jane W. "Marriages in Urban Communal Households: Comparing the Spiritual and the Secular." In Peter J. Stein, Judith Richman, and Natalie Hannon (eds.), *The Family*. Reading, MA: Addison-Wesley, 1977, pp. 409–419.

Ferree, M., and B. Hess. *Controversy and Coalition: The New Feminist Movement Across Three Decades of Change*. Old Tappan, NJ: Twayne Publishers, 1994.

Feuer, Lewis S. (ed.). *Marx and Engels*. New York: Doubleday, 1959.

Fiedler, Fred E. *A Theory of Leadership Effectiveness*. New York: McGraw-Hill, 1967.

Fine, Gary Alan. *With the Boys: Little League Baseball and Preadolescent Culture*. Chicago: University of Chicago Press, 1987.

———. "The Sad Demise, Mysterious Disappearance, and Glorious Triumph of Symbolic Interactionism." *Annual Review of Sociology* 19 (1993):61–87.

———. *The Culture of Restaurant Work*. Berkeley and Los Angeles: University of California Press, 1996.

Fine, Gary Alan, and Patricia Turner. *Whispers on the Color Line: Rumor and Race in America*. Berkeley, CA: University of California Press, 2001.

Fineman, Howard. "Leadership? Don't Ask Us." *Newsweek* (May 11, 1992):42.

———. "Shifting Racial Lines." *Newsweek* (July 10, 1995):38–39.

Fineman, Howard, and Michael Isikoff. "Laying Down the Law." *Newsweek* (August 5, 2002):20–23.

Finke, Roger, and Rodney Stark. *The Churching of America*. New Brunswick, NJ: Rutgers University Press, 1992.

Finlay, Barbara. *Facing the Stained Glass Ceiling: Gender in a Protestant Seminary*. Lanham, MD: University Press of America, 2003.

Fins, Deborah. *Death Row USA*. Criminal Justice Project, NAACP Legal Defense and Educational Fund, Inc., 2002.

Firebaugh, Glenn. "Structural Determinants of Urbanization in Asia and Latin America." *American Sociological Review* 44 (April 1979):199–215.

Fischer, Claude S. *To Dwell Among Friends*. Chicago: University of Chicago Press, 1982.

———. *The Urban Experience*. 2nd ed. New York: Harcourt Brace Jovanovich, 1984.

———. "The Subcultural Theory of Urbanism: A Twentieth-Year Assessment." *American Journal of Sociology* 101 (November 1995):543–577.

Fischer, Claude S., et al. *Inequality by Design: Cracking the Bell Curve Myth*. Princeton, NJ: Princeton University Press, 1996.

Fischer, David Hackett. *Growing Old in America*. New York: Oxford University Press, 1977.

Fisher, Helen E. *The First Sex: The Natural Talents of Women and How They are Changing the World*. New York: Random House, 1999.

Fischer, Joannie. "Scientists Have Finally Cloned a Human Embryo." *U.S. News & World Report* (December 3, 2001):51–63.

Flanagan, William G. *Contemporary Urban Sociology.* 4th ed. Boston: Allyn & Bacon, Inc., 2001.

Fletcher, Max E. "Harriet Martineau and Ayn Rand: Economics in the Guise of Fiction." *American Journal of Economics and Sociology* 33 (October 1974):367–379.

Fletcher, Michael A. "In Wisconsin, Vouching for Vouchers." *Washington Post* (March 20, 2001):A1, A14.

Florida, Richard, and Martin Kenney. "Transplanted Organizations: The Transfer of Japanese Industrial Organization to the U.S." *American Sociological Review* 56 (June 1991):381–398.

Flory, Richard W., and Donald E. Miller (eds.). *GenX Religion.* New York: Routledge, 2000.

Flynn, C. P. "Relationship Violence by Women: Issues and Implications." *Family Relations* 39 (1990):194–198.

Foerstel, Karen N., and Herbert N. Foerstel. *Climbing the Hill: Gender Conflict in Congress.* Westport, CA: Greenwood, 1996.

Fogarty, Thomas A., and Edward Iwata. "Regulators: Firms Traded Coverage for Business." *USA Today* (April 29, 2003):3B.

Fonda, Daren. "National Prosperous Radio." *Time* (March 24, 2003):49–51.

Foss, Brad. "Airlines, Facing Grim Future, Shrink Operations Again." *USA Today* (October 21, 2002). Available online at www.usatoday.com/travel/news/2002/2002–10–21-airline-out-look.htm.

Fouts, Roger. *Next of Kin.* New York: William Morrow, 1997.

Francis, David K. *The Internet in Healthcare: The Final Frontier.* Nashville, TN: J. C. Bradford, 1999.

Frank, Robert H. *Luxury Fever: Money and Happiness in an Era of Excess.* Princeton, NJ: Princeton University Press, 2000.

Frankel, Glenn. "Persian Gulf Region Rife with Rumors Amid News Blackout." *Washington Post* (August 6, 1990):A16.

Frankenberg, Erica, and Chungmei Lee. "Race in American Public Schools: Rapidly Resegregating School Districts." Cambridge, MA: The Civil Rights Project, August 2002.

Fraser, Jill Andresky. *White Collar Sweatshop: The Deterioration of Work and Its Rewards in Corporate America.* New York: Norton, 2002.

Freeman, David M. *Technology and Society.* Chicago: Rand McNally, 1974.

Freeman, Jo., and Victoria Johnson (eds.). *Waves of Protest: Social Movements Since the Sixties.* Lanham, MD: Rowman & Littlefield, 1999.

Freeman, Linton C. "The Sociological Concept of 'Group': An Empirical Test of Two Models." *American Journal of Sociology* 98 (July 1992):152–166.

Freeman, Richard B. *Black Elite.* New York: McGraw-Hill, 1977.

Freedman, Jane. *Feminism.* Buckingham, United Kingdom: Open University Press, 2001.

Freese, Lee (ed.). *Advances in Human Ecology.* Vol. 7. Stamford, CT: JAI Press, 1998.

French, Howard W. "North Korea to Let Capitalism Loose in Investment Zone." *New York Times* (September 25, 2002):A1, A3.

Freund, Peter E., and Meredith B. McGuire. *Health, Illness, and the Social Body: A Critical Sociology.* 3rd ed. Englewood Cliffs, NJ: Prentice Hall, 1998.

Frey, William H. "Lifecourse Migration and Metropolitan Whites and Blacks and the Structure of Demographic Change in Large Central Cities." *American Sociological Review* 49 (December 1984):803–827.

———. "Mover Destination Selectivity and the Changing Suburbanization of Metropolitan Blacks and Whites." *Demography* 22 (May 1985):223–243.

Frey, William H., and Reynolds Farley. "Latino, Asian, and Black Segregation in U.S. Metropolitan Areas: Are Multiethnic Metros Different?" *Demography* 31 (February 1996):35–50.

Friedan, Betty. *The Feminine Mystique.* New York: Dell, 1963.

———. *The Fountain of Age.* New York: Simon & Schuster, 1993.

Friedman, Milton. *Capitalism and Freedom.* Chicago: University of Chicago Press, 1982.

Friedrich, Carl J., and Zbigniew K. Brzezinski. *Totalitarian Dictatorship and Autocracy.* 2nd ed. Cambridge, MA: Harvard University Press, 1965.

Friend, Peter M. "PHO Growing Pains." *Healthcare Executive* (May/June 1996):12–16.

Frieze, Irene H., Jacquelynne E. Parsons, Paula B. Johnson, Diane N. Rable, and Gail L. Zellman. *Women and Sex Roles.* New York: Norton, 1978.

From the Capital to the Classroom: State and Federal Efforts to Implement the No Child Left Behind Act. Washington, DC: Center on Education Policy, January 2003.

Fruch, Terry, and Paul McGhee. "Traditional Sex Role Development and Amount of Time Watching Television." *Developmental Psychology* 11 (1975):109 ff.

Fukuyama, Francis. *The End of History and The Last Man.* New York: Basic Books, 1991.

———. "The Great Disruption." *Atlantic Monthly* 283 (May 1999a):55–80.

———. *The Great Disruption: Human Nature and the Reconstitution of Social Order.* New York: Free Press, 1999b.

———. *Our Posthuman Future: Consequences of the Biotechnology Revolution.* New York: Farrar, Straus & Giroux, 2002.

Fuller, Bruce, and Richard F. Elmore (eds.). *Who Wins, Who Loses?* New York: Teachers College Press, 1996.

Fuller, Bruce, et al. "New Lives for Poor Families?: Mothers and Young Children Move through Welfare Reform." The Growing Up in Poverty Project. Berkeley: University of California, 2002.

Fuller, John L., and William R. Thompson. *Behavior Genetics.* St. Louis: Mosby, 1978.

Furstenberg, Frank F. Jr., and Andrew J. Cherlin. *Divided Families.* Cambridge, MA: Harvard University Press, 1991.

Gall, T. L., D. R. Evans, and J. Howard. "The Retirement Adjustment Process: Changes in the Well-Being of Male Retirees Across Time." *Journal of Gerontology* 52 (1997):110–117.

Galloway, Joseph L. "In the Heart of Darkness." *U.S. News & World Report* (March 3, 1999):18.

Gallup, Alec, and Frank Newport. "Death Penalty Support Remains Strong, But Most Feel Unfairly Applied." *The Gallup Poll News Service* 56 (June 26, 1991):1.

Gallup, George H. Jr. *The Gallup Poll.* Wilmington, DE: Scholarly Resources, 1988.

———. *Religion in America, 1992–93.* Princeton, NJ: Princeton Religion Research Center, 1993.

———. *The Gallup Poll, Public Opinion 1993.* Wilmington, DE: Scholarly Resources, 1994.

———. *Religion in America 2002.* Princeton, NJ: The Princeton Religion Research Center, 2002.

Gallup, George H. Jr., and Frank Newport. "Time at a Premium for Many Americans." *Gallup Poll Monthly* (November 1990):43–49.

Game, Ann, and Andrew W. Metcalfe. *Passionate Sociology.* Thousand Oaks, CA: Sage, 1996.

Gamoran, Adam. "Instructional and Institutional Effects of Ability Grouping." *Sociology and Education* 59 (1986):185–198.

———. "The Variable Effects of High School Tracking." *American Sociological Review* 57 (December 1992):812–828.

Gamson, William. *The Strategy of Social Protest.* 2nd ed. Belmont, CA: Wadsworth, 1990.

Gamst, Glenn, and Charles M. Otten. "Job Satisfaction in High Technology and Traditional Industry: Is There a Difference?" *Psychological Record* (Summer 1992):413–425.

Ganong, Lawrence, and Marilyn Coleman. *Remarried Family Relationships.* Thousand Oaks, CA: Sage, 1994.

Gans, Herbert J. *The Urban Villagers.* New York: Free Press, 1962a.

———. "Urbanism and Suburbanism as Ways of Life." In Arnold M. Rose (ed.), *Human Behavior and Social Processes.* Boston: Houghton Mifflin, 1962b, pp. 625–648.

———. *People and Plans.* New York: Basic Books, 1968.

———. "Reconstructing the Underclass: The Term's Danger as a Planning Concept." *Journal of the American Planning Association* 56 (Summer 1990):271–277.

Ganz, Alexander. "Where Has the Urban Crisis Gone?" *Urban Affairs Quarterly* 20 (June 1985):449–468.

Garbarino, J. "The Incidence and Prevalence of Child Maltreatment." In L. Ohlin and M. Tonry (eds.), *Family Violence*. Chicago: University of Chicago Press, 1989, pp. 219–261.

Gardner, Carol Brooks. "A Family Among Strangers: Kinship Claims Among Gay Men in Public Places." In Don A. Chekki, Spencer E. Cahill, and Lyn H. Lofland (eds.), *Research in Community Sociology*, Supplement 1. Greenwich, CT: JAI Press, 1994, pp. 95–118.

Gardner, David Pierpont. *A Nation at Risk: The Imperative for Educational Reform*. Report of the National Commission on Excellence in Education. Washington, DC: U.S. Government Printing Office, 1983.

Gardner, Lytt. "Deprivation Dwarfism." *Scientific American* 227 (July 1972):76–82.

Garfinkel, Harold. *Studies in Ethnomethodology*. Cambridge, UK: Polity Press, 1984.

Garland, David (ed.). *Mass Imprisonment: Social Causes and Consequences*. Thousand Oaks, CA: Sage, 2001.

Garreau, J. *Edge City: Life on the New Frontier*. New York: Doubleday, 1991.

Garst, Jennifer, and Galen V. Bodenhausen. "Advertising's Effects on Men's Gender Role Attitudes." *Sex Roles* 36 (May 1997):551–572.

Gatewood, James V., and Min Zhou (eds.). *Contemporary Asian America: A Multidisciplinary Reader*. New York: New York University Press, 2000.

Gauntlett, David. *Media Gender and Identity: An Introduction*. New York: Routledge, 2002.

Gay, Peter. *The Enlightenment: The Rise of Modern Paganism*. New York: Norton, 1966.

———. *The Enlightenment: The Science of Freedom*. New York: Norton, 1977.

Geary, David C. *Male, Female: The Evolution of Human Sex Differences*. Washington, DC: American Psychological Association, 1998.

Gecas, Viktor, and Michael L. Schwalbe. "Beyond the Looking-Glass Self: Social Structure and Efficacy-Based Self-Esteem." *Social Psychology Quarterly* 46 (1983):77–88.

Gehrig, Gail. "The American Civil Religion Debate: A Source for Theory Construction." *Journal for the Scientific Study of Religion* 20 (March 1981):51–63.

Geis, Gilbert, Robert Meier, and Lawrence Salinger (eds.). *White-Collar Crime: Offenses in Business, Politics and the Professions*. 3rd ed. New York: Free Press, 1995.

Gelbard, Alene, and Carl Haub. "Population 'Explosion' Not Over for Half the World." *Population Today* 26 (March 1998):1–2.

Gelbard, Alene, Carl Haub, and Mary M. Kent. "World Population Beyond Six Billion." Washington, DC: Population Reference Bureau (March 1999):1–44.

Gellately, Robert, and Ben Kiernan (eds.). *Specter of Genocide: Mass Murder in Historical Perspective*. New York: Cambridge University Press, 2003.

Gelles, Richard J. *Intimate Violence in Families*. 3rd ed. Thousand Oaks, CA: Sage, 1997.

Gelles, Richard J., and Donileen Loseke (eds.). *Current Controversies on Family Violence*. Newbury Park, CA: Sage, 1993.

George, Linda K. "Sociological Perspectives on Life Transitions." *Annual Review of Sociology* 19 (1993):353–373.

Gerber, Jurg. "Heidi and Imelda: The Changing Image of Crime in Switzerland." *Criminology* 8 (March 1991):121–128.

Gersoni-Stavn, Deane. *Sexism and Youth*. New York: Bowker, 1974.

Gerth, H. H., and C. Wright Mills (eds.). *From Max Weber*. New York: Oxford University Press, 1958.

Gest, Ted. "Are White-Collar Crooks Getting Off Too Easy?" *U.S. News & World Report* 99 (July 1985):43.

Gibbons, Tom. "Justice Not Equal for Poor Here." *Chicago Sun-Times* (February 24, 1985):1,18.

Gibbs, Nancy. "Oh My God, They're Killing Themselves." *Time* (May 3, 1993a):27–43.

———. "Angels Among Us." *Time* (December 27, 1993b):56–65.

———. "An American Family Goes to War." *Time* (March 24, 2003):26–33.

Gibson, Margaret A., and John V. Ogbu. *Minority Status and Schooling*. New York: Garland Publishing, 1991.

Giddens, Anthony. *Studies in Social and Political Theory*. London: Hutchinson, 1979.

———. *Social Theory and Modern Sociology*. Stanford, CA: Stanford University Press, 1987.

———. *Runaway World: How Globalisation Is Reshaping Our Lives*. New York: Routledge, 2000.

Giddens, Anthony, and Mitchell Duneier. *Introduction to Sociology*. 3rd ed. New York: Norton, 2000.

Gilbert, Dennis A. *The American Class Structure in an Age of Growing Inequality*. 6th ed. Belmont, CA: Wadsworth, 2003.

Gilbert, Lucia Albino. *Two Careers/One Family*. Newbury Park, CA: Sage, 1993.

Gilbert, Neil. *Transformation of the Welfare State: The Silent Surrender of Public Responsibility*. Oxford, UK: Oxford University Press, 2002.

Gildea, William. "Violence Mars Night of Victory." *Washington Post* (June 21, 2000):D1.

Gilens, Martin. *Why Americans Hate Welfare: Race, Media, and the Politics of Antipoverty Policy*. Chicago: University of Chicago Press, 1999.

Giles, Jeff. "Generation X." *Newsweek* (June 6, 1994):62–72.

Gill, Brian P., et al. *Rhetoric versus Reality: What We Know and What We Need to Know about Vouchers and Charter Schools*. Santa Monica, CA: The Rand Corporation, 2001.

Gillam, Carey. "Kansas Eliminates Evolution from Public School Curricula." *Washington Post* (August 12, 1999):A13.

Gilleard, C., and A. Gurkan. "Socioeconomic Development and Status of Elderly Men in Turkey: A Test of Modernization Theory." *Journal of Gerontology* 42 (July 1987):353–357.

Giroux, Henry A. "Theories of Reproduction and Resistance in the New Sociology of Education: A Critical Analysis." *Harvard Educational Review* 53 (August 1983):257–293.

Gist, Noel P., and Sylvia Fleis Fava. *Urban Society*. 6th ed. New York: Crowell, 1974.

Gittell, Marilyn J. (ed.). *Strategies for School Equity: Creating Productive Schools in a Just Society*. New Haven, CT: Yale University Press, 1998.

Glazer, Nathan. "The End of Meritocracy." *The New Republic* (September 27, 1999):26.

Gleckman, Howard. "Big Spending Wrapped in a Flag." *Business Week* (February 18, 2002):44.

Gleckman, H., J. Carey, R. Mitchell, T. Smart, and C. Pousch. "The Technology Payoff." *Business Week* (June 14, 1993):51 ff.

Gleick, Elizabeth. "Teletubbies Revealed." *Time* (July 20, 1998):60–61.

Gleick, James. *Faster: The Acceleration of Just About Everything*. New York: Pantheon, 1999.

Glick, P., and S. L. Lin. "More Young Adults Are Living with Their Parents: Who Are They?" *Journal of Marriage and the Family* (February 1986):107–112.

Glock, Charles Y., and Robert N. Bellah (eds.). *The New Religious Consciousness*. Berkeley: University of California Press, 1976.

Glock, Charles Y., and Rodney Stark. *Religion and Society in Tension*. Chicago: Rand McNally, 1965.

———. *American Piety*. Berkeley: University of California Press, 1968.

Glueck, Sheldon, and Eleanor Glueck. *Unraveling Juvenile Delinquency*. Cambridge, MA: Harvard University Press, 1950.

Gober, Patricia. "Americans on the Move." *Population Bulletin* 48 (November 1993):1–40.

Goff, D., L. Goff, and S. Lehrer. "Sex-Role Portrayals of Selected Female Television Characters." *Journal of Broadcasting* 24 (1980):466–478.

Goffman, Erving. *Encounters*. Indianapolis: Bobbs-Merrill, 1961a.

———. *Asylums*. Garden City, NY: Anchor Books, 1961b.

———. *Stigma*. Englewood Cliffs, NJ: Prentice Hall, 1963.

———. *The Presentation of Self in Everyday Life*. New York: Overlook Press, 1974.

———. *Gender Advertisements*. New York: Harper and Row, 1979.

———. "The Interaction Order." *American Sociological Review* 48 (February 1983):1–17.

Goldberger, Arthur S., and Glen G. Cain. "The Causal Analysis of Cognitive Outcomes in the Coleman Report." Madison, WI: Institute for Research on Poverty. University of Wisconsin, December 1981.

———. "The Causal Analysis of Cognitive Outcomes in the Coleman, Hoffer, and Kilgore Report." *Sociology of Education* 55 (April/July 1982):103–122.

Golden, Frederic. "Albert Einstein: Person of the Century." *Time* (December 31, 1999):62–65.

Goldfarb, William. "Psychological Privation in Infancy and Subsequent Adjustment." *American Journal of Orthopsychiatry* 15 (April 1945):247–255.

Goldman, Ari L. "Religion Notes: Choosing Jews." *New York Times* (October 23, 1993):8.

Goldscheider, Frances K., and Calvin Goldscheider. "Leaving and Returning Home in 20th Century America." *Population Bulletin* 48 (March 1994):1–35.

Goldscheider, Frances K., and Linda J. Waite. *New Families, No Families?* Chapel Hill: University of North Carolina Press, 1991.

Goldstein, Amy. "'Pro-Life' Activists Take on Death." *Washington Post* (November 10, 1998):A1, A12–A13.

———. "Health Coverage Falls: Uninsured Numbers Up After 2 Years of Decline." *Washington Post* (September 30, 2002):A1.

Goldstein, Amy, and Richard Morin. "Young Voters' Disengagement Skews Politics." *Washington Post* (October 20, 2002):A1, A8.

Goldstein, Joshua R., and Catherine T. Kenney. "Marriage Delayed or Marriage Forgone? New Cohort Forecasts of First Marriage for U.S. Women." *American Sociological Review* 66 (August 2001):506–519.

Goldstein, M., and C. Beall. "Indirect Modernization and the Status of the Elderly in a Rural Third World Setting." *Journal of Gerontology* 37 (November 1982):716–724.

Goldstein, M. et al. "Social and Economic Forces Affecting Intergenerational Relations in Extended Families in a Third World Country: A Cautionary Tale from South Asia." *Journal of Gerontology* 38 (November 1983):716–724.

Goleman, D. "An Emerging Theory on Blacks' IQ Scores." *New York Times Education Supplement* (April 10, 1988):22–24.

Golf World. (July 4, 1997):30, 33.

Golf World 57 (December 19, 2003):A21.

Goo, Sara Kehaulani. "American Airlines Chief Executive Ousted." *Washington Post* (April 25, 2003):E1.

Good, Mary-Jo DelVecchio, Paul E. Brodwin, Byron J. Good, and Arthur Kleinman (eds.). *Pain as Human Experience: An Anthropological Perspective.* Berkeley: University of California Press, 1994.

Goode, Erich. *Collective Behavior.* Fort Worth, TX: Sanders College Publishing, 1992.

Goode, Judith, and Jeff Maskovsky (eds.). *The New Poverty Studies: The Ethnography of Power, Politics, and Impoverished People in the United States.* New York: New York University Press, 2001.

Goode, William J. *World Revolution and Family Patterns.* New York: Free Press, 1970.

Goodman, Ann B., Carole Siegel, Thomas J. Craig, and Shang P. Lin. "The Relationship Between Socioeconomic Class and Prevalence of Schizophrenia, Alcoholism, and Affective Disorders Treated by Inpatient Care in a Suburban Area." *American Journal of Psychiatry* 140 (1983):166–170.

Goodstein, David. "Scientific Misconduct." *Academe* 88 (January/February 2002):28–31.

Goodwin, Jeff, and James Jasper (eds.). *The Social Movements Reader.* Malden, MA: Blackwell, 2003.

Goozner, Merrill. "Forty Acres and a Sheepskin." *The American Prospect* (March/April 1999).

Gordon, Edmund W. "Toward Defining Equality of Educational Opportunity." In Frederick Mosteller and Daniel P. Moynihan (eds.), *On Equality of Educational Opportunity.* New York: Vintage, 1972, pp. 423–426.

Gordon, Greg. "At What Price?" *Minneapolis Star Tribune* (November 8, 2000):23A.

Gordon, Milton M. *Assimilation in American Life.* New York: Oxford University Press, 1964.

———. *Human Nature, Class, and Ethnicity.* New York: Oxford University Press, 1978.

Gorer, Geofrey. *Grief and Mourning in Contemporary Britain.* London: Cresset Press, 1965.

Gorman, Christine. "Blood, Sweat and Tears." *Time* (February 1996):59.

———. "Bleak Days for Doctors." *Time* (February 8, 1999):53.

Gottfredson, Michael R., and Travis Hirschi. *A General Theory of Crime.* Stanford, CA: Stanford University Press, 1990.

Gould, R. "Growth Toward Self Tolerance." *Psychology Today* (February 1975):47–48.

Gould, Stephen Jay. *The Mismeasurement of Man.* New York: Norton, 1981.

Gouldner, Alvin N. "Metaphysical Pathos and the Theory of Bureaucracy." *American Political Science Review* 49 (June 1955):496–507.

Grabosky, Peter N., and Russell G. Smith. *Electronic Theft: Unlawful Acquisition in Cyberspace.* New York: Cambridge University Press, 2001.

"Graduates Take Rites of Passage into Japanese Corporate Life." *Financial Times* (April 8, 1991):4.

Grandjean, B. "An Economic Analysis of the Davis-Moore Theory of Stratification." *Social Forces* 53 (June 1975):543–552.

Granovetter, Mark. "The Strength of Weak Ties." *American Journal of Sociology* 78 (May 1973):1360–1380.

Grapes, Bryan J. *Prisons.* San Diego, CA : Greenhaven Press, 2000.

Graubard, Allen. "The Free School Movement." *Harvard Educational Review* 42 (August 1972):351–373.

Greeley, Andrew M. "Political Attitudes Among American White Ethnics." In Charles H. Anderson (ed.), *Sociological Essays and Research,* rev. ed. Homewood, IL: Dorsey, 1974, pp. 202–209.

———. *Ethnicity, Denomination, and Inequality.* Beverly Hills, CA: Sage, 1976.

———. *Religious Change in America.* Cambridge, MA: Harvard University Press, 1996.

Green, Gary S. *Occupational Crime.* 2d ed. Chicago: Nelson-Hall Publishers, 1997.

Green, Donald P., Laurence H. McFalls, and Jennifer K. Smith. "Hate Crime: An Emergent Research Agenda." *Annual Review of Sociology* 27 (2001):479–504.

Green, Gary S. *Occupational Crime.* 2nd ed. Chicago: Nelson-Hall Publishers, 1997.

Green, Mark. *Selling Out.* New York: HarperCollins Publishers, 2002.

Greenberg, Daniel S. *Science, Money, and Politics: Political Triumph and Ethical Erosion.* Chicago: University of Chicago Press, 2001.

Greenberg, David F., and Ronald C. Kessler. "The Effects of Arrests on Crime: A Multivariate Panel Analysis." *Social Forces* 60 (March 1982):771–790.

Greenhouse, Linda. "No Help for Dying." *New York Times* (June 27, 1997):A1, A15.

Greenhouse, Steven. "The Mood at Work: Anger and Anxiety." *New York Times* (October 29, 2002):G1, G9.

———. "17,000 G.E. Workers Strike Over Higher Health Costs." *New York Times* (January 15, 2003).

Greenwald, Elissa A., Hilary R. Persky, Jay R. Campbell, and John Mazzeo. *NAEP 1998 Writing: Report Card for the Nation and the States.* Washington, DC: National Center for Education Statistics, 1999.

Gregory, Paul R., and Robert C. Stuart. *Comparative Economic Systems.* 3rd ed. Boston: Houghton Mifflin, 1989.

Greider, William. *Who Will Tell the People?* New York: Simon & Schuster, 1992.

———. "Is This America's Top Corporate Crime Fighter?" *The Nation* (August 5, 2002):11–15.

Grescoe, P. *The Merchants of Venus: Inside Harlequin and the Empire of Romance.* Custer, WA: Orca Book Publishers, 1997.

Griffin, John Howard. *Black Like Me.* Boston: Houghton Mifflin, 1961.

Gross, Beatrice, and Ronald Gross. *The Great School Debate*. New York: Simon & Schuster, 1985.

Gross, L., and S. Jeffries-Fox. "What Do You Want to Be When You Grow Up, Little Girl?" In Gary Tuchman, Arlene Kaplan Daniels, and James Benet (eds.), *Hearth and Home*. New York: Oxford University Press, 1978, pp. 240–265.

Grubb, W. Norton, and Robert H. Wilson. "Trends in Wage and Salary Inequality, 1967–1988." *Monthly Labor Review* 115 (June 1992):23–39.

Guillen, Mauro F. *The Limits of Convergence: Globalization and Organizational Change in Argentina, South Korea, and Spain*. Princeton, NJ: Princeton University Press, 2001.

Guinzburg, Suzanne. "Education's Earning Power." *Psychology Today* 17 (October 1983):20–21.

Gur, R. C., et al. "Sex Differences in Regional Cerebral Glucose Metabolism During a Resting State." *Science* 267 (January 27, 1995):528–531.

Gutman, Henry G. *The Black Family in Slavery and Freedom, 1750–1925*. New York: Pantheon, 1983.

Hacker, Andrew. *Two Nations: Black and White, Separate, Hostile, and Unequal*. Expanded and updated edition. New York: Ballantine Books, 1995.

Hadaway, C. Kirk, Penny Long Marler, and Mark Chaves. "What the Polls Don't Show: A Closer Look at U.S. Church Attendance." *American Sociological Review* 58 (December 1993):741–752.

Hadden, Jeffrey K., and Raymond R. Rymph. "The Marching Ministers." In Jeffrey K. Hadden (ed.), *Religion in Radical Transition*. 2nd ed. New Brunswick, NJ: Transaction Books, 1973, pp. 99–109.

Hadden, Jeffrey K., and Charles E. Swann. *Prime Time Preachers*. Reading, MA: Addison-Wesley, 1981.

Haferkamp, Hans, and Neil J. Smelser (eds.). *Social Change and Modernity*. Chapel Hill: University of North Carolina Press, 1992.

Hagan, Frank E. *Introduction to Criminology*. 5th ed. Chicago: Nelson-Hall Publishers, 2001.

Hagan, John. "The Social Embeddedness in Crime and Unemployment." *Criminology* 31 (November 1993):464–492.

———. *Crime and Disrepute*. Thousand Oaks, CA: Pine Forge Press, 1994.

Hagedorn, John M. "Gang Violence in the Postindustrial Era." In Michael Tonry and Mark H. Moore (eds.), *Youth Violence*. Chicago: University of Chicago Press, 1998, pp. 365–419.

Hall, Edward T. "How Cultures Collide." *Psychology Today* 10 (July 1976):66 ff.

———. "Learning the Arabs' Silent Language." *Psychology Today* 13 (August 1979):45 ff.

———. *The Silent Language*. New York: Anchor Books, 1990.

Hall, John. "President's Remarks Leave Settlement Up in the Air." *American Business Review* (September 18, 1997):1.

Hall, Richard H. "Intraorganizational Structural Variation: Application of the Bureaucratic Model." *Administrative Science Quarterly* 7 (December 1962):295–308.

———. *Organizations*. 7th ed. Upper Saddle River, NJ: Prentice Hall, 2001.

Hallinan, Maureen T. "Equality of Educational Opportunity." In W. Richard Scott and Judith Blake (eds.), *Annual Review of Sociology* 14 (1988):249–268.

——— (ed.). *Restructuring Schools: Promising Practices and Policies*. South Bend, IN: University of Notre Dame, 1995.

——— (ed.). *Handbook of the Sociology of Education*. New York: Kluwer Academic/Plenum Publishers, 2000.

Hallock, Kevin F. "Reciprocally Interlocking Boards of Directors and Executive Compensation." *Journal of Financial and Quantitative Analysis* 32 (September 1997):331–344.

Halpern, Diane F. *Critical Thinking Across the Curriculum: A Brief Edition of Thought and Knowledge*. Mahwah, NJ: Erlbaum, 1997.

Hamilton, Brady E., Joyce A. Martin, and Paul D. Sutton. *Births: Preliminary Data for 2002*. Rockville, MD: U.S. Department of Health and Human Services, 51, 2003.

Hamilton, Dona Cooper, and Charles V. Hamilton. *The Dual Agenda*. New York: Columbia University Press, 1998.

Hamilton, Kendall, and Devin Gordon. "Waiting for Star Wars." *Newsweek* (February 1, 1999):60–64.

Hamilton, Peter, and Kenneth Thompson (eds.). *The Uses of Sociology*. Malden, MA: Blackwell Publishers, 2002.

Hamilton, Richard F. *The Social Misconstruction of Reality: Validity and Verification in the Scholarly Community*. New Haven, CT: Yale University Press, 1996.

Hammel, Sara, and Anna Mulrine. "They Get More than Just Game." *U.S. News & World Report* (July 12, 1999):53.

Hammer, Joshua, and Zoran Cirjakovic. "Free at Last." *Newsweek* (October 16, 2000):24–31.

Hammond, Phillip E. *The Sacred in a Secular Age*. Berkeley: University of California Press, 1985.

Hampden-Turner, Charles, and Fons Trompenaars. *The Seven Cultures of Capitalism*. London: Piatkus, 1995.

Hampton, Robert L., et al. (eds.). *Family Violence*. Newbury Park, CA: Sage, 1993.

Hancock, LynNell, and Pat Wingert. "A Mixed Report Card." *Newsweek* (November 13, 1995):69.

Handel, Gerald (ed.). "Revising Socialization Theory." *American Sociological Review* 55 (June 1990):463–466.

Handler, Joel F., and Yeheskel Hasenfeld. *We the Poor People: Work, Poverty, and Welfare*. New Haven, CT: Yale University Press, 1997.

Hannan, Michael T., and John Freeman. *Organizational Ecology*. Cambridge, MA: Harvard University Press, 1989.

Hardy, Quentin. "Family Feud: When One Spouse Embraces the Latest Gadgets, the Other Often Looks on in Horror." *Wall Street Journal* (June 16, 1997):R10–R11.

Harlow, Harry F. "The Young Monkeys." *Psychology Today* 5 (1967):40–47.

Harlow, Harry F., and Margaret Harlow. "Social Deprivation in Monkeys." *Scientific American* 207 (November 1962):137–146.

Harlow, Harry F., and Robert R. Zimmerman. "Affectional Responses in the Infant Monkey." *Science* 21 (August 1959):421–432.

Harrington, Michael J. *The Other America*. New York: Macmillan, 1962.

Harris, Chauncy D. "The Nature of Cities and Urban Geography in the Last Half Century." *Urban Geography* 18 (1997):15–35.

Harris, Chauncy D., and Edward L. Ullman. "The Nature of Cities." *Annals of the American Academy of Political and Social Science* 242 (November 1945):7–17.

Harris, Judith Rich. *The Nurture Assumption: Why Children Turn Out the Way They Do*. New York: Free Press, 1998.

Harris, Kathleen Mullan. "Work and Welfare Among Single Mothers in Poverty." *American Journal of Sociology* 99 (September 1993):317–352.

Harris, Marvin. *Cow, Pigs, Wars, and Witches*. New York: Random House, 1974.

———. "Sociobiology and Biological Reductionism." In Ashley Montagu (ed.), *Sociobiology Examined*. New York: Oxford University Press, 1980, pp. 311–335.

———. *Our Kind: Who We Are, Where We Came From, and Where We Are Going*. New York: HarperTrade, 1990.

Harrison, Bennett. *Lean and Mean*. New York: Basic Books, 1994.

Hart, Robert A., and Seiichi Kawasaki. *Work and Pay in Japan*. New York: Cambridge University Press, 1999.

Hartman, Laura P. "Technology and Ethics: Privacy in the Workplace." Waltham, MA: Center for Business Ethics (February 28, 2000):1–28.

Hasegawa, Keitaro. *Japanese Style Management*. Tokyo: Kodansha International Ltd., 1986.

Haupt, Arthur, and Thomas T. Kane. *Population Handbook*. Washington, DC: Population Reference Bureau, 1997.

———. *Population Handbook*. 4th International ed. Washington, DC: Population Reference Bureau, 1998.

Hauser, Marc D. *Wild Minds: What Animals Really Think*. New York: Henry Holt, 2001.

Haviland, William A. *Cultural Anthropology*. 10th ed. New York: Harcourt, 2002.

Hawking, Stephen W. *A Brief History of Time.* New York: Bantam Books, 1998.

Hawkins, Asher. "Will the Silver State Go Green?" *Newsweek* (July 29, 2002):12.

Hawley, Amos H. "Human Ecology." In David L. Sills (ed.), *International Encyclopedia of the Social Sciences.* New York: Macmillan, 1968, pp. 329–337.

———. *Urban Society.* 2nd ed. New York: Wiley, 1971.

———. "The Logic of Macrosociology." *Annual Review of Sociology* 18 (1992):1–14.

Hawley, W. D. "Achieving Quality Integrated Education—With or Without Federal Help." In F. Schultz (ed.), *Annual Editions: Education 85/86.* Guilford, CT: Dushkin Publishing, 1985, pp. 142–145.

Hawley, W. D., and M. A. Smylie. "The Contribution of School Desegregation to Academic Achievement and Racial Integration." In P. A. Katz and D. A. Taylor (eds.), *Eliminating Racism.* New York: Plenum Press, 1988, pp. 281–297.

Hays, Kim. *Practicing Virtues: Moral Traditions at Quaker and Military Boarding Schools.* Berkeley and Los Angeles: University of California Press, 1994.

Heath, Chip, Chris Bell, and Emily Sternberg. "Emotional Selection in Memes: The Case of Urban Legends." *Journal of Personality and Social Psychology* 81 (December 2001):1028–1041.

Heatherton, Todd F., Michelle R. Hebl, and Jay G. Hull (eds.). *The Social Psychology of Stigma.* New York: Guilford Publications, 2000.

Hechter, Michael, and Karl-Dieter Opp (eds.). *Social Norms.* New York: Russell Sage Foundation, 2001.

Heckert, Druann Maria. "Positive Deviance." In Patricia A. Adler and Peter Adler (eds.), *Constructions of Deviance: Social Power, Context, and Interaction.* Belmont, CA: Wadsworth, 2003, pp. 30–42.

Heckscher, Charles. *White-Collar Blues.* New York: Basic Books, 1995.

Heelas, Paul. *The New Age Movement.* Cambridge, MA: Blackwell Publishers, 1996.

Hefner, Robert, and Rebecca Meda. "The Future of Sex Roles." In Marie Richmond-Abbott (ed.), *The American Woman.* New York: Holt, Rinehart and Winston, 1979, pp. 243–264.

Heilbroner, Robert L. *The Making of Economic Society.* 9th ed. Englewood Cliffs, NJ: Prentice Hall, 1993.

Heineman, Kenneth J. *Put Your Bodies upon the Wheels: Student Revolt in the 1960s.* Blue Ridge Summit, PA: Ivan R. Dee, 2002.

Heclo, Hugh, and Wilfred M. McClay (eds.). *Religion Returns to the Public Square: Faith and Policy in America.* Washington, DC: Woodrow Wilson Center Press, 2002.

Helfer, Mary Edna, Ruth S. Kempe, and Richard D. Krugman (eds.). *The Battered Child.* 5th ed. Chicago: University of Chicago Press, 1999.

Helle, H. J., and S. N. Eisenstadt (eds.). *Macro-Sociological Theory.* Vol. 1. Beverly Hills, CA: Sage, 1985a.

———. *Micro-Sociological Theory.* Vol. 2. Beverly Hills, CA: Sage, 1985b.

"Hello, Dolly." *The Economist* 18 (March 1, 1997):17–18.

Hendricks, Jon, and C. Davis Hendricks. *Aging in Mass Society.* 3rd ed. New York: HarperCollege, 1987.

Henig, Jeffrey R. *Rethinking School Choice.* Princeton, NJ: Princeton University Press, 1994.

Henry, Stuart, and Mark Lanier (eds.). *What Is Crime? Controversies over the Nature of Crime and What to Do about It.* Lanham, MD: Rowman & Littlefield, 2001.

Henry, Tamara. "Poll: Vouchers Lose Support, but Public Schools Gain." *USA Today* (August 23, 2001):1A.

Henry, William A. III. "News as Entertainment: The Search for Dramatic Unity." In Elie Abel (ed.), *What's News.* San Francisco: Institute for Contemporary Studies, 1981, pp. 133–158.

Herbert, Wray. "When Strangers Become Family." *U.S. News & World Report* (November 29, 1999):58–67.

Herivel, Tara, and Paul Wright (eds.). *Prison Nation: The Warehousing of America's Poor.* New York: Routledge, 2002.

Herlinger, Chris. "US Public Confidence in Religious Institutions Tumbles." *Presbyterian News Service* (July 11, 2002). Available online at www.pcusa.org/pcnews/02243.htm.

Herman, Arthur. *How the Scots Invented the Modern World.* New York: Crown Publishers, 2001.

Herring, Cedric. "Is Job Discrimination Dead?" *Contexts* (Summer 2002):13–18.

Herrnstein, Richard J., and Charles A. Murray. *The Bell Curve: Intelligence and Class Structure in American Life.* New York: Free Press Paperbacks, 1996.

Hertz, Rosanna, and Nancy L. Marshall (eds.). *Working Families: The Transformation of the American Home.* Berkeley, CA: University of California Press, 2001.

Herzlinger, Regina. *Market-Driven Health Care.* Reading, MA: Addison-Wesley, 1997.

Hesiod. *Theogony, Works and Days, Shield. Translation, Introduction, and Notes by Apostolos N. Athanassakis.* Baltimore: Johns Hopkins University Press, 1983.

Hewitt, John P.*Self and Society: A Symbolic Interactionist Social Psychology.* 9th ed. Boston: Allyn & Bacon, 2002.

Hilbert, Richard A. "Ethnomethodology and the Micro-Macro Order." *American Sociological Review* 55 (December 1990):794–808.

Hill, Michael R., and Susan Hoecker-Drysdale (eds.). *Harriet Martineau: Theoretical and Methodological Perspectives.* New York: Routledge, 2001.

Hill, Patrice. "Clinton May Trim Some Budget Pork." *Washington Times* (August 5, 1997):A1, A14.

Hill, Steven. *Fixing Elections: The Failure of America's Winner Take All Politics.* Bristol, PA: Taylor and Francis, 2002.

Hillier, Susan, and Georgia M. Barrow. *Aging, the Individual, and Society.* 7th ed. Belmont, CA: Wadsworth, 1999.

Hillis, Danny. "Why Do We Buy the Myth of Y2K?" *Newsweek* (May 31, 1999):12.

Hilt, M. L., and J. H. Lipschultz. "Broadcast News and Elderly People: Attitudes of Local Television Managers." *Educational Gerontology* 22 (1996):669–682.

Himes, Christine L. "Elderly Americans." *Population Bulletin* 56 (December 2001):1–40.

Hirsch, E. D. Jr. *Cultural Literacy.* Boston: Houghton Mifflin, 1987.

———. *The Schools We Need and Why We Don't Have Them.* New York: Doubleday, 1996.

Hirschfeld, Lawrence A. *Race in the Making.* Cambridge, MA: MIT Press, 1998.

Hirschi, Travis. *Causes of Delinquency.* Berkeley: University of California Press, 1972.

Hirschi, Travis, and Michael R. Gottfredson. "Substantive Positivism and the Idea of Crime." *Rationality and Society* 2 (October 1990):412–428.

Hirschi, Travis, and Hanan Selvin. *Principles of Survey Analysis.* New York: Free Press, 1973.

Hobbes, Thomas. *Leviathan.* Edited by Herbert W. Schneider. New York: Liberal Arts Press, 1958.

Hochschild, Arlie R. *The Second Shift.* With Anne Machung, contributor. New York: Avon, 1997.

———. *The Time Bind: When Work Becomes Home and Home Becomes Work.* New York: Henry Holt, 2001.

———. *The Managed Heart: Commercialization of Feeling.* 20th anniversary edition. Berkeley: University of California Press, 2003.

Hodge, Robert W., Paul M. Siegel, and Peter H. Rossi. "Occupational Prestige in the United States, 1925–1963." *American Journal of Sociology* 70 (November 1964):286–302.

Hodge, Robert W., and Donald J. Treiman. "Class Identification in the U.S." *American Journal of Sociology* 73 (March 1968):535–547.

Hodge, Robert W., Donald J. Treiman, and Peter H. Rossi. "A Comparative Study of Occupational Prestige." In Reinhard Bendix and Seymour Martin Lipset (eds.), *Class, Status, and Power.* 2nd ed. New York: Free Press, 1966, pp. 309–321.

Hodson, Randy, and Robert L. Kaufman. "Economic Dualism: A Critical Review." *American Sociological Review* 47 (December 1982):727–739.

Hodson, Randy, and Teresa A. Sullivan. *The Social Organization of Work.* Belmont, CA: Wadsworth, 1990.

Hoebel, E. Adamson. *The Law of Primitive Man.* New York: Atheneum, 1983.

Hoff, Gerard Alan, and Lawrence J. Schneidermann. "Having Babies at Home: Is It Safe? Is It Ethical?" *Hastings Center Report* 15 (December 1985):19–27.

Hoffer, Richard. "Bannister and Hillary." *Sports Illustrated* (December 27, 1999):96.

Hofferth, Sandra. "Did Welfare Reform Work? Implications for 2002 and Beyond." *Contexts* 1 (Spring 2002):45–51.

Hoffman, Karen E. "Internet as Gender-Equalizer?" *Internet World* (November 9, 1998).

Hoffman, Lois W., and Francis I. Nye. *Working Mothers.* San Francisco: Jossey-Bass, 1975.

Hofstadter, Richard. *America at 1750.* New York: Vintage, 1973.

Hogan, Dennis P., and Nan Marie Astone. "Transition to Adulthood." In Ralph H. Turner and James F. Short Jr. (eds.), *Annual Review of Sociology* 12 (1986):109–130.

Hogg, Michael A., and Dominic Abrams (eds.). *Intergroup Relations: Essential Readings.* Philadelphia: Psychology Press, 2001.

Hollingshead, August B. *Elmtown's Youth.* New York: Wiley, 1949.

Hollingshead, August B., and Frederick C. Redlich. *Social Class and Mental Illness.* New York: Wiley, 1958.

Holmberg, Allan R. *Nomads of the Long Bow.* Garden City, NY: Natural History Press, 1969.

Holt, John. *How Children Fail.* Rev. ed. Reading, MA: Perseus Books, 1995.

Hook, Donald D. *Death in the Balance.* Lexington, MA: Heath, 1989.

Hoover, Stewart M. *The Social Sources of the Electronic Church.* Newbury Park, CA: Sage, 1988.

Hooyman, Nancy R., and H. Asuman Kiyak. *Social Gerontology.* 6th ed. Boston: Allyn & Bacon, 2001.

Hope, Keith. "Vertical and Nonvertical Class Mobility in Three Countries." *American Sociological Review* 47 (February 1982):99–113.

Horsley, Richard A. *Jesus and Empire: The Kingdom of God and the New World Disorder.* Minneapolis: Fortress Press, 2003.

Horwitz, Allan V. *The Logic of Social Control.* New York: Plenum, 1990.

Horwitz, Robert A. "Psychological Effects of the Open Classroom." *Review of Educational Research* 49 (Winter 1979):71–86.

Hosenball, Mark. "The Anatomy of a Rumor." *Newsweek* (November 23, 1996):43.

Hotaling, G. T., et al. (eds.). *Family Abuse and Its Consequences.* Beverly Hills, CA: Sage, 1988.

Hougland, James G. Jr., and Willis A. Sutton Jr. "Factors Influencing Degree of Involvement in Interorganizational Relationships in a Rural County." *Rural Sociology* 43 (Winter 1978):649–670.

Houseman, John. "The Men from Mars." *Harper's Magazine* 197 (December 1948):74–82.

Hoyt, Homer. *The Structure and Growth of Residential Neighborhoods in American Cities.* Washington, DC: Federal Housing Authority, 1939.

Hrdy, Sarah Blaffer. *The Woman That Never Evolved.* Rev. ed. Cambridge, MA: Harvard University Press, 1999. Available online at www.tvb.org/tvfacts/tvbacis/basics6.html.

———. *Mother Nature: Maternal Instincts and How They Shape the Human Species.* New York: Ballantine Books, 2000.

Hubbard, L. Ron. *What Is Scientology?* Los Angeles: Bridge, 1998.

Huber, Peter, and Jessica Korn. "The Plug-and-Play Economy." *Forbes* (July 7, 1997):268–272.

Hudak, Glenn M., and Paul Kihn (eds.). *Labeling: Pedagogy and Politics.* New York: Routledge, 2001.

Hudson, Ken. "Duality and Dual Labor Markets." Revision of a paper presented at the 93rd Annual Meetings of the American Sociological Association, 2000.

Huffington, Arianna. *Pigs at the Trough: How Corporate Greed and Political Corruption Are Undermining America.* New York: Crown Publishers, 2003.

Hughes, John A., Peter J. Martin, and W. W. Sharrock. *Understanding Classical Sociology.* Thousand Oaks, CA: Sage, 1995.

Hughes, Michael, and Melvin E. Thomas. "The Continuing Significance of Race Revisited: A Study of Race, Class, and Quality of Life in America, 1972 to 1996." *American Sociological Review* 63 (December 1998):785–795.

Huizinga, David, and Delbert S. Elliot. "Juvenile Offenders: Prevalence, Offender Incidence, and Arrest Rates by Race." *Crime and Delinquency* 33 (April, 1987):206–223.

Hull, J. B. "Women Find Parents Need Them Just When Careers Are Resuming." *Wall Street Journal* (September 9, 1985):27.

Human Development Report 1999. New York: Oxford University Press, 1999.

Human Development Report 2002. New York: Oxford University Press, 2002.

Hummel, Ralph P. *The Bureaucratic Experience.* 3rd ed. New York: St. Martin's Press, 1987.

Humphrey, Craig R., and Frederick H. Buttel. *Environment, Energy and Society.* Malabar, FL: Krieger Publishing, 1986.

Humphreys, Laud. *Tearoom Trade.* New York: Aldine, 1979.

Hunt, Morton M. *The New Know-Nothings: The Political Foes of the Scientific Study of Human Nature.* New Brunswick, NJ: Transactions, 1999.

Hunt, Stephen, Malcolm Hamilton, and Tony Walter (eds.). *Charismatic Christianity: Sociological Perspectives.* New York: St. Martin's Press, 1998.

Hunter, Floyd. *Community Power Structure.* Chapel Hill: University of North Carolina Press, 1953.

———. *Community Power Succession.* Chapel Hill: University of North Carolina Press, 1980.

Hunter, James Davison. *Evangelicism.* Chicago: University of Chicago Press, 1987.

Hurley, Jennifer A. (ed.). *Racism.* San Diego, CA: Greenhaven Press, 1998.

———. *American Values.* Guilford, CT: Greenhaven Press, 2000.

Hurn, Christopher J. *The Limits and Possibilities of Schooling.* 3rd ed. Boston: Allyn & Bacon, 1993.

Iacocca, Lee, and William Novak. *Iacocca.* New York: Bantam, 1984.

Illich, Ivan. *Medical Nemesis.* New York: Pantheon, 1982.

"The Impact of Racial and Ethnic Diversity on Educational Outcomes." Cambridge, MA: The Civil Rights Project, January 2002.

Inciardi, James A. *Criminal Justice.* 7th ed. Fort Worth, TX: Harcourt Brace College Publishers, 2002.

Infante, Esme M. "Panel: Nicotine Addictive." *USA Today* (August 13, 1994):A1.

Inglehart, Ronald, and Wayne E. Baker. "Modernization, Cultural Change, and the Persistence of Traditional Values." *American Sociological Review* 65 (2000):19–51.

Ingrassia, Michele, and Melinda Beck. "Patterns of Abuse." *Newsweek* (July 4, 1994):26–33.

Inkeles, Alex. *One World Emerging? Convergence and Divergence in Industrial Societies.* Boulder, CO: Westview Press, 1998.

Insana, Ron. *Trendwatching: Don't Be Fooled by the Next Investment Fad, Mania, or Bubble.* New York: HarperCollins, 2002.

International Labour Organization. *Yearbook of Labour Statistics: 1996.* 55th ed. Geneva: International Labour Office, 1996.

"Internet 101," 1999. Available online at www2.famvid.com/i101/start.html.

Isikoff, Michael. "Christian Coalition Steps Boldly Into Politics." *Washington Post* (September 10, 1992):A1, A14.

———. "The Secret Money Chase." *Newsweek* (June 5, 2000):23.

Iwata, Edward. "Dynegy to Quit Energy Trading." *USA Today* (October 16, 2002). Available online at www.usatoday.com/money/industries/energy/2002–10–16–dynegy_x.htm.

Jackman, Mary R., and Robert W. Jackman. *Class Awareness in the United States.* Berkeley: University of California Press, 1983.

Jackson, Elton F., and Harry J. Crockett Jr. "Occupational Mobility in the United States: A Point Estimate and Trend Comparison." *American Sociological Review* 29 (February 1964):5–15.

Jackson, Robert M. *Destined for Equality.* Cambridge, MA: Harvard University Press, 1998.

Jackson, Stevi, and Sue Scott (eds.). *Gender: A Sociological Reader.* New York: Routledge, 2001.

Jacob, John E. "Black America, 1993: An Overview." In Billy J. Tidwell (ed.), *The State of Black America 1994.* New York: National Urban League, 1994, pp. 1–9.

Jacobs, H. "Aging and Politics." In R. H. Binstock and L. K. George (eds.), *Handbook of Aging and the Social Sciences.* New York: Academic Press, 1990, pp. 349–361.

Jacobson, Neil S. and John Gottman. *When Men Batter Women.* New York: Simon & Schuster Trade, 1998.

Jacoby, Sanford M. *Modern Manors: Welfare Capitalism Since the New Deal.* Princeton, NJ: Princeton University Press, 1997.

Jaggar, Alison M., and Paula S. Rothenberg. *Feminist Frameworks.* 3rd ed. New York: McGraw-Hill, 1993.

James, James J. "Impact of the Medical Malpractice Slowdown in Los Angeles County: January 1976." *American Journal of Public Health* 69 (May 1979):437–443.

James, Paul. *Nation Formation.* Thousand Oaks, CA: Sage, 1996.

Janis, Irving L. *Groupthink.* Boston: Houghton Mifflin, 1982.

Janofsky, Michael. "Rustic Life of Amish Is Changing but Slowly." *New York Times* (July 6, 1997):7.

Janowitz, Morris. *Sociology and the Military Establishment.* 3rd ed. Beverly Hills, CA: Sage, 1974.

Janssen-Jurreit, Marie Louise. *Sexism.* New York: Farrar Strauss Giroux, 1982.

Jasinski, Jana L. (ed.). *Partner Violence: A Comprehensive Review of Twenty Years of Research.* Thousand Oaks, CA: Sage, 2000.

Jeffreys-Fox, Bruce. *How Realistic Are Television's Portrayals of the Elderly?* University Park, PA: The Annerberg School of Communications, 1977.

Jencks, Christopher. *Inequality.* New York: HarperCollins Publishers, 1981.

Jencks, Christopher, and Meredith Phillips (eds.). *The Black-White Test Score Gap.* Washington, DC: Brookings Institution, 1998.

Jenkins, J. Craig. "Resource Mobilization Theory and the Study of Social Movements." In Ralph H. Turner and James F. Short Jr. (eds.), *Annual Review of Sociology* 9 (1983):527–553.

Jenness, Valerie, and Ryken Grattet. *Making Hate a Crime: From Social Movement to Law Enforcement.* New York: Russell Sage Foundation, 2001.

Jensen, Arthur. "How Much Can We Boost IQ and Scholastic Achievement?" *Harvard Educational Review* 39 (Winter 1969):1–123.

Johnson, Charles Richard, and Patricia Smith. *Africans in America: America's Journey Through Slavery.* San Diego, CA: Harcourt Brace and Company, 1998.

Johnson, Darragh. "SAT Scores Dip Slightly Overall: Widening Racial Gap Is Greater Concern." *Washington Post* (August 29, 2002):A03.

Johnson, David W., and Frank P. Johnson. *Joining Together.* 5th ed. Boston: Allyn & Bacon, 1994.

Johnson, Dirk. "Until Dust Do Us Part." *Newsweek* (March 25, 2002):41.

———. "A Leap of Fate." *Newsweek* (January 20, 2003):34.

———. "The Night the Music Died." *Newsweek* (March 3, 2003):40–42.

Johnson, Haynes. *Sleepwalking Through History.* New York: Norton, 1991.

———. *Divided We Fall.* New York: Norton, 1995.

Johnson, Robert. *Hard Time: Understanding and Reforming the Prison.* 3rd ed. Belmont, CA: Wadsworth, 2002.

Johnson, Troy R. *Contemporary Native American Political Issues.* Thousand Oaks, CA: AltaMira Press, 2000.

Johnson, William B., and Arnold H. Packer. *Workforce 2000.* Indianapolis: Hudson Institute, 1987.

Johnston, David Cay. "Enron Avoided Income Taxes in 4 of 5 Years." *New York Times* (January 17, 2002):A1.

Johnstone, Ronald L. *Religion in Society.* 6th ed. Upper Saddle River, NJ: Prentice Hall, 2000.

Jonas, Steven. *Medical Mystery.* New York: Norton, 1978.

Jones, A. H. M. *The Decline of the Ancient World.* New York: Holt, Rinehart and Winston, 1966.

Jones, Daniel (ed.). *Dylan Thomas: The Poems.* London: J. M. Dent and Sons, 1971.

Jones, James H. *Alfred C. Kinsey: A Public/Private Life.* New York: Norton, 1997.

Jones, James M. *Prejudice and Racism.* Reading, MA: Addison-Wesley, 1972.

———. *Bad Blood.* Expanded ed. New York: Free Press, 1993.

Jones, Robert Alun. *The Development of Durkheim's Social Realism.* New York: Cambridge University Press, 1999.

Kadlec, Daniel. "Everyone, Back in the Labor Pool." *Time* (July 29, 2002):23–31.

Kadushin, Charles. "Mental Health and the Interpersonal Environment: A Reexamination of Some Effects of Social Structure on Mental Health." *American Sociological Review* 48 (April 1983):188–198.

Kalb, Marvin. "Perspective on Journalism: Get Ready for the Really Bad News." *Los Angeles Times* (July 10, 1998):B-9.

Kalisch, Philip A., and Beatrice J. Kalisch. "Sex-Role Stereotyping of Nurses and Physicians on Prime-Time Television: A Dichotomy of Occupational Portrayals." *Sex Roles* 10 (1984):533–553.

Kalish, Carol B. *International Crime Rates.* U.S. Bureau of Justice Statistics. Washington, DC: U.S. Government Printing Office, 1988.

Kalmijn, Matthijs. "Mother's Occupational Status and Children's Schooling." *American Sociological Review* 59 (April 1994):257–275.

Kamarck, Elaine Ciulla. "Family Policy." In Mark Green (ed.), *Changing America.* New York: Newmarket Press, 1992, pp. 443–456.

Kanter, Rosabeth Moss. *Commitment and Community.* Cambridge, MA: Harvard University Press, 1972.

———. *Men and Women of the Corporation.* New York: Basic Books, 1993.

Kantrowitz, Barbara. "A Nation Still at Risk." *Newsweek* (April 19, 1993a):46–49.

———. "The Group Classroom." *Newsweek* (May 10, 1993b):73.

Kantrowitz, Barbara, and Pat Wingert. "Unmarried, With Children." *Newsweek* (May 28, 2001):46–55.

Kaplan, David A. "Bobbitt Fever." *Newsweek* (January 24, 1994):52–55.

Kaplan, Jeffrey, and Helene Loow (eds.). *Cultic Milieu.* Thousand Oaks, CA: AltaMira Press, 2002.

Kaplan, Robert D. "Looking the World in the Eye." *The Atlantic Monthly* (December 2001):68–82.

Karatnycky, Adrian. "Democracies on the Rise, Democracies at Risk." *Freedom Review* 26 (January/February 1995):5–10.

Kardulias, P. Nick (ed.). *World-Systems Theory in Practice: Leadership, Production, and Exchange.* Lanham, MD: Rowman & Littlefield, 1999.

Karson, Jill (ed.). *Cults.* San Diego, CA: Greenhaven Press, 2000.

Kart, C. S. *The Realities of Aging.* Boston: Allyn & Bacon, 1990.

Kasarda, John D., and Morris Janowitz. "Community Attachment in Mass Society." *American Sociological Review* 39 (June 1974):328–339.

Katchadourian, Herant, and John Boli. *Cream of the Crop: The Impact of Elite Education in the Decade After College.* New York: Basic Books, 1994.

Kate, Nancy T. "Two Careers, One Marriage." *American Demographics* (April 1998):28.

Katel, Peter. "The Bust in Boot Camps." *Newsweek* (February 21, 1994):26.

Katsiaficas, George N., and Teodros Kiros (eds.). *The Promise of Multiculturalism: Education and Autonomy in the 21st Century.* New York: Routledge, 1998.

Katz, James E., and Philip Aspden. "A Nation of Strangers?" *Communications of the ACM* (December 1997):81–87.

Katz, Irwin. "Desegregation or Integration in Public Schools?: The Policy Implications of Research." *Integrated Education* 5 (December 1967/January 1968):15–28.

Katz, Michael B. *Class, Bureaucracy and Schools*. New York: Praeger, 1975.

——— (ed.). *The "Underclass" Debate*. Princeton, NJ: Princeton University Press, 1993.

Katznelson, Ira, Mark Kesselman, and Alan Draper. *The Politics of Power: A Critical Introduction to American Government*. 4th ed. Forth Worth: Harcourt Brace College Publishers, 2002.

Kaufman, Philip. *Dropout Rates in the United States: 1999*. Washington, DC: National Center for Education Statistics, November 2000.

———. *The Nation's Report Card: Fourth Grade Reading 2000*. Washington, DC: National Center for Education Statistics, 2001.

Kearl, Michael C., and Anoel Rinaldi. "The Political Uses of the Dead as Symbols in Contemporary Civil Religions." *Social Forces* 61 (March 1983):693–708.

Keen, Judy. "In Capital, Business and Politics Firmly Entwined." *USA Today* (July 31, 2002):1A, 2A.

Keister, Lisa. *Wealth in America: Trends in Wealth Inequality*. Cambridge, UK: Cambridge University Press, 2000.

Kellam, Sheppard G., Margaret E. Ensminger, and R. Jay Turner. "Family Structure and the Mental Health of Children." *Archives of General Psychiatry* 34 (September 1977):1012–1022.

Keller, Amy. "Roll Call's List of the 50 Richest Lawmakers." *Roll Call* (September 9, 2002).

Kelley, Harold H., and John W. Thibaut. "Group Problem Solving." In Gardner Lindzey and Elliot Aronson (eds.), *The Handbook of Social Psychology*. Vol. 4. Reading, MA: Addison-Wesley, 1969, pp. 735–785.

Kelley, Maryellen R. "New Process Technology, Job Design, and Work Organization: A Contingency Model." *American Sociological Review* 55 (1990):191–208.

Kelly, Delos H. (ed.). *Deviant Behavior*. 5th ed. New York: St. Martin's Press, 1996.

Kennedy, Randall. *Interracial Intimacies: Sex, Marriage, Identity, and Adoption*. New York: Knopf, 2003.

Kent, Mary M., and Mark Mather. "What Drives U.S. Population Growth?" *Population Bulletin* 57 (December 2002):3–40.

Kephart, William M., and Davor Jedlicka. *The Family, Society, and the Individual*. 7th ed. New York: HarperCollins, 1997.

Keppel, Kenneth G., Jeffrey N. Pearcy, and Diane K. Wagener. "Trends in Racial and Ethnic-Specific Rates for the Health Status Indicators: United States, 1990–98." *Health People Statistical Notes* 23. Hyattsville, MD: National Center for Health Statistics, 2002.

Kertzer, David I. "Age Structuring in Comparative and Historical Perspective." In David I. Kertzer and K. Warner Shaie (eds.), *Age Structuring in Comparative Perspective*. Hillsdale, NJ: Erlbaum, 1989, pp. 3–20.

Kessler, Ronald C., and Harold W. Neighbors. "A New Perspective on the Relationships Among Race, Social Class, and Psychological Distress." *Journal of Health and Science Behavior* 27 (1986):107–115.

Kessler, Ronald C., et al. "Lifetime and 12–Month Prevalence of *DSM-III-R* Psychiatric Disorders in the United States. Results from the National Comorbidity Survey." *Archives of General Psychiatry* 51 (1994):8–19.

Khouri, Norma. *Honor Lost: Love and Death in Modern-Day Jordan*. New York: Simon & Schuster, 2003.

Kibria, Nazli. *Becoming Asian American: Second-Generation Chinese and Korean American Identities*. Baltimore, MD: Johns Hopkins University Press, 2002.

2002 Kids Count. Baltimore, MD: The Annie E. Casey Foundation, 2002.

Kiechel, W. "How Will We Work in the Year 2000?" *Fortune* (May 17, 1993):38.

Kiecolt, K. Jill. "Satisfaction with Work and Family Life: No Evidence of a Cultural Reversal." *Journal of Marriage and Family* 65 (February 2003):23–35.

Kilbourne, Jean. "Beauty and the Beast of Advertising." In James M. Henslin (ed.), *Down to Earth Sociology*. New York: Free Press, 2001, pp. 391–394.

Kilgore, Sally B. "The Organizational Context of Tracking in Schools." *American Sociological Review* 56 (April 1991):189–203.

Killian, Lewis M. "Organization, Rationality and Spontaneity in the Civil Rights Movement." *American Sociological Review* 49 (December 1984):770–783.

Kilson, Martin L. "The State of African-American Politics." In Lee A. Daniels (ed.), *The State of Black America 1998*. New York: National Urban League, 1998, pp. 247–270.

———. "African Americans and American Politics 2002: The Maturation Phase." In Lee A. Daniels (ed.), *The State of Black America 2002*. New York: National Urban League, 2002, pp. 147–180.

Kimmel, Michael S. *Gendered Society*. New York: Oxford University Press, 2000.

King, Ruth S. "Managed Care 'Freedom'?" *Wall Street Journal* (February 29, 1996):A14.

Kinsey, Alfred C., and Paul Gebhard. *Sexual Behavior in the Human Female*. Philadelphia: Saunders, 1953.

Kinsey, Alfred C., Wardell Pomeroy, and Clyde Martin. *Sexual Behavior in the Human Male*. Philadelphia: Saunders, 1948.

Kinsley, Michael. "In Defense of Matt Drudge." *Time* (February 2, 1998):41.

Kirk, Stuart. *The Selling of DSM: The Rhetoric of Science in Psychiatry*. New York: Aldine de Gruyter, 1992.

Kirkpatrick, David. "Smart New Ways to Use Temps." *Fortune* (February 15, 1988):110–116.

Kissman, Kris, and Jo Ann Allen. *Single-Parent Families*. Newbury Park, CA: Sage, 1993.

Kitano, Harry H. *The Japanese Americans*. 2nd ed. New York: Chelsea House, 1995.

Kitano, Harry H., and Roger Daniels. *Asian Americans*. Englewood Cliffs, NJ: Prentice Hall, 1988.

Kitson, Gay C., and Helen J. Raschke. "Divorce Research: What We Know; What We Need to Know." *Journal of Divorce* 4 (Spring 1981):1–37.

Klandermans, Bert. "Mobilization and Participation: Social Psychological Explanations of Resource Mobilization Theory." *American Sociological Review* 49 (1984):583–600.

Kleiber, Douglas A., and John R. Kelly. "Leisure, Socialization and the Life Cycle." In Seppo E. Iso-Ahola (ed.), *Social Psychological Perspectives on Leisure and Recreation*. Springfield, IL: Charles C. Thomas, 1980, pp. 91–137.

Klein, Stephen P., Susan Turner, and Joan Petersilia. *Racial Equity in Sentencing*. Santa Monica, CA: Rand, 1988.

Kleniewski, Nancy. *Cities, Change, and Conflict*. 2nd ed. Belmont, CA: Wadsworth, 2001.

Klineberg, Otto. "Race Differences: The Present Position of the Problem." *International Social Science Bulletin* 2 (Autumn 1950):460–466.

Klinker, Philip A., and Rogers M. Smith. *The Unsteady March: The Rise and Decline of Racial Equality in America*. Chicago: University of Chicago Press, 1999.

Kluegel, James R., and Eliot R. Smith. *Beliefs About Inequality: Americans' Views of What Ought to Be*. Hawthorne, NY: Aldine de Gruyter, 1990.

Kluger, Jeffrey. "Here, Kitty, Kitty!" *Time* (February 25, 2002):58–59.

Kluger, Jeffrey, and Andrea Dorfman. "The Challenges We Face." *Time* (August 26, 2002):A6–A12.

Knox, David, and Caroline Schacht. *Marriage and the Family: A Brief Introduction*. Belmont, CA: Wadsworth, 1999.

———. *Choices in Relationships*. 7th ed. Belmont, CA: Wadsworth, 2002.

Kohl, Herbert. *Growing Minds*. New York: Harper Touchbook, 1988.

Kohn, Melvin L. "Social Class and Parent-Child Relationships: An Interpretation." *American Sociological Review* 68 (January 1963):471–480.

———. *Class and Conformity*. 2nd ed. Chicago: University of Chicago Press, 1977.

———. "Unresolved Issues in the Relationship Between Work and Personality." In Kai Erikson and Steven Peter Vallas (eds.), *The Nature of Work*. New Haven, CT: Yale University Press, 1990, pp. 36–68.

Kohn, Melvin L., and Carmi Schooler. *Work and Personality: An Inquiry into the Impact of Social Stratification.* Norwood, NJ: Ablex, 1983.

Kohut, Andrew. "Among Wealthy Nations . . . U.S. Stands Alone In Its Embrace of Religion." Washington, DC: The Pew Research Center for the People and the Press, December 19, 2002.

Kohut, Andrew, et al. *The Diminishing Divide: Religion's Changing Role in American Politics.* Washington, DC: Brookings Institution, 2000.

Koon, Jeff, Andy Powell, and Ward Schumaker. *You May Not Tie an Alligator to a Fire Hydrant: 101 Real Dumb Laws.* New York: Free Press, 2002.

Kolbe, Richard, and Joseph C. LaVoie. "Sex-Role Stereotyping in Preschool Children's Picture Books." *Social Psychology Quarterly* 44 (December 1981):369–374.

Konner, Melvin. "Darwin's Truth, Jefferson's Vision." *The American Prospect* (July/August 1999):30–38.

Koppel, Ross. "American Public Policy: Formation and Implementation." In Roger A. Straus (ed.), *Using Sociology,* 3rd ed. Lanham, MD: Rowman & Littlefield, 2002, pp. 265–228.

Korgen, Kathleen Odell. *From Black to Biracial: Transforming Racial Identity Among Americans.* Westport, CT: Praeger, 1998.

Kornheiser, Tony. "You Get No Shot With a Bad Lie." *The Washington Post* (December 15, 2001):D1,D6.

Koss-Feder, Laura. "Providing for Parents." *Time Bonus Section* (March 2003).

Kovner, Christine. "Nursing." In Anthony R. Kovner (ed.), *Health Care Delivery in the United States.* 4th ed. New York: Springer, 1990, pp. 87–105.

Kowalewski, David. *Transnational Corporations and Caribbean Inequalities.* New York: Praeger, 1982.

Kozol, Jonathan. *Death at an Early Age.* New York: Penguin, 1985.

———. *Savage Inequalities: Children in America's Schools.* New York: HarperCollins, 1992.

Kraar, Louis. "The Real Threat to China's Hong Kong." *Fortune* 135 (May 26, 1997):84.

Kraditor, Aileen S. *The Ideas of the Woman's Suffrage Movement 1890–1920.* Garden City, NY: Anchor Books, 1971.

Kraybill, Donald B., and Marc A. Olshan (eds.). *The Amish Struggle with Modernity.* Hanover, NH: University Press of New England, 1994.

Kriesi, Hanspeter. "New Social Movements and the New Class in the Netherlands." *American Journal of Sociology* 94 (March 1989):1078–1116.

Kritz, Francesca Lunzer. "uncertainty@dr-mail.com" *Washington Post* (April 1, 2003):F1, F5.

Kronenfeld, Jennie Jacobs. *The Changing Federal Role in U.S. Health Care Policy.* Westport, CT: Praeger, 1997.

——— (ed.). *Research in the Sociology of Health Care: Changing Organizational Forms of Delivering Health Care.* Vol. 15. Greenwich, CT: JAI Press, 1998.

Krupp, Charla. "The Pajama Game: Not Just for Bedtime." *Time* (March 3, 2003):75.

Kruschwitz, Robert B., and Robert C. Roberts. *The Virtues.* Belmont, CA: Wadsworth, 1987.

Kübler-Ross, Elisabeth. *Death's Final Stage.* New York: Simon & Schuster, 1986.

———. *Death.* New York: Simon & Schuster, 1997a.

———. *On Death and Dying.* New York: Touchstone, 1997b.

Kumar, Krishan. *From Post-Industrial to Post-Modern Society.* Malden, MA: Blackwell, 1995.

Kurlaender, Michael, and John T. Yun. "Is Diversity a Compelling Educational Interest? Evidence from Metropolitan Louisville." Cambridge, MA: Civil Rights Project, August 2000.

Kurzweil, Ray. *The Age of Spiritual Machines.* New York: Viking, 1999.

Kuttner, R. "Competitiveness Craze." *Washington Post* (August 12, 1993a):A17.

———. "The Corporation in America: Is It Socially Redeemable?" *Dissent* (Winter 1993b):35–49.

——— (ed.). *Making Work Pay: America After Welfare.* New York: New Press, 2002.

Labaree, David F. *How to Succeed in School Without Really Learning: The Credentials Race in American Education.* New Haven: Yale University Press, 1997.

Labich, Kenneth. "Can Your Career Hurt Your Kids?" *Fortune* (May 20, 1991):38–56.

Lacayo, Richard. "Wounding the Gun Lobby." *Time* (March 29, 1993):29–30.

Lachman, Margie E., et al. "Images of Midlife Development Among Young, Middle-Aged, and Older Adults." *Journal of Adult Development* (1994):201–211.

Lachman, Margie E., and Jacquelyn Boone James. "Charting the Course of Midlife Development: An Overview." In Margie E. Lachman and Jacquelyn Boone James (eds.), *Multiple Paths of Midlife Development.* Chicago: University of Chicago Press, 1997, pp. 1–17.

Lamanna, Mary Ann, Agnes Riedmann, and Agnes Czerwinski Riedmann. *Marriages and Families: Making Choices and Facing Change.* 7th ed. Belmont, CA: Wadsworth, 2000.

Lamb, Michael B., M. Ann Easterbrooks, and George W. Holden. "Reinforcement and Punishment Among Preschoolers: Characteristics, Effects, and Correlates." *Child Development* 51 (1980):1230–1236.

Lampert, Leslie. "Fat Like Me." *Ladies Home Journal* (May 1993):154 ff.

Landes, David S., and Charles Tilly (eds.). *History as Social Science.* Englewood Cliffs, NJ: Prentice Hall, 1991.

Landry, Bart. *The New Black Middle Class.* Berkeley: University of California Press, 1988.

Lane, Charles. "High Court Faces New Political Foray." *Washington Post* (May 3, 2003):A1.

———. "Ruling Is Landmark Victory for Gay Rights." *The Washington Post* (June 27, 2003):A1,A16.

Lang, Robert E., and Patrick E. Simmons. "'Boomburbs': The Emergence of Large, Fast-Growing Suburban Cities in the United States." Fannie Mae Foundation *Census Note* 6 (June 2001).

Lanier, Mark M., and Stuart Henry. *Essential Criminology.* Boulder, CO: Westview Press, 1997.

Lantz, Paula M., et al. "Socioeconomic Factors, Health Behaviors, and Mortality." *Journal of the American Medical Association* 279 (June 3, 1998):1703–1708.

Lapchick, Richard E., and Kevin J. Matthews. *1998 Racial and Gender Report Card.* Center for the Study of Sport in Society: Northeastern University, 1999.

Lasswell, Harold. "The Structure and Function of Communication in Society." In Lyman Bryson (ed.), *The Communication of Ideas.* New York: Harper and Row, 1948.

Lauer, Robert H. *Perspectives on Social Change.* 4th ed. Boston: Allyn & Bacon, 1991.

Lawrence, Frederick M. *Punishing Hate; Bias Crimes Under American Law.* Cambridge, MA: Harvard University Press, 1999 2002.

Lawrence, Paul R., and Jay W. Lorsch. *Organization and Environment.* Boston: Division of Research, Graduate School of Business Administration, Harvard University, 1967.

Lazarsfeld, Paul. *Survey Design and Analysis.* New York: Free Press, 1955.

Le Bon, Gustave. *The Crowd.* New York: Viking, 1960.

Leach, Edmond. *Social Anthropology.* New York: Oxford University Press, 1982.

Leacock, Eleanor Burke. *Teaching and Learning in City Schools.* New York: Basic Books, 1969.

The Least Developed Countries Report 2002: Escaping the Poverty Trap. United Nations Conference on Trade and Development. New York: United Nations, 2002.

Lee, Alfred McClung. *Toward Humanist Sociology.* New York: Oxford University Press, 1978.

———. *Sociology for People.* Syracuse, NY: Syracuse University Press, 1990.

Lee, Gary R. *Family Structure and Interaction.* Philadelphia: Lippincott, 1977.

Lee, Jennifer, and Frank D. Bean. "Beyond Black and White: Remaking Race in America." *Contexts* (Summer 2003):26–33.

Lee, Ronald, and John Haaga. "Government Spending in an Older America." Washington, DC: Population Reference Bureau, 2002.

Lee, Sharon M. "Asian Americans: Diverse and Growing." *Population Bulletin* 53 (June 1998):1–40.

"Left Out." *Newsweek* (March 21, 1983):26–35.

"Lego: Fighting the Video Monsters." *The Economist* (January 30, 1999):57.

Lehrmann, Hartmut, and Guenther Roth. *Weber's Protestant Ethic: Origins, Evidence, Context.* New York: Cambridge University Press, 1993.

Leibenstein, Harvey. *Inside the Firm.* Cambridge, MA: Harvard University Press, 1987.

Lemann, Nicholas. *The Big Test: The Secret History of the American Meritocracy.* New York: Farrar, Straus & Giroux, 1999.

Lemert, Charles, and Ann Branaman (eds.). *The Goffman Reader.* Malden, MA: Blackwell, 1997.

Lemert, Edwin McCarthy, Michael Winter, and Charles C. Lemert (eds.). *Crime and Deviance: Essays and Innovations of Edwin M. Lemert.* Lanham, MD: Rowman & Littlefield, 2000.

Leming, Michael R., and George E. Dickinson. *Understanding Dying, Death, and Bereavement.* 5th ed. Fort Worth, TX: Harcourt Brace, 2002.

Lemonick, Michael D. "How Man Began." *Time* (March 14, 1994):80–87.

Lengermann, Patricia Madoo, and Jill Niebrugge-Brantley. *The Women Founders: Sociology and Social Theory, 1830–1930.* Boston: McGraw-Hill, 1998.

Lenski, Gerhard E. *Power and Privilege.* Chapel Hill: University of North Carolina Press, 1984.

———. "Rethinking Macrosociological Theory." *American Sociological Review* 53 (April 1988):163–171.

Leonard, Wilbert Marcellus II. *A Sociological Perspective of Sport.* 5th ed. Boston: Allyn & Bacon, 1998.

Leslie, Gerald R., and Sheila K. Korman. *The Family in Social Context.* 7th ed. New York: Oxford University Press, 1989.

Levin, Jack, and James Alan Fox. *Elementary Statistics in Social Research.* 9th ed. Boston: Allyn & Bacon, 2002.

Levin, Jack, and William C. Levin. *Ageism.* Belmont, CA: Wadsworth, 1980.

Levin, Jack, and Jack McDevitt. *Hate Crimes Revisited: America's War on Those Who Are Different.* Boulder, CO: Westview Press, 2002.

Levine, Daniel U., and Rayna F. Levine. *Society and Education.* 9th ed. Boston: Allyn & Bacon, 1996.

Levine, John M., and Richard L. Moreland (eds.). *Group Processes: Essential Readings.* Philadelphia: Psychology Press, 2001.

Levine, Rhonda. *Social Class and Stratification.* Blue Ridge Summit, PA: Rowman & Littlefield, 1998.

Levine, Robert V. "The Kindness of Strangers." *American Scientist* (May–June 2003).

Levine, Sol, and Paul E. White. "Exchange as a Conceptual Framework for the Study of Interorganizational Relationships." *Administrative Science Quarterly* 5 (March 1961):583–610.

Levinson, Daniel J. *Seasons of a Man's Life.* New York: Random House, 1978.

———. "A Conception of Adult Development." *American Psychologist* 41 (January 1986):3–13.

Levitan, Sar A. *Programs in Aid of the Poor.* Ann Arbor: University of Michigan Institute of Labor, 1990.

Levy, Becca R., Martin D. Slade, Suzzane Kunkel, and Stanislav Kasl. "Longevity Increased by Positive Self-Perceptions of Aging." *Journal of Personality and Social Psychology* 83 (2002):261–270.

Levy, Steven. "New Media's Dark Star." *Newsweek* (February 16, 1998):78.

Lewis, Charles. *The Buying of Congress.* New York: Avon Books, 1998.

———. *The Buying of the President 2000.* New York: Avon Books, 2000.

Lewis, Charles, and Bill Allison. *Cheating of America: How Tax Avoidance and Evasion by the Super Rich are Costing the Country Billions—and What You Can Do about It.* New York: Morrow, 2001.

Lewis, David Levering. *W. E. B. Du Bois: The Fight for Equality and the American Century, 1919–1963.* Vol. 2. New York: Henry Holt, 2000.

Lewis, David Levering, and W. E. B. Du Bois. *W. E. B. Du Bois: Biography of a Race, 1868–1919.* New York: Henry Holt, 1993.

Lewis, Gail. *"Race," Gender, Social Welfare: Encounters in a Postcolonial Society.* Malden, MA: Blackwell Publishers, 2000.

Lewis, Gordon H. "Role Differentiation." *American Sociological Review* 37 (August 1972):424–434.

Lewis, I. A., and William Schneider. "Hard Times: The Public on Poverty." *Public Opinion* 8 (June/July 1985):2 ff.

Lewis, Lionel. "Working at Leisure." *Society* (July/August 1982):27–32.

Lewis, Neil A., and Richard A. Oppel Jr. "U.S. Court Issues Discordant Ruling on Campaign Law." *New York Times* (May 3, 2003).

Lexis-Nexis Corporate Affiliations. New Providence, NJ: LexisNexis Group. 2002.

Liazos, Alexander. "The Poverty of the Sociology of Deviance: Nuts, Sluts, and Perverts." *Social Problems* 20 (Summer 1972):103–120.

Lichter, Daniel T., and Martha L. Crowley. "Poverty in America: Beyond Welfare Reform." *Population Bulletin* 57 (June 2002):1–36.

Lieberman, Myron. *Public Education.* Cambridge, MA: Harvard University Press, 1993.

Lieberman, Robert C. *Shifting the Color Line: Race and the American Welfare State.* Cambridge, MA: Harvard University Press, 1998.

Liebow, Elliot. *Talley's Corner.* Boston: Little, Brown, 1998.

Light, Ivan. *Cities in World Perspective.* New York: Macmillan, 1983.

Likert, Rensis. *New Patterns of Management.* New York: McGraw-Hill, 1961.

———. *The Human Organization.* New York: McGraw-Hill, 1967.

Lin, Nan, Walter M. Ensel, and John C. Vaughn. "Social Resources and Strength of Ties: Structural Factors in Occupational Status Attainment." *American Sociological Review* 46 (August 1981):382–399.

Lincoln, James R., and Arne L. Kalleberg. *Culture, Control, and Commitment.* New York: Cambridge University Press, 1990.

Lincoln, James R., and Kerry McBride. "Japanese Industrial Organization in Comparative Perspective." In W. Richard Scott and James F. Short Jr. (eds.), *Annual Review of Sociology* 13 (1987):289–312.

Lincoln, Yvonna S., and Norman K. Denzin. *Turning Points in Qualitative Research: Tying Knots in the Handkerchief.* Thousand Oaks, CA: AltaMira Press, 2003.

Linden, Eugene. "Can Animals Think?" *Time* (March 22, 1993a):55–61.

———. "Megacities." *Time* (January 11, 1993b):28–38.

———. *The Octopus and the Orangutan.* New York: Dutton, 2002a.

———. "The Wife Beaters of Kibale." *Time* (August 19, 2002b):56–57.

Linder, Eileen W. (ed.). *Yearbook of American and Canadian Churches.* Nashville, TN: Abingdon Press, 2000.

Link, Bruce G. "Understanding Labeling Effects in the Area of Mental Disorders." *American Sociological Review* 52 (February 1987):96–112.

Link, Bruce G., and Francis Cullen. "Reconsidering the Social Rejection of Ex-Mental Patients: Levels of Attitudinal Response." *American Journal of Community Psychology* 11 (1983):261–273.

Link, Bruce G., and Bruce P. Dohrenwend. "The Epidemiology of Mental Disorders." In Howard E. Freeman and Sol Levine (eds.), *Handbook of Medical Sociology* 4th ed. Englewood Cliffs, NJ: Prentice-Hall, 1989, pp. 102–127.

Link, Bruce G., Bruce P. Dohrenwend, and Andrew E. Skodol. "Socio-Economic Status and Schizophrenia: Noisome Occupational Characteristics as a Risk Factor." *American Sociological Review* 51 (April 1986):242–258.

Link, Bruce G., Francis T. Cullen, James Frank, and John Wozniak. "The Social Rejection of Former Mental Patients: Understanding Why Labels Matter." *American Journal of Sociology* 92 (May 1987):1461–1500.

Link, Bruce G., and Jo C. Phelan. "Conceptualizing Stigma." *Annual Review of Sociology* 27 (2001):363–385.

Linz, Juan J. "An Authoritarian Regime: Spain." In Erik Allardt and Yrjo Littunen (eds.), *Cleavages, Ideologies and Party Systems.* Helsinki: The Academic Bookstore, 1964.

Lipset, Seymour Martin. "Harriet Martineau's America." In Seymour Martin Lipset (ed.), *Harriet Martineau, Society in America.* Garden City, NY: Doubleday, 1962, pp. 5–41.

———. *American Exceptionalism: A Double-Edged Sword.* New York: Norton, 1996.

Lipset, Seymour Martin, Martin Trow, and James Coleman. *Union Democracy.* New York: Free Press, 1956.

Lissit, Robert. "Internet News: Cybergold or Cybersludge?" *World and I* (October 1998):94–99.

Little, Suzanne. "Sex Roles in Faraway Places." *Ms.* (February 1975):77 ff.

Litwak, Eugene. "Occupational Mobility and Extended Family Cohesion." *American Sociological Review* 25 (February 1960):9–21.

———. "Models of Bureaucracy Which Permit Conflict." *American Journal of Sociology* 67 (September 1961):177–184.

Liu, Melinda. "Party Time in Beijing." *Newsweek* (November 25, 2002):36.

Livernash, Robert, and Eric Rodenburg. *Population Change, Resources, and the Environment.* Washington, DC: Population Reference Bureau, 1998.

Livi-Bacci, Massimo. *A Concise History of World Population.* 2nd ed. Malden, MA: Blackwell Publishers, 1997.

Livingston, E. *Making Sense of Ethnomethodology.* London: Routledge and Kegan Paul, 1987.

Lobo, Susan, and Kurt Peters (eds.). *American Indians and the Urban Experience.* Thousand Oaks, CA: AltaMira Press, 2001.

Locke, John. *Second Treatise of Government.* Edited by C. B. Macpherson. Indianapolis, IN: Hackett, 1980.

Lodge, George C. *The New American Ideology.* New York: Knopf, 1975.

———. *The American Disease.* Washington Square: New York University Press, 1986.

Lodge, George C., and Ezra F. Vogel (eds.). *Ideology and National Competitiveness.* Cambridge, MA: Harvard Business School, 1987.

Lofland, John. "White-Hat Mobilization: Strategies of a Millenarian Movement." In Mayer N. Zald and John D. McCarthy (eds.), *The Dynamics of Social Movements.* Cambridge, MA: Winthrop, 1979, pp. 157–166.

———. *Polite Protesters.* Syracuse, NY: Syracuse University Press, 1993.

———. *Social Movement Organizations.* Hawthorne, NY: Aldine de Gruyter, 1996.

Logan, Charles H. "Problems in Ratio Correlation: The Case of Deterrence Research." *Social Forces* 60 (March 1982):791–810.

Loiz, Adam, and Alison Cassady. *The Wealth Primary: The Role of Big Money in the 2002 Congressional Primaries.* Washington, DC: U.S. PIRG Education Fund, 2002.

Lombroso, Cesare. *Crime.* Boston: Little, Brown and Company, 1918.

"The Longest Journey." *Economist* (November 2, 2002):3–16.

Longmire, Dennis R. "American Attitudes Among the Ultimate Weapon: Capital Punishment." In Dennis Longmire and Timothy J. Flanagan (eds.), *Americans View Crime and Justice.* Thousand Oaks, CA: Sage, 1996, pp. 93–108.

Longshore, Douglas. "School Racial Composition and Blacks' Attitudes Toward Desegregation: The Problem of Control in Desegregated Schools." *Social Science Quarterly* 63 (December 1982a):674–687.

———. "Race Composition and White Hostility: A Research Note on the Problem of Control in Desegregated Schools." *Social Forces* 61 (September 1982b):73–78.

Loo, Chalsa M. *Chinese American.* Rev. ed. Thousand Oaks, CA: Sage, 1998.

Lopreato, Joseph. "From Social Evolutionism to Biocultural Evolutionism." *Sociological Forum* 5 (1990):187–212.

Lorch, Donatella. "Ghosts of Tailhook." *Newsweek* (March 17, 2003):43.

Lord, George F. III, and William W. Falk. "Hidden Income and Segmentation: Structural Determinants of Fringe Benefits." *Social Science Quarterly* 63 (June 1982):208–224.

Lore, Diane. "Physicians Prescribe Unionizing." *Atlanta Constitution* (June 24, 1999):1A.

Lott, Bernice. *Women's Lives: Themes and Variations in Gender Learning.* 2nd ed. Pacific Grove, CA: Brooks/Cole, 1994.

Loury, Glenn C. *Anatomy of Racial Inequality.* Cambridge, MA: Harvard University Press, 2001.

Lowell, B. Lindsay. "Recession Pounds U.S. Hispanics." *Population Today* 30 (April 2002):1, 4.

Lucas, Samuel Roundfield. *Tracking Inequality: Stratification and Mobility in American High Schools.* New York: Teachers College Press, 1999.

Luckmann, Thomas. *The Invisible Religion.* New York: Macmillan, 1967.

Luhman, Reid. *Race and Ethnicity in the United States: Our Differences and Roots.* Forth Worth, TX: Harcourt College Publishers, 2002.

Lüschen, Günther. *The Sociology of Sport.* Paris: Mouton, 1968.

Lüschen, Günther, and G. S. Sage. *Handbook of Social Science of Sport.* Champaign, IL: Stipes, 1981.

Lux, Kenneth. *Adam Smith's Mistake.* Boston: Shambhala, 1990.

Lyman, Stanford M. *Chinese Americans.* New York: Random House, 1974.

Lynaugh, Joan E., and Barbara L. Brush. *American Nursing.* Malden, MA: Blackwell, 1996.

Maccoby, Eleanor E. *The Two Sexes: Growing Up Apart, Coming Together.* Cambridge, MA: Belknap Press of Harvard University Press, 1997.

Machalek, Richard, and Michael Martin. "'Invisible' Religions: Some Preliminary Evidence." *Journal for the Scientific Study of Religion* 15 (December 1976):311–321.

MacIntyre, Alasdair. *A Short History of Ethics.* 1st Touchstone ed. New York: Simon & Schuster, 1997.

Macionis, John J., and Vincent N. Parrillo. *Cities and Urban Life.* 3rd ed. Upper Saddle River, NJ: Prentice-Hall, 2004.

MacKay, Natalie J., and Katherine Covell. "The Impact of Women in Advertisements on Attitudes Toward Women." *Sex Roles* 36 (May 1997):573–583.

MacKellar, F. Landis, and Machiko Yanaghishita. "Homicide in the United States: Who's at Risk." Washington, DC: Population Reference Bureau, 1995.

MacKenzie, Donald A., and Judy Wajcman (eds.). *The Social Shaping of Technology.* 2nd ed. Bristol, PA: Taylor and Francis, 1999.

MacKinnon, Malcolm. "Calvinism and the Infallible Assurance of Grace: The Weber Thesis Reconsidered." *British Journal of Sociology* 39 (1988a):143–177.

———. "Weber's Exploration of Calvinism: The Undiscovered Provenance of Capitalism." *British Journal of Sociology* 39 (1988b):178–210.

Maddox, Brenda. "Homosexual Parents." *Psychology Today* 16 (February 1982):62, 66–69.

Madon, S., L. Jussim, and J. Eccles. "In Search of the Powerful Self-Fulfilling Prophecy." *Journal of Personality and Social Psychology* 72 (1997):791–809.

Magnet, Myron. *The Dream and the Nightmare: The Sixties' Legacy to the Underclass.* San Francisco: Encounter Books, 2000.

Maguire, Brendan, and Polly F. Radosh. *Introduction to Criminology.* Belmont, CA: Wadsworth, 1999.

"The Making of a Skinhead." Simon Wiesenthal Center. 1999. Available online at www.wiesenthal.com/tj/index.html.

Malinowski, Bronislaw. *The Sexual Life of Savages.* New York: Liveright, 1929.

Malthus, Thomas. *An Essay on the Principle of Population.* London: Reeves and Turner, 1798.

Manchester, William. *A World Lit Only by Fire.* Boston: Little, Brown and Company, 1993.

Mandelbaum, Michael. *The Ideas That Conquered the World: Peace, Democracy, and Free Markets in the Twenty-First Century.* New York: PublicAffairs, 2002.

Mann, Charles C. "Homeland Insecurity." *The Atlantic Monthly* (September 2002):81–102.

Mann, Patricia. *Micro-Politics: Agency in a Postfeminist Era.* Minneapolis: University of Minnesota Press, 1994.

Manning, Wendy D., and Pamela J. Smock. "First Comes Cohabitation and Then Comes Marriage?" *Journal of Family Issues,* 23 (November 2002):1065–1087.

Marchese, Aileen A. "What Companies Need to Know about Employment Practices Liability and Layoffs." 2002. Available online at www.plusweb.org/Downloads/Events/NeedtoKnowAboutEPLI&Layoffs-AileenMarchese.doc.

Marger, Martin N. *Race and Ethnic Relations: American and Global Perspectives.* 6th ed. Belmont, CA: Wadsworth/Thomson Learning, 2003.

Markle, Gerald E., and Ronald J. Troyer. "Cigarette Smoking as Deviant Behavior." In Ronald A. Farrell and Victoria Lynn Swigert (eds.), *Social Deviance.* 3rd ed. Belmont, CA: Wadsworth, 1988, pp. 82–91.

Marks, John. "A Hideous Déjà Vu from the Greatest Evil of All." *U.S. News & World Report* (April 26, 1999):30.

Marmor, Theodore R., and Michael S. Barr. "Health." In Mark Green (ed.), *Changing America.* New York: Newmarket Press, 1992, pp. 399–414.

Marmor, Theodore R., Jerry L. Mashaw, and Philip L. Harvey. *America's Misunderstood Welfare State.* New York: Basic Books, 1992.

Marmot, M. G., M. J. Shipley, and Geoffrey Rose. "Inequalities in Death-Specific Explanations of a General Pattern." *Lancet* 83 (May 5, 1984):1003–1006.

Marriott, M. "Afrocentrism: Balancing or Skewing History?" *New York Times* (August 11, 1991):1, 18.

Marsden, Peter V. "Core Discussion Networks of Americans." *American Sociological Review* 52 (February 1987):122–131.

Marshall, Eliot. "DNA Studies Challenge the Meaning of Race." *Science* 282 (October 23, 1998):654–655.

Marshall, Ray F. (ed.). *Back to Shared Prosperity: The Growing Inequality of Wealth and Income in America.* Armonk, NY: M.E. Sharpe, 1999.

Martin, Joyce A., et al. "Births: Final Data for 2000." *National Vital Statistics Reports* 50 (February 12, 2002).

Martin, Karin A. "Becoming a Gendered Body: Practices of Preschools." *American Sociological Review* 63 (August 1998):494–511.

Martin, Lindag. "The Graying of Japan." *Population Bulletin* 44 (July 1989):5–39.

Martin, Philip, and Elizabeth Midgley. "Immigration: Shaping and Reshaping America." *Population Bulletin* 58 (June 2003):1–44.

Martin, Philip, and Jonas Widgren. "International Migration: Facing the Challenge." *Population Bulletin* 57 (March 2002):1–40.

Martineau, Harriet. *Society in America.* Paris: Bandry's European Library, 1837.

———. *How to Observe Manners and Morals.* London: C. Knight, 1838.

Martinelli, Alberto, and Neil J. Smelser. *Economy and Society: Overview in Economic Sociology.* Newbury Park, CA: Sage, 1990.

Martinez, Valerie J., R. Kenneth Godwin, Frank R. Kemerer, and Laura Perna. "The Consequences of School Choice: Who Leaves and Who Stays in the Inner City." *Social Science Quarterly* 76 (September 1995):485–501.

Martocchio, B. "Grief and Bereavement." *Nursing Clinics of North America* 20 (1985):327–346.

Marty, Martin E. "Fundamentalism Reborn: Faith and Fanaticism." *Saturday Review* (May 1980):37–42.

———. *Pilgrims in Their Own Land.* New York: Penguin, 1985.

Marx, Gary T., and Douglas McAdam. *Collective Behavior and Social Movements.* Englewood Cliffs, NJ: Prentice Hall, 1994.

Marx, Karl. *The Poverty of Philosophy.* Translated by H. Quelch. Chicago: Charles H. Kerr, 1920.

Masland, Tom. "A Pageant Turns Ugly." *Newsweek* (December 2, 2002):37.

Mason, Philip. *Patterns of Dominance.* New York: Oxford University Press, 1970.

Massey, Douglas S. "American Apartheid: Segregation and the Making of the Underclass." *American Journal of Sociology* 96 (September 1990):329–357.

Massey, Douglas S., and Nancy A. Denton. "Trends in the Residential Segregation of Blacks, Hispanics, and Asians: 1970–1980." *American Sociological Review* 52 (December 1987):802–825.

———. *American Apartheid.* Cambridge, MA: Harvard University Press, 1996.

Massey, Douglas S., and Mitchell L. Eggers. "The Ecology of Inequality: Minorities and the Concentration of Poverty, 1970–1980." *American Journal of Sociology* 95 (March 1990):1153–1188.

Mathews, Tom. "The Siege of L.A." *Newsweek* (May 11, 1992):30–38.

Mauer, Marc. *Race to Incarcerate.* New York: New Press, 2001.

Maxwell, Gerald. "Why Ascription: Parts of a More-or-Less Formal Theory of the Functions and Dysfunctions of Sex Roles." *American Sociological Review* 40 (1975):445–455.

May, Tim. *Social Research: Issues, Methods and Process.* Buckingham, United Kingdom: Open University Press, 2001.

Maynard, Douglas W. "On the Functions of Social Conflict Among Children." *American Sociological Review* 50 (April 1985):207–223.

Mazaar, Michael J. *Global Trends 2005: An Owner's Manual for the Next Decade.* New York: St. Martin's Press, 1999.

Mazlish, Bruce. *A New Science.* University Park, PA: Penn State Press, 1993.

McCaghy, Charles H., and Timothy A. Capron. *Deviant Behavior.* 6th ed. Boston: Allyn & Bacon, 2002.

McCall, Leslie. *Complex Inequality: Gender, Class and Race in the New Economy.* New York: Routledge, 2001.

McCarthy, John D., and Mark Wolfson. "Resource Mobilization by Local Social Movement Organizations: Agency, Strategy, and Organization in the Movement Against Drinking and Driving." *American Sociological Review* 61 (December 1996):1070–1088.

McCarthy, John D., and Mayer N. Zald. "Resource Mobilization and Social Movements: A Partial Theory." *American Journal of Sociology* 82 (May 1977):1212–1241.

McCarthy, Michael, and Erik Brady. "Privacy Becomes Public." *USA Today* (September 27, 2002):1C, 2C.

McCormack, Gavan, and Yoshio Sugimoto. *The Japanese Trajectory.* Cambridge, MA: Cambridge University Press, 1988.

McCracken, Amber. *Ageism: How Healthcare Fails the Elderly.* Washington, DC: Alliance for Aging Research, 2003.

McDonald, Scott. "Sex Bias in the Representation of Male and Female Characters in Children's Picture Books." *Journal of Genetic Psychology* 150 (December 1989):389–401.

McFalls, Joseph A. Jr. *Population: A Lively Introduction.* Washington, DC: Population Reference Bureau, 2003.

McGinn, Daniel. "A Defeat for Dr. Death." *Newsweek* (April 5, 1999):45.

McGinn, Daniel, and Holly Peterson. "Playground of the Rich." *Newsweek* (November 25, 2002):34.

McGinn, Daniel, and Joan Raymond. "Workers of the World, Get Online." *Newsweek* (February 1998, Winter 1997–98 Special Edition):32–33.

McGrath, Ellie. "Confucian Work Ethic." *Time* (March 28, 1983):52.

McGuire, Meredith B. *Religion, The Social Context.* 5th ed. Belmont, CA: Wadsworth, 2001.

McIntyre, Robert S. "The Taxonomist." *The American Prospect* (November 20, 2000):12.

———. "One for Oil." *The American Prospect* (March 11, 2002):17.

McKinlay, John B. (ed.). *Issues in the Political Economy of Health Care.* New York: Tavistock, 1985.

McLanahan, Sara. "Life Without Father: What Happens to the Children?" *Contexts* 1 (Spring 2002):35–44.

McLanahan, Sara, and Gary Fam Sandefur. *Growing Up with a Single Parent.* Cambridge, MA: Harvard University Press, 1996.

McLaughlin, Steven, and Associates. *The Changing Lives of Women.* Chapel Hill: University of North Carolina Press, 1988.

McManus, Ed. "The Death Penalty and the Race Factor." *Illinois Issues* 11 (March 1985):47.

McNamee, Mike, and Joann Muller. "A Tale of Two Job Markets." *Business Week* (December 21, 1998):38.

McNary, Sharon. "Neighbors Didn't Believe Norco Mother's Tale." *The Inland Southern California Press-Enterprise* (September 11, 1999):B1.

McPhail, Clark. *Myth of the Madding Crowd.* Hawthorne, NY: Aldine de Gruyter, 1991.

McPhail, Clark, and Ronald T. Wohlstein. "Individual and Collective Behaviors Within Gatherings, Demonstrations, and Riots." In Ralph H. Turner and James F. Short Jr. (eds.), *Annual Review of Sociology* 9 (1983):579–600.

McQuail, Denis. *Mass Communication Theory.* 2nd ed. Beverly Hills, CA: Sage, 1987.

McTighe, Kathryn. "Internet-Healthcare: Ultimate Savior of the System?" *Health Management Technology* 20 (December 1999):24–27.

Mead, George Herbert. *Mind, Self and Society.* Chicago: University of Chicago Press, 1934.

Mead, Margaret. *Sex and Temperament in Three Primitive Societies.* New York: Mentor Books, 1950.

Mechanic, David. "Foreword." In John Colombotos and Corinne Kirchner, *Physicians and Social Change.* New York: Oxford University Press, 1986, pp. vii–x.

Mednick, Sarnoff A., William J. Gabrielli, and Barry Hutchings. "Genetic Influences in Criminal Convictions: Evidence from an Adoption Cohort." *Science* 224 (1984):891–894.

Melucci, Alberto. "The New Social Movements: A Theoretical Approach." *Social Science Information* 19 (May 1980):199–226.

Menaghan, E. G., and T. L. Parcel. "Parental Employment and Family Life: Research in the 1980s." In A. Booth (ed.), *Contemporary Families.* Minneapolis: National Council on Family Relations, 1991, pp. 361–380.

Menand, Louis. *The Metaphysical Club.* New York: Farrar, Straus & Giroux, 2001.

Mental Health: A Report of the Surgeon General. Washington, DC: Government Printing Office, 1999.

Merton, Robert K. *Social Theory and Social Structure.* Enlarged ed. New York: Free Press, 1968.

———. *On Social Structure and Science.* Chicago: University of Chicago Press, 1996.

Messner, Michael A. *Politics of Masculinities: Men in Movements.* Thousand Oaks, CA: Sage, 1997.

———. *Taking the Field: Women, Men, and Sports.* Minneapolis, MN: University of Minnesota Press, 2002.

Meuller, Walt. *Understanding Today's Youth Culture.* Wheaton, IL: Tyndale House, 1999.

Meyer, John W., and W. Richard Scott. *Organizational Environments.* Beverly Hills, CA: Sage, 1985.

Michels, Robert. *Political Parties.* New York: Free Press, 1949.

Micklin, Michael, and Dudley L. Poston Jr. (eds.). *Continuities in Sociological Human Ecology.* New York: Plenum Press, 1998.

Miech, Richard, and Robert M. Hauser. *Social Class Indicators and Health at Midlife.* Madison: Center for Demography and Ecology, University of Wisconsin—Madison, 1998.

Milgram, Stanley. "Behavioral Study of Obedience." *Journal of Abnormal and Social Psychology* 67 (1963):371–378.

———. "Group Pressure and Action Against a Person." *Journal of Abnormal and Social Psychology* 68 (1964): 137–143.

———. "Some Conditions of Obedience and Disobedience to Authority." *Human Relations* 18 (1965):57–76.

———. "The Small World Problem." *Psychology Today* 1 (1967):61–67.

———. *Obedience to Authority.* New York: Harper and Row, 1974.

Milkman, Ruth. *Farewell to the Factory.* Berkeley: University of California Press, 1997.

Mill, John Stuart. *On Liberty and Utilitarianism.* New York: Bantam Books, 1993.

Miller, Dale T. (ed.). *Cultural Divides: Understanding and Overcoming Group Conflict.* New York: Russell Sage Foundation, 1999.

Miller, Delbert C. "Sociologists in the Corporate World: Academic, Research, and Practice Roles in Business and Industry." Washington, DC: American Sociological Association, 1994.

Miller, Ellen, and Nick Penniman. "The Road to Nowhere." *The American Prospect* (August 12, 2002):14.

Miller, Joanne, and Howard H. Garrison. "Sex Roles: The Division of Labor at Home and in the Workplace." *Annual Review of Sociology* 8 (1982):237–262.

Millicent, Lawton. "Schools' 'Glass Ceiling' Imperils Girls, Study Says." *Education Week* (February 12, 1992):17.

Mills, C. Wright. *The Power Elite.* New York: Oxford University Press, 1956.

Mills, C. Wright, and Amitai Etzioni. *The Sociological Imagination.* 40th ed. New York: Oxford University Press, 1999.

Mills, Kay. *Something Better for My Children: The History and People of Head Start.* New York: NAL/Dutton, 1998.

Miner, Barbara. "Who's Vouching for Vouchers?" *The Nation* (June 5, 2000):23–24.

Mintz, Beth, and Michael Schwartz. "Interlocking Directorates and Interest Group Formation." *American Sociological Review* 46 (December 1981a):851–869.

———. "The Structure of Intercorporate Unity in American Business." *Social Problems* 29 (December 1981b):87–103.

———. *The Power Structure of American Business.* Chicago: University of Chicago Press, 1985.

Mintz, Morton, and Jerry Cohen. *America, Inc.* New York: Dial Press, 1972.

Mintzberg, Henry. *The Structuring of Organizations.* Englewood Cliffs, NJ: Prentice Hall, 1979.

Miron, Gary, and Brooks Applegate. "An Evaluation of Student Achievement in Edison Schools Opened in 1995 and 1996." Kalamazoo, MI: Western Michigan University Evaluation Center, 2000.

Mirowsky, J., and C. E. Ross. "Social Patterns of Distress." *Annual Review of Sociology* 12 (1986):23–45.

Mishel, Lawrence, Jared Bernstein, and Heather Boushey. *State of Working America, 2002–2003.* Ithaca, NY: Cornell University Press, 2003.

Mishel, Lawrence, John Schmitt, and Jared Bernstein. *State of Working America, 2000–2001.* Ithaca, NY: Cornell University Press, 2001.

Mitchell, B., and E. Gee. "Boomerang Kids and Midlife Parental Marital Satisfaction." *Family Relations* (October 1996):442–448.

Mitford, Jessica. *The American Way of Death Revisited.* New York: Alfred A. Knopf, 1998.

Mizruchi, Mark S., and Linda Brewster Stearns. "Getting Deals Done: The Use of Social Networks in Bank Decision-Making." *American Sociological Review* 66 (October 2001):647–671.

Moe, Terry M. *Schools, Vouchers, and the American Public.* Washington, DC: Brookings Institution Press, 2001.

Moen, Phyllis. "The Family Impact of Maternal Employment: Its Effects on Children." In Sameha Peterson, Judy Richardson, and Gretchen Kreuter (eds.), *The Two-Career Family.* Washington, DC: University Press, 1978, pp. 182–197.

———. *Women's Two Roles.* New York: Auburn House, 1992.

———. *It's About Time: Couples and Careers.* Ithaca, NY: Cornell University Press, 2003.

Moir, Anne. *Brain Sex: The Real Difference Between Men and Women.* New York: Lyle Stuart, 1991.

Moller, David Wendell. *Confronting Death.* New York: Oxford University Press, 1995a.

———. *Death and Dying.* New York: Oxford University Press, 1995b.

"Money Kills." *The Economist* 341 (November 16, 1996):51.

Montagu, Ashley. "Don't Be Adultish." *Psychology Today* 11 (August 1977):46–50, 55.

———. *The Natural Superiority of Women.* 5th ed. Thousand Oaks, CA: Altamira Press, 1998.

———. *Natural Superiority of Women.* Thousand Oaks, CA: AltaMira Press, 2000.

Montero, Darrel. "The Japanese Americans: Changing Patterns of Assimilation Over Three Generations." *American Sociological Review* 46 (December 1981):829–839.

Moody, John. "Street Wars: Youths Vent Their Rage." *Time* (October 14, 1985):51.

Moore, Joan, and Harry Pachon. *Hispanics in the United States.* Englewood Cliffs, NJ: Prentice Hall, 1985.

Moore, Michael. *Stupid White Men.* New York: HarperCollins, 2001.

Moore, Thomas S. *The Disposable Work Force.* Hawthorne, NY: Aldine de Gruyter, 1996.

Moorhead, Gregory, Richard Ference, and Chris P. Neck. "Group Decision Fiascoes Continue: Space Shuttle Challenger and a Revised Groupthink Framework." *Human Relations* 44 (1991):539–550.

Mor, V., et al. (eds.). *The Hospice Experiment.* Baltimore: Johns Hopkins Press, 1988.

Morgan, C. "Adjusting to Widowhood: Do Social Networks Make It Easier?" *Gerontologist* 29 (1989):101–107.

Morganthau, Tom. "IQ: Is It Destiny?" *Newsweek* (October 24, 1994):53–55.

Morganthau, Tom, and Seema Nayyar. "Those Scary College Costs." *Newsweek* (April 29, 1996):52–56.

Morin, Richard. "America's Middle-Class Meltdown." *Washington Post* (December 1, 1991):C1–C2.

———. "Crime Time: The Fear, The Facts." *Washington Post* (January 30, 1994):C1, C14.

Morris, Aldon D. "Black Southern Sit-in Movement: An Analysis of Internal Organization." *American Sociological Review* 46 (December 1981):744–767.

———. *The Origins of the Civil Rights Movement.* New York: Free Press, 1984.

Morris, Charles. *Historical Tales.* Atlanta: Martin and Hoyt, 1893.

Morris, Norval, and Michael Tonry. *Between Prison and Probation.* New York: Oxford University Press, 1990.

Morse, Jodie. "A Victory for Vouchers." *Time* (July 8, 2002):32–34.

Mouw, Ted. "Job Relocation and the Racial Gap in Unemployment in Detroit and Chicago, 1980–1990." *American Sociological Review* 65 (October 2000):730–753.

Mueller, D. P., and P. W. Cooper. "Children of Single Parent Families: How They Fare as Young Adults." *Family Relations* 35 (January 1986):169–176.

Muldavin, Joshua. "Commentary: Market Reforms Breed Discontent." *Los Angeles Times* (June 3, 1999):B9.

Muller, Jerry Z. *Adam Smith in His Time and Ours.* New York: Free Press, 1993.

Mumford, Emily. *Medical Sociology.* New York: McGraw-Hill, 1983.

Munger, Frank (ed.). *Laboring Below the Line: The New Ethnography of Poverty, Low-Wage Work, and Survival in the Global Economy.* New York: Russell Sage Foundation, 2000.

Murdock, George P. *Our Primitive Contemporaries.* New York: Macmillan, 1935.

———. "The Common Denominator of Cultures." In Ralph Linton (ed.), *The Science of Man in the World Crisis.* New York: Columbia University Press, 1945, pp. 123–142.

———. "World Ethnographic Sample." *American Anthropologist* 59 (August 1957):664–687.

Murray, Charles A. *Losing Ground.* 2nd ed. New York: Basic Books, 1994.

Mussen, Paul H. "Early Sex-Role Development." In David Goslin (ed.), *Handbook of Socialization Theory and Research.* Chicago: Rand McNally, 1969, pp. 707–732.

Myers, David G. *Social Psychology.* 6th ed. Boston: McGraw-Hill College, 2001.

Myrdal, Gunnar. *An American Dilemma.* New York: Harper and Row, 1944.

———. *Objectivity in Social Research.* Middletown, CT: Wesleyan University Press, 1983.

Nader, Ralph, and Mark Green. "Crime in the Suites." *The New Republic* (April 29, 1972):17–19.

Nagata, Donna K. *Legacy of Injustice.* New York: Plenum, 1993.

Naimark, Norman N. *Fires of Hatred: Ethnic Cleansing in Twentieth-Century Europe.* Cambridge, MA: Harvard University Press, 2002.

Nakamura, David. "Equity Leaves Its Mark on Male Athletes." *Washington Post* (July 7, 1997):A1, A10.

Nakao, Keiko, Robert W. Hodge, and Judith Treas. "On Revising Prestige Scores for All Occupations." General Social Survey Methodological Report Number 69, October 1990.

Nakao, Keiko, and Judith Treas. "Computing 1989 Occupational Prestige Scores." General Social Survey Methodological Report Number 70, November 1990.

Namboodiri, Krishman. *A Primer of Population Dynamics.* New York: Plenum, 1996.

Nanda, Serena, and Richard L. Warms. *Cultural Anthropology.* 7th ed. Belmont, CA: Wadsworth, 2002.

Nasser, Haya El. "Minorities Reshape Suburbs." *USA Today* (July 9, 2001):1A, 8A.

Nathanson, C. A. "Social Roles and Health Status Among Women and the Significance of Employment." *Social Science and Medicine* 14 (1980):463–471.

———. "Sex Differences in Mortality." *Annual Review of Sociology* 10 (1989):191–213.

National Exchange Club Foundation. "About Child Abuse Frequently Asked Questions." 2000. Available online at www.preventchildabuse.com/abuse.htm.

2002 National Survey of Latinos. Pew Hispanic Center/Kaiser Family Foundation. Los Angeles: University of Southern California, 2002.

"National Survey on Poverty in America." National Public Radio/Kaiser Family Foundation/Kennedy School of Government, 2001.

Naughton, Keith. "Cyber Slacking." *Newsweek* (November 29, 1999):62–65.

Navarro, Vicente. *Medicine Under Capitalism.* New York: Prodist, 1976.

———. *Crisis, Health, and Medicine.* New York: Tavistock, 1986.

Neal, Terry M., and Caroline Daniel. "In Kenosha, Most Voters Have an HMO Story." *Washington Post* (July 19, 1998):A7.

Nee, Victor, and David Stark (eds.). *Remaking the Economic Institutions of Socialism.* Stanford, CA: Stanford University Press, 1989.

Nelan, Bruce W. "More Harm Than Good." *Time* (March 15, 1993a):40–45.

———. "The Dark Side of Islam." *Time* (October 4, 1993b):62–64.

Nelson, Bryn. "Creation vs. Evolution." *Newsday* (March 11, 2002):B-6.

Nelson, Douglas W. "The High Cost of Being Poor." Baltimore, MD: The Annie E. Casey Foundation, 2003.

Nelson, Joel I., and David Cooperman. "Out of Utopia: The Paradox of Postindustrialization." *Sociological Quarterly* (Fall 1998):583.

Nettler, Gwynn. *Explaining Crime.* 3rd ed. New York: McGraw-Hill, 1984.

Neubeck, Kenneth J., and Noel A. Cazenave. *Welfare Racism: Playing the Race Card Against America's Poor.* London: Routledge, 2001.

Neugarten, B. L. "Adult Personality: Toward a Psychology of the Life Cycle." In B. L. Neugarten (ed.), *Middle Age and Aging.* Chicago: University of Chicago Press, 1968, pp. 137–147.

Neuman, William Lawrence. *Social Research Methods: Qualitative and Quantitative Approaches.* 5th ed. Boston: Allyn & Bacon, 2002.

Newman, Katherine S. *Falling from Grace.* New York: Free Press, 1988.

———. *Declining Fortunes.* New York: Basic Books, 1993.

———. *No Shame in My Game: Working Poor in the Inner City.* New York: Alfred A. Knopf, 1999.

"The New World of Work." *Business Week* (October 17, 1994):76–86.

Nie, Norman H., and Lutz Erbring. "Internet and Society: A Preliminary Report." Stanford Institute for the Quantitative Study of Society. February 17, 2000.

Niebuhr, H. Richard. *The Social Sources of Denominationalism.* New York: World, 1968.

Nielsen, Niels C. Jr., et al. *Religions of the World.* 3rd ed. New York: St. Martin's Press, 1993.

Nierenberg, Danielle. *Correcting Gender Myopia: Gender Equality, Women's Welfare, and the Environment.* Washington, DC: Worldwatch Institute, 2002.

Nietzsche, Friedrich. *Beyond Good and Evil.* New York: Penguin, 1987.

Nimkoff, Meyer F. (ed.). *Comparative Family Systems.* Boston: Houghton Mifflin, 1965.

Nisbet, Robert A. *The Sociological Tradition.* New York: Basic Books,1966.

———. *The Social Bond.* New York: Knopf, 1970.

———. *The Present Age.* New York: Harper and Row, 1989.

Nobel, Jeremy J., and Julie C. Cherry. "IT Helps Manage Patients with Chronic Illness." *Health Management Technology* 20 (December 1999):38–40.

Noble, Charles. *Welfare as We Knew It.* New York: Oxford University Press, 1997.

Noguchi, Yuki. "Acterna to Lay Off More as Losses Mount." *Washington Post* (October 31, 2002):E5.

Nock, Steven L. "A Comparison of Marriages and Cohabiting Relationships." *Journal of Family Issues* 16 (January 1995):53–76.

Nolan, Patrick, and Gerhard E. Lenski. *Human Societies.* 8th ed. New York: McGraw-Hill College, 1999.

Noonan, D. "Prescription Drugs: Why They Cost So Much." *Newsweek* (September 25, 2000):53–57.

Nordland, Rod. "Learning the Killing Game." *Newsweek* (September 13, 1999):38–39.

Norris, Pippa. *Digital Divide? Civil Engagement, Information Poverty, and the Internet Worldwide.* New York: Cambridge University Press, 2001.

Novak, Michael. *The Unmeltable Ethnics.* 2nd ed. New Brunswick: Transaction, 1996.

Novak, Viveca. "Carding the Truckers." *Time* (March 3, 2003).

Nursing Homes: More Can Be Done to Protect Residents from Abuse. GAO-02–312. Washington, DC: U.S. General Accounting Office, 2002.

Nydeggar, Corinne N. "Family Ties of the Aged in Cross-Cultural Perspective." In Beth B. Hess and Elizabeth W. Markson (eds.), *Growing Old in America.* New Brunswick, NJ: Transaction, 1985.

Oakes, Jeannie. *Keeping Track.* New Haven, CT: Yale University Press, 1985.

Oakes, Jeannie, and Martin Lipton. In Laura I. Rendon and Richard O. Hope (eds.), *Educating a New Majority: Transforming America's Educational System for Diversity.* San Francisco: Jossey-Bass, 1996, pp. 168–200.

Oates, Stephen B. *Malice Toward None.* New York: Mentor Books, 1977.

Oberschall, Anthony. *Social Conflict and Social Movements.* Englewood Cliffs, NJ: Prentice Hall, 1973.

Obie, Clayton. *An American Dilemma Revisited: Race Relations in a Changing World.* New York: Russell Sage Foundation, 1996.

O'Brien, Joanne, and Martin Palmer. *The State of Religion Atlas.* New York: Touchstone, 1993.

O'Brien, Timothy L. "Officials Worried Over a Sharp Rise in Identity Theft." *New York Times* (April 3, 2000):A1, A19.

O'Connell, Martin. "Where's Papa? Fathers' Role in Child Care." *Population Bulletin* 20 (September 1993):1–20.

O'Connor, Alice. *Poverty Knowledge: Social Science, Social Policy, and the Poor in Twentieth-Century U.S. History.* Princeton, NJ: Princeton University Press, 2002.

O'Dwyer, Thomas. "The Taliban's Gender Apartheid." *Jerusalem Post* (October 27, 1999):6.

Office of Management and Budget. "A Citizen's Guide to the Federal Budget." *Budget of the United States Government Fiscal Year 2000,* 1999. Available online at www.access.gpo.gov/usbudget/fy2000/guide02.html.

Ogburn, William F. *On Culture and Social Change.* Chicago: University of Chicago Press, 1964.

O'Hare, William P. "Can the Underclass Concept Be Applied to Rural Areas?" Population Reference Bureau. Staff Working Papers, January 1992a.

———. "America's Minorities—The Demographics of Diversity." *Population Bulletin* 47 (December 1992b):1–47.

———. "A New Look at Poverty in America." *Population Bulletin* 51 (1996):1–48.

———. "Tracking the Trends in Low-Income Working Families." *Population Today* (August/September 2002):1–3.

O'Hare, William P., and Brenda Curry-White. "The Rural Underclass: Examination of Multiple-Problem Populations in Urban and Rural Settings." Population Reference Bureau. Staff Working Papers, January 1992.

O'Hare, William P., and Judy C. Felt. "Asian Americans: America's Fastest Growing Minority Group." *Population Trends and Public Policy.* Washington, DC: Population Reference Bureau, 1991.

O'Hare William P., et al. *Real Life Poverty in America: Where the American Public Would Set the Poverty Line.* Washington, DC: A Center on Budget and Policy Priorities and Families USA Foundation Report, 1990.

O'Harrow, Robert Jr. "Identity Thieves Thrive in Information Age." *Washington Post* (May 31, 2001):A1, A9.

Ollman, Bertel. *Market Socialism.* New York: Routledge, 1998.

Olson, Laura Katz (ed.). *Age through Ethnic Lenses.* Lanham, MD: Rowman & Littlefield, 2001.

Olson, Steve. *Mapping Human History: Discovering the Past through Our Genes.* Boston: Houghton Mifflin, 2002.

———. *Mapping Human History: Genes, Race, and Our Common Origins.* Boston: Houghton Mifflin, 2003.

Olzak, Susan, and Joane Nagel (eds.). *Competitive Ethnic Relations.* San Diego, CA: Academic Press, 1986.

Olzak, Susan, and Elizabeth West. "Ethnic Conflict and the Rise and Fall of Ethnic Newspapers." *American Sociological Review* 56 (August 1991):458–474.

Omi, Michael, and Howard Winant. *Racial Formation in the United States.* New York: Routledge and Kegan Paul, 1994.

O'Neill, Hugh M., and D. Jeffrey Lenn. "Voices of Survivors: Words that Downsizing CEOs Should Hear." *Academy of Management Executive* 9 (November 1995):23–34.

O'Neill, June E., and M. Anne Hill. "Gaining Ground? Measuring the Impact of Welfare Reform on Welfare and Work." New York: Center for Civic Innovation, 2001.

Opp, Karl-Dieter. "Grievances and Participation in Social Movements." *American Sociological Review* 53 (1988):853–864.

———. *The Rationality of Political Protest.* Boulder, CO: Westview, 1989.

Opp, Karl-Dieter, and Wolfgang Roehl. "Repression, Micromobilization, and Political Protest." *Social Forces* 69 (1990):521–547.

Orecklin, Michele. "Now She's Got Game." *Time* (March 3, 2003):57–59.

Orfield, Gary. "Schools More Separate: Consequences of a Decade of Resegregation." Cambridge, MA: The Civil Rights Project, July 2001.

Orfield, Gary, and Susan E. Eaton. *Dismantling Desegregation: The Quiet Reversal of* Brown v. Board of Education. New York: New Press, 1997.

Orfield, Gary A., et al. "Status of School Desegregation: The Next Generation." Report to the National School Board Association. Alexandria, VA: National School Board Association, 1992.

Orshansky, Mollie. "Who's Who Among the Poor: A Demographic View of Poverty." *Social Security Bulletin* 28 (July 1965):3–32.

Ornstein, Robert, and Paul Ehrlich. *New World New Mind.* London: Paladin, 1991.

Orum, Anthony M. *Introduction to Political Sociology.* 2nd ed. Englewood Cliffs, NJ: Prentice Hall, 1983.

Osterman, Paul, et al. *Working in America: A Blueprint for the New Labor Market.* Cambridge, MA: MIT Press, 2001.

O'Sullivan, Christine Y., Clyde M. Reese, and John Mazzeo. *NAEP 1996 Science Report Card for the Nation and the States.* Washington, DC: National Center for Educational Statistics, 1997.

O'Toole, Patricia. *Money and Morals in America.* New York: Clarkson/Potter Publishers, 1998.

Ouchi, William G. *Theory Z.* New York: Avon Books, 1993.

Pacione, Michael. *Urban Geography: A Global Perspective.* New York: Routledge, 2001.

Paddock, Richard C. "Republic Stirs Debate by Allowing for Multiple Wives." *Los Angeles Times* (August 15, 1999):A1, A27–28.

Page, Benjamin I., and James R. Simmons. *What Government Can Do: Dealing with Poverty and Inequality.* Chicago: University of Chicago Press, 2002.

Page, Susan. "Corporate Credentials Weigh Down Bush's Team." *USA Today* (August 6, 2002):1B.

Palast, Greg. *The Best Democracy Money Can Buy.* Revised American edition. New York: Plume, 2003.

Palen, John J. *The Urban World.* 6th ed. New York: McGraw-Hill, 2001.

Palmore, Erdman. "Are the Aged a Minority Group?" *Journal of the American Geriatrics Society* 26 (May 1978):214–217.

———. "The Facts on Aging Quiz: A Review of Findings." *Gerontologist* 20 (December 1980):669–672.

———. "More on Palmore's Facts on Aging Quiz." *Gerontologist* 21 (April 1981):115–116.

Pampel, Fred. *Aging, Inequality and Public Policy.* Thousand Oaks, CA: Pine Forge Press, 1998.

Parenti, Michael. *Democracy for the Few.* 7th ed. New York: St. Martin's Press, 2002.

Parker, Laura. "Surgeons' Strike Forces Hospitals of Juggle Service." *USA Today* (January 20, 2003).

Parker, Laura, and Peter Eisler. "Ballots in Black Fla. Precincts Invalidated More." *USA Today* (April 6–8, 2001):1A, 3A.

Parker, Richard, and Delia Easton. "Sexuality, Culture, and Political Economy: Recent Developments in Anthropological and Cross-Cultural Sex Research." *Annual Review of Sex Research* 9 (1998):1–19.

Parkinson, C. Northcote. *Parkinson's Law.* Cutchogue, NY: Buccaneer Books, 1993.

Parrillo, Vincent N. *Diversity in America.* Thousand Oaks, CA: Pine Forge Press, 1996.

———. *Strangers to These Shores.* 7th ed. Boston: Allyn & Bacon, 2002.

Parsons, Talcott. *The Structure of Social Action.* New York: McGraw-Hill, 1937.

———. *The Social System.* New York: Free Press, 1951.

———. *Politics and Social Structure.* New York: Free Press, 1959.

———. "A Functional Theory of Change." In Amitai Etzioni and Eva Etzioni (eds.), *Social Change.* New York: Basic Books, 1964a, pp. 83–97.

———. "Definitions of Health and Illness in the Light of American Values and Social Structure." In Talcott Parsons, *Social Structure and Personality.* New York: Free Press, 1964b, pp. 257–291.

———. *Societies: Evolutionary and Comparative Perspectives.* Englewood Cliffs, NJ: Prentice Hall, 1966.

———. *The System of Modern Societies.* Englewood Cliffs, NJ: Prentice Hall, 1971.

———. "The Sick Role and the Role of the Physician Reconsidered." *Health and Society* (Summer 1975):257–278.

———. *The Evolution of Societies.* Englewood Cliffs, NJ: Prentice Hall, 1977.

Pascale, Richard Tanner, and Anthony G. Athos. *The Art of Japanese Management.* New York: Simon & Schuster, 1981.

Passas, Nilos and Robert Agnew (eds.). *The Future of Anomie Theory.* Boston: Northeastern University Press, 1997.

Passell, Peter. "'Bell Curve's Critics Say Early I.Q. Isn't Destiny." *New York Times* (November 9, 1994):B10.

Patchen, Martin. *Black-White Contact in Schools.* West Lafayette, IN: Purdue University Press, 1982.

Patterson, James T. *America's Struggle against Poverty in the Twentieth Century.* Cambridge, MA: Harvard University Press, 2000.

Patterson, James T., and Peter Kim. *The Day America Told the Truth.* Englewood Cliffs, NJ: Prentice Hall, 1991.

Patterson, Orlando. *Rituals of Blood: Consequences of Slavery in Two American Centuries.* Washington, DC: Civitas/Counterpoint, 1998.

Patterson, Thomas E. *Vanishing Voter: Public Involvement in an Age of Uncertainty.* New York: Knopf, 2002a.

———. *The American Democracy.* 6th ed. Boston: McGraw-Hill, 2002b.

Pattillo-McCoy, Mary. *Black Picket Fences: Privilege and Peril Among the Black Middle Class.* Chicago: University of Chicago Press, 1999.

Pattison, Robert V., and Hallie M. Katz. "Investor-Owned and Not-for-Profit Hospitals: A Comparison Based on California Data." *New England Journal of Medicine* 309 (August 1983):347–353.

Payer, Lynn. *Medicine and Culture.* New York: Henry Holt, 1996.

Payne, Richard J. *Getting Beyond Race: The Changing American Culture.* Boulder, CO: Westview Press, 1998.

Pearlstein, Steven, and DeNeen L. Brown. "Black Teenagers Facing Worse Job Prospects." *Washington Post* (June 4, 1994):A1.

Pear, Robert. "Spending on Health Care Increased Sharply in 2001." *New York Times* (January 8, 2003).

———. "Drug Benefit Not Certain for All on Medicare." *New York Times* (February 7, 2003).

Pearson, Patricia, and Michael Finley. *When She Was Bad: Violent Women and the Myth of Innocence.* New York: Viking Penguin, 1997.

Peck, Dennis L., and J. Selwyn Hollingsworth (eds.). *Demographic and Structural Change: The Effects of the 1980s on American Society.* Westport, CT: Greenwood, 1996.

Peck, Jamie, Frances Fox Piven, and Richard Cloward. *Workfare States.* New York: Guilford Publications, 2001.

Pelikan, Jaroslav. *The Idea of the University.* New Haven, CT: Yale University Press, 1992.

Pendleton, Brian F., Margaret M. Poloma, and T. Neal Garland. "Scales for Investigation of the Dual Career Family." *Journal of Marriage and the Family* 42 (May 1980):269–275.

Pennar, Karen. "The Ties that Lead to Prosperity." *Business Week* (December 15, 1997):153–155.

Pennings, Johannes M. "The Relevance of the Structural-Contingency Model for Organizational Effectiveness." *Administrative Science Quarterly* 20 (September 1975):393–410.

Peoples, James, and Garrick Bailey. *Humanity: An Introduction to Cultural Anthropology.* 5th ed. Belmont, CA: Wadsworth, 2000.

Peplau, L. "Roles and Gender." In Harold H. Kelly et al. (eds.), *Close Relationships.* New York: Freeman, 1983, pp. 220–264.

Perlmann, Joel, and Robert A. Margo. *Women's Work? American Schoolteachers, 1650–1920.* Chicago: University of Chicago Press, 2001.

Perrow, Charles. *Complex Organizations.* 3rd ed. New York: McGraw-Hill, 1986.

———. *Organizing America: Wealth, Power, and the Origins of Corporate Capitalism.* Princeton, NJ: Princeton University Press, 2002.

Perrucci, Robert, and Earl Wysong. *The New Class Society.* 2nd ed. Lanham, MD: Rowman & Littlefield, 2002.

Perry, Barbara. *In the Name of Hate: Understanding Hate Crimes.* Bristol, PA: Taylor and Francis, New York: Routledge, 2000.

———. *Hate and Bias Crime: A Reader.* New York: Routledge, 2003.

Perry, Joellen, and Dan McGraw. "In Cleveland, It's a Back-to-School Daze." *U.S. News & World Report* (September 6, 1999):34.

Pessen, Edward. *The Log Cabin Myth.* New Haven, CT: Yale University Press, 1984.

Peter, Laurence J., and Raymond Hull. *The Peter Principle.* New York: William Morrow, 1996.

Peters, E. L. "Aspects of Family Life Among the Bedouin of Cyrenaica." In Meyer F. Nimkoff (ed.), *Comparative Family Systems.* Boston: Houghton Mifflin, 1965, pp. 121–146.

Peters, Tom. "Restoring American Competitiveness: Looking for New Models of Organizations." *Academy of Management Executive* 2 (May 1988):103–109.

Petersen, Trond, Ishak Saporta, and Marc-David L. Seidel. "Offering a Job: Meritocracy and Social Networks." *American Journal of Sociology* 106 (November 2000):763–816.

Peterson, Karen S. "Changing the Shape of the American Family." *USA Today* (April 18, 2000):1D, 2D.

Peyser, Marc. "Tyranny of the Red Ribbon." *Newsweek* (June 28, 1993):61.

Pfeffer, Jeffrey. "The Micropolitics of Organizations." In Marshall W. Meyer and Associates (eds.), *Environments and Organizations.* San Francisco: Jossey-Bass, 1978, pp. 29–50.

———. *Power in Organizations.* Marshfield, MA: Pitman, 1981.

———. *New Directions for Organization Theory.* New York: Oxford University Press, 1997.

Phillips, Derek. "Rejection: A Possible Consequence of Seeking Help for Mental Disorders." *American Sociological Review* 28 (December 1963):963–972.

———. "Rejection of the Mentally Ill: The Influence of Behavior and Sex." *American Sociological Review* 29 (October 1964):679–687.

Phillips, John L. *The Origins of the Intellect.* 2nd ed. San Francisco: Freeman, 1975.

Phillips, Kevin. *Wealth and Democracy: How Great Fortunes and Government Created America's Aristocracy.* New York: Broadway Books, 2002.

Phu, Vu Duy. "Vietnam—Anticipating It." *Vietnam Economic News* (December 14, 1998).

Piaget, Jean. *The Psychology of Intelligence.* Totowa, NJ: Littlefield, Adams, 1981.

Piaget, Jean, and Bärbel Inhelder. *The Psychology of the Child.* London: Routledge, 1973.

Piaget, Jean, and Jane Valsiner. *The Child's Conception of Physical Causality.* New Brunswick, NJ: Transaction Publishers, 1999.

Pichanik, V. K. *Harriet Martineau.* Ann Arbor: University of Michigan Press, 1980.

Pickering, Michael. *Stereotyping: The Politics of Representation.* New York: Palgrave, 2001.

Pickering, W. S., and Geoffrey Walford (eds.). *Durkheim's Suicide.* New York: Routledge, 2000.

Pierre, Robert E., and Karl Lydersen. "Illinois Death Row Emptied." *Washington Post* (January 12, 2003):A1, A11.

Pillemer, K., and D. Finkelhor. "The Prevalence of Elder Abuse: A Random Sample Survey." *Gerontologist* 28 (February 1988):51–57.

Pillemer, V., and J. Suiter. "Will I Ever Escape My Children's Problems? Effects of Adult Children's Problems on Elderly Parents." *Journal of Marriage and the Family* (August 1991):585–594.

Pines, Maya. "The Civilizing of Genie." *Psychology Today* 15 (September 1981):28 ff.

Pinker, Steven. *The Blank Slate: The Modern Denial of Human Nature.* New York: Viking Penguin, 2002.

Pipher, Mary Bray. *The Shelter of Each Other: Rebuilding Our Families.* New York: Ballantine Books, 1997.

———. *The Middle of Everywhere: The World's Refugees Come to Our Town.* San Diego, CA: Harcourt Trade, 2002.

Plath, David. *Long Engagements.* Stanford, CA: Stanford University Press, 1980.

———. "Estasy Years – Old Age in Japan." In Jay Sokolovsky (ed.), *Growing Old in Different Societies.* Belmont, CA: Wadsworth, 1983, pp. 147–153.

Plog, Fred, and Daniel G. Bates. *Cultural Anthropology.* 3rd ed. New York: McGraw-Hill, 1990.

Poddel, Lawrence. "Welfare History and Expectancy." *Families on Welfare in New York City.* Preliminary Report No. 5. New York: The Center for Social Research, City University of New York, 1968.

Pollard, Kelvin. "U.S. Diversity Is More than Black and White." Washington, DC: Population Reference Bureau, 1999.

Pollard, Kelvin, and William P. O'Hare. "America's Racial and Ethnic Minorities." *Population Bulletin* 54 (September 1999):1–48.

PollingReport.com. "Major Institutions." 2002. Available online at www.pollingreport.com/institute.htm.

Pollner, Melvin. "Left of Ethnomethodology: The Rise and Decline of Radical Reflexivity." *American Sociological Review* 56 (June 1991):370–380.

Polonko, Karen A., John Scanzoni, and Jay D. Teachman. "Childlessness and Marital Satisfaction: A Further Assessment." *Journal of Family Issues* 3 (December 1982):545–573.

Pomfret, John. "China Allows Its Capitalists to Join Party." *Washington Post* (July 2, 2001):A1, A14.

———. "Chinese Back New Breed of Capitalism." *Washington Post* (September 29, 2002):A1, A25.

Ponnuru, Ramesh. "Affirmative Reaction." *National Review* 43 (October 13, 1997):52–56.

Pontell, Henry N. *A Capacity to Punish.* Bloomington: Indiana University Press, 1984.

Pope, Liston. *Millhands and Preachers.* New Haven, CT: Yale University Press, 1942.

Popenoe, David. *Life Without Father.* Cambridge, MA: Harvard University Press, 1999.

Popenoe, David, Jean Bethke Elshtain, and David Blankenhorn. *Promises to Keep: Decline and Renewal of Marriage in America.* Lanham, MD: Rowman & Littlefield, 1996.

Porter, Bruce, and Marvin Dunn. *The Miami Riot of 1980.* Lexington, MA: Lexington Books, 1984.

Posner, Richard A. *Aging and Old Age.* Chicago: University of Chicago Press, 1996.

Powell, Gary N., and Laura M. Graves. *Women and Men in Management.* 3rd ed. Thousand Oaks, CA: Sage, 2002.

Powell, Lawrence Alfred, John B. Williamson, and Kenneth J. Branco. *Senior Rights Movement.* Old Tappan, NJ: Twayne Publishers, 1996.

Power, Carla. "Europeans Just Say 'Maybe.'" *Newsweek* (November 1, 1999):53–54.

Power, Samantha. "Bystanders to Genocide." *The Atlantic Monthly* (September 2001):84–108.

———. *A Problem from Hell: America and the Age of Genocide.* New York: Basic Books, 2002.

Pratt, Travis, "Race and Sentencing: A Meta-Analysis of Conflicting Empirical Research Results." *Journal of Criminal Justice* 26 (1998):513–525.

Pringle, Rosemary. *Sex and Medicine: Gender, Power and Authority in the Medical Profession.* New York: Cambridge University Press, 1998.

Prunier, Gerard. *The Rwanda Crisis: History of a Genocide.* New York: Columbia University Press, 1997.

Pryor, Douglas W. *Unspeakable Acts: Why Men Sexually Abuse Children.* New York: New York University Press, 1999.

Puente, Maria. "Election of Governor a Sign of Growing Political Clout." *USA Today* (November 19, 1996):1A.

Pulera, Dominic J. *Visible Differences: How Race Will Matter to Americans in the Twenty-First Century.* New York: Continuum International Publishing Group, 2002.

Purcell, Piper, and Lara Stewart. "Dick and Jane in 1989." *Sex Roles* 22 (1990):177–185.

Putnam, Robert. "Who Killed America?" *American Prospect* (March 1996):26–28.

———. *Bowling Alone: The Collapse and Revival of American Community.* New York: Simon & Schuster, 2000.

Quarantelli, E. L. "Society of Panic." In *International Encyclopedia of the Social and Behavioral Sciences.* Oxford, UK: Pergamon Press, 2001, pp. 11020–11023.

Quarantelli, E. L., and Joseph Cooper. "Self-Conceptions and Others: A Further Test of the Meadian Hypothesis." *Sociological Quarterly* 7 (Summer 1966):281–297.

Quebedeaux, Richard. *The Worldly Evangelicals.* New York: Harper and Row, 1978.

Queen, Stuart A., Robert W. Habenstein, and Jill S. Quadagno. *The Family in Various Cultures.* 5th ed. New York: Harper and Row, 1985.

Quinn, Jane Bryant. "What's for Dinner, Mom?" *Newsweek* (April 5, 1993):68.

Quinney, Richard. *Class, State and Crime.* 2nd ed. New York: Longman, 1980.

Quintanilla, Michael. "Turning Off to Save the Family." *Los Angeles Times* (May 20, 1996):E1–E2.

"Race, Place, and Segregation." Cambridge, MA: The Civil Rights Project, May 2002.

Radin, Paul. *The World of Primitive Man.* New York: Henry Schuman, 1953.

Rae, John. *Life of Adam Smith.* Fairfield, NJ: Augustus M. Kelley, 1965.

Raine, Adrian (ed.). *The Psychopathology of Crime.* San Diego, CA: Academic Press, 1993.

Ramey, Craig T., and Frances Campbell. "Poverty, Early Childhood Education, and Academic Competence: The Abecedarian Experiment." In Aletha C. Huston (ed.), *Children in Poverty.* New York: Cambridge University Press, 1991, pp. 190–221.

Ramirez, Roberto R., and G. Patricia de la Cruz. *The Hispanic Population in the United States : March 2002*. Current Population Reports, P20–545, Washington, DC: U.S. Census Bureau, 2002.

Ranade, Wendy. *Markets and Health Care: A Comparative Analysis*. New York: Longman, 1998.

Rank, Mark R. "As American as Apple Pie: Poverty and Welfare." *Contexts* (Summer 2003):41–49.

Rashid, Ahmed. *Taliban: Militant Islam, Oil and Fundamentalism in Central Asia*. New Haven, CT: Yale University Press, 2001.

Ravitch, Diane. "What Makes a Good School?" *Society* 19 (January/February 1982):10–11.

Ray, Larry J. *Theorizing Classical Sociology*. Levittown, PA: Open University Press, 1999.

Reaves, Jessica. "Human Cloning: A Reason for Rejoicing or Despair?" *Time* (March 13, 2001).

Redfield, Robert. *The Folk Culture of Yucatan*. Chicago: University of Chicago Press, 1941.

Redhead, Steve. *Subcultures to Clubcultures*. Malden, MA: Blackwell, 1997.

Reeves, Tracey A. "DNA Frees Suspect in Md. Slaying." *Washington Post* (June 23, 2000):A1, A14.

Reich, Robert B.. *The Work of Nations*. New York: Knopf, 1993.

———. *Locked in the Cabinet*. New York: Knopf, 1997.

———. *The Future of Success*. New York: Vintage, 2001.

Reichel, Philip (ed.). *Selected Readings in Criminal Justice*. San Diego, CA: Greenhaven Press, 1998.

Reiman, Jeffrey H. *The Rich Get Richer and the Poor Get Prison*. 6th ed. Boston: Allyn & Bacon, 2000.

Reischauer, Edwin O. *The Japanese Today*. Cambridge, MA: Harvard University Press, 1995.

Reiser, Stanley Joel. *Medicine and the Reign of Technology*. New York: Cambridge University Press, 1978.

Reiss, Ira L. *Family Systems in America*. 2nd ed. Hinsdale, IL: Dryden Press, 1976.

Relman, Arnold S. "Dr. Business." *American Prospect* (September/October 1997):91–95.

Rendon, Laura J., and Richard O. Hope (eds.). *Educating a New Majority*. San Francisco: Jossey-Bass, 1996.

Renner, Michael. *Vital Signs 2003: The Trends That Are Shaping Our Future*. New York: Worldwatch Institute, 2003.

2000 Report on Television: The First Fifty Years. New York: Nielsen Media Research, 2000.

Report of the National Advisory Commission on Civil Disorders. Washington, DC: U.S. Government Printing Office, 1968.

Reskin, Barbara. "Sex Segregation in the Workplace." *Annual Review of Sociology* 19 (1993):241–270.

Reskin, Barbara, and Irene Padavic. *Women and Men at Work*. 2nd ed. Thousand Oaks, CA: Pine Forge Press, 2000.

Reynolds, Larry T., and Nancy J. Herman-Kinney (eds.). *Handbook of Symbolic Interactionism*. Thousand Oaks, CA: AltaMira Press, 2003.

Rheingold, Howard. *Smart Mobs: The Next Social Revolution: Transforming Cultures and Communities in the Age of Instant Access*. Cambridge, MA: Perseus Publishing, 2003.

Rhodes, E. "Reevaluation of the Aging and Modernization Theory: The Samoan Evidence." *Gerontologist* 24 (June 1984):243–250.

Richardson, James T. "Conversion, Brainwashing, and Deprogramming." *The Center Magazine* 15 (March/April 1982):18–24.

Richardson, Laurel, and Verta Taylor (eds.). *Feminist Frontiers*. Reading, MA: Addison-Wesley, 1983.

Riche, Martha Farnsworth. "America's Diversity and Growth: Signposts for the 21st Century." *Population Bulletin* 55 (June 2001):1–43.

Ridley, Matt. *The Origins of Virtue*. New York: Viking, 1996.

———. *Nature Via Nurture: Genes, Experience, and What Makes Us Human*. New York: HarperCollins Publishers, 2003.

Riederer, Richard K. "Battered by the World Financial Crisis." *Metal Producing 33* (March 1999):42–45.

Riesebrodt, Martin. *Pious Passion*. Berkeley: University of California Press, 1998.

Riesman, David. *The Lonely Crowd*. New Haven, CT: Yale University Press, 1961.

———. *On Higher Education*. San Francisco: Jossey-Bass, 1980.

———, with Nathan Glazer and Reuel Denney. *The Lonely Crowd: A Study of the Changing American Character*. New Haven, CT: Yale University Press, 2001.

Rifkin, Jeremy. *The End of Work*. New York: Tarcher/Putnam, 1995.

Riley, Nancy E. *Gender, Power, and Population Change*. Washington, DC: Population Reference Bureau, 1997.

Risman, Barbara J. *Gender Vertigo: American Families in Transition*. New Haven, CT: Yale University Press, 1998.

Risman, Barbara, and Pepper Schwartz. "After the Sexual Revolution: Gender Politics in Teen Dating." *Contexts* (Spring 2002):16–24.

Rist, Ray C. "On Understanding the Process of Schooling: The Contributions of Labeling Theory." In Jerome Kagan and A. H. Halsey (eds.), *Power and Ideology in Education*. New York: Oxford University Press, 1977, pp. 292–305.

Ritzer, George. *The McDonaldization Thesis*. Thousand Oaks, CA: Sage, 1998.

———. *Sociological Theory*. 5th ed. Boston: McGraw-Hill, 2000a.

———. *The McDonaldization of Society*. 3rd ed. Thousand Oaks, CA: Pine Forge Press, 2000b.

———. *McDonaldization: The Reader*. Thousand Oaks, CA: Sage, 2002.

Robbins, Richard Howard. *Global Problems and the Culture of Capitalism*. 2nd ed. Boston: Allyn & Bacon, 2001.

Robbins, Stephen P. *Organization Theory*. 3rd ed. Englewood Cliffs, NJ: Prentice Hall, 1990.

Robbins, Thomas, Dick Anthony, and James Richardson. "Theory and Research on Today's New Religions." *Sociological Analysis* 39 (Summer 1978):95–122.

Robinson, Greg. *By Order of the President: FDR and the Internment of Japanese Americans*. Cambridge, MA: Harvard University Press, 2001.

Robinson, J. D., and T. Skill. "The Invisible Generation: Portrayals of the Elderly on Prime-Time Television." *Communication Reports* 8 (1995):111–119.

Robinson, James C. *The Corporate Practice of Medicine: Competition and Innovation in Health Care*. Berkeley: University of California Press, 1999.

Robinson, Simon. "Casting Stones." *Time* (September 2, 2002):36–37.

Rochon, Thomas R. *Culture Moves: Ideas, Activism, and Changing Values*. Princeton, NJ: Princeton University Press, 1998.

Rockett, Ian R. H. "Population and Health: An Introduction to Epidemiology." *Population Bulletin* 49 (November 1994):1–47.

———. *Population and Health: An Introduction to Epidemiology*. Washington, DC: Population Reference Bureau, 1999.

Rodriguez, Gregory. "From Minority to Mainstream, Latinos Find Their Voice." *Washington Post* (January 24, 1999):B1, B4.

Roediger, David R. *Colored White: Transcending the Racial Past*. Berkeley: University of California Press, 2002.

Roethlisberger, F. J., and William J. Dickson. *Management and the Worker*. New York: Wiley, 1964.

Rogers, Adam. "Making a Killing." *Newsweek* (February 1, 1999):64.

Rohlen, Thomas P. *For Harmony and Strength*. Berkeley: University of California Press, 1974.

Romano, Renee Christine. *Race Mixing: Black-White Marriage in Postwar America*. Cambridge, MA: Harvard University Press, 2003.

Roof, Wade Clark. *A Generation of Seekers*. San Francisco: Harper San Francisco, 1994.

———. *Spiritual Marketplace: Baby Boomers and the Remaking of American Religion*. Princeton, NJ: Princeton University Press, 2001.

Roper Organization. *Virginia Slims Poll*. New York: Richard Weiner, 1985.

Rosario, L. "Gray Heirs Will Pick Up the Bill." *Far Eastern Economic Review* (June 1990):20–21.

Rose, Jerry D. *Outbreaks*. New York: Free Press, 1982.

Rosen, Jeffrey. *The Unwanted Gaze: The Destruction of Privacy in America*. New York: Random House, 2000.

Rosenbaum, Jill. "Social Control, Gender and Delinquency: An Analysis of Drug, Property and Violent Offenders." *Justice Quarterly* 4 (March 1987):117–132.

Rosenberg, Debra. "'Civil Unions' Cleared." *Newsweek* (April 25, 2000). Available online at www.newsweek.com/nw-srv/printed/us/na/a19076-2000apr25.htm.

Rosenfeld, Anne, and Elizabeth Stark. "The Prime of Our Lives." *Psychology Today* (May 1987):62–72.

Rosenfield, Sarah. "The Effects of Women's Employment: Personal Control and Sex Differences in Mental Health." *Journal of Health and Social Behavior* 30 (March 1989):77–91.

Rosenthal, Marilynn. "Where Rumors Raged." *Transaction* 8 (February 1971):34–43.

Rosenthal, Robert, and Lenore Jacobson. *Pygmalion in the Classroom.* New York: Irvington Publishers, 1989.

Rosenwein, Rifka. "Why Media Mergers Matter." *Brill's Content* (December 1999/January 2000):92–111.

Rosenzweig, Jane. "Can TV Improve Us?" *The American Prospect* 45 (July/August 1999):58–63.

Rosin, Hanna. "Same-Sex Couples Win Ruling in Vermont." *Washington Post* (December 21, 1999):A1, A14.

———. "Southern Baptists Vote to Ban Female Pastors." *Washington Post* (June 15, 2000):A1, A18.

Rosnow, Ralph L. "Rumor as Communication: A Contextualist Approach." *Journal of Communication* 38 (Winter 1988):12–28.

———. "Inside Rumor: A Personal Journey." *American Psychologist* 46 (May 1991):484–496.

Rossi, Alice S. (ed.). *The Feminist Papers.* New York: Bantam, 1974.

Rossides, Daniel W. *Social Stratification.* Upper Saddle River, NJ: Prentice Hall, 1997.

Roszak, Theodore. *The Making of a Counter Culture.* Berkeley and Los Angeles: University of California Press, 1995.

Rothschild, Joyce. "Alternatives to Bureaucracy: Democratic Participation in the Economy." *Annual Review of Sociology* 12 (1986):307–328.

Rothschild-Whitt, Joyce. "The Collectivistic Organization: An Alternative to Rational Bureaucratic Models." *American Sociological Review* 44 (August 1979):509–527.

Rothschild, Joyce, and J. Allen Whitt. *The Cooperative Workplace.* New York: Cambridge University Press, 1987.

Rotundo, E. Anthony. *American Manhood.* New York: Basic Books, 1993.

Rowe, David C. "Biometrical Models of Self-Reported Delinquent Behavior: A Twin Study." *Behavior Genetics* 13 (1983):473–489.

———. "Genetic and Environmental Components of Antisocial Behavior: A Study of 265 Twin Pairs." *Criminology* 24 (August 1986):513–532.

Roy, William G. *Socializing Capital: The Rise of the Large Industrial Corporation in America.* Princeton, NJ: Princeton University Press, 1999.

Rubenstein, Carin. "Guilty or Not Guilty." *Working Mother* (May 1991):53–56.

Rubin, Lillian B. *Worlds of Pain.* New York: Basic Books, 1992.

———. *Families on the Faultline.* New York: HarperCollins, 1994.

Ruesch, Hans. *Top of the World.* New York: Pocket Books, 1959.

Ruggiero, Vincent Ryan. *A Guide to Sociological Thinking.* Thousand Oaks, CA: Sage, 1996.

Rule, James B. "Rationality and Nonrationality in Militant Collective Action." *Sociological Theory* 7 (1989):145–160.

Russell, Alan. *Guinness Book of World Records.* New York: Sterling, 1987.

Rust, Michael, and Tiffany Danitz. "New Medium Fuels Ancient Passion." *Insight on the News* (March 9, 1998):22–23.

Ryan, William. *Blaming the Victim.* Rev. and updated ed. New York: Vintage, 1976.

Sadker, Myra, and David Sadker. "Sexism in the Schoolroom of the 80s." *Psychology Today* 19 (March 1985):54–57.

———. *Failing at Fairness.* New York: Simon & Schuster, 1995.

Sahlins, Marshall D. *Tribesmen.* Englewood Cliffs, NJ: Prentice Hall, 1968.

Sakamoto, Arthur, and Meichu D. Chen. "Inequality and Attainment in a Dual Labor Market." *American Sociological Review* 56 (June 1991):295–308.

"The 1997 Salary Report." *Working Woman* (January 1997):31–76.

Salholz, Eloise. "A Conflict of the Have-Nots." *Newsweek* (December 12, 1988):28–29.

Salmon, J. Warren, et al. "Corporatization of Medicine: The Use of Medical Management Information Systems to Increase the Clinical Productivity of Physicians." *International Journal of Health Services* 20 (1990a):233.

———. "The Futures of Physicians: Agency and Autonomy Reconsidered." *Theoretical Medicine* 11 (December 1990b):261–275.

Saltman, Kenneth J. *Collateral Damage: Corporatizing Public Schools: A Threat to Democracy.* Lanham, MD: Rowman & Littlefield, 2001.

Sampson, Robert J., and John H. Laub. "Crime and Deviance Over the Life Course: The Salience of Adult Social Bonds." *American Sociological Review* 55 (October 1990):609–627.

Samuda, Ronald. *The Psychological Testing of American Minorities.* New York: Dodd, Mead, 1975.

Samuelson, Paul A., and William D. Nordhaus. *Microeconomics.* 17th ed. New York: McGraw-Hill, 2001.

Samuelson, Robert J. "Welfare for Capitalists." *Newsweek* (May 5, 2003):54.

Sandberg, Jared. "Spinning a Web of Hate." *Newsweek* (July 19, 1999):28–29.

Sanders, Catherine M. *Surviving Grief.* New York: Wiley, 1992.

Sanderson, Stephen K. *Social Evolutionism.* New York: Basil Blackwell, 1990.

Sangree, David H. "Age and Power: Life-Course Trajectories and Age Structuring of Power Relations in East and West Africa." In David I. Kertez and Warner Schare (eds.), *Age Structuring in Comparative Perspective.* Hillsdale, NJ: Erlbaum, 1989, pp. 23–46.

Santrock, John W. *Life-Span Development.* 8th ed. Boston: McGraw-Hill, 2002.

Sapir, Edward. "The Status of Linguistics as a Science." *Language* 5 (1929):207–214.

Sapolsky, Robert. "A Gene for Nothing." *Discover* (October 1997):40–46.

Saporito, Bill. "Getting Teed Off." *Time* (September 9, 2002):50.

Sarat, Austin (ed.). *The Killing State: Capital Punishment in Law, Politics, and Culture.* New York: Oxford University Press, 1998.

Sassen, Saskia. *Cities in a World Economy.* 2nd ed. Thousand Oaks, CA: Pine Forge Press, 2000.

Sawyer, R. Keith. "Emergence in Sociology: Contemporary Philosophy of Mind and Some Implications for Sociological Theory." *American Journal of Sociology* 107 (November 2001):551–585.

Sayles, Leonard R., and George Strauss. *The Local Union.* Rev. ed. New York: Harcourt Brace Jovanovich, 1967.

Scambler, Graham. *Health and Social Change: A Critical Theory.* Buckingham, UK: Open University Press, 2002.

Schaefer, Richard T. *Racial and Ethnic Groups.* 8th ed. Upper Saddle River, NJ: Prentice Hall, 2001.

Schaeffer, Robert K. *Understanding Globalization.* 2nd ed. Lanham, MD: Rowman & Littlefield, 2002.

Schaie, K. Warner and Carmi Schooler (eds.). *Impact of Work on Older Adults.* New York: Springer, 1998.

Scheff, Thomas. "The Labeling Theory of Mental Illness." *American Sociological Review* 39 (June 1974):444–452.

———. *Microsociology.* Chicago: University of Chicago Press, 1990.

———. *Being Mentally Ill: A Sociological Theory.* 3rd ed. New York: Aldine de Gruyter, 1999.

Schell, Orville. "Deng's Revolution." *Newsweek* (August 22, 1997):21–27.

Schellenberg, James. A. *The Science of Conflict.* New York: Oxford University Press, 1982.

Scheper-Hughes, Nancy. "Deposed Kings: The Demise of Rural Irish Gerontocracy." In Jay Sokolorsky (ed.). *Growing Old in Different Societies.* Belmont, CA: Wadsworth,1983, pp. 130–146.

Schiff, Michel, and Richard Lewontin. *Education and Class*. New York: Oxford University Press, 1987.

Schlesinger, Arthur M. Jr. *The Disuniting of America: Reflections on a Multicultural Society*. New York: Norton, 1998.

Schmidley, Dianne A. "Profile of the Foreign-Born Population in the United States: 2000." *Current Population Reports* P23–206 (Washington, DC: U.S. Government Printing Office, 2001):3–9.

Schnitzer, Martin C. *Comparative Economic Systems*. 8th ed. Cincinnati, OH: South-Western College Publishing, 2000.

Schonebaum, Steve. *Does Capital Punishment Deter Crime?* San Diego, CA: Greenhaven Press, 1998.

School Vouchers: Publicly Funded Programs in Cleveland and Milwaukee. United States General Accounting Office (GAO). Washington, DC: U.S. Government Printing Office, August 2001.

Schorr, Jonathan. "Giving Charter Schools a Chance." *The Nation* (June 5, 2000):19–22.

Schrag, Peter. "The Voucher Seduction." *The American Prospect* 11 (November 23, 1999):46.

Schriesheim, Chester A., and Steven Kerr. "Theories and Measures of Leadership: A Critical Appraisal of Current and Future Directions." In James G. Hunt and Lars L. Larson (eds.), *Leadership*. Carbondale: Southern Illinois University Press, 1977, pp. 9–45.

Schuler, Doug. *New Community Networks: Wired for Change*. New York: ACM Press, 1996.

Schuman, Howard. "Sense and Nonsense about Surveys." *Contexts* (Summer 2002):40–47.

Schutt, Russell K. *Investigating the Social World*. 3rd ed. Thousand Oaks, CA: Pine Forge Press, 2001.

Schwartz, Barry. *The Battle for Human Nature*. New York: Norton, 1987.

Schwarz, John E. *Illusions of Opportunity: The American Dream in Question*. New York: Norton, 1997.

Schwarz, John E., and Thomas J. Volgy. *The Forgotten Americans*. New York: Norton, 1999.

Schwartz, Martin D. *Researching Sexual Violence Against Women*. Thousand Oaks, CA: Sage, 1997.

Schweinhart, Lawrence J. "How Much Do Good Early Childhood Programs Cost?" *Early Childhood Education and the Public Schools* 3 (April 1992):115–127.

Scimecca, Joseph A. "Humanist Sociological Theory: The State of the Art." *Humanity and Society* 11 (1987):335–352.

Scott, Alan. *Ideology and the New Social Movements*. London: Routledge, 1995.

Scott, W. Richard. *Organizations*. 3rd ed. Englewood Cliffs, NJ: Prentice Hall, 1992.

Scott, W. Richard, et al. *Institutional Change and Healthcare Organizations: From Professional Dominance to Managed Care*. Chicago: University of Chicago Press, 2000.

Scripps- Howard. "Soon, Most of World Will Live in Cities." *Quad City Times* (May 16, 2002):A12.

Seale, Clive. *Constructing Death: The Sociology of Dying and Bereavement*. Cambridge, MA: Cambridge University Press, 1998.

Seefeldt, Carol, Richard K. Jantz, Alice Galper, and Kathy Serock. "Using Pictures to Explore Children's Attitudes Toward the Elderly." *Gerontologist* 17 (December 1977):506–512.

Seligmann, Jean. "Husbands No, Babies, Yes." *Newsweek* (July 26, 1999):53.

Senna, Joseph J., and Larry J. Siegel. *Introduction to Criminal Justice*. 9th ed. Belmont, CA: Wadsworth, 2002.

Service, Elman R. *Origins of the State and Civilization*. New York: Norton, 1975.

———. *The Hunters*. 2nd ed. Englewood Cliffs, NJ: Prentice Hall, 1979.

Shaller, Elliot H., and Mary K. Qualiana. "The Family and Medical Leave Act—Key Provisions and Potential Problems." *Employee Relations Law Journal* 19 (Summer 1993):5–22.

Shanas, Ethel. *Old People in Three Industrial Societies*. New York: Arno Press, 1980.

Shapiro, Isaac, and Robert Greenstein. *The Widening Income Gulf*. Washington, DC: Center on Budget and Policy Priorities, 1999.

Shapiro, Laura. "Guns and Dolls." *Newsweek* (May 28, 1990):54–65.

Sharot, Stephen. *A Comparative Sociology of World Religions: Virtuosos, Priests, and Popular Religion*. New York: New York University Press, 2001.

Shaw, Clifford, and Henry McKay. *Delinquency Areas*. Chicago: University of Chicago Press, 1929.

, Martin. *War and Genocide: Organized Killing in Modern Society*. Oxford: Polity Press, 2003.

Sheehan, Molly O. "Poverty Persists." In *Vital Signs 2002: The Trends That Are Shaping Our Future*. New York: Norton, 2002, pp. 148–149.

Sheldon, William H. *Varieties of Delinquent Youth*. New York: Harper, 1949.

Shell, Adam. "Wall St. Firms Hit Over '90s Misdeeds." *USA Today* (April 29, 2003):1A.

Shelton, Austin J. "The Aged and Eldership Among the Igbo." In Donald O. Cowgill and Lowell D. Holmes (eds.), *Aging and Modernization*. New York: Appleton-Century-Crofts, 1972, pp. 31–49.

Shepard, Jon M. *Automation and Alienation*. Cambridge, MA: MIT Press, 1971.

Shepard, Jon M., and James G. Hougland Jr. "Contingency Theory: 'Complex Man' or 'Complex Organization'?" *Academy of Management Review* 3 (July 1978):413–427.

Shepard, Jon M., Jon Shepard, and Richard E. Wokutch. "The Problem of Business Ethics: Oxymoron or Inadequate Vocabulary." *Journal of Business and Psychology* 6 (1991):9–23.

Shepherd, William. "Legal Protection for Freedom of Religion." *The Center Magazine* (March/April 1982):30–37.

Sherman, Lawrence. "Defiance, Deference, and Irrelevance: A Theory of the Criminal Sanction." *Journal of Research in Crime and Delinquency* 30 (November 1993a):445–473.

———. "A Brave New Darwinian Workplace." *Fortune* (February 25, 1993b):50.

Sherrid, Pamela. "2000 Retirement Guide: Aging Baby Boomers are Finding an Accommodating Workplace." *U.S. News & World Report* 128 (June 5, 2000):64.

Shichor, David, and Dale K. Sechrest (eds.). *Three Strikes and You're Out*. Newbury Park, CA: Sage, 1996.

Shields, Vickie Rutledge, and Dawn Heinecken. *Measuring Up: How Advertising Images Shape Gender Identity*. Philadelphia: University of Pennsylvania Press, 2001.

Shils, Edward A., and Morris Janowitz. "Cohesion and Disintegration in the Wehrmacht in World War II." *Public Opinion Quarterly* 12 (Summer 1948):280–315.

Sidel, Ruth. *Keeping Women and Children Last*. New York: Penguin, 1996.

Sifford, Darrell. *The Only Child*. New York: Putnam, 1989.

Signorielli, Nancy, and Aaron Bacue. "Recognition and Respect: A Content Analysis of Prime-Time Television Characters Across Three Decades." *Sex Roles* 40 (April 1999):527–544.

Silberman, Charles E. *Crisis in the Classroom*. London: Wildwood House, 1973.

Simmel, Georg. *Conflict and the Web of Group Affiliation*. Translated by Kurt H. Wolff. New York: Free Press, 1964.

Simmons, J. L. *Deviants*. Berkeley, CA: Glendessary Press, 1969.

Simmons, Sally Lynn, and Amelia E. El-Hindi. "Six Transformations for Thinking about Integrative Curriculum." *Middle School Journal* 30 (November 1998):32–36.

Simmons, Wendy W. "Majority of Americans Say More Women in Political Office Would Be Positive for the Country." Princeton, NJ: The Gallup Organization, 2001.

Simon, David R. *Elite Deviance*. 7th ed. Boston: Allyn & Bacon, 2001.

Simon, Herbert A. *Administrative Behavior*. 3rd ed. New York: Free Press, 1976.

Simpson, George Eaton. *Emile Durkheim*. New York: Crowell, 1963.

Sinclair, Karen. "Cross-Cultural Perspectives on American Sex Roles." In Marie Richmond-Abbott (ed.), *The American Woman*. New York: Holt, Rinehart and Winston, 1979, pp. 28–47.

Singer, Max. "The Population Surprise." *The Atlantic Monthly* (August 1999):22–25.

Singh, B. Krishna, and J. Sherwood Williams. "Childlessness and Family Satisfaction." *Research on Aging* 3 (June 1981):218–227.

Singh, Sarban. "Shift in Focus." *Malaysian Business* (September 16, 1998):54.

Sizer, Theodore. *Horace's Hope: What Works for the American High School.* Boston: Houghton Mifflin, 1996.

Sjoberg, Gideon. *The Preindustrial City, Past and Present.* New York: Free Press, 1966.

Skinner, Denise A. "Dual-Career Families: Strains of Sharing." In Hamilton I. McCubbin and Charles R. Figley (eds.), *Stress and the Family.* Vol. 1. New York: Brunner/Mazel, 1983, pp. 90–101.

Skocpol, Theda. "Bringing the State Back In: Strategies of Analysis in Current Research." In Peter B. Evans, Dietrich Rueschemeyer, and Theda Skocpol (eds.), *Bringing the State Back In.* Cambridge, UK: Cambridge University Press, 1985, pp. 3–37.

Skocpol, Theda, and Edwin Amenta. "States and Social Policies." In Ralph H. Turner and James F. Short Jr. (eds.), *Annual Review of Sociology* 12 (1986):131–157.

Skolnick, Arlene S., and Jerome H. Skolnick (eds.). *Family in Transition.* 12th ed. Boston: Allyn & Bacon, 2002.

Skolnick, Jerome H. "The Color of the Law." *American Prospect* 39 (July/August 1998):90–95.

Sloan, Allan. "The Hit Men." *Newsweek* (February 26, 1996):44–48.

Smaje, Chris. *Natural Hierarchies: The Historical Sociology of Race and Caste.* Malden, MA: Blackwell, 2000.

Small, Mario, and Katherine Newman. "Urban Poverty after *The Truly Disadvantaged*: The Rediscovery of the Family, the Neighborhood, and Culture. *Annual Review of Sociology* 27 (August 2001):23–45.

Smedley, Brian D., Adrienne Y. Stith, and Alan R. Nelson (eds.). *Unequal Treatment: Confronting Racial and Ethnic Disparities in Health Care.* Washington, DC: National Academy Press, 2002.

Smelser, Neil J. *Theory of Collective Behavior.* New York: Free Press, 1971.

———. *The Sociology of Economic Life.* 2nd ed. Englewood Cliffs, NJ: Prentice Hall, 1976.

Smelser, Neil J., William Julius Wilson, and Faith Mitchell (eds.). *America Becoming: Racial Trends and Their Consequences.* Vols. 1 and 2. Washington, DC: National Academy Press, 2001.

Smith, Adam. *An Inquiry into the Nature and Causes of the Wealth of Nations.* Edwin Cannan (ed.). New York: Modern Library, 1965.

———. *The Theory of Moral Sentiments.* North Shadeland, IN: Liberty Press/Liberty Classics, 1976.

Smith, Adam, and Edwin Cannan. *The Wealth of Nations.* New York: Random House, 2000.

Smith, Christian. *Christian America.* Berkeley: University of California Press, 2000.

Smith, Christian, and Michael Emerson. *American Evangelicalism: Embattled and Thriving.* Chicago: University of Chicago Press, 1998.

Smith, Heather. "The Promotion of Women and Men." Ph.D. Dissertation, University of Delaware, 1999.

Smith, Lee. "Burned-Out Bosses." *Fortune* (July 25, 1994):44–52.

Smolowe, Jill. "When Violence Hits Home." *Time* (July 4, 1994):18–25.

Snarr, Brian B. "The Family and Medical Leave Act of 1993." *Compensation and Benefits Review* 25 (May/June 1993):6–9.

Snipp, C. Matthew. "Sociological Perspectives on American Indians." *Annual Review of Sociology* 18 (1992):351–371.

———. "A Demographic Comeback for American Indians?" *Population Today* 24 (November 1996):4–5.

Snizek, William. "Survivors as Victims: Some Little Publicized Consequences of Corporate Downsizing." Paper Presented to the Ministerie van Binnenlandse Zaken Den Haag, The Netherlands, July 1994.

Snizek, William E., and Joseph J. Kestel. "Understanding and Preventing the Premature Exodus of Mature Middle Managers from Today's Corporations." *Organization Development Journal* (Fall 1999):63–71.

Snyder, Howard N., and Melissa Sickmund. *Juvenile Offenders and Victims: A National Report.* Rockville, MD: Juvenile Justice Clearinghouse, 1999.

Sobel, Michael E., Mark P. Becker, and Susan M. Minick. "Origins, Destinations, and Association in Occupational Mobility." *American Journal of Sociology* 104 (November 1998):687–721.

Sokoloff, Natalie J. *Between Money and Love.* New York: Praeger, 1980.

Solomon, Jolie. "Smoke From Washington." *Newsweek* (April 4, 1994):45.

———. "Texaco's Troubles." *Newsweek* (November 25, 1996):48–50.

Solow, Robert M., Gertrude Himmelfarb, and Amy Gutmann. *Work and Welfare.* Princeton, NJ: Princeton University Press, 1998.

Sorokin, Pitirim A. *Social Mobility.* New York: Harper, 1927.

Sourcebook of Criminal Justice Statistics Online. "Attitudes Toward the Penalty for Murder." 2002. Available online at www.albany.edu/sourcebook/1995/pdf/t260.pdf.

South, Scott J., and Kyle D. Crowder. "Escaping Distressed Neighborhoods: Individual, Community, and Metropolitan Influences." *American Journal of Sociology* 102 (January 1997):1040–1084.

Spain, Daphne, and Suzanne M. Bianchi. *Balancing Act: Motherhood, Marriage, and Employment among American Women.* New York: Russell Sage Foundation, 1996.

Spencer, Jon Michael. *The New Colored People.* New York: New York University Press, 1997.

Spengler, Oswald. *The Decline of the West.* New York: Knopf, 1926–1928.

Spickard, Paul R. "The Illogic of American Racial Categories." In Maria P. P. Root (ed.), *Racially Mixed People in America.* Thousand Oaks, CA: Sage, 1992.

Spindler, George D., and Louise Spindler. "Anthropologists View American Culture." *Annual Review of Anthropology* 12 (1983):49–78.

Spitz, René A. "Hospitalism." In Anna Freud et al. (eds.), *The Psychoanalytic Study of the Child.* Vol. 2. New York: International Universities Press, 1946a, pp. 52–74.

———. "Hospitalism: A Follow-Up Report." In Anna Freud et al. (eds.), *The Psychoanalytic Study of the Child.* Vol. 2. New York: International Universities Press, 1946b, pp. 113–117.

Spitze, B., and J. Logan. "Sons, Daughters, and Intergenerational Social Support." *Journal of Marriage and the Family* (May 1990):420–430.

Spitze, G. "Women's Employment and Family Relations: A Review." In A. Booth (ed.), *Contemporary Families.* Minneapolis: National Council on Family Relations, 1991, pp. 381–404.

Spitzer, Robert L., Janet B. W. Williams, and Andrew Skodol. "DSM-III: The Major Achievements and an Overview." *American Journal of Psychiatry* 137 (1980):151–164.

Spitzer, Steven. "Toward a Marxian Theory of Deviance." In Delos H. Kelly (ed.), *Criminal Behavior.* New York: St. Martin's Press 1980, pp. 175–191.

Spohn, Cassia C. "Courts, Sentences, and Prisons." *Daedalus* 124 (Winter 1995):119–143.

Spotts, Peter N. "Science Labs, Too, 'Cooking the Books'." *Christian Science Monitor* (July 19, 2002):1, 9.

Spragins, Ellyn. "Does Your HMO Stack Up?" *Newsweek* (June 24, 1996):56–63.

———. "Does Managed Care Work?" *Newsweek* (September 28, 1998):61–64.

Srole, Leo T., et al. *Mental Health in the Metropolis.* New York: McGraw-Hill, 1962.

———. *Mental Health in the Metropolis: The Midtown Manhattan Study.* New York: New York University Press, 1978.

Stacey, Judith. *Brave New Families.* New York: Basic Books, 1990.

Stacey, William, and Anson Shupe. "Correlates of Support for the Electronic Church." *Journal for the Scientific Study of Religion* 21 (December 1982):291–303.

Stafford, Mark. *W. E. B. Du Bois: Scholar and Activist.* New York: Chelsea House Publishers, 1989.

Stangor, Charles. *Stereotypes and Prejudice.* Philadelphia, PA: Psychology Press, 2000.

Stark, David, and Laszlo Bruszt. *Postsocialist Pathways.* Cambridge, UK: Cambridge University Press, 1998.

Stark, Rodney, and William Sims Bainbridge. "Secularization and Cult Formation in the Jazz Age." *Journal for the Scientific Study of Religion* 20 (December 1981):360–373.

———. *The Future of Religion*. Berkeley: University of California Press, 1985.

Stark, Rodney, and Roger Finke. *Acts of Faith: Explaining the Human Side of Religion*. Berkeley: University of California Press, 2000.

Stark, Rodney, Lori Kent, and Daniel P. Doyle. "Religion and the Ecology of a 'Lost' Relationship." *Journal of Research in Crime and Delinquency* 19–20 (January 1982):4–24.

Starr, Mark, and Allison Samuels. "Ear Today, But Gone Tomorrow." *Newsweek* (July 14, 1997):58–60.

Starr, Paul E. *The Social Transformation of American Medicine*. New York: Basic Books, 1983.

———. *The Limits of Privatization*. Washington, DC: Economic Policy Institute, 1988.

Starting Points. The Report of the Carnegie Task Force on Meeting the Needs of Young Children. Carnegie Corporation of New York, April 1994.

"Statement on Race and Intelligence." *Journal of Social Issues* 25 (Summer 1969):1–3.

The State of America's Children: A Report from the Children's Defense Fund. Boston: Beacon Press, 1998.

The State of Food Insecurity in the World 2002. Food and Agriculture Organization of the United Nations. Rome, Italy, 2002.

The State of Our Nation's Youth. Alexandria, VA: Horatio Alger Association of Distinguished Americans, 2001.

The State of the World's Children 2000. New York: UNICEF (United Nation's Children's Fund), 2000.

State of Florida Hospital Cost Containment Board. Annual Reports. Tallahassee: Florida Cost Containment Board, 1981–82, 1982–83.

Stavans, Ilan. *The Hispanic Condition*. New York: Harper Collins, 1996.

———. *The Hispanic Condition: The Future Power of a People*. 2nd ed. New York: HarperTrade, 2001.

Steedley, Gilbert. "The Forbes 500." *Forbes* 153 (April 25, 1994):195–230.

Stefancic, Jean, and Richard Delgado (eds.). *The Latinola Condition*. New York: New York University Press, 1998.

Stein, Joel. "The New View of Pot." *Time* (November 4, 2002):56–66.

———. "The Real Face of Homelessness." *Time* (January 20, 2003):52–57.

Stein, Leonard I. "The Doctor-Nurse Game." *Archives of General Psychiatry* 16 (June 1967):699–700.

Stein, Rob. "Increase in Physicians' Insurance Hurts Care: Services Are Being Pared, and Clinics Are Closing." *Washington Post* (January 5, 2003):A1.

Steinberg, Jacques. "At 42 Newly Privatized Philadelphia Schools, Uncertainty Abounds." *New York Times* (April 19, 2002):A16.

Steinmetz, Suzanne K. *Duty Bound: Elder Abuse and Family Care*. Newbury Park, CA: Sage, 1988.

Steinnes, Donald N. "Suburbanization and the 'Malling of America': A Time-Series Approach." *Urban Affairs Quarterly* 17 (June 1982):401–418.

Stephens, Martha. *The Treatment: The Story of Those Who Died in the Cincinnati Radiation Tests*. Durham, NC: Duke University Press, 2002.

Stephens, W. Richard Jr., "What Now? The Relevance of Sociology to Your Life and Career." In Roger A. Straus (ed.), *Using Sociology*, 3rd ed. Lanham, MD: Rowman & Littlefield, 2002, pp. 347–363.

Stern, Kenneth S. *A Force Upon the Plane*. New York: Simon & Schuster, 1996.

Stern, Philip M. *The Best Congress Money Can Buy*. New York: Pantheon, 1988.

Steward, Julian H. *Theory of Culture Change*. Urbana: University of Illinois Press, 1979.

Stewart, Janet Kidd. "The '99 Salary Survey." *Working Woman* (July/August 1999):46–56.

Stiglitz, Joseph E. *Globalization and Its Discontents*. New York: Norton, 2002.

Stockard, Janice. *Marriage in Culture: Practice and Meaning Across Diverse Societies*. Fort Worth, TX: Harcourt College Publishers, 2002.

Stockard, Jean, and Miriam M. Johnson. *Sex and Gender in Society*. 2nd ed. Englewood Cliffs, NJ: Prentice Hall, 1992.

Stoessinger, John. "The Anatomy of the Nation-State and the Nature of Power." In Michael Smith, Richard Little, and Michael Shackleton (eds.), *Perspectives on World Politics*. London: Croom Helm, 1981, pp. 25–26.

Stogdill, Ralph M. *Handbook of Leadership*. New York: Free Press, 1974.

Stoll, Cliff. *Silicon Snake Oil*. New York: Doubleday, 1995.

Stoltenberg, John, and Hanna McCrum. "The Seventh Annual Working Woman Salary Survey." *Working Woman* (January 1986):73–82.

Stone, Deborah. "Care and Trembling." *The American Prospect* (March/April 1999):61–67.

Stone, Jon R. *On the Boundaries of American Evangelicalism*. New York: St. Martin's Press, 1997.

Stranger, Janice, Nicole C. Batchelder, William Brossman, and Gerald L. Uslarder. "What the Family and Medical Leave Act Means for Employers." *Journal of Compensation and Benefits* 9 (July/August 1993):12–19.

Strasburger, Victor C. *Adolescents and the Media: Medical and Psychological Impact*. Newbury Park, CA: Sage, 1995.

Straus, Murray A., Richard J. Gelles, and Suzanne K. Steinmetz. *Behind Closed Doors*. Garden City, NY: Doubleday, 1980.

Straus, Murray A., David B. Sugarman, and Jean Giles-Sims. "Spanking by Parents and Subsequent Antisocial Behavior of the Child." *Archives of Pediatrics and Adolescent Medicine* (August 1997):761–767.

Straus, Roger A. *Using Sociology*. 3rd ed. Lanham, MD: Rowman & Littlefield, 2002a.

———. "Using Sociological Theory to Make Practical Sense Out of Social Life." In Roger A. Straus (ed.), *Using Sociology*, 3rd ed. Lanham, MD: Rowman & Littlefield, 2002b, pp. 21–43.

Strauss, Anselm, and Juliet Corbin. *Basics of Qualitative Research*. 2nd ed. Thousand Oaks, CA: Sage, 1998.

Strauss, Anselm, Shizuko Fagerhaugh, Barbara Suczek, and Carolyn Wiener. *Social Organization of Medical Work*. Chicago: University of Chicago Press, 1985.

"Street-Level Youth Media." 1999. Available online at http://streetlevel.iit.edu

Strickland, S. S., and Prakash S. Shetty. *Human Biology and Social Inequality*. New York: Cambridge University Press, 1998.

Suarez-Orozco, Marcelo M., and Mariela Paez (eds.). *Latinos: Remaking America*. Berkeley: University of California Press, 2002.

Suchman, Edward A. "The 'Hang-Loose' Ethic and the Spirit of Drug Use." *Journal of Health and Social Behavior* 9 (June 1968):146–155.

Suggs, Welch. "Uneven Progress for Women's Sports." *Chronicle of Higher Education* (April 7, 2000):A52–A57.

Sumner, William Graham. *Folkways*. Boston: Ginn, 1906.

Suro, Roberto. "Mixed Doubles." *American Demographics* 21 (November 1999):57–62.

Sussman, Marvin B. "The Help Pattern in the Middle Class Family." *American Sociological Review* 18 (February 1953):22–28.

Sussman, Marvin, Suzanne K. Steinmetz, and Gary W. Peterson. *Handbook of Marriage and the Family*. New York: Plenum, 1998.

Sutherland, Edwin H. "White Collar Criminality." *American Sociological Review* 5 (February 1940):1–12.

———. *White-Collar Crime*. New Haven, CT: Yale University Press, 1983.

Sutherland, Edwin H., and Donald R. Cressey. *Principles of Criminology*. 11th ed. Dix Hills, NY: General Hall, 1992.

Suttles, Gerald D. *The Social Order of the Slum*. Chicago: University of Chicago Press, 1968.

Swaaningen, René van. *Critical Criminology: Visions from Europe*. London: Sage, 1997.

Swain, Carol M. *New White Nationalism in America: Its Challenge to Integration*. New York: Cambridge University Press, 2002.

Swann, William B. Jr., Judith H. Langlois, and Lucia Albino Gilbert (eds.). *Sexism and Stereotypes*. Washington, DC: American Psychological Association, 1998.

Swartz, Thomas R., and Kathleen Maas Weigert (eds.). *America's Working Poor.* Notre Dame, IN: University of Notre Dame Press, 1996.

Swatos, William H. Jr., and Daniel V. Olson (eds.). *The Secularization Debate.* Lanham, MD: Rowman & Littlefield, 2000.

Sweet, Jones, Larry Bumpass, and Vaughn Call. *National Survey of Families and Households.* Madison: University of Wisconsin, Center for Demography and Ecology, 1988.

Swidler, Ann. "Culture in Action: Symbols and Strategies." *American Sociological Review* 51 (April 1986):273–286.

Swinton, David H. "Economic Status of Blacks." In *The State of Black America 1989.* New York: National Urban League, 1989, pp. 129–152.

Symonds, William C. "For-Profit-Schools." *Business Week* (February 7, 2000):64 ff.

———. "Colleges in Crisis." *Business Week* (April 28, 2003):72–79.

Symonds, William C., and Kathleen Kerwin. "It's Gonna Get Ugly." *Business Week* (October 5, 1998):38–42.

Szasz, Thomas. *The Myth of Mental Illness.* Rev. ed. New York: Harper and Row, 1986.

Szymanski, Albert. "Racial Discrimination and White Gain." *American Sociological Review* 41 (June 1976):403–414.

———. *Class Structure.* New York: Praeger, 1983.

Tannenbaum, Frank. *Slave and Citizen.* New York: Knopf, 1947.

Taylor, Alexander L. "The Growing Gap in Retraining." *Time* (March 28, 1983):50–51.

Taylor, Chris. "Day of the Smart Mobs." *Time* (March 10, 2003):53.

Taylor, Humphrey. "The Canadian and U.S. Health Care Systems Compared." *The Harris Poll* (April 26, 1994):1–6.

Taylor, Humphrey, and Robert Leitman (eds.). "While Managed Care Is Still Unpopular, Hostility Has Declined." *Health Care News.* Harris Interactive Inc., 2002.

Taylor, P. "Therapists Rethink Attitudes on Divorce: New Movement to Save Marriages Focuses on Impact on Children." *Washington Post* (January 29, 1991):A1, A6.

Taylor, Shelley E., Letitia Ann Peplau, and David O. Sears. *Social Psychology.* 9th ed. Englewood Cliffs, NJ: Prentice Hall, 1997.

Taylor, Verta. "Social Movement Continuity: The Women's Movement in Abeyance." *American Sociological Review* 54 (October 1989):761–775.

Teich, Albert H. (ed.). *Technology and the Future.* 6th ed. New York: St. Martin's Press, 1993.

Tellegen, Auke, D. T. Lykken, T. J. Bouchard Jr., and M. McGue. "Heritability of Interests: A Twin Study." *Journal of Applied Psychology* (August 1993):649–661.

Terry, James L. "Bringing Women In: A Modest Proposal." *Teaching Sociology* 10 (January 1983):251–261.

Therborn, Goran. *The Power of Ideology and the Ideology of Power.* London: New Left Books, 1982.

———. "The Prospects of Labor and the Transformation of Advanced Capitalism." *New Left Review* 145 (May/June 1984):5–38.

Thernstrom, Stephan, and Abigail Thernstrom. *America in Black and White: One Nation, Indivisible.* New York: Simon & Schuster, 1997.

Thomas, Cathy Booth. "Called to Account: Guilty of Obstruction, Arthur Andersen Becomes the First Courtroom Casualty of the Enron Collapse." *Time* (June 24, 2002):52.

Thomas, Evan. "The Next Level." *Newsweek* (April 7, 1997):28–36.

———. "The War Over Gay Marriage." *Newsweek* (July 8, 2003):38–45.

Thomas, Evan, and Lynette Clemetson. "A New War Over Vouchers." *Newsweek* (November 22, 1999):46.

Thomas, Evan, and Gregory L. Vistica. "A Question of Consent." *Newsweek* (April 28, 1997):41.

Thomas, Melvin E., and Michael Hughes. "The Continuing Significance of Race: A Study of Race, Class, and Quality of Life in America, 1972–1985." *American Sociological Review* 51 (December 1986):830–841.

Thomas, Melvin E., and Linda A. Trieber. "Race, Gender, and Status: A Content Analysis of Print Advertisements in Four Popular Magazines." *Sociological Spectrum* 20 (July/September 2000):357–371.

Thomas, Susan L. *Gender and Poverty.* New York: Garland, 1994.

Thompson, Kenneth. "The Organizational Society." In Graeme Salaman and Kenneth Thompson (eds.), *Control and Ideology in Organizations.* Cambridge, MA: MIT Press, 1980, pp. 3–23.

Thompson, L., and A. J. Walker. "Gender in Families." In A. Booth (ed.), *Contemporary Families.* Minneapolis: National Council on Family Relations, 1991, pp. 76–102.

Thomson, Randall J., and Matthew T. Zingraff. "Detecting Sentencing Disparity: Some Problems and Evidence." *American Journal of Sociology* 86 (January 1981):869–880.

Thornton, Arland, and Linda Young-DeMarco. "Four Decades of Trends in Attitudes toward Family Issues in the United States: The 1960s through the 1990s." *Journal of Marriage and Family* 63 (2001):1009–1037.

Thurow, Lester C. *Head to Head.* London: Nicholas Brealey, 1994.

———. *Fortune Favors the Bold: What We Must Do to Build a New and Lasting Global Prosperity.* New York: HarperCollins Publishers, 2003.

Tilly, Charles. *From Mobilization to Revolution.* Reading, MA: Addison-Wesley, 1978.

———. "Does Modernization Breed Revolution?" In Jack A. Gladstone (ed.), *Revolutions.* New York: Harcourt Brace Jovanovich, 1986, pp. 47–57.

———. *Coercion, Capital, and European States, AD 990–1990.* Cambridge, MA: Blackwell, 1990.

———. *Coercion, Capital, and European States, AD 990–1992.* Cambridge, MA: Blackwell, 1992.

———. *Social Processes.* Lanham, MD: Rowman & Littlefield, 1997.

———. *Durable Inequality.* Berkeley: University of California Press, 1999.

Tilly, Chris. "Beyond Patching the Safety Net: A Welfare and Work Survival Strategy." *Dollars and Sense* (January/February 1999):14, 36–38.

Tilly, Louise, and Joan Scott. *Women, Work, and Family.* New York: Holt, Rinehart and Winston, 1978.

Tilove, Jonathan. "The New Map of American Politics." *The American Prospect* (May/June 1999):34.

Timms, Duncan. *The Urban Mosaic.* Cambridge, UK: Cambridge University Press, 1971.

Tobias, Sheila. *Sexual Politics.* Boulder, CO: Westview Press, 1997.

Tobin, J. J. "The American Idealization of Old Age in Japan." *Gerontologist* 27 (February 1987):53–58.

Toch, Thomas, and Warren Cohen. "Public Education: A Monopoly No Longer." *U.S. News & World Report* (November 23, 1998):25.

Tocqueville, Alexis de. *The Old Regime and the French Revolution.* Translated by Stuart Gilbert. Garden City, NY: Doubleday, 1955.

———. *Democracy in America.* New York: Bantam Books, 2000.

To Establish Justice, to Insure Domestic Tranquility: A Thirty Year Update of the National Commission on the Causes and Prevention of Violence. Washington, DC: The Milton S. Eisenhower Foundation, 1999.

Toland, John. *Adolph Hitler.* Garden City, NY: Doubleday, 1976.

Tomaskovic-Devey, Don, and Sheryl Skaggs. "Sex Segregation, Labor Process Organization, and Gender Earnings Inequality." *American Journal of Sociology* 108 (July 202):102–128.

Tönnies, Ferdinand. *Community and Society.* Translated and edited by Charles P. Loomis. East Lansing: Michigan State University Press, 1957.

Tonry, Michael (ed.). *The Handbook of Crime and Punishment.* New York: Oxford University Press, 2000.

Tonry, Michael, and Mary Lynch. "Intermediate Sanctions." In Michael Tonry (ed.), *Crime and Justice: A Review of Research.* Vol. 20. Chicago: University of Chicago Press, 1996, pp. 94–144.

Towers Perrin Workplace Index. Boston: Towers Perrin, 1997.

Toynbee, Arnold. *A Study of History.* New York: Oxford University Press, 1946.

Trattner, Walter I. *From Poor Law to Welfare State: A History of Social Welfare in America.* 6th ed. New York: Free Press, 1999.

Treiman, Donald J. *Occupational Prestige in Comparative Perspective.* New York: Academic Press, 1977.

Trigaboff, Dan. "Drudge Begrudged." *Broadcasting & Cable* (June 8, 1998):55.

Trimble, J. E. "Stereotypical Images, American Indians, and Prejudice." In P. A. Katz and D. A. Taylor (eds.), *Eliminating Racism.* New York: Plenum Press, 1988, pp. 181–202.

Trister, Michael. "The Rise and Reform of Stealth PACs." *The American Prospect* (September 25–October 9, 2000):32–35.

Troeltsch, Ernst. *The Social Teachings of the Christian Churches.* New York: Macmillan, 1931.

Troll, Lillian E. "The Family of Later Life: A Decade Review." *Journal of Marriage and the Family* 33 (May 1971):263–290.

Trueba, Enrique T. *Latinos Unidos: From Cultural Diversity to the Politics of Solidarity.* Lanham, MA: Rowman & Littlefield, 1999.

Tucker, Robert C. (ed.). *The Marx-Engels Reader.* New York: Norton, 1972.

Tumin, Melvin M. "Some Principles of Stratification: A Critical Analysis." *American Sociological Review* 18 (August 1953):387–394.

Tumulty, Karen. "Let's Play Doctor." *Time* (July 13, 1998):28–32.

Turnbull, Colin. *The Mountain People.* London: Pimlico, 1994.

Turner, Jonathan H. *American Society.* 2nd ed. New York: Harper and Row, 1976.

———. *The Structure of Sociological Theory.* 5th ed. Belmont, CA: Wadsworth, 1991.

———. *Classical Sociological Theory.* Chicago: Nelson-Hall, 1993.

——— (ed.). *Handbook of Sociological Theory.* New York: Kluwer Academic/Plenum Publishers, 2001.

———. *Face to Face: Toward a Sociological Theory of Interpersonal Behavior.* Stanford, CA: Stanford University Press, 2002.

Turner, Jonathan H., Leonard Beeghley, and Charles H. Powers. *The Emergence of Sociological Theory.* 5th ed. Belmont, CA: Wadsworth, 2002.

Turner, Margery Austin, Raymond J. Struyk, and John Yinger. *Housing Discrimination Study: Synthesis.* U.S. Department of Housing and Urban Development. Washington, DC: U.S. Government Printing Office, 1991.

Turner, Ralph H. "Collective Behavior." In Robert E. L. Faris (ed.), *Handbook of Modern Sociology.* Chicago: Rand McNally, 1964, pp. 382–425.

Turner, Ralph H., and Lewis M. Killian. *Collective Behavior.* 3rd ed. Englewood Cliffs, NJ: Prentice Hall, 1987.

Twaddle, Andrew C. *Sickness Behavior and the Sick Role.* Cambridge, MA: Schenkman, 1981.

Twaddle, Andrew C., and Richard M. Hessler. *A Sociology of Health.* 2nd ed. New York: Macmillan, 1987.

Tyler, Patrick E. "Threats and Responses." *New York Times* (February 17, 2003):A1.

Uchitelle, Louis, and N. R. Kleinfield. "The Price of Jobs Lost." In *New York Times* Special Report, *The Downsizing of America.* New York: Random House, 1996.

United Nations. *Family.* New York: United Nations Publications, 1996.

United Nations Population Division. *World Population Prospects: The 1998 Revision.* New York: United Nations, 2000.

U.S. Bureau of the Census. *Money Income of Households, Families, and Persons in the United States: 1992.* Current Population Reports. Series P-60. No. 184. Washington, DC: U.S. Government Printing Office, 1993a.

———. *1990 Census of Population, Social and Economic Characteristics, American Indian and Alaska Native Areas.* Section 1 of 2. CP-2–1A. Washington, DC: U.S. Government Printing Office, 1993b.

———. *1990 Census of Population. Social and Economic Characteristics, United States.* CP-2–1. November 1993. Washington, DC: U.S. Government Printing Office, 1993c.

———. *School Enrollment, Social and Economic Characteristics of Students: October 1992.* Current Population Reports. Series P-20. No. 474. Washington, DC: U.S. Government Printing Office, 1993d.

———. *Statistical Abstract of the United States: 1996.* Washington, DC: U.S. Government Printing Office, 1996a.

———. *Poverty in the United States: 1995.* Current Population Reports. Series P-60. No. 194. Washington, DC: U.S. Government Printing Office, 1996b.

———. *Voting and Registration in the Election of November 1996.* Current Population Reports. Series P-20. No. 504. Washington, DC: U.S. Government Printing Office, 1997a.

———. "Metropolitan Areas." 1997b. Available online at www.census.gov/population/www/estimates/metrodef.html.

———. "Poverty 1997—Poverty Estimates by Selected Characteristics." *Current Population Survey, March 1998*, 1999. Available online at www.census.gov/hhes/poverty/poverty97/pv97est1.html.

———. "Percentage Change in Metropolitan Population Inside and Outside Central Cities, by Region and Division: 1990 to 1998." 2000. Available online at www.census.gov/population/estimates/metro-city/ma98–c2.gif.

———. *Money Income in the United States: 2000.* 2001a. Washington, DC: U.S. Government Printing Office.

———. "Asset Ownership of Households: 1995—Table 4." 2001b. Available online at www.census.gov/hhes/www/wealth/1995/wlth95–4.html.

———. *Poverty in the United States: 2001.* Current Population Reports, P60–219. Washington, DC: U.S. Government Printing Office, 2002a.

———. *Statistical Abstract of the United States.* 121st ed. Washington, DC: U.S. Government Printing Office, 2002b.

———. "Reported Voting and Registration, by Race, Hispanic Origin, Sex, and Age, for the United States: November 2000." 2002c. Available online at www.census.gov/population/socdemo/voting/p20–542/tab02.pdf.

———. "Historical Income Tables—People, Table P-38." 2002d. Available online at www.census.gov/hhes/income/histinc/p.38.html.

U.S. Bureau of the Census. Poverty in the United States: 2001. Current Population Reports, P60–219. Washington, DC: U.S. Government Printing Office, 2002e.

U.S. Commission on Civil Rights. *Racial Isolation in the Public Schools.* Vol. II. Washington, DC: U.S. Government Printing Office, 1967.

———. *Success of Asian Americans: Fact or Fiction?* Clearinghouse Publication 64. Washington, DC: U.S. Government Printing Office, 1980.

U.S. Commission on Wartime Relocation and Internment of Civilians. *Personal Justice Denied.* Washington, DC: U.S. Government Printing Office, 1983.

U.S. Department of Health and Human Services. *Mental Health: Culture, Race, and Ethnicity—A Supplement to Mental Health: A Report of the Surgeon General.* Rockville, MD: U.S. Department of Health and Human Services, 2001.

———. "U.S. Birth Rate Reaches Record Low." Washington, DC: HHS News, July 25, 2003. http://www.hhs.gov.

U.S. Department of Justice. *White Collar Crime.* Washington, DC: U.S. Government Printing Office, 1987.

———. *Criminal Victimization in the United States, 2000.* Washington, DC: U.S. Government Printing Office, 2001.

———. Federal Bureau of Investigation. *Uniform Crime Reports, 2001.* Washington, DC: U.S. Government Printing Office, 2002.

U.S. Department of Labor. Bureau of Labor Statistics. *Employment in Perspective: Women in the Labor Force.* Report 860. Washington, DC: U.S. Government Printing Office, 1997a.

———. Bureau of Labor Statistics. *Employment and Earnings.* Washington, DC: U.S. Government Printing Office, January 1997b.

———. Bureau of Labor Statistics. "Unemployed Persons by Marital Status, Race, Age, and Sex." 2001a. Available online at www.bls.gov/cps/cpsaat24.pdf.

———. Bureau of Labor Statistics. "Labor Force Participation Trends for Women and Men." 2001b. Available online at www.bls.gov/opub/ted/2001/dec/wk3/art02.htm.

———. Bureau of Labor Statistics. *Highlights of Women's Earnings in 2000.* Report 952. Washington, DC: U.S. Government Printing Office, 2001c.

Ulbrich, Patricia M. "The Determinants of Depression in Two-Income Marriages." *Journal of Marriage and the Family* 50 (February 1988):121–131.

Useem, Michael. *The Inner Circle*. New York: Oxford University Press, 1984.

Vago, Steven. *Social Change*. 4th ed. Upper Saddle River, NJ: Prentice Hall, 1998.

Valdivieso, Rafael, and Cary Davis. "U.S. Hispanics: Challenging Issues for the 1990s." *Population Trends and Public Policy*. Washington, DC: Population Reference Bureau, 1988.

Valian, Virginia. *Why So Slow? The Advancement of Women*. Cambridge, MA: MIT Press, 1999.

Van Ausdale, Debra, and Joe R. Feagin. *The First R: How Children Learn Race and Racism*. Lanham, MD: Rowman & Littlefield, 2001.

Van Biema, David. "The Bishops Get Off the Hook." *Time* (June 24, 2002):18.

VandeHei, Jim, and David S. Hilzenrath. "Hill Leaders Agree on Corporate Curbs." *Washington Post* (July 25, 2002):A-1.

VandeHei, Jim, and Tom Hamburger. "Businesses Prepare New Legislative Wish List for Bush." *Wall Street Journal* (January 8, 2002):A20.

van den Berghe, Pierre L. "Dialectic and Functionalism: Toward a Theoretical Synthesis." *American Sociological Review* 28 (October 1963):695–705.

———. *Race and Racism*. 2nd ed. New York: Wiley, 1978.

Van Der Lippe, Tanja (ed.). *Women's Employment in a Comparative Perspective*. Hawthorne, NY: Aldine de Gruyter, 2001.

Vander Zanden, James W. *American Minority Relations*. New York: McGraw-Hill, 1990.

Van Gelder, Lindsy. "It's Tough Parenting in the High-Tech Age." *PC/Computing* (November 1989):52.

van Gennep, A. *The Rites of Passage*. Chicago: University of Chicago Press, 1960.

Vanhanen, Tatu. *Prospects of Democracy*. London: Routledge, 1997.

Van Zelst, Raymond H. "Sociometrically Selected Work Teams Increase Production." *Personnel Psychology* 5 (Autumn 1952):175–185.

Vasil, L., and H. Wass. "Portrayal of the Elderly in the Media: A Literature Review and Implications for Educational Gerontologists." *Educational Gerontology* 19 (1993):71–85.

Vazque, Francisco H., and Rodolfo D. Torres. *Latino/a Thought: Culture, Politics, and Society*. Lanham, MD: Rowman & Littlefield, 2002.

Veblen, Thorstein. *The Engineers and the Price System*. New York: Viking, 1933.

———. *Theory of the Leisure Class*. New York: Pengiun, 1995.

Verbrugge, Lois. "From Sneezes to Adieux." *American Demographics* (May 1986):35–38, 53–54.

———. "The Twain Meet: Empirical Explanations of Sex Differences in Health and Mortality." *Journal of Health and Social Behavior* 30 (September 1989):282–304.

"Violence Against Women: Research Into Practice." Washington, DC: National Institute of Justice, 2002.

Virnoche, Mary E. "The Seamless Web and Communications Equity: The Shaping of a Community Network." *Science, Technology, & Human Values* 23 (Spring 1998):199–220.

Virnoche, Mary E., and Gary T. Marx. "'Only Connect': E. M. Forster in an Age of Electronic Communication: Computer-Mediated Association and Community Networks." *Sociology Inquiry* 67 (1997):635–650.

Vobejda, Barbara. "The Heartland Pulses with New Blood." *The Washington Post* (August 11, 1991):A1,A18.

Volti, Rudi. *Society and Technological Change*. 4th ed. New York: St. Martin's Press, 2000.

Von Drehle, David. "A Debate on Marriage, And More, Now Looms." *The Washington Post* (June 27, 2003):A1,A16.

———. "Gay Marriage Is a Right, Massachusetts Court Rules." *Washington Post* (November 19, 2003):A1.

Voss, Kim, and Rachel Sherman. "Breaking the Iron Law of Oligarchy: Union Revitalization in the American Labor Movement." *American Journal of Sociology* 106 (September 2000):303–349.

Wade, Nicholas. "Now, the Hard Part: Putting the Genome to Work." *New York Times* (June 27, 2000):A1.

Wagner, Tony. *Making the Grade: Reinventing America's Schools*. Bristol, PA: Taylor and Francis, 2001.

Wagster, Emily. "Many Workers Hurt Abroad by Culture Shock." *USA Today* (December 17, 1993):1B.

Waitzkin, Howard. *The Second Sickness*. Chicago: University of Chicago Press, 1986.

———. *The Second Sickness: Contradictions of Capitalist Health Care*. Lanham, MD: Rowman & Littlefield, 2000.

———. *At the Front Lines of Medicine: How the Health Care System Alienates Doctors and Mistreats Patients . . . and What We Can Do about It*. Lanham, MD: Rowman & Littlefield, 2001.

Waldman, Steven. "The Stingy Politics of Head Start." *Newsweek. Special Issue. Education: A Consumer's Handbook* (Fall/Winter 1990–91):78–79.

Walker, A., C. Pratt, and N. Oppy. "Perceived Reciprocity in Family Caregiving." *Family Relations* (January 1992):82–85.

Wallace, Walter J. *The Future of Ethnicity, Race and Nationality*. Westport, CT: Praeger, 1997.

Wallechinsky, David, and Irving Wallace. *The People's Almanac #3*. New York: Bantam, 1981.

Waller, Margy. "Welfare, Working Families, and Reauthorization: Mayors' Views." Center on Urban and Metropolitan Policy, The Brookings Institution, Survey Series, May 2003, pp. 1–15.

Waller, Willard, and Reuben Hill. *The Family*. Rev. ed. New York: Dryden Press, 1951.

Wallerstein, Immanuel. *The Capitalist World Economy*. New York: Cambridge University Press, 1979.

———. *The Politics of the World Economy*. Cambridge, UK: Cambridge University Press, 1984.

———. *The Modern World-System*. London: Academic Press, 1989.

———. *The Essential Wallerstein*. New York: New Press, 2000.

———. *The End of the World as We Know It: Social Science for the Twenty-First Century*. Minneapolis, MN: University of Minnesota Press, 2001.

———. *Decline of American Power: The U.S. in a Chaotic World*. New York: New Press, 2003.

Wallich, Paul, and Elizabeth Corcoran. "The Discreet Disappearance of the Bourgeoisie." *Scientific American* 226 (February 1992):111.

Wallulis, Jerald. *The New Insecurity*. Albany: State University of New York Press, 1998.

Walsh, Edward. "In Midwest, Bush Calls Gore Obstacle to Reform." *The Washington Post* (October 24, 2000):A12.

Wang, C. T., and D. Daro. *Current Trends in Child Abuse Reporting and Fatalities: The Results of the 1997 Annual Fifty-State Survey*. Chicago, IL: Prevent Child Abuse America, 1998.

Ward, R., J. Logan, and G. Spitze. "The Influence of Parent and Child Needs on Coresidence in Middle and Later Years." *Journal of Marriage and the Family* (February 1992):209–221.

Ward, Russell B. "The Impact of Subjective Age and Stigma on Older Persons." *Journal of Gerontology* 32 (March 1977):227–232.

Ware, Leland, and Antoine Allen. "The Geography of Discrimination: Hypersegregation, Isolation and Fragmentation within the African-American Community." In Lee A. Daniels (ed.), *The State of Black America 2002*. New York: National Urban League, 2002, pp. 69–92.

Warner, R. Stephen. *Public Religions in the Modern World*. Chicago: University of Chicago Press, 1994.

Warner, W. Lloyd. *Social Class in America*. New York: Harper and Row, 1960.

Warner, W. Lloyd, and Paul S. Lunt. *The Social Life of a Modern Community*. New Haven, CT: Yale University Press, 1941.

Warr, Mark. *Companions in Crime: The Social Aspects of Criminal Conduct*. New York: Cambridge University Press, 2002.

Warren, Martin, and Sheldon Berkowitz. "The Employability of AFDC Mothers and Fathers." *Welfare in Review* 7 (July/August 1969):1–7.

Wasserman, Gary. *The Basics of American Politics*. 10th ed. New York: Longman, 2001.

Waters, Malcolm. *Globalization*. 2nd ed. New York: Routledge, 2001.

Waters, Mary C. *Ethnic Options*. Berkeley: University of California Press, 1990.

Watson, C. W. *Multiculturalism*. Bristol, PA: Taylor and Francis, 2000.

Watson, George W., and Jon M. Shepard. "The Disaffected Professional in a Changing Contractual Environment." Unpublished manuscript, 1997.

Watson, Russell, and John Barry. "Tomorrow's New Face of Battle." *Newsweek* (February 1998, Winter 1997–98 Special Edition):66–67.

Watterson, T. "Social Security: Many Widows Don't Get Benefits." *Baltimore Sun* (June 13, 1990):C24.

Watzman, Nancy, James Youngclaus, and Jennifer Shecter. *Cashing In*. Washington, DC: Center for Responsive Politics, 1997.

Wax, Emily. "Across the E-Divide." *Washington Post* (May 17, 2000):G8.

———. "In Times of Terror, Teens Talk the Talk." *Washington Post* (March 19, 2002):A1, A10.

Waxman, Chaim I. *The Stigma of Poverty*. New York: Pergamon, 1983.

Weakliem, David L. "Relative Wages and the Radical Theory of Economic Segmentation." *American Sociological Review* 55 (1990):574–590.

Webb, R. K. *Harriet Martineau, A Radical Victorian*. New York: Columbia University Press, 1960.

Weber, Max. *From Max Weber*. Edited by H. H. Gerth and C. Wright Mills. New York: Oxford University Press, 1946.

———. *The Religion of China*. New York: Free Press, 1951.

———. *Ancient Judaism*. New York: Free Press, 1952.

———. *The Religion of India*. New York: Free Press, 1958.

———. *The Sociology of Religion*. Boston: Beacon Press, 1964.

Webster, Murray Jr., and Stuart J. Hysom. "Creating Status Characteristics." *American Sociological Review* 63 (June 1998):351–378.

Weeden, Kim A. "Why Do Some Occupations Pay More than Others? Social Closure and Earnings Inequality in the United States." *American Journal of Sociology* 108 (July 2002):55–101.

Weeks, John R. *Population*. 7th ed. Belmont, CA: Wadsworth, 2002.

Wehrfritz, George. "Angry at the Yanks." *Newsweek* (January 13, 2003):28–29.

Weinberg, Meyer. *Desegregation Research*. 2nd ed. Bloomington, IN: Phi Delta Kappan, 1970.

Weingart, Peter, et al. (eds.). *Human by Nature: Between Biology and the Social Sciences*. Mahwah, NJ: Erlbaum, 1997.

Weinstein, Jay. "A (Further) Common on the Differences between Applied and Academic Sociology." *Contemporary Sociology* (March 2000):344–347.

Weisbrot, Mark. "The Mirage of Progress: The Economic Failure of the Last Two Decades of the Twentieth Century." *American Prospect* (Winter 2002):A10–A12.

Weisburd, David, and Elin Waring. *White Collar Crime and Criminal Careers*. Cambridge, UK: Cambridge University Press, 2001.

Weise, Elizabeth. "In a Web First, Women Are in the Majority." *USA Today* (August 9, 2000).

Weiss, Gregory L., and Lynne E. Lonnquist. *The Sociology of Health, Healing, and Illness*. New York: Simon & Schuster, 1996.

Weiss, Lawrence D. *Private Medicine and Public Health*. Boulder, CO: Westview Press, 1997.

Weiss, Rick. "Transplant Researchers Clone 5 Pigs." *Washington Post* (March 15, 2000):A1, A16.

———. "Legal Barriers to Human Cloning May Not Hold Up." *Washington Post* (May 23, 2001):A1, A14, A15.

———. "In Laboratory, Ordinary Cells Are Turned Into Eggs." *Washington Post* (May 2, 2003):A1, A12.

Weiss, Rick, and Justin Gillis. "DNA Mapping Milestone Heralded." *Washington Post* (June 27, 2000):A1, A12.

Weiss, Steven. "Money is the Victor in 2002 Midterm Elections." Washington, DC: The Center for Responsive Politics, 2002.

Weisse, Allen B. *The Staff and the Serpent*. Carbondale, IL: Southern Illinois University Press, 1998.

Weitz, Rose. *The Sociology of Health, Illness, and Health Care: A Critical Approach*. 2nd ed. Belmont, CA: Wadsworth, 2000.

———. *The Sociology of Health, Illness and Health Care: A Critical Approach*. 2nd ed. Belmont, CA: Wadsworth, 2001.

Weitzman, Leonore J., and Deborah Eifler. "Sex Role Socialization in Picture Books for Preschool Children." *American Journal of Sociology* 77 (May 1972):1125–1150.

Weitzman, Susan. *"Not to People Like Us": Hidden Abuse in Upscale Marriages*. New York: Basic Books, 2000.

Welch, William M. "Senate OKs Limited Set of Patient Protections." *USA Today* (July 16–18, 1999):1A.

Weller, Jack E. *Yesterday's People: Life in Contemporary Appalachia*. Lexington: University Press of Kentucky, 1980.

Wells, Spencer, and Mark Read. *The Journey of Man: A Genetic Odyssey*. Princeton, NJ: Princeton University Press, 2003.

Werhane, Patricia H. *Adam Smith and His Legacy for Modern Capitalism*. New York: Oxford University Press, 1991.

Wermuth, Laurie A. *Global Inequality and Human Needs: Health and Illness in an Increasingly Unequal World*. Boston: Allyn & Bacon, 2002.

West, Candace, and Don H. Zimmerman. "Doing Gender." *Gender and Society* 1 (June 1987):125–151.

West, Guida, and Rhoda Lois Blumberg (eds.). *Women and Social Protest*. New York: Oxford University Press, 1990.

Western, Bruce, and Becky Pettit. "Beyond Crime and Punishment: Prisons and Inequality." *Contexts* 1 (Fall 2002):37–43.

Westoff, Leslie Aldridge, and Charles F. Westoff. *From Now to Zero*. Boston: Little, Brown and Company, 1971.

Wetcher, Kenneth, Art Barker, and Rex McCaughtry. *Save the Males: Why Men Are Mistreated, Misdiagnosed, and Misunderstood*. Summit, NJ: PIA Press, 1991.

White, Leslie A. *The Science of Culture*. 2nd ed. New York: Farrar, Straus & Giroux, 1969.

White, Lynn. *Medieval Technology and Social Change*. New York: Oxford University Press, 1972.

White, Lynn, and David B. Brinkerhoff. "The Sexual Division of Labor: Evidence from Childhood." *Social Forces* 60 (September 1981):170–181.

White, Ralph K., and Ronald O. Lippitt. *Autocracy and Democracy*. New York: Harper and Row, 1960.

White, Robert, and Robert Althauser. "Internal Labor Markets, Promotions, and Worker Skill: An Indirect Test of Skill ILMS." *Social Science Research* 13 (December 1984):373–392.

Whitehead, Alfred North. *Adventures of Ideas*. New York: Mentor Books, 1933.

Whiten, A., et al. "Cultures in Chimpanzees." *Nature* 399 (June 1999):682–685.

Who Owns Whom 1998/1999. High Wycombe, UK: Dun and Bradstreet Ltd., 1998.

Whorf, Benjamin Lee. *Language, Thought and Reality*. Edited by John B. Carroll. Cambridge, MA: MIT Press, 1956.

"Why Don't Americans Trust the Government?" Menlo Park, CA: *Washington Post*/Kaiser Family Foundation/Harvard University Survey Project, 1996.

Whyte, William Foote. *Street Corner Society*. 4th ed. Chicago: University of Chicago Press, 1993.

Whyte, William H. Jr. *The Organization Man*. New York: Simon & Schuster, 1972.

Whyte, William H., and Joseph Nocera. *The Organization Man: The Book That Defined a Generation*. Philadelphia: University of Pennsylvania Press, 2002.

Wiehe, Vernon R. *Sibling Abuse*. Lexington, MA: Heath, 1990.

Wiesel, Elie. "The Question of Genocide." *Newsweek* (April 12, 1999):37.

Wisensale, Steven K. *Family Leave Policy: The Political Economy of Work and Family in America*. Armonk, NY: M.E. Sharpe, 2001.

Wieviorka, Michel. *The Arena of Racism*. Newbury Park, CA: Sage, 1995.

Wildavsky, Ben. "Grading on a Curve: A New Controversy Erupts Over Race, Class, and SAT Scores." *U.S. News & World Report* (September 13, 1999):53.

Wiley, Norbert (ed.). *The Marx-Weber Debate*. Beverly Hills, CA: Sage, 1987.

Wilkins, David E. *American Indian Politics and the American Political System*. Lanham, MD: Rowman & Littlefield, 2001.

Wilkinson, Doris Y. "Gender and Social Inequality: The Prevailing Significance of Race." *Daedalus* 124 (Winter 1995):167–178.

Wilkinson, Richard (ed.). *Class and Health*. New York: Tavistock, 1986.

Will, George. "The Nature of Human Nature." *Newsweek* (August 19, 2002):64.

Williams, Christine L. *Gender Differences in Work*. Berkeley: University of California Press, 1989.

Williams, J. Allen, Joetta A. Vernon, Martha C. Williams, and Karen Malecha. "Sex Role Socialization in Picture Books: An Update." *Social Science Quarterly* 68 (March 1987):148–156.

Williams, John E., and Deborah L. Best. *Measuring Sex Stereotypes*. Rev. ed. Newbury Park, CA: Sage, 1990.

Williams, Mary E. *Minorities*. San Diego, CA: Greenhaven Press, 1997.

———. *Capital Punishment*. San Diego, CA : Greenhaven Press, 2000.

Williams, Mary E., Karin Swisher, and Brenda Stalcup (eds.). *Working Women: Opposing Viewpoints*. San Diego, CA: Greenhaven Press, 1997.

Williams, Robert L. "Scientific Racism and IQ: The Silent Mugging of the Black Community." *Psychology Today* 7 (May 1974):32–41, 101.

Williams, Robin M. Jr. *American Society*. 3rd ed. New York: Knopf, 1970.

Williamson, Robert C. "A Partial Replication of the Kohn-Gecas-Nye Thesis in a German Sample." *Journal of Marriage and the Family* 46 (November 1984):971–979.

Wilson, Bryan. *Religion in Sociological Perspective*. New York: Oxford University Press, 1982.

Wilson, Craig. "Why People Will Do Almost Anything to Get on TV." *USA Today* (February 25–27, 2000):1A, 2A.

Wilson, Edward O. *On Human Nature*. Cambridge, MA: Harvard University Press, 1978.

———. *Biophilia*. Cambridge, MA: Harvard University Press, 1986.

Wilson, James Q. *The Moral Sense*. New York: Free Press, 1993.

———. *Political Organizations*. Princeton, NJ: Princeton University Press, 1995.

Wilson, James Q., and Richard J. Herrnstein. *Crime and Human Nature*. New York: Simon & Schuster, 1985.

Wilson, Kenneth L., and W. Allen Martin. "Ethnic Enclaves: A Comparison of the Cuban and Black Economies in Miami." *American Journal of Sociology* 88 (July 1982):135–160.

Wilson, Stephen. *Informal Groups*. Englewood Cliffs, NJ: Prentice Hall, 1978.

Wilson, Thomas C. "Urbanism and Tolerance: A Test of Some Hypotheses Drawn from Wirth and Stouffer." *American Sociological Review* 50 (February 1985):117–123.

———. "Urbanism, Migration, and Tolerance: A Reassessment." *American Sociological Review* 56 (February 1991):117–123.

Wilson, William Julius. *Power, Racism, and Privilege*. New York: Macmillan, 1973.

———. *The Declining Significance of Race*. 2nd ed. Chicago: University of Chicago Press, 1980.

———. "The Urban Underclass." In Leslie W. Dunbar (ed.), *Minority Report*. New York: Pantheon, 1984, pp. 75–117.

———. *The Truly Disadvantaged*. Chicago: University of Chicago Press, 1987.

———. "Studying Inner-City Social Dislocations: The Challenge of Public Agenda Research." *American Sociological Review* 56 (February 1991):1–14.

——— (ed.). *The Ghetto Underclass*. Newbury Park, CA: Sage, 1993.

———. *When Work Disappears: The World of the New Urban Poor*. New York: Knopf, 1997.

———. *Visions of Social Inequality: Race, Class, and Poverty in Urban America*. Lanham, MD: Rowman & Littlefield, 2002.

Wilson, William Julius, et al. "The Ghetto Underclass and the Changing Structure of Urban Poverty." In Fred R. Harris and Roger W. Wilkins (eds.), *Quiet Riots*. New York: Pantheon, 1988, pp. 123–154.

Wimberley, Ronald C., and James A. Christenson. "Civil Religion and Church and State." *Sociological Quarterly* 21 (Spring 1980):35–40.

Wimberley, Ronald C., Donald A. Clelland, Thomas C. Hood, and C. M. Lipsey. "The Civil Religious Dimension: Is It There?" *Social Forces* 54 (June 1976):890–900.

Wind, James P., and James W. Lewis (eds.). *American Congregations*. Vol. 2. Chicago: University of Chicago Press, 1994.

Wingert, Pat. "The Sum of Mediocrity." *Newsweek* (December 2, 1996):96.

———. "The Report Card on Charter Schools." *Newsweek* (July 22, 2002):7.

Wingfield, Nick. "Family Planning: The Computer Server Promises to Do to the Home What It Has Already Done for Business." *Wall Street Journal* (June 15, 1998):R18, R23.

Winship, Christopher, and Robert D. Mare. "Models for Sample Selection Bias." *Annual Review of Sociology* 18 (1992):327–350.

Winters, Rebecca. "The Philadelphia Experiment." *Time* (October 21, 2002):64–69.

Wirth, Louis. "Clinical Sociology." *American Journal of Sociology* 37 (July 1931):49–66.

———. "Urbanism as a Way of Life." *American Journal of Sociology* 44 (July 1938):1–24.

———. "The Problem of Minority Groups." In Ralph Linton (ed.), *The Science of Man in the World Crisis*. New York: Columbia University Press, 1945, pp. 347–372.

Wise, William. *Massacre at Mountain Meadows*. New York: Thomas Y. Crowell, 1976.

Wisensale, Steven K. *Family Leave Policy: The Political Economy of Work and Family in America*. Armonk, NY: M.E. Sharpe, 2001.

Witte, John F. *The Market Approach to Education: An Analysis of America's First Voucher Program*. Princeton, NJ: Princeton University Press, 2000.

Wodak, Ruth (ed.). *Gender and Discourse*. Thousand Oaks, CA: Sage, 1997.

Wohl, Andrzej. "Sport and Social Development." *International Review of Sport Sociology* 14 (1979):5–18.

Wokutch, Richard E. *Worker Protection, Japanese Style*. Ithaca, NY: ILR Press, 1992.

Wokutch, Richard E., and Jon M. Shepard. "The Maturing of the Japanese Economy: Corporate Social Responsibility Implications." *Business Ethics Quarterly* (July 1999):527–540.

Wolfe, Alan. "The Moral Meanings of Work." *The American Prospect* (September/October 1997):82–90.

———. *One Nation, After All*. New York: Penguin Books, 1999.

Wolff, Edward N., and Richard C. Leone. *Top Heavy: The Increasing Inequality of Wealth in America and What Can Be Done about It*. New York: New Press, 2002.

Wolfinger, Raymond E., and Steven J. Rosenstone. *Who Votes?* New Haven, CT: Yale University Press, 1980.

Wolman, William, and Anne Colamosca. *The Judas Economy: The Triumph of Capital and the Betrayal of Work*. Reading, MA: Addison-Wesley, 1997.

Wood, James R. *Leadership in Voluntary Organizations*. New Brunswick, NJ: Rutgers University Press, 1981.

Wood, Gordon S. *The American Revolution: A History*. New York: The Modern Library, 2002.

Wooden, Wayne S. *Renegade Kids, Suburban Outlaws: From Youth Culture to Delinquency*. Belmont, CA: Wadsworth, 1995.

The World Almanac and Book of Facts 2002. New York: World Almanac Education Group, Inc., 2002.

The World Bank Atlas, 1999. Washington, DC: World Bank, 1999.

World Development Report 1991. New York: Oxford University Press, 1991.

World Development Report 1993. New York: Oxford University Press, 1993.

World Development Report 1994. New York: Oxford University Press, 1994.

World Development Report 2000–2001. New York: Oxford University Press, 2001.

World Development Report 2003. New York: Oxford University Press, 2003.

World Population Data Sheet. Washington, DC: Population Reference Bureau, 2002.

The World's Women 2000: Trends and Statistics. New York: United Nations Publications, 2000.

"World Without Work." In Robert Staples (ed.), *The Black Family*. Belmont, CA: Wadsworth, 1999, pp. 291–311.

Wortzel, Larry M. *Class in China*. Westport, CT: Greenwood Press, 1987.

Wright, Charles R. "Functional Analysis and Mass Communication." *Public Opinion Quarterly* 24 (Winter 1960):606–620.

———. "Functional Analysis and Mass Communication Revisited." In Jay G. Blumler and Elihu Katz (eds.), *The Uses of Mass Communication*. Beverly Hills, CA: Sage, 1974, pp. 197–212.

Wright, Erik Olin. *Classes*. London: Verse, 1985.

———. *Class Counts*. Cambridge, UK: Cambridge University Press, 2000.

Wright, James D., and Sonia R. Wright, "Social Class and Parental Values for Children: A Partial Replication and Extension of the Kohn Thesis." *American Sociological Review* 4 (June 1976):527–537.

Wright, John W. (ed.). *New York Times Almanac*. New York: Penguin Putnam, 2002.

Wright, Kevin N. *The Great American Crime Myth*. New York: Praeger, 1987.

Wright, Lawrence. "Double Mystery." *The New Yorker* (August 7, 1995):45–62.

Wright, Robert. *The Moral Animal*. New York: Vintage Books, 1996.

Wu, Jean Y., and Min Song (eds.). *Asian American Studies: A Reader*. Piscataway, NJ: Rutgers University Press, 2000.

Wuthnow, Robert. "Recent Patterns of Secularization: A Problem of Generations?" *American Sociological Review* 41 (October 1976):850–867.

———. *Experimentation in American Religion*. Berkeley: University of California Press, 1978.

———. "Review of All Faithful People." *Society* 22 (March/April 1985):87–88.

———. *The Restructuring of American Religion*. Princeton, NJ: Princeton University Press, 1990.

———. *Poor Richard's Principle: Rediscovering the American Dream Through the Moral Dimensions of Work, Business, and Money*. Princeton, NJ: Princeton University Press, 1996.

Yamagata, Hisashi, et al. "Sex Segregation and Glass Ceilings: A Comparative Statistics Model of Women's Career Opportunities in the Federal Government Over a Quarter Century." *American Journal of Sociology* 103 (November 1997):566–632.

Yang, Fenggang, and Helen Rose Ebaugh. "Transformations in New Immigrant Religions and Their Global Implications." *American Sociological Review* 66 (April 2001):269–288.

Yap, M. T. "Singapore's 'Three or More' Policy: The First Five Years." *Asia-Pacific Population Journal* 10 (1995):39–52.

Yinger, John. *Closed Doors, Opportunities Lost: The Continuing Costs of Housing Discrimination*. New York: Russell Sage Foundation, 1995.

Yinger, Milton J. "A Structural Examination of Religion." *Journal of the Scientific Study of Religion* 8 (Spring 1969):88–89.

———. *The Scientific Study of Religion*. New York: Macmillan, 1970.

Yinger, Milton J., and Stephen J. Cutler. "The Moral Majority Viewed Sociologically." *Sociological Focus* 15 (October 1982):289–306.

Yochelson, Samuel, and Stanton E. Samenow. *The Criminal Personality*. Vols. 1 and 2. London: James Aronson, 1994.

Yoo, Grace J. "Racial Inequality, Welfare Reform and Black Families: The 1996 Personal Responsibility and Work Reconciliation Act." In Robert Staples (ed.), *The Black Family*. Belmont, CA: Wadsworth, 1999, pp. 357–366.

Young, Alford A. Jr., and Donald R. Deskins Jr. "Early Traditions of African-American Sociological Thought." *Annual Review of Sociology* 27 (2001):445–477.

Young, Iris Marion. *Justice and the Politics of Difference*. Princeton, NJ: Princeton University Press, 1990.

Young, Michael. *The Rise of the Meritocracy*. New York: Penguin, 1967.

Young, T. R. *The Drama of Social Life*. New Brunswick, NJ: Transaction Books, 1990.

———. "The Sociology of Sport: Structural Marxist and Cultural Marxist Approaches." *Sociological Perspectives* 29 (January 1986):3–28.

Yunker, James A. "A New Statistical Analysis of Capital Punishment Incorporating U.S. Postmoratorium Data," *Social Science Quarterly* 82 (June 2001):297–311.

Zald, Mayer N., and John D. McCarthy. "Introduction." In Mayer N. Zald and John D. McCarthy (eds.), *Social Movements in an Organizational Society*. New Brunswick, NJ: Transaction Books, 1987.

Zamble, Edward, and Vernon L. Quinsey. *The Criminal Recidivism Process*. New York: Cambridge University Press, 1997.

Zaret, David. "Calvin, Covenant Theology and the Weber Thesis." *British Journal of Sociology* 43 (1992):369–391.

Zborowski, Mark. "Cultural Components in Response to Pain." *Journal of Social Issues* 8 (1952):16–30.

———. *People in Pain*. San Francisco: Jossey-Bass, 1969.

Zelditch, Morris Jr. "Role Differentiation in the Nuclear Family: A Comparative Study." In Talcott Parsons and Robert F. Bales (eds.), *Family, Socialization and Interaction Process*. New York: Free Press, 1955, pp. 307–352.

Zellner, William M., and William M. Kephart. *Extraordinary Groups*. 6th ed. New York: St. Martin's Press, 1997.

Zellner, William W. *Countercultures: A Sociological Analysis*. 2nd ed. New York: St. Martin's Press, 1995.

Zellner, William W., and Marc Petrowsky. *Sects, Cults, and Spiritual Communities*. Westport, CT: Praeger, 1998.

Zepezauer, Mark, and Arthur Naiman. T*ake the Rich Off Welfare: The Real Story*. Monroe, ME: Odonian Press, 1996.

Zigler, Edward, and S. Muenchow. *Head Start*. New York: Basic Books, 1992.

Zigler, Edward, and Sally J. Styfco (eds.). *Head Start and Beyond*. New Haven, CT: Yale University Press, 1993.

Zimbardo, Philip G., S. M. Anderson, and L. G. Kabat. "Induced Hearing Deficit Generates Experimental Paranoia." *Science* (June 26, 1981):1529–1531.

Zimring, Franklin E. *Contradictions of American Capital Punishment*. New York: Oxford University Press, 2003.

Zinn, Howard. *Declarations of Independence: Cross-Examining American Ideology*. New York: HarperPerennial, 2003.

———. *A People's History of the United States*: 1492–Present. Twentieth Anniversary Edition. New York: HarperCollins Publishers, 2001.

Zogby, James J. *What Arabs Think: Values, Beliefs, and Concerns*. Utica, NY, and Beirut, Lebanon: Zogby International/The Arab Thought Foundation, 2002.

Zola, Irving K. "Culture and Symptoms—An Analysis of Patients Presenting Complaints." *American Sociological Review* 31 (October 1966):615–630.

Zorbaugh, Harvey. *The Gold Coast and the Slum*. Chicago: University of Chicago Press, 1929.

Zucher, Louis A., and David A. Snow. "Collective Behavior: Social Movements." In Morris Rosenberg and Ralph H. Turner (eds.), *Social Psychology*. New Brunswick, NJ: Transaction Books, 1990, pp. 447–482.

Zuo, J., and S. Tang. "Breadwinner Status and Gender Ideologies of Men and Women Regarding Family Roles." *Sociological Perspectives* 43 (2000):29–43.

Zweig, Michael. *The Working Class Majority: America's Best Kept Secret*. Ithaca, NY: Cornell University Press, 2000.

Zweigenhaft, Richard L., and G. William Domhoff. Diversity in the *Power Elite*. New Haven: Yale University Press, 1999.

Zwerling, Craig, and Hilary Silver. "Race and Job Dismissals in a Federal Bureaucracy." *American Sociological Review* 57 (October 1992):651–660.

Name Index

Subject Index

*Terms in **bold** appear in the glossary.*